# HANDBOOK OF BREADMAKING TECHNOLOGY

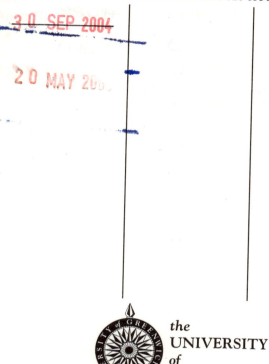

# HANDBOOK OF
# BREADMAKING TECHNOLOGY

Charles A. Stear

*Consultant, Salcombe, Devon, UK*

## ELSEVIER APPLIED SCIENCE
### LONDON and NEW YORK

ELSEVIER SCIENCE PUBLISHERS LTD
Crown House, Linton Road, Barking, Essex IG11 8JU, England

*Sole Distributor in the USA and Canada*
ELSEVIER SCIENCE PUBLISHING CO., INC.
655 Avenue of the Americas, New York, NY 10010, USA

WITH 77 TABLES AND 160 ILLUSTRATIONS

© 1990 ELSEVIER SCIENCE PUBLISHERS LTD

**British Library Cataloguing in Publication Data**

Stear, Charles A.
    Handbook of breadmaking technology.
    1. Bread. Manufacture
    I. Title
    664′.7523

**Library of Congress Cataloging-in-Publication Data**

Stear, Charles A.
    Handbook of breadmaking technology/Charles A. Stear.
        p.   cm.
    Bibliography: p.
    Includes index.
    ISBN 1-85166-394-0
    1. Bread.   2. Baking.   I. Title.
    TX769.S794   1990
    664′.7523—dc20

**Special regulations for readers in the USA**

Printed in Great Britain by Galliard (Printers) Ltd, Great Yarmouth

# Preface

The author's aim in writing this book is to integrate currently available knowledge concerning the basic scientific and technological aspects of breadmaking processes with the diverse breadmaking methods used to manufacture bread in Europe and on the North American continent today. To date, the main technological advances have been in process mechanization, starting with oven development, then dough-processing or make-up equipment, followed by continuous and batch mixing techniques from the 1950s to the present time. On the engineering side, universal emphasis is now being placed on the application of high technology, in the form of microprocessors, computer-controlled equipment and robotization, the long-term objective being computer integrated manufacture (CIM) with full automation within the large chain bakery groups in the capitalist countries and the state-run collectives of Eastern Europe. The application of these key technologies with biotechnology, as yet only applied to a limited degree in food manufacture, coupled with advances in biochemical and rheological understanding of dough as a biomass for breadmaking, should provide us with more expertise and ability to control the processes with greater efficiency. The application of fermentable substrates and industrial enzymes under strict kinetic control should contribute to improving the flavour characteristics of bread. Current trends towards improving the nutritional contribution of bread to the daily diet are improving the competitive edge of bread as a basic food in the market-place.

With the increasing consumer demand for a greater variety of quality oven-fresh baked products, the balance in favour of the industrial sliced-and-wrapped hygienic products is changing. The in-store bakeries of the chain grocers in city-centres and out-of-town hypermarkets provide an improving range of fresh products, owing to their ability to select the best equipment and to train staff to be quality conscious. This swing in favour of traditional fresh products will maintain the status of baking as a craft industry, especially in the UK where its strength has never been as durable as on the continent of Europe. In the East European countries, there is also a change of emphasis taking place in the baking industry, brought about by the need to provide the consumer outside the large cities with a greater choice of baked products which are freshly baked and not transported from the factory-bakeries. This is being achieved by in-stores and by fresh-baked product shops, and the planning of country-production units with staff accommodation which can independently supply products at district level.

The book consists of three parts dealing with mixing-stage technology and international formulation and processing, with yeast and sour-dough fermentation technology, and with bread-baking and preservation processes. It is addressed internationally to cereal chemists, bakery technologists and production-orientated readers, and its contents will contribute to a broader appreciation of the technology of breadmaking. If, in so doing, it stimulates the generation of new ideas which result in the improvement of the quality and variety of bread as the most important source of nutrition for mankind, the collective international expertise of this work will have served the baking industry world-wide.

# Acknowledgements

The author wishes to acknowledge the generous help given by the many bakery machinery manufacturers and their agents in allowing the use of equipment illustrations; and to the scientists and technologists in Eastern and Western Europe, as well as North America, whose valuable collective contributions to research and expertise in the bakery field have enhanced understanding and made this book possible. Also, many thanks to the Editorial and Production team of Elsevier Applied Science for their patient work and expertise in the organisation of this book.

# Contents

*Preface* . . . . . . . . . . . . v
*Acknowledgements* . . . . . . . . . . vii

## PART 1. FUNDAMENTAL DYNAMICS OF THE MIXING PROCESS, AND THEIR IMPLICATIONS FOR DOUGH RHEOLOGICAL BEHAVIOUR, PROCESS CONTROL AND OPTIMIZATION

1.1 Theoretical Model to Explain the Doughmaking Process . . 3
1.2 Application of Fundamental Dough-Mixing Parameters . . 9
    1.2.1. Practical application of dough-mixing parameters for the baker . . . . . . . . . . . 9
1.3 Fundamental Considerations Concerning Dough Rheological Elements and Dynamic Mixing Parameters . . . . . 11
1.4 Water-Binding Capacity of Dough Components and Dough Consistency Control . . . . . . . . . 21
1.5 Effects of Dough Additives . . . . . . . . 27
    1.5.1 Cysteine . . . . . . . . . . 27
    1.5.2 Sodium sulphite and metabisulphite (pyrosulphite) . . 29
    1.5.3 Enzymes . . . . . . . . . . 30
    1.5.4 Ascorbic acid . . . . . . . . . 31
    1.5.5 Fast-acting oxidants . . . . . . . . 34
    1.5.6 Delayed-action oxidants . . . . . . . 35
    1.5.7 Glutathione . . . . . . . . . 35
    1.5.8 Surfactants, strengtheners/conditioners and crumb-softeners 37
    1.5.9 Soy products . . . . . . . . . 41
    1.5.10 Vital wheat gluten and its derivatives . . . . 45
    1.5.11 Hydrocolloids . . . . . . . . . 53
    1.5.12 Starch-based products . . . . . . . 54
    1.5.13 Cellulose and cellulose derivatives . . . . 55
    1.5.14 Salt, acids and metallic ions . . . . . 58
1.6 Chemical bonding during doughmaking . . . . . 60
1.7 Typical Formulation and Process Schedules (including Case Studies) for Wheat and Rye Breads employed in Western and Eastern Europe and North America . . . . . . . . . 86
    1.7.1 France . . . . . . . . . . 86
    1.7.2 Spain . . . . . . . . . . 90

ix

|  |  |  |
|---|---|---|
| 1.7.3 | Switzerland | 92 |
| 1.7.4 | Austria | 94 |
| 1.7.5 | Federal Republic of Germany (FRG) | 96 |
| 1.7.6 | German Democratic Republic (GDR) | 108 |
| 1.7.7 | USSR | 143 |
| 1.7.8 | Hungary | 196 |
| 1.7.9 | Czechoslovakia | 197 |
| 1.7.10 | Poland | 201 |
| 1.7.11 | Collaboration within the COMECON baking industries | 203 |
| 1.7.12 | United Kingdom (UK) | 206 |
| 1.7.13 | North America | 268 |
| 1.8 | Measurement and Control Techniques for Raw Materials and Process Variables | 306 |
| 1.8.1 | Raw materials—chemical and microbiological | 306 |
| 1.8.2 | Process variables | 359 |
| 1.8.3 | Measuring systems based on dough deformation | 371 |
| 1.8.4 | Cleanliness and sanitation (hygiene) | 372 |
| 1.9 | Weigher-Mixer Functions and Diverse Types of Mixers and Mixing-Regimes | 394 |
| 1.9.1 | Weigher–mixer functions | 394 |
| 1.9.2 | Weighing methods | 399 |
| 1.9.3 | Types of mixer and mixing-regimes | 408 |

## PART 2. FERMENTATION OF WHEAT- AND RYE-FLOUR DOUGHS

|  |  |  |
|---|---|---|
| 2.1 | Introduction | 467 |
| 2.2 | Industrial Propagation and Production of Yeast for the Baking Industry | 469 |
| 2.2.1 | Yeast physiology | 471 |
| 2.2.2 | Improvement of industrial yeasts | 475 |
| 2.2.3 | Post-fermentation yeast technology | 476 |
| 2.3 | Chemical Changes in Yeasted Doughs during Fermentation | 479 |
| 2.4 | Wheat- and Rye-Sours and Sour-Dough Processing | 492 |
| 2.4.1 | Process variables | 497 |
| 2.4.2 | Processing stages | 501 |
| 2.5 | Formulation and Processing Techniques for Specialty-Breads | 522 |
| 2.5.1 | Multi-grain breads | 523 |
| 2.5.2 | High-fibre breads | 524 |
| 2.5.3 | High-protein breads | 525 |
| 2.5.4 | *Mecklenburger Landbrot* | 528 |
| 2.5.5 | *Malfa-Kraftma-Brot* | 529 |
| 2.5.6 | Pumpernickel | 530 |
| 2.5.7 | Wheat or rye wholemeal | 532 |
| 2.5.8 | Bran-bread with raisins | 534 |

# PART 3. THE BAKING PROCESS

3.1 Aims and Requirements of the Baking Process . . . . 539
3.2 Elements of the Baking Process and their Control . . . 553
    3.2.1 Influence of the elements on the dough-piece and its components . . . . . . . . . 579
3.3 Energy Sources, Types of Oven and Oven Design . . . . 596
    3.3.1 Energy sources . . . . . . . . . 596
    3.3.2 Types of oven and oven design . . . . . 602
3.4 Control Technology and Energy Recovery . . . . . 620
3.5 Bread Cooling and Setting . . . . . . . 638
    3.5.1 Types of bread cooler . . . . . . . 639
    3.5.2 Moisture movements during the cooling and maturation of bread . . . . . . . . . . 641
    3.5.3 Bread quality changes due to storage . . . . 649
3.6 Dough and Bread Preservation . . . . . . 679
    3.6.1 Retardation of staling processes . . . . 679
    3.6.2 Prevention of microbial infection . . . . 682
    3.6.3 Dough preservation by freezing . . . . 689
    3.6.4 Bread preservation by freezing . . . . . 708
3.7 A Preview of the 1990s and Changes in Product Demand and Supply . . . . . . . . . . . 715

# PART 4. NOTES AND REFERENCES

4.1 Notes and References for Part 1 . . . . . . 721
4.2 Notes and References for Part 2 . . . . . . 745
4.3 Notes and References for Part 3 . . . . . . 764

*Index* . . . . . . . . . . . 837

# PART 1

**Fundamental Dynamics of the Mixing Process, and their Implications for Dough Rheological Behaviour, Process Control and Optimization**

# 1.1 Theoretical Model to Explain the Doughmaking Process

The chemical, physical and biochemical changes which the contents of the dough undergo during the assembly process are very complex. Therefore, model ideas are conceived, which attempt to separate the various phases, although it is accepted that in practice any differentiation in terms of time is impossible. Under production conditions, these phases overlap, irrespective of whether the process is discontinuous or continuous in execution.

*First Phase of Dough Formation*
The first phase of dough formation mainly involves the moistening of the flour particles. The dough water becomes dispersed between the flour particles by the movements of the mixer elements and/or bowl. The hydrophilic properties of the particles adsorb liquid onto their surface, which results in the shearing elements of the mixer developing a high mechanical moment to counter the forces of adhesion. This effect can be measured as the first turning moment maximum, which, in the case of the Brabender Farinograph, is expressed as consistency units. During the first phase of dough formation, the most important material variables are the surface properties of the flour particles, viz. shape, size, and adsorption properties, and the properties of the water, salt, and other additives and their temperature. The process variable is the rate of shear or shear gradient, which is the differential of speed with respect to time, and is determined by the number and geometry of the mixing elements.

*Second Phase of Dough Formation*
During the second phase of dough formation the main processes involved are solubilization and swelling. As the mixing proceeds, the water-soluble flour components, amounting to 3–5% of the total flour weight, become dissolved. About 9% of the protein (albumins and globulins), being water-soluble and on account of their swelling-power, form a colloidal solution by spontaneous peptization. Also, the 2–3% soluble carbohydrates (mono-di- and oligosaccharides) and mineral salts become taken up in the water-soluble phase. The resultant swelling of 75% of the protein (prolamines and glutelins) give a volume increase of 2–3 times, and 10–15% of the total flour weight is transformed into a hydrogel. The starch adsorbs water, the undamaged fraction taking up 30% of its own weight, resulting in a volume increase of about 75%, and the damaged fraction binding 20–40% more water than the undamaged fraction. Of the flour pentosans (1–2% of the flour weight), 25% are soluble, and swell to a level of up to 800% by volume. The swelling ability of the

3

TABLE 1

**Relationship between the processing stages and the changes in the composition and structure of various dough components**

| Process stage | Raw material / Additive / Intermediate product | Process function |
|---|---|---|

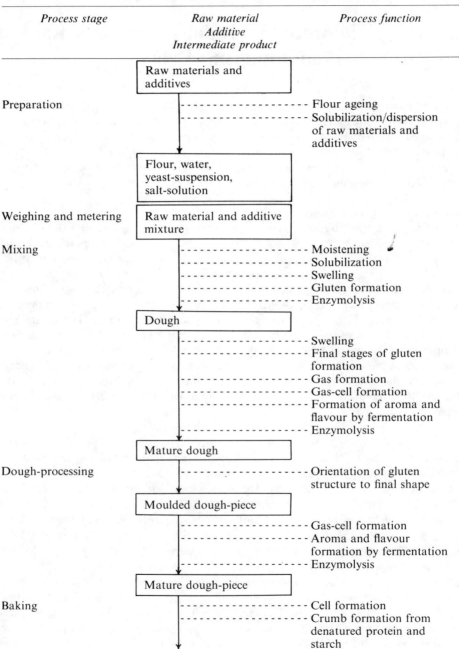

| | Raw materials and additives | |
| Preparation | ------------------- | Flour ageing |
| | ------------------- | Solubilization/dispersion of raw materials and additives |
| | Flour, water, yeast-suspension, salt-solution | |
| Weighing and metering | Raw material and additive mixture | |
| Mixing | ------------------- | Moistening |
| | ------------------- | Solubilization |
| | ------------------- | Swelling |
| | ------------------- | Gluten formation |
| | ------------------- | Enzymolysis |
| | Dough | |
| | ------------------- | Swelling |
| | ------------------- | Final stages of gluten formation |
| | ------------------- | Gas formation |
| | ------------------- | Gas-cell formation |
| | ------------------- | Formation of aroma and flavour by fermentation |
| | ------------------- | Enzymolysis |
| | Mature dough | |
| Dough-processing | ------------------- | Orientation of gluten structure to final shape |
| | Moulded dough-piece | |
| | ------------------- | Gas-cell formation |
| | ------------------- | Aroma and flavour formation by fermentation |
| | ------------------- | Enzymolysis |
| | Mature dough-piece | |
| Baking | ------------------- | Cell formation |
| | ------------------- | Crumb formation from denatured protein and starch |

TABLE 1—*contd.*

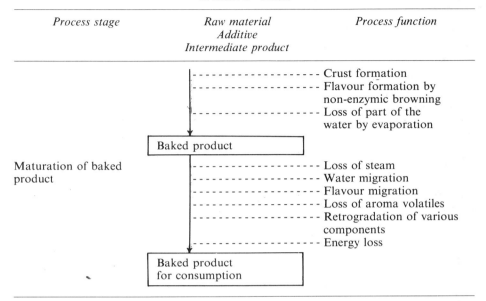

| Process stage | Raw material Additive Intermediate product | Process function |
|---|---|---|
| | | Crust formation |
| | | Flavour formation by non-enzymic browning |
| | | Loss of part of the water by evaporation |
| | Baked product | |
| Maturation of baked product | | Loss of steam |
| | | Water migration |
| | | Flavour migration |
| | | Loss of aroma volatiles |
| | | Retrogradation of various components |
| | | Energy loss |
| | Baked product for consumption | |

Source: *Technology of Industrial Baking*, R. Schneeweiss and O. Klose, VEB-Fachbuchverlag, Leipzig, 1981.

starch fractions is more rapid than that of the proteins. The influence of the water-solubles reduces the turning moment of the mixer, whereas the limited protein, and starch swelling both result in increases in the turning moment.

The dissolution and swelling processes described overlap one another in both time and space, having a complex effect on the rheological behaviour of the dough, which manifests itself in the form of a second turning moment maximum. On reaching this point, the solubility and swelling reactions are complete. Solubility and swelling depend on both raw materials and mixing parameters. The percentage damaged starch, the protein solubility, the pentosan swelling and the activity of flour enzymes, all exert an important influence on dough rheological behaviour. Other influences of importance are: flour particle size and morphology, mineral salt content, and the presence of salt, added enzymes, oxidants, reducing agents and emulsifiers.

The water content of the dough affects the formation of the second turning moment maximum. A reduction in added water increases the cohesion, hindering the swelling of the dough components, and the potential swelling-time. Higher water contents require more mixing to reach the optimum dough consistency.

Increases in the shear gradient (rate of shear) result in corresponding increases in the rate of swelling, but there is a danger that with extremely high mixing intensities protein damage could give rise to reduced swelling capacity.

Temperature exerts a considerable influence on the rheological properties of the dough. Dough temperature depends on the temperature of the raw materials, the mechanical energy absorbed during mixing, and the frictional heat energy dissipated through the dough. The solubilization and swelling processes also generate heat owing to hydration reactions, which are exothermic. Doughs set at about 30°C result in optimal rheological properties for further processing. Temperatures much in excess of 30°C reduce swelling and dough viscosity.

*Third Phase of Dough Formation*
The third phase of dough formation is essentially one of restructuring of the gluten proteins, the swollen flour proteins being changed by the energy input of mechanical mixing into polypeptide chains, which are aligned as more linear film-forming molecules. Progressive energy input results in the initiation of chemical reactions, the most important of which is the breaking of disulphide bonds and their reformation, the formation of hydrogen bonds and hydrophobic bonding. Chemical bonding during doughmaking is discussed in detail in Section 1.6.

As a result of the continuous building of intermolecular and intramolecular disulphide bonds, the gluten structure takes form, and a viscoelastic rheological system emerges. At this stage, enzymic degradation reactions begin to manifest themselves, and dough consistency starts to fall, which is a signal to terminate the mixing operation. The determining factor here is the flour quality, and the content of available swelling-proteins to provide the disulphide groups for cross-linking. The formation of the gluten complex structure can be influenced by oxidants and reducing agents, the former acting as stabilizers and the latter as weakeners of the structure.

The addition of oxidants tends to block the reactive sulphydryl groups after the exchange of disulphide groups at the completion of gluten formation; thus by blocking the residual reactive sulphydryl groups the possible degradation of the gluten with increased shear is reduced. The use of oxidants is only justified either where the flour has insufficient gluten-forming ability or where its qualitative properties are in doubt.

Summarizing, the mixing of wheat-flour, yeast, salt and water into a dough involves the control of the following variables: water-content, flour quality, mechanical energy input, shear gradient, and duration of the shear force on the dough.

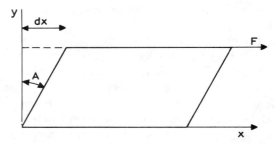

Fig. 1.   Model of shear deformation. Tangential force, $F$. Relative shear deformation, d$x$.
Distance between displaced layers, $y$. Angle of shear, $A$.

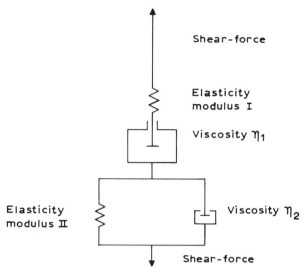

Fig. 2. Combined rheological model of dough behaviour according to Schofield and Scott-Blair. Dough mixing and mechanical manipulation results in an overlap of elastic, viscous and plastic deformation.

Dough is a dispersion of three phases. The solid phase is the gluten network with starch and the other insoluble components embedded in it. The liquid phase consists of the dough water and the water-soluble dough components. The third phase is the entrapped gases adsorbed by the raw materials during the mixing operation. Rheologically, the dough is regarded as a viscoelastic mass with non-Newtonian flow characteristics, although it also develops plastic properties.

The forces acting on a dough during the mixing process and their inter-relationship are as follows (see Figs 1 and 2). Shear is a deformation in which all the layers of an elastic body parallel to a given plane are displaced relative to one another. During shear, the volume of a body remains unchanged. Where $dx$ is the relative deformation due to a tangential force, $F$, a displacement is set up tangentially called the shear displacement, $P$. The relative shear deformation set up is therefore given by

$$S \frac{dx}{y} = \tan A$$

where $y$ is the distance between the displaced layers, and $A$ the angle of shear with the perpendicular.

The differentiation of the relative shear with respect to duration (mixing time), $t$, gives the acceleration, which in this case is the 'rate of shear', or 'shear gradient':

$$Sg = \frac{dS}{dt}$$

According to Newton, the shear displacement force, $P$, is proportional to the speed gradient, viz. acceleration, which in this case is the shear gradient, $Sg$:

$$P = \eta Sg$$

where $\eta$ is the coefficient of internal friction, commonly known as viscosity. The magnitude of the force of internal friction is

$$F_{fr} = \eta \frac{dv}{dl} S$$

where $dv/dl$ is the rate of change of velocities of the layers with the distance separating them, and $S$ the area of contact of the layers. The coefficient, $\eta$, is defined by the force of friction existing between two layers of unit area, with $dv/dl$ ratio equal to unity.

On application of a mechanical shear force to the dough the elastic mode 1 comes into operation, the energy being transferred to the parallel modes, elastic mode 2 and viscous mode $\eta_2$, but owing to the inertia of the viscous mode it absorbs the shear force. As a result of the stretching of the viscous mode $\eta_2$, the parallel elastic mode 2 is activated in the form of an 'elastic after-effect', or retardation. After overcoming the threshold displacement, which hitherto prevented the activation of $\eta_1$, energy is transmitted with the help of this element. As a result of the effect of $\eta_1$, the deformation subsides into the 'relaxation' mode in the elastic element II.

# 1.2 Application of Fundamental Dough-Mixing Parameters

For a given flour quality profile, and formulation, there is a specific mixing intensity which should be imparted to the dough mass to ensure optimal dough development. Optimal dough development will result in the best possible dough viscosity, as a direct consequence of the attainment of the best compromise between the various processes of molecular build-up and breakdown which proceed simultaneously within the dough. The simplest method of achieving this is by adjustment of the mixing time, rather than by utilizing a fixed energy input. Specific mixing intensity in joules per kilogram per second ($J k^{-1} s^{-1}$), is used as the integral parameter for the evaluation of the mixing process. Its value will depend upon the mixing speed, in rpm, and mixing time. The optimal specific mixing intensity is directly correlated with the effective dough viscosity measured at constant rate of shear. A convenient rate of shear for dough-effective viscosity measurements is $0.167 s^{-1}$, at any chosen mixer-rpm. The most suitable type of viscometer for doughs is the rotation viscometer design, available from Brookfield (USA) and Haake (West Berlin), and the Rheotest 2, manufactured in the GDR by VEB Prüfgeräte. Shear rate ranges of $0.1$ to $10 s^{-1}$ are adequate for most dough viscosity measurements.

If the rotational speed is varied, as in the case of the Rotovisco RV-3 of Haake or the Brookfield LVF viscometer, a viscous flow curve is obtained of viscosity *v*. rpm, called a rheogram. In this latter case, readings are proportional to viscosity.

## 1.2.1 PRACTICAL APPLICATION OF DOUGH-MIXING PARAMETERS FOR THE BAKER

The best possible dough properties and potential bread quality relative to mixer-rpm and mixing time, correlates or corresponds with the highest or peak value for effective dough viscosity in SI ($ML^{-1} T^{-1}$) units measured at constant shear rate of $0.17 s^{-1}$. Furthermore, this peak viscosity value corresponds with the specific mixing intensity (in joules per kilogram per second) maximum at any chosen mixer-rpm and mixing time. Therefore, the baker, with a knowledge of the rpm of his mixer, e.g. 150 or 200 rpm, can take dough samples at intervals of 60, 120, 180, 240, 300 and 360 s and establish the optimal mixing time by measurement of the effective dough viscosity at a constant shear rate of $0.17 s^{-1}$. The highest or peak viscosity value obtained in viscosity units will correspond to the optimal mixing time. If the mixer is fitted with an energy recorder, and the specific mixing intensity in joules $kg^{-1} s^{-1}$, is measured at the same time intervals, the maximum or peak specific mixing intensity

9

will correspond with the same mixing time as that previously found to be optimal for effective dough viscosity. Therefore dough viscosity, a fundamental physical variable, is directly related, at any given mixer-rpm, to the specific mixing intensity, when the optimal mixing time is reached. Mixing time is used to optimize the dough properties (see Figs 3, 4, 7 and 9). This relationship holds good at any mixer-rpm within the normal dough-mixing range of 50 to 250 rpm, irrespective of the mixer geometry, dough formulation, method of dough deformation, and flour strength, assuming that the flour is suitable for breadmaking. Data obtained by the author during the 1970s, using the Do-corder and the same experimental design, but not published until 1986 (*Getreide Mehl und Brot* **40**, 10, 1986, pp. 294–297) have confirmed these fundamental dynamics of the discontinuous mixing process. However, in all cases the dough temperature must be maintained constant during the mixing process. A comparison of the graphical relationships established between these fundamental mixing parameters by both the USSR/GDR researchers and the author confirm their validity.

In the test bakery, these experiments can be carried out with the help of a Brabender Do-corder and any suitable rotation viscometer at a shear gradient of $0.17\,s^{-1}$. However, any baker interested in optimizing the properties of his doughs could run these tests on his commercial mixer, providing that it is fitted with an energy recorder and timer. In order to avoid excessive dough handling, it would be preferable to use a suitable viscometer built into the mixer, thus allowing viscosity measurements to be made *in situ*.

# 1.3  Fundamental Considerations Concerning Dough Rheological Elements and Dynamic Mixing Parameters

At the beginning of mixing, the dough formulation components form agglomerates as a result of their mobility, damping, swelling and solubilization. These agglomerates become distributed at a speed dependent on the shear gradient of the mixer. The timing of these processes within the dough mass proceed more evenly, and parallel to one another, when the components become mixed with each other to form the thinnest possible layers as rapidly as possible. This process results in the formation of hydrogen bonds, and other intermolecular cross-linking, and explains why water binding occurs more rapidly with increasing number of turns of the mixing-head, as does the increase in dough effective viscosity. The increase in effective viscosity, using a constant number of revolutions, results in a higher turning moment, or torque, and shear gradient. Since the specific mixing intensity is a function of the number of revolutions, shear gradient and mass of dough, the mixing intensity changes considerably with changes in the revolutions of the mixing-head and shear gradient, with a constant dough mass. Mixers of different geometrical construction as regards mixing-elements and mixer bowl design working at the same rpm will not therefore deliver the same mixing intensity. The commonly used concept of expending a fixed amount of energy on or work input to the dough mass cannot be regarded as a reliable dynamic mixing parameter. The reason is that the energy expended, depending on mixing intensity and mixing time, contributes not only to dough structural build-up, but also to irreversible structural breakdown and rheological deformation of the dough. This latter process leads to energy being transformed and dissipated as heat and cannot be regarded as cost-effective or energy-effective.

The effective utilization of expended mechanical energy, depends on its application and control, which involves the rate of shear or shear gradient delivered by the mixer, the effective dough viscosity, the nature of the deformation and the duration of the regime. The necessary specific work input to achieve optimal dough properties, using shorter mixing times at higher mixer-rpm, is always greater in terms of electrical energy expended.

With increasing mixing intensity, the optimal mixing time decreases and must be strictly controlled to avoid irreversible viscous and plastic deformation of the dough structure. The practical significance of these considerations is depicted in graphical form, from data obtained by USSR/GDR collaborative research work, and by the author (see Figs 3 to 6 and 7 to 10).

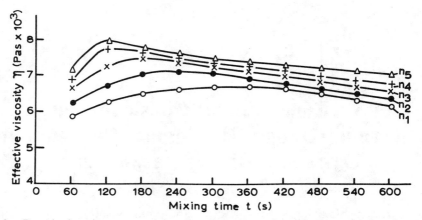

Fig. 3.  Dough viscosity versus mixing time at various mixer rpm (measured at shear-rate $0.167 \, s^{-1}$. $n_1 = 50$ rpm, $n_2 = 100$, $n_3 = 150$, $n_4 = 200$, $n_5 = 250$. (Source: Influence of mixing intensity and mixing-time during discontinuous (batch) mixing of wheat dough, W. J. Tschernych, L. J. Putschkowa, D. I. Ljaskowskij, M. B. Salapin, USSR and H.-D. Tscheuschner, *Bäcker und Konditor* **1**, 1985, pp. 22–4 (in German).)

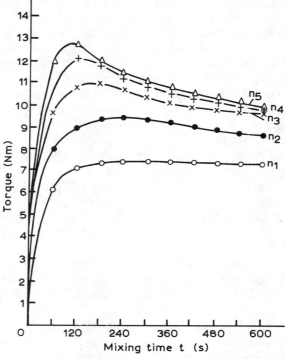

Fig. 4.  Mixer torque versus mixing time (at various mixer rpm). $n_1 = 50$ rpm, $n_2 = 100$, $n_3 = 150$, $n_4 = 200$, $n_5 = 250$. (Source: Influence of mixing intensity and mixing-time during discontinuous (batch) mixing of wheat dough, W. J. Tschernych, L. J. Putschkowa, D. I. Ljaskowskij, M. B. Salapin, USSR and H.-D. Tscheuschner, *Bäcker und Konditor* **1**, 1985, pp. 22–4 (in German).)

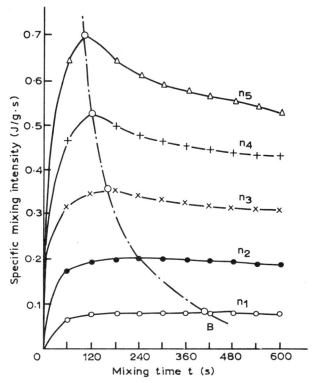

Fig. 5. Specific mixing intensity versus mixing time (at various mixer rpm). $n_1 = 50$ rpm, $n_2 = 100$, $n_3 = 150$, $n_4 = 200$, $n_5 = 250$. (Source: Influence of mixing intensity and mixing-time during discontinuous (batch) mixing of wheat dough, W. J. Tschernych, L. J. Putschkowa, D. I. Ljaskowskij, M. B. Salapin, USSR and H.-D. Tscheuschner, *Bäcker und Konditor* 1, 1985, pp. 22–4 (in German).)

The peak mixing times to attain optimal dough properties, in terms both of viscosity and of specific mixing intensity, utilizing various mixer-rpm, are clearly indicated in the figures. At these mixing times, dough porosity and bread quality scores are found to be optimal. The stages of dough processing at which dough properties become crucial are: dividing, final moulding and proofing. Therefore, it is of fundamental importance to aim for optimal dough properties at the outset, to avoid machining problems, due to 'sticky' doughs, at a later stage.

Recent advances in sensor technology have provided optical inline torque transducers, which now find wide application for inline process control of viscosity. A wide range of these transducers measure both speed and torque of the mixing-element. This enables measurement of the viscosity using either laboratory mixers or full-scale production batches. Available outputs allow the system to be applied to process control situations where ingredients are added to the mix until the desired viscosity is achieved. These optical transducers, available in torque ranges from 10 mN m to 5000 N m, are mounted in line with the mixing element drive shaft.

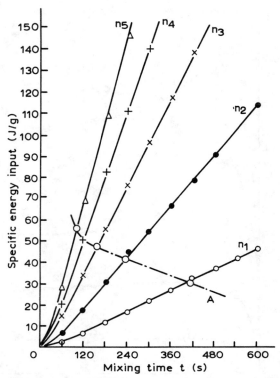

Fig. 6. Specific energy input versus mixing time (at various mixer rpm). $n_1 = 50$ rpm, $n_2 = 100$, $n_3 = 150$, $n_4 = 200$, $n_5 = 250$. (Source: Influence of mixing intensity and mixing-time during discontinuous (batch) mixing of wheat dough, W. J. Tschernych, L. J. Putschkowa, D. I. Ljaskowskij, M. B. Salapin, USSR and H.-D. Tscheuschner, *Bäcker und Konditor* 1, 1985, pp. 22–4 (in German).)

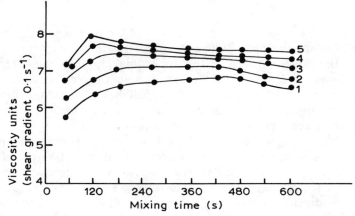

Fig. 7. Dough viscosity versus mixing time (measured at shear rate $0.175^{-1}$). $1 = 50$ rpm, $2 = 100, 3 = 150, 4 = 200, 5 = 250$. (Source: Dynamics of the discontinuous mixing process and its rationalization, C. A. Stear, *Getreide Mehl und Brot* **40**, 10, 1986, pp. 294–7.)

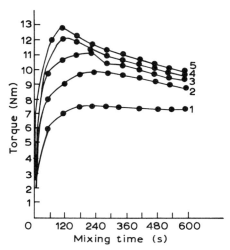

Fig. 8.   Mixer torque versus mixing time. 1 = 50 rpm, 2 = 100, 3 = 150, 4 = 200, 5 = 250. (Source: Dynamics of the discontinuous mixing process and its rationalization, C. A. Stear, *Getreide Mehl und Brot* **40**, 10, 1986, pp. 294–7.)

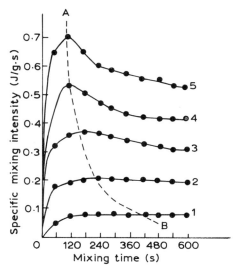

Fig. 9.   Specific mixing intensity versus mixing time. AB, line joining points at optimal dough development (properties). 1 = 50 rpm, 2 = 100, 3 = 150, 4 = 200, 5 = 250. (Source: Dynamics of the discontinuous mixing process and its rationalization, C. A. Stear, *Getreide Mehl und Brot* **40**, 10, 1986, pp. 294–7.)

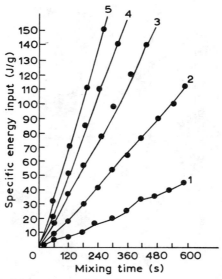

Fig. 10.   Specific energy input versus mixing time. $1 = 50$ rpm, $2 = 100$, $3 = 150$, $4 = 200$, $5 = 250$. (Source: Dynamics of the discontinuous mixing process and its rationalization. C. A. Stear, *Getreide und Brot* **40**, 10, 1986, pp. 294–7.)

Torque variations within a fraction of a revolution can be measured, and transients as short as 1 millisecond may be detected and recorded. Torque, speed and power measurement systems based on the optical rotary torque transducer can provide digital display of these parameters, and differential analogue outputs for recording or logging. Adjustable limit setting on torque, speed and power can provide instant visual, audible and electrical alarm signals for rapid shut-down of the drive source. The extremely fast frequency response of this optical measuring system, enables torque transients as short as 0·1 ms to be monitored and recorded, the peak values being held in memory. The current cost (1989) of such a measurement system is in the region of £40 000 sterling. Basic flour quality parameters must be discussed and agreed between miller and baker; they include: moisture, gluten quantity and quality, damaged-starch content, and alpha-amylase activity, i.e. Falling number of about 250. Percentage water absorption or Farinograph dough consistency at the required dough yield in the bakery is also a desirable parameter to specify.

Dough is a filled-polymer system consisting of starch, protein and pentosans, filled with water, some of which becomes molecularly bonded with the help of small amounts of surface-active lipoprotein and glycolipids formed during mixing, the rest remaining in a free state as layers superimposed on the bonded molecules. These water molecules are an essential solvent for the soluble flour components and dough additives. The degree of freedom of these molecules increases with their remoteness from bonding-force sites within the dough mass. In practical terms, too much free water results in a slack dough, too little in a tight or bound dough. An optimized mixing regime will result in the best possible distribution and water-binding of the structural components of the dough.

Chemical and colloidal properties of these components will determine the relative thickness of the layers of the water molecules and contribute to dough viscosity. Proteins, for example, have a large molecular surface in proportion to their mass, therefore flours with high protein content and good swelling properties give more viscous doughs than those with less protein and limited swelling capacity, when made with the same quantity of water. However, the physical properties of proteins and flour particle size also influence swelling and dough viscosity. In rheological science, viscosity represents an important element, and any changes in dough viscosity are a direct consequence of the water-binding capacity of the dough components at mixing and their structural build-up. During fermentation and processing, the autolytic processes of degradation take over, and progressively reduce viscosity. The other important rheological element is spring, or elasticity, which is described as the Hookean element. Halton and Scott-Blair (1936 and 1937) determined the viscosities and elastic moduli of a large number of wheat doughs, and concluded that these two rheological elements had a profound influence on baking quality. However, they pointed out that these values could not be regarded as constants, since their magnitudes depend on the forces of stress and strain applied to the dough. Both the effective viscosity and elastic modulus values fall with increasing water content, time and temperature. Investigations using a wide range of flour qualities at several water absorptions showed similar elastic moduli; but considerable variations in viscosity were apparent. This close relationship between water absorption and a constant elasticity modulus was confirmed by Halton and Scott-Blair in later work; furthermore, they concluded that the ratio of viscosity to elastic modulus was of fundamental importance for baking performance. They referred to it as 'Maxwell's relaxation time', based on a mechanical model consisting of an elastic element and a viscous rheological element combined in series, together forming the Maxwellian model. In this case, the relationship between the applied force and the displacement or deformation can be expressed as the differential equation:

$$\frac{\mathrm{d}s}{\mathrm{d}t} = \frac{1}{E}F + \frac{1}{n}F$$

where $\mathrm{d}s/\mathrm{d}t$ is the differential of displacement $s$ with respect to time, $E$ is the elastic or spring constant (analogue Young's stress/strain modulus of elasticity), $F$ is the force applied to the elastic element and $n$ is the viscosity.

Lerchental and Müller (1967) also used the mechanical model approach, using the elastic and viscous element to describe the viscoelastic behaviour of wheat-flour doughs. The $n/E$ ratio is closely related to the dough characteristic of 'spring', and for bread of good quality a high $n/E$ ratio is desirable, i.e. a high relaxation time. The baker often refers to this factor as 'work hardening', the capacity of a dough to recover after mechanical forces have been applied to it. Between mixing and baking, as a result of swelling and enzymic breakdown, the rheological properties are in a constant state of change. Any mechanical force applied to the dough results in structural activation, and the resistance to deformation will increase proportionally and the apparent viscosity fall. The change in resistance with time or 'relaxation', which is reversible, is important for dough ripening and fermentation tolerance, passing through all stages from shortness to an optimal value, and then into flow and eventual 'stickiness'. Where a high degree of mechanization is employed at the make-up stage, these changes must be carefully controlled. Stress relaxation was defined by Hlynka and Anderson (1952), as the fall in stress of a stretched piece of dough; whereas structural relaxation describes any loss of resistance which a dough piece undergoes on resting, when subsequently stretched to a defined length, according to Dempster, Hlynka and Anderson (1953). Dough relaxation curves can be determined for both yeasted and non-yeasted doughs by using the Brabender Extensograph. Doughs are prepared in the usual way, and the strips stretched after rest (relaxation) periods of 10, 20, 30 and 40 minutes. In each case, the resistance is read off at a definite extensibility point between 3 and 11 cm, and these values plotted against the relaxation times. Another instrument designed to measure elastic deformation manufactured by Instron of High Wycombe, Bucks, UK, can also be used to measure dough relaxation times. Frazier, Brimblecombe, Daniels and Russell-Eggitt (1979) used it to relate relaxation time to expended energy input and baking properties. They established high correlations between maximum values for energy input (kJ kg$^{-1}$), relaxation time and bread quality, as illustrated in Figs 11 and 12.

Rheological properties form an important part of overall texture evaluation, which is of great importance at all manufacturing stages, and ultimately determines consumer acceptability. Available test instruments range from fundamental to empirical in their units of measurement. Fundamental test properties of materials are established in terms of well-defined physical constants, whereas empirical techniques deliver data that is difficult to define, since it is obtained under arbitrary conditions. Empirical techniques have found wide application as quality control tools. For more basic research applications, the modified Extensigraph with strain-gauge recording system and variable hook-speed made by the Canadian Engineering Research Service, Ottawa, Ontario, Canada, and the Instron universal testing machine with cross-head attachment, of Instron Limited, High Wycombe, Bucks, UK. The Instron can be utilized for stress/strain

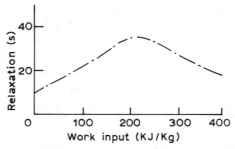

Fig. 11. Relaxation versus work input for mechanically developed doughs. (Source: Comparison of rheological data and baking quality of mechanically developed doughs, Frazier, Brimblecombe, Daniels, and Russell-Eggitt, *Getreide Mehl und Brot* **33**, 10, 1979, pp. 268–71.)

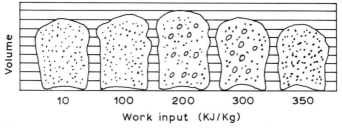

Fig. 12. Volume versus work input for mechanically developed doughs. (Source: Comparison of rheological data and baking quality of mechanically developed doughs, Frazier, Brimblecombe, Daniels, and Russell-Eggitt, *Getreide Mehl und Brot* **33**, 10, 1979, pp. 268–71.)

studies on doughs at different extension rates. The greater accuracy of the Instron stress/strain curves at very low values of strain compared to those obtained from the modified extensigraph allow an evaluation of the curves in fundamental terms at low rates of strain where linear behaviour remains valid, for example, for such studies as flour oxidant treatments and the use of vital gluten and starch additives, where sensitivity is important. At higher rates of deformation, both Instron and Extensigraph curves are suitable for graphical transformation into other forms of stress/strain relationships, e.g. deformation and flow curves, deformation curves relating dough extension to time at constant stress and flow curves relating the velocity of deformation to stress.

Attempts to explain the rheological behaviour of dough and gluten in terms of molecular structure have involved the application of theoretical thermodynamic equations as found applicable to swollen elastic polymers or elastomers, e.g. collagen. In simple tensile mode, applied to wheat gluten, Treloar's equation is:

$$f = \frac{RT\,dv^{1/3}}{M_c}\left(a - \frac{1^2}{a}\right)$$

where $M_c$ is the average molecular weight of the chain between the protein cross-links, $a$ is the elongated length of the swollen polymer divided by its initial length, $d$ is the density of the dry material, $R$ is the gas constant, $T$ the absolute temperature, and $f$ the tensile stress. The volume fraction, $v$, is given by:

$$\frac{W_1 q}{W_2 p - W_1(p - q)}$$

where $W_1$, is the dry weight of the polymer, $p$ its dry density, $W_2$ its wet weight, and $q$ the liquid density. Since two cross-linked chain segments constitute each cross-link, the number of cross-links per unit mass is $1/2M_c$. Since the molecular weight is of the order of 100 000 in the case of gluten, the number of cross-links can be expressed in molecular weight units of $10^5$. The tensile stress operates over the whole cross-sectional area of the dough; considering the cross-sectional area of the gluten which is the continuous phase polymer of the dough system, the relative percentage ratio of gluten to the dispersed water, starch, and pentosan fractions of the dough become important. Of equal importance is the ratio of free to bound water which exists within the dough mass after mixing, and how this water is distributed between liquid phase, pentosans, gluten and starch. Furthermore, Treloar's derived thermodynamic equation for elastomers is only valid for one molecular network, in this case gluten, and although gluten constitutes the dough matrix, the influence of the other dough polymers, and glycoprotein, lipoprotein and glycolipid, and also flour and added lipids, cannot be overlooked. The filled-polymer theories at present advanced cannot be used to explain dough structure with satisfaction, but elastomer reinforcement theories are providing a sound mathematical basis for the explanation of the fundamental phenomena of dough behaviour.

More recent studies on wheat gluten proteins by Lutsishina and co-workers in the USSR (1984), using wide-band proton NMR spectroscopy, have revealed wide differences in the conformational state of protein complexes. Glutens derived from hard wheats show a more compact structure with stronger hydrogen bonding than glutens from soft wheats. Positive correlations between NMR parameters and rheological and technological properties of wheats were established, which serve as reliable indices for wheat quality determinations. Other approaches, using electrophoretic protein spectra, in Hungary by Örsi and Pallagi-Bánkfalvi (1985) have demonstrated that (sodium decyl sulphate) SDS-electrophoresis, combined with densitometric interpretation and a statistical approach, can predict valuable information concerning gluten, dough and baking properties. Such studies depend on the relative intensities of certain spectral-bands compared with the total protein picture or 'finger-printing', and are useful for wheat-flour quality 'typing'.

Selective degradation studies on wheat glutenin by Kawamura and co-workers (1985) under various conditions have shown that glutenin is a linear polymer-protein. All isolated sub-units contain two molecules of inter-polypeptide—SS—links per molecule of sub-unit, and in every case 2–3 molecules of intra-peptide — SS— links. Therefore, the glutenin fraction is made up of many different sub-units, which are linearly linked by —SS— bonds. Moonen, Scheepstra and Graveland (1985) found that any residual cystein-groups appear at the end of the glutenin molecules, but owing to the amino acid sequence analysis, and the reactivity of the residual cystein-groups, only a small number remain exposed on the outside of the molecule.

# 1.4  Water-Binding Capacity of Dough Components and Dough Consistency Control

The physical, colloidal and biochemical processes involved in dough preparation, depending on the mixing intensity and duration of mixing, produce a dough which increases in its viscosity to a maximum and then begins to fall off. At high mixing intensity, this maximum is reached within a shorter time than in the case of a lower mixing intensity, and the maximum value is correspondingly higher at higher mixing intensities than is the case at lower mixing intensities. An important reason for these viscosity changes in dough is the water-binding capacity of the flour components coupled with the processes of structural build-up and breakdown which take place within the system.

During the mixing process, the dough water added becomes distributed between the flour components (pentosans, gluten, starch), and the remainder forms the liquid phase. The structural properties of these components and their water-binding capacity, coupled with the duration and intensity of the mixing regime, will determine the dough properties and the ultimate bread quality.

With increasing mixing intensity, changes in the water-binding capacity of the dough components become more apparent, and the percentage remaining as the liquid phase will progressively diminish. For any dough formulation and mixing regime there exists an optimal water distribution between these four dough fractions at which dough apparent viscosity and relaxation-time, and therefore potential dough properties, become maximized. Any attempt to generalize concerning these water distribution optima would not be realistic, and there is no substitute for a carefully controlled experiment, *in situ*, using the available raw materials and mixing regime.

However, some examples of typical water distribution data obtained at dough moisture 46%, corresponding to a dough water absorption level of approximately 57%, obtained by Jazuba, Siderowa, Putschkova and Tscheuschner (1985) in USSR/GDR collaborative research, will illustrate the fundamental importance of these aspects of dough preparation. These research workers prepared non-yeasted doughs by mixing at the following speeds and times:

  50 rpm for 90, 660 and 900 seconds
 100 rpm for 60, 450 and 660 seconds
 200 rpm for 45, 200 and 480 seconds
 250 rpm for 30, 165 and 360 seconds

Samples of each dough were ultracentrifuged at 26 000 rpm, $10^5$ g for 110 minutes, ensuring equilibrium. Similar techniques having been used by Baker, Parker and

Mize (1946). Since microbiological processes play no part in doughing-up, (Auerman, 1972; and Kosima, 1978), non-yeasted doughs were prepared for these tests. The 21 ml centrifugates yielded the following fractions:

*Fraction 1:* A liquid phase containing lipids, and water-soluble dough components such as sugar, salts and water-soluble proteins.

*Fraction 2:* Pentosans, which form a highly viscous solution or gel.

*Fraction 3:* Hydrated gluten, consisting of proteins in a high state of hydration, form the major component of this fraction. However, this fraction also contains a large amount of starch, which even a centrifugal force of $10^5$ g does not completely separate from the gluten proteins. Therefore, the starch granules must be firmly bound to certain protein components of gluten.

*Fraction 4:* Starch without free water.

The mixing regime, as determined by mixing time and intensity, has an important influence on relative contents of these four fractions in the dough. In this respect, the liquid phase shows the greatest variation. Depending on the rpm of the mixing elements and the mixing time, the liquid phase can range from 0% to 14%, and shows a maximum of 9·5% to 14·0% in the case of doughs which have received in-sufficient mechanical work-input. With increasing mixing time the content of the dough liquid phase diminishes, and at an increasingly rapid rate as the rpm of the mixer is increased. As the mixer-rpm is increased from 50 to 200, a reduction in the percentage liquid phase from 3·4 to 10·1, based on dough weight, takes place. Furthermore, at mixer-rpm rates of 200 and 250, and excessive mixing times, the liquid phase becomes progressively inseparable from the other fractions.

Parallel to changes in the liquid phase during dough-mixing, changes in the contents of the other components occur. The pentosan content of the dough can vary depending on the mixing regime from 6·4% to 14·0%. However, in the case of pentosan, the change is more uniform in character, and increases progressively with increasing mixing time, independent of the rpm of the mixing elements. Nevertheless, the largest increase in pentosan occurs at the highest rpm level at mixing. For example, the amount of the pentosan fraction in the dough, when mixed at 100 rpm for the optimal mixing time of 450 s, and then for an excessive 660 s, increases by 1·3%. Using a mixing-rpm of 250, mixing for 450 and 660 s, the pentosan fraction rises by 4·0%.

Gluten and starch are quantitatively the main components of wheat flour doughs. The gluten fraction of the dough varies from 23·6% to 27·0%, and the starch fraction from 55·0 to 60·1%, and both fractions increase with mixing time and mixing intensity.

These researchers suggest that most of the liquid phase which can be separated by centrifugation, consists of water from the structural network, residing in the capillaries and in the inter-micellar spaces in the free state, which does not change its properties appreciably. Any changes in content of the liquid fraction in the dough at constant dough moisture is due to an increase in the amount of moisture which becomes bound to the various dough components by adsorption or osmosis. This would account for the hydration capacity of the dough components changing during mixing. Considering the dough components in terms of their water-binding capacity, the pentosan fraction, due to its maximal hydration capacity, differs markedly from the other fractions. Under conditions of insufficient dough moisture, viz. 46% moisture, the water-binding capacity of the pentosans amounted to 80·0% to 85·8%. Under the same conditions, the water-binding capacity of the gluten ranged from 46·9% to 56·9%, and the starch from 29·8% to 31·7%. As the research data clearly illustrates, the water-binding capacity of the structural components of the dough depend to a large extent on mixing regime.

With increasing mixing time, at all rpm of the mixing-element, the water-binding capacity of the dough pentosans increase. Mixing at 50 rpm for 90 s, the water-binding capacity of the pentosans is 80·5%. If the mixing time is increased to 900 s, the water-binding increases to 81·6%. When the mixer-rpm is increased the water-binding capacity of the pentosans also increase. Although mixing at 50 rpm did not result in the water binding of the pentosans exceeding 81·6%, the use of 250 rpm raised it to 85·8%.

At the optimal mixing time, the pentosan content of the dough ranged from 8·4% to 8·6% practically irrespective of mixer-rpm. The water-binding capacity of the pentosans at optimal mixing regimes also showed only small variations, from 80·8% to 81·2%. The water-binding capacity of the starch separated from the dough increases with the mixing time during dough preparation. Mixing at 200 rpm, for example, for 45 s results in a water-binding capacity of 30·4%, whereas an extension of the mixing time to 480 s increases water-binding to 31·7%. In addition, the influence of the mixer-rpm is also apparent. Water-binding increases by only 0·3% at 50 rpm, but amounts to 1·5% at 250 rpm.

At optimal mixing time, the starch fraction differed in water binding depending on the mixer-rpm. From 50 rpm to 250 rpm the water binding of the starch increased by 1·0%. This clearly demonstrates the increase in adsorptively bound moisture on the starch granules at higher mixing speeds. This seems to indicate that during intensive mixing a partial structural change in the starch granule takes place, resulting in a marked increase in its total surface area and the number of small capillaries. Therefore, the end effect is a rise in the water-binding capacity.

However, the greatest change in water-binding capacity during doughmaking occurs in the extracted gluten fraction. Mixing at 50 rpm, the maximal water binding amounted to 49·9%; but at 200 and 250 rpm it increased to 56·8% and 56·9% respectively, thus demonstrating that the greatest effect on gluten hydration is the speed of the mixing elements. The effect of mixing time appears to be less dramatic: mixing at 50 rpm for insufficient time the water binding of the gluten was only 46·9%, at optimal mixing time 49·3%, and at overmixing a rise to 49·9% became apparent. The experimental data obtained by these researchers from the USSR and the GDR in collaboration, show very clearly that the water distribution and water binding in dough by its structural components depend not only on the hydrophilic properties of the flour particles, but also on the dough-mixing regime. On the basis of their investigations, the moisture distribution throughout the components of a wheat-flour dough of 46% moisture, and the analysis of the dough fractions after ultracentrifugation are summarized in Tables 2 and 3.

TABLE 2
**Distribution of moisture in the dough components**

| Dough fraction components | Moisture % based on total dough moisture | Optimal moisture level % based on total dough moisture | Optimal specific water-binding g/g |
|---|---|---|---|
| Liquid phase | 25·7...0 | 12·2 | — |
| Pentosans | 11·5...26·7 | 17·7 | 4·2 |
| Gluten | 23·0...34·0 | 29·7 | 1·2 |
| Starch | 36·0...43·0 | 40·4 | 0·5 |

Source: Jazuba *et al.*, *Bäcker und Konditor* **3**, 1985, p. 92.

Technical literature gives large differences in data concerning water distribution and binding in doughs, but most of the studies have been obtained at higher levels of dough moisture in countries where bread moisture levels in excess of 40% are permissible.

The data presented in Table 2 show not only the possible shift in the percentage water content of the various dough components, but also the optimal distribution of moisture in them. The optimal specific water binding in each dough component is defined as the state at which a dough will produce the best possible bread. The data also show the possible ranges of water binding in the various dough components, which are quite large. The largest water content (36·0...43·0%) is represented by the starch fraction of the dough, but the optimal specific water binding of the starch granules amounts to only 0·5g/g. The large quantity of water that the starch binds is explained by the fact it represents by far the largest quantity of all the dough components.

The gluten, under conditions of insufficient dough moisture, was able to bind

TABLE 3
**Analysis of the dough fractions after ultracentrifugation**

| Dough fraction: sensory evaluation | Content % dry matter | | | |
|---|---|---|---|---|
| | Protein | Starch | Pentosans | Rest |
| I Liquid, syrupy clear fraction | — | — | — | — |
| II Viscous, slimy white fraction | 16·0 | — | 60·8 | 23·2 |
| III Viscous, sticky, elastic yellow/brown fraction | 60·3 | 31·4 | — | 8·3 |
| IV Granular white fraction | 5·4 | 91·5 | — | 3·1 |

Source: Jazuba *et al.*, *Bäcker und Konditor* **3**, 1985, p. 92.

23·0...34·0% of the total dough moisture. The optimal specific water-binding for gluten being 1·2 g water for each gram of gluten dry matter, which is considerably higher than the specific water binding of starch.

The largest specific water binding capacity is exhibited by the pentosan fraction with an optimal value of 4·2 g/g. Although the pentosan content of the dough is small, it is capable of binding 11·5...26·4% of the total dough moisture.

In addition to raw material and mixing parameters, there are other external variables which exert an influence on dough consistency and water absorption. The stages of dough processing at which dough consistency is of crucial importance are dividing, final moulding and proofing. Depending on the processing system, this will occur at various times after completion of mixing cycles, e.g., 10–15 minutes for Chorleywood Bread Process (CBP), or 10–40 minutes for the chemically developed Activated Dough Development process (ADD), depending on whether a 10-minute floor (or bulk fermentation) time is allowed. In the case of such short-time processing systems, comparisons of the effect of various dough water additions can be made at appropriate intervals after mixing in order to detect any changes in consistency and dough softening. Relatively small differences in dough water absorption of less than 0·7% are probably insignificant. Where pre-fermentation processing systems are in use, as in North America, whether sponge or brew, adjustments can be easily made at the dough-up stage to final consistency, sponge-doughs tending to work-harden during a floor-time of 20 minutes. Similarly in France, dough mixing is carried out in two stages, final adjustments to water absorption being made at the last stage after a short relaxation period. In France, the salt is also added 5 minutes before the end of mixing to control swelling and facilitate consistency adjustments. Careful control of dough temperature is perhaps the most important single external variable affecting dough consistency during the doughmaking process, irrespective of the process or mixing regime. Dough temperature differences of 10°C, working doughs at 20 and 30°C, using the same flour and formulation, have shown differences in water absorption levels of 65% and 70%.

In general, doughs worked at lower temperatures also give better dough properties and final bread quality than those worked at higher temperatures.

In the case of mechanically developed short-time doughs, in order to maintain a constant dough consistency, reductions in water absorption of 0·3 gallons per 280 pounds of flour and a further 0·2 gallons per 280 pounds respectively, are required as the temperature of the dough water is raised from 80 to 85°F, and then to 90°F, using the same flour.

Since dough consistency is indirectly related to power consumption of the motor driving the mixer-shaft, recording wattmeters or ammeters are frequently utilized for this purpose. Whether analogue or digital, these instruments are often equipped with limit contacts, which either signal any fluctuations in power input which exceed acceptable tolerances, or automatically adjust the amount of water being fed to the mixer. In the case of batch mechanically developed doughs, such as those for the CBP, which are set up to work at a standard work input of 5 W/lb of dough, a consistency control unit is often used to compute consistency. The function of the consistency control unit is to maintain a target dough consistency by automatically adjusting the water addition to the dough whenever the average mixing time deviates from the target mixing time. The latter being used as a measure for target dough consistency.

The prerequisites for the successful operation of this programme with the consistency control unit are:

—A constant dough temperature must be maintained.
—The target mixing time is only valid for a given work input.
—If, in order to maintain dough temperature, it becomes necessary to change the work level, the target mixing time must be adjusted proportionally.

Any vacuum settings fitted to the mixer must be kept constant. The instrument records the mixing times of a series of doughs, and when the average deviates from a predetermined standard by more than a minimum of seconds, the water addition to the subsequent dough is either reduced or increased, according to whether the mixing time was too long or too short. Calibration of the unit is carried out by running at least 25 batches of dough of acceptable consistency at the same water addition, recording the mixing times in seconds. The average mixing time is calculated and fed into the instrument by selecting the appropriate latitude-level setting, which corresponds to the established mixing time range. When the cumulative values of target-mixing-time minus actual mixing time are positive, water is added; when they are negative water is reduced.

An average change in water addition of 1 lb for every 2 s accumulated in the memory of such a consistency control unit, could be found valid for several mixers of the same type using a standard doughmaking procedure.

Common sources of consistency variations in doughs due to raw materials are:

—Fluctuations in weight of flour delivered by automatic weighing units.
—Fluctuations in weight of water delivered by water-measuring devices, volumetric devices being less accurate than gravity-based units.
—Flour compositional variations affecting water absorption and binding, e.g. particle size distribution, physical and chemical structure of the constituent biopolymers as well as their bonding and interactions, uncontrolled enzymic autolysis, extraction rates.

In countries where no legislation exists concerning bread moisture levels, and procedures for the control of percentage damaged starch are undertaken by the miller, appreciable variations in dough water absorption can occur.

The important advantages of consistency control to the baker using any direct one-stage mixing system are:

—Reduction of problems with runny or sticky doughs.
—Uniformity of dough density for presentation to dividers, rounders, and moulder-panners.
—Continuous monitoring of the dough-up process and yields.

# 1.5 Effects of Dough Additives

### 1.5.1 CYSTEINE

Cysteine, used commercially in the form of L-cysteine–HCl, is a highly reactive reducing amino acid which is universally applied in biochemical dough development systems. Its reducing action accelerates the disaggregation by scission of disulphide bonds located between protein aggregates, therefore alleviating the mechanical mixing requirement for optimal dough development, and simultaneously saving energy. The reducing action starts as soon as it comes into contact with flour protein, and unlike the effect of oxidizing agents, appears to be less dependent on the mixing intensity. The free base form of cysteine is more effective in increasing protein solubility, or extractability, than the commercial HCl form, owing to greater reactivity of the SH groups at higher pH. However, the HCl form is more widely available and competitive in price.

Cysteine markedly reduces the energy required for peak dough development, and the critical minimal speed for optimal bread volume. The optimal level or range required depends on the flour protein content, but it also increases tolerance to undermixing where the work level is less than that needed to achieve peak dough consistency.

It has become a very versatile processing aid for modifying dough consistency and structure at the 'clearing' stage in practical breadmaking applications, including very diverse systems, from slow mixing to high-speed mixing conditions, whether batch or continuous and it allows great flexibility in bulk fermentation or floor-times. In warmer climates, it provides the baker with a useful means of controlling excessive dough temperatures.

For short-time mechanically developed batch processing, e.g. CBP, useful reductions in mixing times, that increase the number of doughs per hour, and in total work inputs can be achieved. However, if work inputs fall below about 3 W/lb against the normal 5 W/lb over 170 s, unacceptable gas cell distribution may result. The creation of a large number of small gas cells for later expansion by the yeast largely determines the crumb structure of mechanically developed dough. In such systems L-cysteine–HCl levels of up to 25 ppm (based on flour weight), depending on flour strength, normally within the range of 11–12%, are appropriate.

In the case of Activated Dough Development (ADD) systems, L-cysteine–HCl levels of 35 to 40 ppm can be used with CBP-type mixing equipment and a baker's grade ordinary flour treated at the mill with up to 20 ppm potassium bromate, having a protein content of 11–12% at 14% moisture.

The biochemical activation technique has also been applied to slow-speed batch mixing systems in Australia by P. E. Marston (1971) at the Bread Research Institute,

North Ryde, NSW. Using open-pan mixing, Marston found that doughs could be 'cleared' more quickly by adding L–cysteine–HCl and the salt at approximately two-thirds of the total mixing time, then completing mixing by continued mixing to peak development. The range of L-cysteine–HCl treatment varied from 25 to 75 ppm depending on flour strength. Commercial working tolerances are reported to be exceptionally good with this system, using several types of low-speed mixer of capacities up to 750 lb of flour. Floor-time schedules of 20 to 30 minutes can be used to advantage for bread quality, and prolonged dividing-periods of up to 30 minutes still give uniformly good quality bread. The addition of cysteine results in a structural activation of the dough mass, the presence of more SH-groups causes a fundamental change in the stress relaxation during mixing, reducing both mixing time and expended energy.

For full-scale, continuously mixed bread production with the Do-maker or Amflow units, the addition of L-cysteine–HCl at the developer stage can reduce the necessary developer-rpm from 205 down to 167, a drop of 38 units. Simultaneously dough temperature decreases of about 12°F, and production rate increases of up to 40% can be realized. A convenient method of introducing cysteine into the system was developed by Henika and Rogers (US Patent 3,053,666, 11 September 1962) at the Research and Development and Industrial Division of Foremost Dairies, Inc., Dublin, California. This technique involves the premixing of the cysteine with sweet dry whey. Close or open grain quality of the bread is a function of dough consistency, which in turn is determined by absorption, oxidant and level of flour in the pre-ferment. Since most bakeries prefer a moderately open, slightly irregular grain quality, the average bromate level is increased by 5 ppm to 60 ppm, the iodate level remaining constant at 10 ppm. Levels of flour in the brew below 10%, combined with about 65 ppm oxidant, potassium bromate/iodate ratio 4:1, require water absorption levels of about 70% to prevent tight, small-celled bread textures. An increase in bromate level is necessary owing to the whey solids, in the whey/cysteine mixture; but no change in iodate level is normally required. Since L-cysteine–HCl is a highly reactive, reducing amino acid and since it becomes rapidly consumed when mixed with an oxidant, the addition of whey allows better control and synchronization of the cysteine–flour protein reactions. The presence of the whey increases the mixing tolerance and gas retention characteristics of the cysteine-developed dough. In most bakeries that use the continuous mixing process, the preferred method of adding the mixed product is by means of a dry feeder that meters it into the flour stream to the premixer. Careful control of the particle size and density, in addition to the L-cysteine level, is of extreme importance when volumetric feeders are used. The mixed product is used at 2·0%, replacing all or part of the dairy solids where used. Pre-ferment flour levels using the Do-maker or Amflow units can vary from 0% to 50%, average ferment time being 2·5 hours, set at 80°F, dropping to about 70°F when taken for premixing. In general, pre-ferments without any flour addition tend to become excessively acidic, especially where the buffering effect of non-fat milk solids is absent. This can result in dough and bread structural weaknesses, lack of oven-spring, and lack of flavour and aroma, when compared with the sponge-doughs, which are regarded as the yardstick of quality in

North America. Advantages and improvements derived from the use of the whey/cysteine mixture as a more effective continuous biochemical dough development procedure are:

—A reduction in dough outlet temperature of 12°F, on average, down to 90 to 92°F, which results in significant improvement in dough and bread properties.
—The doughs more closely resemble sponge-doughs both in feel and behaviour. Improved extensibility and excellent pan-flow, lower bake-out loss allowing scaling weights to be reduced by up to 0·5 ounces have been reported by many North American continuous bread bakeries.
—Improved oven spring permits reductions of up to 5 minutes in final proof time, giving standard loaf volume. Loaf symmetry is also improved and side-wall strength is greater, allowing stacking three loaves high on supermarket shelves.

L-cysteine can also be used in combination with the oxidant azodicarbonamide, to reduce mixing intensity and work-input in batch short-time dough processes. Levels of treatment will depend on flour strength, up to about 75 ppm, azodicarbonamide being added at up to 45 ppm. In this case the time sequence of the chemical reactions must be such that the cysteine can perform its function before the oxidizing agent acts on the dough, or directly on the cysteine itself.

L-cysteine reduces the optimal mixing speed (rpm), and the time required to reach critical optimal dough development—as indicated by the mixing curve—and to produce optimal bread quality. Within certain limits, cysteine will increase tolerance to and compensate for undermixing, i.e. where the specific mixing intensity is less than that required to reach optimal dough consistency or effective viscosity. For a given flour and mixing regime there is an optimal range of cysteine addition within which satisfactory dough and bread quality is obtained. Levels in excess of these limits give rise to progressive deterioration in crumb structure, close texture, and, externally, shrinkage and 'green-dough' characteristics.

The functional effect of cysteine is one of protein 'disaggregation', breaking down the large protein gliadin and glutenin molecules into smaller, lower-molecular-weight, sub-units. These changes manifest themselves as the 'clearing' of the dough during mixing, without which good bread could not be produced. 'Clearing' can be seen as the dough changes from a mass with a dull and rough appearance into one with a smooth sheen. At the latter stage, a dough sample may be stretched into a thin balloon-like membrane of near-uniform thickness; whereas, prior to visible clearing, a stretched membrane ruptures owing to heterogeneous lumps and strands of agglomerates.

## 1.5.2 SODIUM SULPHITE AND METABISULPHITE (PYROSULPHITE)

Sodium sulphite and metabisulphite are included in the EC additives list as E221 and E223 respectively, and are classified as preservatives. These two substances cause allergic skin reactions, are dangerous for asthmatics, and reduce the thiamine (vitamin B1) content of cereal foods. However, they are on the permitted list for use in trace amounts as reducing dough-clearing agents in some countries, e.g. the USA and Australia. The normal range of addition is 20–60 ppm, and subjects include pie, cracker and cookie doughs. On a weight for weight basis, sulphites are less functional than L-cysteine, and the floor-time cannot be eliminated when it is added at a late stage in the mixing process.

### 1.5.3 ENZYMES

Enzymes which break down the flour constituents into lower-molecular-weight units, invariably result in changes in dough viscosity and consistency, with time. Simultaneously, the lower molecular weight sub-units stimulate fermentation and gas production on a continuous basis to leaven the dough mass.

Alpha-amylase preparations are added to flour and dough in various forms—flours, viscous liquids, powders, tablets—and in powder-carriers. Concentration levels used in baking vary between 6000 and 12 000 alpha-amylase units per 100 lb flour.

Malted wheat and barley flours give alpha-amylase potencies of about 50 units/g and can be added at the mill or bakery. When used at the rate of 250 mg/100 g of flour, and adjusting to a comparable gassing powder with alpha-amylases from fungal and bacterial sources, they produce optimal loaf volume, grain, texture, external appearance and excellent dough properties.

Alpha-amylase isolates of cereal, fungal and bacterial origin are available in powder-carrier and tablet form, potency of the former is of the order of 50–200 alpha-amylase units/g, and that of the latter about 5000/tablet.

Table 4 shows some important properties of commercial amylases from various sources, as applied to baking.

The relative activity of the alpha-amylase isolates are expressed in SKB units/g, which is a widely recognized assay method for biological materials. A modified

TABLE 4
**Properties of commercial amylases for bakery applications**

| Origin | Type | pH ranges | | Temperature ranges | |
|---|---|---|---|---|---|
| | | Optimum | Stability | Optimum | Effective |
| *Fungal*: | | | | | |
|   *Aspergillus oryzae* | alpha | 4·8–5·8 | 5·5–8·5 | 45–55°C | up to 60°C |
|   *Aspergillus niger* | gluco-amylase | 4·0–4·5 | 3·5–5·0 | 55–60°C | up to 70°C |
| *Bacterial*: | | | | | |
|   *Bacillus subtilis* | alpha | 5·0–7·0 | 4·8–8·5 | 60–70°C | up to 90°C |
| *Cereal*: | | | | | |
|   Barley malt | beta | 5·0–5·5 | 4·5–8·0 | 40–50°C | up to 70°C |
| *Animal*: | | | | | |
|   Porcine pancreas | alpha | 6·0–7·0 | 7·0–8·8 | 45–55°C | up to 55° |

Source: Allen and Spradlin, *Bakers Digest* **48**, (3), 14 (1974).

TABLE 5
**Proteolytic activity of malt and fungal enzyme supplements**

| *Supplement* | *Activity* |
| --- | --- |
| Malted wheat-flour | about 50 H.U./g |
| Dilute fungal enzymes | 50–600 H.U./g |
| Fungal enzyme concentrates | 30 000–100 000 H.U./g |
| Fungal enzyme tablets | 10 000–35 000 H.U./g |

Source: Reference Source 1985, *Bakers Digest.*

Wohlgemuth procedure devised by Sandstedt, Keen and Blish, involves the use of a standardized dextrin substrate, which is prepared by digesting potato starch with beta-amylase. Alpha-amylase activity in SKB units/g is inversely related to the time required to hydrolyse the dextrin to the achromatic point with iodine staining. In practice, dosage according to the same SKB-value is not possible, since European flours require two to three orders of magnitude less than their counterparts in North America. Practical dosage procedures in the mill or bakery usually involve measurement of the reducing sugar produced as maltose, viscosity reduction methods, e.g. Falling Number (Hagberg–Perten), Amylograph, or those based on gassing power determinations. These methods, however, only measure specific aspects of changes in the starch granule, and an exact characterization of the various amylases can only be achieved by gel chromatographic separation of their products of hydrolysis. Quantitative determination of the various sugars produced is performed in a continuous carbohydrate-analyser by reacting with Orcin and sulphuric acid, which gives a yellow to brown colouration, depending on the concentration. Continuous read-out is with a photometer and recorder.

In countries that produce wheat-flour glutens that are hard and difficult for machining, e.g. North America, Argentina, USSR and Australia, proteolytic enzymes are sometimes added to white flour doughs for both conventional and continuous processing. Usually proteases of fungal origin are applied, but papain has proved to be equally effective in reducing dough-mixing times and dough consistencies, resulting in moderately soft doughs which machine well. Concentration levels used in baking are about 50 000 H.U. (haemoglobin units) per 100 lb flour. Table 5 indicates the relative proteolytic activity of cereal and fungal enzyme supplements, as applied in various forms.

## 1.5.4 ASCORBIC ACID

Ascorbic acid is the most generally acceptable compound additive used to improve the structure of bread doughs, being listed as a vitamin. The Food and Drug Administration (FDA) in the USA set no limits on its use in bread, and the range used for good manufacturing practice is 10–200 ppm, based on flour weight. Used as L-ascorbic acid, as a fast-acting oxidant, it behaves quite differently from bromate and iodate, whether it is added to doughs with or without yeast. It causes an immediate reduction in dough consistency, necessitating a reduction of the order of 1·0% in water absorption to maintain it, when added to 70% sponge-doughs at the 120 ppm level. In continuously mixed doughs, when used at 100–200 ppm it acts as a reducing agent, reducing mixer-rpm and work-input. The effect of ascorbic acid on farinogram curves is less apparent at higher levels, which is indicative of the existence of a maximum oxidation effect for any given flour; concentrations much in excess of this level give rise to a reducing effect. The redox reaction kinetics of L-ascorbic acid and L-dehydroascorbic acid in wheat flour doughs has been the subject of much research and postulation for many years. The oxidation of ascorbic acid to dehydroascorbic acid, which is catalysed by ascorbic acid oxidase, necessitates the availability of oxygen in the free form. The source of this oxygen is atmospheric, brought about during dough mixing, and it is the dehydroascorbic acid that effects the improver action by oxidatively converting SH groupings to disulphide, SS bonds. This oxidative effect of dehydroascorbic acid is catalysed by the dehydro-ascorbic acid reductase enzyme system present in the flour, yielding ascorbic acid.

Physiological systems such as dough are characterized by the simultaneous interaction of catalysts and enzymes, and the movement of electrons, since a bridge is formed between the hydrogen ion donor and the acceptor.

In a yeasted dough the yeast's own redox systems are likely to compete with the ascorbic acid oxidase system for available oxygen. Therefore, the reducing effect becomes more predominant at lower levels of ascorbic acid treatment. This balance between a reducing and an oxidation effect of ascorbic acid will depend on the availability and accessibility of suitable coupling systems acting as hydrogen acceptors from the ascorbic acid. The lowering of dough consistency, making slight reductions in water absorption necessary, is more indicative of a reducing action than an oxidative one. Doughs prepared from an unbleached, improver-free, straight-run flour containing 20–30 ppm L-ascorbic acid show rapid oxidation on mixing, about 81% of L-ascorbic acid being oxidized to dehydroascorbic acid within 15 minutes. Average recovery levels of the added L-ascorbic acid, both as L-ascorbic acid and as dehydroascorbic acid, is 87%. Doughs containing up to 30 ppm L-ascorbic acid show reduced extension and increased resistance and relaxation when compared with untreated doughs. The addition of L-ascorbic acid oxidase at the start of mixing does not affect relaxation curves. The reduction of dehydroascorbic acid has been considered to be due to the SH groups of the flour proteins, and the rate of oxidation of L-ascorbic acid does not appear to be a limiting factor in its improver action. Recent studies on the redox reaction kinetics in wheat-flour doughs by Pfeilsticker and Marx (1986), using gas chromatographic and mass spectrographic techniques have shown that the loss of total ascorbic acid (L-ascorbic acid and L-dehydroascorbic acid) after doughmaking was about 30%. Also, only a small amount of L-ascorbic acid is formed when L-dehydroascorbic acid is added; and in this case the loss of total ascorbic acid was about 70%. From this and studies on doughs containing [14]C-labelled L-ascorbic acid coupled with measurements of available thiol groups, it was concluded that L-ascorbic/L-dehydroascorbic acid did not react entirely like a redox system. L-dehydroascorbic acid forms intermolecular condensation products with protein amino acids, which may contribute to improved baking quality.

In all these investigations, the L-dehydroascorbic acid was in its monomeric 3,6-anhydro-2-hydrate form. A recent comparative evaluation of the effect of ascorbic and isoascorbic acid on bread quality by Vangelev and co-workers (1985), in Bulgaria, showed that both forms of the acid, when added to wheat-flour, at the 10 to 30 ppm level, improved the baking quality of gluten, the water-holding capacity of the flour, the rheological properties of the dough, and the bread quality indices, as compared with unsupplemented controls. However, the isoascorbic acid-supplemented bread exhibited a greater improvement in volume and porosity than that containing ascorbic acid. Therefore, the catalytic systems appear to be specific, and to depend on the isomeric form of ascorbic acid. Water extracts of unbleached, improver-free, straight-run flours, in the presence of glutathione (GSH) and dehydro-L-ascorbic acid, molar ratio 1:2, showed that about 90% dehydro-L-ascorbic acid becomes reduced to L-ascorbic acid within 30 minutes at 35°C and pH 5·9, whereas, without the flour extract, only 1·5% becomes reduced. Average molecular ratios of oxidized glutathione to dehydro-L-ascorbic acid in the system are 2:1, and the reaction rate is maximal at pH 7·0. It also appears to be thermolabile, since it is rapidly inactivated at approximately 70% at 60°C and 100% at 70°C. Substituting cysteine for glutathione in the system, without the flour extract, cysteine appears more active than glutathione in reducing dehydro-L-ascorbic acid, and the addition of the extract does not increase their activity. Furthermore, dehydro-D-isoascorbic acid becomes reduced at approximately one-third the rate of dehydro-L-ascorbic acid. Therefore, there would appear to be a selective catalytic system for dehydro-L-ascorbic and glutathione, according to Kuninori and Matsumoto (1963), who carried out all reactions under nitrogen. The specificity for the L-isomer of dehydroascorbic acid as a bread improver would also suggest enzymic catalysis. Whether the conversion of L-ascorbic acid to dehydro-L-ascorbic acid is the result of enzymic activity due to ascorbic acid oxidase and a coupling reaction with free oxygen, will depend on oxygen availability. Where availability is limited, and there is competition for it, possibly from the excessive presence of glutathione in some wheat-flours, the reducing effects of L-ascorbic acid will predominate, especially at the higher levels of addition.

The universal importance of L-ascorbic acid as a dough additive is due to its relative safety and acceptability. In comparison with such additives as potassium bromate and iodate, calcium bromate and iodate, calcium peroxide and azodicarbonamide, it enjoys a high legislative status, the FDA (USA) setting no upper limit for its use in bread. The technological implications in terms of dough rheology and consistency were studied by Prihoda, Hampl and Holas (1971). Using the Hoeppler Consistometer (VEB Prüfgerätewerk, Medingen, Dresden, GDR), they calculated dough consistency as a function of the velocity of fall of a ball through the dough and the applied force. Comparisons of the effects of 0·3% (flour weight) of ascorbic acid and potassium bromate on doughs made from flours of various qualities in terms of consistency changes were evaluated. Additions of ascorbic acid resulted in a decrease in energy consumption for all flour qualities, whereas the addition of bromate caused a decrease for weak flour-doughs and increases for medium and strong flour-doughs, at the mixing stage using a farinograph fitted with a recording amperometer. Decreases in dough consistency shortly after mixing with treated flours were greater for weak than for strong flour, and the differences were accentuated by the presence of both bromate and ascorbic acid. Consistency/stress curves of freshly mixed doughs show quasiplastic behaviour—showing variable viscosity—but after about 45 minutes fermentation period, consistency changed less with stress, and both additives stabilized this effect, since both have a slow reaction rate.

Giacanelli (1972) reported that in continuous dough processing in Italy, 60 ppm ascorbic acid in combination with about 15 ppm cysteine gives satisfactory bread with respect to volume and texture. Mixing energy requirements were also reduced by ascorbic acid in the presence of cysteine, confirming a synergistic mechanism.

### 1.5.5 FAST-ACTING OXIDANTS

Fast-acting oxidants, which include potassium iodate, calcium iodate, calcium peroxide and azodicarbonamide, are not generally acceptable additives under current EC legislation. However, under FDA (USA) legislation they are permissible when used at a maximum level of 75 ppm, and 45 ppm in the case of azodicarbonamide. Estimated usage ranges for processing in the USA are as follows:

| | |
|---|---|
| Potassium iodate | 3–20 ppm |
| Calcium iodate | 3–20 ppm |
| Calcium peroxide | 10–50 ppm |
| Azodicarbonamide | 5–30 ppm |

These oxidants, when carefully used in combination with the slower-acting ones, such as bromates, increase dough consistency almost immediately after mixing and development and commencement of the floor-time. The gluten (coiled molecular complex) becomes stretched and ruptured by the mechanical energy imparted at mixing, followed by enzymic action with the onset of fermentation, and possibly, the action of reducing agents. Bonds become broken and disulphide links combine with hydrogen. The presence of oxidants in adequate concentrations then reform the disulphide links, conferring strength and elasticity to the gluten structure, the process being essentially one of electron transfer. Oxidation requirements increase with milling extraction, owing to location and access to active groups in the wheat kernel. The SH, or sulphydryl groups, occurring in larger amounts in the aleurone (outer) layer and in the germ than in the endosperm (centre). Oxidation requirements are therefore greater to convert the SH groups to SS bonds needed for strength.

Under-oxidized doughs tend to be weak, soft, sticky and extensible, making them difficult to machine and process. The finished loaves have reduced volume, weak crusts, uneven grain and texture, poor break and shred and loss of symmetry.

Over-oxidized doughs are tight, firm, bucky, and difficult to mould, tearing easily. They break on proofing owing to their inelasticity. Bread will be small-volumed, with a rough break and shred, uneven grain and large holes.

## 1.5.6 DELAYED-ACTION OXIDANTS

The bromates fall into the category of delayed-action oxidants. Potassium bromate is the longest-established of all the chemical maturing additives, having been discovered in 1916. Generally, it is the most useful and economic of all the maturing agents, although it has not found acceptance in the majority of European countries. Apart from its acceptance in North America and Australasia, it has found wide application in the United Kingdom and Holland as a flour-maturing agent and component of bread-improver preparations. Currently, it is under consideration by the EC for an 'E' prefix number, the FDA (USA) maximum rate of usage being 75 ppm based on flour weight. The estimated usage range for potassium bromate is 10–75 ppm, depending on flour strength and choice of process, i.e. sponge-dough, short-time or continuous. For example, an average dosage for a sponge-dough might be 22 ppm, and that for a mechanically developed dough about 44 ppm, whether batch or continuous mixing is in use. In the latter case the rapid-acting iodate is also used at up to 20 ppm. Bromated doughs show less immediate fall in consistency at normal water absorptions of 60–70%, and tend to increase in consistency with floor-times of up to 6 hours. In short-time mechanical dough development processing, the use of bromate can permit the elimination of an intermediate proofing stage. Calcium bromate is also permitted by the FDA (USA), and is used as a delayed-action oxidant at 10–75 ppm, being less soluble than the potassium salt.

Between oxygen, oxidants, raw material oxidase and reductase enzyme systems, flour peroxidase and catalase activity, and the mixing intensity, there exists an exceedingly complex series of mechanisms, the kinetics of which remain obscure. The exchange and interchange of these reactions, and their effects on the functional properties of the components of bread-dough, all proceeding simultaneously, are profound.

## 1.5.7 GLUTATHIONE

Glutathione, although not generally used as a dough additive, is naturally present in wheat and flour, since all living cells contain such reducing agents; they are essential to life processes. This low-molecular-weight tripeptide (made up of three amino acids), present in many tissues, is built up from glutamic acid, cysteine and glycine linked through the peptide bond —CO—NH—. For simplicity it is often written in the abbreviated form G—SH, the —SH group belonging to cysteine, which on oxidation goes to G—S—S—G, thus:

$$2\,\text{G—SH} \xrightarrow{\text{oxidation}} \text{G—S—S—G}$$

Glutathione appears to be important in keeping the —SH group of certain enzymes in the active reduced form. Enzymes which have —SH (sulphydryl) groups lose their activity when these are oxidized to —S—S— groups, and regain it when the reverse reduction process brought about by the glutathione or ascorbic acid takes place.

$$R—SH \; HS—R \; \rightleftharpoons \; R—S—S—R + H_2O$$
glutathione or ascorbic acid coenzyme system

Glutathione is introduced into bread dough in the following forms: aleurone cells, which are very strongly bound to the bran; germ cells, which the miller keeps out of his flour, because of their limited keeping properties; and also the added yeast cells. All these cells contain relatively large amounts of glutathione, which is readily assimilable and closely related to the vitality of cell function. This substance, in common with other proteins contain SH groups which can be transformed from consistency-reducing components into gluten-strengthening ones. Jørgensen (1935) suggested how potassium bromate can achieve this by inhibition of the flour-proteolytic enzymes, while other non-improver oxidizing agents fail to do so. Conversely, he found that ascorbic acid, a reducing agent, which inhibited papain, also acted as a flour-improver of the bromate type. Later, Elion (1945) found that other compounds, including glutathione, inhibited papain and acted as flour-improvers.

In a fermenting dough the reaction time or 'floor-time' to which all these complex interrelated series of mechanisms are subjected cannot be overlooked. Although much investigation of this complex subject has been carried out since Jørgensen's work, there appears to be no clear-cut explanation of the pathway of the observed dough-improving effects.

Work done by Coventry, Carnegie and Jones (1972) in Australia, established that the weakening of dough was related to the total glutathione content, and also to total CysS—, which is the summation of free cysteine and cystine as well as cystine which can be bound to the protein at only one end of the disulphide link, i.e. total CysS— = CysSH + CysSSCys + CysSSPr, total CysS being isolated and estimated as cysteic acid. These investigations were done on a Brabender Extensograph, using maximum resistance as the index of weakening at 0, 5 and 15 minutes after balling and shaping. Flours tested were derived from 27 wheat samples of diverse quality, and linear regression analysis of total endogenous wheat glutathione, and maximum resistance in Brabender units, showed linearity at the 0·1% to 5% probability level. Although it has been known for a long time that added glutathione causes a weakening and reduction in dough consistency, the work described indicates the importance of this factor as a determinant in the evaluation of wheat samples for milling and baking by wheat breeders and agronomists, especially if this is related to protein content as well as flour sample weight.

Obviously glutathione causes a reduction in the cohesive molecular forces within the dough, and assuming SS bonds directly confer cohesion, either GSH or GSSG could weaken the dough structure by permanent cleavage or by catalysing disulphide interchange.

Apart from intermolecular bonding, it is possible that SS bonds also help to maintain the macromolecules of gluten proteins in compact conformations with strong hydrogen bonding, as evidenced by wideband proton NMR-spectra parameters. The presence of GSH or GSSG, the oxidized form, could catalyse conformational changes thus giving rise to reduced intermolecular hydrophobicity of the gluten network, and in consequence a less compact dough structure. Typical total glutathione contents of flours vary from 5 to 12 $\mu$mol/100 g. These figures are made up of free reduced GSH, oxidized GSSG, and protein-bound glutathione Pr–SSG. Hard Red Spring quality samples show lower total glutathione contents of up to 3 $\mu$mol/100 g. In using total glutathione contents to assess flour baking quality, it must be appreciated that other variables as well as total glutathione influence dough rheology, which is confirmed by a scatter of points around the regression lines obtained.

Baker's yeast contains approximately 0·65% glutathione on a dry matter basis, and the relationship between dough properties and bread quality as a function of yeast glutathione has been studied by Karpenko (1985) in the USSR. The separation of glutathione from pressed yeast, to avoid its stimulating effect on proteolysis and its deleterious effect on dough properties, did not negatively affect yeast performance (vitality, maltose activity, and fermenting capacity). Rheological properties of gluten and dough, such as elasticity, stability and viscosity, were improved. Dough spread or flow was reduced by 25%, and bread yield increased by 19% compared with non-separated glutathione yeast. Dough adhesion to surfaces is also reduced.

## 1.5.8 SURFACTANTS, STRENGTHENERS/CONDITIONERS AND CRUMB-SOFTENERS

The physical state of the surfactants, strengtheners/conditioners and crumb-softeners group of additives is important for their succesful incorporation and functionality in breadmaking, unless some form of heating is applied to effect a change of state. Being derived from fatty acids, they can be in the form of a plastic solid, hydrate, powder or flakes, and can be incorporated with optimal functionality in a form which is appropriate to the desired processing stage, i.e. sponge, dough, brew-shortening or pre-heated pumpable, which is generally feasible for all stages when agitated.

*Bath Processing*
SPONGE
Plastic form most appropriate, or as a hydrate.
Powder products must be adequately dispersed.
Flakes not feasible

DOUGH
Both the plastic and hydrated forms are appropriate.
Powder products only feasible if easily dispersable, or added to heated shortening.

*Continuous Processing*

BREW SHORTENING

Plastic, hydrated and powder forms are all possible, but foaming results and must be reduced by agitator control or the addition of up to 25% of total dough salt concentration.

Flakes are not feasible.

PREHEATED TANK

Plastic, hydrated, powder and flaked forms are all feasible, but the hydrate will require agitation.

In yeasted doughs, the functionality of surface-active agents is a result either of starch complexing or of protein interaction. However, if shortening and sugars are present, the properties of emulsification and foam stabilization will also be involved. Especially suitable starch-complexing agents are the distilled mono-glycerides derived from fully hydrogenated shortenings with their straight carbon chain, which easily becomes entrapped in the helical structure of the amylose component of the starch macromolecule to form a water-insoluble complex. When water is added to starch and subjected to heat, the two types of carbohydrate, amylose and amylopectin which form the large molecule from basic glucose units, swell and eventually form a gel. This gel then undergoes a phenomenon called 'retrogradation' which is a process of recrystallization of the starch components. Amylose, owing to its molecular size and structure, retrogrades more rapidly than amylopectin, and is partly responsible for the ageing or 'staling' of bread. Therefore, the formation of the amylose/monoglyceride complex (Fig. 13) results in an anti-staling effect on bread and softens the crumb. It also reduces the adhesion of such products as pasta and instant starch-based foods. The saturated, distilled monoglycerides designated E471 under EC legislation, tends to increase extensogram resistance. However, owing to the diverse bonding possibilities—electrostatic interchange, ionic bridge-building, hydrogen bonding, van der Waals interchange, hydrophobic interchange and, in exceptional cases, covalent bonding,

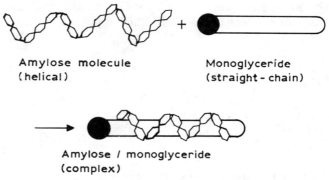

Amylose molecule        Monoglyceride
(helical)               (straight - chain)

Amylose / monoglyceride
(complex)

Fig. 13.   Mechanism of starch-complexing.

all proceeding simultaneously in varying degrees of participation, and in differing media—no generalizations concerning molecular-interchange mechanisms can be made.

Apart from saturated, distilled monoglycerides, ionic surfactants such as diacetyl tartaric acid esters of monoglycerides (E472e), and calcium and sodium stearoyl-2-lactylates (E482 and E481) also show good starch-complexing effects in bread-doughs, pasta and starch-based foods, but to a lesser degree.

The interaction of certain surfactants with protein results in a 'conditioning' effect in yeasted doughs. The viscoelastic properties of the wheat gluten complex are enhanced, resulting in improved gas retention and therefore increased volume and texture. Ionic surfactants such as diacetyl tartaric acid esters of monoglycerides, and calcium and sodium stearoyl-2-lactylates, strengthen the gluten protein structure, enabling it to retain the carbon dioxide produced. These dough conditioners also give the doughs tolerance to mixing and shock during processing.

Other additives with surfactant properties, which are not on the permitted list in EC countries are:

—Ethoxylated monoglycerides and diglycerides, which are very good dough strengtheners, and also give improved texture, when used at the 0·5% level based on flour weight, as permitted under FDA (USA) legislation.
—Succinylated monoglycerides, which act as dough strengtheners and texture improvers, are permitted under FDA (USA) legislation at 0·5% based on flour weight.
—Polysorbate 60, also permitted by the FDA, is used as a dough-strengthener and crumb-softener at the 0·5% level based on flour weight.
—Strongly anionically active surfactants, such as sodium dodecyl-sulphate, sodium soaps, and the dough-strengthener/conditioner types calcium and sodium stearoyl-2-lactylates, used at 0·5% based on flour weight, give rise to a molecular fragmentation. This fragmentation allows exchange reactions of a complex nature to occur, these being hydrophilic and involving the gliadin wheat protein fractions rather than the glutenins.

These interactions bring about certain functionally advantageous effects on bread doughs. Water absorption tolerance can be increased by 1% to 2% without affecting machining qualities, tolerance to under- and over-mixing because of flour quality variations is increased. Owing to improved dough homogeneity and gas retention, proofing times can be reduced by at least 10% in bringing doughs to volume. Bread symmetry and cell structure are more even, and premature staling is prevented.

Recent studies concerning the interactions between selected proteins—casein, fractionated gliadin and glutenin (according to Osborne), and their residual lipids— and a wide range of commercial emulsifiers at the Humboldt-Universität in Berlin, GDR, by Grunert, Möhr and Kroll (1986), using various analytical techniques, have produced interesting results. The general conclusions were that non-ionically-active emulsifiers become more firmly bound to gliadin and glutenin than anionically active types, giving rise to relatively stable complexes. The anionically active types, on the other hand, show complex interactions, which can proceed as far as a fragmentation of the protein molecule. Interactions between glutenin and emulsifiers, compared with those between gliadin and emulsifiers, are less pronounced and are mainly of a hydrophobic nature. Between gliadin and emulsifiers hydrophilic interactions predominate, anionically active emulsifiers and hydrophilic non-ionic emulsifiers being particularly effective substances. Therefore, for exerting a targeted influence on the properties of gluten-containing systems, the use of emulsifier mixtures consisting of anionically active and non-ionically-active emulsifier substances are recommended.

The glutenin proteins are in themselves complexes of lipids, carbohydrates and proteins. The lipid content is variable while the carbohydrate, alpha-glucan, remains consistent at approximately 17%. The alpha-glucan forms specific and stable complexes with protein sub-units to form aggregates. The carbohydrate in the gliadin protein fraction is mainly galactose, in the form of digalactosyl diglyceride complexes, which aggregates the relatively low-molecular-weight gliadins (see Fig. 14).

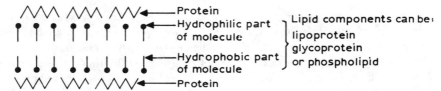

Fig. 14.   Formation of triple-layers of protein–lipid–protein, resulting from ionic or dipole interactions of phospholipids and glycolipids.

When distilled monoglyceride from hydrogenated lard or hydrogenated soya bean oil, i.e. with a monostearin content of 65% to 85%, is added to dough the starch is changed in a two-stage process. Providing the monoglyceride is in a fine crystal form, as a hydrate or extremely fine particulate powder, it becomes absorbed onto the starch surface during mixing. Then, during the baking process, when the temperature reaches about 55°C, the monoglyceride goes into a liquid crystalline state in conjunction with part of the dough water (mesophase). When the starch begins to gelatinize, the monoglyceride comes into contact with dissolved amylose both outside and inside the granule. This allows a helical complex between the straight carbon-chain of the distilled monoglyceride and the helical amylose to form. This complex, being insoluble in water, gives bread with a softer crumb when freshly baked, and retards the subsequent staling process. Saturated distilled monoglycerides are the most effective starch-complexing agents, reducing dough stickiness, impeding gelation and improving dough consistency.

The ionic emulsifiers, such as diacetylated tartaric acid esters (DATA-esters) and the stearoyl-2-lactylates which interact with proteins, are responsible in yeast-raised doughs for a conditioning effect. This effect manifesting itself as an improvement in the viscoelastic properties of wheat gluten, resulting in increased tolerance to processing and bread volume and textural enhancement. The effects of various emulsifiers during fermentation can be measured with the Brabender Maturograph and Ofentriebgerät ('oven-spring apparatus'), the Chopin Alveograph and a controlled baking test, in terms of dough elasticity, stability and gas retention. The improver effect of these additives have also been reported in mixed wheat and rye flour doughs by Lucny (1983) in Czechoslovakia (Czech., CS 210,528), and in Poland by Kawka and Gasiorowski (1985). Recent studies by Moore and Hoseney at the Department of Grain Science and Industry, Kansas State University, Manhattan, Kansas, USA (1986) on the influence of shortening and surfactants on retention of carbon dioxide in bread doughs at higher temperatures indicate rheological differences. Where shortening or surfactants are added, carbon dioxide is lost at a slow rate in the oven during the first 9 minutes, after which it is lost at a greater rate and the rate of expansion slows down. With doughs having no added shortening or surfactant, the rate of carbon dioxide loss increases after only 2 minutes heating, and the rate of dough expansion is correspondingly lower. Using a resistance oven, these workers found that artificially restricting dough expansion tended to increase the rate of carbon dioxide loss. Rheological measurements further revealed that dough without shortening or surfactants undergoes a change at about 55°C, whereas dough with one or both additives does not. It was thus concluded that the increased loss of carbon dioxide at higher temperatures was the result of a rheological change in the dough that restricted expansion rather than increased dough permeability to carbon dioxide gas.

## 1.5.9 SOY PRODUCTS

Various forms of soy-flours have been used in the baking industry in the USA and the UK since the late 1930s. However, most of the work concerning the application of commercial, full-fat soy-flours and defatted soy-flours by Ofelt and Smith (1954) at the Peoria, Illinois, Research Laboratories of the USDA, were directed towards its nutritional value as a supplement for wheat-flour bread, which was deficient in the important amino acid lysine. Loss of bread quality owing to additions of up to 5% soy-flours were overcome at that time by increasing the levels of potassium bromate. Also, American soy-flours were derived from defatted soy, since the extraction of soy-oil was, and remains, of prime importance for the food industry in the USA. Current trends in the USA are towards the use of heat-treated soy-flours at up to 2%, based on flour weight, with added dextrose monohydrate or high-fructose corn-syrup (HFCS) in bread-doughs as replacers for milk solids. An average analysis of an American defatted enzyme-active soy-flour would be: protein 50%, carbohydrate 30%, oil 1%, crude fibre 3% and ash 6%, compared with a European equivalent of: protein 40%, carbohydrate 23%, oil 20%, crude fibre 2%, and ash 4%, both samples having a moisture content of about 7%. As a measure of the degree of heat treatment or denaturation, the Protein Digestibility Index (EDI) is quoted. A PDI of 90 is average, values in excess of 90 are indicative of little or no heat treatment. The UK specification for full-fat enzyme-active soy-flour is designed for use in lean formulated bread-doughs. It is manufactured by milling cleaned dehulled soy beans to a particle size not in excess of 100 mesh. Current usage rates for this product are usually up to a level of 0·7% based on flour weight, either added separately or as a component of a commercially available composite dough conditioner.

The majority of dough conditioners currently available in the UK contain soy-flour, so, unless there is a special reason for using extra additions, no benefit is derived from the use of concentrations above this level. In the USA and Canada, the built-in natural benefit of soy-flour enzymes is marketed in the form of a series of bread-dough-conditioners by the J. R. Short Milling Company of Chicago, Illinois. These products are based on special enzyme-active soy-flour blended with other natural components for both sponge (batch) and continuous processing. In the UK, enzyme-active full-fat soy-flour is marketed as Do-soy by British Arkady Co. Ltd of Old Trafford, Manchester. During the 1950s soy flour, in the full-fat enzyme-active form, marketed under the trade mark 'Diasoy', was a standard ingredient in many large and small bakeries for the production of National white bread in the UK. The inclusion of 2 lb per 280 lb in a straight dough with a floor-time of 3·5 hours would give an increase in dough water absorption of 1 to 1·5 gallons (10 to 15 lb) per 280 lb of flour, and improve tolerance to handling or machining of the dough, as well as improving bread colour, texture, flavour and keeping properties. Therefore, quite apart from the inclusion of soy-flour as a component of composite dough conditioners, full-fat enzyme-active soy-flour merits more consideration as a bread-dough ingredient in its own right. Furthermore, owing to the present desire throughout the baking industry to produce bread requiring no declaration of EC 'E' numbers, and to utilize only unbleached, untreated flour as standard raw material, the functional natural bleaching and improving action of enzyme-active full-fat soy-flour attracts renewed interest. The aim to produce 'clean label' products demands the use of the statement 'made with unbleached flour' and removal of the declaration 'contains flour improver 926' (chlorine dioxide), preferably also excluding benzoyl peroxide, which is currently threatened with a declaration obligation under EC legislation. The sole declaration, 'ascorbic acid (vitamin C) E300, will satisfy most consumers within the EC. In Germany, France, Belgium and Switzerland, ascorbic acid has been the only chemical flour treatment permitted by law for the past 30 years, only organically based substances being acceptable.

The value of enzyme-active soy-flour as a flour-improver depends on its content of the enzyme lipoxygenase. This enzyme acts as a catalyst for the transfer of oxygen to specific flour components via pro-oxidant intermediates produced during the oxidation of polyunsaturated flour lipids (mainly linoleic acid).

The yellow carotenoid pigments and wheat proteins of the flour ultimately reacting with the transient oxygen, produce a bleaching and improving effect.

The improvement in breadmaking properties and crumb colour effected by the use of full-fat enzyme-active soy-flour was the subject of four British and one French patent due to J. Rank Limited, partly in collaboration with J. G. Hay. Known as the Rank and Hay process, this involved the use of small amounts of enzyme-active soy-flour added to a wheat-flour/water batter, into which air is incorporated by intensive stirring or by mixing for 2 to 7 times the normal mixing time (French patent 1.070174). During the 1950s, the Rank patent was applied in the

UK by mixing a 50% flour batter, containing 50% of the total dough salt, 0·7% enzyme-active full-fat soy-flour and about 97% of the total flour water absorption at 350 rpm for a period of 1 minute in excess of the maximum amperometer reading. The final dough was then prepared by adding the yeast and the water absorption balance, with the salt balance. Final dough temperatures were 26°C, and the bulk fermentation time or floor-time 3 hours. Using slow-speed mixers, similar results could be obtained by mixing for four times the normal mixing time.

Another process using atmospheric oxygen and unprocessed soy-flour (enzyme-active full-fat soy-flour) is that due to Blanchard. This process also eliminates the use of flour-bleaching agents and mineral oxidants. A slack sponge is prepared by mixing 75% of the total flour with approximately 94% of the total flour water absorption, yeast, 0·2% malt-flour (total flour weight) and 0·7% enzyme-active full-fat soy-flour (total flour weight) made into an emulsion (1:1) with shortening. This sponge is mixed for 18 minutes at 60 rpm to a final temperature of 26°C and allowed to rest for 45 minutes. The final dough is made by adding the flour balance of 25%, and total dough salt dissolved in the residual 6% of the flour water absorption. The whole being mixed for 15 minutes at 30 rpm. The dough, at 26°C, is bulk fermented for 1·75 hours.

Although these bread-production techniques now belong to a past decade, they demonstrate the versatility of full-fat enzyme-active soy-flour as a dough component. Obviously it is not as effective as chlorine dioxide and benzoyl peroxide as a bleaching agent for flour, but current trends are likely to rekindle technological interest in such patents, in order to satisfy the labelling 'made with unbleached flour'.

Daniels, Frazier and Wood (1971) reported in a study of flour lipids and dough development that a distinct improver effect of mixing in oxygen in the presence of enzyme-active full-fat soy-flour was established. This improving effect was confirmed by marked increases in dough relaxation times and extensograph maximum resistance values (as much as 45·2%). Dough relaxation time as measured with an Instron tensile tester, expressed in seconds, using loads of 180 g to 80 g, for example, will give a reliable indication of gluten strength. These workers concluded that the close relationship between free versus bound lipid and dough relaxation was indicative of a soy–lipoxygenase coupled oxidation of lipid binding sites, and the release of bound lipid, which gives an oxidative improvement of the gluten. Pashchenko and co-workers (1984) in the USSR, also report optimal rheological properties for liquid-dough semiproducts when prepared at 30°C, with the addition of a phosphatide concentration of 0·26–0·29%, coupled with an oxygen saturation rate of 15 mg/litre.

A valuable component of the soy bean is the phosphatide lecithin, or structurally, phosphatidyl choline, the nitrogen-containing choline base forming an ionic dipole with the phosphatidic acid. Phosphatidic acid is structurally a 3-glyceride with phosphoric acid replacing the third —CH— group. Therefore, the lecithin molecule possesses both a hydrophilic element and two hydrophobic elements, and can be classed as an ionic, dipolar compound, which is simultaneously ampholytic, having two centres capable of ionization, giving rise to 'zwitterions'. Lecithin is not to be regarded as the name of a single substance, but rather as a whole group of allied compounds, which may differ (1) in the nature of their fatty acids, (2) in the site of attachment of the phosphoric acid relative to the glycerol framework, i.e. alpha or beta, (3) in optically active forms of the compound, and (4) in the structure of the phosphoryl-choline moiety, which is water soluble, and, owing to the dipolar ion, can react under certain conditions as an ampholyte with either acids or bases.

In the USA, phosphatides in the form of crude lecithins are extracted in considerable quantities as a byproduct in the 'dewaxing' of soy oil, and as early as 1948 some 4000 tons of soy were being extracted. In addition, various derivatives of phosphatides have been prepared, and their surface-active properties examined, and in some cases industrial applications found for them. Modified lecithins are produced by treatment with hydrogen peroxide and lactic acid. Such products find wide application in bakery processing owing to their synergy with mono(1) and di(2)-glycerides and improved hydrophilic properties for oil-in-water emulsions, being marketed in the USA by Central Soy, Chicago, Illinois. The oil-free lecithins, extracted at low temperature, in the form of yellow, waxy, granulates with a bland to nutty flavour and odour are marketed in the USA by the American Lecithin

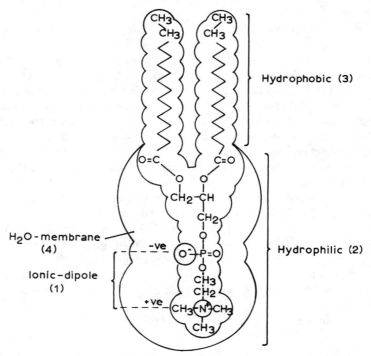

Fig. 15. Model of lecithin molecule. 1. Ionic dipole between positively charged $N^+$ of choline (hydroxy-ethyl tri-methyl ammonium hydroxide) and negatively charged $O^-$ of the phosphoric acid $H_3PO_4$. 2. Hydrophilic (water-liking) element. 3. Hydrophobic (water-hating) elements fatty acid chains. 4. Aqueous membrane enveloping the hydrophilic elements. (Source: *Kleine Enzyklopädie—Natur*, VEB Bibliographisches Institut Leipzig, 1981 (in German).)

Company, Atlanta, Georgia, and in West Germany by Lucas Meyer, Hamburg, who also distribute in France, UK, Belgium, Italy, Netherlands, Spain and USA. The latter company also markets a wide range of soy-flours, protein isolates, grits, and whipping agents, as well as a product 'Emulthin' for wheat-flour improvement.

For dietary breads, soy-flour, protein isolates and bran find wide application. As a source of dietary fibre, soy bran products contain 75% to 99%, and 'clean label' health breads can be made from blends of 100% wholewheat-flour, 100% soy-flour and malted wheat grains. Blends of soy-flour and gluten 60:40 will satisfy human energy and protein requirements, as laid down in a joint FAO/WHO report of 1973. The lysine content being increased from 1·6 to 4·1 g per 100 g. If desired, quite acceptable high-protein bread of 22% protein (dry basis) can be produced by using soy isolates, and a formulation rebalance of sugar, hydrogenated vegetable shortening and oxidant.

## 1.5.10 VITAL WHEAT GLUTEN AND ITS DERIVATIVES

Although many other factors merit consideration, gluten is the principal key to the utility of wheat. Primarily a water-insoluble protein of flour, it confers properties to dough which make wheat unique among cereal grains, breadmaking being at present its major outlet in the world market. However, research is progressing towards the achievement of a broader utility for wheat.

Flour prepared by milling wheat is the source of gluten, being formed as an elastic cohesive complex within the dough, when mixed with water. It can be isolated from dough by washing and kneading in running water, the kneading and washing carrying the starch away, leaving the gluten as a cohesive mass. This procedure, with some modifications, is used commercially. Good-quality clear flours are blended and then mixed into a slack flour-water dough with 80 to 90 pounds of water per 100 pounds of flour. When the flour is thoroughly hydrated, the dough is washed and kneaded several times to remove the starch. This is known as the Martin process, yielding about 15% of the flour as gluten, 60% being recovered as starch. The rest of the flour remains in the wash water as soluble protein, pentosan, very small granule starch and bran. At this stage the gluten is composed of about 30 parts gluten and 70 parts water. In common with other proteins, gluten is readily denatured by heat, especially in the presence of large amounts of water. Therefore the method of drying is very important: a special form of flash drying, assuring a dry gluten with minimum loss of vitality. The dried product is then screened to a uniform particle size. Uniformity of quality is obtained by sampling, testing and blending for baking performance, before packaging in multiwall plastic bags, usually despatched in robust cardboard drums containing 100 pounds net weight. Vital wheat gluten is very stable under normal storage conditions. Samples stored at room temperature for up to 18 months show no adverse chemical or physical changes. The average moisture content of 6·0% does not support insect life, and most of the amylolytic and proteolytic enzymes in flour are removed during the washing process. A good sample of dried gluten should have a uniform light tan colour, have a uniform

TABLE 6

| Protein (N × 5·7) | 75·0% to 80·0% |
| Ether extractable fat | 0·5% to 1·5% |
| Ash | 0·8% to 1·2% |

Moisture contents should be 5·0% to 8·0%, and the water absorption capacity on rehydration 150% to 200%.
Source: Reference Source 1985, *Bakers Digest*, p. 34.

particle size and be free-flowing, 97% passing a US 60 mesh screen. Specifications should conform to the data given in Table 6, expressed on a dry basis.

A typical analysis of a dried gluten sample on an as-received basis is shown in Table 7.

It is very important that the dough should contain sufficient good quality gluten to ensure volume, symmetry and texture in the baked product. Many speciality breads contain large quantities of non-flour ingredients, which dilute and overload the inherent flour gluten-forming proteins, resulting in diminished volume, symmetry and texture. Examples of such products are: multi-grain breads, wholemeal fibre-increased breads, wholewheat breads from flaked wheat, high-protein breads with low slice weight, and also some rolls and buns as well as rye breads. In such products, vital wheat gluten provides an important quality boost. Bakers wishing to standardize on flour types and reduce storage requirements can purchase one basic flour grade, supplementing with 2% to 4% vital wheat gluten for variety bread, roll and bun production. The addition of vital wheat gluten results in a direct increase in flour protein proportional to the amount added. However, the increase in gluten content is actually greater than this direct proportion, since almost all the protein in vital wheat gluten forms gluten, whereas some 20% of flour

TABLE 7

| | % |
| --- | --- |
| Moisture | 6·0 |
| Protein (N × 5·7) | 71·0 |
| Ash | 1·2 |
| Lipid, ether extr. | 1·5 |
| Lipid, acid hydrolysis | 6·5 |
| Carbohydrate | |
| (N-free extract and fibre) | 15·3 |
| Particle size, % passing US | |
| 60-mesh screen | 97 |
| Colour | light tan |
| Form | free-flowing powder |

Source: Hercules Powder Company Inc., Wilmington, Delaware, USA.

TABLE 8
**Effect of vital wheat gluten addition on protein content of flour**

| Protein content of flour before vital wheat gluten addition (%) | Protein content of flour after vital wheat gluten addition (%) | | | |
|---|---|---|---|---|
| | *1*[a] | *2*[a] | *3*[a] | *4*[a] |
| 10·5 | 11·1 | 11·7 | 12·2 | 12·8 |
| 11·0 | 11·6 | 12·2 | 12·7 | 13·3 |
| 11·5 | 12·1 | 12·6 | 13·2 | 13·8 |
| 12·0 | 12·6 | 13·1 | 13·7 | 14·2 |
| 12·5 | 13·1 | 13·6 | 14·2 | 14·7 |
| 13·0 | 13·5 | 14·1 | 14·7 | 15·2 |
| 13·5 | 14·0 | 14·6 | 15·1 | 15·7 |

[a] Parts vital wheat gluten added to 100 parts flour.
Source: Hercules Powder Company Inc., Wilmington, Delaware, USA.

protein is non-gluten protein. Gluten, the structure-forming functional component of the system, confers the desired properties to both dough and the final baked product. Tables 8 and 9 show the effect of vital wheat gluten on flour protein content and water absorption, which in turn results in increases in dough and bread yields. Since bread flours have a narrow mixing tolerance, and since variety bread doughs containing substantial amounts of non-flour ingredients are very sensitive to mixing, the addition of vital wheat gluten improves mixing tolerance in such doughs, allowing quality products to be made under a wider range of mixing conditions.

For each part of vital wheat gluten in the bread formula, the addition of a minimum of 1·5 parts extra water is required. The dough yield is simultaneously

TABLE 9
**Absorption increase from vital wheat gluten addition**

| Parts of flour | Parts of additional water to be added to dough | | | |
|---|---|---|---|---|
| | *1*[a] | *2*[a] | *3*[a] | *4*[a] |
| 100 | 1·5 | 3·0 | 4·5 | 6·0 |
| 200 | 3·0 | 6·0 | 9·0 | 12·0 |
| 300 | 4·5 | 9·0 | 13·5 | 18·0 |
| 400 | 6·0 | 12·0 | 18·0 | 24·0 |
| 500 | 7·5 | 15·0 | 22·5 | 30·0 |
| 600 | 9·0 | 18·0 | 27·0 | 36·0 |
| 700 | 10·5 | 21·0 | 31·5 | 42·0 |
| 800 | 12·0 | 24·0 | 36·0 | 48·0 |

[a] Parts of vital wheat gluten added to 100 parts flour.
Source: Hercules Powder Company Inc., Wilmington, Delaware, USA.

increased, each part of vital wheat gluten giving about 2·5 parts extra dough, and since part of the water is retained in the finished baked product, bread yields are increased in proportion. On average, with each percentage point of added vital wheat gluten, loaf volume should increase by 2·5% to 3·0%, and additions of 2% to 4% to a dough should give further volume increases of 5% to 12%. These increases will, however, only be realized when certain readjustments are made, and preconditions satisfied, viz.:

1.  Elimination of formulation overload with non-flour ingredients which dilute both flour and added vital wheat gluten, protein.
2.  A readjustment of dough oxidants, reducing agents, conditioners, and yeast food may be necessary to optimize the potential benefits of the added gluten proteins.
3.  Provision for adequate gas production by readjustment with malt or enzymes.
4.  Optimization of mixing time to achieve optimal dough viscosity and specific mixing intensity, thus ensuring the best possible water distribution within the dough and final bread quality.
5.  Adjustment of dough water absorption, depending on flour grade used and amount of vital wheat gluten added.

The amounts of vital wheat gluten recommended for various products are expressed on a flour weight of 100 units (always), which is referred to as 'Baker's percent' and is used for all formula ingredients. 'Formula percent' measures the weight of formula ingredients as a percentage of the total formula weight. The sum of all ingredients being always equal to 100%.

   The best procedure for adding vital wheat gluten depends on the breadmaking process.

*Sponge doughs*

Unless the sponge contains coarse materials such as wholewheat-flour, rye-flour, oatmeal or crushed grains, best results are obtained by adding vital wheat gluten at the sponge stage. Where low-protein flour is used, the sponge absorption must be increased by about 1 pound for each pound of vital wheat gluten added, and the water addition at the dough stage reduced by that amount. In this case total mixing time at the dough stage would be the same as that utilized when a high-protein straight-grade or clear flour is employed in the formula. Bromate-type yeast foods or conditioners could be increased from 0·15% to 0·25% of flour weight in order to exploit the presence of the added gluten proteins.

*Straight doughs*

In straight doughs vital wheat gluten is best added with the dry ingredients. The flour is added to the mixer, and the gluten put on top of the flour covering with several scoops of flour. Alternatively, the flour and gluten can be premixed by closing the mixer and turning the blades two or three revolutions before adding the liquid. For the average high-speed mixer, 2% vital wheat gluten requires about 0·5 to 1·0 minute additional mixing, 3%, 1·0 to 1·5 minutes more, and 4%, 1·5 to 2·0 minutes more than doughs without added gluten. Absorption can be increased by about 1·5 pounds for every pound of vital wheat gluten at whatever stage it is added.

## Continuous Bread Processes

Continuous bread processes involve mixing sugars, e.g. high-fructose corn syrup, salt, yeast, milk solids, and often flour with water to form a brew. This is allowed to ferment for 2–3 hours, after which it is fed continuously into the predeveloper, together with oxidant, liquid shortening and the residual flour. After predevelopment the ingredients are pumped continuously into a developer, which is a high-speed dough mixer. The dough is then dropped continuously into bread-pans from the mixer, before proofing and baking in the usual manner. Therefore, it is desirable to add the vital wheat gluten to the brew, since the added gluten then has the benefit of the 2–3-hour fermentation and a more uniform dispersion is assured. Additionally, the dried gluten must be screened into the blending tank while the mixer is in operation; adding it into the mixer vortex close to the head facilitates dispersion. A minimum of 10% of total flour weight in the brew is recommended. Alternatively, the dried gluten may be added continuously with the flour at the predeveloper, in which case a hopper and automatic feeder will be required. In continuous processing, in addition to the usual advantages gained by adding gluten, the sidewalls of the baked bread, which tend to be weaker than in the case of sponge-bread, are strengthened. Sidewall-weakness becomes apparent during proofing, especially when doughs are jostled on transfer to the oven, resulting in excessive 'cripples'. Also, the buffering effect of the milk solids in the brew have a structural weakening effect at subsequent stages, and the added gluten helps to offset this defect. Table 10 illustrates the applications of vital wheat gluten according to baked product.

Owing to the imposition of EC tariffs, in the form of net import charges per tonne of 40–100 pounds sterling, it has become attractive for millers in the UK to replace Canadian wheat with increasing percentages of English-grown wheats and added vital wheat gluten to adjust the breadmaking quality of their grists. However, this is only of financial interest when the price of vital wheat gluten is less than about 2000 pounds sterling per tonne, and the quality of the gluten is such that 1 part Canadian wheat protein can be replaced by 1 part vital wheat gluten. Furthermore, technological implications regarding the functionality of the vital wheat glutens are of paramount importance in the evaluation of any scheme to replace Canadian wheat in mill grists.

The composition and functional performance of vital wheat glutens are as variable as the wheats and flours from which they are derived. An average sample according to Sullivan (1954) gives a typical analysis on a dry basis: protein 85%; lipid 8·3%; starch 6·0% and ash 0·7%. The lipid is held strongly to the protein, only 0·5% to 1·5% being extractable with ethyl or petroleum ether. The residual lipid balance, amounting to 65–75% of total lipid, can only be extracted by acid hydrolysis. Of this 65–75%, some 35–50% is phospholipid. These residual lipids are often referred to as 'bound' lipids, because of the strong electrostatic forces which

TABLE 10
**Wheat gluten applications**

| Baked product | Optimum level | Function |
|---|---|---|
| Wheat bread with bran | 3·0 baker's % ⎫ | Permits use of inert ingredients; |
| Multi-grain bread | 5·0 baker's % ⎭ | prevents degassing |
| Wholemeal fibre-increased | 6·0 parts by wt | Permits inert additions |
| Vienna | 2·0 parts by wt | Improves crispness, volume and texture |
| Hamburger buns | 2·0 bakers % | Improves softness |
| Low slice-weight bread | 30·0 parts by wt | Essential for high volume |
| Wholewheat bread from flaked-wheat (Netherlands) | 10·0 baker's % | Permits use of high levels of inert ingredients; improves volume, grain texture, yield and shelf-life |
| Bran bread (UK) | 2·0 parts by wt | Improves loaf volume, texture, yield, shelf-life; permits use of increased levels of inert ingredients |
| Brown soft rolls (UK) | 2·0 parts by wt | Improves softness of dough; enhances appearance and shelf-life; improves dough machinability |
| Pizza crust | 1·0–2·0 baker's % | Supplements and strengthens the natural flour protein; reduces moisture transfer from sauce to crust; provides chewiness, strength and body |
| Kaiser (hard) rolls (USA) | 2·0 baker's % | Improves structure, shelf-life, moisture retention; enhances appearance and eating quality |
| Salad rolls (Australia) | 4·0 baker's % | Aids required dough softness and crumb strength; improves appearance and keeping qualities |
| Protein-increased bread (Australia) | 5·0 variable | Increases protein level of bread to meet a minimum protein requirement in Australia of 15·4% total protein (N × 7·5) % dry basis; exact usage depends on protein of flour used |

Source: Reference Source 1985, *Bakers Digest*, p. 34.

bind them to the other molecular components of gluten. It is the presence of lipids which determines to a large extent the functional properties of the gluten for breadmaking. Grosskreutz (1960, 1961) showed how the removal of the bound lipids deprives the gluten complex of its elasticity by changing the bonded sheet structure, which can be clearly identified from X-ray diffraction patterns. Also, Hess (1954, 1955) presented evidence to show that gluten does not occur as such in the wheat endosperm, but is only formed by mechanical treatment of flour in the presence of water. The isolation of proteins from flour, freed of starch granules without the use of water results in a protein complex with a swell or water absorption capacity of 25%, whereas vital wheat gluten swells in contact with water 200%. The X-ray crystallographic techniques used by Hess indicate the existence of two distinct protein components in the wheat endosperm: a 'wedge-protein', which fills the spaces between starch granules, and an 'adherent-protein', consisting of a network of fibrils, surface-coated with lipid, which extend over the surface of each starch granule. Rohrlich and Müller (1968) isolated and characterized a lipid–protein complex from wheat-flour with a composition 70% protein:30% lipid, which they described as proteolipid PL1, in order to differentiate it from another proteolipid, PL2, with quite another composition, isolated from wheatgerm. PL1 has a glutamic acid content of 30–32% and a relatively large amount of proline, but small amounts of the basic amino acids resembled the gliadins or 'wedge-protein' in composition. X-ray crystallographic patterns indicate that one part of the protein is closely associated with lipid, i.e. lipophilic and the other part only co-solubilized. Samples of PL1 were isolated from different wheat varieties, and commercial flours, using methyl alcohol/chloroform (1:2). The PL1 sample from Canada Western Red Spring (CWRS) wheat not only gave the highest yield of material but, when added at 1·0% flour weight to commercial flour, brought about a marked improvement in baking performance and final bread characteristics. This effect would confirm the validity of the suggested gluten model put forward by Grosskreutz, namely that a stabilizing double-lipid layer exists between protein platelets (Fig. 16). When this lipid layer is removed by extraction with polar solvents, the stability of the gluten is lost, and also its ability to take up added fat, sugar and other additives. Furthermore, when the polar lipids which are bound to protein (PL1) are extracted from CWRS quality wheat, it loses its 'improver' wheat effect. Controlled baking tests carried out with 1:3 flour mixtures of CWRS and C-grade German wheats before and after removal of proteolipid PL1 have confirmed the importance of this

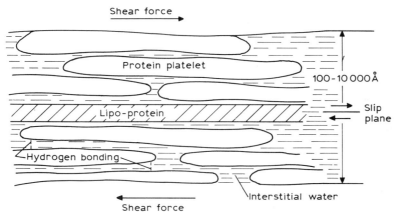

Fig. 16.   Model of gluten sheet structure. (Source: J. C. Grosskreutz, *Cereal Chem.* **38**, 336, 1961.)

complex formation in all wheat glutens, and in particular the connection between the 'improver' effect of CRWS wheat in grists and the special properties of its proteolipid content in developing glutens of outstanding functional and technological importance in breadmaking processes worldwide.

The detailed evaluation of the breadmaking functionality of any gluten sample demands a consideration of the country of origin, the genetic material used in breeding the wheat, the milling process used to mill the wheat into flour, the gluten extraction process, and especially the process employed to dry it. Proteins are very susceptible to heat damage and denaturation, and gluten is no exception. Inefficient drying techniques will reduce both the vitality of the sample and its colour grading. A convenient method of assessing its vitality is by swelling tests using standardized concentrations of dilute lactic acid and a constant weight of re-hydrated gluten tested at zero and 30 minutes after preparation. Good quality glutens show large swelling volumes, when allowed to settle in measuring cylinders of standardized dimensions. This procedure is based on that developed by Berliner and Koopman for measurement of the 'Quellzahl' of glutens washed from wheats and flours, which dates back to 1929, and which still presents valid comparisons between gluten swelling under controlled test conditions. In its original form, 1-g samples of washed out gluten, divided into 30 pieces, are placed in special-shaped and graduated Quellzahl-flasks and incubated at 32°C for 60 minutes. The flasks are then inverted and allowed to sediment, after which the swollen volumes can be read off from the graduations.

Lactic acid concentrations normally used were 0·03N as swelling medium. The original flask graduations gave a swelling-number or Quellzahl range of 0 to 38 units on the scale. In the interpretation of the data derived from swelling tests, it must be noted that drying processes increase the swelling of glutens as well as wheat samples. Differences between swelling after 30 minutes standing of gluten samples and those measured at zero time give an index of the presence of proteolytic breakdown in gluten samples. Gluten softening and tendency to flow can be measured either with the help of a penetrometer, or more simply by the Auerman method. The latter involves the incubation of a 5-g gluten ball on a Petri-dish marked and graduated in concentric circles, the ball being placed in the centre. Samples are incubated at 30°C with adequate humidity for 1 to 2 hours, after which the area enclosed by flow-out of the samples is measured and used as an index of gluten softening.

Rapid and homogeneous hydration are signs of 'vitality' in gluten samples; failure to hydrate is indicative of heat damage during the drying process. Freeze-drying or the use of liquid nitrogen are ideal, since they produce minimal tissue damage due to denaturation, but cost and expediency dictate the more conventional methods of vacuum and flash-drying.

Gluten supplementation of flour is carried out in the mill by continuous metering of gluten powder into the flour stream as required to maintain the protein content. The legal position in the UK regarding flour supplementation is covered basically by the Bread and Flour Regulations 1984 and Food Labelling Regulations 1984, but interpretation problems remain, especially with regard to the declaration of ingredients. Additions to flour are covered by the clause 'small quantities for technological purposes' without the need for changing the name or labelling of the flour, and no exclusion clauses exist for wholemeal flours. However, when added to flour or doughs in the bakery, vital wheat gluten should be declared with the other ingredients. Nevertheless, if vital wheat gluten supplemented flours become designated as 'compound ingredients', then a labelling declaration would become obligatory.

European patent application 134,658 dated 20 March 1985 taken out by Kyowa Hakko Kogyo Company Limited of Japan describes a method of improving the functional and staling-retardation properties of vital wheat gluten by treating the dough before washing with a phospholipase A preparation, thus increasing the monoacylglycerophospholipase content to 72 mol%. The conditions of such treatment were 60 minutes at 30°C using strong wheat flour at 80% water absorption. Improved wetting of vital wheat gluten is obtained by encapsulation with succinylated monoglycerides at 5% to 30%; baking properties are also improved. Acyl amino acid derivatives have been found to improve dough fermentation activity and the acidity of doughs by improving conditions for fermentation microorganisms. Such derivatives are: lauroylasparagine, myristoyl-norleucine and stearoylorleucine. N-Steroyl-alpha-alanine and N-steroyl-glycine improve bread volume, porosity and flavour when added to flour at 0·5%, working synergistically with potassium bromate. Initial work on this group of compounds has been carried out in USSR by Stepanova and co-workers at the Kemerovo Technical Institute, and Konova and co-workers at the Moscow Technological Institute (1986).

## 1.5.11 HYDROCOLLOIDS

The hydrocolloids are a very diverse group of substances both in their source and in their application in the food industry. Their functionality and application depends on their ability to bind water and, in so doing, to effect rheological changes, especially in viscosity, in food processing systems in which they are included. In cereal processing this commences with steps taken by the miller to achieve a reasonable amount of starch damage during the milling process, and continues in the bakery where various hydrocolloids are added as single or composite additives to improve elasticity and shelf-life.

### Damaged Starch
During the process of milling wheat into flour, on average 10% of the starch granules become mechanically damaged by the direct pressure exerted by the reduction rolls at the head of the mill. This percentage varies, depending on the grain structure or hardness, and can be controlled to a degree by roll surface-texture, roll gap, roll pressure, roll-speed differential and feed-rate by the miller. These damaged starch granules, in common with 'pregelatinized' starch, i.e. starch which has undergone water treatment at gelatinization temperature, can become over a thousand times more rapidly broken down than undamaged granules. Therefore, the main source of gas-production, especially in short-time breadmaking processes,

is the quantity of available damaged starch granules. In any case, the beta-amylase, of which there is always an adequate supply in flour, cannot initiate the digestion of undamaged starch granules. Alpha-amylase is capable of attacking the 1,4 links of the starch macromolecule internally once the molecule has been opened up, breaking the molecules into high molecular weight dextrins by random fragmentation. These then become degraded into lower-molecular-weight dextrins and thence into maltose. The straight-chained amylose fraction becomes hydrolysed to within a few residual glucose units, but the branched amylopectin element only becomes degraded to about 64% of the theoretical maltose yield. Since alpha-amylase is synthesized during germination, in contrast to beta-amylase, it is only present in sound, unsprouted wheat flour in trace amounts. Therefore, in order to ensure a progressive fermentable sugar supply throughout the dough fermentation process, supplementation is normally necessary from external sources, e.g. malt or fungal alpha-amylases.

Nevertheless, it is quite possible to find low concentrations of maltose, even in the presence of adequate concentrations of alpha-amylase, owing to the lack of substrate, i.e. inadequate amounts of damaged starch granules. This fact can be confirmed by testing flours with about the same amylase activity, but with different starch damage percentages. The degree of starch damage can be determined by the standard AACC-method, in which the flours are incubated in aqueous suspension with an excess of amylase. The sugar produced is a measure of the starch damage, since the maltose produced in any given incubation period will depend on the amount of available damaged starch granules acting as substrate for the alpha-amylase. However, in practice, a low level of maltose production in the dough can also be due to a lack of flour alpha-amylase. Therefore, a flour supplementation with heat-sensitive fungal alpha-amylases will cover any deficiency. Since fungal alpha-amylases are effective up to temperatures of 60–70°C, they also exert an effect during the oven phase, especially on crumb grain and texture, but care is necessary when they are used in sponge-doughs to avoid difficult machinability. Where short-time dough processing is used, careful control of starch damage and alpha-amylase can provide improved machinability and increases in dough water absorption. Special process optimization techniques involving starch damage, alpha-amylase, protein and dough water absorption were devised by Farrand (1969). Mathematical models involving these parameters were constructed, and their interrelationship established by statistical multiple regression analysis, thus permitting the estimation of flour water absorption on a commercial basis. In addition, an equation was established relating optimal levels of alpha-amylase to starch damage, thus controlling gassing-power and starch rheology within the dough processing system in operation. Data utilized in these calculations were obtained by the Farrand methods for both alpha-amylase and starch damage.

## 1.5.12 STARCH-BASED PRODUCTS

Starch-based products are added to doughs in order to regulate the water distribution or water-holding capacity. Such products include heat-treated flours and starches, which absorb increased amounts of water in the cold as a result of pregelatinization, e.g. pregelatinized wheat flour, potato-flour, and extruder-produced variations. Carob bean flour (*Ceratonia siliqua*) is also used on account of its swelling properties, a common source of Quellmehl or swell-flour often used in Germany to control swelling in rye- and mixed wheat/rye-doughs in combination with lactic acid, especially when excess enzymic activity due to amylases and pentosanases, coupled with susceptible native rye starch, give rise to low amylograph and swelling-curve viscosities. In the case of grain sprout damage, the addition of starch powder or acidulated Quellmehl at 3–5% flour weight inactivates excess starch damage and starch and protein degrading enzymes, and ensures in most cases an acceptable end-product. In the USSR, similar processing effects are achieved by preparing a pregelatinized flour/water (PFW) suspension in various forms for routine addition to production bread doughs. Special machinery, in the form of autoclave cookers, is used to prepare the various types of PFW suspensions. The advantages of using such intermediate products are: a strengthening of the bread crumb-structure, improvement in dough and bread yields, and improvement in the digestibility of bread, especially wheat-bread varieties. The re-processing of up to 3·0% old bread crumb from the same bakery is also permitted in some countries, which also acts as a pregelatinized additive, having similar effects on doughs and bread.

### 1.5.13 CELLULOSE AND CELLULOSE DERIVATIVES

As a result of current trends in the baking industry, the market for low-calorie raw materials and end-products is growing at a tremendous rate. The number of calorie-light bread is an indication of this trend. One major ingredient in this type of product is powdered cellulose, which functions as a non-caloric bulking agent.

Cellulose is the world's most abundant organic chemical, being the structural part of plant cell walls together with hemicelluloses, pectins, gums, mineral components, and a non-degradable substance called lignin.

Chemically cellulose is a molecule composed of many repeating glucose units, bonded together by a beta 1–4 bond. Starch is the same as cellulose, except that it is bonded by an alpha 1–4 bond. This difference in bonding makes starch digestible by humans, whereas cellulose is non-digestible. Commercial food grade cellulose is only about 90% cellulose, the residual 10% being hemicellulose. The hemicelluloses consist of such sugars as xylose and galactose, which are found in close association with cellulose in the native state, and remain as such in the commercial food-grade product. Cellulose is available in various particle sizes from 20 microns to 120 microns, but those falling in the 20–60 micron range are most often used in baked goods. The coarse grades absorb up to seven times their weight of water, and the finer grades about three to four times their weight of water.

The main reason for using powdered cellulose in foods is that it is considered to be non-caloric. In 1984, the FDA (USA) proposed that the amount of dietary fibre in products be excluded from any determination of calorie count for nutritional labelling purposes. Powdered cellulose is 99% dietary fibre and appears on the GRAS (generally recognized as safe) list of the FDA, being designated as alpha-cellulose. To date there is no evidence that alpha-cellulose presents a safety problem when used in food. In order to effect a 33% reduction in calories, cellulose is often used in conjunction with other low-calorie products such as gums and polydextrose. In the case of powdered cellulose and polydextrose, low-calorie products are formulated by partial replacement of flour with cellulose. In so doing, a 75% weight replacement of every weight unit of flour is recommended as a starting point. Although the analytical interpretation of 'dietary fibre' depends on the procedure used, dietary fibre is considered to be the cellulose plus hemicellulose, lignin, gums and pectin found in foods. All these constituents have the characteristics of not being digested before reaching the ileocaecal valve; the colonic microflora may, however, cause some breakdown beyond this point. As pointed out by Professor P. J. van Soest of Cornell University, Ithaca, New York, total dietary fibre contents range from 100–700% of the crude fibre content, depending on the relative proportions of lignin, cellulose, and hemicellulose.

Table 11 shows the relative sources of fibre from various bread, flours, mill products and cellulose isolates marketed under certain trade-marks.

Carboxy methyl cellulose (CMC) is commercially available as the sodium salt, being manufactured from natural cellulose by chemical substitution. Wood-pulp is alkalized, converted to cellulose either by treatment with monochloroacetic acid, and washed free of the byproducts sodium chloride and sodium glycolate in order to convert the technical grade CMC to a degree of purity which can be accepted as food grade CMC. The sodium salt of CMC combines the valuable properties of solubility with the capacity to absorb large amounts of water and swell, forming highly viscous liquid media. The release of the bound water is slow, and on drying out it

TABLE 11
**Dietary fibre contents of some common bakery foods**

| Commodity | Carbohydrate (%) | | Dietary fibre (%) |
|---|---|---|---|
| | Total | Complex starches | |
| Bread, white | 47·5 | 46·1 | 2·7 |
| Bread, rye | 49·1 | 48·3 | 3·0 |
| Bread, wholewheat | 40·4 | 39·2 | 9·5 |
| Flour, white | 72·9 | 70·8 | 3·2 |
| Flour, wholewheat | 59·6 | 57·8 | 9·5 |
| Flour, full-fat soy | 17·3 | 9·1 | 11·9 |
| Wheat bran | 25·1 | 21·9 | 42·4 |
| Soy bran | | | 75·0 |
| Solka-Floc powdered cellulose | | | 99·0 |
| Barley fibre | | | 55·0 |
| Keycel alpha cellulose powder | | | 99·0 |
| Vitacel cellulose powders | | | 99·0 |
| Mexpectin pectin | | | 90·0 |

forms smooth films which possess a certain elasticity. An additional important property is its adsorption capacity, becoming attached to small particles and surrounding them, thus preventing the build-up of larger particles. This property allows the use of CMC as a dispersing agent and inhibitor of crystallization. A decisive factor in the use of CMC in the food industry is its physiological safety status. CMC is not degraded during its passage through the human metabolic tract. Therefore, it can be utilized in dietetic foods, since it only constitutes a part of the dietary fibre, and has no adverse effect on taste or smell. Furthermore, owing to chemical modification of the molecule, it becomes less susceptible to microbial attack. For breadmaking, the optimal level of addition of CMC is 1% based on flour weight or 1 part dry CMC to 100 parts bread-flour.

However, it is not advisable to add CMC in an undispersed state if the full benefit of its properties are to be exploited. Instead, for initial swelling, 1 part CMC is dispersed in 20 parts warm water, ensuring that no CMC particles remain undispersed, by using a high-speed liquid mixer. In this case, 21 parts of dispersed CMC are added to 100 parts flour to give the optimal concentration. This concentration and method of addition is used for both wheat- and rye-flour processing. In the case of the normal dough-souring process for mixed rye/wheat-doughs (80:20) and 60:40), known as the *Sauerteigführung*', 50% of the CMC dispersion is added at the *Vollsauer* or second ripening stage, and the residual 50% at the final doughmaking stage, so that the dough temperature can be kept constant.

CMC is also available in granular form, which can be added to dough in the dry form, but unless the process involves a pre-fermentation stage or stages, as in the case of rye-flour doughs, the swelling time is insufficient for complete water uptake by the CMC. In the preparation of wheat-flour doughs, where the granulated form of CMC is being used, high-speed mixing is essential for complete hydration of the CMC. Where kneading or pressure-action mixers are in operation, initial dispersion and swelling is recommended in all cases.

The use of CMC in mixed wheat/rye-doughs prolongs the shelf-life of the resultant bread by 24 hours, and that of wheat-flour bread by 20 hours. This improvement in shelf-life refers to the use of 1% CMC, based on flour weight, and bread units of 1 kg baked weight. The prolongation of shelf-life in the case of bread rolls, using 1% CMC, is restricted to 6 hours, owing to the relative unit size. The full potential of CMC in baked products has yet to be realized since, in spite of its wide application as a stabilizer in fillings, desserts, dairy, and frozen foods as E466 under EC legislation, it has not yet been included in the bread and flour regulations. Under FDA (USA) legislation, it is given a GRAS listing under section 409. This permits its use in concentrations not exceeding that required to accomplish the intended technical effect in the food product. Also, its purity must be adequate for food use, being not less than 99·5% sodium CMC on a dry weight basis, and with maximum substitution of 0·95 carboxymethyl groups per anhydroglucose unit, minimum viscosity of a 2·0% solution in water at 25°C being 25 centipoises. In view of the advantages for production, retail and the consumer, it became accepted for use in rye, mixed wheat/rye, wheat and speciality breads at 0·75%, based on baked-product weight, in 1969 in the German Democratic Republic. However, in the latter case, all production is on an industrial centralized basis, and under the control of the Grain Processing Institute for the GDR in Potsdam (VEB WTÖZ, and VEB IGV).

## 1.5.14 SALT, ACIDS AND METALLIC IONS

Salt has always been an essential ingredient in any yeasted dough, and although it is necessary for flavour or seasoning, its presence as common salt or sodium chloride has a very important effect on functional properties and fermentation control of all doughs, whether yeasts or sour-dough cultures are being utilized. The technological processing improvements, as expressed by the baker, are an increase in dough stability, firmness and capacity to retain the fermentation gas. The bread-crumb becomes more elastic in the case of wheat bread, and in combination with the correct acidulation and pH adjustment during rye and mixed wheat/rye processing, similar advantageous effects are apparent. The browning process in the baking cycle becomes more even and intensified. Bread flavour progresses from being bland and tasteless to becoming a distinctively aromatic, well-seasoned sensory experience.

Considerable differences exist between the influence of common salt on wheat- and rye-flour doughs. Wheat-flours, which have starch with a high degree of condensation, i.e. compact molecular structure, and low amylolytic activity show progressively high amylograph viscosities of up to 1000 units, even with relatively small additions of salt at the 1·0% level. Maltose production is also progressively reduced with increasing salt concentrations. Starch gelatinization temperatures of wheat-flours without salt addition begin at about 63°C, but the addition of about 1·0% salt raises the temperature to about 69°C. Wheat-flours with a lower degree of starch condensation also respond positively to increasing additions of salt; for example, a flour giving an amylograph peak paste viscosity of 400 units without any salt addition, will give viscosities of over 800 units when 1·0% salt solution is used. Maltose production of the salted paste is correspondingly reduced to 2·0% from 8·0% without any salt addition. Maltose is expressed on flour weight, and amylograph pastes are prepared by adding 60 g flour to 450 ml of water or salt solution as appropriate. Amylograph paste viscosity behaviour simulates the initial phase of the baking process, since flour-water pastes are continuously heated at the rate of 1·5°C per minute for about 20 minutes. Rye-flours also give higher paste viscosities with the addition of salt, but larger concentrations are necessary than in the case of wheat-flours. Rye-flours with initially low paste viscosities due to sprouting usually require salt concentrations in excess of 3·0% in order to effect any increase in paste viscosity. Starch gelatinization temperatures of rye-flours commence at about 54°C for Type 1150 and 997 without salt addition, but the addition of about 1·0% salt raises the temperature to about 60°C. However, the limiting factor in the addition of salt is the taste of the baked bread. The optimal level of salt in most bread, whether it is made from wheat- or rye-flour, is about 2·0% based on flour weight. In the case of rye-flours, the simultaneous adjustment of dough pH is essential for the production of quality bread. This adjustment is traditionally achieved by use of a progressive dough sour, initiated by a lactic and acetic acid fermentation starter culture, which results in bread of optimal quality at pH about 4·5. In addition, however, the dough should show as low a buffering effect as possible, which is usually the case when rye-flour protein is low. Rye-flours with relatively low amylograph peak paste viscosities of about 200 units, require 2·0% salt and approximately 0·7–1·0% dough-acidulant (flour weight), consisting of prepared mixtures of lactic, acetic and citric acid to adjust dough pH to around 4·0–4·5. This results in the best combination of the following rye bread quality characteristics: symmetry, crust colour, texture, crumb colour, crumb elasticity, crumb strength, crumb acidity, pH, and crumb moisture content. This combined effect of salt and acid on the water-soluble proteins of wheat and rye, within the constraints of final bread flavour characteristics, have been studied by Schulz (1962 and 1963), Schultz and Stephan (1960 and 1961) at the Grain Research Institute at Detmold, FRG, in considerable detail. The technological significance of salt and acids were further researched in great detail by Huber (1961, 1962 and 1964), involving extensive types and quantities of wheat- and rye-flours. Huber concluded that salt delays the onset of both the initial and maximum gelatinization temperatures of flours by 5–10°C; and the use of salt and acids allows the optimization of the baking properties of rye-flours in particular, giving them processing properties similar to those of wheat-flours.

Another effect of salt was established by Mecham and Weinstein (1952). These workers found that normal concentrations of salt in doughs, about 2·0%, markedly reduce the total bound lipids of the gluten complex in general and the phospholipids in particular. Fullington (1969) has reported the formation of well-ordered structures between wheat proteins, phospholipid–metal complexes. Such phospholipid–metal interaction provides a manipulative mode for the water-soluble wheat proteins. Depending on the metal ion and phospholipid composition, certain proteins can be removed from solution, including enzymes. Binding of protein to lipid in such cases can involve carboxyl side groups, SH-groups, tyrosine, serine and threonine, in both bound and non-bound wheat water-soluble proteins.

# 1.6   Chemical Bonding During Doughmaking

During dough assembly and mixing, as well as during the fermentation process, a whole series of colloidal, chemical and biochemical processes are set in motion. These processes proceed alongside one another, and in some cases cut across and compete, constituting an exceptionally complex system which cannot be considered as being completely understood. In addition, additives and shortenings are involved in the dynamics of the system, the changes being between low-molecular-weight, soluble or gaseous substances on the one hand, and high molecular substances with defined structure on the other. The type and extent of these reactions also depend on the intensity of the mixing process, as illustrated in Section 1.4.

The structural elements of the main ingredient, flour, consists mainly of broken endosperm fragments, which in turn are composed of basic protein material with embedded starch granules of various sizes. Particles in excess of about 28 $\mu$m consist of endosperm cell fragments of starch embedded in a protein matrix, those falling between 20 and 30 $\mu$m contain mainly starch, and those from 0 to 20 $\mu$m are high protein fines. Variations in composition depend on the grain kernel hardness, and the milling process. Soft-textured wheats yield more free endosperm cell-wall fragments, and traces of germ, which defy even the most sophisticated milling technology.

Chemical changes which take place at an early stage of the mixing process involve the water-soluble dough fraction. The flour lipids change their distribution, and become rapidly oxidized, oxidation being particularly high in the aqueous fraction that contains the bound lipids. The petroleum ether-soluble fraction, containing mainly free lipids, according to Laignelet and Dumas (1984), show less signs of oxidation. After about 10 minutes mixing, the soluble and insoluble glutenin protein and the gliadin also become oxidized. With changes in the lipid content, the protein content of the fractions change their distribution, such changes being considered to be due to the lipoxygenase activity. Phenolic acids, also present in the water-soluble dough fraction, and in particular ferulic acid, tend to migrate out of the soluble phase, forming an adduct with cysteine when the dough is overmixed. This results in a progressive breakdown and loss of dough resilience according to Jackson (1983).

Crucial for the formation of wheat-flour dough is the gradual synthesis from the flour endosperm components of the heterogeneous, rubber-like complex known as gluten. On addition of approximately 60–65% water to the flour, this complex forms as one-third gluten and two-thirds water. Sullivan (1954) gives a typical general analysis of dried wheat gluten as: protein 85%, lipid 8·3%, starch 6·0%, and mineral ash 0·7%. This analysis can vary somewhat with wheat source and mixing intensity.

The lipid is held very strongly to the protein, and is not extractable with either ethyl or petroleum ether. The exact nature of the chemical reactions by which gluten is formed continues to concern cereal chemists, and much work on flour proteins and their behaviour in doughs has been carried out by Dale K. Mecham and co-workers at the Western Regional Laboratory Agricultural Research Service of the USDA, Berkeley, California 94710, USA, during the early nineteen-seventies. In wheat gluten two protein groups dominate, the prolamines and the glutelins, which are the chief proteins of all cereals. In the endosperm of the wheat grain, a prolamine called gliadin and a glutelin called glutenin are present in about equal concentrations. It is the unique presence of gliadin and glutenin which contributes towards the formation of gluten, which is capable of retaining gases and leavening doughs. Bran proteins consist mainly of prolamines with fair amounts of albumens and globulins. The gliadin is soluble in alcohols, highly extensible and has low elasticity, having a molecular weight of less than 100 000, forming predominantly intramolecular bonds. The glutenin fraction is insoluble in alcohols, less extensible but highly elastic, having a molecular weight of over 100 000, forming both intramolecular and intermolecular bonds. In considering the structure of wheat proteins, as with all complex proteins, there are several levels or elements which determine the configuration of the macromolecule. These are generally referred to as primary, secondary, tertiary and quarternary. The primary structure is determined by the amino acid sequence of the polypeptide molecule. The polypeptide chains are held together by peptide bonds, and each polypeptide has a definite sequence in the arrangement of its amino acid molecule residues. The following scheme illustrates the possible amino acid sequence of a polypeptide molecule:

$$NH_2—P—M—M—V—S—P—A—M—P—M—M—A—COOH$$

P proline, M methionine, V valine, S serine, A alanine

The polypeptide chains form the backbone of the protein molecule. Each polypeptide chain is built up from the repeated linking of the two amino acid groups —$CH(NH_2)$—COOH. The residual part of the amino acid molecules form the side chains, which determine the properties of the macromolecule. If the primary structure becomes disrupted, the protein macromolecule undergoes progressive degradation, and loses its natural properties (see Fig. 17(a)). According to Kelesov (1951) the gliadin molecule consists of relatively short-chained polypeptides, formed from 12–15 amino acids. 25 peptide chains, each with a molecular weight of 1080 forming the complete gliadin molecule (Reznichenko, 1951). Methods involving amino acid end-group assay due to Sanger and Edman, will reveal the amino acid which initiates the polypeptide chain, as well as the sequence of the other amino acids adjoining. The number of N-terminal and C-terminal groups discloses the number of complete peptide chains in a protein.

The secondary structure refers to the spatial arrangement of the polypeptide molecule. Polypeptide molecules are not long-drawnout chains, but instead are either folded back onto one another as pleated sheet structures, or form spiral-shaped helices, according to Pauling and Corey. The nature of the folding and the

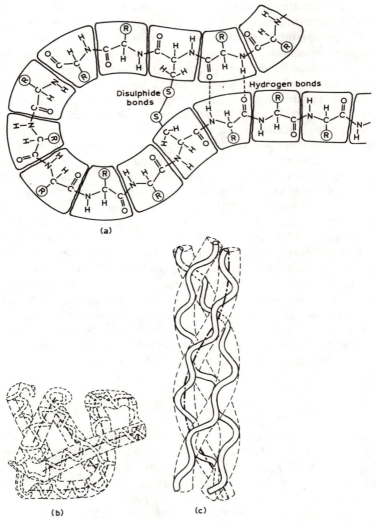

Fig. 17. Protein structures. (a) Primary structure with S—S and H-bond folding. (b) Tertiary structure of a globular protein. (c) Tertiary structure of a fibrillar protein.

distance between the residual amino acids are due to the hydrogen bonding between the O of the C=O groups and the H of the N—H groups in the polypeptide chains. Hydrogen bonding between groups in the same chain (intramolecular bonding) is responsible for the helical structure, and hydrogen bonding between groups on different chains (intermolecular bonding) is responsible for the folded or pleated sheet structures (Fig. 17(a)).

The tertiary structure refers to the distinctive and characteristic conformation or shape of the protein molecule (see Fig. 17(b, c)). The polypeptide chains become folded and re-orientated to form a relatively stable and compact shape, viz. filaments or globular-formations (see Fig. 17(b) and (c)). These spatial structures, in common with the secondary structures, are regulated by the amino acid sequence of the individual polypeptide chains. The tertiary structure is maintained by a variety of interactions between the amino acid side chain groups. These interactions include hydrogen bonding, ionic bonding, disulphide bonding, van der Waals forces and more basic and fundamental bondings.

*Hydrogen Bonding* (Fig. 17(a))
Hydrogen bonding arises because of the tendency of hydrogen, when attached to oxygen or nitrogen, to share electrons with a neighbouring oxygen. The oxygen or nitrogen draws electrons away from the hydrogen, leaving it with a positive bias. Hydrogen bonding is a relatively weak electrostatic linkage of energy levels 2–10 kcal/mol, involving electron sharing instead of proton transfer.

Although weak in terms of bond energy compared with covalent and ionic electrostatic bonds, the occurrence of hydrogen bonds is frequent and their total effect highly significant. They occur between amide and carbonyl groups, tyrosine and carbonyls, or between two carbonyls; and the role of hydrogen bonding in the secondary and tertiary structure of all proteins has received too little attention to date, considering the diversity and total effect of such bonds.

$$=C::\ddot{O}:\quad H:\ddot{O}- \qquad =C::\ddot{O}:H:\ddot{O}-$$

*Ionic Bonding*
Ionic bonding is due to electrostatic coulomb attractions between different charges at energy levels of 10–20 kcal/mol. Occurring when pH conditions are favourable between neighbouring ionized amino and carboxylate groups, present either as end-groups of peptide chains or in the side groups of diamino, e.g. lysine, and dicarboxylic, e.g. glutamic acids. Further examples are the lipid ring bonding mechanisms of the trimethylamine group of lecithins with negatively charged protein groups (see Fig. 11).

$$\vdash NH_3^+ : {}^-OOC\dashv$$

This type of bonding is not as strong as a covalent bond, and can be broken by a change in pH of the surrounding medium.

*Disulphide Bonding*
Disulphide bonds between two neighbouring cysteine units at energy levels of 30–100 kcal/mol, due to electron distribution, are covalent. These play a very important part in protein structure, giving it strength and rigidity. It is possible that the disulphide bridge of cystine is built into the peptide chain directly during synthesis, but it is more often formed in position by oxidation of two adjacent cysteine units.

$$\vdash CH_2 - S - S - CH_2 \dashv$$

As is the case with hydrogen bonding, disulphide bonding can occur between groups in the same chain (intramolecular) or between groups on different chains (intermolecular), thus contributing to the formation of both helical and pleated sheet structures. Electrostatic repulsion between particles of the same charge gives rise to polar groups of amino acid chains, which the sulphur-containing amino acid cystine can cross-link with its two sulphur atoms, forming the disulphide linkage. The reduction of disulphide linkages, followed by re-oxidation leads to different results, in dilute solution weak adhesive products form, whereas in concentrated solution intermolecular disulphide cross-linking gives rise to insoluble cohesive products.

Ewart (1985) has suggested that a measurement of the quantity of glutenin-bound low molecular thiols or end-blockers would permit the number of glutenin molecules to be calculated, and hence also an average molecular weight, which is correlated with glutenin viscosity and bread volume per gram of bread protein. The hypothesis is that glutenin viscosity is inversely proportional to the number of peptide chains, and that the main aim of reducing the native thiol content of flour with additions of oxidizing agents, is to limit or prevent the degradation of the glutenin molecules.

### Van der Waals Forces

Van der Waals forces involve non-polar group interaction at energy levels of 1–3 kcal/mol, e.g. alanine, valine and proline. Proline, which accounts for about 14% of gluten amino acids, can twist peptide chains, thus contributing to the gluten functional properties. Such bonding between amino acids with hydrophobic side-chains during gluten synthesis, is generally referred to as hydrophobic bonding. Longer polar side-chains, e.g. glutamic acid and lysine, could also be drawn into interactions. In addition, repulsion due to van der Waals forces, between non-polar groups in close proximity is widespread amongst all groups.

### More Basic and Fundamental Bondings

More basic and fundamental bondings occur between groups within amino acids, and the elimination of water between the anion $COO^-$ of one amino acid, and the cation $NH_3^+$ of another, forming the peptide bond. This is a covalent carbon–nitrogen bond resulting when water is removed between two amino acids, forming a dipeptide. Like the parent amino acids, dipeptides contain a free amino and a free carboxyl group, which combine further to form tripeptides. Multiple units being known as polypeptides. X-ray studies of such structures show that the side-groups project alternately on either side of the peptide chain.

The secondary and tertiary structure of a protein (Fig. 16(a)(b) and (c)) characterizes its distinctive chain conformation or shape, and involves all the bond interactions described. The so-called quaternary structure describes the shape of the entire complex molecule, and is determined by the way in which the subunits are held together by non-covalent bonds, i.e. by hydrogen bonding, ionic bonding, etc. Many enzymes are also known to have quaternary structures.

Clarification studies concerning the secondary and tertiary structure of wheat proteins have been carried out by Hess (1954), Traub *et al.* (1957, and Grosskreutz (1961). Using physical methods such as X-ray and electron microscopy, a picture of the gluten macromolecule, and its component fractions could be built up. It showed that the gluten protein was made up of folded polypeptide chains within a alpha-helix structure. The beta-sheets showing a strength of 70 angstrom units (7 nm). Since the X-ray plates showed the marked influence of the presence or absence of lipids, the lipoprotein model was suggested by Grosskreutz to explain the viscoelastic properties of gluten.

Vercouteren and Lontie (1954) demonstrated a structural change in the protein molecule at the moment of gluten formation. A gluten film can be produced by dissolving it in dimethyl formamide, dialysing off the solvent, and allowing to dry on a mirrored surface. This film was found to be soluble in water, but became insoluble when moistened with water and stretched. The stretching of the gluten apparently brings about a change in the alpha-helix structure.

The bonds and forces interacting in the formation of the gluten structure are numerous, as illustrated in Table 12.

For the cross-linking of wheat gluten molecule polypeptides, the already-mentioned disulphide (—SS—) bridges are of particular importance. Oxidation of the reactive sulphydryl (—SH) groups of the cysteine within the protein molecule results in a firm bonding, which can easily be split with reducing agents. These

TABLE 12

**Forces participating in the formation of protein structure**

| Type of bond | Mechanism | Energy (kcal/mol) | Interacting groups | Examples |
|---|---|---|---|---|
| Covalent | Electron distribution | 30–100 | C—C, C—N, C=O, C—H, C—N—C, S—S | Internal amino acid, peptide and disulphide bonds |
| Ionic | Coulomb attractions | 10–20 | $-NH_3^+$ $NH^+-COO^-$ $-C$  $NH_2$  $NH_2$ | Lysine, glutamic acid, histidine, arginine |
| Hydrogen | Hydrogen distribution between two electronegative atoms | 2–10 | —N—H...O=C— —OH...O=C— O...HO C— OH...O | Amide-carbonyl tyrosine-carbonyl carboxyl-carboxyl |
| Hydrophobic (van der Waals forces) | Non-polar interaction | 1–3 | Non-polar | Alanine, valine proline, etc. |
| Electrostatic repulsion | Coulomb forces of same charge | $\dfrac{Q_1 Q_2}{r^2}$ | Polar | Polar groups in the side-chains |
| Repulsive (van der Waals forces) | Repulsion between non-polar groups in close proximity | | All groups | All groups |

protein reactions in the keratin of wool and hair were reported by Elsworth and Philipps in 1958, using sodium bisulphite to open up the disulphide bridges to give sulphhydryl groups. Such reactions are also applied to hairdressing technology for the production of permanent wave coiffures. In 1956 Bourdet mentioned a similar reaction carried out on gliadin by Lindley in 1948 in *Ann. Technol.*, **2**, 2, 181–318, 1956 under the title, 'Les protides des Céréales'. This initiated a spate of rheological and technological studies concerning the behaviour of gluten proteins relative to the 'available' cysteine—SH groups, and their ratio to cystine —SS— groups in the macromolecule. Also, the resulting redox potential in aqueous flour/water suspensions in this connection, e.g. Bloksma (1958) and Bushuk (1961). In order to obtain a quantitative measurement of the ratio of SH:SS groups in flour protein, amperometric titration with 0·001N silver nitrate under an inert nitrogen atmosphere to prevent oxidation of the —SH— groups, followed by reduction of the pre-existing —SS— bridges with sodium bisulphite and further titration, is the standard procedure. By this method wheat flours of 65–72% extraction rate give values of 0·28–0·44 SH-groups per 1000 atoms of nitrogen, which are typical. The quantity of disulphide groups shows a certain relationship with flour protein content; for example, according to Axford *et al.* (1962), a flour of protein content 16·4% contained 6·6 —SS— groups and one of protein content 7·5%, 9·1 —SS— groups per 1000 atoms of nitrogen. However, —SS— and —SH— groups are present not only in the gluten proteins, but also in aqueous flour extracts. Depending on the wheat variety, Agatova and Proskurjakov (1962), in the USSR, found 34–72 $\mu$-equivalents of —SH— groups, and 33–89 $\mu$-equivalents of —SS— groups per gram of protein. The ratio of —SS— to —SH— being 0·67–1·53, whereas gluten proteins dispersed in 0·1N acetic acid gave ratios of 3·3–9·9, indicating the potentially higher disulphide content of the latter.

The addition of substances having chemically reactive groups, e.g. maleinimide, which block the —SH— groups in flour, preventing the formation of —SS— bridges, profoundly reduce the baking properties of flours, thus confirming the importance of disulphide bonding.

The presence of oxidants and oxygen in dough result in a shift in the ratio of —SH— groups in favour of —SS— bonding, the required concentration, and relative speed of reaction of the various oxidants varies somewhat. Potassium bromate is less reactive than potassium iodate, calcium iodate, azodicarbonamide and the oxidized form (dehydro-) of ascorbic acid. A radiochemical method for the estimation of SH:SS ratios in wheat gluten was developed by Lee and Samuels (1962).

The important role of hydrogen bonds in dough formation and for ensuring suitable rheological characteristics for subsequent processing, can be demonstrated by adding chemicals which liberate or decompose hydrogen bonds. One mole dimethyl formamide, sodium salicylate at 3·0% plus and carbamide at 10·0% plus concentrations, when compared with distilled water as control for doughmaking in a Farinograph, or other experimental mixer, show marked changes in dough rheological features, independent of basic flour quality. In all cases there is a decrease in dough formation time, a rapid softening after the maximum, and a narrowing of the bandwidth, indicative of a rapid decrease in dough elasticity and resilience. In absolute terms, the largest decrease is experienced in the case of good quality flours. Therefore, the solubility of the gluten proteins in saturated aqueous solutions of such chemicals indicate the role of hydrogen bonds in the structure of the gluten complex and dough formation. These chemicals may split the hydrogen bonds within the molecule, thus influencing the conformation of polypeptide chains, or decomposition of bonds of larger units of the molecules. During the initial period of dough formation, these chemicals accelerate gluten swelling by loosening the structure. Disaggregation of the swollen, formed dough due to the hydrogen bond decomposing chemicals then proceeds at a rapid rate. Doughs with poor quality gluten are less sensitive to such chemicals, since they contain a reduced number of hydrogen bonds, and have a looser gluten structure. Recent studies in the USSR by Lutsishina and Matyash (1984), using NMR wide-band proton spectroscopy, confirmed differences in the conformational state of the gluten protein complex of hard and soft wheats. Wide-band proton NMR-spectrum parameters were higher for hard wheats, since the macromolecules were more compact, with stronger hydrogen bonds. Positive correlations between NMR parameters and rheological and technological properties data were also established.

The rheological and technological effects of hydrophobic bonding, involving non-polar interactions has been reported by several protein research groups. In the field of cereal chemistry, Ponte (1968) demonstrated the functionality of certain aliphatic hydrocarbons of alkane carbon chain length 6 to 8. For example, 0·0086 g-molecules of heptane, added at the dough stage, per 100 grams of flour, resulted in increases in dough formation and stability times measured in the Farinogram, and significant improvement in bread-crumb score, compared with control doughs. Glutens from these doughs showed improved expansion on drying, owing to the production of more stable films. Thus, the addition of aliphatic organic solvents to dough favours protein aggregation resulting from hydrophobic bonding. Non-polar amino acid side-chains of the gluten proteins tend to be attracted towards one another by van der Waals forces, becoming remote from contact with the aqueous phase. Hydrophobic bonding within membranes between lipid and protein is also possible, and the hydrophobic parts of the longer polar side chains (e.g. glutamic acid and lysine) may also enter into interactions. Recent studies by Popineau (1985) using hydrophobic interaction chromatography (polyacrylamide gel electrophoresis, in the presence of sodium dodecyl sulphate), revealed large glutenin aggregates present in several fractions. This suggests differences in their surface hydrophobicities, as is the case with the gliadin components. Differences in sub-unit composition were observed among glutenin aggregates which differed in surface hydrophobicity.

Although glutamic acid content is quantitatively the highest in gliadin, glutenin, and the gluten complex, most of it exists in the protein molecules as the amide glutamine, amounting to about 42% of total gluten protein. The next in predominance is proline, which is found in the greatest amount in the gluten complex, at about 14% of total gluten protein (Fig. 18). The classical Osborne fractions of flour proteins, according to solubility are shown in Table 13. However, owing to the close affinity of some low-molecular-weight wheat protein sub-units

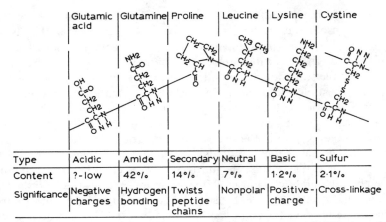

Fig. 18.    Amino acids in wheat gluten protein. (Source: R. J. Dimler, *Bakers Digest* **33**, (1), 52, 1963.)

for endogenous polar lipids, some gluten-forming fractions cannot be classified according to the above scheme of Osborne. Zawistowska and Bushuk (1986) have reported the existence of a low-molecular-weight wheat protein of variable solubility using an electrophoretic procedure.

The relative importance of the amide groups of the gluten complex which provide hydrogen bonding during the breadmaking process, has been demonstrated by Ma Ching Yung, Oomah and Holme (1986). These workers found that deamination led to a progressive degradation of gliadin, giving rise to an increase in low-molecular-weight components. Deamination increases the negative charge and surface hydrophobicity of gluten, reducing loaf volume and dough extensibility. Succinylation of wheat gluten results in a decrease in dough extensibility, but in no significant change in specific load volume. In addition, changes in molecular weight distribution and hydrophobic interaction also influence gluten baking performance, whereas ionic interactions are more likely to be involved in dough mixing and development. The water- and salt-soluble albumins and globulins contain a higher content of ionic functional groups. Table 14 shows the approximate relative distribution of important protein functional groups in the basic classical Osborne

TABLE 13
**Wheat-flour protein fractions according to solubility**

| *Fraction* | *Solvent* | *Percentage of total protein* |
|---|---|---|
| Albumens, non-gluten-forming | Water | 7–10 |
| Globulins, non-gluten-forming | Salt solution | 6–10 |
| Gliadin, gluten-forming | 70% ethanol | 40 |
| Glutenin, gluten-forming | 0·1 N acetic acid | 40 |

TABLE 14

**Distribution of functional groups in Osborne flour fractions (millimoles per 100 g protein)**

| Group | Amino acids | Albumins and globulins | Gliadin | Glutenin |
|---|---|---|---|---|
| Acidic | glutamic and aspartic acid | 91 | 27 | 35 |
| Basic | lysine, arginine, histidine, tryptophane | 100 | 39 | 52 |
| Amide | glutamine, asparagine | 90 | 309 | 266 |
| Sulphydryl and disulphide | cysteine, cystine | 45 | 12 | 12 |
| Total ionic (acid + basic) | | 191 | 66 | 87 |
| Total polar (hydroxy amide) | | 190 | 381 | 365 |
| Total non-polar | | 372 | 390 | 301 |

Source: John Holme, Review of Whear Flour Proteins and their Functional Properties, *Bakers Digest*, **40**, 6, 41, 1966.

fractions, the data being adapted from the original work: J. S. Wall, Proteins and their reactions, Ch. 14 in *Symposium on Foods*, AVI Publishing, Westport, CT (1964). The classical method of separation of the gliadin and glutenin based on 70% ethanol due to Osborne does not yield products which are completely free of contaminants, and improved techniques based on gel filtration and electrophoresis, depending on molecular weight and mobility have to be used. By studying the detailed structure of these purified proteins, and the location of disulphide and the other functional groups in the polypeptide chain, some possible explanations for the resulting dough rheological and processing behaviour can be undertaken. Rheological properties are the summation of the interactions of all the reactive functional groups, and the effect of superimposed dough additives such as reducing and oxidizing agents. By measurements of relative viscosities and the physical behaviour of films of gluten protein fractions in concentrated solution, using appropriate solvents, some ideas to explain why glutenin molecules have a greater tendency to associate than gliadin molecules can be put forward. The relatively small, uniform, compact gliadin molecules do not offer much surface for contact with other molecules. On the other hand, glutenin consists of many large molecules, arranged in random coils, offering numerous opportunities for molecular associations. Such glutenin associations give rise to cohesion and elasticity. These two proteins blend on hydration to give properties intermediate between the separate proteins. This interaction is then further modified by the presence of starch and lipids. Studies by Bushuk, Bekes, McMaster and Zawistowska (1985) on flour and defatted flour glutens involving fractionation into carbohydrate and lipid-containing protein complexes are relevant in this context. These workers isolated a glutenin corresponding to 40% of the flour protein, containing 17% carbohydrate, which was identified as α-glucan, with a molecular weight of 12 000. This formed a highly specific and stable complex with glutenin sub-units and was found to be involved in the formation of glutenin aggregates. The carbohydrate in the gliadin fraction was found to be galactose, present as digalactosyl diglyceride complexes. Low-molecular-weight gliadins aggregated in the presence of these lipids, significant positive correlations being obtained between both contents of polar- and galacto-lipids and loaf volumes.

Hoseney *et al.* (1970), concluded that the ability of gluten proteins to retain carbon dioxide during fermentation depends on the presence of the polar lipids, especially the glycolipids. Free polar lipids are bound to the gliadin gluten proteins by hydrophilic bonds, and simultaneously to the glutenin proteins by hydrophobic bonds, as depicted in Fig. 19(d) below. Using IR and NMR spectroscopy, Wehrli and Pomeranz (1970) studied the interaction between glycolipids and wheat-flour macromolecules. They concluded that gelatinized starch formed a complex with

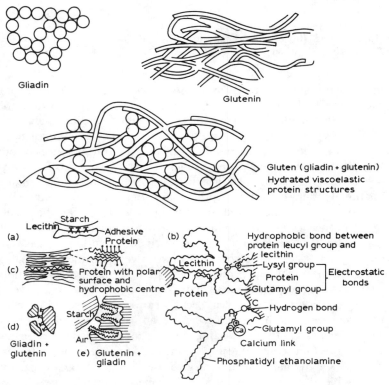

Fig. 19. Molecular bondings in dough formation. Gluten (gliadin + glutenin) hydrated viscoelastic protein structures. (a) Starch–lipid–protein complex of flour, according to Hess. (b) Phospholipid–protein bonding. (Source: M. Kerer, *J. Food Sci.*, **38** (1973) 756–63.) (c) Lipoprotein model, according to Grosskreutz. (d) Gliadin–glutolipid–glutenin complex, according to Hoseney *et al.* (e) Starch–glycolipid–gluten complex, according to Wehrli. (Source: J. S. Wall, Review of wheat protein. *Bakers Digest* **47**, 1, 26, 1973.)

glycolipid, in which case the polar sugar moiety of the lipid is bound to the starch by hydrogen bonds (not extractable with petroleum ether), whereas non-polar lipid side-chains form a layer in which the chains exist mainly in the extended *trans* conformation. Glycolipid can also be bound to gliadin by van der Waals and hydrogen bonds, since no band appears at $720\,\mathrm{cm}^{-1}$, and petroleum ether does not extract the glycolipid. Therefore, in dough, glycolipids can become bound by hydrogen bonds to starch and to gliadin and glutenin. In the presence of water, binding to glutenin is mainly by hydrophobic bonds. In an aqueous environment, hydrophobic groups of protein and lipid will be insoluble when pulled into solution by hydrophilic groups on the same molecule. The hydrophobic groups, not being in their preferred environment, will tend to migrate from the water into zones more akin to their natural groupings. Therefore, in solution the hydrophobic groups of various molecules become forced together to form so-called 'hydrophobic' bonding. Glycolipids are classed as 'polar' lipids since the carbohydrate part of the molecule is water-soluble, and the hydrocarbon part is insoluble and hydrophobic. Generally, lipids range in structure as follows:

| | |
|---|---|
| Non-polar | no asymmetric electronic charge distribution, e.g. triglycerides and waxes |
| Polar | one part of the molecule with its electronic structure differs from the other, e.g. glycolipids |
| Ionic | definite positive and negative charges on part of the molecule, e.g. phospholipids, with a hydrophilic, water-soluble ionizable part, and a hydrophobic part |

Lipid/protein interactions can take many forms: hydrophobic interactions, ionic interactions, hydrogen bonding, and mixed chelate formations. All classes of lipid, by virtue of their hydrocarbon group, are potentially capable of hydrophobic bonding to protein. Other protein/lipid interactions involve the more polar lipids, such as glycolipids, or the ionic lipids, such as phospholipids, which may be in molecular, micellar or lamellar mode.

Depending on conditions such as temperature and concentration, micelles or bimolecular layers may form in solution. In both cases this exposes the polar head groups for protein interaction. The formation of new membranous structures from lipid and protein during dough formation could contribute to such rheological parameters as viscosity, elasticity, plasticity, and extensibility. Analogous is the longstanding Grosskreutz model, which postulates how lipoprotein structure in dough can account for gluten extensibility by providing slip planes in the form of lipid bilayers.

Figure 19(a) illustrates the Hess model, which explains the formation of a starch–lipid–protein complex due to the adherent-protein (Ger: Haftprotein). Hess isolated the adherent-protein by non-aqueous flotation techniques using benzene/carbontetrachloride mixtures of densities between 1·30 and 1·50 g/ml at 20%. According to Rohrlich *et al.* (*Die Mühle*, **106** (1969) 384–6, 'Wedge-protein and Adherent-probe from a new viewpoint', in German), the flour protein is made up of an average of 50% wedge-protein (Ger: Zwickelprotein) and 50% bound to the

starch as adherent-protein (Ger: Haftprotein) at a sedimentation density of
1·37 g/ml at 20°C.

In the interpretation of data from studies concerning lipid extraction, and binding
to other flour and dough components, where solvents other than water are used,
caution is necessary. In view of the multiplicity of possible interactions already
described, the relative polarity of the solvents used can dramatically change the
component distribution in the system under examination. For example, when using
a less polar solvent such as 70% ethanol, more lipid tends to be bound to the gliadin
fraction than to the glutenin, the polar lipids migrating to the gliadin, and the
neutral lipids binding to the glutenin (monoglycerides, di- and triglycerides and free
fatty acids). The same trend applies, whether wheat-flour or dough is subjected to
extraction with the less polar solvent extraction system. The presence of 30% water
reduces the polarity of the ethanol enough to expose hydrophobic aggregates of the
glutenin, but adequate polarity is retained to permit the triglycerides to migrate
towards the more non-polar environment of certain zones of the glutenin. In
general, a mixture of polar and non-polar solvent is necessary to separate lipids from
protein and other tissue, e.g. ethanol: diethyl ether (3:1) or methanol:chloroform
(1:2) (vol./vol.). The common neutral lipids not being particularly soluble in the
polar solvents ethanol, methanol and acetone. Nevertheless, the extraction
temperature also requires control, for example, when acetone is used to separate
phospholipids (acetone insoluble) from neutral lipids (acetone soluble), if the
procedure is carried out at 4°C or below, a significant amount of neutral lipids
(cholesterolesters and glycerides) can be precipitated along with the phospholipids.

Pomeranz (1985), has also reported loss of gluten protein solvency from flours
previously defatted with Skellysolve, benzene or 2-propanol (listed in decreasing
order of ability to remove polar lipids), using 0·01 M sodium pyrophosphate buffer,
0·05 N acetic acid or 3 M urea in pyrophosphate buffer. Losses being greatest in the
case of the urea- or acetic acid-soluble gluten proteins. Furthermore, protein
solvency was restored when the defatted flours were reconstituted with non-polar or
total lipids, but not when restored with polar lipids alone. Data obtained appears to
relate bread volume linearly with free polar lipids, but not to total polar lipids in
flours. Also, flour native polar lipids interact with added shortening inasmuch as,
when absent from doughs containing 3% shortening, reduced loaf volume results.
This would seem to indicate the formation of a barrier effect, which interferes with the
formation of protein/protein complexes. In dough systems where no shortening is
added, the removal of both free and bound lipids from flour enhances the formation
of protein aggregates improving both loaf volume and crumb grain. Such
phenomena exemplify the complexity of the wheat-flour dough system and the
multiplicity of component and additive interactions. It is important to identify
where bond-breaking and solvent fractionation give rise to detrimental functional
changes in the reaction dynamics of dough, which in some cases lead to irreversible
changes in function on reconstitution. MacRitchie (1985) at the Bread Research
Institute, North Ryde, Australia, has outlined a scheme for separation,
fractionation, and reconstitution of flour components with complete recovery of
original flour functionality on mixing and baking.

In Hungary, Pallagi-Bánkfalvi (1983) has studied the changes in wheat-flour proteins during mixing and fermentation of good- and poor-quality flours. Three initial protein fractions were separated: 1 M sodium chloride solubles (albumens and globulins), 0·1 M acetic acid solubles (glutenin) and residual glutenin. Their peptide chains were studied by sodium dodecyl sulphate/polyacrylamide gel (SDS-PAGE) electrophoresis. These techniques were applied to flours, to doughs immediately after mixing, and to doughs after 180 minutes proofing. The quantity of gluten proteins soluble in 0·1 M acetic acid was found to increase in all samples during mixing and fermentation of the dough, but to a differing degree. The lowest increases being found in good-quality flours (7·0%), and the highest in poor-quality flours (28·0%), whilst mixtures of these flours showed increases of about 15–16%. Changes in solubilities of the proteins during mixing and proofing were accompanied by changes in molecular weight and structure of the gluten complex.

Significant changes in the 1 M sodium chloride soluble protein fraction of flours and doughs were not observed. However, the polypeptide chain composition of the gluten proteins extracted with 0·1 M acetic acid changed during mixing and proofing. The amount of polypeptide chains of molecular weight 95 500, 91 000, and 79 000 daltons, representing 12–15% in flours, diminished to 5–6% on mixing. Then, after the proofing stage, about 80% of the polypeptide chains soluble in acetic acid were detected in the molecular weight range 44 000 down to 30 000 daltons. The amount of gluten proteins insoluble in 0·1 M acetic acid diminished according to flour quality. In good-quality flours the reduction approximated 10%, compared with 22% for poor-quality flours. The polypeptide chains, however, did not show any changes in molecular weight distribution by the SDS-PAGE method. These results show that the stability of the gluten structure formed during dough formation does not depend on the original flour protein composition alone. Changes occurring during mixing and proofing, as a result of mechanical treatment and enzyme activity, are of great significance. Studies carried out on 56 samples of 26 soft wheat varieties in the USSR by Bogdanov and co-workers (1985) at the All-Union Scientific Research Institute of Applied Molecular Biology and Genetics, Moscow, using the PAGE method for glutenin sub-unit assay, have yielded interesting electrophoretic correlation patterns. Protein amounts in the peptide band of molecular weight approximately 50 000 correlated well with baking quality of all samples. Also employing the SDS-PAGE technique. Graveland *et al.* (1985) at the Institute of Cereals, Flour and Bread, Wageningen, Netherlands, has postulated a model for the molecular structure of wheat-flour glutenins. They suggest that the high molecular sub-unit fractions are linked via head-to-tail interchain —SS— bonds, forming a linear chain. Whereas, the lower molecular weight subunits form a cluster which is linked via —SS— bonds to two specific large-molecular-weight sub-units of the linear chain.

The important role of wheat-flour lipids in forming functional chemical bonds has been referred to already, but a more detailed consideration of the relative changes in their distribution during mixing and fermentation merits attention. Wheat-flours of extraction rates between 70% and 75% will contain about 1·0–2·0% lipids, of which 0·8% to 1·0% are designated as 'free' lipids. These free lipids include unbound lipids and those held by hydrophobic bonding, which are extractable with non-polar solvents, e.g. petroleum ether. About 0·6% of these free lipids are non-polar components, mainly triglycerides, and about 0·2% are polar components, classified as either glycolipids or phospholipids. The residual 0·6% to 1·0% are referred to as 'bound' lipids, since they cannot be extracted from the proteins with petroleum ether, although essentially polar. Instead, they have to be first removed from the protein surface with the less polar water-saturated butanol, after which they can be taken up in petroleum ether. The bound lipids are a mixture of phospholipids and glycolipids, amounting to 50–55% and 45–50% respectively. Although differences in lipid content and composition of different wheat varieties do not have the same profound effect on baking quality as gluten proteins, their absence on removal with petroleum ether results in loaf volume loss, and loss of crumb texture and grain. Significantly, when gluten is washed from a wheat flour, almost all the polar, and most of the non-polar, lipid components become bound to the gluten complex. As reported by Hoseney *et al.* (1970), the free polar lipids (mainly monogalactosyl and digalactosyl diglycerides) can become bound by hydrophilic bonds to the gliadin fraction, and to glutenins by hydrophobic bonds. When gluten is formed *in situ*, these lipids become integrated into both proteins simultaneously.

Most of the free flour lipids become non-extractable with petroleum ether after the mixing process, and some of the triglycerides and all polar lipids become bound. Binding increases with mixing speed and intensity up to the optimum mixing time. In fact, the glycolipids become firmly bound even at dough moisture levels below normal dough water absorption levels. The influence of lipids on dough properties and bread quality is the result of a physical exchange reaction of lipids with the main components, carbohydrate and protein. This complex formation between lipid and protein, especially the hydrophilic bonding between polar lipids or glycolipids and the gliadins, as well as the hydrophobic bonding between the non-polar lipids and the glutenins, all contribute towards the strengthening of the network of the flour proteins in the dough, and make possible the optimal development of the wheat gluten complex. In addition to the effects of different commercial mixing systems, and mixing intensity, on lipid-binding, the absence of oxygen or replacement by nitrogen during mixing results in a marked increase in bound lipid, according to Daniels *et al.* (1971). However, this phenomenon appears to be interrelated with pigment bleaching, and lipid peroxidation, involving flour or added soy lipoxygenase enzyme systems.

The significance of bonding between glycolipid and wheat-flour macromolecules has been demonstrated by Wehrli *et al.* (1970). These authors established glycolipid bond complexes with starch, gliadin and glutenin wheat proteins by using such methods as: solvent extraction, lipid binding in starch doughs, infrared spectroscopy, NMR, autoradiography and controlled baking tests.

Experiments with synthetic glycolipids have shown that both hydrogen and hydrophobic bonds are necessary for their improving effect. Acetylated glycolipids, for example, will not interact since there is no hydrogen-donating hydroxyl group available. Glycolipids with a disaccharide group have proved to be more effective bread improvers than monosaccharide derivatives, and the fatty acid chain length of the hydrophobic group is also important for optimal function. Commercially available glycolipids of the sucroester type are widely utilized to overcome the adverse effect on loaf volume of high levels of soy flour and protein isolates in nutritionally fortified breads.

The improver effect of available glycolipids increases with increasing HLB (hydrophilic/lipophilic balance), which involves decreasing numbers and chain length of the fatty acids attached to the carbohydrate molecule.

The effects of fermentation on the total lipids and their fractions, non-polar lipids, glycolipids and phospholipids, have been studied in some detail by Tortosa, Ortola and Barber (1985) in Spain. The lipids were extracted from the bread dough with chloroform/methanol (2:1) and fractionated by TLC (thin-layer chromatography). In the non-fermented dough the non-polar lipids were separated into ten fractions, the glycolipids into seven and phospholipids into ten groups. After 6 hours fermentation at 28°C, two new bands were detected, one in the non-polar lipid fraction and another in the phospholipid fraction. Conversely, one band of the glycolipid fraction, possibly including free sugars, digalactosyl monoglycerides and at least one other glycosidic compound, disappeared. The total lipid content of the dough did not change during fermentation. The monoglyceride content decreased from 22·0 to 11·1 mg/100 g dough on a dry basis. The diglycerides, monogalactosyl diglycerides and digalactosyl diglycerides contents increased from 32·3 to 50·3, 11·8 to 28·1, and 56·6 to 79·2 mg/100 g dough dry basis, respectively. No significant changes in lysophosphatidyl choline, phosphatidyl inositide, phosphatidyl choline, *N*-acylphosphatidylethanolamine or lysolphosphatidylethanolamine fractions were detected. Such increases in diglyceride, and in the polar glycodiglycerides at the expense of monoglyceride, coupled with the disappearance of one glycolipid band and the appearance of a new phospholipid band, is indicative of considerable potential interchange of free radicals during fermentation. The known sensitivity of hydrogen bonds towards heat, and increasing stability of hydrophobic bonds at higher temperatures could explain the proposed binding of the hydrophilic part of the glycolipid molecule with the gliadin protein fraction, and the affinity of the hydrophobic part for the glutenin proteins.

On the other hand, interaction of glycolipids with the starch granules in dough has been found, through hydrogen and hydrophobic bonding, both of which are essential for improving bread quality. Tritium labelling of synthesized glycolipids has indicated that some glycolipid becomes associated with the starch in doughs, but during the baking process most of the galactolipid migrates in association with gelatinized starch granules.

Wheat gluten proteins contain about 35% glutamic acid, and 14% proline, and about 40% of all acidic amino acid side chains are amides. Deamidation causes progressive degradation of the gliadin proteins, and an increase in low molecular weight components, but the glutenin proteins are not affected, according to Ma, Oomah and Holme (1986). Deamidation increases the net negative charge and surface hydrophobicity, resulting in reduced loaf volume. It is the amide groups which provide the source of hydrogen bonding during the breadmaking process. Changes in molecular weight distribution and hydrophobic interaction would appear to affect the baking performance of gluten, but ionic interaction, although involved in dough development, is less critical in controlling overall baking performance. Hydrophobicities of gliadin and glutenin derived from amino acid composition data, according to Woychik, Boundy and Dimler (1961) are 1109 and 1016 calories respectively, based on an average amino acid residue. Gliadin contains about twice as many alpha-helices as glutenin, which are stabilized by the hydrophobic amino acids, as confirmed by Cluskey and Wu (1966) and also Lenard and Singer (1966). However, glutenin has more non-polar amino acids available for intermolecular hydrophobic binding with lipids than does gliadin. Methylation of the amide groups of both the gliadin and glutenin fractions results in intrinsic viscosity loss. Whereas, on using deuterium oxide instead of water for doughmaking, gluten strength and elasticity increase (Vakar *et al.*, 1965).

The effects of hydrogen bonding on grain polymers have been clearly established by Kretowich and Vakar in a series of studies carried out at the Academy of Sciences of the USSR during 1964–65, which were described at the ICC Conference in Vienna in 1972. Following work on the deuteration of proteins carried out earlier by Lindström-Lang, these workers prepared freeze-dried gluten samples and mixed them with $D_2O$ instead of $H_2O$ in a Farinograph. The resultant dough became firmer and extensibility was reduced by 51–77%. Similar effects were found with rye protein prepared from rye wedge-protein (according to Hess). The viscosity of starch gels are reduced whether the $D_2O$ is added before or after gelatinization. These experiments clearly confirm the earlier assumption that hydrogen bonding plays an important role in the swelling of protein and starch and in their rheological behaviour. Gluten containing 98–99% protein in dry matter, and retaining all its native properties, was prepared by Vakar and co-workers by the Hess flotation technique, using a chloroform/benzene mixture. Gluten from strong flour was found to be less soluble than gluten from weak flour in 12% sodium chloride and 0·05N acetic acid. Dispersions of strong glutens in 12% sodium chloride had a lower intrinsic viscosity than weak gluten. Additions of urea increased intrinsic viscosity, the effects being greater on strong gluten. Thiol contents of weak and strong glutens were not significant; however, strong glutens contained more disulphide groups. Vakar and co-workers concluded therefore that gluten quality is determined by cross-linking of gluten proteins by hydrogen, disulphide or other types of bonds as yet undefined.

Changes in chemical bonding are also induced by various additives, as discussed in detail in Chapter 1.5. However, a discussion of the reaction mechanisms involved, which manifest themselves in rheological properties and baking performance, is appropriate here. Probably the most important is the sensitivity to oxidation and reduction. These well-established effects, have propagated numerous hypotheses concerning the relative importance of thiol and disulphide groups for wheat gluten structure. These have been well chronicled and summarized by Bloksma (1975) and Ewart (1977), (1978) and, more recently, (1985). The effect of reducing substances, e.g. cysteine, glutathione and sulphites, are considered to be due to the reduction of disulphide bonds, which give rise to a fall in molecular weight of the gluten proteins in general, and the glutelin fraction in particular. Conversely, the effect of oxidizing substances and sulphydryl blocking agents is not primarily due to increases in molecular weight, as a result of the cross-linking of intermolecular disulphide bonds between protein molecules. Instead, it is the oxidation of low-molecular-weight thiol groups to corresponding disulphides, i.e. the prevention of losses in molecular weight as a result of thiol–disulphide exchange reactions between high-molecular-weight proteins and low-molecular-weight thiols. Tests with a Canadian Hard Red Spring sample carried out by Tsen and Bushuk (1963) confirmed that mixing in air gives a firmer dough (high resistance/low extensibility), than when mixed under nitrogen. According to Sullivan *et al.* (1963), mixing in air, the concentration of — SH— groups quickly becomes reduced by about 50% of the initial flour concentration. Evidence shows that response to oxidation depends to a large extent on the wheat variety. The concentration of —SH— and —SS— groups in the total protein of hard wheat varieties tends to be smaller than is the case with softer varieties. However, owing to the relatively large difference in their protein contents, the converse becomes the fact on milling into flours.

Mamaril and Pomeranz (1966) during studies on the isolation and characterization of wheat-flour proteins, also included the effects of dough mixing and oxidants on the proteins. Mixing in the Farinograph, they found that the proteins dispersible in 3M urea increased from 65% to 85%, this increase correlating positively with length of mixing. Protein dispersibility was highest in doughs mixed with an excess of potassium iodate, and lowest in the original flours used. Doughs mixed in air, in a nitrogen atmosphere, or in air with excess potassium bromate showed intermediate amounts of extracted protein. The 3M urea extracts of doughs mixed to optimum consistency were further fractionated on Sephadex G-100 into a minimum of six fractions, giving three distinct peaks. Protein distribution in the peaks was measured by Kjeldahl-N, Biuret, and absorbance at 280 $\mu$m. Mixing in air, but not in nitrogen atmosphere, substantially increased the amount of proteins with a molecular weight above 150 000, and decreased the amount of low-molecular-weight proteins. The effects of mixing bromated doughs were similar to mixing untreated flour doughs in air, but the chemical modifications in bromated doughs were much more drastic. No such changes were observed in iodate-treated doughs. Similar changes in distribution of protein fractions were observed when the gross protein extracts, or Sephadex-fractionated proteins were studied by starch-gel electrophoresis in the presence of 3M urea. More recently, these effects of oxidants and reducing agents on flour protein molecular weight distribution were confirmed by Nguyen-Brem, Kieffer and Grosch (1983). Oxygen has the effect on all wheat gluten proteins of changing the molecular weight distribution in favour of the very-high-molecular-weight fractions. The extent of these changes depends strongly on the wheat variety. Varieties with good baking qualities show high molecular weight increases, of the order of 30–40%, in the high-molecular-weight fraction, as a result of the presence of atmospheric oxygen. With increasing hydrophobicity a marked increase in the prolin content at the expense of the polar amino acids and glycine in terms of molecular percentage of total amino acid composition becomes apparent. The varieties with poor baking qualities, however, show no such changes in amino acid composition of the high-molecular-weight protein fraction. These changes are paralleled by changes in gluten firmness (changes in high-molecular-weight protein amino acid composition, due to oxidation, resulting in increasing resistance to extension of the washed out glutens at the expense of their extensibility, as measured by stretching tests). The maturing effect of flour appears to follow the same reaction mechanism, since doughs from freshly milled wheat are difficult to process, yielding low-volume bread with an inelastic crumb structure. On the other hand, long-storage flours often result in short, bucky doughs and reduced-volume bread. Such changes in baking behaviour are attributed to the influence of atmospheric oxygen on the formation of high-molecular-weight proteins in the dry state, i.e. in the absence of lower-molecular-weight compounds. This would appear to favour the formation of additional closed disulphide bridges between certain gluten components of intermediate molecular weight.

Since the yields of wet and dry gluten show no change as a result of oxidation with such substances as potassium bromate and L-ascorbic acid, the improvement in gluten firmness and tenacity experienced with good quality wheat-flours cannot be attributed to an improvement in protein hydration capacity. Instead, all evidence is indicative of the observed effects being due to change in composition of the high-molecular-weight protein fraction. Furthermore, any possible role of the lipoxygenase enzyme system in these reaction mechanisms would not appear to be of any significance, since the inhibition of lipoxygenase with an excess of L-ascorbic acid (about five times the optimal dosage level, according to Walsh *et al.*, *Cereal Chemistry*, **47**, 119, 1970) does not change the molecular weight distribution of unoxidized glutens. Additionally, glutens washed out under a nitrogen atmosphere by Dirndorfer, Kieffer and Belitz (1986), and subsequently oxidized, also showed increases in the isolated high-molecular-weight fraction in the presence of ascorbic acid, to the same extent as untreated glutens, in the case of good-quality EC wheat-flours. The poorer baking quality wheat varieties by comparison showed smaller increases (7–11%) in high-molecular-weight fractions of their gluten proteins on subsequent oxidation. Furthermore, their rheological response to oxidation is insignificant, showing little sign of cohesion and dryness or resistance to extension. The resultant bread is of poor structure and lacking in symmetry. Typical examples of such EC wheats are the varieties Clement and Maris Huntsman.

Various amino acid derivatives, and the gluten/emulsifier complexes have been prepared and patented for use as processing aids. In addition, acyl derivatives of starch, and various hydrocolloids have shown useful improver interactions in bread doughs. These innovations clearly indicate that the chemical bonding or complex formations possible have yet to be probed and applied. The presence of such substances, often in concentrations of less than 1%, can improve the economic utilization of wheat varieties which would otherwise be unsuitable for breadmaking purposes. Ideally, compounds of organic origin, or those synthesized from naturally occurring organic groups would be preferred, since they have more chance of current legislative approval. Examples of derivatives include Lauroylasparagine, myristoylnorleucine, and stearoylorleucine. All acylated amino acids improve the quality of bread, and accelerate dough fermentation, but lauroylasparagine in particular improves both dough fermentation activity and titratable acidity by producing conditions favourable for fermenting organisms. *N*-Steroyl-α-alanine and *N*-steroyl-glycine are reported to improve bread volume, porosity and flavour when added to flour at 0·5%, and works synergistically with potassium bromate. Complexes formed between vital wheat gluten and hydrated emulsifiers include, for example, mixtures of sodium stearoyllactylate hydrate reacted at about 50°C with vital wheat gluten, freeze-dried and powdered. Complex formation between certain emulsifiers and starch, e.g. distilled monoglycerides, and with proteins, e.g. diacetyl tartaric acid esters (DATA-esters) have found wide commercial application in the baking industry.

Gluten-improving properties for bread doughs can also be enhanced by digestion with certain enzymes. Phospholipases added to strong wheat-flour doughs for digestion for periods of about an hour at 30°C, prior to gluten isolation by washing and freeze-drying, results in increases in the monoacylglycerophospholipid content from about 30 mol% to over 70 mol%. Gluten modification based on this procedure improves its functionality in breadmaking by increasing loaf volume and resistance to staling, compared with unmodified vital wheat gluten.

Basically, all proteins are polymers composed of amino acids. Each amino acid contributes to the chemical and physical properties of the macromolecule. Gluten is a typical example of this relationship, and owing to the diversity of its amino acids it has great potential for structural modification to improve functionality. Although insoluble in water at pH 7, it is soluble in acidic and basic aqueous solutions of low ionic strength. Its 70% ethanol-soluble fraction, in the hydrated state, called gliadin, has the consistency of syrup, whereas hydrated insolubles, named glutenin, form a rubbery mass. Together these form a cohesive/elastic protein polymer. As already mentioned, native gluten proteins contain inter- and intramolecular cystine—disulphide bonding which influences its physical properties. Gliadin bonding is intramolecular disulphide, whereas glutenin shows intramolecular as well as intermolecular disulphides, which link protein chains into high molecular weight polymers. Thus, differences in the properties of gliadin and glutenin can partly be attributed to molecular size. The high concentration of non-polar amino acid residues, such as proline and leucine, contribute to gluten insolubility in aqueous solution. Also, a scarcity of readily ionizable amino acids such as lysine or glutamic acid reduces solubility at neutral pH. Gluten cohesion is not restricted to the hydrated solid state; in solution the proteins associate into larger molecular entities, as evidenced by intrinsic viscosity and molecular weight determinations. Hydrolytic cleavage of gluten proteins yields polypeptides, a more uniform-molecular-sized product. Alternatively, protein functional groups can be transformed into other groups, as an amide–ester conversion. Gluten protein can also serve as a substrate for graft polymerization, yielding a protein-synthetic polymeric material. Mild hydrolysis of gluten proteins, yielding polypeptides, which contain more ionizable functional groups, are more water-soluble, and can be modified more easily.

In the case of rye bread manufacture, protein bonding does not play such a crucial role in determining bread structure. Rye proteins are more water-soluble than wheat proteins. They are not able to form a gluten complex on hydration, owing to the presence of the pentosans, and the weak electrostatic forces between the protein molecules, according to Berliner and Koopmann (1929/30). However, protein can be extracted and precipitated from rye-flour with dilute lactic acid, which exhibits a certain weak elasticity similar to a soft wheat-flour (Kosima, 1958). According to Golenkov (1965), a wedge protein can be precipitated from rye-flour, which on hydration also yields a gluten-like mass. The significance of wedge protein for rye-dough formation has been emphasized by Vones, Podrazy, Simova and Vesely (1964). The wedge protein influences the mechanical properties of the dough, especially its coherence and water absorption capacity. This significance for dough formation probably lies in an interaction between protein fractions and soluble high-molecular-weight gums. Recent work by Anger, Dörfer and Berth (1986) concerning molecular weight/viscosity relationships on the basis of the Mark–Honwink ratio, have thrown new light on the composition of the rye hemicelluloses. Rye hemicelluloses affect dough rheological properties, therefore, the pentosan, arabinoxylan, component was isolated and studied. A 1% aqueous solution showed a relative viscosity of 5·6. Its pentosan content was found to be 74·2%, and the protein, inseparable by proteolysis and dialysis, amounted to 5·7%. In addition, a glucose content of about 5% indicated incomplete separation from the $\beta$-D-glucans. The pentosan was further fractionated on Sepharose 2B/Sepharose 4B, and about 60% of the fraction with molecular weight less than 160 000, had a viscosity–molecular weight relationship of $(n) = 0.003\,47$ (mol.wt)$^{0.98}$. A highly branched high-molecular-weight fraction, amounting to about 40% of the preparation, showing a viscosity–molecular weight relationship of $(n) = 29.5$ (mol.wt)$^{0.23}$. According to Neukom and Markwalder (1978), in the structure of the pentosans and protein–pentosan complexes, hydrophenols form the covalent links, especially over ferulic acid. This linking gives rise to an increase in viscosity in aqueous suspension, and is considered to be an oxidative gelling effect. However, during dough formation, consistency is determined by the cumulative effect of more bonding forces. The generalized term 'swelling-substances' or '*Quellstoffe*' (German), is used to describe a whole series of components and complexes, which can be water-soluble or only partially water-soluble.

According to studies carried out by Holas *et al.* (1973), the functional components of rye-flour are polysaccharides, which do not belong either to the starches or celluloses, glycoprotein complexes and proteins. The components which are soluble in either water or aqueous alcohol form a mucilaginous mass, the amount and viscosity of which in solution are parameters for rye-flour quality. The extraction of rye-flour with aqueous alcohol yields a gliadin, which is similar to wheat gliadin, with a high glutamic acid content, but, like the prolamine of wheat, it is formed by an association of relatively short polypeptide chains of 11 to 13 amino acids, according to Kolesov (1951).

The glutenin-type proteins remaining after aqueous alcohol extraction have yet to be defined.

The centrifugation of an aqueous suspension of rye-flour yields two distinct layers, the lower one being starch and the supernatant consisting of strongly water-binding components with small starch granules embedded. This viscous fraction contains proteins, hemicelluloses and amylodextrin, together with lignin and pectins. The hemicelluloses consist mainly of pentose units, and are often referred to as pentosans. In the literature, the term 'pentosans' is often used to describe the water-soluble components of rye-flour, and the term 'hemicelluloses' the insolubles. However, both these fractions contain hexoses and hexuronic acids as well as pentoses. Rye, unlike wheat, contains the polysaccharide homoglycan graminin, which is a grain-fructane, consisting of ten molecules of fructose (decafructosan).

Structural characterization of the various pentosan fractions by acetylation or DEAE-cellulose procedures by Perlin (1951), Montgomery and Smith (1955), Cole (1967), and Medcalf, D'Appolonia and Gilles (1968) have yielded the following information. Both the water-soluble pentosans, and those associated with the 'sludge' pentosan fraction are mixtures of free arabinoxylans and polysaccharide–protein complexes. The endosperm monomers are D-xylose, L-arabinose, D-glucose, D-galactose, and glucouronic acid; other grain hemicelluloses from the outer layers also contain methyl-D-glucouronic acid. The basic chain consists of xylose units which are $\beta$-1,4 linked. Single unit L-arabinofuranose branches are attached to carbon atom 3 of the xylose units, or to the 2 and 3 carbon atoms as double-branching. The branching or substitutions occur mainly alternately on the xylopyranose basal units.

Data obtained concerning the fractionation of the water-soluble pentosans on DEAE-cellulose show the average characteristics given in Table 15.

The functionality of the pentosans and hemicelluloses and their processing properties for the manufacture of rye bread have been mainly researched in the USSR by Kosima and Holas (1974) in Czechoslovakia, Schmieder (1977) in the GDR, and Drews and co-workers (1973), (1977) and (1979) in the FRG. Apart from the pentosan content and their viscosity (which depend on their degree of substitution, branching or polymerization) the effect on macromolecules and their complexes of the degrading enzymes, e.g. pentosanases, cellulases, and proteases, must be taken into account. For example, although it has been well established that

TABLE 15
**DEAE-cellulose fractions of water-soluble pentosans**

| Material (wholewheat meal) | Yields (%) | Protein (%) | Arabinose:xylose ratio (fractions) | Arabinose:xylose ratio (total) | Molar weight |
|---|---|---|---|---|---|
| US HRS | | | | | |
| Fraction 1 | 20·4 | 1·8 | 1:1·8 | 1:1·5 | 96 000 |
| Fraction 2 | 18·0 | 1·4 | 1:1·2 | | 109 000 |
| US soft | | | | | |
| Fraction 1 | 17·1 | 1·8 | 1:2·0 | 1:1·9 | 76 000 |
| Fraction 2 | 23·8 | 1·5 | 1:1·8 | | — |
| US rye | | | | | |
| Fraction 1 | 22·0 | 1·7 | 1:1·7 | 1:1·5 | 120 000 |
| Fraction 2 | 19·6 | 2·0 | 1:1·2 | | 133 000 |

Source: Medcalf, D'Appolonia and Gilles, *Cereal Chemistry*, **45**, 6, 546, 1968.

the water-soluble pentosan content correlates with the baking properties, high enzymic activity can nullify the positive effect of this component of a rye-flour. Also, the soluble components are more susceptible to enzymic breakdown, and a more rapid fall in molecular weight results. In general, molecular weight of the rye pentosan fractions are significantly higher than those of wheat. Differences in degree of branching of the pentosans could affect the type and extent of their interaction with the proteins. Molecular weight differences would also affect these interactions, profoundly changing dough water absorption. Hydration capacities of all pentosan fractions are exceptionally high compared with other flour components, being within the range of 80% to 100%. The extent of any subsequent buildup of soluble high-molecular-weight pentosans from the insoluble components has not been fully researched.

There is considerable overlapping of the effects of the pentosan–pentosanase complex and the starch–amylase complex, which have an important influence on the baking results. No absolute values for these variables can be laid down in the light of present knowledge. Relatively low pentosan contents, coupled with low pentosanase activity, tend to give good rye bread quality. High pentosanase activity, in spite of relatively high pentosan content, results in inferior quality bread, owing to excessive reduction in dough viscosity. Processing problems are also encountered when the pentosan content is high, giving high initial viscosity, and enzymic activity is inadequate to bring about a subsequent fall to an optimal viscosity. Rye-flours with excessively high degree of starch condensation resistant to amylase activity can also lead to inferior bread quality. The pentosans determine the dough consistency, crumb texture and elasticity, bread volume and symmetry and also its keeping properties. The insoluble pentosan fraction is more effective in stabilizing the shelf-life than the soluble fraction. The insoluble pentosans control the retrogradation of the amylose and amylopectin, and the soluble pentosans inhibit amylose retrogradation. According to Neukom and Markwalder (1978), hydrophenols, especially ferulic acid, form covalent linkages contributing to the structure of the pentosans and protein–pentosan complexes. In aqueous suspension an oxidative gelation occurs, which raises viscosity and contributes to dough consistency. The enzymic reactions result in a fall in dough consistency during proofing, and excess enzyme activity causes the stability, gas-retention properties, and oven spring of the dough-pieces during final proof to diminish. Resultant bread quality deficiencies are: uncontrolled flow-out of the dough-pieces, cracks in the crust and uneven crumb texture and often sticky-crumb, as well as a parting of crust from crumb in extreme cases. Such effects involve the following variables: pentosans, proteins and starch as well as the enzymes pentosanases, proteases and amylases. The consequences of these variables for processing purposes are measured rheologically with Brabender Amylograph or Rotation viscometers. The Quellkurve (swelling-curve) according to Professor E. Drews, Detmold, FRG, measures the initial viscosity and fall in viscosity under a programmed regime of time and temperature of an aqueous rye-flour suspension. The Amylograph monitors the initial viscosity of an aqueous suspension of rye-flour, and the changes it undergoes during progressive heating at a constant rate, including the gelatinization temperature. The

Quellkurve, also carried out on the Brabender Amylograph, measures the consistency potential of the pentosans and the changes which they undergo as a result of pentosanase activity at temperatures below the temperature of swelling and gelatinization of the starch. The Amylogram also provides a guide to the pentosan–pentosanase effect during processing in the form of the initial consistency and how it changes with gelatinization. Specially modified rotation viscometers give similar information when carefully programmed. Supplementary information concerning contents of total pentosan, and soluble pentosan are obtained by chemical hydrolysis at acid pH, yielding furfuraldehyde, which gives colour reactions with either aniline acetate or orsin, and is measured colorimetrically. Recent work on the chemical and physical characteristics of rye pentosans by Meuser *et al.* (1986) in heat-treated and untreated flours give additional information. No significant differences between pentosans soluble in hot or cold water are apparent. The insoluble pentosans of rye differ from those of wheat in containing galactose. The water-insoluble arabinoxylan of untreated flour is more highly branched than is the case with heated flour, the mean molecular weight of the arabinoxylan in the heated flour being threefold that in untreated flour. Heat-treated flour had 2·1% water-soluble pentosan, 82% arabinoxylan and 18% arabinogalactan. Water-soluble pentosans amounted to 40% of total flour. Bran water-soluble pentosans amounted to about 14%, were more highly branched and were of higher molecular weight. Polyphenol content of the arabinoxylan-containing glycoprotein was slightly higher in the flour than in the bran, being about fivefold that in wheat-flour. Also, compared with wheat-flour, rye-flour water-soluble pentosans showed a higher mineral ash content. Relative viscosity of the rye arabinoxylan–glycoprotein complex was greater than that of wheat-flour. Average arabinoxylan:arabinogalactan ratios were 4·5:1 in rye- and 2·3:1 in wheat-flour.

Fractionation and reconstitution studies by Kühn and Grosch (1986) involved the initial separation of the dark-coloured non-starch polysaccharide supernatant layer, from the white-coloured starch by centrifugation at 900 g and 4°C. After further similar purification, the soluble fraction was heated with lactic acid at 100°C and pH 4·6, and then precipitated with 50% and 70% ethanol. The soluble fraction represented in total 15% of the flour, most of which was contained in the 70% ethanol precipitate. The non-starch polysaccharide (dark-coloured supernatant) amounted to 44% of the flour and contained secondary starch, insoluble pentosan and protein. On passing through a 0·04-mm sieve, the retained non-starch polysaccharides contained higher levels of pentosans. The various fractions were recombined and used to produce bread with normal characteristics, and enzymic treatments of the fractions before recombination gave useful indications of the effects on baking of the various flour components. The effect of cellulase treatment on the whole rye-flour, total non-starch polysaccharide, water-soluble pentosan, and the insoluble, high-non-starch polysaccharide pentosan fractions could also be demonstrated.

For the control of rye grain mixtures in the mill, the viscometric Falling number method in the form of the Liquefaction number due to Hagberg-Perten, has given the best linear correlation for the use of optimization nomograms. When evaluating rye-flours, e.g. types 997 and 1150 in the FRG, a more complete potential baking quality picture would be obtained from the following set of tests: Falling number, Liquefaction number (Hagberg-Perten); Amylogram maximum viscosity, maximum temperature (Brabender); Quellkurve log viscosity fall-off, final viscosity (Drews); maltose % dry matter; sour-dough baking test, including an assessment of the following criteria: dough yield, dough texture, crumb elasticity, crumb structure, and bread yield per 100 g flour, giving an overall quality score index. In general, rye grain mixtures are extremely sensitive to the addition of even small percentages of sprouted, enzyme-active grain. Considerable variations in the potential baking quality of rye grain, as measured by the above testing-scheme, are common in all the central European areas from year to year, owing to climate and location. Therefore, processing adjustments have to be made according to bread variety, by flour blending and/or adjustments to the following: sour-dough percentage; wheat-flour percentage (rye/wheat-flour mixed breads); dough yield maintenance or increase by the use of improver additives containing hydrocolloids, e.g. xanthan gum, deacetylated xanthan, acidulated starches, or cellulose derivatives—the deacetylation of xanthan increases the flexibility of the side-chains attached to the inner mannose residues, allowing better control of dynamic viscoelasticity with increasing temperature at various concentrations; enzyme-active additives containing fungal pentosanases (Veron HE, Röhm Technical Inc., New York), which releases the insoluble pentosan–protein complex, thus permitting optimization of the water-binding capacity of the dough. Where amylase activity is deficient, the fungal amylases Veron AV/AC can be added, since these fungal enzymes are inactivated at rye starch gelatinization temperature.

It is not possible to specify analytical norms for breadmaking potentials, and attempts to hold the miller to meaningless narrow units of measurement without meaningful discussion and collaboration is time-consuming and retrogressive. Nevertheless, one can venture to quote some basic data (see Table 16) as an approximate guide, providing one bears in mind that the criteria are interdependent and will depend for their success in practice on the processing schedule and set-up in operation.

TABLE 16

**Typical analytical data for commercial rye-flour types 997 and 1150 as used in the Federal Republic of Germany**

| Method | Typical data range |
| --- | --- |
| Falling number (Hagberg-Perten) | 120–300 s |
| Liquefaction number (Hagberg-Perten) | 120–150 AU |
| Amylogram (Brabender) Viscosity maximum | 320–700 AU |
| Temperature at maximum | 63–70°C |
| Quellkurve (Drews) Log viscosity fall-off | 90–240 |
| Final viscosity | 120–400 AU |
| Dough yield/100 g flour | 164–172 g |
| Bread yield/100 g flour | 250–350 g |

Glutaraldehyde, a difunctional aldehyde, often used as a cross-linking reagent, tends to exert the same effect on gluten, forming covalent intermolecular cross-linking and intracross-linking between the amino groups of lysine and terminal alpha-amino acids, according to Simmonds and Orth (1973), and also between lysine and tyrosine residues, as reported by Ewart (1968).

Differences in cross-linking mechanisms are often reflected in stress relaxation behaviour, and this has yet to be fully understood. Bohlin (1984) has studied such effects using a special relaxation quality control instrument (relaxometer), which provides an analysis of the flow pattern of doughs in the relaxation mode. From such tests, it is apparent that the influence of potassium bromate in the relaxation mode differs from that of ascorbic acid. The effect of ascorbic acid is the same as gluteraldehyde. As a colloidal system, dough consists of a continuous aqueous gluten (protein/lipid) gel phase with dispersed phases of starch granules with other components, together with gas cells. The hydrophobic zones of the system, consisting of one phase finely dispersed in another, has a large proportion of high-energy interfacial region and is inherently unstable. Solid particles in an emulsion or suspension tend, on impact, to stick together, thereby reducing the energy of the system. The speed of this process depends on the characteristics of the interfacial region, notably the electrical properties of the particle surface and its surrounding electrical double-layer. This parameter, referred to as the zeta-potential, can now be rapidly measured by the technique of laser doppler microelectrophoresis, which permits the calculation of the particle mobility spectrum and hence zeta-potential.

# 1.7 Typical Formulation and Process Schedules (Including Case Studies) for Wheat and Rye Breads Employed in Western and Eastern Europe and North America

In order to satisfy national consumer taste as regards bread texture and flavour and achieve varying degrees of processing rationalization, a very diverse range of mixing techniques and fermentation procedures is in use throughout Europe and the North American continent. Fermentation or floor-times have been progressively reduced by the application of technology and mechanization, but the quality of the end-product as an eating experience has not always improved. Although wheat-flour bread has always dominated the bread market globally, consumer tastes are changing in favour of bread varieties with more 'flavour-token', which at the same time make an improved contribution to a healthy diet. Although an acquired taste, rye-flour bread has distinct advantages over wheat since it has a more intense flavour-token and is more satisfying on a weight for weight basis, its acidic crumb often complementing the eating experience of other foods. Its keeping qualities are unquestionably superior to wheat-flour bread varieties.

## 1.7.1 FRANCE

In France particular attention is paid to the wheat at the breeding stage as regards the quality criteria, the protein quantity being of secondary interest. Wheat variety classification has always been the preferred scheme, rather than any classification according to protein quantity. Further categorization is according to their corrective capacity in blending, or variability from year to year, and growth location.

Such varieties as Prinqual and Florence Aurore from the south-eastern region are quality wheats comparable with North American Spring wheat in the Amylograph $W$ value, and superior in terms of hydration, flour yield and ash content. Other long-established wheats having an improver or corrective capacity are Rex and Magdalena. Category 1 varieties include Hardi, Capitole and Capelle; the commonest variety cultivated, being of irregular quality, is Fidel. Standard bread-flours amount to more than 85% of all milled flours, and belong to the type 55, with ash between 0·50% and 0·60%. This flour is milled from 100% French wheat to a yield of 75–77%, and a relatively fine particle-size granulation (75–85% through on a 100 $\mu$m, and a 1–3% residue on a 160 $\mu$m sieve) (Calvel, 1972). Alveograph $W$ values are 130 to 160 according to region, and protein contents vary from 10% to 11%. To obtain the best results, a French wheat-flour from a French mill of type 55 is recommended.

French bread is made ideally from the basic ingredients flour, water, salt and yeast into a soft dough, which is allowed to hydrate progressively. This ensures a good tolerance to processing under the practical techniques normally used into oven-bottom or hearth bread. Important dough characteristics are:

—dry enough not to adhere to mixing-bowl or surfaces after mixing;
—extensible and elastic;
—development of firm stability on standing;
—good fermentation activity.

After cooling for at least one hour, the bread must show the following characteristics:

—crisp, golden-brown crust, without excessive hardness or toughness;
—the transverse section on cutting should be clean and of uniform shape and good volume;
—the crumb colour should be a creamy-white, and its texture fine, but not too regular, and well-aerated with no tendency to closeness or rubbery texture.

Bread is traditionally eaten by breaking, rather than cutting, and eaten fresh. In Paris, bread is baked the whole day, and is usually bought twice daily on average. Whereas, in the regions and in the countryside, bread is baked in the morning, and bought once daily. The large 500-gram units, generally known as *'flûtes'* or *'Parisien'* in Paris, amount to over 50% of the white bread consumption. The smaller 250-gram units called *'baguettes'*, about 40%; and the 125-gram *'ficelles'* or small *flûtes* make up the rest of white bread sales. Normally the same dough is used to produce these bread varieties. A typical formulation is:

| | |
|---|---|
| white flour, type 55 | 100 parts |
| water | 61–64 (approximate Farinogram consistency, 500) |
| yeast | 2 (fast-fermenting) |
| salt | 2·2 |

Procedure:
1. Add the finely crumbled yeast to most of the hydration water, tempered to give a final dough of 25°C, in the mixing bowl.
2. Using a mixer with cyclic movements, e.g. Artofex-type, add the flour and mix for 3 minutes at low speed (about 50 rpm). Continue mixing at fast speed (about 70 rpm) and introduce the salt during the 12 minutes that is the duration of this mixing stage.
3. Allow the dough to rest for 3 minutes.
4. Mix for a further 5 minutes at fast speed, adjusting the consistency to optimum with additional water as necessary.

This procedure gives a total mixing time of 20 minutes, in three stages, whilst using both mixing-arms throughout doughmaking, amounting to 1400 contacts with the dough.

Using an oblique-shaft, fork-type mixer (Loiselet), the total mixing time would be 24 minutes in three stages, amounting to about 2100 contacts with the dough, assuming that the speed of rotation of the bowl is set at 13 rpm in both slow and fast speeds.

Procedure:
1.    3 minutes on slow speed (60 dough-contacts).
2.    16 minutes on high speed (90 dough-contacts).
3.    3-minute rest-period.
4.    5 minutes on high speed (90 dough-contacts).

Where more intensive mixers are in use, e.g. Spirel or Wendel types, the flour quality must be such that the doughs possess adequate fermentation tolerance after the second fermentation period.

*Initial Fermentation Period ('Pointage')*
This operation is carried out at 27°C and at relative humidity 70% in a covered bowl. The average duration is 15 minutes, but will depend on the regional characteristics of the wheat used to mill the flour. A better general guide would be a 2·5 volume increase in the dough mass after mixing.

*Intermediate Proof (Façonnage)*
The dough is divided into pieces of desired weight, e.g. 500, 350 or 250 grams, and rounded, then allowed to rest for 15 minutes. The rounded pieces are then mechanically moulded by correct setting of the moulding-machine to give *baguettes* of 70 cm length, or are moulded by hand into *bâtards* about 35 cm long.

*Final Fermentation Period ('Apprêt')*
The moulded dough-pieces are proofed in a cabinet of relative humidity 70% at 27°C for a period of 2 to 4 hours, depending on regional practice and raw material characteristics. The long final proof time is typical for French bread, and contributes to the excellent flavour-token of the products.

A more reliable assessment of the optimal final proof time is obtained by taking a dough sample of about 25 grams, moulding round by hand, and placing at the bottom of a graduated cylinder (maximum of five divisions) placed within the proofing cabinet. When the dough piece reaches the 4·5 level, the batch is ready for baking. This simple test can be modified to accommodate regional variations in flour quality and procedures.

This tolerance to final proof is extremely important for the successful production of French bread, and is achieved through flour quality, with small adjustments made by using certain additives. Malted wheat or flour can be added to adjust the Falling number to between 250 and 300 s, about 0·1% to 0·25% being used, often added to bean-flour as a carrier.

The use of bean-flour, especially where speed-mixing is used, at about 1·0% effects an oxidative colour improvement and contributes to dough tolerance, but detracts from bread flavour and aroma. Ascorbic acid at about 20 ppm, especially with speed-mixing, improves tolerance at final proof, and contributes to an open and improved volumed bread. Loaf-shape must be rounded at the base without any splitting at the sides.

### Baking ('Mise au Four')

*Baguettes* are cut about seven times, and *bâtards* three before baking. Oven temperature should be 260°C on setting, and the oven chamber saturated with steam, being used to full capacity. *Baguettes* are baked for about 20 minutes, and *bâtards* about 25 minutes. In the south of France, a slightly shorter baking-time results in a paler crust colour, and a tendency to larger volume, whereas in Paris and northern France a golden-brown crust is appreciated by the consumer. In southern France the drying effect of the *mistral* on bread must be countered in some way. Although *baguettes* have now become a common sight in other EC countries, their crumb structure is often too compact, and the crust tough and leathery by comparison.

In the south of France and certain regional areas, e.g. Brittany, a special sour-culture is used called '*levain*', which is perpetuated for continuous use. This can be used with white flour or wheat or rye meals. In Brittany bread made by this procedure can be found on sale in the market-places of Finisterre in the Morlaix

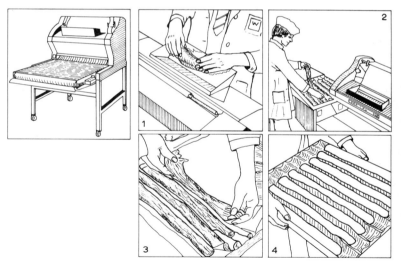

Fig. 20. Use of the mobile long-roller B 750 of Werner and Pfleiderer with extendable conveyor for *baguettes*. 1. Folding and elongation input. 2. Withdrawal of the elongated dough-pieces. 3. Setting the dough-pieces on racks with folded cloths for the proofing-cabinet before baking in a deck-oven. 4. Setting in hanging compartments, instead of folded cloths, for wheel-in proofing and baking in a rack-oven, thus avoiding transfer of the dough-pieces after final proof. (Source: Backtechnick-Langroller B 750 published by Werner and Pfleiderer, D-7000 Stuttgart 30, FRG.)

area. These products are very aromatic and have a longer shelf-life, but have a distinctly bitter after-taste. The San Francisco French sour-dough process, so popular in the USA, is closely related to the '*levain*' procedure.

The long final proof period of 2 to 4 hours is ideal for the dough retardation process (*la fermentation dirigée*), which is quite widely used.

Mechanization of the rolling process is achieved with a special long-roller, e.g. Langroller B 750 of Werner and Pfleiderer (Fig. 20) This allows the elongated dough-pieces to be either placed on racks in folded cloths, or on specially made hanging-compartments for proofing. In the former case the proofed pieces can be set by hand-paddles, or loaded into the oven directly where setter-cloths and setter-ovens are used. In the latter case, wheel-in racks for both proofing and baking in rack-ovens are generally used. Where space is limited, baking cabinets with rotating-racks, e.g. Rototherm RE of Werner and Pfleiderer, offer more baking space per square metre of floor space, and the options of manual, automatic or computer control of the baking operation.

## 1.7.2 SPAIN

In Spain, baking potential is determined by the Alveograph (Chopin) P/L ratio, $W$ value and protein content. Many French wheat varieties are grown, e.g. Florencia-Aurora, Provence-Rex, Hardi and Capitole, alongside Spanish varieties, all of which are classified into four types according to quality. However, owing to extreme climatic conditions of this part of the Mediterranean, large variations within and between varieties is inevitable. Nevertheless the quality and enormous selection of bread varieties, over 50 in number, confirm the ingenuity and innovation of the craftsman baker, and his ability to satisfy a wide diversity of consumer taste, and bread culture. Broadly speaking, the cultural zones may be divided into Castille, Andalusia, and Catalonia, but additional, more local, variations exist.

Depending on the final proof time used in the various regions, the Alveograph $W$ value of the flour will usually need to increase from about 100 to 160, even if the protein content is about 10% to 11%, on average. Therefore, as in France, protein quality becomes more important than quantity for the prevailing processing techniques. In Castilian Spain, especially in the Madrid area, final proof times of only 30 to 60 minutes are used, producing bread of intermediate volume but plenty of crust. In Andalusia, final proof times tend to be up to about 2 hours in duration, in common with those employed in the provinces of Galicia, Asturias, Estramadura and the Canary Islands. Whereas in Catalonia and Valencia, 3–4 hours is regarded as normal, especially in the Barcelona area. Crumb textures become increasingly more open with longer final proof times, which also demand greater tolerance in flour quality. In all cases crispness of crust, adequate volume and an aromatic crumb are the requirements of the Spanish consumer. This objective is achieved by the use of special wheat sours (*masa madre panaria*) which are carefully cultured and maintained by individual bakeries from a starter-culture. The starter-culture being a blend of various yeasts and lactobacilli, which produce the optimal control of

TABLE 17

| Masa madre pre-ferment | Dough: masa panaria | |
| --- | --- | --- |
| | *Short final proof* | *Long final proof* |
| Wheat-flour 54·0% | Castilian, e.g. Madrid | Catalonia, e.g. Barcelona |
| *Masa madre* 21·0% | Wheat-flour 100 parts | 100 parts (strong) |
| Water 25·0% | Pre-ferment 28 | 21 |
| | Water 55 | 66 |
| | Yeast 8 | 3 |
| | Salt 2 | 2 |
| Temperature: | 24°C | 24°C |
| Fermentation in bulk | 5 minutes | 5 minutes |
| Intermediate proof | 5 minutes | 5 minutes |
| Final proof | 45 minutes | 2–5 hours |
| Average baking time | 30 minutes | 40 minutes |
| (400 g units) | | |

pH and dough acidity during fermentation, resulting in the distinctive aroma and taste of Spanish bread. These cultures are stored at 4 to 5°C for average periods of up to 8 hours in the form of tight flour/water doughs, the acidity developed depending on the storage temperature and time. Perpetuation of the sour culture is achieved by taking material from the pre-ferment rather than the final dough, which contains salt and baker's yeast, etc.

The family-bread (*pan de familia*) accounts for approximately 70% of bread production in Spain, and typical short and longer fermentation schedules would be as shown in Table 17.

In Spain conventional-type mixers are favoured, since they exert a mixing regime which slowly develops the dough into a mass with a wide processing tolerance, and produces less heat-dissipation during dough-mixing at ambient temperatures in excess of 24°C. Speed-mixing is stimulating interest, but requires appropriate changes in technology, which may not contribute to bread quality and acceptance.

The microflora of the *mass madre panaria* has been studied in some depth by Barber (1983) and (1985), and co-workers. They identified the yeasts as *Saccharomyces cerevisiae, Saccharomyces fructuum, Pichia polymorpha, Hansenula subpelliculosa* and *Trichosporon margaritiferum*.

Total yeast counts for three industrial *masa madre* samples varied within the range $10^4$ to $10^8$ cells/gram. The type *S. cerevisiae* was the most active, the various strains showing significant variation in gas and acid production. Likewise the *Lactobacilli* also vary widely in their gas and acid production capacity, according to species and strain. The homofermentative *Lactobacilli* predominated in the three *masa madre* samples, and the ratio of homofermentative to heterofermentative species profoundly affect the properties of the final bread. The homofermentatives produce enough acidity to give adequate dough elasticity, and the hetero-fermentatives impart the distinctive flavour to the bread. The heterofermenting *Lactobacillus brevis*, common to all three *masa madre* samples studied, is also found

in other wheat sours, e.g. the Italian '*panettone*' (Galli and Ottogalli, 1973) and
Sangak bread (Azar *et al.*, 1977). Lactic acid bacteria generally produce much higher
acidities than the yeasts, and the species *Lactobacillus plantarum* var. *plantarum* in
particular produces most of the acidification. Optimal conditions for yeasts and
bacteria vary widely as regards temperature, time, pH and consistency of the
medium, and simultaneous optimal adjustment is not possible. Either the yeasts or
the bacteria of the sour predominate. Yeast concentrations of more than about $10^8$
cells/gram tend to inhibit acid production of both homo- and heterofermentative
*Lactobacilli*. Whereas, if the *Lactobacilli* are allowed to predominate at around
30°C, the resulting acid build-up will inhibit the yeasts.

### 1.7.3 SWITZERLAND

The three cultural influences in the French, German and Italian cantons of
Switzerland have made a decisive contribution to the processing methods and bread
varieties, as is apparent in most areas of the economy. In the west of Switzerland, in
the Geneva and Lausanne area, the influence of French baking methods becomes
apparent; in Ticino, south of the Alps around Bellinzona, Locarno and Lugano,
typical Italian bread varieties are widespread. Further east, in Canton Graubünden,
known as Bündnerland, the Italian influence is also present in the village bakeries in
such resorts as Davos, St Moritz and Pontresina, in Engadin. In the German-
speaking cantons of Bern, Basel, Luzern, Zürich, Schwyz, Zug, Glarus, Uri,
Unterwalden, Aargau, and Solothurn, in central Switzerland, and the northeastern
cantons of Thurgau and St Gallen, some variations in consumer taste are apparent,
but there is more evidence of uniformity in product variety than elsewhere. The
external character of most bread is rustic with a thick crust, the crumb uneven and
open, but at the same time firm and with a good sheen.

Although dough yields are relatively high at around 180, based on 100 parts of
flour, the baking process is intensive, the crust remaining crisp for some time and the
crumb moist and elastic. The standard 1 kg national loaf is either made from
*Halbweißmehl* (semi-white flour or *mi-blanche*) or *Ruchmehl*. The corresponding
breads being referred to in Swiss-German Dialect as '*helles*' or '*dunkles*'.

The wheat breeding programme in Switzerland has been successful over many
decades owing to the federal government policy of self-sufficiency for defence
reasons, and government reserve grain is controlled and turned over regularly to the
mills, being replaced from each harvest. Wheat quality varies considerably within
varieties owing to extreme differences in altitude and climate as well as location.
Class 1 wheat varieties with a hard structure, e.g. Probus, have wet gluten contents of
24·0–26·0%, although protein contents are 13·0–14·5% in dry matter, and the gluten
tends to be short but elastic. Other newer varieties such as Zenith, Zenta and Kolibri
have shown increasing acceptability. Swiss mills work with two main wheat
mixtures (grists), one for *Halbweißmehl* (semi-white) of about 72% extraction, which
is divided to produce 42% *Halbweißmehl* (semi-white flour) of mineral ash content

0·68–0·70% in dry matter, and 30% white flour (*Weißmehl*) of mineral ash content 0·42–0·45% in dry matter. The other wheat mixture or grist is used to produce *Ruchmehl* (a wheatmeal-type flour) of about 82% extraction, which is likewise divided to produce 42% *Ruchmehl* of mineral ash content 1·00–1·05% in dry matter, and 40% white flour (*Weißmehl*) of mineral ash content 0·42–0·45% in dry matter. In the French-speaking cantons around Geneva and Neuchâtel, the semi-white flour is referred to as *mi-blanche* and the wheatmeal-type as *bise*, the white flour being designated *farine* fleur. In these regions the production of the white-flour bread tends to be greater than in the German-speaking cantons, and the particle-size distribution of the *mi-blanche*, *bise* and *farine* flour is somewhat finer on a 35 μm sieve.

Any adjustments at the mill are made by adding appropriate amounts of ascorbic acid to the various divided mill-streams.

Two breadmaking processes are in general use, depending on regional choice. A directly fermented straight dough of relatively soft consistency of about 180 dough yield per 100 grams of flour for *Halbweißmehl*, which can be mixed with an Artofex-type twin-armed mixer with cyclic movements, or the Spiral- and Wendel-type mixers of more recent design. Alternatively, the more traditional sponge-dough, referred to as '*Hebel*', is widely accepted; it can be processed after standing overnight or after 2–4 hours fermentation. The advantages of this method are (1) improved swelling and dough yield of about 185/100 grams of flour, and better flavour and aroma build-up in the final bread, (2) improved aeration of the dough mass, giving a rustic, somewhat irregular crumb, which is moist, elastic and of improved sheen. Typical short and long fermentation schedules are given in Table 18.

One kilogram units are divided either by hand or machine, given a short intermediate proof, and then elongated through a long-roller, adjusted for 1 kg

TABLE 18

|  | Straight dough Basel bread | Sponge (Hebel) St Gallen |
|---|---|---|
| *Halbweißmehl* (semi-white flour) | 100 parts | 40 parts |
| Water (approximate) | 80 | 35 |
| Yeast | 4 | 1 |
| Salt | 1·8 | Rested for |
| Suitable dough improver | 1·0–2·0 | about 3 h. |
| Temperature | 24°C | 25°C |
| Fermentation in bulk | 1 hour | 1 hour (dough) |
| *Dough* | | |
| *Halbweißmehl* | | 60 parts |
| Water | | 50 |
| Yeast | | 2 |
| Salt | | 2 |
| Suitable dough improver if required | | 1·0–2·0 |

dough-pieces. The pieces are then proofed fully on racks with folding-cloths to separate each dough-piece, cut obliquely, and baked strongly on the oven-bottom with steam injection in the traditional manner. Hand-manipulation can be reduced by using setter-ovens or rack-ovens, according to normal practice for all 1 kg hearth-bread varieties.

In the more commercialized cantons of Basel, e.g. *Basler Brot*, and around Zürich, the more direct straight-dough procedure is used, but in the more rural cantons and alpine resorts the sponge or *Hebel* is preferred, e.g. St Gallen *Bürli Brot* and, in Canton Bern, *Berner Brot*. Also, south of the Alps, in Ticino, the *Hebel* is preferred for many bread varieties. Certain characteristics of bread-culture in Switzerland can be compared with those existing in Spain, perhaps on account of the strong regional and localized single-mindedness of the Alpine inhabitants of Switzerland and the language differences, which are comparable to the differences in Spain between the Castilian culture and that of Catalonia. Just as in Spain, where regional differences in processing techniques exist between Madrid and Barcelona, and result in bread of different texture and flavour, so in Switzerland, owing to inherent differences in language and culture, and more localized differences within cultures, a very diverse range of bread and small bread-rolls are to be found throughout the Confederation. During the last centuries before Christ, most of Switzerland was inhabited by Celts, who in their turn mixed with an earlier race that had migrated into areas now designated as the cantons of Graubünden, Glarus and St Gallen. This race originated from the sea-faring peoples of Homer's Greece, of Etruscan blood, and their descendants, known as Raetians, inhabit parts of these cantons today. Their language, Raeto-ramansch is derived from the Latin, with some Germanic influence. The Germanic and Roman cultures were only a result of later migrations, and the Celtic culture was never completely overwhelmed by them, owing to the seclusion of the Grison mountains.

## 1.7.4 AUSTRIA

In Austria the Germanic influence from the north dominates the bread culture, and rye bread varieties are to be found everywhere. Apart from a few wheat bread varieties in the regions of Lower Austria and Styria, the main wheat bread variety is the *Kaisersemmeln*, a small bread-roll with folds in the form of a rosette which is both traditional and widespread. Although the gluten content of flour used to produce *Kaisersemmeln* is normally over 30%, the water absorption levels used are relatively low, usually being below 160 dough yield, based on 100 parts of flour. The flour is known as type 700, of mineral ash 0·66–0·70%, and about 13·0% protein in dry matter, of relatively coarse particle-size. Any adjustments to Extensograph balance between extensibility and resistance is done by adding ascorbic acid at the mill. In general, apart from special bread-type doughs where pre-ferments are often used, doughs for rolls are short, direct processes of bulk fermentation times of not more than 45 minutes. A typical fermentation schedule for *Kaisersemmeln* would be the following:

| | |
|---|---|
| White flour, type 700 | 100 parts |
| Water (appropriate) | 57 |
| Yeast | 6 |
| Salt | 1·8 |
| Suitable improver | 2–3 |
| (malt-lecithin-ascorbic acid basis) | |
| Temperature | 24–30°C |
| Intermediate proof | 10 minutes |

Procedure:
Speed-mixing requires careful adjustment of water temperature, to ensure that the dough temperature does not rise above 30°C in summer.

Although machines are available for marking the flattened dough-pieces, the best results are obtained by hand-pinning and folding into the traditional rosette-form. This ensures a symmetrical *Semmel* with sharply defined folds and edges, which appear as well-defined, golden-brown, prominent radial folds on baking.

Quality standards are very demanding: the appearance of the *Kaisersemmel* is ensured a permanent place in the craft-bakery. In the tourist resorts around Salzburg, Tirol and Vorarlberg the *Semmeln* tend to be well-finished and of good volume and light texture, whereas further east a more compact, smaller but better-flavoured *Semmel* is preferred.

On hand-pinning or machining the dough-pieces are given a final proof time of 10–15 minutes. The *Semmeln* are baked traditionally in a deck-type oven with a sloping-sole, whereby when the steam is injected before setting, the whole of the oven-sole will be covered right to the door. In this way, all goods will be at once covered with steam, and by gradually lowering the door as the oven is filled to the mouth, the steam will envelop the goods as they are set. This is impossible in a flat-sole oven, no matter what improvisations are employed. Such ovens are referred to as 'Vienna' ovens. Steam is injected at the rate of 3000/100 kg of goods on a weight for weight basis, allowing sufficient residence-time to effect gelatinization of the surface starch, which on drying out gives a characteristic glaze to the *Semmeln*. During the final stages all steam must be removed if a hard, crisp crust is desired. Ordinary deck-ovens or setter-type ovens are also used, but the results in terms of crust quality will become apparent on comparison. Temperature in the oven-chamber should be 230°C falling to 220°C for 18 to 25 minutes according to size of the *Semmeln*. By spraying with water immediately after withdrawal from the oven the sheen of the crust can be increased. The crust should amount to about 35% of the baked *Semmeln*.

Other forms of *Semmeln*, apart from *Kaisersemmeln*, are *Sternsemmeln*, *Mohnsemmeln*, *Kärntner Semmeln* and *Langsemmeln*, as well as other folded, pressed and cut varieties. These varieties can be produced by hand manipulation, or with the help of semi-automatic and fully-automated roll-plants, depending on market and output. Typical machine-aggregates for dough-pieces ranging from 30 to 75 g, and dough yields up to 165/100 g flour, with hourly outputs of 4000 to 18 000 and more units are manufactured by Werner and Pfleiderer, Stuttgart, FRG.

## 1.7.5 FEDERAL REPUBLIC OF GERMANY (FRG)

The West German baking industry offers its consumers the greatest choice in baked goods of any country, amounting to at least 200 different types of variety breads and, up to 60 different types of rolls. The quality, within a free market economy, is maintained by voluntary participation of industrial and smaller craftsman bakeries in the bi-yearly quality examination and competition sponsored by the German Agriculture Cooperative (DLG). Qualified teams of expert judges examine competitors' products and award a grand prize, silver prize and bronze prize as appropriate. Methods of evaluation and scoring are being standardized and introduced into commercial bakery practice both for bread and confectionery. Success in the DLG bakery-product competition earns the coveted insignia of the Central Marketing Organization (CMA), which is a quality status symbol for home and export sales of 'Food from Germany'.

In the Federal Republic wheat varieties fall into three classes, viz. A, B or C according to protein percentage dry matter, sedimentation value, and potential baking quality, as loaf volume, obtained in the rapid mix test (RMT) expressed in millilitres. In order to estimate the potential loaf volume in the RMT, regression equations and nomograms have been computed for the three wheat classes by Lein (1972) and Bolling and Zwingelberg (1972, 1973). The RMT loaf volume in millilitres per 100 gram of flour is the most important criterion, and volumes between 520 and 600 ml are considered good, those between 600 and 700 ml very good, and those over 700 ml are improver wheats, which are capable of carrying the other wheats in mill blends (grists). In comparing wheats on this basis, it is essential that the $\alpha$-amylase activity is standardized; therefore, the Falling number is adjusted to 250 before baking by adding an appropriate amount of malt flour. Current popular varieties, depending on location, grown throughout the FRG are: Kanzler(A6), Okapi(B4), Caribo(B4), and Rektor(A9). Potentially, the highest RMT-volumes are obtained from: Monopol(A9), Urban(A8), Kormoran(A6), Rektor(A9), Okapi(B4), Kanzler(A6) and Kraka(A6). Class A wheats averaged 13·5% protein in dry matter and sedimentation value 44 for the 1986 crop year, giving RMT-volumes of about 690 ml. Class B wheats averaged 12·3% protein in dry matter, and sedimentation value 27, averaging 600 ml RMT-volume. Cultivation of A-class wheats for the 1986 crop on a national basis amounted to 52·7%, and that for B-class wheats 42·1%. Of the A-class varieties, Kanzler was the most widely cultivated at 23·0%, and Okapi the most popular B-class wheat at 13·1%. Locations for best quality are Schleswig-Holstein, Baden-Württemberg, Rheinland-Pfalz and Bayern (Detmold, Quality of German wheat crop 1986, *Die Mühle und Mischfuttertechnik*, **42**, 572–575, 1986).

The baking quality of the rye crop is less dependent on variety and nitrogen fertilization compared with wheat. Instead, location, coupled with climatic conditions, is the determining factor. In the FRG, the rye crop at over 1 million tonnes per year is very important for the producer, miller and consumer alike. Important quality criteria for rye are the Amylogram and Falling number data.

Amylogram maximum viscosity values between 400 and 500 for flours of about 82% yield. and maximum gelatinization temperatures within the range 64 to 73°C allow normal processing into bread. Correspondingly, Falling numbers between 80 and 200 s are within the normal range. As a measure of dough texture and crumb elasticity, the Quellkurve or swelling-curve, according to Drews, is a useful indicator. Amylograph viscosity at 30°C of 350–400, 300–350 at 42°C, falling to about 260 after holding at 42°C for 30 minutes, provide satisfactory dough stability and crumb elasticity. Such relatively high initial viscosities, followed by a gradual fall-off slope are indicative of intact, undamaged starch and pentosan structure, which is resistant to enzymic breakdown. In order to produce the many types of flour in use in the FRG, and satisfy the legal tolerances for mineral ash content, the various flour-types are prepared in the mill by a system of 'divides'. The wheat-mixture or 'grist' is milled to an average yield of 78–79%, but the various mill-streams or 'passages' are used to produce two to three types of flour at the required ash content. The following milling schemes are given as examples.

Mill I:    45% wheat-flour type 405 for household use
              12% wheat-flour type 550 for specialty breads
              22% wheat-flour type 1050 for mixed wheat/rye bread

Mill II:    60% wheat flour type 550 for specialty bread rolls
              15% wheat flour type 405 for household use
               3% wheat flour type 1050 for mixed wheat/rye bread

Mill III:   45% wheat flour type 405 for household use
              34% wheat flour type 812 for specialty breads or rolls

Other type 550 flours produced from appropriate selection of mill-passages include a type 550 gluten-rich flour for specialty products, and a type 550 toast flour especially for panned toast-bread manufacture. The various flour type numbers correspond to the average ash content expressed in mg%, in dry matter. Minimum and maximum ash contents in dry matter have been laid down for each flour-type by the federal government in Bonn under the 17th Grain-regulation for all milled-products (1961). This regulation covers 22 wheat- and rye-flours and meals. The most important of these are dealt with in the following sections.

## Wheat-Flours

Type 550 is for general bakery use, mainly for the manufacture of the many types of white bread-rolls, e.g. *Weißbrötchen*, Hamburger *Rundstücke*, Berliner *Schrippen*, or Bavarian *Semmeln*. A typical analysis of such a flour is as follows: ash 0·58% dry matter, moisture 14·5%, protein 12·0% dry matter, wet gluten 31·0%, Falling number 280, maltose 1·6%, ascorbic acid added at the mill 1 to 3 g/100 kg flour, giving a dough yield of 158/100 g of flour. Some regional qualitative differences in type 550 exist, and in some areas a type 550 gluten-rich flour of slightly higher gluten content is preferred for bread-roll production. A typical fermentation schedule for the manufacture of bread-rolls from this type of flour is the following:

| | |
|---|---|
| Flour type 550 | 100 parts |
| Water (approximate) | 58 |
| Yeast | 6 |
| Salt | 2 |
| Malt-based improver | 2–3 |
| Dough temperature: | 26–27°C |
| Proof in bulk: | 10–15 minutes, depending on mixing intensity |
| Final proof: | 10 minutes |

Procedure:

Adjustment of dough temperature and proof time depends on type of mixer and mixing-intensity,

Dough-pieces pressed at 45 grams/unit.

Processing is either by semi-automatic roll divider/moulder with about 30 divisions at a dough yield/consistency of 158, pressing/dividing at 45–160 grams, for example using a Werner and Pfleiderer bun and roll divider/moulder Rota H 30/120, or the automatic roll-plant aggregates '*Schrippenanlage*' of Werner and Pfleiderer, based on an appropriate combination of machine-aggregates. For example, divider, moulder and intermediate proofer ZNK with Longroller ELR 680, or the intermediate proofer ZGL 5/6 bridged with a final-proofer and tunnel-oven. On this basis, roll-plants with hourly outputs of up to 18 000 units or more are possible.

Where deck-ovens are in use, a better quality product can be produced by injection of steam at about 3000 kg/100 kg of dough mass during the initial baking phase to effect surface gelatinization of the starch and impart a glaze to the rolls. All traces of steam are removed during the final baking phase, thus ensuring a crisp, short crust. Oven temperature should be about 230°C, falling to 220°C for 18 to 25 minutes depending on roll size. Since the average shelf-life of such products is limited to 4–5 hours for crispness of crust, the craftsman-baker must adjust his production schedules as far as possible so that the consumer can purchase the rolls within this time-span. However, since demand and supply do not always coincide, other innovations have been devised to ensure continuity of supply. One possibility is the use of ready-mix concentrates for bread-roll production, but this may not be expedient enough to satisfy demand. Fortunately, modern processing technology allows the separation of doughmaking and fermentation in both time and place from the baking process. Thus, the following techniques can be considered.

1. Partial baking of the rolls in one location, followed by subsequent bake-off as required in another location, for which purpose a microwave can be used for the partial-baking phase. This technique is known as 'brown-and-serve', as originated from the USA in the early 1950s.

2. Dough-retardation methods and the provision of an inventory of moulded dough-pieces at refrigeration temperatures, which can be brought up to ambient temperature and baked-off as required. The reduction of dough yeast content is necessary in this method. In fact, this procedure, involving a prolonged fermentation phase, has a positive effect on the flavour of the rolls.

3. Deep-freezing techniques involving the freeze/store/thaw cycle, which can take place in the same unit. The oven-fresh rolls are fresh-air cooled to about 35°C, followed by rapid cooling of the centre of the crumb to − 15°C by blast-freezing at a speed of about 3·5 m/s. The critical temperature of − 15°C must not be exceeded to avoid separation of crumb from crust. The deep-freeze time is noted by recording when the centre of the crumb reaches − 8°C, which requires a chamber temperature of about − 30°C. A chamber temperature of − 17°C at air speed about 0·5 m/s will maintain the crumb at the critical temperature during storage. The freezing time and storage temperature are critical for product quality. The freezing time for rolls under the above conditions would be 1–2 hours without packaging. Thawing can be carried out automatically by using a timing device and warm air unit. When the chamber temperature rises to zero, the humidity control system is actuated in two stages of about 5 minutes, maintaining a relative humidity of 50% to 70% at air speeds of 0·5 to 2·0 m/s. This control of relative humidity helps to maintain crispness and the correct hardness of the crust during defrosting, while the temperature is being raised to 60 to 65°C as rapidly as possible. A rapid defrosting process is recommended for lean-formulated bread rolls to prevent the onset of starch-retrogradation. Although such a system requires special conditioning equipment and thermostatic control, the whole cycle can take place in the same unit. Each compartment consisting of two adjacent stacks of steel-wire baskets mounted on wheeled-frames. This set-up eliminates night-work, and makes better use of an 8-hour day, baking being completed by 07.30 hours daily, including Saturdays. Freezing commences immediately, and the thaw-cycle can be actuated by 05.00 hours or whenever desired to provide bread rolls for sale from the previously frozen stock within 4 hours. This system is suitable for the craftsman-baker or in-store bakery where the sales point is adjacent to production, but in the case of larger industrial bakery groups, with scattered distribution points, another organizational system must be effected. In the latter case, the bread rolls can be transported in the frozen state in freezer-vans, and either defrosted in transit or on arrival at the point of sale. For special deliveries to hotels, restaurants, and canteens, refrigeration or thermo-insulated vans are necessary. In emergency situations, bread-rolls can be defrosted at about 170°C in the oven within about 5 minutes, but steam injection must be used. At sales points, infrared is often used for this purpose. If, for economic reasons, defrosting is carried out at ambient temperature, e.g. 20°C, shrinkage and the separation of crust from crumb can be inhibited by shortening the baking time by about 20%, and increasing the baking temperature by about 15°C. Certain additives such as invert sugars, lecithin and distilled monoglycerides are also helpful in preventing shrinkage.

Other wheat flour types for bakery use are the following:

Type 630 ash content 0·60–0·70% dry matter
Type 812 ash content 0·75–0·87% dry matter
Type 1050 ash content 1·00–1·15% dry matter
Type 1600 ash content 1·55–1·75% dry matter
Type 1700 ash content 1·60–1·90% dry matter (wheatmeal)

The flour types 812 and 1050 are mainly used for the production of mixed wheat/rye bread, but can also be blended with other flours for special biscuit and confectionery formulations.

Most of the products manufactured from wheat-flour are formulated from: wheat-flour type 550, *Brötchenmehl* (bread-roll flour) or a gluten-rich version of the same type, water, yeast, salt, improver based on malt–ascorbic acid–lecithin, as an emulsified hydrate.

The larger 500-g and 1-kg bread units are either hearth bloomer-type, cut open-pan or special toast-bread. In the case of the latter, a special higher-protein toast-flour is recommended to ensure a shelf-life of 4–6 days, and the inclusion of corn-syrup, vegetable oil and milk-powder coupled with intensive mixing at about 1400 rpm for 60 s produces a better product.

Bread-rolls are distinguished from one another by their shape, viz. *Spitzbrötchen* (*Schnittbrötchen*), which are very popular in northern and western (Nordrhein-Westpfalen) areas. They break open on baking, and have a crisp, open/split ('*ausgebunden*') appearance, i.e. '*Ausbundgebäck*', being ellipsoidal in form.

*Schrippen* are a typical Berlin roll, but are now popular far beyond the bounds of the city, and can be prepared either by 'pressing' with machine or 'cutting' longitudinally.

*Semmeln* or *Sternsemmeln* are round, and cut to form a cross, being of oriental origin (*samidu* = wheat-flour), a simplified form of the distinctive *Kaisersemmel* of Austria. The addition of 3–5% rye-flour improves dough manipulation and flavour of this product. Best results are obtained with this product when they are set in the oven slightly under-proofed with a full-oven, thus allowing the consequent fall in temperature to produce a good volume and open appearance (*ausgebunden*).

A similar group of products, known as '*Weizenkleingebäck mit Zusätzen*' (small wheat-rolls with other additions), are distinguished from the leaner-formulated bread-rolls (*Brötchen*) in that they may contain up to 10% (based on dough weight) of shortening and/or sugar. In addition, the weight of the baked product must not exceed 180 g. Depending on the dough fermentation used, the yeast quantity would be at least 3·0%, and the salt up to 2·0% of the flour weight. The most important products in this group are: *Knüppel-, Mohn-, Kümmel-* and *Salzbrötchen, Brezeln* and *Hörnchen*. The raw material preparation and doughmaking procedures are very similar to the lean bread-roll formulations, apart from additions of margarine, milk powder and malt. The softened margarine is worked-in at the last stage of mixing, and the dough given 2 to 3 knock-backs or punches during its 30-minute rest in bulk. Examples of formulation variations are given in Table 19.

*Knüppel* originated in Berlin as '*Berliner Knüppel*, but are now more widely known as '*Tafelbrötchen*' (table-rolls). The dough should contain milk-solids in some form. The rounded pieces are moulded at the slightly under-proofed stage, flattened, and joined end-to-end. They are then passed through the long-roller, the blunted ends being characteristic for this product. The pieces being set for final-proof with the fold underneath. On setting in the oven, the pieces are turned over and set end-to-end, thus giving the blunted ends after baking.

*Mohnbrötchen* can be made in the form of rosettes, star-shapes or as elongated strips, being either pressed or cut with scissors in the rounded-form. The shaped

TABLE 19

|  | *Knüppel* | *Mohn* | *Kümmel* |
|---|---|---|---|
| Wheat-flour type 550 | 50 parts | — | — |
| Wheat-flour type 550 (gluten-rich) | 50 parts | 100 parts | 100 parts |
| Yeast | 4·0 | 4·8 | 3·2 |
| Salt | 2·0 | 0·7 | 1·5 |
| Margarine | 5·0 | 12·0 | — |
| Ground sugar | 1·0 | 9·8 | — |
| Milk powder | 2·0 | 2·5 | — |
| Water (approximate) | 56·0 | 49·0 | 53·0 |
| *Kümmel* (caraway seed) | — | — | 3·6 |
| *Mohn* (poppy seed) | — | 5·2 | — |

pieces are moistened with water or milk and either sprinkled or placed in *Backmohn* (bakery poppy seed) before setting in the oven.

*Kümmelbrötchen* can be prepared either by flattening the dough-pieces, rolling and forming into crescents, or leaving the rolled-up pieces in the elongated form. In both cases, the pieces are moistened and either sprinkled with *Kümmel* (caraway-seed) or placed in it before setting in the oven. The seeded top of this product gives an attractive appearance and herbal flavour to the baked rolls.

*Final-proof*
The length of time and degree of the dough-ripening have an important influence on the structure and porosity of the crumb of all the products described. It commences at moulding and ends when the pieces are set in the oven; it can be carried out in a dry or humid atmosphere. This will depend on whether the dough-piece is to be inverted before oven-setting. Humid final-proofing has the advantages of imparting increased extensibility to the dough-pieces, which allows uniform oven-spring and improved volume, and the formation of dextrin, which imparts a glaze to the product and a smooth surface. Final-proof can be divided into two stages, which becomes essential where pressing, folding and long-rolling are involved. Where continuous processing lines are in use, this is obligatory in the form of an intermediate and final proofing unit. However, this regime is not possible when the dough-pieces are subjected to diverse shaping procedures before baking.

The duration of final-proof will depend on the moulding and processing conditions as shown in Table 20.

TABLE 20

| *Moulding conditions* | *Extended final-proof* | *Shorter final-proof* |
|---|---|---|
| Rounding-up intensity | Mechanized | Manual |
| Initial proof during moulding | No bulk ferment | Bulk ferment |
| Diverse moulding procedures | Mechanical pressing, folding, rolling, etc. | Long-rolling, manual pressing, cutting, etc. |

In addition, final-proof will depend on the raw materials and the addition of fermentation-promoting improvers, and dough-mixing and fermentation technology.

Both temperature and relative humidity at final-proof should be measured and regulated as follows. For products with *Ausbund* (open/split appearance) fermentation temperatures should not exceed 40°C, and in the case of cooler doughs of 24 to 28°C, the upper limit is about 30°C in order to avoid excessive condensation. Relative humidities for products with '*Ausbund*' should fall between 50% and 60%, and between 70% and 80% for other products.

The correct judgement of the final-proof before setting in the oven is critical, and depends on oven-temperature and process used. Mistakes made in this judgement will result in defective products. Cutting of the dough-pieces is best done at the slightly under-proofed stage (on the 'green' side of ripeness) manually. Baking of 45-g dough-pieces commences at 220°C for about 18 min, using steam at the rate of about 3000 litres/100 kg product.

The production of *Brötchen* and *Weizenkleingebäck* represents the technical skill of the German baker. The variety, in texture, flavour and shape, gives a representative picture of German bakery-culture and its traditional quality standards. Many of these varieties are intended to complement quality wines and good beer. Although the individual consumption of wheat bread and rolls has shown a small decline over the past three decades in favour of special and dietary breads, the bread-roll sector of the market remains an important one for craftsman-baker and the industrial-baker alike.

### Rye Flours
In the FRG, all rye-flours and rye-meals for bakery use are classified according to mineral ash content in the same manner as the wheat-flours. The most important of them are as follows:

Type 815 ash content 0·79–0·87% dry matter
Type 997 ash content 0·95–0·07% dry matter
Type 1150 ash content 1·10–1·25% dry matter
Type 1370 ash content 1·30–1·45% dry matter
Type 1800 rye-meal ash content 1·65–2·0% dry matter

These ash ranges were laid down in the 17th Grain-regulation for all milled products (1961) by the federal government in Bonn, as described for the wheat-flour types. Although the consumption of rye bread and rolls tended to decline over the previous two decades, its popularity in the 1980s gradually increased. This is due to its valuable contribution to the diet, which include certain components which are complementary to the consumption of products made only with wheat-flours. Its amino acid spectrum for the following important acids is higher than for wheat: valine, lysine, threonine, and histidine. The biological value of rye-flour at comparable extraction rate of 98%, according to the method of Osborne and Mendel is 54 compared with wheat-flour at 47, the index being based on whole-egg as the reference food at 100. Furthermore, the rye proteins have a lower molecular weight and higher water-solubility than wheat proteins, which is due to protein–pentosan complex formation. Also, the latent swelling capacity of the lignin and hemicellulose components of rye bread, coupled with its sour-dough, bacterial-derived lower pH, favour good peristaltic movement and digestion in the intestinal tract, resulting in high volume:weight ratio stools. The high source of fructose from the graminin-decafructosan is also a useful dietary supplement. Type 997 corresponds to an extraction rate of 79–82% and type 1150 to extraction rates between 84% and 87%, depending on grain structure, milling-diagram, and choice of machinery. The manufacture of bread from rye-flour requires different conditions and processing technology to that for wheat-flour. Rye-flours, owing to their composition, can only be processed into good bread with the help and control of acids and salt. The starting-point, as the source of necessary yeasts and bacteria, is the sour-dough starter-culture. Its careful preparation and composition will ensure rye bread without structural faults and of appetizing flavour. Once the pure starter-culture is used to prepare the initial sour-dough stage, useful and harmful microorganisms manifest themselves, and the purpose of the choice of fermentation conditions is to propagate the useful ones and suppress the harmful ones. This process is best controlled by fermenting in stages at optimal temperatures, thus ensuring the necessary balance between bread flavour, crumb-structure and loaf symmetry. By using defined dough temperature and consistencies at each stage of the technological process, the desired microbial growth results, as follows: sour-dough yeasts 25–27°C, sour-dough lactic acid bacteria 35–40°C, and sour-dough acetic acid bacteria 20–25°C. Although the temperature is initially adjusted by the dough water, it is maintained by the fermenting bacteria, and controlled surroundings. Progressive swelling of the rye-flour components and optimal dough yield at each stage depend on the maintenance of a controlled regime. Table 21 gives

TABLE 21
**Summary of sour-dough control variables**

| Variable | Starter (*Anstellgut*) | Initial-sour (*Anfrischsauer*) | Basic-sour (*Grundsauer*) | Full-sour (*Vollsauer*) | Dough (*Teig*) |
|---|---|---|---|---|---|
| Flour | — | rye | rye | rye | rye/wheat |
| Standing-time (h) | — | 5 | 6 | 3 | 10–25mins |
| Temperature (°C) (range) | — | 25–27 | 22–27 | 28–34 | 28–32 |
| Dough yield (%) (flour weight) | — | 180–200 | 150–160 | 160–200 | 150–165 |

| | | |
|---|---|---|
| Short process | 10·0% of rye-flour content | Depending on: Sour-temperature |
| Long process | 0·5% of rye-flour content | Sour-quantity |
| | | Ambient temperature |

| Aeration requirement | | high | modest | very high | according to mixing regime |
|---|---|---|---|---|---|

some optimal ranges for the more important controllable variables of the various stages of sour-dough manufacture.

### Starter-culture (Anstellgut)

This is the 'seed' for the manufacture of rye-doughs, and must be carefully maintained according to the bakery processing conditions, within the limits of the controllable variables, e.g. temperature, dough-consistency, standing-time, choice of flour substrate, reproduction rate, aeration requirements. For the perpetuation of the multi-stage sour-dough process, the starter-culture is taken from the centre of the mature full-sour, where the microorganisms are most vigorous and concentrated. Its pH has been raised somewhat by the addition of flour during the doughmaking process. It should be stored in a cool place in containers of poor thermal conductivity, e.g. wood. The required amount of starter-culture will depend on weight of the dough to be produced, on average 0·5% of total rye-flour weight to be acidified. Higher concentrations are possible depending on type of process, flour-type and desired sour-dough percentage of total dough mass. For doughs containing 80–100% rye-flour, 35–40% sour-dough, based on total flour weight is to be recommended, and in the case of mixed wheat/rye doughs an average of 50% of the rye-flour should be soured.

The most important sour-dough bacteria in the starter-culture are, the homofermentative species *Lactobacillus delbrückii*, which produces lactic acid, and the heterofermentative *Lactobacillus fermentum*, which also produces acetic and other volatile acids.

### Initial-sour (Anfrischsauer)

As a result of the low consistency of this culture-stage (180–200), the microorganisms can reproduce freely, the yeasts in particular dominate at 25–26°C, and the production of lactic and acetic acids are suppressed over the 4–5-hour rest period.

### Basic-sour (Grundsauer)

At the higher temperature of this stage, and the higher consistency, the lactic acid build-up is favoured. The longer resting-time prevents an excess of acidity.

### Full-sour (Vollsauer)

For this stage, the controllable variables are adjusted so that both the sour-dough yeasts and bacteria can reach peak activity. The resultant strong build-up of lactic acid peptidizes the proteins, providing lower-molecular-weight units on which the sour or added yeasts can feed.

### Dough (Teig)

Rye- or wheat-flour is added, depending on the type of bread being produced. A mixture of rye and wheat, 50:50, is popular, and is known generally as '*Mischbrot*'.

In practice, the successful application of the four-stage sour-dough process described demands careful planning and organization. The prime consideration is

TABLE 22
**Sour-dough process using an overnight '*Grundsauer*' (basic-sour)**

| Stage | Schedule (h) | Sour (kg) | Flour (kg) | Water (litres) | Total | Dough yield (%) | Temperature (°C) Start | Temperature (°C) Finish |
|---|---|---|---|---|---|---|---|---|
| Starter | 10.00– | 1 | — | — | — | — | — | — |
| Initial-sour | 10.00–16.00 | 1 | 2·6 | 3 | 6·6 | 213 | 23 | 25 |
| Basic-sour | 16.00–24.00 | 6·6 | 15 | 6 | 27·6 | 150 | 22 | 26 |
| Full-sour | 24.00–03.00 | 27·6 | 30 | 25 | 82·6 | 175 | 28 | 30 |
| Dough | 03.00–03.15 | 81·6[a] | 52 | 20 | 153·6 | 154 | 29 | 30 |

[a] 1 kg starter taken from dough for next batch.

when the dough is required (time factor) and the process technology in operation. The controllable variables are adjusted and applied accordingly. Since it is not possible to culture all the microorganisms simultaneously at their optimal level of growth, the four-stage process allows each specific group of microorganisms to reproduce preferentially. At the doughmaking stage, there must be an optimal balance between sour-dough yeasts and sour-dough bacteria.

In the small and medium-sized bakery, in order to produce enough dough for daily requirements, using the four-stage process, each stage must be proportionally increased starting with the starter-culture (*Anstellgut*). Table 22 illustrates a typical four-stage sour-dough process, based on an overnight '*Grundsauer*' (basic-sour).

*Preparation of Initial-sour ( Anfrischsauer )*
At 10.00 hours 1 kg starter-culture, 2·6 kg rye-flour and 3 litres water are made into a soft dough of dough yield/consistency 213 and temperature 25°C. The dough is covered with a little flour and set aside for 6 hours to ripen. After which it is processed into the basic-sour (*Grundsauer*). In the initial-sour (*Anfrischsauer*), the sour-dough yeasts can reproduce freely. The temperature of 25°C is optimal, and the soft dough contains dissolved nutrients for yeast growth, and during fermentation the dough temperature will rise by 1 to 2°C. With each successive sour-dough stage the gas-retention capacity of the dough increases.

Although the above procedure for sour-dough preparation and processing into bread explains the basic technology, it is very complicated and time-consuming, as well as difficult to rationalize. Therefore, technical institutes, and technical people in the industry have devised numerous modifications which separate the sour-dough preparation in time from dough preparation, thus simplifying the whole operation.

The principle of one of these schemes is to prepare a larger amount of the basic-sour (*Grundsauer*). The basic-sour, being the most important biologically for the souring process, and the one which requires the longest resting period, provides the best stage for bulk sour-dough reserve for subsequent processing into bread. For example, a bulk basic-sour can be divided into two equal parts by weight, and each of these sub-divided into four equal parts by weight for the preparation of four full-sours (*Vollsauers*), thus allowing four doughs to be prepared simultaneously. A further shortening of the traditional multi-stage process is achieved by elimination of the initial-sour (*Anfrischsauer*) and instead using a larger amount of starter-culture with increased flour and water. The large amount of basic-sour (*Grundsauer*) thus produced can be stored for 16 to 40 hours, owing to the self-conservation effect of the acid produced. About 18–20% starter, based on flour weight, is made into a basic-sour of dough consistency/yield 154. On ripening, this is then used to produce up to 10 full-sours and subsequent doughs, as required.

The following processes for rye bread production have been well defined, and give the baker a wide choice to suit his individual operations.

Detmold one-stage process (*Detmolder Einstufenführung*, DEF) (Stephan, 1957). A one-stage process for preparation at midday, and doughmaking the following day between 04.00 and 10.00 hours.

Berlin short-sour process (*Berliner Kurzsauerführung*, BKF) (Pelshenke, 1941). A one-stage sour of standing time 3 hours at 35°C.

Salt-sour process (*Salzsauerführung*, SSF) (von Stein, 1956). By adding salt to the sour, the standing-time can be from 1 to 3 days, allowing the dough to be made at any time.

Detmold two-stage process (*Detmolder Zweistufenführung*, DZF) (Stephan, 1956). This involves the preparation of a basic-sour (*Grundsauer*) to stand for 15 to 24 hours; the full-sour (*Vollsauer*) is prepared the following morning. Or, alternatively, the full-sour can be prepared in the evening, and the doughmaking take place at 04.00 hours the next morning.

Applying the two-stage Detmold process, with a 3-hour final-sour (*Vollsauer*) ripening time, the following is a description of the procedure employed for the manufacture of mixed wheat/rye bread (*Mischbrot*), which is the largest market of the rye sales sector.

Basic-sour (*Grundsauer*)

| | |
|---|---|
| Starter-culture (*Anstellgut*) | 0·800 kg |
| Rye-flour type 1150 | 9·600 |
| Water | 4·800 litres |
| Total basic-sour | 15·200 kg |

| | |
|---|---|
| Sour temperature: | 20–23°C |
| Dough yield: | 150 |
| Resting time: | 15–24 hours |

In order to calculate the amount of flour required to prepare the basic-sour, the amount of starter-culture is multiplied by a factor of 12, i.e. $0·800 \times 12 = 9·600$ kg. To determine the amount of water, the same figure is multiplied by a factor of 6, i.e. $0·800 \text{ kg} \times 6 = 4·800 \text{ kg}$.

*Final-sour (Vollsauer)*

| | |
|---|---|
| Basic-sour (*Grundsauer*) | 15·200 kg |
| Rye-flour type 1150 | 30·400 |
| Water | 30·400 litres |
| Total final-sour | 76·000 kg |

Sour temperature:  28–31°C
Dough yield:  188
Resting time:  3 hours

In order to calculate the amount of flour and water for the final-sour, the weight of the basic-sour, 15·200 kg, is multiplied by a factor of 2, which gives 30·400. This fixes the dough yield at 188.

*Dough preparation*

| | |
|---|---|
| Final-sour (*Vollsauer*) | 75·200 kg (76·000 − 800 kg mature full-sour for starter) |
| Rye-flour type 1150 | 40·000 |
| Wheat-flour type 1050 | 20·000 |
| Water | 24·800 litres |
| Total dough weight | 160·000 kg |

Dough temperature:  28–29°C
Dough yield:  160
Resting time:  10 minutes

In addition, 1·800 kg (1·8%) salt, and 0·500–1200 kg (0·5–1·2%) yeast are added to the dough.

The resting-time of the basic-sour can vary between 15 and 24 hours, which does not adversely affect bread quality. This scheme is also suitable for various ratios of rye: wheat flour, but the percentage of soured-dough should not fall below 40% based on flour weight, otherwise crumb-elasticity, and bread flavour become downgraded. With increasing proportions of wheat-flour, more yeast is added to the dough, and a souring of more than 50% of the flour is recommended to improve bread flavour. However, as a general rule, the sour-dough quantity is usually based on the amount of rye-flour being processed, and not on total dough water, total flour weight or total dough weight, since in the latter cases sources of error creep in. Wheat-flour type 1050 is considered by most bakers to be the best partner for rye-flour, owing to its higher extraction, having compatible colour and doughmaking properties. Attempts to replace the natural dough-sour with patented dough-acidulants, produce bread of inferior quality, and standards have now been laid down for minimal sour-dough contents for rye breads in the FRG. Dough-acidulants (*Versäureungsmittel*) are useful additives when used as intended, for the correction of abnormal rye-flour quality due to adverse conditions of rye-grain cultivation, but they are not intended to replace sour-doughs.

A more detailed consideration of sour-dough technology will be undertaken in Chapter 2.3.

## 1.7.6 GERMAN DEMOCRATIC REPUBLIC (GDR)

The bread product range in the German Democratic Republic is similar to that in the FRG, offering the consumer a wide choice of appetizing and nutritionally excellent cereal-based products. A very diverse range of special and dietary breads have been developed, which have shown a steady rise in consumption, especially during the past two decades, in common with similar trends in the capitalist countries. Although flour confectionery sales remain buoyant, white bread and rolls, and to a lesser extent rye bread and rolls, have both shown slight falls in per capita consumption over the past two decades. This trend is also apparent in most countries with rising living-standards. Nevertheless, with quality-maintenance, and explanation to the consumer of the value of rye products, appreciable falls in consumption should be prevented. In the GDR, all quality standards are decided by free discussion, involving all aspects of scientific, technical and economic importance. This requires close cooperation between production specialists, quality-control personnel, scientific and state institutions, and appropriate ministries.

The resulting standards, designated by assigned TGL numbers, cover raw materials and finished products, and their texture and analytical control methods. Similar systems are in operation covering machines and machine-aggregates, for both processing and packaging. In addition, sensory standards have been fixed for raw materials, additives and finished products in the form of ASMW–VW reference numbers. Since 1958, the GDR has been a member of the International Association for Cereal Chemistry (ICC), and has been active in working-groups seeking to standardize quality-control methods for cereals and bread on an international basis.

Since 1977 the quality standards for bread and rolls have been revised and updated. Bread standards divide bread varieties into three categories, viz. basic, special and dietary breads, each of which is defined. The definition of basic bread varieties as laid down under Standard TGL-3067 is the following: a bread utilizing rye- and/or wheat-flours, possibly also rye- or wheat-meal, with partial souring of the rye-flour, with or without the addition of salt and drinking-water, yeast, and improver, which, after dough preparation, is manufactured by baking.

Since 1979, a minimum baking-time has applied to all types of bread, and specific setting and end-temperatures are recommended in order to achieve these baking-times, e.g.

—bread-rolls, 45 grams baked weight, 18 minutes at 220°C, giving an average baking-loss of 16–20%;
—mixed rye/wheat (*Roggenmischbrot*), scaling weight 1500 kg, for hearth-bread, 60 minutes (maximum deviation: 10 minutes) at 280°C falling to 210°C, giving an average baking-loss of 10–12%.

Every specific bakery product is assigned standards to cover raw materials, additives, weight and weight-loss, volume and sensory qualities, labelling, packaging, storage and transport.

In all large bakeries and cooperatives (*Kombinate*) in the GDR, there is a Technical Control Organization (*Technischen Kontrollorganisation*) known as TKO. The number of personnel engaged in this operation depends on the output of the production unit. Thus, for a bakery producing over 9000 t of bread and rolls, or over 2000 t specialty products and confectionery per annum, the TKO team would consist of one TKO manager (specialized engineer) and one TKO co-worker (qualified baker) for each shift, and, additionally, one TKO assistant, recruited from the foremen of each shift—the most experienced and reliable in the appropriate area of production. As further support for the TKO, additional TKO assistants are appointed as honorary members of the team. The laboratory, in addition to analytical work, also carries out any routine measurements for the TKO. The TKO manager being directly responsible to the director or general manager of the production unit. The TKO manager is responsible for the collation of results of checks carried out during the various production stages by the TKO assistants, e.g. raw materials, preparation of starter-sour, initial-sours and basic-sours, moulding-operations, baking-process, storage and despatch, including laboratory analytical data and final product assessments. Final compilation of findings and conclusions is made in report form for the production manager and general manager. A centralized system of information flow, allowing the comparison of planned objectives with those achieved, ensures that any deficiencies within departments, or between them, are communicated rapidly, involving all the workforce in responsible collaboration for the benefit of the community as a whole. In socialist societies, the effective implementation of complex measures aimed at improving productivity and the quality of work is of paramount importance to all members of society. The basic concept for achieving these goals was developed in the USSR as the Saratower system, and has been applied and developed by other socialist societies. The most important preconditions for good quality work and quality assurance are the socialist work-ethic, a healthy attitude towards work, and a disciplined approach to work problems. Additionally, a high degree of technical training and skill is essential for all personnel involved in the production process.

The necessary stimuli for achieving a high standard of work and simultaneously minimizing mistakes must be directed to the moral and material aspects of human nature. The results of the quality control are published, with mention of personnel who have achieved both good and poor work standards. Good quality work is recognized by the award of appropriate certification and medals. Material rewards are provided in the form of the dependence of a percentage of salary on the quality of workmanship. The same system applies to salaries of both foremen and experienced and qualified bakery personnel. In addition, individual consistency in work quality is recognized by the payment of yearly premiums to all members of the production unit, depending on the surplus of income over outgoings. TKO assistants are rewarded for good control work, performed on an honorary basis, by additional premiums. Exceptional achievements of planned targets or initiative in quality control and development work are rewarded either individually or collectively by premium payments. Any serious damage to product quality, due to neglect or breach of quality control procedures, is subject to material fines, and registration with the Ministry for Standardization, Measurement and Product-Control—Foodstuffs (ASMW).

Another very important function of the TKO in any bakery production unit is the close cooperation with TKO of raw material suppliers. For example, there is open discussion of analytical and other data of flour and shortening deliveries with the supplier TKO, in order to agree on supply specifications. There is free discussion of any quality problems encountered in processing the raw materials with the existing machinery and technology. Contracts can be withdrawn in the extreme event of failure to meet required quality standards. There is regular exchange of information with product-distribution outlets regarding product-quality and consumer complaints.

Although the methodology applied to the control of production unit operations is based on the same statistical approach as that employed in many other countries, the successful implementation of these techniques always depends on the motivation of the workforce. It is the different socio-economic order compared with that of capitalist societies, which allows a more comprehensive integration of technology and human resources. It is the lack of a collective approach to production problems, which often hinders the best use of resources and the maintenance of product quality standards.

In the GDR, flours are classified basically on optical reflection rather than mineral ash content, since the latter depends on the flour extraction rate. Furthermore, any quality evaluation of flours will depend on the end-use and the quality aspects necessary to satisfy the technological processing involved. For all flours and meals, moisture content must not exceed 15%, soil content must not exceed 0·1% in dry matter, the content of iron in metallic form must not exceed 3 mg per kilogram. The grain from which it is milled must be free of all insect-contamination or microbial growth, as laid down in TGL 27424/03. The same regulation covers semolina, which must not exceed 0·55% ash, 15·5% moisture, and 4·5 acidity-index. The soil and metallic iron contamination maxima are the same as for flours and meals. The diverse technology used in dough processing and the various product-types demand specific dough properties. Therefore, the deliveries must be laboratory-tested at regular intervals so that the doughmaker is aware of the correct flour:water ratio, and mixing time, as well as water-temperature adjustment.

The following aspects of flour quality are tested: moisture, optical-reflection (%), falling-number, gluten-content, and gluten-extensibility. As a result of these tests, small changes in dough formulation and/or the use of certain improver additives can be undertaken. Table 23 shows the quality requirements of flours and meals in the GDR, according to TGL 27424/01.

A visual colour-contrast definition is also given for each group of milled products, as follows:

Wheat-flours—white with a yellow tone
Rye-flours—white with a greenish-grey to blue-grey tone
Wheatmeals—cream with a reddish-brown tone
Ryemeals—white with a greenish-yellow to bluish-grey tone

The percentage reflected-light index (brightness) is the basis for flour classification, and no form of flour-bleaching is permitted in the GDR; also, with the exception of ascorbic acid, no flour additives can be used.

Wheat bread and smaller wheat products (*Weizenklein-gebäck*) are classified according to their weight and composition.

## Wheat Breads

Wheat breads can vary in baked weight from 250 g to 1500 kg, depending on type and form.

White bread units of 500 g and 1 kg are made from gluten-rich wheat-flour, in various lengths and baked as hearth-breads on the oven-sole. Panned bread units of 1·5 kg are baked in long, narrow pans and cut lengthwise.

## Wholewheat Breads

Wholewheat breads are either made with wholewheat flour or wholemeals in units of 500 g, 1 kg and 1·5 kg, mainly as panned bread.

## Special Breads

The special breads group include the various Rehbrücker Spezial-Toast products, viz. Spezial-Toast, Spezial-Toast with cheese, Spezial-Vitamin-Toast, Spezial-Pußta-Toast, and Spezial-Toast with milk-protein, being made from wheatflour of optical-reflection not less than 55% (*Weizenauszugmehl*) and ash content 0·38–0·48% dry matter. These formulations are enriched with shortening, milk-solids, and sugar, and, in certain cases, additions of cream-cheese, tomato-paste, and paprika. Conventional mixing techniques for periods in excess of 12 minutes using traditional mixers are recommended for these doughs, set at about 30°C. Two remixes are then given at 30-minute intervals, followed by 40 minutes final-proof in pans. Baking is at 240°C falling to 210°C for 30 minutes in the case of 500-g units, and 35 minutes for 750 gram units. Adequate cooling is essential before slicing and packaging, 24 hours at ambient temperature with good ventilation, or accelerated cooling at reduced temperature over about 2 hours.

*Grahambrot* is a long-established special wheat bread made with a specially treated wholewheat flour or wheatmeal. The special Graham-flour being processed without salt or yeast; instead, a spontaneous souring process, initiated with a starter, rye-flour and water, provides the texture and distinctive taste. The spontaneous-sour depends on progressive culture of specific acid producing bacteria and yeasts which reproduce in symbiosis. By choice of resting-times and remixing with further additions of flour and water, an optimal balance of carbon dioxide development and acidulation results. *Grahambrot* is usually produced in bakeries which specialize in this type of bread, in the form of panned-bread units of 1·5 kg.

# TABLE 23
## Quality standards of flours and meals, TGL 27424/01

| Flour type | Mineral ash (%) DM | Optical brightness (%) | Acidity-index maximum | Particle size (mm) | Falling number minimum (s) | Gluten-content (%) |
|---|---|---|---|---|---|---|
| Wheat-flour | 0·38–0·48 | not less than 55 | 2·6 | under 0·160 90% | 200 | not less than 21 |
| Wheat-flour, gluten-rich | 0·58–0·68 | 51–54 | 2·6 | under 0·160 90% | 200 | 25–33 |
| Wheat-flour, low-gluten | 0·58–0·68 | 51–54 | 2·6 | under 0·160 90% | 200 | 19–24 |
| Wheat-flour, for bread | 0·57–0·87 | 47–50 | 2·9 | under 0·160 90% | 200 | — |
| Whole wheat-flour | up to 2·00 | 30–33 | 4·5 | under 0·160 90% | 200 | — |
| Rye-flour I | 1·12–1·25 | 31–34 | 4·0 | under 0·160 90% | 110 | — |
| Rye-flour II | 1·35–1·58 | 28–30 | 4·3 | under 0·160 90% | 110 | — |
| Whole rye-flour | up to 2·00 | 24–27 | 5·0 | under 0·160 90% | 110 | — |
| Wheatmeal fine | up to 2·00 | 30–33 | 4·5 | 100% under 1·25 40% max. under 0·4 | 200 | — |
| Wheatmeal medium | up to 2·00 | 30–33 | 4·5 | 100% under 2·0 20% max. under 0·4 | 200 | — |
| Wheatmeal coarse | up to 2·00 | 30–33 | 4·5 | 100% under 4·0 10% max. under 0·4 | 200 | — |
| Ryemeal fine | up to 2·00 | 24–27 | 5·0 | 100% under 1·25 40% max. under 0·4 | 110 | — |

TABLE 23—*contd.*

| Flour type | Mineral ash (%) DM | Optical brightness (%) | Acidity-index maximum | Particle size (mm) | Falling number minimum (s) | Gluten-content (%) |
|---|---|---|---|---|---|---|
| Ryemeal medium | up to 2·00 | 24–27 | 5·0 | 100% under 2·00 20% max. under 0·4 | 110 | — |
| Ryemeal coarse | up to 2·00 | 24–27 | 5·0 | 100% under 4·00 10% max. under 4·0 | 110 | — |

Source: R. Schneeweiss and O. Klose, *Technologie der industriellen Backwarenproduktion*, VEB-Fachbuchverlag, Leipzig, 1981, p. 394, Table 66.

*Driftbrot* is another special type of bread which is made from a flour of special composition called '*Driftmehl*'. *Driftmehl* is a hydrothermally gelatinized mixture of 80% wheat bran and 20% fine bran-flour (*Nachmehl*), which has a moisture content of 15%, mineral ash 7%, and fibre 12%. It is reddish-brown in colour, and must not be musty, clumpy or caramelized (Schneeweiss and Klose, *Technology of Industrial-baking*, in German). *Driftbrot* is an energy-reduced bread with a high protein and fibre content, eminently suitable for the diet of those who are both overweight and carry excessive adipose-tissue. A typical *Driftbrot* formulation is as follows:

| | |
|---|---|
| Wheat-flour, gluten-rich | 100 parts |
| Rye-flour I | 20 |
| *Driftmehl* | 80 |
| Yeast | 4 |
| Salt | 3 |
| Water (approximate) | 138 |
| Emulsifier-hydrate | 3 |

Procedure:
*Driftbrot* is prepared by a short-process direct dough procedure. The rye-flour is best added in the soured state, using a three-stage sour process, adding the resulting full-sour (*Vollsauer*) to the other dough-components.

The final dough is given 60 min fermentation in bulk, with a remix after 20 min. The baking-time must be at least 80 min for a 1·5 kg unit, setting at a temperature of 280°C, falling to 210°C (W. Schwate and U. Ulrich, *Spezielle Verfahren Bäckereiwaren* VEB-Fachbuchverlag, Leipzig, 1986).

Another special flour is that used for the manufacture of *Malfa-Kraftma-Spezialbrot*, which is prepared from malted grain (summer brewing-barley); milled to a uniform particle size, the residue on a 0·4 mm sieve being not more than 25%, and the moisture maximum 10%. Because of the rye-flour content, the dough is prepared with a three-stage sour-process, and the *Malfa-Kraftma* special flour added at the doughing-up stage, after being allowed to swell in part of the dough water. The total dough composition is as follows:

| | |
|---|---|
| Rye-flour I | 76·5 parts |
| Wheat-bread flour | 13·5 |
| *Malfa-Kraftma* special flour | 10·0 |
| Water (approximate) | 62·0 |
| Salt | 1·6 |

Procedure:
The dough is prepared with a traditional-type mixer for more than 12 min, and the 1·5 kg dough-pieces are given minimal final-proof in pans. The minimal baking time for 1·5 kg units is 80 min (W. Schwate and O. Ulrich, *Spezielle Verfahren Bäckereiwaren*, VEB-Fachbuchverlag, Leipzig, 1986, p. 94).

A high-protein bread of similar formulation and preparation is the *Spezialbrot mit Soja* (special soy-bread). Total dough composition is as follows:

| | |
|---|---|
| Rye-flour I | 76·5 parts |
| Wheat-bread flour | 13·5 |
| Full-fat soy-flour | 10·0 |
| Water (approximate) | 62·0 |
| Salt | 1·6 |

Procedure:
The rye-flour content is soured in the usual way, using a three-stage process, and the soy-flour is added to the dough in the form of a 1:2 paste with part of the dough water. The dough is mixed with a traditional-type mixer for more than 12 min. The dough-pieces are given a minimal final-proof in pans.

For 1-kg units the baking time is 65 min, and for 1·5-kg units 80 min, at a setting temperature of 250°C falling to 210°C (W. Schwate and O. Ulrich, *Spezielle Verfahren Bäckereiwaren*, VEB-Fachbuchverlag, Leipzig, 1986, p. 94).

The *Hagenower Spezialbrot* is a recently developed special bread with advantageous nutritional components to meet the increasing demand for better-quality food. The total dough composition is as follows:

| | |
|---|---|
| Rye-flour I | 47·7 parts |
| Wheat-flour, low-gluten | 47·8 |
| Whole dried egg | 0·5 |
| Whey powder | 4·0 |
| Water (approximate) | 65·0 |
| Salt | 1·6 |

Procedure:

The rye-flour is soured with a three-stage process, and the dried egg powder and whey powder sieved before adding to the rest of the dough components. Mixing is with a traditional-type mixer for more than 12 min. This type of bread is only produced as 1·5 kg hearth-bread, requiring a baking time of 60 min, setting at 280°C falling to 210°C (W. Schwate and O. Ulrich, *Spezielle Verfahren Bäckereiwaren*, VEB-Fachbuchverlag, Leipzig, 1986, p. 95).

Pumpernickel has long been established as one of the most popular special bread varieties, and is made from ryemeal, water and salt using either a dough-sour or 7-hour pre-swelling, yeast-free sponge known as a '*Quellstück*', made with 25% of the ryemeal and water. The bread has a characteristically dark crumb and aromatic smell, requiring a long baking time. A typical formulation is as follows:

| | |
|---|---|
| Ryemeal (fine, medium or coarse) | 100·0 parts |
| Salt | 1·5 |
| Yeast | 0·6 |
| Water (approximate) | 60·0 |

Procedure:

Dough-sour method: prepare a three-stage sour with 25% of the meal. *Quellstück* method: prepare a sponge with 25% of the meal with warm water at 22°C, allowing it to swell for about 7 hours.

Prepare a dough with the rest of the meal, water, salt and yeast, adding the swollen sponge. Assemble the dough and mix intensively to yield a dough of 160%.

The dough-pieces are shaped and moulded, being placed in greased pans and final-proofed in regulated humidity.

The baking process is important and requires the regulated use of saturated steam in a specially constructed oven-chamber for about 16 hours at 100°C. The baked units are stored to cool, sliced, and carefully sealed, ideally under sterile conditions. Pasteurization is carried out by using a shrinkable packaging material for the sliced 250-g units, which reside for a period of at least 2 hours at 90°C. If higher temperatures are used, exposure must be reduced to about 10 s. As a result of the heat-treatment, the packaging material shrinks, and the packs are sealed to prevent any further microbial contamination, thus prolonging shelf-life.

### Dietetic Breads

Dietetic breads are defined as being special breads which, on account of their composition and properties, are suitable for the nutrition of people with special health conditions, environmental problems, or age-associated requirements. In the GDR they fall into three types: low-sodium, gluten-free, and energy-reduced dietary breads.

Low-sodium bread has been long-established, and is important for people with certain kidney, bladder, stomach or heart conditions. In order to reduce the effects of the absence of sodium on bread quality, potassium chloride is used as a substitute.

The following is a formulation for rye/wheat mixed bread (*Roggenmischbrot*):

| | |
|---|---|
| Rye-flour | 85 parts |
| Wheat-bread flour | 15 |
| Water (approximate) | 65 |
| Potassium chloride | 0·3 |

Procedure:
The processing procedure followed is basically the same as for *Roggenmischbrot*, apart from small reductions in bulk and final proof-times.

Baking times for 1 kg and 1·5 kg are 50 and 60 min respectively for hearth-breads, and 65 and 80 min in the case of panned units (W. Schwate and O. Ulrich, *Spezielle Verfahren Bäckereiwaren*, VEB-Fachbuchverlag, Leipzig, 1986, p. 96).

Gluten-free dietetic bread must contain no trace of flour, and all raw materials should be stored separately, and all equipment thoroughly cleaned before use. The following formulation is typical for bread made without flour:

| | |
|---|---|
| Maize starch | 44·0 parts |
| Potato starch | 9·0 |
| Carob bean (locust bean) flour† | 1·0 |
| Sugar | 2·1 |
| Marinea (shortening) | 6·2 |
| Salt | 1·4 |
| Yeast | 4·0 |
| Water | 43·7 |
| Calcium propionate as preservative | 0·1 |

† For the preparation of a darker-colour bread, a higher extraction carob bean-flour can be substituted for the white bean-flour.

Procedure:
A soft dough is prepared, placed in pans and given a final-proof time of about 30 min. Baking is at 200°C with steam-injection for about 30 min in the case of bread with a baked weight of 750 g (W. Schwate and O. Ulrich, *Spezielle Verfahren Bäckereiwaren*, VEB-Fachbuchverlag, Leipzig, 1986, p. 96).

Another type of specialty-bread (*Spezialbrot*) using wheat-flour is linseed-wholewheat bread (*Leinsamenvollkornbrot*), which is sold as sliced-bread units of 250 and 500 g. It is made from a mixture of 75% wheatmeal and 25% ryemeal with added linseed. Linseed is a valuable aid to the proper function of the digestive tract, containing 25 g/100 g dietary fibre, of which some 9·0 g is mucins. In addition it contains about 28 g/100 g of the polyunsaturated fatty acids, both α-linolenic and linoleic. Ideally, the seed should be cultivated with 100% organic material, and without the use of pesticides.

Most of the specialty bread types described are also available in the Federal Republic of Germany; but in the GDR in recent years, an intensive development programme for such breads has yielded many unique innovations to improve the dietary intake of the people as a whole.

**National Bread and Rolls**

The main types of bread manufactured in the GDR, and the flour types used to produce them are as summarized in Table 24. Wheat breads include the following, which weigh more than 180 g each: wheat bread, wheat bread with shortening, *Kaviar* bread, tea bread, poppy-seed bread, poppy-seed twist, wheat bread with sultanas, *Knüppel* bread, and small *Kaviar* bread units. The most important small white bread units are the rolls (*Brötchen*), which are sometimes referred to as '*Semmel*' or '*Schrippe*', all of which may not exceed 180 g in weight, flour being the main ingredient. The minimum legal weight of a roll is 22·5 g, but the 45-g unit is generally accepted as being the standard. They are either made from wheat-flour or whole wheat flour, and as improvers, the following can be added: malt-flour, emulsifiers in hydrated form, soy-flour, or sugar, at up to 3·0% flour weight. Doughs for the manufacture of rolls (*Brötchen*) must conform to the Standard TGL 26972 and contain: wheat-flour, gluten-rich or wholewheat flour, yeast 0·75, up to at least 3·0% of flour weight for a direct straight dough, salt concentration being 1·2–1·9%

TABLE 24

| Bread type | Flour type | Remarks |
|---|---|---|
| Whole-rye bread (*Roggenvollkornbrot*) | Whole-rye flour 98–99% extraction 100 parts | Baked as hearth or panned units |
| Mixed rye/wheat bread (*Roggenmischbrot*) | Rye-flour I 84–86% extraction 85 parts Wheat-flour for bread 25% divide or 76–78% straight flour 15 parts | Baked as hearth or panned bread |
| Mixed wheat/rye bread (*Mischbrot*) | Rye-flour I 84–86% extraction 60 parts Wheat-flour for bread 25% divide or 76–78% straight-flour 40 parts | Baked as hearth or panned bread |
| Rye bread (*Roggenbrot*) | Rye-flour I 84–86% extraction 100 parts | |
| Wholewheat bread (*Weizenvollkornbrot*) | Wholewheat flour 98–99% extraction 100 parts | Baked as panned bread only |
| Wheat bread | Wheat-flour for bread 25% divide 100 parts | Baked as hearth or panned bread |
| Wheat bread | Wheat-flour for bread 25–73% divide 100 parts | Baked as hearth or panned bread |

depending on the process employed. The only other ingredient is water. A typical composition of such a roll (*Brötchen*) dough is the following:

| | |
|---|---|
| Wheat-flour, gluten-rich | 78·0 parts |
| Yeast | 2·5 |
| Salt | 1·2 |
| Malt-flour | 1·6 |
| Water (approximate) | 41·0 |

The required fermentation time determines the quantity of yeast, dough temperature and dough consistency/yield per 100 parts flour. For a 1–2-hour dough, 3–4% yeast, based on total flour weight, is necessary, setting the dough temperature at 28–30°C, giving a dough yield of 160% (flour weight). Similarly, a 7–8-hour dough would require only 1·0% yeast maximum, setting the dough temperature at 22–23°C, giving a reduced dough yield of 150%.

In general, the fermentation time is kept as short as possible, within the limits of an acceptable product flavour and texture. Prolonged fermentation schedules can result in excessive dough-softening owing to enzymic action, and higher fermentation losses are unavoidable.

Legislation requires that rolls (*Brötchen*) on removal from the oven must show an average weight of 45 g/unit, sampling 20 rolls at random. Therefore, 20 rolls should give a total weight of at least 914 g in order to be within the legal limit in the GDR. Furthermore, after 8 hours, the loss in weight must not exceed 3%, and after 12 hours must not exceed 5% of the statutory 45-g baked weight (W. Schwate and U. Ulrich, *Spezielle Verfahren Bäckereiwaren*, VEB-Fachbuchverlag, Leipzig, 1986, p. 207).

Working on an average baking-loss of 16–20%, the baking time is laid down at 18 minutes at 220°C for rolls.

Wheat bread and rolls account for an important part of the diet and energy-requirements of the population of the GDR, the individual per capitum consumption per annum in kilograms having risen from 24 000 in 1960, to over 28 000 in 1985. By comparison, the per capitum consumption of specialty and dietary breads per annum has increased from 4300 in 1970, to 7600 in 1985 (W. Schwate and U. Ulrich, *Spezielle Verfahren Bäckereiwaren*, VEB-Fachbuchverlag, Leipzig, 1986, p. 215, Tables 40 and 41).

Rolls which contain up to 10% shortening or sugar, and possibly also contain whey powder with seeds on the dough-surface, e.g. poppy-seed (*Backmohn*), are classed as 'Rolls with other ingredients' (*Weizenkleingebäck mit Zusätze*) and have a potentially longer shelf-life. However, the sought-after crispness of the crust, crumb-solubility and excellent flavour of the leaner rolls made without these additions, is lacking. Prolonged crispness of crust is achieved by allowing the rolls to cool in a forced-air environment, allowing the moist air to freely evaporate.

An excessive variety of different bread-roll products reduces the output efficiency of the plant. Therefore, with increasing full-mechanization and automation of roll-production, standardization of roll-lines to accommodate *Knüppel*, *Rosenbrötchen*, *Hörnchen*, *Zöpfe*, and *Brezeln*, the most popular 'Rolls with other ingredients' in the

TABLE 25
**Important quality criteria for rolls and how to maintain them**

| Quality criteria | Measures to ensure acceptable quality |
|---|---|
| Open/split appearance (*Ausbund*) | Dough-pieces must be pressed through at the centre longitudinally, at the optimal stage of final-proof. Bake in adequate steam, at 220°C for 18 minutes. |
| Crisp crust | Use a flour with adequate gluten quantity and quality, i.e. gluten-rich flour. Avoid too much bottom-heat at baking. Cool in adequate forced-ventilation. |
| Crumb-elasticity | According to available flour quality add an improver which improves swelling and gas-production. |
| Volume | Use a gluten-rich flour. Increase the yeast addition. Extension of bulk fermentation and/or final-proof. Adequate humidity during fermentation and baking. |
| Crumb-porosity | Sieve flour thoroughly. Work ingredients at uniform temperature. Check yeast for fermentation-power. Work dough soft, and allow adequate bulk fermentation and final-proof. |

GDR, gives the best compromise between rationalization and consumer satisfaction.

Table 25 shows the most important quality criteria for rolls in the GDR; they are also valid in the FRG and other countries where value is placed on quality products.

## Production Lines in the GDR

*Processing of Bread-rolls (Brötchen)*

Doughs used on this plant can be prepared by batch (discontinuous) or continuous-mixing procedures, using sponges/pre-ferment systems, and medium and high-speed, intensive, mixing-machines. Conventional or traditional slow-speed mixers are not recommended for wheat-flour doughs.

An advantageous mixing-unit is the Type IMK 150, this high-output, versatile-machine is of the batch-type, and is in the high-speed mixer category. It can also be used for rye and mixed rye/wheat-doughs. The hydraulic and electrical controls, as well as the mixing-elements are all housed within a double-panelled enclosure. The mixing-bowls are equipped with castors, and where the bread-roll line has an oven of the type BN 50, four machine-bowls are necessary; but if a total of eight are available, the mixer can be used at peak output of 1900 kg/h, feeding two lines with the BN 50 oven on a continuous basis. Weight and temperature of the dough-water, as well as the mixing time, can be controlled at the IMK control-panel. The sequence: raising of the bowl, water dosage, mixing, lowering of the bowl is programmed, and the water mass and mixing-time pre-set. Manual operations include dosage of all dry ingredients, and the yeast can be added in the unsuspended state. In the case of direct, straight doughs, the doughs are rested for 20–30 minutes in the machine-bowls. Since the IMK cannot provide an extrusion facility, provision of a pair of rollers before feeding to the fully automated divider–moulder is an option.

The IMK has rotary mixing-elements, one arm being positioned on either side of the central drive-shaft, fixed with a downward slope, one being near the top of the shaft, and the other at the lower end close to the floor of the mixing-bowl. The drive-shaft rotates at a speed of 415 to 420 rpm within a machine-bowl of internal volume 290 litres, which can accommodate dough masses of 50 to 170 kg, the actual mixing duration being 120 to 130 s. The total mixing time cycle, involving nine work-operations is 300 s, based on an actual mixing duration of 120 s. Manual dosage-operations involving flour, salt, yeast, improver, sugar, and shortening, including bowl transport time, require on average 160 s.

Maximal throughput using four machine-bowls and the BN 50 wire-band, continuous tunnel oven is 1700 kg dough per hour. The temperature increase during mixing with the IMK is about 10°C for wheat-flour doughs, and optimal dough-temperature is approximately 33°C; therefore the water temperature must be adjusted accordingly. The power output of the motor is 60 kW.

The swelling capacity of the dough is increased, and the structure of the rolls improved when air is fed into the mixer, whereas, in the case of toast-bread manufacture the use of a vacuum improves dough yields and gives a finer texture to the bread.

For the processing of bread-rolls, using the A12 assembly with the BN 50 wire-band continuous oven, 5–6 batches per hour is the normal throughput, one batch being processed in about 12 minutes. The bulk resting time is adjusted accordingly. Using a transfer-aggregate, the doughs are fed continuously to the processing-line. The line, together with the BN 50 oven unit, permits the manufacture of elongated, pressed rolls. The various processing stages: are:

—Feeding of the dough to the divider–moulder, known as the 'Vollautomatische Teil- und Wirkmaschine (VATW)'. Weight 45–60-gram units.
—Moulding operation.
—Proofing of the pieces. 13–18 minutes intermediate and final-proof times.
—Shaping of the pieces and orientation. Stamped/pressed lengthwise.
—Transport to the oven.

The necessary machine-aggregates are:

—Fully automatic dough divider–moulder (VATW 515).
—Long-rolling machine.
—Depositing device.
—Intermediate proofer.
—Stamping or pressing unit.
—Final proofer.
—Transfer bridge.
—Control panel.
—BN 50 wire-band continuous tunnel oven.

Using the BN 50 wire-band tunnel oven the technological parameters are:

Throughput: 15 000 rolls per hour
Proof 1: 13–18 minutes
Proof 2: 13–18 minutes
Weight: 45–60 grams each
Shape: stamped lengthwise
Number of dough-pieces per row:   15

Service personnel:
1 Doughmaker
1 VATW 515 operator
1 Operator to sort cripples
1 Experienced chargehand

Two lines imported from Czechoslovakia known as the t 940 and the t 985, used in combination, are also suitable for the manufacture of elongated bread-rolls of the various finishes and with the additions described. The t 940 aggregates form the dough-preparation operations, and those of the t 985 the dividing and processing functions. With a mixing-time duration of 5 minutes, the dough throughput is up to 700 kg/h, producing 7000 to 9000 rolls per hour. The mixing unit is of the continuous intensive design, with two horizontal mixing-screws, rotating within the elongated housing in opposite directions at variable controllable speeds. The exact mixing time depends on the dough mass (maximum 35 kg), mixer-rpm, and desired throughput. The resident dough-mass is controlled by the opening of a sliding-outlet. However, although the mixer-rpm is adjustable between 21 and 78, the output of the motor is not adequate at 78 rpm at maximal throughput. An amperometer measures the current rating.

The salt solution and water-tempering units are separate sections of one basic unit. The water-tempering tank has a capacity of 45 litres, with automatic levelling device, outlet and overflow, as well as connections to hot and cold water. The latter is adjusted according to the desired dough temperature. The salt solution is covered with a grating, the salt being added manually. This section has a capacity of 300 litres, and a pump forces water from below through the salt layer, ensuring the continuous presence of a saturated solution of about 25% concentration, which is monitored by a salinometer.

The preparation section for the liquid components consists of six containers for shortening, malt, sugar and yeast. Each container is fitted with a stirrer, a cover and an outlet with an easily replaceable filter-membrane. The capacity of each of these containers is 220 litres. The shortening container is double-walled, and can be steam-heated to melt the solid shortening. For the solubilization of the salt and yeast there are two containers, and one container for the sugar.

The following mixing ratios are provided for: malt extract/water 1:2, sugar/water 1:1, yeast/water 1:1. The water necessary for the solution/suspension of these components is provided by a water-mixing and dosage unit, which is part of the preparation section. All connections to the dosage unit are provided by hoses, and a heated pipeline in the case of the shortening. The dosage unit is equipped with eight pumps, which are centrally driven. The dosage-quantities of the various components are set by manually operated wheels.

The shortening and the mixture of the other liquid components are conveyed to the mixer by separate pipelines. The flour-hopper has a capacity of 150 kg; vibrators feed the flour to the scales and out of the hopper weighing-unit at the rate of three oscillations per minute, the weight delivered by each oscillation being adjustable. The scale is emptied by means of a hose direct into the mixer, any blockages being signalled both optically and acoustically.

The travelling-band leading to the dough-ripening channel moves the extruded dough from the mixer, in the form of a strip, upwards at an angle to the dough-ripening channel, the band being plastic-coated. The dough-ripening tunnel which is positioned on top of the t 985 line, consists of a closed housing within which two travelling-bands, covered with plastic, run on top of one another. At the end of the upper band is situated a power-driven roller which serves to give the partially ripened dough a 'knock-back'. The dough is conveyed as a wide strip along the upper band through the whole housing and back along the lower band. The dough ripening time is adjustable by changing the speed of the band. The climatic conditions are regulated by injecting steam and air-conditioning. The 'knock-back' and moulding operation is carried out below the dough-ripening canal. It consists of a funnel within which is a pair of rollers. Over both rollers an endless band is positioned. The fermentation gases are pressed out of the dough, leaving a flat dough-band. The throughput of this set-up is quite independent and directly adjustable. The dough-moulding and transport consists of a band over which vertical and horizontal rollers are positioned, which produce a dough-band of uniform cross-section. Warm air is fed in from all sides to produce a dry dough-surface, preventing it from sticking to the rollers. The positioning of the rollers, and the rate of feed can be adjusted to suit the requirements of the divider–moulder. The return-conveyor is also a belt and allows the return of 'cripples' in the feed-hopper of the 'knock-back' and moulder' The control-panel regulates the plant; the line consists of the following processing machine-aggregates: divider–moulder, conveyor, proofer, bridge-conveyor, which together make up the t 985 line.

The divider–moulder resembles the divider–moulder VATW, but produces six moulded pieces in each row. The conveyor set-up, which serves the function of a spreader-band, consists of six small band-conveyors, fitted with 6-cm wide belts, which are driven by the divider–moulder.

The depositing operation takes place with a fixed rhythm; the whole mechanism rotates around a fixed point and takes up three different positions. Since in each case six moulded-pieces are deposited, rows with 18 moulded pieces are therefore built up. The proofer consists of five band-conveyors arranged on top of one another. At the inlet a felt-covered roller, rotating over the band, elongates the dough-pieces, the length of the pieces depending on the gap between the roller and the band. A device consisting of rods and rollers turns the pieces through 90 degrees. At the end of the top band-conveyor there is an alignment device, and this is followed by a stamping-device equipped with 18 different stamping-rollers. The rollers have different profiles, which permit various shapes for the dough-pieces. The pressed pieces are then conveyed to the next band-conveyor, travelling over a special three-roller/endless-belt, transit-device. The placement of this device results in the inversion of the dough-pieces. On reaching the lower band-conveyor, the dough-pieces are turned again, so that the stamped profile is uppermost. That is immediately followed by another band-conveyor fitted with dough-piece separators, which conveys the dough-pieces upwards at an angle and into the wire-band continuous tunnel-oven. The band-conveyor can be folded upwards when the plant is non-operational. The proofing time can be regulated by changing the rpm of the drive-motor, and the proofer is equipped with a humidity and warm-air facility. The band-conveyor feeding the oven has a moistening and sprinkling device mounted above it. The surface of the dough-pieces are moistened by a brush-roller, which rotates in a water-bath. The sprinkler-device consists of a funnel containing bakery poppy-seed, caraway-seed, salt etc., which is activated by a powered dosage-regulator. Surplus seed is collected in a box below and re-used. The control-panel regulates the line-functions, and all units of the dough-preparation and processing plant are synchronized, together with all transport-elements, in order to ensure the continuous smooth running of production. For the baking operation, a wire-band continuous tunnel-oven of bandwidth 2·1 metres, and a baking-surface of 25 m² can be utilized.

Technical parameters of the plant:

| | t 940 | t 985 |
|---|---|---|
| Length | 26 000 mm | |
| Width | 3 000 mm | |
| Height | 4 510 mm | |
| Weight | 20 000 kg | 9 000 kg |
| Energy input | 14 kW | 11 kW |

Technological parameters of the plant:

| | | |
|---|---|---|
| Throughput | up to 700 kg dough per hour | 7000 to 9000 rolls per hour |
| Mixing time | about 5 minutes | |
| Proof 1 | 1 to 7 minutes | |
| Proof 2 | 15 to 40 minutes | |
| Weight | 30–60-g or 60–90-g units | |
| Shape | elongated with various shapes | |
| Number of dough-pieces per row | 18 | |

Service personnel:
1 Doughmaker
1 Experienced chargehand
1 Operator feeding the divider–moulder
1 Operator to sort cripples

(Information source: R. Schneeweiss and O. Klose, *Technologie der industriellen Backwarenproduction*, VEB-Fachbuchverlag, Leipzig, 1981 edn, pp. 309–318.)

The most significant difference between the A12 roll-processing plant and the one imported from Czechoslovakia is the method used to prepare the dough before processing. The A12 allows a wide option of mixers and fermentation schedules, and either a batch or continuous mixing procedure could be utilized. Whereas, the t 940 combined with the t 985 forms a continuous roll-line, which is capable of producing rolls of various shapes and finishes at reduced rate of output. The A12, comprising the IMK 150 intensive mixer and the VATW 515 is ideal for the production of rolls of the pressed variety, i.e. *Schrippen* and *Knüppel*.

### Bread-processing Lines (Brotlinien)

The following is a description of bread-processing line which can be used to manufacture rye bread, mixed rye/wheat bread (*Roggenmischbrot*) made with up to 50% wheat-flour, and specialty breads (*Spezialbrote*) with up to 50% wheat-flour. In addition, the line is suitable for specialty breads with enrichment which are cooled and packed for a 7-day shelf-life in polyethylene foil.

The mixed rye/wheat breads are of the hearth (oven-bottom) type of various weights. The baked products are stored under hygienic conditions at 20–25°C and a relative humidity not under 50%.

Technical parameters:

| | |
|---|---|
| Maximum throughput | 935 kg of baked products |
| Length | 50·0 metres |
| Width | 4·0 metres |
| Height | 3·3 metres |
| Energy input | |
|   Electroenergy | 500 kW approx. |
|   Gas | 90 metres$^3$/h |
|   Steam | 160 kg/h |
|   Degree of automation | 90% |

Service personnel:
1 Experienced doughmaker
1 Experienced person to take change of the processing-line

The processing can be divided into four phases:

### Phase 1: Dough-preparation

The continuous doughmaking plant, type KVT 1000 (Czechoslovakia) consists of the following aggregates:

—2 mixing and dosage aggregates for flour
—1 mixing and dosage-tank for water
—1 salt solution and dosage-tank
—sour-dough mixer
—sour-dough ripening container with 12 dough-ripening compartments
—sour-dough divider and carbon dioxide remover
—continuous intensive-mixer
—pipelines for salt solution, water and dough
—endless band-conveyor to transport dough to processing phase

*Phase 2:* Dough-processing
—1 bread divider–moulder (BTW)

*Phase 3:* Dough-ripening (proofing)
—Proofing-cabinet with bridge and tipping set-up for the moulded dough-pieces from the BTW, and transfer set-up with specialized pressing/stamping station.
—oven-transfer bridge

*Phase 4:* Baking
—wire-band continuous tunnel-oven (BN 50)

The flour-dosage aggregates, consisting of a container with mixer and weighing cells, are filled with flour from the daily-flour silo-station. The full-level of the container is regulated by a membrane-switch set-up, and the flour is automatically dosed-in accompanied by an optical and acoustic signal. The salt-solution aggregate consists of two tanks fitted with stirrer-units. In the first tank an appropriate amount of salt is added, with stirring, to tempered water until a saturated solution results. The saturated salt solution is then pumped from the first tank into the second one. The concentration of the salt solution can be determined with a salinometer, which measures the density. The necessary amount of salt solution, according to the formulation, is added to the mixer-unit by operating the control-panel. In the same manner the correct amount of tempered water is added to the sour-dough mixer.

To start the sour-dough ripening process, flour, water and mature full-sour (*Vollsauer*) are mixed in a sour-dough mixer to a pumpable sour-dough consistency for filling the sour-dough ripening container with the 12 dough-ripening compartments. The sour-dough mixer, a container with mixing-head, mixes the continuously added components to a homogeneous, pumpable mass. The sour-dough mixture is then pumped by means of a cogged-wheel-pump into one of the 12 dough-ripening compartments of the sour-dough ripening container. The compartments rotate according to a predetermined interval within the sour-dough ripening container. In the base of the sour-dough ripening container are openings, which serve to fill and empty the compartments. In this way, the appropriate compartment is filled with sour-dough, or the mature full-sour is drawn off by vacuum. When the full-mark of a compartment is reached, the ripening-container is automatically turned, so that a continuous flow in and out of sour-dough is maintained. The time of rotation of the sour-dough ripening container can be regulated, the maximum time being 180 minutes, which corresponds to the required ripening-time. The mature full-sour is pumped by cogged-wheel-pumps into the carbon dioxide remover, which consists of a container equipped with a mixer. The mature full-sour is thus degassed, in order to facilitate more exact dosage measurement. About 40% of the gas-free full-sour is pumped by means of cogged-wheel-pump, into the sour-dough mixer, and simultaneously the residual 60% is pumped into the continuous intensive-mixer. The full-sour in the mixer can take up to 42% flour (based on total flour content of the dough). The components, full-sour, flour, water and salt solution, are continuously fed to the mixer and, depending on the formulation, yeast and stale-bread suspension can also be added. The mixing process is intensive; two screw-shaped mixing-shafts rotating in opposite directions convey the dough-mass towards the outlet, the residence-time being 4–5 minutes. The continuous band of dough from the mixer is conveyed by a transport-band to the divider–moulder, the time-lapse involved having a very advantageous effect on the dough-ripening process.

*Dough-processing*

The continuous band of dough is divided by volume and then moulded in the divider–moulder. The dough is moved by two worm-feeds into a cylinder, the dough being pressed through the cylinder and cut off on emergence from the cylinder by a knife. The length of the dough-strip determines the weight of the dough-piece. The machine is powered by a continuously adjustable drive. The elongated dough-pieces are then fed to the tipper-band.

*Dough-ripening*

The tipper-band serves to collect and orientate five moulded pieces placed equidistant in a row, and to transfer them to the dough-pockets of the proofing-cabinet. After each transfer, the transport system of the proofing-cabinet is intermittently set in motion by a control-system. After depositing the five dough-pieces in dough-pockets, the pocketed trays move forward, so that after each transfer, an empty pocketed-tray stands ready at the tipper-band. The throughput-time of the proofing-cabinet, which determines the ripening-time, can be adjusted from 36 minutes to 61 minutes. The adjustment is carried out by altering the length of the chain-drive, which results in the change in the ratio of filled trays to empty trays on return. In order to maintain an appropriate climate in the proofing-cabinet, heating-elements and humidifiers are installed within the unit. The necessary climatic conditions are:

Temperature        33–35°C
Relative humidity 70–80%

The proofing-cabinet is also provided with appropriate temperature and humidity measurement instrumentation.

The proofed dough-pieces are then tipped from the pockets onto the transfer-bridge band-conveyor. On their path to the oven, cutting or stamping-aggregates are positioned as appropriate. The oven is a wire-band continuous tunnel-oven BN 50 with a baking-surface of 50 m². The cut, and correctly positioned, proofed dough-pieces then pass through the tunnel-oven according to the appropriate technological parameters (see Table 26). This processing-line can also be used in combination with other continuous-mixers, e.g. FTK 1000 UM, manufactured in Hungary, or batch-processing units, such as conventional kneaders, spiral or Wendel types. The GDR bakery-machinery manufacturers also build lines suitable for rye bread and wheat-rolls, the bread line being above the roll line, either line being used in combination with a wire-band continuous tunnel-oven. (Information source: R. Schneeweiss and O. Klose, *Technologie der industriellen Backwaren-produktion*, VEB-Fachbuchverlag, Leipzig, 1981 edn, pp. 323–326.)

TABLE 26

| | Wt | Form | Temperature (°C) | | Baking time (minutes) |
|---|---|---|---|---|---|
| | | | Start | Finish | |
| Mixed rye/wheat (*Roggenmischbrot*) and | 1 kg | Hearth-bread (oven-bottom) | 280 | 200 | 50 |
| Mixed wheat/rye (*Mischenbrot*) | 1·5 kg | Hearth-bread | 280 | 210 | 60 |
| | 2 kg | Hearth-bread | 280 | 210 | 65 |
| Whole-rye bread (*Roggenvollkornbrot*) using rye-flour | 1 kg | Hearth-bread | 250 | 210 | 50 |
| Whole-rye bread using ryemeal | 1 kg | Hearth-bread | 250 | 210 | 60 |
| | 1·5 kg | Hearth-bread | 250 | 210 | 70 |
| | 2 kg | Hearth-bread | 240 | 210 | 80 |

Continuous lines for the manufacture of rye and mixed rye/wheat (*Mischbrotteigen*) bread-doughs are quite different in conception to those intended for wheat bread-dough manufacture owing to the incorporation of the sour-dough process. The following continuous lines are in use in the GDR, KVT 1000, KVT 1500 and KVT 1800 (imported from Czechoslovakia), and the FTK 1000, FTK 1000 U, FTK 1000 UM and FTK 1500 (imported from Hungary), all require a continuous sour-dough process. These sour-dough processes must be so arranged that a mature full-sour (*Vollsauer*) is always available for dough-processing. The mature sour is continuously fed to the mixer. The lines are designated by definite numbers, i.e. KVT '1500', which indicates a throughput of 1500 kg/h, and the FTK '1000' similarly indicates 1000 kg/h throughput.

The sour-dough process used in continuous processing lines is basically a production schedule which runs from one full-sour to another, which ideally runs uninterrupted for the whole working-week from Monday to Friday. The important parameters for sour-dough management are temperature, dough-consistency and maturing time, thus maintaining optimal microbial growth. These parameters must be so chosen that the acidity-index (*Säuregrad*) is kept within well defined limits.

Considering the KVT 1500 schedule as a popular example of such a system, the production begins by adding the starter-culture (*Anstellgut*) to the initial-sour (*Anfrischsauer*) and mixing discontinuously. The source of the starter-culture is normally the full-sour. Conventional mixers are used to mix the initial-sour, resulting in a discontinuous propagation of the starter-culture at the start of production. The sour-dough scheme for the KVT 1500 is as shown in Table 27.

The basic-sour (*Grundsauer*) is then prepared by the addition of flour and tempered water in the sour-dough mixing machine, which is then pumped by cogged-wheel-pumps into the ripening compartments, leading to a continuous propagation of the initial-sour over a 3-hour period. After the addition of more flour and water the mass is pumped back into the ripening compartments for a further 2·75 hours to form the first mature full-sour. At this stage, the *full-sour* is

TABLE 27

| Stage | Standing-time (h) | Sour (kg) | Flour (kg) | Water (kg) | Total (kg) | Temperature (°C) | | Dough yield (%) |
|---|---|---|---|---|---|---|---|---|
| | | | | | | Start | End | |
| Starter-culture | 17 | 37·2 | — | 93·3 | 130·5 | Cool-storage | | — |
| Initial-sour | 30 | 130·5 | 148·5 | 102·0 | 381·0 | 26 | 28 | 235 |

The acidity-index should be 16–18, and the pH 3·5–3·6. The mature initial-sour is added manually to the mixer, and continuously fed to the sour-dough mixing unit. (Information source: W. Schwate and O. Ulrich, *Spezielle Verfahren Bäckereiwaren*, VEB-Fachbuchverlag, Leipzig, 1986, 2nd edn.)

divided into three equal parts, one-third being returned to the ripening compartment for continuous propagation, and the residual two-thirds proceeding to the continuous mixer. Stale bread is often added at the doughmaking stage. The preparation and division of full-sours is repeated (see Table 28).

The dosage-time required to fill the first ripening-compartments is 8 minutes and in the case of compartments 2–11 only 4 minutes. The 8 minutes necessary for the first compartment is due to the fact that the pipelines are empty, and the initial-sour would be inadequate for an accurate dosage. On starting up the plant, the dosage-time and compartment-transfer time are variable, since the basic-sour and first full sour must first attain their full maturity. In order to allow for this, an extra ripening-time for the basic-sour and the first full-sour is given, which correspondingly reduces the standing-time during the subsequent production phase. The twelfth compartment is kept free to receive the next full-sour for the production phase. During the dosage period and the filling of the compartments, each compartment is programmed for filling, in order to ensure that mature full-sour cannot mix with unripened full-sour.

The compartment-transfer operation is pre-set with a time-relay. The necessary

TABLE 28

| Stage | Dosage time (minutes per compartment) | Transfer time | Standing time (h) | Sour (kg/min) | Flour (kg/min) | Water (kg/min) | Total (kg/min) | Temp. (°C) | Dough yield (%) |
|---|---|---|---|---|---|---|---|---|---|
| Basic-sour | 4 | 17 | 3 | 4·7 | 6·34 | 8·56 | 19·6 | 26/28 | 235 |
| First full-sour | 13 | 15 | 2·75 | 7·2 | 6·34 | 8·3 | 21·84 | 28/30 | 235 |

The acidity-index of the basic-sour should be 10–12, and the pH 3·8–3·9. The acidity-index of the first full-sour should be 9–11, and the pH 3·8–4·0.

Information source: W. Schwate and O. Ulrich, *Spezielle Verfahren Bäckereiwaren*, VEB-Fachbuchverlag, Leipzig, 1986, p. 23.

dosage and compartment-transfer times are controlled by the time-relay system, and in order to be able to control the dosage and compartment-transfer times, the operator must be conversant with the time-intervals of the relays.

The floor-space required by the KVT 1500 is only about 4 metres × 8 metres. The machinery required for the preparation of the various sour-dough stages is as follows:

*Starter-culture (Anstellgut)*
—Taken from mature full-sour (*Vollsauer*) of the continuous production.
   *Initial-sour (Anfrischsauer)*
—Conventional mixers of various design and operating speeds.
*Basic-sour (Grundsauer)*
   —Mixer, sour-dough beating-machine
   —Flour and water measurement and metering equipment
   —Cogged-wheel pumps
   —Fermentation equipment, comprising a fermentation-container and fermentation-compartments
   —Time-relays
*Full-sour (Vollsauer)*
   —Sour-dough mixer
   —Flour and water measurement and metering equipment
   —Cogged wheel pumps
   —Sour-dough beating-machine
   —Fermentation equipment (as for basic-sour)
   —Time-relays
*Production phase of the full-sour*
   —Mixer
   —Flour and water measurement and metering equipment
   —Sour-dough beating-machine
   —Cogged-wheel pumps
   —Fermentation equipment (as for basic- and full-sour)
   —Time-relays

The total process involves the continuous and progressive preparation and dividing of the full-sour (*Vollsauer*). One-third is used to perpetuate the continuous full-sour production in the ripening-compartments, and two-thirds of the mature

TABLE 29

| Stage | Sour (kg/min) | Flour (kg/min) | Salt-soln (kg/min) | Stale bread paste (kg/min) | Total (kg/min) | Starting temp. (°C) | Dough yield (%) | pH value | Acidity-index |
|---|---|---|---|---|---|---|---|---|---|
| Dough | 14·46 | 10·92 | 1·46 | 1·16 | 28·0 | 30 | 162 | 4·3–4·6 | 7–9 |

Information source: W. Schwate and O. Ulrich, *Spezielle Verfahren Bäckereiwaren*, VEB-Fachbuchverlag, Leipzig, 1986, p. 38.

full-sour passes on into the continuous-mixing unit. Hence the name 'full-sour/full-sour process'. The dough preparation scheme for the KVT 1500 is given in Table 29.

*Continuous Dough Mixing with the Matured Full-sour (Vollsauer)*
The feeding of the various raw materials takes place from the various aggregates, as follows:

    —from the sour-dough mixer—the mature full-sour (*Vollsauer*)
    —from the flour-measuring and dosage plant—the sieved and mixed flour
    —from the salt-solution and tempering unit—the tempered salt-solution
    —from the water-tempering and metering unit—the necessary dough water
    —for the full-sour preparation—stale bread paste

The outlets of the various feed-in pipelines have diameters which are appropriate to their respective feed-in mass. Abnormalities which occur in connection with the raw materials feed-in are immediately signalled by built-in flowmeters.

    The actual mixing takes place in a closed housing in the form of a horizontal cylinder. An intensive mixing and kneading action takes place by means of two screw-shaped mixing elements rotating in opposite directions. The mixing-chamber is continuously filled by the constant feed-in of the raw materials and, because of the spiral shape of the mixing elements, the dough is conveyed towards the outlet-orifice. The diameter of this outlet-orifice can be controlled from 0 to 75 mm by means of two sliding apertures, which in turn control the rate of extrusion and dough-throughput. An increase in residence-time increases the mixing-time, the average time required for converting raw materials into dough being 4 minutes.

    The throughput is 1400 to 1500 kg/h rye bread or mixed rye/wheat bread. Any abnormalities during mixing result in automatic power switch-off. The mixing-barrel can be opened for ease of access for cleaning. The extruded dough is transported on a band-conveyor, during which time it is undergoes proof, to the bread-dividing and -moulding machine (Brotteigteil- und Wirkmaschine, BTW).

    Therefore, after mixing with the KVT 1500, dough-proofing takes place on a special fermentation/proofer band. The duration of this fermentation can be varied by means of a time-relay, and is about 25 to 30 minutes. The fermentation/proofer band is 1500 mm wide, hollow-shaped, forming a trough, and covered; the distance travelled is adjustable within about 20 cm. No provision is made for a knock-back or punching operation, and this operation is performed by the transport worm of the divider and moulder (BTW).

    Fermentation in bulk is important in the case of rye and mixed rye/wheat doughs, since it has a great influence on swelling, solubility, the resultant acidity-index and the final-proof time. Doughs made with whole-grain flours and meals require more mixing, and a longer fermentation time than doughs containing, for example, rye-flour I of lower extraction rate.

    Owing to the increased product demand on Mondays and Fridays in certain larger bakery-cooperatives where the KVT 1500 is in use, the workload necessary prior to starting-up production must be rationalized and reduced where possible.

For example, in the bakery-cooperative (Konsum-Backwarenkombinat) in Karl-Marx-Stadt, Dipl.-Ing. U. Fiedler and his co-workers, as a result of many years experience with discontinuous (batch) sour-dough production using a basic wort-seasoning to bridge production interruptions of up to 72 hours, devised a new sour-dough schedule. This new schedule eliminates the time-consuming sour-dough preparation over several stages with the KVT 1500. The plan is initiated by preparing a wort-seasoning at the end of production at about 14.00 hours on a Friday, consisting of 60 kg mature full-sour and 90 kg water at 6–15°C. After standing for up to 72 hours, covered, in a machine-bowl (during the summer months storage is in a cool-room at 6°C), the broth increases by a factor of 2. The wort-seasoning is then further enriched on the following Sunday at 17.00 hours with whole-rye flour and water at 23–25°C to give a yield of 326% as follows:

| | | | |
|---|---|---|---|
| Wort-seasoning | 60 kg | Standing time | 2 hours |
| Whole-rye flour | 25 kg | | |
| Water | 40 kg | | |
| Initial-sour (*Anfrischsauer*) | 125 kg × 5 | | |

Procedure:
The wort-seasoning is manually stirred and 60 kg is added to 25 kg flour.
Mixing takes place in a fast mixer DK 150, adding the water in stages.
Then, at 19.00 hours, the basic-sour (*Grundsauer*) is prepared as follows:

| | | | |
|---|---|---|---|
| Initial-sour (*Anfrischsauer*) | 125 kg | Standing time | 2 hours |
| Whole-rye flour | 35 kg | Temperature | 20–28°C |
| Water | 40 kg | Dough yield | 277% |
| Yeast | 3·2 kg | | |
| Basic-sour (*Grundsauer*) | 203·2 kg × 5 | | |

At 21.00 hours, the basic-sour can be fed to the KVT-mixer as follows:

| | | | |
|---|---|---|---|
| Basic-sour (*Grundsauer*) | 4·64 kg/min | Feed time | 17·5 min |
| Rye-flour Ia | 4·68 kg/min | | + 2·5 min |
| Water | 5·40 kg/min | Temperature | 27–29°C |
| | | Dough yield | 232% |
| Full-sour (*Vollsauer*) | 14·72 kg/min | Acidity-index | 8–11 |
| | | pH-value | 3·8–4·2 |

This operation completes the Sunday preparation work.
On Monday, at 01.00 hours, dough production commences, providing full-sour (*Vollsauer*) for continuous production.

| | | | |
|---|---|---|---|
| Full-sour (*Vollsauer*) | 4·64 kg/min | Feed time | 17·5 min |
| Rye-flour Ia | 4·29 kg/min | | + 2·5 min |
| Water | 5·80 kg/min | Temperature | 28–30°C |
| | | Dough yield | 234% |
| Full-sour (*Vollsauer*) | 14·73 kg/min | Acidity-index | 9–12 |
| | | pH-value | 3·5–3·8 |

Further rationalization and reduction of manual work are achieved by: using a self-activated suction-pump to transport the sour-dough at the rate of 5 m³/h from a depth of 2 m into the KVT-mixer; using a larger amount of starter-culture, thus reducing the sour-dough standing-time; and adding 10 kg of fine ice-crystals to the initial wort-seasoning to save transport to the cool-room. Additionally, the high degree of mechanization of the KVT 1500 can also be utilized to produce specialty-breads for the Monday peak delivery programme, as well as producing sour-dough for certain batch processing requirements.

The following sour-dough plan illustrates how the preparation period for the KVT 1500 can be reduced to 3 hours by using the wort-seasoning, initial-sour and basic-sour.

Wort-seasoning prepared after the completion of a production-run:

| | |
|---|---|
| Full-sour | 60 kg |
| Water | 90 kg |
| Ice | 10 kg |
| Wort-seasoning | 160 kg × 6 transport to cool-room unnecessary |

The factor ' × 6' indicates the reproductive capacity of the medium, in this case six-times. A definite quantity of fermentable sugar $(Qt_{n-1})$ in the starter-culture (*Anstellgut*) is used up by fermentation, and becomes replaced by more fermentable sugar converted from the starch. Thus, the situation $Qt_{n-1}$ becomes $Qt_n$. From a quantity of sour-dough organisms $(Qw_{n-1})$, a larger number of sour-dough organisms $(Gw_n)$ develop within a definite standing time and appropriate temperature, when nutrient (from flour) and water are added. During the sour-dough process the following variables can be changed: time, temperature, dough yield and the reproductive capacity, all of which are interrelated. As a general rule, the following approximation forms a useful guide:

$$M_{s2} = M_{s1} \times t_{s2}$$

where $M_{s2}$ represents the weight of flour in the second stage, s2, $M_{s1}$ represents the weight of flour in the first stage, s1, and $t_{s2}$ represents the standing-time of s2. For example, where $M_{s1} = 10$ kg and $t_{s2} = 5$ hours, $M_{s2} = 10 \times 5 = 50$ kg flour are required for the next stage of the sour-dough process.

The standing time is the first to be laid down in practice, owing to the organization of the work-schedule. Then, by regulating temperature, dough yield, and the reproductive capacity, the development of the sour-dough is adjusted to reach maturity at the desired standing time. The above example illustrates that if a 5-hour standing time is chosen, the following sour-dough stage will require five times the quantity of flour used in the previous sour-dough stage. Although such processing variables as mixing also constitute the characterization of any process, the reproductive rule described has proved a reliable practical guide. When standing times are reduced, the reproductive factor must be reduced, and when prolonged standing times are chosen, the reproductive factor must be correspondingly increased. The flour extraction-rate must also be considered, flours of relatively lower extraction cannot sour as well as those of higher extraction rate, owing to reduced enzyme-content. Therefore, in the latter case a higher reproductive factor must be chosen. Fuchs (*Qualität von Brot und Kleingebäck*, VEB-Fachbuchverlag, Leipzig, 1968) has established in trials, that a high tolerance exists between the reproductive factor of basic-sour (*Grundsauer*) and full-sour (*Vollsauer*).

Having prepared the wort-seasoning, the initial-sour (*Anfrischsauer*) is prepared on Sunday at 17.00 hours as follows:

| Wort-seasoning | 80 kg | | |
| Whole-rye flour | 48 kg | | |
| Water | 50 kg | | |
| Yeast | 2 kg | | |
| | | | |
| Initial-sour (*Anfrischsauer*) | 180 kg × 12 | Dough yield | 278% |

At 19.00 hours, the initial-sour is pumped into the KVT-mixer, using the following feed-rates:

| Initial-sour (*Anfrischsauer*) | 11·92 kg/min | Standing time | 3 h |
| Rye-flour Ia | 3·88 kg/min | Feed time | 12 min |
| Water | 3·40 kg/min | | + 2·5 min |
| | | | |
| Basic-sour (*Grundsauer*) | 19·20 kg/min | Dough yield | 235% |

Then, at 22.00 hours on the same day the full-sour preparation for continuous production commences as follows:

| Basic-sour (*Grundsauer*) | 4·64 kg/min | Standing time | 4 h |
| Rye-flour Ia | 4·29 kg/min | Feed time | 17·5 min |
| Water | 5·80 kg/min | | + 2·5 min |
| | | | |
| Full-sour (*Vollsauer*) | 14·73 kg/min | Dough yield | 235% |

In addition to enough full-sour (*Vollsauer*) being provided for the continuous production throughput of 1058 kg/h, the dough-mixing aggregates can also be utilized. The mature basic-sour (*Grundsauer*) can also provide enough full-sour (*Vollsauer*) for the production of 3·6 t specialty-bread (*Spezialbrot*), in which case 4·3 kg yeast/t specialty-bread is also added, providing a throughput of 1000 kg dough/h.

Owing to the increased demand for such bread varieties as *Roggenbrot*, *Bauernbrot* and *Mecklenburger Landbrot* for the Monday delivery programme, certain quantities of bread must be produced on the previous Friday, or produced during the night-shift on Sunday. In the latter case, the processing is only partially mechanized, and the number of personnel required for this operation is high— baking with a wire-band continuous tunnel-oven type BN 40. Therefore, the objective of the management has been directed to making more effective use of the high degree of mechanization of the KVT 1500, especially for producing additional sour-dough for the rationalization of batch processing.

The wort-seasoning is prepared at the end of production as already described, and used to manufacture enough basic-sour (*Grundsauer*), which is treated with appropriate proportions of rye-flour Ia and water to produce batches of full-sour (*Vollsauer*). The standing times of full-sour batches are adapted to the throughput of the wire-band oven BN 40, for example, 1·5, 2·0, 3·0 and 4·0 hours respectively.

Adaptation of the formulation of the full-sour process to suit the throughput of the wire-band oven BN 40:

| | | | | |
|---|---|---|---|---|
| Basic-sour (*Grundsauer*) (kg) | 90 | 60 | 40 | 30 |
| Rye-flour Ia | 17 | 33 | 44 | 50 |
| Water (litres) | 13 | 27 | 36 | 40 |
| | — | — | — | — |
| Full-sour (*Vollsauer*) (kg) | 120 | 120 | 120 | 120 |
| Standing time (h) | 1·5 | 2·0 | 3·0 | 4·0 |

Sunday, 17.00 hours preparation of the initial-sour (*Anfrischsauer*)

| | |
|---|---|
| Wort-seasoning | 80 kg |
| Whole-rye flour | 48 kg |
| Water | 50 kg |
| Yeast | 2 kg |
| | — |
| Initial-sour (*Anfrischsauer*) | 180 kg × 15 |

Sunday, 19.00 hours: pumping of the initial-sour into the KVT-mixer and commencement of the basic-sour (*Grundsauer*) production using the KVT dough-developer unit

—sour-dough side as previously described
—adjustments for sour-dough and flour on the dough-side are made according to appropriate values for following specialty-bread (*Spezialbrot*) production.

| | | | |
|---|---|---|---|
| Initial-sour (*Anfrischsauer*) | 8·88 kg/min | | |
| Rye-flour Ia | 6·30 kg/min | Temperature | 26°C |
| Water | 1·91 kg/min | Standing time | 3 h |
| | — | | |
| Basic-sour (*Grundsauer*) | 17·09 kg/min | Dough yield | 180% |

From the discharge-outlet of the KVT dough-developer unit, the fermentation/proofer band is utilized to fill 25 machine-bowls with 120 kg each of basic-sour (*Grundsauer*), at intervals of 7 minutes. This amount represents the limit of the baking capacity available in the wire-band oven BN 40. Following this procedure, with the use of a second line, 6·5 t of bread could be produced within 3 h before 03.00 hours on Monday. Where, however, only one wire-band oven BN 40 is available as additional oven capacity, and the transport path to the wire-band oven is only available until 20.30 hours, only 3·8 t bread can be produced on a batch system, using the following procedure:

Basic-sour (*Grundsauer*) deposited in 15 machine-bowls

19.00 hours:       $3 \times 120\,kg = 3 \times 7\,min$

$3 \times 90\,kg = 3 \times 5\,min + 16\,s$

$3 \times 60\,kg = 3 \times 3\,min + 31\,s$

$3 \times 40\,kg = 3 \times 2\,min + 20\,s$

$3 \times 30\,kg = 3 \times 1\,min + 45\,s$

20.00 hours:       completion of depositing, and transport of the bowls under the bridge-conveyor of the bread-roll line

21.20 hours:       commence with the batch sour-dough and dough preparation, according to the scheme described above under 'Adaptation of the formulation of the full-sour process to suit the throughput of the wire-band oven BN 40', combined with a detailed 'time-plan schedule' for all sour-doughs and dough preparations, including any yeast additions as appropriate. The plan will cover the period 21.00 to 02.40 hours.

The development of this shortened, combined starting-up phase in this factory will reduce the necessity of the pre-baked bread inventory by 2 t. (Information source: U. Fiedler, Rationalization of sour-dough preparation with the KVT 1500 dough-line at the beginning of the week (in German), *Bäcker und Konditor*, **4**, 102–104, 1985.)

These actual case-studies with the KVT 1500 illustrate how the sour-dough mixing and dough-development aggregates can be used to overcome peak demand periods by shortening the sour-dough ripening stages, and combining this with batch processing techniques, especially during 'start-up' phases. However, adequate reserve baking-capacity must be available at the right time to take full advantage of these innovations. Reductions in the necessity for producing bread too much in advance of the despatch time reduces wastage, and satisfies the consumer demand for fresher baked products.

Other methods have been worked out to cope with planned and unplanned interruptions in production with the KVT 1500. These have been directed towards dealing with the following emergency situations:

—Interruption of the machine-aggregates on the dough-side of the KVT dough-developer without production-stoppage.

—Interruption of the machine-aggregates on the dough-side of the KVT 1500 bread-processing-line/oven, with production stoppages of up to 5 h.

—Interruption of the machine-aggregates on the dough-side, bread-processing line/oven, without production stoppage by changing the full-sour (*Vollsauer*) production to suit the capacity of the wire-band oven BN 94 (U. Fiedler, Various methods of interrupting the sour-dough production using the continuous sour-dough and dough-line KVT 1500, *Bäcker und Konditor*, **5**, 136, 1985).

In the case of planned interruptions of sour-dough production, various processe have been devised to reduce the over-souring (acidulation) of the full-sour dough. For example, prolonged standing times of 5·5 hours can result in acidity-index figures in excess of 13. The Institute for Grain-processing, Potsdam-Rehbrücke, has developed a method of improving the quality of sour-doughs during production interruptions (E. Köhler and G. Träger, Process for improvement of sour-dough quality during production-interruptions, *Bäcker und Konditor*, **31**, 266–267, 1983). According to this method, the excess acidity of the full-sour dough is treated with a carefully measured amount of potassium carbonate solution, and loss of gas-production supplemented by yeast addition. Depending on the sour-dough process and standing-time, practical instructions regarding dosage of potassium carbonate and yeast solutions based on the full-sour have been worked out, and found to give satisfactory end-product results for both planned and unplanned interruptions on the sour-dough side in the Konsum Backwarenbetrieben (bakery cooperatives) of Karl-Marx-Stadt and Hagenow. Permanent inventories of potassium carbonate are maintained enabling all shifts to cope with sour-dough stoppages of up to 6 hours.

These technical innovations for scheduled and unscheduled stoppages in sour-dough production make an important contribution to the economic use of material resources, reducing the wastage of rye-flour.

In the Karl-Marx-Stadt Bakery Cooperative, when the KVT 1500 continuous-doughmaking plant was commissioned in May 1981, the production of two basic bread varieties with a three-shift system was envisaged. Alongside hearth (oven-bottom) varieties, a large amount of panned-bread was planned—97% of the specialty-breads being panned.

However, the demand for panned-bread has steadily declined, and from 1981 to 1982 the demand for specialty-breads (*Spezialbrot*) in the panned form also fell from 22% to 19%. This situation could only be reversed by producing all the specialty-bread varieties as hearth (oven-bottom) bread. Owing to the quality and especially the mild flavour of the basic bread varieties produced with the KVT 1500, which are appreciated by the population, the demand for the basic varieties showed an increase to 105% from 1981 to 1982. However, in 1983, for nutritional reasons, bread varieties made with high-extraction rye-flours were given first priority, which meant the manufacture of 30% specialty-bread (*Spezialbrot*). Therefore, as a result of a research and development programme, a process for the production of two specialty-breads and two basic bread varieties with the continuous sour-dough and dough development plant KVT 1500 was developed. In so doing, the process for producing the specialty-breads was adapted to the most important technological parameters of the specification-standards for the basic bread varieties. Thus minimizing the changeover operations, and simultaneously any adverse changes in end-product quality. (Fiedler, Sour-dough and dough preparation using the KVT 1500 depending on the production requirements and flour-quality in the Karl-Marx-Stadt Bakery Cooperative, *Bäcker und Konditor*, **30**, 230–232, 1982.)

*Specialty-bread Production with the KVT 1500:*
Continuing the case-study of the Karl-Marx-Stadt Bakery Cooperative, the following improvements were completed in order to meet the production requirements:

(1) The silo-storage capacity was increased by joining-up with two unused sugar-silos.

(2) Separation of the control-system of the daily-silo-pairs, thus allowing one

daily-silo to be filled with rye-flour Ia, another with rye-meal for the sour-dough preparation. For dough preparation, one daily-silo each is filled with rye-flour Ia, mixed-flour for mixed rye/wheat (*Mischbrot*), mixed-flour for *Mecklenburger Landbrot* (ML) and mixed-flour for *Bauernbrot* (farmhouse-bread).

(3)   Increase the daily-silo capacity before the necessity to refill by blocking the 'empty-levels', thus ensuring that the night-shift have all the flour they need without manning the control-room.

(4)   Improvement in the uniformity during mixing of the various flour components in the daily-silos, achieved by frequent movement of the flours according to a set procedure.

(5)   The installation of a small cogged-wheel pump complete with a motor with controllable-output for the dosage-aggregate T 416·0 which feeds the KVT 1500. Thus allowing a dosage-range from 1·1 to 4·6 kg/min for media in paste-form.

(6)   Modification of the flour-dusting unit of the divider-moulder, by enlargement of the grill, and provision for regulating the outflow.

(7)   Building-on of a mechanized labelling-machine to the divider–moulder.

After discussion with the Supply and Demand Organization, the following programme was laid down:

(1)   *Mecklenburger Landbrot* and *Bauernbrot* despatched daily to all sales points according to order.

(2)   The number of full-sour compartments necessary for dough-preparation in the fermentation-equipment-aggregate, taken as the batch-weight for raw-material call-up.

(3)   *Bauernbrot* and *Mecklenburger Landbrot* being labelled with brown and orange-coloured bread-stamps for effective promotion.

(4)   Setting-up of a seminar for bakery-sales personnel, led by production-management and Supply and Demand Organization colleagues.

According to the formulation developed by the Institute for Grain-processing for *Mecklenburger Landbrot*, the following shows the feed-rates for 17·25 kg dough/min for the KVT 1500:

| | |
|---|---|
| Full-sour (*Vollsauer*) | 2·19 kg/30 s |
| Water | 2·74 kg/30 s |
| Rye-meal | 1·36 kg/30 s |
| Full-sour | 13·94 kg/60 s |
| Full-sour | 4·44 kg/30 s |
| ML-suspension | 0·60 kg/30 s |
| Salt-solution | 0·39 kg/30 s |
| Mixed rye/wheat-flour | 1·93 kg/20 s |
| Water | 0·30 kg/30 s |
| Dough | 17·25 kg/60 s |

Preparation of the ML-suspension is shown in Table 30.

TABLE 30

| Compartments full-sour (290 loaves from each) | Loaves | Caramel (kg) | Sugar (kg) | Caraway (kg) | Stale bread crumb from ML (kg) | Water (litres) |
|---|---|---|---|---|---|---|
| 1 | 290 | 1·49 | 0·69 | 0·160 | 4·66 | 15·9 |
| 2 | 580 | 2·98 | 1·39 | 0·319 | 9·31 | 31·7 |
| 3 | 870 | 4·47 | 2·08 | 0·478 | 13·97 | 47·6 |
| 4 | 1 160 | 5·96 | 2·77 | 0·638 | 18·62 | 63·5 |
| 5 | 1 450 | 7·45 | 3·74 | 0·798 | 23·28 | 79·4 |
| 6 | 1 740 | 8·94 | 4·16 | 0·957 | 27·94 | 95·2 |
| 7 | 2 030 | 10·43 | 4·85 | 1·117 | 32·59 | 111·1 |
| 8 | 2 320 | 11·92 | 5·54 | 1·276 | 37·25 | 127·0 |
| 9 | 2 610 | 13·42 | 6·24 | 1·436 | 41·91 | 142·9 |
| 10 | 2 900 | 14·91 | 6·93 | 1·595 | 46·56 | 158·7 |
| 11 | 3 190 | 16·40 | 7·62 | 1·754 | 51·21 | 174·6 |
| 12 | 3 480 | 17·89 | 8·32 | 1·914 | 55·86 | 190·5 |

The organization of the daily production-schedule is such that dough-making operation starts at about 22.00 hours. Therefore, the sour-dough daily-silos must be filled with rye-meal, and the dough daily-silos with mixed rye/wheat-flour by the early-shift.

The flour-dosage sequence for 1·25 t mixed rye/wheat flour for *Mecklenburger Landrot* (ML) is shown in Table 31.

Depending on requirements, the caramel, sugar, and caraway are also weighed out by the raw materials department during the early-shift. At about 17.00 hours the KVT chargehand starts to assemble the raw materials. In the second container of the dosage-aggregate T 416.0 water is metered to the desired level. Sugar and caramel are dissolved by starting the stirrer. Three hours before commencement of production, dried bread-crumbs and chopped caraway seed are suspended in the solution. Four hours before the doughmaking process, the souring of the rye-meal is started with sour-dough from the daily sour-dough silos. The technological parameters: temperature, dough yield, and feed-time remain according to the working specification standards for the basic bread varieties.

Owing to the longer baking-time for specialty-breads, the throughput for full-sour preparation must be reduced from 14·73 kg/min to 13·94 kg/min (see p. 131 Monday 01.00 hour-schedule for full-sour).

TABLE 31

| Flour type | Dosage-sequence | | | | |
|---|---|---|---|---|---|
| | 1 | 2 | 3 | 4 | 5 |
| Rye-meal | | | | 1 | |
| Whole-rye flour | 2 | 3 | 2 | 2 | 3 |
| Rye-flour Ia | 1 | 1 | 1 | 1 | 1 |
| White-bread flour | 1 | 2 | 1 | 2 | 1 |

The production-order determines the number of compartments of the KVT fermentation-equipment which must be filled. When demand is high 10 to 16 compartments are in use.

At about 21.30 hours, the changeover from rye-bread to *Mecklenburger Landbrot* begins. The dough daily-silo is set to withdraw mixed-flour (*Mischmehl*) for *Mecklenburger Landbrot* production. The throughput of the raw materials being reduced according to the formulation from 17·64 kg/min to 17·25 kg/min. Having closed-down all dosage-aggregates, the actual dough preparation for *Mecklenburger Landbrot* commences. The first 20–30 kg of mixed rye/wheat dough are returned for discontinuous (batch) production. A mechanized labelling-machine is loaded with bread-stamps, and after long-rolling each dough-piece is marked on the sealed side. The flour-dusting aggregate of the divider–moulder is set in motion after filling with fractionated rye-meal. Then the dough-pieces pass through the fermentation-cabinet (final-proofer) at a chain length of 25·7 metres in 38 minutes. The wire-band continuous oven BN 50 is set for a baking-time of 48 minutes for 1 kg units, and a temperature in Zone 1 of 300–320°C falling. After the throughput of *Mecklenburger Landbrot*, a production-run of rye-bread follows until about 11.00 hours.

Within a time-span of only five months the demand for this bread variety increased to 140%, the continuous manufacturing process contributing to this in no small measure. *Mecklenburger Landbrot* is very popular with the inhabitants of Karl-Marx-Stadt, and in terms of raw material, a saving of 14 g of yeast per tonne of bread coupled with an increase in dough yield of about 4%, relative to the specified raw material input, results in a gross increase in raw material yield. Productivity is higher than in the case of any discontinuous (batch) dough processing.

Another specialty-bread variety called *Bauernbrot* (farmhouse bread) has been produced on the KVT 1500 in the same bakery since 1983, by a similar process and without excessive preparation work for the changeover. In this case a *Bauernbrot* suspension is prepared 3 h before doughmaking. This involves the preparation of a salt solution to which potato-semolina is gradually added in a machine T 635.0 fitted with a cutter-blade. After about 15 minutes the suspension is transferred to a second container of the dosage-aggregate T 416.0 by starting the stirrer, which pumps it in. On account of the high water-binding capacity of the potato-semolina, the processing of stale bread-crumb is not included for *Bauernbrot*.

The souring of the rye-flour Ia is carried out according to the technological parameters of the working specification standards for the basic bread varieties, and a reduction in the throughput as in the case of *Mecklenburger Landbrot*. For dough preparation, the only machine-aggregates required are the mixed-flour weighing unit, full-sour weigher and suspension-dosage unit. Rye-flour Ia is used for dusting through the divider–moulder, and the dough-pieces are labelled with brown bread-stamps after long-moulding. The dough-pieces pass through the final-proofer in 38 minutes. Baking is carried out in the wire-band continuous oven BN 50 for 45 minutes at 280–300°C with a falling temperature-regime.

By producing the two specialty-breads described, alongside the basic varieties, the consumption of specialty-breads were increased from 19% in 1982 to 26% in 1983, this upward-trend continued during the first-half of 1984, reaching a level of 32%.

(Fiedler, U. Production of Specialty-breads using the KVT 1500, *Bäcker und Konditor*, **33**, 167–168, 1985.)

Supplying the population with the daily basic necessities at stable uniform prices is the fundamental objective of government in the GDR. The task of the baking industry is to provide as large a choice of products of improved quality as possible from available resources. With regard to quality, the oven-freshness of the baked products is of paramount importance. Bread-rolls (*Brötchen*) are particularly prone to quality deterioration, and lose their crispness even after 4–6 hours, depending on storage-conditions. Therefore, between 1976 and 1983 the VEB Bakery Collective in Potsdam opened four *Kaufhallenbäckereien* (in-store bakeries) in order to improve the oven-fresh service for the customer. These in-store bakeries were equipped with the following machine-aggregates:

—Two cloth-silos, each with a capacity of 5·5 t flour, pneumatically filled and manually discharged
—Mixer Type MB III
—Mixer type HLK 50
—Full-automatic divider–moulder Type VATW-4
—One to three electrically heated deck-ovens with forced-convection

The doughs are made discontinuously (Batch-process), worked off manually and fed into a fully automatic divider–moulder Type VATW-4.

The moulded pieces are transferred from the delivery-band onto tipping frames by two operators, and stamping or pressing is also done manually. The moulded and pressed dough-pieces are then placed in a final-proofing room, situated between the two deck-ovens. To facilitate the setting operation with the deck-ovens, two wheeled setter-tables are provided.

The baked bread rolls are placed in specially constructed transport containers for intermediate storage. Apart from rolls, part of the production includes *Kaviarbrote* and *Potsdamer Weizenstangen* as 250-g units, the latter being similar to *baguettes*. All products go oven-fresh for immediate sale into the Kaufhalle Bake-shop open shelves. The production output of the Kaufhallen-bakeries is determined by the throughput of the deck-ovens, and the demand for products in the store. It was established on opening the Kaufhallen-bakeries, that the weekly sales of fresh bread-rolls increased fivefold, and that the demand for fresh bread-rolls after midday greatly increased. The important role of the in-store (*Kaufhallenbäckereien*) in the supply politics is to complement the line-production of the industrial bakeries in areas where new apartments have been built, and where no craft-bakeries exist, simultaneously improving the oven-fresh service. Depending on the population, e.g. 5000 up to 30 000, the daily-production of rolls would be 11 300 rising to 41 100 units, requiring a baking surface of from 10 up to about 30 m$^2$. In addition to bread-rolls, other product lines produced include, *Kaviarbrot*, croissants, pound-cakes, cut-cake, poppy-seed rolls, gateaux, fruit-and-cream desserts, and choux-pastries. By January 1984 the fourth in-store bakery in the Potsdam district was in production, following one becoming operational in Schwerin in April of 1983 with a floor area of 350 m$^2$, and a weekly bread-roll production of 234 000. Of the 350 000 bread-rolls sold daily in Potsdam, 120 000 are baked in the in-stores, and about 90 000 in craft-bakeries, reaching the sales points oven-fresh. The remainder are produced in the industrial bakeries. In common with experiences in capitalist countries, where wheat bread and rolls are important sales products, the proximity of production and sales is essential to satisfy the increasing consumer demand for oven-freshness. During the past 20 years, a network of more than 100 modern industrial bakeries and over 6660 cooperative and private craft-bakeries have grown up; but in larger towns without any industrial baking capacity, either more industrial bakeries or *Kaufhallenbäckereien* are necessary, situated close to the consumer. In rural areas smaller production units are planned supplying several shops, including accommodation for the master-baker. For these and the in-stores, suitable machine-aggregates must be designed to give increased mechanization and automation for outputs of 5000–6000 bread-rolls per hour.

In recognition of the 11th Party-anniversary of the Socialist Unity Party of the GDR, a new bake-shop was opened in the Kaufhalle Leipzig-Lößnig. The main feature of which was the newly developed electrically-heated rack-oven, which was designed and built by the engineers of the Rationalization Department of the VEB Bakery Cooperative in Leipzig. In this store the new oven was used to bake-off partially baked, shock-frozen bread-rolls. The bread-rolls had been made in the City's Industrial Bakery on a bread-roll line, and baked in a wire-band continuous tunnel-oven for 12 minutes. After a short cooling-time, the bread-rolls were frozen at $-30°C$. Delivery to the Kaufhalle store was in a thermally insulated van. In the Kaufhalle the bread-rolls were placed in a deep-freeze compartment until required. The rolls being baked-off as required in the rack-oven with strong steam-injection. Using this procedure, 3200 oven-fresh rolls per hour could be prepared. However, the application of the rack-oven for this purpose is a special case. The basic purpose of the oven is to bake various types of wheat-flour rolls, rye and wheat breads and confectionery products. The main components of the rack-oven are:

—oven-chamber housing with steam-injection boiler
—power-block with cylindrical heating-elements
—outer-housing and insulation-material
—electrical and control components
—Racks on wheels with flat-edged sheets

The baking characteristic of this type of oven is that it takes place by forced convection of hot-air. The intensive heat-transfer results in a 10% reduction-minimum in baking-time. Further advantages are the reduced space and standing-area, coupled with the lightness of construction and ease of loading and unloading of the products. It gives particularly good results with wheat-flour products. The air in the baking-chamber is forced by a fan over the cylindrical heating-elements and is conveyed from the hot-air channel through vents into the baking chamber. To achieve an even heat-distribution over the products, the rack is rotated about its axis lengthwise. The necessary steam is introduced from the side into the hot-air channel from the boiler. The dosage of the necessary water is by means of a timed magnetic valve.

Technical parameters:

| | |
|---|---|
| Oven-chamber material | Stainless-steel |
| Dimensions | |
| (width × height × depth) | 1700 mm × 2750 mm × 1300 mm |
| Weight | 1500 kg |
| Power ratings | 30 kW for heating (wheat products) |
| | 40 kW maximum (rye products) |
| | 1·5 kW for the fan |
| | 0·18 kW for powering the rotation of the racks during the baking cycle |
| Water consumption | Up to 50 litres/h for steam-generation |
| Baking temperatures | Up to 320°C according to type of product |
| Throughput for bread-rolls | About 1800 units/h |
| | Baking at 210°C for 17 minutes |
| Wheat-bread 0·5 kg | 120 units/h |
| | Baking at 220°C for 30 minutes |
| Special toast bread 0·75 kg | 120 units/h |
| | Baking at 220°C for 35 minutes |
| Rye-bread 1 kg | 75 units/h |
| | Baking at 300°C, falling to 210°C |
| Baking-surface | 5·2 metres$^2$ using baking-sheets 420 mm × 620 mm |
| Racks, on wheels | |
| width × height × depth | 500 mm × 1750 mm × 700 mm |
| weight | 50 kg |
| number of racks | 20 for small units |
| | 10 for large units |

The Rationalization Department of the VEB Bakery Cooperative in Leipzig manufactured four of these ovens during 1986, and during the second half of 1987 the VEB Metal-works in Leipzig has taken over the mass production, producing about 100 rack-ovens per year. A gas-heated version is also planned as an alternative. The construction of the rack-oven has provided the baking-industry in the GDR with a very effective tool for rationalization of small production-scales, especially for oven-fresh products, thus complementing the work of the mass-production centralized industrial bakeries (Löscher, VEB Backwarenkombinat, Leipzig, *Bäcker und Konditor*, **34**, 109, 1986).

The actual industrial case studies described demonstrate the value of targeted integration of scientific, technical and economic political planning within the industry, in order to make the best possible use of manpower and resources. The decision of the Politbüro (Council of Ministers of the GDR) in 1985 to integrate all scientific, technical and economic strategy for the Baking Industry into a centralized unit, resulted in the formation of the WTÖZ (Scientific–Technical–Economic Centre), which, in the case of the baking branch of the food industry, is the 'VEB Institute for Grain-processing' in Bergholz-Rehbrücke.

The function of the WTÖZ is to ensure the conception and implementation of a consistent scientific, technical and economic policy for long-term research and top-level decision-making. This includes planning and coordination of the various research and development work done in the many cooperatives and contractual research with universities, technical high schools (polytechnics) and trade-schools. Important areas of research and development include: problem case-studies, objective basic research, applied research, process-research, building of machines and process-lines, computer application, issuing of licences and patents, and the working out of Standards for raw-materials, products and processes. In the foreground are certain key technologies such as extruder-technology, fluidized-bed drying, continuous-foaming and biotechnology, which allow the rationalization of the manufacture of traditional products with improved material and energy economy. Additionally, processing costs are reduced, and a new range of products can be created. The directive for the 5-year-plan 1986–1990 emphasizes the need for increasing intensification, in order to increase productivity and improve its effectiveness at the same time. This involves better use of existing conditions by increasing the number of shifts, thus increasing the output of machinery, and increasing the baking-surface of wire-band continuous ovens from $50\,m^2$ to $80\,m^2$, whilst utilizing the same floor-space. The key role of CAD-engineers in the design of flour-silos, ovens, etc., and the increasing use of CAM engineers to improve efficiency of production in large bakeries is of paramount importance to the economy. In the medium-sized cooperatives and private craft-bakeries, the crucial role of their own rationalization-workshops, on site, for improvements and innovative new construction is to provide the following services:

—diminishing requirement for the import of replacement parts
—support in introducing the new technologies
—rationalization projects
—improvements in the working conditions of production personnel.

In common with other branches of industry in the GDR, the Food and Baking industry is fully committed to the rapid application of the latest scientific and technological ideas, involving the fullest participation of all its citizens, with the total support of the SED (Socialist Unity Party).

### 1.7.7 USSR

In pre-revolutionary Russia, the baking industry was in the hands of about 140 000 small unmechanized bakeries; and residual, manually operated, small bakeries from these feudal times were still in evidence in many Russian cities at the beginning of the twentieth century. At the beginning of the second-half of the previous century, under existing capitalist production conditions, some concentration of bakery production and sales into the hands of larger companies took place. However, this did not lead to any significant degree of mechanization of production since there were vast reserves of unemployed bakers and a plentiful supply of cheap labour. Any larger, partially mechanized bakeries which did exist were equipped with imported machines and ovens.

In the early years after the Great October Revolution of 1917, up until 1920, the nationalization of the baking industry took place, and production became concentrated in larger and better-equipped bakeries. Nevertheless, during the period of peaceful reconstruction of industry from 1921 to 1925, with the new political economy, the re-emergence of the small entrepreneur-baker became apparent. Some of these small bakeries remained independent and others joined their cooperatives. During the same period, the previously nationalized bakeries were handed over to the cooperatives and this led to a struggle for the improvement and mechanization of production, which then led to the ousting of private capital from bread manufacture generally. According to Central Committee for Food Industry Union statistics for 1925, the bakeries were divided as follows: under state ownership 3·5%, cooperatively organized 38·7% and privately owned 57·8%. However, the distribution of the workforce was such that 79·2% belonged to the state and cooperative bakeries, and only 20·8% to privately owned bakeries. The average number of workers in each state or cooperative bakery amounted to 16, and in each private bakery to only 3. Production capacity was mainly in the medium and smaller-sized bakeries at 94·6%, a mere 3·6% coming from the bread-factories.

In 1925, the Ministry of Work and Defence decided to push ahead with a total industrialization programme, which included the baking industry. A bakery machinery construction industry was set up and mechanized; bread-factories were erected, which transformed the baking industry and set the pattern for its development into the twentieth century. In October 1931, the Central Committee initiated a plan for improved food supply for the Soviet people. Mechanization of bread production was thus fully completed by 1935, starting in the large towns and industrial centres and in the Urals, as well as in Moscow, Leningrad and the Don basin. This resulted in 58% of bread production being in bread-factories, 16·8% in mechanized bakeries and only 25·2% in small bakeries. At that time the Soviet Union was already leading in the degree of mechanization of the baking industry on a global basis. The number of newly constructed bakeries amounted to about 100 large units, which supplied over 0·5 million people with bread. At that time bread-factories were regarded as models of the success of Socialism. Maxim Gorky, who worked as a baker in pre-revolutionary Russia at two Leningrad bakeries, summed up the achievements on re-visiting the new factory in 1929 as follows: 'This factory is the most amazing experience in Leningrad … nothing speaks more forcibly about the Revolution in daily life than the bread-factory, which has made light of hard work'.

The great achievement during these years was the work of Soviet project-engineers and co-workers in constructing bakeries fitted out with machines and ovens made in Soviet machine-shops. The Soviet Engineer G. P. Marsakov was the first to build a bread-factory working on a ring-transport system.

In 1935 the baking industry in the towns and cities was removed from its dependence on the cooperatives and placed under the direction of the People's Commissariat of the Food Industry of the USSR. Between 1935 and 1941 the new bakeries that were built were fitted with modern Russian-made conveyor-ovens, mixers and processing machines. By the beginning of 1941, 77% of all bread production was in bread-factories or mechanized medium-size bakeries.

During the Great Patriotic War of 1941–1945 on Russian soil, the bakeries in areas under occupation by the enemy were almost completely destroyed. With the gradual liberation of these areas the factories were re-built, and new mechanized bakeries built in cities and new industrialized areas in the eastern and central regions of the country. At the end of 1947, the production capacity of the bakeries reached a level 17% more than at the beginning of 1941. During the post-war years, building of large and medium-sized bakeries continued, and as regards degree of mechanization of processing, the baking-industry has become one of the leading branches of the food industry of the USSR.

In more recent years, modern dough-mixing aggregates, which are suitable for both discontinuous and continuous processes without the use of machine-bowls, have been developed and manufactured in large quantities, as have dough-processing machine lines and wire-band continuous tunnel-ovens, which allow fully mechanized bread production. The use of sacks for transport, storage, and transport within the bakeries has been replaced by tankers, silos and pneumatic transport systems, in common with developments in Western Europe.

In addition, continuous processes have been devised, not only for bread but also for specialty breads, bread-rolls of various shapes, crispbread and rusks. Mechanization of bread-cooling, storage and despatch are also well advanced. All the bakeries in the USSR are under the direction of different systems of organization:

(1) The Ministry for Food of the USSR, which is all-union, covering the republics, to which 60% of the bakeries belong.
(2) The various cooperatives, which are united within the Zentrosojus, accounting for 30%.
(3) The Ministry for Travel and other subsidiary activities.

The total production of bread and baked-products from these craft and industrial bakeries is progressively increasing, the present level being a little over 30 million tons. The total number of bakeries amounts to approximately 23 000, which employ some 0·5 million people. Bakeries belonging to the Food Ministry are almost entirely bread-factories with a relatively small number of mechanized small units. The population of the cities and industrial centres are completely supplied with bread and flour confectionery from these production units. A more complex problem is the supply of the country-people of the USSR. However, the bakeries which belong to the system of the Zentrosojus (*c.* 30%) have made great headway in recent years in solving this complicated problem.

In 1964, the total number of country-bakeries exceeded 17 000, of which 423 were bread-factories and about 5500 smaller mechanized units. By 1969, the total number of bakeries belonging to the Zentrosojus system had been reduced to 14 000. Included in this figure are about 800 bread-factories, and about 7500 mechanized smaller units, which together produced about 70% of the total bread and confectionery manufactured in the countryside. The number of small bakeries had been reduced to less than 6000.

The population in country areas nearer to the towns are supplied by the city bread-factories which belong to the Food Ministry system. Taking this into consideration, by 1969 around 70% of the country population were being supplied with bread from industrial bakeries, self-sufficiency through household-bakeries being restricted. Further mechanization and rationalization of country bakery production and supply of the whole Soviet people with baked products of uniform quality is the task of the cooperative system. The introduction of in-store bakeries and production units based on rack-oven baking is considered to merit increasing attention to supply the country people with oven-fresh baked products in a rational and efficient system. Thus complementing the work of the large industrial production units.

The future objectives of the baking industry in the USSR is to constantly improve the production and distribution of bakery products, especially for the people in the countryside. In such a vast country, with diverse climatic conditions, nationalities and cultures, the best balance must be found between industrial city bakery production and those more suited to the country population. This situation demands the unification of the energies and skills of all those engaged in design-offices, machine-construction, bakery-production, and scientific research and development. The collective effort of these groups of people, and the critical appraisal of the results of scientists and technologists in other countries, form the basis for the further development of progressive rationalization of production methods and processes.

For breadmaking purposes, wheat and rye flours of various types are used. Barley, maize and oat flours as well as other flours are frequently added to wheat and rye flours in amounts laid down in the state processing-standards, known as GOST. During the late 1960s the percentages of the various grains cultivated in the USSR were 50% wheat, 13% rye and the rest maize and other grains. Since that time annual yields have been greatly increased in favour of wheat at the expense of rye and other cereals, and the long-term objective is for self-sufficiency. With the constant improvement in the availability of chemical fertilizers, pesticides and herbicides, and investment in increasing mechanization, these objectives can be reached. The limiting factors are climatic conditions and the vast areas of permafrost. Wheat quality is generally above average for continental Europe, wet gluten contents ranging from 26% up to 45%, the average falling between 30% and 32% and of good quality.

In the USSR wheats are classified according to their blending potential as so-called 'improver wheats'. For evaluation purposes wheats are milled to extraction rates of between 70% and 78%, the latter being the industrial milling average. Controlled baking tests are carried out both with individual varieties, testing their response to chemical and biochemical improvers, i.e. potassium bromate and malt, etc., and with blended flours from two wheat varieties. A typical test-baking formulation is as follows:

Flour        600 parts
Yeast         18
Sugar         30
Salt           8
Water, variable to obtain normal dough consistency
(Source: V. L. Kretovich, Biochemistry of grain and breadmaking (in Russian), *Biokhimiya zerna i Klebopecheniya.*)

The straight dough is mixed for 5 minutes for blended samples, but when strong 'improver wheat' varieties are being evaluated a more intensive mixing is necessary coupled with a complementary fermentation period. Dough temperature is set for 30°C and controlled at this temperature and relative humidity 75–85%. The first punch is at the 90-minute stage, and the second at 150 minutes, moulding being carried out 30 minutes later. The dough is divided in half, moulded and placed in pans greased with vegetable oil measuring $10 \times 15$ cm at the base, and $12 \times 17$ cm at the top, the height of the pans being 8·5 cm. Proofing is to standard height at the same temperature and relative humidity as for fermentation. Baking is in an electrically heated oven with humidified chamber and rotating floor at 230°C for 25 min. The baked loaves are cooled in a cupboard and evaluated 16 hours later. Evaluation of bread quality is from weight and volume, external appearance, porosity, colour, condition of the crumb, elasticity, aroma and taste. Basic indices are combined to form a 5-point system of evaluation. 4·6–5·0 points—excellent; 4·0–4·5—good; 3·5–3·9—above average; 3·0–3·4—average; 2·5–2·9—below average; less than 2·5—poor. Breadmaking strength is defined as the capacity of a flour to produce, under standard conditions, a loaf of large volume, good appearance, and satisfactory crumb porosity.

To establish an 'improver-wheat' effect, mixtures are made by mixing from 25% up to 50% of a potentially strong wheat-flour with a weak wheat-flour. In this way new varieties of strong, 'improver' wheats can be identified. Generally, when two wheat-flours are blended, one with inferior quantitative and qualitative indices; the quality of the blend corresponds to a weighted mean for most indices. The addition of 25% flour from improver wheats with wet glutens below 36% to weak wheat-flours does not produce blends of adequate gluten content, being in all cases less than 30%. However, a 50% addition usually effects a marked rise in gluten content to about 30%. Improver wheat varieties normally contain at least 15–16% protein in dry matter, and not less than 33–35% of wet gluten of good resilience, normal extensibility, producing first-grade flour. Medium-strength wheats can often produce satisfactory bread, but such wheats are not efficient improvers for weak wheats. Wheats requiring no improver wheats, but not strong enough to be efficient improvers of weak varieties in the breadmaking sense are termed 'filler-wheats'. Wheats of medium breadmaking flour strength form the bulk of raw materials for breadmaking, the gluten content of weak wheat-flours being either too low, but of satisfactory quality, or fairly high and of poor quality.

In addition to the direct method of baking tests, indirect tests for breadmaking strength of wheat varieties are used. These include the Chopin Alveograph *P* value, a measure of dough resilience and stability, the $L$ value, a measure of extensibility, and the specific deformation energy $W$ in ergs per gram of dough. Water absorption and the valorimeter mixing-index are measured on the Farinograph of Brabender. Gluten quality is determined according to the Standard GOST 13586 1-68 method, using the apparatus IDK-1, which measures its response to a pressure deformation between two discs. This apparatus is standard equipment in silo and mill laboratories in the USSR, and is similar in measurement principle to the penetrometer, having been developed by the All-union Institute for Grain and Grain-processing Research (WNIIS) in Moscow. By carrying out sample baking tests of the breadmaking strengths of flours and indirect tests, new varieties of strong 'improver' wheats can be recognized. For example, in addition to the well-known Sarrubra, Cesium 111, and Erythrospermum 841 strains, improver varieties for spring wheat include the following:

Bezenchukskaya 98
Produced by the Kuibyshev State Research Station by hybridizing, Lutescens DS-11-22-44 (USA) with Erythrospermum B-47, being recommended for this region. A variety with large vitreous grains, red in colour, with good milling and baking qualities. Alveograph specific deformation energy values range between 158 and 452 ergs, with an average of 322 ergs. Dough resilience and stability is high at 94 mm, and it has a good resilience:extensibility ratio ($P:L$) of 0·83.

Dal'nevostochnaya
An improver variety, showing an average Alveograph specific deformation energy value, $W$, of 336 ergs, very high resilience and stability of 111 mm, and an excellent $P:L$ ratio of 1·3.

Lutescens 758
An improver variety, showing an average Alveograph $W$ value of 338 ergs, $P$ value of 99 mm, and $P:L$ ratio of 0·82.

Saratovskaya 29
An improver variety, showing an average Alveograph $W$ value of 356 ergs, $P$ value of 111 mm, and $P:L$ ratio of 1·09.

Apart from the established winter wheat varieties, Ukrainka, Kooperatorka and Krymkas, the following newer varieties have been used as improvers for winter wheats:

Bezostaya 1
A successful improver variety with an average Alveograph $W$ value of 315 ergs, $P$ value of 110 mm, and $P:L$ ratio of 1·2. This wheat gave good yields and breadmaking qualities when introduced into Hungary in 1959 under ideal conditions of growth (Pollhamer, 1980). However, the Hungarian research workers established that Bezostaya 1, in common with Hungarian improver wheat variety Bankuti 1201 showed quality deteriorations by the 1970s, owing to excessively intensified husbandry (seed-sowing density, intensive fertilization and use of weedkillers). It is emphasized that such quality changes are not due to the

breeding, but rather to the reaction of the cultivar to environmental changes, owing to changes in husbandry methods. Generally, balanced growth and avoidance of 'distress' conditions are best achieved by an N:P:K ratio of 1:1:1, avoidance of drought and excessive application of mineral fertilizers. Optimal levels of plant nutrients and humic acids are essential for breeders' material.

Bezostaya 4
Another improver cultivar with an average Alveograph $W$ value of 325 ergs, $P$ value of 99 mm, and $P:L$ ratio of 0·93.

Novoukrainka 83
A variety of outstanding deformation energy, with Alveograph $W$ values of 480 erg/g dough, $P$ value 128 mm, and $P:L$ ratio of 0·98.

Summarizing the classification of varieties from the 1957 harvest, controlled baking tests of varietal wheat flour-blends gave the following results:

—Strong wheats: Dal'nevostochnaya, Sarrubra, Novoukrainka 83, Bezostaya 4.
—Medium strength wheats of average or above average quality (filler varieties): Odesskaya 3, Odesskaya 16.
—Weak wheats of below average quality: Agropyron hybrid 1, Shortandinka.
(Source: *Biochemistry of Grain and Breadmaking*, V. L. Kretovich, Moscow, Publication of Academy of Sciences of USSR, 1958.)

In common with researchers in other countries, Soviet researchers have concluded that any meaningful differentiation between the potential breadmaking strengths of various wheat-breeders' samples can only be achieved by subjecting the doughs to intensive mechanical work-inputs. Whether this energy input can be optimally performed at the initial mixing stage or in two stages, as with the 'Remix' method developed in 1960 by the Canadian Grain Research Laboratory, is open to debate. The inclusion of a direct source of sugar, and about 0·0015% potassium bromate based on flour weight, is essential; and the inclusion of an inorganic source of nitrogen and phosphate as yeast nutrient may be desirable. The water-absorption addition should be such that dough viscosity is at a maximum for the chosen mixer-rpm and mixing-time.

The progressive decrease in grain purchase on the world market by the USSR from year to year is indicative of the success of the breeding programme and methods of husbandry. Furthermore, the FAO confirm worldwide increases in national grain production over national consumption, therefore the pressure on the world price of grain currently continues unabated, resulting in a buyer's market. At the time of writing, traditional importers of grain, e.g. India, have become exporters. However, much of the high yielding, mass-produced wheat is of the animal-feed variety, and for breadmaking the emphasis is on quality improver-wheat varieties, which command a price premium. The current trend being a closer cooperation between the bread wheat producer and the large national milling and bakery processor.

Much research effort has been put into finding reliable and rapid methods of identifying improver-wheat quality cultivars, enabling the breeder to screen at an early stage a large number of samples of minimal sample size. In addition to methods based on the quantitative isolation of specific protein fractions using reversed-phase, high-performance liquid chromatography (RP-HPLC), for correlation with controlled baking test scores, sub-atomic spectral techniques have shown promising results according to Soviet research. Lutsishina and Matyash (1984), using wide-band proton NMR-spectroscopic methods, have established higher parameters in NMR-spectra for the harder improver-quality wheat varieties than for the weaker soft varieties. This is due to the gluten protein macromolecules being more compact with stronger H-bonding. These workers found positive correlations with rheological and baking data.

Wheat-flours produced for bakery use in the USSR are the following:

| | |
|---|---|
| Coarse flour | maximum ash content in dry matter 0·60% |
| Top-grade | maximum ash content in dry matter 0·55% |
| 1st grade | maximum ash content in dry matter 0·75% |
| 2nd grade | maximum ash content in dry matter 1·25% |
| Wholewheat-flour | not less than 0·07% of the ash content of the grain used to produce it, before the cleaning process |

Additional State Standards for flour include the following:

—Colour, taste and smell
—Granularity of the flour particles (particle-size analysis using various sieves to assess percentage retained maximum, and percentage passing through-minimum)
—Moisture 14–14·5%
—Wet-gluten content and quality, measured by pressure deformation using discs, and extensibility tests
—Freedom from extraneous matter (insect particles and metal particles)

The mills provide a quality guarantee for each delivery, including ash and moisture content.

TABLE 32
**Milling-plan and flour types produced from them in USSR**

| Flour type | Milling-plan | | | | | | | | | | | | |
|---|---|---|---|---|---|---|---|---|---|---|---|---|---|
| | Three-flour-types | | | | | | | Two-flour-types | | | One-flour-type | | |
| Coarse flour | 10 | — | — | — | — | — | — | — | — | 10 | — | — | — |
| Top-grade | — | 15 | 15 | 15 | 10 | 10 | 10 | 40 | — | — | — | — | — |
| 1st grade | 35 | 30 | 35 | 40 | 35 | 40 | 45 | — | 45 | 60 | 72 | — | — |
| 2nd grade | 33 | 33 | 28 | 23 | 33 | 28 | 23 | 38 | 33 | — | — | 85 | — |
| Wholewheat-flour | — | — | — | — | — | — | — | — | — | — | — | — | 96 |
| Total % extraction | 78 | 78 | 78 | 78 | 78 | 78 | 78 | 78 | 78 | 70 | 72 | 85 | 96 |

Source: *Technology of Breadmaking*, L. J. Auerman, Moscow, 1977.

In the USSR a combined system of flour-milling is used, in common with most continental European countries. This involves the production of composite flour divides, which can be made up of one, two or three different components, average total flour yield being 78%; varying according to milling-plan and type of flours to be produced, from 70%, up to 96% for a single component milling of wholewheat-flour.

The three-flour-type milling-plan is the average, producing top grade, 1st grade and 2nd grade from the same milling with a total flour extraction rate of 78%. The percentage yields of each flour are summarized in Table 32 which shows possible one-, two- and three-flour-type milling-plans.

The average milling plan for wheat in a milling unit with an output of 1000 to 1500 tonnes per day is a combined system of continental milling, producing three flour types:

—15% top grade flour of 0·55% ash in dry matter (type 550) at an energy consumption of about 115 kWh/t.
—35–40% 1st grade flour of 0·75% ash in dry matter (type 750) at an energy consumption of about 66 kWh/t.
—The percentage yield of 2nd grade flour depends on the required percentages of the two higher grades, averaging 30% of production, with an ash content of not more than 1·25% in dry matter (type 1250), at an energy consumption level of about 22 kWh/t.

On the basis of experimental data, the specific energy consumption for a 78% total extraction rate, producing three types of flour from wheat in any mill of known daily production capacity can be computed. The wheat prepared for the milling process is brought to a moisture of about 16·5% by automated control before being subjected to first break rolls, a process known traditionally as 'conditioning' or 'tempering'. The typical modern mill with pneumatic transport of mill-stocks, consists of 84 pairs of rolls, each of length 800–1000 mm, 250 mm diameter, rotating at speeds of 6 m/s, average feed rates to the rolls being of the order of 125 kg/cm per day (24 hours). The break rolls have a differential of 1:2.5, and 8 flutes/cm, and the scratch rolls working at a differential of 1:1·5 with 9–10·5 flutes/cm in combination with the reduction system. The sifting and sorting operations consisting of 64 plansifters and 32 semolina-purifiers, the sifting surface being 109 kg/m$^2$.

Measurement and control systems are well developed, and centralized on one control-panel. Laboratory control methods are designed to produce data which adequately describe processing quality parameters, and a 10 t/h pilot mill is available for test-milling.

Wet gluten percentage is determined according to the State Standard GOST 9404-60 method, using 10-g weight samples for flour of low extraction rate (type 550), and 25-g samples for second grade higher extraction flours (type 1250). Gluten quality evaluation consists of measurement of firmness, extensibility and hydration capacity. Firmness being measured with the Plastometer AB-1, which is a specialized type of viscometer which measures the extrusion-time in seconds of a fixed amount of gluten sample under standard conditions of temperature, after a

1–3 h rest period in water. The URK-2 apparatus measures the gluten extensibility in cm/min under standardized conditions. An inverse relationship exists between firmness and extensibility, which is expressed mathematically by a correlation coefficient of $-0.533$.

An average storage time for flour, before delivery to the bread factories is 10 days, and all grain and flour silos are regularly gassed. Grain silo capacities are of the order of 72 000 t, and flour silo about 10 000 t (compartments holding 80–100 t each). In the flour mills, production continues without interruption 6 days a week. In each factory, generally, the workforce is divided into four brigades, of which three work whilst the fourth rests. Basically the system operates on three 8-h shifts, with two breaks of 30 minutes for each shift. Monthly working hours amount to 160–180. Women earn the same as men, salaries varying between 60 and 130 roubles per month. Average earnings being about 90 roubles per month, an engineer earning about 110 roubles per month. Premiums are paid when production norms are exceeded. When targets have been reached or reached before fixed deadlines, premiums of 20–40% are paid to all workers, based on their salaries. In flour mills norms are based on tonnage produced, but in the bakeries product quality predominates, since the machines and degree of mechanization/automation dictate the work-speed. The state provides 25% of the salary sum in the form of subsidies for accommodation, health and pensions. Flour prices vary somewhat according to region, and the luxury character of the products produced from it. For example, assuming a basic average wheat price in Moscow of 60 roubles per ton, and flour prices, according to type, of 75 to 90 roubles/t; Leningrad wheat-flour prices rise to 120 roubles/t; and in Kiev the norm is 60 to 80 roubles/t. Assuming an average wheat-flour price for breadmaking of 75 to 85 roubles/t, bread prices vary between 200 and 300 roubles/t. Therefore, the baking margin is exceptionally high compared with the milling one. Bread is both the best value food commodity and the cheapest; therefore the average consumption remains high at about 150 kg/head per year, in spite of the rising standard of living. Regional variations in preference for wheat and rye bread exist, and the manufacture of specialty breads of improved nutritional value are gaining in acceptance. Great emphasis is placed on bread reaching the consumer as fresh as possible, and in the large capital cities, most sale-points are delivered bread from the bread-factories four to five times a day. Table 33 shows the

TABLE 33
**Milling-plan for rye-flour types produced in the USSR**

| Flour type | Milling-plan | | | |
|---|---|---|---|---|
| | Two-flour-types | One-flour-types | | |
| Light grade | 15 | 63 | — | — |
| Medium grade | 65 | — | 87 | — |
| Whole-rye flour | — | — | — | 95 |
| Total % extraction | 80 | 63 | 87 | 95 |

TABLE 34
**Important quality standards for wheat- and rye-flours in USSR**

| Flour type | Max. ash (%) DM | Min. gluten (%) wet | Particle-size distribution | | | |
|---|---|---|---|---|---|---|
| | | | Residue on sieve no. | Max. (%) | Throughs on sieve no. | Min. (%) |
| *Wheat* | | | | | | |
| Top grade | 0·55 | 28 | 43 | 5 | — | — |
| 1st grade | 0·75 | 30 | 35 | 2 | 43 | 75 |
| 2nd grade | 1·25 | 25 | 27 | 2 | 38 | 60 |
| Wholewheat | + + | 20 | 067 | 2 | 38 | 30 |
| *Rye* | | | | | | |
| Light grade | 0·75 | — | 27 | 2 | 38 | 90 |
| Medium grade | 1·45 | — | 045 | 2 | 38 | 60 |
| Whole-rye | 2·00 | — | 067 | 2 | 38 | 30 |

Source: *Technology of Breadmaking*, L. J. Auerman, Moscow, 1977.

milling-plan for the rye-flour types produced in the USSR, and Table 34 the most important quality standards for wheat- and rye-flours for breadmaking.

The assortment of baked products produced in the USSR amounts to several hundreds, and includes the following main groups:

—Wheat and rye breads from 0·5 to 3 kg, baked as panned and hearth types.
—Small hearth breads and rolls in units of up to and including 500 grams, of many different formulations and shapes.
—Yeasted rolls containing various amounts of sugar, shortening, egg and jam.
—Products for prolonged shelf-life, made from a stiff dough scaled at 6 to 100 g each. These are shaped into either 'rings' or 'strips' from doughs of moisture contents of 33–36%, the baked products being baked-out to moistures of between 9% and 27%, depending on the desired shelf-life of the product.
—Rusks of various shape made from yeasted doughs containing sugar, butter and margarine, for consumption of both adults and infants. Bread-rusks, made from rye-, mixed wheat/rye- or wheat-flours, which are dried to 10% moisture and packed as slices. Such products have the advantage of extended shelf-life, and are unaffected by ambient temperature variations from sub-zero up to 30°C.
—National-breads produced in the various republics according to traditional ethnic formulations and methods.
—Diet and health-breads, which are specially formulated so that their composition benefit the health and nutrition of specific groups of adults and infants, according to their requirements.

Over the past 10 years, the trend has been increasingly in favour of the smaller units at the expense of the larger units of 1 kg and over. In the case of the larger units, the trend is away from panned varieties, in favour of hearth varieties. The number of new bread and small unit varieties produced in the many republics increase yearly. However, in the interests of collective control aimed at increasing productivity, efficiency and quality of baked-products, some degree of rationalization must be imposed centrally.

A full-description of the many formulations and methods of production of baked products in the USSR is not possible within the scope of this book, and for this the reader is referred to more specialized sources, viz.

—*The Production of Baked-products—a Technological Handbook*, Roiter, I. M., Kiev, published by Technika, 1966 (in Russian).
—*Recipe-collection for National Small Baked-products*, Iljinskaja, T. N. and co-workers, published by Ekonomika, Moscow, 1967 (in Russian).
—*The Modern Technology of Industrial Bread Production*, Roiter, I. M., Kiev, published by Technika, 1971 (in Russian).

However, in view of the lack of information available in the English language concerning baking technology, some attempt will be made to outline the basis of technology and methods in general use in the USSR.

The Central Bureau of Statistics of the USSR have compiled a list detailing the nomenclature of the most important bread varieties, which forms the basis for statistical records for the baking industry of the USSR. This list is presented in Table 35 as an initial guide. More detailed descriptions of the listed types, and the raw materials used to produce them are as follows.

### Rye Bread from Ryemeal

Rye bread from ryemeal is made in pan and hearth form, the weight depending on whether it is to be sold in parts or as whole loaves. The former can be from 2–5 kg, and the latter 500-g or 1-kg units. In addition to ryemeal bread, there is also: rye bread prepared from a 'soak' (*Brühstück*) with 5% red rye-malt and 0·1% caraway seed; Moscow rye bread made with 7% red rye-malt and 0·1% caraway seed; and Borodin-bread consisting of 80% ryemeal, 15% wheat-flour 2nd grade, 5% red rye-malt, 6% sugar, 4% syrup, and 0·5% coriander.

### Rye Bread from Medium and Light Grade Rye-flours

This group also includes bread varieties which are made with the replacement of rye-flour by wheat-flour.

Rye bread is made from both medium-grade and light-grade rye-flours. These are produced in both the panned and hearth forms, and are sold according to weight or as whole loaves.

Ukrainian bread is prepared from 20–80% medium-grade rye-flour, and 80–20% wholewheat flour, according to the desired blend. A newer version of Ukrainian bread is produced by blending medium-grade rye-flour with 2nd grade wheatflour.

Minsk-bread, which is also sold according to weight or as whole loaves and in the panned or hearth form, is made from 90% light-grade rye-flour and 10% 1st grade wheat-flour with the addition of 2% syrup and a 'soak' (*Brühstück*) of 2% white malt.

TABLE 35
**Official designation of types of bread into groups for statistical records in the USSR**

| Group | Product description |
|---|---|
| 1 | Rye bread from whole-rye flour. |
| 2 | Rye bread made from medium and light rye-flour with wheat-flour types, including bread from flour mixtures, produced as hearth-bread. |
| 3 | Mixed rye/wheat bread. Wheat/rye types made from wholemeal flours, including hearth-bread. |
| 4 | Wheat bread from wholemeal flour as hearth-bread. |
| 5 | Wheat bread from 2nd grade wheat-flour, made into bread of weight exceeding 500 g/unit, as hearth-bread. |
| 6 | Wheat bread from 1st grade wheat-flour of weight exceeding 500 g/unit, as hearth-bread. |
| 7 | Wheat bread from top-grade wheat-flour of weight exceeding 500 g/unit, as hearth-bread. |
| 8 | Small products made from 2nd grade wheat-flour of weight 500 g/unit and below. |
| 9 | Small products made from 1st grade wheat-flour of weight 500 g/unit and below, including: <br> (a) Long white loaves, either hearth or panned, and cut (slit) lengthwise. <br> (b) City-rolls (*gorodskije bulki*). |
| 10 | Small products from top-grade wheat-flour of weight 500 g/unit and below. |
| 11 | *Baranotschnye isdelija*, which are made from a tight, low-moisture dough (10–36%), the dough-pieces weighing from 7 to 100 g, and shaped into rings or boat-shapes. |
| 12 | Small yeasted products such as: croissant-type rolls, salt-covered rolls, Moscow *rosantschicks* (Rosette-shaped rolls or *Semmeln*), Leningrader *kalatsche* and other types containing 7% shortening and sugar (minimum) and other types of raw materials. This group also includes various types of bread e.g. Donezker-bread, top-quality bread made from top-grade wheat-flour. <br> Various crispbread, e.g. Leningrad crispbread. <br> Special small units of 50–200 g, which are either plain '*Ljubitelskij*' or shaped '*Wyborger*'. Top-quality flour is used for the '*Wyborger*', and '*Ljubitelskij*'. |
| 13 | A wide range of rusks made with different grades of wheat- and rye-flour, enriched with diverse raw materials, and made into various shapes and sizes. |
| 14 | Various lean and enriched formulated yeasted products. |

Adapted from Table 46, *Technology of Breadmaking*, L. J. Auerman, Moscow, 1977.

Riga-bread consists of 85% light-grade rye-flour, 10% 1st grade wheat-flour, 5% white malt and 0·4% caraway seed, using a 'soak' to provide hydrolysed sugars as substrate. This variety is made in the hearth form and sold as whole loaves.

Kaunas-bread, also made as hearth bread is made from 92% medium-grade rye-flour, 5% 2nd grade wheat-flour, 3% red rye malt and 0·4% caraway seed.

Orlow bread is made in the panned form, and sold as whole loaves, being made from 70% medium-grade rye-flour, 30% 2nd grade wheat-flour, and 6% syrup.

## Mixed Wheat/Rye Bread from Wholemeal Flours

These are prepared both as panned and hearth loaves, and sold both by weight and as whole loaves. Their composition varies from 60% whole-rye flour and 40% wholewheat flour which is called mixed rye/wheat bread, to 60% wholewheat flour and 40% whole-rye flour. Normally a 'soak' (*Brühstück*) with 5% red rye-malt is prepared to provide a substrate.

*Wheat Bread*
Wheat bread includes the hearth and panned breads sold by weight and as whole loaves made from wholewheat flour (see Group 4, Table 35). Ordinary white bread, which must weigh more than 500 g, can be made from 2nd grade wheat-flour (Group 5, Table 35), 1st grade wheat-flour or top-grade wheat-flour (Groups 6 and 7, respectively).

Transbaikal-bread is made from 50% wholewheat flour and 50% 2nd grade wheat-flour. In addition, from 2nd grade wheat-flour the following panned and hearth breads are produced: Krassnosselsk-bread, Ukrainian *paljanytza* bread, Kiev *arnaut* bread, New-land bread, milk bread and fisherman's bread. Tea-bread is made from 90% 2nd grade wheat-flour, 10% light-grade rye-flour with red rye-malt 2·5%, syrup 10% and coriander 0·2%. Karelisch-bread is made from 85% 2nd grade wheat-flour, 10% light-grade rye-flour, 5% red rye-malt, sugar, coriander and raisins.

From 1st grade wheat-flour the following bread varieties are produced: normal white bread in panned or hearth form, mustard bread which is made using mustard-oil, Krassnosselsk-bread, household-bread, milk-bread, Saratower *kalatsch*, Ukrainian *paljanytza*, Kiev *arnaut*, etc.

From top-grade wheat-flour another series of bread varieties are made: normal white bread in panned or hearth form, white bread with raisins, milk bread, Saratower *kalatsch*, etc.

*Small Products*
Small products include those listed under Groups 8, 9 and 10 in Table 35, hearth-baked products weighing up to 500 g, including city long-loaves, twists, croissants and Moscow *kalatsch*, made with less than 7 kg shortening and 7 kg sugar per 100 kg flour. Certain types of small products, e.g. country-rolls, and city white-bread do not contain either sugar or shortening. Some typical examples of these products are illustrated in Fig. 21.

For more detailed technological information and formulations of small bakery products in the USSR, the reader is referred to the following specialized literature:

—*Recipe Collection for Small Wheat-flour Products*, Glawchleb M.P.P., Moscow, Pistschepromisdat, 1947 (in Russian).
—*Bread and Baked Products in the USSR and Abroad*, ZINTIpistscheprom, Moscow, 1967 (in Russian), by Grischin, A. S., Selman, G. S. and Iljinskaja, T. N.
—*Production of Small Baked Products and Fancy Goods on Mechanized Lines*, by Grischin, A. S., ZINTIpistscheprom, Moscow, 1969 (in Russian).

*Small Specialty Products*
This group of fancy goods are also yeasted, and are listed under Group 12 in Table 35. They include shaped products such as croissants, salt-covered rolls, Moscow

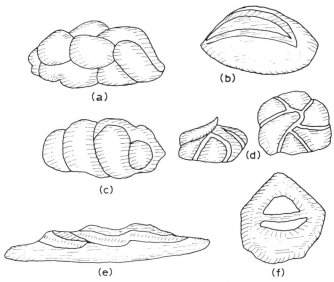

Fig. 21.   (a) Plait or twist. (b) City roll. (c) Plaited roll. (d) *Rosantschik*. (e) City long white loaf. (f) Moscow *kalatsch*. (Source: *Technology of Breadmaking*, L. J. Auerman, Moscow, 1972 (in Russian); Leipzig, 1977 (in German).)

*rosantschiks* (rosettes), Leningrad *kalatsch* etc. and products containing 7% or more of shortening and sugar per 100 kg flour. In this category, certain types of bread are also included, e.g. Donezker bread, made from top-grade flour, and crispbread, e.g. Leningrad crispbread. The various rolls and smaller units, of weight 50, 100 and 200 g are divided into plain '*Ljubitelskij*' products, and '*Wyborger*', which are shaped, both being made from top-grade flour. The leanest of these products contains 10% sugar, 7% butter, and 3·6 kg egg per 100 kg flour. The *Ljubitelskij* contain 17% sugar, 13% butter, 8·8 kg egg and 0·042% vanilla per 100 kg flour. *Wyborger* are more highly enriched, containing 20–25% sugar, 2% syrup, 7–10% butter, 4 kg egg, 12% fruit-conserve, 0·05% vanilla per 100 kg flour, being sugar-glazed. The various Danish-type pastries also fall into this group, which are layered with butter.

### Ring-shaped, Low-moisture Products
These products are listed in Group 11 of Table 35, and include such products as: *Ssuschki, baranki* and *bubliki*.

   *Ssuschki* are small thin rings of 6–12 g each, with moisture contents of not more than 9–13%.

   *Baranki* are somewhat larger and thicker, being made to weigh 25–40 g each, having a slightly higher moisture content of 14–19%.

   *Bubliki* are even larger units weighing 50–100 g each, shaped thicker, the moisture content range being 22–27%.

   *Baranki* and *Ssuschki*, owing to their low moisture, have a long shelf-life.

   *Bubliki*, owing to their higher moisture content, are eaten freshly baked. All the described products are prepared from a firm dough of average moisture 33–37%. These ring-shaped products are prepared either with a perpetual sour, or using a sponge with yeast. Owing to the low dough moisture content, the production process is divided into two main operations:

(1) Kneading in a special kneader, specially constructed for firm doughs.
(2) Mechanical development by means of a fluted roller, in order to ensure a homogeneous dough structure and texture.

After mixing, the dough is given a period of recovery and fermentation of between 30 and 60 minutes. The dough then passes through a special type of divider–moulder, the dough-pieces emerging ready-shaped. The All-Union Research Institute for the Baking Industry, WNIIChP, has developed a series of divider–moulder–shaper machines especially designed for *ssuschki, baranki* and *bubliki*, and also universal units with interchangeable working elements, allowing the processing of all three types of ring-shaped products. A well-known machine for the shaping of the dough-pieces for *bubliki*, the BM-2, was developed by F. P. Karpenko. The shaped pieces receive the required resting-time (30–90 minutes in the case of machine-shaping) for final proof. The pieces are then boiled in boiling water with added syrup or sugar to improve oven-browning later. Alternatively, the boiling-process can be achieved in an autoclave at a pressure of 40–50 kN/m² (0·4–0·5 atm). The duration of the cooking-process varies from 0·5 to 3·0 minutes, depending on the product-size. As a result of cooking, the volume of the ring-shaped products increases greatly, their weight remaining relatively the same; therefore the cooked pieces rise to the surface. The centre of the dough-pieces reaches 60°C, and the surface about 70°C. The cooking results in the gelatinization of the starch, and a denaturation of the protein, which proceeds at a more rapid rate on the dough surface-layer. This causes the surface of the product to acquire a shine on baking. After cooking the dough-pieces are dried by blowing a current of hot-air over them. The baking process is done in a wire-band tunnel oven of the type BK, the baking-time, depending on the type and size of the product, is approximately 10–20 minutes. The baking process for these products is quite different from that for bread, owing to their rapid temperature increase; moisture, even in the centre of the product, rapidly relocates in the form of steam to the surface. The temperature of the central layers at the end of the baking process reaches, 104°C to 106°C for *bubliki*, 107–108°C for *baranki*, and 110–112°C for *ssuschki*. The actual process is a combination of baking and drying. Radiation-heat is important for this product to achieve a rapid drying-speed of 1·6–1·8% per minute, which improves its texture. Using a wire-band oven, the shape, texture and surface colour are enhanced by setting the top-heat at 300–350°C for 1–2 minutes initially, and the bottom-heat at 250–280°C. The optimal baking regime for these three products has been researched in the Moscow Technological Institute of the Food Industry by Auerman, Ginsburg, Kulikow, and Lylowa, *MTIPP*, **3**, 26, 1954. The separation of glutathione from baker's yeast also improves the appearance of these products where used, especially *bubliki*, and decreases dough adhesion to baking surfaces whether steel or fluoroplastic, Karpenko (1985). Sweet *baranki* and the leaner *ssuschki* are made with 2nd grade wheat-flour. Grade 1 wheat-flour is used for *bubliki* (Ukrainian *bubliki, bubliki* with poppy-seed, caraway or sesame); milk *bubliki, baranki*, both lean and sweet; *baranki* with mustard-oil, and *ssuschki* both with and without salt. From top-grade wheat-flour, the more enriched *baranki* containing poppy-seed, lemon essence, vanilla, and *ssuschki* with poppy-seed, lemon essence, vanilla and mustard-oil, are manufactured. A detailed description of the machinery used to manufacture these ring-shaped products on a factory scale can be found in the following original literature: *Technological Equipment for Bread Factories*, Moscow, Publisher: Pistschewaja promyschlennost, Author: Saizew, N. W., 1967 (in Russian).

It should be emphasized that the manufacture of these ring-shaped products, known in Germany as '*Kringeln*', would be difficult by hand and very time-consuming for the general craftsman-baker. In the USSR, the development of complex mechanized and automated production-lines for *Kringeln* by Soviet construction engineers, led to a technical revolution, in this, until comparatively recently, backward branch of the baking industry.

## Rusks

Rusks are listed under Group 13 of Table 35. These products in many cases have moisture contents as low as 8–12%, and are therefore long-shelf-life products. The finest quality rusks are made from wheat-flour, either 1st or 2nd grade, and, in the case of infant-rusks, top-grade flour is always used.

Formulations for quality rusks contain sugar and shortening, in the form of butter or margarine, in various concentrations, and also, in some cases, eggs, almonds and vanilla. In formulations with the highest levels of sugar and shortenings, the levels of yeast are also the highest. In the case of manual methods of production, for each type of rusk the formulation, shape and weight is specified, and for full details the reader is referred to the following literature:

—*The Production of Bakery-products*—a technological handbook by I. M. Roiter, Publishers: Technika, Kiev, 1966 (in Russian).
—*The Production of Fine Rusks* by Pawperowa, N. A., Moscow, Pistschepromis-dat, 1954 (in Russian).

Infant-rusks are made up of 200–230 pieces/kg, and the city and traveller-rusks of 40–50 very large pieces.

Rusk doughs are prepared from a sponge and dough process, and the sugar and shortening must be added in a finely divided state. After full-proof, the dough is shaped into strips of the desired shape and size required. These strips must have small, even, thin-walled cells, this necessitates adherence to the following procedure:

(1) Dividing of the doughs into small pieces.
(2) Long-rolling of these pieces into 'sausage-shapes', the length being in proportion to the width.
(3) Placement of the sausage-shapes onto trays in rows, close to one another.
(4) Final shaping of the sausage-shapes by vertical pressure into the desired external profile.

In the large-scale production plants, as in Moscow Bread Factory No. 2, a machine has been developed which presses or extrudes the dough through dies of the required profiles, the shaped strips being placed lengthwise onto sheets. However, the porosity and texture of the baked strips, is inferior to that obtained by hand-moulding of the sausage-shapes. WNIIChP (the All-Union Research Institute) has developed improved machines in recent years for preparation of the sausage-shapes for baking.

The final-proof of these sausage-shapes is done on sheets in a proofer or travelling-proofer at 35–40°C and RH 75–85% for 40–120 minutes, depending on the formulation used and flour quality. The dough-pieces are washed with eggwash, and sprinkled with crumbs in some cases. Baking takes place in various types of oven at 120–250°C for 7–20 minutes, depending on the formulation and the dimensions of the product. The injection of steam during baking is not appropriate. The baked pieces are cooled for about 20 minutes on the sheets, and each piece is turned over after about 7 minutes has elapsed.

The cooled baked strips are then placed on special wooden shelves and stored for 8–24 hours. The storage-room must be dry and well-ventilated, the temperature being 12–15°C and RH 65–70%. The use of optimal storage regimes is important, in order to ensure that the products are in the optimum condition for slicing. Special slicing machines are used for cutting the pieces to the right thickness, according to the type of end-product. Certain types of rusk are egg-washed after slicing and dressed with sugar, poppy-seed, grated-nuts, or blends of each.

In all cases, the final production phase is the drying and roasting, which is done in an oven at temperatures between 165 and 220°C for 12–35 minutes, again dependent on the formulation and the product size. The products must be dried down to the required moisture of 8–12%. The roasting should give a brown colour to the surface of the edges, due to radiated heat from the oven-surfaces, which cannot be obtained from the warm-air convection currents of drying chambers. The finished rusks are cooled for 2–3 hours, sorted and packed in moisture-proof material.

*Bread-rusks*
Bread-rusks are prepared by drying conventional rye, mixed wheat/rye and wheat breads after slicing down to a moisture of 10%, thus producing products with prolonged shelf-life. The great advantage is that such products can be used as food under any temperature conditions. For example, at temperatures of −20 to −30°C and under permafrost conditions, conventional bread freezes, whereas bread-rusk remains edible.

For the manufacture of bread-rusk, bread of the following types can be used: whole-rye bread, mixed whole-rye/wholewheat, and wholewheat. However, rye or mixed wheat/rye are usually used as the starting bread. The bread must have a total moisture within the range 42–44%, corresponding to a crumb-moisture of 47–50%. In this context it must be made clear that in drying technology it is customary to express the moisture of the product to be dried as a percentage of the weight of dry matter which the product contains. This moisture figure is then termed the 'absolute moisture' (Wa). The absolute moisture of the bread-rusk must be between 72% and 78·5%, and in all other aspects of quality, the bread-rusk must conform to the normal standard of the respective bread from which it is dried. Of particular importance is the crumb structure and its even pore-size. Usually bread intended for bread-rusks is baked in pans weighing 1·5 to 2 kg per unit.

It has been established by L. J. Auerman and co-workers at the Moscow Technological Institute for the Food Industry, first in 1936, and later in research work carried out in 1956 and 1958, that direct electro-contact baking is appropriate for the drying of bread-rusks. This produces bread without crust, better porosity and even texture, by generating heat within the product, which is induced by electromagnetic-wave frequencies of chosen magnitude. This effects a rapid drying and more even shrinkage in the cross-section. Bread for the manufacture of bread-rusks is stored for 12–18 hours after baking at least, often up to 24 hours, before slicing. The storage-period is critical for attainment of the optimal crumb properties, thus reducing excessive waste. However, the storage-period can be shortened somewhat by storage temperature control, and use of a slicing-machine of improved action and blade efficiency. Later, in 1973, H.-D. Tscheuschner and co-workers at the TU Dresden, GDR, developed a slicing process whereby fresh bread can be sliced without storage, the resulting slices being of good quality and free from permanent deformation.

The bread is cut to a thickness of 20–25 mm, 22 mm being ideal. This allows a rapid drying of the slices, and results in bread-rusks which are firm and undistorted. The slices are then placed vertically in metal cassettes and dried in a tunnel-drier, which removes 350–400 kg water from every 1 tonne rusk. For this process, conventionally, convection is used; important parameters are temperature, air RH, and air speed relative to the product to be dried. In order to dry the slices down to a moisture level of 10%, at an air speed of 4 m/s and RH 5%, a drying-time of 20 hours at 40°C is required, 10 hours at 60°C, 4 hours at 90°C, and only 2 hours at 130°C. Higher temperatures result in undesirable browning-reactions and a fall in general quality of the rusks. For further details concerning the thermodynamics of the drying process, the reader is referred to the standard reference book in the Russian language, *Technology of Breadmaking*, by L. J. Auerman, Moscow, 1977. Further contributions to the technology of the process are described by Gupta, N. K. and Tscheuschner, H.-D. in *Bäcker und Konditor*, (24)11, 1974, p. 330.

The convection type of tunnel-drier used for drying rusks works on the principle of a counter-current system, air being used as the heat conductor, which is heated by means of a system of pipes. Dry air is passed over the material to be dried, and part of the exhausted air is recirculated.

The process begins with the drying of the bread-slices with an air temperature of 70–80°C and RH 25–35%; whereas the actual drying of the rusks, which represents the second phase, is at 120–130°C and RH 5–10%, at an air speed of 3·5 m/s, which completes the process within 7–8 hours. Detailed studies of this technology with experimental results can be found in more specialized sources, i.e.

—*The Production of Rusks*, by Auerman, L. J., Moscow, Pistschepromisdat, 1943.
—*The Drying of Food Products*, by Ginsburg, A. S., Moscow, Pistschepromisdat, 1960 (in Russian).
—*The Drying of Rye and Rye/Wheat Rusks*, by Maslow, I. N., Moscow, 1942 (in Russian).

The dried rusks are cooled by wheeling the racks with cassettes into a cooling-room. They are then tipped onto a sorting-table where they are checked for quality before packaging. Important quality parameters during sorting are: burnt and cracked slices, misshapen and slices not conforming to standard dimensions, partially dried slices, also the amount of large and small broken-pieces occurring, and how these relate to fixed standards for production. Laboratory tests include, moisture, acidity-index, and other sensory procedures designed for rusk quality assessment.

The rusks are then packed in multi-layered strong paper bags, in tight rows to avoid damage. The packaged rusks must be stored in a clean, dry insulated room with temperature and RH control to avoid changes in weight on prolonged storage. The equilibrium moisture content (ERH) of rye-rusks will depend on the relative humidity of the surrounding atmosphere. The hygroscopic moisture of rye-rusks at 100% relative humidity is about 25% (*Technology of Breadmaking*, L. J. Auerman, Moscow, 1977, p. 386–390).

*Additional Types of Bread and Small Products in USSR*
Apart from the bread varieties described, which are listed in the USSR Bureaux of Statistics in Table 35, in the Soviet Baking Industry there also exists a further collection of two recipe and product groups, viz. regional (ethnic) bread and small products, and health and dietary products.

The regional ethnic products produced in the southeastern and southern parts of the USSR (which include: Armenian SSR (capital: Yerevan), Azerbaijan SSR (Baku), Kazakh (Alma-Ata), Kirghiz SSR (Frunze), Tajik SSR (Dushambe), Turkmen SSR (Ashkhabad), Uzbek SSR (Tashkent)) are prepared mainly from wheat-flour, top-grade, 1st grade or 2nd grade, and are known as '*lepjoschka*' and '*Lawasch*' (thin, ring-shaped dough-pieces, baking-off into air-pockets), '*Sangjak*' (thin ring-shaped dough-pieces, baked on a hearth of round stones), and Armenian '*dogik*', similar to *lepjoschka*, with a hole in the centre. Some of these products are sold by weight, most are sold as loaves of various dimensions using diverse recipes. These recipes and how to produce the products can be found in the following literature:

  —*Collection of Recipes for National Small Baked-products*, by Iljinych, K. E., Faradshew, M. R., Saakjan, R. W., Cheniaschwili, R. I., Moissejenko, T. T. and Babajew, J. G., Moscow, Ekonomika, 1967.
  —*Uzbek Bread—Technology and Recipes*, by Machkamow, G. M., Pogosjanz, A. I. and Swinkin, S. N., Tashkent, 1961.

Most of the recipes for these products contain: flour, salt and hops to prepare a liquid hop-yeast or sour. The Uzbekian *lepjoschka*, termed '*schirmai*', and Kazakian '*Damda-nan*' contain sugar, shortening, and often mutton-tallow. The basis of the recipes is, liquid hop-yeast or sour and some ripened intermediate product from the last batch. For special types of *lepjoschka* a multi-stage sour is made from ground-peas, aniseed, onion-pieces, meat-bouillon or sour-milk.

The preparation of these doughs with home-made sours is an old tradition, as in many other surviving cultures, especially in remote country areas. Industrial bakeries in these regions use liquid yeast for the manufacture of ethnic-bread varieties. The remaining small country bakeries still use their prototype ovens, similar to those used by their ancestors of 3000 years ago, to bake the *lepjoschka* and *lawasch* types of bread. The dough-pieces are stuck to the sides and crown of the oven, and baking takes place during the burning of the fuel within the oven-chamber. For industrial production newer versions of the old ovens have been designed and built, and projects for specialized modern mechanized production of these products worked out (Agababjan, R. J., Nikoladse, W. I., Gamsachurdija, R. R., in *Chlebopekarnaja i Konditerskaja Promyschlennost*, **12**, 7, 1968).

Such bread varieties as *lepjoschka, lawasch, tschurek, sangjak* and variations of them, have been in existence for many centuries, forming part of the traditional bread-culture in the respective republics of the Soviet Union. Similarly, other republics have their own national identity in bread-culture, viz. Byelorussian SSR (Minsk), Latvian SSR (Riga), Lithuanian SSR (Vilnius), Moldavian SSR (Kishinev), Estonian SSR (Tallinn) and the RSFSR (Moscow). In latter years the bread varieties with locational identity have increased strongly.

Health and dietary breads are placed in a category of their own, depending on the patients for which they are intended: protein-rich breads for diabetics; salt-free and chloride-free for kidney-sufferers; gluten-free for coeliac-disease; iodated-bread for thyroid deficiency; bread with added calcium and bran with reduced or increased energy-values; bread with increased protein, mineral and vitamin contents; bread with reduced acid content, etc. In addition there are a series of health and diet breads for children, the recipes for which are to be found in the following publications:

—*Dietary Bread and Baked-products*, by Maslow, I. N. and Nikolajew, B. A., Moscow, Rosgismestprom, 1950 (in Russian).

—*New Varieties of Small Baked Products with Increased Nutritive Value and Medical–Dietetic Properties*, by Patt, W. A and Stscherbatenko, W. W. Moscow, ZINTIpistscheprom, 1964 (in Russian).

—*The Production of Bakery Products—a Technological Handbook*, by Roiter, I. M., Kiev, Technika, 1966 (in Russian).

Current trends in consumption in the USSR have shown a fall in the production of panned bread in favour of bread baked on the hearth or oven-bottom. Bread sold in the form of whole loaves has increased at the expense of those sold by weight in the form of parts of larger units. The increased production of small, yeasted-dough products, i.e. rolls, rusks and the ring-shaped low-moisture products, known in German as '*Kringeln*'. The number of national-ethnic varieties produced in the various Socialist Soviet Republics (SSR) have increased yearly. However, in the interests of quality control and rationalization of production, some form of regulation to ensure increased mechanization and automation of the manufacture of these products is essential. The nutritional contribution of bread to the national (USSR) diet is being increased by the development of specialty-breads with improved digestive properties and sources of protein, carbohydrate, minerals and vitamins. These are similar to those described in the appropriate section concerning the GDR.

Within the Socialist community, the aim of the Food Industry, and its duty under the framework of the Constitution, is to provide all its citizens with bread of the highest possible nutritional and sensory quality. This dictates production of bread of controlled moisture content, which means the correct balance between bread-crumb and bread-crust, the former having significantly higher moisture content. Any reduction in the crust-proportion of bread reduces its energy-value in kJ/100 g of product. A wholemeal loaf baked as panned bread of average crumb moisture content of 46%, has an average energy-value of 908 kJ/100 g product, whereas the same weight and type of bread baked as a hearth or oven-bottom loaf of average crumb moisture 43% would show an average energy value of 955 kJ/100 g product. This difference in energy is due to a cumulative increase in protein, carbohydrate and fat energy in the case of the latter.

Any determination of gross energy content of a food product demands a knowledge of its content of protein, carbohydrate and fat, and their maximal energy content. For this reason, the Ministry of Health and Food of the USSR has published tables showing the chemical components and nutritive value of foods. For this purpose, gross energy-values for protein and carbohydrate of 17·1 kJ, and 38·9 kJ for fat per gram are taken as the basis. Cellulose-material contents of limited digestibility are not considered for this purpose. The chemical composition and energy content of the main bread varieties are detailed in Table 36.

Table 36 shows clearly that the gross energy content of any bread or baked product increases with decreasing water content, and increases with the amount of shortening added in the recipe. In order to calculate the energy content of bread and baked products it is necessary to consider their water content and total chemical composition. Most importantly, as far as water (moisture) content is concerned, the crust as well as the crumb must be taken into consideration. Bread-crust has much less moisture content than bread-crumb, therefore any reduction of the proportion of crust will reduce the gross energy content of the bread. Hence it is of paramount importance that the baking process is strictly controlled as regards temperature/ baking-time for the whole baking cycle to deliver the optimal crust/crumb balance and baked bread water content. These considerations are in sharp contrast to those practised in some capitalist countries, where some milling/baking corporations which manufacture sliced-bread in the panned form expend much technological ingenuity in finding ways of maximizing crumb moisture to within 45–48%. In some countries there are legal maximum limits for bread-moisture levels, e.g. USA, Australia and New Zealand. In the UK, there is no legal standard for bread-moisture, therefore bread energy-contents will vary, and are generally lower than in those countries where strict control of the baking-cycle is exercised, e.g. USSR and GDR. In order to determine the real energy-content of bread and bakery products, the digestibility coefficients of protein, carbohydrate and fat for the product must be known. It has been established in the USSR and other countries, that the human organism fed on a mixed diet can digest 92–98% of bread or baked product carbohydrate, 85–93% of the fat, and only 70–85% of the protein. The digestibility of carbohydrate, fat and especially protein depends on the flour extraction-rate. The higher the extraction-rate the lower the digestibility of protein, fat and

## TABLE 36
**Chemical composition and energy content of various bread and baked products in the USSR**

| Product | Flour | Water (%) | Protein (%) | Fat (%) | Carbo-hydrate (%) | Cellu-lose (%) | Salts (%) | Gross (kJ/100 g product) |
|---|---|---|---|---|---|---|---|---|
| Panned rye bread | Whole-meal | 45·5 | 5·9 | 1·1 | 44·5 | 1·0 | 2·0 | 908·36 |
| Hearth rye bread | Whole-meal | 43·1 | 6·2 | 1·3 | 46·3 | 1·0 | 2·0 | 954·40 |
| Hearth rye bread | Light-grade | 42·0 | 6·2 | 0·8 | 49·0 | 0·5 | 1·6 | 975·33 |
| Ukrainian hearth wheat bread | Whole-meal | 40·2 | 7·8 | 1·3 | 47·5 | 1·0 | 2·2 | 1 000·45 |
| Panned wheat bread | Whole-meal | 43·1 | 7·0 | 1·6 | 45·1 | 1·2 | 2·0 | 954·40 |
| Ordinary white bread | 2nd grade wheat | 35·8 | 9·0 | 1·3 | 51·4 | 0·7 | 1·8 | 1 088·36 |
| White rolls | 1st grade wheat | 31·7 | 9·4 | 2·0 | 55·1 | 0·2 | 1·6 | 1 184·63 |
| *Wyborger* | Top-grade wheat | 35·0 | 7·6 | 5·5 | 50·7 | 0·2 | 1·0 | 1 213·94 |
| Rye-rusks | Whole-meal | 11·0 | 11·4 | 1·4 | 70·6 | 1·9 | 3·7 | 1 460·91 |
| Full-milk rusks | Top-grade wheat | 11·0 | 9·4 | 5·5 | 72·3 | 0·2 | 1·5 | 1 619·98 |
| Ordinary rings (*Kringeln*) | 1st grade wheat | 17·0 | 10·6 | 1·2 | 69·0 | 0·2 | 2·0 | 1 414·86 |

Source: *Technology of Breadmaking*, L. J. Auerman, Moscow, 1977, p. 397, Table 50.

## TABLE 37
**Digestibility coefficients for the components of USSR flour types dependent on flour extraction-rate**

| Bread and baked-product from | Digestibility coefficient for | | |
|---|---|---|---|
| | Protein | Fat | Carbohydrate |
| Wholemeal flour | 0·70 | 0·92 | 0·94 |
| *Wheat-flour* | | | |
| 2nd grade | 0·75 | 0·92 | 0·95 |
| 1st grade | 0·85 | 0·93 | 0·96 |
| Top-grade | 0·87 | 0·95 | 0·98 |
| *Rye-flour* | | | |
| Medium-grade | 0·75 | 0·92 | 0·95 |
| Light-grade | 0·80 | 0·93 | 0·96 |

Source: *Technology of Breadmaking*, L. J. Auerman, Moscow, 1977, p. 398 Table 51 (in Russian).

TABLE 38
**Real energy-content of various baked-products in the USSR**

| Product description | Flour type | Real energy-content (kJ/100 g product) |
|---|---|---|
| Panned bread | Whole-rye flour | 828·82 |
| Hearth-bread | Whole-rye flour | 870·68 |
| Hearth-bread | Light-rye flour | 920·92 |
| Panned bread | Wholewheat flour | 916·73 |
| Bread-rolls | 1st grade wheat-flour | 1 117·66 |
| *Wyborger* | Top grade wheat-flour | 1 079·98 |
| Full-milk rusks | Top grade wheat-flour | 1 561·37 |
| Ordinary rings (*Kringeln*) | 1st grade wheat-flour | 1 335·33 |

Source: *Technology of Breadmaking*, L. J. Auerman, Moscow, 1977, p. 389 Table 52 (in Russian).

carbohydrate, therefore the digestibility coefficients of these components also varies with the flour extraction-rate. With a knowledge of the chemical composition of each type of bread or baked product (Table 36), and the digestibility coefficients (Table 37), the real energy-contents can be calculated. Table 38 shows the real energy-contents of various bread and baked products in the USSR. The highest real energy-contents are from products made from low extraction flours of low moisture (crumb and crust) and high fat content, i.e. rusks, rings (*Kringeln*) made from top-grade wheat-flour or 1st grade wheat-flour. Full-milk rusks made from top-grade wheat-flour, and a crumb moisture of 11%, have a real energy-content of 1561·3 kJ/100 g product.

For any meaningful determination of real energy-content or real energy values of foods, the following information is required:

(1) Complete chemical composition of the food, including: moisture %, protein %, carbohydrate %, fat %, cellulose %, mineral salts %.
(2) In the case of bread and rolls, the ratio of crust to crumb is important, since the crust has much less moisture than the crumb. Therefore, any reduction of crust will reduce the real energy value of the product.
(3) The digestibility coefficients for the flours used to produce the bread and rolls in terms of protein, fat and carbohydrate, as detailed in Table 37.

It is clear from Table 38, that the real energy values increase considerably in the case of products with a high proportion of crust *v*. crumb, and lower moisture content.

In quoting the final real energy values of products, it is usual to round up the values in kJ to the nearest unit, since energy values cannot be obtained with such precision as two decimal digits would indicate.

**Dough-processing in the USSR**
Traditionally, in USSR bakeries wheat-flour doughs are made from the various flour types already described using the indirect method with pressed or liquid yeast or a

mixture of both. The indirect process takes from 4 to 6 hours, and requires a large number of machine-bowls for the sponges and doughs, as well as a large amount of space in the dough-preparation department. Transport of the machine-bowls is a manual operation, unless the bread-factory has the more highly mechanized systems, e.g. that of G. P. Marsakow, or other patent.

The Marsakow system was developed in 1927, and involves a full mechanization, whereby the machine-bowls are arranged on a carousel. The two-stage discontinuous (batch) process of sponge and dough, including transport of the machine-bowls, being fully mechanized. For a fuller description of this process, the reader is referred to the following:

—*Machines and Equipment for Dough Preparation*, by Goroschenko, M. K., Moscow, Pistschepromisdat, 1963 (in Russian).
—*Technical Equipment for Bread-factories*, by Saizew, N. W., Moscow, Pistschewaja promyschlennost, 1967 (in Russian).

Machinery according to the Marsakow system is still working in several Moscow and Leningrad bread-factories. In 1932 A. A. Pawperow invented another fully mechanized set-up, consisting of machine-bowls mounted on a conveyor covering four levels, which was developed originally as a mobile field-bakery, for the two-stage discontinuous dough production (*Machines and Equipment for Dough Preparation*, by Goroschenko, M. K., Moscow, Pistschepromisdat, 1963 (in Russian)).

A further development by A. A. Pawperow and B. I. Tschernjakow using a closed conveyor to transport the sponges and doughs mechanically was installed in the Moscow bread-factory No. 12 (*Handbook for the Bakery Technologist*, 2nd edn, by Micheljew, A. A., Kiev, Technika, 1966 (in Russian).

Other attempts at mechanizing the doughmaking process could not solve the problem of batch preparation and fermentation in machine-bowls. However, the system devised by W. F. Gatilin did eliminate the use of more than one machine-bowl, and also the transport-problem, but it remained a batch-process. Nevertheless, the so-called bunker doughmaking-aggregate developed by Gatilin found wide application in the bread-factories of the USSR from about 1946. It was developed to use in combination with the AZCh oven for the daily production of 100 t of rye-bread from a firm sour-dough. However, it was later applied in the Sverdlovsk Bread-factory for the preparation of wholewheat bread-doughs from a firm wholewheat sour-dough, and was also adapted to produce doughs from 2nd grade wheat-flour by using a sponge and dough system. The design of the machine-aggregates is such that the handling of an intermediate product, i.e. sour-dough or sponge is assumed. For use in bakeries with less oven-capacity, a smaller bunker-aggregate was developed, the BAG-20, for 20 tonnes/day of wheat bread by the indirect sponge method, and 30 tonnes/day of rye bread by the firm sour-dough procedure.

This type of bunker-aggregate completely eliminates the manual transport of machine-bowls with the intermediate products (sours and sponges) and the final doughs, and requires less floor-area, although the production remains a batch system. This bunker-aggregate concept has also been coupled with a continuous mixing set-up, thus allowing reduced fermentation times before work-off. The KVT 1500 described in the section dealing with the GDR is based on the same principle. The original basic principle of the Gatelin batch process is shown in Fig. 22.

The rye sour or wheat sponge is mixed in the mixer (1), which can be turned to coincide with any of the five compartments of the bunker (2). The mixed intermediate product is deposited into one of the five compartments of the bunker through the outlet-vent in the base of the mixing-bowl. After one compartment is

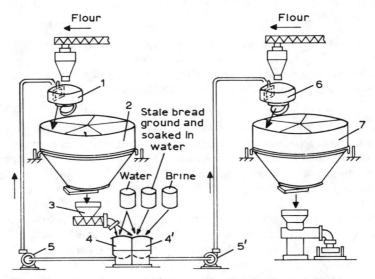

Fig. 22.    Bunker system of doughmaking (N. F. Gatilin). 1. Batch mixer with movable bowls with discharge vent for wheat-flour sponges or rye-flour sours. 2. Rotating bunker with five sections for sponges/sours. 3. Dosage hopper for sponges/sours. 4. Mixers. 5. Pump. 6. Machine bowl. 7. Bunker. (Source: *Technology of Breadmaking*, L. J. Auerman, p. 193, Moscow (in Russian), 1977.)

filled, the machine-bowl is rotated to coincide with the other compartments in turn, each being filled with the appropriate intermediate product on completion of mixing. On completion of these operations, the first intermediate product is completely fermented, and can be discharged into the hopper and worm-feed (3), from whence it is fed to the two adjacent mixers (4) and (4′). In the case of rye doughs, one part is deposited into mixer (4), and the rest into mixer (4′). The sour in mixer (4) is diluted with water, and then pumped by pump (5) up into the machine-bowl (1). The main part of the sour in mixer (4′) is diluted with water, and the soaked stale bread and salt solution also diluted and pumped into the machine-bowl (6) by pump (5′).

In the case of wheat doughs, the sponge is deposited into mixer (4′) and after mixing with water and salt solution, it is pumped by pump (5′) into the mixing-bowl of mixer (6). The compartments of bunker (7) are then successively filled with the batch mixed doughs. The finished dough being discharged either into the appropriate compartment of the bunker (7), or directly into the hopper and worm-feed and thence into the divider. (Diagram and information source: *Technology of Breadmaking*, L. J. Auerman, Moscow, 1977, pp. 193–194 (in Russian).)

The scheme for the continuous bunker-aggregate used to produce wheat doughs, as installed in the Moscow Bakery No. 12, has an output of 30 t of city long white loaves per day. The doughs are prepared from one large sponge, which contains about 70% of the total flour, and two-thirds of the total water. The resting-time for the sponge is 4·5–5 hours at 28°C, and that for the dough 20–25 minutes in a stationary bunker set at 32°C before the dividing operation. The control of the aggregate is automatic. Another aggregate, the L4-ChAG, is based on a similar principle, having universal application for both wheat and rye doughs with an output of about 15 t/day.

Apart from dough-preparation aggregates with a rotating bunker, there are also aggregates with stationary bunkers. For example, in 1958, the Suchumier Bread-factory No. 2 installed an aggregate designed by W. I. Dshalagani. This aggregate was designed for the production of wheat-flour doughs from 1st grade and 2nd grade wheat-flours using a liquid brew and liquid yeast. The dough is fermented in a vertical rectangular container, remaining stationary, inside which are fitted gratings at various heights to prevent the dough from rising excessively as it expands. A detailed description of the Dshalagani system can be found in the following literature:

—*Working Experiences in the Abchaskier Bakery*, Moscow, ZINTIpistscheprom, 1965.
—*Technological Equipment for Bread-factories*, by Saizew, N. W., Moscow, Pistschewaja promyschlennost, 1967 (in Russian).

In 1959 a new doughmaking-aggregate designed by A. M. Chrenow was installed in the Moscow Bread-factory No. 10 for the manufacture of firm rye sours and doughs. In this aggregate, fermentation takes place in a rectangular bunker, which is slightly inclined to the horizontal. The bunker is fitted with baffle-plates on its base and cover in such a manner that they do not quite reach the opposite side of the bunker. It is thus divided into a series of sections into which the sours or doughs are successively introduced. Regulatory sliding of the plates of the sections allow the fermentation times of the sours and doughs to be controlled. The Chrenow system has also been modified to manufacture wheat doughs on a two-stage sponge and dough process, and in the Moscow Bread-factory No. 3, it has been adapted to produce wheat doughs by using liquid-brews. Details of these innovations can be found in the following literature:

—*Handbook for Bakery Technologists*, 2nd edn, Kiev, Technika, 1966 (in Russian) by Micheljew, A. A.
—*Chlebopekarnaja i konditorskaja promyschlennost* (Journal), **10**, **11** & **19**, 1966 (in Russian) by Kogan, M. A.

In 1946 I. L. Rabinowitsch designed a continuous system of doughmaking, which was built in 1949 for the continuous one-stage manufacture of wheat bread, this aggregate was known as the ChTR. The standard additions for this process were: 1·5% pressed-yeast and 5–10% liquid-yeast, or 50–60% liquid-yeast. The fermentation time was 4·5–5 hours at 27–29°C.

Experience with the ChTR using the one-stage direct method, however, produced bread of inferior quality, especially in respect of flavour and aroma, when compared with bread made by the traditional indirect sponge and dough procedure. In the USSR, in common with most continental European countries, volume and external appearance is not enough to make bread acceptable; the flavour and eating experience being of paramount importance. Chorleywood bread is judged as unacceptable on flavour and textural grounds by Soviet bakery experts. It is also of relevance to note here that many North American bakers of experience consider that sponge doughs are superior to any continuously manufactured doughs, whether made with zero or 50% flour in the pre-ferment. It is also the author's own experience that a fermentation of the flour either in the form of a firm sponge or in bulk as a straight dough remains difficult to replace where the overriding quality parameters are flavour and firmness of crumb. Until technology can simulate the flavour-token of ripened, fermented dough, in addition to shortening the processing-time, the various short-time processes, whether mechanical or chemically designed, will remain a compromise in terms of bread quality. Crumb-structure is also dependent on an adequate gluten quantity and quality profile, which no amount of mechanical or chemical development for a short time appears to successfully transform into quality bread. In view of these negative influences on quality experienced with the ChTR and direct processed doughs, the All-Union Research Institute for the Baking Industry in Moscow constructed a series of modifications which could be used for the direct processing of doughs. As an example, the ChTR System of I. L. Rabinowitsch, which was mass-produced for the manufacture of wheat-flour doughs by the indirect sponge-dough procedure, a description of which follows (refer to Fig. 23).

The aggregate is divided into two sections, section A being designed for the fermentation of the sponge, and section B for the dough. The sponges are mixed in the mixer (1) and pass into section A of the round-shaped fermentation-channel, which is separated from section B by a holed-partition (2). The B-section is for the dough fermentation. A short, removable partition (3) is also provided, and the prongs (4) fixed to the central shaft are for the mixing of the fermented sponges. The movement of the sponges in section A is provided for by the slope of the fermentation-channel, assisted by the propeller-like blades mounted on the central shaft at the inlet of section A (5). The ripened sponge is then pumped by means of a dosage-worm pump (6) through the pipeline (7) into the dough-mixer (8). The mixed dough then passes for fermentation into section B, where it is moved by propeller-like blades (9), also mounted on the central shaft. On completion of ripening, the dough is discharged from section B through an outlet positioned at the lowest point at the extreme end of the fermentation-channel. The size of the aperture can be controlled by the movement of a hand-operated wheel (10).

The aggregate ChTR is designed to produce 25 t of small roll-type products per day by the indirect dough procedure. Other variations of the aggregate ChTR are

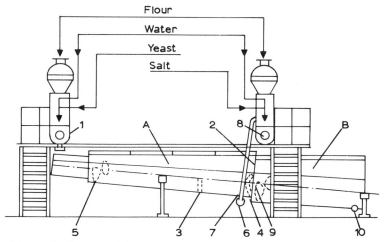

Fig. 23. Schematic diagram of doughmaking unit ChTR with an indirect sponge process: section A for sponge fermentation; section B for dough fermentation. (Source: *Technology of Breadmaking*, L. J. Auerman, Moscow, p. 197 (in Russian), 1977.)

also in use. The complete equipment for the ChTR aggregate includes the Ch-12, a dough-mixer unit which functions continuously. Using these mixers, production people became aware that the specific work intensity for wheat doughs of moisture level 42% (based on total dough weight) was not intensive enough at only 6·8 J/g. Therefore, as a result of collaboration between the Odessa Institute for Food Machinery Automation, production people, and other specialists in research, construction and design engineers, construction modifications were undertaken to increase the specific work intensity of the Ch-12. In the case of wheat dough processing it was recognized that the intensity of the mechanical work-input at the mixing stage was critical for the reduction of the necessary fermentation time from the moment of mixing to processing or work-off, always within the constraints of bread quality being acceptable in terms of texture and flavour.

With the above considerations in mind, W. M. Dontschenko developed a continuous dough-mixing aggregate, known as the ChTU-D. This was made up of the above-mentioned Ch-12 mixer unit, beneath which is mounted a worm, the dough from the Ch-12 outlet feeding directly into the worm, which serves the dual function of conveyor and dough mechanical developer. Thence the dough passes through a pipeline either directly into the hopper of the divider, or into a dough-bunker/or machine-bowl of appropriate capacity. If desired the dough can be conveyed instead to another unit for the continuous fermentation, the ChTU-D combination of the Ch-12 with a worm-developer finding wide application in bread-factories in the USSR. A schematic representation of the ChTU-D is presented in Fig. 24.

In the USSR, special equipment is used for the preparation and fermentation of liquid intermediate products. This equipment consists of stationary containers fitted with water-jackets, mixer-elements, and instrumentation, e.g. thermometers. Such

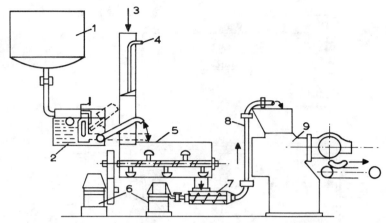

Fig. 24. Doughmaking unit ChTU-D. 1. Liquid pre-ferment. 2. Dosage unit for pre-ferment. 3. Flour-feed. 4. Flour measuring unit. 5. Mixing aggregate Ch-12. 6. Motors. 7. Worm-feed pump. 8. Pipeline. 9. Dough-divider. (Source: *Technology of Breadmaking*, L. J. Auerman, Moscow, p. 198 (in Russian), 1977.)

equipment can be used either for discontinuous (batch) preparation of intermediate products or for continuous processing, in which case additional containers are necessary to provide a series of containers of the product. A typical example is the ChSM-300, which is also used to prepare a gelatinized flour/water suspension. This heated flour/water suspension, owing to the resultant conversion of starch to sugars and proteins to lower-molecular-weight sub-units serves as a substrate for yeasts and lactic acid bacteria. These lower molecular substrates are utilized by a process due to A. I. Ostrowski to prepare liquid-yeast and liquid-sour, which are also classed as liquid intermediate products. The first stage in the process is the fermentation of a gelatinized flour/water suspension at 48–54°C with thermophilic lactic acid bacteria. This is followed by cooling the fermented mash to 28–30°C, which now has a high lactic acid content. The latter material is set aside for further use as substrate for the propagation of yeast. Similarly, other liquid intermediates prepared from gelatinized and hydrolysed flour/water suspensions at 28–30°C propagate both non-thermophilic lactic acid bacteria and yeast, and are referred to as liquid-sours.

During the course of production, a definite amount of the liquid-yeast or liquid-sour is taken to prepare the dough, and an appropriate amount of nutrient added. After a fixed fermentation time, further amounts of liquid-yeast or liquid-sour are taken, for dough preparation and more nutrient added to perpetuate the propagation-cycle.

Wheat-flour doughs can be made with liquid-yeast and liquid-sour according to the direct straight-dough, or indirect sponge-dough processing procedures. The amount used depending on the flour strength and the process, as well as the fermentation power of the yeast or sour. If a low-extraction flour, and the yeast or sour of potent strength, are available, correspondingly less must be added. For the indirect sponge process, the amount of liquid-yeast or liquid-sour would be 15–25%

of total flour weight per batch, less than that used for the direct straight-dough procedure. Liquid-yeast which has been acidulated with the thermophilic bacteria *Lactobacillus delbrückii* can be ideally used for preparing doughs from top-grade and 1st and 2nd grade wheat-flours, owing to the high acidity-index which results. Liquid-sour, which is prepared without the initial acidulation with the thermophiles is used for making doughs from wholemeal-flours. Pressed-yeast can also be added to wheat-flour doughs together with liquid-yeast and liquid-sour, in which case appropriate concentration adjustments must be made. Assemblies for the continuous preparation of liquid-intermediate products have also been developed, and in this connection the All-Union Soviet Research Institute in Moscow (WNIIChP) has applied this liquid-sour technique to a process for preparing dough with a strongly reduced fermentation period before dividing and work-off. After detailed analysis of the research data concerning the biochemical and colloidal changes during dough preparation, the WNIIChP developed a new technological approach to breadmaking, and translated their concepts into practice. The basis of this technology is an intensification of the dough-ripening processes aimed at reducing the total duration of the breadmaking process including baking, normally 7–8 hours, to 2–5 hours. This involves the reduction of the stages of fermentation of intermediate firm sponges before the dividing process to a minimum. This is achieved by using liquid intermediate products without salt addition, and intensive mixing of the dough. Energy-inputs ranging from 25–46 J/g, depending on the gluten quality of the flour, effect better swelling of the protein, and gluten hydration. According to this process, the liquid intermediate product is continuously prepared, but discontinuously processed. The main part is used to prepare the dough, and the small residual of the ripened intermediate product is returned to the first section of the fermentation-container, where it acts as a carrier of microflora for subsequent intermediate product. In addition, to stimulate fermentation enzymic-ferment preparations from *Aspergillus awamori* are added, together with non-ionic surface-active agents, which improve the properties of the gluten. Depending on flour quality and extraction-rate, certain organic acids, e.g. lactic and citric acids, can be added at the mixing stage to adjust dough acidity. A typical formulation for white-bread with sugar and shortening found to be optimal was: flour 100 parts, dried-yeast 2·0 parts, salt 1·3 parts, citric acid 0·246 parts, ethanol 2·2 parts, water addition as required. The dough is mixed intensively, or conventionally for 20 minutes, and moulded into the required form for proofing. After a final-proof of 60–65 minutes, the 1 kg Bread-units are baked for 50 minutes. The resultant bread being considered to be organoleptically and physicochemically as good as the conventional process.

Compared with the normal process of dough manufacture, this procedure achieves a fully-ripened dough, which cannot be differentiated either in dough properties or in final bread quality from its conventionally prepared counterpart. For a full description of this rapid method for manufacturing wheat bread, the reader is referred to the original publication by Stoljarowa, Tschtscherbatenko, and Lure in *Chlebopekarnaja i Konditerskaja Promyschlennost*, **1**, 5–10, 1963 (in Russian).

The influence of the processing method used in making rye and wheat bread on the rapidity of staling has been thoroughly researched in the USSR during the early 1960s by I. M. Roiter, A. J. Kowaljenko, N. I. Bersina, and E. W. Ljach. Staling intensity was measured at 17, 41 and 65 hours after the baking process by three methods, viz. compressibility of the crumb, swelling ability of the crumb, and breakdown of the crumb by beta-amylase. Increasing the amount of flour in the sponge for wheat bread, or in the sour in the case of rye, retards the staling process. The addition of 0·7% salt to the sponge or sour (based on total sponge or sour weight) also retards the staling process. The use of a liquid-sponge or sour is recommended, since it permits a shortening of the fermentation-time, reduced loss of carbohydrate, and the possibility of controlling the liquid-state fermentation by automation.

Equipment for the continuous manufacture of liquid intermediate products (liquid sours) for processes with greatly reduced fermentation periods up until work-off, have been developed by the WNIIChP Technological Laboratories in Moscow and include the following: Tschtscherbatenko, W. W., Tschishowa, K. N., Schkwarkina, T. I., Lurje, T. S. *Chlebopekarnaja i Konditerskaja Promyschlennost* (ChKP) **1**, 1, 7, 1957; Tschtscherbatenko, W. W., Tschishowa, K. N., Schkwarkina, T. I., Lurje, T. S., Pronin, S. I., *Research Work of WNIIChP*, **7**, 14, 1958. Tschtscherbatenko, W. W., Lurje, T. S. Work collection: *Rye and Wheat Dough Manufacture using Continuous-Aggregates*, 26 and 52, Moscow, GosINTI, 1961 (all Russian original references).

Another innovative variation and technical solution for dough preparation was devised in the Leningrad bakery-factory 'Badajew'. The mixer being serviced by a series of machine-bowls arranged in a circle on a conveyor travelling in a circular path. The aim of this set-up was to avoid the manual handling of the machine-bowls for the ChDSch-mixer, and simultaneously to make use of them as fermentation-containers. This set-up was developed in two versions:

(1)  For the preparation of wheat doughs with sponges, using two circular conveyors, one with 12 machine-bowls for the sponges, and one with four machine-bowls for the dough.

(2)  For wheat doughs prepared from liquid intermediate products, the liquid intermediate product is prepared in tanks, the doughs being mixed in the ChDSch-mixer and fermented in the ChDSch machine-bowls of the small circular-conveyor, upon which the four machine-bowls are arranged.

The original reference, describing this application, due to P. M. Miljukow, is published in *ChKP*, **11**, 10, 13, 1967.

Traditionally in the USSR, as in the USA and many other countries, wheat-flour doughs are manufactured by the indirect sponge/dough method, in which case, between mixing and work-off the dough must be allowed to ferment.

In the USSR, the fermentation process is initiated by using pressed-yeast, liquid-yeast, or a combination of both, the salt being added at the dough stage.

However, between 1960 and 1970 many new processes and process variations for both wheat- and rye-flour processing were developed. The basis of these processes involves the preparation of an intermediate liquid product (liquid-sour) with a water content of about 65% for wheat-flour, and about 70% for rye-flour. Fermentation of the dough up until work-off or processing is carried out in stationary bunker-type containers positioned above the divider, for periods of 15–30 minutes. An intensive mixing process has also been recommended, the intensity of which depends on the flour type and quality. The liquid intermediate product is prepared by a continuous process in a specially constructed liquid-sour aggregate, part of the ripened-sour being returned to the first section to perpetuate the process. A typical formulation, based on 100 kg flour, and the process-schedule for wheat-flour is given in Table 39.

Between 1956 and 1959, a new plant was developed in the Research Bakery of the WNIIChP in Moscow. The liquid intermediate product and dough being mixed in a continuous mixer type Ch-12. The new concept of the process was not the continuous production of the intermediate product or pre-ferment, but its application in reducing the fermentation-stage, which would normally take place in a bunker-type container above the divider. An additional element of the concept was the increase in mechanical mixing intensity expended during mixing, which was achieved after the close collaboration of the Research Bakery WNIIChP and technical people involved in industrial bakery production.

The idea of a stepwise addition of salt during wheat-flour bread manufacture was originated by W. M. Dontschenko, W. W. Kusmenko and co-workers at the Krasnodar bakery. Up until 1956, the classical sponge-process, using liquid-yeast, sponge and dough was still in general use. However, the salt was often added at all stages of manufacture of the intermediate product, i.e. 1 part to each of the following: pre-gelatinized and acidified flour suspension; liquid-yeast; sponge and dough. The total concentration of salt in the dough remained unchanged. In the Krasnodar Bread-factory No. 3 wheat-flour doughs were made with the ChTR

TABLE 39
**Formulation and process-schedule for wheat-flour**

| *Raw material and parameter* | *Liquid intermediate-product (pre-ferment or brew)* | *Dough* |
|---|---|---|
| Flour (kg) | 30 | 70 |
| Pressed-yeast/liquid-yeast (kg) | 0·8–1·5 | — |
| Salt (kg) | — | 1·3 |
| Water (%) | 65 | 44–46[a] dependent on bread type |
| Initial temperature °C | 27–30 | 28–30 |
| Final acidity-index | | |
|   1st Grade Wheat flour | 4·5–5·0 | — |
|   2nd Grade Wheat flour | 5·0–6·0 | — |

[a] Adjusted final dough moisture percentage content.
Source: *Technology of Breadmaking*, L. J. Auerman, Moscow, 1977, p. 200, Table 21 (in Russian).

aggregate, without sponges to improve bread quality. The processing scheme was liquid-yeast or liquid yeast/sour-dough. The salt was added in stages over all processing phases. The liquid-yeast is continuously produced by progressive culture with increasing salt concentration, whilst the salt-containing yeast-phase is completely used up during the dough preparation stage in a ChTR doughmaking aggregate. For a full description of this particular case study, the reader is referred to the original Russian text, *A Rationalized Wheat-flour dough preparation process, which has been applied in the Bread-factories of the Krasnodar Trust of the Baking Industry*, by Dontschenko, W. M. and Kusmenko, W. W., published by Sowjetskaja Kuban, 1958. In some bread-factories, such as those in Krasnodar and Tuapse, doughs were also prepared by progressive stepwise addition of salt using the scheme: liquid-yeast/yeast-sour/sponge/dough. Nevertheless, this process variation did not eliminate the traditional ripening of the dough before work-off. Additionally, in the late 1950s, further variations were devised, in which doughs were processed from liquid-yeast containing salt, and sponges containing salt, the latter also being in liquid-form. In such cases, no water is added at the dough-up stage, the water content of the liquid-sponge, containing salt, being 70–72%. The liquid, salted, sponge is fermented for a 4-hour period, but the final dough only for 15–25 minutes.

For the preparation of salted liquid sponges containing up to 30% of total flour, the mixer ChSM-300, used for the preparation of gelatinized flour–water suspensions, is frequently used. Both batch and continuous processes have been developed for dough processing by the method described, involving the use of salt-containing liquid pre-ferments. For continuous processing, a shortened version of the ChTR aggregate is frequently used, as described in the original Russian literature reference: *Preparation of Rye and Wheat-doughs with Continuous Aggregates*, by Dontschenko, W. M., Vol. 78, Moscow, GostINTI, 1961. In 1965, Dontschenko developed another process based on the use of salt-containing liquid pre-ferments in the Armawir and other bread-factories, which required no fermentation prior to dough work-off, and for this process the ChTU-D doughmaking unit was used. The salt-containing liquid pre-ferment for these processes is prepared from 25–30% of a salt-containing liquid yeast or a slightly salted liquid sponge (yeast-sour), and 70–75% of a flour suspension of 71–73% water content. After 4 to 4·5 hours fermentation, the sponge or pre-ferment reaches an acidity-index of 6·5 to 8·5. The total salt, apart from that in the liquid-yeast or yeast-sour is present in the salt-containing liquid pre-ferment. The dough is mixed without any further water addition, and can be immediately worked-off or placed in a suitable container.

W. W. Kusmenko, the co-developer of this process, in his evaluation of the new process in 'New Technology of Doughmaking with Liquid Intermediate Products', *Bulletin No. 18*, Moscow, ZINTIpistschepromisdat, 1965, states 'Bread quality is the main criterion for the choice and evaluation of any breadmaking process. Working with the liquid intermediate product, followed by no further fermentation of the dough until work-off; and subjecting the dough to an intensive mechanical mixing through the worm of aggregate ChTU-D, followed by passage through a ChTR aggregate with shortened residence-time of 30–35 minutes fermentation only, a bread of completely satisfactory quality can be obtained'.

In this connection, and in evaluating new process developments for wheat-flour doughs, both in the USSR and from other countries, the following considerations must be thoroughly tested and proved satisfactory for application in the Soviet baking industry:

(1) In addition to being of acceptable density, about 0·34 g/ml, good symmetrical shape, adequate volume and small even crumb-porosity, bread must exhibit a fully developed flavour and taste token on consumption, and a good aroma. These properties must not be inferior to bread produced by an established breadmaking process with fully acceptable quality standards.

(2) Considerable success and acceptance of the progressive mechanization and automation of breadmaking processes, which reduce manual transport of machine-bowls and dough have been achieved, and research and development will continue in this direction. The application of such key technologies as biotechnology, microprocessor-technology and biosensor-technology, combined with computer-aided design (CAD) and computer-integrated manufacturing (CIM) techniques, when successful in the industrial factory-bakeries, should improve quality control and work efficiency.

(3) Wheat-flour dough processes which allow reductions or the elimination of fermentation between mixing and work-off are only of practical application where a fully developed flavour token and aroma akin to established processes prevail.

(4) The bread-crust must be thin and the crumb must be moist but remain elastic.

(5) The consumer is orientated towards fresh bread, especially in the capital cities, e.g. Moscow, where most sale-points receive 4–5 deliveries from the bread-factories daily.

In addition, other process variations for wheat dough using liquid intermediate, and liquid oxidation-phases (LOP) have been developed and applied in the Soviet baking industry.

In the Lugansk Bread-factory No. 2, a mechanized line for the production of dough from 2nd grade wheat-flour was developed. In this process, the dough is made from salted liquid pre-ferment with a water content of 70%, salt being added to both pre-ferment and dough. A schematic diagram of the equipment-layout in the Lugansk factory is illustrated in Fig. 25.

The salted liquid pre-ferment is prepared continuously according to the illustrated scheme, although the pre-ferment is still mixed on a discontinuous (batch) basis. However, in the Kiev Bread-factory No. 6, the pre-ferment is mixed in a Ch-12 aggregate, thus making the whole process continuous.

Initially, doughs made with the salted liquid pre-ferment were mixed by using the ChTU-D aggregate, and fermented in the sloping fermentation-channel; but since 1965 the ChTU-D aggregate has been used, and the dough deposited into the hopper above the divider. Similarly, in the No. 2 Bread-factory of the Donezk

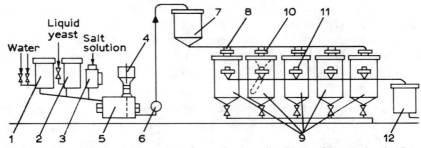

Fig. 25.    Equipment-layout for preparation of salted liquid pre-ferment in the Lugansk Bread-factory No. 2. 1, 2 and 3. Dosage units for liquid components. 4. Automatic flour dosage unit. 5. Pre-ferment mixer (ChSM). 6. Pump. 7. Holding tank for freshly prepared pre-ferment. 8. Control unit for measuring amount of freshly prepared pre-ferment. 9. Fermentation containers for pre-ferments. 10. Hopper and discharge pipe for deposit of fresh pre-ferment into bottom of container. 11. Pipeline for removal of fermented pre-ferment. 12. Collection tank for fermented pre-ferment. (Source: *Technology of Breadmaking*, L. J. Auerman, Moscow, p. 202, 1977.)

Baking-collective, another machine-assembly is used for wheat-flour doughs using salted liquid pre-ferments, but in this case-study, part of the salt is added to the dough. The salted liquid pre-ferment is prepared from liquid yeast in the case of wholewheat-flour and 2nd grade wheat-flour, and from a mixture of liquid- and pressed-yeast for 1st grade wheat-flour, the mixing being completed in a ChSM-300 machine. The water content of the salted liquid pre-ferment is 69% for 1st grade wheat-flour, and 73–74% for 2nd grade wheat-flour. The pre-ferments are prepared discontinuously in this bakery, and the doughs mixed in the ChTU-D aggregate, passing directly into the divider unit. For the original source literature on this case study, the reader is referred to Ionowa, W. W., *Chlebopekarnaja i konditerskaja promyschlennost*, **9**, 6, 27, 1965.

Many other bread-factories in the Soviet Union manufacture wheat-flour doughs from salted liquid pre-ferments, without fermentation of the dough before work-off, achieved by collaboration with the engineers from the Krasnodar, Lugansk and Donezk Bakery-collectives. Most of these prepare the salted liquid pre-ferment from liquid yeast or, in the case of 1st grade wheat-flour, using a mixture of liquid- and pressed-yeast. The Leningrad bread-factories found that for the manufacture of bread-rolls from 1st grade wheat-flour, and top-grade wheat-flour, pressed-yeast is preferable. For this reason the Leningrad WNIIChP developed a process for wheat-flour doughs using salted liquid intermediate products (pre-ferments) containing only pressed-yeast. Pre-ferments made from 1st grade wheat-flour are prepared from 27–30% of the total flour, with a water content of 64%. All the yeast and salt are added to the pre-ferment, which is set at a temperature of 32–33°C. Within a 5-hour fermentation period, the pre-ferment reached an acidity-index of 3–4 units. In some Leningrad bakeries, the salted liquid pre-ferment ripens within 2 hours, and in another the pre-ferment is in the form of a sponge, containing only 43–45% water and pressed-yeast. In the latter case, however, the dough is mixed for 25 minutes, and dough-fermentation only takes place in the holding-hopper above the divider.

In 1963 and 1965, the MTIPP developed a process based on the action of the enzyme lipoxygenase for the improvement of bread and roll quality from wheat-flours. An Inventor's certificate was taken out under the number 164860; w.e.f 8.7.1963 by Auerman, L. J., Kretowitsch, W. L., and Polandowa, R. D., under the title *Process for Improvement of the Quality of Wheat-bread* with the State Committee for Inventions of the USSR; and in 1965 Auerman, L. J., Kretowitsch, W. L., and Polandowa, R. D., published an article in the journal *Prikladnaja Biochimija Mikrobiologija*, **1**, 1, 66, 1965. According to this process, a liquid oxidation-phase is prepared (LOP). After laboratory-scale preparation of LOP by the MTIPP, trials were carried out in the WNIIChP test-bakery and the Moscow Bread-factory No. 4 under production conditions by discontinuous processing using machine-bowls. Using the experience and results gained in these trials, a production-scale aggregate for the preparation of LOP was built in the Orechowo-Sujewsker Bread-factory, which was linked with a ChTR aggregate for continuous dough-processing. The procedure for the preparation of LOP is to take: in the case of a direct (straight dough) procedure, 50–75% of the total dough water (or all the dough water for an indirect (pre-ferment) procedure), 0·3% (flour weight) of freshly milled, full-fat soy-flour as a source of lipoxygenase, and a vegetable oil emulsion made by using phosphatide-concentrate as the emulsifier, consisting of 90% water, 5% vegetable-oil and 5% phosphatide-concentrate prepared in a hydrodynamic vibrator. An amount of this emulsion is added to the LOP which corresponds to 0·05% of total flour weight in the dough, and 20–25% of total flour used to prepare the dough. Therefore, the LOP consists of a batter/emulsion composed of water, vegetable-oil, and concentrated phosphatides dispersed in a 20–25% flour batter, with 0·3% full-fat enzyme-active soy-flour as the source of enzyme lipoxygenase. The emulsion of water/vegetable-oil/phosphatide concentrate acts as a substrate for the lipoxygenase. This LOP concept is similar to that devised by G. Blanchard in the UK, as published in *Milling*, **147**, 24, 519, 1966, 'The Blanchard dough process for breadmaking'. Details of this process are described earlier in section 1.5.9. Depending on the dough or pre-ferment process in use, the LOP can be applied on a batch or continuous basis, the equipment being designed to produce enough LOP for output capacities of 15 to 18 t bread-rolls per day. For a 24-hour production-line, with an output of 17·3 t wheat-flour bread made from 1st grade wheat-flour, two batches of LOP must be prepared per hour. The composition of a batch of LOP is as follows: water 87·6 litres, full-fat enzyme-active soy flour 0·8 kg, 2·65 kg of a 10% emulsion of 90% water/5% vegetable-oil/5% phosphatide concentrate, and 52·1 kg wheat-flour. Since only a small amount of emulsion is required, enough material is prepared for at least one shift or more, and stored in a holding-tank for LOP preparation as required.

The equipment-layout necessary for the preparation of the liquid oxidation-phase is illustrated in Fig. 26. Using the set-up illustrated, the LOP preparation process is a batch one. The prepared LOP is pumped into the holding-tank (7) which holds two batches, and is then used to prepare the sponge, liquid pre-ferment or straight-dough.

The main part of the assembly is the high-rpm mixer (5), in which the preparation

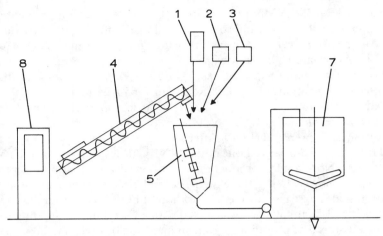

Fig. 26.    Equipment-layout for preparation of liquid oxidation-phase (LOP). 1. Water-mixing and dosage unit AWB-100. 2. Dosage unit for soy-flour. 3. Dosage pump for emulsion. 4. Dosage worm for wheat-flour. 5. High-speed agitator for preparation of LOP. 6. Pump. 7. Collection tank prepared LOP. 8. Control-panel. (Source: *Technology of Breadmaking*, L. J. Auerman, Moscow, p. 204, 1977.)

of the LOP takes place. The mixer container is conical in shape with a capacity of 200 litres, the inner surface having baffle-plates extending vertically. The mixer-vortex is inclined at an angle of 10 degrees to the vertical and three propeller mixing elements rotate at 1400 rpm. This not only results in an intensive mixing of the contents, but also a saturation of the mass with air-cells. The presence of atmospheric oxygen is essential for the oxidative action of the lipoxygenase within the LOP. The various components of the LOP are metered into the mixer (5) by means of appropriate dosage units, the water over the AWB-100 (1), the soy-flour over dosage-unit (2), and the emulsion over dosage-unit (3). The wheat-flour can either be fed to the mixer by a worm, (4), or over an automatic flour-weigher MD-100 direct into the mixer. The whole set-up is automatically controlled from the control-panel (8).

The detailed procedure and chemical mechanisms for the preparation of the LOP are as follows:

*Stage 1*
Water is metered at a maximum temperature of 35°C, using the dosage-apparatus AWB-100, into the mixer (5) and the mixer-motor is switched on, the appropriate amount of soy-flour being added by the dosage-apparatus (2). The water/soy-flour mixture is stirred for 1 minute, in order to solubilize the lipoxygenase of the soy-flour. The requisite amount of emulsion is then added using dosage-apparatus (3), and the mixture stirred again. This results in the polyunsaturated fatty acids of the vegetable-oil and the phosphatide-concentrate being oxidized by the lipoxygenase with the help of the trapped atmospheric-oxygen due to the mixing-action. The fatty acids form peroxide compounds, mainly hydro-

peroxides, which function as active oxidants. It is the formation of hydro-peroxides that is the main objective of this first stage of the LOP preparation.

*Stage 2*

After the emulsion has been added to the mixer, the appropriate amount of wheat-flour is added over a 2-minute period, and after mixing for a further 4 minutes the preparation of the LOP is complete. It is during this second stage that the peroxide-compounds of the fatty acids further oxidize the —SH— groups of the protein/proteinase complex of the wheat-flour which has been added to the LOP. This results in an improvement of the physical properties of doughs made with LOP, i.e. gas-retention, form or symmetry, an increase in water-binding energy. The resultant bread shows improved volume, symmetry and crumb characteristics. Simultaneously the peroxide-compounds of the fatty acids oxidize and bleach the carotinoid flour-pigments, which is immediately apparent from the improvement in crumb brightness and colour.

The prepared LOP batch is pumped into the holding-tank (7), from which it is taken to prepare the liquid pre-ferment or direct, straight-dough as required.

As a result of the use of LOP, the physical (rheological) properties of the dough are significantly improved, the moisture content of bread-roll doughs can be adjusted to correspond to the legal norm for crumb-moisture. In this connection more recent work carried out by Pashchenko, L. P., Mazur, P. Ya., Serbulov, Yu. S., and Zolotykh, I. N. (1984), at the Voronezh. Technological Institute, USSR, has shown that optimal rheological properties of liquid-pre-ferments are obtained at 30°C, with the addition of the phosphatide-concentrate or the dough improver known as 'volzhskiï' at 0·26–0·29% and simultaneous saturation of the pre-ferment with 15 mg oxygen/litre. Further work by Mazur, P. Ya. (1986) established an improvement in binding-energy of moisture in flour and dough when they are enriched with oxygen, the tightly bound water increasing by 5%, corresponding to a water-binding energy increase of 19 kJ/mol in thermodynamic terms. Auerman also points out from earlier experiences, that owing to the improvement in rheological properties of the doughs, an increase in the final-proof time is recommended. Also, the ripening of the dough is speeded up, especially the accumulation of the fermentation acids (acidity-index). Another point of practical significance in the application of the LOP-process, is the marked improvement in bread quality achieved with immature flour. The net improvement is equivalent to an in-sack storage of about 2 months (Ambrosjak, S. W., Auerman, L. J., *NTInf.*, **5**, 7, 1967), but this does not mean that the application of the LOP-process can eliminate flour maturation. The storage and maturing of any flour will instead improve the quality of the bread still further.

Basic technological instruction in the USSR for the preparation of wheat-flour doughs generally, and bread-roll doughs in particular recommends the use of indirect processes rather than direct, straight-dough techniques. This breadmaking philosophy is founded on the national standard requirement for high-quality wheat-bread, and the importance of bread quality standards generally in the diet of the Soviet people. A new process is only acceptable for industrial application, when the bread quality can be judged to be in no way inferior to that produced from the indirect method. The indirect-process produces a high-quality product of superior taste and aroma, with a better-developed and thin-walled crumb porosity. The taste and flavour result from a larger accumulation of fermentation products, and the improved structure and crumb physical properties from the unchecked swelling and peptide-formation from flour proteins. The relatively rapid build-up of acids in the pre-ferment results in a more complete and intensive flavour using less yeast, compared with direct processes. Additionally, owing to the presence of flour in the pre-ferment, the process can be better controlled according to flour quality, water content and temperature, and fermentation-time of both pre-ferment and dough can be optimized for bread quality. Although the indirect method is less economical, takes longer, and gives a bread yield of about 0·5% less than a direct method, the bread quality is much better than any direct procedure yet devised, and less yeast is necessary.

This is also the experience of the author, where bread quality is the overriding factor, and also applies to direct, straight dough procedures with reduced floor-times when compared with 3-hour floor-times, as practised in the UK in the 1950–60 decade and earlier. Specialists from the USSR baking-industry had the opportunity to taste bread made by the Chorleywood process in the UK. Their opinion was that, although the bread was of high volume and had a fine thin-walled crumb, the taste and aroma was inadequate. The most important aspect of the Chorleywood process in technological terms is the fact that a minimum energy-input of 40 J/g over a period of 4–5 minutes results in enough specific mixing-intensity to effect sufficient structural changes, mechanically, to the dough without the need for a floor-time or bulk-fermentation period. However, the presence of other additives, viz. ascorbic acid and/or potassium bromate, special shortening, extra yeast and extra water are also essential to the process. The spread in popularity of the Chorleywood process in the UK, uniquely linked with the development of the new Tweedy dough-mixers, was mainly due to its bakery industrial-machinery sponsors. The vested interest of the machine-manufacturer with an organized sales force was in a strong position to spread the gospel.

Prompted by the public concern and publicity surrounding the treatment of flour in the 1950s with Agene (nitrogen trichloride), as an improver, G. Blanchard, a master-baker from Nottingham pioneered a non-chemical treatment process of breadmaking. His process was based on the use of untreated flour, and a diverse range of conventional mixers in commercial use at that time, including a speeded-up version of the Baker Perkins High Speed Mixer. A 75% flour sponge-batter was made with water and a small amount of unprocessed full-fat enzyme-active soy-flour as a source of the enzyme lipoxygenase, along with yeast. The batter was beaten to a stage of total hydration and physical development of the gluten, thus replacing normal fermentation. This can be achieved in an ordinary horizontal-bar mixer if speeded up from 75 to 98 rpm. The residual 25% flour together with the salt and shortening if required are then added and thoroughly blended into the developed batter. The resultant bread was a well-developed plant-bakery loaf, the texture and crumb-sheen and brightness as well as the eating quality being judged to be very good. This fact being confirmed by the subsequent expansion of G. Blanchard's business and bread sales, which employed over 1000 workers at the time of the sale of the business to Spillers. Dr Coppock and the staff of the then Research Association also pronounced the bread of very good quality. The Blanchard process has many advantages over the Chorleywood bread process, viz.

—Conventional high-speed mixers can be used, thus eliminating the extra expenditure of a mechanical-developer.
—Less energy is required for physical development and therefore less electrical power than for the CBP.
—Heat dissipation and temperature rises due to frictional heat are less, and doughs can be run at lower temperatures.
—Any oxidation requirements are lower than with the CBP.

In the Estonian SSR, the Blanchard process is used for the processing of bread containing milk-products with the use of extra yeast, and with the problem of flour treatment back in the headlines in the UK, the Blanchard method could well provide the answer to manufacturing untreated bread with no chemical additives.

Complementary to the use of liquid intermediate products, there are two other aspects of wheat-flour dough processing in the USSR, which have been designed to improve bread quality by fermentation control, viz. pre-activation of the yeast, and the use of liquid-yeasts.

**Pre-activation of the Yeast**
Research work has established that the economics of both direct and indirect processes of breadmaking using yeast can be significantly improved by pre-activation of the yeast.

During manufacture the yeast-cells are propagated by using an intensive aeration of the culture medium. Therefore, the internal structure of the yeast and its enzyme complex becomes adapted to an aerobic environment. In pre-ferments or doughs, the yeast finds itself under almost anaerobic conditions, and must adapt from using oxygen as the hydrogen-accepter (breathing) to using organic substances as hydrogen-accepter (fermenting). As shown by the work of M. N. Meissel, detailed in *The Functional Morphology of Yeast-microorganisms*, Moscow, AN USSR, 1950 (in Russian), this transition of environment results in considerable changes in the internal structure of the yeast-cell. The enzyme-system of the cell must change, and adjust to the new environment, this change to fermentation requires an induction-period, and demands specific conditions. With the aim of speeding-up the fermentation process of the pre-ferment or dough, a small amount of soluble substrate was used at a concentration considered as optimal for the activation of the onset of the fermentative processes. In this way, the yeast passes out of the passive resting state and into active state, adjusting to the dough fermentation process by increasing its maltase enzyme activity. The research work on the activation of yeast before dough-preparation was carried out in the Faculty for the Technology of Baking of the MTIPP under the direction of A. G. Ginsburg. In collaboration with the co-workers of the Moscow Bread-factory No. 4, a process for yeast activation was researched, and the necessary pilot-plant designed.

The process of yeast-activation involves the preparation of a nutrient solution, to act as substrate, the homogeneous distribution of the yeast cells in this medium, followed by an incubation period of 1–2 hours at 30–32°C. The nutrient solution consists of a pre-gelatinized flour/water suspension made from wheat-flour with the addition of active white-malt to the hot, 50–60°C, flour/water suspension. This is followed by the further addition of wheat-flour with a little soy-flour. After thorough mixing of the mixture, and subsequent cooling down to 30–32°C by the addition of an appropriate quantity of cold water and constant stirring, the crumbled pressed-yeast is added and thoroughly suspended by further stirring.

The formulation of the flour/water suspension is as follows:

| | |
|---|---|
| Wheat-flour (kg) | 1·3–2·0 |
| Water at 95–98°C (litres) | 4·0–6·0 |
| White-malt (kg) | 0·2 |

The preparation of the nutrient solution of yeast-activation phase then proceeds as follows:

| | |
|---|---|
| Pre-gelatinized flour/water suspension (kg) (prepared as above) | 5·5–8·2 |
| Cold water (litres) | 5·5–5·7 |
| Wheat-flour (kg) | 1·3–2·0 |
| Soy-flour (kg) | 0·5 |
| Yeast (pressed) crumbled | total amount used in the dough recipe |

Source: *Technology of Breadmaking*, L. J. Auerman, Moscow, 1977, p. 184 (in Russian).

The experience of many Bread-factories has indicated that the use of yeast activated in this way results in a 20–40% reduction in the use of pressed-yeast, and a simultaneous reduction of the sponge or pre-ferment fermentation-time by about 30 minutes. Yeast cell propagation does not take place during the 'yeast-activation phase', and the lower the amount of yeast in the dough, and its fermentation-strength, the greater the 'activation' effect. This effect is greater for direct processes of doughmaking than for indirect methods.

Special equipment for the preparation of the 'activated pressed-yeast', such as the UAD-60 (for daily production of 60–90 t of wheat bread), and the UAD-15 (for 15–25 t daily capacities), have been developed and produced on a mass-production basis. The use of ultrasonic vibration equipment or mixers with rpm of approximately 2800, for mixing the water-suspension will greatly improve the fermentation-activity of the yeast, owing to better distribution of the yeast, and the breakdown of cellular-conglomerates present in most yeast/water suspensions.

The technological significance of the use of pre-gelatinized flour/water suspensions, depends on the fact that the flour starch, in a gelatinized state, is rapidly converted to sugar by the amylases. These pre-gelatinized flour/water suspensions are used extensively for bread production as nutritive substrates for the growth of yeast and acid-forming bacteria, and also for the preparation of liquid-yeast and liquid-sours. They can also be added to sponges, pre-ferments and doughs to improve bread quality in the Soviet Union. In the case of certain types of bread, the use of pre-gelatinized flour/water suspensions are essential, quite apart from the amylase-activity of the flour.

Normally a pre-gelatinized flour/water suspension (PFWS) is made from flour and water in the ratio 1:3 to 1:2. For use as a bread improver, 3–5%, maximum 10%, of the total flour is used to prepare the PFWS. The process has been rationalized by using the machine ChSM-300. This machine consists of a horizontal cylindrical container fitted with screw-like elements mounted on the horizontal drive-shaft. The capacity of the chamber is about 300 litres, being fitted with a water-jacket. Heating of the flour/water suspension is effected by four steam-injection pipes passing through the cover, throughput being approximately 200 kg of PFWS per hour. The heating process is controlled by use of a thermometer, which shows the gelatinization temperature required. Cooling of the gelatinized mass follows by passing cold water through the water-jacket. The heating process for each batch takes about 90 minutes, and the cooling cycle about 50–60 minutes, using 580 litres of water at 15°C. The output of the drive-shaft motor is rated at 2·2 kW. Larger units tend to be of vertical design in larger bread-factories, in which case the heating and cooling cycle takes a total of 120 minutes, allowing 90 minutes for cooling which requires 1500 litres at 15°C.

In 1960, as the result of collaboration between researchers at the MTIPP and the co-workers of the Moscow Bread-factory No. 5, an automatic system for the production of PFWS was developed and accepted for production use. The introduction of this equipment allowed the preparation of a PFWS with a higher flour content (a flour:water ratio of 1:1·3) using electro-contact heating, within 22–25 minutes. Rapid cooling is effected by mixing the PFWS with an additional quantity of cold water, thus reducing the total preparation-time for a batch of PFWS to 36 minutes. The energy requirement for a tonne of flour into PFWS is 39 kWh. These automatic electro-contact heated units for PFWS are used in Moscow Bread-factory No. 5 for wheat-flour, but they were also introduced into the Leningrad Bread-factories Nos. 10 and 11, for the preparation of PFWS from rye-flour. For the actual case-study details of the application of automated PFWS preparation, the reader is referred to the following original reports:

—Ostrowskij, J. G., Auerman, L. J., Schmain, M. M., Czech. Journal *Prumysl potravin*, **3**, 11, 1961 (in Czech.).
—Jegorowa, A. G., Kasanskaja, L. N., and co-workers, *Manufacture of Rounded Hearth-bread from Medium Grade Rye-Flour in Bread-factories using the Marsakow-system*, Moscow, ZINTIpistscheprom, 1965.
—Saizew, A. D., Michailow, W. N., Tschumakow, W. F., *Complex Mechanization and Automation of the Bread-factory*, Moscow, ZINTIpistscheprom, 1962.

The application of these automated units releases four operators from PFWS preparation in its original form. Furthermore, trials in Moscow Bread-factory No. 5 have established that PFWS prepared with the new electro-contact heating system, produces a gelatinized-mash which when soured with *Lactobacillus delbrückii*, gives a high level of water-soluble nitrogen and an intensive acid build-up. This produces a good, stable liquid-yeast when used as substrate for the preparation of liquid-yeast.

The various types of PFWS used in the Soviet baking industry can be classified as follows: (a) hydrolysed PFWS, (b) non-hydrolysed PFWS, (c) salted PFWS, (d) fermented or soured PFWS.

Hydrolysed PFWS is formed as a result of the amylolysis of the gelatinized starch in the flour. However, the extent of this will depend on the treatment of the PFWS. If saccharification has taken place as a result the action of the amylases during the cooling-down process from about 70°C down to 35°C, after gelatinization, the conversion will be comparatively mild. Whereas, if the PFWS is treated with an appropriate amount of enzyme-active white-malt, or held at 62–65°C for 2–4 hours in the newer automated cookers after gelatinization is complete, the end-product will be fully saccharified and in the hydrolysed form. Initially, since the saccharified PFWS was not only used as a nutrient substrate for the preparation of liquid-yeast or liquid-sour, but also as a bread-improver for direct addition to sponges and doughs, the degree of saccharification was not critical. Subsequent research work confirmed that the saccharification of PFWS for use as a bread-improver was unnecessary, since the gelatinized starch of the PFWS is easily saccharified by the flour amylases in the dough during fermentation and baking. For the preparation of non-hydrolysed PFWS, 3–10% of the total flour is used, and 2·5–3·0 times that amount of water added, and in the case of graded wheat-flours, temperatures of 63–65°C must be reached for gelatinization, and 70–73°C in the case of wholewheat-flours. The heated and carefully mixed mass is immediately cooled down to 35°C, and can be used for sponge or dough preparation.

Some authors recommend the use of salted PFWS, in which case the flour is heated with a salt solution instead of water, the salt concentration corresponding to the total amount added to the dough.

### Liquid-yeast and Liquid-sours

As already described, the technological processes for the manufacture of wheat-flour doughs with any type of liquid intermediate product must involve the use of wheat-flour as a source of nitrogen and vitamins for yeast-cell growth.

The other technological characteristic of process schemes in Soviet bakery is the wide application of liquid-yeast, either as a substitute for, or complementary to, pressed-yeast. Such special features of processing result in the build-up of certain organic metabolites during fermentation, which appear in the intermediate products, doughs and bread.

The starting material for the preparation of liquid-yeast and liquid-sours is PFWS. After cooling, pressed-yeast, lactic-acid bateria, or fermented dough are added, and allowed to ferment for several hours. A typical procedure may be described as follows.

A saccharified PFWS is cooled down to 30–32°C, and 0·8–1·0% pressed-yeast crumbled and evenly suspended (based on flour weight used to prepare the PFWS). This is allowed to ferment for a period of 3·0 to 3·5 hours. The fermented product can then, if desired, be used instead of a pre-ferment or sponge.

Another technique for souring the PFWS, is to add liquid-yeast, fermenting for 12–16 hours, until the acidity-index reaches 16–17. This method is considered particularly suited to the processing of flours with weaker-glutens. PFWS soured with the thermophilic lactic acid bacteria is especially appropriate for the manufacture of dough without pre-ferments or sponges.

For souring the PFWS, another guideline is to add 25% by weight of fermented dough; after a further 2 hours fermentation, the soured PFWS can be used for the preparation of the dough instead of a pre-ferment or sponge. This latter technique using a PFWS soured with the thermophilic lactic-acid bacteria, *L. delbrückii* instead of a sponge was devised by A. I. Losa, as early as 1939, before the designation 'biotechnology', as a branch of the biological sciences had emerged. Considerable research into the biochemistry of liquid-yeast and liquid-sour fermentation processes has been done in the USSR, as well as that of liquid intermediate products in general. The results of their work will be reported in Chapter 2.3.

Recent attempts to improve the biological activity of culture media for the propagation of lactic-acid bacteria for dough souring, and the fermentation power of yeasts, in order to perform better in specific doughmaking processes, have also been published. Chernaya, L. S., Vydrina, O. A., Polandova, R. D. and Petrash, I. P. of WNII Khlebopekarn. Prom., USSR, have reported in *Khlebopek. Konditer Prom.-st*, **5**, 28–29, 1986, a biologically active mixture, named BIAKS, containing pressed baker's yeast and other ingredients for improving dough-fermentation and bread-quality. A modified yeast suspension is dried from a moisture content of 30% down to 10% without significant loss of fermentation-power, being higher in this respect than control preparations. Its application reduced the period of dough-fermentation by 50–80 minutes. Bread-quality was improved, volume increasing by 17%, porosity by 3%, and shape resistance increased from 0·45 to 0·53 units, compared with controls, providing the BIAKS was fresh. Although prolonged storage of 10% moisture BIAKS for about 3 months slightly reduced its activity, it remained superior to control preparations.

Another approach to the improvement of the performance of pressed-yeast by removing the glutathione under vacuum, is described by V. I. Karpenko of the Voronezh. Tekhnol. Institute, Voronezh, USSR, in *Khlebopek. Konditer Prom.-st*, **5**, 42–44, 1986. Since yeast glutathione negatively affects the physicochemical properties of dough, it was effectively removed by holding a pressed-yeast suspension at 35°C and 0·5–0·6 atm vacuum for 4–5 minutes. The water containing the dissolved glutathione is then centrifuged off, leaving a glutathione-free yeast for dough-fermentation. The vacuum process has no negative effect on gassing or maltase activity.

The use of starch-syrup, obtained by the acid hydrolysis of potato or wheat starch instead of molasses as a substrate for the manufacture of baker's pressed-yeast with added supplements, has also been found to be satisfactory by L. D. Belova and co-workers at the WNII Khlebopek. Prom., Moscow, USSR, as reported in *Khlebopek*.

*Konditer. Prom.-st*, **4**, 40–41, 1985. The resultant yeast is stable and has a high maltase and zymase activity. The starch substrate being more compatible with flour used in bread-production.

Work done at the WNIIChP allows the following facts concerning the use of the various types of PFWS for improving bread-quality to be stated:

(1)  The use of PFWS improves the physical properties of the wheat-flour dough, the salted PFWS producing the greatest improver effect in this respect. This is due to increased ability of the dough to form aqueous colloids and so increase water-binding, and the resulting thermal effect on the flour-proteins.

(2)  The loss of dry matter due to fermentation when using PFWS is in general not greater, but when using soured and fermented PFWS with yeast, a small additional loss may be apparent, compared with doughs without PFWS.

(3)  When the flour used has a low amylase-activity, all types of PFWS increase the volume of the bread per 100 g flour. However, in most cases there is a slight fall in the volume yield of bread, which enhances bread-quality in general by increasing crumb density, elasticity and porosity. Most importantly, the flavour and taste of PFWS-containing bread is vastly improved without exception, which is reason enough for its use in the USSR.

(4)  The bread-crust and crumb are enhanced without exception. The crust, becomes darker owing to the increased sugar content, and the crumb elastic and of fine porosity. When salted PFWS is used, the crumb becomes drier and of more open porosity.

(5)  Without exception, all PFWS preparations result in bread with slower staling characteristics compared with bread without their inclusion.

(6)  As a result of the increase in binding of colloidal-water owing to the use of PFWS, the baking loss and loss on cooling of the bread are reduced, and the weight yield of bread correspondingly increases.

The complete fulfilment of the people's requirements for baked products of high quality and wide variety is one of the most important objectives of the food industry in the USSR. The ongoing solution to this problem includes, among other things, the scientific and technological work in the development of new progressive technologies, and plant-machinery development. The Soviet baking industry has accumulated much experience in the development of systems for the control of machinery, and automatic systems in many factory-bakeries situated in the larger cities. With a high degree of mechanization, an increase in output capacity is possible, without increasing waste and specific energy-consumption. By the reconstruction and re-equipping of bakeries, with early preparation for progressive automation, the capacity is increased, manual work reduced, and production quality improved. For example in the Bakeries No. 2 and No. 4 in Odessa, the former, which was built in 1932, with a capacity of 160 t/24 hours relied on four oil-fired ovens which were out-of-date. Reconstruction of the bakery in the early 1980s, included provision for a bulk-flour silo capacity of 500 t powered by four compressors. For doughmaking, with a liquid-pre-ferment, worm-pumps and dosage-aggregates with large throughputs were installed. For dough-mixing, specially developed continuous aggregates were used, followed by worm-pumps feeding the mixed-doughs into dough-ripening containers. The ripened-dough is fed to a 'Sotscha' Divider unit, from which the dough-pieces pass through a conveyor-proofer unit for recovery. On leaving the conveyor/recovery unit, the dough-pieces pass through a cone rounder 'sabotin'. Three such lines as described feed into wire-band tunnel ovens each of $100\,m^2$ baking surface, for baking of round hearth-bread. The most significant reconstruction work in this case-study was the automation of the flour silos and sieve-station, the compressor-station and the mixing-station. The centralization of the control of these sections of production by the silo-manager is completed by an audio-communication system. The scheme for silo automation consisted of remote-control of the silo-compartment filling process, automated remote-control of the transport of flour into the daily-silos and into the continuous-mixer hopper, the weighing of the flour in the daily-silos, and a control of levels (a control of levels in the discharge-units, and in the containers before and after weighing), air-pressure control in the feed-mechanisms of the pneumatic-transport, and a safety-device to prevent blocking of the pipelines. Automation of the compressors ensures automatic control of the station during start-up, operation, and switch-off of the compressors in a fixed sequence, thus ensuring provision for the necessary amount of pressurized-air at each stage, the use of 1–3 compressors being optional. The working sequence of the compressors can be changed depending on the time-programme of the run. The design allows for an emergency switch-gear for the compressors, when the air-supply to the compressor-cylinder and cooler fails, when the oil-pressure falls, or when the temperature or air-pressure at the compressor outlet exceeds a predetermined value. Additionally, there is provision for the automatic 'blowing-out' of the oil/water filter of the cooler, and the air build-up within the pipeline, at each switch-on, and after 2 hours continuous operation. The water-supply and circulation-system is provided by a cooler GPN-25, with breather and pump for the supply of warm and cooled water, which is automatically controlled, and linked to the function of the compressors.

This first phase in the reconstruction of Bakery No. 2 in Odessa increased production and efficiency, allowing the release of personnel for other work, reduced flour losses and energy consumption. The Minel production-lines, made in Yugoslavia, with a height of only 3·5 m, allow the cooling and despatch areas to be increased, which in turn made room for the mechanization of the despatch operation, and delivery of the bread on trays loaded directly by hydraulic-manipulators through the sides of transport-trucks, for ease of withdrawal at the sales-points. At the point of sale, the 8-shelved trolleys and trays serve also as sales containers for the various bread varieties, since they can be housed in sales-cabins of standardized dimensions with open fronts, which are labelled by product and price for consumer clarity. These tailor-made transport-systems with full mechanization, loading and unloading, have reduced costs by 25–30%. In the Moscow Bakery No. 10 the same system has achieved a saving of 100 000 roubles per annum (*Progressive Technology of the Movement of Merchandise—Bread and Baked-goods*, by Kanevskij, published in the Soviet journal, *Promyshlennyj transport*, No. 5, 1983).

The mechanized, continuous technological complex, from the receipt of raw material to despatch of the finished products, forms the basis for comprehensive data-processing aimed at directing the technological processes with the aid of microprocessor techniques. The prime objective is the control of each technological phase with the help of microprocessor control units, which are linked to one another by data-channels, enabling them to control, direct and communicate with the despatch-service of the combined production, and also to pass on orders and information concerning the fulfilment-status to a centralized computer data-bank. Information concerning the actual raw material inventory, current production, and the quantity of products for various outlets should all be immediately accessible from measurement-points for production, stores and despatch. This facility is only possible with mechanized processes. The ultimate aim remains the synchronization of the flow of material and information, thus ensuring their scheduling and accuracy.

This case-study demonstrates the development of continuous bakery production-lines *v.* automation as applied in the Soviet Union. For the processing of rye-flour doughs, a large number of procedures exist in the Soviet Union. They can vary depending on the number of processing-stages, formulation, technological regime of the processing stages, and scheduling of the doughmaking cycle. For detailed descriptions of these processes for the various types of rye and mixed rye/wheat bread, the reader is referred to the following technological sources of reference:

—*The Production of Baked-products—Technological Handbook*, by I. M. Roiter, Kiev, 'Technika', 1966 (in Russian).
—*ZNIIChP Technological Instructions for the Manufacture of Small Baked-products*, Moscow, Pistschepromisdat, 1960 (in Russian).

However, the most important processes for manufacturing rye bread in the USSR can be categorized, and summarized in general form for practical purposes, as follows. Dough made for the production of rye bread from whole-rye flour is normally either made with a sour of firm consistency, known as '*golowka*'; or a relatively soft sour, in which case the water-content is from 1·5–2·0% higher than the '*golowka*' and is known as '*Kwaß*'. Therefore, the rye doughs are either made with '*golowka*' or '*Kwaß*', both of which must be matured stepwise to produce a full-sour before being used to prepare the final-dough. The method for the processing of the firm sour (*golowka*) requires the complete propagation-cycle, which includes the following stages: (1) initial-sour, (2) intermediate-sour, (3) basic-sour, (4) full-sour. A typical formulation, based on 100 kg of full-sour, and the regime of the sour-dough cycle, originate from the two reference works of I. M. Roiter and ZNIIChP quoted above, and are referred to by L. J. Auerman in his reference book, *Technology of Breadmaking*, Moscow, 1977, p. 225, Table 23. A reproduction is presented in Table 40.

The water content of the full-sour is approximately 50%. The shortened production-cycle for dough-preparation consists of two stages only, i.e. full-sour (*golowka*) and final dough. The full-sour stage is continuously maintained, part of the mature full-sour being used as starter-culture, and the rest to prepare the dough. In the case of discontinuous, batch processing of rye doughs in machine-bowls, the

TABLE 40
**Formulation and regime for the preparation of a firm sour (*golowka*)**

| | Initial-sour | Inter.-sour | Basic-sour | Full-sour |
|---|---|---|---|---|
| Starter (*golowka*) | | | | |
| (from previous batch) | 1·0 | — | — | — |
| Initial-sour | — | 6·5 | — | — |
| Inter.-sour | — | — | 18·0 | — |
| Basic-sour | — | — | — | 56·0 |
| Flour | 2·8 | 6·5 | 22·2 | 68·0 |
| Water | 2·6 | 5·0 | 15·8 | 46·0 |
| Baker's yeast | 0·1 | — | — | — |
| Initial temperature (°C) | 25–26 | 26–27 | 27–28 | 28–30 |
| Fermentation time (h) | 3·5–4·5 | 4–4·5 | 4–4·5 | 3·5–4·0 |
| Final acidity-index | 9–11 | 11–13 | 13–15 | 13–16 |

full-sour (*golowka*) is normally divided into three or four parts. From one-third or one-half a fresh sour is prepared by the addition of an appropriate amount of flour and water. The residual two or three parts are used to prepare two or three batches of dough.

The formulation in kg per 100 kg flour into dough, and the regime of the production-cycle using firm full-sour, are shown in Table 41.

In the case of the soft sour (*Kwaß*), the propagation-cycle is made up of the following stages: (1) initial-sour, (2) intermediate or half-sour, (3) full-sour (*Kwaß*). The recommended formulation in kg per 100 kg flour in the full-sour, and the regime for the production-cycle using the soft sour, originate from the reference sources quoted, due to I. M. Roiter and L. J. Auerman, and appear in Table 42.

For the production-cycle, two thirds of the full-sour is used to prepare the dough, and one-third for the preparation of a fresh batch of sour. The formulation, expressed in kg per 100 kg flour into dough, and the regime of the production-cycle using the soft full-sour (*Kwaß*) is detailed in Table 43. Attention is drawn to the fact

TABLE 41
**Formulation and regime for the preparation of dough from firm sour**

| | Full-sour | Dough |
|---|---|---|
| Full-sour | 15 | 46 |
| Flour | 18 | 74 |
| Water | 13 | as required |
| Salt | — | 1·5 |
| Initial temperature (°C) | 28–29 | 30–31 |
| Fermentation time (h) | 3·5–4·0 | 1·5–1·75 |
| Final acidity-index | 13–16 | 10–12 |

Source: *Technology of Breadmaking*, L. J. Auerman, Moscow, 1977, p. 226, Table 24.

TABLE 42
**Formulation and regime for the preparation of a soft sour (*Kwaß*)**

|  | *Initial-sour* | *Inter.-sour* | *Full-sour* (Kwaß) |
|---|---|---|---|
| Starter (*Kwaß*) | | | |
| (from previous batch) | 2·0 | — | — |
| Initial-sour | — | 15·0 | — |
| Inter.-sour | — | — | 50·0 |
| Flour | 7·0 | 20·0 | 72·0 |
| Water | 6·5 | 15·0 | 54·0 |
| Baker's yeast | 0·15 | — | — |
| Initial temperature (°C) | 27–28 | 27–28 | 28–29 |
| Fermentation time (h) | 4–4·5 | 3·5–4·0 | 3·0–3·5 |
| Final acidity-index | 10 | 10–11 | 11–12 |

that the water content of the softer sour (*Kwaß*) is approximately 1·5–2·0% higher than the water content of the firm sour (*golowka*). Furthermore, in many bread-factories, in order to reduce the acidity of the bread flavour and facilitate the transport of sour through the pipelines, a full-sour with even higher water-content and softer consistency is prepared.

The amounts of firm and soft sour quoted in Tables 41 and 43, for preparing the doughs must be regarded as guidelines, and can be varied depending on raw material, processing and environmental parameters. In practice, the amount of full-sour used to prepare the dough will depend on such factors as: time of year, temperature, fermentation-activity of the sour, flour-quality, and dough-scheduling for processing.

A process has been developed using a liquid-sour S-1 by E. A. Gladkowa and co-workers of the Saratow Union for the Baking Industry. The application of this liquid-sour, known as '*Saratowskaja perwaja*', is detailed in *New Technology for the Manufacture of Rye Bread with the Liquid Sour Saratowskaja perwaja (C-1)*, E. A. Gladkowa, Moscow, Pistschepromisdat, 1955. This process involves the continuous

TABLE 43
**Formulation and regime for the preparation of dough from soft sour (*Kwaß*)**

|  | *Kwaß* | *Dough* |
|---|---|---|
| Kwaß | 20·0 | 76·0 |
| Flour | 32·0 | 56·0 |
| Water | 24·0 | as required |
| Salt | — | 1·5 |
| Initial temperature (°C) | 28–29 | 28–30 |
| Fermentation time (h) | 3–3·5 | 0·8–1·2 |
| Final acidity-index | 11–12 | 9–12 |

Source: *Technology of Breadmaking*, L. J. Auerman, Moscow, 1977, p. 277, Table 26.

propagation of the liquid-sour by adding pure cultures of lactic-acid bacteria in the form of those classified by the Seliber-system as Group B, Tribe 1, 2 and 7. During the production-cycle, one-half of the mature sour is used for doughmaking, and the residual is treated with the following nutrients: saccharified PFWS 40%, water 51%, and flour 9%, made up as a concentrate, and added at an appropriate concentration. The sour is then matured for a period of 60–75 minutes at 33–35°C, during which time it reaches an acidity-index of 8–11. As before, one-half is used to prepare the sour, the latter being mixed in stationary containers. At doughmaking, 50% sour, based on flour weight is taken and added to the flour, water and salt solution. The dough is mixed at an initial temperature of 34–35°C, and fermented for 120–140 minutes. During this time, a whole-rye flour dough reaches an acidity-index of 8·5–9·0.

In the Jaroslawer Bread-factory No. 1, and in other bread-factories, the ChTR-aggregate, previously described and illustrated, has been used extensively for doughmaking with the Liquid-sour S-1. Another Liquid-sour process without using PFWS, was devised at the Leningrad Technological Institute for the Food Industry by P. M. Plotnikow, M. I. Knjaginitschew and S. T. Schmidt as early as 1950, published in *LTIPP Report*, **3**, (11), 37, 1953. This publication was followed by another in 1961 from Leningrad entitled, *Ways of Speeding-up Rye-bread Production*, edited by P. M. Plotnikow. The basic concepts of this process can be summarized as follows. A sour is prepared from lactic-acid bacterial cultures of Group A (Tribe 8 and 27), and the yeast *Saccharomyces cerevisiae* (Tribe 90). The liquid-sour is prepared at a water-content of 75%, using a flour/water mixture as substrate. Fermentation is at 33–35°C for 120–145 minutes, resulting in a final acidity-index of 10–11. The dough is also prepared at a temperature of 33–35°C, 43·8 kg of liquid-sour being added to 100 kg of flour. After 2 hours fermentation, the dough reaches an acidity-index of 9–9·5. A detailed description can be found in the following practical reference handbooks by I. M. Roiter:

—*The Production of Baked-products—a Technological Handbook*, Technika, Kiev, 1966 (in Russian).
—*Modern Technology of Doughmaking in Bread-factories*, Technika, Kiev, 1971 (in Russian).

The rye-dough processing methods developed and recommended by the WNIIChP, the All-Union Research Institute for the Baking Industry, are published in the ZNIIChP *Technological Instructions for the Manufacture of Small-products*, Moscow, Pistschepromisdat, 1960. These recommendations are based on: (1) separately prepared liquid-yeast and lactic-acid sour, and (2) continuously added liquid intermediate products, resulting in a shortening of the dough-fermentation period before processing.

There are also a number of rye-dough processes using liquid-sours to which a part of the salt has been added.

An aspect of production common to all rye dough processing methods is the souring process cycle, whether this medium be firm or soft/liquid in form. This requires the addition to the flour/water suspension of either part of the full-sour, derived from the previous production, often containing some yeast, or a previously cultured and propagated pure culture of lactic-acid bacteria or yeast. The maintenance of the sour, using pure cultures of the fermentation microflora was devised in 1937 by the Central Laboratories of the Leningrad Union of the Baking Industry, and is detailed in the technological instruction manual last quoted. In recent years this has been accomplished in Leningrad bread-factories by using pure cultures of lactic-acid bacteria of the Tribe A-63, B-5 and B-78, and the yeast *Saccharomyces minor* Race No. 7.

Traditionally, discontinuous batch processes were employed for rye dough-making, using sours of firm or soft consistency with the help of machines with hand-manipulated machine-bowls. However, during the past 20 years, increasingly mechanized equipment has been applied. The previously described bunker-doughmaking aggregate, operating on the system of N. P. Gatilin has found wide application for rye doughs as well as for wheat doughs. This system involves the preparation of the full-sour in one bunker, and the dough in another. The full-sour when mature is metered and diluted before being pumped into the doughmaking-aggregate, an appropriate amount of the mature full-sour being kept back for the preparation of the fresh sour. In many bread-factories a firm sour is made using the doughmaking-aggregate and system of A. M. Chrenow, also described previously, using stationary bunkers for the sour and dough-fermentation.

For rye dough processing with liquid intermediate products, viz. liquid-yeast, liquid-sour, liquid salted intermediate products, etc., stationary containers are also employed, and the sour is transported either by pumping or by pipeline and gravity. For the purpose of mixing the liquid intermediate products, machines of the type ChSM-300, normally used to prepare PFWS, have been successfully applied.

Liquid intermediate products for rye dough preparation are made continuously using the type Ch-12 mixer, and transported via trough-shaped equipment for fermentation out of ChTR aggregates. In a few bakeries, the rye dough is discharged either direct from a Ch-12 aggregate or via an additional mechanical working through a worm, directly into a staionary dough-hopper above the divider.

The application of the machine-aggregates mentioned, which are illustrated in schematic diagrams featured earlier concerning wheat dough processing, offer solutions to the desire to reduce or eliminate time-consuming manual handling of doughs. More recent research into the intensive-mixing of liquid-pre-ferments by V. Kovbasa and co-workers, published in *Khlebopek. i Konditersk. Prom.*, Moscow, **28**, No. 3, 23–26, 1984, has resulted in the filing of a patent under SU-US 854349. This concerns the construction of a circulation-mixer for pre-ferments, the mixing-element rotating at 800–2400 rpm, which mixes a batch of pre-ferment within 45–60 s. Comparative tests with 3-hour pre-ferments, resulted in an improvement in wheat bread quality and simultaneous reduction of mixing times.

In 1984, a communication edited by L. Kazanskaja, of the Leningrad Department of the All-Union WNIIChP, published in a catalogue by the Light and Food Industry entitled *Technological Instructions for the Baking Industry*, details the future trends for wheat and rye bread processing in the USSR. These technologies include the application of the following: activated yeast, amylolytic enzymes, whey milk-products, potassium bromate, starch modification, ascorbic acid, emulsifiers, liquid intermediate products, magnetization of water, gelatinized flours, ultrasonic and high-frequency treatment.

Another subject of research is the use of monosaccharide high-fructose syrups, and 50% sucrose solutions added both separately, and in combination with liquid whey and yeast. The latter method shows some improvement in bread-dough quality when added at specific concentrations. In general, 50% sucrose solutions are considered necessary when doughmaking with Grade 1 wheat-flour in order to produce maximum volume bread, depending on the flour gluten-content, e.g. a flour with 30·5% gluten requires the addition of 3% of a 50% sucrose solution, and a flour of lower gluten-content correspondingly less.

Mathematical modelling techniques are also being applied to the process of autolysis and dough-fermentation, by measuring the aerobic and anaerobic oxidation-reduction potential and pH.

In the USSR, the variety of bread available is considerable, in the three largest cities on average 10–20 main types of bread, small wheat-flour products, e.g. bread-rolls, numerous prolonged shelf-life products, e.g. '*Kringel*' (a bagel-type product), in many shapes and sizes, numerous types of rusk, crispbreads, salted-beer-snacks and various plaited-breads, confectionery and pastry products similar to those available in EC countries. Both rye and wheat breads are available in panned and hearth forms, the latter in long and round shapes. Also, a large number of specialty-breads and dietary-breads, many of rustic origin and formulated to improve the nutritional contribution to the daily diet. Typical original specialty breads include, mustard-seed-oil bread with a strong flavour of its own, a bread with a strong seasoning for consumption with caviar. Top-quality wheat bread is formulated with sugar, margarine and milk, and is much favoured by the consumer. Apart from that, wheat-flour breads are differentiated according to the types of flour used to make them, Top-grade (0·55% ash), 1st grade (0·75% ash), 2nd grade (1·25%). Most breads have a relatively thin crust, and a moist but elastic crumb structure.

In the USSR, bread is both the best value and the cheapest of foods, hence the high consumption of about 150 kg/head per year. In Moscow the preference for rye-bread has fallen in favour of white goods, but in Leningrad the consumption is more evenly balanced.

By way of comparison, the average industrial worker would have to work for about 20 minutes for 1 kg rye-bread, 30 minutes for 1 kg wheat bread, 1·5 hours for 1 litre of full-fat milk, 2·5 hours for a 500-g pound-cake, and about 4 hours for 1 kg of butter-creme torten/gateau. In the large cities, bakery shops are open from 08.00 to 23.00 hours and on Sundays from 09.00 to 21.00 hours, and in many shops the most popular breads are sold from specially constructed automatic vending machines, which attract queues of people. Baked-products are generally only sold in bakery shops, and only in the biggest stores, in special departments, can bread be found. Moscow has more than 700 bakery shops, and bread is delivered from the bread-factories 4–5 times a day, great importance being placed on fresh bread and baked-products reaching the consumer.

## 1.7.8 HUNGARY

Visitors to Hungary confirm the high quality and freshness of the bread and other national specialty baked-products. This is due not only to the craftsmanship and technical expertise of those engaged in bakery production, but also to the success of those scientists and agronomists in the grain-breeding institutes who have maintained the baking-quality of new varieties of wheat and rye. The quality and unique processing characteristics of Hungarian wheat were well known in the 1930s, both in Western Europe and Britain, when Hungarian wheat was exported. For example, the wheat variety Bánkuti 1201 which was grown for 42 years up until 1972, showed excellent breadmaking qualities and was maintained by careful husbandry, thus avoiding degeneration due to excessively intensive environmental conditions of breeding and growth.

There are about 17–20 main types of wheat and rye bread, and various types of 'Semmeln' made from Vienna-type doughs, and also 'Kipfel', which are made from a shortened-layered dough similar to croissants, Since Hungary is a very popular tourist area, especially around the Lake Balaton, bakery production increases considerably during the summer season. Bakeries working a 2–3 shift system produce daily during the summer season about 160 t of bread and on average 1 million Semmeln and Kipfeln, which are delivered once or several times daily to an average of 2000 points of sale, which include in-store bakeries (Kaufhallen), hotels, guest-houses, holiday chalets, and factories.

Many bakeries are equipped with machines and production-lines made in the GDR, which have a good reputation for reliability, e.g. bread-roll lines producing 64 000 bread-rolls during an 8-hour shift. For the maintenance of the machines and production lines, every large bakery has 2–10 mechanics, and a repair workshop with about 30 trained technicians and craftsmen for repairs and the construction of new machinery for production-rationalization and other innovative work.

Quality and freshness are maintained by setting norms, which are controlled by the operative personnel during the production process, if these specification norms are not realized, financial deductions are incurred. Every bakery unit is set a planned target objective, and the economic results are measured according to the surplus of income over expenditure.

As a typical case-study of an Hungarian bakery consider the State Bakery of the Komitat (district) of Veszprém, which is one of the largest in the Republic, catering for the local inhabitants and tourists on the northern bank of Lake Balaton. This bakery was built in 1961, and employs 140 personnel including the administration, being one of the eight largest bakeries within this bakery-group. The bakery was re-equipped in 1984, and the most important technological equipment consists of the following: one bread-roll line with a five-in-a-row, fully automatic divider–moulder, and a wire-band tunnel oven type BN 25, manufactured in the GDR; one bread line with wire-band tunnel oven with 52 m$^2$ baking-surface, and throughput of 1 t/hour, manufactured in Poland; one wire-band tunnel oven, type PTC-24, with 24 m$^2$ baking-surface, manufactured in Poland, for the production of various products e.g. French sticks or *baguettes*: a doughmaking station equipped with Hungarian high-speed mixers. The erection of a flour-silo is planned.

The production of *Kipfeln* is carried out on the bread-roll line, with a *Kipfel* rolling-machine (manufactured in the bakery-workshop) placed between the divider–moulder and proofer. The dough-pieces proceed direct from the spreader-band into the aggregate, and after the rolling-up operation, are positioned for a three-in-a-row outlet by a special mechanism. The throughput of this line is 4000 pieces/hour. The production-output of this bakery with a two-shift system is approximately 165 t bread, 64 000 rolls and 3000 *Kipfel*, daily. In addition, on three days per week, 4000 *baguettes* are produced on each day.

Typical rustic products produced in this district are '*turosbatyũ*' and '*tepertös pogácsa*', a type of small-bread with a shallow-cut top surface, being round in shape.

Nine different varieties of bread are produced, which includes the popular 'Bakonyer-bread', which is a mixed rye/wheat formulation containing 50% rye-flour. This bread is produced three times per week, 12 t of which per week is delivered to the capital city, Budapest. The confectionery department, which operates on a single-shift system, is housed separately, and is equipped with an 80-litre cake-mixer, a high-speed dough-mixer HLK-50, dough rolling-machine, croissant rolling-machine, dough-nut production unit, etc. On average, 80 different articles are produced, which include the following: Danish and puff-pastry varieties, strudel, choux-pastry for eclairs and cream-buns, Hamburger-rolls (120-gram units) for the snack trade, and the unsweetened local products *turos-batyû* and *tepertös pogácsa* using quark and dripping-with-crackling.

The delivery area covers a radius of 30 km, and all products are despatched fresh by 12 lorries owned by the bakery. (The Várpalota Bakery—a modern production unit in the District of Veszprém in the Peoples Republic of Hungary, *Bäcker und Konditor*, **1**, 11–12, 1987.)

Hungary also exports continuous doughmaking aggregates such as the FTK 1000 UM, which is used in the GDR for the manufacture of sliced-bread varieties, e.g. rye-bread, mixed wheat/rye bread, specialty-breads, wheat bread and toast-breads. A similar larger unit is the FTK 1500, which also requires a continuous sour-process for rye-doughs, providing a mature full-sour continuously for a working-week. Part-lines are also exported for the manufacture of wheat-bread, which includes the Divider S-70 and Band-rounder S-64.

### 1.7.9  CZECHOSLOVAKIA

Characteristically, the baking industry in Czechoslovakia manufactures a large assortment of traditional baked-products, which are made from old recipes either of Czech of Slovakian origin and include various types of bread. However, many new specialty-breads have been developed in recent years, in particular breads designed to provide an improved nutritional-balance according to current dietary requirements. These include in particular specialty-breads containing a higher level of dietary-fibre and other dietary supplements.

In addition to the standard wheat- and rye-flour types, the milling industry in Prague and Bratislava manufacture special and composite ready-mixed flours for the housewife, e.g. *Knödeln* (dumplings), bread, and various cake-mixes.

The standard types of bread of 500 g and 1 kg, are similar to those produced in the other central European countries, based on wheat- and rye-flours and mixtures of these flours sold as mixed-bread (*Mischbrot*). Most of the bread and baked-products are packed in foil or packaging during manufacture on line-production systems.

Czechoslovakia is a leading exporter of machinery for the food industry in general and the baking industry particularly. Technoexport (Prague) and Technopol (Bratislava) and the national manufacturer Strojobal, also based in Prague, are all leading exporters. Continuous sour-dough and doughmaking lines for rye and mixed wheat/rye bread, are assembled from the flour-weighing aggregate T 437.0, Mixing-aggregate T 457.0 coupled with the continuous sour-dough and processing aggregate T 995.0. All these aggregates are manufactured by Topos of Sluknov. This centrally controlled system is made up of a total of 18 aggregates and has a throughput capacity of 1800–2500 kg dough/hour. The doughmaking-lines KVT 1000, KVT 1500 and KVT 1800, widely used in the GDR, are based on a continuous sour, whereby a mature full-sour must always be available for processing into the final dough.

Wire-band tunnel-ovens manufactured in the PPC Type series by the milling machinery group in Pardubice, are a popular choice for many bakeries in the GDR and elsewhere, when increased capacity and reconstruction is necessary. Heating is indirect with oil or gas, offering a throughput of 450–2000 kg bread or bread-rolls/hour. The ovens are manufactured in six sizes, three with a working-width of 2·10 metres and three with 3·00 metres. The baking-surface choice is from 25–108 m². A common assembly of bread production-lines is the doughmaking line KVT 1500, and the wire-band oven PPC 381.12, providing 81 m² of baking-surface. The PPC series ovens have an improved energy-efficiency.

Complete bread-roll lines manufactured in Czechoslovakia are also widely accepted, the two lines t 940 and t 985 producing 7000–9000 bread-rolls/hour, at a dough throughput rate of up to 700 kg/h. Such lines require four operators: for doughmaking, machine-control, divider–moulder, and sorting of the dough-pieces as necessary.

These lines comprise the following aggregates (refer to Fig. 27): salt-solubilizer

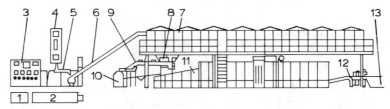

Fig. 27. Bread-roll production line manufactured in Czechoslovakia. 1. Salt-solution and water-tempering tank. 2. Preparation assembly for ingredients. 3. Dosage aggregate for liquids. 4. Flour dosage aggregate. 5. Mixer. 6. Band-conveyor to dough-fermentation unit. 7. Fermentation unit. 8. Dough remix and shaping assembly. 9. Dough-strip shaper and conveyor. 10. Divider–moulder and spreader-aggregate. 11. Proofer. 12. Oven-transfer bridge. 13. Wire-band tunnel-oven. (Source: *Technology of Industrial Baking*, R. Schneeweiss and O. Klose, VEB-Fachbuchverlag, Leipzig, 1981, p. 314 (Fig. 148).)

and water-tempering tank, which is bifunctional. The water-tempering section has a capacity of 45 litres, an automatic levelling device, inlet and outlet with hot and cold water feeds, which regulate the temperature after setting. Salt is added manually, this section holding a capacity of 300 litres. A pump circulates the water from below up through the salt-layer, ensuring a saturated solution of about 25% concentration controlled with a salinometer.

The preparation-station for the liquid components consists of six containers for shortening, malt, sugar and yeast, each container having a stirrer, cover and easily changeable sieve. The holding capacity of each container is 220 litres, and the shortening container double-walled and heated with steam. The malt and yeast are provided for by two interchangeable containers, with one for the sugar. The mixing ratios are: malt/water 1:2, sugar/water 1:1, and yeast/water 1:1. Water for the solution or suspension of the components is provided for by a mixing and metering apparatus, which is part of the preparation-station. Hoses are used for connections to the preparation-station, and electrically heated pipeline for the shortening.

The preparation-station for the liquid components is equipped with eight pumps, which are controlled centrally. The metering of the various components are set by hand-operated wheels. The metered shortening and the other mixed liquid components are added to the mixer by separate pipelines. The flour-weighing station consists of a hopper with a capacity of about 150 kg, vibrators being installed to feed the flour to the weigher and discharge it from the container. The vibrator giving three deliveries/minute, the weight of a delivery being pre-set. The weigher is emptied through a pipe direct into the mixer, any malfunctions being registered both optically and acoustically.

The mixer is a continuous high-speed design, within the horizontally mounted barrel two screw-like shafts rotate in opposite directions, which mix and develop the dough, and extrude the finished mass. The energy input is about 4 Wh/kg, the actual mixing-time depends on the volume of dough relative to the volume of the mixer-barrel, mixer-rpm and the throughput, being on average 5 minutes. The maximum capacity of the mixer-barrel is 60 litres, which corresponds to approximately 35 kg of dough, the volume of dough resident in the mixer-barrel can be regulated by the opening and closure of a sliding-vent at the discharge-end. The rpm of the mixing-elements can be regulated from 27 to 78, but the motor-output at maximum throughput and maximum-speed is inadequate. An ammeter records the current consumed.

A band-conveyor transports the extruded dough-ribbon upwards at an angle of about 40 degrees to the horizontal, into the dough-fermentation channel, the band being coated with plastic. The dough-fermentation channel, which is built on top of the line t 985, consists of a closed housing, inside which two band-conveyors (plastic-coated) are built, one above the other. At the end of the uppermost band-conveyor is a driven-roller, which pushes out the partially ripened dough. The dough then passes as a wider ribbon onto the uppermost band-conveyor, through the whole housing, and is conveyed back onto the lower band-conveyor. The dough maturing-time can be regulated by changing the band-conveyor speed. The internal-climate can be adjusted by steam-injection and forced-air control.

The dough-extrusion and degassing equipment is located under the fermentation-channel. It consists of a hopper with a pair of rollers mounted inside. The fermentation gases are pressed out of the dough, and a dough-band is formed. The throughput rate of this set-up is separate and can be controlled directly.

The dough-band shaping and conveyor set-up is a travelling-band over which vertical and horizontal rollers are mounted, resulting in a dough-band with approximately the same cross-section being formed. The dough-band is subjected to warm-air currents from all sides, in order to ensure the retention of a dry surface. The positioning of the rollers and the speed of the travelling-band can be regulated to suit the feed-rate of the divider–moulder. The return conveyor is also a travelling-band, and allows the return of misshapen dough-pieces into the hopper and thence to the dough-band shaping and conveyor set-up.

All control-functions and switches are actuated at the control-panel. The line for the processing of wheat doughs t 985 consists of the following:

Dough divider–moulder
Spreader-aggregate
Proofer
Depositing-device
Control-panel

The dough divider–moulder produces six units in a row (a six-pocket configuration). The spreader-aggregate comprises six small band-conveyors, fitted with 6-cm width belts, which are driven by the divider–moulder. The spreading operation is effected by timing and synchronization; the whole aggregate rotates about one point and takes up three different positions. Since in each case, six dough pieces are set down, rows of 18 dough-pieces are formed. The proofer consists of five band-conveyors positioned one over another. At the inlet, there is a felt-covered rotating roller positioned above the band-conveyor, which elongates the dough-pieces by flattening as they pass through. The length is controlled by the distance between the roller and band-conveyor. An assembly consisting of rods with rollers, rotates the dough-pieces through 90 degrees. At the end of the uppermost band-conveyor is spreader-band, followed by a stamping-unit equipped with 18 stamping-rollers. These rollers have various profiles, thus allowing a wide choice of shapes. The shaped dough-pieces then pass to the next band-conveyor, where they are inverted. Adjacent to the lower band-conveyor, on which the dough-pieces are deposited with the stamped-profile uppermost, is a transfer-bridge which transports the dough-pieces at an upwardly inclined angle into the wire-band oven. The transfer-bridge can be raised when the line is not in use. The proofing time can be regulated by changing the rpm of the motor. The proofer can be humidified and warm-air passed in as required. The transfer-bridge from proofer to wire-band oven consists of a conveyor with rollers, over which is mounted a humidifying and sprinkling device. The surface of the dough-pieces are moistened by means of a brush-roller fed with water by another roller rotating in a water-bath. For sprinkling of poppy-seed, caraway or salt, a hopper fitted with a dosage-oscillator is used. Surplus material is collected in a box under the rollers and re-used. The control-panel serves to provide all activation and control functions.

All components of the doughmaking and dough-processing lines are synchronized and unified by appropriate transport-elements, thus ensuring a smooth production flow. For baking, a wire-band oven with a band width of 2·1 metres, and a baking-surface of 25 m² can be utilized. The technological parameters of the line can be summarized as follows:

| | | |
|---|---|---|
| Throughput | up to 700 kg dough/hour | 7000–9000 bread-rolls/hour |
| Mixing-time | about 5 minutes | |
| Floor-time | 60–180 minutes | |
| 1st proof | 1–7 minutes | |
| 2nd proof | 30 minutes | |
| Scaling-wt. | 30–60 g and 60–90 g | |
| Shape | long, with various design-finish | |
| Dough-pieces | | |
| per row | 18 | |
| Personnel | 4 operators | |

At the 1985 Leipzig Exhibition, the Czechoslovakia export-organization, Technopol Bratislava, displayed their latest modifications to the continuous doughmaking unit, KVT Type 955.0 for rye and mixed rye/wheat doughs, which has a throughput of 1800–2500 kg dough/hour. A new flour-weighing unit T 437.0 was introduced, together with a new technical solution by TOPOS, Sluknow, for the sour-dough ripening containers. Instead of the large compartments formerly fitted, seven compartments have been introduced. This reduces the quality differences that occur owing to density variations of the mature full-sour. In addition, the new vertical conveyor T 725.0 transports the dough between two plastic-coated bands vertically into the proofer, thus reducing the required floor-space considerably. The improved flexibility of the part aggregates due to the 'building-block' construction-principle, increase in throughput up to 2500 kg dough/hour, energy saving of about 35%, and reduction in required floor-space area, all offer new advantages for the baking industry.

### 1.7.10 POLAND

In common with the other countries within the Socialist community, the population in or near the capital and larger cities are supplied with fresh baked-products from large industrial bread-factories. Also, as practised in the other countries, all raw materials and the full-range of finished products are subjected to strict quality-control regimes under the direction of the factory laboratory. The basic product varieties are:

—standard-bread wheat/rye 60:40 produced in 800-g units
—wheat bread varieties produced as 400-g units, which are popular, and distributed to over 200 points-of-sale

—long bread-rolls and *Semmeln* or star-shaped bread-rolls
—*baguettes*, known as '*bagietky*', which, since their introduction in Poland in the late 1970s, have shown a steady growth in demand. In 1980, following the success of the manual production of *bagietky* by smaller bakeries, a special bakery was equipped with a French Pavailler plant in the centre of Warsaw for the mechanized production of this increasingly sought-after bread.

As a typical bread-factory case-study in Poland, one can consider the large bakery and confectionery situated in the Krakowiak Street near the Warsaw airport. This bakery was built in 1976, but the project is not yet completely finished, since more production-lines will be installed, and a final personnel complement of about 200 is planned. The cooling-plant and despatch areas are not yet installed, and the current level of personnel is about 50 trained bakers, the scheduled recruitment of personnel being difficult to achieve, technically trained personnel being in short supply.

Bread production started in December 1982, and is made up of silos and heating-oil storage with workshops outside the production area. Within the production-hall are six production-lines and wire-band ovens adjacent to a despatch-area. Two laboratories, administration and facility rooms are on the first floor. The confectionery building, not yet completed internally, will have a production-hall, social and administration suites, including suites for the bakery management.

The production programme for the confectionery-hall will include amongst the usual pound-cakes, Danish and puff-pastries, gateaux, and filled and decorated products, such specialities as the *panettoni* (of Italian origin), *babka*, biscuits and sweet bread-rolls amounting to about 20 t/day. Also projected is a bakery-shop adjacent to the main entrance, facing onto the street. This new bakery is, in effect, a prototype, originally intended for a capacity of 60 t/16 hour, and after project-changes increased to 70 t/16 hour, for the supply of baked-products to the Ochota and Mokotów areas of Warsaw.

Most of the plant and machinery are of Polish manufacture, built in the machinery works 'Spomasz' (however, two continuous sour-dough and dough-making lines KVT 1800 and 1300 came from Czechoslovakia): dough divider–moulders, transfer-assemblies, proofers, band-conveyors with photo-cells, together with six doughmaking-stations for discontinuous dough preparation. The six production-lines are equipped with two wire-band ovens of $72 \, m^2$ baking-surface, two wire-band ovens of $52 \, m^2$ baking-surface, and two wire-band ovens of $25 \, m^2$ baking-surface. Although the bakery was intended for two-shift working, only three production-lines using continuous dough-preparation were in use during 1985, owing to the necessity for more exhaust-gas collectors to be installed. Originally, environmental experts had planned for all exhaust-gas channels to be connected to one exhaust-gas collector, but this led to gas-currents causing interference in the exhaust-gas circulation, and consequent heating-burner ignition failures. On completion of these modifications, and the additional recruitment of 70–80 more personnel, the other three lines will become operational, thus meeting the target demand for bread and bread-rolls. During 1985, at times only nine operators were servicing the three lines during the night-shift.

Daily outputs of 22 000–40 000 loaves of bread of 800 g each is the average, and on the peak delivery days, viz. Mondays and Fridays, 30 000 and 40 000 respectively are baked.

Owing to the increasing demand for *baguettes* (*Bagietky*) in Poland, it became necessary to rationalize the production process, which hitherto had been performed manually. Therefore, a machine was developed by the Polish Research and Development Centre for Grain and Feed Industries 'Spomasz' in Bydgoszcz. The machine became mass-produced in 1983, being available in three versions, viz.

— type O2WAO, which is a table-model requiring manual feeding and removal of the elongated dough-pieces, the function of the machine being that of a long-moulding operation;
— type O2WA1 in the form of transportable unit mounted on a stand with wheels, but still manually fed and removed;
— type O2WA2 as a transportable unit with extending conveyor-belt for use with production-lines, but also with manual feed option.

This long-moulding machine, is also suitable for other baked-products, e.g. croissants, *batons* and Wroclawer large-rolls.

In the case of the long-moulding machine type O2WA2, the moulded dough-piece falls onto the extending conveyor-belt. The machine can be rotated through 180 degrees, so that the feeding and removal operation can be completed from the opposite side of the extending conveyor-belt, e.g. for manual operation, as desired.

Technological parameters:
Dough-piece weight (after baking)                    0·1–1·0 kg
Maximum elongation of dough-piece                    750 mm
Maximum throughput at minimum dough weight:
—with manual feeding                                 2 400 pieces/hour
—with mechanical feeding                             5 000 pieces/hour
Motor-output                                         0·55 kW
Motor-output of extending conveyor-belt (type O2WA2)  0·25 kW
Weight of various models:
            type O2WA0 170 kg, type O2WA1 194 kg, type O2WA2 285 kg
(Source: *Przegl. Piek. i Cukiern.*, **31**, 3, 12–14, 1983.)

## 1.7.11 COLLABORATION WITHIN THE COMECON BAKING INDUSTRIES

Since the 1950s, considerable research and development work has been carried out in the field of bakery industrial mass-production within the Socialist community countries (RGW or COMECON). Much of this effort has been directed to the continuous production of bread and bread-rolls, which has involved machine construction development to meet the demands of new processing technologies. In the Socialist baking industry the ultimate aim is to provide the community with the quantity, quality and assortment of bread, small yeasted products and confectionery required, by using the best possible input of available materials and manpower.

Society and technology demands from the technical bakery worker, as co-owner and operator of modern mechanized and often automated equipment, a high degree of technical ability, political responsibility, moral and character qualities.

In the construction of wire-band tunnel-ovens, the VEB Kyffhäuserhütte Artern, GDR, produced a versatile and flexible series of ovens, which constituted an advance in construction and heating-technology. This type of oven is the most widely used in Socialist industrial bakeries, and includes the BN 25 (25 m² baking-surface), BN 50 (50 m² baking-surface), and since 1976, the BN 72, with a baking-surface of 72 m² and working-width of 3 metres. These ovens can be controlled from either the right or left side, as desired, and the heating-options are, city-gas, natural-gas or heating-oil. The divider–moulder type 612 (Habämfa), produced in the GDR, was for many years the only dough-processing machine, and was used throughout the Czechoslovak baking industry.

From Hungary, the divider–moulder type S 70 has become accepted in all countries of the Socialist community. This machine meets the diverse requirements of rye, rye/wheat (up to 40% wheat-flour content), wholewheat and wheat dough processing. It can be used for various scaling-weights and long or rounded bread shapes.

Also from Hungary are the universal wheat and rye processing aggregates FTK, the FTK 1000 having a throughput of 1000 kg rye dough per hour, or 810 kg wheat dough per hour. Other aggregates include the FTK 1000 U, FTK 1000 UM and the FTK 1500, all of which require a continuous sour-process, organized in such a way that a mature full-sour is constantly available for continuous processing into dough. This processing-line is sophisticated in conception and design, and is controlled from a central control-panel, with additional control-units built into the component-aggregates. The central control-panel has the following blocks of control-units: energy-distribution, programme-control, measurement and recording facilities, switches for the various line-functions, malfunction alarm systems, switches for manual or automatic control, safety-switches for the electrics, illuminated schematic-flow board. The various components of the complete aggregate are as follows: yeast-suspension unit, water-mixing and metering unit, flour-weighing assembly, sour-dough homogenizer and pump, sour-dough fermentation container with 12 compartments and filling and emptying assembly, sour/sponge cooler, liquid dosage assembly, e.g. salt and ascorbic acid, water-jacketed mixing aggregate with flattened rods mounted on the shaft and horizontal barrel. The movement of the mixing dough through the barrel relies on the pressure exerted by the following dough, thus intensifying the mixing process. A buffer-plate mounted between the end of the shaft and the outlet orifice enhances mixing, and also provides steady conditions for measurements.

Dough consistency depends on the energy absorbed by the dough during mixing, the maximum being 6–6·5 Wh/kg. Doughs which are too firm often show energy absorptions of 8 kW or more, in which case the doughmaker will adjust the reading to 7 kW, finding that the dough consistency is now optimal. However, in fact, the net energy absorption is 6 kW, which corresponds to a dough which is normally too soft for the divider–moulder. This discrepancy is due to residual dough of too high a viscosity sticking to the outlet-cone, thus giving a false energy reading. Such examples illustrate the necessity for the doughmaker to gain experience concerning the behaviour of a machine. Biological material changes with time, and compensatory measures to correct for such changes must frequently be made, hence the provision for manual correction in most automated production lines. The dosage-pump feeding sour or sponge to the dough-mixing aggregate is powered by a motor of synchronized, adjustable variable-speed design, allowing the rpm to be varied depending on the instrument readings. When the dough consistency is too firm, the rpm increases so that more sour/sponge liquid pre-ferment is added, thus adjusting to optimal consistency. Mixing-time is standard at 40 s. Nevertheless, the state of maturity of the sour/sponge added as liquid pre-ferment will affect its fermentation-power, and this variable cannot be corrected by automation.

Czechoslovakia has contributed the continuous sour-dough and dough-processing lines type KVT, which includes the KVT 1000, KVT 1500, and KVT 1800 with throughputs of 1000, 1500, and 1800 kg dough/hour, and the KVPT lines for the production of small baked products. All these lines have proved themselves in the baking industry of the GDR for many years. As a result of specialization arrangements within the Socialist community (RGW), Czechoslovakia has mass-produced these lines for the other partners. The processing aggregates are similar to those of the FTK series, the dough-fermentation container also being divided into 12 segments or compartments. The mixing-aggregate interior is readily accessible, since the top cover around the mixing-elements can be opened on hinges. The dough components are mixed with two elements rotating in opposite directions, fitted with spiral-shaped working-strips wound around each drive-shaft (similar in design to a cylinder-type grass-cutting machine). Power transmission from the motor is by shaft and triplex chain-drives to the main mixer-shaft, and thence by cog-wheels to the parallel mixer-shaft. An ammeter measures the current-uptake of the motor. Since this is proportional to the resistance the dough offers to the mixing-elements, the amount of current is a function of the mixing-intensity. The latter variable can be regulated by a screw, which controls the area of the outlet-orifice from 0 to 75 cm$^2$.

Bulk-flour storage technology from the GDR has also been successfully applied in Czechoslovakia and other RGW-countries.

These exchanges of technologies include the respective Research Institutes and Grain-processing Institutes which consult one another on common solutions to problems and scientific and technical information. Also, the technical press, and specialized trade and technical high schools maintain exchanges of ideas and technical personnel. Industrial bakery collectives also exchange visits with those in the USSR and their partners, thus gaining processing and new product ideas, which introduce new baking-cultures on a mutually profitable basis; such contacts are drawn-up as 'friendship-contracts', between neighbourly Socialist countries.

This philosophy strengthens the growth of technology, and avoids unnecessary reliance on Western technology, simultaneously economizing on hard-currency payments for Western machinery.

The participation of Socialist community countries at the international level within the framework of the International Society of Cereal Chemists (ICC) dates back to the late 1950s, and at the Discussion Conference of the ICC in Vienna in June 1964, the following Socialist countries sent delegates: USSR, GDR, Poland and Hungary, as well as Cuba and Yugoslavia.

In 1966, the study-group 'Documentation' edited a *Multilingual Glossary* for the ICC of about 1000 special terms used in the analysis and processing of cereals. The languages included: English, French, German and Russian, which was a firm start for more efficient use of international literature. This was then supplemented by subsidiary booklets in Czech, Danish, Dutch, Finnish, Hungarian, Italian, Norwegian, Polish, Portuguese, Serbo-Croatian, Spanish and Swedish. These glossaries were produced under the Direction of Professor R. Schneeweiss, Director of the Institute for Grain-processing of the GDR, Potsdam-Rehbrücke, in collaboration with many specialists from Capitalist and Socialist countries. The latter Institute also organized a Conference at Potsdam, entitled 'International Problems in Modern Cereal Chemistry and Processing'. The 6th Conference held in September 1975, which included papers from a very broad spectrum of countries comprised the following programme-sections: dietetic-foods, additives, grain-quality, yeast-quality, line-process, rationalization in the baking industry. At this conference, the Research Institutes of USSR, USA, UK, Canada, FRG, Poland, Hungary, Bulgaria, GDR, Czechoslovakia, France and Belgium were all represented, as well as some private industrial corporations.

## 1.7.12  UNITED KINGDOM (UK)

During the post-World War II years, 1945–50, the war-time extraction rate of 85% remained mandatory for the miller. However, he succeeded in milling a flour of reasonable quality, and not just a coarse meal, by milling at lower moisture and performing more work than usual on break-rolls to remove as much endosperm as possible, followed by the scratch-system. The aim was to remove the bulk of the endosperm in the form of chunks or nodules, referred to as 'semolina' or 'middlings'. When the endosperm is in this form, it can be 'purified' from adhering branny particles by sieves and air currents, thus producing a flour of good colour and free of branny-specks. The storage life of these high, or 'long-extraction' flours, as they were then termed, was restricted to 4 weeks maximum, owing to the high germ content. The official labelling at that time was 'National flour', and in 1950 the National flour extraction rate was reduced to 80%. All these flours described were 'straight-run' flours, i.e. 80% extraction rate, which means, from 100 parts of wheat, 80 parts National flour and 20 parts offal (low-grade flour stock and bran). At that time UK mill wheat-mixtures or 'grists' were made up of over 50% imported, third-country (non-European) 'hard' wheats, mainly of North American origin. From 1945 up until the early 1950s, wheats were allocated to the miller by the Ministry of Food and Agriculture, and National flour contained about 11·0–12·5% dry-gluten. During this era, many large bakeries and some smaller ones, purchased English wheat-flour from a local miller and blended it with an imported strong flour milled from Manitoban wheat, thus producing a 'tailor-made' breadmaking-flour at lower cost. The best English wheat variety for such blending at that time was Yeoman, containing up to 11·0% dry-gluten. The addition of 5% gluten, obtained from Manitoban wheat, when blended with an English wheat-flour, would also produce a comparable breadmaking-flour. Two proprietary brands of high-protein/low-starch breads, marketed under the trade-names 'Procea' and 'Nutrex' were also produced from soft-flour/gluten blends.

Owing to their high-gluten, low-carbohydrate balance, these breads had a much improved shelf-life compared with ordinary National bread, made with National-flour. The National-flour was marketed in hessian (jute) bags containing 140 pounds (lb) each, the standard unit of flour measurement in UK bakeries was 2 bags or 1 sack (280 lb). Only the large 'plant-bakeries' had bulk-flour storage facilities, the average bulk-flour intake being a 12 ton (2240 lb × 12) tanker, which could be unloaded in about 1 hour, and pneumatically transported to the appropriate processing-station in the bakery at the rate of 1 ton within 9–10 minutes. At the doughmaking-station, the flour was received in a twin-hopper, which could be discharged by a worm into the weight-sensitive lower-hopper, pivoted on two weight-cells, having been pre-set to the requisite weight. Typical automatic flour-weighing aggregates were manufactured by Dietrich Riemelt of Frankfurt-am-Main, who are specialists in bulk-flour-handling equipment and silos. Doughmaking was completed in Baker Perkins 'Viennara' single-arm, rotating-bowl mixers of 2 sack (560 lb) dough capacity. The mixer-arm (T-shaped) rotating in a vertical plane, at right-angles to the rotating-bowl. Average mixing-times were 15 minutes, after which the bowls were transported on wheels to the fermentation-room, the bowls or pans acting as containers for return to the mixer for remix, known as a 'knock-back' or 'punch-back', at the appropriate fermentation stage. Such mixers were generally known as 'open-pan mixers', being at that time the favourite form of dough-mixer, especially in large bakeries. However, the high-speed, stationary-bowl-horizontal-bar mixer, already the most popular high-speed machine in the USA, and under manufacture in the UK by Baker Perkins, were gaining in popularity, and the larger chain-bakeries were standardizing on this mixing unit. These mixers were powered by a 30-horse-power motor, which necessitated the mixing-chamber to be enclosed in a water-jacket. However, the geometry of the mixing-elements on the centrally mounted drive-shaft, rotating about a horizontal-axis, was less severe in mixing-intensity than the American-version. Nevertheless, the mixing-time was critical, and was restricted to: 1·5 minutes at slow-speed, and 3·0 minutes at fast-speed, making a total mixing-time of 4·5 minutes. The sliding front-panel door was electrically operated with safety-sequence, facilitating the discharge of the mixed dough.

Irrespective of the type of mixer used, the favoured processing procedure was a straight-dough of 3·5 hours fermentation in bulk, a typical formulation being as follows:

| | | |
|---|---|---|
| Straight-run flour (protein 11–12% at 14·0% moisture) | 100·0 parts | 560 lb |
| Water | 57·0 | 320 lb |
| Yeast | 1·1 | 6 lb |
| Salt | 1·6 | 9 lb |
| Yeast-food, bromated and malted | 0·2 | 1 lb |
| Glycerol monostearate hydrate | 0·7 | 4 lb |
| Enzyme-active full fat soy-flour | 0·9 | 5 lb |

Procedure:

| | |
|---|---|
| Dough temperature | 78–82°F |
| Remix/knock-back/punch at | 2·25 h |
| Work-off after a further | 1·25 h |
| | 3·50 h |

Alternatively, the same formulation was often used, but without adding the salt, which was instead added about 2 h after mixing the dough, blending it in with the open-pan Viennara mixers for about 10 minutes.

The final dough was then rested for the remaining 1·5 h, when it was worked-off. This processing technique is known as the 'delayed-salt process', and was often utilized both by the larger chain-bakeries and the medium to small craftsman-baker alike. Normally, in the large-chain bakeries the doughs were wheeled into the fermentation-room in the dough-mixer pans for controlled fermentation in bulk at 66–75% relative humidity and 80–82°F. Under such conditions of controlled temperature and humidity, the average loss in weight of the doughs over 3·5 h in the fermentation-room was 0·44%, whereas the same weight of dough left to bulk-ferment in the bakery under uncontrolled conditions showed average weight losses of 0·68% over 3·5 h, although covered with linen-sheeting (author's unpublished data, 1959).

The standard baked weight of UK National bread units was, and remains to date, 28 ounces and 14 ounces, although completion of harmonization with the other EC countries into 1-kg and 500-g standard bread units, and units of 60 g up to 200 g for small bread and rolls, will materialize with time.

In order to comply with the legislation, the baker must adjust his scaling-weights to allow for evaporation losses resulting from proofing, baking and cooling. Since this will depend on the processing conditions being used, and the proofing, baking and cooling regimes, only data-ranges can be cited as guidelines. Additional variations are introduced owing to pan size and shape and moulding methods, i.e. one-piece, four-piece/eight-piece moulded-round and flattened, placing face-to-face in the pans, thus allowing the bread to be cut with the grain instead of across it. This moulding-process is known as the 'Supertex', due to Baker Perkins Limited. Similar results are obtained by twisting two pound-sized dough-pieces passed through a long-shaper, which originated in the USA.

In the large chain-bakeries 2·5 to 3·75 ounces were allowed for evaporation losses in the standard 28-ounce National unit baked in a 9-inch open-pan. Approximate National 28-ounce unit yields per sack (280 lb) of flour were 214–220. Owing to cell-density variations and those due to the various pan-shapes and sizes and hearth-bread varieties, followed by moulding-techniques, careful consideration as to the baking-time must always be given in order to avoid circumstances whereby divider weight-settings are producing short-weights after baking.

The following was a typical schedule for a large bakery capable of throughputs of 10 sacks/h 14-ounce National bread units, or 14 sacks/h 28-ounce units:

*Dough-room*
Open-pan 'Viennara'; mixing-time, 15–20 min; dough-temperature, 78–82°F
Baker Perkins
Horizontal-bar: mixing-time 1·5 min slow-speed
High-speed                              3·0 min high-speed
Baker Perkins
Atmospheric temperature 82°F; relative-humidity 54%
Doughs prepared discontinuously every 15 minutes
Delayed-salt addition 10 minutes
2 sack (560 lb) flour per charge, yielding 428–440 units at 28 ounces

*Fermentation-room*
Relative humidity 66–75%; temperature 80–82°F
Residence-time 2 h 50 min to 3 h 30 min average range

*Divider*
Average processing-times: National 14-ounce units, 15–20 minutes; National 28-ounce units, 10–15 minutes for 1–2 doughs.
Cleaning-time interval for changeover from National white bread to wholemeals, 10 minutes.

*Rounder (Umbrella-moulder)*
Average throughput time for each 2 sack (280 lb per sack) dough as 28-ounce units, 8–10 minutes.

*First/intermediate or recovery proofer*
Consists of an enclosed cabinet with travelling-pockets, four in a row (powered by a 3 h.p. motor) into which the rounded dough-pieces fall. The speed of travel of the pockets can be adjusted and synchronized with the speed of the divider and rounder. It is a short-travel proofer of about 12 minutes duration, serving only to allow the dough-pieces to recover and gain gas before final shaping.

SPINDLE-MOULDER OF STRAIGHT-THROUGH BASIC DESIGN WITH A DRUM
The commonest design was made up of three sets of plastic-spindles, which are set to rotate alternately clockwise and counter-clockwise. The gap between the spindles can

be pre-set by three manually operated wheels, which set the adjustment of each pair or set of spindles. The final-roller is fluted and of large diameter, rotating in a clockwise direction, ejecting the moulded pieces. Moulding is effected from the centre of the dough-piece outwards, thus degassing more efficiently and securing a better texture. This type used to be in common use in large plant bakeries. The surface of the drum is fluted or scored in order to 'work' the dough-piece along and against the steel pressure-plate. This moulder design is more compact than the horizontal designs, but a minimum of pressure is necessary to elevate the dough-piece from the bottom of the drum to the discharge point. The dynamic forces acting on the dough-pieces in this machine are: sheeting, curling, and extending.

Sheeting involves the degassing of the dough-piece by gradual reduction of the gap between the three sets of spindles or rolls. Dusting is unnecessary, since the machine is fitted with non-stick PTFE-covered rolls and a controlled air-blast.

The efficient flattening effect of triple sheeting rolls extrudes some free-moisture from the dough to the surface, which assists in producing a good union of the curled piece. The length of the sheeted dough-piece gives a curl of about 3·5 spirals, compared with 1·5 spirals of lamination for smaller machines.

The curling-roll consists of a heavily fluted roll of large diameter revolving against the direction of the travel of the leading edge. The edge being held by a groove, turned back, thus initiating the curling. On formation of the roll, it drops down into the spacing between the drum and the steel pressure-plate.

The final moulding process is considered of great importance in the industrial mass-production breadmaking process, in which the panned, sliced and wrapped 28-ounce unit has become the accepted National loaf. In the past, the sliced and wrapped unit has dominated the marketplace, especially in the large grocery chain-stores in the UK; but the 1980s have presented a changing situation. The public's increasing awareness of healthier nutritional habits, has resulted in increasing demand for the higher fibre breads and baked products. In an effort to redeem the drop in sales of sliced bread, the large chain bakeries have introduced the 'soft-grain' loaf with added soaked wheat, rye or other grains, to increase the fibre content, and still appeal to the palate of the younger generation.

In the UK, where the bread-crumb has been ideally defined as a mass of thin-walled cells, the nature of the cell determines in large measure the quality of the bread. Therefore, the effect of different moulding-machines exert a significant change in cell-structure, and their design and operation remains an important factor in any large chain-bakery schedule. The various 'Supertex' moulding techniques already mentioned also confirm this fact.

FINAL AUTOMATIC PROOFER

An essential part of any automated plant-bakery, the travelling-proofer moved the rows of pans, about 200 rows of eight pans, very slowly through the insulated cabinet, often in an up-and-down motion in a steamy and warm atmosphere. A humidifier unit maintained the relative humidity at about 53% at the inlet-zone, where the temperature was 90–101°F, and the average levels of relative humidity and temperature at the outlet-zone were approximately 46% and 108–110°F respectively. Average throughput times being 40 to 52 minutes, at an average speed of 0·1 ft/s.

The commonest type in breadmaking-lines during the 1950 decade was the travelling-tray oven. Each tray is permanently attached to a chain which pulls it from the front of the oven to the back. Where the oven-tunnel had two openings at either end, the swinging-trays remained on the same track, moving very slowly around the oil-fired burners, mounted in the centre of the tunnel, being unloaded at the opposite end. If the oven-tunnel had only one opening, the trays would be transferred to a lower track, which returned to the front of the oven, where the bread is unloaded. In the case of the Baker Perkins 'Uniflow' of the swinging-tray design (Fig. 28), each tray or cradle held five sets of 3 × 28-ounce units each. Baking the 14-ounce units, four sets of 4 × 14-ounces each was the loading. The oven capacity for 28-ounce National bread units was therefore 1350, baked in 38 minutes, corresponding to a throughput of 2131 units/h. Baking the 14-ounce National bread, the oven capacity was 1440, baked in 36 minutes, giving a throughput of 2400/h.

The 'wholemeal' bread made at that time contained 25% National straight-run white flour and 75% wholemeal-flour, enriched with shortening at the rate of about 2 lb/280 lb flour.

Baking of the 28-ounce 'wholemeal' bread units was also carried out with five sets of three units each per tray, making a total of 1 350 units. In this case the baking-time was 39 minutes, corresponding to a throughput of 2080 units/h. The baking-loss of a 28-ounce unit baked in a 9-inch open-pan was approximately 2·5 ounces. Using a scaling weight of 31·75 or 31·25 ounces, this is equivalent to a baking-loss of 7–8%.

The average baking-temperature for 28-ounce National bread units with the Uniflow oven was 560°F, but the heat-retention under solid-heat conditions was such that during an uninterrupted long baking-run, two oven-loads or 'batches' could be baked without the use of the burners. Average fuel consumption of these ovens was approximately 5 gal/h.

The outputs of these ovens could be increased if desired by either increasing the oven-speed, or increasing the number of pans per tray. For example, a speed increase on a 90-tray Uniflow from 40 to 30 minutes, using 16 pans per tray would increase the sackage/h from 9·8 to 13·1, thus increasing the hourly output by 3·3 sacks (1 sack = 280 lb) of flour into dough or 34%. Similarly, increasing the number of pans per tray from 14 to 16 would increase the sackage/h by 0·14 or 14%.

Later models such as the Baker Perkins 'Turbo-radiant' oven (Fig. 29) were a travelling-hearth design with a baking-surface of steel segments moving through the baking chamber on conveyor chains. Loading and discharge were at opposite ends of the tunnel, allowing maximum flexibility. Individual pans, straps of several joined pans (sets), or even unpanned units could be placed on the hearth. The average speed of travel of the hearth was 0·1 ft/s, with average baking-times of 40 minutes for 28-ounce National bread units. The throughput-capacity was 10 sacks/h or 5 × 2-sack doughs (560 lb). The ovens were fuelled by diesel oil (Derv grade),

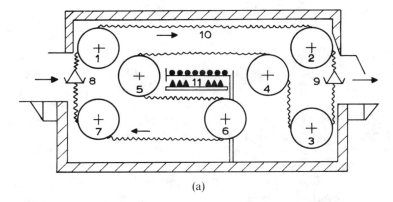

(a)

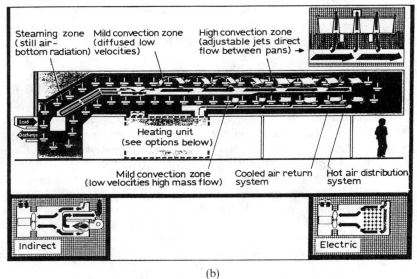

(b)

Fig. 28.    (a) Schematic of Baker Perkins 'Uniflow' oven with swing-trays. 1–7. Cog wheels. 8. Swing-tray loading. 9. Swing-tray unloading. 10. Chain drive. 11. Burners heating Perkins' steam pipes. (b) Series 440 swing-tray oven heat pattern zones. (Source: APV–Baker Peterborough, UK, 1989.)

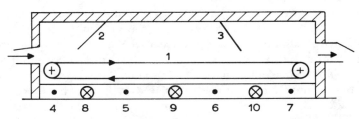

Fig. 29.    Schematic of Baker Perkins 'Turbo-radiant' oven. 1. Endless steel-band or wire-mesh. 2 and 3. Baffle plates. 4–7. Oil-burners. 8–10. Circulating fans.

average consumption being 5 gal/h. The heat generated by the burner-jets was dissipated within the oven-chamber by three circulating-fans. Lateral vents controlled the top and bottom heat, and baffle-plates fixed to the oven-crown could be adjusted at various angles, thus effecting 'heat-zoning' within the chamber (a setting of 7·5 on the scale being the norm). With this oven, owing to improved heat-transfer and dissipation, baking-times of 26 to 28 minutes could be achieved using a top temperature of 375°F, and a bottom temperature of 425°F. In some bakeries 28-ounce National bread units were scaled at 31·25 ounces, allowing only about 3·25 ounces for baking and cooling-loss due to evaporation. Therefore there was a very narrow safety margin *v.* short-weights, and the control of baking-times was very critical, and also the humidity control of the cooler, in order to keep the cooling-loss to a minimum. Under such stringent production conditions, the baking of a 'crusty' loaf was out of the question.

**Bread-cooling and Wrapping**

In the larger bakeries properly designed bread-cooling plants were installed. This is important when bread is to be wrapped, especially during the summer months, to avoid the onset of mould, or bacterial e.g. 'rope' contamination. As an extra deterrent, it was also normal practice to add acid or a mould inhibitor during doughmaking in the summer. Acid calcium phosphate or calcium di-hydrogen pyrophosphate added at 0·5%, or calcium propionate at 0·1% were the usual choice. The bread-cooling plant was housed separately from the production area adjacent to the despatch-station. The cooler consisted of spacious steel-girder construction housing a cog-and-chain-driven continuous-conveyor occupying about five levels. The conveyor was formed from wooden sticks protruding in V-formation, positioned closely enough to form elongated cradles for the loaves of bread. The temperature of the atmosphere was maintained at 76–78°F by an air-conditioning plant; humidity is controlled, so that the cooling-loss due to evaporation is minimized. Such 'humidity-controlled coolers' maintained a controlled saturation vapour-pressure throughout the cooling-cycle.

When the optimal steady-state conditions in crumb temperature and humidity had been reached, the bread was sliced and wrapped immediately to avoid excess loss by evaporation. Average residence time for the bread inside the cooler was 2 h 47 min, and the average time from loading to unloading was about 3 h 3 min. By comparison, uncontrolled bread cooling on racks took 4 to 4·5 h, and in this case scaling-weights must be set high enough to allow for higher losses due to evaporation on cooling. Already by the middle 1930s, it had become common to slice and wrap bread before leaving the bakery, which was encouraged by the ever-increasing popularity of panned and sandwich loaves. Although held in abeyance during World War II, the hygienic practice soon became normal in the post-war large chain-bakeries, and shelf-life was extended at ambient temperatures for up to a week. The sales of sliced-and-wrapped bread increased so dramatically nationwide, especially in the growing self-service grocery stores, that the small baker experienced great difficulty in maintaining the sales of his crusty, oven-fresh, unwrapped and unsliced product. Failure to diversify the product range resulted in the closure of many small bakeries between 1948 and 1960, as a result of the sliced-bread revolution in the UK.

After cooling to 76–80°F over 4–5 h, the bread was ready for slicing. Fully automatic machines enveloped and sealed the loaves in waxed or 'cellophane' wrappers, made attractive as a good form of advertisement. Where 'cellophane' was used, it was usual first to surround the loaf with an attractive paper-band.

The principle of the machine used for slicing employed multiple blades, resembling hacksaw blades, arranged in parallel, which oscillate and cut through the loaf as it is brought to meet them on a mechanically propelled carriage.

Waxed-paper was the first choice in packaging material at that time, the wax-coating being a relatively high melting-point wax of paraffin hydrocarbon composition. The 'cellophane' used was of medium gauge, manufactured by British Cellophane, and used for certain specialty products only, e.g. American-type breads for the United States Armed Forces, and farmhouse-bread for the UK market. Polythene as a wrapping material for bread was unproven at that time owing to its tendency to stretch and become elastic under stress, although its advantages, viz. robustness and tear-resistance were recognized. Slicing-machines were usually powered by 1·5 horse-power motors, one driving the slicing-unit and oscillating the knives, and another driving the conveyor-belt. Power transmission was by belt, the moving cams and levers being chain-driven. The loaves of bread were fed by the conveyor-belt into the knives, which vibrate or oscillate in a vertical plane, being set in a frame, two such frames oscillating in opposite directions at the same time. The number of knives required always exceeds the number of bread-slices by one unit. The knives oscillate continuously in a vertical plane at right-angles to the path of the loaf of bread. On passing through the knives or blades, the sliced-loaf is mechanically moved upwards into a loop of waxed-paper which envelops it, followed by a folding and tearing action. The loaves then passed by means of the conveyor-belt through an adjustable channel to the heat-sealer, with temperature setting 200–500°F according to the melting-point of the wax used on the paper. The wrapped and sealed loaves then passed through the refrigerant-panels, which rapidly solidified the melted-wax, thus effecting a seal *in situ*. The cellophane material, owing to its chemical composition, does not require sudden cooling to effect the seal. At that time such wrapping materials as microperforated polypropylene, PVC, or laminated films were not applied to bread-wrapping. The 28-ounce National loaf was then divided into 19 slices and wrapped in waxed paper of 13·5 inches width, 28-ounce Farmhouse loaves were cut into 20 slices and wrapped in 15-inch width cellophane. Any standardization of the cutting-process with these machines has always been difficult, since the rapidity of the cut continuously changes similar to a sine-function curve, and the resulting variable cutting-speed does not meet the requirements of the crumb rheology. In order to ensure an efficient cutting process, the speed at which the loaves of bread approach the blades, and their angle of approach is important. The blades must be continuously freed of adhering crumb, and can only be used for a limited time before requiring re-sharpening. When the cutting-life of the blades is exceeded, the quality of the sliced surface deteriorates, and the amount of adhering crumb increases. The regular cleaning of slicing-machines has a direct effect on the efficiency of the machine. Various blade-profiles are available, made of special steel-alloys for strength, and versatility. The blade-teeth are either the concave waveform or zigzag, depending the crumb properties of the baked-product.

During the 1950s, a typical formulation for a 'wholemeal' dough in a large chain or plant-bakery, using the 'delayed-salt process' was as follows:

| | | |
|---|---|---|
| Wholemeal-flour | 75 parts | 420 lb |
| Straight-run National-flour | 25 | 140 lb |
| Water | 58 | 325 lb |
| Yeast food, bromated and malted | 0·18 | 1 lb |
| Shortening | 0·7 | 4 lb |
| Yeast | 0·98 | 5·5 lb |
| Salt | 1·78 | 10·0 lb |

Procedure:
Dough temperature: 80–82°F
Mixed for 12 minutes in the open-pan Viennara mixers
Bulk fermentation in the fermentation-room for 1·5 h
Salt (10 lb) added, and mixed in with the Viennara mixers for 8 minutes.
After a further 45 min rest in the fermentation-room, the dough was worked-off.

Scaling the dough at 32–33 ounces per unit, the yield was about 216 units at 28 ounces. The throughput rate for a 90-tray 'Uniflow' Baker Perkins oven was about 2080 units/h based on a baking-time of 39–40 minutes.

Wholemeal-flour was produced by grinding the whole of the grain to varying degrees of fineness and, if stone-ground, it contained practically the whole of the bran and germ. Although the actual composition of the 'wholemeal' depended on the wheat used to produce it, the average wholemeal bread contains more protein, minerals and fibre, but less carbohydrate, than National straight-run flour. There was, at that time, little difference between 'wholemeal' and 'wheatmeal', some wheatmeals were finer than the coarser wholemeals, being preferred by the relatively small numbers of brown bread eaters. Wheatmeal was produced in various degrees of fineness, but was generally not as coarse as true 'wholemeal', containing less of the coarser bran. It was usually made by blending together lower-grade white flour with various proportions of sharps or middlings. At that time it was a common practice for bakers to blend a proportion of strong white flour with the meals, and so improve the volume and processing tolerance of the brown breads. However, such practices depended on local customer tastes. In general, unless stronger white flour and/or shortening was added to the dough, and adequate salt used, the crumb was coarse and of dry eating-quality. At least one large grocery-chain plant bakery produced high quality 'brown breads', some of which could have been described as specialty-breads, containing, in addition to strong white flour and shortening, milk-powders, sugar, honey and yeast food. Many of these variety breads were manufactured especially for the United States Armed Forces in Europe, under contract. The addition of dry or wet gluten was also used to improve 'wholemeal' and other types of 'brown bread'.

Protein-enriched 'gluten bread' was produced by adding 40 lb of wet-gluten, or about 13 lb of dry-gluten, 5–5·5 lb yeast, 5 lb salt, 4 lb milk-powder, and 4 lb of shortening to 280 lb of National-flour. Enough water was added to give a slightly slacker dough than for National bread dough. To improve the flavour, either 0·5–2·0 lb of malt or 2 lb of sugar were added, which also improved the bloom. Doughs were worked at 78–80°F, and mixed longer than normal, to ensure even distribution and optimal development. A bulk fermentation of 2 h, with a remix or knock-back at 1·5 h was normally sufficient. Scaling was 16 ounces for a 14-ounce baked unit, followed by a 15-min recovery-period, and final moulding and tinning-up. A fairly full-proof was given, baking as for ordinary bread. For diabetic-patients full 'gluten bread' was made by adding two parts gluten to one part soft flour, making 280 lb composite-flour to prepare a dough using the same quantities of other ingredients as quoted above. Moulding was done through heavy rollers or a 'brake', after scaling at 13 ounces, the loaves being set in the oven at three-quarters of pan height, at about 380°F.

Legislation during the 1950s in the UK was such that in the case of 'protein bread', no special labelling was required for breads with less than 16% protein. Bread containing more than 16% but less than 22% could be called 'gluten bread', and breads with more than 22% had to be described as 'high-protein bread', but without any specific reference to protein-type, e.g. 'contains $x\%$ protein'. No bread could be described as 'wheat-germ bread', unless it contained at least 10% added, processed wheat-germ calculated on a dry-basis on the baked bread.

The purpose of flour treatment at the mill was, and still remains, (a) to improve flour colour and (b) to improve its baking properties. Up until the early 1950s, the most widely used chemical for improving flour colour, viz. 'bleaching-agent', was nitrogen trichloride or 'Agene'. However, some feeding-trials on dogs by Sir Edward Mellanby, using heavily treated flour made into bread, resulted in the formation of certain amino acid derivatives with nitrogen trichloride, which affected the nervous system, inducing fits. Public reaction at the time rapidly forced the Ministry of Food to prohibit its use as a flour bleaching-agent. Therefore, the only remaining alternatives were benzoyl-peroxide, marketed under the trade names 'Novadelox' and 'Chefarox', a solid substance, and chlorine dioxide, which is a yellow gas. The first patent for the application of chlorine dioxide to flour treatment was due to Bühler Brothers of Uzwil in 1929 (Deutsche Reichspatent 41023 Kl 53c, *Chem. Zeitung*, 93, 904, 1929). Since it is both fat and water-soluble, it acts simultaneously as a flour-improver and bleaching-agent.

Substances which improved the baking properties included the following: potassium bromate, potassium iodate, ammonium persulphate, calcium peroxide and ascorbic acid. Although the working mechanism of these substances vary somewhat, the final result is about the same, inasmuch as they all either transfer their own or atmospheric oxygen to substances present in flour which are harmful to its baking properties, or to those formed on doughmaking, e.g. glutathione. The quantity of each required is of the order of 1 to 5 g/100 kg flour.

The structure of the UK milling industry during the 1950s was such that, in addition to the existence of three large milling/baking complex groups, viz. Allied (Weston group), Rank-Hovis, and Spillers, a large number of independent millers supplied the medium and small baker with specialty-flours. Many of these regional 'town-mills', which supplied local small bakers, went out of existence with the increasing concentration and centralization of the milling and baking industries into larger public groups. Apart from the three large milling/bakery complexes already mentioned, there were a number of medium-sized bakery-chains which remained independent for some time, mainly located in the UK Midlands and North, Wales, Scotland and Northern Ireland. In addition, the UK Cooperative movement had its own mills and bakeries nationwide, and J. Lyons and Company, based at Cadby Hall, London W14, manufactured bread and baked-products on a large and mechanized scale. The latter company supplied their large hotel chain and restaurants in central London and the home-counties, as well as private retail shops, supermarkets and the US Forces in Europe.

The quality standards of UK mass-produced bread using 3–3·5 h straight-doughs, fermented under controlled conditions in bulk, was generally good. Produced from a relatively 'lean formulation', the standard 28-ounce National loaf, had an even texture and good crumb-elasticity and colour. Whether unsliced and unwrapped, or in the sliced and wrapped form, the crumb had a good sheen and aroma, and a flavour-full crust, within the constraints of processing and baking parameters of plant bread manufacture.

It is the writer's experience, that the replacement of pre-processing fermentation completely, and reliance on high-energy mixing levels, coupled with chemical dough development, has not hitherto produced a baked product with the all-round, general standards of quality of the 3–3·5 h bulk-fermented straight-dough produced in the 1950s. The main deficiencies are crumb-structure, flavour and aroma. Unless new technological innovations can produce bread of at least the all-round, general quality standards of existing technology, including crumb-structure and taste/flavour and aroma, no technological breakthrough has been achieved. Any compromise only represents the subordination of final-product quality to economy of production and monetary considerations. The net result is a progressive reduction in the quality standards of UK commercial bread, compared with those experienced in other European countries with both capitalist and Socialist societies. Nevertheless, it must be acknowledged that the UK consumer has shown considerable apathy in allowing this to happen. At the time of writing, an improving nutritionally educated public, and the increasing desire for non-additive 'clean-labelling' of foods, and even biologically produced food, could reverse the downward trend in quality standards. Furthermore, the desire for 'oven-fresh' baked products is re-emerging, favouring the smaller retail outlet generally.

Turning to the production-scene of the small baker during the 1950s, and taking a predominantly rural area of the UK as a case-study, the following manufacturing schedule prevailed in the author's own family-bakery located in the South Hams district of South Devon, UK, founded in 1924.

The enterprise was based on the bakery, restaurant, tea-rooms, and retail shop marketing bread, confectionery and grocery products, and two delivery-vans for town and country wholesale and retail distribution. The bakery, restaurant and tea-rooms were a self-contained unit, located near a water-front with car-parking facilities.

Flour and other raw materials were purchased according to price quotation or contract, depending on the quality performance of the merchandise. Flour-storage was in jute sacks in a well-ventilated loft above the bakery, flour being fed into the bakery below and into the dough-mixer by means of a wooden hopper, fitted with brushes, driven by the dough-mixer power unit. The dough-mixer, a double-arm open-pan type had a capacity of 1·5 sacks or 420 lb of flour into dough. The pan rotated about its axis, and the two arms described a lifting cyclic/scissors movement in the same path as the dough. The pan could be removed on wheels, but this facility was not utilized, since only one pan was used. During the mixing process, one arm could be immobilized by use of a gear-box, thus changing the mixing-action. The mixer was manufactured by T. Collins Ltd of Bristol, UK. The mixer was driven by a 1·5 h.p. Lister petrol-engine, housed in an engine-house situated a short distance outside the bake-shop. Power transmission was by belt onto a fly-wheel, which rotated a shaft mounted at roof-level, under cover, through to a small diameter wheel, and thence power was transmitted by belt to mixer over a fly-wheel. This source of power was economic, and rendered the doughmaking process immune to electricity power-cuts.

The normal daily production schedule was $2 \times 420$ lb flour batches during the winter months, and often $3 \times 420$ lb batches during the July–August months to meet vacational demands of hoteliers, campers, summer-residences and lettings.

The first batch of the day was prepared at about 17.00 hours, being bulk-fermented for about 12 h in a long, narrow U-shaped wooden container cladded with galvanized metal sheeting for ease of cleaning. The dough was covered with galvanized metal cladded top boards, which also served as additional work-tops. The formulation used was the following:

| National straight-run flour | | |
|---|---|---|
| (protein 12–13% at 14·0% moisture) | 100 parts | 420 lb |
| Water | 55·3 | 232·5 lb |
| Yeast | 0·9 | 4·0 lb |
| Salt | 1·9 | 8·0 lb |
| Enzyme-active full-fat soy-flour (Bredsoy) | 0·7 | 3·0 lb |

Procedure:
Dough temperature                          74°F
Mixing-time                                15–20 min
Place in fermentation-container for approx. 12 h
Worked-off at approx. 05.00 h scaling at 2 lb 3·5 ounces (2·22 lb)
Rounded-up by hand, giving 10–15 minutes intermediate-proof, covered with linen cloths.

Final shaping and moulding was also by hand, into a wide variety of shapes including the following 1-lb and 2-lb units: lidded-pan, open-pan, canister, cottage, round and open-batch, and coburg. The 4-lb units, known as quarterns, were baked in lidded pans as long-sandwich bread (long-fours).

The dough water absorption and resultant yield varied according to flour quality, 190–200 × 2-lb units from 280 lb flour being the average range. Dough consistency was adjusted by the baker, after adding 14 gal/sack (280 lb) initially, at the dough assembly-stage. Town-mill flour quality invariably showed greater performance differences than the National-mills, owing to the lack of laboratory-testing facilities.

All the panned bread was given a final-proof on the work-tops, often without being covered; whereas the wholemeal, germ-bread, malted-bread and all the hearth-breads (baked on the oven-bottom), as well as the soft-rolls and sweetened/shortened morning products (buns), were proofed in the in-built proofer under the oven.

By 07.00 hours the overnight-dough had been set in the oven, together with the soft-rolls (Devon-tuffs), and buns, the work having been completed by the chief-baker and two trained helpers.

At this point, a short break was taken, after which the chief-baker prepared the second batch, according to the following formulation:

|  | % | Weight (*lb*) |
|---|---|---|
| National straight-run flour | 100 | 420 |
| Water | 57·1 | 240 |
| Yeast | 1·2 | 5·44 |
| Salt | 1·2 | 5·0 |
| Sugar | 2·5 | 10·5 |
| Enzyme-active full-fat soy-flour | 0·7 | 3·0 |

Procedure:

Dough temperature                     84°F
Mixing-time                         15–20 minutes

Covered with linen cloth and bulk ferment in the mixer-pan for 2 h, 07.30 to 09.30 hours.

Worked-off at 09.30 hours, scaling at 2 lb 3·5 ounces (2·22 lb)

Rounded-up by hand, giving 10–15 minutes intermediate-proof, covered with linen cloths.

Final shaping and moulding by hand into the same bread varieties, according to time available.

Final-proof of panned-bread on the work-tops, and specialty-breads in the proofer. Proofing to constant height, approx. 45–60 minutes.

The second batch would be ready for setting in the oven by about 12.00–12.30 hours.

Allowing for a fermentation loss of about 10 lb, a weight equivalent to the amount of added sucrose (which is rapidly fermented), the sack yield was about 204 × 2-lb units.

The oven was of the internally heated side-flue design also manufactured by T. Collins of Bristol, UK. These ovens were of very solid construction, being of brickwork and filled around the chamber with clay or sand, which is an efficient insulator, and the baking-process is one of stored, solid heat. The oven structure is integral with the building itself, and the furnace positioned on the right-hand wall of the oven. By opening the damper, the flame and hot gases are drawn around and across the baking chamber. An additional draught was created by the use of a blower, a curved steel, convex closure for the mouth of the oven, and the air-vent positioned outside the oven-door in the roof of the oven-mouth. The blower was only used during the firing of the oven, to get the 'solid-heat' of the chamber to a temperature of 550°F, which was measured by a pyrometer situated on the left of the oven-mouth, mounted through the front-wall. However, the pyrometer did not measure the solid-heat baking capacity of the oven or 'stored-heat', and was only reliable when conditions of steady-state had been reached by previous firing. Since the oven was fired daily at 05.00 hours, when work commenced, the steady-state conditions were attained without much difficulty.

The solid fuels used were coal and coke, the former, since it contains a high percentage of hydrogen, gives a high calorific-value when burnt in ample air, forming steam and about 20 000 calories. Coke did not have as high a calorific-value, requiring 112 lb, compared with 100 lb of coal, to achieve the same heating-capacity. Coke was a good complementary fuel for coal, since it could be used to produce a smokeless flame, after reaching the required temperature with coal, thus removing any sooty deposits.

The stoking-operation of solid fuel ovens is a skilled exercise, the fire-bars must be properly raked out, to ensure a plentiful supply of air for efficient combustion. It also prevents the formation of clinker, and a minimum of ash. The fire is fed with fuel in small quantities, and spread evenly over the surface. Frequent use of the rake and care exercised in banking up the fire overnight maintained as 'solid' a heat as was possible.

Burnt-out fire-bars had to be replaced to prevent the others from buckling with the heat. The aim was always to use a falling temperature for baking.

When the solid temperature, steady-state conditions have been reached, the oven is scoured-out with a 'scuffler', which is a wet cloth or jute-sack attached to a long pole by a length of chain. The oven was then ready for loading with a 'peel', a flat board with a sharpened edge, attached to a long pole, which reached to the back of the oven-chamber. On setting the oven with the complete batch of products, the temperature would fall to about 500°F, providing the stoking and firing operations had been carefully executed, and conditions of a solid steady-state between heat-transfer and heat-retention thus achieved.

For a small mixed trade of bread and confectionery, the oven could be worked economically, by arranging production so that the products reach the oven in descending order of their baking temperatures. Thus, bread is baked first, followed by sweetened/shortened morning-sale products, then pastry, sponge-products, pound-cakes and genoese, and, last of all, rich fruited cakes or slab-cakes. Even then, almond goods and meringues could be fitted in if required, and all the day's production was baked on the heat reserve from the first batch and the lesser re-firing for the second batch.

The baking-time for a batch of bread, about 1 hour, resulted in a baked product with a relatively dense crumb-structure, moist but also of good elasticity. The flavour and aroma of both crumb and crust was such that it was a pleasant experience to eat without any butter, largely owing to the fermentation-schedule and the method of baking. The loaves had a 'break-shred', which refers to the lift due to oven-spring of a dough-piece as it expands in the oven, stretching and breaking at the side in the case of pan-bread.

The crumb-stability and moistness (versus moisture) were a function of the colloidal state of the baked polymer-complex, brought about by the influence of raw materials, fermentation, manipulation and the baking process. In the case of the side-flue oven the heat-transfer during baking is due to heat-radiation, heat conduction from the flames of gas-ignition to oven-sole and thence to dough bottom surface, and convection currents around the top and side surfaces. The quantity of heat energy necessary to fully bake a 2-lb or 1-kg loaf of bread is approximately 500–550 kJ. The absolute total crumb-moisture of a baked loaf may vary from 35% to 42%.

These two sharply contrasting case-studies of two bakery operations of the 1950s both confirm that good product quality standards can be achieved in both a large- and a small-scale enterprise, although the structure and eating-experiences of the products are very different. The advantages of the mass-produced plant loaf of bread were uniformity, convenience and prolonged shelf-life, and, for the manufacturer, a rationalized, mechanized and continuous production schedule, allowing an acceptable return on capital investment. The smaller-scale, rurally based enterprise, with a basic and limited financial investment, provided a more personalized service within a limited area with a scattered, relatively low-density population. The diversity of the enterprise compensated in some measure for the smaller bakery production output, and distribution costs, which could not be covered by extra charges at that time. The merits of the bread and baked-products so produced were the freshly baked appetizing eating-experience, with ample crust formation, a firm but elastic crumb-structure, and excellent flavour and aroma. Such products were consumed readily in large quantities by the local population, forming an integral part of a staple diet for the manual and the more sedentary consumer. Also, a door-to-door delivery service was provided daily or three times per week, depending on location, free of charge. In contrast, the plant bakery operation could not compete on freshness and eating quality or delivery service and personal attention. Therefore, an analysis of the two operations reveals that the consumer receives less value for money from the plant bakery. The chief advantages of the mass-production enterprise is that it is a rational method of production, and is more cost-effective for the manufacturer on account of the following:

—High level of mechanization allows a larger output per man-hour.
—Product yields are higher owing to processing and product standardization.
—Machinery depreciation and plant costs can be written-off more quickly owing to continuous shift-production working systems.
—The plant-baker invariably operates in or near large cities, so that the distribution lines are relatively short, and confined to wholesale outlets, thus reducing indirect costs.

Although much of the baking-process technology used during the 1950s is no longer applicable today, it should not be regarded as being of little value. Since bread quality in the UK has depreciated in no small measure, in spite of process innovations, it is quite possible that processes no longer in use may be re-introduced, in order to comply with new legal directives or consumer demand. Many examples of the re-cycling of past technologies are already in evidence in baking technology, e.g. the need to avoid additives, and fermentation process modification to produce products with 'new' flavours and textures, etc.

As a result of the 'Agene' (nitrogen trichloride) publicity, which hit the headlines in the early 1950s, considerable interest was already being aroused in the possible use of air as an improver instead of chemical bleaching and oxidizing agents in the flour mill and bakery. In 1951, A. Schulz, working at the Cereals Research Institute at Detmold, FRG, published an article entitled 'The influence of dough-aeration on baking-quality' in *Brot und Gebäck*, **5**, 10, 145–146. In the UK and in other countries, the significance of the mixing-intensity, with or without the addition of substances which induce oxidation, was gaining increasing attention.

During 1955, the topics under test were: (1) the use of air instead of chlorine bleaches or mineral improvers, and (2) the use of air and unprocessed soy-flour. Under topic (1), several techniques were being tested involving the preparation of 'batters', which consisted of 50% of the total flour weight, containing 50% of the salt and about 90% of the dough water. Various types of mixer were used to prepare the batters, from heavy-duty cake-mixers to high-speed horizontal-bar mixers. A patent taken out by J. Rank Ltd and Hay, used untreated flour. 50% of the flour, containing a sufficient supply of the oxidase enzyme (lipoxidase), responsible for the natural bleaching of flour, was mixed with all the water and the yeast in a high-speed mixer at about 350 rpm. Then, after about 4–8 minutes, the salt and the remainder of the flour were mixed in at normal speed. The bread resulting from such a dough showed as good a quality as that made with the same flour chemically treated. Although the process was used commercially, as long as chemicals were permitted there was little incentive to change. In fact, the process performs well with flours of high or long extraction, i.e. 80–85%; but with flours of 70–76% extraction, additional oxidase becomes necessary. This phenomenon explains the introduction of topic (2), whereby, the extra oxidase was supplied by an addition of about 4 ounces (112 g) per sack (280 lb) of flour of non-processed soy-flour. At that time it was general practice to bulk-ferment doughs for 3–3·5 h after completion of the mixing process. The subject was also studied by Todd, Hawthorn and Blain (1954), and Hawthorn and Todd (1955).

Numerous variations of these techniques, and suggestions, such as increasing the mixing-time and the amount of soy-flour, in order to reduce mixing-speed, were all evaluated. However, these mixing-systems are oxidative rather than physical in character, involving the action of air on the dough at high-speed (compare the 'liquid oxidation-phase' and the Blanchard processes, discussed in section 1.7.7).

During the 1950s, the then British Baking Industries Research Association recommended that a similar improving effect could be obtained by using four-times the normal mixing time in a conventional mixer.

The author's own baking-test evaluation of the relative merits of these 'batter' processes, with and without unprocessed soy-flour were conducted as follows.

Slack-doughs were made with 420 lb untreated National-flour, 6 lb yeast, 3 lb salt, 1 lb malt-flour, and about 290 lb water, (1) without unprocessed soy-flour, and (2) with 1·5 lb unprocessed soy-flour. This was mixed for 17–18 minutes at 60 rpm at a dough temperature of 78°F, after which it was rested for 45 minutes. The final dough was made by adding the balance of 140 lb flour, 3 lb salt, and 20 lb water. This was mixed for 10–15 minutes at 30 rpm at a dough temperature of 78°F, and rested for 1·75 h. These test-doughs were compared with a control-dough of 2 h bulk-fermentation, containing 420 lb untreated National-flour, 6 lb yeast, 6 lb salt, 1 lb malt-flour and 310–320 lb water, and, as a second control-dough, a 420 lb mill-treated National-flour, containing the same raw materials. In the case of test-doughs (1) and (2) the yeast was added by suspension in the residual water at the dough-stage, and the salt added in two-stages, 3 lb to the batters, and 3 lb to the doughs (total salt 6 lb). The control-doughs were given a knock-back or punch at the 1 h stage. All doughs were scaled at 1 lb 2 ounces (18 ounces) each piece, and passed through a Mono-moulding machine, after rounding-up and allowing 10 minutes recovery. Final pan-proof was in a humidified proofing-cabinet for 30 minutes. The loaves were set in a solid oven at 500°F. After overnight storage the bread was judged as shown in Table 44. The untreated-flour sample gave a loaf of lower oven-spring and volume than the mill-treated National-flour of 80% extraction-rate. The loaf had a somewhat bound appearance, the texture being close and inelastic compared with the mill-treated flour, lacking in sheen and resilience. The crust of the untreated-flour bread had a thick leathery texture, deficient in bloom, being generally below the standard of normal bread made from the mill-treated commercial National-flour of that period.

TABLE 44

| Parameter | Untreated control National-flour | Treated control National-flour | Batter | Batter + soy-flour |
|---|---|---|---|---|
| Height at setting (inches) | 4·2 | 4·3 | 4·4 | 4·2 |
| Height after baking (inches) | 4·9 | 5·6 | 6·1 | 5·7 |
| Oven-spring | 0·7 | 1·3 | 1·7 | 1·5 |
| Volume (10) | 8·0 | 10·0 | 11·0 | 10·0 |
| Shape (5) | 4·0 | 5·0 | 5·0 | 5·0 |
| Bloom (10) | 8·0 | 10·0 | 10·0 | 9·0 |
| Crust-colour (5) | 3·5 | 5·0 | 5·5 | 5·5 |
| Softness (10) | 8·0 | 9·0 | 10·0 | 10·5 |
| Texture (10) | 8·0 | 9·0 | 11·0 | 13·0 |
| Sheen (10) | 7·0 | 9·0 | 11·0 | 12·0 |
| Crumbliness (10) | 7·0 | 9·0 | 10·0 | 10·0 |
| Eating-quality (10) | 8·0 | 9·5 | 11·0 | 11·0 |

The overall effect of the oxidative-action of mixing the untreated-flour batter even at 60 rpm for 17–18 minutes, had a dramatic effect on the baking performance of the flour, with or without the addition of unprocessed soy-flour. Oven-spring and volume approximated to that obtained with the mill-treated National-flour. Crumb softness, elasticity, sheen and colour showed significant improvement in both the batter and batter + soy-flour test-doughs. The addition of the soy-flour resulted in the best improvement in the internal parameters of the bread, giving improved texture, crumb-colour and sheen which surpassed that of the mill-treated National-flour.

The beneficial effects of direct or indirect flour oxidation were not fully exploited using the only mixing regime available at that time. The use of higher specific mixing-intensities would have resulted in much higher levels of trapped atmospheric-oxygen. Since that time, the mechanism of non-chemical oxidation of wheat-flour doughs has been researched and elucidated. The soy-flour provides a source of soluble lipoxygenase enzyme, as well as vegetable oil polyunsaturated fatty acids, which become converted into peroxide-compounds, mainly hydro-peroxides, by the lipoxygenase and the entrapped atmospheric-oxygen. These hydroperoxides function as highly reactive donors of 'nascent' oxygen, which, on contact with the flour proteins oxidize the —SH— groups of the protein/proteinase complex. The net result is an improvement in physical dough properties, manifested as gas-retention and water-binding energy. Therefore, the resultant bread shows improved volume, symmetry and crumb characteristics. Simultaneously, the peroxide-compounds of the fatty acids oxidize and bleach the carotenoid flour-pigments, immediately apparent as an improvement in crumb brightness and colour (compare 'liquid oxidation-phase' discussed in section 1.7.7). The process developed by George Blanchard of Watnall in Nottinghamshire, known as the 'Blanchard process' depended on these reactions described. He had already developed a method of making bread with untreated flours in the early 1950s, as already described in earlier sections 1.5.9 and 1.7.7. This procedure, developed in his own bakery, involved the intensive mixing of an untreated-flour batter made with 75% of the total flour and all the water. The remaining flour was then added, and the dough worked-off in the usual way after an appropriate bulk-fermentation period. Blanchard used speeded-up Baker Perkins high-speed mixers, but he claimed this was not essential, and that the process could be adapted to any type of mixer. Although the process was being used in his own bakery during the early 1950s, it received little publicity until 1965–66, when two articles appeared in *Milling* entitled, 'The Blanchard batter process—no-time dough with less power', *Milling*, **146**, 520–521, 1965; and 'Blanchard batter process for bread', *Milling*, **147**, 519, 1966. The renewed interest during the 1965–66 period was the elimination of the necessity for bulk-fermentation, which Blanchard devised during 1963. The basis of the process is the same as he devised for use with unbleached flour in his own bakery, incorporating a two-stage mixing method, based on an initial batter-stage. The batter is again made from 75% or three-quarters of the total flour and the whole of the water, including the yeast and oxidant. This batter is beaten to the stage of thorough gluten hydration, so that it acquires enough physical development to replace that produced during conventional fermentation processes. An important advantage of this process is that conventional high-speed mixers can be used, thus avoiding the need for an expensive mechanical-developer. Blanchard suggested the use of the ordinary high-speed horizontal-bar mixer, which produces optimal results when speeded up from the fast-speed normal rpm-rate of 75 to 98. This batter process achieves complete gluten physical development at a lower energy requirement and considerably less electrical power-input than that required by the Chorleywood bread process. Also, dough temperature rises are considerably less, owing to less dissipation of frictional heat-energy, therefore doughs can be worked cooler than is the case with mechanical-development methods. Oxidation requirements are correspondingly lower than with the CBP.

However, the British Baking Industries Research Association was already fully committed to their own innovation, 'the Chorleywood bread process', in collaboration with the pioneer mixer development work of Tweedy of Burnley, Lancashire, UK. The vested interest of the machinery-manufacturer, with the necessary marketing expertise succeeded in selling the process and mixing-machinery to the large chain milling/bakery groups, and some regional chain-bakeries in the UK, as well as achieving considerable export success within European and Asian countries.

Nevertheless, the technology of the Blanchard process with its various aspects and diverse potential application should not be regarded as a technology of the past. On the contrary, whenever the topic of 'short-process/no-time methods' or 'clean-label' bread, made with unbleached/untreated flour is under discussion, the basic technological concepts of Blanchard cannot be overlooked. Indeed, wherever the final bread eating-qualities must be balanced against any production economics derived from mechanical dough-development, the Blanchard process merits serious consideration. The Blanchard process is described in Soviet literature and, in spite of the fact that continuous processes with firm and liquid pre-ferments, making use of various special machine-aggregates, find increasing application, in one of the Republics the Blanchard technique is applied. In the Estonian SSR of the USSR, the process is used for the manufacture of baked-products containing various milk-products. The batter is intensively mixed and an increased amount of yeast is also added. In 1963, the Moscow Technological Institute for the Food Industry (MTIPP) reported a 'Process for the improvement of wheat-flour bread' under the State Committee for Inventions of USSR No. 164860, with priority from 7 July 1963, in the name of Auerman, L. J., Kretowitsch, W. L., and Polandowa, R. D. This process was an enzymatic one, involving lipoxygenase, for the improvement of the quality of wheat-flour bread and rolls from the various flour grades. For this purpose, a liquid oxidation-phase (LOP) was prepared, which is described in detail in section 1.7.7.

The underlying conception and mechanism of the Blanchard process and the Soviet patent shows general similarities. The soy-flour provides the source of lipoxygenase, vegetable oil polyunsaturated fatty acids (which can be supplemented by adding soy-oil). These components, when mixed at high intensity of up to 1400 rpm, form hydroperoxides owing to entrapped atmospheric oxygen. The hydroperoxides then function as reactive donors of 'nascent' oxygen, which contacts with the exposed areas of the flour proteins, oxidizing their —SH— groups, in particular those —SH— groups associated with the protein/proteinase complex, e.g. the tripeptide glutathione.

With the commercial acceptance in the USA of the continuous 'Do-Maker' and 'Amflow' mixing units in 1953, increasing interest in continuous doughmaking units spread to the UK. Certain aspects of these processes, however, produced a type of bread quite different from that produced conventionally in the UK. The crumb texture and grain being very fine and uniform, the crumb lacking in strength and resilience. The density of the final product is much less than either UK or European bread in general. This fundamental difference was the result of certain raw-material and processing aspects such as:

(1)   the use of leaner formulations in the UK, i.e. lower levels of shortening, no milk or sugar solids;
(2)   the use of low flour-content pre-ferments in the USA, or 'liquid-brews', varying from 0 to 50% of total dough flavour;
(3)   differing flour-quality profiles in the UK compared with those in the USA, owing to the use of more diverse wheat mixtures (grists) and differing selection of mill-streams in the UK milling industry.

Although the type of bread and the crumb could be regulated within limits by processing variations, such as, degree of premixing, developer throughput, dough temperature, pressure in the developer, speed of the mixing-elements in the developer, and developer gas-injection, these processes did not gain acceptance in the UK. In the UK, the most successful continuous mixing system was that manufactured by E. T. Oakes Limited, which was used in conjunction with the Chorleywood bread process (CBP), capable of outputs of 1800 to 6750 pounds of dough/hour. It consisted essentially of a horizontal barrel with fixed pegs on its inner surface. A central shaft, also with projecting pegs, was powered by a 30 h.p. variable speed motor, which delivered the standard 0·4 h.p. per minute per pound of dough mass. The ingredients were worm-fed directly into the main barrel, which was water-jacketed and secured by tie-bars. Inside and between each stator were rotors keyed to the central-shaft. The design of stator and rotor are optimized to give minimum resistance to flow and correct mixing action for the texture required. The dough from the mixer/developer is extruded as a continuous ribbon. It is then put through a specially designed divider for continuous doughs accurate to ±0·14 ounce. A recording wattmeter allows constant control of work input at 5 watt-hours per pound of dough for any chosen throughput, there being aggregates produced in five different capacities. The system allowed an intermediate proofing stage if desired, and represents the only continuous system which can be counted as a success in UK breadmaking. The Baker Perkins–Ivarsson continuous sponge and dough system, used during the period 1962–65, produced good bread, but showed the following disadvantages:

(1)   the sponge process gave reduced yields compared with straight dough processes;
(2)   it necessitated the purchase of expensive plant.
(3)   it required more personnel than bulk-fermentation processes.

In general, there has been a distinct lack of enthusiasm for a continuous breadmaking revolution in Western Europe; and although at one time, about the late 1960s, more than 40% of panned white bread in the USA was produced by the continuous process, many bakeries still prefer the quality of sponge and dough bread. Those bakeries which used the liquid pre-ferment or brew, found that 50% of total flour when added to the brew, strengthened crumb-structure and improved bread quality.

In the UK, the British Baking Industries Research Association at Chorleywood became more interested in a process for producing bread by a 'no-dough-time' method. In 1961 they introduced the Chorleywood bread process (CBP) to the UK baking industry. The principle of this process is much simpler than continuous methods, being similar to the hitherto conventional straight-dough method. However, in the case of the CBP, bulk fermentation is eliminated, the dough going direct from mixer to divider. The dough is ripened by intense mechanical development in specially constructed high-speed mixers instead of by fermentation. By 1967, six years after its commercial introduction, more than one-third of all bread in the UK was made by the CBP. The large plant bakeries were the first to change, and over 50% of all plant bread became produced by CBP-type processing systems, which was a revolution for the UK baking industry.

The first mixer to be commercially applied to the CBP was the Tweedy 280, capable of mixing 280 pounds of flour (1 sack). The output of this machine was about 10 doughs or a total of 4600 lb of dough/hour. Various capacities are available with throughputs of 300–3000 kg flour/hour. Owing to the extremely short mixing cycle, these mixers can be also used in conjunction with continuous-processing lines. The Tweedy 6600, introduced in 1973, could be used in conjunction with a 15-sack processing-line. Continuous production incorporating a batch or discontinuous mixing stage is quite feasible, providing the batch mixing cycle is short enough. Adaptation is achieved by automatic feeding of flour, water, fat and minor ingredients to the mixer. Automatic sequencing of all mixing operations, including, ingredient call-up, mixing control within limits, dough-discharge from the mixing-chamber and transport to the divider-hopper at the required time. This synchronization of dough-mixing with the make-up plant is often referred to as 'scheduling'. The Tweedy-160 Automatic, for example, has a minimum charge/mix of 50 lb (22 kg) and a maximum charge of 275 lb (125 kg)/mix, providing for a maximum dough output/hour of 3025 lb (1375 kg). The Tweedy mixers fall into the category of 'intensive-mixers', allowing the energy input to be expended on the dough within the shortest possible time, 90–120 s for 8–11 Wh/kg. However, normal commercial practice is about 11 Wh/kg (40 J/g) over 200 s. Particularly advantageous is the programming facility for the mixing cycle, which includes loading, energy-input, and vacuum for crumb-structural adjustment if desired. This demands a degree of technical expertise from the operator, in order to optimize the technological and economic aspects of the mixing process. The watt-hour counter setting is based on a 1 unit scale reading for 10 watt-hours, assuming a desired work-factor of 5 watt-hours/lb of total ingredients, or 0·4 h.p. min/lb. Also, given a fixed flour temperature, the doughmaker must seek to optimize dough temperature by using the cooling-jacket in order to achieve a target consistency *v.* dough water absorption. On a basis of 11 Wh/kg energy-input for UK panned bread, unless chilled water is available at the mixer below 4°C, the flour temperature cannot exceed 27°C for a dough temperature of 30°C with the Tweedy mixer. Mains cold water supply is usually at about 14°C, therefore a considerable amount of chilled water may be necessary to adjust CBP dough temperatures, owing to heat dissipation. Popular Tweedy high-speed mixer models for the small to medium-size bakery are the 35, 70 and 140.

The typical Tweedy mixer-head design consists of an octagonal-plate, with four tooth-like blades and a propeller attachment in the centre to draw the dough down into the mixing zone. The plate rotates at about 400–500 rpm, being driven by a 60 or 75 h.p. motor. The mixing chamber is cylindrical, and has prism-shaped protrusions, which prevent the dough spinning around on top of the plate, and the resistance offered by these elongated prism-shaped protrusions increase the mixing-intensity. It is also possible to achieve different technological effects by changing the plate for others with another blade-shape, a different number of blades and/or different positioning of the blades. The chamber is fitted with a lid, and mixing can be carried out under various partial-vacuums of the order of $0.05 \, \text{N mm}^{-2}$, thus permitting a degree of control of dough-porosity and bread volume up to about 3·6 ml/g.

Bread in the UK is of relatively low specific volume 3·7 ml/g, compared with 4 hour sponge-and-dough in North American formulations, which is of the order of 6 ml/g.

Since in the CBP dough-ripening relies on intense mechanical development in special high-speed mixers, certain formulation modifications as well as processing modifications are essential, which include the following.

The mixer is of the intensive high-speed discontinuous or batch type, capable of imparting the necessary energy level of 11 watt-hours/kg or 40 J/g within 4–5 minutes (35·7 kJ/800 g unit).

Flour-quality variations demand energy adjustments, and since the mixer is adjusted to stop automatically when the required energy has been imparted to the dough, the mixing time is used as the control variable for optimal development. From the addition of the dry ingredients and water to discharging the dough into the divider-hopper normally takes 5–6 minutes.

Shortening is essential for good quality bread by the CBP, of which some 5% must be of high-melting-point (38°C), and a slip point higher than the dough temperature, in which case 0·7% is sufficient.

Water absorption should be increased by 2–4% above that used for conventional processing, 3·5% being an average increase. This is due to more flour-solids being retained compared with conventional bulk-fermentation, also, autolytic enzyme reactions have less time to present themselves, thus reducing dough softening. Any further increases in absorption can only be achieved with low-protein, Special CBP-flours of about 11·0% protein (14% moisture basis), such as 'T-blend' or 'Challenge', where processing adjustments at the mill have increased the 'damaged-starch' content to about 25–30 Farrand-units, and regulated the alpha-amylase content at about 5 Farrand-units (250 s Falling number). However, in order to avoid a damp, soggy-crumb at the slicing-stage, it is sometimes necessary to reduce the work-input from 5 watt-hours/lb dough to 4·5 watt-hours, which also reduces dough temperature to about 84°F (28·9°C). Average water absorptions of bread-flours for the CBP during 1974 were 167–173 lb/280 lb flour (1 sack). Alpha-amylase levels of 7 Farrand-units (200 s Falling number) result in slicing problems, owing to dextrin build-up, at the 11·0% flour protein level. The correct relationship between flour-protein, flour alpha-amylase and flour damaged-starch levels expressed in Farrand-units must be adhered to, as emphasized by E. A. Farrand, in order to ensure that CBP doughs pass through the divider and moulding-equipment without sticking. In addition, it must be stated that protein-quality is also important, and increasing emphasis is being placed on named quality wheat varieties, e.g. 'Avalon', with 11·0% protein and 250 s Falling number.

Flour quality variations can occur, owing to the increasing use of UK-grown wheat varieties of variable quality, and the addition of dried-gluten of variable quality at the mill in order to fortify breadmaking flours and raise the protein content. This latter practice has become widespread in the UK milling industry, aimed at reducing the percentage of protein derived from expensive imported Canadian and other third-country (non-EC) wheats in the breadmaking grist. The Bread and Flour Regulations 1984 and Food Labelling Regulations 1984, does not classify flour as a 'compound' ingredient for labelling purposes, therefore the miller is at liberty to add 'small quantities of gluten for technological purposes' without having to make a declaration to that effect. At the time of writing (1987), flour is not legally classified as a 'compound ingredient' for labelling purposes, and the gluten-fortified flour requires no special designation. The economics of gluten addition depend on its market-price and functionality in replacing Canadian Western Red spring (CWRS) or other quality wheats in the mill-grist, e.g. American Northern spring DNS 14%. Furthermore, if the gluten required could also be derived from UK-grown wheat, an extra demand for some 500 000 tonnes of UK home-grown wheat surplus would be created. The functionality potential of the gluten derived from UK-grown wheats would need continuous evaluation.

Oxidation levels, as with continuous processes, are high and costly compared with conventional bulk-fermentation processes. It is essential to use a fast-acting oxidant, and in the UK ascorbic acid at 75 ppm (based on flour weight) is now generally used. The use of ascorbic acid at 75 ppm in combination with 25–30 ppm potassium bromate can give better results according to Meredith (1966). If a vacuum is applied, the use of potassium bromate is essential, since it works without the presence of oxygen. Potassium bromate addition at the mill was being voluntarily fixed at 15 ppm, in common with agreed levels in commercial bread improvers. Therefore, provided the baker used any mixed improver at the recommended level, there was little danger that his bread would contain more than the maximum permitted level of potassium bromate set out in the Bread and Flour Regulations 1984, which was 50 ppm. Having been added to flour by millers to a maximum of 15 ppm; and by bakers in improvers containing not more than 30 ppm, E-924 Bromate, has been used in USA, UK, Canada, Australia, New Zealand, and South Africa since 1915. However, by 1989, only the UK was using it in Europe. Therefore, the MAFF Food Advisory Committee reconsidered the status of bromate at its September 1989 meeting. In view of EC Legislation harmonization, on evidence of some residues when used in excess of 75 ppm, this potentially carcinogenous oxidant is now banned in the UK with effect from 1 April 1990. The two accepted oxidants are now azodicarbamide and ascorbic acid. The use of these, combined with vital-gluten fortification and fungal alpha-amylase, all added at optimum concentrations with permitted emulsifiers, should produce products of acceptable baked quality. Since ascorbic acid requires available oxygen to provide the maximal improver effect, the use of open-bowl mixers, especially Spirals and Windels, represent the most efficient incorporators of air into the dough. However, the closed mixer-chambers of plant bakeries, operating under partial vacuum would inhibit oxidation. Low-speed mixing with 2-arm mixers, also do not incorporate as much air as the Spiral-mixers for Short-time doughs.

Since UK flours have become increasingly dependent on weaker European wheat varieties in recent years, the use of specialized composite improvers have increased. In addition, 70–80% of the flour now used for bread production is unbleached and untreated. At this time, maximum levels for ascorbic acid is 200 ppm and for azodicarbonamide 45 ppm.

The Bakery and Allied Trades Association (BATA) suggested the use of untreated flours only, so that the baker could use the correct improver concentration in compounded form for the various processes and products desired. Presently, azodicarbonamide added at the flour mill is limited to 10 ppm, with the improver suppliers limiting their product concentrations below 35 ppm azodicarbonamide for use with treated flours. Corresponding agreed levels of ascorbic acid are 50 ppm for flour at the mill, and 150 ppm in compound improvers. Compound improvers formulated for untreated flours can contain the maximum permitted levels of the oxidants.

Dough temperatures with the CBP are difficult to maintain much below 86°F or 30°C, owing to frictional heat dissipation, even using chilled water, which is always a limiting factor *v.* water absorption.

Yeast levels must be increased by a factor of 1·5–2·0, otherwise longer proof times will be required.

Other intensive high-speed batch mixers suitable for the CBP process which were in competition with the Tweedy during the 1960–70 decade included the 'Supertex', which had a rotating blade which continuously pinched off dough-pieces and kneaded them around the bottom of the bowl. The 'Cresta Doughmaster' had a completely different design, the bowl rotating, and the dough being pushed through revolving knife-edged blades located in the mixer-lid. The 'Cresta' is manufactured and exclusively exported by Rosedowns of Hull, UK. The most popular CBP continuous unit was the Oakes mixer already described, but the batch-mixing systems dominated the mixing processes in the UK. During the 1960–70 period, at least 10 UK manufacturers were marketing batch and continuous mixers for the CBP, with batch capacities of 300–600 lb of dough, capable of outputs of up to 6500 lb/hour.

Mechanically developed doughs in general, which are further processed by direct panning, tend to produce bread with the fine, even crumb porosity and sponge-like texture of those produced by a continuous process. Whereas, if the dough is divided, rounded, and given an intermediate-proofing stage before final-moulding and panning, the crumb porosity and texture will be more uneven and open, appearing like conventional bulk-fermented bread. Basically, the main points of difference between the continuous processes (Amflow/Domaker) and CBP continuous systems is the elimination of the pre-ferment, and use of the conventional make-up stages. The most important advantages derived from the use of the CBP process can be summarized as follows:

(1) Short-process-time of about 2 hours from start of dough-mixing to withdrawal from the oven, compared with 4–7 hours for continuous, straight and sponge-and-dough processes.
(2) Greatly reduced processing-area.
(3) Lower protein flours of 11·0–11·5% protein, containing larger amounts of UK-grown wheats, at lower costs, can be processed into bread of acceptable commercial UK standard—a cost-effective situation in a competitive market of discount-conscious supermarket chains and marginal profits.
(4) Process control is facilitated, and fermentation-losses are eliminated.
(5) Water-absorption levels can be increased, giving average yields of 180 × 807 g units at 39% moisture from 100 kg flour at 14·0% moisture.

The main disadvantages of the CBP are the large capital investment required for mixers, and the high energy-input levels necessary to replace bulk-fermentation. Additionally, although by 1976 some 80% of all bread baked in the UK was produced by the CBP, the adoption of the CBP technology in other countries has shown a very mixed pattern. The type of bread which the CBP produces is not acceptable as the national loaf in Western or Eastern continental European countries, where panned bread is regarded as a speciality product. In these countries the benefits of intensive mixing technology have been well-researched, and suitable mixing aggregates developed by their own bakery machinery engineers. In Western Europe (continental), the baking industry is very much in the hands of smaller bakers, whereas in the UK the large plant groups dominate, for whom the CBP is ideally suited. However, owing to the emergence of new sectors, such as in-store bakeries and hot-bread shops, as well as 'on the premises bake-offs', it is predicted that both the craft-baker and the plant groups are likely to lose appreciable market-shares before 1990. The number of plant bakery establishments has fallen by about 50%, and the number of master-bakers by about 55% between 1976 and 1987. In France, panned Dutch-type toast-bread has made headway, two three-shift plants having been installed by Baker Perkins for national distribution near Paris and in Lyons.

The other main developments in UK breadmaking, the activated dough development process (ADD), which is based on the use of L-cysteine, and the ADA, based on the use of azodicarbonamide, were both adapted from USA baking technology. Both these maturing additives became permitted in the UK under the Bread and Flour (Amendment) Regulations of 1972. L-Cysteine is permitted at levels of up to 75 ppm in bread other than wholemeal, and azodicarbonamide at levels not exceeding 45 ppm. In 1980, L-cysteine also became permitted in the Federal Republic of Germany at levels not exceeding 100 ppm for bread and small goods, and up to 150 ppm for fine baked goods. The ADD is primarily intended for the small baker, which gives him many of the advantages of the CBP, without the large equipment requirement of the latter. The necessary L-cysteine, potassium bromate, ascorbic acid and shortening is added in the form of a composite proprietary improver concentrate, marketed by most bread-improver suppliers. Almost any bread formulation can be adapted to ADD, being mixed in a conventional low-speed mixer to a final dough temperature of 85–87°F (29–30°C). The quality of the bread so produced should be equal to that made by bulk-fermentation in most respects, apart from crumb elasticity, flavour and aroma, providing the baker uses a flour suitable for bulk-fermentation, such as baker's grade flour of average protein content 11·5–12·0%, e.g. Sovereign, Radiant and Baker's Choice. Where flours intended for the CBP are used, e.g. T-Blend and Challenge of average protein content 11·0%, bread quality may well be below standard. This situation does not present itself when changing from bulk-fermentation to a mechanical dough-development process like the CBP. The following formulation would be typical for a small baker using a baker's grade flour, mill treated with 14–20 ppm potassium bromate:

| | |
|---|---|
| Baker's grade flour (11·0–12·0% protein, 15 ppm KBrO$_3$) | 280 lb |
| Yeast | 6 lb |
| Salt | 5·5–6 lb |
| Water, approximate | 168–170 lb |
| Shortening (high-melting-point) | 1–2 lb |

Composite ADD improver containing:

| | |
|---|---|
| L-cysteine–HCl monohydrate | 35–40 ppm (or 27 ppm L-cysteine) |
| potassium bromate | 25 ppm |
| L-ascorbic acid | 50 ppm |

(All additive concentrations in ppm are based on total flour weight.)

Some composite ADD improvers also contain the optimal amount of shortening together with the L-cysteine–HCl, KBrO$_3$, and ascorbic acid. The addition of the specified amount of improver to the basic ingredients is all that is required of the baker.

Procedure:

Mixing should be extended, depending on the type of conventional mixer in use, e.g. Artofex 25 minutes, RVK 25–30 minutes.

The dough is immediately divided, giving a first proof-time of about 10 minutes, after work-off and rounding.

Final-proof is the same as with bulk-fermentation, but moulders need to be adjusted for a denser dough.

The use of the ADD improver, permits work-input levels of 3 to 4 Wh/lb. The same technique can be made use of to reduce the work input levels of mechanically developed doughs, e.g. CBP. In such cases, where L-cysteine–HCl is added, the work-input can be reduced from 5 watt-hour/lb for 170 s, to 3 watt-hour/lb. By so doing, dough temperature rises during mixing are reduced by at least 5°F, and watt-counter settings reduced from about 235 to about 180–200 on a Tweedy 280, 1-sack mixer (280 lb).

When converting bulk fermentation doughs to the ADD, water-absorptions are adjusted up by 0·75–1·0 gallons/sack to compensate for dough-softening due to bulk-fermentation.

The use of azodicarbonamide (ADA) as an alternative to the oxidizing agents used in the CBP, bulk-fermentation and ADD, is possible without any major changes in dough-processing. However, it is a faster-acting oxidant, and can be used to eliminate intermediate proof when added at the 25 ppm level in conjunction with 35–40 ppm L-cysteine–HCl with baker's grade flour of protein content 11–12%. The cysteine level is less critical, but adjustments might have to be made for large differences in flour protein, in which case, a higher level of ADA may also be necessary.

The reaction rate of ADA is rapid, being comparable to potassium iodate, and produces a very fine cellular structure in the bread. It has been used in the USA as a flour-maturing agent marketed as 'Maturox' up to the maximum permitted level of 45 ppm, since 1962. Tsen reported in 1963, that oxidation of all —SH— exposed by mixing was effectively complete within 2·5 minutes, and further resting or mixing did not change residual —SH— content. According to Tsen, the formation of the —SS— bridges owing to ADA oxidation is between the polypeptide-chains. The higher the flour extraction rate, the greater the amount of ADA required for optimal treatment. ADA starts to react as soon as the mixing is complete, improving the dough structure and increasing water-absorption. At the correct dosage concentration, it increases loaf volume; when used in excess the over-oxidation results in a loss of gas-retention. Forty-five minutes after water-addition, it has completely decomposed and cannot be analytically detected.

In the application of ADA as the oxidant in chemical dough-developers, the relative reaction time sequence of the oxidizing agent with which it is paired must be considered. Barret and Joiner (1967) recommend the use of ADA and potassium bromate in the ratio 1:2, especially for continuous dough-processing. Such a combination is particularly effective with intensive-mixing, the authors suggesting the combined addition of 20–30 ppm ADA and 40–60 ppm potassium bromate.

In the development of improved 'short-time' processing systems to simulate the effects of pre-process fermentation, the critical factor is a balanced system of dough-improvers to put the dough into optimal condition for processing, compensating for lack of fermentation, dough properties and bread quality. The major reactions in the cycle can be categorized as follows.

GAS-PRODUCTION

Yeast activity must be accelerated by using improved strains or higher levels, with the help of growth-stimulants and substrates.

GLUTEN-PROTEIN MODIFICATION

Gluten-proteins must be rapidly modified by using the optimal balance of permitted oxidants and reducing agents, in order to ensure adequate gas-retention properties, and enough extensibility to expand under internal pressure of the carbon dioxide from the yeast. The use of proteolytic enzymes to modify the mixing and dough-handling properties of certain flours could also be beneficial.

POLYSACCHARIDE MODIFICATION

In the case of wheat-flour doughs, the application of amylolytic enzymes, e.g. alpha-amylases from various fungal and bacterial sources, amyloglucosidase, and glucose-isomerase is a possibility. The manufacture of low-molecular-weight glucose polymers from flour/water substrate media, by first heating the media to 95°C, cooling, then adding alpha-amylase to allow hydrolysis to proceed for about 2 hours. Residual bread-crumb could also be recycled by adding a proportion to the media. On cooling the media down to about 60°C, amyloglucosidase and glucose-isomerase can be added at appropriate substrate/enzyme concentrations, together with a trace-source of magnesium. The stirred media are reacted for about 10 hours, after which the media is sterilized at 95°C, and used to prepare the bread. The purpose of such media, is to provide a low-molecular-weight hydrolysate for the yeast, and to produce a flavour-token.

The use of pentosanases could also be applied to rye, ryemeal and wheatmeal doughs. The analogous manufacture of lower-molecular-weight amino acid and glucose monomers from residual rye and wheat crumb could be carried out by acid hydrolysis, using sulphuric acid at pH about 1·0–2·0 at 100°C for 1 hour. After appropriate neutralization with ammonium hydroxide, the 80% hydrolysate, containing amino acids, reducing sugar and glucose, can be used as a stimulant for the microbial rye starter medium, adding at about 20%. The same hydrolysate can be added to yeasted wheatmeal doughs.

The proteolytic and amylolytic enzymes extracted from *Aspergillus oryzae* and *Aspergillus awamori*, used at concentrations of the order of 0·002–0·05% (flour weight) are particularly effective in improving volume, porosity, colour, taste and aroma of bread. This is due to the increased levels of reducing sugars and free amines from enzyme action. The maltose disappears, and glucose, fructose and other hexoses appear.

ACID/SALT BALANCE

The effect of salt on the activity of the amylases depends on the concentration and pH of the medium. At pH around 7 to 8, with increasing salt concentration amylase activity is stimulated, but it decreases as the pH falls from 6·0 down to 4·5, becoming completely inactivated below 4·0. The addition of 0·2% citric acid to dough-improvers/conditioners used to prepare 'short-time' doughs, also lowers the pH at the mixing stage.

The following is an example of a recently patented Australian dough-conditioner, attributed to Mauri Bros and Thomson Pty Ltd, filed under NL 8301,974 2/1/85: potassium bromate 1·20 parts, ascorbic acid 4·0, malt 30·7. fungal enzyme 6·7, sodium thiosulphate 1·2, sodium metabisulphate 0·4, and wheat-flour 55·8. 0·25 parts of this mixture are added to flour 100 parts, salt 2·0, shortening 2·0, yeast 3·0, and water 60·0. Mixing is for 25 minutes at 27–30°C on a conventional-type mixer; ferment for 0–45 minutes; then proof for about 40 minutes before baking.

However, it is very doubtful whether the acidity of this dough would fall enough to provide the structure and eating-qualities required by most continental bakers, and their consumers. Where eating-quality standards take priority, the elimination of some form of pre-ferment in liquid or solid form, remains difficult.

The introduction of mechanical dough development and short processing times, resulted in the mills producing specialty types of flour and the inclusion of increasing percentages of UK home-grown wheats in the mixture (grist).

For a bulk-fermented dough, a baker's grade flour of average protein 11·5–12·0%, marketed under such trade names as 'Sovereign', 'Radiant' and 'Baker's Choice' is recommended. These flours were treated at the mill with a maximum of 20 ppm potassium bromate, the bleaching agent benzoyl peroxide, and the bleaching and improving agent chlorine dioxide. Commercial bread improvers also had their potassium bromate contents limited by agreement, so that the final bread would not contain more than the maximum permitted bromate level of 50 ppm, as set out in the Bread and Flour Regulations of 1984.

Baker's grade flours are also appropriate for the ADD and ADA chemical dough-development processes. However, since flours can now be supplemented with gluten at the mill, it is important that the baker checks the quality of the flour gluten proteins as well as their quantity, since quality variations can now arise from both the UK home wheat percentage and the quality of the gluten the miller adds to his grist.

The lower protein CBP flours, marketed under the trade names 'T-blend' and 'Challenge', amongst others, have protein contents of about 11·0%, and when processing such flours, the average water absorption will fall by about 5 lb/280 lb flour compared with a flour of 11·5% protein, i.e. 170 lb water/280 lb flour reducing to 165 lb/280 lb flour. Such adjustments, coupled with reduced work-inputs of 4·5 watt-hours/lb, instead of 5 watt-hours/lb are often necessary with flours of higher damaged-starch levels in order to avoid the symptoms of a damp, coarse-textured bread. The consequent fall in dough temperature also favours dough consistency at the mixing stage.

Bread-slicing problems can arise, owing to dextrin formation when the alpha-amylase activity of the flour exceeds about 250 s Falling number (Hagberg) or 5 Farrand-units of cereal alpha-amylase. However, these values must always be related to damaged-starch level and the protein quality (flour gluten and added gluten). Frequently, when alpha-amylase levels rise to 7–11 Farrand-units (200 s Falling number falling to 150 s), sticky-crumb texture make slicing very difficult to impossible. An average analytical profile of a CBP grade flour in the UK is as follows: moisture 14·0–14·5%, protein (as is) 11–12%, damaged-starch (Farrand-units) 25–30, alpha-amylase (Farrand-units) 7–10, Falling number (Hagberg) 200–150 s, Farinograph absorption (as is) 56–57%, colour-grade (Kent-Jones) 2·0–3·0, 10–15 ppm potassium bromate added at the mill (pre-1989).

When L-cysteine became a permitted additive for all flours other than wholemeal in 1972, various mixed dough-improvers and conditioners came on the market in the UK, based on the chemical ADD or ADA techniques. Such compositions could contain, L-cysteine monohydrate or HCl, L-ascorbic acid, potassium bromate, unprocessed soy-flour, emulsifying agents, e.g. diacetyl tartaric acid esters (DATA) and ADA, with or without specially selected shortenings for their solids, liquid fat index and plasticity. The British Arkady Company developed the following products for use during the 1970s: 'Crystal' for bread and fermented products generally, 'Sprial' and 'Super Vencrust', containing L-cysteine, oxidants and emulsifers. A product known as 'FAD', developed by Arkady for the Republic of Ireland contained L-cysteine, L-ascorbic acid, and potassium bromate as well as shortening and the enzyme-active full-fat soy-flour 'Do-soy' (Arkady). The purpose of FAD was to facilitate the preparation of doughs by the 'no-dough-time' method. The small baker could use 3 lb FAD/280 lb flour, and instead of bulk-fermenting his doughs for 1–2 hour by the conventional process, he could eliminate bulk-fermentation. The necessary changes were:

(1)   An increase in mixing-time of 20–25% irrespective of the mixer type.
(2)   An increase in the yeast concentration is obligatory, but can vary according to conditions, a maximum of 6 lb/280 lb flour being the guideline.
(3)   An increase in water-adsorption is also obligatory, and a good guideline in this case is approximately 0·5 gal/280 lb.
(4)   Increase dough temperature to about 82–83°F.

For the small baker a typical formula could be as follows: Baker's grade flour 280 lb, salt 5 lb, yeast 6 lb, FAD 3 lb, water 166 lb. The working-off procedure of dividing, rounding-up, moulding and final-proof including baking remain the same as for conventional bulk-fermentation. Mixing for 25–30 minutes conventionally at a dough temperature of 83°F. Divide at 28 or 14 ounces immediately, giving a 10-minute first proof before moulding. Final-proof is 50–60 minutes at 106°F and 80% relative humidity.

Following this schedule, fermentation-losses are reduced, and an extra 5 lb/280 lb water is possible, owing to the absence of the dough-softening of bulk-fermentation. Dough consistency is improved, and the density at the dividing-stage is constant.

The procedures for the Arkady 'FAD' and 'Crystal' are similar. Where 'FAD' is used, the use of shortening is optional, since 'FAD' contains the required amount of appropriate type of shortening. Current traditional Arkady products marketed for the Chorleywood system include 'Dynarex' for white bread, and 'Tolerance' for bread-roll production. For conventional breadmaking processes based on bulk-fermentation and sponge-and-dough, 'Opal' or 'Diamond' are the recommended conditioners.

In 1971, the structure of the UK milling and baking industry changed as a result of the merger between Spillers-French, Cooperative Wholesale Society and J. Lyons companies. This left the bread market divided between: Rank Hovis McDougall, 25%, Associated British Foods 24%, and Spillers-French 20·5%, leaving the smaller independent and small bakers with a combined slice of 30%. The 'big-three' total production capacity in flour-milling and baking then analysed as follows:

|  | *Mills* | *Bakeries* |
|---|---|---|
| Associated British Foods (ABF) | 22 | 53 |
| Rank Hovis McDougall (RHM) | 20 | 100 |
| Spillers-French | 20 | 61 |

The imposition of graduated minimum import prices for all imported cereals, accompanied by the variable levy system was a major handicap for the UK milling industry, which depended so much on North American and other imported third-country (outside the EC) grain. This concentration within the industry was indicative of a policy of integrated bulk-buying resources, and although pressure for a price escalation of flour and bread seemed imminent, this all happened against a background of a fall in demand over the previous 15 years. At that time breadmaking wheats were a very minor factor in UK agriculture, Kolibri, a spring-wheat variety having the largest acreage at 54%, followed by the soft winter-wheats Joss Cambier at 30%, and Cappelle at 20%. Therefore, the milling industry, backed by their own experience and technical resources, initiated their own individual 'producer-incentive bonus schemes'. For example, Allied Mills (ABF) offered top-premiums on a tonnage basis for the winter-wheat varieties 'Maris Widgeon' and 'Bouquet', and bonuses for the varieties 'Cama' and 'West Desprez', also winter-sown. However, this group still required some 400 000 tons (now expressed as 'tonnes') of UK home-grown wheat with protein content less than 10·0% to replace imported wheats. RHM Agricultural Division also set up a 'premium milling-wheat scheme' offering top bonuses for Class I wheats from the 1974 and 1975 harvests. Their list had been increased to include eight wheat varieties comprising: Flinor, Maris Freeman, Maris Widgeon, Sappo, Bouquet, Kolibri, Kleiber and Maris Dove. This scheme only applied to RHM seed sown in the autumn (winter varieties) and the following spring (spring varieties), the delivery period extending from the 1975 harvest to June of 1976. Grain merchants recommended farmers to grow Cappelle, Cama, West Desprez, Bouquet, and Champlein for the bread-flour trade, giving the crop a good May/June nitrogen dressing. Considering that marginally higher yielding varieties did not have the potential as bread-flour wheats. The variety Chalk was considered ideal for the milling of cake-flours on account of the flour-colour-grade, flour yield, and disease-resistance in the field.

In September 1974, the National Seed Development Organization devised a new winter-wheat cropping plan, which included the high crop-yielding variety Maris Huntsman, this wheat making up 50% of the total crop. Maris Huntsman being at that time top of the National Institute of Agricultural Botany (NIAB) list, on account of its high crop yield. However, its breadmaking properties were considered to be very poor, being graded down to 4 on the FMBRA grading system published at the end of 1974. The low protein contents of the 1974 harvest had resulted in a shortfall of UK and EC wheats of about 200 000 tons.

The planting of Maris Huntsman proved to be a nightmare for the milling industry, both in the UK and the Federal Republic of Germany, due to the 'sticky', softening doughs it produced. An EC test had to be devised to enable the miller to identify and segregate this wheat variety, and a similar variety named 'Clement' planted in the FRG. The best milling/breadmaking varieties in 1974 were: Atle, Cama, Bouquet, Drabant, Elite Lepeuple, Flinor, Janus, Maris Widgeon, Maris Ploughman, Sappo, Saxon, Sirius, and West Desprez, according to the FMBRA home-grown wheat assessment of December 1974. After circulation to the various trade associations of a document for discussion in April 1974, the Home-Grown

Cereals Authority launched a Wheat Classification Scheme, which operated from January 1975. The prime objective of this scheme was to make home-grown wheat more attractive to the miller, and a better competitor with imported wheat, thus reconciling the interests of growers, users and the nation generally. To merit classification, a wheat variety must be suitable for milling and breadmaking, and must meet a specified standard for protein content, alpha-amylase level, specific weight (kg/hl), screenings and impurities, moisture content and general condition. The particular class into which a wheat is placed depends on the variety and protein content. The scheme provided for two classes of wheat, A and B, according to the milling and breadmaking characteristics of the variety. Both classes being divided into three sub-classes of protein content on a dry-matter basis, as follows:

Class A: A1 13·5% protein (dry basis)     Class B: B1 13·5% protein
         A2 12·5%                             B2 12·5%
         A3 11·5%                             B3 11·5%

Basic standards: minimum 90% germination, minimum specific weight 72 kg/hl, maximum moisture 16%, maximum screenings 2%, damaged grains maximum 2%, maximum ergot 0·001%, alpha-amylase 180 Hagberg.

The varieties listed in each class depend on the FMBRA-assessment for milling and breadmaking properties, which constitutes an aggregate score of 3 for eligibility for Class A. A comprehensive list of all the FMBRA varieties is constantly updated, and it also includes qualifying varieties of limited commercial availability. The scheme was administered by the Home-Grown Cereals Authority, which appointed registered agents to carry out wheat classification, and approve laboratories, where certain tests can be done. Since the correlation between milling and breadmaking value and variety is of paramount importance, indisputable evidence of the identification of samples must be provided by the grower, including seed-purchase documentation. In problem cases, samples can be identified by the 'finger-printing' of the electrophoretic-pattern of the protein in a single wheat kernel, or the sample can be 'grown-on' in order to confirm its variety. It is also the agent's duty to ensure that the commodity is being stored under the correct conditions, and delivered to the buyer up to specification. Classified wheat premiums relate partly to the comparative value to the miller, and partly to the incentive needed by the grower to grow and market classified breadmaking wheats.

It is important to appreciate that every sample of wheat within the individual varieties is a variable commodity, depending on soil, climate, location and plant-husbandry, and as such must be judged on its own merits.

There will be continuing efforts to increase the level of home-grown wheats in breadmaking grists. The use of wheat-gluten to supplement the protein quantity and quality of grists containing increasing levels of these wheats, is likely to be the subject of ongoing research. Gluten-quality testing methods are gaining increasing attention, in order to supplement the relatively limited information gained from mere protein-content.

New strains of wheat require about 15 years from the initial crossing of genotypes to the commercial availability of the seed. Unless low alpha-amylase strains for high-rainfall/high humidity climates at the ripening time can be bred, quality will continue to be unpredictable from year to year. UK grain crops are very prone to fungal-disease, owing to the humid, maritime climate, but the controlled use of fungicides help to contain the detrimental effects of fungal enzyme systems on the wheat endosperm.

Often, over-intensive husbandry, i.e. planting-density, excessive N:P:K ratios greater than 1:1:1, give rise to distress-growth conditions, and result in varietal degeneration. Symptomatic of these effects are imbalances in the enzymic metabolic activity of the grain.

The structure and development of wheat breeding, production and quality classification before marketing to the milling industry has important consequences for the baking industry, and the lack of communication and flow of information between these various sectors within the food-chain of a private-enterprise system frequently does not allow the best use to be made of manpower and resources. The decision to include increasing percentages of home-grown wheats in mill-grists is a valid one, where the quality and availability of the varieties is ensured by integrated forward planning from a vertical organizational standpoint. Bread quality standards are at first defined by the baking industry, both plant and craftsman's specifications, for flours to meet their respective production tolerances. The milling industry then formulates a grist to meet the specification, and informs the grain-supplier, whether under state or private control, of the wheat varieties which he needs to meet the baker's standards. The grain-supplier would then collaborate with the Wheat-breeding Institute, and grower's cooperative in order to plan grain production over a 5-year period.

Where planning, or lack of it, emanates from the agronomist, crop yields will always be the dominating criterion, and the desired quality varieties will not be cultivated. The non-availability of wheats of the requisite varieties inevitably leads to weaker grists and flour specifications, and a fall in bread quality standards. Under such circumstances the miller and baker will increasingly seek additives to improve his products, or modify his processing methods, or both. This pattern would seem to fit the development of the UK grain, milling and baking industries over the period from 1960 to the present day. A comprehensive range of dough-improvers have appeared on the market, variously defined as: 'compound-improver', 'compound-dough-additive', 'compound-dough-conditioner', owing to the critically stringent raw material requirements of the CBP version of mechanical dough development from about 1960, and the ADD version of chemical dough development from the early 1970s.

Treatment begins at the mill with benzoyl peroxide, chlorine dioxide, potassium bromate (now ascorbic acid and azodicarbonamide) and dried-gluten, according to flour requirements, based on laboratory tests. For the baker, the essential ingredients in the form of a compound improver contain: soy-flour, oxidants— ascorbic acid and/or additional azodicarbonamide—a suitable shortening and yeast foods. This product being added at a level of 3 lb/280 lb of flour to ensure the correct final concentration in the dough, is effectively a mixed concentrate containing the essential CBP functional-components. Similar products were developed for the ADD, containing: soy-flour, L-cysteine–HCl, oxidants—ascorbic acid and/or azodicarbonamide—the correct type of shortening and yeast foods. This product also is added at the rate of 3 lb/280 lb of flour. Therefore, by the early 1970s, three types of compound-improvers became available, one for the bulk-fermentation processes, one for the CBP, and another for ADD. However, not all contained soy-flour, and were instead in paste form, using a base of shortening and emulsifier. From these, a second generation of product was developed containing mixed emulsifiers to improve volume using the same three processes. This application was used for bread, bread-rolls, Vienna-bread and rolls, and *baguettes*. Other innovations were mixtures for fermented buns and cookies, containing emulsifiers, shortening, sugar and milk-powder. Both these latter products are added at the 2% level, based on flour weight. A general-purpose mixture, suitable for the CBP, no-dough-time, doughs mixed on low-speed or spiral mixers and short bulk-fermentation processes, often contains: soy-flour, ascorbic-acid, emulsifier-blends, glucose, malt-flour or fungal alpha-amylase, yeast foods and sometimes L-cysteine. Such a mix can be applied to all fermented products by variation of the usage level between 1% and 2% of flour used. Metrication was introduced in the late 1970s, which made some reformulation necessary, and all usage levels became expressed in percentages. Dosage weights were than quoted as kg/100 kg flour on all bags. These general-purpose mixtures were designed for the hot-bread shop, in-store bakery and family bakery, owing to the wide range of products and mixing and processing conditions involved. The emulsifier-blends used consist of diacetyl tartaric acid esters of the monoglycerides, having good gluten-complexing properties, and sodium and calcium stearoyl-2-lactylates which complex with both gluten and starch. The chemical composition and physical state, as well as the ratio of these blends are important for functionality in the dough.

Special mixed improver concentrates are available for hamburger plants, containing special blends of emulsifiers and oxidants, providing dough-stability for processing, adequate flow for pan-proof, and a good volume, fine-textured product.

Some bakers use ADD improver mixes in high-speed mixing in order to reduce energy-input levels or shorten mixing-times to achieve lower dough temperatures, when processing with high flour and high minimum water temperatures.

A range of concentrates is also marketed, containing all the essential functional ingredients and additional shortening, and sugar where required, as well as the correct salt concentration. In such cases, only concentrate at the correct level, flour, water and yeast are necessary to prepare the dough. These complete mixes are available for bread, fermented morning products and the numerous special-breads.

Spillers Premier Products Limited, a part of Spillers Milling, also market a wide and increasing range of concentrates for the baking industry. The main group of bread improvers/conditioners include products for the CBP, no-dough-time processes and those using traditional bulk-fermentation. Similar to the British Arkady Company, they also market a range of special products for plant bakers, small family bakers and in-store bakeries, and a range of bakery concentrates for standard and high-fibre products. The other main bakery range of Spillers Premier Products (SPP) is the 'Slimcea' and 'Procea' bread premixes, which Spillers bought from Cavenham Foods in the 1960s. SPP was formed from Soya Foods, British

Soya Products, and Slimcea as a result of a succession of strategic commercial fusions to form part of the Dalgety-Spillers Group of companies. The 'Countryman' concentrates are formulated by SPP for whole-grain, brown high-fibre, white high-fibre and Muesli breads. This type of product is of particular interest to the smaller baker, enabling him to compete in the growing specialty-bread market without the capital expenditure of new product development and product-launching. The supermarket in-store bakeries also find such products useful, selecting the type of concentrate product most appropriate to their own ideas, in order to market under their own label. The basic raw material of most of these premixed improvers and concentrates marketed by such companies as British Arkady and SPP is the raw soybean. The raw beans are cooked and milled to a fine flour, without any extractions or additions. The full-fat enzyme-active flour is used for addition to yeasted doughs, and the processed material, marketed by both British Arkady and SPP, after appropriate heat-treatment, as nutritional supplements, and for confectionery manufacturing purposes.

At the BBex 1987 Baking Industry Exhibition held at Birmingham, UK, SPP launched a new generation of tea and savoury bread formulation ideas under the registered trade mark 'Gemini'. By formulating with a Gemini Flour Base, tea-loaf concentrate, and various Gemini fruit and nut or spiced mixtures, the baker can create various tea loaves, e.g. Swiss muesli, Caribbean, citrus fruit, sultana and malt and spiced varieties. Formulating with the Gemini flour base, bread concentrate and various Gemini savoury mixtures, the baker is able to create such savoury breads as, for example, mushroom bread, pizza bread, cheese and onion bread, eastern spice bread and celery and apple bread. Formulations are provided, and all the baker has to add at mixing is yeast and water, processing within 1 hour of mixing.

For the preparation of high-fibre bread and other products, SPP launched a wholemeal with 10% added fibre, and more than four times the wheatgerm of standard wholemeal. This composite flour called 'Utopia', when made into bread contains about three times the fibre content of ordinary white bread.

UK wholemeal flour, as the name suggests, is milled from the whole of the wheatgerm, and can either be produced by stone-grinding, 'stone-ground wholemeal', or conventional roller-milling. Wholemeal flours are marketed by national milling-groups and small individual mills supplying local bakers. In the past, wholemeal-flours produced a dry, dense crumbly loaf with a poor shelf-life. However, in recent years this situation has improved, bread of better volume, crumb texture and shelf-life being possible because of the following changes:

(1) Use of stronger mill-grists containing higher proportions of imported and home-grown strong wheats, resulting in improved dough-stability and gas-retention.

(2) The changes in the Bread and Flour Regulations of 1984, permitted the addition of dry-gluten and ascorbic acid to whole-meal-flour at the doughmaking stage.

(3) The availability of compound bread-improvers formulated especially for wholemeal bread, containing DATA-ester emulsifiers, shortening, malt-flour or fungal alpha-amylase, and/or gluten, also ascorbic acid, with flour as a carrier.

(4) Improved processing technology through better control of such variables as: yeast level, mixing method, dough temperature, proofing temperature and time.

(5) Greater choice of wholemeal-flours, with different particle size distributions to suit individual baker's requirements. Stone-ground wholemeals from organically grown grain without any chemical additives are also available from small, independent mills.

Brown flour, formerly designated as 'wheatmeal flour', must contain a minimum of 0·6% fibre, based on dry matter, which approximates to a flour extraction rate of about 78%. Often brown flours contain much higher levels of fibre, and the bran particles can be coarse, medium or fine, according to the choice of the baker.

Germmeal is a blend of brown flour and a minimum of 10% wheatgerm, which is stabilized by heat, and can also contain salt. Typical examples of these are the proprietary brands Hovis of RHM Ltd, and Vitbe of Allied Mills, both established national products, famous for their nutritional value and good flavour.

There are also numerous brands of multigrain brown flours, which may contain wholemeal-flour, brown flour, rye-flour, malted wheat flakes and/or grains, as well as rice and barley and, in some cases, oats. Also, fibre-enrichment such as wheat-bran, soy-bran, pea-bran, shredded citrus-fruit peels, and brewers by-products are widely applied for the preparation of high-fibre specialty-breads. Complete-mix grain breads are also on the market, some, for example 'Granary', having national brand labels, backed by advertising, requiring only the addition of yeast and water. 'Granary' flour is a registered trade mark of RHM. Some concentrates contain wheat-bran, blended with other ingredients, to add flavour and improve dough development, for addition to white and brown flours to create individual types of specialty-breads.

Rolled or cracked wheat-flours are produced by passing wheat through a roller set to crush into flakes or simply crack the grain open. However, hydration of this material by normal doughing methods is impossible since most of the endosperm is still locked inside the cracked grains. Therefore, such flours have to be mixed in two stages; the material is first mixed on low speed with the dough water for about 30 minutes. This allows the material to 'soak', separating the endosperm from the bran, and a cohesive gluten structure to form. At this point, the remaining dough ingredients are added, which includes an appropriate dough improver/conditioner containing dried gluten to improve dough stability and volume. Alternatively, the material can be soaked batchwise overnight in containers, especially where high-speed mixers are in use, since the speed is too high for a 'wet-milling' effect to take place. The bread produced with this material and technique has a smoother, and moister crumb with large flakes, giving a good eating quality and shelf-life. A popular nationally produced example of such a bread is the 100% Rustic wholemeal marketed by Tesco in-store bakeries, made in the form of panned bread, cobs, *baguettes* and rolls. Providing 13·5 g/125 g-serving, the dietary-fibre contribution is even higher than ordinary wholemeal bread (Tesco Consumer Advisory Fact

Sheet—'Bread'). This type of bread is very popular in Holland and Belgium, and continues to gain in sales in the UK since its introduction.

Nevertheless, the baker can develop his own specialty-breads by balanced addition of specific ingredients. A high-bran loaf can be made by adding wheat-bran, soy-bran or pea-bran to a strong white flour. Using a North American spring wheat flour (protein 13·0% basis 14·0% moisture), 5–10% bran-flakes can be added, which will reduce the flour content of the dough by that amount, i.e. 95 or 90 parts flour: 5–10 parts bran-flakes. Since the gluten content of the dough has been reduced, a rebalance is necessary in the form of vital wheat gluten to restore dough structure and final loaf volume. Also, because of the water absorption capacity of the high-cellulose bran-flakes, it is important to increase the water content of the dough. These two factors are the most important in formulating this type of dough. In the case of white bread with 5–10 parts bran-flakes, 3 parts vital wheat gluten should be sufficient, where 95 or 90 parts of North American spring flour is used. Dough water absorption could be as high as 73–75%, depending on flour quality and the particle-size of the bran. The yeast concentration in the dough must be increased from 2·0% to 4·0%, and the addition of 0·25% DATA-ester will increase bread volume. If higher volume bread is desired, the use of a cellulase enzyme preparation is recommended. It is possible to prepare bread from a mixture of 100 parts white flour and 30 parts bran, in which case the correct balance between the addition of dried gluten and cellulase enzyme preparation must be established. Further improvements to the recipe re-balance can be made by adding vegetable-shortening at 1–2%, and high-fructose syrup at 1–2% flour weight. For the manufacture of bread-rolls with added bran, shortening levels of up to 5% flour weight should be added. Dough temperatures should be about 28–29°C, and mixing times will depend on the type of mixer in use:

—Spiral: 2 minutes slow, 8 minutes fast speeds.
—Conventional: 20–25 minutes.
—High-speed: 2·5 minutes or 11 watt-hours/kg dough.
—Using cake-mixers: 2 minutes first speed, 10 minutes second speed.

For the UK bread market it is desirable to maintain loaf volume as far as possible when re-balancing recipes for fibre-enrichment. The formulation of specialty-breads is discussed in more detail in Chapter 2.

Other structural changes took place in the milling and baking industries of the UK during the late 1970s and early 1980s. Spillers terminating their baking operations about 1978, which resulted in the disappearance of the 'Wonderloaf' labels, which dated back to the 1950s. Then RHM sold their Agricultural Division to Dalgety in 1984, resulting in a cash realization of £25m sterling, and an important release from external borrowings of £40m. During 1981, RHM had reduced its British Bakeries chain from more than 60 bakeries down to 40 production units, in order to rationalize manufacturing, and move the Baking Division from a loss-making situation slowly into a profit-making one. The trading position had been made more difficult owing to plant becoming obsolete, and the savage discount price-war within the baking industry. In 1985, bread prices were increased by 1 and 2 pence a loaf, and

Allied Bakeries (Weston Group) in common with RHM, took steps to restore bread margins, thus suggesting an end to the long-running price discount war for the valuable supermarket business. Whether this situation can prevail and stabilize will depend on market volatility, and the effects of supermarket in-store bakeries. Meanwhile, the milling industry wheat-prices have fallen, reducing raw material costs, which make up about 80% of total operating costs. This situation, however, will depend on the harvest and the availability of sufficient hard, home-grown breadmaking varieties for sale on the home market. Recent flour-price increases and the proximity of mills to ports are important factors for profitability, since approximately 50% of the wheat requirement is imported from North America and other EC countries, and transport is the second largest cost for the miller. Against a background of slowing down in aggregate profit growth from milling and baking during 1986/87, diversified expansion is indicated and long-awaited, as an alternative to attracting takeover bids. The likelihood of the latter in RHM's case, remained, while S. and W. Berisford held a 15% strategic stake in the company, and the Australian foods group, Goodman Fielder, held another key stake in the company. In recent years RHM have been one of the food sector's longest-running takeover candidates. The main competitor, Garfield Weston's Associated British Foods, also have a strategic holding in the commodities group S. and W. Berisford, as a long-term investment. Both ABF and RHM have balance-sheet provisions of £200m and £100m sterling respectively for takeover moves. Moreover, the third group, Dalgety, which comprises Spillers Milling, Foods and Agriculture, has concentrated its interests in the food and agriculture sector by selling off its non-food companies during 1987, also realizing about £200m sterling. On the acquisition side, Dalgety has bought Golden Wonder snacks, another snack company in The Netherlands and a commodities company Gill and Duffus. To date, unlike ABF and RHM, Dalgety is not directly involved in the manufacture of bread.

In 1984, the plant bakeries had 70% of national bread sales, and the craft sector 27%. Over the past 10 years, the number of master bakers has declined by about 50%, and during the same period the number of plant bakery manufacturing establishments also fell by about 50%. New market sectors in the form of supermarket in-store bakeries, hot-bread shops and 'bake-off on the premises' establishments have emerged. Current trends indicate that the in-stores will continue to capture an increasing share of the baked-products market, including freshly baked bread, bread-rolls, *baguettes*, fermented goods and flour-confectionery. However, the bake-offs are poised to take a greater share of the bakery market. Fresh bread and other baked products available all day, as is normal practice on the continent, is the popular consumer requirement of the 1990s. The lack of freshly baked quality bread and baked products in the UK, has resulted in the annual progressive decline in bread consumption compared with the rest of Europe. As already stated on more than one occasion, the eating-quality standards of UK bakery products, leaves much to be desired compared with other EC countries generally, and with the Federal Republic of Germany in particular. The same statement can also be made with confidence when comparing the UK with bakery products produced in most of the countries of the Socialist community, and in particular those found in the German Democratic Republic. The practical skills and expertise applied in the German-speaking countries stems from a well-conceived and organized apprenticeship training-scheme leading to the '*Meister-prüfung*', and only a person with this standard of education can operate a bakery business and produce baked products for the general public. Individual product standards are maintained by the German Agricultural Cooperatives' annual competitive examination, which is conducted by groups of recognized craftsmen-bakers led by trained and qualified group-leaders.

A comparison of factory or plant bakery products in the German-speaking countries with those produced in the UK, also reveals that UK standards of production technology, and in-depth knowledge of how to produce appetizing baked products is sadly lacking. Nevertheless, even German factory-baked products cannot compare with those of the private craftsman baker for freshness or quality, although, as in the UK, the price of the mass-produced products is usually lower.

Preoccupation of the consumer with health foods and diet is now acknowledged as a major factor influencing the European food market of the 1980s and will continue well into the next century. In the UK, the government NACNE and COMA reports have strengthened the case for increased dietary fibre, and more discriminatory dietary habits. The dietary contribution of bread remains excellent for growing children and for adults at any reasonable level of intake. A comparison of the composition of wholemeal and white flour of approximately 70% extraction shows that wholemeal differs from white in containing a little more protein (13·6% *v.* 12·8%), more cellulose (2·0% *v.* 0·2%), less available carbohydrate (69·0 *v.* 77·0 g/ 100 g), more iron (3·8 *v.* 2·1 mg/100 g), more magnesium (141 *v.* 27·0 mg/100 g), more calcium (28·0 *v.* 12·6 mg/100 g), although phytic acid interferes with its absorption, and more B-group vitamins, biotin (5·0 *v.* 0·8 μg/100 g), also folic acid (35 *v.* 14 μg/ 100 g). However, any differences only become apparent if it can be demonstrated that they affect the growth rate, or general wellbeing of man when the resultant breads are eaten as part of a normal diet. The most conclusive evidence concerning this long-standing topic was obtained in Germany during the food shortage following World War II. Children in orphanages within the age group 5–15 were given unlimited amounts of bread, deriving their calorie intake at the 75% level from various types of bread made from extraction rates from 70% up to 100%, but all were enriched with calcium. The subjects were divided into five groups, obtaining their calories at 75% from five bread varieties. The duration of the trial was 52 weeks. The results of these trials were quite revealing, since white flour had long been decried by scientists and less scientific allies, so much so, that bread generally was regarded as being of low nutritional value compared with milk, meat and vegetables. The results of this gruelling test showed that average weight-gains of the five groups, all getting about 75% of their calories from five different breads, were very similar. The 70% extraction flour bread gave the most rapid weight gains, the 85% extraction flour bread the slowest, and the 100% extraction-flour bread weight-gains intermediate between the other two. Further tests were carried out, in which the bread diet of one group was supplemented with milk, and growth compared with an unsupplemented control group, over a period of 28 weeks in this case. The result of this experiment showed that the two groups grew equally well, as measured by both weight-gain and height-gain. Therefore, one can but conclude that wheat-flour bread in any form, whatever the extraction rate, represents an excellent food for growing children and adults, at any reasonable intake level. In general, bread is the main source of trace elements, followed by potatoes and vegetables. In particular, the daily requirements of the important trace elements are contributed by bread in the following percentages: copper 40%, chromium 27%, manganese 80%, cobalt 29%, iron 29%. With the reversion to lower extraction-rates after the war years, many countries fortified their flour with various vitamins and minerals. As the experiments described show, this policy is not a necessity unless the national diet as a whole fails to supply certain important vitamins, growth factors or trace elements. Any decision on this matter demands a detailed knowledge of the composition of the national diet, particularly the amounts of bread, milk, fats, sugar, meat and vegetables, in the fresh form.

In order to determine the part played by bread as a source of energy and nutritional requirements in the national diet the following information must be available:

(1)  The daily requirements of the people concerning energy and nutrients, depending on their occupations.
(2)  The energy content of bread, and its various nutrients.
(3)  The daily consumption of bread and baked products.

In the case of (1), only average figures covering such consumer categories as: sex, age, working-intensity, etc., can be given. Assuming an average energy requirement of 3000 kcal (12 558 kJ) per day, the daily requirement for some of the most important nutrients would be as follows: Water 1750–2000 g, protein 80–100 g (of which 50 g should be animal-protein and include the following amounts of amino acids: tryptophane 1 g, leucine 4–6 g, isoleucine 3–4 g, valine 4 g, threonine 2–3 g, lysine 3–5 g, methionine 2–4 g, phenylalanine 2–4 g, glutamic acid 16 g, arginine 6 g, tyrosine 3–4 g, alanine 3 g, proline 5 g, cysteine 2–3 g, aspartic acid 6 g, histidine 2 g), carbohydrate, including starch, 400–500 g, sugar 50–100 g, cellulose and pectin 25 g. Fat coverage should comprise 25 g vegetable fat, 3–6 g unsaturated fatty acids, 0·3–0·4 g cholesterol, and 5 g phospholipids. Mineral intake should cover calcium 800–1000 mg, magnesium 300–500 mg, iron 15 mg, zinc 10–15 mg, phosphorus 1000–1500 mg, copper 2 mg, sodium 4000–6000 mg, potassium 2500–5000 mg, manganese 5–10 mg, selenium 0·5 mg, chromium 2–2·5 mg, chloride 5000–7000 mg, fluoride 0·5–1·0 mg, iodine 0·1–0·2 mg, molybdenum 0·5 mg. Vitamin intake requires vitamins C 70–100 mg, A 1·5–2·5 mg, B1 1·5–2·0 mg, B2 2·0–2·5 mg, B3 5–10 mg, B6 (pyridoxine) 2–3 mg, B12 (cobalamine) 0·005–0·080 mg, biotin 0·15–0·30 mg, D in various forms 0·04 mg, P (rutin) 25 mg, B9 (folic acid) 0·2–0·5 mg, E (various forms) 2·6 mg, K (various forms) 2·0 mg, choline 500–1000 mg, inositol 0·5–1·0 mg. For calculating the gross energy values of protein and carbohydrate, 17·1 kJ (4·1 kcal) for proteins, and for fats 40 kJ (9·4 kcal) per gram can be utilized.

The gross energy-value of bread per 100 grams will vary depending on the bread variety, and its moisture and fat content. Products with higher levels of cellulose of low digestibility, generally have lower energy-values, e.g. white-bread 1088 gross kJ/100 g of product (260 kcal), wholemeal bread 1000 gross kJ/100 g of product (239 kcal), bread-rolls 1184 gross kJ/100 g of product (282 kcal). The latter relatively high gross energy-value is due to the lower moisture content of bread-rolls (*c.* 32%), compared with 35–40% in the case of most panned breads.

The daily consumption of bread and baked products cannot be quantified generally, but in order to satisfy the daily requirement list of nutrients specified above, it would be necessary to consume about 300 g of rye wholemeal bread and about 200 g of bread-rolls made from white flour.

People on slimming diets of about 1200 calories/day, consume an average of 56–84 g/day, of bread. This calorie cut, however, reduces the economic intake of large amounts of protein, iron, vitamins trace elements, vitamins, vegetable lipids and the valuable digestive-aid cellulose for the intestinal flora. White bread contains about 2·7% dietary fibre, and wholewheat bread about 9·5%. The milling of wheat into refined white flour removes considerable amounts of major elements, i.e. calcium 60%, magnesium 81%, phosphorus 70%, potassium 77%, sodium 78%, and many trace elements essential to general wellbeing, i.e. iron 76%, zinc 78%, cobalt 86%, manganese 86%, copper 68%, molybdenum 47% and folic acid 60%, to quote average data. Enrichment programmes for flour return about 960 mg/lb calcium, 2·9 mg/lb thiamin, 1·8 mg/lb riboflavin, 24 mg/lb niacin, and 13·0–16·5 mg/lb iron. Although the iron is added in a form which is not efficiently absorbed in the human tract. Flours with increasing extraction rate, e.g. 70% to 92% contain increasing amounts of phytin, which forms phytic acid, which in turn reacts with the calcium and iron in the flour, forming calcium and iron phytates, from which compounds the two elements are not readily absorbed. Therefore, the relatively small increases in the calcium and iron contents of bread made from the higher extraction flours, amounting only to about 3 mg/ounce of calcium, and 0·5 mg/ounce of iron, could well become insignificant. Ironically, wholemeal flour is not fortified with minerals in the UK, since it is claimed that these are present naturally.

With the increasing realization in the UK that bread and baked products must be eaten oven-fresh, at their crusty best, an increasing number of consumers will become convinced of the value of bread as the most complete food next to milk, and an excellent complement to foods rich in fat and animal protein, since, unless fortified with certain amino acids, e.g. lysine, methionine, threonine and tryptophane, its protein balance shows a deficit as a first-class protein. Rye-flour bread gives a better amino acid spectrum than wheat-flour. In the grain endosperm, and in flour, the concentration of the amino acids glutamic acid and proline is higher than in the grain as a whole. Baker's yeast contributes about 1·22 g of lysine/100 g of product, and skimmed milk powder about 2·95 g/100 g of product. Milk protein isolate contains about 7·8 g lysine/100 g of protein.

In view of the importance of calcium in bone-building, maintenance of nervous tone, and blood-clotting, the supplementation of flour, ensuring a daily intake of 1 g, as carried out by the UK Ministry of Food during the war-years and post-war, ensures a safe dietary minimum.

Current bread varieties produced in the UK, and the flour types used to make them are as follows.

White breads are made from fortified white flour, providing about 0·33% of the daily requirement of calcium, and about 3 g of dietary fibre/125 g consumed daily. Common National versions of this type of bread are detailed as follows. Farmhouse, a panned bread, sprinkled with flour and cut lengthwise to give a crisper top-crust. Bloomer, a hearth or oven-bottom variety, moulded into a long loaf, cut diagonally across the top along its whole length to give the traditional 'London-bloomer' appearance. Baked on the oven sole, the crisp, light crust is very popular. Danish, is often sold as a sliced-loaf, containing a small amount of shortening and sugar for enrichment. Popular as a toast-bread, becoming crisp on the outside, and soft internally. *Baguettes*, and *batons*, are long, narrow crusty loaves made in the traditional way, preferably from French-milled flour. The dough is given a first and second proof, after which the dough-pieces are baked-off with adequate steam-injection to give a light texture and crisp light crust. The same dough can also be used to make crusty continental breakfast-rolls. The basic white bread dough can also be used to make poppy- and sesame-rolls, garlic and onion breads and plaited-breads. Baps, muffins and hamburger rolls are made with a soft dough, containing a small amount of shortening, milk-solids, and sugar. High-fibre breads, are made from white-flour fortified with vegetable cellulose material to increase its fibre content to approximately 10 g/125 g of bread. This satisfies those who prefer the lighter colour and texture of white bread but who wish to benefit from the higher dietary fibre levels of the darker breads.

The soft-grain breads were introduced by the large plant chains in order to recapture and hold their share of the market in sliced–wrapped bread. However, a soft-grain mix is also available for the smaller baker to add to his white bread flour at the 15% level, in the form of a type of concentrate. This soft-grain mix can also be added to baps and muffins, as well as to other specialty-breads, either alone or in combination with other fibre-rich materials and new flavour innovations.

The RHM Mothers Pride 'Champion' label consists of a blend of wheat, rye, and maize. Allied Bakeries were the first in this field, owing to their Australian connection. In Australia, soft-grain came onto the market in about 1983, in order to increase the limited range of bread varieties in that country, and took a massive 20% share of the market. Allied started their 'Mighty White' label soft-grain launch in the North of England in early-1986, followed up by the Midlands, and by late-1986 it became a national label. 'Mighty White' contains kibbled wheat and rye grains, increasing the fibre content of the bread by 30% compared with the standard white loaf. Allied claim that the 'Mighty White' sales figures have originated from former white bread buyers, rather than from wholemeal or brown bread sales. In addition to the standard 800-g medium and thick-sliced wrapped loaves, both Allied and British Bakeries have introduced a 400-g unit. The soft-grain idea has filled a market gap between a desire to protect children's health and to remain traditional in terms of taste and flavour. Obtaining a price premium to cover the extra time and labour involved for soft-grain bread is difficult in the large supermarkets, where the lowest price is the aim. For the plant baker, the long soaking-time, longer mixing-time, and labour-intensity, all add to production costs, and although an average price premium of 3 pence/loaf is the target, this can become eroded on 'own-label' contracts where competition is sharp. RHM's British Bakeries blend the grains, and then soak them for 24 hours, after which, they are fed to the mixer at 15% of flour weight. Allied Mills developed a soft-grain bread flour for the master baker craftsman and in-store bakeries, and supplies recipe cards for panned-bread, crusty and soft rolls. In addition, the soft-grain bread flour can be applied in ways to suit the baker's own formulation ideas, with the aid of the company's own computerized recipe service. The soft-grain bread flour is prepared at the mill by putting the kibbled wheat and rye grains through the soaking process prior to blending. In this way, the baker only has to soak the mixture for another 2–5 minutes before adding the other ingredients and mixing in the normal way.

The independent bread plants now also produce their own soft-grain labels, the Family Loaf Bakery, Avonmouth, having developed several variety breads including a soft-grain under the 'Countryman' brand name. Soft-grain is the best-seller within the range, claiming about 7% of all wrapped-bread sales. Associated Family Bakeries at Byfleet and High Wycombe have also launched 'Countryman' soft-grain.

The large plant bakery groups are beginning to realize that something has to be done to improve the quality and status of their bread to stimulate sales. In order to compete with the in-stores, hot-bread shops, bake-offs and the craftsman baker, bread has to be presented as a food in its own right, and not just as a carrier for meat and cheese. Furthermore, the whole baking industry is in competition with the other sectors of the food industry, dairy, meat, fish and fresh fruit and vegetables, and must convince the consumer that bread is a better buy as a source of nutrition. Bread must be marketed as an appetizing fresh product, straight from the oven, available throughout the day, and even over 24 hours in the future. Specialty-breads should be marketed on their nutritional merits, to meet the needs of all ages and professions. The longer shelf-life, sterilized breads, mainly imported from continental Europe, represent another hitherto untapped market and challenge for the baking industry.

The manufacture of many baked-products in the UK requires reformulation, and the introduction of better processing methods in order to build-in more quality and add value to the final result. Process development since the 1950s has tended to have a negative effect on bread quality in general, in spite of acknowledged achievements in production rationalization.

Brown bread, is made from flour which has had 10–15% of the wheat-grain removed during the milling-process. The flour was formerly called Wheatmeal-flour, which was a misleading description. Brown flour must contain a minimum of 0·6% fibre in dry matter, which is equivalent to a flour of approximately 78% extraction. Many brown flours contain much higher levels of fibre than the minimum, and are available in various degrees of fineness to suit the baker and his customers. The flour is light brown in colour, containing about 50% of the fibre of wholemeal-flour. 125 g of brown bread will contain about 6 g of fibre, and provide more than 33% of the daily requirement of calcium, and about 25% of the daily requirement of iron, as well as useful contributions of the B-group vitamins. By adding extra bran to the brown flour, the fibre contribution of the 125 g of brown bread can be increased to at least 8 g.

Wheatgerm or germmeal-bread is made from a blend of brown flour and a minimum of 10% wheatgerm, which has been stabilized by heat. This latter process may involve the addition of salt. Proprietary brands of germmeal breads have been on the market for many years, Hovis, an RHM label, and Vitbe of Allied Bakeries are advertised nationally on the basis of their nutty flavour and health value. Germ is a valuable addition, since it contains significant amounts of the B-group vitamins.

Mixed-grain and malted mixed-grain bread, is made from brown flour to which a wide variety of other products may be added, viz. wholemeal flour, crushed wheat and rye grains, rye-flour, malted wholewheat, malted whole-rye, and/or flakes of malted wheat and rye, oat products, rice and barley products, maize products, wheat-bran, soy-bran, pea-bran, dried citrus-pulp sacs, apple-fibre, oat-fibre, 99% cellulose dietary powders, pectins, buckwheat products, linseed, brewer's residues, sesame-seed, sunflower-seed, dried banana, and honeys. The application of these raw materials to specialty-bread formulation is discussed in Chapter 2.5.

Many complete mixed-grain bread premixes are available, viz. Granary of RHM, and a Countryman mix marketed by Spillers Premier Products. These latter products require only the addition of water and yeast. Mixed concentrates are also available, containing flavour and nutritionally added-value components and dough-conditioners and developers, which can be added to white or brown flours to create new bread types, e.g. SPP 'Gemini' series and British Arkady products, and others. Many independent milling companies market excellent quality malted grain and soft-grain and flaked flours for the craftsman baker, e.g. Smith's mills at Worksop, Notts., Walsall, Staffs., and Langley Mill, Notts., and now also at Nafferton Mill, near Driffield, North Humberside, and Heygate's Mill, Bugbrooke, Northampton.

A basic formulation guide for mixed-grain or malted mixed-grain bread would be as follows. Mixed-grain of malted mixed-grain flour 100 parts, shortening 2·0, yeast 2·5–3·0, salt 2·0, water approximately 55·0 (as required). Mix for about 25 minutes in a traditional slow mixer, or 8–10 minutes using a spiral mixer. These doughs can be either bulk-fermented for about 30 minutes, or worked-off immediately, and proofed to pan-height.

Wholemeal bread, is made from flour that contains all of the wheat grain, as the name logically states, being either milled by modern roller-milling or by crushing the grain between millstones in the case of traditional stoneground wholemeal flour. With all the outer layers of the grain included in the flour, wholemeal bread is much coarser and more crumbly than brown breads. It contains more fibre, 125 g (4·5 ounces) containing about 11 g, but is not fortified with vitamins or minerals. Wholemeal flour is used for large 800-g and 400-g bread units and rolls.

Rolled or cracked wholemeal bread, is made from crushed whole wheat grains in the form of flakes rather than a flour. The wheat grains are passed through one set of rollers, which simply cracks the grain open. The flakes are soaked for about 30 minutes to allow complete hydration of the endosperm components, mixing at low-speed. The final dough is then mixed by adding the other ingredients in the usual way to make a smoother, moister-textured dough. A suitable dough-conditioner or improver is also often added, containing dried gluten to aid dough stability and increase bread volume. This type of bread originated in The Netherlands, where it is very popular, and is gaining ground in the UK, owing to its rustic character. It is better than ordinary wholemeal bread, being less crumbly and having a full nutty flavour. As a source of dietary fibre, it is even higher than wholemeal bread, providing about 13 g/125 g (4·5 ounces). The rolled or cracked wholemeal flakes can be made into rolls, panned bread, cobs and *baguettes*.

Pitta bread, traditionally of Greek origin, has become popular in the UK generally since its introduction by the ethnic population. It can be made with white or wholemeal flour, and only contains a small amount of yeast. After a short fermentation period, the dough is divided, and the pieces rolled out to de-gas at about 6–7 inches (16 cm) diameter. After proofing for 10 minutes, the 90-g pieces are baked at 230°C or higher for 5–10 minutes. On removal from the oven, the dough-pieces are inflated with steam, but this leaks out leaving the flat, oval shape. Average yields per kg flour are 12 pieces, the yeast, salt, flour and water being mixed to a stiff dough. Plants are marketed at about £60 000 sterling, made in the Middle East, to produce at capacities of about 6000 per hour. Although demand for pitta bread tends to be concentrated within areas of the UK where certain ethnic groups have settled, there is an increasing general demand from the student population in the UK. Production and distribution has been restricted to 2–3 bakeries, but since the large supermarket chains are finding increasing demand, other bakeries may break into the market. Pitta is ideal for grill-and-serve dishes, or it can be sliced in half to make a pocket for filling with roast or grilled meats and salads. In Greek and Cypriot restaurants, it is served as an accompaniment to shishkebabs.

Pizzas are also a useful diversification for craftsman bakers who run a take-away section within the shop for the lunchtime-trade. The pizza can be sold cold or re-heated in a microwave-oven. Large sheets can be made and cut into squares according to demand and size. The pizza base is produced from a yeast-raised dough, or a chemically leavened dough, typical formulations being as follows: flour 100 parts, yeast 3·0, salt 2·0, shortening 6·2, milk-powder 3·0, egg 15·0, water 50·0. The dough is mixed to a smooth clear consistency at 25°C, and allowed to ferment in bulk for 30 minutes. In the case of the chemically leavened dough: flour 100 parts, baking-powder 5, salt 2·0, shortening 6·2, milk-powder 3·0, egg 15·0, water (cold) 50·0. The flour, baking-powder and milk-powder is sieved together, and the shortening rubbed in. The salt and liquids are mixed, added, and worked into a clear dough.

For the production of round pizza bases, one method involves sheeting the dough to a specific thickness, and cutting out discs with a cutter. In this case, the dough must be shaken-out after sheeting to remove and prevent excessive tension and shrinkage after the cutting operation. The problem with this method is the re-working of the cuttings into the dough without any toughening, especially with the chemically leavened dough.

An alternative method is to scale the dough into pieces and mould round. For this purpose, a moulding machine with a 12-piece cutting and moulding head for dough-pieces of 110–170 g is required. After moulding, the pieces are proofed for 10–15 minutes, and pinned-out to the required size, although pinning-machines are also available. For sheeted pizza doughs, the dough-sheet is laid on the baking-sheet and trimmed to size. Some bakers lightly bake the bases before filling in order to obtain a crisp, sealed base and a moist filling, which can be treated in a number of ways to meet demand. A variety of stock fillings can be made up and stored under refrigeration to be used for filling into the lightly baked bases, the whole being baked-off as required.

Alternatively, the filled bases can be sold with the filling raw or baked. A shop with a bake-off section can bake-off the pizzas to customer demand, requiring about 5–10 minutes only, compared with about 15–20 minutes when baked from a raw dough from scratch. The bases are baked after proofing for 20–30 minutes in the case of the yeasted dough, or resting for 30 minutes after cutting in the case of the chemically leavened dough. Baking is at 215°C for 5–10 minutes, aiming to produce a moist lightly baked base which is not dried out.

A wide variety of fillings can be prepared, using a tomato-purée gel base, consisting of pre-gelatinized starch, seasoning, pepper and salt. This mixture will carry the various cooked meats, fish and vegetable additions, which can be either mixed or placed on top of the filling, depending on the type of pizza. Other ingredients include grated cheese. Popular meats are ham and salami. Suitable fish are anchovies and sardines, and vegetables are mushroom, onion, paprika, maize corn and black olives. A typical filling mixture is: italian canned tomatoes 100 parts, tomato-purée 3·4, salt 0·6, pepper 0·01, oregano 0·28, pre-gelatinized starch (as used for pie-fillings) 2·3. Break down the tomatoes and prepare a smooth paste, allowing to stand for 1 h at least to swell and form a gel. To prepare a 20-cm diameter pizza, pin out 112 g dough-pieces, which have been rounded-up and rested for 10–15 minutes, placing on lined baking-trays. Brush the dough-pieces with olive oil, and spread with 154 g of filling, adding the meat/fish and vegetable trimmings.

A pizza-sheet is prepared by rolling the dough to a thickness of about 10 mm to cover the lined tray (silicone-paper lining), and trimmed. Brush the dough-sheet with olive oil, and spread with 2 kg filling/30 inch × 18 inch tray, dressing with meat or fish and vegetables. The simplest dressing is 1 kg diced onion and 500 g of grated cheese.

The dressed pizzas are proofed for 30 minutes, or relaxed in the case of the chemically leavened variety. Baking times are 10–15 minutes for the 20-cm diameter units at 230°C, and 20 minutes for the pizza-sheets.

A large range of pizza equipment from cutters to special ovens is available from Crest Catering Equipment Limited of Croydon, Surrey CR0 3EB, UK. Examples of machines for production rationalization are: dough-rollers with a two-pass front operation, two-pass side operation or single-pass side operation.

Dough retarders are useful in restaurants for the storage bank of unbaked pizzas ready-filled for dressing and bake-off as demand requires. This allows the bases to be prepared the previous afternoon or early in the morning, the topping or dressing being done during the following day or evening, thus providing space, staff and equipment free for the dressing and baking operations. Although small travelling-ovens are ideal for rapid heat-transfer and a short bake-time, resulting in a crisp crust and moist filling, small deck-ovens with a ceramic-tile lining give similar results.

Turning to the structural trends within the UK baking industry for the 1980s and 1990s, large-scale production will further concentrate into the hands of Allied Bakeries (ABF-group), Dalgety Foods (Spillers Milling and Spillers Premier Products, etc.), and another group at present represented by RHM's British Bakeries, which, during 1989, was becoming increasingly taken-over by the Australian food group Goodman Fielder who at one stage owned 20% of the equity. At this time (1990) the Anglo–French financier, Goldsmith, holds about 30% of the equity in the RHM group. Goodman Fielder is the biggest milling company in Australia, and has bread-baking and other food operations, which it intends to expand in the form of overseas joint-venture deals. Independent regional plant bakery groups also exist, which operate in certain specific regional towns and cities, but are not national companies in the same sense as the 'big-three' mentioned. Dalgety, originally an international agricultural group, have disposed of their non-food companies worldwide, and are concentrating on the food sector. Any takeover ambitions will be in the foods field.

Bread consumption statistics are eagerly awaited by all bakers, from the 'big three' to the one-shop craftsman. At present, sales of white bread fluctuate from 51% to 54%, and those for wholemeal from 15% to 17% each quartal-year. Any loss of sales for the 'big three' means gains for the in-stores, hot-bread shops, bake-offs, or craftsman-baker, and the latter sectors are in the ascendancy. The craftsman-baker with an in-shop take-away section is in the unique position to capitalize on hot or cold croissants, with or without a choice of fillings, with the help of a commercial 1 kW microwave-oven. *Crêpes* and waffles are another growth area, with the aid of twin hot-plate and waffle-maker, the customer being allowed to choose fillings from a hot/cold display unit.

However, the pizza is the most popular of all take-away foods, and can be made up and sold in a relatively small area of the shop, provided the dough is already prepared and divided. An insulated storage section for toppings, and a double-deck oven with a specially designed firebrick base to take the pizza-bases (about eight units of 20-cm diameter), allows the pizzas to be made on demand rather than held warm. Such a set-up would require a floor-space of about 6 m², and an equipment capital cost of about £6600 sterling. Nutritionally, an ordinary tomato/cheese pizza supplies about 33% of the recommended daily amount (RDA) of protein, and the amounts of protein, fat and carbohydrate are well-balanced. 100 g of tomato/cheese pizza contains about 10 g protein, 12 g fat and 34 g carbohydrate, and supplies about 270 calories. One of the easiest ways for a baker to retail pizza is to batch-bake the bases, top with the dressings and freeze complete. This provides him with two options, he can sell them in the frozen form, or the pizzas can be heated in a microwave-oven as required for sales. The microwave-oven should be a commercial one, costing about £500 to £1000 sterling to cope with the workload demanded.

This type of set-up could produce better pizzas than local pizza houses, for supplying bars, clubs and other instant-food outlets. The consumer's current

TABLE 45
**Nutritional information printed on UK national breads (per 100 g)**

|  | Brown bread with oatmeal | Vitbe | Windmill white, high-fibre | Sunblest white | Mothers Pride white |
|---|---|---|---|---|---|
| Fat (g) | 3·8 | 3·7 | ·1·8 | — | 2·2 |
| Protein (g) | 8·7 | 10·4 | 7·7 | 7·2 | 7·9 |
| Dietary fibre (g) | 5·1 | 4·8 | 8·6 | 2·7 | 2·7 |
| Carbohydrate (g) | 45·2 | 43·5 | 47·2 | 49·3 | 47·9 |
| Energy (kcal) | 231 | 223 | 230 | 230 | 230 |
| Thiamin B1 (mg) |  | 0·52 |  | 0·15 |  |
| Niacin (mg) |  | 3·90 |  | 1·60 |  |
| Iron (mg) |  | 2·20 |  | 1·80 |  |
| Calcium (mg) |  | 125·00 |  | 110·00 |  |

The trade names 'Windmill' and 'Mothers Pride' belong to British Bakeries Limited (RHM group), and 'Vitbe' and 'Sunblest' to Allied Bakeries (ABF group).

preoccupation with diet, brought about by the communication media and 'opinion-formers' is having a great impact on the food industry, and baked products are being critically analysed on a value-for-money basis. The large plant bakery groups and the supermarket chain 'own-labels' all declare detailed nutritional information on their packages, expressed as g/100 g bread. Some current examples of such labelling are given in Table 45.

Following Dr D. Burkitt's well-balanced lecture on the significance of dietary fibre in the early 1970s, there was considerable scepticism concerning his concepts and considerable research from the baking industry. This was due to the industry being put on the defensive as a result of implications that white bread was detrimental to the nation's health by certain 'health-food cranks' at that time. Gradually, the baking industry modified its standpoint since: (1) the evidence presented by Dr Burkitt of the beneficial effects of a higher fibre diet seemed convincing, and (2) there seemed to be an interesting potential market for improved brown breads, which contrasted sharply with the falling white bread sales; therefore, the financial investment for development and marketing was justified.

The traditional dense, dry texture of brown breads had long been a deterrent to selling, since the consumer generally was accustomed to a lighter, moister, and more elastic crumb, and better shelf-life. Therefore, any new product development had to embrace these qualities as well as the nutritionally beneficial dietary fibre aspect. Accordingly, there has been a great improvement in both the range and quality of brown bread flours available from the milling industry, and the range of breads offered by the baker. These include the following categories: high-fibre white breads, using various sources of fibre to increase the dietary fibre level; muesli breads, containing diverse composite mixtures of flakes and seed material; mixed grain or multi-grain breads; brown-breads with malted grains; cracked-wheat breads, etc. In an attempt to improve the nation's health, the Government has already implemented various new items of food legislation, and certain new proposals regarding the declaration of fats and fat content of most baked products. This latter aspect will present difficulties in implementation for the industry, especially if the composition of the fat in terms of percentage saturated, trans and polyunsaturated fatty acids, is also included in the package-declaration. Nutritional labelling is becoming standard with the large bakery groups and supermarket chains, and although no legal obligation has yet been laid down, standard guidelines for the nutritional labelling of foods will mean the use of the same terms and format, to enable the consumer to make comparisons of the value of different foods.

The current aspects of a healthy diet which are attracting particular attention are the following:

(1) An increase in dietary fibre intake.
(2) A reduction in fat, in particular animal fats.
(3) Replacement of these fats with fats containing high amounts of the polyunsaturated fatty acids.
(4) Sugar levels generally reduced, which only provide energy in the form of 'empty' calories.
(5) The level of sodium intake should be reduced, in order to reduce the danger of hypertension, or high blood-pressure levels.
(6) Alcohol intake should be reduced to a minimum, in order to avoid obesity and addiction.

Foods such as red meat, milk, and butter, traditionally regarded as important to a healthy diet, are now to be consumed with caution. Whereas, potatoes, bread (in particular wholemeal bread), and fresh fruit and vegetables are now recommended, and their consumption should be increased.

The reason for this shift in emphasis is the progressive rise in living standards, which has resulted in the proliferation of those diseases associated with 'civilized' urbanized existence. Refined and convenience-foods cannot provide enough fibre, which is essential 'ballast' material for the gut and bowels generally, acting as a carrier which has valuable swelling properties, thus increasing the surface-area of nutrients to be exposed to enzymic metabolism, and therefore more efficient absorption across the gut. Typical 'civilization-diseases' are: high blood-pressure, varicose veins, piles, gall-bladder conditions, appendicitis, diverticular disease and, in extreme cases, bowel-cancer. The primitive life-style of the tribal native does not expose him to such hazardous eating habits, therefore he is less prone to these 'diseases-of-civilization'. The general term 'dietary fibre' is somewhat ambiguous, since many of the substances which comprise it, such as hemicelluloses and pectins, are non-fibrous and form polysaccharide colloidal gels. Plants depend on these fibrous and non-fibrous substances within the mature parenchyma tissue to form sclerenchyma cell wall tissue which is thickly lignified (woody), providing structural support for the stems. The actual 'fibrous' fraction consists of lignins and celluloses which are analytically defined as 'crude fibre', being the non-digestible residue remaining after a food has been treated with strong acid and alkali. It is this material which is determined in brown bread in order to ensure that the 'crude fibre' content exceeds 0·6%, the legal minimum. The term 'roughage' is based on crude fibre.

This is the reason why wholemeal flour has 2% crude fibre, but approximately 9% dietary fibre. Since a large proportion of 'fibre' is, in fact, non-fibrous, it is a problem finding a term which will accurately describe it, the term 'roughage' being clearly inappropriate. The analysis of dietary fibre is difficult, since the heterogeneous nature of the substances involved demand not only one determination, but a system of determinations run in parallel and in succession with one another. Not only the insoluble, but also the soluble dietary fibre components must be isolated and determined. For this reason acid or alkaline hydrolysis methods are unsuitable. The best results are obtained by using combined methods, which initially precipitate the soluble components, enzymically break down the digestible food components, and determine the indigestible fraction by difference gravimetrically. Crude fibre determinations only determine a quantitatively undefined fraction of the dietary fibre complex, and only in isolated cases is crude fibre data constantly related to the total dietary fibre complex of a food product. The Southgate method differentiates each component of the cell wall, therefore, the analytical results are useful in evaluating nutritive values. This method is, however, complicated and takes some days to complete, being a chemical determination. The Asp method is simpler, and useful for the analysis of many samples, but does not differentiate each component of the cell wall, being an enzymic method. Moreover, the general definition of the dietary fibre complex of a food product is a group of components which are either not digested in the small intestine, or only partially digested. These substances proceed into the large intestine, and can there be broken down by the intestinal flora, or become excreted in an undigested form. Often in healthy people, 10% of the starch ingested can reach the end of the large intestine, and the baking process also forms a so-called 'resistant-starch', which is not broken down in the large intestine, and can therefore raise the dietary fibre complex content found in the flour by as much as 30%. Therefore, any meaningful interpretation of effective total dietary-fibre-complex components and their function within the alimentary tract must be regarded as difficult to quantify in practice.

Internationally, owing to the physiological and technological importance of the total dietary fibre complex, methods involving amylolytic and proteolytic enzyme pretreatment with the help of detergents have been developed (Menger, A., *Getreide Mehl und Brot*, **31**, 48, 1977, and **32**, 13, 1978).

Sir Robert McCarrison, the well-known research worker in nutrition, and founder of the McCarrison Society, demonstrated repeatedly as early as 1920 in humans and animals how food, deprived by over-processing of vitamins fibre, minerals, and other fresh-food factors hitherto unidentified, resulted in lower resistance to infection, poor stature, and exposure to degenerative disorders when consumed continuously by living subjects. McCarrison's work in connection with deficiency diseases also indicated that foodstuffs grown in soil treated with animal manure had a higher nutritional value than that treated with inorganic chemicals, in spite of possible higher yields in the case of the latter. The nutritional value of crops appears to be enhanced by biological rather than chemical processes in the soil.

In addition to the British patents top-grade flour selling at about £33 sterling per 100 kg bag currently, Baker's ordinary at £32 sterling, and English baker's at £28 sterling, there are other non-standard flour grades available to the baker. A Canadian basic is marketed at £35 sterling per 100 kg bag, but such prices are negotiable within limits, and discounts for prompt payment may also be offered. Many bakers wishing to produce better quality bread with an improved crust and elastic crumb structure would choose the latter for variety bloomer and Vienna type breads. French wheat-flour blends, often quoted below a British patent (as much as £115 sterling per tonne cheaper), will produce better crusty bread than standard gristed flours with soft English wheat supplemented with 2% and more dried gluten. For wholemeal bread, an 80% Canadian flour will produce bread with an improved crust and internal crumb structure. The large national millers, e.g. Allied Mills, market a French flour under the 'Flèche d'Or' label, and in common with Spillers Milling, also market a wide range of unbleached, untreated flours.

The majority of Independent Millers such as Bowman and Sons Limited of Hitchin, W. Marriage and Son Limited of Chelmsford, Smiths Flour Mills of Worksop, William Nelstrop and Company Limited, all market specialty-flours, e.g. unbleached and untreated flours, malted wholegrain flours, heat-treated, pre-gelatinized, and stabilized brown wholemeal flours, also composite flour mixtures and flours with extended shelf-life. Some independent mills have their own farms for the cultivation of organically grown wheat. Doves Farm Flour grow wheat to organic grade I standard for stone-grinding, and also market a 100% fine wholemeal, a brand leader for making pastry and cakes. Their range also includes an unbleached strong white flour with no added improvers, a gluten-free brown rice flour, and a slightly brown barley flour for people on a gluten-free diet. Specialist blending and milling contracts to manufacturer's specifications as well as 'own-label' products, are on offer.

St. Nicholas Mill also has its own farm, St. Nicholas Court Farm, and is in the position to use its own English wheat. Stone-ground wholemeal is produced from this wheat.

Pinhill growers and millers have farmed organically since 1949, and produce Muesli-based products, 100% wholemeal flour, an 85% extraction flour, breakfast oats, three grades of oatmeal and wheat flakes, cracked-wheat and crushed grains. In the case of the 100% wholemeal, the value of the whole-grain remains by grinding slowly in one operation, the germ oil becoming evenly distributed throughout the flour, thus avoiding overheating of the germ oils. There is a definite trend in favour of natural, untreated foods from cultivation through to the consumer, thus achieving a 'clean' label product. If a wheat grower/miller collaborates with a craftsman-baker, they could well find a place in the market for biologically pure bread made from organically grown wheat, containing no pesticide or herbicide residues or bleaching agents and chemical improvers. Such products would be the ultimate in the health-bread sector, and would compete with industrially produced baked products.

In the UK, the limiting factor for the use of home-grown wheats is the climatic conditions prevailing at the time of harvesting, and motivating the grower to plant varieties with good milling and breadmaking properties. For example, owing to the wet harvesting conditions in most regions, with the exception of Devon, the 1987 harvest left a shortfall of about 1 million tonnes of breadmaking wheat, mainly because of the poor quality of the crop in East Anglia, the largest grain-growing area. This shortfall has to be made good by importing EC wheat from France at about £128 sterling/tonne, or from Spain at £170 sterling/tonne. These prices have to be compared with average English milling-wheat ex-farm of £138–140/tonne, or American Northern Spring (DNS 14%) selling at £93–94 sterling/tonne, which attracts a daily non-EC import levy of £121 sterling/tonne, i.e. a net price of £214 sterling. For the miller the calculation is clear under such circumstances: use French wheat and/or English wheat, and supplement with dried-gluten, providing the gluten price is about £1000 sterling/tonne.

Under such circumstances, a miller might be prepared to pay an extra few pounds per tonne premium for wheat samples with the minimum of 72 kg/hl specific weight, 11·0% protein, 180 Hagberg, and 2% shrunken grains at 16% moisture.

Providing the price of dried gluten remains below £2000 sterling/tonne, the miller could save about 25% of quality third-country wheat in his grist by replacement with an EC wheat of 10% protein and adding 1–2% gluten of protein content 75%.

However, the addition of gluten cannot produce the same bread quality as an equivalent amount of quality wheat in the grist. The addition of gluten with higher levels of soft wheats, gives rise to an increasingly 'spongy' texture, and an increasingly pallid, soft crust, similar to high-protein gluten-enriched breads. These effects can be offset to a degree by rebalancing with sugar solids, preferably high-fructose corn-syrup, and reducing dough pH below 6·0, either by bulk fermentation or the addition of lactic or acetic acid. The latter acidulation of dough also progressively reduces alpha-amylase activity from pH 6·0 down to pH 4·5, pH 5·5 being an optimal target value for bread doughs. Bakers in the UK should demand flour specifications from millers which include the following quality-profile parameters: colour-grade, protein content, gluten quantity and quality, Hagberg, Farinograph (water absorption, consistency changes during mixing), and Extensograph. Information on treatment with ascorbic acid and azodicarbonamide is also useful. Even a bakery without testing facilities can perform a gluten-washing test and Pekar-colour check to compare flour deliveries.

Since the emphasis is now on fresh, crusty bread to counter balance the sale of 70% longer shelf-life sliced-and-wrapped bread, it is essential that the baker is supplied with a flour of sufficient breadmaking quality to produce a better standard of appetizing fresh, crusty products, at a competitive price.

Modern refrigeration techniques have also been applied in the UK by the baking industry generally in order to make better use of working-hours. Production peaks and troughs can be overcome, and very early starts can be eliminated, making life less hectic and allowing a wider range of products to be marketed according to consumer demand. It is particularly suitable for higher profit margin products which are normally labour-intensive to produce. Some examples of the uses of retarder/proofers are the following. The overnight filling of a retarder/proofer with bread-rolls and *baguettes*. This will enable the baker to get one or more oven loads onto the shop shelves for the morning trade. Microprocessor controls allow the proofing schedule to be timed to suit oven capacity. Bread and rolls can be stored over the week-ends or over holiday periods by switching to a low temperature retard setting at about $-5°C$ instead of the normal $+3°C$.

Doughs can be left overnight independent of time. For example, rolls can be left for 4–5 hours at 20°C to condition, thus improving product flavour, without any risk of temperature, time or humidity variations. Blast-freeze storage of gateaux allows them to be kept without losing their shape and texture or going soggy. If desired the blast-frozen products can be transferred by tray or rack to an adjacent cold-store, thus allowing the blast-freezer to be utilized for another batch.

For in-shop baking, a retarder cabinet can be installed for about £5000 sterling, for thawing products under controlled conditions, the price including a convection oven/proofer to bake-off a limited range. The level of skill required is low, space minimal, and the shop area more active with an attractive aroma. For bakers who do not want to freeze their own product, a wide range of quality frozen Danish pastry and other products can be bought in for baking-off. Retarder/proofers have distinct benefits over a separate system. Proofing takes place before production begins each morning, with fully proofed products being ready for the oven right away or soon after production commences, thus allowing a later start in the morning. The proofing time, as well as the defrost and recovery time, where appropriate, can take as long as required, since it takes place before production begins and does not delay setting in the oven. Whereas, in the case of separate units, a long recovery and proof time could cancel out the benefit of a retarder. Oven utilization is also better organized, since it can be loaded with products as soon as production begins, products from the retarder/proofer being baked before the freshly prepared products are ready for the oven. On removal of the initial load, the retarder/proofer can be used to proof freshly produced doughs until required again for the retarding operation.

In-store bakeries and hot-bread shops make use of retarder/proofers to defrost overnight, recover and proof frozen-doughs bought in from specialist manufacturers; this range includes dough-pieces for 400-g pan and hearth breads, *baguettes*, baps, croissants, etc. The various types of retarder/proofers available and their applications are discussed in more detail in Section 3.6.

The introduction of dough-retarders and retarder/proofers, especially for low-temperature retarding of large dough-pieces, demands the maintenance of consistent dough temperatures to reduce fermentation prior to and during retardation. Lower dough temperatures are desirable, compared with immediate processing. Since this is difficult using a high-speed mixer, which produces frictional-heat increases of about 15°C, necessitating the use of a chilled water supply, many bakers have changed to spiral mixing.

The spiral mixer became very popular on the continent during the late 1960s, and appeared on the UK market in the early 1970s. This introduction coincided with the acceptance of the activated dough development (ADD) process, and the addition of L-cysteine to the UK permitted list of bread additives. Bakers were able to use either traditional low-speed mixers or spirals, the spiral offering several advantages over the traditional low-speed mixer. Mixing times were reduced to about 8 minutes, 3 minutes slow speed and 5 minutes fast speed, compared with 20–30 minutes.

Timing is automatically controlled in sequence, and the arm and bowl design requires little scrape-down.

High-speed mixing is universally used in plant bakeries today, apart from certain small-scale production operations, and this practice remains with the progressive development of automated and computer-controlled mixing systems.

With the increasing use of spirals, a new generation of bread-improver appeared on the market, which was being used on the continent with spirals. These contained shortening, special permitted emulsifiers, ascorbic acid and fungal alpha-amylase. These improvers, which are marketed in the UK by both UK and continental manufacturers, permit 'no dough-time' methods to be used, and are especially formulated for use with spirals. This range of improvers is both versatile and more acceptable where ascorbic acid is the component, harmonizing with other EC country legislation versus oxidant. The usage level can be varied according to the product produced, e.g. bread 1%, 1·5% for enriched fermented products, and 2% for bread-rolls and products of relatively high volume. Yeast levels are varied according to process schedules.

Using spiral mixers result in smaller dough temperature rises, being on average 6–7°C only, compared with average rises of 15°C for high-speed mixing units. Although the spiral is a discontinuous (batch) processing unit, it can be used in conjunction with a bowl-tipper or other dough-removal system. A system for continuously producing doughs with the spiral mixer has been devised by All Foods Engineering Services Limited of Newcastle, Staffs ST5 6NR, UK. This system is computer-controlled, programmable with memory, being a top of the range mixing plant. For versatility, special interchangeable mixing arms for a wide range of product is possible, retaining simplicity in engineering for dependability and maintenance. The complete plant comprises ingredient-loading, mixing, resting and tipping, and gives a throughput of 3 tonnes/hour. Mixer-arm speeds are variably controlled from zero to the maximum rpm required, and operation can be fully automatic, with a manual override and vacuum mixing as necessary.

The range of spirals on the market is enormous, and certain design points have to be considered before making a purchase decision on type and size. The incorporation of a centre column or plinth top-bearing gives improved bowl stability and positive drive. Spiral drive through the centre column, permits a smaller head size and improved accessibility. Capacity is always geared to bakery-output, and may vary from 32 kg flour to 128 kg flour; dough capacities will be approximately 0·625 of quoted flour capacities. Some machine reference numbers relate to kg-dough capacities. Most of the spirals on the UK market are of continental origin, e.g. Dierks 'Diosna' of Osnabrück, dough capacities 24–240 kg with the option of six fixed, and four removable, bowl models; Boku mixers, also of FRG, equipped with a stainless-steel bowl of dough capacity 265 kg (165 kg flour); Eberhardt of FRG, with a unique spiral, bowl-revolution and energy-input, can be used for bread, pie-crust and Danish pastry doughs. The Eberhardt Maximat-Spiralkneter can be used for all doughs, and has a high mixing energy due to special power transmission and broad spiral; Sottoriva of Italy range of spirals also have good reliability and dough development, being available with fixed or removable bowls, two speeds and a timer, with dough-capacity range of 25–330 kg; Dahlen of Sweden market the Girrek spirals with capacities of 20–350 kg dough; Artofex model S-100 has a stainless-steel forged mixing element with twin support, and centre cutting column. The stainless-steel mixing bowl is driven independently by a separate, reversible motor for reversing bowl action. Two separate timers are linked for programming to individual needs, e.g. 3 minutes slow, 5 minutes fast speeds; Thomas Collins of Bristol, UK, market heavy-duty spirals with automatic controls, two-speed mixing, reversible stainless-steel bowls and timing device. Dough capacities range from 6 to 300 kg, with unloading facility at any desired height, i.e. work-table or hopper-feed.

Another mixing technique designed by Dierks of Osnabrück, FRG, which is a multi-purpose mixer for bread and biscuit doughs is the 'Diosna' Wendel Kneter. It can also be used for CBP doughs, ADD-processing, straight doughs, sponge doughs, no-dough-time procedures, and pastry and biscuit doughs. It is available in three sizes of upright design, with outputs ranging from 2200 lb to 7000 lb dough/hour. A Hi-tech in-line version has been developed for UK plant bakeries with advanced automated-production systems. Two mixing elements rotate at a selected speed in opposite directions, keeping the dough away from the rotating-bowl to minimize temperature rises. The two elements are machined to give optimal mixing efficiency. The exact mixing dynamics and regime of this Wendel Kneter are discussed in Section 1.9.3.

The Baker Perkins range of spirals combine high/low speed, and bowl-reverse suitable for fermented, ADD, straight dough, sponge dough and pastry. A central kneading-bar is said to improve mixing efficiency. The automatic head-lifting gear incorporating shock-absorbers to reduce stress on the machine when the dough stiffens, is fitted to the removable-bowl version.

The new 'Kemper' of Austria range of spirals, have ovecome machine mobility problems by introducing a range of Kemper wheel-out bowl mixers with heavy-duty removable bowls, but no longer having fixed base plates. This SP-ALH range has flour capacities of 100 to 250 kg. Location and securing of the bowl is automatic at the start of the mixing sequence, being programmable. Fixed bowl spirals for 32 kg bags of flour, operating manually or automatically, have a two-speed mixing cycle, creep-button for facilitating unloading, and bowl reverse for pastry or fruit addition. For monitoring dough temperature, an optional display can be provided, thus allowing water temperature adjustments to be made as dough temperature changes with mixing conditions.

In making a machine purchase decision, the baker should see as many spiral or other mixiers in operation in other bakeries as possible, and gain the opinions of their owners/operators, in order to be informed about the strengths and weaknesses of each. Important points are: ease of maintenance, after-sales service, and availability and cost of spare-parts. Metal-fatigue and drive-bearings give defects. Operational aspects are: choice of the optimal capacity for production requirements, differentiating between flour capacity and dough capacity expressed in kg. The installation of too small a mixer, e.g. 32 kg flour capacity, may result in too many doughs having to be produced *v.* daily output, thus overloading and increasing labour costs. On the other hand, the installation of too large a mixer *v.* existing processing equipment capacity downstream, will result in over-fermented doughs during a production run. Therefore, a good compromise must be struck with enough reserve doughmaking capacity. The other aspect is the choice of a fixed or removable bowl machine, the latter choice giving much greater flexibility for transport to various processing lines, e.g. bread, fermented-goods scaling-table, etc. A bowl-elevator can be used to deposit the dough into the divider-hopper, or a dough can be given a period of bulk-fermentation in the bowl. Whilst one or more doughs are fermenting in bowls, another bowl can be used to mix a fresh dough. Another mixing innovation was launched in 1984 by Baker Perkins, the latter company having since merged with APV in 1987, forming the APV-Baker group. This Peterborough, UK based group introduced the 'Uniplex' and 'Biplex' mixing processes. The resulting series of horizontal-fixed-bowl (HFB) automatic batch-mixers for industrial bakeries, include four output capacities, from 1600 to 4000 kg dough/hour.

Both the UniPlex and Biplex systems of mixing can be applied to: Mechanically Developed Doughs (MDD), and Activated Dough Development (ADD) using L-cysteine and/or other ingredients. However, bulk fermented (BF) doughs require the application of the BiPlex system, as well as sponge doughs. Both systems are applied to bread and bread-roll products, and fruited doughs, but wheat/rye-flour doughs with up to 50% rye-flour content require the application of the BiPlex mixer. The BiPlex and UniPlex are two distinct mixing processes. The UniPlex features a simple rotary motion of the spiral-shaped mixing-element about a fixed axis. The BiPlex process consists of two stages. The first involving a planetary and rotary motion, the second a high-speed rotation about a fixed axis. The BiPlex being a new process in its own right, comparable with the MDD (e.g., CBP), ADD, bulk fermentation, sponge and pre-ferment procedures. They are available with three levels of control:

Level 1:   Mixer only, no weigher, thus integrating with bakery individual weighing equipment.
Level 2:   With flour and water weigher, for full automation, but installed for use with manually set weights and energy control.
Level 3:   Flour, water and optional small ingredient weigher. Water temperature computation and control. VDU/operator interface, with recipe storage, printer, and diagnostic aids.

Therefore, these mixers can operate with no manual intervention, feeding all ingredients automatically, and cycling at a rate in synchronization with the processing line. The mixer is closely integrated with the ingredients system upstream, and the processing equipment downstream, thus playing a vital role in process control and automation. The UniPlex features intensive mixing with the spiral blade giving efficient ingredient dispersion and uniform dough development. Consistent doughs are ensured by accurate load cell weighing of flour and water, ingredient metering, energy and dough-temperature control.

Mixing-times are on average 2·0–3·25 minutes, the screw action of the beater giving positive and complete discharge of the mixed dough with a mixing-cycle time of about 5 minutes.

All the horizontal-fixed-bowl (HFB) mixers feature a dough-consistency-monitor, and a dough consistency control unit is available as an optional extra. The various model numbers, 1600, 2200, 3000 and 4000 refer to the respective maximum dough output in kg/h, based on 2·0–3·25-minute mixing-times. Doughs requiring longer mixing-cycles will reduce output accordingly. The respective maximum and minimum dough batch weight in kg for the four models are: 133/55, 183/55, 250/100, 333/100.

The BiPlex mixer, coupled with the 'BiPlex-process' is a new process for producing mechanically developed dough for immediate work-off, using one-third less than the energy-input required for the CBP, made by Baker Perkins of Peterborough, UK. Since the energy-input is one-third less than with the CBP, the temperature rise during mixing is correspondingly one-third less. The BiPlex mixing process is made up of two separate mixing actions.

(1) A planetary movement of the special spiral beater for 2·5–3·0 minutes allows full hydration of the flour components, efficient mixing and dispersion of all ingredients, and incorporation of oxygen into the dough.

(2) Basically a different beater action consisting of a high-speed rotation about a fixed shaft, which completes development of the dough within 0·5–1·0 minute.

The advantages of mechanical dough-development are retained, and in addition the following innovative improvements are obtained:

(1) Reduced dependence on the use of chilled water. Normal CBP energy-input levels of 11 Wh/kg for UK panned-bread will not accommodate flour temperatures in excess of 27°C for doughs at 30°C, without the application of chilled water. Using the BiPlex mixing system, flours with temperatures up to 40°C can be processed, since the optimal energy-input level is reduced to 7·6 Wh/kg. With mains water at about 14°C, and flour temperature 26°C, no chilled water is necessary with the BiPlex. Although higher temperatures require the use of chilled water, the amount will be considerably less than with the CBP.

(2) Lower concentrations of yeast can be used, or a higher volume product obtained, or lower scaling weights for the same volume product.

(3) Additive cost is also reduced, since oxidant requirements are less. Since the process mechanically incorporates oxygen into the dough, the best type of oxidant is one which reacts in conjunction with oxygen present, e.g. ascorbic acid, or azodicarbonamide. If, however, the use of a vacuum is desired, then an oxidant such as potassium bromate, which reacts without oxygen will be obligatory. The fact that potassium bromate can be eliminated, is a distinct advantage for clean-labelling purposes, and even the use of ascorbic acid can be reduced to a level of 40 ppm based on flour weight. Additive cost can be reduced by as much as 67%.

(4)    Dough water absorption can be increased and the same dough-machin-ability obtained, or in the case of hearth-breads the same water-absorption can be obtained without any loss of dough-machinability. Typical examples when changing from the CBP are increases from 59% to 62%, and 64% to 65%, depending on flour quality; when changing from traditional bulk-fermented doughs to the BiPlex-process, dough water absorptions can be increased from 57% to 61%, often also allowing the use of a less expensive flour grade with a 4% lower raw material cost; when changing from the ADD process, dough water absorptions can be increased from 58% to 60%, in addition to the reduced additive cost.

(5)    When processing bread-rolls with the BiPlex-process, unit scaling weights can be reduced by about 4 g in some cases when changing from the CBP.

As is the case with the CBP mechanical development process, the use of a high-melting shortening at 0·7% flour weight, either separate or as a mixed improver, is recommended.

The BiPlex mixers can be either used as a high-speed mixer for bulk-fermented doughs, in which case only the first planetary-action is utilized, or as a mechanical dough-developer for immediate dough work-off, in which case the second higher-speed single-rotation mode is also utilized, which develops the dough in a short time period. Its unique planetary, sweeping action eliminates any unmixed dough pockets around the mixing-bowl entirely. Therefore, the BiPlex can be used as a dual-purpose mixing aggregate, providing two distinct mixing regimes, the first a high-speed planetary kneading action, and the second being a high-energy dough-development. Typical formulations with these mixers are as follows:

FORMULATION 1 Wheat bread
FORMULATION 2 Mixed wheat/rye as used in Germany and Scandinavia

Generally, when using these mixers, mixing-times are reduced by 25%. The APV-Baker HFB mixer, shown in Fig. 69, represents a flexible solution to dough-mixing technology in general, and automated Industrial mixing aggregates in particular.

The APV-Baker series of UniPlex and BiPlex mixers, launched in 1984, have become market leaders for industrial bakery mixers in the UK and Europe, and sales are progressing in North America and Australasia. Their market share increased from 17% to 45% for automated aggregates, and from 4% to 15% for all industrial mixers worldwide.

The UniPlex mixing-chamber or bowl is positioned at about 15 degrees to the horizontal, with a powered opening door for ease of discharge and cleaning. Consistent doughs are ensured by accurate load-cell weighing of flour and water, exact ingredient proportioning, energy measurement and dough-temperature controls. The weighing hoppers, suspended from load-cells are mounted on a structure isolated from the mixer frame. The flour-hopper is fitted with a vibrator to ensure complete discharge, and feeds for any number of flours can be connected.

The floor-mounted motor reduces vibration levels to a minimum, driving a spiral

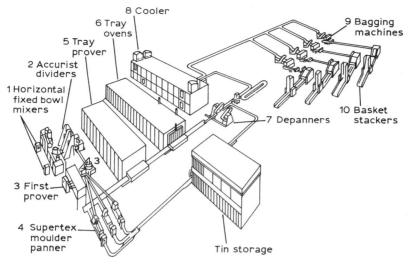

Fig. 30. (a) Pan bread production system—2000–8000 + loaves per hour. Tray-plant system (proofing and baking in pan-straps on trays). Main features include: maximum fuel efficiency; integration in-oven lidding to each tray within oven; proofer, oven and cooler can be sited in weatherproof enclosures outside the main production building; compact, straightforward layout reduces personnel and maintenance. 1. Horizontal fixed bowl mixers—UniPlex. 2. Accurist dividers—best possible scaling accuracy. 3. Conical moulder and first-proofer. 4. Supertex moulder-panner mounted over pan-conveyor. 5. Tray-proofer with double- or triple-deck swing-trays. 6. 440 swing-tray oven with 'in-oven' lidding (gas, oil or electric). 7. Suction-depanner (flexible, quiet and clean in operation). 8. Rack-cooler with biological air-filtration and controlled climate. 9. Synchronized slicing and wrapping machines. 10. Microprocessor-controlled chain-conveyor and automatic stacker.

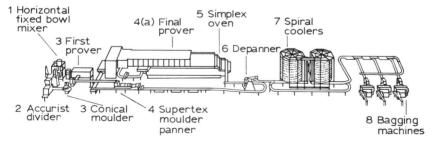

Fig. 30. (b) Hearth or variety-bread production system—2000–8000 + loaves per hour. (Hearth baking of bread and rolls in a Tunnel oven.) Main features include: versatility and flexibility for hearth, sheet or pan baking; rapid temperature and humidity adjustment (microprocessor); automatic loading and unloading systems. 1. Horizontal fixed bowl mixers—UniPlex. 2. Accurist dividers. 3. Conical moulder and first-proofer. 4. Supertex moulder/panner. 4(a). Overhead or spiral proofers. 5. Simplex 2000 tunnel oven. 6. Suction-depanner. 7. Spiral-coolers. 8. Synchronized slicing and bagging machines. (Source: APV–Baker, Peterborough, UK.)

blade, giving efficient ingredient dispersion and uniform dough development. The screw-like action of the element provides a clean discharge of the mixed dough, giving batch times of about 5 minutes. The choice of mixing regimes with control systems allows a wider range of product-mixing applications, operating either as a high-speed or high-energy mixer. Thus, it is suitable for mechanically developed doughs, in addition to ADD or bulk-fermented doughs. Fig. 30(a) shows a schematic of an up-to-date APV-Baker tray plant system for the production of 2000–8000 loaves per hour into panned bread. The production of such bread represents about 70% of the total UK bread production output (1989). The schematic Fig. 30(b) is an APV-Baker variety bread production system for the production of 2000–8000 loaves per hour of variety-bread, being more European in orientation.

The design of these mixing aggregates have made provision for three levels of automated control and lower energy consumption, and have taken account of the fact that no single mixing regime can be optimal for all processes or bakery products.

It is the task of all bakery enterprises, irrespective of the size and type of ownership, to meet the market demands for an increasing product range, and product quality with simultaneous regard for products of dietary-value-added, and limited or no additives in the form of processing aids. The raw material suppliers, especially the miller, is also involved in these legislative transitions. In turn, the manufacturer of bakery equipment, plant and machines must develop new aggregates and processes to meet the rationalization demands of the baking industry. Since the trend over the past 20 years has been the erection and equipping of large plant chains, the present trend is the equipping of medium-sized units, including in-stores, fresh-baked product shops, and country bakeries. This latter trend is not restricted to countries with free-enterprise social orders, but also extends to the Socialist countries.

Breadmaking technology in the UK has increasingly relied on the application of 'bread-improver' functionality. Sometimes referred to as 'dough-conditioners', to distinguish them from 'flour-improvers', these products have acquired a certain 'designer' status, being formulated to suit the doughmaking-system for which they are intended. It is, however, important that flour and dough treatments are both compatible and complementary for final breadmaking functionality. During the 1975–85 decade, the introduction of the EC tariff barriers have led to an economic disincentive in the use of hard wheats from USA (Dark Northern Springs), and Canada (Canada Western Red Springs) in milling grists. Instead, economic pressures have dictated the use of more home-grown or European wheats, the majority of which are soft in endosperm structure, and inferior in gluten-forming properties. Therefore, research and development has been directed towards finding methods of incorporating increasing amounts of soft wheats into the bread flour grists, without any consequent devaluation of final bread quality. In 1974, a typical bread flour grist would have contained up to 70% CWRS or DNS wheats, providing a flour of about 12% protein on a 14% moisture basis. By 1975, flour proteins had fallen to about 10·5–11·0% and, without an improvement in improver functionality, bread quality would have deteriorated in volume, shape, softness, and rate of staling. Such criteria are those most critical for the UK bread consumer, the housewife's 'squeeze-test' for freshness being the bottom-line.

The active components of these improvers are oxidants, high-melting-point shortenings, emulsifiers, yeast-foods, soy-flour, and fillers, the final composition being adjusted to suit the processing system used (designer-improvers). These currently include: Chorleywood Bread Process (CBP), activated dough development (ADD), traditional bulk-fermentation, or a combination of all three systems. In addition, with the introduction of American burger buns into the UK baking industry, for such companies as MacDonalds, the requisite American type Pan-o-mat Roll Plant demanded extensible doughs with adequate sponge-and-dough-type flow-out properties. This has been achieved by a further development of high volume CBP-improvers, the principal components being an oxidant system to give a close, even texture, and the use of DATA esters for stability, gas-retention and even porosity. Since high-speed mixers with relatively longer mixing-times are replacing high energy/short mixing-time CBP systems, these short-time or no-time processes produce doughs with greater floor-time sensitivity. After mixing, 10–15 minutes is the maximum before deterioration sets in. Therefore, the doughs require an extended tolerance to handling at work-off. This is achieved by using an improver with a slow acting oxidant system, coupled with DATA-esters to provide dough stability, and an even crumb-porosity with a biscuit-like crust. In terms of functionality, all improvers represent a compromise between baking performance (volume, softness, symmetry), and tolerance to a range of processing conditions and abuse to which the dough may be subjected. Cost-effectiveness is the final compromise in both cases. A CBP-type bread improver for standard panned-bread, being used with highly automated bread-lines with tight control, will be required to produce a soft loaf of high volume. Such process variables as water-absorption, dough-temperature, energy-input, dough-hopper residence-time, moulding stresses, proof-time, proof-temperature and humidity will be reasonably under control. Therefore, the improver only has to correct for small variations in flour quality. Whereas, a general-purpose improver, finding application in craft, and in-store bakeries, will be required to perform over a wide range of tolerance, e.g. various mixers, energy-input rates, mixing-times, flour profiles, water-absorptions, floor-times and dough-temperatures. The criterion for an optimum compromise is the ability of the mixed improver to produce a wide range of products using a specific processing system. This is achieved by finding the best possible combination of oxidants and emulsifiers for a desired set of bread characteristics or dough tolerance. In practice this would require millions of baking tests with limited reproducibility. Instead, by utilizing statistical techniques, coupled with the computer, the task can be facilitated. In this way, all ingredients and variables involved can be considered simultaneously for an improver formulation, and correlated with any desired bread-quality profile/dough-tolerance profile. Bread-quality profile parameters applied in the UK are: specific volume, loaf height, external softness, crumb softness, crumb elastic recovery, appearance, staling rate, and sensory qualities. Unfortunately, only the first two can be easily quantified, and physical methods have to be relied upon, coupled with subjective assessments. It is also possible to incorporate process variables and mechanical abuse effects into the program, in order to optimize performance to tolerance aspects.

The performance and properties of currently permitted oxidants show considerable differences in terms of dough relaxation time and energy-input response, which parameters can be related to bread quality and floor-time. Azodicarbonamide is capable of producing the highest bread-quality profile, but this can only be achieved over a narrow band of energy-input or dough floor-time. The level of azodicarbonamide used is also critical for optimal results. Ascorbic acid, although not capable of producing such a high quality-profile, shows stability over a wider latitude of either energy-imput or floor-time. Soy-flour, in the full-fat enzyme active form, as a source of lipoxygenase, is capable of an even wider tolerance latitude. Of the permitted emulsifiers, DATA-esters seem to be the most effective for tolerance latitude in doughs, owing to their complexing capacity with proteins and starches. A typical CBP bread improver would contain more azodicarbonamide, owing to the narrow tolerance range possible, as a result of the narrow processing limits inherent in this system. Whereas, a general-purpose improver would be based on ascorbic acid and enzyme-active soy. However, this is an oversimplification, since many ingredients used in bread improvers are synergistic, interacting with one another in a non-additive manner. For example, when ascorbic acid is superimposed on an improver system of azodicarbonamide and potassium bromate, the loaf-volume computer contours show a shift in pattern, two possible combinations of the oxidants producing maximum volume increases.

Loaf-volume shows a significant increase generally, due to ascorbic acid inclusion, and the corresponding levels for both azodicarbonamide and potassium bromate, for a given target-loaf-volume widen in their tolerances, thus illustrating the complexity of the interactions of these three oxidants used simultaneously.

In order to save time and avoid weighing errors, the use of improvers has been extended to form premixes or concentrates, which include, in addition to the oxidants, high-melting-point shortening, salt, emulsifiers, yeast-foods, and often yeast. These premixes are used for the CBP process, bun and roll doughs, doughnuts, high-fibre breads, and rye breads. Emulsifiers currently permitted in the UK are lecithin, monoglycerides, DATA-esters or DATEMS, stearyl tartrate, and the stearoyl-2-lactylates. Stearyl tartrate is often included to safeguard against so-called 'fat-failure' in the CBP; lecithins are not widely used in the UK, owing to cost and handling problems in the viscous liquid form. The DATEMS are the most important group applied in the UK baking industry. Currently permitted oxidants in the UK are: ascorbic acid 75 ppm, azodicarbonamide 45 ppm, and L-cysteine 75 ppm. It is possible, however, that some of these may be removed from the list as a result of EC legislation harmonization.

## 1.7.13 NORTH AMERICA

Owing to the extreme climatic conditions existing on the North American continent, full use is made of the different growing requirements of spring and winter wheats.

The spring wheats, which are planted in early spring, are of two types: Hard Red Spring and Durum. They grow continuously, reaching maturity during the summer, and are harvested in late summer. The Hard Red Spring subclasses are: dark northern, northern, and red. In Europe the best-known example imported is Dark Northern Spring (DNS 14% protein), which is widely used by millers as an improver wheat in mill-grists. The ideal climatic conditions for these wheats are severe winters, followed by dry hot summers, such as those found in the north-central states that have a deep, rich soil, viz. North Dakota, Montana, Minnesota and South Dakota. The high-grade, high-protein flour milled from these wheats are used in the production of Hamburger buns, pizza crusts, and ethnic foods. The other type of spring wheat, Durum with the well-known subclasses, hard amber, amber and durum, are milled into semolina, which is used for spaghetti, macaroni and other pasta products, the wheat kernels being hard and vitreous. These wheats need timely rains from April to July, being naturally resistant to rusts. Growing areas are: North Dakota 75%, and Montana, Minnesota, South Dakota, California, and Arizona.

The winter wheats, which are planted in the autumn in the case of the Hard Red Winters (subclasses, dark red, hard, yellow hard), require frequent dry periods, with less than 25 inches annual precipitation, and sub-zero temperatures in winter. Such areas are Kansas, Oklahoma, Nebraska, the Texas panhandle, and Colorado. These wheats are milled into flour for bread and all-purpose or family flours, since it does not have the high protein or gluten strength that the spring wheats have. Harvesting of these wheats is late spring to early summer.

Another type, Soft Red Winter, grown in the Mississippi river valley, northeast towards Ohio, and southeast towards the Atlantic coast and Gulf states, requires a sub-humid climate, which does not produce wheat with that hard kernel texture. This wheat is milled into flour for cakes, pastries, doughnuts, cookies, and crackers. The other type of winter wheat, called white wheat (with subclasses, hard white, soft white, white club, and western white), require a more varied climate, generally in the north with a more plentiful rainfall. Such areas are: Washington, Oregon, Idaho, Michigan and western New York.

This wheat is of low protein, and gluten-forming quality and is ideal milled into flours for products of a crispy nature, such as pie-crusts, and shredded and puffed breakfast cereals.

Rye is planted in early spring and harvested in late summer. Wheat-flours are either referred to as 'straights', i.e. straight-run, or straight-grade flour, which is a composite flour, milled from either hard or soft wheat, made up of various percentages of first clears and second clears (flours obtained by screening off the break-roll products, as plansifter-residuals, and passing through purifiers and reducing rolls before collecting as throughs of another plansifter pass), and a percentage of patent-flour (purest and best mill-stream throughs), or 'patents', which can be an all-purpose or family flour quality, or baker's bread flour patent. Patent flours always contain purest and best mill-stream throughs, and the only additives are the enrichment vitamins and minerals if required by Federal law, plus bleaching and maturing agents, bromate and malt (if appropriate). Summarizing, the white wheat-flours are:

|                                      | Protein | Ash    |
| ------------------------------------ | ------- | ------ |
| Straight, hard wheat                 | 11·8%   | 0·46%  |
| Straight, soft wheat                 | 9·7%    | 0·42%  |
| Patent, bread enriched               |         |        |
|     non-enriched                     | 11·8%   | 0·44%  |
| Patent, all-purpose or family enriched |       |        |
|     non-enriched                     | 11·8%   | 0·44%  |

Other wheat-flours include: wholewheat from hard wheats, 80% extraction from hard wheats, cake or pastry flour, gluten-flour (45% gluten, 55% patent flour), and enriched self-raising flour. Average moisture contents of white flours are 12·0%.

Rye-flours of various extractions are available with the following ash contents:

|                   | Moisture | Ash   | Fibre |
| ----------------- | -------- | ----- | ----- |
| 60% extraction    | 15%      | 0·5%  | 0·2%  |
| 75% extraction    | 15%      | 0·7%  | 0·5%  |
| 85% extraction    | 15%      | 1·0%  | 0·8%  |
| 100% extraction   | 15%      | 1·7%  | 1·6%  |
| Rye grain         | 15%      | 1·7%  | 1·7%  |
| Rye-bran, fine    | 15%      | 4·1%  | 4·9%  |
| Rye-bran, coarse  | 15%      | 3·7%  | 9·2%  |
| Ryegerm           | 15%      | 4·7%  | 3·3%  |

(Source: E. J. Pyler, *Baking Science and Technology*, 1973.)

In the United States there are mandatory enrichment laws for bread in most states and territories, therefore, most flours are enriched at the mill. When enrichment is applied, the product must be labelled or advertised as being 'enriched', and must comply with the enrichment standards. In addition, the states of California, Washington, Massachusetts and the territory of Puerto Rico, require the use of enriched flour in all products which contain flour. Enrichment Standards are as given in Table 46. The main sources of calcium-enrichment are: USG Terra Alba (chalk), calcium propionate, monocalcium phosphate, non-fat dry milk, Arkady-type yeast food double strength and Fermaloid-type yeast food. Nutritional labelling was introduced in the United States in 1973, providing a set of standards

TABLE 46

| Enrichment          | Enriched flour (mg/lb) | Enriched bread (mg/lb) |
| ------------------- | ---------------------- | ---------------------- |
| Thiamin             | 2·9                    | 1·8                    |
| Riboflavin          | 1·8                    | 1·1                    |
| Niacin              | 24·0                   | 15·0                   |
| Iron                | 13·5–16·5              | 8·0–12·5               |
| Calcium (optional)  | 960·0                  | 600·0                  |

upon which the consumer can base his choice of what to eat. The practice of putting nutrition labels on packaged food products, including baked products is voluntary. Only under specific circumstances does the law require nutrition labelling on packaged food. Such mandatory labelling situations are, when:

—adding a nutrient to enrich a food, e.g. enriched bread,
—advertising the nutritive value of a product,
—printing on the packaging any nutritional information.

To facilitate the communication of information in a clear and concise manner, the form, language and position of the label must conform to rigid regulations.

In the interests of consumers and manufacturers, a yardstick of nutritive values was devised, called the US Recommended Daily Allowance (USRDA). This data, expressed in percentage on package labels, is used to inform how good that food is as a source of a particular nutrient. The original source of this data was the Recommended Dietary Allowances published by the National Academy of Sciences, being based on the amounts recommended for men in most cases, but in the case of iron, the standard is based on women. Although the 'adult USRDA' is the most widely used, USRDA standards for children under 4 years, and pregnant women have also been ascertained.

Not all nutrients in the USRDA standards must be included on the label; required listings are: protein, vitamin A, vitamin C, thiamin, riboflavin, niacin, calcium, and iron. Optional listings are: vitamin D, vitamin E, vitamin B6, folacin, vitamin B12, phosphorus, iodine, magnesium, zinc, copper, biotin, pantothenic acid.

There are also regulations governing the relative accuracy of the percentages declared. When a nutrient is present at more than 50% of USRDA, its percentage is given to the nearest 10%. Nutrients between 10% and 50%, to the nearest 5%. Nutrients below 10% to the nearest 2%, and nutrients present at less than 2%, are listed either as 0% or designated as 'containing less than 2% of USRDA for these nutrients', the nutrients being then named. In the case of products normally consumed with other foods, e.g. breakfast cereals with milk, two lists are quoted, the first referring to the USRDA percentages in the cereal, and the second reports the percentages in the complete dish.

In addition, nutrition labels carry other information, such as: energy content listed in calories, protein as total amount per serving, and also the extent to which it meets the USRDA. Protein quality is also important, being divided into two classes, using casein as the reference, with a 'protein-efficiency ratio' (PER) of 2·5. If the food has a PER equal to or exceeding 2·5, its percentage is listed based on a USRDA of 45 g. Where the rating lies between casein and 20% casein, the protein content is listed with the USRDA set at 65 g. If the PER is less than 20% of casein, the protein cannot be listed.

Information concerning fat and cholesterol content is also allowed on labels. However, only total fat per serving is listed, any listing regarding type of fat is optional. No information is necessary for foods with less than 2 g/serving. Where optional information on the type of fat is shown, it must be accompanied by the statement: 'Information on fat content provided for individuals, who, on the advice of a physician, are modifying their total dietary intake of fat'. The same statement is required when cholesterol content is declared, cholesterol being shown in mg/ serving, or mg/100 g of food.

Declaration of sodium content is optional, unless a nutrition label is used, in which case it becomes mandatory. Sodium is declared as mg/serving, or mg/100 g food. Salt is 40% sodium.

The USRDA for nutrition labelling with reference to adults and children over 4 years are as follows: protein 45/65 g, vitamin A 5000 IU, vitamin C 60 mg, thiamine 1·5 mg, riboflavin 1·7 mg, niacin 20 mg, calcium 1·0 mg, iron 18 mg, vitamin D 400 IU, vitamin E 30 IU, vitamin B6 2·0 mg, folacin 0·4 mg, vitamin B12 6 mcg, phosphorus 1·0 g, iodine 150 mcg, magnesium 400 mg, zinc 15 mg, copper 2 mg, biotin 0·3 mg, pantothenic acid 10 mg.

The milling industry in the USA is broader-based and less concentrated than in European countries, due in part to the area of the country and the Federal State system, but also the prevailing spirit of free-enterprise.

There are at present 217 wheat mills, 12 durum mills and 10 rye mills, these mills together have a daily output of 1 126 758 cwt. The USA mill outputs are expressed in sacks (1 cwt)/24 hours, 1 cwt = 50·80 kg. The total daily output of the 217 wheat mills, 12 durum mills and 10 rye mills can be analysed as follows: bread flour 824 396 cwt, soft flour 245 733 cwt, wholewheat flour 26 195 cwt, and in the case of combined grain mills 30 434 cwt. The daily output of the durum mills is 82 300 cwt, with an additional 6000 cwt in the case of combined bread wheat/durum mills. The cumulative daily output of the mills baking-flours and durum-semolina is 1 209 058 cwt, and together with the rye mills, the total daily capacity is 1 218 162 cwt. In terms of size, the classification is: 23 mills producing less than 200 cwt, 23 mills producing between 200 and 399 cwt, 17 mills producing between 400 and 999 cwt, 63 mills producing between 1000 and 4999 cwt, 49 mills producing between 5000 and 9999 cwt, and 42 mills producing 10 000 cwt and more. The largest US mills are: ConAgra, Inc., 22 mills (196 000 cwt), Cargill Inc., 13 mills (143 000 cwt), The Pillsbury Co., 10 mills (140 900 cwt), and Archer Daniels Milling Co., 15 mills (133 500 cwt). For the manufacture of variety and specialty-breads, bases, mixes and concentrates are available. The high-profit specialty-bread market including such varieties as: hard rolls, hoagies, Italian, French, rye, bagels, pumpernickel, wholewheat, hamburger buns and English muffins.

White bread can be made with either enriched or non-enriched flours, containing the following levels of non-fat dry milk solids: 1–2%, 3–4% and 5–6%.

American rye bread is made from 15% light or low extraction rye-flour and 15% dark or high extraction rye-flour 75–100%, the rest being a clear flour; fermentation is with yeast. Foam soft-rye is made with 25% of a 75% extraction rye-flour, the rest being wheat-flour, with all the dough water added to a pre-ferment with yeast.

Pumpernickel and rye breads made with sours are also marketed in the USA.

Italian bread is made with white flour, but contains, in addition to yeast, some old-dough to provide a wheat-sour.

Common sour rye bread is made from 6% light rye, 21% medium and 8% dark rye flour, the rest being white flour. A three-stage sour and yeast is used, caraway seed being added at about 0·3% flour weight. Combination rye bread contains about 20% light rye, 8% dark rye flour and 36% wheatmeal with a balance of white flour. This is a yeasted dough.

Swedish rye bread is made using 35% medium rye flour with white flour, adding 12% molasses, fermenting as a yeasted dough. Garlic is another added flavour for rye bread, originating from Russia and Central Europe during the 1930s. Most specialty-breads in the USA have their roots in the early immigrant populations from Germany, Sweden, Poland, Czechoslovakia, Austria, Russia, Italy, Ireland, Greece and Turkey, etc. Bagels, for example, originated from Russia, Poland and the Balkans, coming to New York's lower East-side basement bakeries, spreading across to Brooklyn. They have now become big business nationwide, being non-sectarian and unionized.

On the West Coast, in the San Francisco area, early settlers brought their French-sours. Doughs made with two sponges and old mature dough using sour milk. The distinctive flavour of this bread is cherished from coast to coast.

It is difficult to generalize about the American baking industry. Various systems are employed for bread production, having been adapted to accommodate a wide range of tastes from state to state, to flour quality profiles and to type of water available. However, the vast majority of bread produced in the USA utilizes a pre-ferment system in one form or another.

Flour protein quality and quantity profiles will vary between bakeries, depending on where the wheat is grown. In general, protein levels are higher than European bread flours, being on average 12·0%. The gluten formed is much tougher and more elastic, and requires adequate development, compared with any European bread flour. Flour blends of Red Spring and Red Winter are frequently utilized. Although water is usually treated before reaching the bakery, it can vary in pH from below 6·8 on the acidic or 'soft' scale to more than pH 8·0 on the alkaline or 'hard' side. Equally important is the mineral content of the water, which places it in a subclass within a type, viz.

—Soft, with less than 120 ppm, requires a regular bromated type of yeast-food, added at a level above the normal. Additionally, salt can be added to the sponge, and calcium sulphate added in extreme cases.
—Hard, with 180–1211 ppm, requires a little less than normal of regular bromated yeast-food, and malt can be added to the sponge in extreme cases.
—Soft, but with a pH above 8·0 requires a regular yeast-food with added calcium monohydrogen phosphate or acid, and extra $CaHPO_4$ as alkalinity demands.
—Hard, and with pH above 8·0 requires an acidic yeast-food, with the use of extra malt in combination with lactic or acetic acid in extreme cases.

Owing to the effect of pH on fermentation-rate, particularly where brews are used, the American baker must pay special attention to final brew pH and total titratable acid (TTA).

Pre-fermentation of part of the flour results in a positive modification of the gluten proteins, which aids the development of the final dough, due to improved gassing, gas-retention properties, and yeast nutrition. Since liquid sugar solids are about 10%, based on flour weight, the yeast must have reached its optimal gassing-power by the doughing-up stage. The three main fermentation systems utilized are as follows:

## Sponge-dough

This traditional process is the most popular and widely used in the United States, accounting for some 60% or more of pan bread produced. Sponges are popular since they yield loaves of good volume, fine grain and texture and desirable flavour. Doughs are pliable and extensible with excellent machining properties. 65–70% of the total flour, usually 50% spring: 50% winter wheat flour, is mixed to a tight dough with 40% of the total water, including all the yeast and a yeast-food. The sponge stage requires about 2 minutes mixing in a horizontal-bar type mixer, of a total mixing time of 10–14 minutes. The resultant sponge is given a normal 4–5 hours fermentation under controlled conditions of temperature and humidity, during which, the dough temperature rises by about 10°F (5°C) from an initial 78–80°F (26–27°C) to 88–90°F (31–32°C). When the fermentation is complete, the remaining ingredients are added, and the whole remixed for 10–12 minutes to produce a fully developed dough. After a floor-time of 15–30 minutes, the dough is divided, and given a 10-minute overhead proof before moulding.

The dough is allowed to rise in the pan for 50–55 minutes, and then baked for about 18 minutes. The total time required being approximately 5 hours and 45 minutes duration, for a 4-hour sponge system, and can be as much as 7–8 hours, from start to finish, in the case of longer systems.

This long processing-time, coupled with the high production costs in space and equipment of the sponge-dough process, have long motivated researchers to seek out viable alternatives.

## Flour Pre-ferments in Liquid Form ('Brews' or 'Broths')

Since the introduction of the continuous mixer in about 1959, attempts have been made to replace the sponge-mixing and fermentation steps with a more liquid sponge, which could be blended, stored and transported by simple equipment using less labour. By 1974, the continuous-process accounted for about 35% of total US bread production. Today, flour-pre-ferments are used not only in conjunction with continuous processes, but also in conventionally mixed doughs. In the latter case, a proportion of the flour, from 20% to 60%, the current trend being to put at least 25% flour in the pre-ferment, is mixed with the water in the average ratio of 1·0:1·1 parts. This pre-ferment or brew contains all the yeast, yeast-food, a low concentration of sugar (about 2·5%) in the liquid form, and about 0·25% salt to control the fermentation rate (all percentages based on flour weight). Since the brew is essentially a liquid, it can be blended by simple mixers of low power, and transferred by pumps. The brew is normally fermented in large vertical tanks, compared with traditional solid sponges, which are fermented in 1000-pound capacity troughs, which must be moved to and from the fermentation-room either manually or by special conveyor systems. These liquid pre-ferments are fermented for 2–3 hours, during which time, a heat rise of 10°F (5°C) is desirable, the pre-ferment having been set at 78–80°F (26–27°C). When fermentation is complete, the pre-ferment is chilled down to 38–50°F (3–10°C), and held at this temperature to be used as required. For continuous-mixing procedures, the pre-ferment is pumped to the incorporator, where the other ingredients are added, after which the dough is pumped to the mixer head. Here, the mixing-time is only a few seconds. The fully developed dough is panned directly, and given a normal 50–55-minute floor-time before baking for 18 minutes at 425°F (218°C). In the case of the continuous pre-ferment system, the total processing time is approximately 4 hours. The trend in recent years has been to include up to 50% flour in the pre-ferment broth, since without enough flour, the bread flavour is very bland and not up to the standard of sponge-dough bread, which consumers have come to accept. However, the Rogers Fermitech Company, a division of Rogers Group Inc., have developed a concentrated dry pre-ferment marketed in 50-pound bags as 'Fermitech-101', manufactured by Rogers Fermitech of Oberlin, Ohio 44074, USA. This product contains: wheat-flour, natural fruit flavourings, fermented light-cream, yeast, whey, cottonseed/soybean oil, malt and corn syrup.

Whether the baker is using sponge-dough or brew systems, equipment costs, operating costs, maintenance costs, and fermentation-time can be reduced, and flavour, aroma, texture, and shelf-life improved. This product is used in conjunction with the 'Fermitech-pre-ferment process'.

Nutrient slurries are an important component of these pre-ferment processes, in order to ensure a rapid propagation of the yeast and ensure adequate gassing-capacity at the make-up stage. Yeast-foods can be either of the regular bromated type, often referred to as 'Arkady-type yeast-food', or specially formulated to compensate for acid/alkali imbalance in the water supply. A typical regular bromated yeast-food formulation could be: ammonium chloride 9·4%, potassium bromate 0·3%, calcium sulphate 30%, sodium chloride 35·5%, with starch as filler at 24·8%. Where the water supply pH exceeds 8·0, and is simultaneously deficient in acidic mineral content, calcium monohydrogen phosphate is a typical addition. In cases of extreme hardness, and pH above 8·0, an acidic yeast-food with added malt in combination with lactic or acetic acid, is necessary.

As a case-study of an automated flour-pre-ferment system, that recently installed by the Flowers Baking Company at their Miami plant, continuously producing pre-ferment for roll and bread manufacture, will clarify such an operation.

The operation is started by producing a yeast slurry, and nutrient slurry in separate stainless-steel tanks of 450-gallon capacity, equipped with high-speed mixers.

The yeast-slurry consists of 2 parts yeast: 1 part water, and the nutrient-slurry contains high-fructose corn syrup (HFCS), dough-strengthener (sodium stearyl lactylate), yeast-food and water. After preparation, the yeast-slurry is held at 38°F in a cold-wall tank. The nutrient-slurry is held in a glass-fibre tank, which maintains the temperature at 78°F.

The pre-ferment brew is prepared in a continuous 'sponge-maker', a small chamber fitted with a high-speed/high-shear mixer. Yeast and nutrient slurries are pumped into the unit at feed-rates of 7 lb/min in each case, and flour of about 12·0%

protein content is fed in from an overhead-hopper via a rotary-valve, and tempered water is also metered into the unit. The resultant mixture, which contains 50% of the total flour in the final dough, is then pumped at a feed-rate of 96 lb/min into a fermentation-tank of 5-tonne capacity. The horizontal, cylindrical tank is divided internally by baffle-plates into four compartments to ensure the correct residence time of about 2 hours. A reel moves the mixture from compartment to compartment, and finally to the discharge pipe. It is then pumped through a plate heat-exchanger, where the circulation of cold water cools the pre-ferment down from 86°F to about 45°F. It then passes into a storage-tank. Flowers Bakery utilizes four dough-mixers, one for bread-rolls, and three for bread.

To prepare a batch of dough, the operator presets a digitally operated weighing system, which adds the liquid ingredients in the correct quantity and sequence into a batch weighing container mounted on load-cells. Water, pre-ferment brew, HFCS, and vegetable oil shortening are all batched-up in this container. By actuation of a button, the operator discharges the contents into the mixer together with the flour. The cycle is repeated by actuation of another button to refill the batch weighing container.

Bread-roll doughs are mixed for 18 minutes in a horizontal-bar mixer of 1300 pounds capacity, operating at ultra high-speed. The dough then passes at about 78°F into an auger-type degasser, after which it passes into the divider hopper automatically. 400 dough-pieces/minute are deposited across a floured zigzag table positioned at the end of the roll-machine. After make-up (sheeting, panning, racking and proofing), the rolls are baked for 7·5 minutes at 440°F.

A battery of three 1600-pound capacity horizontal-bar mixers, two for white pan bread and one for variety-bread, feeds 5 tonnes dough per hour to the bread-line. White pan bread doughs are mixed for about 13 minutes at high speed, and variety-bread doughs receive an initial mix of 1 minute at low-speed, before the 13 minutes at high-speed. After a floor-time of 10–15 minutes, the dough is deposited by a hoist into a six-pocket divider-hopper, which weighs 114 dough-pieces/minute. Dough-pieces are then deposited into pans in straps of 5 across. The automatic rack-loader then loads 40 five-strap pans onto a monorail-rack system. The proof-box tender then loads the racks into a 32-rack monorail proofer for a 50-minute proof-time. The proofed pan loaves are automatically unloaded from the racks onto a table-top conveyor feeding the oven.

Baking is carried out in a five-zone direct-fired, single-lap oven, which bakes the loaves in 19 minutes at 440°F. The baked loaves are vacuum depanned, and cooled on a continuous conveyor for 45 minutes.

After cooling down to an internal temperature of 110°F (43°C), the loaves pass through a metal-detector before entering the bread-wrapping station.

There are two separate bread packaging lines, the primary bread-bagging line consists of four band-slicers and four reciprocal baggers with wire-tie closures. A secondary wrapping-line is also equipped with four band-slicers, which are mounted on tracks for movement to the right or left, thus allowing the slicers to be set-up to feed either paddle-baggers or overlap wrapping-machines. Pre-ferments containing no flour are referred to as 'water or yeast brews', and are more commonly used for bun-doughs. This brew consists of water, all the yeast, a small amount of sugar, and a buffer-salt in the form of calcium carbonate. These 'lean' brews usually contain about 45% water (expressed as formula per cent), and are fermented for 1–2 hours, depending on the amount of liquid-sugar and brew-buffer used. The temperature is set at 78–80°F (26–27°C), and a temperature rise of 10°F (5°C) is optimal. The brew is then cooled, and held at about 40°F (4°C) to inhibit fermentation. Since no flour is included, there is no flour protein development or development of flavour components, and if used to produce bread the resultant product is bland in flavour and has a reduced shelf-life.

**'Short-time' dough processes**
In recent years, several 'short-time' or 'no-dough-time' processes have been introduced on the North American continent, using the traditional 'rich-formulations', aiming to achieve the physical characteristics, flavour and aroma of sponge-and-dough processed bread. These will be discussed and compared with conventional processing techniques. The conventional processing formulations detailed in Table 47 are typical of current procedures in use in the USA.

Liquid pre-ferments or brews were developed to fit in with such continuous mixing units as the Domaker and Amflow, but flour-brews are now being used in conjunction with conventional batch mixers as described in the case-study of the Flowers Miami bakery. However, large bakeries turning out white pan bread usually use the sponge-and-dough method. The standard mixer is the horizontal-bar high-speed type, which is used for mixing sponges and doughs.

The sponge stage of the sponge-and-dough process requires about 2 minutes mixing of a total mixing-time of 10–20 minutes. The latter depends on flour-strength, quantity of flour used in the pre-ferment, and type of mixer. Mixing-times increase as the amount of flour pre-fermented is reduced. A typical horizontal-bar high-speed mixer can process 1000–1300 pounds of dough/batch, and these mixers are ideally suited to both the liquid pre-ferment and sponge-dough process systems.

Whereas, sponges have to be fermented in 1000-pound capacity 'troughs', which have to be transported either manually or by conveyor-system, liquid pre-ferments can be stored in tanks and metered into batch-weigher systems for direct feeding into the mixers. Sponges require 4 hours fermentation (before a remix of 10–12 minutes on adding the remaining ingredients), followed by a 15-minute floor-time for the final dough. Then, after dividing, the dough-pieces are given a 10-minute overhead proof before moulding and panning. Pan-proof is carried out at 110°F (43°C) at 92% relative humidity for 50–60 minutes, and baking is at 425°F (218°C) for 18–20 minutes. Therefore the total time required is 5–6 hours. Nevertheless, the sponge-dough process accounts for approximately 60% of total bread production in the USA. As already stated, repeated attempts have been made to introduce short-time or no-dough-time processes, eliminating high-energy inputs, floor-times and/or solid-sponges with mechanical or chemical dough-development techniques.

In Canada, Kilborn and Tipples (1968) reported attempts to produce mechanically developed Canadian bread with the same quality characteristics as sponge-and-dough process bread. These workers, demonstrated under laboratory

TABLE 47
**Conventional pre-ferment formulae for white bread in the USA**

| Pre-ferment | Water-brew (%) | Flour-brew (%) | Sponge-dough (%) |
|---|---|---|---|
| | *(based on percentage of total flour)* | | |
| Flour | — | 50·0 | 70·0 |
| Water | 16·0 | 55·0 | 40·0 |
| Yeast | 2·5 | 2·5 | 3·0 |
| Yeast-food | — | 0·5 | 0·5 |
| Buffer | 0·1 | — | — |
| Calcium sulphate | — | 0·6 | 0·6 |
| Liquid sugar (HFCS) | 2·0 | 2·5 | — |
| Salt | — | 0·25 | — |
| Fermentation time | 1·5 h | 2·5 h | 4·0 h |
| Setting-temperature | 78–80°F | 78–80°F | 78–80°F |
| *Dough:* | | | |
| Flour | 100·0 | 50·0 | 30·0 |
| Water | 44·0 | 5·0 | 20·0 |
| Salt | 2·0 | 2·0 | 2·0 |
| Liquid sugar (HFCS) | 8·0 | 8·0 | 8·0 |
| Non-fat dry milk (whey) | 3·0 | 3·0 | 3·0 |
| Soy-flour | 1·5 | 1·5 | 1·5 |
| Dough conditioner (sodium stearyl lactylate) | 0·5 | 0·5 | 0·5 |
| Crumb softener (0·25% distilled monoglycerides) | 0·5 | 0·5 | 0·5 |
| Vegetable shortening | 3·0 | 3·0 | 3·0 |
| Calcium sulphate | 0·6 | — | — |
| Yeast-food | 0·5 | — | — |
| Mould-inhibitor (calcium propionate) | 0·125 | 0·125 | 0·125 |
| Floor-time | 20 min | 20 min | 20 min |
| Dough temperature | 78°F | 78°F | 78°F |

conditions, that it was possible to produce by a mechanical development no-dough-time method, bread which was indistinguishable from that made by the sponge-and-dough process in all respects. These results were obtained with a typical North American formulation, and high specific loaf volumes, using a range of flours of 12·8–16·0% protein content. Oxidant levels were, on average, 44 ppm bromate, and 16 ppm iodate for the mechanically developed dough, and 22 ppm bromate only for the sponge-and-dough procedure as control. In 1974, J. A. Johnson, published the Kansas State University 'Flavol short-time dough process', which eliminated primary or pre-fermentation, and controlled flavour-intensity of the bread by small additions of amino acids, and organic acid salts in ratios found in 4 hour sponges. Elimination of the lengthy primary fermentation step greatly reduces production time, and bread flavour intensity is under control. The Flavol-process can also be applied to the straight-dough method with the elimination of primary fermentation. When applied to a continuous mixing system, total time for bread production is 70–80 min, compared with about 4 hours 10 minutes when a 3-hour floor-time is in use.

Applied to the straight-dough procedure, the total processing-time amounts to 80–90 minutes, compared with about 4 hours 32 minutes. Furthermore, these total process-times must be set against a cumulative total process-time of about 5 hours 45 minutes in the case of sponge-and-doughs with 4-hour floor-times.

Comparisons of loaf volume, grain, and texture have shown that bread made with 0·2% and 0·4% Flavol are almost identical with white pan bread made by the sponge-and-dough process. White pan bread is the largest sales item produced by the US baking industry. The ingredients of Flavol are approved by the FDA in Washington, DC, and temporary permits for interstate shipment of Flavol-bread were granted. The process was used in 18 commercial bakeries for various types of baked products, viz. American rye bread, French hearth-bread and dinner-rolls. Most products were judged as not significantly different by the 16 semi-trained judges for sweetness, sourness, over-all flavour, and over-all aroma. French bread, and American rye bread made with 0·2% Flavol and no fermentation-time were judged superior to controls, and no bread made by the Flavol-process gave scores lower than the control sponge-and-dough process bread.

Further research into short-time doughs was reported by Ponte (1985) at Kansas State University, in order to simplify pan bread processing. Attention was drawn to the long process-times of up to 7–8 hours, from start to finish for sponge-doughs, high production costs, and equipment and space demands. Hitherto, the main application of short-time (no-time) doughs has been for hearth-baked bread and rolls for short-term consumption within 1–2 days. However, recent work suggests that short-time doughs may be applied to white pan-bread. The continued popularity of sponge-doughs is due to their pliable, extensible machinability, yielding loaves of good volume (approx. 6·8 ml/g), fine grain, texture and flavour. The key requirement is a correctly balanced dough-improver system, to offset the fermentation deficiency. Yeast-activity and gassing-power must be accelerated by the use of higher concentrations, and provision of yeast stimulants and substrates, and the gluten proteins must be rapidly modified with the optimal balance of oxidants and reducing agents. This will ensure that the dough attains optimum gas-retention properties, and become extensible enough to expand as a result of increasing internal pressure of the carbon dioxide from the yeast. The short-time formulation suggested is as follows: flour 100 parts, water 62, yeast 4·5, salt 2·0, vegetable-shortening 3·0, sugar 6·0, non-fat dry milk 3·0, dough conditioner 1·0. The dough conditioner is one especially designed for pan-bread, and has the following composition: calcium stearoyl lactylate, soy-flour, monoglycerides, oxidizing and reducing agents and enzymes. The type of conditioner used in this process is marketed by the Puratos Corporation, Pennsauken, NJ 08110, USA. Similar products are available from Foremost Whey Products, a division of Wisconsin Dairies, Baraboo, WI 53913, USA, under the 'Reddi-Sponge' label. Reddi-sponge rapid dough-developer simplifies the baking process by eliminating the time-intensive process of dough-fermentation or sponging, cutting up to 3 hours from the processing of yeasted white-pan, variety bread and hard-rolls. Similar results can be achieved with Danish pastry, doughnuts, English muffins, soft-rolls, and brown-and-serve frozen items.

A typical make-up schedule for such a 'short-time' would be: dough temperature

82°F (28°C), dough floor-time 20 minutes, intermediate (overhead) proof 15 minutes, final proof at 110°F (43°C) and 92% relative humidity 60 minutes, bake 425°F (218°C) for 20 minutes.

Dry concentrated ferments are also marketed by Rogers Fermitech Co., as already mentioned under 'Flour pre-ferments in liquid form'. High-quality white pan-bread can be produced with 'Fermitech' in about 2 hours, compared with 6–8 hours of normal production time.

Dough-conditioners are also available in tablet form, which makes dosage accurate and expedient. A typical balanced conditioner would contain: mono-calcium acid phosphate, ammonium sulphate, ascorbic acid, fungal alpha-amylase and possibly a proteolytic enzyme, which creates an ideal environment for proper yeast metabolism, The functionality of each of these components is as follows:

—Monocalcium acid phosphate assists the adjustment of pH.
—Ammonium sulphate acts as an inorganic nitrogen source to encourage vigorous yeast-cell reproduction.
—Ascorbic acid restructures the gluten proteins and improves their gas-retention properties.
—Fungal alpha-amylase improves water retention, giving more pliable doughs, and a softer crumb.
—Proteolytic enzymes, e.g. papain, to reduce excessive dough elasticity, increase dough extensibility, and reduce mixing-times. Both papain, and another plant proteinase, bromelin, are classed as 'endopeptidases', which act on the interior peptide-bonds of the gluten proteins.

Average usage levels of these balanced combinations in tablet form are: 7–10 tablets/cwt of flour (100 lb or 90·7 kg). Enrichment, oxidation, enzymic and yeast-food functions are all available either in powder or tablet form.

The introduction of continuous breadmaking processes in the USA dates back to 1953–54, with the Baker 'Do-maker', and 'Amflow' plants. The original concept of bread production by continuous mixing, differed widely from the conventional sponge-dough systems. When first introduced, fermentation took place in a liquid pre-ferment, which contained no flour, the sugar and other ingredients supplying the substances for yeast activity. However, it was soon found that fermentation of flour produced more desirable end results in the finished bread. Thus increasing amounts of flour began to be added to the liquid pre-ferment. In some instances, amounts of flour in excess of 50% were added, and the pre-ferment soon became a liquid-sponge, thus reverting to conventional baking. The main difference remaining is the length of the fermentation-time, being considerably shortened in continuous processing and much more vigorous than in batch processing.

In the case of the Do-maker process, the dough is prepared from a flour-free liquid pre-ferment. The dough is continuously mixed, and proceeding to the processing line without any fermentation or floor-time. The plant layout sequence is as follows.

Three vertical tanks for the liquid pre-ferment, connected via a distribution-valve to a heat-exchanger unit to regulate pre-ferment temperature. Other tanks for liquid ingredient mixing and storage included, one for oxidant equipped with a dosage-pump, and two for handling the shortening. One is equipped with a steam-jacket to melt and mix the shortening, the other is a heated holding-tank for the liquefied shortening, the two being interconnected via a pump. A second dosage-pump metering the required amount. The pre-ferment, oxidant and shortening pipelines, together with the flour silo and dosage-aggregate feed into the continuous premixer. Here the ingredients are homogeneously blended to form a raw, rough-textured mass. A dough-pump conveys the mass into the developer, where the developer-head imparts a large amount of mechanical energy to the dough, whilst confining it within a relatively small area, under pressure. Since the ratio of energy to mass of dough is large compared with batch mixing, the gluten is quickly developed sufficiently to produce desirable properties. Nevertheless, where no flour is added to the pre-ferment, hydration problems could occur at the premixer stage where the residence time of the pre-ferment and flour was only 3–6 minutes, and as production or throughput rates increase, this tended to become even shorter.

From the developer, the dough was fed directly into the divider, and thence into the pans for final-proof and baking.

In the case of the Amflow process, developed by the American Machine and Foundry Company, the basis was also a liquid pre-ferment, a two-stage continuous mixing of the dough, and immediate dividing and panning after leaving the dough-developer. However, in this process, 10–16% flour is added to the pre-ferment, making it a flour-brew. The preparation of the liquid pre-ferment commences as a batch operation. In the mixing-tank yeast, yeast-food, water and part of the mature pre-ferment from the previous batch are thoroughly mixed. After 1-hour fermentation, salt, non-fat milk solids, and the 10–16% of total dough flour are added. The mixed flour-brew is then pumped into the first of two fermentation-tanks maintained at 30–32°C for 1 hour. Meanwhile, the second batch of pre-ferment is prepared and pumped into the second fermentation-tank. The pre-ferment from the first fermentation tank is then pumped into a horizontal continuous-fermentation container divided into three compartments for completion of ripening. In the first compartment, the sugar-solids are introduced, and the complete flour-brew is allowed to ripen for a further 1 hour. Most of the ripened flour-brew is then pumped into the batch-weigher, and thence through a heat-exchange cooler, after which it proceeds into the premixer. The remaining part of the ripened flour-brew is returned to the initial mixing-tank to act as starter for the next cycle. The cooled flour-brew, liquid vegetable shortening, and oxidant in solution, are then mixed in the premixer, before being pumped into the developer-unit for intensive mechanical development. The dough is then immediately divided and panned for final-proof.

Certain aggregates of the Amflow-plant were also applied in other variations, in order to rationalize and contribute to production flexibility. For example, in the case of bread-roll production, a simplified version of the Amflow liquid-pre-ferment stage has been utilized for the batchwise preparation of pre-ferment for mixing of doughs in conventional mixers, with transportable bowls. This eliminated the preparation of sponges in machine-bowls and their manual transport. A continuous dough-developer unit was also built, described by Freed (1963). In this unit, the dough was transported through a barrel-like aggregate, which was directly coupled with a unit which divided and panned the extruded dough-pieces. The special features of this process were that the doughs were made on a straight-dough system in conventional mixers with transportable-bowls, using higher dough-temperature of 33–36°C. Chemical development was achieved either by the addition of 44–45 ppm potassium bromate, or lipoxygenase-active soy-flour at 0·9% flour weight. After a floor-time of 1·5 h in the mixer-bowls, the dough was deposited into the hopper of the developer-unit, and divided and panned immediately for final-proof at 40°C for 50 minutes. In this process, the developer-unit did not eliminate the floor-time, but instead the need for a divider, rounder, intermediate or overhead proofer-aggregate and final-moulder.

In summary, the most important aspects of the Do-maker and Amflow continuous dough-processing plants in the USA are:

(1) The use of a liquid pre-ferment of water, yeast, sugar and yeast-food with added non-fat milk solids and salt; and with the Amflow an added percentage of flour.

(2) Mixing takes place in two stages, viz. premixing and developing. The dough is intensively mixed with shortening and oxidant in solution in the developer-unit.

(3) Immediate work-off of the dough from the developer into pan for final-proof.

As a result of the 1966 Conference of the American Society of Bakery Engineers concerning continuous-doughmaking, certain ways of improving the quality of the end-product were put forward, which were incorporated into the Amflow procedure, viz.

(1) The inclusion of up to 50% flour in the pre-ferment, to improve the taste and aroma of the bread, allowing a reduction of the sugar and oxidant requirements.

(2) The use of liquid-shortening with the optimum ratio of solid:liquid phase (dilatation or solid–liquid index), the solid fats being either in the form of crystals or flakes. The inclusion of surfactants were also recommended. The melting-point of the solid fat phase must be higher than the dough-temperature after mixing and fermentation, i.e. above 40–42°C.

Research done by the manufacturers of the Do-maker was directed towards simulation of the crumb-structure, flavour and aroma of sponge-dough bread. Nitrogen, oxygen, air, and carbon dioxide were injected between the premixer and developer stages. The use of oxygen and atmospheric oxygen both improved the crumb and flavour and aroma of the bread, but the inert gases only improved the crumb-structure of the bread. Control of the following process variables will determine the type and crumb-structure of the bread obtained:

(1) Premix residence-time.

(2) Developer throughput rate.

(3)  Developer-speed.
(4)  Developer-pressure.
(5)  Dough-temperature.
(6)  Type and concentration of gas-injection.

In spite of subsequent improvements to the processes, due to ingredient and process modifications, most bakers consider that something basic is missing when compared to the sponge-and-dough process. Typical disadvantages encountered with Do-maker and Amflow plants were as follows.

Excessively high dough temperatures, a need to increase scaling weights, deficient oven-spring, weak-sidewalls, lower crumb resilience, inferior eating-quality, and lack of the conventional sponge-and-dough flavour. Increasing the amount of flour in the pre-ferment approximating to that of sponge-doughs was beneficial in some bakeries, but was not the complete answer to the problem. Setting batch pre-ferments did not fit in with a continuous process. Initial work done by Henika and co-workers during 1965–66 suggested that under-development of the doughs was the reason for these defects. They proposed the use of sweet dry whey, in combination with L-cysteine. Production trials were carried out in 18 bakeries, located in the East, deep South, Central and Western States, 12 of which operated Do-maker units, and the remaining 6 Amflows. Flour levels being used in the pre-ferments were from 0% to 30%. The amount of whey/cysteine blend added was 1·5–2·0% of flour weight, being metered by means of a dry-feeder into the flour stream on entering the premixer unit. Control dough developer-speeds of 205 rpm, could be reduced to about 167 rpm for whey/cysteine doughs, a reduction of 38 units. This reduced the power requirements of the developer-unit, and increased the production rate by an average of 5–40%, depending on reserve proof-box and oven-capacity of the lines. These rpm reductions were more apparent when pre-ferment flour-contents were zero to 25%. At 50%, rpm and power requirements are normally reduced. Dough outlet temperatures must, however, be maintained high to give sufficient extensibility and pan-flow.

Before using the whey/cysteine product, developer outlet temperatures of 104–110°F were normal with continuous mixing, owing to the mechanical energy input necessary to reorientate the gluten proteins becoming dissipated as heat energy. These high temperatures reduced development-times to about 100 pounds/minute with an average output-motor and developer-head size. Decreasing premix-temperatures in order to lower dough outlet temperatures, only increased developer requirements, and throughputs had to be reduced or the developer-rpm increased. This inevitably led to even higher temperatures.

The presence of whey/cysteine at 1·5–2·0% flour weight reduced average developer outlet temperatures from 104°F to 92°F (12°F). Reducing pre-ferment temperatures also helped to adjust to lower dough temperatures. Average dough temperatures of 92–94°F resulted in significant changes in dough and bread properties. The dough became more extensible, and overmixing had to be carefully controlled. Pan-flow was excellent within the dough-temperature range 84–97°F. Bake-out loss was much lower, and dough scaling weights had to be reduced by 0·1–0·5 ounces, and final bread moisture checked carefully. In so doing, scaling weights approximate to sponge-dough scaling weights, and dough behaviour is very similar. In some bakeries scaling-weights were reduced from 27·75 to 27·5 ounces with no change in baked-weight or bread moisture levels. In other bakeries, the average baked-weight increased from 23·9 to 24·9 ounces, with no change in water absorption or scaling weight. Other dough work-off variables required only minor corrections. Proof-time to normal height remained the same, but the extra oven-spring demanded a reduction of the proof-time by about 5 minutes to give the same loaf volume.

Owing to improved flour protein development, oven-spring and loaf volume were in the hands of the production superintendent once more. Volume was controlled by striking a balance between developer-speed and the amount and choice of oxidant, and/or increasing flour strength. The use of whey/cysteine allows the baker to use the flour best suited to his market. Improved bread characteristics included:

(1)   Loaf symmetry was much improved.
(2)   Side-wall strength increase enabled loaves to be stacked about three high on supermarket shelves.
(3)   Crumb-softness and bread resilience, typical of sponge process bread, were markedly improved.

Close or open texture can be controlled by water absorption, oxidant, and level of flour in the pre-ferment. For example, stiff doughs, containing about 65 ppm oxidant at the bromate/iodate ratio of 4:1, and flour levels of 10% or less in the pre-ferment, give small-celled crumb-grain in the final loaf. Most bakeries prefer a moderately open, slightly irregular crumb-grain. Average bromate levels were increased by up to 5 ppm to compensate for the non-fat milk solids from the whey/cysteine product, no increases in the iodate levels were normally necessary. When finer, whiter grain was required, the oxidant level was raised by 8–10 ppm, and/or the bromate:iodate ratio adjusted; alternatively, the pre-ferment-time could be reduced. Normally, where the flour level in the pre-ferment is above 30% and total oxidant level not more than 50 ppm, only the bromate level is adjusted. Working at dough temperatures below 95°F necessitated an adjustment to the shortening composition, since the melting-point was too high. The amount of lard or vegetable flakes in the blend were therefore reduced from 10% to 4% in steps of 2%. Crumb colour was improved along with a tighter grain when the lard flakes were reduced.

Where greater resilience, side-wall strength, and flavour is desired, the dough outlet temperature was reduced by steps of 5°F by cooling the pre-ferment. Optimum developer-rpm was then increased by increments of 5 rpm. This resulted in stiffer doughs, improved gas-retention, oven-spring, shorter proof-times and finer-grained loaves.

When dough temperatures fell below 92°F, the water absorption normally had to be increased by 1–2%, and developer speed adjusted by increments to produce a dough with optimum extensibility and pan-flow, on relaxing for a few minutes in the pans. Having achieved this condition, the developer-operator could set the optimum rpm by feel.

Many bakeries were so dissatisfied with continuous mixing and bread quality compared with conventional sponge-dough bread, that they were prepared to sell their continuous plant and revert to sponging, before the use of whey/cysteine and cooler doughs. The cost of the additive was more than compensated for by the increases in bread yield.

The subsequent use of these cooler doughs, for variety bread and bun doughs, e.g. approximately 92°F, made possible the continuous mixing and extrusion of such doughs. In which case, the addition of the cysteine/whey product was added with a dry-feeder into the flour stream, as the flour entered the premixer unit. The use of a pre-ferment step in continuous or batch processing conditions the yeast cells and stimulates gas-production, providing metabolites which assist in dough development, maturing and flavour enhancement. Any controlled biochemical activation of these processes by the addition of carefully selected additives such as whey/cysteine, will shorten pre-ferment-times and enrich the fermentation substrate.

A typical schematic diagram (see Table 48) for a continuous breadmaking process is represented by the Amflow plant of the American Machine and Foundry Company.

Flour types used in the American baking industry are derived, for breadmaking, from blends of spring/winter wheats 50:50, representing over 60% of US bakeries' basic breadmaking raw material, for both sponge-dough and continuous processing.

Sponge-dough oxidation levels may only be 15–20 ppm, whereas continuous processing may require amounts approaching the legal upper limit of 75 ppm.

Conventional sponge-doughs have two-thirds of their flour content left to soak and hydrate for up to 4 hours in the sponge. Then, at the remix stage, the remaining one-third, plus water and other ingredients are added, and the complete dough

TABLE 48

| *Mixing-tank* ⟶ | *Pre-ferment fermentation-tanks* ⟶ | *Horizontal tank* |
|---|---|---|
| Water | Constant temperature 30–30°C | *with compartments* |
| Liquid sugar | insulated | Additional liquid- |
| Yeast | | sugar and water |
| Yeast-food | | added for |
| up to 50% flour | | continuous |
| NFMS | | fermentation of |
| Salt | One-third mature pre-ferment | pre-ferment |

| *Batch-weigher tank* ⟶ | *Heat-exchanger* ⟶ | Liquid-shortening |
|---|---|---|
| | cooler | oxidant-solution |
| | | added from |
| | | holding-tanks |

| *Premixer* ⟶ | *Developer* ⟶ | *Divider and panner* |
|---|---|---|
| Dough assembly | Intensive | *unit* |
| stage | mechanical | |
| | energy-input | |
| Flour balance | | |
| added | | |

mixed for 10–14 minutes. It is then given a floor-time of 20 minutes or more before make-up and proof.

In continuous processing, the time between contact of flour with water, and the beginning of proofing is only a matter of 2–3 minutes. Therefore, there is very little time for hydration. For this reason, the incorporation of substantial amounts of flour in pre-ferments results in improved continuous processed bread, viz. improved hydration. Flours which are processed at the mill to give rapid hydration, should help to alleviate this time-dependent processing variable.

If, however, nearly all the ingredient water is added to a pre-ferment containing 60% flour, it becomes in effect a 'fluidized' sponge, which can be readily pumped, agitated and metered. Fluidized sponges ferment faster, and mild agitation creates new surfaces for additional yeast activity, therefore producing similar fermentation end-products in less than half the time as a conventional sponge.

In addition to white panned bread, the American Baking Industry offers the consumer a wide range of variety breads, the most well-known of which are the following: wholewheat, raisin, raisin-cinnamon, cheese, rye, cinnamon, French, Italian, honey-wheat, honey-raisin, cracked wheat, multi-grain, potato, Wisconsin-rye, petite loaves, rolls, muffins, bagels, molasses-bread, and Pullman-toast.

In the production of rye breads, owing to the lack of gluten-forming proteins, the amount of rye-flour used in a formulation is limited to a certain percentage depending on the rye-flour type and rye bread being produced. One of the most critical stages in rye bread production is mixing, rye doughs requiring only about 50% as long as white-flour doughs. Over-mixing leads to loss of loaf volume and splitting. Dough temperatures should be about 80°F, doughs being made on the stiff side, with absorptions a little over 60%. Prepared sours are available, and the dough is given a floor-time of about 35 minutes. The rounded doughs are given an intermediate proof of about 8 minutes before final moulding. Cross-grain type moulders with minimal sheeting are applied, the moulded pieces being placed in perforated-baskets for proofing for 58–60 minutes in a rack-type proof-box. Proofing temperature is 114°F, and relative humidity is kept to a minimum. Typical formulations are as follows.

*Light rye bread*
Rye-flour 100 parts, clear flour 100, yeast 4·0, salt 4·0, water 120.
*Raisin-bread*
White patent flour 100 parts, water 65, NFMS 10, liquid-sugar 8, shortening 8, salt 2·5, yeast 4·0, seedless raisins 50, yielding 220 pounds baked bread/100 pounds flour.
*Wholewheat white bread*
White patent flour 50 parts, wholewheat-flour 50, water 64, NFMS 6·0, liquid-sugar 6·0, shortening 6·0, salt 2·5, yeast 4·0, yielding 165 pounds of bread/100 pounds of flour.
*Wheat bread with soy and wheatgerm*
White patent flour 97 parts, full-fat soy-flour 3·0, water 65, processed wheatgerm 3·0, NFMS 6·0, liquid-sugar 8·0, shortening 6·0, salt 2·5, yeast 4·0, yielding 170 pounds bread/100 pounds flour.

### High vegetable-protein bread

Sponge:    white patent flour 65 parts, water 34·6, yeast 2·25, yeast-food 0·5, shortening 3·5, set for 4 hours at 78–82°F

Dough:    white patent flour 35 parts, water 33·6, vegetable-protein isolate 9·4, sodium carbonate 0·6, liquid-sugar 6·0. floor-time 20 minutes, overhead proof 15 minutes, final-proof 1 hour. Bake 425°F for 25 minutes.

### Wheat bread with soy

White patent flour 96 parts, NFMS 6·0, full-fat soy-flour 4·0, water 65·0, liquid-sugar 7·5, shortening 7·5, salt 2·5, yeast 4·0, yielding 170 pound bread/100 pound flour.

### Molasses-rye bread

Sponge:    hard wheat-flour (springs) 60 parts, water 40, yeast 3·0, yeast-food 0·5, set for 3 hours at 80°F.

Dough:    hard wheat-flour (winter) 30 parts, light rye-flour 10, dehydrated molasses (molasses solids 60·0%, partially gelatinized wheat-flour 37·0%, water 3·0), shortening 4·0, light brown sugar 2·0, NFMS 3·0, salt 2·0, soda 44·0 g, floor-time 20 minutes, final-proof 1 hour, bake 425°F for 25 minutes.

The annual valuation of the US baked-products market is about $23 666m, based on 1985 published statistics. Sales of bread and rolls showed total increases of 2·1%, and within this total, a 4·1% increase in variety bread sales. This strong performance of the variety bread sector is explained by the in-store bakeries initially capturing the market share from wrapped white bread, and then introducing new product varieties. The plant bakeries then fought back by increasing their own product ranges, based on fibre and other nutritional value upgrading. Such products as wholewheat and cracked-wheat breads, which are much favoured by the health-conscious middle or intermediate age groups, should maintain a strong annual growth of about 5%, which has held for the past 8 years. New nutritional specialty-breads are centred on the following enrichment areas: dietary fibre, protein, calcium, sodium-reduction, dietetic, and breads formulated to satisfy specific dietary requirements. This latter area is expected to show particularly strong growth, and new formulations are being aimed at the specific nutritional requirements of the various age groups in society. However, it is often difficult to reconcile nutritional needs with acceptable sensory properties in such products. Therefore, ways of improving texture and flavour of some of these specialty-breads have to be further researched.

The frozen-dough industry, which began to appear in the 1950s, has, after periods of fluctuating fortunes, emerged with a steady growth rate since the 1960s. Ownership of this industry has always been diverse, with wide variations in products manufactured, sales outlets and promotional programmes. The main channels of distribution have been through franchise operations or brokers into the chain-grocery stores. Frozen-dough going to institutional and baking outlets is baked at the outlet before it is sold to the consumer. Chain stores with in-store bakeries produce frozen-dough in their own centrally located plants for bake-off in these bakeries, or retail outlets. The number of firms producing frozen-dough exclusively for chain-store bake-offs and grocery supermarkets continue to show progressive growth. Typical frozen-dough product ranges are white bread dough, wholewheat, raisin, honey-raisin, multi-grain, cracked-wheat, raisin-cinnamon, cheese, rye, Wisconsin-rye, cinnamon, French, honey-wheat, petite loaves, rolls and muffins. Dinner rolls, Danish pastry and fruit pies are additional popular items.

Owing to the steady decline in per capita consumption of bakery products during the period 1952–67, frozen-dough was introduced as a more efficient, low-cost method of producing and distributing bakery products. Frozen-dough does not include any products that are baked or partially baked before freezing.

Partial baking techniques, known as 'two-stage baking' or 'brown-and-serve' were introduced in the early 1950s. This technique was applied to both lean and enriched formulations, the aim being to achieve maximum volume without colouring the crust. Oven temperatures between 275 and 300°F, resulted in internal dough temperatures of about 170°F, which gave maximum rigidity. Baking times of 10–15 minutes at these temperatures were found to be adequate for this purpose. For this purpose, the dough is mixed without excessive dough development to a firm consistency, set for 90°F. Floor-times were limited to about 15 minutes, followed by three-quarter final-proof with a minimum of steam at 98–105°F. After cooling for 20 minutes, without forced-air, the products had a mould-free shelf-life of 3–5 days. Mould inhibitors such as sodium diacetate of calcium propionate could extend the mould-free shelf life of these products significantly.

The growth of frozen-dough for bake-off operations or on-premises baking in supermarkets and in-store bakeries, is due to the following market factors:

(1) Consumer demand for one-stop shopping.
(2) The desire for fresh baked products with aroma and eye-appeal.
(3) Providing one supermarket-chain with a competitive-edge over the competition.
(4) Removing the baking industry from fixed time and place of production, and allowing production versus shelf-life planning.
(5) Ensuring round-the-clock sales, without unsocial working-hours.
(6) Reduction of distribution costs and problems.

The most significant factor in the performance of frozen-dough is the amount of fermentation before freezing, stability of the dough being inversely related to the floor-time before freezing. For direct freezing, allow the dough just enough floor-time to recover from mixing, about 15 minutes. Then divide and round using conventional equipment, giving a minimum overhead proof before shaping. Doughs processed in this way prior to freezing should have a stability of 3–4 months.

A typical straight-dough formulation for frozen-dough preparation would have the following composition: white patent flour 100 parts, water 65, dry yeast 4·0, yeast-food with bromate 0·5, salt 2·0, dextrose 7·0, shortening 9·0, sweet whey 5·0.

Procedure: all ingredients are metered into a high-speed mixer fitted with a refrigeration-jacket, and mixed to a fully developed dough. Dough temperature is maintained at 65–70°F (18–21°C), and 10–15 minutes floor-time is allowed during the make-up stage. Freezing is by blast/contact at −20 to −25°F (−30 to −32°C), using an air-flow rate of 60 ft/min. Slabs of bread dough in units of 1 lb when unwrapped require a freeze-time of 2·75 hours, single-wrapped units require about 3·25 hours, and double-wrapped units (bag and wrapper) require a minimum of 3·5 hours. Slabs of bread dough prepared in this manner should show good performance after 16 months storage below ±10°F (−12°C). Two-pound slabs of bread dough would require 3–4 hours freeze-time, depending on the use of wrapping and its thickness, to reach +10°F (−12°C). Final temperatures for most yeasted bread and roll doughs (lean or rich formula) should reach −20°F (−30°C). One-pound slabs of bread dough require, when double wrapped, 3·5 hours thawing-time, rising from 0°F to 70°F at a defroster temperature 100–120°F. Whereas, 2-lb slabs of bread dough would require 3·75 hours to thaw under the same temperature conditions, using a standard air-flow speed of 150–200 ft³/min. For comparison, if a 1 lb slab of bread dough is packaged in a Poly-bag, identical thawing conditions would require 4·5 hours duration.

Dinner-rolls packaged in a plastic bag would need about 1 hour to freeze down to 10°F (−12°C), and 1·25 hours to thaw between 0°F and 70°F, using the same defroster temperature and air-flow. Obviously, the time factor will depend on the size and package, but ideal thawing-conditions can be summarized as follows: thawing-temperature 100–150°F, relative humidity range 50–60%, at air velocity 200–500 ft/min.

Popular products for frozen-dough production are: bread- and dinner-rolls, cinnamon-rolls, fruit pies, Danish pastry, read-to-serve products. The ideal type of packaging varies with the product, viz.

| | |
|---|---|
| —bread and rolls | single/double wrapped, or bag and wrapper |
| —dinner-rolls | plastic bag or foil pan |
| —cinnamon-rolls | plastic bag or foil pan |
| —fruit pies | foil pan or box and overlay |
| —Danish pastry | foil pan or sealed and overwrapped box |
| —ready-to-bake products | foil pan or box and overlay |
| —frozen bread dough (rich formula) | plastic bag |
| —yeast-raised dough (rich formula) | foil-lined, fibre-wound can. |

When removed from the freezer, the products are proofed at 95°F (35°C) with 80% relative humidity for 70–80 minutes. Baking is at 425°F (218°C) for an average of 17–20 minutes. Specific volumes of the baked products should not be less than 4·5 ml/g, when less than 4·0 ml/g the product becomes compact and heavy.

For up to about 8 weeks, the dough gives a normal final-proof time on thawing, but after about 21 days storage, rolls are usually unacceptable in quality. When final-proof times exceed 3 hours, specific volume is low and product structure poor. Minimal fermentation before freezing is essential to avoid fall-off in final-proofing with storage time.

A whey/cysteine/bromate formulation is ideal for processing frozen-doughs, owing to the direct biochemical development of the flour proteins instead of floor-time fermentation. This process results in a doubling of the shelf-life stability of the dough, and allows a reduction of dough outlet-temperature down to 65–70°F (18–21°C).

The following formulation is based on the Reddi-sponge process of R. G. Henika, Foremost Foods, Dublin, California, USA.

Flour 100 parts, water 62, yeast 4·0, yeast-food 0·5, sugar 8·0, salt 2·0, shortening 3·0, emulsifier (mono- and diglycerides) 0·4, Reddi-sponge 3·0, total oxidant 70–80 ppm.

Procedure:
Mix all ingredients at low and high-speed to full development.
If the total mixing-time exceeds 12 minutes, withhold the yeast, and add during the last 4–5 minutes of mixing.
Adjust the water and jacket temperature for a final dough temperature of 70–75°F (21–24°C).
Allow the dough to relax for 10 minutes.
Divide, round, give minimum overhead proof, and mould into shape.
Freeze in a blast-freezer at −10°F (−24°C) or lower.

Liquid nitrogen blast-freezing reduces the temperature of items very quickly, and in this way they retain their shape and structure without becoming soggy, retaining maximum freshness on thawing. After blast-freezing, the trays or entire racks of products are transferred to adjacent cold-stores, making the blast-freezer free for another batch. This type of operation is designed for long storage periods, ideal for the large-scale frozen-dough corporations who distribute on a wholesale basis coast-to-coast nationwide, through brokers, franchises or frozen-food distributors. With this scale of operation, tight product inventory dating and stock rotation is essential.

The products pass under a nitrogen spray, which evaporates on contact. Fans are used to move the cold evaporated-gas flow in a countercurrent direction. A state of equilibrium at −18°C is achieved by retaining the products in the tunnel. Tunnel lengths of up to 12 m, and bent widths of 1·8 m, with throughputs of up to 1000 kg/hour are considered economic.

Chemically leavened products containing baking powder mixtures can destabilize during the freezing cycle; in such cases the sodium bicarbonate component must be coated. Bake-off or in-store operators who purchase frozen-dough products for storage and subsequent proof and bake-off, should follow the supplier's recommendations concerning storage temperature, as this is critical. The average optimum temperature range is −18 to −20°C, but some manufacturers regard −15°C as the optimum level. Therefore, products should not be stored at −26°C with ice-cream. A box of dough is similar to a carton of ice-cream in that it thaws from the outside, but unlike ice-cream it contains a living organism, yeast. Yeast enzymes are sensitive to temperatures approaching −30°C, and become inactivated at −40°C. On the other hand, yeast begins to ferment when the storage temperature rises much above −18°C, and once the defrost process starts it must not be reversed. The bake-off operator must also take care to use his stock of frozen products in strict rotation, and not allow older supplies to become buried in the freezer. Apart from conventional in-store bake-off operations, retarder/proofing facilities are available to suit any retailer, however small his initial turnover. The diverse choice of retarding and/or proofing equipment can control retard modes from 2 to 4°C, −5 to +4°C, −10 to +4°C or −18 to +4°C. At the higher temperature retardation-ranges, no defrost sequence is built-in, but from −5 to −18°C defrost and recovery sequences are required. Final-proofing temperatures are adjusted as required from 20–43°C.

More detailed information concerning frozen-dough techniques and bake-off equipment, which is a very important sector of market growth for the baking industry worldwide, is discussed in Chapter 3.6, 'Dough and bread preservation'.

A somewhat rare bread variety for the North American continent is the San Francisco sour-dough French bread, which has been made in the San Francisco region of North California for over 100 years. Sour-dough French bread is made only from flour, water, salt and the all-important starter-sponge, which provides both the leavening and the souring action, which is maintained commercially by rebuilding every 6–8 hours. Surprisingly little has been published concerning the mechanics and microbiology of this dough system. However, Kline, Sugihara and McCready (1970 and 1971) have made considerable contributions to knowledge concerning both aspects of this dough system. Their research work has identified the active yeast type as *Saccharomyces exiguus*, and the sour-dough bacterium as *Lactobacillus sanfrancisco* (suggested name). In this system, the yeast and bacteria coexist within the pH range 3·8–4·5. The bacteria utilize maltose as a source of carbon, whereas the yeast does not, the two microorganisms living in symbiosis.

Instead, the sour-dough yeast utilizes the flour sugars, glucose, fructose and glucofructans and also sucrose. Since the sour-dough *Lactobacillus* only utilizes about 56% of the total maltose, the rest contributes to crust-browning. The yeast does not contribute at all to the acidity of the system, the *Lactobacillus* producing the lactic and acetic acids.

The flour used for preparation of the sour or mother-dough is a high-gluten Montana spring wheat flour of 14·0% protein (as is). In practice, the starter or mother-dough used is one carried through from successive transfers from previous batches, such transfers being carried out with the following formulation: Previous sour-dough 100 parts, high-gluten flour 100, water 53. Incubation is for 7–8 hours at 80°F (27°C), during which time the pH falls from 4·5 to 3·9. Sour-dough starters should be stored at 10°C or less between transfers.

The final bread dough is prepared as follows:

Bread patent flour of 12% protein (as is) 100 parts, water 60, salt 2·0, starter or mother-dough (as prepared above) 18.

Procedure:
The final dough is given a floor-time of 1 hour after mixing.
After scaling and moulding, the dough-pieces are either proofed on canvas or in perforated metal-pans for 7–8 hours at 86°F (30°C), during which time the pH

value of the dough drops from about 5·0 to 4·0. Baking is for about 45 minutes at 375°F (190°C), using low-pressure steam bled in for 50% of the baking-time. Suitable culture media for the propagation of the sour-dough yeast cake (B-98), and bacterial cake (B-1) in pure culture are as follows:

Sour-dough yeast: 100 ml of substrate
Glucose 2·0%, 1·0% trypticase, and 0·5% yeast extract. Inoculate and allow to grow for 18–24 hours, centrifuge and wash with sterile 1·0% saline, using the packed cells (cake) immediately.

Sour-dough bacteria: 100 ml of substrate
Glucose 2·0%, trypticase 1·0%, yeast extract 0·5%, monopotassium phosphate 0·6%, ammonium citrate 0·2%, sorbitan monooleate 0·1%, sodium acetate hydrate 2·5%, magnesium sulphate 0·06%, maganese sulphate 0·01%, ferrous sulphate 0·003%, agar 1·5%.
Prepare by adding 100 ml of distilled water, and 0·1 ml of acetic acid. Heat with agitation, and boil for 2 minutes, use without autoclaving, and cool to room temperature.

By comparison with conventional dough and bread, the French sour mother-dough, final bread dough and bread contains approximately 50% of total acidity as acetic acid and 50% as lactic acid, whereas, the conventional dough and bread contains 60–75% acetic acid and considerable amounts of propionic acid.

The United States baking industry's 17 000 commercially operated bakeries produce some 20 million loaves of bread a day, apart from large amounts of cakes, pastries, pies and cookies, as well as such specialized items as bagels, croissants, pizza crust, high fibre and dietary breads. Required outputs have necessitated an ever-increasing process mechanization and automation.

Sophisticated marketing techniques, and revolutionary methods of product distribution, such as the in-store bakery, found in the supermarket chains, and franchised cookie shop bake-offs have all contributed to the application of basic science and new technologies increasingly within the industrial corporations. The total current size of the market for wholesale, retail and in-store operations has now exceeded $30 billion per year, and growth continues at the time of writing.

Cereals and baking research in the USA is carried out both on a national and on a Federal-state basis. The USDA Agricultural Research Service, has the Northern Utilization Research and Development Division located at Peoria, Illinois 61604, and also the Western Regional Research Center, 800 Buchanen Street, Albany, California 94710. Research is also carried on at Fargo, North Dakota, in collaboration with North Dakota State University, Department of Cereal Chemistry and Technology. Non-state organizations which often sponsor research and national conferences include: Great Plains Wheat Inc., and affiliated state agencies, Millers' National Federation, National Association of Wheat Growers, state wheat commissions, Western Wheat Associates, USA, Inc., and affiliated state agencies.

In addition, the large baking corporations sponsor research at the American

Institute of Baking, located at Manhatten, Kansas 66502, which also organizes bakery educational courses for the baking industry worldwide, on a non-profit basis. The Institute is also in close proximity to the Department of Grain Science and Industry at Kansas State University, the USDA Grain-marketing and Research Center, and Wheat Quality Council. The KSU offers the only fully degreed grain science programme in the world, and attracts many international visitors and students.

The American Society of Bakery Engineers, based in Chicago, Illinois, offers production-orientated members contact with bakery technologists internationally, annually, thus allowing the continuous exchange of new ideas, technical information, and machinery updates. At the annual meetings, specialists give papers on diverse aspects of the technology of baking. The American Association of Cereal Chemists (AACC) based at St. Paul, Minnesota, offers professional membership to the cereal chemist worldwide. The association's main journal, *Cereal Chemistry*, publishes research papers monthly, submitted by corporate and non-corporate members on topics of interest to the cereal-processing and baking industry generally. The AACC publishes monographs and textbooks written by leading specialists, and organizes an annual convention in the USA for members and affilitates.

There are also a number of independent consultants who offer their experience and technical services to the baking industry and ingredient manufacturers. This deep reserve of technical expertise and flow of information, has placed the American baking industry in particular, and the American food industry generally in the forefront of technological innovation for many decades.

For the baker, quality control begins with the basic ingredient, flour, and in the USA flour standards and processing parameters can be summarized as follows.

Standards requiring definition for the fulfilment of a buying contract include: flour strength, flour tolerance and quality, protein quality and starch quality. The objective is accurate prediction of baking results. Typical US standard flour specifications for function are:

Chemical analysis:
Protein $11.5\% \pm 0.2$, basis $14.0\%$ moisture
Moisture $14.0\%$ maximum
Ash $0.46\% \pm 0.02$
Functionality:
Agatron colour 57–60, amylograph $450 \pm 50$ BU, falling number $250 \pm 20$ s, gassing-power $470 \pm 30$ at the fifth hour; Farinograph: absorption $62\% \pm 1.5\%$, peak time $7.0 \pm 1$ min, stability $13 \pm 1.5$ min, protease and amylase activity, damaged starch, functional standards for isolated gluten (using Glutomatic or other gluten-washer) and testing and setting functional performance standards.

In general, recognized methods of testing are those included in *Cereal Laboratory Methods* published and constantly updated by the AACC.

Important parameters for bakeries using pre-ferments are: pH and titratable acidity. For measurement of baked product quality, the Instron, Penetrometer, Volan-Stevens and Bakers Compressimeter can be utilized. Other physical testing on doughs are often carried out with the Brabender Extensograph, Swanson Mixograph, and possible use of the Chopin-Alveographe is being studied. The speed and versatility of near-infrared spectroscopy (NIR) has found wide application in the milling industry, and in soybean and barley processing, for the analysis of moisture, protein, fibre, and ash. More recently, it has also been applied for categorizing wheat varieties into hard and soft. In the baking industry, NIR has been used for the measurement of moisture in white-bread dough, white bread, crackers, cake mixes, doughnuts, biscuits, snack-food, and raw materials, e.g. egg blends, milk-powders, and other components such as total sugars and sodium stearoyl lactylate. Dough taken from the hopper with a minimum of handling, and placed in the standard reflectance cup of a Gardner/Neotec 3000 or 7000 unit will measure dough moisture within the range 40·01–41·93% as the first derivative at wavelength 2045, with a standard error of 0·43 within 1 minute. This compares with 6 hours laboratory time, using a vacuum oven at 100°C to constant weight, and shows a correlation coefficient of 0·88 with the reference method.

Baked bread is judged according to the following characteristics: volume, colour of crust, symmetry of form, evenness of bake, character of crust, break and shred, grain, colour of crumb, flavour, aroma, texture, mastication, and crumb softness, each characteristic being assigned a maximum of 10 points. Other items quantified are: bread weight, height, number of slices and specific volume. Packaging criteria such as: loose, tight, open-bottom, end labels off-centre, are often also scored in the case of sliced and wrapped bread, and slicing defects such as rough or wavy penalized.

Canada can, with justification, be described as the world's most consistent exporter of quality wheats. All classes of wheat grow on the Eastern and Western Prairies and are renowned for their consistently good quality from one crop year to another. This is due to the successful collaboration between breeder and producer, and the handling and grading of the grain, monitored by the Canadian Wheat Board. This authority thoroughly tests all crop classes of wheat, and publishes yearly *Bulletins*, which include quality data on wheat, flour, and bread. Comparisons are also made with both previous crop years, and average data presented covering the previous 10 years quality profile. On the world market, the following wheat classes are offered:

—Canada Western Red spring wheat (CWRS), a hard summer wheat with top quality milling and baking properties, which is available with various guaranteed protein contents. The 1986 leading variety was Katepwa, comprising 34·4% of the crop, with Neepawa in second place at 32·5% for the summer wheat crop. A total of 25·22 million tonnes were planted on 11·44 million hectares. Average data for CWRS 1 for the 1986 crop year were: kg/hl 82·0, protein 15·8% in dry matter, Falling number 380, flour yield 75·4%. Milled flours show wet gluten 39·0% at 0·54% ash in dry matter, and test-bake water absorptions of 64%, bread volume being on average 855 ml with consistently good external and internal characteristics.

—Canada Western Red winter wheat, is a hard red winter wheat, which is suitable for a diverse range of products from French and other specialty-breads to certain types of noodles.

—Canada Prairie spring wheat, is a new class of summer wheat of medium hardness, protein content and gluten strength. This is also a versatile wheat, suitable for French, other specialty-breads, crackers and Eastern bread varieties.

—Canadian Utility wheat, is a hard red summer wheat with a particularly strong gluten, suitable for blending with weaker varieties, and for panned and hearth-breads. Protein content is about 14·0% in dry matter, Falling number 300+, wet-gluten 33·0% for a 75% flour extraction at 0·70% ash, test-bake water absorption being 61%, yielding a bread volume of about 600 ml.

—Other classes are either durum, or soft wheats for non-breadmaking applications.

For the Canadian home market an average mill grist (wheat-mix) would contain about 13·6% protein on a 14·0% moisture basis. Extraction rates for white flours are limited to about 75%, since the quality of the flour from high-grade wheat slowly falls off between 65% and 70%, and rapidly at 70–75%. A straight-run or straight-grade flour is a blend of all the flour streams produced in the mill, the target ash level being 0·50%, and the flour must also meet a colour specification, measured on a colour grader, an optical instrument. A straight-run flour would have a protein content of about 13·0%, on the above basis.

Apart from straight-run grade, there are several others which are prepared from four primary blends. No. 1 blend consists of 60% of the total millstreams, No. 2 blend 30%, No. 3 blend 5% and No. 4 blend 5%. Often, various additives are incorporated into the blends for bleaching, maturing and enrichment. For bleaching, 50 ppm benzoyl peroxide is added to the flour stream with a carrier in powder form. Maturing is achieved by adding potassium bromate at between 5 and 25 ppm for the home market, or up to 75 ppm for other markets. In addition, either chlorine dioxide at 0 to 20 ppm, depending on flour grade, or azodicarbonamide at up to 200 ppm are added. The use of these additives is self-limiting, and over-dosage produces adverse baking results. An average Canadian baker would have a hard time, using conventional systems, to utilize flours containing bromate levels of 60 ppm.

Most flours in Canada are enriched with three vitamins and iron. Legal requirements, 1 lb (454 g) of enriched flour must contain: thiamine 2·0–2·5 mg, riboflavin 1·2–1·5 mg, niacin 16–20 mg, and iron 13·0–16·5 mg. The primary blends are fed with a mix of the four nutrients according to legal specification. Powder feeders are used to meter all additives apart from chlorine dioxide, which is a gas. After thorough mixing in agitators, and final sieving, the primary blends can become commercial flours, but more often they are cross-blended in various combinations to form various commercial flours. Primary blends can also be divided by valves in the flour streams, allowing the use of either a whole stream or only a part of it. These various percentages are known as 'divides', and allow the miller more flexibility in achieving target levels for ash and colour-grade.

When two flours are produced from the same run, the better grade is known as a 'patent', and the lower grade as a 'clear'. A commercial flour for bread production could be made up of the following primary blends: medium patent comprising primary blend No. 1 and No. 2, amounting to on average 90% of total flour produced on a mill run. Such a flour would be about 12·5% protein with an ash of 0·44%. Alternatively, it could be a long patent comprising primary blend Nos. 1, 2, 3, plus 1·6% of primary blend No. 4, amounting to an average 96·6% of total flour produced on a mill run. In this case, the flour would be about 13·0% protein with an ash of 0·47%. By comparison, a first clear comprising primary blends Nos. 2, 3, and 4, amounting to 40% of the total flour produced on a mill run, would be suitable for mixed rye/wheat bread production. The remaining 60% of this latter mill run would then be a short patent or primary blend No. 1 with an ash content 0·36%, and protein 11·9%, which is suitable for household use.

A straight-run commercial flour for bread production would be made up of primary blends 1, 2, 3, and 4, representing 100% of total flour produced on a mill run. Such a flour would have an ash content of 0·50%, and protein 13·1%.

If the miller wants to produce a baker's long patent and a short patent for household use, he would take primary blend No. 1, comprising 25% of total flour produced, with an ash content of 0·36% and protein about 12·0%, as the household flour. The baker's long patent would then consist of 35% of blend No. 1 and No. 2, making up about 65% of total flour produced. Thus producing a flour of 0·47% ash and 13·1% protein. Blends No. 3 and 4 being used to produce a 2nd Clear flour, amounting to 10% of total flour with an ash content of 1·04% and 16·1% protein, which is used for general industrial use.

Flours milled from high-grade wheats are generally deficient in alpha-amylase activity, therefore, malted barley flour is added as a source to act on the damaged starch to ensure an ample supply of fermentable sugars. The addition of all additives are under laboratory supervision, the baking performance of any flour depending largely upon the grade and class of wheat from which it was milled, protein functionality, and appropriate maturing and malt treatments.

In common with most industrial plant bakeries in the USA, the Canadian industrial bakeries still prefer the results obtained from conventional sponge-and-dough systems. The small baker traditionally uses a straight dough procedure, but additional use has been made of the advantages of whey/cysteine/bromate systems coupled with mechanical dough development.

Work done at the Grain Research Laboratories, by Kilborn and Tipples (1968), succeeded in producing Canadian bread from mechanically developed doughs comparable with that of the sponge-and-dough process on a laboratory scale. Physical properties, flavour and aroma were judged indistinguishable from those of traditional sponge-doughs, when using flours of protein contents 12·8% to 16·0%, and North American rich-formulations with high specific loaf volumes (5·0–6·6 ml/g). Additionally, straight-grade flours of 75% extraction were used, with protein ranges 12·8–13·7%. All flour grades are derived from Canadian Hard Red spring wheat. Typical Canadian formulations and processing schedules are given in Table 49.

TABLE 49

| | Sponge-and-dough method | | Straight dough (small bakers) |
|---|---|---|---|
| Flour | 70 parts | 30 parts | 100 parts |
| Yeast | 2·0 | — | 3·3 |
| Salt | 0·2 | 2·2 | 1·9 |
| Ammonium phosphate | 0·14 | — | 0·1 |
| Sugar | — | 5·0 | 3·0 |
| Shortening | — | 3·2 | 2·3 |
| Milk-powder | — | 2·0 | 2·0 |
| Water | 44·0 | 20·0 | 60·0 |
| Malt syrup 250 Lintner | 0·3 | 0·1 | 0·2 |
| *Variable additives:* | | | |
| L-cysteine/bromate/whey | — | — | 0%, 1% or 2% |
| Potassium bromate | | 30 ppm | 10, 20 or 30 ppm |
| Fungal enzyme | 0%–0·6%[a] | | |
| *Processing conditions:* | | | |
| Mixing | 2·0 min high-speed type mixer | 11·0 min horizontal-bar | 15·0 min McDuffee 60 rpm |
| Dough temp. | 77–79°F | 84–86°F | 86°F |
| Fermentation | 4·5 h | 20 min floor-time | 1–2 h |
| Intermediate proof | | 10 min | 25 min |
| Pan proof | 70 min (100°F) | | 55–60 min |
| Baking 420°F | 30 min, 1 lb unit | | 35 min, 1 lb unit |

[a] Proteinases of fungal or bacterial origin act on the intramolecular peptide bonds of the gluten proteins, resulting in reduced mixing times, and improved dough extensibility. Most widely used are papain and bromelin, both of plant origin.

For mechanical development, the following formulation has produced comparable results: flour 100 parts, yeast 3·0, salt 2·4, ammonium phosphate 0·1, sucrose 4·0, shortening 3·0, skim milk-powder 2·0, malt syrup 250 Lintner 0·08, water 63·0, potassium bromate 40 ppm, potassium iodate 15 ppm (Kilborn and Tipples, 1968).

In Western Canada, which includes the provinces of Ontario, Manitoba, Saskatchewan and Alberta, the culture and language is English. The panned white bread in these provinces, and many hearth and variety breads, are similar to those produced in the USA. Maple Leaf Mills Limited has a production enterprise in three provinces, employing between 1000 and 2500 people; Ogilvie Mills Limited is also represented in three provinces and employs over 5000 personnel. Another Canadian group is Robin Hood Multifoods Limited, which has a large manufacturing unit in Montreal, also employing over 5000 workers.

The American Corporation is represented by General Mills Canada Limited in Ontario, operating a general food and grocery division. In the province of Quebec, in addition to the presence of Maple Leaf Mills, and Ogilvie Mills, there are a number of smaller flour mills, e.g. Farines Phenix Flour Limited in Montreal, and La Meunerie du Moulon Bleu Inc. at Assomption. The culture and language of

Quebec is predominantly French, and baked products in the province retain their traditional French character. Canada's leading manufacturer and distributor of baked goods is Weston Bakeries.

During 1987, Maple Leaf, the Canadian flour and bakery business was purchased by Hillsdown. This is Hillsdown's largest-ever take-over move, and follows a small but significant move into the British baking industry with a take-over of the Kara company. This constitutes 50 new business acquisitions during 1987, representing an investment of £600m, and a return of £35m from the sale of small business assets.

With the advent of closer trading relations with the USA, increasing merger activity in the baking and food industry will continue during 1989 and beyond to 1992.

Canadian companies are also very active in the bakery ingredients industry. Ogilvie Mills Limited, from their mills in Montreal, Quebec, market such products as: 'Aytex' starches, 'Provim' vital wheat gluten, 'Bioblend' vitamin/mineral blends, 'Honi-bake' dry honey replacement, 'Dri-mol' dry molasses amongst others which are marketed in the United States through six outlets from coast to coast. The largest US milling corporation, ConAgra Fibers Division, market specialty fibres under the brand name 'Canadian Harvest' from Ontario. Lallemand of Toronto market a wide variety of yeast products such as: compressed, liquid, active dry, Instaferm, and malt products; mineral yeast food 'Yeasko'; fermentation buffer for pre-ferment pH buffering; rye flavour and pan-release agents.

The most successful sector of the North American baking industry is franchising, which is by no means a new concept. In 1948, when the MacDonald brothers started their success, doughnuts were already being sold in franchised bakeries. The continued success of franchising is a direct function of the fast-moving society, the growth of the city malls, and city centre development, providing high-traffic catchment areas for franchises. Even in high-rent areas, enough sales can be generated from a $500\,\text{ft}^2$ site to ensure success, typical products being: muffins, cinnamon buns, sweet and savoury bagels, croissants filled with ham or cheese, or fruit and cream, white bread and buns, and doughnuts. Hillsdown Holdings, a UK-owned company, now controls Buns-master, the most successful bun franchise on the North American continent, with more than 120 outlets. In the Pacific North-West and Canada, Hillsdown's franchises revolutionized sandwich-making with the introduction of the giant kaiser-bun. Where product lines are limited, relatively unskilled labour is used, the baking process being simplified by corporate guidelines for personnel, involving standardized times and temperatures. Muffins, the peaked-bun, is a popular franchise item, since it is versatile in presentation, and undemanding in equipment. They can be made without sugar, allowing the production of both sweet and savoury items. Classic plain muffin varieties are: cornmeal, wholewheat, bran, blueberry, and banana nut; dietary varieties include: oat-bran, sugar-free bran, blueberry, prune-bran and sugar-free. The oat-bran was the top-selling variety during 1989, owing to its medical research link with serum cholesterol reduction. Plain varieties, such as cornmeal can be filled with cheese, carrot and tomato, or cheese and onion, making them a breakfast and lunch variety. Being formulated with fibre, natural fruit sugars, and canola oil, muffins have become the priority item with the yuppie-cult.

Following a dramatic sales increase of muffins in 1976, they have shown continued growth, owing to increased home consumption and their general acceptance in the fast-food sector. The texture of the average muffin is tough and chewy and has medium to large size holes, honeycombed throughout. The flavour is bland with acidic traces, and the product is round and flat in shape, about 10 cm in diameter and 2·5 cm in height, weighing about 56 g. The side walls are straight and light coloured, the edge between the flat darker brown top crust rounded rather than sharp. The product is grilled on one side, turned over, and grilled on the other side. They are made using a sponge-dough, straight dough or brew process. Equipment has been developed to automate production, producing 250 to 300 units per minute. Such systems, however, demand special formulations and processing procedures with strict control of processing parameters. High-absorption doughs for automated handling (but fluid enough to flow out in the griddle-cup and provide the steam-vapour leavening at the crucial time in the griddle) are essential. Short-time straight dough procedures have been developed to satisfy these requirements, and most high-speed systems now use them. Muffin doughs are basically cold, slack doughs with very short fermentation times, proofed in a hot, semi-humid climate, and baked on a griddle. A premix base has been developed for a no-time dough by Thompson (*Proc. Am. Soc. Bakery Engrs.*, 141, 1981), requiring the addition of only flour, water, and yeast by the baker. The flour should be a high-protein spring wheat, or spring/winter wheat patent blend, or straight grade, with a protein content of 13%, basis 14% moisture. A high protein quality is also necessary to carry a high water-absorption of over 80%, and to provide a gluten network to retain the gas during proofing, and in the griddle, also to ensure the essential chewy texture. This can also be supplemented with 1–2% vital wheat gluten, on a flour weight basis, each 1% gluten increasing the protein content of the flour/gluten blend by approximately 0·6%. Since leavening is a combination of fermentation gases and steam liberated by water vaporization in the griddle, sufficient free-water must be available to produce the open porous structure or blow-holes. Water absorption ranges are often 83% to 87%, based on flour weight. In order to achieve this level, and ensure that the dough can be handled in the divider and rounder, the dough must be very cold on discharge from the mixer (20–21 °C); if it is only a few degrees higher, the dough will be sticky on make-up, and become soft and sticky in the proofer, making depositing in the griddle cups a difficult operation. The correct dough temperature is obtained by using ice-water and mixer refrigeration. The water is added in two stages, followed by the salt, thus allowing efficient hydration, and avoiding prolonged mixing-times. Typical formulae for a straight dough and a no-time dough are the following:

*Straight dough:*
flour (13% protein) 100 parts, water 80, yeast 1·0, salt 1·75, sugar 2·25, shortening 1·0, calcium propionate 0·625, fungal protease 1·0 (Noel, Sr., *E. M. Proc. Am. Soc. Bakery Engrs.*, 128, 1971).

*No-time dough:*
flour (13% protein) 100 parts, water 76, vinegar 4·0, yeast 6·5, salt 1·75, sugar 2·0, shortening 1·0, calcium propionate 0·5, fungal protease 1·0 (Jackal, S. S., *Bakery Prod. Marktg.*, **18**, 2, 115, 1984).

The dough is scaled at 2·25 ounces for 2-ounce units, using special dividers for muffin production, or conventional roll or bun dividers. The scaled and rounded dough-pieces are dusted with rice flour and fed into an overhead proofer at 43°C, where they remain for about 30 minutes. At the proofer discharge end, they are inverted by a mechanism, so that the dusted bottoms come to rest in the griddle-cups, which travel on an endless conveyor through the oven. As they enter the oven, a mobile cover assembly places a lid on the cups to retain the steam and leavening gases for maximum expansion. At the oven discharge end, the muffins are turned over onto a mobile flat hearth for grilling on the other side. On receiving the dough-pieces, the cups are at about 136°C, which is maintained for about 2 minutes, after which the temperature is raised to 200°C for the remaining 4 minutes to complete the grilling of the bottom surface. On transfer to the lower hearth, the muffins are again exposed to 200°C for about 3 minutes. The effective grilling-time is 7·5 minutes, whereas the total residence-time on the griddle, including warm-up time, is 9 minutes. On completion, the muffins are coloured on the top and bottom surfaces, and the sides remain white and firm. The grilled muffins are cooled on a conveyor for 1 h, and on reaching ambient temperature are ready for packaging, after being sliced horizontally.

The other important franchise item is the hamburger bun and hot-dog roll, which are made from formulations similar to those for white bread, but with shortening and sweetener enrichment. Sweetener levels range from 10% to 12%, compared with 7% in white bread; and 4% to 6% shortening, compared with the 2–3% in panned-bread. Additional ingredients included are: dough strengtheners, oxidants, crumb-softeners, yeast-food, and mould-inhibitors, in accordance with good manufacturing practice (GMP). A typical soft roll formulation is the following: flour 100 parts, water 60, sugar 10, shortening 5, non-fat dry milk 2, yeast 3, salt 2.

For large-scale production, a liquid pre-ferment with continuous mixing is preferred, but for optimal results 50% of the flour, at least, must be included in the pre-ferment. This ensures good volume, strong crumb, and lower dough temperatures. Otherwise a straight dough or sponge can be utilized, with appropriate adjustments.

Liquid pre-ferments, after fermenting at 30°C, must first be cooled to 15–18°C through a heat exchanger, before entering the incorporator so that the final dough-temperature from the mixer is maintained. Bun production is now a fully integrated and automated process, comprising: divider, rounder, overhead proofer, moulder and panning unit, a standard design being the Pan-o-mat designed by Winkler. This original design, dating back to the 1950s was used in conjunction with a conical moulder, small-unit divider and travelling intermediate-proofer. The Winkler machine orientating the dough-pieces four abreast, flattening them through a roller, depositing them at a lower level into the pockets of the baking pans, which accepted 24 units/pan. These then passed into the final-proofer, and ultimately into the oven. Hamburger or soft-roll doughs should be worked off in a slightly under-developed state to allow for any energy input resulting from mechanical processing before final-proof. Flour used for bun and roll production, should have a protein content of 12–13% (14% moisture) and contain a high proportion of spring wheat to ensure bloom and colour in the product. In the case of hearth roll varieties, the dough-pieces emerging from the intermediate-proofer pass into a roll stamping or cutting machine, imparting the desired design, such as the star or kaiser roll impression. The Winkler Derby, available as a single machine, or as a semi-automatic or automatic roll and bagel production plant, has a maximum capacity of 38 000 dough-pieces/h, scaled at 1–4 ounces (28–112 g), and 2–7 ounces (56–196 g). The unique design is suitable for kaiser rolls, tortillas, bagels, hamburger buns, pittas and raisin doughs, and is marketed in the USA, the UK, and continental Europe. This unit provides for manual feeding or automatic feed, accurate divider, efficient round-moulding, and conveyor to final-proofer. These machines are available as 4, 5, 6, or 8-pocket models, corresponding to capacities of 8000 to 16 000 h.

In franchising operations, the use of pre-mixes limits errors, and provides consistent results.

Another expanding franchise of long standing is the doughnut, the daily doughnut and coffee being part of the American way of life. Modern production units are compact and can be operated by one person, carefully controlling water addition, temperatures, mixing-times, and recording mix absorption. The frying fat has tight specifications, and is filtered daily, product standards often being unmatched by skilled bakers from in-stores and retail outlets. New varieties of doughnut include: a lighter calorie-reduced product and oat-bran and wheat-bran varieties, which contribute to a healthier diet. Doughnut complete premixes constitute about 16% of all complete mixes utilized in supermarket combination in-store bakeries ('combos') in North America. Combination in-stores make use of the scratch, the premix, and the frozen-dough/batters for bake-off, techniques, thus maximizing the use of both skilled and unskilled labour on a cost-effective basis. Glass panels have also been introduced, as in the French hypermarkets, to reveal glimpses of the bakers at work.

Doughnuts are of two types: cake doughnuts and yeast raised doughnuts. Their quality, however, depends as much on the quality and cleanliness of the frying medium (fat) as on the ingredients used in their formulation. Doughnuts can be either formulated from scratch, weighing out the individual ingredients, or prepared by the use of prepared premixes or bases, requiring only flour and water and optional minor ingredients, or prepared from commercially available complete premixes, requiring only water to prepare. The use of premix bases or complete premixes offer complete uniformity and economy in labour. Cake doughnuts are made from batters similar to layer-cake, being cut into ring-shaped pieces weighing from 0·5 to 2 ounces (14 to 56 g). These are deposited into shortening heated and maintained at 183–193°C for about 2 minutes until cooked through. The finished product must have a golden-brown exterior colour, not over-browned, have a crisp, dry crust, and have an internal structure resembling a baked product, about 25% residual moisture eventually migrating into the crust on storage making it softer. A representative example of a complete doughnut mix is the following: cake flour 65 parts, bread flour 35 parts (total 100 parts), fine granulated sugar 40, dextrose 3·50, non-fat dry milk 10·00, egg yolk solids 7·0, defatted soy-flour 7·0, shortening 5·75, salt 1·50, mace 0·75, sodium bicarbonate 1·70, sodium acid phosphate 2·30, vanilla to taste (source: Wheeler, F. G., and Stingley, D. V., *Cereal Science Today*, **8**, 120, 1963). The batter is prepared by mixing an appropriate amount of water to prepared mix at a temperature within the range 24–26°C to avoid final product variations. Mixing-times are limited to 1–3 minutes to prevent excessive toughening of the batter, and inadequate fat uptake, whereas under-mixing produces friable batters with excessive fat absorption. This phenomenon is due to the degree of protein hydration, a 50% reduction in mixing-time increasing fat absorption from 1·3 ounces (36·4 g) to 2·4 ounces (67·2 g) per 12 units. An average rate of fat absorption for 14 ounces of dough (12 doughnuts) is 1·65 ounces/12 units, or 392 g dough absorbing 44·8 g fat. A floor-time of 10–15 minutes normally gives adequate batter hydration, and leavening-system activation, but this period is crucial, and must be optimized to reduce such defects as poor volume, product symmetry, excessive fat absorption, or split crusts. The batter then proceeds to the hopper, from which it is deposited in preset amounts by the depositor, which is gravity-fed, air-pressure operated, or of the vacuum type. The first two designs depend on the gravity flow from the hopper into feed-tubes, fitted with plungers, the lower end of which acts as a cutter. A sequenced opening and closing of the cutters extrudes the batter-rings for deposit into the hot fat. In the case of pressure-head depositors, a rotary gate forces the batter from the hopper into a chamber, which serves as a manifold for more depositor-tubes. The batter is forced into the tubes by air-pressure, from which it is extruded as the cutters open and close in a synchronized cycle. Doughnut weight is controlled by the air-pressure adjustment, the duration of cutter remaining open, the dimension of the opening, and batter viscosity. With the vacuum-mechanical type head, metering of the batter is volumetric. The individual plungers create a vacuum, drawing the batter from the open hopper into the cutter cylinders. On accepting a predetermined amount of batter, the cylinders become closed off, the measured batter is then extruded in ring form through the preset opening. This volumetric plunger/cutter head provides good accuracy in controlling the volume of batter extruded. In addition, a double-chambered doughnut extruder is available, which allows the rational production of filled ring-shaped cake doughnuts. The extruder being made up of an annular-shaped chamber, set within the batter chamber, serving as a reservoir for the filling. This second chamber is provided with flanges to direct the filling flow into the surrounding batter on extrusion. The proportion of filling to batter, and weight/volume of the doughnuts, are adjustable within limits.

The rings of batter are deposited directly into the hot frying medium, in which they remain for 1 to 2·5 minutes, depending on the size and variety, and the temperature of the fat medium, which should be maintained within the range 183–193°C with the help of the independent controls at the feed and discharge ends of the fryer. When deposited into the hot fat, the doughnuts sink below the surface, but should rise again to the surface within about 3–6 seconds provided the leavening level, and state of proof is correct. On deposit, the raw batter takes longer to fry the initial side, due to the conduction-time; the reverse side, on turning, requiring a shorter residence-time in the fryer. Fat absorption by the doughnuts appears to depend on several factors, associated with the frying medium and the formulation and structure of the batter. A certain amount of fat absorption is necessary for doughnut quality, contributing to flavour and eating quality and extending shelf-life, provided that the frying medium is kept clean and not allowed to decompose. Frequent filtration of the medium, coupled with adherence to a replacement of one-third of the total fryer capacity per day, should maintain the frying medium in a satisfactory condition. Fresh fats do not reach their peak performance until about 48–60 h, depending on their composition, coinciding with free fatty acid (FFA) increases to about 0·35% from an initial value of 0·05%, as a general guideline. This 'conditioning' of the frying medium during use is brought about by the driving off of steam during frying, which also takes with it many volatile reaction products that have strong odours. These are formed at high temperatures and tend to build-up in the kettle. Fat quality is best maintained by filtering at the end of each day's production, removing all burnt particles that contribute to its deterioration. For this purpose, filters coated with diatomaceous earth are found to be most efficient. A weekly clean with a detergent ensures that the kettle itself is kept free of deposits and residues. Although hydrogenated fats are generally considered suitable for frying, vegetable oils seem to be the popular choice. Vegetable oils giving average doughnut absorption levels of 11·75% or 1·64 ounces/12 units (45·92 g/12 units). Properly hydrogenated and stabilized animal fats, with added antioxidants can, however, give equally good performances and exceed the vegetable oils in smoke point and foaming resistance. Hydrogenated lard shows average doughnut absorption levels of 11·9% or 1·67 ounces/12 units (46·76 g/12 units). The most important properties of a frying fat are: a bland flavour, low smoke and foam generation, and clear in appearance, free from burnt residues. A specification guideline for frying fat is the following: free fatty acids (% oleic) max. 0·05%; smoke point min. 400°F (205°C); peroxide value meq/kg max. 0·5; solid fat index (SFI) 10°C 37–43, 21°C 25–29, 27°C 20–24, 33°C 15–17, 40°C 10–12; crystallization rate % cooled to 33°C for 30 minutes 1–7%, 120 minutes 13–16%: cooled to 27°C for 30 minutes 14–17%, 120 minutes 16–19%. Such a specification would ensure that no fat adsorption onto the doughnut surface, due to set-up, could occur before draining was complete.

Yeast-raised doughnuts, being made from a fermented dough rather than a chemically leavened batter, and requiring a final-proof before frying, demand specific formulation, ingredients, processing methods and equipment. A typical formulation for yeast-raised doughnuts is the following: flour 100 parts, water 55–62, sweetener (sucrose and/or dextrose) 10, shortening 12, non-fat dry milk 3, egg yolk 5, yeast 5. This basic formulation is usually optimized by the addition of yeast food, flavours, emulsifiers and dough conditioners of the types generally applied to fermented doughs. Both pre-ferments, including sponge-and-dough, and straight dough procedures can be utilized, but an extended fermentation period gives the dough improved elasticity and proof. The best results are obtained by using a sour-dough starter, added at about 25% to each successive dough. Production procedures vary from semi-automated to fully automated. In semi-automated systems, the dough is first sheeted, and the doughnuts cut out on the make-up table. The dough-pieces are then placed on proofing-cloths or boards for proofing in a cabinet at 35 to 45°C, and 65% RH. The proofed dough-pieces then pass on an automatic conveyor to the infeed-conveyor, which deposits them into the conveyor-type fryer. Fully automated systems utilize an automatic stamping device, made up of a hopper which extrudes a band of dough onto a belt conveyor, feeding the dough into a batch of sheeters, which roll the dough to a preset width and thickness. The dough sheet then passes through the stamper stencil fitted with a set of 6 to 8 cutters. A short conveyor length rises, pressing the dough sheet against the cutters, cutting the dough-pieces from the sheet. The cutter-rings are provided with a vacuum from within, which holds the dough rings, the centres being ejected by the application of air pressure. Simultaneously, the short conveyor length becomes lowered again, transferring the dough trimmings onto a lateral band-conveyor for return to the dough hopper. The cutters, mounted on a track, are then relocated above a proofer tray, onto which the doughnut rings are released from the vacuum-hold, being directly transferred to an automatic continuous proofer. Cutters can be either round or hexagonal in design, ranging from 2·75 to 3·0 inches in diameter. It is claimed that the hexagonal design reduces trimmings from 40% down to 20%, compared to the round design.

It is also possible to produce yeast-raised doughnuts by extrusion, similar to that used for cake doughnuts. This method utilizes either air-pressure or vacuum extrusion techniques. Pressure extrusion is achieved by depositing the dough in a pressure chamber, from which it is extruded through a battery of cutter-dies, depositing the cut pieces directly onto the feed conveyor of a continuous proofer. This technique involves a batchwise refilling of the dough-chamber, and is therefore not completely automated. The vacuum version utilizes an open dough-hopper, permitting a fully continuous operation, providing the hopper is kept filled with dough. In this case, the dough is drawn in by vacuum from the hopper into the cutter tubes, the dies cutting and depositing the dough-pieces directly onto the automatic proofer-conveyor. Although these extrusion techniques are highly automated and labour-effective, flexibility is limited, and stricter control on dough preparation becomes critical. Doughs under pressure become sensitive to floor-times, and when processed too immature ('green'), their response to final-proof is limited, producing products of reduced volume. In contrast, over-mature doughs result in poor product symmetry and dough-piece uniformity. In general, sponge-doughs fermented at 26°C for 2·0 to 2·5 h appear to be best suited to extrusion techniques. This involves mixing the doughs to full development at absorption levels high enough to give adequate dough extensibility. Floor-times can then be limited to less than 5 minutes, and the dough processed over about 25 minutes.

Proofing of yeast-raised doughnuts demands a comparatively dry, warm proofer, maintained within the temperature range 35 to 45°C and 35–40% RH, allowing a slight surface skin to form, taking about 30 minutes to attain optimal proof. The newer automatic-proofers provide zonal control, thus permitting adjustment to a moist zone at the feed-end, followed by a dry zone at the discharge-end, favouring stability and expansion as well as reducing dough adhesion.

The temperature of the frying medium should be kept higher than in the case of chemically leavened cake doughnuts, within the range 205 to 215°C. Residence-times are normally 50 to 60 s for each side, according to size. In the USA it is customary to apply a glaze during the finishing of yeast-raised doughnuts. An example of a glaze formulation is the following: powdered sugar 60 parts, granular sugar 40, hot water 22, gelatine 1·5, corn syrup or honey 10, glycerine 1·5, hard fat flakes (m.p. 54–60°C) 1·0. This should be applied at a temperature of about 40°C, the inclusion of fat flakes minimizes moisture release from the coating after packaging. The application of microwave energy to the proofing process has been researched, since proofing-times can be reduced from 30 minutes to about 4 minutes. However, at this time it has not achieved wide commercial acceptance.

In the USA, most in-store supermarket bakeries were traditionally either producing products from scratch, or utilizing frozen dough/bake-off systems. This situation has now changed in favour of combination ('combo') bakeries, making use of both techniques, employing both skilled and unskilled labour, offering more variety and quality at lower unit cost. Production areas have diminished in prime locations owing to lease costs, and better use is being made of available space by using rack-ovens and freezers. The production area typically accommodates a reversible sheeter, small bread and roll plant, and a spiral mixer, reducing manual labour considerably. By keeping fridge and freezer capacities well stocked, heavy demand peaks can be accommodated. Many in-store bakeries now bake later in the day to meet the heavy demand for fresh products between 16.00 and 18.00 hours, instead of filling the showcases at 09.00 hours. In-store growth is based on fresh products, but the products must also be as near fresh as possible for eating after 24 h has elapsed, therefore formulation, raw materials, and packaging are all important areas for constant research and innovation. A typical combo in-store bakery product programme is the following:

Frozen-dough baked/ready for bake-off:
Pies 6%, danish pastry 5%, bagels 3%, croissants 3%, puff pastry 3%, total 20%.
Complete mixes:
Yeast-raised and cake doughnuts 16%, variety-breads 13%, buns and rolls 9%, sweet products 3%, total 41%.
Frozen/refrigerated batters:
layer cakes 9%, cookies 6%, muffins 4%, total 19%.
Baking from scratch:
white and brown breads 10%, crusty breads and buns 9%, Pullman squares 2%, total 21%.

Most in-store bakeries in the USA feature discounts over a week each month, and wholesale bakeries 20% discounts for a week each month. In general, the in-store bakeries hold the competitive edge for most bakery products.

# 1.8  Measurement and Control Techniques for Raw Materials and Process Variables

## 1.8.1 RAW MATERIALS—CHEMICAL AND MICROBIOLOGICAL

**Flour**

The basic quality and performance potential of any flour depends on the skill of the wheat-breeder, and the miller in formulating his wheat mixture or 'grist', and upgrading it by careful selection and control techniques during the milling process. The process of wheat-breeding takes about 10 years from initial crossing of cultivars to commercial production of a new variety. This work involves close collaboration of the expertise of the plant-physiologist, geneticist, entomologist, agronomist and cereal chemist, not forgetting the valuable advice from the miller and baker at the market end. The method for developing new varieties proceeds in stages. The first step is the establishment of 'true-breeding' (homogeneous and stable) varieties by careful selection and crossing, followed by detailed evaluation of the milling and baking aspects of potential varieties. The final step being one of purification, propagation and distribution of a new variety.

Small-scale screening methods for breeder's samples have been developed for reliable evaluation of the breadmaking potential. The sodium dodecyl sulphate (SDS) sedimentation test has shown the most promising potential to date. This is due to the sample size, sample throughput, and high correlation with baking results, location, and inherent genetic differences in protein quality. The method is similar to the Zeleny test in principle, a 6-g sample of wholemeal being dispersed in an SDS-lactic acid solution in a graduated cylinder, and the sedimentation volume read off after 20 minutes standing. The method is also suitable for evaluating flour quality, being a reliable measure of the gluten colloidal response, and swelling-capacity. Owing to the choice of reagents, it is superior to the Zeleny for protein resolution. Another proven method involves a measurement of the glutenin fraction of wheat-flour proteins. Often referred to as the 'residual-protein' test, it predicts the doughmaking properties of flours. The flour is extracted with 0·05M acetic acid once only using only 2·0 g of flour sample. The 'percentage residual-protein' remaining in the flour correlates highly with Farinograph mixing-stability and resistance to extension tests (maximum resistance). Using this method, wheat-flours can be segregated into their suitability for breadmaking or other applications.

As a practical control method, the measurement of both gluten content and quality are fundamental. However, they cannot be determined in wheatmeal or wheat-flour, since neither contain gluten, but they do contain the protein fractions prolamine and glutelin as storage proteins. Both are insoluble in water but have

306

swelling properties. Only when the flour is mixed into a dough can the two fractions form 'gluten', The gluten content is therefore expressed as a wet or dry percentage of 100 g of the flour. The determination is carried out by preparing a dough from 10 g flour, the gluten is then isolated by washing out with a 2% salt solution for 5 minutes or until the salt solution is completely starch-free to iodine solution 0·001 N. The salt solution should be within the temperature range 18–20°C, and the isolated gluten rinsed in running water to remove the salt. The wet gluten is then weighed, after pressing out the unbound moisture with a gluten-press. The gluten-press consists of two rough-glass-plates (sand-blasted) set in a frame which is hinged on one side. The clearance between the two glass-plates should be about 2 mm. The press is clapped-closed about 5–8 times, moving the gluten piece each time to a dry zone on the glass-plates. The plates are then dried with a cloth, and the operation repeated from 20–40 times until no moisture is visible on the plates. Flour wet-gluten contents over 27% in Europe are considered high, 20–27% average, and under 20% below average. For speed and convenience, the Glutomatic 2100 programmed gluten-washer is recommended. The sample weight is 10 g flour, and the dough is prepared and washed with a 2% salt solution according to ICC-Standard No. 106 using potassium dihydrogen phosphate and disodium hydrogen as buffer. In this case the doughing-up time is only 20 s, and the washing-time 5 minutes. The amount of wash-water can be limited to 300 ml. The gluten is then centrifuged in the Glutomatic centrifuge 2011, at 6000 rpm for 1 minute. The gluten is then weighed, and the amount as a percentage calculated, based on the flour sample weight. As qualitative parameters of the washed-out gluten, the extensibility, elasticity and swelling-capacity must be determined. The extensibility is determined by stretching the ball of gluten until it breaks, and the length of the gluten-strip is measured. Gluten which can be stretched up to 5 cm, is classed as 'elastic and of good extensibility' and designated 1st Class. Gluten which can only be stretched to 2 cm is classed as 'elastic with adequately extensible', being designated as 2nd Class. Gluten which breaks below 2 cm, and has a typical ragged structure is designated as Class 3, having elasticity but deficient in extensibility. Only a small number of samples fall into Class 4, which can be described as inelastic and flowing. The reproducibility of the quantitative determination of gluten by this method is ±0·2%. An experienced operator can carry out the qualitative evaluation of gluten manually with adequate precision. However, various apparatuses have been developed for this purpose. The most recent of these is the Brabender Gluten-tester, in which the washed out gluten is placed between two discs with a fluted-surface. The upper disc is rotated by a constant shear force against the lower disc. The shear-angle resulting from tangential shear force is 0 to 40 degrees, which is translated into a scale reading of 0–1000 Brabender Units onto a horizontal pen and chart recorder. Gluten samples are relaxed for 90 s in buffered 2% salt solution before measurement. The gluten sample is placed in the working-head and deformed until the pen reaches 800 BU, and then released. The curve build-up to the peak at 800 BU represents the response to the shear force and the fall-off corresponds to the relaxation behaviour of the gluten. A 1st Class gluten sample of good elasticity and extensibility shows a rapid rise to 800 BU within 9 s, and a smooth deep fall-off as relaxation. A sample with elasticity and only adequate extensibility, shows a slower rise to the 800 BU line and a shallow steep fall-off during the relaxation mode. Earlier work by Tscheuschner and Auerman demonstrated in 1964 how the Penetrometer AP-4/1 could be used to evaluate gluten and dough properties, as well as other raw materials, thus replacing expensive apparatus import. By this method, the gluten is placed in a round holder, and force necessary to penetrate the gluten-ball taken as an index of gluten quality. The exact procedure is as follows.

A 4-g gluten sample is formed into a ball and placed in a dish of water at 20°C to rest for 15 minutes. The sample is then placed in the centre of the level cylinder. By pressing a button, the pressure-disc is released, and allowed to exert pressure on the gluten-ball for exactly 5 seconds. The pressure is then automatically terminated, and the depth of fall of the pressure-disc read-off on a scale mounted behind glass at eye-level. Since the distance between the pressure disc and the cylinder is scaled at 20·0 mm, which is equivalent to 200 Penetrometer units (PU), the gluten-strength-index is given by:

$$200 - d = Hd$$

where $d$ is the depth of fall with the gluten sample in position. This index ranges from over 85 PU for a strong gluten down to less than 40 PU for a weak sample. Later versions were automated including the Mark PAU-1 of Soviet construction, and the AP-4/2 from the GDR, as well as the laboratory penetrometer III Type OW-204 manufactured in Hungary. These penetrometers were also provided with additional components which enabled them to be used for other measurements on wheat- and rye-flours, including elasticity. The All-Union Research Institute for Grain and Grain-processing (WNIIS), Moscow, USSR, during the period 1967–69, developed a new version, the PEK-3A for measurement of gluten quality. The measurement principle in this case is to exert a programmed 30-s pressure deformation on the gluten ball in the relaxed mode. A timed relay-switch after the 30 s stops the deformation, and the reading can be read off on the instrument scale, graduated from 0 to 120 units, as shown by a pointer which moves from left to right. This measurement is known as the 'Hpek', and represents the degree of travel of the upper pressure disc or plate during the pressure deformation. This value also corresponds to the degree of pressure deformation of the gluten ball in terms of height. This differs from the gluten-strength-index, as determined with the Penetrometer AP-4/1, which measures the height of the gluten ball after deformation. The stronger the gluten, the less the compression, and the value 'Hpek' in scale units. In the case of very strong samples, the value of 'Hpek' can be less than 40 units, and for very weak glutens more than 100 units. For mill laboratories, the 'Hpek' parameter is the most appropriate for wheat gluten evaluation, whereas for factory bakeries the $Hd$ (gluten-strength-index) is preferred, since the Penetrometer can also be used for many other determinations. Between the various physical properties of gluten, viz. extensibility, flow (viscosity), elasticity and resistance to various deformations, there exists a law of proportionality or inverse proportionality. For example, the weaker the gluten, the greater the extensibility and flow, and inversely, the lower the elasticity and its resistance to deformation by stretching, on compression, penetration by projectiles, or extrusion from Plastometers. The correlation coefficient between 'Hpek' and *Hd* is linear $r = -0.988 \pm 0.0034$, which allows the calculation of one value from the other. The physical properties of doughs and the strength of wheat-flours depend both on the quantity and quality (physical properties) of the gluten which forms.

Therefore, the gluten content or protein alone is inadequate to define the quality profile of a flour for breadmaking. Instead, the choice of parameters which define both quantity and quality of gluten in a flour, are generally regarded as being most reliable as a basis for potential flour strength determinations. Although dry gluten contents correlate better with flour protein contents, the time taken to dry samples is not necessary for routine quality control purposes. More important is a detailed evaluation of quality. This is, in fact, carried out in the following countries: USSR, Austria, Poland, Czechoslovakia, Hungary, GDR, Switzerland, Italy, and FRG, as well as in many other countries. A complete gluten evaluation is achieved by awarding points on merit for both quantity (50 points maximum), and quality (50 points maximum), making a possible total score of 100. Dry gluten contents of 6% and less per flour dry matter, receive no points. Gluten contents of between 6·1% and 20% are given an appropriate graduation, but values over 20% receive no more than the 50-point maximum. As quality parameters, either the 'Hpek' range of 116 to 30 units, or Penetrometer *Hd* range 34 to 92 units, are awarded an appropriate number of points from 0 to 50.

The summation of the points for dry gluten content and the quality evaluations give the final assessment. The nearer it is to 100 points, the greater the potential flour strength. This system correlates much better with dough physical properties and baking results than individual values for gluten quantity and quality. The correlation coefficient between 'Hpek' and *Hd* are so high ($r = 0.978 \pm 0.007$) that these values are of equal merit in expressing the potential flour strength (Auerman, 1972). A detailed description of the complete gluten evaluation method described, using the 'Hpek' and *Hd* parameters can be found in *The Laboratory Handbook for Baking Technology* by L. I. Putschkowa, Moscow, 1971, published by Pistschewaja promyschlennost, which includes the appropriate calculation-tables.

Dough constitutes a complex polydispersed aqueous colloidal system, the internal structure of which changes its physio-mechanical properties with time. It shows elasticity, viscosity, plasticity, and the ability to relax with a retardation effect (elastic after-effect). Depending on the mode of deformation, its speed, magnitude, and time over which it acts, dough can behave as an elastic, viscous, viscoelastic, and plastic material.

For this reason, a complete understanding of the structural and mechanical response of dough is of prime importance, both from a scientific and a processing viewpoint.

An important contribution to the early development of the theory and methods of measuring the structural–mechanical properties of materials is due to the work of the Soviet scientists P. A. Rebinder, M. P. Wolarowitsch and other workers. From their basic research, methods involving time-dependent deformation curves on doughs were devised. Measurements being made from a specific deformation kinetic from a defined force during and after its application, the force remaining constant throughout. Various apparatus design evolved, the one due to Weiler–Rebinder was based on the principle of a tangential shear force, applied to the product between two plates (cf. Brabender Gluten-tester described in this section). The apparatus due to Schwedoff and Wolarowitsch uses coaxial cylinders with the test product placed between; that due to Tolstoi uses a dough layer between parallel solid surfaces without any movement of the dough at the interfaces, applying a shear-force. The latter represents a shear deformation in which all layers of the material parallel to a given plane are displaced relative to one another. The planar displacement is defined as the 'absolute-shear', and for small displacements, unit shear is the ratio of absolute shear: the vertical distance of the shear force from the pivot, i.e. the tangent of the angle between them. Hooke's law for shear can be written in the form:

$$\text{stress, } p = G \times \text{angle } a, \qquad \text{where } G = \text{shear modulus}$$

In general terms, when dealing with material such as dough, with elastic, viscous and plastic properties, and which relax and retard under defined conditions, three types of curve depicting the deformation-kinetics are obtained depending on the magnitude of the applied constant stress. When the constant stress applied is less than the lower elastic-limit, the curve of deformation $v$. time shows ideal elastic behaviour, throughout the period of applied stress. On removing the stress, the entire elastic deformation reverts to zero. In the case of structured systems of high molecular weight, like dough, the lower elastic-limit is almost zero, hence a curve of the deformation-kinetics of dough is difficult to obtain.

When the constant stress applied exceeds the lower elastic-limit, and the upper elastic-limit of the material exceeds the lower elastic-limit, the deformation-kinetics of dough take a distinct configuration. Under these conditions, the test material reaches the initial deformation within 1 s, after which the period of delayed elastic deformation begins, and proceeds until the upper elastic-limit is reached and remains at that level until the constant stress is terminated. The elastic after-effect, known as retardation, is a function of the heterogeneous nature of the structure and composition of the material, resulting in the elastic deformation within different zones of the material at different rates. The retardation effect is a characteristic of elasto-viscous materials. On termination of constant stress, the deformation falls abruptly (elastic), and then with deceleration, the delayed elastic deformation. Under these conditions, the deformation can be described as ideal elasticity, since deformation is completely recovered.

When the constant stress applied exceeds the upper elastic-limit of the test material, the elastic deformation commences instantaneously at almost zero time. However, since the applied constant stress exceeds the upper elastic-limit, the curve of the deformation-kinetics at constant stress not only reflects the delayed elastic deformation, but also the constant speed build-up of the irreversible deformation, which proceeds at a constant rate in a straight line. The velocity of this irreversible deformation remains constant for the material under test, and the stress regime. When this point is reached, only irreversible deformation takes place, and the material goes into a state of flow. The total reversible elastic deformation is shown in

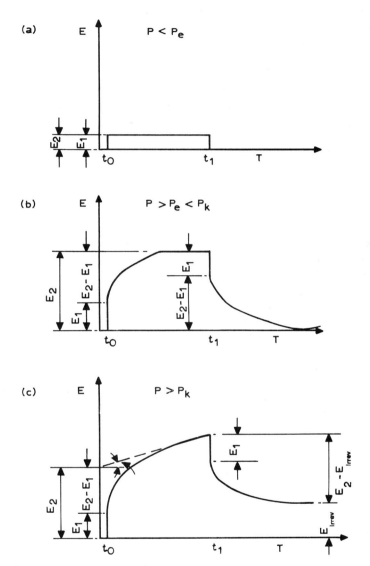

Fig. 31. Kinetics of shear deformation at constant stress, *P*, parallel to the horizontal plane (viscoelastic/plastic props). (a) $P < P_e$. ($P_e$ = elasticity threshold) ideal elastic behaviour, deformation *e* reached instantly (0·5–1 s) in time $t_0$; remains constant under stress; on release at time $t_1$, deformation returns to zero. Dough shows poor response to these conditions. (b) $P > P_e$ and $P_k > P_e$. $P_k$ = elastic limit (viscoelastic retardation). (c) $P > P_k$ gives irreversible deformation, e.g. dough (viscoelastic/plastic). (Source: *Technology of Breadmaking*, L. J. Auerman, Moscow, 1972.)

Fig. 31 as $E_2$ on the ordinate axis, the portion intersected by the dotted line. The angle subtended between this dotted line and the line drawn at right-angles to the ordinate axis remains constant, independent of the duration of the test. The portion $E_2 - E_1$ on the ordinate axis represents the magnitude of the delayed elastic deformation. By termination of the stress at time $t_1$, the momentary deformation $E_1$ (elastic) falls off almost immediately, and after that the elastic after-effect or retardation diminishes with reducing velocity.

However, since during these tests irreversible deformation occurs, after subsidence of the elastic deformation, this also remains after the system is destressed. Irreversible deformation is given by the expression:

$$t_1 \left( \frac{dE}{dT} \right)_{\text{Irrev.}}$$

The complete graphical description of the deformation-kinetics involved at constant stress is given in Figs 31 and 32. Under this regime, the constant stress exceeds the upper elastic-limit of the material under test (dough).

In order to confirm the relative importance of the gluten-complex, compared with starch during stress deformation, B. A. Nikolajew produced three stress curves of the deformation-kinetics of wheat-flour, gluten and starch-paste, using the same dimensions of time $v$. deformation. The resulting curves showed clearly that gluten determined the elasticity of the dough, and starch its plasticity. Water contents of the materials tested were as follows: wheat dough 46%, gluten 61·8%, and starch-paste 40%. These levels were optimal for each material, and required stress deformation forces of 580·4, 1241, and 1987 dynes/cm$^2$ respectively. Methods used in flour mills and bakeries to control the daily strength, physical properties of flours and doughs do not normally measure rheological constants in absolute terms as fundamentally as depicted above by Rebinder, Wolarowitsch, Schwedoff, Tolstoi, and Nikolajew. However, for practical purposes the relative empirical measurements derived from

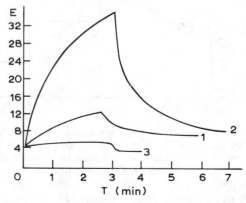

Fig. 32.   Shear deformation curves. (1) Wheat-flour dough. Moisture 45%; $P = 580·4$ dynes cm$^{-2}$. (2) Gluten. Moisture 61·8%; $P = 1241$ dynes cm$^{-2}$. (3) Starch paste. Moisture 40%; $P = 1987$ dynes cm$^{-2}$. (Source: *Technology of Breadmaking*, L. J. Auerman, Moscow, 1972.)

mixing and stretching tests world wide are quite adequate to characterize the strength of a flour.

In bakeries where dough-testing apparatus is limited or non-existent, the stability of a dough can be quite easily assessed by making 100 g of the flour into a dough with an appropriate amount of water, which always remains constant (dough moisture about 46%). The dough-piece is rounded up and placed on a glass plate at 30°C for 60, 120, and 180 minutes. The average diameter of the contour of the piece is then measured at these time intervals and expressed in millimetres. The weaker the flour, the quicker and more extensive the flow-out at the quoted resting-times. This simple test was described by Auerman in 1937, and could even be applied today, using appropriate processing variables. More sophisticated methods based on consistency measurements are penetrometer methods, e.g. Penetrometer MOCKIP, AP-4/1, AP-4/2, etc., which measure the depth of penetration of a spherical or other shaped projectile into the dough at a fixed termperature, force, time of resting, and deformation-time. Mixers with variously designed elements are used, e.g. double-Z-blades, static and moving-pin connected to electromotor-dynamometers, or electronic sensors with analogue or digital read-out. Typical examples are the Brabender Farinograph, Mixograph, Grain Research Laboratory (GRL) mixers made in Canada, Voisey pin-mixer, Brabender Do-corder, Valorigraph (Hungary). These types of apparatus measure: water absorption, dough development, stability, softening and mixing tolerance in general. Apparatus designed to measure response to stress deformation in the form of extensibility, resistance to extensibility, deformation-energy, and elasticity in empirical units are Brabender Extensograph, Chopin Alveograph, Simon Extensimeter, all of which have specific empirical measurement parameters which correlate adequately with some fundamental rheological constants. The sedimentation test, according to Zeleny (ICC-Standard 116), measures both the protein quantity and quality at the same time. However, this method can only be used for low extraction flours, with mineral ash content below 0·6%, since it is based on a swelling and sedimentation rate measurement. The principle is that the quantity and degree of swelling of gluten agglomerates in lactic acid/isopropanol solution determines the speed of sedimentation of the flour suspension. The method has also been modified for the determination of the protein quality of rye-flours in the GDR, by measurement of the sedimentation speed (TGL No. 22292/17), introduced in 1978 as a technological standard for industrial bakeries with line-production systems.

The 'Quellzahl' is also a measure of the change in volume of wet gluten by swelling in lactic acid, and since its conception by Dr Berliner in the 1920s, it has maintained its reputation as a reliable index of gluten quality, especially in the German-speaking countries, i.e. FRG, GDR, Austria, and Switzerland. In the GDR it has been assigned TGL No. 8029 since 1963, as a standard method. This method involves a measurement of the relative difference in gluten solubility in 0·01 N lactic acid of 0·5 g salt-free washed gluten after standing of the dough for zero time and 30 minutes. The gluten samples are incubated and rotated in a thermostat for 40 minutes at 27°C. After standing for 5 minutes in the sample test-tubes, the turbidity is measured using a photometer. The relative enzymatic breakdown of the gluten from zero time to 30 minutes is expressed as a percentage of the value for zero time. For flour samples with the average wet gluten content of 30·0%, Quellzahl values above 20 units indicate excessive breakdown, 10–20 average, and below 10 low. When comparing values at zero time and 30 minutes, 40–47% is regarded as highly enzymic, 15–40% average and 8–15% below average.

Where an accurate measure of protein-nitrogen of a flour sample is required, the standard Kjeldahl wet analytical method is appropriate, using $N \times 5\cdot7$ for conversion to percentage protein, expressed in some countries on a 14·0% moisture and in others on dry matter. The next flour component in order of importance is the starch. The structure and resistance of the flour starch granule depends on the granularity of the flour, size of the starch granule, and degree of mechanical damage incurred during the milling process. During development and ripening of the wheat-grain, the size and molecular structure of the starch-grain is laid down, and this manifests itself in the physicochemical and colloidal properties of the flour starch. Starch from large wheat grains has high molecular weight, high blue-staining values with iodine and good swelling capacity in warm water. Starch from smaller wheat-grains give a high viscosity and hydroscopicity, but are readily attacked by the beta-amylases. The decisive factor is the exposed surface of the granule and pieces of the granule, which allow adsorption of degrading enzymes onto it. The smaller the flour particles, the smaller the granules, and greater the damage during milling, which increases their vulnerability to beta-amylase degradation. In unsprouted wheat-flours, there is beta-amylase available in a free and active state, and in more than adequate concentration. Therefore, sugar build-up of a normal wheat-flour is not limited by beta-amylase, but instead by the availability of adequate degradable substrate, i.e. flour starch, which has been adequately damaged. The optimal temperature for beta-amylase in wheat-flour dough at pH 5·9 is 62–64°C, and that of alpha-amylase 70–74°C. Beta-amylse is completely inactivated at temperatures of 82–84°C, but alpha-amylase activity can extend up to 97–98°C. Even in the centre of the baked bread alpha-amylase can retain activity, and inactivation of beta-amylase during the baking cycle will depend on the rapidity and duration of heat to the various zones of the bread crumb. The water concentration prevailing at the time of processing will also influence the effect of these enzymes.

In the case of rye-flour doughs, the influence of lower pH and acidity-index have a great influence on the inactivation temperatures of both beta- and alpha-amylases. Beta-amylase is totally inactivated during baking at acidity-indices 10–11, which correspond to pH range 4·3–4·6, and on reaching 60°C; also at acidity-indices 4·6–6·3 (pH 4·7–4·9) on reaching 73–78°C. Rye-flour alpha-amylase becomes totally inactivated at acidity indices 10·5–11·5 (pH 4·3–4·2) and a temperature of 71°C. At acidity-index 4·4 (pH 5·0), the alpha-amylase in the centre of the loaf-crumb remains active for the whole baking cycle, or until a temperature of 96°C is exceeded.

Under comparable conditions of temperature, pH and enzyme/substrate concentrations, gelatinized wheat starch as a starch substrate will produce more than 300 times more maltose when digested with wheat beta-amylase for a fixed incubation period, than will undamaged native wheat starch. The effect of starch particle-size is readily demonstrated by grinding wheat starch to various particle-sizes, and reacting with adequate cereal beta-amylase. The amounts of maltose produced under standardized conditions shows increases of up to 15 times that of the original coarse starch. Therefore, evaluation of the starch complex is important for both grain and flour quality to the miller and baker. The two aspects of this are the properties of the granule (swelling, gelling and sensitivity to enzymic attack), and the activity of the amylases. The Falling-number method of Hagberg-Perten (ICC-Standard 107) is a measure both of starch granule properties and amylase activity. It makes use of the flour starch as substrate for the flour alpha-amylase. The Falling number constitutes the time in seconds necessary to stir a flour/water suspension, together with that required for the same stirrer/viscometer to fall a fixed distance through the hot gelatinized suspension. The suspension is gelatinized in special tubes immersed in a boiling water-bath, after which, the alpha-amylase liquefies the gelatinized starch. For wheat-flour 7 g, and in the case of rye-flour 9 g, of sample is suspended in 25 ml water at 20°C in the special viscometer tubes. Falling numbers 200–250 in the case of wheat-flours, are average and regarded as ideal for starch properties and alpha-amylase activity. Those below 150 signify the presence of sprouting in the wheat from which the flour was milled. Flours with Falling numbers above 300, tend to be deficient in alpha-amylase, and should be adjusted to 250 by adding malt flour or fungal alpha-amylase preparations.

The safe minimum Falling number of rye-flours is about 100 for flour types 997 and 1150, but in good crop years values of 200–300 are normal, and values in excess of 300 indicate a 'hard', high degree of starch polymerization.

The Brabender Amylograph, another indirect viscometric method, is more frequently used for rye-flour evaluation, but is often applied to wheat-flours for special uses. 60 g of wheat-flour or 90 g in the case of rye are made into a suspension with 450 ml distilled water and placed in the rotating cylindrical bowl, the vertically pivoted but non-rotating mixing-pins fixed to the measuring-head are then locked in position. The suspension is then progressively heated at 1·5°C/min, having set the top mounted horizontal stylus and chart for the start of read-out. The mercury-contact thermometer is programmed by setting at 25°C and the heating-cycle started simultaneously. When the suspension temperature reaches about 65°C, depending on the gelling properties, the viscosity steadily increases with gelatinization of the starch complex. This reaches a peak value, measured in Brabender units, up to about 90°C, and then falls off.

In the case of rye-flours, both the maximum temperature and viscosity are important data. Gelatinization temperatures below 62°C for flour types 997 and 1150, are indicative of starch corrosion. Viscosity maxima below 300, are less tolerant to processing and can give loss of symmetry in the final bread cross-section.

Gelatinization temperatures between 64 and 70°C, and viscosity maxima between 400 and 500 can be processed without difficulty. In evaluating rye-flours, another useful parameter for processing into bread is the water absorption at 350 Farinograph consistency units, or, inversely, the consistency reached at 60% water (160 dough yield) after 2 minutes and 10 minutes mixing in the Farinograph. Such methods are empirical internal bakery methods based on processing experience. Similar methods can be applied to wheat-flours, using the appropriate water absorption (say) 64% (164 dough yield), and noting the consistency reached after 5 and 15 minutes mixing. The standard 300 g of flour is used in all cases.

Another type of viscosity curve, using the Amylograph, for rye-flours, devised by E. Drews, involves the measurement of the consistency changes of rye-flours stirred at 30°C and 42°C in water suspension. The relative fall in consistency at these temperatures over a fixed time correlates with baking tests. Alpha-amylase activity can be specifically determined by the ICC-Standard 108, which involves the measurement of the speed of hydrolysis of a limit-dextrin substrate. The speed of hydrolysis is monitored photometrically as the fall in colour intensity of a dilute potassium iodide–iodine solution. However, this method lacks some precision and selectivity, and is operator-dependent. It can also give falsely high results at high levels of alpha-amylase. Rapid breakdown of the beta-limit dextrin, allowing beta-amylase attack, resulting in loss of the red iodine/beta-limit dextrin colour. Accurate assay of alpha-amylase in the presence of beta-amylase, which is present in large amounts in both ungerminated and germinated cereal grains, is possible when ether-bonded starch substrates are used, these being resistant to beta-amylase. Such dye-labelled cross-linked soluble starch is marketed by Pharmacia AB under the trade mark 'Phadebas'. W. C. Barnes and A. B. Blakeney applied this substrate for a rapid sorting method for cereal alpha-amylase. 5 g of flour are extracted with a mixed solution of 5 g sodium chloride and 0·2 g calcium chloride per litre in distilled water, using a mechanical shaker for 5 minutes. 5 ml of clear supernatant are then transferred to a test-tube in a water-bath at 70°C, and allowed to equilibrate. A timer is started on the addition of a Phadebas tablet with dispersion by shaking. Digestion is for 15 minutes, with a hand-shake every 5 minutes. The reaction is terminated by the addition of 1 ml of 0·5M sodium hydroxide, made up to 10 ml, filtered, and absorbance read at 620 nm with a cuvette for 1 cm light-path. If absorbance exceeds the scale of the spectrophotometer, dilute accordingly. A maximum absorbance of 5 may be read, as beyond this the substrate is limiting. Conversion of absorbance into milli-enzyme units per gram of material is as follows. Using the calibration chart supplied by Pharmacia with each batch of tablets, the absorbance reading is converted into milli-enzyme units per gram (mEU/g) by first reading off the units/litre, and then multiplied by a conversion factor of 0·04 to give mEU/10 ml of coloured solution.

For the determination of the mechanical starch-damage of a flour during the milling process, the amylose content is determined. The amylose reacts with a formamide-sulphosalicilic acid to give a colour, which is measured colorimetrically and compared with a copper tetra-amine/hydroxide solution as colour standard. Alternatively, it can be determined enzymically by digesting the flour with 2500 SKB bacterial alpha-amylase at 30°C in a buffer solution pH 5·9–6·1 for 17 minutes. The starch damage value is expressed in gram-equivalent of maltose/100 g flour, per cent, on a dry basis. The maltose being determined iodometrically by the potassium ferricyanide (alkaline)/sodium thiosulphate procedure, using soluble starch as indicator. This method is due to Donelson and Yamazaki at the Soft Wheat Quality Laboratories, ARS, USDA, Wooster, Ohio, USA. This method was designed for use on 1·0-g flour samples, being standardized against autoclaved wheat starch, simulating a maximum starch-damage content in flour of about 12%. It is based on the rapid hydrolysis of cold, gelatinized (damaged) starch. This level of testing is sufficient for soft wheat-flours, and is also applicable to most hard wheat-flours. If flours have damaged-starch levels greater than the 12% limit, sample size is reduced first down to 0·1 g, which will cover the range up to and beyond 30% on a dry basis. Flour sample weight of 0·50 g is suitable for the range 10–20%, and 0·25 g for 20–30% damaged-starch on a dry basis. The method due to E. A. Farrand, is based on the same enzymic hydrolysis principle and iodometric titration procedure, but differs in the following aspects:

(1) Starch-damage is measured in arbitrary units (Farrand units), expressed as a percentage of total starch content of the flour.
(2) Sample flour weight is adjusted according to the summation of flour protein content and moisture, varying within a range 4·5–5·5 g.
(3) The source of alpha-amylase is an extract of a 10 000 unit alpha-amylase malt flour, extracted with a mixture of sodium chloride/calcium acetate and pH 4·6–4·8 sodium acetate/acetic acid buffer (1:1).
(4) Digestion time is 1 h at 30°C.
(5) Tables used to read off starch-damage per cent from titration figure.
(6) Choice of arbitrary unit enables flour water absorption to be computed from protein, moisture and starch-damage data, based on a mathematical model.

The purpose in designing this type of method is to set up an optimization system, which enables the miller to measure and control starch-damage. Thus enabling him to offer the baker a flour with a controlled, consistent level of water absorption. This topic has already been discussed in Section 1.5.11. This concept is of particular value in the UK, for the control of flours for the CBP-process. Most CBP flours requiring starch-damage levels in the range 25–30, which maximizes water absorption to a safe level, without the danger of excessive water levels, where flour protein is limiting factor. This variable is particularly important in short process systems, where there is limited time for swelling and water-uptake, often resulting in a grey, waterlogged crumb after baking.

The maltose build-up due to the effect of the flour amylases can be determined by incubating the flour-in-water suspension for 1 h at 27°C. It is expressed in grams of maltose per 100 g flour. The maltose can either be determined by titration, using potassium ferricyanide and 0·1N potassium permanganate according to Rumsey-Ritter, or the more straightforward method of Dr Berliner. The latter simply requires the incubation of 10 g of flour with 50 ml distilled water at 27–30°C for 1 h in a rotating-thermostat. After filtration, 15 ml of clear filtrate are treated with 5 ml 1N sodium hydroxide in a test tube, mixed and placed in a rack in a boiling water-bath for exactly 5 minutes. It is then withdrawn, and placed in a cold water-bath to cool and allow turbidity to sediment. The absorption is then measured, using a photometer with a green filter, and the percentage maltose read off using Dr Berliner's maltose-tables.

Commercial wheat-flours normally fall within the range 1·7–3·0% maltose up to 75% extraction rate, and rye-flours types 997 and 1150 within the range 1·5–3·0%.

The Dr Berliner method can be made more accurate by calibration with known amounts of maltose-hydrate. The particle-size or granulation of a flour has an influence on the properties of its components, and its behaviour during mixing and processing. Therefore, some standard should be specified for processing in the bakery. For most industrial baking processes wheat-flours should give a 90% fall-through on a 0·160 mm sieve; in some EC countries 100% fall-through occurs at 0·125 mm and 0·230 mm. depending on the degree of industrialization of the processing methods used.

50 g of flour are mechanically sieved for 10 minutes at 260 rotations/minute, on a cluster of sieves bolted into a frame, on the Laboratory Plansifter Bühler-Miag MLU-300. The appropriate sieves are: 10 xx (0·125 mm), 15 xx (0·075 mm), 25 Prima (0·055 mm), and 240 Taffet silk (0·035 mm). The results are then plotted on millimetre graph-paper, the percentage residuals and throughs appearing on the vertical axes, and the mesh size in millimetres (0·001 mm = 1 micron) along the horizontal axis. If particle sizes under 40 microns are to be measured, sieves are no longer appropriate, instead, sedimentation or air-classification methods must be used. Flour particle-size has a profound influence on flour composition and processing properties, when milled from the same wheat-mixture (grist), and is dependent on the skill of the miller and his equipment.

All flours are made up of particles with distinct shapes, sizes, and therefore composition and properties. The smallest particles, to 10 microns in diameter (compared with human hair, 70 microns), are small starch granules or secondary starch; those up to 12 microns, consist of free wedge-protein. These particles constitute approximately 5–8% each, of total flour. Particles from 12–30 microns make up the bulk of the flour, and consist of small endosperm chunks, viz. protein and starch attached during growth, amounting to 60–65% of total flour. Particles from 30 to 55 microns, making up 5–8% of total flour are in the main large starch granules or prime starch. The residual percentage of total flour, which can be up to 200 microns, consists of large endosperm-chunks. If these endosperm-chunks are subjected to an intensive further milling with a pin-mill or hammer-mill at high-speed, the wedge-protein and starch granules can be separated. In a normal commercial flour, the amount of free wedge-protein will depend on the structure of the wheat grain. The hard wheats liberate little free wedge-protein, whereas the softer ones liberate more. However, the adherent-protein which is attached to the starch granule cannot be separated in this way. Flour particle-size distributions vary from country to country, depending on the mill-diagram, and the breadmaking processes employed. French and Swiss wheat-flours tend to be milled to finer overall particle-size distributions than Spanish flours, the latter often being subjected to long fermentation periods either in bulk or during final-proof.

The bakery laboratory mechanical sieve analysis method, is to place 50 g of flour sample onto the various sieves: 10xx (125 microns), 15xx (75 microns), 25 Prima (55 microns), and 240 Taffet (35 microns), each with firm metal trays separating them to collect the throughfalls. The sieves and their trays are stacked and bolted into the mechanical sieve MLU-300 (Bühler-Miag), and sieved for 10 minutes at 260 rotations/minute. The throughfalls are expressed as a percentage by multiplying the respective throughfalls in grams from each sieve by a factor of 2.

Although no strict norms can be specified, the following throughfalls are typical for a breadmaking flour: 125 micrometres 100%, 75 micrometres 85%, 55 micrometres 55%, and 35 micrometres 30%.

The measurement of flour colour is carried out in the flour-mill on a continuous on-line basis by optical methods, which monitor flour 'brightness'. Brightness is a measure of the percentage reflected light from the flour sample, which has been made into a paste, placed in a glass-cuvette, and placed in the path of a beam of monochromatic-light of chosen wavelength. 530 $\mu$m in the green range is a useful choice, since it measures the effect of impurities and bran particles in flour, without being sensitive to the flour's natural yellow pigment, which is a wheat property, and has nothing to do with the milling process. If the effect of bleaching is to be measured, then the wavelength 440 $\mu$m in the blue range is chosen. As colour standard for 100% reflection, magnesium oxide powder, which reflects over 97% monochromatic light is often used. Laboratory versions of flour-colour meters are also on the market, viz. Kent-Jones Color grader (Henry Simon Ltd, Cheadle, UK), Leukometer (VEB Carl Zeiss, Jena), and remission meter (Bühler-Miag, Switzerland).

Before a reliable measurement of flour-brightness was possible, flour mineral ash content was used as the only index of flour extraction-rate. In many countries, commercial flours are still typed according to ash content, which is legally binding. Where no instruments are in use, flour colour can be quite usefully monitored by the Pekar test. 5–10 g of flour are heaped onto a wooden-board, 20 × 6 cm, with a spatula or thick-glass sheet, pressed firmly and compared with acceptable control samples. The bran specks and background pigmentation can be further contrasted by immersing the pressed sample carefully in water for a few seconds, and allowing to stand in the open.

Mineral ash is determined by first burning the sample in air, and then mineralizing in a furnace at 600–900°C, until the weight of the residue remains constant (ICC-Standard 104). The earth-content of a flour is determined by treating the ash residue with 10% hydrochloric acid, allowing to dissolve, filtering, and weighing the insoluble residue as earth content.

The flour moisture content is determined by weighing at least 5 g into a corrosion-proof metal dish with an accuracy of $\pm 1$ mg, and placing in an oven at 130–133°C, electrically heated and with adequate ventilation, for 2 h. The dried sample is then placed in a desiccator fitted with a metal perforated plate, and filled below with phosphorous anhydride ($P_2O_5$) or calcium sulphate granules inpregnated with cobalt chloride as indicator, when water-saturated. The loss in weight is expressed as a percentage of the sample weight. This method (ICC-Standard 110) is accurate to within 0·15%.

The acidity of flours and meals are determined by the Schulerud method. This involves stirring 10 g of flour with 50 ml 67% ethanol in a 100 ml beaker at 20°C for 5 minutes, meals requiring 15 minutes extraction. The extract is then filtered, using a folded Sleicher and Schüll No. 588 or equivalent, covering the extract with a watch-glass to prevent evaporation. Collect at least 30 ml of filtrate, pipette off 25 ml into a 100 ml Erlenmeyer, add 3 drops of 3% alcoholic phenolphthalein-solution, and titrate with 0·1 N sodium hydroxide dropwise to a distinct rose colour tone. The

colour sequence transition is yellow/greenish-yellow to rose, and the amount of
0·1 N sodium hydroxide required is read to an accuracy of 0·2 ml. This amount is
multiplied by a factor of 2 to give the acidity value. For acidity values up to 3·0,
discrepancies of 0·2 units are the upper limit, for duplicates.

Average acidity values are as follows: wheat-flour, ash 0·55% 1·7, values above 2·4
being excessive; rye-flour type 1150, 3·0, values above 3·7 being excessive. In the case
of wheatmeals and ryemeals, values of 3·0 and 4·0 can be regarded as normal.

Flour pH can be measured either with a pH-meter, or by treating with methyl-red
indicator; a red colouration indicates a pH below 4·5, which is abnormal.

A useful qualitative, quick test for the presence of such mineral improver
additives as potassium bromate or iodate is to perform a Pekar test, and moisten
with a mixture of 2% potassium iodide and dilute sulphuric acid. In the presence of
oxidative treatment, dark-red to brownish-black spots or streaks appear on the
surface on standing.

For the determination of added L-ascorbic acid to flour on a quantitative basis,
which is permitted in most countries, 2 × 30 g of flour are suspended each in 150 ml
of 2% metaphosphoric acid solution free of all clumping. After 5 minutes, the
solutions are filtered through a folded (fluted) Sleicher and Schüll No. 597·5 filter-
paper, diameter 15 cm. The first 10 ml of filtrate 1 and 2 are rejected, and about 60 ml
collected. Alternatively, one can first centrifuge and filter the supernatant.

Take 50 ml of filtrate 1 and adjust to pH 3·8 with 20% sodium acetate solution;
add 100 ml of copper-free distilled water. Divide the whole into two equal volumes
in 400 ml beakers (A and B). Titrate briskly with the prepared solution made up
from: 0·1 g 2,6-dichlorophenol-indophenol dissolved in 40 ml boiling water in a
250 ml volumetric, by transfer, rinsing and cooling. After cooling, bring to the 250-
ml mark, mix well and filter. Store in a brown bottle in a refrigerator.

The titration proceeds until the colour is identical with the DI-colour-standard
solution, which is a rose colour. This standard DI-solution is prepared as follows. In
a 400-ml beaker add 0·2 ml DI-titration solution to 70 ml water and 5 ml of 10%
acetic acid.

When titrating, make sure that the two 400 ml beakers A and B have the same
colour-tone at the same layer-depth.

The 50 ml of filtrate 2 are also treated with the 20% sodium acetate solution to
adjust the pH to 3·8, treated with 25 ml formaldehyde, and allowed to stand at room
temperature for 15–20 minutes to block the L-ascorbic acid. Then proceed as for
filtrate 1. The calculation is then completed thus:

mg L-ascorbic acid in 100 g flour
$$= \text{titre of filtrate } 1 - \text{Titre for filtrate } 2 \times \text{DI-titre} \times 10$$

Titre of filtrate 1 is the DI required for the flour sample. Titre of filtrate 2 is the DI
required for the formaldehyde blank 1 mg of L-ascorbic acid/100 g flour is the lower
limit of detection. The formaldehyde solution for the blank is 30–40%
concentration. Standardization of the titre is achieved by titrating against 0·02%
L-ascorbic acid test solution, which has been previously standardized against 0·01 N

iodine solution, 1 ml 0·01 N iodine solution = 0·8805 mg ascorbic acid, and 1 ml DI solution is equivalent to 0·2 mg ascorbic acid.

An ascorbic acid test solution for checks is prepared by dissolving exactly 0·0500 g of crystalline L-ascorbic acid in 2% metaphosphoric acid in a 250 ml volumetric, and filling to the mark. This solution is stable for about 3 weeks in a refrigerator. In view of the importance of oxidation and reduction processes—Redox-systems—in dough, the significance, among other enzyme classes, of the oxidoreductases merit consideration and evaluation in bread flours. The oxidoreductases can be subdivided into two sub-classes, the dehydrogenases, which catalyse the dehydrogenation process involved in oxidation, and the oxidases, which catalyse reactions involved in the direct use of oxygen. Of the oxidases, peroxidase, polyphenoloxidase and catalase have been determined in wheat, wheat-flour, and by-products. Significant correlation has been found between polyphenoloxidase activity of flour and the crumb-colour of the resultant bread.

The presence in the dough of free tyrosine, which is the polyphenoloxidase substrate, gives rise to a darkening of the dough, and a deterioration of the bread-crumb. This phenomenon often occurs when two to three types of flours are produced from a mill-run with a total flour yield of 75–77%. In such cases, the white flour divides often show increased protein solubility in 10% salt solution, i.e. the non-gluten-forming globulins, albumins, peptides and amino acids.

All these enzymes can be separated by electrophoresis on polyacrylamide gel, each fraction showing enzyme activity. Polyphenoloxidase (tyrosinase) catalyses the oxidation of tyrosine, liberating melanoidine, which builds up in the dough as a dark-coloured compound, causing a darkening of the dough as well as the bread-crumb. The content of free tyrosine in the flour, or the total non-protein, amino-nitrogen is as good a method of indexing the darkening during processing as the actual measurement of the polyphenoloxidase activity. In measuring the latter, the change in optical density of a 0·01 M pyrocatechol solution induced by 1 ml of enzyme extract per minute, is the unit of activity. Another unit of enzyme measurement is the Anson unit, defined as the amount of enzyme, which, under test conditions liberates 1 mmol of folin-positive amino acids (tyrosine) per minute. These darkening effects from flour to dough can be monitored by using the flour-colour photometers mentioned earlier, used for routine reflectance or brightness measurements.

Peroxidase activity in flour or cereal products is determined by its reaction with pyrogallol, with which it forms a purple colour. 5 g of flour are sampled, and 0·5 g sub-samples ground in a mortar and pestle with sand. Transfer quantitatively to an Erlenmeyer, and add 0·5 g pyrogallol, 2 ml distilled water, and 3 ml of 1% hydrogen peroxide. The whole is vigorously shaken, when the reaction-mixture slowly becomes red. After 1 hour at room-temperature, the peroxidase reaction is terminated by the addition of 2 ml of 10% sulphuric acid. For comparison, a sample blank is prepared by treating 0·5 g of flour sample with 0·5 g pyrogallol and 5 ml distilled water. The purple colouration formed in the presence of the hydrogen peroxide is extracted five times with ether. The coloured extracts are unified in a 100 ml volumetric flask, and brought to the mark. The colour intensity is measured in a photometer, and the activity read off from a previously prepared calibration curve.

Another possibility is to add an appropriate amount of flour extract, to a mixture of ascorbic acid, phosphate/citric acid buffer, *o*-Toluidine and hydrogen peroxide. After a specific reaction-time, the peroxidase is inactivated with 2N sulphuric acid, and the residual ascorbic acid titrated with 0·004N iodine solution. The reaction catalysed is: $DH_2 + H_2O_2 = D + 2H_2O$ ($DH_2$ = leuko dye, $D$ = dye).

The catalase activity in flour can either be determined on the basis of the enzymatically decomposed hydrogen peroxide, or the liberated oxygen. 50 ml 0·01 M hydrogen peroxide in pH 6·8 phosphate buffer, are cooled to 0°C, and treated with 1 ml of the flour extract. After rapid shaking, 5 ml are withdrawn, and the catalase destroyed by adding 5 ml 2N sulphuric acid. The residual hydrogen peroxide is then titrated with 0·2N potassium permanganate solution. After 3, 6 and 9 minutes further samples are taken and treated in the same manner. From the required amount of potassium permanganate solution at zero time and time *t*, the catalase activity is determined. The decomposition of the hydrogen peroxide by the catalase, can be monitored more accurately electrometrically. The reaction catalysed in this case is:

$$H_2O_2 + H_2O_2 = O_2 + 2H_2O$$

The significance of the enzyme systems of cereal flours have not received much attention in the past, since white flours contain relatively low levels of the oxidases. However, with the increasing demand for higher extraction and wholewheat and wholemeal mill-fractions, the location of these oxidases is gaining in significance. For example, wheat-bran contains higher levels of polyphenoloxidase than the wheat from which it is derived. The tail-end streams from the reduction system, being particularly rich in polyphenoloxidase. Apart from wheat varietal variations in the content of this enzyme, summer wheats appear to be richer endowed with this material than winter wheats in general. In the case of wholemeal flour most of the polyphenoloxidase tends to remain in the solid residue after aqueous extraction, and the addition of a detergent with swelling properties to improve extraction such as, lactic acid or SDS may be necessary. Depending on the concentration, the addition of whey, sucrose, yeast and salt can all interact with wheat-flour peroxidase and cause darkening during bread manufacture. Present information on this topic of interactions between flour and dough components and the many flour enzyme systems is sparse.

Any mineral-contaminants in flour can be readily detected by treating a 5-g representative sub-sample with 50 ml chloroform in a separating funnel, shaking well. The dense material settles at the bottom, and the rest either floats on the chloroform or becomes suspended. The dense gravitated material can be run off into a weighed dish, and reweighed after drying.

Animal contaminants, in the form of hair and insect fragments and even moulds, are resistant to chemicals and must be interfaced with oil treatment. This is known as the 'Filth-test'. About 50 g of representative sample are defatted with petroleum ether several times using a centrifuge. The extract is saved for later filtration. The defatted residue is removed quantitatively from the centrifuge tubes into a glass-dish, and carefully dried in two stages for about 2 hours at about 35°C, and finally up to 100°C. The dry residue is then transferred quantitatively into a 500-ml beaker, and treated with 200 ml of boiling 0·5N HCl, with stirring with a glass-rod. The contents are simmered on low heat for about 40 minutes while covered with a watch-glass. After cooling, the contents are transferred quantitatively into a 1-litre separating-funnel and topped up to about 600 ml, adding 20 ml of liquid paraffin (density 0·880) and well shaken. After allowing to settle for 30–40 minutes, the aqueous layer is tapped off into another separating-funnel and treated with another 20 ml of liquid paraffin. After a further 30-minute settling time, the separation is completed to within a few millilitres. The paraffin layers in both funnels, are then washed out twice with water, the second time with the addition of 50 ml petroleum ether. After tapping-off the washings to within a few millilitres, the residual layer is filtered through a 9-cm diameter filter fitted into an 8 cm diameter glass funnel, thus allowing the side walls to be covered with an overlap to prevent leak. The contents of the second separating-funnel are filtered first, followed by that of the first, and the collective petroleum ether extracts. The funnels are then washed with water twice and once with ethanol. The filter is then dried for about 1 hour, sandwiched between two watch-glasses at 105°C, and moistened with a few drops of paraffin to make the whole surface more transparent. The moistened paper is then observed at a magnification of 60 times with a microscope, and the hair and insect-fragments counted and identified by comparing with sample-photographs. In 227 g of flour sample, a maximum of 2–3 rodent-hairs, and 20 insect-fragments is tolerable.

The presence of mites in flour can best be observed microscopically. Living mites can be detected by placing a tablespoonful of flour on a piece of paper and pressing a glass-plate to flatten the sample surface. After standing at 30°C for about 24 hours, the presence of mites is confirmed by the presence of tunnelling and heaps. In the case of meals, 100 g are placed in a 250 ml Erlenmeyer, spread over the bottom surface and the flask stoppered. After a few hours at 30°C, the mites emerge, and can be observed with a magnifying-glass.

Vegetable contaminants in flour can be detected by the following initial spot-test: 2 g flour are vigorously shaken in a test-tube with 10 ml of a mixture of 95 ml 70% ethanol and 5 ml HCl (density 1·19), and placed in a warm water-bath. The colouration of the supernatant liquid is, according to A. E. Vogl indicative of specific weed and grass contamination from the field.

Orange-yellow: *Agrostemma githago, Lolium temulentum*
Rose-red, violet or purple-red: *Vicia sativa* ssp. angustifolia
Blue-green or green: *Melampyrum arvense, Alectorolophus major Rhinanthus alectorolophus*
Bright-red: Spore sacs of the parasitic mould *Claviceps purpurea*, known in German as '*Mutterkorn*'. Hosted by about 600 plants including grain. Shape and size of spore-sacs vary with host, viz. rye, triticale, winter-wheat, barley.

These undesirable weed seeds and mould spore-sacs belong to the grain contaminants known under the grain-intervention EC guidelines as '*Schwarzbesatz*' or black-contaminants, which should not exceed 3·0% in total. Within this total, not more than 0·05% *Claviceps* (*Mutterkorn*), and a maximum of 0·1% undesirable weed seeds, are permitted.

The distribution of these grain-contaminants depends on the intensity of cultivation of land, and the extent of the usage of herbicides, both of which are in imminent decline rather than increase for political reasons. Also, current trends towards biologically cultivated grain, will result in an increase of weed-seed contaminants. Their removal at the mill during the cleaning process can be achieved to a high degree, but this always represents an additional cost in equipment and energy Therefore, for the consumer who purchases grain and does his own milling to ensure that he has biological raw material for breadmaking, there is a potential toxicity hazard from weed-seed contamination. This is also the case with any mill which produces flour from biologically grown grain, efficient cleaning-room equipment in the form of sieves, *trieurs*, aspiration and washing is essential to remove potentially hazardous weed and grass seeds, which contain toxic alkaloids and glycosides.

*Agrostemma githago* (corn-cockle, *Kornrade, nielle des blés*) are liberated during the thrashing process as dull-black, kidney-shaped seeds. 3–5 g constitute a strong poison for humans and animals. It affects the nervous-system, common symptoms being irritation of the mucous-membrane, and giddiness. The influence on flour taste is caustic and 'scratchy', giving the flour and bread a bluish-green colour-tone. They can be removed during the mill-cleaning process with *trieurs* fitted with 4·5-mm slit-aggregates for wheat, and 5·5-mm for rye. The poison is a low hazard when present at less than 0·1%. The procedure due to H. Medicus and H. Kober will detect flour contamination of less than 1·0%, when large enough samples are taken. Not less than 20 g of flour sub-sample, representative of the delivery batch, are initially extracted with petroleum ether, followed by a warm extraction with 55 ml chloroform, and 25 ml ethanol, the solvent being drawn off as warm as possible. The solvent residual is then concentrated on a steam-bath. The residue is then taken up in a little warm distilled water, filtered and concentrated again. On addition of a few drops of concentrated sulphuric acid, the *Agrostemma githago* containing flour develops a yellow colour, becoming brown-red within a few minutes. By comparison, a sulphuric acid treated pure wheat-flour will remain colourless under the same treatment for up to 2 hours. Contents of 1·0% in flour are clearly detected, and lower levels by increased sample size. Detection based on its haemolytic effect are also possible.

*Claviceps purpurea* (ergot, *Mutterkorn*), which represents the spore-sacs of this mould are present as attached mycelium. The shape and size depends on the host, and is very difficult to remove. Reliance has to be placed on washing, scouring, sieves or *Tischausleser*. Apart from the qualitative detection of claviceps by the Vogl reaction described earlier, a semi-quantitative determination of its alkaloids colorimetrically is possible by use of the Limes reaction (*Chimie analytique*, **36**, 33, 1954), or the paper-chromatographic method of Stoll and Bouteville (*Helv. Chim. Acta*, **37**, 1725, 1954).

In the GDR, the new grain contaminant-regulations (Ackmann, A., *Getreidewirtschaft*, **20**, 4, 93–95, 1986, TGL-Besatzbestimmung) has been made more exact by the inclusion of *Sinapis arvensis* (charlock, *Ackersenf, moutarde des champs*),

*Raphanus raphanistrum* (wild radish, *Hederich, ravenelle*), and *Lathyrus* spp. (chick-pea, *Platterbsen, gesse*), alongside *Vicia sativa* ssp. angustifolia (vetch, *Wicke, vesce cultivée*).

Charlock, when present in flour at 0·2% level makes the dough slightly darker, 1·0% visibly darker without any increase in smell, but a pronounced mustard-like taste, and 2·0% gives a strong mustard-like odour and taste. It can be almost completely removed in the mill by the use of sieves, aspiration, and the *trieur*. Wild radish gives a bitter taste in flour and a scratchy taste to the dough, but is reduced during baking. It can be removed in the mill by *Tischausleser* or sieves smaller than 2–3 mm.

When correctly stored, flour for breadmaking purposes should not require any detailed examination of its microbiological aspects. The most important requirements for safe storage is moisture content, which should not exceed 15·0%, temperature, approximately 25°C, adequate aeration in order to assist and promote maturation, a 6 hour aeration using 2–3 m³ of air/tonne at 25°C, as suggested by N. J. Sosedow is ideal for the optimum development of free fatty acids, reduction of proteolytic-activity, polymerization of the pentosans. These changes result in a flour pH optimum at processing of about 5·0, instead of about 6·2 freshly milled. Storage-tolerance will depend on lipid content, i.e. flour extraction rate. Wholemeal flours should not be stored for more than 10 days, wheat-flours maximum 2–3 months, and rye-flours 2–4 weeks. During the pneumatic transport of flour, the air can increase the temperature of the flour to about 30°C within 40 s, thus contributing to ripening. An organized preparation of flour as the most important ingredient of the dough, by blending, aeration and elimination of clumping and 'bridge-building', prior to processing cannot be over-emphasized. All storage silos should be regularly gassed. Should a microbiological examination ever be necessary, the first step would be to evaluate a total bactierial and mould plate count, using the following nutrient-agar media as substrates:

bacterial counts, tryptone–yeast–glucose agar, containing:
bacto-tryptone 5·0 g, yeast extract 2·5 g, dextrose 1·0 g, agar 15 g, made up by adding 1 litre distilled water, pH 7·0; available from Difco Laboratories, Detroit, USA under the trade-name 'Plate-count Agar'.
mould counts, malt–salt–agar, containing:
malt extract 20·0 g, sodium chloride 75 g, agar 20·0 g, made up by adding 1 litre distilled water, pH 6·8.

These media are prepared by dispensing 10 ml quantities into test-tubes, and autoclaving for 15 minutes at 1 atm, and 120°C in order to sterilize.

10 g of a representative sub-sample of flour, taken under aseptic conditions, are weighed into a four-sided Breed-Demeter type sterile bottle with 10 g sterile quartz-sand and 100 ml 0·1% peptone solution or saline, closed and shaken for 20 minutes in a shaker. 10 ml of this suspension are transferred aseptically with a pipette into another sterile bottle containing 90 ml of 0·1% peptone solution, and likewise shaken to homogenize. 10 ml of this latter suspension is aseptically pipetted into a third sterile bottle containing 90 ml of 0·1% peptone solution. Therefore, from the original flour sample, three dilutions have been prepared, 1:10, 1:100 and 1:1000.

Now, using special, sterile 1·1 ml Demeter pipettes, 1·0 and 0·1 ml from each prepared dilution are pipetted aseptically into labelled sterile petri-dishes, and the appropriate liquid nutrient agar, cooled to 48–50°C, poured in from the pre-pared test-tubes. Gently rotate the dishes to homogenize the mix until clear, and cover; allow to solidify. The culture-plates are then placed in an incubator, in the case of the bacterial plates for 2 and 5 days at 30°C, and in the case of the mould plates 5 days at 25°C. After the incubation periods have elapsed, all plates showing less than 20, and more than 300 colonies are rejected. The remainder are evaluated, by counting the number of colonies on each plate, using a magnification glass with 6–10 times magnification. At least three parallel determinations are necessary, the average calculated, and the final result reported in colonies per gram of flour, taking into consideration the dilutions used. The accuracy of these counts is about $\pm 10$–20%, but this can be raised to $\pm 5$–6% by carrying out more parallel series of plates. Once a guide is obtained concerning the number of colonies, certain dilutions can be omitted, only 5–10 plates being inoculated with the appropriate suspension. Flour bacterial counts can vary from 200 g to 8000/g, being about 70 times less than the parent wheat or rye, whereas mould counts range from 90/g to 8000/g, being often more than in the parent grain by a factor of 3–5. Therefore, there is often mould growth and pick-up during milling. Most of the bacteria found in flour are Gram-negative, belonging to, or related to the genera *Flavobacterium*, and *Paracolobacterium aerogencides* which constitutes about 30% of the total bacterial flora of flour. Another *Aerobacter* species, which represents about 15% of total flour population is *Aerobacter cloaceae*. However, these genera are normal grain microflora. Alongside these native inhabitants of grain and flour, *Escherichia coli*, *Staphylococcus aureus*, both dangerous pathogens, have been found at about 1% of the population. If these do present themselves, they indicate a secondary infection, often due to grain contamination during transit, or infections in the mill-silos and/or handling equipment. Typical examples of moulds found in flour are: *Aspergillus candidus*, *A. flavus* and the rest of the *Aspergillus* and *Penicillium* genera, when the moisture content reaches 16% to 17%. Another troublesome inhabitant is the 'potato-bacillus', the mesophilic bacillus species *subtilis mesentericus*, the causative agent of 'ropiness' in bread. These spores occur on the surface of grain, and can be found in flour. When a flour suspension is pasteurized at 80°C for 10 minutes and plated quantitatively, a count of 20 or more spores per 100 g flour is considered as suspect. In such instances, quantitative dilutions of the flour in nutrient-broth, followed by incubation at 37°C for 48 hours, will reveal the presence of the *Bacillus* species as a greyish-white surface pellicle on the broth surface. Factors contributing to the general hygiene and sanitation of grain, and especially milled flours are:

(1)  Sanitary status of the wheat intake.
(2)  Wheat cleaning operations and equipment.
(3)  Sanitizing agents used before milling.
(4)  Mill equipment sanitation.
(5)  Use of bleaching and maturing agents if permitted.
(6)  Storage conditions imposed on the flour before processing.

Thermal treatment of flour is very effective in reducing the microflora of flour for special applications. According to Weisblatt, counts of from 300/g to 2400/g can be reduced from zero to 2·4/g in the case of *E. coli, A. flavus, S. aureus* and *B. subtilis*, in decreasing order of efficiency.

The cereal chemist/technologist must decide for himself which of these analyses described are meaningful in their interpretation for the processing of flour into bread. Determining factors here are available staff and facilities, processing methods, degree of mechanization/automation, quality control at the flour-mill and, above all, his own accumulated experience.

The final summation of flour quality is the controlled baking test under standardized conditions. The method used for evaluating wheat-flour in particular varies considerably from one country to another. Typical differences can be enumerated as follows:

(1)  Dough formulation: water addition according to dough moisture or dough consistency. Yeast and salt concentrations vary slightly, although an average of 2·0% yeast on flour weight is most common. Some countries, e.g. USA, use sugar and milk-solids, a range of fast- and slow-acting oxidants, and various dough-conditioners. In Canada, the addition of a malt/phosphate yeast-food is established, and in Continental Europe malt/ascorbic acid/emulsifier-based improvers are popular.

(2)  Processes: these can be divided generally into indirect (sponges, or liquid pre-ferments with various amounts of flour content, as practised in the USA, USSR, Spain, Switzerland, Canada, Austria, GDR, some parts of France and elsewhere) and direct (as practised in UK, Holland, parts of France and Switzerland and elsewhere.

(3)  Mixing and remixing of doughs: the use of laboratory mixers of various design and construction working for various mixing-times, and delivering diverse mixing-intensities, e.g. spirals. Wendels, two-arm kneaders, Z-blade, pin-type, and orbitals.

(4)  Processing conditions and length of fermentation: initial dough-temperature, temperature of proofers and relative humidities climates. Length of intermediate, floor-time and final proof-times.

(5)  Type of product, shape and scaling-weight.

(6)  Moulding method used: manual, long-roller, sheeting, curling, dragnet, straight-through horizontal or drum, cross-grain, and reverse-sheeting actions.

(7)  Conditions and duration of final-proof: temperature and humidity of the atmosphere, proofing to constant height or a fixed time.

(8)  Baking conditions: type of oven, temperature and amount of steam injected, baking-time.

(9)  Conditions for quality evaluation: time elapsed between baking and evaluation, criteria and scoring procedure, depending whether pan- or hearth-type bread.

Evaluation procedures for the baked bread also vary widely from country to country, depending on the relative importance of the various sensory aspects and structure of crust and crumb. Conventionally, subjective assessments are made, based on a points system concerning: shape or symmetry, volume, oven-spring, crumb-porosity, crumb-elasticity, crumb-structure, crust-colour, crust-texture, flavour and aroma. However, considerable research effort is being made to quantify and automate measurement of the porosity and structure of both the dough and the resultant bread, using optical-electronic analysis (Wünsche and Tscheuschner, 1984). Other attempts to measure structural dough and baked-product parameters objectively include Bloksma (1981), Carlson and Bohlin (1978), Kuzminskij, Scerbatenko and Vassin (1977); Zimmermann and Schmidt (1977).

Attempts to evolve an international standard baking-test procedure would require an equipment standardization (mixer, moulder, fermentation-cabinet, oven, and bread-volume measurement apparatus), and may not be of practical significance, owing to the diversity of processing requirements from flours worldwide.

In the USSR, according to the Laboratory Wheat-flour Baking-test GOST 9404-60 Standard, paragraph 55–64—Flour and Bran Test Methods, Moscow Isdatelstwo Standardow, 1963, and the Laboratory Manual for Technology of Baked-products by Putschkowa, L. I., Moscow, published by Pistschewaja promyschlennost, 1971, the procedure is a direct or straight-dough procedure, consisting of the flour to be tested, yeast and salt. An amount of flour equivalent to 966 g dry matter is mixed with an amount of water which is standard for each flour type (top-grade, 2nd grade, 1st grade). The method and regime for mixing is not laid down. Dough temperature is fixed at 32°C, at which temperature the dough is fermented in a thermostat enclosure for 170 minutes. At 60 and 120 minutes after mixing, the dough is remixed or punched. The dough is divided into two pieces, one for panning, and the other shaped for a hearth or oven-bottom loaf. Final-proof of the pieces proceeds to the 'full-proof' stage, the panned piece being given 5 minutes longer than the other. Baking is in Oven ZNII ChP-P6-56, at 220–230°C, which is a fixed time for panned and hearth bread, depending on flour type. After 4–24 hours, the bread quality is judged according to volume-yield in ml/100 g flour at 14·5% moisture, and the relationship between height, $H$, and diameter, $D$, for the hearth bread. Sensory evaluation includes: shape, colour, character of crust, elasticity and porosity of crumb, and flavour.

All characteristics are reported as verbal descriptions, instead of on a points system.

The standard method for assessment of the baking quality of bread flour, by the sponge-dough, pound-loaf method in USA, according to AACC Method 10–11, can be summarized as follows:

Apparatus:
(1)  Electric Hobart A-120 or A-200 mixer, equipped with constant temperature jacketed mixing bowl and fork.
(2)  Red-spirit filled thermometers.

(3)  Fermentation-cabinet, maintained at 86°F (30 ± 1°C) and 85% RH.

(4)  Moulder: National Sheeter and Moulder, 6-inch rolls.

(5)  Proofing-cabinet, maintained at constant temperature 96°F, 35·5 ± 1°C, and 92% RH.

(6)  Aluminium fermentation bowls No. 9045 Wear-Ever, top diameter 9·5 in, bottom 4·75 in; depth 4·5 in; capacity 4 quarts.

(7)  Baking pans; 4 × tinplate, unglazed; dimensions, top inside 10 × 4$\frac{3}{8}$ in; bottom outside, 9·5 × 3·75 in; depth inside 2·75 in. Official pans obtainable from either of two suppliers: Ekco Engineering Co., Division of Ekco Products, Chicago, Illinois: Stock No. 60 (Shur-Bake Conditioned).

(8)  Scales 1 kg capacity, accurate to 0·1 g.

(9)  Flour containers 1·5 quart capacity, with tight-fitting lids.

(10)  Oven: rotating reel or plate, capable of maintaining constant temperature of 425°F (220°C).

Procedure:

Formula: for 700 g flour-dough

Sponge: flour (14·0 moisture) 420·0 g, water 252·0 (variable), compressed yeast 14·0, yeast-food 3·5 (ammonium chloride 9·4%, potassium bromate 0·3%, calcium sulphate 30·0%, sodium chloride 35·5%, starch 24·8%).

(1)  Add yeast to mixing-bowl as suspension in portion of ingredient water.

(2)  Add the yeast-food in dry form with the flour, do not dissolve yeast-food in yeast suspension.

(3)  Mix sponge 0·5 minutes in first speed, and 1 minute in second speed, or until smooth. Temperature of sponge after mixing: 80°F (26·5 ± 0·3°C).

(4)  Place sponge in fermentation-bowl, and set aside in fermentation-cabinet for 4 h.

Dough: flour 280·0 g, water 168·0 (variable), sucrose 35·0, salt 14·0, shortening (hydrogenated vegetable) 21·0.

(1)  Place ingredient water in mixing-bowl, add dry ingredients and shortening, start mixer in first speed, add the sponge in three approximately equal proportions, at 15, 25 and 35 s mixing time. Continue mixing in first speed until 1 minute has elapsed from beginning of mixing.

(2)  Shift mixer into second speed, and mix to optimal development, noting the total time dough was mixed. Dough temperature out of mixer 81°F (27·0 ± 0·3°C).

Make-up:

(1)  Recovery-time: round up dough lightly as it comes from the mixer, place in lightly greased fermentation-bowl, place in fermentation-cabinet, and leave for 30 minutes.

(2)  Scaling: after 30 minutes rest in fermentation-cabinet, remove dough and divide into two 500-g dough-pieces.

(3)   Intermediate-proof: round each piece lightly, and place in fermentation-cabinet. Allow 12–15 minutes in cabinet for relaxation before moulding.
(4)   Moulding: pass dough through National Sheeter twice. For first pass, set rolls at $\frac{5}{16}$ in; for second pass, set rolls at $\frac{3}{16}$ in. Curl 'as a ribbon', seal, and elongate to 9·5 in, using National Moulder. Place in lightly greased baking pan, seam down.

Proof:
Place panned dough in proofing-cabinet at 96°F (35·5°C), and 92% RH.
Proof to template of $\frac{3}{8}$ in above centre of pan.
Bake:
Bake 25 minutes at 425°F (220°C).

Score:
(1)   Measure volume of loaf within 10 minutes, after removal from oven.
(2)   When loaf has cooled to internal temperature of 90°F (32°C), wrap in waxed paper, or place in polyethylene bag and close bag.
(3)   Score for external characteristics and internal characteristics 18 h after baking.

For measurement of dough response to certain variables, e.g. mixing time/mixing intensity, fermentation oxidation, dough-conditioners, dough-developers or other additives, a straight-dough procedure can be employed based on 100 g flour at 14·0% moisture (86% flour solids).

For this purpose, it is convenient to prepare:
(1)   Yeast suspension: 12 g fresh refrigerated yeast in water, and make up to 100 ml, freshly prepared for each series of tests.
(2)   Salt/water solution: Dissolve 4 g sodium chloride and 20 g sucrose and make up to 100 ml.

Mixing:
Add 25 ml of salt/sugar solution to a quantity of flour equivalent to 100 g flour to be tested/standard control-flour, based on 14·0% moisture (86% flour solids). Add sufficient additional water to bring dough to desired standard consistency after mixing. Start mixer, and mix 2 minutes in a Swanson-type pin-mixer until a standard consistency is obtained. Final dough temperature should be 86°F (30°C).

Fermentation:
Round-up by folding 20 times with the hands, put in a fermentation-bowl, and place in a cabinet. Ferment in bulk for 180 minutes, giving a first punch after 105 minutes, and a second punch after an additional 50 minutes, moulding after a further 25 minutes.

Punching by machine:
Remove dough from fermentation-bowl, lightly seal over the wet side of the dough by drawing the dry edges together with thumb and fingers. Elongate dough-ball slightly, pass once through sheeting-rolls with a clearance of $\frac{9}{32}$ in.

Roll up dough sheet slightly without rounding or sealing the edges, replace in bowl and return to cabinet.

Moulding method:
Sheet the dough, lightly roll up the piece, seal the seam, roll lightly under the palm of the hand, and place seam down in the pan. The dough-piece should not exceed the length of pan prior to final rolling-out.

Proofing:
Proof at 86°F (30°C), and at least 75% RH.
Two methods of proofing are recommended:
(1)  to constant time, usually 55 minutes, or
(2)  to constant height, say 9·5 cm.
Choice will depend on purpose of test, and interpretation to be placed on results.

Baking:
Bake for 25 minutes at a temperature of 446°F (230°C), as determined by an oven-thermometer placed at the level of the baking-pans (top), and 5 cm distant from them, on the side next to the axis of rotation of the shelf. Precise control of temperature is essential. An open pan of water should be placed in the oven, and to provide more uniform oven conditions, bake a series of dummy loaves just prior to and after the experimental series. (AACC Method for ascertaining the baking-quality of bread-flour: wheat—Cereal Lab., Methods.)

In order to determine the 'response-reaction' of a flour or dough to a set of conditions, known or specific stimuli, it must be established by baking it in comparison with a control. The established response may be positive, neutral, or negative. Unless otherwise qualified, any differences are understood to represent differences in loaf volume, expressed in ml/100 g flour. Other responses, e.g. texture, colour or flavour may be required. To measure 'mixing-response', the dough is mixed for varying lengths of time, keeping final dough temperature as near to 86°F (30°C) as possible, noting the changes in general bread quality. Using this procedure, optimal mixing-time and mixing-tolerance can be established. To measure 'fermentation-response', at any chosen dough temperature, floor-times (ferment-ation in bulk), intermediate (first proof), and final-proof can be varied to determine: (1) optimum fermentation-times, and (2) fermentation-tolerance under specified conditions.

The 'gassing-power response' of a flour is determined by its starch-damage level at any given alpha-amylase activity, which may have been supplemented with malt-flour or fungal alpha-amylase at the mill. Additions of sucrose or high fructose corn-syrup (HFCS), provide spontaneous gassing-power.

Flour 'oxidation-response', to give maximum loaf volume and crumb-grain with any given processing system, ranging from 0 to 75 ppm flour weight, can be ascertained by adding various increments and noting the effects.

Oxidant, shortening and surfactant responses are particularly important in short-time processes.

Flour 'baking-response' is best measured by oven-spring and loaf-symmetry.

The three criteria which determine the potential 'strength' of a flour are: oven-spring and loaf volume at optimum response, protein quantity and quality, and the ability to carry low grade or 'weak' flours (blending value).

In Canada, in order to evaluate their Red spring flours, the so-called 'remix baking-test' is used. This involves a remix operation on doughs made from strong Canadian wheat-flours, which was developed as early as 1960. The procedure followed is as follows. A direct straight-dough is prepared from: flour 100 g, water, salt 1·0%, sucrose 2·5% and yeast 3·0%, with the following additions: malt-extract 0·3%, potassium bromate 0·0015%, ammonium hydrogen phosphate 0·1%. Flour water absorption is variable, according to the Farinograph and the flour tested. Mixing is with the GRL Laboratory mixer (pin-type) at 130 rpm, initially for a duration of 3·5 minutes, then relaxed for 165 minutes in a thermostat at 30°C. The dough is then mixed for 2·5 minutes in the GRL pin-mixer instead of the conventional punch, and relaxed for another 25 minutes. The dough is then rolled-out, folded, and long-rolled using a laboratory moulder. Final-proof takes place in pans of the same measurements as those quoted in the AACC method already described, at 30°C in a thermostat for 55 minutes. The total fermentation-time, including final-proof is 4 hours 8 minutes. Baking is in a rotary-oven, electrically heated at 220°C for 25 minutes. Bread volume is measured 1 hour after baking, and sensory evaluation the following day. This procedure is ideal, since it gives a linear relationship between bread volume and flour strength over a broad spectrum of flour quality profiles, from very weak to very strong. The combined effect of adequate mechanical development, with added sucrose, malt, phosphate and bromate, permits the full baking potential of stronger flours to be realized. Thus producing better correlations with dough-rheological and gluten-evaluation methods, especially in terms of loaf-volume. The production baking-test is usually carried out in full-batch capacity of the mixer employed in the actual production process and worked-off on the normal plant equipment, followed by production proofing and baking. Objectives of such test-trials could be to evaluate recipe or formulation changes, process parameters, floor-time or proofing parameters, baking parameters, or the evaluation of a flour delivery or flour blend proposed for a purchasing contract from a flour-mill for an advance period. Typical variations in wheat-flour bread parameters are in volume, crumb-porosity, and darkening of the crumb during processing; they can arise from one flour delivery to another. But in the case of rye-flour bread, the crumb properties, its degree of stickiness, elasticity, relative dryness or moistness, as judged by sensory feel, are of paramount importance. Crumb colour of rye-flour breads are only important for the light rye-flours, e.g. types 997 and 1150. In the case of darker rye-flours, the bread crumb is always darker, less open-pored, and more sticky.

Quality differences in rye-bread are due to certain specific reactions within the carbohydrate-amylase, and protein-proteinase of the rye-grain and resultant flours, often because of unfavourable weather conditions.

In the evaluation of wheat-flour, in addition to the sensory judgement of the bread, scaling-weights, bread-volume, bread-yield, volume-yield and baking-loss should all be calculated, based on 100 g flour at 14·0% moisture, as follows:

$$\text{Dough-yield} = \frac{\text{dough weight} \times 100}{\text{flour-weight}}$$

$$\text{Bread-yield} = \frac{\text{bread-weight} \times \text{dough-yield}}{\text{dough scaling-weight}}$$

$$\text{Volume-yield} = \frac{\text{bread-volume} \times \text{dough-yield}}{\text{dough scaling-weight}}$$

$$\text{Baking-loss} = \frac{(\text{dough scaling-weight} - \text{bread-weight}) \times 100}{\text{dough scaling-weight}}$$

MacRitchie has suggested the calculation of the loaf-volume index (LVI), which is expressed as follows:

$$\text{LVI} = \frac{\text{loaf-volume (ml)} \times 200}{\text{flour protein \% } \times \text{ weight of flour (g)}}$$

The value 200 is arbitrarily chosen so that extra-good performance flours give values close to 100.

For the evaluation of rye-flours, doughs are prepared from flour, salt and either sour-dough or lactic acid, for European processing conditions.

The Detmold sour-dough baking test is the closest to industrial baking practice. Using the Berliner short-sour process, the following scheme is employed:

Short-sour:
Starter culture (either taken from a mature short-sour or from a bakery) 90 g, rye-flour 450 g, water 405 g.
Dough-yield: 190, dough-temperature: 35°C, resting-time: 3 hours.

Dough:
Short-sour from above 855 g, rye-flour 550 g, yeast 10 g, salt 15 g, water approximately 195 g.
Dough-yield: approximately 160 dough-temperature: 29°C, resting-time: 20 minutes.

The sour represents 45% of total flour, and the starter-culture 20% of the flour to be soured. Using 1 kg flour in total, 450 g are mixed with 90 g starter culture and 405 ml water to form a short-sour with a dough-yield of 190. The short-sour temperature must be fixed at 35°C, and maintained at that temperature for the 3 hours resting-time. The final dough is prepared by adding the residual flour, 10 g yeast, and 15 g salt to the mature short-sour, together with enough water to adjust the dough-yield to approximately 160, and the dough-temperature to 29°C exactly. The dough resting-time must not exceed 20 minutes. The dough-piece is then shaped and proofed to the full-proof stage in a special proofing-basket without steam. Baking is at 250°C for 5–10 minutes setting-time, followed by 1 hour at 220°C to complete bake-out. After an initial 1-minute treatment with steam, the vent should be opened for 5 minutes.

The lactic acid baking-test for rye-flour is designed to ascertain the crumb characteristics of the resultant bread, in particular its elasticity. Other properties which can be judged are shape and colour, and any storage damage deduced from flavour and aroma defects. The procedure for a 1 hour direct process is as follows:

Rye-flour 1000 g, lactic acid (DAB 6, density 1·206–1·216) 6·5 or 8·0 ml, yeast 10 g, salt 15 g, water approximately, 600 to 650 ml.
Dough-temperature: 29°C, resting-time: 60 minutes.
The addition of 0·65% or 0·80% lactic acid is equivalent to 40% or 50% full-sour in technological processing terms.

Criteria for the judgement of rye-bread include the following: symmetry and volume, crust-colour, crust-texture, crumb-openness, evenness of crumb porosity, flavour (according to process), crumb-elasticity, acidity-index of the crumb. The maximum number of points awarded is 20. In evaluating the processing-value of a rye-flour, the most important criteria are: crumb-elasticity (maximum 3 points), and flavour (maximum 5 points), crumb-openness and evenness of porosity (maximum 5 points), symmetry and volume (maximum 2 points), crust-colour (2 points maximum), crumb-structure (maximum 2 points), acidity-index of the crumb (maximum 1 point). A rye-flour quality score of 19–20 points is classed as very good, one scoring 16–18 points as good, one scoring 15 points as adequate, and one with fewer than 15 points requires the addition of an 'improver-quality rye-flour' for commercial processing.

Common crumb-faults are: lack of crumb-openness, dense crumb, and in more extreme measure, unbaked damp crumb-zones. Common crust defects are separation of crust from crumb and splitting of the crust. Flavour and aroma deficiencies vary from lack of flavour and aroma to bitterness and contamination effects (mouldy and storage off-flavours). The ideal rye loaf should have an evenly browned, elastic but crisp crust; have even symmetry and oval cross-section; have an evenly pored, but aerated, crumb of even colour and good elasticity; have an aromatic, mild acid taste due to the souring of at least 50% of the rye-flour used, and adequate ripening or maturation of the sour-process used. A hearth (oven-bottom) rye loaf should have at least 5 mm of crust surrounding the crumb, depending on variety, and addition of wheat-flour. Compared with wheat bread, it is more satisfying of the appetite, especially when eaten only spread with butter. Nutritionally, rye-flour bread provides 3–5 g/100 g of insoluble-fibre (B. Thomas procedure), compared with 0·9–1·0 g/100 g in the case of wheat breads. As a dietary source of the minerals iron, phosphorus and potassium rye bread is 1·5–2·0 times richer than wheat bread in mg/100 g bread. However, wheat-flour bread is about twice as rich a source of calcium (mg/100 g bread) as rye-flour bread. The use of properly matured sour-dough and yeast (50% of total rye-flour content used to prepare the basic-sour) ensures a distinctive, appetizing and satisfying taste, and a vastly superior shelf-life compared with that of wheat-flour bread.

In the USA, where rye-flours are classed as light, medium and dark, flour test-bakes are carried out with these flours blended with a clear wheat-flour 50:50, 60:40, and 30:70 respectively. Doughs are made by adding 2% yeast and salt (flour basis), and 60–68 ml water/100 parts flour. First remix or punch is at 75 minutes after mixing, the second 45 minutes later, and the dough is moulded into shape after a further 30 minutes rest-period. The baked bread is scored accordingly regards volume, weight, general appearance, crust-colour, grain, texture, odour and taste, 25 points being awarded for general appearance, and 25 for odour and taste; 20 points each for grain and texture, and 10 for crust-colour, i.e. a maximum total of 100 points. A typical American rye bread is made from 15% light rye-flour, 15% dark rye-flour and 70% wheat-flour using a yeasted dough-process. A dark-rye bread is made from 30% dark rye-flour and 70% wheat-flour; a light-rye bread from 20–25% light rye-flour and 75–80% wheat-flour; and a raisin-rye bread from 10% dark rye-flour, 5% light rye-flour and 85% wheat-flour and 35% raisins (flour weight), all doughs being predominantly yeast-dough processes. For the evaluation of wholewheat-flours the following procedure is appropriate.

Wholewheat-flour 100 parts, yeast 2·0, sucrose 3·0, salt 2·0, shortening 2·0, water (approximately) 68·0.

Bulk-ferment for 1 hour at 85°F (30°C), remix or punch-down and ferment for a further 1 hour. Scale at 20 oz (560 g), and proof finally for 1 hour. Bake for 45 minutes at 435°F (225°C) in a fan-assisted hot-air oven.

The hemicelluloses and hemicellulase-activity (pentosans and pentosanases) of flours are significant on account of their effect on the water-binding capacity. This is particularly important in the processing of rye-flour bread. The dough-forming properties depending on the content, polymerization and substitution of the pentosans, i.e. degree of branching. The depolymerization effect of the hemicellu-lases (pentosanes and glucanases), as a result of grain-sprouting, can lower the processing value of flours. The pentosan concentration, after acid hydrolysis, is determined colorimetrically with aniline, according to Cerning and Guilbot (*Cereal Chemistry*, **50**, 1973, p. 176), or Petzold and Volkmer (*Bäcker und Konditor*, **28**, 1980, p. 282).

## Yeast

Pressed or baker's yeast is manufactured by culture on sugar containing or saccharified substrate as the genus *Saccharomyces cerevisiae*. The minimum dry matter content should be 25%, and it should not contain any starch, flour, brewer's yeast or other additions. Fresh yeast has a light-grey to yellowish-white colour, with a pleasant, characteristically fresh, faintly acid taste and smell. Dry, crumbly, brownish samples are indicative of autolysis, with accompanying yeast-extract odours.

For the identification of dead yeast cells, a suspension of the cells is mixed with the same volume of a solution containing: methylene blue 200 mg, potassium dihydrogen phosphate 27·2 g, disodium hydrogen phosphate 0·07 g per litre of distilled water. In contrast to live cells, the dead cells stain red, allowing counting under a microscope.

The use of molasses as a source of energy in the commercial manufacture of baker's yeast means that the main carbon sources are sucrose with a small amount of invert-sugar. These conditions of growth give the biochemical properties of the yeast a special emphasis, and demands specific measures to ensure high quality and efficient dough aeration. The technological role of yeast in wheat-flour doughs is a strong alcoholic fermentation with a large carbon dioxide liberation. The gassing power of pressed-yeast depends on the activity of the zymase enzyme-complex of the yeast-cells, and the fermentable carbohydrate available.

The gassing-power of a yeast for breadmaking purposes is evaluated by the following procedure.

A standard production flour with acceptable analytical specification is sampled by weighing 280 g and tempering to 35°C. 160 ml of a 2·5% salt solution, prepared with tap-water, are brought to about 55°C, and placed in a laboratory mixer. When the temperature falls to 35°C, 5 g (approximately 1·8%) yeast are stirred in, the prepared flour added, and the whole mixed for 5 minutes. The dough is then immediately placed in a greased baking-pan measuring 8·5 cm high, with base measurements 9 cm × 14 cm, and top measurements 10 cm × 15 cm. The panned dough-piece is then placed in a thermostat at $35 \pm 1°C$ to ferment. When the dough has risen to the height of a marker positioned 7 cm above the bottom of the pan, the time required is noted, and the dough removed and mixed for 1 minute intensively. It is then replaced in the pan and returned to the thermostat to ferment further. The second rise to the 7 cm marker represents the second proof-time, the gassing-power being expressed as the proof-time in minutes. Proof-times are always measured from the commencement of mixing. The first or main proof-time should not take more than 80 minutes, and the second proof-time not longer than 50 minutes. The signalling of the time to reach the 7 cm marker can be accomplished either electro-optically or electro-acoustically. (Standard test-method for the Baker's Yeast Industry—'proof-time', TGL-25111/03, GDR, May 1971.)

Although the principle of the above method has been used in many countries for a long time, its validity for industrial bakery production must be subjected to the following critical analysis. Particularly where direct straight-dough processes are in use, it is more than likely that the same yeast when used under production circumstances, would show a lower gassing-power during the first or main proof-time in most cases, and often also during the second proof-time. This difference in gassing-power of yeast, as determined by the above standard method for yeast, and its performance under production conditions is due to the low maltase activity of yeast commercially grown on molasses. Unlike the zymase complex, maltase is an adaptable enzyme, which the living cell only produces when the substrate contains maltose, and even then only after a period of induction. During the relatively short fermentation period for the above standard method, the pre-existing sugars in the flour (sucrose, glucose and certain polyfructosides) are sufficient to provide enough gassing-power. However, when the same yeast is added to a production dough, often made from a flour containing less readily fermentable sugars, the first or main proof-time will take very much longer. After using up all the available sugar, the yeast cells lack fermentable carbohydrate, since they cannot use the maltose in the dough. For this reason, Jelitzki and Semichatowa in the USSR, developed a production-based method in which the time required to produce a predetermined volume of carbon dioxide from both a glucose and maltose solution was measured. As a quantitative yardstick for the relative activity of the maltase and zymase complex the time is recorded in minutes. For this special method of measuring the maltase activity alongside the zymase activity, Jelitzki devised a Microgasometer of special design, to stand in a thermostat at 30°C. The maltase activity and zymase activity of pressed-yeast varies considerably, and the All-Union Research for the Baking Industry of USSR have suggested the scheme shown in Table 50 for the evaluation of the enzyme activity and quality of yeast samples. These values are intended as guidelines

TABLE 50

| Activity in minutes | | Yeast quality |
|---|---|---|
| Zymase | Maltase | |
| 30–40 | 50–80 | good |
| 41–60 | 90–120 | satisfactory |
| 61–70 and above | 121 and above | unsatisfactory |

only, as there is no direct relationship between zymase and maltase activity, and a high maltase activity does not always result in a high yeast gassing-power. In the case of direct straight-dough processing methods, the use of a yeast sample with low maltase-activity will have a negative influence on the dough-ripening process. In such processes, the gassing-power of the yeast must start immediately after mixing. And, although in the case of indirect sponge-doughs, and pre-ferments this factor can be less critical owing to substrate adaptation, a general shortening of any total process will result from the use of yeast with a high maltase activity. Dough leavening is an entirely anaerobic fermentation, depending on the rate at which the organism can push hexose sugars through glycolysis. Sucrose is rapidly converted to glucose and fructose, owing to the activity of the yeast enzyme invertase. Invertase activity can be measured from the amount of reducing sugar formed within 10 minutes from a standard 15% sucrose solution at pH 5·2 at 60°C, by a yeast suspension containing 6·2 mg dry cells/ml. Although both glucose and fructose are the first sugars to become fermented, when both are present glucose disappears at a more rapid rate than fructose. Once these hexose sugars have been consumed, the yeast enzyme system has to adapt to maltose, which results in a sudden reduction in dough gassing-power. However, during indirect pre-ferment processing systems, no sudden reduction in maltose conversion is observable. The limiting factors for progressive maltose availability would be the percentage of damaged-starch granules and the presence of sufficient alpha-amylase. In countries where sucrose or corn-syrups are standard raw materials, their additions will delay the onset of maltose fermentation, and where sufficient is added, eliminate it entirely.

The shelf-life of yeast can be subjectively determined by cutting three 2-cm slices from a packet, and wrapping in greaseproof paper. The wrapped pieces are then placed in an incubator at 35°C for at least 96 hours at a relative humidity of 90%. No signs of excessive softening should occur within this time. More objective methods requiring specialized apparatus involve the use of penetrometers, or an electrometric measurement of pH and redox-potentials (Bergander and Bahrmann, *Nahrung*, **1**, 1957, 74, and **2**, 1958, 500; standard test method TGL-25111/04, GDR, Shelf-life of baker's yeast).

The water content of yeast is determined by direct drying; 2 g yeast accurate to within 0·5 mg, are weighed into a dish, and 1 ml ethanol stirred in. After drying for 3 hours at 105°C, the percentage weight loss is calculated. Moisture content should not exceed 75% by weight. The acidity-index of yeast is determined by suspending 20 g in distilled water in a 100 ml volumetric, adjusting to the mark. Filter, collecting 50 ml of filtrate. Titrate against 0·1 N sodium hydroxide, using phenolphthalein as indicator.

The acidity-index is expressed as the number of millilitres of sodium hydroxide needed to neutralize the 50 ml extract from 100 g yeast. 1 ml 1 N sodium hydroxide corresponds to 0·09 g lactic acid.

Although most bakeries only determine the gassing-power or fermentation power of yeast deliveries, there are certain properties of great processing significance which are usually overlooked. One of these is the glutathione content of yeast, which is water-soluble and which acts as a proteolysis-activator in dough. Of the polypeptides of the yeast-cell, glutathione (glutaminyl–cysteinyl–glycerine), a tripeptide, has a special significance technologically for the breadmaking process. It exists in two forms; in the oxidized form it contains a disulphide group, and in the reduced form a sulphhydryl group. It becomes involved in the oxidation/reduction reactions within the living yeast-cell, and can represent up to 1·0% of yeast dry-matter; and when yeast-cells die off, the glutathione is released into the dough-mass in the reduced form, having a negative effect on dough proteins. Dried yeast releases higher levels of glutathione into the dough than pressed yeast, depending on the drying process used to prepare it. It can be removed from yeast by mixing with water in a suspension at 35°C for 12 minutes, and holding under vacuum at 0·055 MPa (0·55 atm) for 5 minutes. The water with dissolved glutathione is then centrifuged, yielding a glutathione-free yeast. Vacuuming has no effect on the yeast rising strength or maltose activity. Its removal improves dough rheological properties (elasticity, viscosity, stability), bread yield increases, and dough-spread and adhesion to baking-surfaces are reduced, whether steel or fluoro-plastic.

Glutathione can be either qualitatively detected or quantitatively determined in the vacuumed centrifugate by treating with a denaturing agent such as 3 M urea to open up the bonds to expose the free thiol or —SH— groups, including cysteine and glutathione. The mixture is then alkalized with ammonium hydroxide, and enough sodium nitroprusside added to develop the purple colouration. Denaturation may require the addition of a high concentration of urea to maintain the colour. To quantify, measure the absorption at 280 $\mu$m, using a calibration curve prepared by using standard concentrations of glutathione, and the same reagents.

Baker's yeast should be a technically pure culture of *Saccharomyces* yeast-cells, and the percentage contamination with wild or other genera (e.g. *Candida, Torulopsis,* etc.) should be limited to less than 20% of cell dry weight. Their presence may increase the yield of the industrial yeast product, but have a negative effect on the fermentation power and storage-life of the yeast. More attention should also be paid to the maltose activity of yeast, and this property standardized and controlled by the baker.

Dry yeasts are available in the form of granules or pellets. Such products should not contain more than 10% of fine 'dusty' material. Top quality products should not have moisture contents in excess of 8·0%, or fermentation or gassing-power of more than 70 minutes, using the standard method described earlier. Second-grade dry yeasts could have moisture contents of up to 10%, and take up to 90 minutes to attain the required gassing standard. Storage-life would be correspondingly reduced to about 5 months compared with about 12 months for top-quality products. Dry yeasts must be appropriately packaged, have moisture levels of 7–9% and be stored at up to 15°C during transport and in the bakery.

For testing the gassing or fermentation power of dried yeasts, a reduced amount of 2·5 g must be taken, and be premixed with a small amount of flour/water mixture at 35°C for a period of 30 minutes to activate before testing in the manner described for pressed samples. The actual amount of carbon dioxide produced in a dough by sugar fermentation amounts to about 70% of the theoretical quantity indicated by the chemical equation for alcoholic fermentation. This is due to the fact that part of the sugar is required by the yeast cells for energy and growth processes. Other factors which affect the relative speed of alcoholic-fermentation in dough are the following:

*Temperature:* Raising dough temperature from 20 to 30°C will almost double the fermentation-speed, and the gassing-power within the dough.

*pH:* The optimum range for fermentation and respiration is pH 4·0–6·0. Flours giving dough pH of 5·5 are regarded as ideal.

*Vitamins and growth factors:* The most important of these are thiamine, biotin, pantothenic acid with others.

*Mineral compounds:* The addition of various salts, in the form of yeast-food used in many countries, speeds up yeast fermentation. Such compounds include: calcium phosphate, ammonium sulphate, ammonium phosphate, ammonium chloride, calcium sulphate. The ammonium salts are particularly useful as a source of inorganic nitrogen for flours of lower extraction-rate.

*Amino-nitrogen compounds:* The presence of protein degradation products such as peptides and amino acids will greatly increase the fermentation-speed. Bread-crumb or waste bread hydrolysates prepared either by acid or enzymic hydrolysis, producing reducing-sugars, glucose and amino acids, will both stimulate fermentation and improve flavour and aroma of the baked bread. These have been the subject of several patents for both wheat and wheat/rye-flour bread in several European countries.

For the quality control of active dry yeast, for comparative tests, the following parameters are suggested: glutathione-content, fermentation-potential, total carbohydrate content, trihalose, and amount of low-molecular material released on rehydration at 8°C. The latter correlating well with fermentation-power changes, according to D. Josić.

## Sour-dough Starter Cultures and Sour-dough

Sour-dough starter cultures (German: *Anstellgut*) for the preparation of rye sour-doughs consist of specially selected pure cultures of lactobacilli (lactic acid bacteria), or mixed cultures of various species selected for their specific properties, either with or without selected yeasts. The lactic acid bacteria are homo- or hetero-fermentative types of the genus *Lactobacillus*, having specific growth requirements, depending on the degree of acidification of the substrate and fermentation process. For the factors pH, acidity-index, and lactic/acetic acid ratio, the technological importance of lactic acid bacteria in the sour-dough starter culture vary with the fermentation process. The hetero-fermentative *Lactobacillus brevis* var. *lindneri* is capable of producing adequate acidity over the widest range of processing conditions, and conclusions concerning processes must always be made with reference to the specific starter-culture used. This, in turn, is a function of the spectrum of microflora which it contains. The amount of starter used depends on the weight of dough to be produced, but a guideline would be approximately 0·5% of total flour weight. Larger amounts of up to 20% are possible, depending on type of process, type of flour or meal, and eventual sour-dough flour percentage based on total rye-flour content. Mixed wheat/rye doughs require at least 50% of the rye-flour content to be acidified to ensure adequate crumb firmness and flavour (Stephan, 1982). Most starter-cultures also contain yeast microflora, which belong to the *S. cerevisiae* genera, but have larger round-to-oval cells, which adapt to the higher acid medium and exist symbiotically alongside the bacteria. Schulz (1941 and 1954) described this Tribe as P-14 or 'sour-dough yeast'. In a rye-flour dough these sour-dough yeasts produce about two-thirds more carbon dioxide than an equal quantity of pressed yeast over a 3-hour fermentation-time. These yeasts can tolerate lactic acid concentrations of up to 1·0% without loss of activity and gassing. By careful cultivation, these yeasts retain their properties after separation from sour-doughs for culture on molasses and isolation in the pressed form. The manufacture of sour-dough yeasts in a pressed form could speed up and simplify the preparation of rye-flour doughs. However, their cell-size and gassing-power diminishes with prolonged culture at temperatures much in excess of 38°C. In addition, certain wild yeasts, in particular the species *S. minor* which has smaller round-to-oval-shaped cells, fermenting glucose, fructose and sucrose, but not maltose, also inhabits rye sour-doughs. These cells show optimum growth at about 25°C, and an acidity-index about 10. Such starter-culture products are added to rye-flours and water with the objective of initiating a lactic acid/sour-dough fermentation in symbiosis with a yeast-fermentation with simultaneous high cell-propagation within a relatively short time.

In order to standardize sour-dough production and facilitate its preparation on an industrial basis, various products are marketed as dried concentrates, in which the life-cycle of their constituent bacteria and yeasts have been temporarily interrupted. Such products make possible the preparation of rye bread without a multi stage sour process, using instead a direct one-stage procedure. By carefully drying pure sour-dough cultures using special freeze-drying techniques, fine dry-powders are produced, which maintain the activity of the bacteria *L. brevis lindneri* for a limited period when stored at refrigeration temperature. However, in order to reproduce at the normal rate, the culture must be fresh, enabling it to build up lactic and acetic acids, alcohol and carbon dioxide to leaven the bread, as well as aroma precursors.

Freeze-dried sour-doughs are prepared by a traditional three-step method, each step differing in consistency, rate of yeast and bacterial development and temperature from the previous one. The dough is cultured with the appropriate homo- and hetero-fermentative lactobacilli, which form the optimal balance of lactic and acetic acids, reaching a pH of about 3·8. The dough is then deep-frozen and dried in a high vacuum to remove the ice crystals by sublimation. The dried sour-dough is then milled to the desired particle-size. The protective nature of the freeze-drying process prevents any damage or denaturation of the bacteria, yeasts and aroma precursors. Other drying techniques are fluid-bed/air-lift, using high temperatures of 70–80°C at the inlet, utilizing the cooling effect of moisture evaporation to reduce the moisture content to about 35%, rapidly without cell damage. A temperature of 35–40°C is then used to reduce moisture to 4–8% at the second stage of drying. This process produces cylindrical particles of 0·2–0·5 mm diameter, and 1·0–2·0 mm in length. The rehydration rate of these particles even when added directly to flour and water is high. Another drying method used is roller spray-drying.

Single-stage production of rye sour-dough bread is achieved within 3 hours, instead of the 3 days fermentation required to obtain the flavour intensity of acidified bread conventionally. Furthermore, there is no variation in quality standards from batch to batch, giving consistent quality bread, time after time. A typical analysis of dried sour-dough starter used in this way would be: moisture 8·0%, pH 3·8, acidity-index 40–44, ash 1·5%. Other dried sour-dough starters are used to produce a basic-sour dough (*Grundsauer*), which stands for 12 hours at 25°C, with a dough yield of about 175/100 g flour. Usually prepared before the end of a working day, and taken the following morning. The ratio of the pure or mixed sour-dough starter to flour for the basic-sour is from about 1:3 to 1:10. Using this system, a weight of basic-sour equal to that of the added starter is retained to prepare the next batch of basic-sour. This process is repeated for one week's production programme, and a fresh batch of starter used to prepare a fresh basic-sour over the week-end. Normally, these starter products are formulated to allow standing-times of up to about 40 hours without excessive acid build-up.

Another type of sour-dough preparation consists of a two-component system, component 1 is a dry, flour-like powder and component 2 a liquid. There is often a choice of product variations, differing in acid content according to the baker's requirements. Using this system, most of the dough acidification is derived from the inclusion of organic acids: lactic, acetic and citric, added in either the crystalline or the liquid form. Such products are often used in combined sour-dough processes, whereby about 10% of the total flour is acidified with a conventional basic-sour of 24 hour standing-time, to ensure a fully rounded aroma-profile in the finished bread. A typical formulation for a rye-wholemeal or mixed rye/wheat-flour dough is as follows: flour 100 kg, basic-sour 15·0, component 1 1·2, component 2 0·2, salt 1·8, yeast 1·0 at a dough yield about 155 kg.

Alternatively, a direct one-stage process can be used to produce a bread within about 3 hours in an emergency situation, as follows: flour 100 kg, component 1 1·4 kg, component 2 0·5 kg, salt 1·8, yeast 1·5. Although this procedure is a useful standby, the bread aroma is not comparable qualitatively with a traditional uninterrupted sour-dough.

In order to maintain a quality standard in bread production, irrespective of the type of sour-dough or process system employed, the sour-dough must be tested at each production stage. In a small bakery the sour-dough is subjected to sensory testing. A sour-dough stage has reached maturity when fermentation has reached a peak, and the acid-forming bacteria have produced a balanced acidic aroma. Dough-ripeness can be judged by the volume increase and the stability of the dough. If an initial-sour (*Anfrischsauer*) or basic-sour (*Grundsauer*) is sprinkled with flour on the surface in the machine-bowl, the appearance of cracks on the surface indicates that the dough is ready for the next processing stage. The maturity of a full-sour (*Vollsauer*) can be judged by the collapse of the dough when punched in the machine-bowl, and the appearance of large blisters (gas-pockets). A mature aroma is indicated by a balanced smell, which is neither too pungent from acetic acid, nor too mild from lactic acid build-up. However, the presence of these acids can only be detected if the gassing-power of the yeast is sufficient to carry the aroma in the liberated carbon dioxide (acting as a carrier-gas effect).

In industrial-bakeries equipped with a laboratory, facilities are available for chemical and physical tests on doughs and finished products, which facilitates the control of the sour-dough process and provides information for judgement of the bread. The acidity-index (*Säuregrad*) can be determined either in an aqueous dough extract, according to Neumann, or by using a method due to Schulerud involving a filtration stage. According to Neumann, 10 g of sour-dough are pasted with 100 ml distilled water, and homogenized with a mixer. 5–10 drops of phenolphthalein indicator solution are immediately added, within a maximum of 3 minutes, and the suspension titrated with 0·1 N sodium hydroxide. Titrate to the first appearance of a permanent red colouration of the suspension. The number of millilitres of 0·1 N sodium hydroxide required is taken as a direct measure of the acidity-index of the sour-dough. The change in acidity-index of the sour-dough during the ripening period will depend on the following parameters: biological composition of the sour-dough starter; type of process; temperature; dough-yield (dough-firmness). A conventional basic-sour of dough-yield 155 at 24°C, after 24 hours standing, could reach an acidity-index of 17–18, compared with an acidity-index of about 4 at the outset. During this time, the pH value of the basic-sour will fall to 4·0 and remain constant, and providing the temperature is kept at 24°C or below, basic-sours can be used for up to 5 days' bread production. The acidity-index measures the total water-soluble acidity of the sour-dough, whereas the pH is an indication of the degree of dissociation or ionization of the dough acids. There is no valid direct relationship between these variables, and only a loose relationship, based on general empirical data gathered from practical experience, is possible. Temperature has a marked influence on the acidity-index data, but relatively little on that of pH. Practical guidelines from experience give the following ranges:

| *pH value* | *Acidity-index* |
|:---:|:---:|
| 3·5 | 16–18 |
| 4·0 | 9–12 |
| 4·5 | 6–8 |

The relative ratio of lactic to acetic acids, depending to some extent on the process system and whether batch or continuous, is an important variable for bread quality. A correctly matured sour-dough should contain 80–85% lactic acid, and 15–20% acetic acid, based on total acid content. Lactic acid and acetic acid favourably influence flavour and aroma, and also suppress the sour-dough yeasts from microbial contamination, and any enzymic breakdown of the carbohydrates. Lactic acid encourages dough cohesion, giving the rye-flour its baking properties by peptidizing the water-soluble proteins and inhibiting potential enzymic breakdown leading to dough-instability. Starch granules swell, hindering potential crumb dextrinization. If the acetic acid content rises too high the sour-dough yeasts become inhibited, giving the bread a pungent smell and unacceptable flavour.

The quantitative determination of both acetic and lactic acids in a fermented sour-dough can be carried out according to a standard procedure for the vegetable volatile organic acids. The acetic acid is separated from the lactic acid by steam-distillation, ensuring that 200 ml of distillate is collected within 50 minutes of passing steam into the sample, the steam used being free of carbon dioxide. The distillation procedure must be such that when a 0·1 N acetic acid solution is steam-distilled, 99·5% must be recovered; and when a 1 N lactic solution is distilled a maximum of 0·5% is found in the distillate. The test sample distillate is collected and just brought to boiling-point, when a few drops of phenolphthalein indicator solution are added. The solution is immediately titrated with 0·1 N sodium hydroxide. The calculation is as follows: 50 ml aliquot boiling distillate required $x$ ml 0·1 N sodium hydroxide. Then, 1 litre of distillate contains $0.12x$ grams of volatile acetic acid. Taking into consideration test-sample weight and dilutions, the percentage can be calculated.

Lactic acid remains in the residue after the acetic acid has been distilled off, since in dilute solution it is not very volatile. The residue is rinsed into a porcelain dish with a small quantity of distilled water (quantitatively), treated with 5 ml 10% barium chloride solution and about 2 g of finely powdered barium carbonate, mixed, and with constant stirring, concentrated down to about 10 ml. This residue is taken up in 15 ml distilled water in a 100 ml measuring cylinder fitted with a ground-glass stopper. The volume at 20°C is brought to 25 ml with quantitative rinsing with water. After bringing to the 100 ml mark with ethanol, the cylinder is stoppered, and set aside after shaking, and periodically shaken over 2 hours. The mixture is then filtered into a clean 100 ml measuring cylinder, and 90 ml of filtrate collected, or, in case this is not reached, collect exactly 81 ml for ease of calculation later (the end result can then be multiplied by a factor of 1·1). Transfer the measured volume of filtrate into a platinum or quartz dish, and carefully concentrate to dryness. The dry residue is carefully carbonized on an asbestos plate, and subsequently ground with a glass pestle. The ground residue is then moistened with a few millilitres of water, evaporated to dryness on a water-bath, and carbonization completed. If necessary, the treatment with water can be repeated. The pure white residue is then treated with 0·5N sulphuric acid, normally 5 ml is adequate, to give a sharp end-point using bromothymol blue indicator (the quantity required being (say) $a$ ml). Dilute with about 50 ml hot water, and heat for a few minutes on a water-bath to remove all carbon dioxide, placing a watch-glass on top. After adding 1–2 ml of 0·1% bromothymol blue solution, the solution is titrated with 0·5N sodium hydroxide (the quantity required being (say) $b$ ml). Then,

$$\text{grams lactic acid/litre} = \frac{a-b}{2} \times 20 \times 0\cdot1$$

(*Z. Lebensmittel-Unters, u. Forsch.*, **113**, 1960.)

This method is a selective one intended for wine organic acids, but for a rapid approximation, a straight titration of the water extract of the sour-dough with 0·1N sodium hydroxide, using phenolphthalein as indicator, or in the case of turbidity or strong coloration from rye-flour, potentiometric titration using the glass electrode and pH is more reliable.

For the manual titration method, 25 ml of carbon dioxide free water extract of the dough are titrated with the 0·1N sodium hydroxide and the respective amounts of acid/1 ml 0·1N sodium hydroxide calculated as follows: acetic acid 6·0 mg, lactic acid 9·0 mg, citric acid 7·0 mg and malic acid 6·7 mg. Having determined the acetic acid present in the dough by the method of steam-distillation described, the lactic acid can be deduced from the difference between total titratable acidity and an accurate acetic acid analysis.

For the determination of dough-acidulants in combination with Quellmehl (pregelatinized starch or flour) for use as biological sour-dough replacements, Gerstenberg has recommended an enzymic method for the detection of citrate and/or malate in the baked bread (Detection of dough-acidulants in bread from the citric acid content, *Z. Lebensmittelchemie und gerichtlichen-Chemie*, **32**, 1978, pp. 125–126). Other useful methods of defining standards for these intermediate sour-dough additives have been contributed by Rabe (AGF Detmold) entitled 'Organic acids in breads from various processes', Part I: Methods for the determination of the mode of acidulation (*Getreide Mehl und Brot*, **34**, 1980, 4, pp. 90–94), Part II: Various methods for the determination of the type of acidity (*Getreide und Brot*, **6**, 1981, pp. 146–150), Spicher, G. and Stephan, H. (1987). In *Handbook of Sour-dough Biology, Biochemistry and Technology*. B. Behr Publishers, Hamburg (in German).

TABLE 51

| | Intensity of acidification | | |
|---|---|---|---|
| | *Mild* | *Average* | *Strong* |
| Acidity-index | 8–10 | 10–13 | 13–20 |
| pH value | 4·2–4·1 | 4·1–4·0 | 4·0–3·6 |

In discussing the various sour-dough processes and the resultant rye-bread quality, Stephan (1982) gives some guidelines with regard to 'Intensity of acidification' of sour-doughs depending on the amount of rye-flour, as a percentage of total flour, which has been 'soured' (Table 51).

In general, the shorter the sour-dough process, the higher the requirement of sour-dough and starter as a percentage of total flour, e.g. Berliner short-sour process requires 60% of total rye-flour processed to be 'soured', and 20% starter in the sour-dough (flour weight). The maturing-time being 3–4 hours at 35°C with a dough-yield 190%. This represents the shortest biological sour-dough process, developed by Pelshenke and Schulz in 1941 (*Mühlenlaboratorium*, **11**, 11, 1941, p. 105). The sour-dough of this process produces about 85% lactic acid and 15% acetic acid, the higher lactic acid content providing a relatively mild bread flavour. Average acidification reaches an acidity-index of 10–13.

Another technique for controlling the intensity and type of acid build-up, is to feed at the beginning of each sour-dough stage—initial, basic and full-sours—with air by using a special beater-aggregate. The continuous sour-dough process plant KVT 1000 of Czechoslovakian manufacture makes use of a similar technique. The process parameters for this 'foam-sour process' is a soft consistency, dough-yield 200–250%, and temperatures of 22–25°C. These conditions suppress acid formation and encourage yeast reproduction, which gives the dough increased gassing-power, and the bread a mild flavour.

A short-sour process for a mixed rye/wheat bread dough (*Mischbrot*) 70:30 of 3–4 hours standing-time would be as follows:

Short-sour: starter 10 kg, rye-flour 50 kg, water 45 kg at 35°C.
Dough: Short-sour 95 kg, rye-flour type 997 or 1150 20 kg, wheat-flour 30 kg, water 15 kg at 30°C, yeast 2 kg, salt 1·5 kg. Standing-time: up to 15 minutes; 10 kg of short-sour is retained as starter for the next batch.

In contrast, a continuous sour-dough process using the doughmaking plants KVT 1000, 1500 or 1800 from Czechoslovakia, or the FTK 1000, 1000 U, 1000 UM or 1500 from Hungary, demands control and synchronization of such process parameters as: temperature, dough-consistency, maturing-time, and acidity-index of each sour-dough processing stage. Such processes must be so organized that a mature full-sour is available for processing, as described earlier. The mature full-sour is continuously fed into the mixing-aggregate. The process is effectively a full-sour to full-sour process, operating over a working-week. For this system, the initial-sour would be ripened when an acidity-index of 16–18, and a pH value of

3·5–3·6 has been reached. At this stage it is aerated by feeding into the sour-dough beater-aggregate, after preparation of the next sour-dough stage, the basic-sour. The basic-sour is mature on reaching acidity-index 10–12, and pH value 3·8–3·9; and the full-sour, on reaching an acidity-index of 9–11, and a pH value 3·8–4·0.

All baked products made with more than 10% rye-flour must be acidified, in order to render the baked products acceptable from a technological and flavour viewpoint. The classical biological sour-dough processes normally applied include: a three-stage process, Detmolder two-stage process, Detmolder one-stage process, Berliner short-sour process, and the Monheimer salt-sour process.

Most of these processes can be applied as batch or continuous with certain process-modifications, but it is generally accepted that there are distinct technological and sensory advantages in employing a classical biological sour-dough process. Even when processing sprouted rye-flours, with Falling number (Hagberg) between 80 and 100, it is possible to produce acceptable bread using purely biological sour-dough processes. This is achieved by reducing dough yield in the initial sour-dough stages, and then adjusting the third or final sour-dough stage for a higher temperature at a higher dough-yield, and for a slightly longer standing-time. Thus inducing a sharp fall in pH value.

As already mentioned, sour-dough acidulants are marketed and used for direct and combined sour-dough/acidulant processes. The acidity-index of these products is usually about 100 to 1000, and the dosage rate is adjusted accordingly.

These preparations are formulated from organic acids and Quellmehl (pre-gelatinized starch or wheat-flour). Crystalline acids are added to the Quellmehl, liquid acids are often added separately but in combination with the Quellmehl/acids blend. Acids used are: lactic acid (E 270), acetic acid (E 260), citric acid (E 330), malic acid (E 296), fumaric acid (E 297), tartaric acid (E 334), adipic acid (E 355), and succinic acid.

Typical of such a product is the Teltoma-sour, manufactured by the VEB Backwarenkombinat, Potsdam, GDR. This product is used for the production of rye- and mixed rye/wheat-flour doughs, and consists of two parts. One part is Quellmehl, blended with organic acids, and the other part is 80% acetic acid, the latter being used in a dilute form. The dry blend (part I) has a maximum moisture content of 8·0%, and an organic acid content of at least 6·0%. In the FRG, similar products are available from Böhringer of Ingelheim/Rhein, Ireks-Arkady, 8650 Kulmbach, amongst others. Because of the processing-time and processing-loss in raw material solids, incurred with the conventional processes, the use of these products has led to more direct processes. These replace the acid build-up from the microorganisms by direct acidulation; but owing to legislation and bread-quality effects, this technique is only used to replace some of the biological souring. Usually, 10–20% of the rye-flour content of the mix is matured to the basic-sour stage, and the following process parameters adhered to: temperature 25°C, dough-yield 155%, standing-time 15–24 hours (increasing the amount of flour to 25–30 times that used in the starter).

Where legislation permits, and no biological souring is used, pressed-yeast must be added, and a fermentation time of 3 hours given to activate the yeast. Direct processes generally, using acidulant preparations with yeast, require a longer fermentation time for gassing and acid build-up than is the case with conventional biological sours. Average times are 60–180 minutes for the former against 15 minutes in the case of the latter.

Normally, the relatively short maturing-times of rye-flour doughs eliminate the need for remixes/punches/knock-backs. Final-proofing times average 50 minutes at 35°C at 60–70% RH.

## Wheat Sour-dough

Wheat sour-dough represents the oldest method of producing fermented bread. A sour-dough is employed to develop acetic acid, thus creating optimum conditions for swelling of the flour components in the dough, for structure for baking, and for aroma and flavour precursors. Spoilage organisms are inhibited, and an excellent flavour profile and balance obtained.

Traditionally, the production of wheat or rye sour-doughs is accomplished by a three-step process before the final mixing stage. This empirical process optimizes the conditions which are essential for the microorganisms to produce the type and quantity of acids leading to dough ripening or maturity.

Wheat-flour and water are initially mixed with a small amount of dough-starter to give a dough yield of 200%. The dough is then allowed to ferment for about 6 hours at a temperature of 25–26°C. This allows the yeasts to propagate, and acidity to develop. Then, by adjusting the dough yield to 170%, and giving a further fermentation-time of 8 hours at 24–27°C, the dough matures by developing a strong acidity and aroma. This is due to the activity of the homo- and hetero-fermentative lactobacilli, which form lactic and acetic acids.

During the third and final stage, optimum aroma and fermentation is ensured by maturing for 3 hours at a temperature of 28–31°C, and readjusting dough yield to 190%. The dough is considered to be mature when it shows a pH of 3·8. Acid formation being a direct function of flour mineral-ash content, and the starch hydrolysing enzyme content of the dough.

Producing a good quality sour-dough demands an exact control of acidity (acidity-index, i.e. total titratable acidity (TTA) = *Säuregrad*), and temperature at each fermentation stage. Even in a modern industrial bakery, exact control of these process parameters can prove difficult. Any quality deviations in the matured dough, can often only be discovered in the baked bread. To achieve rationalization of the production process, and the production of consistently high-quality breads, dried sour-doughs have been developed. The best drying method is freeze-drying, the matured sour-dough being deep-frozen and dried in a high vacuum to remove the ice crystals more easily by sublimation. On drying down to a moisture content maximum of 8·0%, the dried product is milled to a standard particle-size. The freeze-drying process preserves the activity of the yeasts and lactic acid bacteria, as well as the aromatic flavour components.

Other drying techniques such as fluid-bed/air-lift, roller spray-drying; and the use of liquid nitrogen spray by direct contact through a tunnel, using a countercurrent system, can also be applied. The final product is a fine white powder, with a typical acidic flavour and aroma. A typical analysis of this type of product is: moisture maximum 8·0%, pH 3·9–4·0, acidity-index 20–25, ash 0·55%, containing typical sour-dough bacteria with or without sour-dough yeasts. For the manufacture of white wheat bread, the following formulation is appropriate:

> Wheat-flour 100 parts, dried wheat sour-dough 10, salt 2·0, yeast 4·0, water 65 (variable). Dough temperature 27°C.
> Procedure:
> Allow 20 minutes recovery after mixing, scale, make-up and shape for final-proof. Final-proof time 30 minutes, baking at 230°C. When formulating for smaller units, e.g. baguettes or bread-rolls, 2–4% dried wheat-sour dough (flour weight) is usually sufficient to ensure the intensive, full flavour characteristics of acidified bread. Quality is uniform from batch to batch, with an excellent crumb elasticity and shelf-life.

Detailed information on the use of wheat sour-doughs is sparse, but there remains considerable scope for research and application of this technology, aimed at improving the crumb characteristics and flavour of neutral, bland-tasting white bread manufactured in the Anglo-Saxon countries.

The microflora of wheat-sours have not been studied in depth, but those which have become established for certain specialized baked products are detailed here.

The yeast *Saccharomyces exiguus*, present in the sour-dough French bread process of San Francisco, which is distinct from *S. cerevisiae*, in that it is not inhibited by high acidity, does nor ferment maltose, but consumes virtually all the flour-hexoses. This yeast does not, however, contribute to dough acidity, which is provided by a sour-dough bacterium named *Lactobacillus sanfrancisco*. The yeast and bacteria coexist within a pH range 3·8–4·5; the bacteria utilizes maltose as a carbon source.

For the production of the Italian speciality '*panettone*', the sour also contains *S. exiguus*, alongside diverse species of lactobacilli, including *L. brevis*, *L. cellobiosus* and *L. plantarum*, as well as *Enterobacter* and *Citrobacter*.

Another type of bread called '*sangak*', also made with a sour-dough, contains various species of the genus *Leuconostoc*, which are non-pathogens, forming acids, and useful aromas. *Leuc. mesenteroides* is osmotolerant, and can grow in 60% sucrose solutions. Other species include the *Lactobacilli brevis* and *plantarum*, and *Pediococcus cerevisiae* (= *Ped, damnosus*), which is also found in sour beer-wort. Although unwelcome in the latter case, they have useful application in the manufacture of *sauerkraut* and salted gherkins, and in starter-cultures for raw-sausage ripening. Two yeasts present in the *sangak* sour-dough are *Torulopsis candida*, and *T. colluculosa*.

The composition of wheat-dough sours, widely used in Spain, for the preparation of the '*masa madre panaria*' (mother-dough for bread), is of great industrial significance. The microflora of these sour-doughs have been studied in some depth by Barber, Benedito de Barber, Martinez-Anaya, Báguena and Torner at the Cereals Laboratory of the Institute for Agricultural Chemistry and Technology of Foods (CSIC), Valencia 10, Spain. These workers studied three industrial sour-doughs of wheat-flour bread, identifying the microflora, and their functional properties. Yeast cell counts/g varied from $2.2 \times 10^4$ to $1.3 \times 10^8$, and bacterial counts/g from $1.9 \times 10^5$ to $1.5 \times 10^8$. All doughs contained *S. cerevisiae* as the main yeast, but the three samples each contained another species of yeast in the form of *Pichia polymorpha*, *Trichosporon margaritiferum* and *S. fructuum*, and also *Hansenula subpelliculosa*. No other data recording the existence of these yeasts in wheat-sours is known, but Spicher and Schöllhammer (1977), Lorenz (1983) and Spicher *et al.* (1979) have mentioned other species of the genus *Pichia*, and *Trichosporon*, i.e. *P. saitoi* and *T. penicillatum*. The physiological characteristics of these yeasts explain their utility and functionality in wheat-sours for industrial breadmaking. *S. cerevisiae* ferments glucose, maltose, galactose, sucrose, and one-third of the triose, raffinose molecule, and is also capable of assimilating the same sugars completely; *S. fructuum* ferments glucose, galactose, raffinose, and sucrose, but in addition to assimilating these sugars it can also assimilate maltose; *P. polymorpha* assimilates glucose, maltose, galactose, raffinose, lactose, sucrose and ethanol, but does not ferment any of these sugars; *Hansenula subpelliculosa* assimilates glucose, maltose, galactose, raffinose, sucrose, ethanol and nitrite, and is capable of fermenting glucose, raffinose, sucrose; *T. margaritiferum* assimilates glucose, maltose, galactose, raffinose and sucrose, but not lactose or ethanol; and it does not ferment any of these mentioned sugars.

The relative gassing-power of these yeasts, expressed as millilitres of gas/$10^7$ microorganisms, over 3 hours, varied considerably, *S. cerevisiae* producing 500 ml, *H. subpelliculosa* 75 ml, and *T. margaritiferum* and *P. polymorpha* practically none under similar conditions. There are also differences in gassing-power between the various races of *S. cerevisiae*. Furthermore, in the anaerobic pathway of alcoholic fermentation, yeasts cannot metabolize as many types of sugar as is possible in the case of the aerobic pathway. The differences in abilities to assimilate carbohydrates and other organic substances is important for the systematic classification of yeasts in general.

Lactobacilli identified in all sour-dough samples were *L. brevis* and *L. plantarum*, including the *L. plantarum* var. *plantarum*, and *L. plantarum* var. *arabinosus*, all of which differed in fermentation activity according to species and strains. *L. brevis* and some facultative homo-fermentative strains of *L. plantarum* generated only little gas, and *L. plantarum* var. *plantarum* produced the largest changes in total titratable-acidity. The hetero-fermentative *L. brevis* produced generally much less gas under the conditions of the test. Hetero-fermentative lactic acid fermentation proceeds via a pentose as the intermediate stage, decarboxylation of carbon-atom 1 from a hexose forming carbon dioxide. The various stages proceeding from glucose through glucose-6-phosphate, 6-phosphogluconate, to pentose-5-phosphate. Then the pentose becomes split into a 3-carbon and a 2-carbon degradation product. From the 3-carbon product glyceraldehyde-3-phosphate, lactic acid is formed in the same way as with a fructose-1,6-diphosphate pathway. From the 2-carbon product acetylphosphate, acetylaldehyde and then ethanol is formed. Thus explaining how hetero-fermentative lactic acid bacteria, when fermenting glucose produce lactic acid, ethanol and carbon dioxide. *L. brevis* can produce acetic acid from acetyl phosphate instead of ethanol during hetero-fermentative fermentation.

Homo-fermentative lactic acid fermentation is characterized in that the product of metabolism of the lactobacilli from the carbohydrate fermentation is almost entirely lactic acid. This is in sharp contrast to the hetero-fermentative lactobacilli, which take the Embden–Meyerhof–Parnas pathway, and produce, alongside lactic acid, carbon dioxide and alcohol.

The same parameters control the efficiency of wheat sour-doughs for breadmaking, as those discussed for rye-sours, i.e. the type and composition of the microflora, their capacity to produce gas, and their effect on dough pH and total titratable-acidity (TTA) (expressed as ml $0.1$N NaOH/$10^7$ yeast cells, and ml $0.1$N NaOH/$10^{11}$ bacteria).

In order to control flavour and aroma of the baked bread, a determination of the ratio of lactic:acetic, and possibly other dough acid by-products of fermentation such as succinic, malic, tartaric, citric and formic acids, although these only contribute up to 10% of total dough acidity.

Conventional yeasted wheat-flour doughs fall in pH from mixing to maturity from about $6.0$ down to about $5.5$–$5.0$, which corresponds with the isoelectric point of the wheat proteins. The formation of lactic and acetic acids in wheat doughs is due to the hetero- and homo-fermentative lactobacilli present in flour, but commercial yeast also contains a number of acid-forming bacteria. In addition, the carbon-dioxide produced by the yeast forms carbonic acid in the flour/water suspension, and contributes to the rise in the acidity-index (total titratable-acidity) of the dough.

For the control of pre-ferments, the acidity-index becomes a very important parameter, since it is an index of the degree of ripeness or maturity of the intermediate product. A rise in acidity-index increases the rate of swelling and peptidization of the dough-proteins, and controls the activity of the enzymes. Bread flavour and aroma are built up as a result of the accumulation of acids and their reaction products with other dough components such as alcohols. In the case of bread, flavour is determined not only by quantity of acids, but also their composition. Lactic acid gives bread a pleasant taste, whereas acetic and other volatile acids give it a less pleasing flavour. The acidity-index of flour will depend on the type of flour, or more specifically its extraction-rate. Wheat-flours range from $1.5$ to $2.5$ acidity-index for a patent flour, $2.8$ to $3.2$ for a straight baker's grade, and $3.5$ to $4.5$ for a wholemeal-flour. Sponges fermented at normal fermentation temperatures of 28–30°C rise in acidity-index, depending on flour and standing-time, from $1.5$ to $5.0$, and from $3.5$ to $6.0$ in the case of wholemeals. Straight doughs will also approximate to these acidity levels, but only if given sufficient standing-time. The use of high pre-ferment yields, in the form of liquid pre-ferment and higher temperatures 35–40°C will induce acid-formation, but even higher temperatures of 48 to 55°C will encourage the activity of the flour thermophilic lactobacilli, e.g. *L. delbrückii*, thus inducing an acid fermentation. A spontaneous fermentation can be initiated by leaving a flour/water dough to stand for 1–2 days in the open air. Then adding more flour and water to promote further fermentation. During the 1–2 days, bacteria predominate the system, but on adding more flour and water the yeasts take over.

Compared with the mesophilic yeasts, the thermophilic yeasts ferment at temperatures of 39–40°C much faster, the mesophiles dying off at these temperatures. These thermophilic yeast cultures are of great importance for the production of wheat bread sponge doughs in Armenia. Tests have shown that the biological specificity of this culture is for a temperature of 40°C, and the ability to develop in an acid medium. At temperatures of 50–55°C a greater volume of carbon dioxide is liberated.

The Spanish wheat-flour sours produce acidity-index values per $10^7$ yeast cells of various species of *S. cerevisiae*, ranging from 19 to 29, the higher values being obtained from '*masas madre*' (mother-doughs) fermented at ambient temperature for 2·5 hours. Samples direct from storage in a refrigerator at 4–5°C for 22 hours gave acidity-index values from 9 to 26 per $10^7$ yeast cells. The fermentation activity of the lactobacilli in the '*masas madre*', measured as acidity-index values, tended to be lower (11–67) in the sample fermented for 2·5 hours, and higher (30–250) in the case of the samples taken direct from refrigeration, especially the species *Lactobacillus plantarum* var. *plantarum* (Barber, S., Báguena, M. A., Martinez-Anaya and Torner, M. J., Microflora of sour-dough of wheat-flour bread. I. Identification and functional properties of microorganisms of industrial sour-doughs. *Rev. Agroquim. Tecnol. Aliment*, **23**(4) (1983) (in Spanish)).

According to Cavel, R. (1980), Fermentation au Levain naturel 1er Partie (Fermentation with natural sour, Part I), *Industries des céréales*, **7** (1980) 27–35 (in French) sour doughs can be prepared by a multi-stage process system at 28°C and RH 85%. The process is carried out by using 5 sour-dough stages, the initial-sour being produced over a period of 22 hours from: flour (100 parts) bran extract (50), salt (0·5), which is added as starter material to water (43 parts), salt (0·5) and flour (100) (Stage 2). Successive stages 3–5 are perpetuated by feeding-back the preceding sour-dough stage with an equal amount of fresh flour as follows:

*Sour-dough stages*

|  | 1 | 2 | 3 | 4 | 5 |
|---|---|---|---|---|---|
| Flour | 100 | 100 | 100 | 100 | 100 |
| Bran extract | 50 | — | — | — | — |
| Water | — | 43 | 43 | 43 | 43 |
| Salt | 0·5 | 0·5 | 0·5 | 0·5 | 0·5 |
| Saccharose | — | — | 0·5 | — | — |
| Previous sour | — | 100 | 100 | 100 | 100 |
| Fermentation time (h) | — | 20 | 23 | 20 | 19 |

*Final bread dough*

| Flour: | 100 parts |
|---|---|
| Dry yeast: | 0·37 |
| Water: | 63 |
| Salt: | 2 |
| Sour dough: | 30 |
| Dough fermentation time: | 4–5 h |

Both yeast and lactic acid bacterial counts increase progressively from 1 to 5, being larger from 1 to 3 than between 3 and 5. Microbial counts for yeasts can be monitored on Wort or Dextrose Agar at 30°C for 3 days; and for lactic acid bacteria on MRS Agar at 30°C for 3 days. Yeast counts in the fermented sour-doughs are always lower than lactic acid bacterial counts, yeast growth showing an adaptation phase during initial fermentation. However, at each sour-dough stage both yeast and bacterial counts increased. During the final dough stage yeast counts fall slightly, whilst bacterial counts remain constant. Sour-dough pH values fall progressively from 5·5 down to 3·8 (stage 1 through 5); the largest fall being from stage 1 to stage 3, parallel with the sharp increase in lactic acid bacterial counts.

The full fermented bread doughs vary in pH from 5.3 to 4.3. The lactic:acetic acid ratio of the sour doughs increase parallel with increases in lactic acid bacterial counts 1·77 to 5·59 (stage 1 to stage 5; the ratios in the final doughs rising in favour of lactic acid in all cases, the highest ratio being at stage 3 make-up. Overall, however, bread made with sour-dough stage 5 produced the best quality bread, having highest volume, best symmetry, and crumb texture.

The Danish company Chr. Hansen's Laboratorium A/S, 10–12 Bøge Allé, Hørsholm, Denmark, now market a series of Starter Cultures for bread under the 'Flora Pan' label. These are being marketed in the UK by Chr. Hansen's Laboratory Ltd, Reading, Berks, UK.

These products are freeze-dried starter cultures of cultivated bacterial strains selected for acid, gas and flavour production, with a uniform high cell count. Using these fresh cultures, both rye sours and wheat sour doughs and sponges can be produced with uniformity. Flora Pan L-62, based on *Lactobacillus brevis*, can be used for all types of bread, producing an acetic:lactic acid ratio of 1:4. Flora Pan L-73, based on *Lactobacillus plantarum*, is suitable for specialty-breads, providing a piquant, spicy flavour token with rapid acidification. Flora Pan L-22, based on *Lactobacillus delbrückii*, provides a mild flavour to bread producing L(+)-lactic acid, the natural isomer. It is also useful for the acidification of rye flours with low falling numbers.

The procedure for rye breads is to sprinkle the culture into the dough water when adding the flour, the sour-dough being allowed to mature for 16–20 hours at 25–30°C, reaching an acidity-index of 14–17 at pH 3·6–3·9. Bread production on day 2 is completed by adding the residual flour, water, yeast and salt, allowing the final dough to mature, and receive final proof before baking. Over 48-hour weekend periods, the sour-dough maturation can be delayed by the addition of 2–3% salt at the lower temperature of 20–22°C.

For wheat breads, Graham bread or mixtures, up to 20% mature sour will not retard yeast leavening, the culture is sprinkled into the water at doughmaking, the sour-dough being allowed to mature for 16–20 hours at 25–30°C. On day 2, the residual flour water, yeast, salt-shortening, sugar etc., are added to the liquid mature sour. The final dough is matured, final-proofed and then baked. Over 48-hour weekends, sour-dough ripening is delayed by the addition of 2% salt at 20–22°C.

Under static or discontinuous culture conditions, within a closed system, growth proceeds in six stages: lag-phase, acceleration-phase, exponential-phase, delay-phase, stationery-phase and lethal-phase. This growth curve, according to

Monod, applies to all dormant single-celled microorganisms on inoculating into an ideal substrate under ideal growth conditions. Each microorganism having a specific 'generation-time', defined as the time required to double the cell count.

Apart from *S. cerevisiae*, and its sour-dough mutants, other yeasts potentially suitable for wheat-sours are: *S. minor (S. paradoxus)*, *Candida* species, e.g. *C. crusei*, *Torulopsis holmii* (ferments all sugars except lactose), *Pichia saitoi*, and *Hansenula subpelliculosa*. Some species ferment glycerine, ethanol and lactic acid. Of the lactobacilli, *L. brevis* var. *lindneri* is probably the most tolerant to diverse processing conditions. The homo-fermentative (strepto-bacteria) produce almost entirely lactic acid, which is adequate for acidification, but the presence of the hetero-fermentative (beta-bacteria) produces lactic, acetic acids, alcohol, carbon dioxide, and other aroma precursors which complement bread flavour. The starter-culture, and the selection of suitable yeasts and lactobacilli for it, must be carried out with extreme care to exclude contamination of the culture media with unwanted wild yeasts or bacteria. Cells are selected under a microscope, and inoculated into a sterile wort medium of optimum composition for optimum growth. The pure culture cells are then inoculated aseptically into a larger fresh, sterile volume of culture medium in a fermenter unit with built-in control functions, the process of growth and propagation of the pure strains being perpetuated. Using equipment of ever-increasing size, and exercising the strictest control at every stage, especially in wort media preparation and during transfer of the cultures, a stage is reached when enough starter seed is available.

As with all groups of microorganisms, standardized culture media are available for the isolation, cultivation, identification and maintenance of yeast and lactobacilli strains.

For the efficient growth of yeasts, and maintenance of leavening activity, an exact balance between nitrogen and phosphate content of the yeast cells is important. Molasses do not contain sufficient phosphates for yeast growth. The ammonium phosphates are the preferred source of nitrogen and phosphate (mono-, di, or tri-ammonium), but aqueous ammonia or urea can be utilized. In addition to a source of carbon, nitrogen and phosphates, growth factors and trace minerals are essential for growth; these include: biotin, pantothenic acid, inositol, thiamine, pyridoxine, niacin, thiamine accelerating fermentative activity in particular; the inorganic elements required are Na, K, Zn, Fe, Mg, and Cu, but molasses are normally rich in these. Control of pH and temperature at 4·5 and 30°C respectively, can be rigidly controlled with the help of microprocessors/computers coupled with precise instrumentation. Oxygen transfer can be economically achieved by using a bubble-column fermenter instead of a mechanically agitated fermenter. Traditionally, industrial yeasts have been differentiated from one another by a series of test parameters, viz. giant colony morphology, bios typing, and carbohydrate/nitrogen utilization spectra. Other parameters used, in combination with carbohydrate-fermentation, have been: sporulation-frequency, percentage of 4-spored asci, and the use of desthiobiotin and mating-type segregation in meiotic spore clones. However, such information must be supplemented with the functional response of the organism in the model dough systems for which it is intended.

Culture media for yeasts should always be buffered, since during yeast fermentation a rapid fall in pH will occur. Two useful media are the following:

Culture medium 1: sucrose 100 g, ammonium sulphate 0·5 g, magnesium sulphate ($7H_2O$) 0·7 g, potassium dihydrogen phosphate 1·0 g, calcium chloride ($6H_2O$) 0·4 g, ferrous sulphate ($7H_2O$) 0·1 g, wort 10 ml in 1 litre of distilled water.
Culture medium 2 (Reader's medium): glucose 10 g, ammonium sulphate 3·0 g, magnesium sulphate ($7H_2O$) 0·7 g, potassium dihydrogen phosphate 0·1 g, sodium chloride 0·5 g, calcium nitrate 0·4 g in 1 litre of distilled water.

Buffer for culture media: citric acid 8·4 g, malic acid 6·96 g, succinic acid 7·08 g, lactic acid 10·8 g per litre of medium. The required pH is fixed by adjustment with 1 N sodium hydroxide. pH optima depend on genera and species, e.g. *S. cerevisiae* 4·2–4·8 *S. ellipsoidus* 4·5. pH adjustments can be made down to pH 3·5 by aseptic addition of buffer solution.

Bacto-Wort Agar is a medium prepared in the dehydrated form by Difco Laboratories Inc., Detroit 1, Michigan, USA, which is for the cultivation of yeasts, the reaction being adjusted to give a pH after sterilization of 4.8, being near to the optimum for most yeasts. Most bacterial growth is inhibited at this pH. The formulation closely duplicates the composition of wort.

For quantitative examination, prepare a $10^{-1}$ dilution of the starter-culture in quarter-strength Ringer's solution, setting up a breed-smear preparation. Examining several fields under the microscope, count each chain or clump of cells as one, ascertaining the average number for each field, then, using the microscope-factor, calculate the total numbers of organisms per millilitre of the starter-culture. Enumeration of viable starter microorganisms is achieved by preparing dilutions of the starter-culture in the quarter-strength Ringer's solution up to $10^{-8}$, or according to the breed-smear population. Plate out 1 ml each of the last three dilutions, using the appropriate media suitable for the isolation of the particular groups of microorganisms. Examine the developed colonies, and select plates with between 30 and 300 colonies for evaluation of the colony count per millilitre of starter culture. The technique for the isolation and maintenance of pure cultures of starter microorganisms is to obtain isolated colonies of the organism from the colony count method. Select a suitable isolated colony, in the case of bacteria, e.g. lactobacilli, and prepare a Gram's stain. Subculture the colony into yeast–glucose–lemco broth, incubating at 30°C for 3 days. The culture obtained is then re-examined by Gram's method. Rate of lactic acid production by the isolate, and ability to produce acetylmethylcarbonyl are determined by such biochemical tests as the Voges–Proskauer and the activity-test, by adding 1 ml of starter culture to 100 ml of sterile skim milk, and incubating at 30°C for 6 hours. The acidity developed is determined by adding 1 ml of 0·5% phenolphthalein solution to 10 ml of culture, and titrating with 0·1 N sodium hydroxide until a pink colouration is obtained.

Maintenance of the pure culture can be realized by inoculation into yeast–glucose–chalk litmus milk, and incubating at 30°C for 24 hours. The culture can be stored at room-temperature and subcultured every 3 months.

The sour-dough microflora consist of yeasts, and two groups of lactic acid bacteria (lactobacilli), all of which live in cohabitation or symbiosis, their metabolism influencing the progress of the sour-dough process. Counts of yeasts and bacteria in spontaneous and pure culture sour-doughs vary from: $1 \times 10^6$ to $3 \times 10^8$ yeasts per gram of sour-dough, and $1.4 \times 10^8$ to $6.1 \times 10^9$ bacteria per gram sour-dough. Depending on their enzyme systems, the lactobacilli are subdivided into homo-fermentative and hetero-fermentative. The homo-fermentative lactobacilli ferment glucose by the fructose–diphosphate pathway to over 90% into lactic acid, whereas the hetero-fermentative ferment glucose by the hexose–monophosphate pathway into 50% lactic acid, carbon dioxide, ethanol, acetic acid and other organic acids. Spicher has isolated about 260 species of lactobacilli from various spontaneous and pure sour-dough cultures. The following summary shows the lactobacilli of a spontaneous sour-dough:

Homo-fermentative lactobacilli constituting about 54%:
*L. plantarum, L. casei, L. farciminis, L. alimentarius, L. acidophilus.*
Hetero-fermentative lactobacilli constituting about 46%:
*L. brevis, L. brevis* var. *lindneri, L. fermentum, L. fructivorans,* and *L. buchneri.*

International research work has indicated that strains of the lactobacillus types *L. brevis, L. fermentum, L. plantarum,* and *L. casei* give good dough acidification and aroma formation, and can be utilized without reservation in starter-cultures. Information concerning the exchange effects between yeasts and lactobacilli in sourdoughs is sparse. While the lactobacilli produce mainly acids and aromatic substances, the yeasts produce carbon dioxide, which is essential for aeration, breadvolume and crumb-structure. The liberation of carbon dioxide in the sour-dough provides conditions favourable to facultative aerobes, which aids lactobacilli metabolism and reproduction. Spicher has isolated to date more than 40 strains of yeast, belonging to the following families: *S. cerevisiae, Pichia saitoi, Candida krusei,* and *Torulopsis holmii.*

The microflora of sour-doughs vary according to the production methods and processes from one bakery to another, and within samples. This applies to both quantitative and qualitative compositions of the microorganisms. The efficiency of a starter will depend on the number and physiological condition of the microflora it contains. If processing parameters such as: temperature, dough-yield, their growthrate and metabolism remain reasonably constant. This enables the maturing-time to be regulated by adjusting the amount of starter used. The maturing-time depending on the Monod growth-curve described earlier, depicted in six stages.

The lactobacilli are characterized mainly by their physiological properties. They form Gram-positive rods, which are non-acid fast, non-spore-forming, dividing in one plane only, forming pairs or chains, especially in liquid media. They are with few exceptions non-motile, and show the following growth characteristics: convert carbohydrates into lactic acid; have complex and discriminative nutritional requirements; are aerobic (microaerophilic) but catalase-negative; show limited colony size.

Lactobacilli can only develop in carbohydrate-containing media which have been enriched for optimum growth, viz. trace elements (manganese), vitamins (lactoflavin, biotin, thiamin, pantothenic acid, folic acid, and nicotinic acid), amino acids or peptides, purines and pyrimidines. Some species use lactose, and on account of their special nutritional requirements they grow on such complex substrates as milk, whey, tomato and vegetable juices and farinaceous substrates such as doughs, yeast water and brewer's worts. Excessive lactic acid build-up can lead to toxic concentrations, which sometimes have to be neutralized with the addition of calcium carbonate to the medium, or potassium bicarbonate in the case of sour-doughs.

Lactobacilli belong to the facultative aerobe group of bacteria, they can grow in the presence of air, but do not utilize it. They derive their energy from lactic fermentation, and grow ideally within the medium rather than at the surface. Their rate of growth is relatively slow, and in solid agar media only form small colonies of about 1 mm diameter of a brownish-yellow colour. The lactobacilli are powerful lactic acid-forming bacteria, converting carbohydrate-rich, mildly acid media, into up to 2·3% acid. They are subdivided into two groups according to their metabolism, the homo-fermentative mainly lactic acid-forming types, and the hetero-fermentative types which also produce considerable amounts of acetic acid, alcohol and carbon dioxide, alongside lactic acid. Their temperature growth range is from 5 to 53°C, but most types have an optimum between 30 and 40°C. Orla-Jensen divided the various lactobacilli into three groups, according to both metabolism and optimum temperature.

The streptobacterium: homo-fermentative, with low optimum temperature, e.g. *L. casei* and *L. plantarum.*
The thermobacterium: homo-fermentative, with a high optimum temperature, e.g. *L. acidophilus* and *L. bulgaricus.*
The betabacterium: hetero-fermentative, with either a low or high temperature optimum, e.g. *L. brevis* or *L. fermentum.*

To identify lactobacillus isolates, first determine to which subgroup it belongs. The ability to produce gas from glucose in Gibson's semi-solid medium indicates whether the fermentation is homo- or hetero-fermentative.

Gibson's medium: yeast extract 2·5 g, *d*-glucose 50·0 g, tomato-juice pH 6·5 100 ml, hydrated skim milk powder 800 ml, nutrient agar 200 ml (medium pH 6·5).

The ability to grow on de Man, Rogosa and Sharpe (MRS) broth at 15 and 45°C will indicate the temperature response of the organism, as mentioned above. The ability to produce ammonia from the amino acid arginine, in a medium containing 2% glucose is a supplementary test detected by Nessler's reagent. This arginine test is also useful for the differentiation of hetero-fermentative lactobacilli from the leuconostocs, since the coccal forms of lactobacilli are sometimes difficult to distinguish from the coccal forms of leuconostocs. A suitable medium is MRS broth containing 0·3% arginine, ammonia liberation being detected with Nessler's reagent. The composition of MRS broth, for the culture of lactobacilli is the following:

> Peptone (Oxoid) 10·0 g, Lab-Lemco meat extract (Oxoid) 10·0 g, yeast extract (Difco or Oxoid) 5·0 g, *d*-glucose 20·0 g, Tween 80 1·0 ml, dipotassium hydrogen phosphate 2·0 g, sodium acetate 5·0 g, triammonium citrate 2·0 g, magnesium sulphate hydrate 200 mg, manganese sulphate hydrate 50 mg, distilled water 1 litre. The components are dissolved by steaming, and the pH adjusted to 6·2–6·6. Dispense and sterilize at 121°C for 15 minutes.
> Final pH should be 6·0–6·5.

The same medium can be used for fermentation studies with lactobacilli, by omitting the glucose, and meat extract, adjusting to pH 6·2–6·5 and adding 0·004% chlorophenol red as indicator. Dispense into tubes. Prepare 10% suspensions of test-substrates aseptically, and add to the modified broth medium to give a final concentration of 2% (de Man, Rogosa and Sharpe, 1960).

The amount of lactic acid built up by any starter-culture by homo-fermentative lactate fermentation can be ascertained at any desired time during the fermentation process by a simple formula. In order to achieve this, only a measurement of the actual pH value is required, the empirical function being read off from a prepared calibration-curve. The graphic relationship between medium pH and lactic acid concentration is curvilinear, the lactic acid concentration being read off in ml/100 ml substrate. If more accuracy is desired mathematical modelling and regression analysis can be used. In making use of these mathematical models, it must be understood that the pH of the initial unacidified medium will depend on the concentration of the fat-free dry matter in the medium. This control technique is specially designed for the evaluation of homo-fermentative lactate-producing organisms, which convert the carbon-source into up to 90% lactic acid (lactate). Kinetic models for other strains of lactose-fermenting streptococci both in batch and continuous culture can be devised.

The optimum control of pH is very important for the activity of the sour-dough bacteria; with a progressive fall in pH, the bacteria restrict their acid production. Whereas, in acid media the lactobacilli produce lactic acid, in neutral media they produce acetic and formic acids. Any undesirable change in sour-dough pH can be prevented by adding either inorganic buffering agents such as alkaline phosphates and calcium carbonate, or mineral water and whey solids. The latter method of buffering is preferable, since it delays the fall in dough pH, but allows adequate acid build-up for successful breadmaking. However, the effect of mineral water and whey on sour-dough fermentation depends on the species and strain of bacilli in the starter-culture. Especially when adding whey solids to the dough liquid, there is initial retardation of bacterial reproduction. In general, a well-balanced biological system of yeasts and sour-dough bacteria will ensure a bread of good quality and flavour. The natural vitality of the microorganisms is also very important, since autolysed cell material will both reduce fermentation efficiency and produce off-flavours. A yeast which is completely free of all bacteria will not produce a well-flavoured loaf of bread.

Increasing application of biotechnological techniques will improve the control of breadmaking as a biotechnological process by: (1) adaptation of dough processes to the starter-cultures of yeasts and bacteria, (2) control and exploitation of the natural enzyme systems of grain. (3) Genetic manipulation of yeasts and bacteria for starter-cultures with more desirable and controllable properties.

## Water and Dough-liquid

Since water constitutes more than 50% of the flour weight during doughmaking, its composition merits attention both as a natural raw material and as a vehicle for other water-soluble dough ingredients.

The natural 'hardness' of water is expressed in the European countries in 'grades of hardness', the units varying from one country to another. The German hardness-grade (degrees d) corresponds to a content of 10 mg/litre of calcium oxide; French degrees of hardness are quoted as 'degrees f', and English as 'degrees e'.

The various types of hardness are referred to as carbonate hardness, non-carbonate hardness, and total hardness. Total hardness is determined as the calcium and magnesium content, and carbonate hardness by titration with acid, using 100 ml of the water sample. Non-carbonate hardness is then determined by the difference between total hardness and carbonate hardness. Inorganic impurities can be determined quantitatively by such determinations as: nitrite, phosphate, ammonia, aluminium, iron, lead, copper, arsenic, etc. Organic impurities are usually determined either by titration with potassium permanganate, or gravimetrically by difference between the residue after evaporation, and ashing.

Tap-water contains carbonic acid and bicarbonate, the ratio of these determining the buffering effect. Moderately hard water with a total hardness of 10 degrees dH (carbonate hardness 7·3 dH; 1 dH = 10 mg/litre calcium oxide), would have a weak buffering effect, whereas mineral water of dH 55·0, owing to its high bicarbonate content, would show a strong buffering effect, in ionic exchange with a larger amount of carbonic acid. The significance of the difference in these two water samples is the acid required to adjust their pH value to 4·0, the former requiring only 5·0 ml of 0·1 N lactic acid/100 ml, and the latter 42·5 ml/100 ml. Buffering capacity is at a maximum when acid and salt is in equilibrium (1:1). The buffering capacity of whey solids in dough water has the most dramatic effect, owing to its lactates, citrate, phosphate, carbonate and bicarbonate presence. In this situation, an equilibrium exists over a wide range of pH, ensuring a buffer effect.

This buffering effect has more significance for rye-flour processing than for wheat-flour, since in the case of the former, there is a delay in the pH fall to 4·0 owing to the presence of ions, in spite of bacterial acidulation.

In the USA, where the composition of water varies widely, adjustments are made by adding mineral salts or acidic substances with the yeast-food. For example, an alkaline water with pH greater than 8·0, classed as 'moderately hard/hard' would be treated by adding an acid type yeast-food, probably containing an appropriate acid calcium phosphate; and in extreme cases of hardness, additional acid is added in the form of lactic or acetic acids. In the latter case, extra malt additions are often necessary, and in water containing more than 180 ppm minerals, malt is added to the sponge, and water absorption adjusted accordingly. Acidic waters with pH below 6·8, containing less than 120 ppm of mineral matter, require the addition of calcium sulphate, with a percentage of the dough salt added to the sponge.

Where liquid pre-ferments are in use, water composition is very important, in order to control brew pH and total-titratable acidity.

As already discussed, the optimum relationship between free and bound water, and therefore total water absorption of any dough, is a basic for any dough system. Water acts as a solvent for nutrient, organic or inorganic additives, and mineral salts, and is essential for the diverse enzymic hydrolysis of the flour components. Adequate water is also essential for the optimal metabolism of both yeast and sour-dough bacteria, and as a vehicle for buffer salts.

For doughs prepared with sour-dough cultures, water additions should be between 65% and 75%, depending on the flour quality and other ingredients added. This corresponds to dough-yields of 165 175/100 g flour. Thus allowing the flour to fulfil its functions of providing nutrients, active additives, and mineral salts with their buffering action. Although hearth or oven-bottom breads allow little scope for variations in water absorption, improved bread flavour in panned bread, especially in the case of rye/wheat-flour mixtures, is obtained at dough yields of 170%.

## 1.8.2 PROCESS VARIABLES

### Temperature

Temperature is the most important process control variable for the fermentation of yeasted and sour-dough schedules. Choice of optimum dough-temperature will depend on the formulation and maturing-time employed, but, once chosen, it must be maintained from leaving the mixer, right up until final-proof. In so doing the rate of flour component modification, and progressive changes in pH and gassing with time will be synchronized. In the case of yeasted wheat-flour doughs, the temperature range of processing is generally narrower than is the case with sour-doughs. Wheat-flour doughs are normally set at temperatures between 26 and 30°C, although liquid pre-ferments may be set at higher temperatures for batch and continuous short process schedules. For the various process and stages for rye-flour doughs processing temperatures can range from 24 to 38°C. However, this is a special case, since dough-yield is adjusted in conjunction with temperature to provide optimal conditions for the yeasts or bacteria to reproduce in regulated stages.

For tempering of the dough water, every bakery uses an aggregate which combines the three functions of temperature measurement, water-mixing and metering. There are two basic design principles. One is based on a container of 20–200-litre capacity into which hot and cold water is fed, controlled by valves which are opened or shut by a magnetic control mechanism, which is in turn controlled by a preset contact-thermometer consisting of an expansion-rod sensor located at the base of the container. The floater, which controls the water level, is also controlled by a magnetic switch within the contact-thermometer circuit. These aggregates operate with a water pressure of about $0·09\,N\,mm^{-2}$ ($0·9\,atm$), and deliver tempered-water with an accuracy of about 0·5 to 1 litre. Temperature operation range is 20 to 60°C with an accuracy of ±2°C.

The second basic design type depends on the flow principle. Hot and cold water are fed in at the base of a horizontal, insulated container, controlled by twin valves in the form of pistons, positioned on a rod moving in a horizontal plane. Their movement within the narrow cylinder section of the container, being controlled by the setting of a bimetallic rod via a temperature scale. This function is achieved by the lateral movement of the rod, which actuates the movement of a sliding-jet laterally nearer or farther away from a membrane. The equilibrium is taken up by a pressure-spring, which acts as a pressure-transducer, moving the rod and twin-valve pistons to the right or left, thus controlling the inflow of both hot and cold water. The required temperature is set by screw setting, which fixes the position of the bimetallic strip, and hence the jet, relative to the membrane. The screw-setting for temperature is provided with a calibrated scale, which is located near the outlet-valve, which delivers the tempered-water.

Calculation of the 'friction factor' or machine allowance:
This is the factor or value required to compensate for the heating-up of the dough during mixing. No machine is 100% efficient, and one energy loss is heat-energy due to friction. It must be determined for each individual mixer by using water at ambient temperature, and noting the temperature rise between water temperature before and after mixing a dough, substituting known values in the following formula:

$$\begin{aligned}
\text{friction factor (FF)} = &(4 \times \text{actual dough temperature}) \\
&- (\text{bakery temperature} + \text{flour temperature} \\
&+ \text{water temperature} + \text{shortening temperature}) \\
\text{water temperature (WT)} = &(4 \times \text{desired dough temperature}) \\
&- (\text{bakery temperature} + \text{flour temperature} \\
&+ \text{water temperature} + \text{shortening temperature})
\end{aligned}$$

The number of ingredient temperatures factored into the equation for bread dough would be three, viz. bakery temperature, water temperature, and flour temperature, but shortening temperature can also be included, depending on the formulation. Therefore, the general equation for calculating water temperature is:

$$\begin{aligned}
\text{water temperature (WT)} = &(3 \times \text{desired dough temperature}) \\
&- (\text{bakery temperature} + \text{flour temperature} \\
&+ \text{machine frictional factor})
\end{aligned}$$

for a lean bread dough, limiting the number of temperature factors to three.

If bakery ambient temperature is too high, and the machine frictional factor does not allow tap-water to be used, the use of ice is essential, and the weight of ice necessary is calculated as follows:

$$\text{weight of ice} = \frac{(\text{weight of water}) \times (\text{tap-water temperature} - \text{calculated water temperature})}{\text{tap-water temperature} + 112}$$

where temperatures are in degrees Fahrenheit and weights in pounds. Allowances for frictional heat of the mixer are important, and will depend on the mechanical action of the machine and its capacity. Assuming a target dough temperature of 30°C, which is a good average, using a high-energy input mixer delivering 11 W h kg$^{-1}$, flour temperature cannot exceed 27°C for a dough at 30°C, since in this case the water temperature at the mixer would need to be 4°C. Chilled water, however, is not normally available at the mixer at 4°C and below. By contrast, another type of mixer with an energy input of 7·6 W h kg$^{-1}$ would allow flour temperatures of up to 40°C to be handled, provided the dough water could be chilled to 4°C. Using the latter mixer, with flour temperature at 26°C, no water chilling would be necessary, since the mains cold water is usually at about 14°C. These examples illustrate the effect of high-energy mixers on dough temperature control.

The heat of hydration of flour can also be significant if flour dries out on storage, requiring a water temperature adjustment. Flour granularity also has an influence on frictional heat development during mixing, and brown flours and meals require water a few degrees cooler than white flours on account of the extra friction due to the bran particles.

Flour temperatures in bakeries vary considerably, depending on storage conditions, bulk-handling system, and aeration in the silos at the bakery after bulk delivery. Normal storage temperature for flour in bulk should be about 20°C, but pneumatic transport can result in temperature rises to about 30°C within 30–40 s. Therefore, it is desirable to have a facility for checking the temperature of flour on delivery by the tanker, during external silo-storage at the bakery, and after transfer to the daily-silos within the bakery building.

Intensive batch mixers such as the Tweedy, working at 400–420 rpm and with work-inputs of 8–11 W h kg$^{-1}$, lead to dough temperature rises of 15°C and more in some cases. Therefore, a cooling of the mixing-room and/or cooling of the dough water is imperative. Since the Tweedy mixer is programmable as regards loading, energy-input, vacuum, and cooling, temperature control should be attainable. Used for wheat-flour, rye/wheat flour mixtures and rye-flour doughs, mixing-times range from 1·5 to 5·0 minutes, with an average of 2·0 minutes, but energy inputs are from 6 to 8 times higher than in the case of conventional mixers.

Another very efficient intensive mixer, the IMK 150, manufactured in the GDR by the VEB Bäckereimaschinenbau, Halle, which is used for both batch and continuous processing, also gives dough temperature rises of about 10°C for wheat-flour doughs and about 3–5°C for rye-flour doughs. According to the time of year, water temperatures are adjusted for dough and flour temperatures within the following ranges:

|                   | *Winter*  | *Summer* |
|-------------------|-----------|----------|
| Wheat-flour doughs | 17–25°C  | 12–17°C  |
| Rye-flour doughs   | 25–40°C  | 18–25°C  |

This mixing-aggregate has a motor-output of 55 kW, and provides 415 rpm, giving work-inputs of 8–10 W h kg$^{-1}$ within 2–3 minutes. Similar dough temperature rises occur with the horizontal-bar high-speed mixers used in the USA, but these are provided with a cooling-jacket.

As mentioned earlier, where no cooling facility is provided, the water temperature cannot be higher than 4°C if normal dough temperature is to be obtained (28–30°C).

For measurement of the volume of the dough water various measurement principles are utilized. The Woltman counter is based on a winged-wheel, which rotates about an axis parallel with the direction of water flow, the rpm of the wheel being proportional to the velocity of water flow; and velocity of flow is also proportional to quantity of water per unit time.

Another device consists of two specially toothed cog-wheels mounted on axes, rotating within a chrome–nickel steel spherical housing, which is part of the pipeline. The inflow of the water results in a force, $F$, equal to pressure/area, which resolves itself into two force components $F_1 = p/A_1$ and $F_2 = p/A_2$, one acting on the lower cog-wheel, and the other on the upper one. As a result, the upper wheel rotates in a clockwise direction, and the lower one in a counterclockwise direction. The volume of the water passing through is determined by

$$\frac{\text{volume of the spherical housing}}{\text{rpm of the cog-wheels}}$$

The rpm can be translated on the tachometer principle into units of volume per unit time on a calibrated scale, or by using a digital read-out. In addition, other facilities, such as a continuous registration of the volume of throughput, or the metering of a preset volume of water over a valve on reaching a preset number of rpm, are also possible.

Optimal drinking-water hardness-grade for breadmaking is between 10 and 24 dH (1 dH = 10 mg/litre calcium oxide).

**Flour-preparation**

Apart from analytical standards, all flour must be subjected to a controlled, mechanised preparation for processing into dough. This involves delivery, storage and the preparation procedures. Bulk-flour delivery involves linking the flour-tanker with the external storage-silos by means of a flexible pipeline. The compressed-air is provided by the tanker's compressor and the flour is pneumatically transported to the appropriate storage-silo of capacity 5–60 t. The construction of these silos is either circular, and made of steel, with an average diameter of 2200 mm, or rectangular, measuring about 2200 mm × 4000 mm, their height varying according to location and requirements. Each silo is provided with a fluidized-bed to allow pneumatic transport out of the storage-silos as required. Since the various types of bread require appropriate flours, daily-silos must also be provided to allow accommodation for the daily or shift production programme, and the preparation of flour-blends. These daily-silos are located within the bakery, and are provided with conical, vertical worms mounted on the axis of the silo, to serve as mixing/blending units.

From the external storage-silos, the required flour types are transferred pneumatically into the daily-silos via a continuous weighing-station, which is controlled automatically by presetting on the control-panel. The weighed flours then pass through a mechanical sieve, through a pipeline into a daily-silo. According to demand, the flour is then transferred by dosage-worms into a weigher unit which has been preset for a defined weight of flour. On discharge from the weigher, the flour passes through a control sieve either directly or pneumatically to the dough-room. An average silo-capacity for a medium-size bakery could be, eight external storage-silos, each with a capacity of 35 t of flour, and four internal storage-silos each with a capacity of 8 t of flour. Where sack deliveries are also received, these are emptied, sieved and blended at a preparation-station before being pneumatically despatched to the daily-silos. Often for the preparation of flour-blends, pneumatic-mixers are preferred to the traditional worm-mixers. Sieves are an important part of flour preparation, removing impurities and aerating the flour. The types of sieve design utilized are: centrifugal, rotational, and vibrational.

For the storage of meals and brown-flours for specialty-breads, storage-silos on a bulk-flour basis also cut handling to an absolute minimum. The less the movement the less the separation of the blend components; any gravitational fall-out is returned to the pneumatic conveyor line, rather than to the silo.

Electronically controlled weigh-stations are made of stainless steel, generally operating within the 50–250 kg range, but larger units are available. They vary from single call stations, to totally controlled, proportioning weighing-systems handling multiple weighings and accurate blending, for product uniformity. Since raw material consistency is a prime requirement of the baking industry, 'in-line' monitoring of flour protein, moisture, damaged-starch/granularity, utilizing NIR-reflectance spectroscopy, in the form of the Technicon 'Infra Alyzer', after blending and sieving, would be a possibility. In addition, NIR-data from a large number of flour probes after sieving and blending and prior to mixing, will allow an estimate of flour water absorption from protein and damaged-starch percentage to be computed using multiple regression analysis.

## Dough-yield

Dough-yield is a very important process variable, essential for economic, rationalized, industrial bread production. Low dough-yields result in very large annual losses of raw materials. During batch processing an optimum dough-yield can be ensured by adhering to the weights laid down in the formulation, and dough consistency control relative to flour water absorption. During continuous dough manufacture, dough-yields need to be checked every 2 hours to ensure that the feeding of raw materials from the metering-aggregates is uniform. This operation is necessary for both wheat-flour pre-ferments and sponges, as well as rye-flour full-sours and doughs. The classical oven-drying method takes too long owing to the prolonged drying-time for production interpretation. Therefore, in the GDR, a group of students working at the Food Engineering School at Dippoldiswalde were given the task of developing a plate-dryer with standardized temperature/heat-transfer and time control. This equipment, known as the 'TKO-instrument for the determination of dough-yield', has fulfilled the requirements of the National

Standard TGL-29066. The apparatus comprises two component aggregates, the drying-plate and a control unit. The drying-plate is mounted firmly on top of the control unit for compactness and stability. Two measuring systems for temperature feed back information from the drying-plate to the control-system. The control-unit houses the electrical components, and energy source. The procedure is to first allow the plate to heat up, whilst preparing the dough samples. 5 g of dough are weighed to an accuracy of 1 mg by means of a special stencil. Four samples can be dried in one batch, the drying process requiring 15 minutes at 170°C.

In calculating the weight loss, the flour moisture must be taken into consideration. The weight loss being only a measure of dough moisture. Production samples are taken at 2-h intervals.

Example:

| | |
|---|---|
| Weight of dough sample in plate-dryer | 4·895 g |
| Weight of dried dough | 2·268 g |
| | ——— |
| Weight loss | 2·627 g |

$$\text{Dough moisture} = \frac{2 \cdot 627 \times 100}{4 \cdot 895} = 53 \cdot 67\%$$

The basic theoretical definition of dough-yield is the amount of dough prepared from 100 parts of flour. Owing to variations in the water-binding capacity of flours, however, from the same amounts of the various flours, different amounts of dough are produced, assuming that the baker works to a constant dough consistency. With the help of this value, the amount of flour and water necessary to produce a required amount of dough can be calculated beforehand. However, this calculation only gives exact values when the dough is prepared from flour and water, and hardly any other raw materials.

In the case of bread and lean bread-roll doughs this assumption is valid, dough-yield being expressed as a percentage.

Experience has shown that an optimal ratio of flour to water for doughmaking with the various flour types can be given as a guideline for the preparation of doughs of acceptable consistency. These are as follows:

| | |
|---|---|
| Rye-flours (depending on the process) | 155% to 170% |
| Rye wholemeal | 170% |
| Wheat-flours (depending on process) | 158% to 162% |
| Wheat wholemeal | 165% |
| Mixed rye/wheat-flour (70:30) | 158% |

In calculating actual dough-yield, given the following:
Flour weight (basis 100%), dough weight, then,
Flour weight: Dough-weight = 100%: dough-yield.

In a lean flour/water dough, the water distribution is bound as follows: 40% by the swollen proteins, 40% by the swollen starch and 20% either bound to the swollen pentosans, or remaining as free water. It must be emphasized that the above guidelines are at best approximate, and during production the dough-yield must be constantly controlled. During industrial bread production, when the dough-yield rises by 1%, the economy in flour usage will run into 1 to 2 million units of currency per annum. Owing to the presence of higher concentrations of non-starch carbohydrates, viz. pentosans and pectins, especially the former, rye-flours are capable of achieving much higher dough-yields than wheat-flours. This can be realized without either excessive loss during the baking process, or excessive loosely bound water retention within the crumb. The latter gives a 'doughy' crumb, and reduced flavour and palatability. These characteristics are symptomatic of industrially produced wheat-flour breads in some countries.

During continuous dough-processing, the energy output curve of the mixer-motor can be utilized to control the dough-yield. Variations can be detected, and corrected by adjusting flour content, although throughput, mixer-rpm, residence-time, etc. also exert an influence. The mixing-curve indicates that an increase in energy-input, to an optimal level, results in corresponding dough-yield optimization.

Since the relationship between flour-weight and dough-weight in kilograms is fairly linear, a graph relating flour-weight and dough-yield is useful for bread dough production, using a lean formulation. In this way, the amount of flour required to produce a required amount of dough can be read off to the nearest kilogram. Such a calibration graph can also be utilized as a nomogram by plotting the number of dough-pieces, baked units, etc. produced onto a second ordinate ($x$) axis parallel to the first one, carrying dough-weight values in kilograms, thus producing a monogram relating flour:dough:dough-pieces, allowing read-out facilities shown in Table 52.

The use of such nomograms, relating process product yields to raw materials, facilitates daily production scheduling and control. Dough-consistency during processing can be measured with a penetrometer. This method involves taking a sample of dough from the production-line, allowing a relaxation period, then placing in the penetrometer-cup. The penetration-head is then allowed to penetrate the dough-piece for a period of 5 s. The depth of penetration is taken as an index of dough consistency, the smaller the depth of penetration, the higher the dough consistency.

An alternative method for consistency is to take a 300 g dough sample from the

TABLE 52

| *Given* | *Read-out* |
| --- | --- |
| Number of dough-pieces at a required scaling-weight | Necessary flour- and dough-weight to produce them |
| Flour-weight (kg) | Dough-weight yield obtained (in kg) and number of dough-pieces |
| Dough-weight (kg) | Necessary flour-weight (in kg) and number of dough-pieces |

production-line and mix in a Farinograph bowl; other consistency-measuring instruments such as the Valorigraf, Mixograph, and GRL-mixer could also be used.

**Mixer and Mixing-regime**
The significance of this process variable has been discussed in depth in Chapter 1. The specific type of mixing unit is chosen to suit the range and variety of baked-products being produced. Certain other process variables already discussed, such as temperature and dough-yield, can be influenced by the mixing-regime imposed on the dough, e. g. specific mixing-intensity. This factor can be optimized by variation of the mixing-time, without changing dough temperature at the mixing stage. If dough-yield is showing a fall of up to 1%, in spite of the accurate metering of the flour: water ratio, and a standard flour quality profile, then the chances are that the specific mixing intensity is sub-optimal. Under these conditions, the water-binding capacity of the component raw materials has not been fully exploited, many particles remaining unmixed and inadequately swollen. It should also be mentioned that similar effects are obtained when the amount of dough is either too large or too small for the capacity of the mixer.

If the specific mixing intensity has been exceeded, the doughs show signs of sticking, and this process variable must be reduced. In both instances, this is best achieved by adjusting mixing time, upwards in the former case, and reducing it in the latter. If this corrective measure proves inadequate, the addition of dough-improvers based on swelling-improvement can be considered. Nevertheless, assuming the flour-quality profile is satisfactory, the addition of more flour should be resisted, since this reduces dough-yield and increases raw material costings, particularly in a large industrial bakery.

In summary, as far as 'process-variables' are concerned, one is confronted with a series of interrelated variables, which form a cycle of corrective measures. Initially, the baker has fixed ideas concerning the dough-consistency properties he considers essential to process the dough into a specific type of product. When these properties deviate from the 'norm', he makes use of the process variables to exert a corrective influence on the process. During continuous breadmaking processing, these corrective measures are automatically taken care of by the process-control-system. Load-cells mounted on ingredient blending-tanks retain narrow tolerance-limits for all ingredients, thumb-wheel dials on the control panels of the blending-tanks program the number-code of each ingredient into the computer. The control technique sequence is 'measure–control–regulate', based on 'target-weights' *v.* 'actual-weights' ('*Soll-Wert*' *v.* '*Ist-Wert*'). Water enters a batch-weighing tank, the control-system checking the weight of water against prefixed tolerances. If the weight falls within these tolerances, the system acknowledges that the quantity of water is acceptable, and signals the computer to advance to the next ingredient. When all liquid-ingredient delivery-cycles are complete, dry ingredients are added manually to a slurry-tank, one by one, and acknowledged. The system going through its weight-checking process, confirming that each ingredient-weight falls within the set tolerances. On completion, the computer generates a print-out in the operations control-room, showing 'target-weights' of all ingredients against 'actual-weights' delivered in each batch.

Pre-ferment-slurries, containing water, yeast, oxidizing agents and dough-conditioner are blended in the same way. Water enters the batch-weigh tank first, the system checking the weight; yeast and oxidants are then added manually.

According to the type of machine and geometrical design of the mixing-elements, the time necessary to impart 5 W h of energy to a fully loaded machine can vary from 160 to 200 s. The same mixer, loaded at 50% capacity, to impart 5 W h, could require a mixing-time of 140 to 160 s. Therefore, the effect of machine-loading *v.* machine-capacity, as regards energy-input, does not follow any predictable pattern. 140 weight-units of flour mixed in a machine of 280-weight-unit capacity, will not logically require half the mixing-time. Dough-consistency is the most important single practical control parameter for the production process, and the response of bread quality to it often overrides the effects of raw-material variables. The tighter the dough, the lower the mixing-time, since energy can be imparted more efficiently to a dough offering greater resistance. However, extreme tightness is neither economic or desirable for loaf-volume and crumb characteristics. Extremely slack doughs, apart from becoming unprocessable, will often eliminate any desirable effects of a flour with good quality proteins and swelling-capacity. It is therefore of paramount importance to define the process-system, and optimize consistency for it.

During Chorleywood bread-processing, deviations in dough consistency over a 1 h production period, requiring changes in sack (280 lb) water absorption of 1 lb, necessitate mixing-time adjustments of on average 2 s. Various control techniques, e.g. the Rank–Pullen meter, were designed to control consistency more accurately. The basis of this method is to measure mixing-time variations over a statistically significant number of batch-mixings, and set tolerance limits for dough-consistency. This type of control-system can also be computer-linked. However, many bakeries mix to a constant energy-input of 5 W h lb$^{-1}$ dough, using a single predetermined water level, thus relying on the consistent quality of their flour deliveries. For the CBP, when a baker calculates how much development his dough requires, the procedure is to total all formulation ingredients to the nearest pound or kilogram, then multiply by a factor of 5 W h lb$^{-1}$ or 11 W h kg$^{-1}$ as appropriate. This energy-input is expended over a mixing-time of 120 to 180 s; the shorter the time taken to expend this level of energy, the more efficient the mixer. The geometry of the mixing-element and the presence of prism-shaped elongated baffle-elements fixed at an appropriate angle, will increase the mixing-intensity. A watt-hour meter is used in the case of the Tweedy mixer to measure the amount of mechanical work done by the working head. This instrument having been calibrated to eliminate the 'zero error' watt-hour output when the machine runs empty.

The calibration of the readout is such that 1 unit corresponds to 10 W h, therefore the watt-hour counter setting is determined by dividing the total dough ingredient-weight by 2.

*Electric-motors as Power-sources*
The drive-shaft of dough-mixers are driven by the turning-moment or torque initiated by a 3-phase, 50 Hz, alternating current (AC) energy source. The magnitude of this induced force is given by:

$$T = c \times \phi \times i \cos A$$

where $T$ = turning-moment or torque, $c$ = motor constant, $\phi$ = inductance, $i$ = current in amperes and cos $A$ = cosine of the angle between the field direction and the vertical force component to winding-surface, i.e. electromagnetic flux.

Most AC motors are capable of delivering two or more rpm rates, achieved by switching polarity. When, for example, half of the pole-couples are cut out, the shaft and mixing-element must rotate at twice the number of rpm to keep pace with the AC, and the rpm is then given by:

$$\frac{60 \times \text{frequency}}{\text{number of pole-couples}}$$

Technically, this problem is solved by winding some of the pole-couples in parallel. The resulting increased current flow, at the same mains voltage, demands an increased output from the mains. Such motors are labelled on the data-plate in the form of two or more output-ratings, e.g. 2·5/3·5 kW, according to rpm graduation. Thus, by doubling the rpm, the energy consumption increases by a factor of about 1·5. If such motors are overloaded, a thermal contact will break the circuit.

For intensive mixers like the Tweedy and the IMK 150 (VEB Bäckerei-maschinenbau, Halle, GDR), which works at 415 rpm, various starting-procedures can be selected for 'star' (*Stern*) or 'delta' (*Dreieck*, $\Delta$) use. In the case of star ($\curlywedge$)-switching, only one-third of the current flows compared with delta. Hence, on starting the motor, not only is the initial current reduced, but the torque is reduced, giving a gentle start-up for a period of about 4 s. Then the motor is switched to delta, at which point measurement of energy expended in mixing the dough commences. The watt-hour meter is corrected to take into account losses due to the motor and bearings. When the preset amount of work has been delivered, the motor stops. In the field of measurement and control techniques in industrial baking, the key technologies, microelectronics, biotechnology, and information technology will play an increasing role, depending on their application for the various phases of production.

For raw material handling the aim is the control and transport of flour from silos over sieves, blending and weighing, computer-aided balancing of silo-contents, and the construction and control of recipes.

For dough-processing and make-up, the aims will be a continuous micro-processor-controlled pre-ferment and sour-dough preparation, electronic-control of dough-mixing machines, complete control of dough temperature at mixing, and complete electronic control of dough make-up and maturing ready for baking.

As far as the baking-process is concerned, the aim is control of the energy consumption in the oven, and energy recovery.

The stabilization and maintenance of the quality of production by the determination of process-data at each production phase requires adequately trained personnel, instrumentation and proper facilities. Trained and qualified personnel also require constant up-dating. Organizational prerequisites are the fixation of responsibilities, work-flow, and information-flow. Proper coordination regarding problems involving measurement is achieved by the formation of 'working-groups' within one or more production-units of an organization (company, cooperative or

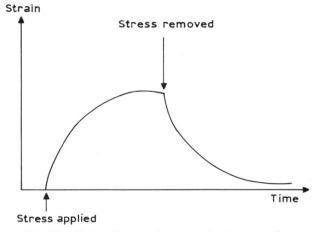

Fig. 33.   Relaxation mode—viscoelastic material.

corporation). Such groups constitute a consultative organ for Production and General Management.

The objective application of the two complementary information technologies 'computer-aided design' (CAD), and 'computer-aided management' (CAM), with the ultimate aim of Computer-integrated management (CIM) will continue, and eliminate much of the manual aspects of the manufacturing control system. CAD makes use of the computer to design and construct an optimal product, and some examples of its application are in the design of silos, flour-preparation plant, ovens, mixers and other bakery equipment. CAD/CAM is capable of organizing the conception, development, planning, raw material preparation and production of an automated manufacturing process.

For in-line testing of the rheology of dough during processing, oscillatory and relaxation operation modes can be applied. The Bohlin Rheometer System marketed by Bohlin Reologi AB, S-22370, Lund, Sweden, is computer-controlled, and includes an integrated temperature-control unit, allowing the operator to make measurements with a high degree of automation. In addition to performing temperature or time-programmed flow-curves or frequency-sweep responses, the system can handle the automatic linking of programmed runs, measuring in the following operation modes: viscometry, oscillation, and relaxation (see Figs 33–36). The integrated

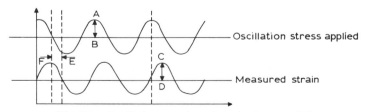

Fig. 34.   Oscillatory mode—viscoelastic material.

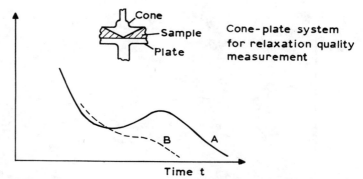

Fig. 35.   Relaxation-spectrum—wheat glutens. A. Good-quality gluten (sharp peak). B.
Good-quality gluten (weak peaking).

system comprises: rheometer, temperature-control unit, rheometer control and
system-interface, central computer for program and data storage with console, high
resolution colour-graphics monitor and four-colour digital plotter. A similar
system, the Rheo-Tech VER, marketed by Rheo-Tech International Limited, Forest
Gate, London E7 0EW, UK, can be connected to any computer through a suitable
interface. However, existing software is written for the IBM system; nevertheless,
other software systems are also available. A program-suite provides rheometer
control for measurement of strain and strain-rates for both dynamic and oscillatory
deformation. Controlled stress, creep, yield, oscillatory, and shear-gradient
programs are run directly from the control-panel. Computer control allows
program variation and data analysis. Measuring heads with the following
geometries are provided: cone and plate, parallel plate, concentric cylinder and
double-gap concentric cylinder. Shear-stress, shear-gradient and viscosity are
calculated and displayed.

Material property testing using a small-amplitude oscillatory shearing motion,
provides information on the ability to both store (elastic) and dissipate (viscous)
energy. The sample is subjected to a sinusoidal shear stress at preset frequencies or
frequency-sweep, ranging from about 0·001 Hz to 10 Hz, under both time- and
temperature-controlled conditions. The phase difference and amplitude-ratios
between the input sine wave and the measured strain wave characterizes the

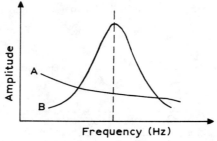

Fig. 36.   A. Oscillation—stress applied. B. Resonance—curve output.

viscoelastic behaviour of the material under test (Fig. 33). This information supplements creep tests, since the time lag is shorter for stored, elastic energy, and dissipated viscous, energy. For longer time-scale tests, the relaxation or creep-recovery mode is a much faster method of obtaining a mechanical spectrum than the oscillation mode. The shear-stress is instantly applied, maintained constant, then, after a period of time, suddenly removed. If recoil occurs, it is defined as creep-recovery (relaxation). During relaxation, retarded elasticity can be isolated from viscous flow. In the relaxation mode, temperature, strain, stress, time and ratio stress:initial-stress all appear on the print-out.

## 1.8.3 MEASURING SYSTEMS BASED ON DOUGH DEFORMATION

Modern rheometers are designed on the following measurement principles:

—cone-plate and parallel-plate systems
—concentric cylinder systems
—double-gap concentric cylinder systems
—high shear-rate concentric cylinder systems

Typical shear-rates range from $0.001$ to $10\,000\,s^{-1}$, delivering shear stresses of $10^{-3}$ to $10^{-5}$ Pa. Viscosity measurement ranges are from $10^{-4}$ to $10^{-8}$ Pa s.

In the oscillatory mode frequencies range from $10^{-3}$ to 20 Hz. The 'in-line' and 'on-line' system R2 of Rheotec AB, S-22370 Lund, Sweden, operates in the oscillatory mode, at frequencies and amplitudes chosen to obtain optimal signal resonance for the material and process being measured (see Fig. 36).

The application of IR-absorption and NIR techniques also show considerable scope for the structural study of biologically significant differences in complex molecules.

IR-absorption bands showing hydrogen-bonding can be used to study changes in wheat and flour during storage (ripening) under specific conditions of time, temperature and humidity. NIR-transmission spectra of whole grains of wheat, with appropriate calibration for temperature, weed-seed contamination, crop-year and moisture, measure protein and moisture of 150 to 250 g samples of wheat within 50 s at wave-lengths between 900 and 1050 nm.

The superconducting electromagnetic fields of NMR are capable of separating and identifying important minor structural differences in complex molecules, polymers, e.g. composition, tacticity, chain-branching, and residual monomer-content. High resolution pulsed proton magnetic resonance can be used to measure spin-lattice ($T_1$), and spin-spin NMR relaxation-times. This allows determination of free and bound water in samples, and water mobility on storage. Thus, changes in state of water in colloids with time can be followed under specific storage conditions, e.g. starches, wheat proteins, and non-starch polysaccharides. Chemical bonding energy levels are consistent with hydrogen-bonding. Such wide-band proton NMR-spectrum parameters as dH, $M_2$ and S are higher for hard-wheat flours than for soft-wheat-flours, reflecting differences in the conformational state of protein complexes, consistent with stronger or weaker hydrogen-bonding. Correlation with rheological data and technological function of wheat-flour proteins is high.


### 1.8.4 CLEANLINESS AND SANITATION (HYGIENE)

Cleanliness and sanitation are very important, and comprise the most exacting operations in every food-processing facility, regardless of size. They represent the heart of every quality-control programme, commencing with the receipt and handling of raw materials and extending through processing, packaging, handling, storage and final delivery to the customer.

Two short maxims summarize the basics of sanitation, which is little more than good housekeeping, viz. 'prevention is better than cure' and 'clean twice and disinfect once'.

Food safety is the number one concern of the food industry, and plant management has the responsibility to see that the consumer receives foods that are safe and free from contamination or adulteration.

A prime contribution to a sanitation programme is the cleanliness of floors, walls, and ceilings. These surfaces must be smooth, resist dirt and bacteria, be easily cleaned and long-lasting. Ceilings and support-columns are best initially power-sanded to expose all cavities, which are then filled, before applying a catalysed filler/sealer with spray and squeegee, making sure all holes are filled. Two coats ensure an even base. The surfaces are then finished with coats of catalysed polyamide-epoxy enamel with an airless spray gun. The result is a smooth, highly reflective surface, which should last for 15 years' exposure to cleaning solutions.

The sanitary design of machines and equipment involves construction in a style or manner which protects the product from contamination, results in equipment that is readily cleaned, and prevents the harbouring of food and insects in dead spots where bacterial growth can be promoted. The use of stainless steel or other non-corrosive material must be complemented by the arrangement of all surfaces in contact with the food so that they can be either readily cleaned 'in-place', or disassembled for cleaning. The degree of sanitary design required will depend on the nature of the process and the susceptibility of the food material processed. Milk and meat, for example, require specialized equipment design and handling, since they are a potential microbiological hazard. Equipment for handling dry ingredients and foods must be designed to prevent contamination by airborne contaminants, insects or rodents.

The American Baking Industry Sanitation Standards Committee (BISSC) (which is made up of representatives from the American Bakers Association, bakery-equipment manufacturers, the American Society of Bakery Engineers, Associated Retailers of America (Bakers), Biscuit and Cracker Manufacturers Association, American Institute of Baking, with consultants from nationally well known sanitarians) and the FDA publishes lists of equipment and manufacturers certified as meeting BISSC standards. Among the equipment for which standards have been published are the following: flour-handling; dough troughs; intermediate-proofers;

horizontal and vertical mixers; conveyors; dividers; rounders; bread-moulders; air-conditioning equipment for fermentation; proofing; cooling and retarding; ingredient water-coolers and ice-makers; bread-slicing, wrapping and bagging machines; mechanical ovens; pan-greasers; liquid ferment and continuous-mix processing equipment; dough-chutes; hoppers; trough-hoists; automatic dough-trough pumps; depanners and delidders; floor ingredient-scales; racks; pan-trucks; skips; pallets; liquid-measuring systems; facilities for handling and storing refined-liquid and dry sweetening products; facilities for handling and storing bulk edible shortening; electrical motors; distribution cabinets and containers; coolers; baking-pans.

Food safety is the main concern of the food industry, and the number one risk of contamination and unsafe foods is from microorganisms or their toxins (toxic products).

In 1971, at the Denver National Conference of Food Protection, the concept of 'hazard analysis and critical control point system' (HACCP) was generally accepted in the USA by the FDA as the main procedure for food process inspection (FDA Compliance Program Guide Manual). Food and Drug law clearly states 'a food may be considered adulterated if it has been prepared, packed, or held under conditions whereby it may have become contaminated'. Industry is therefore expected to prepare and market foods which are as free from contamination as good manufacturing practices (GMPs) will allow. Regulatory agency inspectors look for unsanitary conditions and potential contamination hazards of any sort. In the USA, Congress gave the FDA additional powers of inspection to examine quality-assurance programmes and records, to make a better assessment of a food producer's daily operations. Under Section 402(a) of Chapter IV of the Federal Food and Cosmetic Act, a processor is required to identify his control points, and associated hazards of contamination from possible unsanitary environment, and accept total responsibility legally for its entire programme. The issue of 'good manufacturing practices' for the food industry, sets out guidelines as to how a clean plant is maintained, and is intended to act as an 'umbrella'. Therefore, the industry has to respond by demonstrating to the public, government, and regulatory agencies that it is producing clean, wholesome and safe foods.

Control-points are defined as observations or measurements of the manufacturing operation and environment by which production is controlled to avoid contamination from possible sources. Many of these may be non-critical, but are part of good-housekeeping. Examples of control-point areas are as follows.

**Environment and Buildings**
Condition of external areas and pest harbourages; sealing-up of the building; pest absence; pest-control programme; water-supply quality and bacteriological status; guarding of light-sources. Each of these should have recorded observations or measurements confirming that these operations are being handled thoroughly. Pest-control should include plant locations of all pest-control devices, storage of pesticides, frequency of checks on rodent traps and insect-electrocuters, details of insect species found. Specific needs could be the condition of equipment, e.g. a

detector or sifter, frequency and mode of checking, condition of electronic insect-killers, rodent bait stations. Checking of floor level and pallets for presence of insects or rodents.

### Raw materials

Supplier guarantees and specifications; sampling and quality checks; condition on arrival; clear identification for each raw material. The critical control point system should be applied to the plant programme, i.e. measurement or observation frequency; name of individual responsible; acceptable limits; comments and action to be taken. Each food plant must define its own control-points and set up its own programme. Most importantly, it must have management backing, and all personnel must be properly trained and motivated. Some form of checklisting is necessary to ensure that the programme has been carried out.

### Processing Equipment

Protection from the environment; operating times and temperatures; inspection of filters, sifters, magnets and detecting devices; checking for cleanliness, dead-spaces; waste-disposal.

### Packaging Materials

Specifications; quality assurance approval; protection from the environment; check FDA acceptability of polymer films.

### Ventilation

Air movement between processing and storage areas must be restricted; air-filtration and condition of filters; checks on the airborne microflora in storage, production, dough-room, fermentation, and personnel facility areas.

### Personnel

Regular medical checks for all personnel handling and associated with the food processing operation, including stool-analysis, and issue of 'health-passports'; training in hygienic habits; hygienic equipment in toilets and washrooms; clothing and headgear. In the USA, the approach to sanitation of two large food corporations can be cited. General Mills define their critical control-points by a twofold analysis, 'critical-point flow analysis' using flow-sheets to identify hazards, and a 'functional' analysis involving the operators as responsible for a clean operation. Pillsbury's approach involves four steps: (1) hazard identification, (2) identification of the points at which the hazard can be controlled, (3) fixation of control measures, and (4) method of monitoring the control measures taken.

### Equipment Aids to Sanitation Control

Swab test-kits for testing surface-sanitation, to check on cleaning and disinfection efficiency; counts: bacteria, yeasts and moulds (Millipore Corporation, Bedford, MA 01730, USA).

Electronic insect killers for airborne insects (Insect-o-cutor, Inc., Stone

Mountain, GA 30083, USA), or Gardner Zap (Gardner Mfg Co., Horicon, WI 53032, USA).

Ultraviolet liquid sterilizer, using quartz jacketed lamps, running the entire length of the unit; the liquid to be sterilized, e.g. liquid sugar, is passed under pressure, flowing perpendicularly to the lamps through the entire unit. The unit provides optimal germicidal UV energy at 2537 Å within a thin-sheet flow chamber, maintaining a uniform liquid flow-sheet of less than 0·25 in as it flows through the radiation chamber (PWS Sterilizer Bulletin from Pure Water Systems, 343 Boulevard, Hasbrouck Heights, NJ 07604, USA).

Instant hot-water makers for cleaning requirements, providing mixes of steam and cold water, or hot and cold water, equipped with 50 ft hose and water-saver spray-nozzle (Strahman Valves, Inc., 3 Vreeland Road, Florham park, NJ 07932, USA).

Birco Nocorrode foamer for foam-cleaning of floors unaffected by detergents, chlorine or acidic substances (Birco Chemical Corp., PO Box 1315, Denver, CO 80201, USA).

Morton Biocidal System instantly produces dilute solutions of the non-irritating, non-hazardous sodium hypochlorite from only salt, water and low voltage electricity. This provides the lethal effect of gaseous chlorine without the hazard. Compared to the high pH of commercial sanitizing solutions, the Biocidal's low pH (below neutrality) is much more effective, and is more cost-effective than most commercial biocides. Morton also markets over 35 portable and stationary high-pressure washers and cleaning and sanitizing equipment (Morton Salt Company, Division of Morton Norwich Products Inc., 44 Dock Street, St. Louis, MO 63147, USA). A combined cleaner/disinfectant–sanitizer–viricide is a quarternary-based formula for single-step use. Marketed by Robert Langer Co. Inc., it is effective against *Staphylococcus aureus*, *Salmonella choleraesuis*, and *Pseudomonas aeruginosa*. It is also viricidal against *Herpes simplex*, *Vaccinia*, Influenza A2, and *Adenovirus* type 2, and effective against pathogenic fungi, *Tricophyton interdigitale*. This product is accepted by the USDA.

A measurement of 'water-activity', in addition to moisture content, will help to predict whether certain products will support bacterial growth (American Instrument Company, Division of Travenol Laboratories Inc., Silver Spring, MD 20910, USA).

For automatic insect control in warehouses or plant, strategically mounted aerosol units with a master-timer controlling up to 60 units will spray concentrated pyrethrins to eliminate insects during the mobile, airborne stage of their life-cycle. This system is USDA-approved, and marketed by Virginia Chemicals, Dept 116, Portsmouth, VA 23703, USA.

For efficient sanitation, high-pressure, low-volume, compact, stationary or mobile cleaning-station systems are marketed by Chemidyne Corporation, 8679 Freeway Drive, Macedonia, OH 44056, USA, under the 'Economiser' trademark.

To scare away rodents from warehouses, plants, mills and granaries, ultra-high-frequency sound devices, operating on compressed-air can be applied, which is non-poisonous, maintenance-free and low in cost. Such devices are marketed by Rat-X,

325 W. Huron, Chicago 60610, USA, who are International leading rodent-control specialists.

In applying the 'critical control point' concept in food processing, the total-systems approach is to consider each unit operation, viz.

> Research: ingredient selection, process design, nutrition, product-stability and packaging, consumer tests, product quality.
>
> Materials: Ingredient and packaging specifications, quality assurance, supplier's quality assurance programme audit, legal clearance.
>
> Plants: incoming ingredient inspection, definition and monitoring of critical processing points, document sanitation programmes for plants and personnel, establish plant and corporate sanitation inspection procedures to ensure enforcement and improvement where required, all legal agency inspections and findings immediately reported to corporate level with corrective action taken.
>
> Shipping and Warehouse: carrier inspection, warehouse selection approval, audit warehouse food-protection programmes.
>
> Customer: well-documented consumer complaint procedure, continuing review of complaint patterns at headquarters and plant level.
>
> Recall: All products code dated, documented and tested procedures.
>
> (Source: Reference Source, *Bakers Digest*, 1986, p. 40.)

In Europe, an increasing impetus towards sanitation or hygiene importance in the Food Industry became apparent during the late 1960s, owing to the increasing presence of American-based corporations. These corporations purchased their raw materials locally, and according to headquarters practice demanded tight specifications for microfloral counts, as well as other analytical data. Thus, European flour millers and other raw material manufacturers were made aware of plant sanitation or hygiene as a new quality-control discipline.

Apart from the necessity for pest-control and plant clean-up, the additional need for microbiological testing for mould, bacterial counts, and checks for the absence of such food pathogens as *S. aureus*, *Escherichia coli*, *Salmonella* species, and *Candida albicans* (a pathogenic yeast) became apparent. Sanitation control teams were recruited to work in close collaboration with the company technical control organization in the laboratory. Their task was to limit the microbial contamination of the manufactured products to a minimum by intensive routine cleaning and disinfection, followed by planned microbiological checks of the efficiency of this work.

By means of planned cleaning and disinfection measures, any contamination chain in the processing plant can be interrupted, and its source eliminated. The presence and reproduction of pathogenic microflora result in the build-up of toxic metabolites or 'toxins', which contaminate food and constitute a public health hazard. Therefore, it is of paramount importance that their numerical presence be kept to an absolute minimum.

To detect the source of unwanted microbiological contamination, whether from a raw material or by contamination within the production process, or to control the relative efficiency of the cleaning and disinfection work, microbiological methods and techniques of process control are becoming increasingly important. The following bakery departments must be included in any biological process control system:

—Raw materials, compound improvers, concentrates, water-supply, especially material for nutritional enrichment, e.g. bran, germ, gluten, rolled or broken cereal kernels and cellulose concentrates.
—Machines, machine-aggregates, process-lines, and all equipment for the production process.
—Production and storage areas.
—Packaging materials.
—All transport-equipment and conveyors.
—Aspects of the hygiene of personnel, and provision for facilities to maintain them.

Apart from the standard methods for the microbiological testing of raw materials and finished products, the control of the process is of equal importance. The detection of hygienically critical points and sources of contamination fall into this category. The proper interpretation of these results allow conclusions concerning the efficiency of the sanitation work, or changes in cleaning and disinfection planning or routines.

The following is a description of the sampling methods and procedures used during biological process control.

*Direct Surface Contact*

The solid plate-count agar sterile medium in a petri dish is pressed against the surface under test, the lid replaced, and incubated at several incubation temperatures, 20–25°C giving viable counts of saprophytic organisms, 37°C for parasites and commensals of homothermic animals, 0–10°C for refrigeration temperature psychrotrophs and psychrophiles, and 55°C for thermophilic counts. This use of incubation temperatures will provide useful information about the spectrum of microflora present within the bakery generally. A non-selective nutrient medium such as plate-count agar (tryptone glucose yeast-extract agar) is used since this allows the growth of streptococci, which may be found frequently in food. This method is simple in application, but is only applicable to flat surfaces.

*Indirect Procedures*

Material is removed from the surface under test by scraping or cutting with a sterile scalpel and forceps. The material is then homogenized in quarter-strength sterile Ringer's solution (essentially 0·2% saline) to obtain an initial 1:10 dilution. Closed containers and similar equipment can be tested by rinsing with 0·2% saline, allowing the rinsings to stand in the containers for about 5 minutes before rinsing again and pouring the final rinse into the sterile sample container for a 1:10 dilution. Rinses should be tested as soon as possible within 6 hours of sampling. Mix by inverting the sample container slowly about three times. Pour 1 ml and 0·1 ml plates using plate-count agar and Yeastral milk agar, incubating at 30°C for 72 hours. To test for the presence of coliforms, inoculate 1 ml of rinse into each of three tubes of MacConkey's broth, incubating at 30°C. Record as colony count/1 ml rinse, and also record the presence or absence of coliforms. Colony counts should be less than 10 000/1 ml of rinse, for unwashed surfaces, and less than 200/1 ml after washing. For closed containers, such as bowls and measures, not more than 1 colony/ml of capacity after cleaning is a good sanitation standard.

The use of cottonwool or alginate-wool swabs is a more positive and versatile method of sampling. Cottonwool swabs are prepared from cottonwool wound to a length of 5 cm and a thickness of 2 cm onto wooden sticks or stainless steel wire. The cottonwool should be secured with thread, and placed in a large boiling-tube with 25 ml of 0·2% saline or Ringer's solution containing 0·05% sodium thiosulphate. The tube is then plugged with cottonwool, covered with aluminium foil and sterilized by autoclaving at 121°C for 20 minutes. To swab, press the swab against the side of the tube with the handle to remove excess liquid. Remove the swab, and using heavy pressure rub back and forth over an area of about 30 cm × 30 cm, using parallel strokes in two dimensions, rotating the handle to absorb as much sample material as possible. After swabbing the surface examined, break-off the swab aseptically into a screw-capped bottle containing 10 ml of a 1% solution of sodium hexametaphosphate in quarter-strength Ringer's solution. Replace the cap and shake the bottle vigorously. This causes the wool to disperse and dissolve, giving a suspension of the bacteria on the swab. Prepare plate counts from 1 ml and 0·1 ml amounts, or further dilutions as necessary, using appropriate media. Any slight bacteriostatic effect of the hexametaphosphate can be nullified by using tryptase–soy–agar (BBL) or plate-count agar, and 20 ml of medium added to each plate. Counts are reported per square centimetre of surface swabbed.

It is advisable to swab more than one area of any piece of equipment, with particular attention to points difficult of access for cleaning. The recommended standard is not more than 1 bacterial colony per squate centimetre after thorough cleaning.

For use in bakery production areas, the swabs can be prepared in aluminium alloy or rigid cardboard tubes instead of boiling-tubes to avoid breakages.

In setting standards, quantitative counts must not be interpreted in the same way as chemical or physical tests, but rather viewed as being semi-quantitative data and interpreted statistically over a specific time period. Surface testing of equipment and plant is intended as an index of cleaning/disinfection efficiency, and as long as reasonably high standards are set and maintained, this source of food contamination can be eliminated. Inaccessible edges and corners of plant are the critical points.

Another direct technique which is effective, but somewhat time-consuming for a bakery sanitation control team is to pour the liquid medium directly onto the object or surface to be tested, using a frame to limit flow-out. After setting, the medium is removed, and the adherent microflora incubates.

For this type of work, it is often more economic to purchase plastic screw-cap

'Falcon-swubes' with swab sticks affixed, which are packed in a sterile condition; plastic petri dishes are also available from the international media manufacturer Baltimore Biological Laboratories, a division of BioQuest, under the 'Falcon' label, located at Cockeysville, MD, USA.

During all microbiological control work, the laboratory worker must be protected from the accidental production of bacterial aerosols (vapour-borne bacteria). All cultured material should be treated as 'pure-cultures', and rendered harmless by collective autoclaving before final disposal as garbage. For the determination of bakery atmospheric-counts, various techniques can be applied. Most require specialized equipment which collect airborne dust with the help of a fan-assisted sample-collecter, or the use of special filters. However, a simplified technique is to expose media plates and rely on the exposed sedimentation surface for a fixed time duration (15 minutes). The plates are then incubated and the counts expressed in terms of sedimentation surface for a given time. Since particles less than about 10 $\mu$m do not sediment, and sedimentation generally depends on available air-turbulence, this technique can only give an indication of air contamination.

For the routine bakery sanitation (hygiene) control, non-selective media are usually adequate to provide a general picture of the hygiene standard. Total plate-counts of the aerobic mesophilic bacteria, yeasts, moulds and coliforms, being the basic recommendation.

However, for the surveillance of the raw materials, and personnel hygiene, more selective media for the isolation and identification of specific groups and species are necessary.

In order to keep the mould population in proofing-aggregates down to a minimum, especially in the case of roll-plants and variety, and specialty bread lines, the use of polyamide gauze material instead of any natural fibre such as cotton is essential. The installation of UV light cannot be relied upon to eliminate mould growth under conditions of proofing humidities. It is important to reduce the growth of xerophilic moulds of the type *Aspergillus*, and *Penicillium*, since these produce mycotoxins. There are about 200 mould species which are capable of producing mycotoxins under specific conditions when growing on cereal products. Primary contaminations can occur from flour, wholegrain products and bran, as bakery raw materials, and secondary contamination is possible from baked bread on storage. Amongst the mycotoxins found in baked products are aflatoxin, patulin, penicillic acid, citrinin and ochratoxin A, and also zearalenon in bread, according to Spicher. With the increasing use of crushed cereals, and highly enriched concentrates for health-breads the risk of contamination with aflatoxins B1, B2, G1 and G2 has increased. In such products the maximum content of aflatoxins B1, B2, G1 and G2 in total must not exceed 10 $\mu$g/kg, and the concentration of aflatoxin alone must not be higher than 5 $\mu$g/kg. The possibility of a health risk for the consumer is difficult to quantify, in spite of the amount of work done concerning mycotoxins, but the safest course to follow is the elimination of all mould occurrence within the bakery on sight. Bread-slicing areas in particular must be maintained at low humidity and temperature, since the growth of mould mycelia and aflatoxin B1 and G1 build-up is rapid at 95% RH, and 30°C. Bread-slicing knives and slicing-machines must be cleaned and disinfected against mould growth due to crumb accumulation.

The use of preservatives such as propionic acid, propionates, and sorbic acid, do not guarantee freedom from mycotoxins. The presence of the mycotoxin zearalenon originates from the genus *Fusarium*, which occurs on grain in the field, therefore, this would suggest a primary contamination of a cereal raw material, viz. storage-grain before or at the mill.

All nutritional grain products for the preparation of specialty or variety breads, are a potential source of mycotoxin infection, especially when stored for long periods at moisture contents in excess of 15·5%. These products include the following: wholegrain meals, grain-brans, grain-grits, malted-grains, grain-germs, crushed whole grains, grain-flakes, muesli, oil-seeds and extrudates. The most important sensory tests for these products are smell and taste, followed by plate-counts for moulds and bacteria. When the latter are excessive, usually the material must be rejected on sensory tests in any event. Product appearance and colour is also important but takes second place to smell and tests. In the FRG, top quality products are awarded DLG-prizes according to merit by the German Department of Agriculture, which can be used for advertising purposes.

## Raw Material Sanitation

To detect the presence of insect eggs or larvae, random samples of flours and all grain products should be incubated in bottles with fine-gauze closures at 27°C. After 3–5 days the presence of any insect pests will present themselves as silk threads clumping the particles of the material together and/or the presence of the worm or caterpillar stage. Common pests include: the flour-moths or *Ephestia*, the species *Kuehniella* from the Mediterranean region; *Plodia interpunctella* (*Dörrobstmotte* = dry-fruit moth); *Tenebriones* (Meal-worms); *Tyroglyphus farinae* (flour-mite). Grain-pests include *Calandra granaria* (wheat-weevil); Australian grain-borer (*Rhizopertha dominica*). Bakery pests include, *Blattaria* (cockroaches); *Lepisma saccharinum* (silverfish) and the *Mus domesticus* (house-fly).

In carrying out incubation tests for storage-pests (*Vorrats-schädlinge*), the incubator humidity must be 60–80%, otherwise the eggs will not hatch, even under favourable temperatures. For some life-cycles, temperatures of 18–25°C are optimum.

Fumigants in common use for such pests are methyl bromide, or phosphine, marketed as 'Phostoxin' in tablet form. On moistening, these tablets disintegrate, releasing phosphine from the aluminium phosphide. The legal upper limit of phosphine residue is about 50 mg/t for safety, as enforced in the FRG. Ethylene oxide is used in USA, and is very effective against weevil. For grain, epoxide concentrations in the vapour-space of 0·5 g/litre at 120°F, over a 2-hour period, also reduce the milled flour microbial counts to less than 500/g. Propylene, also used in USA, at 1·5 g/litre concentration in the vapour-space at 120°F, reduces microbial counts to a minimum, mould growth being completely destroyed. The commonest contact-insecticide, which is relatively safe with regard to residue, but has no lasting effect, is based on pyrethrum piperonylbutoxide. In selecting a disinfectant, one which can be diluted many times and remain efficient in killing bacteria, leaving a minimum of residue, is the obvious choice. The dilution factor is not only a cost, storage and transit consideration, but also allows strong dilutions to be used when sporing-organisms require disinfection. Although efficient cleaning with surfactants is normally carried out with agents containing substances of the 'Tween' or 'Span' series (Atlas Powder Company), or by means of quaternary-ammonium compounds, retention of efficiency in the presence of organic matter is essential. A disinfectant depending for its efficiency on oxidation, will waste itself on dead organic matter before it attacks living protoplasm. Another prerequisite is its ability to form either an emulsion or solution, and complete homogeneity on standing. Stability is also an important factor over as wide a temperature range as possible, as is freedom from toxicity at all dilutions.

Few disinfectants are compatible with soaps and surfactants, apart from some coal-tar-based products, many of which are too toxic for routine use. Therefore, the cleaning operation is best tackled separately, and provided an adequate contact-period is allowed, many organisms are killed by detergent-soaps, e.g. streptococci and pneumococci, but the Gram-negative bacteria will survive. Certain classes of synthetic cationic detergents are, however, actively germicidial against both Gram-positive and Gram-negative bacteria, even at dilutions of 1:3000 up to about 1:30 000, by disrupting the bacterial cell wall or membrane. Halogen-based substances, viz. chlorine, bromine and iodine, are used in solution, but have to be used in an excess proportionate to the amount of residual organic matter present. Typical examples are: chloride of lime ($CaOCl_3$), sodium hypochlorite (chloros), chloramine and hermitine, all sources of 'available-chlorine'. The latter is prepared by the electrolysis of a solution of sodium chloride and magnesium chloride, sodium hydroxide being added to fix free hypochlorous acid, also acting as a preservative. Although not as commonly used in the food industry as chlorine, iodine used at about 2% in solution with potassium iodide or with alcohol to form iodoform, is a very long-lasting producer of fresh iodine. Iodine is liberated on contact with the enzymes of the microbial contaminants; otherwise no iodine is released.

The liberated 'nascent' or atomic oxygen is the active agent in many disinfection procedures. The halogens are examples of such oxidizing agents. The action of hypochlorites depends on liberated chlorine, but the process is one of oxidation. Atomic oxygen and ozone ($O_3$) are also powerful disinfectants, and in Japan have been applied to food sterilization at high speed. Ozone combined with ultrasound and ultraviolet irradiation with optimum regulation can provide an efficient and toxin-free disinfection and sterilization as required. For the treatment of wheat and wheat products, clearance has been obtained from the Atomic Energy Agency in Vienna by USSR, Canada, Spain, Brazil, Chile, Thailand and Bangladesh; also rye-bread may be treated in USA and Netherlands (1988), according to the *Official Journal of European Communities*, No. C99, 13:4:87. These clearances refer to ionizing radiation in the gamma form. Present UK legislation (1988) is based on the Food (Control of Irradiation) Regulations Amendment 1972, which states that food for general consumption may only be irradiated up to a level of 50 rad (0·5 Gy or 0·005 kGy), with the exception of sterile diets for hospital patients.

About 90% of all food-poisoning is caused by microorganisms, the remaining 10% arising from intoxication of non-microbial origin, e.g. pesticide residues, and arsenic, antimony or zinc compounds. Owing to the wide distribution of food, this commodity acts as a carrier or transport medium for pathogenic (disease-causing) microorganisms. In such cases, no reproduction of the organism in the food is necessary, since the consumption of a few living cells of these pathogens constitute a potential health hazard. Examples of such types of disease are: brucellosis, cholera, typhus, paratyphus, and certain viruses and parasites.

Other types cause illness owing to their high content in food as living pathogens. Depending on the state of health, age, stomach acidity, etc., a cell count of $10^4$–$10^6$/g of food can be enough to cause a food poisoning. However, since such a high primary contamination of food with pathogenic bacteria is ruled out, reproduction of the bacteria is necessary to cause illness. Typical examples of such bacteria are: the salmonella, which give rise to gastroenteritis, and those caused by such facultative, spore-forming pathogens as *Clostridium perfrigens* and *Bacillus cereus* during the summer months.

A third group, which cause food poisoning, are also microbial, and due to bacterial reproduction within the food, give rise to certain chemical components called 'toxins', which are biosynthesized from specific chemical substances present in the food. In this case, illness is caused by the mere consumption of the toxin-containing food, and ingestion of the living cells of the causative agent is not a prerequisite. Examples of these are: botulism and staphylococcienterotoxis, caused by bacterial toxins. These also include the recently identified mycotoxicoses, caused by various types of mould. This choice of groupings is important in planning measures to prevent outbreaks of illness caused by food organisms. In the case of infections caused by microorganisms from which an increase in cell-count is not a prerequisite for a potential hazard, the only measure which is meaningful is the avoidance of contamination during processing and handling. Food-poisoning caused by microbial contamination, as a result of reproduction, can be reduced or stopped by appropriate conservation or storage measures. Health hazards are not only to be found in the underdeveloped countries, but also in countries with highly industrialized societies with relatively organized sanitation and hygienic services. Incidences of food-poisoning due to *Salmonella enteritidis* (Gärtner's bacillus) have recently shown increases in certain countries, owing to the increasing affluence and trend towards eating out in hotels, restaurants and take-aways, and the increasing international movement of raw materials and animal-feeds. Quite apart from the potential health hazard and the relatively curable instances, fatal cases occur in spite of the advances of modern medicine. Therefore, the consequences for society of microbial food-poisoning must be drawn to everyone's attention, especially those responsible for the processing, handling and sale of food products.

The commonest symptom of food-poisoning from bacteria is diarrhoea, which is symptomatic of gastroenteritis, enteritis and dysentery. Other intestinal illnesses are due to *Klebsiella* and *Shigella*, often associated with dairy products, and *Proteus*, found in meat, fish, and eggs, forming a toxic lipopolysaccharide. A more recent gastroenteritis organism is *Yersinia enterocolitis*, which originates in pig-excreta.

*Escherichia coli*, is a normal human-intestinal inhabitant, but some species are facultative pathogens, forming endotoxins, which are coded by plasmids. In this way plasmids from *E. coli* are carried over to other Enterobacteriaciae such as *Shigella*, *Salmonellae*, and *Pseudomonas aeruginosa*, as well as spore-forming *Vibrios*.

It is likely that cases of food-poisoning caused by entero-pathogens of the *E. coli* group (enteritis-coli), arise more often than generally recognized, since often in cases of intestinal infections, only *E. coli* can be identified. The symptoms show the same pattern as gastroenteritis, i.e., fever, stomach-ache, diarrhoea, and vomiting, with incubation times of up to 24 hour. Infections occur by direct contact with a source in the first instance, typical foods becoming infected being meat, milk, and dressed salads.

The facultative pathogenic species of *Staphylococcus aureus*, which form enterotoxin in the stomach and intestinal canal can also be found on filled-baked products, sandwiches and cakes. Contamination is by direct or indirect contact with personnel, who could be carrying the bacteria on skin, wounds, nasal discharges and saliva, discharged into the air by coughing and sneezing.

However, not all *S. aureus* species produce enterotoxin, and there is to date no simple test to ascertain the pathogenic species; the ability to form acids from glucose and mannose anaerobically, and the formation of coagulase are typical biochemical confirmation tests; but some negative species have been reported to cause illness. An additional biochemical confirmation test for this pathogen is to test its ability to break down a desoxyribonucleic acid (DNA) containing substrate, due to the presence of desoxyribonuclease (DN-ase). The symptoms, after a 3–6-hour incubation period, are giddiness, vomiting, stomach-ache, diarrhoea (often with haemolytic discharge).

The toxin formation usually occurs in the food, and is consumed with the food, but it has also been found that the enterotoxin-former can reproduce within the tract, in which case acute illness occurs. These enterotoxins affecting the central nervous system. The toxin produced by this organism is very difficult to destroy ($121°C$ for 20 minutes at $15 \, lb/in^2$), and is described as heat-stable. Food poisoning from *S. aureus* has been found on a wide spectrum of food products, attributable to human infection sources. Therefore, some rules for the prevention of staphylococcal food-poisoning are: reduce food-handling to a minimum, keep all foods capable of growing bacteria under refrigeration, and ensure that all persons handling food are free from infections, and that they practice good personal hygiene.

Additional pathogens are *Clostridium perfrigens*, a Gram-positive, spore-forming anaerobe, which produces many types, and more than 10 types of toxin. Sources of infection are: soil, water, dust, human and animal faeces; *Bacillus cereus*, a Gram-positive, spore-forming anaerobe, has occurred in puddings which have not been refrigerated after preparation. It has strong associations with cereal products, but mainly sweetened, wet-processed products.

All raw materials received into the bakery-store inventory could be sampled and tested for mould and bacterial content according to the procedure described below. However, where laboratory facilities or time is limited for microbiological testing, resources should be better utilized in concentrating on process plant, utensils, and personnel sanitation and hygiene.

Important raw materials for checks are: egg (fresh, liquid and dried), milk-powders and all dairy-based materials, cereal enrichment products, viz. bran, crushed-grains, wholemeals, germ, oil-seeds, soaked-grain products, malted-grains and malt-products; desiccated coconut can also be a source of infection. Breadmaking flours do not normally require microbiological examination.

*Procedure for Raw Material Microbiological Examination*
The accepted method, known as the 'plate-count', for water and foods is based on the assumption that each microorganism will give rise to one colony after incubation in a suitable medium. A measured amount of sample, or dilution of the sample is mixed in a sterile petri dish with melted agar. After incubation the number of colonies growing in the agar are counted to give an estimation of the bacterial or mould population of the sample material. Although not a very exact method, the technique is simple and gives satisfactory results for routine testing. For the preparation of dilutions of the sample under test, 0·9% saline, peptone-water or quarter-strength Ringer's may be used. Weigh out 10 g of sample aseptically into a sterile wide-mouthed Erlenmeyer flask fitted with a ground-glass stopper. Add 90 ml of sterile 0·9% saline, and shake the flask about 25 times using up-and-down movements of about 30 cm, within 7 s.

Sterile disposable plastic containers and test-tubes, such as those marketed by Falcon Plastics in the USA can reduce preparation-time, but where glassware is in use the standard procedure is to autoclave all equipment, solutions and media at 121°C for 20 minutes, under pressure.

The initial sample extract represents a 1:10 dilution. Transfer with a sterile pipette 1 ml of the extract into a sterile petri dish, and introduce 15–18 ml of the appropriate sterile melted agar medium at 45°C. Mix the medium with the inoculum by gently rotating the dish, first in one direction, and then in the opposite direction, taking care not to splash the mixture up to or over the edge of the dish. Cover, and allow the medium to solidify on a level surface. Incubate the plates in the inverted position for 24 to 48 hours.

Transfer a further 1 ml of the 1:10 extract dilution to a sterile test-tube containing 9 ml of sterile saline, mixing by rotating between the palms of the hand. This produces a 1:100 dilution of the original sample. Again transfer 1 ml of this dilution to a sterile petri dish, preparing the agar plate as already described. Using the same procedure, prepare a 1:1000 dilution of the sample and plate out as before. Further dilutions can be prepared as necessary, depending on the microbial population.

To evaluate the incubated plates, select plates from each dilution which contain between 30 and 300 colonies. View the plates against an illuminated background, counting the number of colonies. Multiply the number of colonies by the dilution of the sample. For example, if the 1:100 dilution plate shows 50 colonies, then 50 × 100 = 5000; the estimate of the bacterial population of the original sample would be 5000 microorganisms per gram, and is reported as the Standard Plate-count per gram. To express the numerical results in line with the precision of the technique used, the recorded number of bacteria per gram should not include more than two significant figures. For example, a count of 153 is reported as 150, and a count of 165 as 170; but a count of 55 is expressed as 55. This method can be used when the sample is believed to contain such a high and mixed population that any differentiation of species is either difficult or impossible. The dilution technique is not used to facilitate enumeration, but rather to reduce the number of organisms, thus allowing the development of discrete colonies so that their individual reactions may be studied.

Since dilution can result in only the predominant organism being detected, pathogenic agents present in only small numbers remaining undetected, appropriate selective media should be included as well as general-purpose media for each raw material sample.

The use of aseptic techniques and sterile equipment is essential for all procedures. Pipettes whether glass or disposable plastic, if pipetted by mouth must be cotton-plugged, or ideally the use of a rubber-bulb is recommended for dispensing dilutions for safety reasons where pathogens are being handled. Inocula are always allowed to flow into tubes and petri dishes, and not blown, to avoid contamination and the production of bacterial aerosols. Covers of petri dishes are raised just enough to allow the introduction of pipette, inoculum and the culture medium, thus reducing exposure to the environment.

Standard Plate-counts should be incubated at 32°C and 35°C for 48 hours. Stacking of plates in incubators should be such that each pile is separated from adjacent piles, and the top and sides of the incubator by at least 4 cm. The inverted piles of plates are placed directly over each other on successive shelves to encourage even circulation of air and temperature in the incubator.

Appropriate selective media should include those for *Staphylococci, Streptococci*, and *E. coli*, the latter acting as a 'faecal-indicator' for the possible presence of other faecal contamination, with such human or animal intestinal pathogens as: *Salmonella, Shigella* and *Clostridium perfrigens* and others.

For the identification of streptococcal groups, and *Staphylococcus aureus*, surface inoculation of the appropriate plated media can be carried out. This involves introducing a measured amount of the inoculum onto the surface of the medium with a pipette or bacteriological loop, and spreading the inoculum uniformly over the surface with a sterile bent glass-rod as 'spreader'.

For the identification of haemolytic *Streptococci*, single or double-poured blood agar plates can be used. Prepare tubes containing 12 ml trypticase soy agar medium, and pour into sterile petri dishes to form a base-layer, allowing to solidify. The base layer is then either streaked or spread with the sample inoculum, and immediately overlaid with about 12 ml of sterile melted blood agar, rotating once or twice to mix, and allowing to solidify. The blood agar is prepared by rehydrating a suitable base medium such as trypticase soy agar according to BBL label instructions. Fill into tubes and sterilize in the autoclave, and allow to cool to 45°C, which is essential before adding about 5% sterile defibrinated sheep or rabbit blood (5 ml blood per 100 ml of medium). Mix thoroughly by rotating the tubes, but avoid bubble formation.

If single poured plates are used pour 15 to 18 ml into the sterile petri dishes, avoiding bubble formation, and allow to solidify on a flat surface. The plates are then ready for inoculation by surface-spreading and can be stored at refrigeration temperature, away from heat and light; but freezing must be avoided, thus allowing plate preparation before use.

Haemolytic colonies of *Streptococci* are identified on blood agar plates as black colonies with transparent surrounding zones. Identification of the haemolytic colonies is carried out by selecting and transferring to 1 ml of Todd-Hewitt Broth (a BBL medium based on beef-heart infusion) which is used to group and type the beta haemolytic *Streptococci* for epidemiological studies. After a 5-hour incubation at 35°C, fluorescent antibody staining is used for the detection of Group A *Streptococci*. The technique is a one-step fluorescence inhibition procedure. Staining consists of a mixture of anti-Group A *Streptococcus* globulin, labelled with fluorescein isothiocyanate, unlabelled anti-Group C *Streptococcus* globulin or serum, and unlabelled 'normal' globulin. BBL test and negative control reagents are prepared by rehydration with 5 ml of sterile distilled water. Incubated broth smears are air-dried, and fixed in 95% ethanol for 1 minute, or by dry-heat with a bunsen burner. One smear is covered with one drop of the diluted test reagent, which is spread over the entire smear with an applicator-stick, in the horizontal position. The other smear is then covered with one drop of the diluted negative control reagent in the same manner. After allowing the smears to stand for 30 minutes at room-temperature, in a moist atmosphere, excess reagent is absorbed with a tissue. They are then rinsed with buffered saline, and then fixed by allowing to soak for 10 minutes in a buffered saline. Excess saline is then removed by momentary rinsing with distilled water. The smears are then carefully blotted, one drop of alkaline buffered glycerol saline being placed on each, and coverslips placed over them. Examine with a fluorescence microscope, using the oil-immersion objective (97 ×) and appropriate filters. Smears stained with the test reagent are examined for the presence of Group A, which stain bright yellow-green with well-defined rims. The inhibition procedure employed results in Group C and G *Streptococci* and *Staphylococcus aureus* appearing non-fluorescent to dull yellow-green with ill-defined rims. For comparative purposes, known strains of these organisms, including strains of Group A can be tested together, thus allowing experience to be gained in distinguishing the bright fluorescence of Group A *Streptococci* with well-defined rims from the low-grade fluorescence occurring with some strains of Group C and G *Streptococci* and *Staphylococcus aureus* (Public Health, USA). For the isolation of *Staphylococci*, dilutions are prepared from the sample as described earlier, depending on the raw material and the suspected potential bacterial population. Complete surface inoculation of the plates, using a bent glass-rod as a spreader, is carried out on Vogel and Johnson agar, which permits early detection of coagulase-positive and mannitol-positive colonies of *S. aureus*. Incubation is at 37°C for 24 hours, but if no colonies appear, reincubate for a further 24 hours. *S. aureus* forms colonies of 1·0–1·5 mm diameter, which are black, shiny and convex. They show a narrow white entire surround with clear zones of 2–5 mm extending into the otherwise opaque medium. An alternative proven medium for the enumeration of coagulase-positive *Staphylococci* in foods, is Baird-Parker's medium, which is enriched with egg-yolk, potassium tellurite and 20% glycine and 20% sodium pyruvate.

After incubation, a number of all of the colonies typical of *S. aureus*, can be examined by Gram-staining, and final identification carried out, either by the coagulase test or the DN-ase test. In the first case, coagulase plasma is inoculated with a loopful of an emulsified colony, or an 18-hour culture in trypticase soy broth, taking 0·05 ml. Place the tubes in a water-bath at 37°C to incubate. Any signs of clotting is a positive reaction.

A coagulase mannitol agar actively fermenting with coagulase-positive *Staphylococci* will produce large, yellow opaque zones, whilst the medium surrounding the growth remains a clear blue-purple. DN-ase or desoxyribonuclease activity is identified on DN-ase test agar. The suspected organisms are streaked on one plate. After incubation at 35°C to establish good growth, flood the plate with dilute HCl, positive cultures show a distinct clear zone around the streak. If the plate is flooded with toluidine blue, the plate becomes generally blue where the DNA has not been broken down, with bright rose pink zones around colonies producing DN-ase.

Staphylococcal food contamination results in an intoxication depending on the ability of the food to support growth of coagulase-positive *Staphylococci* producing the toxin. They grow and produce toxins best at pH 7·5, at moderate temperatures. Although normally requiring oxygen, the organisms can adjust to anaerobic conditions. The toxin produced by *Staphylococci* is heat-stable, Type B enterotoxin being even difficult to decompose by autoclaving at 121°C for 20 minutes. The consumption of about 20–25 $\mu$g of pure enterotoxin B from *S. aureus* is sufficient to cause enterotoxicosis. Most species produce enterotoxin A, which is the main cause of illness, but cases of enterotoxin B and combinations of Types A and B are quite common. To give rise to enterotoxicosis, bacterial cell counts in food of the order of $10^5$ bacterial cells per gram are necessary. The enterotoxins are released during the exponential and stationary growth phases into the substrate, being detectable after about 6 hours. The enterotoxins of *S. aureus* are water-soluble proteins, consisting of 18 different amino acids, of which, asparagine, glutamine, lysine and valine are the most important.

Serologically, toxins are classified into seven types: A, B, C1, C2, D, E and F. They are resistant to most proteolytic enzymes, viz. papain, trypsin and rennin, but are decomposed by pepsin at pH 2. *Staphylococci* are very tolerant to drying, high salt concentrations, and such toxic substances as disinfectants and antibiotics. Whereas the coliforms can live in symbiosis with *S. aureus*, the lactic-acid-forming *Streptococci* (beta-bacteria), *Lactobacilli* and *Pseudomonas* tend to exert an antagonistic effect on growth.

To date there are no international standards set for the various foods for coagulase-positive *Staphylococci* content, and quoted guidelines vary from less than 1 up to 1000 per gram of food. A realistic target specification would be about 100 coagulase-positive *Staphylococci* per gram of food.

In order to avoid contamination with *S. aureus* in the bakery the following points should be noted:

(1) The prime source of *S. aureus* contamination is humans, who either directly or indirectly come into contact with the food.

(2) Enterotoxin formation usually results from faulty conservation, packaging, storage and/or transport of raw materials.

(3) Incorrect and prolonged reduction of temperature of potentially hazardous raw materials is often the cause of infections, especially in bulk. Temperatures below 6°C do not allow growth or toxin build-up.

For routine checks on the level of coliforms, especially *E. coli*, the 'most probable number' technique can be used. Fermentation tubes containing inverted Durham-tubes are filled with Brilliant Green Bile Broth 2%, sterilized and inoculated with appropriate volumes of sample extract (say), five tubes each of two volumes, 10 ml and 1 ml, and incubate for 48 hours at 32°C. Examine for gas-formation, and report as 'Coliforms, Most Probable Number/ml' (Coliforms MPN/ml).

To confirm a test on liquid medium, loopful volumes are streaked from positive tubes onto Levine Eosin Methylene Blue Agar or Endo Agar and incubated for 24 hours at 32°C. Examine for the presence of colonies, and inoculate a Nutrient Agar slant and a Lactose Broth fermentation tube from one or more typical colonies.

Incubate the slant for 24 hours, and the broth tube for 48 hours at 32°C. Gas-formation within 48 hours in broth tubes, coupled with the presence of Gram-negative, non-spore-forming, rod-shaped bacteria on the slant is a positive-test confirmation.

Coliforms are common inhabitants of any normal environment, as well as the intestinal-tracts of man and animals. Although not often pathogenic, their presence in dairy products is taken as an indicator of the general state of sanitation, in particular of the pasteurization process. These members of the genera *Escherichia* and *Aerobacter* contaminate dairy products at source, but if found in heat-treated milk-products, a secondary contamination is indicated.

Coliforms can also be evaluated on solid medium, such as Violet Red Bile Agar or Desoxycholate Lactose Agar, in which case precise counts are obtained; but the use of liquid medium allows a larger sample volume, and is thus more likely to give positive results.

With liquid medium, actual numbers are estimated by multiple-tube examination, and probability-tables used to interpret bacterial numbers from the number of positive tubes at each dilution. The Most Probable Number (MPN) of bacteria is expressed per 100 ml of sample or initial dilution.

For soliform evaluation on solid medium, transfer 1–4 ml of a 1:10 dilution of the sample to sterile glass or plastic petri dishes. Add 15–20 ml of melted, cooled (45°C), Violet Red Bile Agar or Desoxycholate Lactose Agar; mix the sample and agar medium by tilting and rotating the dish, and after solidification, add a 3–4 ml overlay of plating medium to prevent spread of surface colonies. Invert and incubate the plates for $24 \pm 2$ hours at 32°C. Examine the plates for the presence of dark-red colonies measuring 0·5 mm or more in diameter on uncrowded plates (fewer than

150 colonies/plate), and report as coliform colonies per millilitre or gram of raw-material sample. Verify typical colonies by transfer to fermentation-tubes of Lactose Broth, and incubate for 24–48 hours at 32°C, examining for gas production. Transfer a loopful from positive tubes to fermentation-tubes of Brilliant Green Broth 2%, and inoculate for 24–48 hours at 32°C. If gas is produced, this represents a positive verification of coliforms. From time to time flours become contaminated with a field microflora of the thermoduric type, which survives dry heat at 120°C, forming ovoid spores about 1·2–0·06 μm in size. These become activated by the proofing and baking process and liquefy the structure of the bread-crumb. The species of *Bacillus*, which is the causative agent is *B. mesentericus*; it belongs to the same aerobic spore-forming *Bacilli* as *B. subtilis*, according to Bergey's Manual.

To evaluate, prepare a 1:10 suspension of the flour, or enrichment cereal product (bran, meal, crush-grain, etc.), pasteurize at 80°C for 10 minutes, and inoculate a sterile petri dish with 1 ml of cooled suspension (40°C). Fill with 15–18 ml of Plate Count Agar, rotate in both directions on a flat surface to mix and cover. If necessary, prepare a quantitative dilution at 1:100 and plate out as described. Invert the plates and incubate at 34–37°C for 48 hours. A count of 20 or more spores per 100 g raw material is considered excessive, and liable to 'rope' (German: *Fadenziehung*) development or dough structural-breakdown during the baking cycle.

If 1 ml of a 1:10 pasteurized suspension at 80°C develops a grey-white pellicle on the surface when inoculated into a tube of nutrient broth and incubated at 37°C for 48 hours, this is indicative of a *B. subtilis* type species. During breadmaking the symptoms of dough liquefaction can be prevented by adding propionic acid, and if dough pH is adjusted between 3 and 4 with acetic, lactic or citric acids, the molecule will remain more than 90% undissociated, thus increasing its anti-microbial effect.

For the routine evaluation of flours, and other enrichment-cereals for mould and yeast counts (especially the former which produce mycotoxins as secondary metabolites), the following basic procedures can be applied.

Prepare a 1:10 sample suspension, and inoculate 1 ml into a sterile glass or plastic petri dish, adding 15–18 ml of a suitable medium such as malt–salt agar at pH 6·8 (prepared by adding 20 g malt-extract broth and 20 g agar to 750 ml of distilled water; dissolve 74 g sodium chloride in 300 ml distilled water. Autoclave both solutions at 121°C for 15 minutes, mix, and cool the mixture to about 45°C). Mix the inoculate well with the medium by rotating in both directions, cover, invert and incubate at 23°C for 3 days. Count all plated with 30–300 colonies, and report as 'yeast and mould count per millilitre or gram'. Other suitable media are potato–dextrose agar at pH 5·6, and dextrose–salt agar at pH 6·6.

Following the discovery in the 1960s of the aflatoxins, and the later work on mycotoxins and their link with serious illness, known as 'mycotoxicosis' in man, this group of substances has drawn increasing attention. Much information has been revealed in various countries concerning links with the consumption of foods containing mycotoxins and certain forms of illness. These include clinical, epidemiological, and food biological studies. Although to date no fatal cases of mycotoxicosis have been recorded, attention has been drawn to the health hazard of moulds, and their toxic metabolites. Mycotoxins can occur in grain, flour, bran and wholemeals, resulting from mould-growth, and these are carried through into the baked products as a primary contamination. On the other hand, a secondary contamination can occur when the baked products develop mould, and mycotoxins are deposited within them. Laboratory studies have shown that the variety and amount of mycotoxins are greater in baked ·products which have become spontaneously contaminated with mould as secondary infections. In spontaneously mould-contaminated bread and baked products, the following mycotoxins have been found: aflatoxin B1 and G1, patulin, penicillic acid, citrinin, ochratoxin A and zearalenon.

Of the various species of *Aspergillus* and *Penicillium* found in the bake-shop and bread-store atmosphere, Spicher (1969) established that 6% were capable of producing aflatoxins when growing on rye bread. Earlier work by Frank (1966) showed that the pathogenic mould *Aspergillus flavus* was capable of forming aflatoxins in milled products and bread. Constraints on the exploitation of the groundnut as a source of cheap protein in 1966 were due to *A. flavus* producing the aflatoxins B1, B2, G1, and G2 within the nuts. The aflatoxins are primarily liver poisons, and are acutely toxic to most species. As early as 1966, the FAO/WHO advisory group laid down a maximum level of aflatoxin in food at 30 µg/kg (0·03 ppm). *A. flavus* grows on groundnuts with moisture content exceeding 9·0%, and prompt harvesting and drying, followed by dry/cool storage reduces aflatoxin contamination to a minimum. This is also valid for grain and cereal products. In 1968 Frank and Eyrich found direct evidence of aflatoxin in mould-contaminated wholegrain bread. Bösenberg and Eberhardt reported in 1969, that from 42 mould-contaminated loaves in a supermarket, one loaf contained aflatoxin B and G. From tests carried out by Spicher in 1970 on 91 heavily moulded loaves of various types, 16% of the samples contained aflatoxins, mainly G1 and B1. More recent studies by Spicher (1984), on 110 mould-contaminated sliced-loaves of various types, stored at 25°C and 70% RH, for 7 days, revealed 18 loaves (16·4% of the total) with a positive mycotoxin response. The mycotoxins found were: citrinin (11 instances), ochratoxin A(4), zearalenon (5), and in two instances ochratoxin A, and zearalenon within the same sample.

The use of mould-inhibitors such as: sorbic acid, propionic acid or propionates does not prevent the build-up of mycotoxins.

Identification of the types of mould responsible for the various mycotoxins is at an early stage of research, but initial work has shown that *Penicillium citrinum* is responsible for citrinin, as is *Penicillium expansum*; in the case of ochratoxin, no definitive producer has been established, although in the case of *A. ochraceus*, which is also a microflora of flour, the contamination can be carried over from flour to bread as a primary contaminant. Zearalenon is associated with various varieties of *Fusarium*, which are habitually field-microflora. Therefore, its presence is indicative of carry-over from grain into flour at the mill, rather than a spontaneous bread contaminant.

In view of the attention drawn to this topic in recent years, and the toxic nature of mould metabolites, the best advice for industrial and craft baker alike, for consumer protection and preservation of good name, is where no mould is allowed to grow there can be no mycotoxins.

The best fungicide for bread-slicing, storage and despatch-stations is a dilute solution of sodium hypochlorite, and where infections have occurred, a 40% formaldehyde (formalin) solution in liquid and vapour form is an effective room sterilizer. A 4% solution of sodium hypochlorite is also effective in the detoxification of aflatoxin and other mycotoxins.

Mycotoxins are very heterogeneously distributed in food samples; therefore, an adequate sample size is essential if the analytical data are to be meaningful. In sampling groundnuts or grain, a statistical approach is recommended. Samples must then be finely ground and thoroughly mixed to form the 'master-sample'. Extreme care must be taken in handling mycotoxins, and when weighing solid reference mycotoxins, both mouth and nose must be masked from aerosols, and preferably carried out under an extraction-canopy. As is the case with the analysis of pesticide residues in cereals, a specially equipped laboratory is essential both from the safety viewpoint, and because the methods involve sensitive micro-techniques. The basic techniques used in the quantitative estimation of mycotoxins are chromatographic. Thin-layer (Gorst-Allmann and Steyn, *J. Chromatography*, 444 **175** (1979), 325; Duracková, Betina and Nemec, *J. Chromatography*, **116** (1976), 141; Howell and Taylor, *J. Assoc. Off. Analyt. Chemists*, **64** (1981), 1356). Among the high pressure thin-layer chromatography methods used is that of Lee, Poole and Zlatis, 'Simultaneous multi-mycotoxin analysis by HPTLC, with continuous multiple-development and two solvent systems of different polarity', *Instrumental HPTLC*, Bertsch and Kaiser (eds), Hüthig Publishing, New York, 1980, see pages 263 to 273. High pressure liquid chromatographic methods include that of Scott, 'Liquid chromatography in the analysis of mycotoxins', in *Trace Analysis*, Vol. 1, Lawrence, J. F. (ed.), Academic Press, 1981, see pages 193 *et seq.*

Stoloff gives a useful survey of the analysis of mycotoxins in the following: *Clinical Toxicology*, **5** (1972), 465; reports on mycotoxins of *J. Assoc. Off. Analyt. Chemists*, **65** (1982), 316; *Official Methods of Analysis*, 13th edn, Washington, 1980, Chapter 26; Castegnaro and O'Neill, I. K., *Environmental Carcinogens—Selected Methods of Analysis—Some Mycotoxins*, Vol. 5, Lyon, IAAC, 1982.

The Mycotoxins are extracted by shaking the ground sample with a polar solvent, usually chloroform. The extract must then be purified by column-chromatography on kieselgel, gel-filtration on Sephadex LH-20, or liquid–liquid partition.

The acidic mycotoxins, viz. ochratoxin A, penicillic acid and citrinin, can be separated from the remaining mycotoxins by shaking with saturated sodium bicarbonate, thus effecting removal from the chloroform extract. By adjusting the pH to 13, zearalenon also migrates into the aqueous phase as a resorcin derivative. Where thin-layer chromatography, or HPLC, is used the mycotoxins are identified by using either a fluorescence or a UV detector, and also in the case of gas-chromatography after derivative-preparation, confirmation by mass-spectrometry, UV- or fluorescence-spectra, or preparation of derivatives is required.

Chloroform is a suitable solvent for aflatoxins B1, B2, G1, G2, M1, but for ochratoxin A, benzene/acetic acid (99:1 v/v), zearalenon methanol and patulin, ethanol are necessary. The determination of the concentration of standard solutions is done by UV-spectra, the aflatoxins at 361–362 nm, ochratoxin A at 333 nm, zearalenon at 274 nm, and patulin at 275 nm.

As practical analytical methods for mycotoxins become more accurate, increasing legislation will follow to safeguard the potential hazards to consumers generally, and to infants and children in particular.

A good example of the application of mycotoxin analysis is in the monitoring of the abuse of irradiation by using it to kill bacteria on contaminated food. This practice passes the food off as saleable, when in fact it may contain potentially dangerous toxins. However, when correctly applied, ionized irradiation of the gamma type can reduce contamination and wastage in raw materials, providing dosage/decay-time is controlled, and no chromosome-damage can occur.

For 'air-sampling sanitation control' the following plating media are recommended for the identification of the specific organisms:

| *Microorganism* | *Media for plating* |
| --- | --- |
| Enteric bacilli | Desoxycholate Lactose Agar |
| Haemolytic streptococci | Trypticase Soy Blood Agar |
| Staphylococci | |
| Staphylococci (coagulase + ve) | Vogel and Johnson Agar |
| Total aerobes | Eugonagar or Trypticase Soy Agar |
| Total anaerobes | Trypticase Soy Blood Agar, in anaerobic jars |
| Total aerobes with reference to detergent use | Trypticase Soy Agar + Lecithin Polysorbate 80 |

For the examination of raw materials, process, and personnel, the following media are appropriate:

| *Organism* | *Media* |
| --- | --- |
| Total aerobes | Standard Methods Agar |
| | Dextrose Salt Agar |
| Coliforms, e.g. *E. coli* | Brilliant Green Bile Broth 2% |
| | Desoxycholate Lactose Agar |
| Salmonella | Desoxycholate Lactose Agar |
| | Brilliant Green Agar |
| Staphylococci (total) | Mannitol Salt Agar |
| Coagulase + ve | Vogel and Johnson Agar |
| Yeasts and moulds | Potato Dextrose Agar |
| Enterococci | Azide Dextrose Broth |
| e.g. *Shigella, Krebsiella, Proteus* and *Yersinia* | M-Enterococcus Agar (University of New Hampshire) |

*Disinfectant testing*:

For the general evaluation of the efficiency of disinfectants, the following media can be used:

Nutrient Broth (FDA Agar BBL)
Cystine Trypticase Agar
Letheen Broth AOAC
Routine testing of disinfectants for the determination of phenol coefficients of disinfectant products containing cationic surface-active materials.

The topic of sanitation and hygiene is of crucial importance to all food producers, and the responsibility *vis-à-vis* national and international food legislation, designed to protect the consumer, remains with the producer. The purpose of all government-controlled sanitation agencies is to protect the public from unhygienic production operators. Therefore, the best advice for the baker is to practise good-housekeeping by ensuring all working areas are cleaned up at the end of each day or shift, delegating the Operator-in-charge of a group as being responsible for this work. Typical trouble-spots are: under work-tables, machine-aggregates, the top of and underneath deck-ovens, raw-material stores, refrigerators, bread-slicing and despatch stations, refuse-disposal daily to external disposal-areas. In bakeries: ventilation, reduction of humidity and temperature, wherever possible, and adequate natural lighting is very important. Planned positioning of washing and cleaning facilities by zoning, and staff training and total cooperation from all personnel is essential.

For the industrial bakery, sanitation is usually part of the technical-control organization, but where private capitalization is involved, there is a disincentive to expend resources on sanitation unless some enforcement agency provides the required external pressure.

The craft-baker, with much more limited capitalization, is often hard-pressed by government sanitation enforcement agencies. However, if he can establish a cooperative working relationship with Health Department Inspectors, and act on their advice in terms of good-housekeeping and sanitary production processes, he will find most inspectors are, in reality, doing him a service by indirectly maintaining the quality and good name of his products. Product weight and ingredient-labelling is also a controversial issue; where there are arbitrary baked weight minima of (say) 800 g and 400 g per bread unit, the baker has two procedures open to him

(1)  in order not to be prosecuted for baked bread 20 g under weight: he can err on the side of overweight, thus losing about 52 loaves per week, or
(2)  he can exercise critical-baking-control of baking temperature and time, in which case the production of a reasonably crusty loaf can be ruled out.

In some countries a minimum crust-thickness has been stipulated, which is a consumer-oriented regulation, and encourages bread sales.

# 1.9   Weigher–Mixer Functions and Diverse Types of Mixers and Mixing-Regimes

### 1.9.1 WEIGHER–MIXER FUNCTIONS

Efficient production begins with raw material intake and preparation, and most of these materials are biologically active, and therefore are prone to deterioration with time. This can be due to microorganisms, as a result of water content/water-activity or chemical oxidation resulting in rancidity. However, in the case of the most important raw material, flour, a definite storage-period is essential for the biochemical ripening-process to take place. This demands from the technological viewpoint, that at all times enough mature flour is available for processing. In this connection, the important parameters of the flour are: gluten quantity and quality, water content, temperature, and colour/brightness. In the case of sack-storage, the maturity of flour takes about 4–5 weeks, whereas in the case of bulk-transport and pneumatic-handling, this is reduced to 10–12 days. This is brought about by the process of aeration, both in the external silos, during transport of the flour from external silos to the daily-silos within the bakery, and by passage through mixing-worms and sieve machines (Figs 37 and 38). Another important function of the external silos is the maintenance or adjustment of the moisture between 14% and 15%, which demands a storage-temperature of 20°C and a relative humidity of 60% to 65%, depending on the properties of the flour on delivery. This fundamental procedure of flour-storage and preparation by oxygenation is very important for its baking performance, and it also eliminates storing, stacking, lifting, moving and emptying of the bags and boxes of raw materials in the bakery.

However, bulk-handling is only feasible where usage volume is high enough. Some flour millers have adopted delivery systems whereby bulk deliveries as low as 1 tonne can be accommodated, thus including the craft-baker. This allows the negotiation of lower flour prices, reduces wastage and contamination, often making use of dead-space at roof or ceiling level. Internal daily-silos should be located close to the weigh-hopper and doughmaking-station, and long pipeline-runs from intake points avoided to save unloading-time. When installing, the extra cost of a 12-tonne silo compared with an 8-tonne silo is minimal, and allows for future increases in production capacity, whilst being operated at a lower capacity until required. Rigid silo materials are coated-steel, steel-laminate, glass-fibre, aluminium or stainless-steel. Very efficient internal silos are non-rigid, and usually of woven polyester material, e.g. Trevira, which is suspended on a support-framework. Interior or daily-silos can be square, rectangular or round, but external silos are

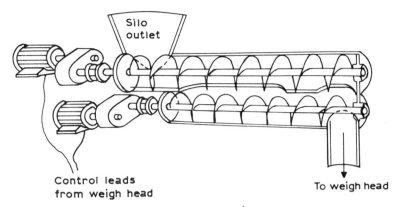

Fig. 37.   Worm-conveyor (silo to weigher). (Source: *Machine Handbook* (*Baking*), 2nd edn, B. Schramm, VEB-Fachbuchverlag, Leipzig, 1978.)

usually cylindrical. Large storage-silos of 5–60 tonnes capacity are usual, which can be constructed as either internal or external units. Common capacities for daily-silos are 0·5 to 5 tonnes, and weigh-hopper throughputs of up to 10 t/h. The average throughput of a pneumatic system, depending on pipeline length and number of curves, is between 4 and 10 t/h. For low pressure systems of 0·01–0·02 N mm$^{-2}$ (1 N mm$^{-2}$ = 10 atm) the flour/air mixture ratio is 10:1, and for a high-pressure system of 0·1 N mm$^{-2}$ the flour/air ratio is 50:1.

Owing to the advantages of efficient use of space, oxygen enrichment of the flour, smooth flow, relatively low energy consumption, and minimal flour-temperature rise, the installation of this type of system is increasing in popularity. The energy requirement for a pneumatic system varies according to pipeline length, junctions and pipeline resistance, from 5 to 17 kW.

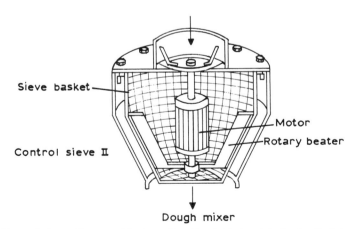

Fig. 38.   Control sieve. (Source: *Machine Handbook* (*Baking*) 2nd edn, B. Schramm, VEB-Fachbuchverlag, Leipzig, 1978.)

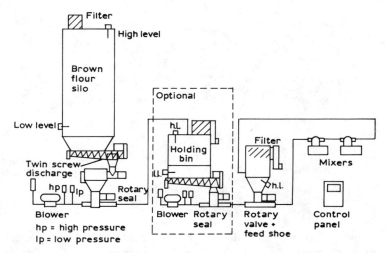

Fig. 39.   Bulk-handling of brown flours. System of Kerry Handling Limited, East Grinstead, Sussex, UK.

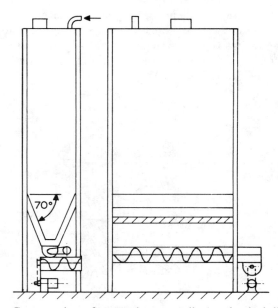

Fig. 40.   Cross-section of external storage-silo (mechanical discharge).

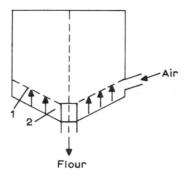

Fig. 41.  Pneumatic fluidized-bed principle. 1. Perforated base-plate. 2. Air cushion.

Two additional important functions of flour-preparation aggregates are:

—the preparation of flour blends for various products (mixed wheat/rye breads, specialty breads, etc.),
—the removal of flour contaminants (textile-fibres, insect-larvae, flour-moth, etc.).

Amongst the typical machine aggregates used in flour preparation are the following:

|  | *Power-rating (kW)* |
|---|---|
| Fluidized-bed or base for (say) 6 silos | 6 |
| Sifter and conveyor | 4 |
| Internal daily-silos with distributor and metering-aggregates for weighers (depending on capacity) | 2–4 |
| Weigher with discharge-worm and valve | 2 |
| 1 storage-silo with pneumatic discharge including distributor | 1·5 |
| Sieve and blender for small and medium-size bakeries | 1·5 |
| Sack-elevator (according to height) | 1·5–2·5 |
| Centrifugal sieve aggregate | 0·5–1·0 |
| Automatic weigher | 0·7 |

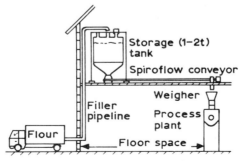

Fig. 42.  Spiroflow automatic system. (Source: Spiroflow (UK) Machinery Ltd., Clitheroe, Lancashire.)

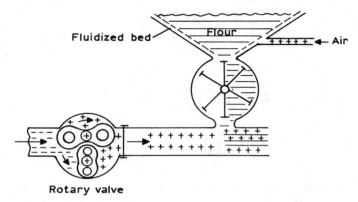

Fig. 43.    Pneumatic system. Positive system using high, medium or low pressure.

Flour can be discharged or transported from a silo either pneumatically or mechanically. Since flour is difficult material to move, the problem is removing it from the silo, and high extraction flours or mixtures can cause special problems (Fig. 39). Mechanical discharge can be achieved by one of the following procedures:

—the use of a vibrator fitted at the base of the silo, the vibrations directly applied preventing 'bridge-building', and ensuring a free flow, in order of silo-loading sequence,
—a U-shaped screw conveyor, fitted along the base of the silo transporting the flour to an outlet point.

The mechanical transfer operation is then achieved by worm conveyors, usually mounted horizontally, transporting the flour to the point of use (Fig. 40).

In the case of pneumatic discharge, the flour at the base of the silo must be 'fluidized', by blowing air through perforated plates. This allows the fluidized flour to flow freely from the outlet point (Fig. 41).

Pneumatic transport is much more flexible than mechanical transfer, and can circumvent most obstacles from silo to usage point (Fig. 42). The fluidized flour is transported by being either blown or sucked from the silo discharge point to the point of use, the air used for transportation of the flour being filtered out at the usage point. This method of transportation allows the use of an efficient weighing

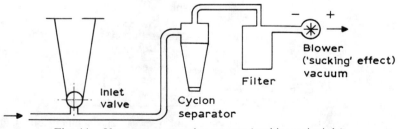

Fig. 44.    Vacuum or negative system (sucking principle).

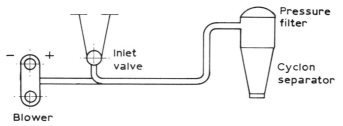

Fig. 45.   Pneumatic system (pressure principle).

aggregate, thus eliminating the need for manual weighing by operators (Figs 43–45). Several methods of weighing are applied.

## 1.9.2 WEIGHING METHODS

### Beam Weigh-scales with Dial-scale or Digital-setting
In this case, the weigh-hopper acts as a balance, which reacts to the weight inside. The required weight is set on the scale-dial, and the flour conveyed into the hopper, the indicator rotates and stops the flour flow on reaching the preset weight. The dial-type scale has been increasingly replaced by a digital setting and read-out. The required weight is keyed into the system, then activated; flour is fed into the weigher and an electronic signal transferred to the unit, thus allowing the required weight to accumulate. On reaching the required weight, the system automatically switches off.

### Metering System
This depends on the timing of the flow of flour into the hopper, rate of flow and time determining the quantity fed into the hopper. This system no longer finds wide application.

### Load Cells
Load cells are sensitive to the weight/load acting on them, and react electronically. The method finds increasing use, since it can be linked to other electronic systems and microprocessor controls.

In the case of small silo-systems, the pneumatic transport capacity is less important than precise weighing, since small amounts of flour are more sensitive to weighing inaccuracies, on a percentage basis, than larger batches.

The electronically controlled digital weigher, which provides absolute weight data, can be used for other dough components, and will soon replace the dial-scale as standard equipment in the bakery.

The advantages of electronics for weighing functions can offer a host of applications. In using multicomponent weighers, recipes for various types of products can be stored. The dosing of ingredients, and the tempering of dough liquids beforehand, based on any batch size, is no problem for electronics. Errors in calculation, and memory are impossible with a correctly programmed computer. The storage capacity for information means that the computer can be used for the silo, mixing and baking.

In the area of the application of electronic and microprocessor based control systems, Dietrich Reimelt, KG, 6074 Rödermark-Urberach, FRG, and cooperation between Reimelt and Atlas Equipment (London), Limited, has resulted in a special 'in-house', SP-80 system for the baking industry (Fig. 46(a)–(g)). More than 100 of these systems have been installed worldwide, some of which control not only the batching of ingredients but also the production line. The SP-80 systems are tailormade, and no two systems are identical, although certain core-elements of software and hardware are the same, e.g. batching control, accuracy checking, and optimization. These represent the basic core-of-intelligence, to which is added the extra intelligence necessary in each individual case. The SP-80 can interpret from an input of the number of finished products required, the appropriate number of ingredient batches. The commonest SP-80 software is formulation-based, so that costing information can be integrated into the programs, thus giving weight of ingredients, and cost/unit material, as well as total production cost. Operation parameters are accessible via a 'menu' series, allowing the user familiarity with the system in minutes. Hardware has been designed to be resistant to the dust and humidity of the production-floor. The complete programme of various flours, ingredients and liquids are pre-programmed, automatically weighed and recorded, and the silo contents determined. Other functions include optimization by correcting, checking tolerances, dosage checks and dialogue. The weighing system may be linked to a continuous recording system, recording flour consumed on a meter or print-out, which balances flour consumed with deliveries. Obviously, the flour transport system and weigher-head must be geared to flour consumption. The weigher-head may feed into two adjacent machine-bowls, using

Fig. 46.    (b) Fluidizing bed discharge.

Fig. 46.    (a) Outdoor storage silo.

Fig. 46.    (c) Indoor storage silo with fluidizing bed discharge.

Fig. 46. (d) Ingredient silos for salt, baking aids and other ingredients.

Fig. 46. (e) Raw material centre.

Fig. 46. (f) Computer-controlled ingredients discharge. (Source: Dietrich Reimelt KG, Rödermark-Urberach, FRG.)

Fig. 46. (g) Flour is discharged through the fluidizing bed and pneumatically conveyed through an integrated sifter to the scales. (Source: Dietrich Reimelt KG, Rödermark-Urberach, FRG.)

a Y-shaped outlet. However, a better set-up is a weigh-head mounted on a rail, allowing movement from one machine to another. Where removable bowls are in use, the bowls can be wheeled to the weigh-head discharge point. The height of the latter should be decided upon when installed, and if desired, the silo can be designed to feed into several weigh-heads. Dietrich Reimelt were installing pre-set automatic weigher-hoppers using load-cells in the 1950s, and at least one large bakery chain in West London were using this system.

In transporting flours and meals containing large amounts of bran and other cereals of varying density, for special-bread production, the pressure exerted on the material is critical to prevent separation when handling pneumatically. Such material, with a static bulk density of $5776 \, kg \, m^{-3}$, could become as low as $232 \, kg \, m^{-3}$ when aerated, and in this state the lighter materials separate easily. One solution to this problem is to modify the receiving silo intake-pipe, ensuring even dispersion over the whole silo area; since fluidization only encourages separation, the material can be discharged from the silo by twin-worms with tapered screws, thus forming a small mixing-chamber at the point of discharge. In addition, the air velocity must be reduced to reduce pressure, and the point of discharge can be changed, creating a P-loop receiving hopper, thus avoiding any separation of material.

There is plenty of scope for new innovation in pneumatic conveyor design, apart from velocity and air-pressure variations, and flour-to-air ratios. Sometimes raw materials can be more economically transported either by mechanical aggregates or 'negative' handling involving a vacuum. For the storage of brown flours and grain-meals, insulated silos of about 40 tonnes maximum capacity are ideal, since this reduces the danger of internal condensation at the top of the silo. The simplest form of silo load assessment is by glass-portholes located at various levels in the silo. Alternatively, continuous level indicators, and high and low level sensors are fitted to most silos to prevent overfilling or the flour level dropping too low before a new delivery. Some internal silos or daily-silos are mounted on load-cells, and the net flour weight electronically assessed and displayed on digital read-out and print-out.

Bulk-handling equipment for other raw materials is mainly suitable for large bakeries. Sugar is stored in silos similar to flour, being available for both internal and external installation. The main differences from flour silos are a steeper-angled base-cone and mechanical rather than pneumatic discharge. Pumpable shortenings are stored in liquid form by temperature controlled jackets, and can meter direct into mixers. For medium-size bakeries, pumpable shortening is delivered in containers, which are exchanged when empty for a full replacement.

With production lines getting larger and more specialized in most bakeries, the trend is towards a 'one-stop' collection of raw materials, whereby they are weighed together and moved as a batch (Fig. 47(a)–(f)). This system requires a mainframe computer to handle the raw material batching, checking and sequencing operations. The one-step weigh and mix procedure could be a reality within the next five years. This system allows considerable personnel reductions, since the complete operation is handled by one program-controller. For bakeries using bagged raw materials and small batch processing, the 'lift and sift' system marketed by Russel Finex Limited, London WC2H 7EQ, UK, is a useful innovation. The operator simply tips the sacks into a hopper which stands at floor level in a framework, eliminating high-level lifting. The material is then conveyed by a 'negative' vacuum system hygienically, and can be sieved at any height or position to suit the process stage. The sieving unit can be positioned over a mixer-bowl, being mounted on a bridge framework. This system leaves no residues, input weight balancing with discharge weight. Therefore, pre-weighed ingredients are sieved and delivered net to the mixer.

Where horizontal high-speed mixers are in use, arranged in line in a doughmaking station, a flour-blending hopper with up to four compartments, followed by a sifter, worm-conveyor and elevator for the one-floor-bakery is a long-established and reliable set-up. The flour passes from the elevator via a worm-conveyor into a travelling-weigher, which services all mixers.

Using the Tweedy mixing system, a master control panel, which indicates the schematic flow of the entire plant, allows the operator to move flour from any silo and discharge into the weigh-hopper of the mixer. When the mixer is in automatic sequence, flour is pneumatically transferred from the silo through cyclone-separators and sifters and thence into the mixer hopper. In the case of the Tweedy 300 (delivering 11 batches of 55 to 270 kg charges, with maximum dough output of 3240 kg/h), the mixer hopper fills in 2·5 minutes. Where two flours are used, as for brown-bread dough, 420 kg of brown flour and 140 kg of white flour respectively, are preset on the flour weigh dial by first dialling in the 420 kg brown flour, followed by the total weight of brown and white flour at 560 kg. A signal from the diaphragm valve, indicates that the weigher is ready to receive flour, and the brown flour is moved from the bulk-flour system to the mixer-hopper. Brown flour continues to discharge into the mixer-hopper until the weigher indicator passes the intermediate air-switch, the brown flour feed then stops with a signal from a diaphragm valve. The mixer control then discharges white flour to the mixer-hopper, the weigher indicator advancing until the final air-switch is passed, and a signal from the diaphragm valve results in the termination of white flour discharge. Both flours are now ready to be fed into the mixing-chamber for the mixing-cycle.

Continuous dough-mixing systems use a scaling system in which all the ingredients specified by a formula are continuously scaled and conveyed to the mixer feed-hopper by an automatic scaling system (Fig. 47(a)–(f)). Each ingredient or blend is provided with its own scaling device. For dry ingredients, such as flour or dry-milk, an electronically regulated volumetric twin-screw feed is appropriate with a scaling accuracy of $\pm 0\cdot5\%$. Liquid ingredients such as dough water, yeast, and salt solutions are fed to the mixer by impeller-pumps. Pumps are electronically regulated, in combination with line pressures, and minimum level tanks, and deliver with accuracies of $\pm 0\cdot5\%$. Any solutions and slurries are prepared in vertical dual-tank systems. The ingredients for solution or dispersion being placed in the lower tank, and the required amount of water is metered into the upper tank. When the outlet valve of the upper tank is opened, an agitator and circulating pump is actuated in the lower tank to disperse the solids uniformly. Slurry tanks have a maximum capacity of 130 gallons (650 litres), and have a reserve compartment containing enough old slurry to keep the mixer supplied without interruption of production. Usually, provision for cleaning is provided *in situ* for the entire liquid preparation and transfer system. Spray-heads installed in the tanks allow low-pressure cleaning and sanitizing of tank interiors, pipelines and pumps right up to mixer inlets, without dismantling any equipment. Plasticized ingredients, such as shortenings, and sour-ferments are metered into the system by geared-pumps, which can be easily dismantled for cleaning. These pumps are integral with the console located under each thermostatically heated ingredient container, and are driven by electronically controlled motors also housed in the console. An agitator within the containers ensures an even feed rate of the paste material into the pump, thus ensuring accurate dosage. Automatic weight-controllers monitor the levels in the containers, and activate a refill system as necessary. Otherwise, this function can be done by a control unit based on weight-loss. The scaling system for both plastic and paste-like ingredients is also provided with an *in situ* cleaning and sanitizing facility.

Fig. 47.   (b) Preparation of pre-weighed
and pre-mixed dry and liquid ingredients
for dough mixers.

Fig. 47.   (a) Dry and
liquid ingredients being
fed to a high capacity
mixer.

Fig. 47.    (c) Automatic weighing, mixing and pressure conveying for the preparation of flour
and other ingredients.

Fig. 47. (d) Central dry ingredients batching and in-line blending station, showing flour and minor ingredient scales and pneumatic blenders with integrated dense phase conveying system for the premixes.

Fig. 47. (e) Major and minor dry ingredient batch receivers and liquid weigh scales above high speed dough mixers. Computerized operation with touch-sensitive graphic display and prompting station for special and micro ingredients handling.

Fig. 47. (f) Computerized liquid ingredient preparation room for liquid brew, fermentation, yeast slurry, honey, calcium propionate, ascorbic acid and ammonia liquefying. Includes fully integrated CIP system. (Source: Dietrich Reimelt KG, Rödermark-Urberach, FRG.)

Electronic controls for the various scaling operations are housed separately within a console which also has the individual digital controllers for the formulae and hourly throughput. Instant checks of scaling accuracy are carried out with the aid of electronically regulated sampling devices. By pushing one control button, all ingredients being scaled at that moment are channelled off into separate containers, if the scaling operation is on target, each container will hold the exact amount of each respective ingredient specified in the formula. Any deviations from the specified feed rates can be adjusted over the control panel.

Although water presents no storage problems, its handling in terms of treatment, tempering and metering can involve certain extra preparation procedures. For example, in some processes, e.g. the CBP, it is necessary to use chilled-water, which requires a refrigeration plant. Water can be measured either by weigh-head or by volumetric metering.

As already described, semi-automatic tempering tanks are used generally for the batch-type mixing processes. Water is metered either direct from the mains supply, or via a water-chilling unit into a batch water-tank, located over, or adjacent to, the mixer bowl or chamber for gravity feeding.

This method, however, is not suitable for the automatic feeding of minor ingredients in solution where digital meters are in use for each solution, and the density of solutions would have to be determined at various temperatures in order to make calibration possible. Therefore, on automatic aggregates, a different technique is applied. Instead of volumetric metering, the water is weighed. The required amount of water is preset by a final air cut-off switch located on the water-weighing dial, and when the mixer calls for water during the cycle, a valve in the supply line is actuated, and water flows into the weigh-tank located above the mixing-chamber, until the weighing-dial pointer reaches the cut-off switch. The water flow is then terminated, and the weighed water held in the tank ready for the next mixing-cycle. This set-up can be easily adapted to the automatic weighing of brews, slurries or premixes in suspension form.

For example, either two tanks of 500 to 5000 litres capacity or four tanks of 250 litres capacity, depending on hourly throughput, can be used. Where four tanks are in use, two are used for salt solution and two for yeast suspension, and where two large capacity tanks are available, one is used for salt solution and the other the yeast suspension. Other minor ingredients such as yeast-foods, soy-flour, mould-inhibitors, malt-flours etc. must then be batched in a compatible manner with either the salt or yeast for dispensing into the mixer.

Where a four-tank system is in operation, two could contain water 400 parts, yeast 50 parts and yeast-food 30 parts, and the other two: water 400 -arts, salt 60 parts, soy-flour 20 parts and mould-inhibitor (calcium propionate) 4 parts. For each dough, 50 parts of the yeast-containing premix is pumped from one tank, and 50 parts of the salt-containing premix from the other. Water is dispensed to each tank by a joint-pipeline from the chilled-water unit, the other ingredients being added to the appropriate tank manually. The contents of each tank being then discharged into the water-weighing tank by displacement-pumps. The water balance is then pumped into the water-weighing tank in the normal way, and the appropriate amounts of salt solution and yeast suspension automatically weighed on top of the water balance by second and third settings on the water-weighing dial. When one yeast-containing and one salt-containing tank reaches the low-level mark, the flowlines are switched to the second pair of tanks, and the first pair are refilled again. This set-up is intended for short runs only, viz. roll and variety-bread lines, or where enriched formulations are in use. The larger two-tank set-up, is ideal for longer production runs with one type of product. In such cases, one brew would be prepared containing the yeast and all the other minor ingredients in a concentrated premix form, apart from mould-inhibitors. One tank-full is enough for one shift, the concentration depending on the formulation and the amount of salt it contains. The second tank is also filled with brew and used as a reserve supply, while the first is being replenished. Examples of brew concentrates are:

Brew A: water 300 parts, yeast 50 parts, salt 50 parts, improver or conditioner 30 parts.
Brew B: water 550 parts, yeast 50 parts, salt 60 parts, improver or conditioner 30 parts, soy-flour 20 parts, and malt-flour 6 parts.

Since Brew A is in a more concentrated form, it would provide for a longer production-run, but it could not carry the higher salt, and enrichments of Brew B without retarding yeast activity. Brew temperatures would depend on time-duration of the production-run. The appropriate amount of brew from the tank is pumped by the constant displacement pump into the water-weigher on top of the dough-water balance by using the dual marker weigh-dial.

The doughmaking components for mixing a 280 part flour batch would be assembled as follows:

Flour 280 parts into the flour weigh-head from the pneumatic system.
Water 140 parts—balance from the water-chilling unit to water weigh-head.
Salt             5 parts ⎫
Yeast           5 parts ⎬ from brew-tank to water weigh-head.
Improver       3 parts ⎪
Water          30 parts ⎭
Shortening 2 parts—delivered to mixer from liquid shortening container.

This latter system of automatic batch cumulative-weighing is typical of that used in conjunction with a high-speed batch mixer of the Tweedy type for the production of bread dough.

Weigher/mixer functions in general tend to fall into specific categories as follows.

Simple manual weighing is where the flour is weighed into the mixer bowl by using a scale with a circular dial and pointer or digital readout, water metered in from a tempering-tank, and the yeast and salt dispersed separately in aliquots of the dough water.

Centralized flour weighing can result in traffic jams, and there is no provision for minor ingredients. The dough liquid target cannot be set unless there is some form of communication between the central dry weigher and the local liquid weigher, the latter being only a simple volumetric water-meter, which is incapable of any compensation for errors.

Dedicated weighing at the mixer allows automation where appropriate, but product control is maintained at the production point (mixer). In combination with a batch system, this keeps all responsibility for the mixing process within the mixer. All weights, temperatures, energies (or better, specific mixing intensities) are known and readily available within one control centre, and any necessary compensations for errors can be implemented.

Electronically controlled digital weighing, coupled with 'one-stop' collection of raw materials and batching, linked to a mainframe computer, checking and sequencing operations, is the future trend in weigher/mixer function technology.

## 1.9.5 TYPES OF MIXER AND MIXING-REGIMES

All doughmaking machines, irrespective of design, must perform the following functions:

—Mixing of the dough component raw materials, flour, water, salt, yeast-suspension, liquid-shortening, emulsifiers and oxidants into a cohesive mass with the help of adhesive, capillary and the chemosorption forces set up during hydration. Chemosorption involving the dipole properties of the water molecule, mainly onto the —OH— groups of the starch molecule.

—Kneading of the raw materials involving the unification or homogenization of the solids with one another, and/or with the liquid components to form a plastic mass, mainly under the influence of shear-forces.

—Indirectly initiating, as a result of mixing and kneading, the important processes of solubilization and swelling, which are essential for the formation of the gluten protein complex, the activity of the microflora (yeasts and sour-dough bacteria), and texture formation.

—Incorporation of air, and the expulsion of excess carbon dioxide during remixing, punching or 'knock-backs'.

The more completely these technological effects are achieved, the better the processing properties of the dough.

Since the design and mixing action of mixers have a considerable influence on dough development owing to the swelling and solubilization processes, an overview of the diverse types and their geometry is appropriate.

Important factors which determine the working-life of a mixer are: daily-cleaning, weekly cleaning and lubrication by the user, and regular servicing by the manufacturer, or 'in-house' engineers, who have been familiarized with its mechanics and electrics.

Although research has shown that specific mixer design is suited to certain types of dough, few machines are exclusively constructed for one type of dough. Processes seem to have had the greatest influence on mixer design to date, and the obvious classification is as follows: discontinuous mixers, continuous mixers, and intensive mixers, which can be either discontinuous or continuous. The other distinguishing design factor is whether they operate vertically or horizontally in the dough container.

## Discontinuous Mixers

Typical of discontinuous mixers is that the doughs are prepared in 'batches', and usually have mixing elements both in shape, and in movement which imitate the traditional technique of hand-mixing. Single-arm versions have either a 'pushing' or 'pulling' action on the raw materials, brought about by pressure and shear forces on the raw material particles, forcing them together. Then, during the second and third phase of doughmaking, these same forces form the gluten network and dough development. These machines operate at 15 to 50 rpm, and by specific positioning of the mixing-element, by the shape of the bowl, and by the path of the element relative to the inner surface of the bowl, the mixing effect and intensity can be varied. In practice, the highest possible mixing-intensity is the aim. The schematic diagrams show the possible positioning of the mixing-element, mixing-bowl, and path of the element relative to the bowl. The mixing times with these conventional mixers is 15 to 20 minutes, but since the bowls can be removed and several bowls made available, according to output, the time interval between the completion of one mixing and another is not much longer than the actual mixing-time. If desired, the one bowl can be emptied, and refilled with raw materials for the next batch.

The characteristics of these conventional mixer designs are a slow and gradual mixing and kneading action, resulting in a limited binding of flour and dough water; much of the solubilization and swelling processes, and hence gluten-formation takes place after mixing. After a mixing time of 7 to 15 minutes, doughs from these mixers exhibit a dry, silky surface; a soft, elastic consistency; and an open porous structure. Mixing can be completed in stages, and water and salt added after a 10-minute relaxation period, thus allowing better water absorption control and structural control for variety bread production, i.e. baguettes, Vienna doughs (Semmel-teig), *flutes* (*parisien*), and *ficelles*.

The single-arm mixer design, which depends on almost elliptical movements of the mixer-arm, as the bowl rotates in a counterclockwise direction, depends on a pushing or pressure action on the down-stroke, and a lifting/stretching action on the up-stroke. Some mixer-bowls are flat in curvature at the base, whereas others are curved to form a central peak at the centre of the bowl base (see Figs 48 to 53). This latter design (Fig. 53) ensures that the arm moves in close proximity to the bowl surface, thus ensuring a maximum of pressure/unit area on the down-stroke in particular. These machines are used in both medium and large bakeries in Europe, and because of their robust design, are suitable for firm doughs of lower water absorption from both wheat- and rye-flours, as well as mixed wheat/rye-flour doughs and their meals.

The power-rating of these mixers is about 5 kW, with a bowl capacity of 520 down to 390 kg. The power transmission is from the motor over a drive-belt to a shaft-worm, which in turn drives a cog-wheel. From this a swinging-disc drives the mixer-arm, and simultaneously a cog-wheel and chain arrangement drives the shaft-worm,

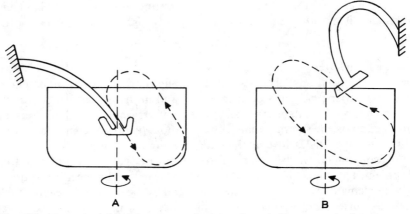

Fig. 48. Working principle of the one-arm impact mixer. A. Down-stroke (impact pressure). B. Up-stroke (stretching).

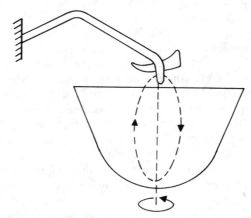

Fig. 49. Working principle of the rotary mixer (bent-arm type).

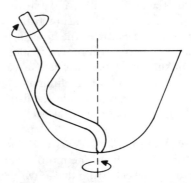

Fig. 50. Working principle of the rotary-mixer (spiral-shaped arm).

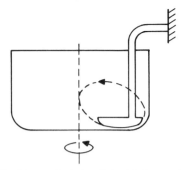

Fig. 51.  Working principle of the high-speed one-arm mixer.

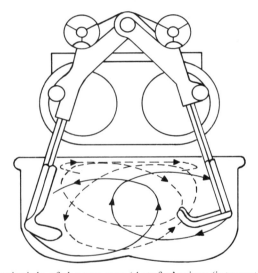

Fig. 52.  Working principle of the two-arm 'Artofex' mixer (intersecting elliptical orbits).

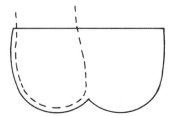

Fig. 53.  Overlapping spherical bowl design for the one-arm impact mixer.

which moves the machine-bowl in counterclockwise rotation. The specific movement of the mixer-arm is achieved by eccentric attachment to the swinging-disc, and by providing a lever-joint at the end of the mixer-arm, which is also pivoted at its other end. The swinging-disc, on rotating, results in the mixer-arm always describing the same angle, but traversing different distances along its path. However, since the rpm of the motor is constant, at different points along its path the mixer-arm is travelling at different velocities. This constant acceleration and retardation demands a robust construction of all moving parts, and the belt-drive prevents any instantaneous overloading of the motor, since it can slip if necessary. The path of the mixer-arm also influences the shape of the base of the mixer-bowl. The overlapping spherical shape, forming a central peak is necessary in order to ensure that all raw materials are reached during mixing, and there is no dead-space in the centre of the bowl. The specific positioning of the moving parts ensures that the velocity of the mixer-arm is greatest when the kneading action is not required. The actual velocity within the bowl itself falls, so that the resistance the mixer-arm offers to the mass of raw materials during mixing is reduced. This protects the moving parts, and also reduces the formation of flour-dust clouds at the beginning of the mixing process. The layout and working principle of this type of single-arm, pressure-lifting-stretching action machine is shown in Figs 48–53.

In operating these machines, it is important to make sure that the drive-worm of the machine is correctly keyed into the teeth of the cog-wheel which drives the bowl. If this proves difficult, withdraw the bowl, and rotate a little before re-engaging. Servicing involves control of the tension of the drive-belts and chains. Chains and greasing-nipples should be regularly greased, after removing all flour deposits beforehand. General cleaning and protection measures against corrosion should be followed, but isolation of the power source from the machine is the *first* step before commencing. The rotary-mixer of the bent-arm type (Figs 54, 55) is mainly used in craft-bakeries, as well as in the meat industry, and chemical industry. It is suitable for medium firm to soft doughs. The figure shows the external appearance of the machine. The torque and rpm are transmitted from the motor of power-rating

Rotary mixer (bent-arm type)

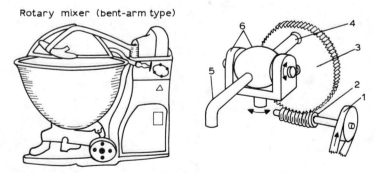

Fig. 54.   Layout and working principle of rotary mixer (bent-arm **type**). 1. V-shaped drive-belt. 2. Shaft-worm. 3. Cog wheel. 4. Ball-joint. 5. Mixing-arm. 6. Moving joint. (Source: *Machine Handbook* (*Baking*), 2nd edn, B. Schramm, VEB-Fachbuchverlag, Leipzig, 1978.)

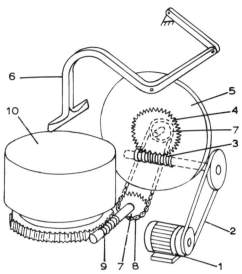

Fig. 55. Layout and working principle of one-arm impact mixer. 1. Motor. 2. V-shaped drive-belt. 3. Shaft-worm. 4. Cog-wheel. 5. Swinging-disc. 6. Mixing-arm. 7. Chain cog wheel. 8. Chain. 9. Shaft-worm. 10. Mixing-bowl. (Source: *Machine Handbook* (*Baking*), 2nd edn, B. Schramm, VEB-Fachbuchverlag, Leipzig, 1978.)

1·6–2·0 kW over a belt-drive onto a shaft-worm, and then to a large diameter cog-wheel, which gives a mechanical advantage of 40:1, where the cog-wheel has 40 teeth and the shaft-worm rotates once. This results in a large reduction in the rpm rate, and a corresponding increase in the turning-moment or torque. The power-transmission can only occur from the shaft-worm to the cog-wheel, which is controlled by making the cog-angles less than 15 degrees. The mixing-arm is connected to the large diameter cog-wheel over a ball-bearing joint. The figure shows the layout and working principle of the internal moving parts of the rotary-mixer of the bent-arm type. The path of the mixing-arm, which is bent downwards, results from it being connected to a joint which can move vertically and horizontally (see Fig. 54). Usually, the machine-bowl is driven from the shaft-worm over V-shaped drive-belts onto another shaft-worm, which keys into the cog-wheel at the base of the mixer-bowl. The machine-bowl can be disengaged, by activating a hand-lever, which functions by disengagement of the shaft-worm. A hand-brake is also provided to anchor the machine-bowl. The disengaged machine-bowl would otherwise continue to rotate, owing to the influence of the resultant force and inclined path of the mixer-arm. The magnitude of this resultant force depends on the distance of the central turning axis from the base of the bowl. Most of this torque force is expended in kneading the dough, and the magnitude of the torque at any time will vary, depending on the location along the mixer-arm orbit, magnitude and direction of the resultant force, and distance from central axis or pivot.

The mixing-intensity will depend on the movement of the bowl, being reduced when the bowl is either free, moving in the same direction as the mixer-arm, or anchored by the brake, and increasing when it moves in the opposite direction to the mixer-arm.

In operating these machines, the following points must be observed. When positioning or removing the bowl, the mixer-arm must be in its highest position, achieved by using the hand-wheel. The machine-bowl must be positively engaged with the drive-worm. The machine should only function when the protection-guard is in closed position, otherwise the switchgear must be checked out. On no account should the machine be emptied during operation. Bowl removal should only be possible when the motor is switched off, and the clutch disengaged. The drive-belts must be protected from flour dust and well guarded during operation. The tension of the drive-belts must be regularly checked, and adjusted or replaced as necessary. All grease-nipples and worms and cog-wheels must be greased at regular intervals, or according to frequency of operation. Machine-bowls must *always* be pushed, and never pulled, to prevent foot and leg injuries. When cleaning, *always* isolate from the mains supply.

The high-speed mixer (Fig. 56) can also be built with one arm, and operated in the vertical position, as opposed to the general conception of the high-speed mixer, as used in the USA and UK, being of the horizontal-bar type. Typical examples of such mixers are the HLK-50, S-125, and S-250, manufactured by the VEB-Bäckerei-maschinenbau, Halle, GDR, and a similar patent by Diosna of Osnabrück, FRG. These mixers operate at two speeds, the higher speed being at 70 to 100 rpm. The machines are extremely versatile, being used in large and small bakeries for all types of dough. The HLK-50 is also used extensively in medium and large restaurants and hotels; the meat industry also use this type of mixer for meat processing. Figure 56 shows the external appearance and internal working principle of the high-speed mixer Type S-125. The power-rating is 2·3 to 3·0 kW, and the bowl capacity 215 litres, holding 125 kg of dough. Referring to the figure for the internal mechanisms: the motor (1) transmits power over a V-shaped drive-belt (2) and shaft-worm (3), which drives the cog-wheel (4). As a result of the reduction of the rpm, owing to the belt-drive and shaft-worm, the turning-moment or torque increases to levels which deliver adequate mixing intensity. In the cog-wheel (4) is a ball-joint (5) and a connecting-piece (6), which is flexible, and positioned so that it can be moved in an axial direction, but is clamped to the oscillating-rod (7). The oscillating-rod (7) can move vertically within the guide (8). Attached to the rod is the actual mixer-arm, which can be moved towards or away from the bowl perimeter and locked in any desired position. The path of the mixer-arm will depend on the relative position of the cog-wheel/ball-joint/connecting piece. This movement results in a combined 'lifting' and 'swivelling' mixing action. This is a unique mixing action, and because of the work-load on the components, the moving parts must function in an oil-bath. The mixing-bowl is driven simultaneously from the cog-wheel by means of a crown-wheel and pinion (10) and (11). In the case of the mixer Types S-125 and S-250, the mixing-bowl is removable. This is achieved by using a spring-loaded metal carrier-plate and lever mechanisms, which holds the bowl in place with retention-pegs.

Before positioning the mixer-bowl, the mixer-arm is moved aside by unscrewing

## High-speed mixer S-125

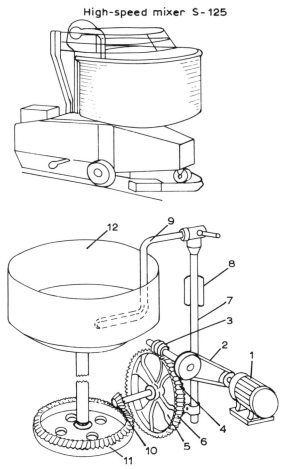

Fig. 56. Layout and working principle of high-speed mixer type S-125/250 (GDR). 1. Motor. 2. V-shaped drive-belt. 3. Shaft-worm. 4. Cog wheel. 5. Ball-joint. 6. Connecting-piece. 7. Oscillating-rod. 8. Guide for rod. 9. Mixing-arm. 10. Pinion. 11. Crown wheel. 12. Mixing-bowl. (Source: *Machine Handbook* (*Baking*), 2nd edn, B. Schramm, VEB-Fachbuchverlag, Leipzig, 1978.)

the hand-key at the top of the oscillating-rod. The bowl carrier-plate is then lowered as far as it will go by using the foot-pedal, and the bowl firmly anchored in the retention-pegs. The mixer-arm is then clamped in the working-position with the hand-key. The dry and liquid ingredients are added, the guard frame pulled down, and the timer set for the mixing-cycle, which can be anything up to 22·5 minutes. After the mixing time has elapsed, wait for the mixer-arm to remain motionless, remove the guard-frame by lifting to the vertical position, and remove the mixed dough and/or mixer-bowl as required.

The greasing and electrics of this machine should be serviced according to the handbook, and only by an engineer. All safety measures must be kept in working order.

This type of mixer is very popular both in the GDR and FRG owing to its versatility for all types of dough, and unique high turning-moment or torque and reduced rpm. This is achieved by the combination of belt-drive and the worm-shaft, the central cogged fly-wheel providing the torque for both the vertically oscillating rod (lifting and swivelling action), and the crown and pinion driven mixer-bowl.

Using this mixer, the dough assembly or 'mixing-phase' can be completed within 1·5 minutes, and the dough fully developed without excessive increases in dough temperature, and use of chilled-water. It has become the accepted mixer for rye, mixed rye/wheat, and continental-roll doughs, as well as pastry and other enriched doughs for a mixed baking-trade.

The one-arm discontinuous 'batch' mixers hitherto described will appeal to a baker who prefers to work with fermented doughs, and who is convinced that, in spite of the time economy of short mixing-cycles, the quality of his baked products demand this type of production system. This type of true craftsman would probably also rely extensively on hand-moulding, at least for variety- and hearth-type breads.

The unique action of the twin-arm batch dough-mixer is acknowledged by all true craftsmen the world over. The two powerful arms working during the entire dough assembly and kneading processes, with their two arms describing intersecting elliptical paths. These are driven by a motor of 2·2 to 5·5 kW power-rating, corresponding to 3 to 6 hp, depending on whether the mixer is intended for 63 kg (140 lb) or 190 kg (420 lb) of flour. For a 2 sack (560 lb) or 254 kg of flour, the power-rating would be 11·0 kW, corresponding to 15 hp.

This mixer design dates back to the 1930s, when Thomas Collins of Bristol, UK, un-veiled a heavy-duty two-armed mixer at the National Bakery Exhibition in London. This mixer, used two twin-pronged mixing arms of about 6 to 7 cm diameter, the prongs being positioned similarly to those of a hay-fork. These arms were mounted at the perimeter of two rotating circular plates, and moved in a vertical plane in intersecting elliptical paths within the mixer-bowl, which simultaneously rotated in a clockwise direction. The arms first moving towards the centre of the bowl in close proximity to the base, and then backwards towards the bowl perimeter in opposite directions. This cyclodynamic action, based on sound technological principles, produces doughs of excellent quality, and without overheating and consequent dough-temperature rises. This enables the required dough temperature to be maintained without any allowances having to be made for frictional temperature increases by the use of chilled-water. There are no possibilities of localized temperature rises within the mixing dough, allowing the maximum possible water absorption at the calculated target dough temperature. Since the optimum mixing-intensity, using any mixer, does not only depend on the mixer-rpm, the unique shaking-in, lifting, folding, stretching and kneading movements produce a dough which has a silky, dry feel, although the mixing-time cycle required to achieve this is greater. Experience has convinced the author that these silky, dry doughs produce bread with an excellent crust colour and bloom, well-aerated texture, elastic crumb, and much-improved eating-experience.

The early Thomas Collins version of this type of mixer, installed in the author's family bakery in 1930, produced at least three batches of bread per day, each of 1·5 sacks of flour (190 kg) for six days per week up until 1962. This machine was powered by a 2 hp Lister petrol-engine, housed in an engine-house 12 metres from the mixer. The transmission was a system of fly-wheels and flat-belts, over a shaft to the mixer fly-wheel. The internal drive cog-wheels of this machine were housed within a large chamber forming the bulkhead of the mixer, which also served as an oil-bath. A clutch was provided, thus allowing one arm to be engaged or disengaged at will. The mixer-bowl was driven by a crown-wheel and pinion.

One surviving mixer of this design is the Artofex, which has certain refinements compared to the original Collins machine. Instead of both arms being in the form of two-pronged forks, one acts as a shovel, moving the contents towards the centre of the bowl, while the other arm carries out the lifting and shaking-in operation. Owing to this action, the Artofex can completely mix a dough within 6 to 8 minutes, depending on flour properties. Cleaning and greasing of the bowl is not required, due to the very close movement of the arms to the sides and bottom of the bowl. These mixers are ideal for doughs of all types and consistencies, and fit in well with delayed salt addition methods, 'dry-salt' processes and sponge-doughs.

The bowl is of heavy-gauge steel, thickly coated internally and externally with tin; the arms are of hand-forged steel also well coated with tin. The gears are helical machine-cut and precision finished.

Spiral mixers have largely replaced the low-speed mixers, and many high-speed mixers installed after the introduction of the CBP in the UK. The introduction of 'short-time' or 'no doughtime' processes did not require a high-speed mixing, with excessive dough temperature rises. Therefore, spirals and traditional low-speed mixers became appropriate for this purpose.

The spiral mixer offers a simple and compact mixing process, being automatically timed over two speeds to deliver a mixing-time of 8 to 10 minutes. Dough removal is easy and cleaning is a simple task. The spiral can be used for quasi-continuous systems where more bowls are in use, and machine-bowl hoists can deposit into a high-level divider-hopper if required. Where dough-retardation is employed, and lower dough temperatures are appropriate, the spiral is also ideal, since no chilled-water is needed, as is the case with high-speed mixers.

The current trend towards making fermented products without any mineral improvers, e.g. potassium bromate, and without additives which carry E-numbers for labelling declaration, also favoured the use of the spiral. Using ascorbic acid as the only additive, combined with unbleached flour and enzyme-active soy-flour, the unique mixing and blending action of a spiral continuously incorporates air and oxygen into the dough, thus activating the bleaching effect, and ascorbic acid interaction with the flour proteins.

The common disadvantage to the craftsman and medium-size bakery, is that the dough has to be cut and lifted from the machine-bowl, unless a removable-bowl machine is used with an expensive bowl hoist or tipper-system. For the large bakery, it can be difficult to integrate into a fully automatic plant process, owing to the longer mixing-time or cycle, and removal of the dough from the bowl. However, at least one company in the UK, All Foods Engineering Services Limited of Newcastle, Staffordshire ST5 6NR, UK, has devised a system for continuously producing doughs with the spiral-mixer. This system is computer-controlled, programmable with memory, being a top of the range plant. Using special inter-changeable mixing arms, a wide range of products can be mixed, combining versatility with simplicity in the engineering, which is essential for a dependable machine. The complete plant comprises ingredient loading, mixing, resting and tipping, giving a throughput of 3 tonnes/h. Bowl rotation and mixing-arm speeds are variable from zero to the maximum rpm required, operation being automatic with manual override; vacuum mixing is an optional extra. The machine-bowl has a full-length centre column, and non-wrapping spiral, allowing efficient dough development. The incorporation of a centre-column top-bearing, gives improved bowl stability and drive efficiency. The tipping model can tip into a divider-hopper, onto a bench or into bins with casters. Although this system is only a quasi-continuous set-up, and mixing is a batch process, the short mixing-cycle allows doughs to be worked off on a line system. It also offers versatility, and conventional make-up for hearth and variety bread production.

Spirals have become increasingly popular in the EC countries, owing to their reliability, versatility and dough development. They are supplied with fixed or removable stainless-steel bowls, two speeds and a timer, for mixing capacities of 50 to 250 kg, powered by 3 to 15 hp motors.

Because of the particular shape of the spiral and rotation speed, which is synchronized to the bowl speed, a homogeneous dough can be produced within a short time. Timers operate on both first and second speed, and control-panel push-buttons allow automatic and manual actuation, as well as reversal of bowl rotation. A typical automatic mixing-cycle is as follows:

(1)  Set pre-kneading and kneading times by means of the two timers.
(2)  Pre-kneading is performed at first speed.
(3)  Kneading follows on second speed.
(4)  Conversion from first to second speed is carried out automatically.
(5)  As soon as the working cycle is completed, the machine stops, and the guard-head automatically lifts so that the carriage can be drawn out. Push-button operation at first and second speeds for manual and automatic sequence are provided on the control-panel, and additional buttons for reversal of the direction of rotation of the bowl.

Some spirals have a spiral arm only, while others have a central pillar or plinth, against which the dough can be mixed. Both types are satisfactory, provided adequate mixing is obtained over the mixing time, with no dead-corners requiring a scrape down.

The other important requirement is that the mixer design allows a wide range of dough sizes to be fully mixed. Capacity has to be geared to a baker's output, processing plant and oven capacity. A 32 kg flour capacity machine is adequate for many small bakers, while a medium-sized operation will require a 128 kg or larger capacity. Capacities are quoted in kilograms of dough or kilograms of flour, and sometimes the model reference number refers to the kilograms of dough capacity. An approximate conversion from dough capacity to flour capacity is to multiply by the factor 0·625. The two extreme capacity dilemmas are:

(1)   Too small a capacity mixer which becomes overworked.
(2)   Too large a capacity mixer, which produces large batches, which cannot be processed or scheduled efficiently into the make-up line or oven, resulting in over-fermented doughs at the end of a run.

Removable-bowl machines are much more expensive than fixed-bowl ones, especially when extra bowls are necessary. Apart from the need for a removable-bowl machine where a bowl elevator is necessary to deposit the doughs into the divider-hopper, other valid reasons are the following:

(1)   The movement of doughs to various processing stations, e.g. variety-bread line, bread-roll line etc.
(2)   Where doughs are fermented in the bowls.

Typical examples of spirals are the following: the 'Diosna' range of spirals manufactured by Dierks & Söhne GmbH, 4500 Osnabrück, FRG, who market about ten high-technology models with dough capacities from 24 to 240 kg, with fixed or removable bowls (Fig. 57).

The Kemper range includes a range of wheel-out bowl mixers, with heavy-duty removable bowls, but without the necessity for fixed base-plates. Their SP-ALH range offers capacities of flour from 100 to 250 kg. The location and locking-in of the bowl is automatic, being the start of the mixing sequence, which is fully programmable. Special fixed-bowl spirals for the UK standard 32 kg bags of flour, can be operated manually or automatically, having a two-speed cycle with 'inching' facility button for ease of bowl-unloading. The bowl-reverse facility is desirable for the pastry production and fruit addition. For the progressive monitoring of dough temperature, a display is provided, so that water temperature adjustments can be made as doughmaking proceeds.

The Thomas Collins range of spirals covers 6 to 300 kg dough capacity. All are heavy-duty machines, and all the larger models are fitted with automatic controls, two-speed mixing, reversible stainless-steel bowls and timers. A standard safety device included enables the operator to by-pass the raising-device as required. Optional extras include an unloading facility for any height to working-table or feed-hopper; computerized models are also obtainable.

Sottoriva, Marano Vicentino, Italy, manufacture a range of spirals which are very popular within the EC countries, owing to their performance and reliability. These are available with fixed or removable bowls, two speeds and timers, and bowl capacities ranging from 25 kg up to 330 kg. The larger capacities have trolleys for ease of transport.

Another Italian manufacturer, Technoplast, produces a compact spiral, with a capacity of 32 kg, which is the new UK bag size. This is a useful size for the small operator, reducing mixing times to about 12 minutes compared with 25 minutes in the case of some more conventional types of mixer. This company also market larger spirals, with separate bowls.

The APV Baker (formerly Baker Perkins), offer a wide choice of spirals in their Master Baker range, suitable for fermented and short-time doughs, and also for pastry. The machines offer high-speed, low-speed and bowl-reverse within one programme, providing the best performance for each product. The kneading-bar mounted in the centre of the bowl improves mixing efficiency, and the machine is well guarded. Removable bowls are available in 120, 160 and 250 kg capacities, and the fixed-bowl type in 25, 50, 80, 120, 160 and 250 kg capacities.

The Sancassiano spiral range includes machines with dough capacities of 25–300 kg on fixed-bowl models, and 140–500 kg with removable bowls. These mixers include bowl-reversing and automatic bowl-location facilities for the 200–500 kg removable-bowl models. Bowl-tippers are available, and the fixed-bowl spirals tip the dough onto a scaling-table by raising the complete machine and inverting. The fully automatic removable-bowl models are ideal for bakeries producing bulk-fermented doughs, including roll and variety-breads.

Werner and Pfleiderer market a range of spirals, including a unique type of twin-spiral design, more akin to a single 'Wendel' mixing element. This works air into the dough without overheating, producing good results even with French bread. Flour capacities of 80 kg and 120 kg are available, the mixing arm operating at two speeds, and the bowl rotating in both directions, computer-controlled if required. This series is known as the UC series. G. L. Eberhardt GmbH, D-8032 Gräfelfing, Munich, FRG, manufacture the Maximat-Spiralknetautomat, complete with automatic tipper, allowing deposit of doughs direct into dividers or onto work-tables. Thus they meet the technical requirements of most craft-bakeries. The Diosna (Fig. 57), and Werner and Pfleiderer (Fig. 58) spirals are illustrated; the latter could be described as a twin-arm spiral format or a single 'wendel' element.

For the craft-bakery which is seeking a high performance multi-functional mixer, the Diosna Wendel mixer (Fig. 59(a)) meets specifications. The Diosna Wendel (Fig. 59(b)), provides an advanced mixing-action which draws the dough or batter

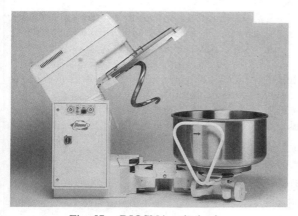

Fig. 57.	DIOSNA spiral mixer.

Fig. 58.	Werner and Pfleiderer Model UC-80. (Source: Stuttgart, FRG.)

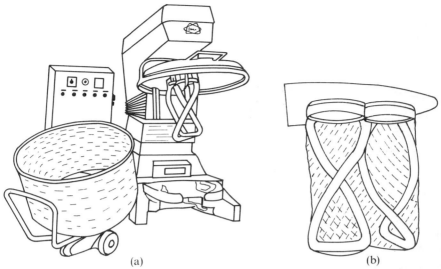

Fig. 59 (a) DIOSNA Wendel-mixer. Hydraulic mixing-head. Removable mixing-bowl. Programmable control panel. (b) Kneading elements drawn to represent them moving in slow motion.

between the two wendels, rotating in opposite directions forming a complex system of wave formation. This results in the gluten being stretched and compressed at short, regular intervals, which is depicted in the illustration in slow-motion (Fig. 59(b)), thus forming a three-dimensional gluten structure. In so doing, the wendel kneader binds in a combination of air and oxygen, giving the dough a light, aerated structure and improved colour. This action is ideal for the use of untreated flour in combination with enzyme-active soy-flour and ascorbic acid, thus producing fermented products with a 'clean' label, suitable for consumption under any legislative regime. The machine is equally efficient for cake batters and biscuit doughs. For normal wheat-flour doughs, mixing times are 5–6 minutes, and 3–4 minutes for mixed wheat/rye-flour doughs and rye-flour doughs. When mixing coarse meal doughs, much saving in hydration-time can be realized. This machine is an advanced up-market design for specialized bakeries and all food production plants, and has a high-technology performance.

The Boku Spiral high-speed machine, marketed by Boku-Maschinenfabrik GmbH, 7000 Stuttgart 30 (Feuerbach), FRG, is capable of mixing all doughs from 1 to 240 kg within 4–8 minutes, producing complete hydration and maximum dough-yield. This machine offers complete machine mobility and anchorage where required, mobile removable bowl, and fully automatic mixing-cycle.

The Oase-Pumpen, Wübker Söhne GmbH, Maschinenfabrik, 4446 Hörstel-Riesenbeck, FRG, market the Fanal-Spiralkneter, which is also mobile and anchorable, full-automatic emptying, with removable bowl.

Two types of discontinuous mixers, with adjustable, eccentrically positioned mixing-elements from the Ukrainian SSR of the USSR exhibited at the Leipzig

Fig. 60.   USSR: Type A2-ChTB. Ukrainian SSR (Leipzig Trade Fair 1987).

Spring Fair 1987, are the A2-ChTB (Fig. 60), which mixes a maximum of 100 kg flour/batch, at a power-rating of 5·5 kW, and the smaller Type A2-ChTM for 45 kg of flour/batch with a power-rating of 1·5 kW. This type of mixer is illustrated as the A2-ChTB. The mixing-time can be preset, the mixing-bowl being fixed, but completely removable and transportable on tyred wheels.

From Yugoslavia, the GOSTOL of Nova Gorica, produce and market a large, heavy-duty spiral-kneader of the Type SM85, with a maximum dough charge/batch of 135 kg. The mixing-bowl, 'plough-share' and mixing-arm are of non-rusting steel. The vertically mounted, spiral-shaped mixing-arm rotates within one half of the machine-bowl at speeds of 110 and 220 rpm. The 'plough-share', mounted within the machine-bowl axis, ensures an even dough distribution. The machine-bowl rotates in the opposite direction to the mixing-arm, and is independently driven by a 2·2 kW motor. During mixing, the direction of rotation becomes reversed. The mixing and kneading times can be preselected and proceed automatically; manual operation is also provided for. After completion of mixing, the mixing-arm, 'plough-share' and drive are raised electromechanically, so that the bowl can be removed. The machine type SM 125, 200 kg of wheat-flour dough maximum capacity, uses a mixing-time of 1·0–1·5 minutes, and a kneading-time of 5–8 minutes. This delivers an energy-input of 11 to 15 kW, whereas the Type SM85 delivers 6·8 to 11 kW.

*High-speed or Intensive Discontinuous Mixers*
These machines are used in medium and large industrial bakeries in conjunction with various bread and roll lines owing to their flexible quasi-continuous capability resulting from a short mixing-cycle, giving minimal delay between batches. To accommodate each dough-batch, either extra bowls or mobile dough-bins can be used.

In the UK, high-speed mixers are almost universally used by industrial plant bakeries using the CBP process. However, some in-store and small bakeries still use their high-speed Tweedy 140, 160, or 200, their Gilbert 35 or 70, or their Mono 100, installed at the time of the introduction of the CBP during the 1960s. The features of these machines are energy-inputs of 8 to 11 W h/1 kg dough over 90 to 120 s, working at 400 to 420 rpm, using a plate with jagged-teeth driven through the base of the closed vertical mixing-chamber. Bread volume can be controlled by the use of vacuum at 0·05 N mm$^{-2}$, which produces wheat-flour bread of specific volume up to 3·6 ml/g. This compares with the high-speed mixer S250, already described and illustrated in this section, which is capable of producing wheat-flour bread volumes of up to 3·0 ml/g, at lower energy-input and higher mixing-intensity, owing to its design by the VEB-Bäckereimaschinenbau, Halle, GDR. This is a classic example of the difference between energy-input and specific mixing-intensity. During mixing with the Tweedy mixer, much heat is dissipated as heat-energy as dough temperature rises. In recent years, developments have been directed towards: computer-control of the total ingredient weighing and feeding system, mixing-cycle control, dough discharge, and line throughput operations. Visual display units (VDUs) are available with many models, displaying which function is operating at any time, and recipes and processing requirements are programmed with the help of microprocessors.

The popularity of continental-style emulsifier dough-conditioners, the pressure to produce 'clean' label products with fewer or only organic additives, coupled with the national health campaigns in favour of wholemeals, brown and specialty-breads, have resulted in changes in processing for some products. Reductions in energy-inputs and oxidants, with the increasing use of oxygenated conditions during mixing, in combination with the use of ascorbic acid as the only oxidant, and in conjunction with sources of lipoxygenase such as enzyme active soy-flour, have been reawakened. Most of these techniques were researched and applied during the 1950s with success, but were discarded owing to more generalized use of chemical flour bleaching and more liberal national legislation concerning additives. The UK had to harmonize its legislation with that of the other EC countries, most of whom have been producing bread with a limited number of organic additives such as: malt, lecithin and ascorbic acid for decades, and using unbleached flours.

Although spirals have replaced high-speed mixers in the craft sector, the latter are still widely used by many master-bakers and supermarket-chain in-stores.

Both fixed-bowl, horizontal tipping, and removable bowl upright models are on the market, usually complete with a timed mixing sequence, and energy-input control is an optional extra. Mixing-times are about 2 to 3 minutes, compared with 8–12 minutes with the spirals, therefore more batches can be produced in any given time. Owing to a short mixing-cycle, these mixers can be used for continuous line-production, and according to mixer capacity, can provide hourly throughputs of 300 up to 3000 kg of flour into dough. Another advantage of the high-speed mixer is the rapid and increased water-uptake, and the use of lower-grade flours. Apart from the Tweedy arch-type batch mixer made famous by the CBP from the 1960s, the Mono Equipment company, of Swansea SA5 4EB, Wales, UK, manufacture a stainless-steel clad range of high speeds in three capabilities. These machines having tipping facilities, operated from buttons positioned on the mixer-cradle. The 70 16 model, is a popular size for the craft-baker.

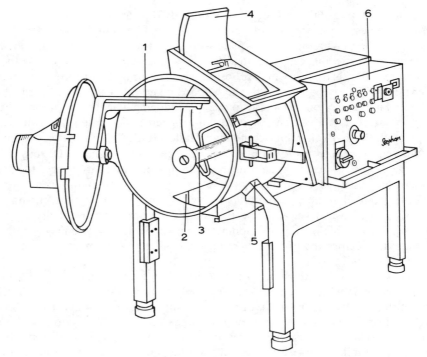

Fig. 61. Stephan TK 150. 1. Mixing-baffle. 2. Hydraulic discharge slide. 3. Kneading-element (removable). 4. Transparent Plexiglass loading window. 5. Rubber vibration-pad mounting. 6. Control panel. Features: automatic timer; digital thermometer; automatic water-dosage.

Another high-speed machine suitable for the smaller operator is that marketed by A. Stephan und Söhne GmbH, Stephanplatz 2, D-3250 Hameln 1, FRG. Stephan TK-150 (Fig. 61) is a small, fast, high-speed mixer with a horizontally positioned mixing-chamber, the ingredients being fed in through a top loading-hatch with a transparent Plexiglass window, allowing visual inspection during mixing. Batch sizes from 5 kg to 28 kg can be mixed within 2 minutes. Various mixing heads can be exchanged for mixing pizza-dough and pastry, cake-batters and grinding of stale-bread for crumb. The horizontal mixing-chamber can be tilted and locked from the adjacent control-panel, which also incorporates an automatic-timer, digital thermometer and automatic water-dosing. Mixing speeds are 750, 1500 or 3000 rpm, according to product. Double-jacket heating and cooling, and a mixing-baffle, which assists in mixing and kneading and dough-discharge makes this small fast high-speed mixer ideal for bakeries producing breads, rolls, cakes and cookies to keep their lines supplied with doughs and batters. Short kneading-time, high dough yield and fast unloading allow dough throughputs of about 800 kg/h. The TK 150 is capable of coping with 10 dough batches per hour, each made with 50 kg flour. The mixing-chamber or drum capacity of the TK 150 is 150 litres, with a maximum dough loading per batch of 90 g, the power-rating being 15·5 kW. The space required for this unit is a mere 2100 mm × 1200 mm × 1300 mm, and a hydraulic discharge-slide will discharge direct into a trolley or onto a conveyor.

It should be pointed out that this type of high-speed mixer, irrespective of design, are only suitable for the processing of wheat-flour doughs, since rye-flour doughs depend mainly on the development of the sour-dough for success, and cannot be appreciably improved by mechanical influences.

The construction of the Stephan and Collette (Fig. 62) machines, which are suitable for medium-size operations, with a wide product range, requiring flexibility, are illustrated.

Collette of Belgium were one of the pioneers in the application of high-speed mixers to all sectors of the baking and food production industries during the 1970s. The Collette SM range of seven high-speed mixers is intended to accommodate all types and mix capacity from 5 to 600 litres. They are engineered in close-grain cast-iron, with stainless-steel removable-bowl, mixing-arm and blades. The wheel-in bowl eliminates handling and tipping, and if desired continuous operations can be carried out with the use of extra bowls. The SM system is an extremely flexible one, accommodating a manually operated small machine and a completely automated mixing cycle with batch, emptying. Floor space is saved with the elimination of

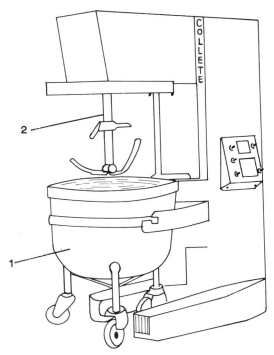

Fig. 62. Collette SM. 1. Stainless steel wheel in bowl (5–600 litres). 2. Mixing-arm and blades.

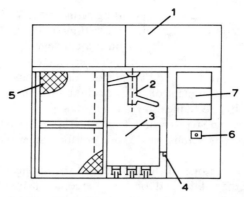

Fig. 63.   Intensive mixing and kneading machine, IMK-150. 1. Double-standing housing. 2. Mixing-element. 3. Mixing-bowl. 4. Mixing-bowl locking. 5. Sliding safety-guard grill. 6. Mains switch. 7. Function controls (see Fig. 64). VEB-Backereimaschinenbau, Halle, GDR. (Sources: *Machine Handbook* (*Baking*), 2nd edn, B. Schramm, VEB-Fachbuchverlag, Leipzig, 1978; *Technology of Industrial Baking*, R. Schneeweiss and O. Klose, VEB-Fachbuchverlag, Leipzig, 1981.)

fermentation-bowls, and loss of dough solids due to fermentation reduced to a minimum.

An extremely heavy-duty intensive mixing and kneading machine is the IMK 150 (Figs 63, 64), engineered in the GDR by VEB Bäckereimaschinenbau, Halle, GDR. This is a universally applied discontinuous mixer, which is based on the high-speed mixer principle. In addition to wheat-flour doughs, rye- and mixed wheat/rye–flour doughs can be successfully prepared with this machine.

It consists of a twin-standing housing, with the mixing-bowl located between them, within the arched recess. The hydraulics, electrics, transmission and mixing-elements are all within the housing, and during operation the mixer is enclosed by a sliding grill, which acts as a safety-guard.

The power transmission from the 60 kW motor is achieved directly onto the mixing-elements via ten drive-belts. This reduces the rpm from 1470 down to

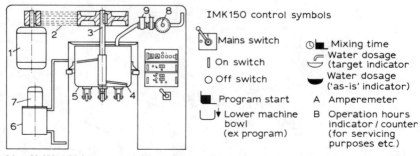

Fig. 64.   IMK-150 layout and function. 1. Motor 60 kW. 2. V-shaped drive-belts. 3. Drive-shaft. 4. Mixing-elements. 5. Machine-bowl. 6. Hydraulics. 7. Motor for hydraulics. 8. Water-dosage indicator. 9. Sieve filter and magnetic valve.

415–420, and the residual net energy at 55 kW. This allows an energy-input of 8–10 W h/kg of dough to be applied within 2–3 minutes. The energy-input is measured by a recording amperometer, and when this registers more than 100 A, the motor is overloaded. If this occurs, and the dough consistency is not reduced, a thermal safety mechanism will automatically break the motor circuit. The raising and lowering of the mixer-bowl functions hydraulically, the oil-pump of which is driven by a separate 3 kW motor.

The metering of the dough water is achieved with a circulatory-pump complete with measuring device, the unit being constructed of chromium-nickel steel, which is suitable to withstand excess pressures of up to $1.6 \, \text{N mm}^{-2}$. A magnetic-valve and sieve to act as filter completes the water-dosage unit, which delivers up to 100 litres of water. All machine functions, apart from movement of the machine-bowl, are controlled from a well laid-out panel mounted on the right-hand side of the twin-standing. Figures 63 and 64 show the layout of the IMK 150, together with the symbols of the control-panel. The working procedure is programmed as follows:

(1) Preset mixing-time and amount of dough water required.
(2) On pressing the starter-button, the following functions occur automatically:
   —elevation of the mixing-bowl, and closing the safety-guard grill (30 s)
   —delivery of the programmed water (50 s)
   —mixing (120 s)
   —immobilization of the mixing-elements (20 s)
   —lowering of the mixing-bowl (20 s)
   —machine-guard grill-door opening, and cleaning of mixing-element (20 s)
   —removal of machine-bowl (20 s)

The average duration for the manual dosage of flour, salt, yeast and water, including dough-conditioner, sugar and shortening, also machine-bowl transport, is 160 s. The programme-duration is about 220 s. Mixing-times can be chosen in 10 s intervals up to 300 s. Before presetting the quantity of dough water required, it is important to ensure that the automatic zeroing of the meter has occurred; otherwise it must be zeroed manually.

According to the time of year, type of dough and flour temperature, the following water-temperatures are recommended:

|                     | *Winter*   | *Summer*   |
|---------------------|------------|------------|
| Wheat-flour doughs  | 17–25°C    | 12–17°C    |
| Rye-flour doughs    | 25–40°C    | 18–25°C    |

In common with all high-speed machines, the high energy-input during mixing results in a temperature rise of about 10 K for wheat-flour doughs, and about 3–5 K in the case of rye-flour doughs. The above-quoted figures already make allowance for these corrections. The large panels and enclosed surfaces of the IMK 150, coupled with the relatively maintenance-free construction, reduces the service requirements of this machine. Servicing at operating-hour intervals comprises: checks on the screw-unions, belt-tensions, magnetic-valve, sieve and filter, and the oil-level of the hydraulic-system; and adjustment of the flow-control valves for perfect balance, thus avoiding unequal stress, and inclination of the machine-bowl to the vertical. The machine will not function until the safety-guard grill is closed.

The dough capacity of the IMK 150 ranges from 50 to 170 kg, and the number of batches per hour is about 10, giving a maximum hourly throughput of 1700 kg. The actual mixing duration is 120–130 s, and the full machine working-cycle takes about 6 minutes.

Mixing wheat-flour doughs, with a frictional temperature rise of about 10 K, the optimal dough temperature is about 33°C.

Doughs from this machine can be either processed manually or fed into any process line, and unlike many intensive high-speed machines, it can be also used for rye- and mixed wheat/rye-flour doughs. Manual operations include the addition of yeast, salt and other ingredients, also the fixing, removal and transport of the mixing-bowls. The machine can be operated by one person. The application of this mixer for wheat-flour doughs, e.g. toast-bread and bread-rolls, allows throughputs of up to 3000 kg of dough per hour, depending on the bowl-capacity, and the required mixing-time. Therefore, on account of the short mixing-cycle and high throughput, this mixer can feed continuous processing lines, only the mixing being a discontinuous or 'batch' operation.

The IMK 150 permits maximum flexibility for the industrial production of wheat-flour doughs for bread-rolls in the GDR, and W. Schwate and U. Ulrich in *Special processes for baked-products*, VEB Fachbuchverlag, Leipzig, 1986, report the following advantages in its application:

—In feeding a bread-roll line, equipped with a BN 50 (50 m² baking-surface) tunnel-oven, the IMK 150 would require four machine-bowls; and where eight machine-bowls are available, one IMK 150 machine at maximum output (1900 kg/h) is capable of feeding two lines continuously, using the BN 50 tunnel-oven.

—Ingredients, apart from water, are added manually, and yeast can be added in the dry form.

—All dough-conditioners or improvers can be added without special preparation beforehand.

—Water weight and temperature, also mixing-time, can be preset and delivered on the IMK 150.

—When using a direct (straight) dough process, the doughs can be given a floor-time of 20–30 minutes in the machine-bowls, the floor area being no more than in the case of continuous-mixers.

—The short mixing-time allows an energy-saving.

The only disadvantage with the IMK 150 is that a remix, punch or knock-back is not possible, but a pair of rolls positioned before the dough is fed to the moulding-machine could serve this purpose.

For large-scale batch-processing of about 1000 pounds (454 kg) of dough, especially for sponge-and-dough systems as traditionally carried out in the USA and Canada, the horizontal-bar and drum-type mixers (Figs 65, 66) are standard equipment. They consist of a heavy-duty metal housing firmly fixed to the floor. The

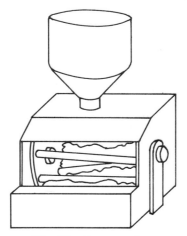

Fig. 65.   Sponge-mixer.

mixing-chamber or drum is of stainless-steel, mounted horizontally to form a drum-shape. The front section being open to allow discharge of the dough after mixing, by opening the sliding, electrically operated door, which moves up and down in a vertical plane. The flour and water are fed in at the top of the mixer from a staging or platform from which the operator can also weigh the minor ingredients.

Usually, several of these mixers are monitored in line at the doughmaking station (Figs 67, 68), the platform allowing access to the ingredient feed, and one flour-hopper mounted on rail-hangers servicing all mixers by sliding along. The flour is either blended and sieved at a lower level and passed through a worm-conveyor via

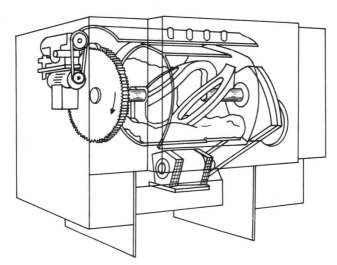

Fig. 66.   Horizontal high-speed mixer. 1000 pound dough capacity. (Source: APV–Baker, Peterborough, UK.)

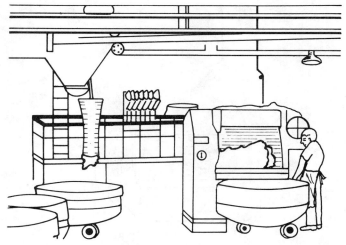

Fig. 67.   High-speed mixer with staging service-platform.

an elevator and additional worm-conveyor into the mixer flour-hopper, or fed into the mixer-hopper via a pneumatic system from above. In either case the hopper should function as a weighing-head, being mounted on load-cells.

The original beater-design of these mixers of which the Baker Perkins was the archetype, of 1950s vintage, consisted of a large beater-arm, with roller-shaped arms, which, on revolving, carried the dough around with it. Thus the dough becomes stretched instead of being torn. Mounted inside the drum were two additional rollers, against which the dough was thrown by the beaters, thus helping to develop the gluten by a stretching action. There were two working speeds, 35 rpm and 70 rpm, the machine being preset for a specific mixing-program. Working at 35 rpm, these mixers will mix a dough in 4–6 minutes, depending on flour strength and formulation. Thus it is possible to produce 6 doughs per hour, equivalent to output of 600 pounds or about 2724 kg dough per hour. For long production runs of

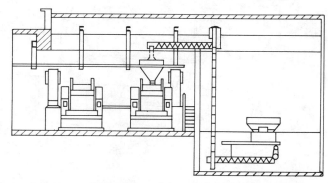

Fig. 68.   Schematic layout of blender, weigher and high-speed mixers at doughmaking station.

a standardized bread product, where oven capacities are of the order of 9 to 12 sacks per hour or 2520 to 3360 pounds flour made into bread per hour, these mixers are ideal. The flour should be of good quality and colour, capable of forming a resilient, elastic gluten network. Flour of extraction levels up to 75% are the best raw material for these mixers, since their potential can be exploited to the full in terms of dough hydration, and a more resilient bread texture and crumb colour.

These mixers are used extensively on the North American continent for the traditional sponges (Fig. 65), which are fermented in 'troughs' of 1000 pounds capacity, that are moved to and from the fermentation-room manually or by special conveyor-systems. Where liquid pre-ferments are now being used, these are fermented in large vertical tanks and contain up to 25% of the total flour, yeast, sugar-solids and yeast-foods. These can be blended and agitated at controlled temperature, and added to the residual flour and ingredients in the horizontal high-speed mixers, thus saving the fermentation-room floor-space.

Modern horizontal high-speed mixers are fitted with two elliptical-shaped beaters, driven from either ends of the mixing-chamber. The constructional layout of the current horizontal-bar high-speed mixer is shown in Fig. 66, together with the arrangement of the blender, weigher and platform for servicing of a battery of these mixers at the doughmaking-station (Figs 67, 68). These mixers are fitted with a cooling-jacket, owing to average temperature rises of about 10 K. The high safety level and trouble-free operation of these machines have made them the accepted system of mass production for standard bread throughout the North American continent for several decades. Baker Perkins Limited (UK), now known as APV-Baker, have a sister company at Saginaw, Michigan 48605, USA, which manufactures these mixers; and similar horizontal high-speed mixers are marketed by the American Machine and Foundry Company of Richmond, Virginia 23227, USA. In the case of North American flours with greater gluten-formation capabilities, longer mixing times are required, unless proteolytic enzymes such as papain or bromelin are used. During the late 1950s, one large bakery company based in London adopted these mixers as standard for producing 3 hours straight-doughs for the manufacture of white bread. In spite of the widely held view that these mixers were only suitable for glutinous North American flours, with appropriate mixing-time adjustments, excellent bread was produced, the overall quality of which was superior to current white bread quality in the UK.

It is the author's experience that for the large-scale production of white bread, whether lean or enriched formula, the dough development capabilities of these mixers, coupled with a fermentation or floor-time in bulk, produces bread-quality superior to any short-time doughs relying on chemical dough development techniques. The sponge-and-dough technique, traditionally used in the USA for white bread production, or its latter-day counterpart the 25% flour pre-ferment or brew, is more flexible than the European straight-dough procedure, since adjustments can be made more easily in scheduling doughs through production lines. Furthermore, it is still open to question whether a pre-ferment in liquid form, containing less than 25% of total flour, can successfully replace a bulk-fermentation or floor-time, with regard to final product quality.

In 1984, Baker Perkins introduced a new generation of horizontal fixed-bowl dough mixers for bread-dough, called the 'UniPlex' and 'BiPlex'. These resulted in the APV-Baker market share in automatic industrial bakery mixers increasing from 17% to 45%, and from 4% to 15% for all industrial bakery mixers worldwide. These mixers include a microprocessor-based controller and colour-VDU, operating without manual intervention, the feeding of all ingredients being automatic, and cycling at rates to suit production line requirements. Every mixer being integrated with the ingredients system upstream, and process equipment downstream. The UniPlex vertical (Fig. 69) and horizontal fixed bowl (HFB) (Fig. 70) are classed as high-speed mixers, suitable for mechanically developed doughs, ADD, and bulk-fermented. It can be applied to bread and bread-roll production, fruit-containing doughs, and wheat/rye-flour doughs. It is available in four sizes: Model 1600, with a maximum dough output of 1600 kg/h; Model 2200, with a maximum dough output of 2200 kg/h; Model 3000, with a maximum dough output of 3000 kg/h; and Model 4000, with a maximum dough output of 4000 kg/h. Minimum and maximum dough batch sizes (kg) are: 55/133, 55/183, 100/250,

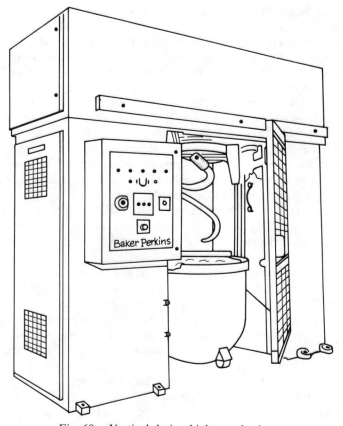

Fig. 69.   Vertical design high-speed mixer.

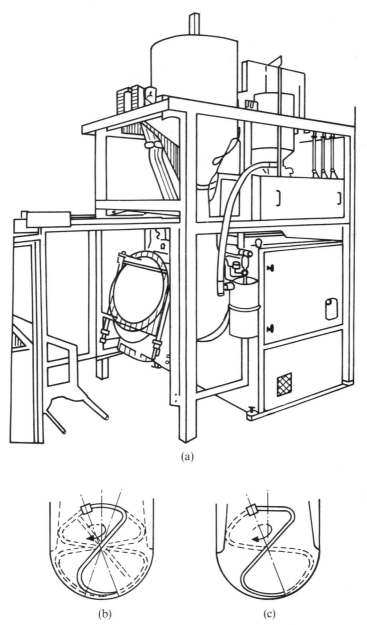

(a)

(b)　　　　　　　　　(c)

Fig. 70. Horizontal fixed bowl design. (a) Schematic diagram; (b) planetary-kneading mode; (c) dough-development mode.

100/333 respectively. The electric-motor power rating of the UniPlex mixers are 35, 45, 55, and 75 kW respectively. Quoted output figures are based on average mixing-times of 2·75 to 3·25 minutes, output decreases with increased mixing-cycle times, according to ingredients, formulation or process applied. The four machine sizes are all available complete with flour and water-weighing, optional hoist, and two machine-bowls. However, each mixer is obtainable with three alternative levels of automation (see Fig. 71):

—mixer only, suitable for integration with existing weighing equipment;
—with flour and water weigher, suitable for full-automation, but with manual weight-setting and energy levels, etc.;
—flour, water and optional small ingredient-weigher; water temperature calculation and control. VDU interface, recipe-storage, print-out and diagnostic aids are also added.

Operating features include:

—intensive mixing with a spiral-blade gives effective ingredient dispersion and dough development;
—consistent results, due to accurate load-cell weighing of flour and water, minor ingredient precise proportioning, energy measuring, and dough-temperature control;
—the screw-like action of the spiral-blade gives a rapid dough discharge, allowing short batch mixing-times or cycles of about 5 minutes;
—vacuum control is optional for control of loaf-texture;
—hoist with automatic oiling system either vertical, angled, in line with or at right-angles to the mixer.

Mechanical features are:

—stainless-steel mixing-chamber, with powered door-opening for ease of discharge and cleaning;
—weigh-hoppers suspended from load-cells, mounted on separate structure, and isolated from the mixer frame;
—separate control-panel, with power panel incorporated on mixer;
—flour-hopper with vibrator to ensure complete discharge, three feed option connections for flour types; additional flour options unlimited;
—liquid hopper with thermal-insulation to prevent condensation; additional feeds for yeast, sugar etc., also possible;
—optional ingredient hoppers for precise weighing of improvers/conditioners;
—floor-level motor mounting to reduce vibration levels;
—all mechanical and electrical components enclosed behind locked door-panels;
—access area in the centre of the machine is enclosed by safety-guards, operated from main control-panel, access only being possible when the machine is safe.

The BiPlex is similar in constructional layout to the Stephan TK 150 described earlier; however, the BiPlex is intended for the larger operator requiring larger batch capacities and more sophisticated ingredient-feed systems.

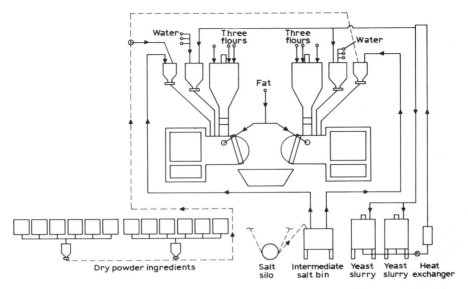

Fig. 71.  Ingredient flow diagram for HFB mixing systems. (Source: Bakery mixing systems, APV-Baker, Peterborough, UK.)

The BiPlex machine incorporates a new mixing concept and process, having two separate mixing actions. The first is a planetary action, which progressively incorporates the ingredients, and draws air into the dough. Thus, when this action is used alone, it provides an effective high-speed mixing action for doughs to be subsequently bulk-fermented. The second stage mixing-action represents a higher-speed single rotation, which develops the dough within a short time. The net result is a dough produced with a lower temperature rise, and a reduced requirement for additives. The most effective oxidants are those which are synergistic towards oxygen, i.e. ascorbic acid and azodicarbonamide (if permitted). Provided no vacuum is applied, the use of potassium bromate becomes unnecessary, and the ascorbic acid concentration can be limited to 40 ppm, based on flour weight. The choice of dough improver/conditioner will, however, depend on the flour specification, product being produced, and the process schedule. The BiPlex 200 Bread-mixer has a bowl capacity of 2 litres, allowing dough batches of 106 kg (64 kg flour). Maximum throughput is 15 batches per hour, equivalent to 1590 kg dough per hour. The main motor has a 22 kW power-rating, and the planetary motor 4 kW. The machine being capable of producing bread by the new 'BiPlex' process, CBP, or fermented doughs generally where high-speed mixing is desirable.

The sweeping action of the beater-blade in the planetary mode, eliminates any unmixed pockets of dough as it sweeps the bowl-surface. The mixing element is mounted vertically within a single recessed standing, with a sealed gearbox and life-long lubrication. Additional machine-bowls of either tinned mild steel or stainless-steel are optional extras. Safety standards are such that access to moving parts are eliminated by guarding, the machine-bowl remaining in the upper position as long as the beater rotates, and the beater cannot be set in motion unless the machine-bowl is clamped in the upper operating position.

The BiPlex 500 (Fig. 69) is similar to the IMK 150, its forerunner from the GDR. The 500 litre mixer-bowl is removable and positioned within the recess between a twin standing housing, its maximum output performance at 10 batches/h giving 2650 kg dough/h is also similar in magnitude to the IMK 150. However, the BiPlex has the two separate mixing actions common to both the BiPlex 200 and 500 models, incorporating both the planetary and high-speed rotary modes. The BiPlex 500 is powered by either a 55 or 75 kW motor, and the transmission is through a sealed gearbox with lifelong lubrication. For the industrial line-production of bread and fermented products, the BiPlex 500 is usually used in conjunction with a machine-bowl hoist, feeding dough into an 'Accurist' APV-Baker (Baker Perkins) dough-divider. In addition APV-Baker market a Biplex 100 mixer of 100 litre machine-bowl capacity for the smaller operator. The new 'BiPlex-process' patented by Baker Perkins, now merged into the APV-Baker machinery group, produces a mechanically developed dough to be divided immediately after mixing. The energy-input, at approximately 7·6 W h/kg dough, and hence the resultant dough temperature-rise during mixing, amounts to about one-third less than the optimum for the Chorleywood bread process (CBP). Since much of the energy-input expended in the CBP becomes dissipated as heat-energy and dough temperature rises, accurate temperature measurement can give an index of energy expended.

For short-time mechanical dough development processes, e.g. the CBP, the recognized advantages are retained, and certain additional rationalization advantages gained by applying the BiPlex-process, viz.

(1) Less dependence on chilled-water. When using the CBP energy-input level of 11 W h/kg, in order to obtain a dough at 30°C, flour temperatures cannot be much in excess of 27°C. Since mains water is at about 14°C, chilling down to 4°C and below is often necessary using the CBP. Using the BiPlex-process, flour temperatures of 26°C can accommodate dough-water temperatures of 14°C.

(2) Lower yeast concentrations. CBP levels of 1·8% to 3·1% can be reduced to 1·5% to 2·7%.

(3) Improved machinability of doughs at the same water-absorption, or the same machinability at increased water-absorption. Depending on basic flour quality in use, from 59% to 62%, or 64% to 65%.

When changing from the chemically developed 'activated dough development' process (ADD), considerable economy in chemical additive costs of up to 67% can be realized. Simultaneously, dough water absorption increases of up to 2% are possible.

Owing to the active oxygen incorporated into the dough, organic oxidants perform best, and often 40 ppm ascorbic acid (flour weight) is adequate.

Where a bulk-fermentation or floor-time was in use, often a cheaper, lower protein flour of about 11·0%, instead of 11·5–12·0% (14·0% moisture basis) can be used for panned-bread and soft bread-roll doughs.

If a vacuum is used to improve texture, potassium bromate is a necessary additive.

Improvers/dough-conditioners for panned bread, in common with the CBP process, should contain 0·7% high-melting-point shortening based on flour weight, in addition to the 40 ppm ascorbic acid. Alternatively, the shortening can be added separately. For a white-bread dough typically 1·0% improver would be added, and about 2·0% for other fermented doughs.

These mixers are fitted with two timers (BiPlex 100 and 200), timer no. 1 registers the total mixing-time, and timer no. 2 the first stage mixing-time. For small 32 kg flour mixes for white-bread dough, using a baker's grade flour of 11·5–12·0% protein (14·0% moisture), total mixing-times are 160 to 180 s, mixing with the first stage planetary mode for 130 to 150 s. Using the automatic BiPlex 500, these mixing times are reduced by 25%, owing to a higher rate of energy input.

For certain types of dough and fermentation schedules, the BiPlex is used in the planetary mode only, in which case both timers are set at the same time, thus eliminating the high-speed beater-action mode.

In the case of bulk-fermented doughs, the planetary-action mode only is used, with both timers set for 3·0 minutes, and the water temperature raised by about 7°C versus when mixing in the higher energy-input mode of 7·6 W h/kg. When mixing mixed wheat/rye-flour doughs, only the planetary action mode is used, and the mixing-time is limited to 150 s. For puff-pastry, only the planetary mode is employed, and the dough assembled over 30–40 s, the pastry-margarine is then distributed over 15–20 s. In the case of short-crust pastry, 50–60 s is adequate to assemble all the ingredients.

The two-stage BiPlex dynamic modes proceed as follows:

(1) The spiral-shaped beater blade, designed with the help of CAD/CAM, describes a planetary orbit, simultaneously sweeping the entire machine-bowl surface, typically for a period of 2·5 to 3·0 minutes, this allows: complete hydration of the flour macromolecules, efficient mixing and dispersion of all ingredients, considerable incorporation of oxygen into the dough mass.

(2) An entirely different dynamic-path, involving a high-speed rotation about a fixed shaft, which stretches and develops the dough, typically for a period of 30–60 s.

The universal popularity of the discontinuous or batch-type mixer is due to its great flexibility for the production of a wide range of products, to the ability of high-speed batch mixers to feed several production lines on a quasi-continuous basis where either a number of machine-bowls are available or mobile-troughs are provided, and to the possibility of correcting errors during formulation or mixing on a relatively small quantity of dough.

In selecting a mixer, it is important to bear in mind that the mixing-time, mixing-effect and mixing-intensity depends on the type and dynamics of the machine. The objective in machine construction is to seek the optimal application of the available dynamic force modes to the type of dough being produced, and the downstream process schedule desired.

**Continuous Dough-mixing Systems**
With continuous doughmaking, there is a demand from the operator for systematic and exact working procedures, since no corrective measures can be undertaken during the mixing stage. The dough is assembled, mixed and developed in distinct phases, and leaves the horizontal cylindrical mixing-chamber in the form of a continuous dough-strip. The processing-system consists of two stages: raw material preparation and dosage, and continuous dough-mixing. The sieved and blended flour is moved into the daily-silos and fed into the weighing-head of the mixing-aggregate. The other dough ingredients, such as: yeast, salt, shortening and dough improver/conditioner (rapid-solubility type) are prepared either as slurries or solutions at the preparation-station using special containers and agitation aggregates. These are then fed to the mixing-chamber through the appropriate dosage-aggregates. The required amount of yeast and salt are automatically metered according to the flour weight measured by the weighing-head. If the weight of flour is changed, a synchronization-device adjusts all liquid ingredients accordingly. Nevertheless, the operator must check this function to ensure that the formulation is adhered to, and correct as necessary before delivery. Formulation changes or changes in dough-yield (based on flour-weight) demand an adjustment of the synchronization-device.

During mixing, the dough temperature rises so much to frictional-heat dissipation, that either the added liquid must be cooled before addition, or the mixer-chamber is fitted with a cooling-jacket. The most important components of all continuous mixers are: feed-hopper, worm, drive-shaft, mixing-elements or shear-discs, and extrusion-outlet.

The continuous doughmaking plant is made up of the following machine-aggregates: raw material containers, liquid-ingredient preparation equipment, dosing or metering-equipment, dough-mixing and kneading machine, and switch and control-panel. The construction of the continuous mixing aggregates show a wide variation in conception and design. Some depend on two shafts rotating in opposite directions, fitted with worm, shovel, shear or wendel-type attachments; others have the mixing-worm and shear-discs mounted on separate shafts, thus allowing each aggregate to rotate at different speeds.

The speeds of continuous mixing-aggregates can usually be regulated between 25 and 170 rpm, and depending on the rpm and extrusion-outlet adjustment, the dough residence-time in the mixer is about 2 to 4 minutes. The mixer housing is usually either cylindrical or spherical in shape, and sometimes baffle-plates or pins are mounted on the inner surface to increase the resistance during mixing. In the first section of the chamber the raw materials are mixed, and in the second section the more intensive swelling, solubilization, and dough-development processes take place.

The dough throughput depends on the chosen rpm, and approximately 500 to 1500 kg dough/h can be processed on average. However, for larger operations units for 1800 to 3000 kg/h are available. The concept of employing an active pre-ferment, comprising part of the ingredients before the actual mixing process resulted in a practical solution to the feasibility of continuous dough-mixing. The two major continuous dough-mixing systems which evolved in the USA, depended on a high-speed developer unit to impart the necessary energy input. For the production of white pan-bread only, the texturizing-effect of high-speed mixing on bread crumb-structure was considered desirable, resulting in a softer type of bread. In spite of modifications to this process during the middle 1950s, to improve flavour, texture and specific volume, the ultra-high-speed mixing proved essential, thus making these units unsuitable for bread products not requiring a fine grain and soft texture (high specific volume).

The European baker, being faced with different consumer demands, required another approach. Most bread varieties tend to be of the hearth or variety type, utilizing a diverse range of flour blends and quality profiles. These also often contain both wheat- and rye-flours of different extractions. The American high-speed versions of continuous mixers are unsuitable for such breads. Therefore, continuous mixing development became adapted to the conventional kneading, folding, stretching, pressure and shearing actions of low-speed batch mixing, but with improved efficiency, and over a shorter mixing-time.

The solution involved the use of cylindrical mixing chambers within which, revolving shafts fitted with discs, cutters, blades or paddles assemble the ingredients, and move the mass towards the discharge point. These proved reasonably efficient for the mixing of one specific type of dough, i.e. wheat-flour, rye-flour or biscuit-dough, but were restricted in application. Therefore, a multi-purpose continuous mixing system, capable of producing hearth-type breads, bread-rolls, rye-bread, mixed rye/wheat breads, and special-breads was developed by a team of German bakery engineers for Werner and Pfleiderer of Stuttgart, FRG. This development required several years of experimentation, and the first large-scale unit of this mixing system was installed in a commercial bakery in 1968. Since that time many of these systems, called the 'Konpetua' were shipped throughout the world. The main features of this continuous mixing system are its flexibility and ease of adaptation to changing production requirements. The constructional layout and mode of operation of this unit will be discussed in detail later. Another successful continuous mixing aggregate is the Kontinua, also manufactured by Werner and Pfleiderer, which has been installed in countries where a large output of more or less standard bread-types are produced. The construction and internal profile of the mixing-elements of the Kontinua are illustrated in Figs 72–75. It is suitable for the processing of wheat-, mixed wheat/rye-, and rye-flours, and has been installed in some socialist-community countries, where district-collectives are re-equipping or building new bakeries. There are two types of unit, with the following outputs and power-ratings:

| Type | Throughput (kg dough/h) rye and mixed wheat/rye doughs | Wheat doughs | Power-rating (kW) |
|---|---|---|---|
| FMS 400 | 1 000–2 000 | 600–1 500 | 42·5 |
| FMS 500 | 1 700–3 500 | 1 000–2 400 | 41·5 |

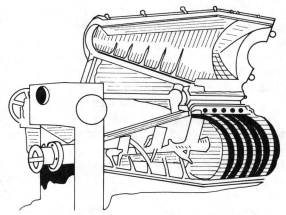

Fig. 72. External layout of mixing aggregate.

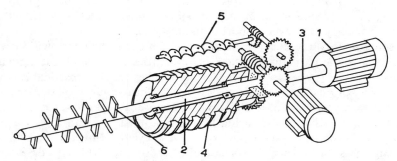

Fig. 73. Internal layout of mixing aggregate. 1. Motor. 2. Shaft. 3. Motor. 4. Kneading elements. 5. Mixing-worm. (Source: *Machine Handbook (Baking)*, 2nd edn, B. Schramm, VEB-Fachbuchverlag, Leipzig, 1978.)

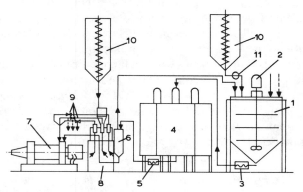

Fig. 74. Schematic of plant. 1. Mixing tank for full-sour preparation. 2. Agitator. 3. Spindle pump. 4. Fermentation tank. 5. Spindle pump. 6. Holding tank. 7. Mixing-aggregate. 8. Dosage-aggregate. 9. Magnetic valves. 10. Silos. 11. Automatic weigh-head. (Source: *Machine Handbook (Baking)*, 2nd edn, B. Schramm, VEB-Fachbuchverlag, Leipzig, 1978.)

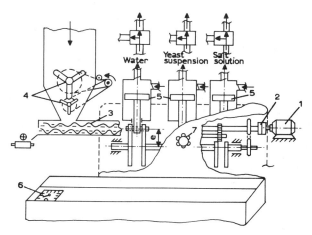

Fig. 75. Simplified schematic of dosage-aggregate. 1. Motor. 2. Coupling. 3. Feed-worm. 4. Three-bladed rotors. 5. Piston-pumps. 6. Rev-counter. 7. Hand-wheel. (Source: *Machine Handbook (Baking)*, 2nd edn, B. Schramm, VEB-Fachbuchverlag, Leipzig, 1978.)

This mixing-aggregate has the mixing and kneading sections separated. In the worm section, the dough components are assembled, and then pass immediately into the section where paddle-shaped resistance elements on both shaft and housing impart an intensive kneading action.

On actuation of a switch on the control-panel, the raw materials are fed into the first section. Control-lamps indicate the correct functioning of the aggregate. The first section of the mixer-housing is cylindrical, where the ingredients are assembled, and conical in shape where the dough is kneaded. The illustration shows the opened worm and kneading sections, together with the respective mixing-elements.

The mixing-intensity is determined by the shape of the chamber, the mixing-elements, and the cross-sectional area of the extrusion outlet of the aggregate. Mixing-intensity and throughput are geared to production requirements by means of a variable-speed motor, delivering between 600 and 2200 rpm, and arrangement of the resistance elements, which can be adjusted. Figures 72–75 illustrate the schematic layout of the Kontinua mixer, and the various upstream machine-aggregates which feed it (Fig. 75).

As an example of the application and flow of the Kontinua, the production of rye and mixed wheat/rye doughs is described. The process commences with the mixing-aggregate for the full-sour preparation (see (1), Fig. 74). The mature basic-sour is the starting-point of the fermentation process.

The appropriate amount of basic-sour, water at the correct temperature from an automatic mixing and measuring unit, and rye- or wheat/rye-flour blends are fed into the mixing-aggregate. These components are mixed with the agitator (2) (Fig. 74). A spindle-type pump (3) pumps the prepared full-sour into one of the six dough-ripening containers (4). After a period of 2 hours the full-sour is fully mature. It is then pumped by another spindle-type pump (5) into the holding-container (6). One-third of this mature full-sour is returned to the full-sour mixing-aggregate (1) where a full-sour to full-sour processing system is in operation, thus acting as starter (*Anstellgut*). The residual two-thirds proceeds into the mixing-aggregate (7).

Into the dosage-aggregate (8), the blended flour from the silo (10) is conveyed by a worm, and together with salt-solution, yeast-suspension and water is accurately metered, before being fed to the mixing-aggregate (7) through a hose-pipeline. Formulation control is adjustable over magnetic-valves, and allow samples to be taken during a production-run. A special feature of the dosage-aggregate, is that when the throughput rate is changed, the ratio of the ingredients in the formulation are automatically corrected to the preset-status. A cog-wheel-type pump is used to feed stale-breadcrumb-paste to the mixing-aggregate. This material, when hygienically prepared from stale bread from the same bakery, acts as a fermentation stimulator, and the pregelatinized polysaccharide content improves crumb-structure and bread shelf-life. The dosage-pumps servicing the yeast and salt solutions, and also the flour-worm and cog-wheel pump, can all be powered by one motor, which is of reversible-polarity and variable speed. The spindle-pump (5) is also powered by a motor of reversible polarity and of variable speed. Since, at that point in time when the one-third full-sour is pumped back into the dough-ripening container (4), the motor is automatically switched to a higher rpm, an adequate supply of sour-dough is always available for the mixing-aggregate. The actuation is triggered by a constant-level sensor in the holding-container (6), the relevant switch being mounted on the dosage-aggregate (8). The simplified illustration (Fig. 75) in schematic form shows the most important technical functions of the dosage-aggregate. The motor (1) drives the two dosage-worms (3) for the flour, the flour being discharged from the blending-silo by the two three-bladed rotors (4). The quantity of flour delivered depends directly on the rpm of the worms. The motor (1) also drives the cogged-wheel pumps for stale-breadcrumb-paste, and the sour-dough, in addition to the piston-pumps (5) for dispensing the liquid ingredients.

Although the illustration does not show all the intermediate moving parts, it is evident that, by adjusting the motor-rpm, the throughput is changed, and the ratio of the raw material dough components become automatically adjusted immediately. The rev-counter (6) indicates the chosen rpm of the feed-worms, and the synchronization of the component mechanisms, including the feed-rates of the whole dosage-aggregate. The piston-stroke, and hence the amount of liquid dispensed, depends on the eccentric adjustment, indicated by 'e'. This is adjustable with the hand-wheel (7) when the aggregate is at a standstill, and operates continuously when in motion.

Preparation of the salt-solution and yeast-suspension is carried out in a group of containers comprising one preparation and one holding-tank for each component. During the preparation of the solutions/suspensions, the preparation tank can be isolated from the holding-tank in each case by valve closure, thus avoiding concentration variations. Floats, which actuate valves in the lower holding-tanks, prevent overfilling. The liquids are withdrawn from the holding-tanks by pumps which are built into the dosage-aggregate.

Figures 72 and 73 show the external and internal layout of the Kontinua mixing-aggregate in schematic form. The motor (1) drives the shaft (2), which is fitted with shovel-like segments and pins at the dough extrusion end. A second motor, of lower power-rating, (3) drives the cylindrical-shaped mixing-element (4), which is machined to form a spiral-worm, thus ensuring an intensive mixing of the dough components. The small spiral-worm (5), also driven by motor (3) at the same rpm, serves to prevent residual raw materials from lodging in the space directly above the mixing-element. Both mixing and kneading sections of the aggregate are accessible for cleaning and repairs by means of the hinged-housing. The surface of the housing and mixing elements are coated with 'Teflon' (PTFE) to reduce friction due to adhesion. The high degree of mixing-intensity is achieved by the positioning of the resistance-elements on the inside of the conical-shaped housing, together with the shaft-segments. Since the rpm-rate can be adjusted from 600 to 2200, and the resistance-elements can be re-orientated at will, mixing-intensity and throughput can be adapted to production requirements within a wide range of tolerance. The mounting of the shaft only through the mixing-element results in a considerable load on the bearings, resulting in considerable wear with time. The space within the aggregate, relative to the interplay between the small spiral-worm and the larger mixing-element, give rise to uneven wear of the worms and cogs. Although the physical demand in operating the Kontinua is minimal, the responsibility in operation, maintenance, and servicing is considerable for an output of about 3·5 t/h of dough maximum. This type of plant also demands exact adjustments, and an operator with a trained sense of observation, response to unusual sounds from motors, transmissions, etc., temperature rises of motors or bearings, etc.

Dough consistency changes, and correct settings of the dosage-aggregate are among the most important functions to control, and during operation constant setting, re-setting and checking is essential. The high investment cost of continuous doughmaking aggregates justify constant maintenance and treatment with anti-corrosion coatings as and where necessary, to prevent valuation losses.

The main features of continuous doughmixing *v.* discontinuous or batch mixing are:

—reductions in the physical work requirement from personnel
—reduction in fermentation or floor-times
—increased productivity
—more uniformity in the production process

However, in most capitalist countries from the USA to Western Europe and the EC countries, these techniques have been applied with only limited success. In the USA, the Domaker and Amflow processes have not replaced the use of batch-mixing with high-speed horizontal mixers, although some sponge-and-dough users have converted to liquid pre-ferments to save on transport and floor-space when re-equipping or modernizing. In Western Europe, the following continuous mixing-aggregates. Domaker, Amflow, Buss-Ko System List, Baker Perkins-Ivarson, Simon-Strahmann and E. T. Oakes were all installed in various industrial bakeries with only limited success. In the UK, the Oakes continuous-mixer was the only unit which could be counted as a limited success. The 8D 30 range of mixer/modifiers gave throughputs of 523 to 3060 kg flour/h (1150 to 6740 lb/h), powered by a 30 hp motor, fan-cooled with variable speed. The internal design was similar to the Kontinua, a screw-feed mixing the ingredients, and a system of stators and rotor-blades completing the kneading and dough-development. Average shaft speeds were 200 rpm, but in practice were varied until optimum dough-consistency was obtained. The rpm being utilized to control energy-input levels.

The main advantage found with the continuously mixed dough was the improved accuracy in dough-dividing. This is due to the inertness of mechanically developed dough, combined with a constant head of dough in the divider-hopper, which gives a dough of relatively constant density. Changes in scaling-weights could be carried out whilst the divider was in operation.

Nevertheless, the many disadvantages of continuous-mixing, eventually phased out one system after another. Bakeries in the UK even preferred to use the CBP on a batch-system, although it was used successfully with the Oakes continuous 8D 30 range for throughputs of up to 3060 kg flour/h.

The important economic disadvantages of continuous-mixing aggregate, compared with batch-mixers, are the following:

(1)  The difficulty in making major changeovers rapidly from one type of bread to another, and from white-bread to wholemeal.

(2)  It was found necessary to use higher levels of yeast for the Oakes continuous mixer/modifier doughs compared with those for GBP batch-mixers. This was due to shorter first or intermediate-proof times, making use of belt-conveyors instead of conventional intermediate proofers. First proof-times were 30 s to 2 minutes instead of 15 to 20 minutes in a proofer.

(3)  Lower levels of dough water absorption were necessary with the continuous-mixer, amounting from 0·25 to 1·0 gallons/280 pounds of flour. At the time this was considered to be due to the shorter dough residence-time in the mixer, 1·25 to 1·75 minutes, compared with 2·5 to 3·25 minutes in batch CBP mixers.

Both high-speed (Tweedy-type) and Intensive (IMK 150-type) discontinuous (batch) mixers will produce doughs of similar characteristics to continuous dough development mixing-aggregates, assuming a short-time processing technique is used with no bulk-fermentation or floor-time. However, the finished loaf-texture in the case of the continuously mixed dough tends to be more spherical in porosity cross-section than its batch-mixer counterpart.

When using the Oakes continuous mixer for standard panned white bread, the procedure is to run the dough water into the liquor-tank until the level reaches the agitators, any dough improvers/conditioners being then added and dispersed. This is then followed by salt, the yeast being added in granular form for ease of handling after the salt has been dispersed. Other ingredients, such as malt-flour and milk-powder, must be pre-mixed to a slurry in a small amount of measured water, which must be included in the total dough absorption water. Any shortening is added directly into the mixing-aggregate, by a constant displacement pump.

On leaving the divider, the dough-pieces traverse a conical-moulder or rounder in the normal way. A much shortened first proof-time of either 30 s using open conveyor-belts, or 4 minutes in a conventional first proofer-cabinet, is appropriate. Since no vacuum can be applied with this mixing-aggregate, textural control must be exercised by variation of first proof, dough temperature 28 to 32°C, mixer-throughput, and configuration of the mixing/kneading elements.

In the socialist-community countries (Comecon), continuous mixing-aggregates have been applied with greater success, owing to the necessity of supplying large populations in cities and surrounding areas with bread of standardized quality, and good keeping-qualities. As already detailed, the USSR pioneered the technology of continuous mixing during the 1950s, and have built many types of mixing-aggregates for liquid pre-ferment dough-processing systems of diverse throughput capacities. A more recent continuous mixing-aggregate, exhibited at the Leipzig trade fair 1987, intended for average throughputs of 1300 kg/h with a power-rating of 3 kW, is the aggregate Type A2-ChTT from the Ukraine. In the other Comecon countries, including the GDR, owing to special trade arrangements, continuous mixing-aggregates are manufactured in Czechoslovakia, or Hungary. For the continuous manufacture of rye-flour and mixed wheat/rye breads, the complete doughmaking plant KVT 1000, KVT 1500 or KVT 1800 are used, corresponding to throughputs of 1000, 1500 and 1800 kg/h respectively (see Figs 76–78). The complete plant assembly is made up of the following aggregates, which can also be utilized for production work which is not of a 'line' nature. Two flour weigh-heads; salt-dissolving and water-tempering equipment; liquid dosage equipment; sour-dough beater/mixer; fermentation drum-shaped container which can be rotated, and is divided into 12 segments; sour-dough agitator; mixing-aggregate; stale-breadcrumb dosage unit. The KVT 1500 and 1800 have the fermentation-aggregate equipped with remix-beaters, thus providing a greater throughput within the same dimensions. The following is a description of the KVT 1000 (refer to Fig. 76).

The total throughput can vary within the range 850 to 1000 kg dough/h, at a total

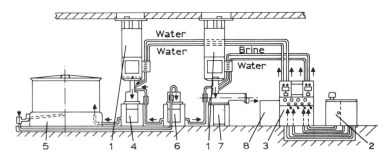

Fig. 76. KVT-1000 doughmaking plant (Czechoslovakia). 1. Flour-dosage units (2). 2. Salt-solution and water-tempering units. 3. Liquid-dosage equipment. 4. Sour-dough beater/mixer. Fermentation-container divided into 12 segments. 6. Sour-dough mixer. 7. Mixing-aggregate. 8. Stale-breadcrumb dosage unit. (Source: *Machine Handbook (Baking)*, 2nd edn, B. Schramm, VEB-Fachbuchverlag, Leipzig, 1978.)

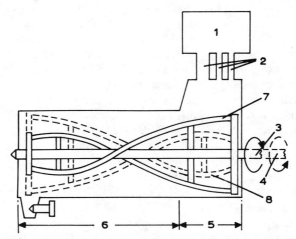

Fig. 77.   KVT-1000 mixing-aggregate. 1. Dosage-aggregate. 2. Raw-materials feed. 3. Front-shaft. 4. Back-shaft. 5. Mixing-zone. 6. Kneading-zone. 7. Wendel-type mixing-element (front). 8. Wendel-type mixing-element (back). (Source: *Basic Processes in Baking*, Hirsekorn and Nehrkorn, GDR.)

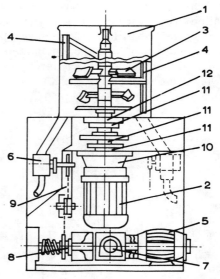

Fig. 78.   KVT-1000 sour-dough beater/mixer. 1. Container. 2 and 5. Motors. 3. Beater-rotor. 4. Scraper. 6. Cog-wheel (gear-type) pump. 7. Variable-speed transmission. 8. Torque-control device. 9. Chain drive. 10. Reduction gears. 11. Bearings. 12. Seal. (Source: *Machine Handbook (Baking)*, 2nd edn, B. Schramm, VEB-Fachbuchverlag, Leipzig, 1978.)

power-rating of 12 kW. The illustration (Fig. 76) shows the overall schematic layout of the most important aggregates which make up the plant. There are two flour-dosage units, one feeding the sour-dough beater/mixer, and the other the mixing-aggregate. The flour passes out of the silo over a sieve and holding-container into the top of the flour-dosage unit. A motor-driven agitator mixes the flour, and conveys it towards the outlet point, the movement of flour being controlled by two membrane-switches positioned at the bottom and top of the holding-container. The flour then proceeds through a vibrating-pipe into the weigh-head. The flour-level from the vibrator is determined by positioning of a sliding-vent. When the required flour weight approaches the preset amount, the frequency of vibration is reduced to give a fine-dosage. An electromagnet opens the bottom of the weighing-container, and the contents fall into a hopper and thence into the sour-dough beater/mixer, or mixing-aggregate. The weighing-mechanism of the flour-dosage unit depends on changing weights, and the effective-length of a chain, the effective weight being registered on a scale. The weights can be changed as required by actuation of a turret-switch, and a hand-operated wheel sets the effective chain-weight via a cogged-wheel for the fine weight-adjustment.

The salt-solution and water-tempering equipment consists of a tank with a grill, through which salt is added manually. Water is added and regulated by a valve, which, in turn, is actuated by a float. A circulation pump ensures intensive mixing and solubilization of the salt, the brine then passes into a holding-tank, from which it is conveyed by pipeline into the liquid-dosage aggregate. The water-tempering aggregate is divided into two parts, and involves the mixing of hot and cold water. The flow of both hot and cold water is controlled by the opening and closing of guillotine closures, which receive a signal of water-temperature from a resistance-thermometer. The water-tempering tank is connected to the dosage-aggregate by a pipeline, and the full-level controlled by a floating-valve. The main components of the dosage-aggregate are the piston-pumps, which can be adjusted from zero to maximum by changing the rpm. The control-panel for the dosage-aggregate also carries the regulation and control mechanisms for the brine and water-tempering equipment.

Sour-dough from the sour-dough mixer (7), and flour and water from the appropriate dosage equipment enter the cylindrical-shaped sour-dough beater/mixer (4). These components are intensively mixed by a 2·2 kW motor, which powers beater-rotors and a rotary-scraper (Fig. 78). The freshly mixed sour is then pumped by a cogged-wheel (gear-type) pump, powered separately by a motor, into the drum-shaped fermentation-container. The output of this pump can be variably regulated from 30 to 180 rpm. It is protected from overloading by a torque-limit coupling, and a motor-protection switch.

The Fermentation-container (Fig. 79) (1) is divided into 12 segments, which can be rotated inside the outer-mantle of the drum-shaped container. This rotation can be either continuous or periodic according to preset adjustment. These segments are sealed with rubber from top to bottom. The container is driven by a two-stage worm (2), powered by a special motor (3). The sour-dough is fed in at the base, and discharged by cogged-wheel (gear-type) pump (6) driven by the motor (4) via a

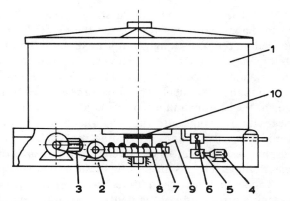

Fig. 79.   KVT-1000 fermentation container. 1. Container divided into 12 segments. 2. Worm transmission. 3 and 4. Motors. 5. Variable adjustable transmission. 6. Cogged-wheel (gear-type) pump. 7. Cams. 8. Driven rotating cogged-wheel. 9. Switch for drive-motor via timer. 10. Seal. (Source: *Machine Handbook (Baking)*, 2nd edn, B. Schramm, VEB-Fachbuchverlag, Leipzig, 1978.)

variable speed transmission (5). The control, under normal working-conditions (periodic rotation), is achieved by the cams (7) on the cog-wheel (8), which actuate a switch which controls the motor by means of a relay. After the preset time has elapsed, the motor is switched on again, and moves the segment-partitions, and with them the contents of the segments. This procedure is repeated continuously. In the base of the fermentation-container are outlets and inlets for loading and discharging the sour-dough. Each compartment or segment is filled with freshly prepared sour-dough. This is sucked out as ripened full-sour. When the sour-dough reaches the maximum level of a compartment, the fermentation-container rotates automatically, thus ensuring a continuous loading and discharge of sour-dough. The rotation period of the fermentation-container can be regulated up to a maximum period of 180 minutes, which is sufficient for maturity. The ripened full-sour is pumped by means of a cogged-wheel (gear-type) pump into the sour-dough mixer (6), which serves to degas the dough and make it more accurate to divide. About 40% of the gas-free full-sour is returned to the sour-dough beater/mixer (4) (Fig. 76), and 60% proceeds via another cogged-wheel (gear-type) pump to the mixing-aggregate (7). The two cogged-wheel (gear-type) pumps are powered by separate motors, with variable-speed drives, which allow a variable control of sour-dough quantity. How the division is carried out will depend on the process. The construction of the sour-dough mixer (6) (Fig. 76) is similar to the sour-dough beater/mixer (4). The detailed illustration of the sour-dough beater/mixer (Fig. 78) also shows the second pump and outlet as dotted-lines.

   The dough components sour-dough, flour, water, and salt-solution are continuously fed into the mixing-aggregate (Figs 77 and 80), together with yeast and stale-breadcrumb-paste. They become intensively mixed by two mixing-elements which rotate in opposite directions, being fitted with spiral-shaped blades similar to a grass-machine cylinder. The illustration shows the schematic design of the twin

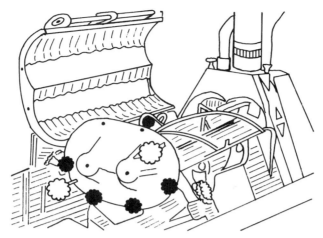

Fig. 80. KVT-1000 mixing-aggregate showing mixing-elements. (Source: *Machine Handbook (Baking)*, 2nd edn, B. Schramm, VEB-Fachbuchverlag, Leipzig, 1978.)

wendel-type elements. The main-shaft is driven by a motor via transmission onto a chain-drive, thence onto cog-wheels, which also drive the parallel mixing-element. The two kneading-rotors move the dough mass towards the outlet, being driven over cog-wheels mounted in the end-plate of the mixer-housing. The mixing process takes about 4 to 5 minutes. Access to the mixer internally is facilitated by the upper housing cover being hinged on the left-hand side. An amperometer measures the current consumed by the motor, which is proportional to the resistance which the dough offers to the mixing-elements. Hence, the ampere-reading is simultaneously related to mixing-intensity. This can be regulated by a variable-shutter, which adjusts the area of the outlet-orifice between zero and 75 cm² actuated by a screw. The output-curve of the KVT mixing-aggregate is used as an objective control for dough-yield, allowing corrections to be made, depending on throughput, mixing-time, etc. For example, mixer-outputs of 1·2–2·3 kW correspond to dough-yields of 164–176% (based on 100 parts of flour), depending on the particular flour-type being processed. For example, rye-flour Ia would given an average yield of 164%, whereas a rye-wholemeal would give an average yield of 170%.

The endless-band of mixed dough is transported by the band-conveyor to the divider and final moulding-machine, which allows the dough an interval of recovery and maturity-time. This can be varied by using a time-relay device, but is on average 25 to 30 minutes. The band-conveyor is about 1500 mm wide, being U-shaped in cross-section and covered.

After completion of mixing, the dough floor- or maturity-time is allowed. The duration of this process will vary, depending on the type of rye-flour and/or the amount of wheat-flour blended with it for mixed rye/wheat-flour doughs (*Mischbrot*), mixing-intensity, and sour-dough process utilized. Floor- or maturity-time and final-proof times are generally considered to be interdependent and complementary for most dough-processes. A longer floor-time requiring a

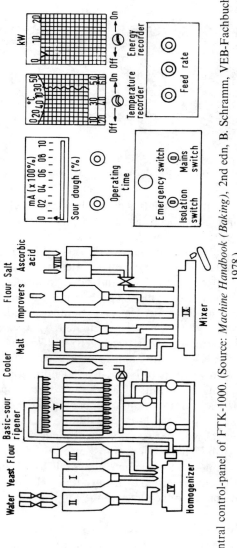

Fig. 81.   Central control-panel of FTK-1000. (Source: *Machine Handbook (Baking)*, 2nd edn, B. Schramm, VEB-Fachbuchverlag, Leipzig, 1978.)

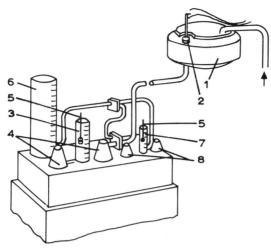

Fig. 82. Liquid dosage equipment. 1. Brine holding container. 2. Float. 3. Measuring cylinder. 4. Magnetic valve. 5. Electrode. 6. Holding cylinder. 7. Measuring cylinder. 8. Magnetic valve. (Source: *Machine Handbook (Baking)*, 2nd edn, B. Schramm, VEB-Fachbuchverlag, Leipzig, 1978.)

shortened final-proof time, and a short or no floor-time demanding a prolonged final-proof time. If this basic procedure is not followed, the quality of the resultant bread will be inferior. This topic will be discussed in greater detail in Part 2.

Another continuous doughmaking plant suitable for the preparation of wheat-flour and rye-flour doughs is the FTK 1000, manufactured in Hungary, which is used in other Comecon countries, including the GDR (see Figs 81–88). This is a well-engineered plant, capable of outputs of 1000 kg of rye-flour or 810 kg of wheat-flour dough/h. The whole plant is controlled from a central control-aggregate (Fig. 81), complemented by further control units on the individual aggregates. On the central control-aggregate are the following control-mechanisms in the form of grouped

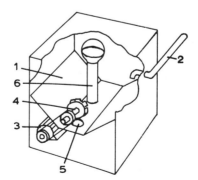

Fig. 83. Yeast-suspension tank. 1. Container. 2. Water-inlet pipe. 3. Motor. 4. Mixing propeller. 5. Outlet opening. 6. Measuring cylinder. (Source: *Machine Handbook (Baking)*, 2nd edn, B. Schramm, VEB-Fachbuchverlag, Leipzig, 1978.)

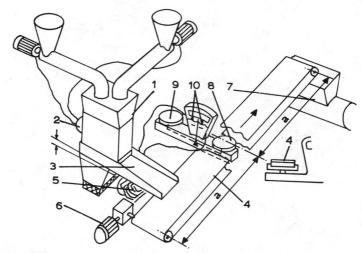

Fig. 84.   Flour-dosage equipment. 1. Holding container. 2. Maximum level device. 3. Vibrating-tray. 4. Conveyor-belt. 5. Electromagnet. 6. Motor. 7. Homogenizer. 8 and 9. Weight-sensitive discs of the scale. 10. Contacts. (Source: *Machine Handbook (Baking)*, 2nd edn, B. Schramm, VEB-Fachbuchverlag, Leipzig, 1978.)

building-elements: energy-distribution; programming; instrumentation; switches and preset facilities; malfunction warning-lights; facilities for switchover from manual to automatic-control; safety devices for the electrics; illuminated display of all functions. The automatic elements all use transistors, and thyristors for the control of the motors. The illustration of the central control-panel (Fig. 81)

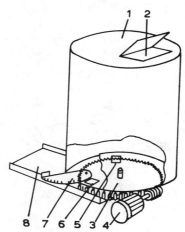

Fig. 85.   Dosage-container for malt-flour and improvers. 1. Container. 2. Filler-vent. 3. Disc. 4. Motor. 5. Window in disc. 6. Window in container floor. 7. Sliding vent. 8. Vibrating-tray. (Source: *Machine Handbook (Baking)*, 2nd edn, B. Schramm, VEB-Fachbuchverlag, Leipzig, 1978.)

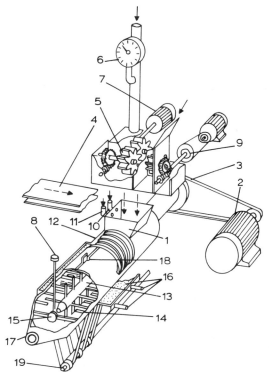

Fig. 86.   Mixing aggregate of the FTK-1000. 1. Feed hopper. 2. Motor. 3. V-shaped fan belt. 4. Conveyor band. 5. Dosage equipment. 6. Manometer. 7. Synchronized motor. 8. Multi-purpose measurement unit. 9. Geared belt drive. 10 and 11. Connection points for pipelines. 12. Mixer element. 13 and 14. Mixer pins. 15. Pressure plate. 16. Water cooling. 17. Extrusion orifice. 18. Rapid closure. 19. Shaft. (Source: *Machine Handbook (Baking)*, 2nd edn, B. Schramm, VEB-Fachbuchverlag, Leipzig, 1978.)

numbered in sequence indicates the construction and schematic layout of the FTK 1000 plant (Figs 81–88).

The addition of yeast is done manually (*c.* 35 kg/100 litres of water) in a square tank fitted with a mixing-propeller, driven by a motor over a belt-drive (Fig. 83). As result of the water circulation, the yeast forms a suspension, which is drawn off at the base, proceeding to the dosage-aggregate. A measuring cylinder within the tank also serves as an overflow in case of production interruptions. The water-tempering tank and suspension dosage-aggregate together make up one unit. In the water-tempering section cold and hot water are blended according to the preset temperature of a resistance thermometer, which controls two magnetic-valves one for the cold-water inlet, the other for the hot-water inlet. The maximum and minimum levels of the tank are maintained within fixed limits by a level-regulator consisting of three electrodes adjusted to give alarm-signals at preset levels. Depending on the water-level, the measuring-beakers collect a larger or smaller

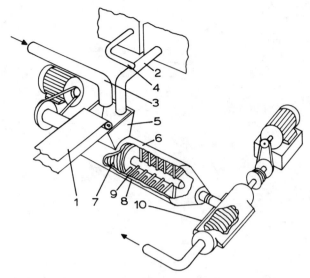

Fig. 87. Homogenizer and pump. 1. Conveyor-belt for flour. 2. Water pipeline. 3. Pipeline for sour-dough. 4. Pipeline for yeast suspension. 5. Hopper. 6. Homogenizer. 7. Feed-worm. 8. Resistance-elements. 9. Mixing-elements. 10. Worm-type pump. (Source: *Machine Handbook (Baking)*, 2nd edn, B. Schramm, VEB-Fachbuchverlag, Leipzig, 1978.)

volume of water, which is registered on a calibrated disc at the preset point. The tempering of the water is continuous, as water is withdrawn and replenished.

From the yeast-suspension-tank, the yeast-suspension is fed continuously through a pipeline into both a large holding-tank, and a smaller intermediate-tank within the holding-tank. The liquid-level is controlled within the intermediate-tank by an adjustable vent, which allows more or less liquid outflow. A rotating ladling-device, equipped with two measuring cylinders transfers the yeast-suspension into an outflow-gutter, which feeds into a pipeline, and from thence into the homogenizer. These dosage-aggregates are powered by a motor. The torque and rpm are transmitted to the rotating ladling-device via a worm and chain-drive onto cog-wheels, which drives the shaft of the ladling-device. This eliminates the variations in rpm often encountered with drive-belts.

The flour-dosage aggregate (Fig. 84) is fed by two flour-hoppers, one for rye-flour and the other for wheat-flour. The level of the holding-tank is maintained at the same level by a constant-level measuring device. This ensures that a constant amount of flour flows onto the vibrating delivery-tray, which is inclined at about 15 degrees, and at right-angles to an endless conveyor-belt, onto which the flour is continuously deposited. The vibrations of the delivery-tray are maintained in oscillation by an electromagnet, on the Wagner-hammer principle. The flour thus proceeds along the motor-driven conveyor-belt into the hopper of the homogenizer. The conveyor-belt moves over the measuring plate of a two-plate weighing scale, a weight of the required magnitude being placed on the other scale-plate. The scale

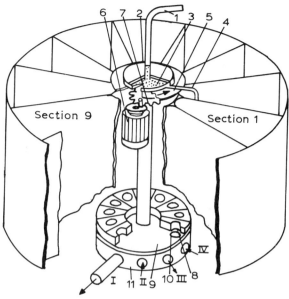

Fig. 88. Sour dough fermentation (ripening) container showing loading and emptying equipment. 1. Inlet pipeline. 2. Distributor. 3. Lower part of distributor. 4. Rotating pipe for filling Sections (1–12). 5. Guidepins. 6. Motor. 7. Maltese drive. 8. Outlet. 9. Rotating disc. 10. Outlet hole. 11. Holding chamber. I. Connection to cooler, thence to mixer. II and IV. Sour dough feedback openings. III. Connection to measuring tank, thence to homogenizer. (Source: *Machine Handbook (Baking)*, 2nd edn, B. Schramm, VEB-Fachbuchverlag, Leipzig, 1978.)

registers a centre-zero read-out when correctly metering the flour, if it deviates in a plus or minus direction, the electrical limit-contacts give an electrical-impulse to a regulator, which in turn adjusts the amplitude of vibration of the delivery-tray until the centre-zero position of the scale is restored.

The homogenizer (Fig. 87) is fed with the following raw materials: flour from the conveyor-belt; tempered-water from the water-mixing and dosage-aggregate via a pipeline; in the case of rye-flour dough preparation, one-third full-sour from the sour-dough maturing-container via the basic-sour measuring-container and pipeline; yeast-suspension from the dosage-aggregate via the pipeline for wheat-flour doughs. These components are fed into the homogenizer via a hopper, and conveyed by a feed-worm into the mixing-chamber, where the firmly mounted resistance-pins on the inside of the chamber-housing, coupled with the rotating pins, give the sour-dough or pre-ferment-sponge an intensive homogenization. The product is then conveyed by a screw-pump into the maturing-container. It reaches the container through a pipeline, passing first through a distributor-unit, from which the viscous dough mass is discharged through a pipe, which can be rotated to deliver into each of 12 sections of the maturing-container (Fig. 88). The rotation of the filler-pipe is achieved by two pegs fixed to a rotating-disc, which is driven by a motor via a system of gears and a maltese-cross, giving intermittent drive only. The control impulse for the function of the motor is housed in the central-control aggregate in the form of a timer-switch, its setting determining the maturing-time of the sour or pre-ferment in the 12 sections.

At the base of the cells, which taper towards the centre of the maturing-container, are outlet holes which lead into a rotating-disc, which is also driven by the intermittent maltese-cross system previously described. Within the rotating-disc there is an outlet so positioned, that the contents of the first section or cell to be filled, having completed a full-rotation and reached maturity is discharged as full-sour. The holding-tank positioned underneath collects the sour or pre-ferment, and it is then pumped through the connecting pipeline into the cooling-unit, and then into the mixing-aggregate, or through another pipeline into the measuring-unit and thence into the homogenizer (one-third of the total weight). The average sour-dough yield on entering the maturing container is 220% to 230%.

Two additional connecting pipelines to the holding-tank from the mixer holding-tank and the measuring unit allow excess sour-dough to be returned for recycling if the levels become too high. An additional pipeline from the holding-tank of the sour-dough maturing-container passes the mature full-sour via a motor-driven spiral-pump into a cooling-unit insulated with polyurethane-foam. This unit cools the sour-dough down to prevent overheating at the mixing stage. Providing the pressure in the dosage-aggregate above the mixing-aggregate is not excessive, the sour-dough is fed directly through the cooling-unit. If this is not the case a return-pipeline is provided for feedback of the sour-dough into the holding-tank. Another pipeline connection to the holding-tank allows the removal of one-third full-sour by means of a worm-type pump into a sour-dough measuring-unit. If this is over-filled, an additional return-pipeline is provided for its return to the distributor-unit at the top of the sour-dough maturing container. It is essential that the worm-type pumps do not operate under 'dry' conditions, therefore each of these suction-pipelines are provided with a funnel and tap, which must be filled with water.

The dosage-container for dough-conditioners/improvers and malt-flour (Fig. 85) is positioned above the vibrating delivery-tray of the flour-dosage aggregate, which feeds into the mixing-aggregate downstream. The additive-material is introduced into the cylindrical container through an opening at the top. At the base of the container is a round plate, which is driven from below by a motor over a worm and large diameter cogged-wheel. The rotating-plate, and the base of the container both have small apertures in them, the one in the base of the container being equipped with a shutter which can be preset at will. Depending on the shutter-setting, when the two apertures coincide during rotation, more or less material passes into the flour-stream. The liquid-dosage aggregates make provision for dispensing salt-solution, and ascorbic acid as required. The salt-solubilization equipment (Fig. 82), is similar in principle to the yeast-suspension aggregate, the salt-solution passing into a holding-reservoir. The liquid level is maintained by a float, which actuates the pump-motor, thus ensuring a constant discharge volume. The salt-solution passes through a pipeline into a measuring-cylinder over a magnetic-valve and into the mixing-aggregate. The amount required is adjusted with the help of an electrode.

The ascorbic acid is conveyed from a holding-cylinder and magnetic-valve into a mixing-chamber, where it is mixed with the salt-solution. The mixture then flows by pipeline into the inlet-hopper of the mixing-aggregate. The timing of the opening of the magnetic-valves is controlled by programmed timer-switch in the central-control unit. The inlet-hopper of the mixing-aggregate has a feed-intake which is driven by a motor via several V-shaped belt-drives.

The flour and flour-treatment additives pass over the weigher onto the conveyor-belt and then into the mixing-aggregate. Sour-dough or pre-ferment-sponge is fed through a cogged-wheel (gear-type) dosage unit, which functions as a pump, from the holding-container. Its static pressure being registered by a manometer in-line. The pump being driven by a variable-speed synchronized motor, the rpm of which can be increased or decreased, depending on a multipurpose measuring instrument (readings) mounted near the mixer-outlet above the buffer-plate. When the dough-consistency is too firm, the rpm is increased; thus more liquid in the form of sour-dough or pre-ferment is added to reach the desired consistency. Within the mixing-aggregate, the mixing-intensity is determined by the consistency of the dough. Mixing-intensities of $6.0$–$6.5$ W h/kg of dough can be achieved, the relationship between mixing-intensity and dough-consistency being the basis for automatic dough-processing. However, production interruptions cause problems, since dough-consistency only is used to control the addition of sour-dough or pre-ferment-sponge and not their respective fermentation-power. Fermentation-power of the same amount of liquid dough component will vary depending on the maturing-time in the fermentation-container. These variations, within a continuous automatic system lead to loss of production continuity and product quality, which are difficult to correct manually.

Stale-breadcrumb-paste or sponges can also be added using the same cogged-wheel, gear-type pumps, changes in the quantity added being accommodated by the use of a flat-belt drive, and various pulley diameters. The use of polyamide cogged-wheels ensures a long life for the dosage equipment and quiet operation. Both containers used for these pumps are constructed of rust-free steel sheeting, the pump-motors being controlled by measurement-probes.

The liquid dough components, salt-solution, mixed with ascorbic acid, and yeast-suspension are conveyed to the mixer-hopper inlet-connections by flexible hoses. All dough components are initially mixed by the mixing-element (12) of the mixing-aggregate (Fig. 86), and then moved into the kneading-section of the unit. Here, a shaft rotates with flat mixing-pins attached (13), and the mixer-housing is also fitted with these pins (14). The dough is transported by following dough from the mixing section, which makes the mixing process more intensive. A buffer-plate (15), helps the mixing process, and ensures that the multi-purpose measuring instrument (8) has steady-state conditions. Average mixing time is 40 s, and the unit is cooled with water (16). For cleaning, the mixing-barrel housing can be opened after removing the conical-shaped extrusion-end (17). When the screw (18) is removed, the housing can be removed for thorough cleaning. The illustration clearly shows the high degree of automation of this plant and the sophistication of the engineering. It demands a thorough knowledge of processes involved in dough preparation, the influence of raw materials, the construction and function of the various aggregates, and the essential measurement and control techniques employed for the various process-parameters.

The value of the plant and the large output capacity justify diligent operation and regular maintenance. The operator must constantly check all control functions. These include: the energy-input of the mixing-aggregate, changes in vibration amplitude of the flour-dosage delivery-tray and checking the flour feed-rate over the conveyor-belt, salt-solution dosage and bread flavour. Critical parameters must not be allowed to exceed or fall below defined tolerances, e.g. dough consistency depends on energy-input and the intensity with which it is imparted. If the dough is too firm, the energy-input can exceed 8 kW, in which case the operator will adjust down to 7 kW. However, the registered 7 kW corresponds to dough consistency which in reality should register 6 kW, although an energy-input of 6 kW is normally too soft for the divider/moulder to handle. Such measurement errors are the result of residence of residual dough in the conical exit part of the aggregate, giving excessively high viscosity, leading to false readings. Other aspects which require consideration are the fact that dough is a biologically active material, which is undergoing continuous change. This demands compensation-measures, therefore, the facility for manual operation of the control mechanisms is built into each aggregate. Production changeovers also demand readjustments of control functions, and often involve initial manual operation of certain aggregates. The working sequence of operation of the FTK 1000 doughmaking plant can be summarized as follows.

The sour-dough is prepared in the sour-dough mixing aggregate by adding flour, water, and matured full-sour to give a dough-yield of about 220% to 230%. The prepared sour is then pumped into the sour-dough fermentation/maturing container, which is divided into 12 compartments or segments. This allows a continuous supply of mature sour-dough, as one compartment is emptied and refilled with fresh sour-dough. The time interval between filling and emptying of a compartment can be varied, but is on average 180 minutes. Two-thirds of the mature full-sour being fed to the mixing-aggregate, the remaining one-third being retained for continuous culture for the next mixing. The mixing-aggregate is equipped with dosage aggregates for flour, dry additives, and liquid components such as, salt-solution, shortening, yeast-suspension, etc. An energy-input of 6 kW gives most types of dough an intensive-mixing, which parameter allows the control of dough consistency (variations). True consistency however, is controlled by the quantity of sour-dough fed into the final dough mixing-aggregate. Mixing-times vary from about 35 to 45 s, and throughputs from 800 to 1200 kg dough/h, depending on the type of dough. Where an assortment of rye- and wheat-flour bread is produced with bread-slicing involved, and frequent changeovers, two separate dough-lines are recommended. This allows the standstill time of the line to be minimized, especially when changing over from rye- to wheat-flour processing.

The FTK 1000 plant can be used to feed either continuous or discontinuous (batch) processing lines. From the dough mixing-aggregate, the dough passes via a belt-conveyor into a dough-divider, which gives an hourly output of 800 to 1000 pieces, depending on scaling weight (500 g to 2400 g units), and dough structure and density. The divided pieces then proceed through a rounder (conical or umbrella moulder), before proceeding through an adjustable long-roller, the Type S 76 manufactured in Hungary being frequently used in FTK 1000 continuous bread-lines. The divider, Type S 70, also from Hungary, working on the chamber/portioning exerts a protective-structuring of the dough-pieces.

The dough-pieces then pass through a synchronically regulated final pan-proofer fitted with pans transported on a travelling-conveyor. These are usually orientated 16 abreast covering a band-width of about 2100 mm. The dough-pieces from the long-roller pass through the pan-filling station, the pans having been first passed through the pan-release spray equipment. The filled pans then pass vertically into the proofing enclosure, maintained at 33–35°C and 70–80% RH. The proofed panned dough-pieces in the grouped/bound pans then proceed into the wire-band tunnel-oven. On leaving the oven, the loaves are depanned by a depanning-unit, and the empty pans conveyed over the top of the oven, the conveyor doubling as a pan-magazine for return to the oven-feed end.

Therefore, the most important aggregates comprising, and making possible, a continuous production system are the continuous dough-making system and the continuous tunnel-oven output capacity. The intermediate processing-line is synchronized accordingly. The depanned loaves are conveyed into the store-room for sorting, and then dispatched by a chain-linked conveyor into the cooling-plant. The cooling-plant consists of a labyrinth of straight and curved transport paths. Flexible wire-band conveyors are utilized, the straight and curved transport stretches being independently driven, with 10-metre straight stretches. Vertically the transport stretches are spiral, and horizontal and vertical diversions are provided. The aim of the cooling-system is to cool the bread down to 40°C, before it proceeds to the slicing-machine, travelling 500 m. In the GDR, for slicing warm, 40°C bread-crumb, the band slicing-machine is utilized, flat blades being used with a wave-form cutting profile. The blades are driven by a 4 kW motor over a shaft and rotating-drum. The average speed of the slicing blades is 10 m/s, and the average feed-time 0·5 m/s, with average hourly throughputs of 700 loaves. Most slice-bread lines utilize two such machines in parallel. The slice-portioning equipment is linked to the slicer and synchronized with it. The packaging-machines pack the apportioned number of slices in polyethylene (high-pressure type) foil, which is transparent, water and steam-proof, and temperature-stable from −50 up to +80°C. The packs are then often passed into a shrinking-tunnel, which involves a heat-treatment, which prolongs their mould-free shelf-life.

The continuous doughmaking system described, used on a three-shift basis, will provide about 1100 kg dough/h or 1000 kg bread/h, providing a selection of rye/wheat-flour mixed bread, wheat-flour bread and specialty-breads. Each shift requires eight personnel, using a floor surface area of about 550 to 700 m$^2$, depending on the length of the cooler. Total electro-energy usage is about 170 kW, and natural-gas consumption 90 m$^3$/h by the BN 50 wire-band oven, of 50 m$^2$

baking-surface. The final-proofer takes a maximum of 75 grouped-pans/h, each group consisting of 16 pans.

In the FRG, mechanization and automatic control of sour-dough acidification and perpetuation processes are described by D. Thörner of Düsseldorf (*Getreide Mehl und Brot*, **2**, 1982, pp. 44–47), patented as the 'Isernhäger natural dough-souring process'. This process can be applied to the production of a complete range of rye-flour bread from type 997 to type 1800. This system provides a uniform sour-dough supply for a whole working week (Monday to Saturday), involving a simple and flexible technique with reproducible quality. This process is divided into three distinct stages:

(I)   Preparation of the sour, involving mixing of the 'starter' with the flour and water.

(II)  Development and ripening of the sour, starting at 36°C, allowing an intensive production of lactic acid, after which an automatic cooler ensures the production of acetic acid at 20°C.

(III) The ripened sour-dough is then transferred through a hose pipeline, in paste-form into the dough-mixer.

For operation I, a tank fitted with a reversible stirrer is required. For operation II, an automatic levelling device is required to cope with overflowing when the stirrer is switched on, and the cooling-unit complete with tank-jacket and polystyrene-foam for temperature control.

For operation III, a dosage-pump and pipeline is necessary, which delivers the required amount of sour-dough (about 36 kg), as registered on the dial of the control-panel, directly into the mixer. The exact amount of sour-dough for a given flour mix is read-off from tables, when using the 'salt-sour process'. After all the sour-dough is used up, a new batch is prepared which again requires 18 hours to ripen. If the mature sour-dough cannot be processed within 6 hours, it is transferred to the starter-machine, and after 24 hours it flows into the dosage-tank, which contains enough sour-dough for the whole day's production. After the starter-machine has been emptied, it is immediately refilled, and the new sour-dough is ready for use within 24 hours. The combined plant is built for 1500 kg sour-dough, but can be stretched to 2000 kg if necessary. If two starter-machines are coupled with one dosage-tank, a daily-output of up to 4000 kg sour-dough is possible.

In Austria, A. Foramitti of Vienna (*Getreide Mehl und Brot*, **2**, 1982, pp. 47–50) describes a patented system for the preparation and storage of natural, progressive sour-dough, according to a joint design and construction by VNI and Riemelt of Frankfurt, FRG. This consists of a two-stage fermentation, a fermentation-tube and a tank. The first stage can be stored for 4 days under control, without excess acid build-up or damage to the fermentation organisms, allowing production stoppages at any time. The second stage gives a uniform quality sour for bread production. By selection of suitable process-parameters, the fermentation-power of the sour-dough yeasts can be maintained to give a balanced lactic:acetic acid content. The first plant of this type is in operation at VNI in Vienna, one of Europe's largest bakeries, producing up to 4·8 t sour-dough/h. The operation of this fully mechanized, continuous, automatic plant requires one person only. The conversion from manual sour-dough systems was achieved in 1979, with no bread-quality problems. This new process guarantees homogeneous sour-dough, and constant bread quality.

The growth in popularity of latter-day processes based on 'no dough-time', or 'short-time' doughs, whether totally mechanical or by a combination of chemical and physical means (activated dough development), is based on the elimination of bulk-fermentation or floor-time, and the use of a flour of somewhat lower quality. Since the flour does not have to respond to an extended floor-time, the quantity and quality of its protein and other flour components can be lower for the production of bread of an acceptable quality. However, these techniques are not acceptable for wheat-flour bread quality in many countries, since the flavour, and crust and crumb structure leaves much to be desired; and for rye-flour bread production the sour-dough is the most important criterion.

Furthermore, in the developing countries, the high capital cost of mixer/developers, and lack of power have precluded their adoption. In Central and South American countries, e.g. Peru, as well as in parts of Spain, south east Asia and some parts of Africa, alternative methods of developing doughs involving repeated passage of doughs through sheeting-rolls are not new. And, although it is labour-intensive because of manual operation in these countries, it is very efficient in terms of mechanical energy required to develop dough. For the production of bread from composite-flours, e.g. wheat/maize, wheat/sorghum, wheat/millet, wheat/triticale in various ratios, low-power pressure sheeting as a form of dough-development has proved very successful. Using a high-quality Canadian Hard Red spring (Manitou) wheat-flour as the control and carrier base-flour, Bushuk and Hulse (*Cereal Science Today*, **19**, 8, 1974, pp. 424–427), produced satisfactory bread with various composite flours containing up to about 20% of non-wheat flour. Acceptability is defined as having adequate minimal volume, and adequate grain, texture and colour. Higher levels of non-wheat flour could be acceptable, depending on regional bread quality criteria. The procedure used in the above studies were as follows: 60 parts flour, 4·0 parts yeast, 1·0 part salt, 2·5 parts sugar, 1·0 part shortening, 0·3 parts malt syrup (250 degrees Lintner), ammonium phosphate 0·1 parts, potassium bromate 1·5 mg, ascorbic acid 7·5 mg and 0·5 sodium stearoyl-2-lactylate, were mixed in a butter-churn with enough water to form a uniform batter. The remaining 40 parts flour were then added and the mass kneaded into a dough by hand, then placed in a bowl, covered with a damp-cloth to rest for 30 minutes. Ten small dough-pieces of equal weight were prepared, and developed by passing through sheeting-rolls as follows: 10 times through rolls set 7/36 inches apart, 10 times at 3/16 inches, and 10 times at 5/32 inches, the rolls being operated either manually or electrically. The dough-pieces were given a first proof of 10 minutes at 35°C and 80% RH, sheeted by passing between rolls set 11/32, 3/16 and 1/8 inches apart in that order, moulded and panned. Moulding was either by hand or mechanical moulder. Final-proof was for 55 minutes at 35°C and 80% RH. Baking at 221°C for 25 minutes. Loaf volumes were determined by rapeseed displacement. After cooling, the loaves were sliced and examined for crumb characteristics.

The same workers compared the baking results obtained from five different flours by the method described with the same flours processed by the Chorleywood Bread Process (CBP), and the remix-procedure (Irvine and McMullan, 'The remix baking-test', *Cereal Chem.*, **37** 603, 1960). They concluded that simple dough development by sheeting, without the use of any high-speed mixers, gave results comparable with those obtained with the CBP procedure.

Further work by Kilborn and Tipples at the Grain Research Laboratory, Winnipeg, Canada (*Cereal Chemistry*, **51**, 5, 1974, pp. 648–657), using flour milled from an average sample of CWRS 1 wheat of protein content 12·5% (14·0% moisture), ash 0·43%, and starch-damage (Farrand) 33 units, confirmed the efficient work done by sheeting-rolls at much less energy-input than high-speed mixers. The procedure used in this case was as follows: flour 600 parts, water 64 parts, sugar 2·5 parts, salt 1·0 parts, shortening 1·5 parts, yeast 3·0 parts, malt syrup (250 degrees Lintner) 0·3 parts, ammonium phosphate (monobasic) 0·1 part, ascorbic acid 75 ppm, and potassium bromate 45 ppm, were premixed for 1·5 minutes at 45 rpm in the GRL-1000 laboratory pin-mixer (*Cereal Chemistry*, **51**, 500, 1974), at 74°F. The 1034 g of dough obtained were then divided as follows: one half (517 g) of the premixed dough (minimum mixing-capacity) was returned to the mixer for further mixing at 105 rpm until a gross energy-input of 6·4 W h/kg (or 5·8 W h/kg net) were registered by the GRL energy-input meter. The mixer-developed dough was then divided into three pieces, and dough make-up and processing procedure for all samples according to the following schedule.

After development and scaling, doughs were rounded, and given 25 minutes intermediate-proof at 95°F, mechanically moulded, panned and proofed 55 minutes at 95°F, followed by baking 25 minutes at 420°F. The second half of the dough was sub-divided into three pieces, one piece being made-up, proofed and baked without any further mixing or development. The remaining two pieces of premixed-dough were passed in the same direction first through sheeting-rolls set at 7/32 inch gap, and then through rolls set at 5/32 inch gap. The pieces were then folded in half, rotated through 90 degrees, and re-passed through the same roll-settings as before. Thus representing four sheetings and one fold at this stage. Subsequent folds were followed by two sheetings, and the total number of sheetings calculated by 2 × number of folds + 2. Up to 15–20 folds produced the best results, and the total time taken to work the dough was about 45 s, expending a net energy level of only 0·78 W h/kg, which is approximately 15% of that required for peak development using a conventional pin-mixer. The rate of energy-input by the sheeting-rolls at about 0·5 W h/min was only slightly less than that imparted to the dough by the conventional GRL-1000 pin-mixer rotating at 105 rpm. This implies that the type of work done on dough with sheeting-rolls is both useful and efficient. Doughs given more than 20 folds begin to show entrapped bubbles of air, which increase as sheeting continues. Over-sheeted doughs show small gas-holes in the bread crumb, and the symptoms of an under-oxidized dough with sharply defined pan-crust edges.

A 20-fold treatment produces over a million layers in theory, each being about 0·0003 $\mu$m in thickness (compared with smallest starch-granules in flour at about 5 $\mu$m diameter). Furthermore, since the dough is rotated through 90 degrees after each fold, the layers tend to become cross-hatched, giving rise to a two-dimensional, rather than a unidirectional network of sheet-like structures within the dough.

Compared with the 6·4 W h/kg dough required for acceptable bread using the GRL pin-mixer, the energy level of approximately 1·0 W h/kg expended by the sheeting-procedure in producing bread equal to the optimum bread from the mixer, represents an energy saving of the order of 70%.

The aim of this technique is to impart a maximum of useful work on the dough within the shortest time interval without tearing the dough-surface.

These experiments have demonstrated the feasibility of producing 'no-bulk-fermentation' bread without a dough-mixer. However, unless the sheeting-operation could be mechanized and automated, this technique could not be considered a commercial proposition in any industrialized country. Sheeting doughs through large steel-rollers or 'dough-brakes' is common practice in countries where adequate manual labour is available, e.g. the Philippines.

Owing to the more efficient utilization of energy by sheeting compared with high-energy mixing for mechanical dough development, further work was carried out by Kilborn, Tweed and Tipples (*Bakers Digest*, **55**, 1981, pp. 18–31). In these experiments, a line of rolls (four sets) were positioned between a continuous-premixer, imparting about 25% of the required energy-input, and a dough-divider, thus allowing more process control to be exercised.

The whole field of conventional mixing of bread-dough to produce bread and the efficient use of energy in so doing merits considerable research. The importance of the fundamental dynamics of the mixing-process has already been discussed in depth, and their effect on dough rheological behaviour for process-control and optimization. Nevertheless, these studies concern measurements made in the rotational mode, involving the effects of shear-forces. The effects of controlled pressure, folding to form a sheet-like structure in two dimensions (cross-hatching), followed by periods of proof-relaxation, involves other modes of dynamic deformation.

The extremely diverse systems of mixer design described in this section clearly demonstrate that new modifications are constantly being devised to improve mixing-efficiency. The application of computers, in the form of CAD has proved a useful tool in this field. It is possible that the time has come for a radical re-think concerning the technology of breadmaking, involving the optimal integration of the various process-phases, starting with raw materials, then mixing, through dough-maturation (fermentation and proofing) to the application of the various forms of energy to the baking-process itself. Although the commercial objective of industrial-baking is to produce an acceptable quality product with the minimal input of energy, raw material and labour resources, the author ventures to suggest that apart from product-freshness, the dough-maturation phases merit special attention to produce a flavour and structure which will increase the desirability of bread as a basic food commodity. As a cereal-based product, bread must compete with meat and meat-products, and vegetable and vegetable-products in the market-place.

In recent years, the emphasis has been on short-time breadmaking processes, eliminating bulk-fermentation and utilizing instead, either mixers of high speed and high mixing efficiency to satisfy the optimum mixing-intensity, or combinations of mechanical and chemical dough-development additives (where legally permitted). In countries where wheat-flour-, panned- or variety-breads predominate in the market, these processes have resulted in an appreciable loss of bread quality in terms of crumb and crust structure, flavour, and general desirability of bread as a dietary-component. Bread sales statistics have reflected these trends, and given rise to some concern in large industrial baking chains.

# PART 2
## Fermentation of Wheat- and Rye-Flour Doughs

# 2.1   Introduction

The application of biotechnology to the production of baked products has been practised for centuries, using yeast and sour-dough cultures. However, with the industrialization of breadmaking, specific microbiological cultures have been selected and maintained systematically for the manufacture of wheat- and rye-flour bread. Their selection and application in mixed or single culture (yeast for wheat-flour bread) is based on the activity of their enzyme-systems.

The baker's yeast used for fermenting, and the gas-texturization of most wheat-flour doughs, and shortened rye-flour sour-dough systems is a 'top-fermenting' race of yeast, being cells of the type *Saccharomyces cerevisiae*. This yeast can metabolize the following mono- and disaccharides: glucose, galactose, sucrose, maltose and one-third raffinose. The pentoses cannot be utilized, since the requisite enzyme is not present in the cell. Also, starch cannot be degraded by the yeast-cell, since it contains no amylases. Therefore, initial breakdown into disaccharides by the flour amylases or amylase-supplements is necessary. Salt and excessive build-up of alcohol inhibit yeast activity, but a low acidity-index stimulates gassing, its optimal pH range being 5·0 to 5·5. The optimal temperature range for growth and reproduction is from 24 to 30°C, but vitality diminishes above 35°C.

The commonest form of baker's yeast is pressed-yeast, which has a moisture content between 73% and 75%, of which some 20% is extracellular. According to moisture content, and cell-size, 1 g of pressed yeast should contain from $8 \times 10^9$ to $13 \times 10^9$ cells. In the pressed form the cells have no nutrient and must survive on their cell contents, a process known as 'autolysis'. The relative speed of this process depends on the activity of the cell protease, and determines its shelf-life. Therefore, the storage temperature must be below 35°C, and protection from microbial contamination ensured.

By extruding the pressed-, or cake-yeast as it is sometimes called, through a perforated-steel plate of 0·5 to 3·5 mm diameter openings and drying, 'active dry yeast' (ADY) is produced. The extruded yeast strands are broken into lengths of about 1·5 mm to 3·0 mm, and dried on a tunnel- or belt-type dryer at about 38°C to a moisture of 8–10%. The dried material can be coated with a protective glaze, packed in air-tight packaging or gas-packed, thus prolonging its shelf-life for 1 year or more. ADY requires rehydration before application in dough, which can be a labour-intensive operation in a mechanized bakery. Rehydration is ideally performed in an atmosphere of steam, otherwise in water at 40°C with the addition of sugar. If water alone is used, about 30% of the cellular components as dry matter become leached out, and gassing-power of the yeast is reduced. Temperatures in excess of 45°C result in progressive inactivation of the yeast cells.

In recent years, improved drying techniques have been applied to the preparation of dry-yeast. The introduction of the newer dry-yeast products, called instant active dry yeast (IADY) has overcome the rehydration problem, and IADY can be directly incorporated into doughs without prior treatment.

# 2.2 Industrial Propagation and Production of Yeast for the Baking Industry

Work done between 1915 and 1920 by the Danish scientist Sak, and the German scientist Hyduck led independently to the introduction of a process in which sugar solution was fed slowly to an aerated yeast-suspension. This procedure, known as the '*Zulaufverfahren*' or 'incremental-feed process', became widely accepted by the fermentation industry. However, the source of sugar changed from the traditional grain-mash to corn (maize), malt, malt-sprouts and then molasses sugar during the 1920–1930 decade. More recent research by Belova, L. D., and co-workers at the VN II ChP (All-Union Research Institute for the Baking Industry) in Moscow, have shown that starch-syrup derived from the acid hydrolysis of potato or wheat starch, can also be used as a substitute for molasses in the production of baker's yeast. However, the culture medium is supplemented with a source of biotin, calcium pantothenate and trace elements, giving biomass yields within the range 60·7–62·8%. The yeast showed good stability with high maltase and zymase activity, giving bread of high quality.

At present, compressed-yeast (pressed-yeast or cake-yeast) is the form preferred by large bakeries, although many institutional bakeries and consumers have taken to ADY and bulk liquid yeast.

For comprehensive reviews describing the progress of the yeast industry, the reader is referred to the following: '*The Yeasts*', by S. Burrows, in A. H. Rose and J. S. Harrison (Eds), Vol. 3, Academic Press, New York, 1970, pp. 349–420; *Progress in Industrial Microbiology*, by J. S. Harrison, in D. J. D. Hockenhall (Ed.), Vol. 10, Elsevier Publishing Company, 1971, pp. 162–163; '*Microbial Technology*' by H. J. Peppler, in D. Perlman and H. J. Peppler (Eds), Vol. 1, Academic Press, New York, 1979, pp. 157–185; *Prescott and Dunn's Industrial Microbiology*, by G. Reed, in G. Reed (Ed.) 4th edn, AVI Publishing Company, Westport, CT, 1982, pp. 593–633; N. B. Trivedi, E. J. Cooper, and B. L. Bruinsma, *Food Technology*, **38** (1984); N. B. Trivedi and G. Jacobson, 'Recent advances in Bakers yeast'. In *Microorganisms in the Production of Food*, ed. M. R. Adams. Elsevier Applied Science, 1986, pp. 45–71.

In common with most fermentation processes, seed culture and propagation is of fundamental importance to yeast manufacture. The initial stages of the process are carried out in a set batch-mode in sterilizable pressurized fermenters to minimize microbial contamination. Fermenters using simple sparged-air by bubble-column are more efficient in terms of oxygen transfer and economy than mechanically agitated fermenters. The final fermentation stage is carried out in a sanitized vessel for the trade. The process is a batch incremental-feeding procedure, in which sugar is

*Handbook of breadmaking technology*

progressively supplied at a concentration of less than 0·1% to avoid the 'Crabtree-effect' (excess sugar accumulation, resulting in ethanol formation and lower yields). Trade fermentation processes are run at 30°C with the pH adjusted at 4·5 ± 0·5, for 12–18 hours depending on the fermenter design and desired final cell-concentration. Aeration must be maintained at a high level, and never below 1 vvm. The fermentation process is rigidly controlled by microprocessors/computers coupled with accurate instrumention. Instead of basing the rate of fermentation on yeast-cell bud formation, recent trends utilize either a measurement of the ethanol present in the exit-gas of the fermenter, or the dissolved oxygen level which is limited to about 12% of saturation. If the latter falls below the set level, the molasses feed is automatically reduced until the dissolved oxygen level is restored to the target value. Monitoring the respiratory quotient (RQ), which is defined as the rate of carbon dioxide evolution divided by the oxygen uptake, is also used to control feeding. For an efficient fermentation, the oxygen uptake and carbon dioxide evolution should be evenly balanced, giving an RQ of between 0·95 and 1·04. This should give a yield of 0·5 g yeast-solids/gram sugar consumed at RQ 1·04 (Wang *et al.*, *Biotechnology and Bioengineering*, **19** (1977), 69–86). However, of more importance to the baking industry is the leavening activity of the resultant yeast, rather than maximization of biomass. Leavening power depends on efficient growth, which in turn depends on an optimal balance of nitrogen and phosphate content of the cells. Molasses is deficient in phosphates, therefore, supplementation with mono-, di, or tri-ammonium phosphate, which supply both phosphate and nitrogen is necessary. Other important growth supplements are: biotin, pantothenic acid, inositol, thiamine, pyridoxine, niacin, and the inorganic trace elements Na, Mg, Zn, K, Fe and Cu (present in molasses).

The propagation of baker's yeast is a highly aerobic process, requiring for aeration 1 volume of air per 1 volume of media per minute. The biochemical equation reflecting the quantitative conversion of sugar into yeast-cell solids is:

200 parts sucrose + 132 parts ammonia + 100·44 parts oxygen + 705 parts mineral ash → 100 parts yeast-cell solids + 140·14 parts carbon dioxide + 78·12 parts water.

which corresponds to approximately 22·7% conversion to yeast-cell solids, based on total raw material weight input.

Beet molasses gives the product a light colour, but the nitrogenous component betain is not utilized by the yeast.

Yeast quality, in terms of leavening ability depends on the strain. Whereas earlier hybridization was used to improve yeast strains, protoplast fusion and rDNA are the main techniques currently used in their construction.

P. Gélinas, G. Fiset, A. Le Duy, and J. Goulet in *Applied and Environmental Microbiology*, **38** (1989) 2453–59 have studied the effects of conditions of growth and trehalose synthesis on the cryotolerance of baker's yeast in frozen-doughs. They concluded that fed-batch cultures were much better than batch cultures, and that strong aeration enhanced cryoresistance in both cases, within the freezing-rate range, 1–56% min$^{-1}$. Loss of cell viability in frozen-doughs was related to the dissolved-oxygen deficit duration during fed-batch growth. Strongly aerobic fed-batch cultures, when grown at a lower average specific rate ($\mu = 0.088\,h^{-1}$ $v$. $0.117\,h^{-1}$) also exhibited greater trehalose synthesis and improved frozen dough stability. Lower growth temperatures ($20°C$ $v$. $30°C$), coupled with a dissolved oxygen deficit, reduces both fed-batch-grown yeast cryoresistance and trehalose content. This cryoprotective effect of trehalose in baker's yeast is, however, extremely sensitive to momentary removal of excess oxygen in the fed-batch growth medium.

## 2.2.1 YEAST PHYSIOLOGY

Baker's yeast strains of *S. cerevisiae* are single-celled fungi, which reproduce by budding, bearing their sexual spores in an 'ascus' or sac, which is derived from the cell wall. The organisms have two mating-types, classified as 'heterothallic' or 'homothallic'. Homothallic strains do not have stable haploid-cell lines, and after germination of the homothallic-spores, the cells of one type can convert to the opposite mating-type (Fig. 89). The mixed population of cells which accumulate will mate and re-establish the diploid cell-line. Heterothallic strains lack the mating type conversion system; stable haploid heterothallic spore-clones, capable of being isolated, have been constructed from mated haploid cell lines for commercial application.

The rate of cell division tends to be regulated by cell growth-rate, cell-size probably depending on cell-protein concentration. On attaining the predetermined size, the cell-division cycle begins, which is more rapid than that of cell maturity. The cell of the offspring is smaller than that of the parent, and requires a longer growth period than the parent to reach maturity and cell-division status. However, trade yeast fermentations are well synchronized, since substrate or feed schedules are pitched accordingly. Within about 8 hours, at least two budding-crops occur, but

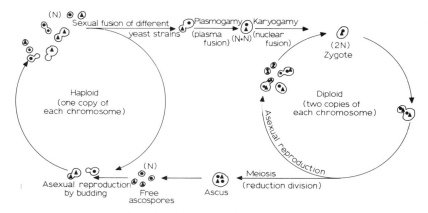

Fig. 89. Life cycle of baker's yeast.

during the ninth hour the maturation phase takes place in which cell-mass is built-up as trihalose and glycogen.

Yeast physiology is important to its success as an industrial microorganism, strains varying in their ability to convert sugar into biomass. Additionally, the selected organisms must exhibit a competitive performance in the market place. Major producers must meet clients' requirements in terms of yeast activity, stability and product consistency, quite apart from distribution and customer service. Specific aspects of yeast physiology which have some bearing on competitiveness are: carbohydrate metabolic pathways, the repression of catabolism, and membrane synthesis.

**Carbohydrate Metabolism**

Both cane and beet molasses are the commonest industrial substrates for yeast production. As waste products of the sugar industry, their composition is variable, but they contain about 50% residual sucrose, up to 5% raffinose, and the inversion products fructose and glucose. Sucrose undergoes extracellular hydrolysis to glucose and fructose by the enzyme invertase. Raffinose acts as a poor substrate for invertase, and becomes hydrolysed randomly to melibiose and fructose. Melibiose is not a substrate for baker's strains of *S. cerevisiae*. Invertase can be encoded by any of the six genes of the *SUC* polymeric system, in the presence of both a cytoplasmic and a periplasmic enzyme. The exoenzyme is a glycolysated octomer of molecular weight 800 000 daltons. The gene is sequenced, and encodes both a 1·9 kbp (kilo base pair), and a 1·8 kbp mRNA, the larger 1·9 kbp mRNA is regulated by glucose and encodes the exo-invertase. This differential regulation ensures a more or less stable endo-invertase concentration, while the exoenzyme activity can vary greatly, depending on the conditions of culture. The products of the actions of invertase, glucose and fructose, are passed into the cell by diffusion, but a phosphorylation step is considered to be involved. The number of *SUC* genes in industrial yeast strains is unknown, but gene-dosage effects on activity have been reported by Grossman and Zimmermann in *Molec. Gen. Genet.*, **175** (1979), 223–229.

Maltase activity is important in lean dough fermentation, and could be linked to cell osmosis. Maltose metabolism is also controlled by a polymeric gene system, for which, five unlinked genes are identified. However, as with the *SUC* system, a given strain does not carry all of these genes. Cohen *et al.* (*Mol. Gen. Genet.*, **196** (1984), 208–216) consider that three factors govern maltose fermentation: a regulatory gene, maltase itself (alpha-glucosidase), and maltose permease. The permease and maltase being induced by the presence of maltose, and repressed by glucose (acting as catabolite). Added glucose results in: a decrease in enzyme activity, followed by enzyme inactivation, and eventual repression of enzyme synthesis.

Baker's yeast strains are also capable of fermenting galactose, and although the use of galactose is not currently of industrial importance, the ability of a particular strain to use this sugar could become significant. The ability to utilize alpha-methyl glucoside, is also of interest, since the cleavage enzyme involved also shows activity with the alpha 1–6 bond of isomaltose, a sugar often present in lean dough systems.

The hexose molecule enters the glycolytic pathway intracellularly by phosphorylation with the help of two hexokinases and a glucokinase. Glucokinase is glucose specific, whereas hexokinase will phosphorylate glucose, fructose or mannose. One hexokinase, named hexokinase P II or hexokinase B, according to Entain and Frohlich (*J. Bacteriol.*, **158** (1984), 29–35), is multifunctional involving catabolic repression. According to Carlson *et al.* (*Molec. Cell Biol.*, **3** (1983), 439–447), the high concentration of these enzymes amount to 30–65% of total cellular protein. Glycolytic rates are controlled by the enzymes phosphofructokinase and pyruvatekinase. Phosphofructokinase (PFK) is partially regulated by fructose-6-phosphate and ATP. Pyruvatekinase is activated by fructose 1,6-biphosphate, the product of the PFK-reaction. The enzyme pyruvate dehydrogenase feeds carbon into the tricarboxylic acid cycle (TCA) by converting pyruvate into acetyl coenzyme-A and carbon dioxide. However, the enzyme pyruvate dehydrogenase becomes repressed under the conditions of fermentation, and the pyruvate becomes decarboxylated, forming acetaldehyde in the presence of the enzyme pyruvate decarboxylase. Acetaldehyde is then reduced to ethanol by one of the three alcohol dehydrogenases present in yeasts.

This pathway is the natural metabolic pathway of yeast, but during industrial manufacture, this route is suppressed by feeding sugar at appropriate rates and regular aeration. The build-up of any ethanol is assimilated by the yeast via alcohol dehydrogenase, which is glucose repressible and induced by oxygen, forming acetaldehyde. Acetaldehyde then enters the TCA cycle with the help of acetaldehyde dehydrogenase and acetyl coenzyme-A synthetase. This suppression of alcoholic fermentation in the presence of oxygen is called the 'Pasteur effect', and can be quantified by comparing the hexose fermented under anaerobic and aerobic conditions. The resulting coefficient, known as the 'Meyerhof coefficient', is expressed as the difference in the amount of carbon dioxide liberated, divided by the quantity of oxygen made available.

This Meyerhof coefficient, is a measure of the inhibition of alcohol formation, in the presence of oxygen, and varies with type, and strain. *S. cerevisiae*, in the presence of oxygen, inhibits alcoholic fermentation by about 66% compared with anaerobic conditions.

The main storage carbohydrate of yeast is trehalose, accounting for up to 20% of cell dry weight, but it also stores glycogen. The function of these compounds is to supply energy to the cell during periods of adaptive growth, e.g. oxidation, sporulation, germination, starvation, change of growth conditions. Storage carbohydrates are built up as a result of nitrogen, sulphur or phosphate deficiency, and when carbon is deficient, glycogen is the first energy source to be drawn on. Cell vitality does not fall unless the trehalose level falls to about 5% of cell dry-weight. Trehalose accumulation is associated with the presence of maltose in the substrate.

High trehalose content is an advantage in pressed-yeast, since high moisture content (70–72%), coupled with low storage temperatures does not completely prevent metabolism, but in the case of dried-yeasts this is less critical. Storage carbohydrates are necessary for both cell survival, and adaptation to the various dough systems in which it is incorporated. In addition, both the storage-carbohydrates trehalose and glycerol have a role in the osmo-tolerance of yeasts.

Osmo-tolerance, or tolerance to diverse osmotic pressures within doughs is demanded by the various sugar concentrations added to doughs. The precise interrelationship and functions of trehalose and glycerol are not yet fully clarified, but research indicates that they both contribute to the osmo-stability of the yeast cell. In the case of dry-yeast products, trehalose stabilizes the cell membrane during the dehydration process, thus preventing excess damage due to desiccation.

## Repression of Catabolism

This is the term applied to the inhibition of the various metabolic pathways by the various fermentable sugars. Of these, glucose is the most repressive, and its presence inhibits the enzyme systems necessary for the metabolism of other sugars. An identification of the critical glucose concentration at which it is converted to ethanol and carbon dioxide in the presence of excess oxygen is an important commercial parameter. However, values obtained vary with strain, nitrogen content, medium phosphate and vitamin supplementation levels. Toda and Yale (*Biotechnol. and Bioeng.*, **21** (1979), 487–502) established ranges of 35 to 280 mg/litre. Once the critical concentration is exceeded, fermentative growth proceeds irrespective of aeration levels. Nevertheless, fermentation behaviour is also influenced by other factors.

The activity of the respiratory enzyme systems, including the TCA cycle, and glyoxylate by-pass is determined by the rate of cell growth; and apart from the hexose galactose, all sugars which support rapid growth rates give rise to a maximum repression of the oxidative-enzyme systems. Without repressive levels of sugar, oxidative enzyme systems become induced by the presence of oxygen. Without the presence of oxygen, isocitrate lyase cannot be synthesized, and the mitochondrial enzyme malate dehydrogenase activity is doubled when the oxygen content of the air supply increases from zero to 0·2% only. The majority of oxidative enzymes undergo maximum induction at oxygen levels of about 20% in the air supply, but levels of 50% and above often inhibit enzyme synthesis in many cases.

## Membrane Structure

Membrane structure and synthesis also contribute to yeast activity. The cytoplasmic membrane acts as an osmotic barrier, controlling the entry and exit of solubilized material for the cell. Apart from its physical function in holding the cellular protoplast, it could also act as a foundation for cell-wall synthesis. The membrane is made up of sterols incorporated into a lipid double-layer, which includes phosphatidyl-choline, -ethanolamine, -inositol, and -serine. The main phospholipid being phosphatidyl-choline, with the predominant fatty acids in the lipid molecules being the C-16 and C-18 acids: palmitic, stearic, and derivatives, palmitoleic and oleic acids.

Membrane composition will dictate how the yeast type or strain reacts towards drying and rehydration of activated dry-yeast, the shelf-life of compressed yeasts, and leavening ability in the various dough-systems. Actual membrane composition is determined by: growth-temperature, level of aeration, substrate composition and ethanol concentration within it.

## 2.2.2 IMPROVEMENT OF INDUSTRIAL YEASTS

The industrial yeast geneticist is confronted with diverse problems in attempting to construct improved baker's yeast strains. In the first place, there is no detailed knowledge concerning the genes responsible for the specific properties of industrial yeast strains. Although certain conclusions can be drawn from biochemical and genetic studies at laboratory level, the reaction of various industrial strains in applications cannot be correlated with data concerning gene copy number and allele of the gene (change in its internal organization due to mutation, yielding another form of the same gene), e.g. inducible or non-repressive enzyme-synthesis. Industrial strains usually show poor sporulation, the resulting spores having low viability. Therefore the genetic content of such strains is difficult to recover in meiotic spore clones both for study or breeding new hybrid yeasts.

Additionally, no reliable screening tests exist at petri-dish level capable of evaluation of the potential of a new strain for industrial application. The only reliable test is evaluation in model small dough systems, which is tedious work in a yeast genetics laboratory producing large numbers of new strains.

In order to systematically classify industrial yeasts a large number of physical, biochemical and serological tests are necessary, but at present these are only applicable to scientific work, and are not suitable for industrial practice. Instead one has to resort to the organism's performance in a variety of model dough systems. Recombinant DNA technology has opened up new possibilities in industrial yeast genetics, but since it is suited to single gene traits, or those regulated by a small number of genes, its application is at present limited. Current recombinant DNA technology has been directed towards creating yeasts which produce specific compounds of commercial value, e.g. insulin, interferons, hormones, antigens, vaccines, and other protein and peptides. Typical large recombinant DNA companies are: Zymogen, Integrated Genetics, Cetus and Chiron. The enzyme chymosin, used in cheese manufacture has been cloned and integrated into yeast.

For baker's yeast manufacture, rapid growth and high biomass yields on molasses, coupled with good leaven and bake activity, are the predominant requirements. Such properties depend on genes and gene-systems which feed carbohydrates into glycolysis, the glycolysis process itself and the tricarboxylic acid cycle (TCA). Also, genes involved in nitrogen metabolism, and storage-carbohydrate synthesis are essential. The flow rate of carbon through the glycolytic pathway is important for both yeast manufacture and baking application. However, considering the relatively short leavening times compared with yeast propagation for biomass (2–4 hours *v.* 18 hours), and that leavening is an entirely anaerobic fermentation, it is the rate at which the yeast organism can feed hexose sugars through gycolysis that is most important for the leavening process.

The European market demands 'quick' strains, with good leavening activity in lean dough formulations. The 'slower' working strains have optimal activity in enriched sweet doughs, and relatively poor activity in the leaner doughs.

A patent taken out by D. T. Rogers and J. W. Szostak of Genetics Institute, US Application 796,551 dated 8 November 1985 (PCT Int. Appl. WO 87,03,006), details a method of increasing the rate of carbon dioxide or ethanol production and decreasing biomass production by yeast. The glycolysis is increased by reducing the intra-cellular ATP levels. This was achieved by stimulating the operation of a futile cycle, thus increasing ATP consumption. A leaderless acid phosphatase is produced, which cannot be excreted, and yeast cultures are treated with dinitrophenol, which acts as an uncoupler. Yeast transformants containing plasmid BA 601 (which contained the fructose 1,6-diphosphatase gene, mutated to remove the phosphorylation, mediated inhibition of the encoded enzyme under control of the glyceraldehyde-3-phosphate dehydrogenase promoter) increased carbon dioxide production towards the end of the growth cycle, and produced 25% more carbon dioxide than a control. Therefore, baker's yeast containing this plasmid had substantially increased leavening power.

**An Example of how a 'Fast' Yeast Strain can be Constructed**
New industrial baker's strains have also been constructed by the hybridization of meiotic spore clones of appropriate types. Spore clones often being derived from existing strains with the sought-after properties. As an example of this technique, the French yeast manufacturer Société Industrielle Lesaffre (UK Patent Specification 1,593,211,1979) use 'quick' and 'slow' strains as genetic material. Reflecting the European preference for 'quick' strains for leaner doughs. The spore clones from these strains were mutagenized using either ethanemethyl sulphonate, or $N$-methyl-$N'$-nitro-$N'$-nitrosoguanidine before mating. Hybrids are selected for testing according to growth-rate; in the presence of inhibitory concentrations of acetic acid; maltose adaptability in the presence of glucose; low invertase content; and the time lag before the onset of growth in the presence of 20% sugar. Crosses are carried out by mass-mating techniques. The spore clones are not always haploid, clones isolated often showing a sporulating ability. Crosses of such clones can give hybrids which are more than diploid in chromosome number. Many successful commercial strains have been constructed using the classical hybridization techniques.

## 2.2.3 POST-FERMENTATION YEAST TECHNOLOGY

The final stage in the separation process yields a 'yeast-cream', which varies in colour depending on whether it is derived from cane or beet molasses, the former being darker, owing to impurities. This cream generally contains 18–20% yeast solids, and passes either through a filter-press or a rotary vacuum filter to remove excess water, resulting in yeast-cake. In the USA, this is referred to as 'cake-yeast' (CY), but in Europe the common name is 'pressed or compressed yeast'. This product is either bagged in a crumbled form or shaped into pound-sized blocks of moisture content 70–75%. By extruding cake yeast through a perforated plate or die, and drying in a tunnel or belt-dryer at about 38°C, 'active dry yeast' (ADY) was first produced. However, ADY needed rehydration before incorporation in the dough, which was labour-intensive in mechanized bakeries. In the late 1960s, Gist-Brocades introduced 'instant active dry yeast' (IADY), which overcame that problem, rehydrating easily owing to the larger surface area of the cylindrically shaped particles, and their porosity. The application of genetics resulted in strains with higher leavening activity, which could be dried in fluid-bed/airlift dryers. This drying technique can take advantage of the cooling of moisture evaporation, so that, even when the inlet-air temperature reaches 70°C, the cells do not become damaged. The higher temperature and shorter drying time allows removal of moisture down to about 35% without any adverse effects. The second stage of drying at about 40°C air temperature reduces the moisture content of the product to 4–8%. The yeast particles are cylindrical with a diameter range of 0·2–0·5 mm and of average length 1·0–2·0 mm. This imparts an instant characteristic to IADY, and the high activity of the hybrid strain provides quicker leavening activity. Membrane structure is considered important for yeast strain activity. The presence of ergosterol and zymosterol in varying concentrations in the yeast cell membrane, depending on such environmental conditions of growth as substrate, aeration level and temperature, determine how the organism reacts to drying and rehydration of ADYs, and the storage of cake-yeast. Leavening activity in various dough-systems is a function of all these inter-related parameters.

Another approach to improving the biological activity of yeast and wheat-flour bread is reported by Chernaya *et al.* of WNIIChP (All-Union Research Institute of USSR) in *Klebopek. Konditer Prom.-st* 5 (1986), 28–29 (in Russian). A biologically active mixture, named BIAKS, containing pressed yeast and other ingredients to improve dough fermentation was made by modifying and drying a friable BIAKS preparation of 30% moisture content. This preparation could be dried down to 10% moisture in an expanded-bed, decreasing its leavening activity only slightly. Storage of the preparation for up to 3 months showed higher activity than control preparations over the same period, and its use reduced the period of dough fermentation by 50–80 minutes, increased bread volume by 17%, porosity by 3%, and shape-resistance from 0·45 to 0·53 units.

Various patents are also described involving the hydrolysates of wheat-flour and wheat-flour bread crumb, producing lower molecular glucose polymers in the dough, and amino acids, which act as yeast substrate.

Similar patents exist that use waste rye-bread crumb/rye-flour hydrolysates which also act as rye starter culture stimulants.

In the Soviet baking industry, liquid, pressed and dried yeasts are utilized. Liquid yeasts are prepared directly in the bakery, and can be regarded as the first phase of doughmaking. Liquid yeast is an intermediate product prepared by a rationalized process developed by A. I. Ostrowski. Initially, a gelatinized flour/water suspension is prepared, which is fermented at 48–54°C with thermophilic Lactobacilli. The second stage involves the cooling of the mash, which then has a high lactic acid content, to about 28–30°C. It is then transferred to another container, and used as substrate for yeast propagation. For the doughmaking process, a definite amount of mature liquid yeast is taken, some nutrient substrate having been added beforehand. After a fixed fermentation period, another portion of the mature liquid yeast is taken for doughmaking, and more substrate added. The propagation cycle for liquid yeast begins with the addition of a culture of thermophilic Lactobacilli, then, after cooling to 28–30°C, various yeast cultures are added.

For pressed-yeast (cake-yeast), quality standards must conform to the Soviet specification GOST 171–69, which includes a sensory evaluation for colour, consistency, smell and taste. Physical-chemical data include: moisture (maximum 75%), time to reach a standard 70 mm height maximum 75 minutes, and the acidity-index and shelf-life at 35°C (not less than 48 hours). Also, it can only contain technically pure cultures of *Saccharomycetes*. In bakery laboratories only the fermentation-speed is determined. Soviet researchers have paid considerable attention to the glutathione content of baker's yeast, which activate the proteolytic enzymes, and methods have been researched to remove this substance from yeast. (This matter is described in Section 1.5.7.) Contamination of baker's yeast with wild or other yeast types, e.g. *Candida* or *Torulopsis*, although increasing the yield of biomass (by up to 30%), considerably reduces its fermentation activity. Maltase activity is considered to be important in dough ripening, and should be both controlled and normed.

As early as 1960 the researchers at the WNIIChP in Moscow devised a production-scale process for large bakeries whereby yeast from molasses fermentation was separated in the form of a concentrate, with a fermentation activity to the standard 70 mm dough height within 45–52 minutes. This product is now produced in large quantities for the large industrial bakeries.

In the USSR, dried-yeast products are only applied where pressed- or liquid-yeast cannot be produced or stored. According to Ministerial Directive MRTU 18/121-66 for Food Industry Technical Standards, two types are manufactured, depending on their physical-chemical parameters, designated as 'top-quality' and 'first quality'. Dry-yeast is either produced in pellet or granular form, with an upper limit of 10% dust material. Top-quality dry yeast has a maximum moisture limit of 8%, a fermentation activity to the standard 70 mm height within 70 minutes, and a storage-life of not less than 12 months. The first quality product has a moisture maximum of 10%, a fermentation activity for the 70 mm standard within 90 minutes, and a storage-life of not less than 5 months. In general, according to research by Plewako, shelf-life is optimal within the moisture 7–9% for dry-yeasts.

# 2.3 Chemical Changes in Yeasted Doughs during Fermentation

Dough fermentation commences during mixing, proceeds through dough-processing or make-up, and only ends during the final baking phase. This includes the processes of dough-dividing, shaping, proofing stages of the dough-pieces and during the early baking phase. However, in practical production terms, the fermentation or floor-time refers to the time elapsing from the end of mixing through to dividing.

The objective in fermenting a sponge, pre-ferment, sour-dough or final-dough is to convert a hitherto dense, inelastic dough-mass into a physical state in which it can be leavened by the gaseous products of fermentation. Equally important is the formation and build up of flavour and aroma substances, which are an integral part of correctly prepared and fermented bread-dough. The final formation of an aerated and porous crumb structure, is accomplished during the final-proofing stage, before setting in the oven.

The total summation of the various process phases of fermentation and remixing (punching and knock-backs) necessary to bring the dough to a condition for processing (make-up) and baking is referred to as 'dough-ripening' or 'dough-maturing'.

A properly processed and matured dough demands the following prerequisites in terms of properties, the achievement of which is the result of a series of complicated processes which work simultaneously and are interdependent:

—gas development immediately after dough preparation must proceed with adequate intensity;
—the physical properties of the dough must be optimal for dividing, rounding and final moulding, gas-retention and stability being optimal for final-proof and baking;
—some unfermented sugars and products of protein hydrolysis must be left over, sufficient for normal colouration of the bread-crust;
—enough similar residual products of fermentation must also be present which will form the essential compounds for the specific flavour and aroma of the bread.

In considering the processes which take place during fermentation and ripening of a chemical nature, those involving the yeast enzymes and the formation of carbon dioxide and alcohol will be considered first.

Since doughmaking is essentially an anaerobic process, the yeast cell utilizes the carbohydrates in the absence of oxygen to produce energy, alcohol and carbon dioxide being the end-products. The process proceeds via a series of intermediate stages, in which many enzymes take part. The scheme for the anaerobic breakdown or fermentation of carbohydrate is known as the tricarboxylic acid cycle (TCA).

The zymase-enzyme-complex of the yeast enzymes is responsible for the breakdown of the monosaccharide molecule into alcohol and carbon dioxide. One molecule of hexose sugar forms two molecules of alcohol, and two molecules of carbon dioxide. Baker's yeast can ferment all the main types of sugar in dough, viz. glucose, fructose, saccharose (sucrose) and maltose. Glucose and fructose become fermented immediately, saccharose first being converted to glucose and fructose by the enzyme saccharase (invertase). This latter process occurs very rapidly, and a few minutes after mixing the dough all saccharose molecules have been converted into glucose and fructose, even when the dough contains up to about 7·0% saccharose (based on flour weight). The maltose molecules with the help of the yeast enzyme maltase, become hydrolysed to two molecules of glucose. The yeast cell is capable of fermenting, whether in the sponge or dough, all the flour sugars, the maltose formed by the amylolytic enzymes from the starch molecule, and also any added sugar. The flour sugars are only of importance during the first fermentation stage. Glucose and fructose are fermented first, the former disappearing more rapidly than the latter, and maltose will only be fermented by the yeast after all the built-up fructose and glucose has almost been depleted. This changeover from fermenting the hexose sugars to maltose represents a readjustment for the yeast enzyme system, and results in an induction period during which the speed of gassing is noticeably reduced. After this adjustment, gassing is restored up until the supply of maltose nears depletion. Where saccharose (sucrose) is added to the dough, the fermentation of maltose becomes postponed, and when the amount of added saccharose is plentiful, the maltose may not be fermented by the yeast cell.

In the case of indirect dough processes (pre-ferments) the yeast adapts itself to maltose fermentation in spite of fresh amounts of glucose and fructose as well as saccharose from the addition of fresh flour during doughmaking.

The amount of carbon dioxide produced in a dough by sugar fermentation amounts to about only 70% of the theoretical amount indicated by the chemical equation; this is explained by the fact that part of the sugar is used for energy and reproduction of the yeast cells within the dough. The gas production in dough results from alcoholic fermentation, in which the yeast cells, with the help of their specific enzymes, convert the hexose sugars (mainly glucose) into alcohol and carbon-dioxide gas. The simplified equation is:

$$\text{glucose} \rightarrow \text{alcohol} + \text{gas} + \text{energy}$$
$$C_6H_{12}O_6 \rightarrow 2C_2H_5OH + 2CO_2 + 234\,kJ$$

The leavening process depends on adequate gas-production and gas-retention within the dough. The former depends mainly on the availability of soluble, fermentable sugars, and the latter on the amount of gluten formed and its rheological properties. However, the reactions involved in the TCA-cycle in fact amount to about 11 successive chain reactions, depending on: temperature, pH, and yeast nutrients in the form of vitamins, and essential mineral compounds, e.g.

ammonium chloride, ammonium phosphate, calcium phosphate, calcium sulphate, the addition of which speed up the fermentation, especially when using flours of lower extraction-rate. The ions K, Mg, $NH_3$, $SO_4$ and $PO_4$ have a profound effect on the process, phosphate being essential to the formation of the initial stage in the process, viz., glucose-6-phosphate, which is formed by phosphorylation of the glucose molecule with the help of adenosine triphosphate (ATP). This reaction is catalysed by the yeast enzyme glucokinase (or hexokinase).

The second stage is the conversion of glucose-6-phosphate into fructose-6-phosphate, which is brought about by the yeast enzyme exoisomerase.

Stage three involves the conversion of fructose-6-phosphate into fructose-1,6-diphosphate, which is brought about by ATP and the yeast enzyme phosphohexokinase. This completes the preparative breakdown stage (anaerobic and aerobic) of the sugar.

The next stage is the decomposition of fructose-1,6-diphosphate into the two molecules, dioxyacetonphosphate and (D)-3-phosphoglyceraldehyde, brought about by the yeast enzyme aldolase (zymohexase). The presence of the enzyme hexokinase makes this reaction reversible, therefore dioxyacetonephosphate can react with other aldehydes as well as (D)-3-phosphoglyceraldehyde, giving rise to pentoses, hexoses, and certain polysaccharides. The reduction of glyceraldehyde to glycerol is also a possibility.

The next stage involves the isomerization of the phosphotrioses by the enzyme phosphotriose-isomerase, a point of equilibrium being reached when a content of 3-(D)-phosphoglyceraldehyde and 97% dioxyacetone-phosphate exists.

The following stage is the dehydration of 3-phosphoglyceraldehyde and its phosphorylation to 1,3-diphosphoglyceric acid. The 1,3-diphosphoglyceric acid is converted to 3-phosphoglyceric acid by the enzyme phosphate-transferase, the phosphate radical being transferred to ADP, thereby generating ATP, thus storing the second moiety of energy.

Then through the action of the enzyme phosphoglyceromutase, 3-phosphoglyceric acid is converted into 2-phosphoglyceric acid. The resulting 2-phosphoglyceric acid is then changed into phosphoenolpyruvic acid through the agency of the enzyme phosphotransferase. However, the 'enol' form of the pyruvic acid is rapidly converted to the 'keto' pyruvic acid form. The reaction, catalysed by the enzyme enolase (phosphopyruvate-hydratase), requires $Mg^{2+}$ (magnesium ions) for activation.

The final stage is the conversion of pyruvic acid to carbon dioxide and acetaldehyde, catalysed by the enzyme pyruvate-decarboxylase, which is also a yeast enzyme containing magnesium. The acetaldehyde then becomes involved in an exchange reaction with the co-enzyme of the dehydrogenase ($NADH_2$), which is nicotinic acid-amide-adenine-dinucleotide in the reduced form, releasing ethanol. The final step is therefore achieved by first decarboxylation, and then hydrogenation.

Yeast cells are very sensitive to temperature, and a rise in dough-temperature from 25°C to 35°C will almost double the speed of fermentation and gas development.

The fermentation product-balance is also dramatically changed by changes in medium pH. Optimal for yeast fermentation and respiration is an acidic medium from pH 4·0 to 6·0. This fits in well with wheat-flour dough average pH values of 6·0 down to about 5·0. The products of alcoholic fermentation vary considerably with pH; pH values in excess of 6·0 result in increasing amounts of such products as glycerol, acetic acid and succinic acid, with a reduction in ethanol and carbon dioxide production. Whereas a medium pH of around 3·0 gives rise to about 50% less glycerol, lower concentrations of organic acids, but more ethanol and carbon dioxide per 100 $\mu$mol, of fermented glucose. Glycerol yields are increased considerably by the addition of bisulphite, since this forms an additional compound with acetaldehyde, and eliminates this as an intermediate product. In the case of dry-yeast cells, glucose becomes converted to pyruvic acid and glycerol. Certain osmophilic yeasts will produce in addition to ethanol, mannitol, *d*-arabitol, erythritol and glycerol. The energy effect derived from the anaerobic utilization of carbohydrates by yeast cells is relatively small, therefore they must ferment a large amount of sugar. Under optimal conditions of temperature (30°C), and substrate composition, 1 g of pressed- or cake-yeast will ferment 1 g of sugar in 1 hour. Important ions for yeast fermentation are: K, Mg, $NH_3$, $SO_4$, and $PO_4$, the latter being essential for the formation of the phosphoric acid esters of glucose. The presence of the B-group vitamins increase the intensity of the fermentation by about 25%. Amino acids are equally essential for cell activity (and their amides), in particular aspartic acid and its amide, which are twice as well assimilated as even ammonium sulphate. Their addition results in a large increase in carbon dioxide release. This acidic amino acid, with its two carboxyl groups and one basic amino group is essential to the glycolytic breakdown products of the carbohydrates for oxidation in the TCA-cycle. Using ammonium sulphate as a source of nitrogen, which is expressed in terms of 100 parts utilized (absorption and assimilation), both *d*- and *l*-aspartic acid and *l*-asparagine are rated at 200 parts utilization under identical fermentation conditions of temperature, time, medium composition and pH. The kinetics of glucose fermentation follows the Michaelis–Menten equation, the Michaelis dissociation constant, $K_m$ being calculated from the graphical plot of the reciprocals of enzyme reaction rate, $1/V$ v. substrate concentration, $1/S$. The linear relationship so obtained, known as the 'Lineweaver–Burk plot', when produced backwards to cut the horizontal axis ($1/S$), gives $-1/K_m$ as the negative intercept.

Within the sugar concentration range 0·2 to 6·0%, the Michaelis constants ($K_m$) for glucose are 0·010 67, and for fructose 0·0225, which is indicative of a predominantly glucose fermentation. Since $K_m$ is a dissociation constant, the greater the affinity of the enzyme for the substrate the lower the value of $K_m$. Saccharose (sucrose) becomes rapidly converted to glucose and fructose by the yeast invertase enzyme system, and provided the speed of inversion by the yeast-cell enzymes exceeds that of the fermentation-rate of the inversion-products, the two reactions become synchronized to a degree. Therefore, saccharose, as base material does not restrain the fermentation process.

Invertase activity can be assessed from the amount of reducing sugar formed within 10 minutes from a 15% saccharose solution at pH 5·2, and 60°C by a yeast-suspension containing 6·2 mg dry-cells/ml. Although, as previously mentioned, a temperature rise from 25°C to 35°C results in approximately twice the rate of fermentation, dough-temperature rises from 35°C up to about 40°C, corresponding to those found in proofer-units, are virtually unknown in their effects on the fermentation process.

The thermophilic yeasts, compared with the mesophilic yeasts, to which baker's yeast strains belong, have fermentation optimums at temperatures of 39–40°C, and thrive in acid media. The thermophilic yeast strains are of practical significance for wheat-flour dough sponges in Armenia, where their ability to ferment at 40°C in an acidic sponge-dough, and produce carbon dioxide and alcohol at a rapid rate, represent a remarkable technical innovation. According to Saruchanjan, such strains as Armenija and Kirowokan show much reduced proofing-times compared with normal strains fermenting at 25–30°C (mesophils).

Depending on whether the growth medium contains oxygen or not, yeast can initiate either an alcoholic fermentation or an aerobic oxidation of the carbohydrates. In either case, part of the carbohydrate is utilized for cell growth, for fermentation about 10% of total carbohydrate is used, whereas under aerobic conditions 50–60% is used for cell growth.

The yeast cell is capable of assimilating almost all the amino acids present in grain. It is also capable of synthesizing amino acids and protein with the help of inorganic nitrogen sources and carbohydrates. Glucose serves not only as an energy source, but also as a source for intermediate metabolic products, e.g. pyruvic acid is a precursor for such amino acids as alanine, valine, and leucine, and also pentoses. Thus there is an exchange relationship between carbohydrate and nitrogen metabolism of the yeast cell. Yeast cells also contain a considerable polypeptide balance, which changes between about 10% and 25% of total cell nitrogen. Changes in cell physiology give rise to corresponding changes in the amounts of the various amino acids in the polypeptide chain. The polypeptide glutathione is of special importance for the cell redox-system (oxidation–reduction reactions). This tripeptide exists in two forms, in the oxidized form it contains a disulphide group, and in the reduced form a sulphydryl group. However, when the yeast cell dies, the glutathione is discharged into the substrate, which is often apparent when dry-yeast is used. In which case its reducing properties can have a negative effect on the dough protein complex. Glutathione contents can reach as much as 1·0% of yeast cell dry matter.

Yeast reproduction in dough increases in rapidity according to the percentage (of flour weight) added initially. The average amount of yeast used in doughs is about 2% (flour weight), in which case over a relatively short fermentation period of up to 3 hours, very little reproduction can take place. Since 1 g of yeast can contain from 8 to 20 million cells, the average dough, containing 2·0% yeast, 1·5% salt, and 60% water (based on flour 100 parts), would contain about 120 million cells/g of dough. If the bulk fermentation time extends long enough to exhaust all available sources of hexose sugars, apart from any available maltose which can be digested by yeast maltase (alpha-glucosidase), starch is the only untapped source for gas-production.

During the average modern milling process, about 10% of the starch granules become mechanically damaged by the direct pressure exerted by the reduction-rolls at the head of the mill. These damaged granules, in common with so-called 'pregelatinized' starch granules, are 1000 times more rapidly broken down than undamaged granules. Therefore, the main source of gas production in all short-time breadmaking processes is the quantity of available damaged-starch granules. In any event, the beta-amylase, of which flour contains an adequate supply, cannot initiate the digestion of undamaged granules. As already described in Section 1.5.3, any alpha-amylase dough supplement, either as pure enzyme or as malt-flour, attacks the 1,4 linkages of the starch molecule. This breaks the polymer into high-molecular-weight dextrins by random fragmentation. These becoming degraded to lower molecular weight dextrins and then maltose. The straight-chained amylose fraction of the polymer is hydrolysed to within a few residual glucose units, but the branched amylopectin part of the starch molecule is only degraded to 64% of the theoretical maltose yield. The carbohydrate/amylases complex is in a state of continuous change during dough fermentation, producing more or less total fermentable sugar as required. The process is kept under control by adjustment of the starch-damage and alpha-amylase treatment at the mill, with due consideration of the dough-process and schedule, and the type of bread for which the flour is to be utilized. The approximate level of fermentable residual sugar necessary for an adequately intensive first and final-proof of the dough-pieces, and crust colouration during baking of wheat-flour bread should not be less than 3·0%.

The high-molecular-weight pentosans of the flour also undergo partial hydrolysis by the enzyme pentosanase, and contribute to the colloidal properties of the dough and waterbinding, by virtue of their gelling tendencies. Rye-flour doughs depend very heavily on the pentosan/pentosanase complex for their optimization of functionality in processing into bread.

The availability of sufficient natural organic nitrogen-containing substances to act as food for the yeast, will depend on the flour extraction rate. Wheat-flours with ash-contents of about 1·0% and above, and extraction-rates from about 80%, will provide ample supplies of amino-nitrogen by proteolysis for rapid yeast fermentation. Wheat-flours of lower extraction rates and ash-contents 0·55% to 0·75% usually require supplementary additions of ammonium salts in the form of phosphates, chlorides, or sulphates to ensure adequate supplies of nitrogen for rapid yeast growth and fermentation. Yeast reproduction is stimulated by the enrichment of the substrate (dough) with mineral salts, and small amounts (up to 1·8% flour weight) of sodium chloride also stimulate reproduction.

During the fermentation process an increase in the pre-ferment and/or dough acidity-index takes place, owing to the accumulation of substances with an acid reaction. The titratable acidity of the pre-ferments and doughs rise, and simultaneously pH value changes from about 6·0 to about 5·0. This acidification process is essential for the production of quality bread of good flavour and structure. Although fermented doughs can contain lactic, acetic, succinic, malic, formic, tartaric and citric acids amongst other acids, in the case of yeasted doughs about 66% of the acidity is contributed by lactic acid. Acetic acid accounts for about 24%, and the remaining acids contribute about 10% only. However, most of the lactic acid is formed by the homofermentative lactic acid bacteria, but also the heterofermentative bacteria present in wheat-flours. Commercial yeasts also contain certain amounts of acid-producing bacteria, but alcoholic fermentation using yeast generally does not give rise to very appreciable quantities of succinic and other organic acids. The main factors in determining the acidity-index of pre-ferments and doughs are the flour extraction-rate and the temperature. Pre-ferments set at 28–30°C depend on the non-thermophilic bacteria for acid build-up, which actually have temperature optimums of about 35°C, whereas the thermophiles, e.g. *L. delbrückii* with temperature optimums of 48–54°C cannot make a significant contribution. Nevertheless, the change in acidity-index of wheat-flour doughs during fermentation is very important. All swelling and peptization processes of proteins within the dough are speeded up by increases in acidity-index and the resultant favourable pH. Enzyme processes are optimized, and flavour- and aroma-forming components made possible by the accumulation of acidic reaction products, resulting from their interaction with dough alcohols. Hence, it is not surprising that the final acidity-index of pre-ferments and doughs is taken as an indicator of the 'degree of ripeness or maturity' of the product. The acidity-index of baked bread is also a quality parameter in many countries, and is included in the Bread Standards. Bread flavour is influenced by both the quantity and composition of the dough acids. In the case of wheat-flour, it is the lactic acid which gives the bread a pleasant taste, whereas acetic and other volatile acids tend to give the bread an unpleasant flavour profile. The adsorptive water-binding which takes place during mixing by the dough colloids—protein, starch and pentosans—progresses again during the fermentation process. A reduction in the compaction of the protein molecules within the dough, as a result of proteolytic degradation and the osmotic swelling processes, increase their surface area for further adsorptive binding of water. Similarly, the hemicellulose and pentosan polymers undergo peptization and swelling according to their solubilities. The increase in dough acidity and accumulation of alcohol increase the hydrophilic properties of the dough colloids. The controlled processes of swelling and peptization (depending on the doughmaking process and time/temperature regime) progressively reduces the liquid phase within the dough, thus improving its physical (rheological) properties. Where this process proceeds uncontrolled, the liquid dough phase components progressively increase and the physical properties depreciate. Apart from the two process variables of time and temperature, the fundamental factor controlling the relative speed of the reactions described is the flour 'strength' (properties of its components). In the case of a 'strong' flour, the reactions proceed slowly, reaching a maximum at the end of a prolonged fermentation process. Whereas, in the case of a 'weak' flour the swelling of the protein and other components proceeds relatively rapidly, and the peptization and proteolytic reactions will lead to a rapid build-up of the liquid dough phase. This in turn is reflected in a more rapid depreciation in dough rheological properties, manifested as dough-softening. This difference in flour performance is often referred to as 'fermentation tolerance'. This inherent difference in 'tolerance' is also reflected in mechanical action during fermentation (remixes, knock-backs, dividing and moulding). Strong flours react positively to work-hardening, and weak flours show a structural weakening under identical mechanical dough treatment.

The various types of chemical-bonding occurring during processing are described in detail in Section 1.6.

The role of air and carbon dioxide gases, and their response to temperature and pressure during the processing stages are also significant. Carbon dioxide becomes absorbed by the liquid dough phase, most of which is adsorbed by the flour particles. Water at 20°C is capable of dissolving, per litre, at $10 \times 10^5 \, N/m^2$ (760 mm mercury) pressure, about 920 ml of carbon dioxide and only 20 ml air at full saturation. Temperature increases result in the release of more air from the dough, and at a more rapid rate than is the case with carbon dioxide. During proofing the dough-pieces adsorb carbon dioxide into the liquid dough phase, as well as containing the gas in the gaseous state. The presence of carbonic acid, together with the lactic and acetic acids formed in the dough exert a reversible peptization of the gluten proteins, the gluten complex being reformed when the peptizate is neutralized during dough remixing, dividing and moulding. In a biological system such as dough, respiration involves the catalytic transfer of hydrogen from primary substrates to carriers. However, this hydrogen is in the atomic state, i.e. a proton and an electron, $H^+ + e'$. Owing to its reactive nature, neither atomic hydrogen nor free electrons exist in biological systems, but instead are taken up by a sequence of carriers, leading to a final hydrogen acceptor. In anaerobic respiration, the final acceptor is an organic compound such as pyruvic acid, whereas in aerobic respiration the final acceptor is molecular oxygen. Bread dough is an example of an oxidation–reduction (redox) system, and during fermentation the redox-potential, or rH, constantly changes as a result of the action of a group of enzymes known as the 'desmolases', which control the various redox-reactions. These are classed as flavoprotein enzymes and contain a prosthetic group, the 'hydrolases', which control hydrolysis and resynthesis of saccharides, amides, esters and proteins simply consist of proteins alone.

The significance of electron transfer in a filled-polymer biomass such as bread dough is important for the formation of an even-textured aerated dough of low density prior to baking. The redox-system depends on the fact that when a substance loses one or more electrons, it exists in an oxidized form. Conversely, when it gains electrons (one or more) it becomes reduced. The involvement of hydrogen and oxygen in these reactions is possible but not a necessity, but the two reactions must occur simultaneously, since electrons can only be lost by a substance if another substance is available in the system capable of accepting them and vice versa. The extent to which a substance tends to lose or gain electrons is a measure of its oxidizing or reducing power. The migration of electrons in redox systems can therefore be measured in terms of electrode potentials. In physiological work, where relatively small changes in pH occur, the rH scale gives a broader spread for the meaningful interpretation of biological data. When dealing with multienzymic systems such as are encountered in panary fermentation, although certain simplifications have to be made, the redox process, involving the simultaneous transfer of electrons for ionic or covalent bonding, is an appropriate and valid conception. On the rH scale, 0–15 signifies very strong reducing properties, and 25–41 very strong oxidizing properties. Wheat-flours have rH values within the range 15–19, and those containing the outer portion of the wheat-grain show lower values of the order of 15·0–15·5. Wheatgerm is strongly reducing, with rH values between 6 and 7. Rye-flours fall between 17 and 18. After about 2 hours fermentation, the rH can fall to between 16 and 17, as the system becomes progressively reduced. In a reversible oxidation–reduction system, the oxidized or reduced forms are ionized. Their effective concentrations therefore depend on the degree of ionization. Thus pH or hydrogen ion concentration is of great importance.

If an unattackable platinum electrode is placed in a reversible oxidation–reduction system, a potential difference is set up. The more positive the potential, the greater the oxidizing power of the system. This potential difference, or redox-potential is measured with a potentiometer set-up, the electrode potential in volts relative to the Standard Hydrogen Electrode being designated Eh. The Eh value will depend on the relative amounts of oxidized and reduced components in the system rather than their actual quantities. The Standard Hydrogen Electrode consists of a platinized platinum electrode two-thirds immersed in sulphuric acid with a pH of 1, saturated with hydrogen gas at 1 atm pressure. Platinum is a good absorbent of hydrogen, and forms a chemical state of equilibrium between the adsorbed hydrogen and the hydrogen ions in solution, i.e.

$$H_2 = 2H^+ + 2e$$

The sole purpose of the platinum electrode is to form a 'solid' hydrogen electrode by adsorption of hydrogen gas onto its surface, this serves to draw off the electrons. The potential of the Standard Hydrogen Electrode is taken as zero for all temperatures. When a system containing both oxidizing and reducing agents is in equilibrium, an electrical potential (the redox-potential) is developed, the magnitude of which depends on the pH and temperature. It can be measured in terms of the hydrogen electrode potential (Eh). The relation between Eh and pH is complex, but a simplified version at 30°C is:

$$Eh = 0·03(rH - 2pH)$$

The redox intensity of a biological system can be expressed in rH units in a manner analogous to that of pH units.

By thermodynamic reasoning, a simple relationship between rH, pH, and Eh can be deduced as follows:

$$rH = \frac{2Eh}{0·06} + pH$$

where 0·06 represents the value of $2·303RT/F$ at fermentation temperature, $R$ being the gas constant $8·315 \, J \, K^{-1} \, mol^{-1}$, and $T$ the absolute temperature. $F$ is the Faraday unit, the amount of electricity carried by 1 gram-equivalent of an ion with a single charge (96 500 coulombs). The multiple 2·303 represents the ratio of the logarithm of any number to the base 10, to the log of the same number to the base e. When only hydrogen is present at 1 atm pressure.

$$rH = - \log 1 = 0$$

In an atmosphere of oxygen, without hydrogen at pH 7, the Eh is found to be 0·83 V. When these values are substituted in the rH–Eh–pH equation, rH = 41. Therefore, the extreme limits of rH at atmospheric pressure, within which an rH scale can be constructed analogous to pH, is made possible.

This technique can be applied to form the basis for the formulation of mathematical models to simulate the autolytic dough-fermentation process. The rate of autolysis can be followed by measurement of both the aerobic and anaerobic oxidation–reduction potential and pH of the dough system.

In formulations where sugar is added to dough, e.g. the addition of sugar hydrolysates such as HFCS (high-fructose corn syrup), it is desirable to relate the amount of sugar needed to produce bread of maximum volume to the quantity of gluten in the flour. This is an example of the application of mathematical regression techniques to exert economic control over the dough-fermentation process. Another desirable objective would be to obtain objective measurements for the control of dough-ripening by monitoring the changes in quantity and quality parameters of the gluten complex during the fermentation period from mixing through to baking.

The automated control of the fermentation process by continuous measurement of such parameters as: dough-viscosity, temperature, pH, rH, gassing-rate, and changes in gluten-complex properties with time, would be an important contribution towards a CIM (computer integrated manufacturing) situation in the industrial baking industry. The application of biotechnological techniques in the baking industry appears very appropriate, since baking, in common with brewing, pioneered the application of biotechnology to industrial processing. For centuries man has made use of sour and yeast cultures by attracting them to grow on suitable cereal-based substrates, from the wild. However, apart from the specific microbiological cultures utilized, the processes involved are enzymic. The cultures have been selected for the specific enzymes which they contain, and for their response to temperature and pH. They begin with the addition of water to flour, as a result of the flour enzymes, and extend through the processing stage to the baking process.

The preparation of freshly baked products from wheat- and rye-flour is achieved on an industrial-scale according to the following general schemes.

*Wheat-flour bread*
Wheat-flour + salt + water + yeast + conditioners, mixed 2–20 minutes.
Fermentation for 40–70 minutes, including bulk, dividing, first moulding (rounding), first-proof, moulding and shaping, and final-proof.
Baking 20–40 minutes.
Cooling, processing (slicing) and packaging 1·5–2·5 hours.
*Rye-flour bread*
Rye-flour + salt + water + sour-dough culture, mixed 1–10 minutes.
Ripening processes for 40–70 minutes, including maturing stages, dividing, rounding/resting, shaping, and final-proof.
Baking 45–90 minutes.
Cooling, processing (slicing) and packaging 5–6 hours.

In practice, every bakery utilizes its own process schedule both for wheat and rye bread production. Conventional processes, whether direct (straight-doughs) or indirect (pre-ferments/sponges) take several hours to bring to completion. Furthermore, a large area is necessary for containers (mixing-bowls, dough-bunkers, etc.). For this reason ways have been progressively sought which reduce or at best eliminate the time required for dough-fermentation and -ripening. Although these innovations have been applied in most countries in one form or another, one cannot overlook the quality parameters of the dough to be processed. In most cases they produce doughs which are deficient in one or more of the following basic requirements:

—necessary content of fermentable sugars,
—optimal rheological properties,
—smooth throughput of the dough-pieces in the rounder and long-roller,
—production of bread of large volume, elastic crumb, good shape and with a crisp crust of good colour,
—production of baked bread with good taste, aroma and smell.

One method of shortening the dough ripening or maturing time is to speed up the fermentation process. This can be achieved by various procedures:

(1)  Increase of the quantity of pressed- or liquid-yeast used in the straight-dough or pre-ferment. By using continuous processes with liquid pre-ferments or brews, larger quantities of yeast cells are introduced at the mixing stage.

(2)  The use of activated yeast in hydrolysed substrates for preparing the pre-ferment or dough.

(3)  Use of biologically activated mixtures containing yeast cells, either as a concentrate or in liquid form, e.g. whey/sugar solution treated with 0·05% pressed-yeast. 50–60% sucrose solution is dissolved in whey at room temperature, which can be stored for up to 3 days without yeast, and not more than 1 day with 0·05% added yeast cells. This material can be concentrated down to 30% or 10% moisture content to increase its storage time. The addition of 3–4%, depending on concentration, decreases the dough-fermentation period by 50–80 minutes.

(4)  Selection of active types and strains of microorganisms for the addition to liquid yeast, and pre-ferments for batch or continuous processing.

(5)  The addition of 'yeast-foods', containing ammonium and calcium salts of hydrochloric, sulphuric and phosphoric acids, as widely used in North America.

(6)  Increasing the temperature of dough or pre-ferment to suit the process schedule.

(7)  Use of dough conditioners, containing stearoyl-2-lactylates, monoglycerides, oxidizing and reducing agents, enzymes and soy flour. This combination can be balanced to compensate to a degree for lack of fermentation, and is used for short-time processing.

Although used for the production of wheat-flour doughs for many years, the quality parameters of the doughs and baked bread do not satisfy the basic criteria detailed earlier. Therefore they have to be judged on their merits by consumer and trade standards. Another approach utilized to expedite dough-ripening depends on chemical agents, which reduce the amount of energy required to mix the dough, and the following fermentation or 'floor-time'. This basic conception was pioneered in the USA, as described in Section 1.5.1. A mixture of cysteine, whey powder and potassium bromate and/or ascorbic acid, was first used to prepare doughs by the direct- or straight-dough procedure instead of preparing sponges. However, it was later found that it could be used in conjunction with Amflow and Domaker continuous plants in the USA to reduce energy inputs and expedite dough-ripening.

This idea was then taken up in the UK when the 1972 Bread and Flour Regulations amendment permitted the use of L-cysteine–HCl. Known as the activated dough development (ADD) process, used in combination with 11–12% protein flours, 35 ppm L-cysteine–HCl, 25 ppm potassium bromate and 50 ppm ascorbic acid, allows the floor-time to be reduced to about 10 minutes before dividing. The inclusion of a plasticized shortening of melting-point 38°C appears essential for this process to succeed. This treatment also allows lower mixing-intensities, and both high-speed or conventional mixers can be utilized.

Another innovation is the use of the fast oxidant azodicarbonamide in combination with L-cysteine–HCl. Using flours of 11–12% protein, up to 75 ppm L-cysteine–HCl and 40 ppm azodicarbonamide based on flour weight, with about 0·7% high-melting-point shortening in the formulation, also reduces the floor-time to 10 minutes before dividing. However, to date, this additive is not permitted widely within Europe.

A more generally acceptable approach would be to simply increase the yeast quantity by 2% to 7%, and employ a more intensive or prolonged mixing-time at a higher temperature (32–33°C). Conventional mixers working at low mixing-intensities would demand mixing-times of 30 minutes for wheat-flour doughs of up to 76% extraction. For wholemeal-flour doughs 20 minutes should be adequate. The dough can be divided, shaped and proofed to the required volume, immediately on completion of mixing.

Depending on flour analysis, the addition of fungal amylases/proteases and selected emulsifiers in combination with adjusted levels of potassium bromate, are recommended for top-patent flours. Since insufficient acid build-up can take place without bulk fermentation, low-extraction wheat-flours require acid supplementation with either 0·2% or 40% lactic acid or 0·13% crystalline citric acid, and 0·05% of 80% acetic acid. This method of the expedition of wheat-flour doughs was originated by the technological laboratories WNIIChP in Moscow, due to Stoljarowa, Stscherbatenko, Lurje, and Berjosnizkaja in 1962–63. After mixing, the dough is divided and moulded into the required shape for proofing. Final proof for 1 kg units is 60–65 minutes, before baking for 50 minutes. Bread made by this procedure was considered structurally and organoleptically as good as the normal bread by Soviet bakery experts, in accordance with the national GOST-standards.

In balancing formulations for short-time dough systems, there is a pronounced interaction between yeast concentration, fermentation-time, proof-time and oxidation requirement. Reduction in bulk fermentation-times require increases in yeast concentration and a combination of 10 ppm potassium bromate and 80 ppm ascorbic acid. Using this technique, systems can be balanced for 45-, 60- and 120-minute fermentation-time schedules. Organoleptic deficiencies are apparent, although structural characteristics often remain acceptable.

The alternative approach for reducing bulk-fermentation or floor-time is to increase the specific mixing-intensity to an optimal level, which is referred to as 'mechanical dough development'. This optimal level is often specified in terms of energy input in either J/g or W h/kg. Weak flours requiring 15–25 J/g, medium-strength flours 25–40 J/g, and strong flours 40–50 J/g.

The Chorleywood Bread Process (CBP) is one example of this approach. The necessary energy-input in this case is 40 J/g or 11 W h/kg expended over 4–5 minutes, which is 5–8 times greater than that expended by a conventional batch mixer which cannot impart intensive mechanical energy to the dough. Other modifications are:

—the addition of 75 ppm ascorbic acid (flour weight),
—the addition of 0·7% shortening, containing about 5% high-melting-point glycerides (38°F),
—increase yeast concentration by a factor of 1·5–2·0,
—increase water absorption by about 3·5%.

Doughs prepared by this procedure, whether on a batch or continuous basis, are passed directly into the divider, receiving a first proof after rounding, final moulding and shaping before final proofing and baking.

Many countries have produced special high-speed/intensive mixers for both batch and continuous processing, based on the principle of mechanical dough-development. These have been described under the individual countries in Section 1.70.

Most countries prefer the use of pre-ferments with added flour, and liquid pre-ferments containing up to 50% of total flour have proved to be the most successful from the viewpoint of process rationalization and rate of gas-production. By adding nearly all the ingredient water in the pre-ferment, the ferment can be agitated, pumped, and metered. In this way a fluid 'sponge' containing 50% flour, yeast, yeast-food and almost all the dough-water, will produce the same fermentation end-products in less than half the time for a conventional sponge. The process eliminates fermentation-rooms and equipment.

The fermentation cycle is complex and difficult to duplicate. However, the development of a combination of ingredients and a process which simplifies and shortens the baking operation remains highly desirable. Certain reactions of oxidants, enzymes, and starch modifiers can be formulated to simulate fermentation changes to a degree.

Similar trends have taken place in the production of rye-flour bread types, aimed at rationalization of the process-flow and improvement of reproducibility of sour-dough production.

# 2.4   Wheat- and Rye-Sours and Sour-Dough Processing

Although yeast in a compressed, dried or liquid form has become standard raw material for industrial baking worldwide, wheat-flour-based sour-cultures remain popular in many countries, e.g. Spain, USA (San Francisco), Central and Eastern European countries, and Scotland. In fact, the oldest method of making fermented bread was to use wheat and rye sour-dough. Sour-doughs were utilized to produce acetic acid, which created the optimal conditions for swelling and baking of the flour, simultaneously preventing spoilage organisms fermenting in the dough, and imparting excellent flavour to the baked bread.

To produce a standardized acidified dough, the ripening or maturing method will depend on sour-dough development. It begins with the quality of the sour-dough produced during the first step, and the selection of raw materials and starter-cultures.

To produce a good sour-dough, demands an exacting control of acidity from bacterial reproduction. Temperature, time and dough-yield (dough-firmness) and type and species of microflora will determine the quality of the sour-dough and baked bread. Even in a modern industrial bakery, control of these parameters can prove difficult and lead to quality variations in the ripened dough, which often is only discovered in the baked bread.

Wheat sour-dough cultures depend on various species of Lactobacilli and yeasts contained in very firm flour/water doughs, or 'mother–doughs', referred to in Spain as 'masa madre' and in German-speaking countries as 'Mutterteige'. The contents of these starter-cultures, Lactobacilli and various species of yeast, are carefully guarded secrets in many bakeries. Their preparation, control and perpetuation has been carried out with great care and individualism by idealists over generations.

The starter-culture is made into a stiff dough with water, stored overnight at 2–8°C for up to 8 hours. The degree of acidulation will depend on the storage-time and temperature.

The application of sours to improve the quality and flavour-token of wheat-flour breads was never more appropriate than it is today, where rationalization and the elimination of bulk-fermentation has reduced many mass-produced breads to the status of a napkin-like carrier food.

Traditionally, a typical wheat-flour sour-dough production process involves a three-step process prior to the final mixing stage. Normally this would extend over 3 days fermentation to obtain the intensive flavour associated with acidified breads. In order to produce an acidified flour dough and optimize the conditions necessary for

492

the microorganisms to produce the type and quantity of developed acids, the following procedure is a typical empirical system.

*Stage 1*
Wheat-flour and water are mixed with a small quantity of starter-dough to give a dough-yield of 200%.
The dough is then allowed to ferment for 6 hours at 26°C.
During this time the yeast cell content increases, and some acidity develops.
*Stage 2*
Adjust the dough-yield to 170% by adding more flour. Ferment for 8 hours at a temperature of 24–27°C, during which time the dough ripens by developing considerable acidity and aroma.
*Stage 3*
This stage allows an optimal aroma development within 3 hours at a temperature of 28–32°C, adjusting the dough-yield up to 190% with water.

The final pH of the Stage 3 dough should be about 4·0, and the acidity-index (*Säuregrad*) 22–24, the mineral-ash content of the wheat-flour used is about 0·55% (flour type 550).

Yeast fermentation contributes little to dough acidity in the form of lactic and acetic acids, but the sour-dough bacteria, depending on type and species, are capable of converting citric and malic acids of flour into lactic and acetic acids. This induces the development of aromatic and acceptable flavours.

The San Francisco sour-dough French bread has been made continuously for over 100 years, and the microbiology of the starter-cultures used were only characterized during the 1970 decade. The yeast, identified as *Saccharomyces exiguus*, and the sour-dough bacterium *Lactobacillus sanfrancisco* coexist in the pH range 3·8–4·5. The bacteria preferentially utilize maltose as a carbon source, whereas the yeast does not utilize maltose at all, thus avoiding any competition for the same carbohydrate in the dough. The Scottish 'barms', used to produce the Scottish square bread, are leavened by preparing a barm starter-culture consisting of a hop-infusion, malt-flour and some old-barm, which is allowed to ferment for 3–4 days. The finished barm is added to a 25% overnight-sponge, or a 20% flour pre-ferment, being processed into a final dough, scaled into 2 lb 3 oz pieces for box-proofing and hearth-baking in one sheet. The loaves are separated from one another by greasing before proofing, and on separation have only a top and bottom crust. Scottish square bread formulations are traditionally lean, consisting only of flour, water, salt and yeast. Both wheat and rye sour-dough can be freeze-dried, and then milled to give wheat and rye sours in a convenient powder form for a single-step production of sour-dough breads.

The naturally fermented wheat sour-dough produces a fine white powder with a typical acid flavour and odour, containing sour-dough bacteria, which can be reactivated. Such products have maximum moisture contents of 8·0%, a pH 3·7–4·0. Acidity-indices are 20–24 for wheat sour-dough, and 40–44 in the case of rye sour-dough. These products allow the production of sour-dough bread within 3 hours, instead of a 3-day maturing using a traditional three-step fermentation schedule. A

3-day ripening procedure is normally required to obtain the intensive flavour characteristic of acidified breads. With the dried sour-dough, no variation in quality is apparent, thus standardizing production batch after batch. The resultant bread gives an improved taste, good crust, good crumb, and easier cutting and prolonged freshness.

For the preparation of sour-dough white bread, 10% dried sour is used, based on flour weight, 4% yeast, 2% salt and approximately 65% water. The dough-temperature is set for 27°C. After mixing for 6–10 minutes depending on the type of mixer, the dough is given a floor-time of 20 minutes. The dough is then divided, shaped and given a 30-minute final-proof, before baking at about 220°C.

The quality and consistency of dried powdered sours depends on the type, species and blend of the microorganisms, and the efficiency of the drying process. Freeze-drying techniques offer the best protection for the microorganisms, since heat-damage is eliminated. A 'sour' is defined as the product obtained when a mixture of white/rye-flour and water is fermented by lactic and acetic acid bacteria, i.e. biologically produced.

In the USA, Caravan Products Company, Inc., of 100 Adams Drive, Totowa, New Jersey 07512, offers sours for: San Francisco-type sour bread, white breads, brown-and-serve bread and rolls, Italian bread, rye breads, pumpernickels, and English muffins. In the FRG, Dr Otto Suwelack GmbH & Co., of 4425 Billerbeck, Böcker Sauerteig-Produkte of Minden 4950, Ulmer-Spatz of Ulm (Donau), Böhringer and Söhne of Ingelheim-am-Rhein, amongst others, market sours mainly for rye/wheat-flour mixtures.

In the USA, specialized sours are marketed, depending on whether the doughs are to be baked and sold fresh or for frozen-doughs subjected to a frozen shelf-life of 8 months or more.

In German-speaking countries, sour-dough is defined as a dough containing microorganisms in an active state, whose life-cycles have never been completely interrupted.

Depending on the type of bread to be produced and the preconditions existing in the bakery, many individual methods of sour-dough processing have been developed. By choosing appropriate process parameters the microbial metabolism of the sour-dough microflora can be controlled to give a balanced acid build-up for aroma, yeast reproduction and fermentation efficiency. Between the microbial composition of the sour-dough, the formation of metabolic products, and the process parameters of the sour-dough process, there is a close interrelationship. Research continues in the FRG, USSR and Poland to clarify the basic composition of sour-dough microflora. Sour-doughs contain two main groups of microorganisms: lactic acid bacteria and yeasts, both existing in a state of symbiosis, influence the character of the baked bread as a result of their metabolism under controlled process conditions. Average yeast and Lactobacilli cell counts in starter-cultures are: $1 \times 10^6$ to $3 \times 10^8$ and $1 \cdot 5 \times 10^8$ to $6 \cdot 1 \times 10^9$ respectively. The Lactobacilli are classified as either homofermentative or heterofermentative, depending on their respective enzyme systems. The homofermentative Lactobacilli ferment glucose by the fructose diphosphate pathway, converting it to yield over 90% lactic acid. The heterofermentative Lactobacilli ferment over the hexose monophosphate route forming about 50% lactic acid, carbon dioxide, ethanol, acetic acid and other organic acids. Spicher and co-workers isolated about 260 Lactobacilli strains from various starter-cultures. An average starter composition is the following:

Homofermentative lactic acid bacteria approximately 54%:
*Lactobacillus plantarum, L. casei, L. farciminis, L. alimentarius*, and *L. acidophilus*.
Heterofermentative lactic acid bacteria approximately 46%:
*L. brevis, L. brevis* var. *lindneri, L. fermentum, L. fructivorans*, and *L. buchneri*.

However, the most important strains of Lactobacilli used internationally, which give good dough-souring and flavour build-up in starter-cultures are: *L. brevis, L. fermentum, L. plantarum* and *L. casei*. Little information is available concerning the exchange processes between yeasts and Lactobacilli in sour-dough.

Whilst the Lactobacilli produce mainly acids and flavour compounds, the yeasts produce carbon dioxide for leavening and the formation of loaf structure and volume. The presence of carbon dioxide creates conditions favourable to facultative anaerobes, which promote the metabolism and reproduction of the Lactobacilli. Spicher has identified over forty yeast strains, which belong the following four types: *Saccharomyces cerevisiae, Pichia saitoi, Candida krussei*, and *Torulopsis holmii*.

Sour-dough microflora show variations depending on the production conditions, and the process utilized from one bakery to another. This variation applies to both the quantity and composition of the sour-dough microflora. Apart from the indigenous flour microflora, the sour-dough microflora will depend on the following variables: initial microflora of the starter-culture (*Anstellgut*) and the processing parameters of the chosen dough-process. Assuming that the process parameters, temperature and dough-yield, are controlled, the rate of reproduction and metabolism within the dough should remain constant over a given time period. Therefore, by increasing or decreasing the amount of starter used, the ripening- or maturing-time can be decreased or increased accordingly. The ripening- or maturing-time of the sour-dough depends on the rhythm of the growth curve of the microflora. The growth curve can be divided into six growth-phases, by plotting the logarithm of the cell-count *v.* time (h), as follows.

Phase I—induction phase, during which the microorganisms adapt themselves to the substrate, and take in nutrients for reproduction and metabolism. The duration of the induction phase is inversely proportional to cell count of the inoculate (starter concentration). The lower the amount of starter used, the longer the induction phase. The average duration of the initial-phase is about 1 hour.

Phase II—acceleration phase, which lasts for about 2 hours.

Phase III—exponential reproduction, during which time the microorganisms reach their maximum growth potential. A doubling of the cell-count is achieved in minimal time (generation-time) during this phase. The total duration of this phase is about 2 hours.

Phase IV—when approximately 4 hours have elapsed the exponential growth lapses into a period of delayed growth.

Phase V—At about the fifth hour, the so-called stationary phase is reached, at which point a state of equilibrium exists between newly formed cells and dying cells.

Phase VI—after about 6 hours, the lethal-phase sets in, whereby the availability of nutrients diminishes, and the products of metabolism concentration reaches such a level that the cells become damaged and cell-autolysis can occur.

In the case of the continuous sour-dough process, the sour-dough reaches the necessary quality parameters in terms of pH, acidity-index, gas-development and cell-count, within about 3 hours ripening-time. The process is so designed that at this point the mature sour is taken, and two-thirds used for final dough-preparation, the residual one-third serving as fresh starter. Were this sour to be taken earlier, there would be a deficiency of microorganisms and metabolic products, resulting in inadequate dough acidulation and leavening. The resultant baked bread would manifest an unbalanced, weak aroma, and a poor crumb-structure. Also, if an immature sour-dough is used as starter, the ripening of the newly prepared sour becomes delayed. On the other hand, if the sour is taken after more than 5 hours, excessive dough acidulation takes place, the acidity resulting in inhibition of and damage to the sour microflora. Furthermore, if this material is further processed as starter, quality depreciation result, and the resultant bread quality is inferior. If the starter parameters are incorrect, this can have a negative effect on any subsequent generations of sour-dough.

In sour-dough, bacteria and yeasts coexist, but each group of microorganisms have different optima regarding temperature, time, pH-value, and dough-yield (consistency of medium). Therefore, the optimal conditions for both bacteria and yeasts cannot be simultaneously provided. Either bacteria or yeasts predominate, and one group can have a negative influence on the other. The optimal temperature for the bacterial acid production is between 30 and 40°C, but differs according to type of bacteria. The function of the sour-dough bacteria is to utilize the carbohydrates, mainly maltose, and the protein degradation products to produce lactic and acetic acids, and other volatile acids for flavour, and liberate carbon dioxide. Acid production in the sour-dough continues down to a pH of about 4·0, below this level the bacteria become inhibited by their metabolites. The Lactobacilli can be differentiated according to the metabolites which they produce, the homofermentatives ferment the carbohydrates to lactic acid, whereas the hetero-fermentatives produce lactic, acetic acids and carbon dioxide. The term Lacto-bacillus for the lactic acid bacteria, due to Beijerinck, no longer corresponds to the current classification system. Krassilnikow has introduced the designation Lactobacterium, which links them with the asporogenous bacteria. The lactic acid bacteria are non-motile, Gram-positive bacilli, forming no spores, the rod-shaped cells (2 μm to 35 μm) reproduce by cell division.

All sour-dough bacteria ferment glucose and maltose, and saccharose (sucrose) is also fermented with the exception of *Lactobacterium brevis*. Lactose is also fermented except by *L. delbrueckii*, and *L. leichmanii*. *Lactobacterium plantarum* and *L. brevis* also ferment pentoses, especially arabinose. The generation-time (time for a doubling of the cell-count) for lactic acid bacteria ranges from 70–200 minutes, depending on species, temperature of culture, and viscosity of the substrate. Every microorganism demands certain optimal conditions for both its reproduction and its metabolism, and these factors are important for all sour-dough processes.

### 2.4.1 PROCESS VARIABLES

**Temperature**
By regulation of the sour-dough temperature, the development of the desired microorganism can be selective to a degree. This is the reason for the sour-dough stages, so characteristic of rye-flour processing. The controllable variables must be used in coordination in order to optimize the chosen technological process. Optimal growth and development temperatures are the following:

| | | |
|---|---|---|
| Sour-dough yeasts | 25–27°C | —gas-production and leavening |
| | | —produce alcohols, aldehydes and organic acids for flavour and aroma |
| | | —Antibiotic to wild Gram-negative bacteria |
| Heterofermentative Lactobacterium | 20–30°C | —favour lactic acid formation, producing a ratio lactic:acetic of 80:20 |
| | | —aromatic bread flavour |
| Homofermentative Lactobacterium | 30–40°C | —favour acetic acid formation |
| | | —milder, balanced flavour. |

The baker can either inhibit or encourage the development of the sour-dough microorganisms to suit his production requirements by adjustment of the temperature of the sour-dough stages as follows.

Starter-sour (*Anstellgut*):
On average, amounts to 0·5% of total flour weight for the batch, together with 1–2% of total rye-flour content, and 2–3% water. This serves as the 'seed' for the dough process, on which the control variables, temperature, dough firmness, ripening-time, added amount of flour at each stage, aeration, and choice of flour must be made to exert an influence, and produce good bread.

Initial-sour (*Anfrischsauer*):
(This stage is set at 24–27°C to stand for 4–5 hours, adding the above percentages of rye-flour and water to form a soft dough, mixing well to aerate as much as possible. The soft consistency of this sour-stage allows the yeasts in particular to reproduce at the set temperature. Any lactic or acetic acid formation is minimized.

Basic-sour (*Grundsauer*):
This stage is set at 22–28°C to stand for 4–8 hours (overnight), adding a further 14–16% of the total rye-flour content, and another 10% water. These conditions also allow the reproduction of the bacteria; higher temperatures favour the

formation of lactic acid over a shorter standing-time, and lower temperatures for a longer standing-time the formation of alcohol and acetic acid. Excessive acid build-up is prevented by the firmer dough of 145–160% dough-yield, compared with that of 180–200% used for the initial-sour.

Full-sour (*Vollsauer*):
This stage is set at 30–34°C to stand for 1·5–4·0 hours, adding a further 32–35% of the total rye-flour content, and 55–60% more water, mixing well for good aeration. These conditions are so chosen that the yeasts and acid-forming bacteria are both activated. Lactic acid formation is favoured, which also peptizes the protein. Dough-yield is increased to 160–220%, compared with 145–160% for the basic-sour.

Dough (*Teig*):
The dough is set at 28–32°C, to stand for 5–20 minutes only. According to bread-type, 48–52% more rye or wheat-flour, and 40–45% more water together with 1·6–1·8% salt are mixed for about 20 minutes using a conventional-mixer for rye/wheat doughs, and 30 minutes for rye-meals and rye-flours. Using high-speed/intensive mixers, 6/2 minutes are sufficient, and up to 20 minutes for rye-meals or -flours. Dough-yield is re-adjusted down to 145–160%. The same control variables applying to the final-dough as to the sour stages for dough-ripening.

This is a typical example of a three-stage (excluding the starter stage) process schedule, using an overnight basic-sour, illustrating how temperature control is successively utilized during the sour-dough stages within the range 23 to 32°C. During each sour-dough stage, one group of microorganisms are propagated at the expense of another. Thus ensuring that an optimal ratio of yeast:bacteria are available at the end of the sour-dough process. This is a typical discontinuous or batch sour-dough process used in a small- to medium-sized bakery operation.

In order to produce more batches during a working-day, the quantities of each stage are scaled-up accordingly, starting with the starter-sour, to meet the requirements. By preparing enough basic-sour, this can be divided on maturity into enough material for four full-sours, which in turn, on reaching maturity, each produce four batches of dough. The basic-sour is the best stage to use for propagation, owing to its superior acidification and fermentation performance. Also, basic-sour can be stood without becoming excessively acid.

**Dough Consistency, Substrate Viscosity and Dough-Yield**
All these variables are determined by the ratio of flour to water in the sour-dough. Ratios of flour:water 1:1 to 1:2 produce a more intensive acid build-up than 1:0·7 to 1:0·8 at temperatures over 30°C. Optimal acidification being obtained at 35°C, and flour:water ratio 1:0·9, according to Schneeweiss and Klose.

Both yeasts and sour-dough bacteria develop better in softer doughs. This is due to the nutrients' solubility, less resistance, and improved distribution at mixing. The homogeneous distribution of the nutrients in the softer doughs with yields 200–300, results in an acceleration of microbial activity of the homofermentative lactic acid bacteria. This gives the bread a mild, harmonious flavour-note, which is distinctive.

**Dough Substrate Nutrients**

Apart from the physical factors, the sour-dough contents are very essential to growth, i.e. assimilable forms of carbon and nitrogen, trace-elements, growth-factors and vitamins. This important complex of variables is identified by the baker as the 'reproductive-factor' (*Vermehrungsfaktor*). This requirement can be estimated for each sour-dough stage with a knowledge of the weight of flour used in the previous stage, and the necessary standing or maturing-time of the following stage.

$$W_2 = W_1 \times t$$

where $W_2$ = weight of flour for next stage, $W_1$ = weight of flour of previous stage and $t$ = standing-time for next stage. Given $W_1 = 10$ kg flour and $t = 5$ hours, then $W_2 = 50$ kg flour.

Using a 5 hour standing-time, the next stage will require the addition of 50 kg flour. If the standing-time is reduced, the 'reproductive factor' must also be reduced, and in the event of prolonged standing-times an increase in the 'reproductive factor' becomes appropriate. Standing-time and the 'reproductive factor' are inter-dependent. Flour extraction-rate is also involved in this relationship, lower extraction flours producing less acid build-up than the higher ones, this being due to differences in enzyme activity. Thus, when processing rye-flours of lower extraction-rates, a higher 'reproductive-factor' is required than in the case of higher-extraction flours. K. Fuchs (*Bread and Small Baked-Products* (in German), VEB-Fachbuchverlag, Leipzig 1968), established a considerable degree of tolerance between the 'reproductive factor' of basic-sour and full-sour, as follows:

| *Initial-sour: Basic-sour* | *Effect* |
|---|---|
| 1:2 | Optimal volume, heavily cracked surface. Irregular crumb and coarse texture with cracking |
| 1:3·5 | Fewer signs of surface cracks, texture regular but coarse with cracking |
| 1:7 | Good quality |
| 1:30 | Round cross-section, close texture |

Fermentation-times, acidity-index and bread-flavour showed no significant differences.

| *Acidity-index: Full-sour* | *Dough behaviour* |
|---|---|
| 1:2 | Increase of fermentation time by about 10 minutes |
| 1:2·5 | Normal bread quality |
| 1:3·5 <br> 1:5·5 | Bread quality consistent |

Aeration of the sour-dough stages encourage the reproduction of the sour-dough yeast cells, which ensure good leavening.

**Standing-time**

Each sour-dough stage requires a definite standing-time, and good bread quality results from allowing the various sour-dough stages time to reach maturity.

Standing-times vary considerably since they depend on the other process variables: dough consistency, temperature, and flour characteristics. The average range is from 2 to 5 hours, with the exception of overnight sours, in which case excessive souring is prevented by adding more flour, giving a firmer consistency, thus slowing down the activity of the microorganisms.

Since standing-time is the first and most important determinant for the production process, in common with wheat-flour processing, research is always directed towards finding ways of reducing standing-time to a minimum without loss of bread quality. Many bread-faults result from standing-times that are too short and immature doughs; fewer instances of over-ripe faults due to prolonged standing-times are encountered.

**Flour Quality**
The dough-yield and water absorption of a rye-flour is determined by its protein and pentosan quantity and quality and the properties of starch laid down during grain-ripening. However, superimposed upon the composition are the autolytic enzyme systems of the flour amylases and pentosanases, as well as proteinases, which are activated and reduce dough viscosity with standing-time.

It is the baker's management of flour composition, using the optimal combination of controllable process variables, and the addition of salt and other additives of organic origin that determine bread quality.

Apart from the flavour token of salt addition, it is a vital control raw material, acting in unison with acid-formation or addition to inhibit enzymic activity. Salt and acid together reduce enzyme solubility owing to a 'salting-out' process. At comparatively low acidity-indices, in the presence of salt, enzyme activity becomes inhibited. Enzyme activity in sour-doughs normally increase up to acidity-indices of 10–12, whereas such high acidities only normally occur at the basic-sour stage. Furthermore, such high levels in the final bread-crumb are restricted to rye-meal breads and pumpernickel (12–14). Therefore, the optimal baking properties of a rye-flour cannot be achieved by acidification alone, but only with the help of salt at about 1·8% flour weight. The presence of salt also delays the rye-starch gelatinization by 5–10°C, which places the alpha-amylase temperature optimum before gelatinization takes place, thus helping to preserve the structure at baking.

This double-action between dough acidification and salt is fundamental for the baking properties of rye-flour doughs. Controlled acidification results in a delayed peptization of the proteins, giving rise to swelling and an increase in dough viscosity. In an acid-free water dough, the protein would rapidly go into solution, lowering viscosity and triggering off an enzymic breakdown of both the pentosan material and the starch. Similarly, low salt concentrations would increase protein solubility, and favour enzymic activity. Only relatively higher concentrations, of the order of 1·8–2·0% flour weight, exert the desired inhibitory effect on the enzymes.

Rye-flour quality testing involves the following parameters: mineral-ash content to determine the flour-type, i.e. type 997, type 1150, etc.; sieve-analysis to check that the particle-size distribution falls within the limits of a flour-type. Changes in viscosity with temperature in the dough-stage are measured by the '*Quellkurve*' (swelling-curve), using the Brabender Amylograph. This consists of measuring the fall in viscosity of a standard flour/water suspension, first from 30°C to 42°C, and then on holding the temperature at 42°C for 30 minutes. Viscosity being measured in Brabender units, differences calculated, and expressed in logarithmic terms.

Dough behaviour during the 'hot' oven stage is measured by the standard Brabender method, using a flour/water suspension, 80 g:450 ml water. Significant parameters are: the Amylograph viscosity maximum, which is an indicator of the degree of polymerization of the starch; and the gelatinization temperature, which indicates the degree of swelling/gelatinization, and resistance to alpha-amylase. About 63°C is an optimum temperature maximum; flour samples below this level require appropriate blending or the use of an enzyme inhibitor, e.g. 0·2% lactic acid. Samples with temperatures in excess of 68°C will give a dry bread-crumb unless blended with another flour, or given appropriate enzyme supplementation with fungal or bacterial amylases, proteinases and/or pentosanases.

Average Falling-number (Hagberg–Perten) values for rye-flours range from about 80 to 200, depending on crop-year and extraction-rate. Types 997 and 1150 should not have values less than 115, as a general processing guide.

Rye-flour water absorptions, measured at 350 Farinograph units, range from about 64 to 73, depending on crop-year and extraction-rate.

Test-bakes should be carried out in the bakery, evaluating the baked bread in terms of volume (ml/250 g), loaf-symmetry (height/width), and sensory evaluation of crust, crumb structure, colour and flavour.

## 2.4.2 PROCESSING STAGES

Having discussed how the various process parameters can influence the souring and leavening of a rye-flour dough, it should be possible, irrespective of the raw material, flour, to provide the necessary conditions for the diverse bread-varieties. By processing in more than one stage, the following objectives can be achieved:

—sour-dough yeast reproduction,
—sour-dough bacterial reproduction,
—inhibition of wild microflora,
—achievement of the optimal pH and acidity-index,
—controlled swelling of part of total flour.

Using discontinuous or 'batch' processing, and several process-stages, changing temperature, dough-consistency, standing-time and the reproductive-factor to suit the various microorganisms, a large quantity of dough can be leavened and acidified with a small amount of starter.

For accurate control of these process parameters in the small bakery, a fixed sour-dough schedule, thermometer, scales, and clock are basic equipment. In addition, a facility for testing the acidity-index of each sour-dough stage is desirable to ensure consistency in bread quality.

Although many bakeries no longer use the multi-stage processes, the process demonstrates an exact knowledge of the procedures involved with sour-doughs, and a correct application of other shorter processes in production schedules.

Taking a three-stage sour-dough process as an example of a multi-stage schedule, fermenting the basic-sour overnight:

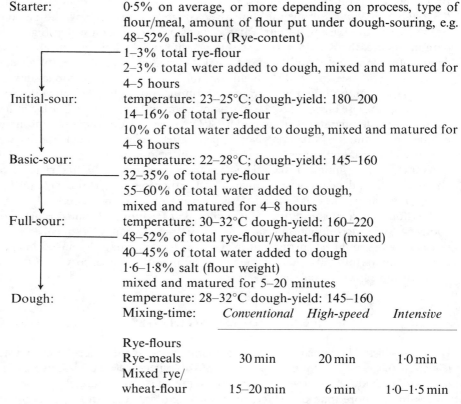

| | |
|---|---|
| Starter: | 0·5% on average, or more depending on process, type of flour/meal, amount of flour put under dough-souring, e.g. 48–52% full-sour (Rye-content) |
| | 1–3% total rye-flour |
| | 2–3% total water added to dough, mixed and matured for 4–5 hours |
| Initial-sour: | temperature: 23–25°C; dough-yield: 180–200 |
| | 14–16% of total rye-flour |
| | 10% of total water added to dough, mixed and matured for 4–8 hours |
| Basic-sour: | temperature: 22–28°C; dough-yield: 145–160 |
| | 32–35% of total rye-flour |
| | 55–60% of total water added to dough, mixed and matured for 4–8 hours |
| Full-sour: | temperature: 30–32°C dough-yield: 160–220 |
| | 48–52% of total rye-flour/wheat-flour (mixed) |
| | 40–45% of total water added to dough |
| | 1·6–1·8% salt (flour weight) |
| | mixed and matured for 5–20 minutes |
| Dough: | temperature: 28–32°C dough-yield: 145–160 |

| Mixing-time: | Conventional | High-speed | Intensive |
|---|---|---|---|
| Rye-flours Rye-meals | 30 min | 20 min | 1·0 min |
| Mixed rye/ wheat-flour | 15–20 min | 6 min | 1·0–1·5 min |

The amount of rye-flour put under dough-souring (full-sour content) is an important technological parameter, and varies with the type of bread being produced as follows: rye breads, 40–50%; mixed rye/wheat breads, 40–60%; mixed rye/wheat-flour (70:30), 50–100%.

During the processing of a 100% rye-flour dough, the normal range for the acidity-indices are as follows: initial-sour, 6–8; basic-sour, 10–13; full-sour, 7–10; dough, 7–9. The acidity-index (according to Neumann) is measured by mixing 10 g of sour-dough with 100 ml distilled water using a mixer. Within 3 minutes add 5–10 drops of phenolphthalein-solution, and titrate with 0·1N NaOH. Titration is continued until a red colouration develops. The number of millilitres of NaOH required gives the acidity-index. The method according to Schulerud, using filtration, gives values on average one unit higher than that of the Neumann.

Acidity-index guidelines for the various types of bread are given in Table 53. The acidity-index is a measure of the total acidity of the dough, with no indication of the content of the various acids. In a well-controlled sour-dough the ratio of lactic to acetic acids should fall within the range 80–85% lactic: 15–20% acetic.

TABLE 53

|  | Acidity-index | |
|---|---|---|
|  | Neumann | Schulerud |
| Whole rye-bread | 7–9 | 8–10 |
| Mixed rye/wheat 70:30 | 6–8 | 7–9 |
| Pumpernickel | over 12 | over 14 |
| Wheat bread | 4–6 | 5–7 |

Both acidity-index and pH give valuable information during the processing of flour into bread, and are essential for optimal management of the process-schedule. There exists an approximate relationship between acidity-index and pH, but pH depends on the buffering-capacity of the flour and its mineral content. Therefore, the acidity-index is a more reliable scale of measurement. Approximate equivalent values are given in Table 54. The use of a pH-meter allows measurements to be made either in aqueous suspension or in more solid substrates. One unit on the acidity-index scale is equivalent to 0·09% lactic acid, therefore acidity-index 10 indicates a lactic acid content of 0·9%. The acidity-index figure is, however, inclusive of mineral acidic phosphates present in flours.

A typical schedule for a multi-stage overnight basic-sour, based on 100 kg rye-flour, the full-sour containing 48% of total flour is given in Table 55.

The final-dough mixing times in the case of rye-flour and rye/wheat-flours tends to be shorter than for wheat-flours alone, since most of the raw materials have already been mixed at the basic/full-sour stages, and only a small amount of water, salt, with a larger amount of flour needs to be worked into the available mature sour-dough (full-sour, or sometimes the basic-sour). Dough-yields from rye-flours are on average 6 to 10 units higher than from wheat-flours. Mixing is complete when all flour particles, water, and sour-dough from the previous stage are homogeneously integrated into a plastic dough.

Rye-flour has about 40% soluble-proteins, compared with wheat-flour which contains only about 25% soluble-protein fraction. The soluble-proteins are mainly gliadins, having considerable swelling capacity. The presence in rye-flour of the high-viscosity pentosans prevents the integration of the proteins into a gluten-type complex formation. Instead, the acidified soluble-proteins have swelling properties,

TABLE 54

| Acidity-index | pH |
|---|---|
| 6 | 4·7 |
| 8 | 4·5 |
| 10 | 4·1 |
| 12 | 4·0 |
| 14 | 3·8 |
| 20 | 3·6 |

TABLE 55

| Sour-dough stage | Time (h) | Sour-dough (kg) | Flour (kg) | Water (litres) | Total weight (kg) | Yield (%) | Acidity index |
|---|---|---|---|---|---|---|---|
| Starter | 06·00–12·00 | 1 | — | — | — | — | — |
| Initial-sour | 12·00–18·00 | 1 | 3 | 3 | 7 | 200 | 6–8 |
| Basic-sour | 18·00–03·00 | 7 | 15 | 6 | 28 | 150 | 10–13 |
| Full-sour | 03·00–06·00 | 28 | 30 | 25 | 83 | 175 | 7–10 |
| Dough | 03·00–06·20 | 82[a] | 52 | 20 | 154 | 154 | 7–9 |

[a] 1 kg final-dough retained as starter for next batch.

the lower dough pH inhibiting the amylases, proteinases, and pentosanases, thus allowing the dough mass to become leavened throughout. During the stepwise maturation of the sour-dough stages, both the protein and starch fractions progressively swell, taking up about three times and one-third respectively of their weight of water. The insoluble cellulose fraction, especially in rye-meals or grits store large amounts of water onto the surface pores (surface-action), which unlike colloidal swelling, becomes lost during baking, giving rise to a moist crumb. Therefore, differences in rye-flour composition must be taken into consideration during mixing. The temperatures quoted above for the individual sour-dough stages and final dough are valid for the tabulated basic-sour schedule given.

In the small bakery, the process begins with the manual preparation of the initial-sour. The starter (mature basic-sour) is broken down in the measured water at about 25°C, and the pre-weighed amount of flour evenly distributed. Small quantities of this mass are then spread over the larger dough mass, until the dough has an even texture, including all dough scrapings from the sides of the container. Further dough mixing is completed using mixers of various design (rotary; pressure and lifting with an elliptical orbit in one plane; Wendel-type with two cone-shaped elements mounted in the cover, rotating; one-arm high-speeds, with arm vertically mounted, oscillating elliptically from bowl-perimeter towards centre and upwards; spirals, adjustable at different heights, rotating about own axis; intensive-mixers with two elements mounted on one vertical rotating shaft at high speed). Mixing-times quoted on p. 502 under the headings: conventional, high-speed, and intensive, are intended as guidelines for rye-flour and mixed wheat/rye-flour doughs.

Final-dough discontinuous 'batch' or 'charge' mixing is completed by mixing: flour, water, full-sour or basic-sour, salt-solution, dough-conditioner (as necessary), and yeast to a smooth dough with a dry surface. For convenience a 25% salt-solution is used instead of adding dry-salt, the amount, in litres, of this solution added to give a desired dough salt concentration, being deducted from the required total amount of dough water.

Stale bread, dried and milled to flour granularity, from the same bakery, of hygienic and microbiologically acceptable quality, can be added to the dough for reprocessing at up to about 3% (flour-weight). This material is soaked in water at about 24°C, and added at the final doughmaking stage.

An alternative three-stage sour-dough process would be to run the full-sour overnight, the full-sour acting as the main reproductive factor, standing for 8 h to provide a reproductive factor of 10 to 12. In this case, the basic-sour would be run for only 5 hours, at a reproductive factor of about 5 only.

This illustrates how the process control variables of temperature, standing-time, dough-yield (consistency), and reproductive-factor can be varied to suit production requirements. The variable standing-time being laid down first, and the other three variables regulated in such a way as to ensure dough maturity at the fixed time. These four variables, together with the mixing and work-off operations constitute a specific process.

For the following reasons, science and technology have worked to simplify the multi-stage sour-dough processes:

—reduction of sensitivity to flour quality, temperature and weather,
—shortage of skilled labour,
—time saving,
—reduction of sour-dough faults due to under- or over-ripeness,
—less fermentation-loss; higher yields,
—savings in fermentation-containers and space.

During the period 1950–60, many diverse short sour-dough processes were introduced for batch schedules. The basis of all these processes is to separate the various functions of sour-doughs, viz. leavening efficiency, formation of adequate acids for optimal pH and total-acidity, and for formation of flavour and aroma components. One such process, known as '*Grundsauer ohne Vollsauer*' (G.o.V.), i.e. 'basic-sour without full-sour', dispenses with the initial- and full-sour stages, which serve mainly for yeast reproduction and fermentation. Instead the process is progressively continued from one basic-sour to the next in one stage, the starter material being taken from each mature basic-sour for the following one.

A basic-sour is prepared with a dough-yield (consistency) of 150, i.e. 2 kg rye-flour/1 litre of water. The amount of starter necessary is a minimum of 5%, based on flour weight; but this quantity can be increased. For example, if 50% starter is used, the basic-sour matures within 4–6 hours, in which case the residual sour from each day can be utilized as starter. Peak maturity is reached when the bacteria are producing enough acid, which is ascertained from the volume increase of the dough-mass, and the cracks or crevices which appear on the dough surface when covered with flour. Also the experienced baker is able to detect progress of fermentation by smell. A sharp, pungent smell being characteristic of excessive acetic acid formation, and a milder, more aromatic smell, indicates the desirable lactic acid formation. A state of maturity is also indicated by the fermentation performance of the yeast. The use of 500 g starter/10 kg flour in the sour, should provide a basic-sour with a pure aroma and flavour over a standing-time of 11–12 hours, maintained at about 24°C. The basic-sour can provide better acid-formation and fermentation performance within a temperature latitude of 20–30°C, although 24°C is optimal since it is nearer to the ambient bakery temperature. Furthermore, it is self-conserving, inasmuch as after 12 hours standing, the acidity-index remains comparatively constant, inhibiting any undesirable fermentations. A basic-sour can stand all day or remain over a weekend without enrichment, since after 12 hours the sour reaches an acidity-index of about 12, and after 48 hours, total acidity content approaches about 2%, further acid formation ceasing. However, the acid-forming bacteria retain their full biological vigour for the next sour-dough.

Making use of the 'basic-sour to basic-sour' scheme, the process can be maintained over months and years by taking starter from the mature basic-sour. Choice of a relatively firm consistency and moderate temperature, over a long time produces acetic acid from the alcohol up to about 25% of total dough acids. This ensures, depending on flour type and mixture, an acetic acid content in the final bread of 10–15% of total acids, which is adequate for bread aroma.

The basic-sour process can be used to produce all types of bread containing rye-flour, any acidity adjustments, according to sour-dough acidity-index, being made by the addition of a commercial brand of dough acidulant, containing organic acids (lactic, citric, adipic acids) or their esters. The basic-sour provides the aroma substances, and dough-acidification being achieved by final adjustment as necessary. Leavening is provided by adding about 0·8–1·5% baker's yeast (flour weight), depending on flour mixture and size of dough.

The following process scheme gives an indication of how a basic-sour can be used to produce three different types of bread, in either hearth (oven-bottom) or small rounded shapes.

*Flour mixtures based on 100 kg total flour weight:*
Using rye-flour types 997 or 1150, and wheat-flour types 550 or 1050
Flour mixtures: rye-flour/wheat-flour 50:50, 70:30, 100% rye-flour
*Basic-sour:* 24°C; standing-time 12 hours
Starter taken from mature basic-sour    1800 g

| | |
|---|---|
| Rye-flour | 35·0 kg |
| Water | 17·5 kg |
| | ——— |
| Basic-sour | 52·5 kg |

| Raw materials | Mixed-bread 50:50 | Mixed-bread 70:30 | Rye-bread 100% |
|---|---|---|---|
| Basic-sour | 15 kg (10%)[a] | 15 kg (10%)[a] | 22·5 kg (15%)[a] |
| Rye-flour | 40 | 60 | 85·0 |
| Wheat-flour | 50 | 30 | — |
| Water (approx.) | 60 | 60 | 60·0 |
| Yeast | 1 | 0·840 | 0·875 |
| Salt | 2 | 2·100 | 2·200 |
| Dough-acidulants: depending on dough acidity-index | | | |
| Dough-temperature | 26°C | 27°C | 29–30°C |
| Proof-time | 25 min | 35 min | 45 min |

[a] Percentage-sour used to prepare dough.

The basic-sour reaches a pH of about 4·0 within 10–12 hour standing-time, and remains constant at that level.

Both rye and mixed rye/wheat breads made by the basic-sour scheme show large volume, fine/even crumb-porosity, and a brighter crumb-colour than is the case with multi-stage sour-dough processes. The desired crumb-elasticity, and acidity-index for each type of bread can be adjusted with a commercially available dough-acidulant which also allows certain local preferences in bread flavour to be met, from mild to strongly aromatic.

Process-time and loss of dough-solids have led to other 'direct' processes, whereby the acid formation of the microorganisms are replaced to varying degrees by the addition of organic acids, such as: lactic, acetic, citric, adipic acids, amongst others. The crystalline acids are mixed with carrier-substances, such as *'Quellmehl'* (a heat-treated potato-starch) to add to the dough. The liquid acids are added in a diluted form. However, these processes usually also involve the use of a minimum of biologically produced sour-dough, usually basic-sour (e.g. the G.o.V. process previously described), in combination with acidulants.

Since, in these 'direct' processes no sour-dough yeasts can be produced, baker's yeast is added, allowing a 3-hour ripening-period to activate the yeast. The inclusion of a biologically produced sour-dough component for aroma-enhancement is achieved by processing 10–17% of the rye-flour into a basic-sour, the processing parameters of which are:

| | |
|---|---|
| Dough-yield (consistency) | 155% |
| Temperature | 24°C |
| Standing-time | 16–24 h |

This procedure involves the increase in flour content by a factor of 32, compared with that in the starter.

Further pressures for rationalization have led to a reduction in the function of the various sour-dough stages, instead several single-stage processes have emerged, dough-yields (consistency) and temperatures being so adjusted that an adequate amount of acid is developed. Since the sour-dough yeasts cannot develop under such conditions, baker's yeast is added at the doughmaking stage. The following is a summary of the well-known single-stage processes.

## Berlin Short-sour Process *(Berliner Kurzsauerteigführung)*

The process parameters are so chosen that the heterofermentative bacteria can produce an adequate amount of acid (acidity-index 10–13) within 3 hours. This ensures an optimum ratio of acetic/lactic acid of 30:70 for bread aroma. The essential processing parameters are:

| | |
|---|---|
| Dough-yield | 180–200% |
| Temperature | 35°C |
| Standing-time | 3 hours |

Increase in flour content five times that in the starter.

## Detmolder Single-stage Process

This process is designed to give a longer standing-time (maturing or ripening-time), with average fermentation tolerance. An improved fermentation tolerance is achieved by lowering the dough-yield (consistency) to 160–180%, and the temperature down to 20–28°C, and increasing the amount of flour subjected to the souring-process. Processing parameters are:

Dough-yield                    160–180%
Temperature                    20–28°C
Standing-time                  15–20 hours
Increasing the flour content five times that in the starter.

## Salt-sour Process

The salt-sour process gives both a longer standing-time, and high fermentation tolerance, owing to the addition of 2–5% salt to the sour-dough. Salt inhibiting the metabolism of the microorganisms, thus prolonging the maturing-time during the period of adaptation. The acidity-index reaches about 10, on maturity, after which acid-formation is retarded, giving a high fermentation tolerance. Processing parameters are:

Dough-yield                    200–220%
Temperature                    35–38°C
Standing-time                  20–72 hours
Flour content increased six times that in the starter.

A more recent modification of this process, known as 'the new Monheimer salt-sour process', due to E. vom Stein (1974) gives an improved bread flavour. The salt addition is reduced from 5% down to 2% of sour-dough weight, in addition, the initial high sour-temperature of 35°C, set at the beginning of the standing-time, is allowed to fall to 20–25°C. These changes result in less loss of bread-aroma, but demand the inclusion of a higher level of starter (approximately 20% of the flour in the sour-dough). The acetic/lactic acid balance in the sour gives an improved mildly acid taste to the final bread.

A reliable guideline for the production of mixed wheat/rye-flour doughs, irrespective of the choice of process, is to subject at least 50% of the total rye-flour content to the sour-dough process. In this way adequate acidification for crumb firmness and flavour is ensured, and a medium acidic intensity (pH 4·1–4·0, and acidity-index 10–13) is produced in the sour-doughs.

Where sour-doughs are stored for several days, a stirring-aggregate with a refrigeration facility is recommended, using temperatures in the range 5–15°C.

Table 56 gives a typical scheme for the production of mixed rye/wheat-flour bread by the short-sour process.

## Quasi-continuous Process

This process is often described as 'continuous', assuming that the further processing of the sour-dough into the final dough proceeds continuously. In fact, it is quasi-

TABLE 56

| Stage | Standing time | Sour (kg) | Flour (kg) | Water (kg) | Total (kg) | Dough yield (%) | Temperature (°C) | Dough addititives |
|---|---|---|---|---|---|---|---|---|
| Starter | — | — | — | — | 9·5 | 190 | — | — |
| Short-sour | 3–4 h | 9·5 | 50[a] | 45 | 104·5[c] | 190 | 35 | — |
| Dough | 15 min | 95 | 20[a]/30[b] | 15 | 160 | 160 | 30 | 2 kg yeast 1·5 kg salt |

[a] Rye-flour type 997.
[b] Wheat-flour type 550 or 1050.
[c] 9·5 kg starter retained.

continuous since the final dough is prepared on a 'batch' or 'charge' basis, although the sour-dough preparation and maturing-system operates continuously.

Typical examples of quasi-continuous sour-dough processes are the KVT 1000, 1500 and 1800, producing 1000, 1500 and 1800 kg/h of rye-flour or mixed rye/wheat-flour bread (*Mischbrot*), respectively. This process is characterized by the following process-parameters:

Dough-yield of full-sour          230%
Temperature                            28–29°C
Standing-time                           3·5–5 hours
Starter flour-content increased 6–10 times.

These plants, together with the FTK 1000, are described in detail in Section 1.7.6. The following illustration shows the process-schematic for the KVT 1000 continuous doughmaking plant of Czechoslovakian origin.

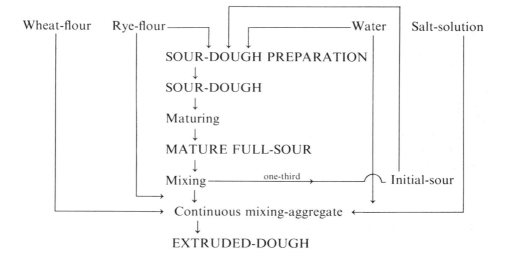

TABLE 57

| Stage | Sour (kg/min) | Flour (kg/min) | Salt- solution (kg/min) | Stale- bread paste (kg/min) | Total (kg/min) | Temperature start (°C) | Dough- yield (%) |
|---|---|---|---|---|---|---|---|
| Dough | 14·46 | 10·92 | 1·46 | 1·16 | 28·0 | 30 | 162 |

Acidity-index 7–9; pH 4·3–4·6
Average output: 1500 kg/h rye-flour or rye/wheat-flour dough

Source: *Special Bakery-processes*, 2nd edn, by Werner Schwate and Udo Ulrich, VEB-Fachbuchverlag, Leipzig, 1986, p. 38.

Table 57 gives the feed rates for the dough-preparation scheme for the KVT 1500.

During continuous processing, the dough-yield is measured every 2 hours, both for the full-sour and the final dough. This ensures that the flow of raw materials from the dosage-aggregates are in order. For this purpose, 10 g of dough are placed between the plates of an electroplate-drier, and the loss in weight calculated as a percentage of the 10 g weigh-in. The standard drying procedure being 26 minutes at 180°C (GDR/TGL 29 066). Dough-yield is defined as the mass of dough obtained from 100 parts of flour, and is expressed as a percentage, which also expresses binding-ratio (absorption) or flour:water. 100 parts flour: 60 parts water, means a dough-yield of 160% and a water-absorption of 60%. Dough-yields vary depending on flour-type and type-quality. Average values, as guidelines, are as follows:

| Flour-type | Average dough-yield |
|---|---|
| Rye-flour types 997/1150 | 166 |
| Whole-rye flour | 170 |
| Wheat-flour type 550 (patent) | 160 |
| Wheat-flour type 550 gluten-rich | 158 |
| Wheat-wholemeal | 165 |
| Rye/wheat-flour: | |
|    Rye-flour type 997/1150 70% | |
|    Wheat-flour type 550 | |
|    gluten-rich 30% | 158 |

Using the KVT, the energy-output curve of the mixer-motor can, to a degree, be utilized as an objective control of dough-yield. Variations can be recognized, and adjustments made, bearing in mind other mixer covariables such as feed-rates, mixing-times, residence-times. Registered mixer-curve energy levels within the range 1·2 to 2·3 kW correspond to dough-yields of 164% to 176% for the KVT mixing-aggregate.

These sour-dough processes must be organized in such a way that a mature full-sour is always available for continuous dough preparation. These continuous plants utilize a full-sour to full-sour process, organized on the basis of a weekly production schedule. The important process parameters are: temperature, dough-yield/consistency, maturing-time, and acidity-index, to ensure optimal development of the microorganisms, and balanced acid-production. The process is initiated by

TABLE 58

| Stage | Standing-time (h) | Sour (kg) | Flour (kg) | Water (kg) | Total (kg) | Temperature (°C) Start | Temperature (°C) End | Dough-yield |
|---|---|---|---|---|---|---|---|---|
| Starter | 17 | 37·2 | — | 93·3 | 130·5 | store | cool | — |
| Initial-sour | 30 | 130·5 | 148·5 | 102·0 | 381·0 | 26 | 28 | 235 |

Source: *Special Bakery-processes*, 2nd edn, by Werner Schwate and Udo Ulrich, VEB-Fachbuchverlag, Leipzig, 1986, p. 21.

preparing an initial-sour, which involves adding a starter on a 'batch' or discontinuous basis. Subsequently, the starter is then always taken from the mature full-sour.

Table 58 gives the sour-dough preparation scheme for the KVT 1500.

The acidity-index of the mature initial-sour reaches 16–18, at a pH 3·5–3·6. When mature, the initial-sour is added to the mixer/agitator, from which it is fed together with flour and water through pipelines into the sour-dough beater/mixer. This forms the basic-sour, being a continuously maintained reproductive development of the initial-sour. This is pumped into one of the compartments (sections) of the fermentation (ripening) container to mature for 3 hours, when it reaches an acidity-index of 10–12, at pH 3·8–3·9.

The matured basic-sour is then remixed with more flour, the consistency adjusted (if necessary) to about 235%, having been conveyed by pipeline to the sour-dough beater/mixer. It is then returned to the appropriate section of the fermentation-container via cogged-wheel type pumps automatically controlled by a central time-relay system. After a further residence-time of 3 hours, the mature full-sour is produced, with an acidity-index of 9–11, at pH 3·8–4·0. The mature full-sour is then divided, two-thirds going to the final dough mixing-aggregate, and the residual one-third returned to the fermentation-container for continuous propagation of the full-sour. The sections of the container rotate at a predetermined time-interval within the container in carousel-fashion. The KVT plant is 90% automated, requiring only two trained operators for doughmaking and line-control. The maximal rotation-time of the fermentation container is 180 minutes, corresponding with the necessary maturing-time. When the maximum level of sour-dough in a section is reached, the section is automatically rotated, so that the filling and emptying of sour-dough is continuous. The average dosage-time to fill each section is about 6 minutes, but the dosage-time on 'start-up' of the plant varies considerably, since the basic-sour and the first full-sour first have to reach maturity. This extra standing-time is no longer required during continuous production. Number 12 section is kept free during production for the following full-sour, and during the dosage-periods and filling of the sections, the appropriate section stands under filling-orders. This is very important to ensure that the mature full-sour does not become mixed with the immature full-sour. All filling and emptying operations of the sections are under the control of the time-relay system, which is under the direction of the line-operator.

Mature full-sours are pumped into the sour-dough beater/mixer, where it becomes degassed for accurate metering. About 40% of the gas-free dough, approximately one-third, goes to the sour-dough-mixer, and the remaining 60% to the final dough-mixing aggregate via cogged-wheel pumps. The mature full-sour which goes to the final dough-mixing aggregate contains up to 42% of the total flour content of the final dough. Mature full-sour, flour, water and salt-solution are continuously fed into the mixing-aggregate, yeast and stale-bread paste also being added, according to the type of bread being produced. The aggregate is capable of intensively mixing the dough within 4–5 minutes, the dough leaving the mixer as a continuous extruded band. It is then transported by conveyor-belt to the divider-moulder, the transport-band conveyor serving as a travelling-proofer. This proofer-band, which is about 1500 mm wide is U-shaped and also covered, and can be set to the desired time-interval, normally 25–30 minutes. A remix or 'knock-back' is taken over by the transport-worm of the divider-moulder.

Rye and mixed rye/wheat-flour are sensitive to proof-times, since it determines the dough-swelling and acidity. Therefore, treatment depends on flour/flour blend rye:wheat, mixing-times, and sour-dough process utilized.

Processing with the KVT, this distance to the divider-moulder has a very positive effect on dough-ripening.

The dough-band continuously enters the divider, and is divided by volume, the dough being fed into a cylinder by two worms, forced by pressure through the cylinder and the dough-length so formed being cut off by a knife. The length of the dough-strip thus determines the weight of the dough-piece. The long-rolled dough-pieces which emerge then pass via a spreader-band into the pockets of the transport system of the final-proofer, which accommodates the dough-pieces in rows of about five. By adjusting the chain length, the residence-time, and hence the proofing-time can be varied from about 30 minutes up to 60 minutes. The necessary climatic conditions within the proofer are provided by heating-elements and an atomized steam-spray, being maintained at 33–35°C, and 70–80% relative humidity. The proofed pieces are then tipped out of the proofing-pockets onto the oven-conveyor, over which is positioned a cutting/stamping aggregate, before entering the wire-band oven of average baking-surface 50 m².

**Special Make-up and Work-off Procedures for Rye-flour Doughs**
Relative to wheat-flour doughs, the structure of rye-flour doughs tends to be less stable, which is the reason for the extra need for care and skill during processing of the latter. With increasing wheat-flour content processing becomes increasingly the same as that employed for wheat-flour doughs. When the wheat-flour content increases to 50%, in common with wheat-flour doughs, a resting-period (relaxation) is recommended between the first rounding-up (deformation) and final elongation or long-rolling of the dough-pieces, amounting to 5–10 minutes.

The dough surface of rye-flour doughs also 'stick' more than wheat-flour doughs. Therefore, during make-up far more 'dusting' of the dough-surface with flour is required. This flour-film will, however, become moistened and integrated into the dough-pieces on final-proofing.

The techniques for shaping and proofing of rye-flour doughs are also different from those for wheat-flour doughs. There are three ways of setting rye doughs in the oven: as hearth or oven-bottom bread, either rounded or elongated, by long-rolling for proofing in special round or elongated 'proofing-baskets'; coated with fat proofing as a batch, to a height below the top of a pan, referred to as '*angeschobene Brote*' varieties; normal rectangular panned bread, proofed to full pan height with no disinct 'break and shred'. More exact definitions of the various baking-processes used follow.

*Hearth or Oven-bottom Varieties ('freigeschobene Brote')*
The dough-pieces are placed in the oven alongside one another, so that each piece can acquire an even thickness of crust over its entire surface. Typical rye-bread varieties which fall into this category are: rustic types such as '*Landbrote*' and '*Bauernbrote*', i.e. *Landbrot* either rounded or elongated; mixed rye/wheat-flour varieties such as those produced by industrial bakeries using continuous processing, e.g. KVT 1000, 1500 and 1800 in the GDR, viz. 1 kg 100% rye-flour bread, and mixed rye/wheat-flour breads, known as '*Mischbrote*'.

*'Angeschobene Brote'*
The dough-pieces are placed close to one another in the oven with the sides touching, fat, oil or oil-in-water emulsions being used to prevent sticking. These doughs must be made up to a firmer consistency than oven-bottom varieties. The breads are without crust along the sides, but have a well-developed top crust. Typical examples are: *Paderborner Bauernbrot*, *Gegerstertes Brot* (Hannover and Lower Saxony), and *Kommissbrot*, all produced in the FRG.

*Panned Varieties*
This mode of setting is not widely used for rye-flour doughs in general, being more appropriate for less firm doughs made from wheat-flour or wheat meal; the amount of crust is less than in the case of oven-bottom varieties. Apart from panned white-bread and toast-bread, other varieties baked in the pan include: ryemeal types, *Rhinelander* and *Holsteiner*, *Steinmetzbrot*, *Simonsbrot* (wheatmeal type), *Keimbrote* (germ-breads made from wheat and rye mill-products).

## Interactions between Methods of Processing, Bread-shape, Bulk-ferment or Floor-times, Equipment and Make-up Procedures

In the case of hearth, *angeschobene Brote*, and panned varieties (as appropriate), the dough is shaped by long-rolling either manually, or by machine. The dough-pieces acquire during the final-proof stage a change in shape, according to the shape of the pan used. The objective of a bulk-ferment or floor-time for a rye-flour or mixed rye/wheat-flour dough is:

—to activate the sour-dough microorganisms,
—to improve the structural-lattice of the dough,
—to promote the processes of swelling and solubility,
—to increase the dough acidity-index, and lower the pH to 3·8–4·0 to control enzymic activity.

The bulk-ferment and final proof-times are, in general, interchangeable. A prolonged bulk-ferment permits a shortened final proof-time, and a shortened, or no, bulk-ferment demands a prolonged final-proof. The duration of the bulk-ferment will depend on the flour type, and/or the ratio of rye-flour to wheat-flour, the mixing-intensity, and the choice of sour-dough process. In the case of light rye-flours, i.e. type 815 and 997, only a short, or perhaps no, bulk-ferment is required Water-binding is rapid owing to the relatively low content of pentosans (the active swelling component in rye-flour); there is also no delayed swelling. In contrast, rye-wholemeals and ryemeals require a longer mixing-time, but in spite of this the swelling and solubilization processes will not have been complete. However, after a resting-time of about 25–30 minutes, these processes will have been completed, indicated by a drying-out of the dough-surface and an improved stability.

When longer bulk-ferments are employed, a remix or knock-back is recommended to degas the dough, rendering its density more uniform for weighing and shaping/moulding.

Short bulk-ferment is appropriate in the following circumstances:

—processing light rye-flours (lower extraction-rates) type 997, and 100% rye-flour doughs,
—when longer mixing-times are used,
—when longer sour-dough processes are in use (three-stage process).

Longer bulk-ferment is appropriate in the following circumstances:

—processing rye-wholemeals and ryemeals of various degrees of fineness,
—when mixing-times are short,
—when processing lower extraction rye-flours, blended with wheat-flour in the ratio 70:30,
—when using shorter sour-dough processes (*Kurzsauer* or two stage processes).

When processing higher extraction rye-flours and course meals, a remix or knock-back is essential for about 5 minutes, which acts as a second mixing-stage, allowing the late-swelling process to be exploited.

Technology and equipment varies somewhat, depending on whether the processing methods are craftsmen-orientated/discontinuous, or continuous line-production. The craftsman's method is to proof the dough-pieces finally on flat boards covered with cloths, which are folded to separate and partially cover each piece. Both cloths and dough are well dusted with dusting-flour or starch-powder to prevent sticking. This technique is appropriate for 1 kg and smaller 500 g hearth (oven-bottom) units made from rye-flour and wheat/rye-flour blends. Shaping and moulding would either be done by hand-moulding or with the help of a long-rolling machine, fitted with an adjustable pressure-board for setting the required length of the dough-piece. The manual process make-up sequence is: weighing—rounding—long-rolling—final-proofing—baking.

A general guide to final-proofing parameters is as follows:

| | |
|---|---|
| Temperature | 35°C |
| Relative humidity | 70% |
| Time | 50 minutes |

Average ferment or floor-times, depending on the choice of process for rye doughs are:

| | |
|---|---|
| Indirect, multi-stage sour-dough processes | 10–15 minutes |
| Direct processes, using acid-preparations and yeast additions | 180 minutes |
| Direct processes, combined with a partial biological souring of the flour | 50–60 minutes |

Panned bread varieties are baked in elongated pans either greased or coated with silicon, with or without covers.

Continuous line-production involves final-proofing in hanging containers moving within a controlled climate. Hearth breads are dusted with starch-powder, and panned varieties are placed in silicon-coated pans. Closely batched varieties (*Angeschoben*), e.g. *Paderborner, Kommissbrot* and *Gegerstertes Brot* must be made up firmer than hearth varieties, and the sides greased with fat/oil or an oil-in-water emulsion, whether craftsman or line-processed. A uniquely designed and engineered system for producing rye sours progressively for continuous automated industrial production, is the result of a joint design and construction patent due to Vereinigte Nahrungsmittelindustrie (VNI) and Reimelt of Frankfürt/Main, FRG. The first plant of this type, producing 4·8 t sour/h, was installed at the VNI Bakery in Vienna in 1979, one of Europe's largest bakeries. Conversion to this fully mechanized, continuous, automatic plant from a manual sour-dough system has been achieved with no bread-quality problems. Its operation requires one person only. It is basically a two-stage sour-dough process, utilizing a sealed fermentation-tube for the first stage, and a cylindrical fermentation-tank for the second stage. The first stage can be stored for four days under technological control without excess acid build-up or damage to the fermentation organisms, thus allowing production stoppages at any time. The second-stage provides a sour of uniform quality for bread production. By careful selection of the process parameters, the fermentation power of the sour can be maintained to give a balanced lactic/acetic acid content of 75:25%. This process provides considerable saving in raw material costs, no acid-preparations or use of baker's yeast are required. Also, no periodic inoculation with fresh pure-cultures are necessary. Hitherto, such advantages were only possible with a classical three- or four-stage process or the 'full-sour to full-sour' system previously described. Even these latter processes have distinct limitations for large-scale industrial production, such as:

—uniformity of sour-dough cannot be guaranteed after prolonged shut-downs, e.g. weekends, vacation periods;
—changes in demand with time, and the production of a diverse product range using several production-lines can be difficult.

The important technological parameters of the process are:

| | |
|---|---|
| Dough-yield | 190% or more |
| Temperature | max. 30°C |

This ensures:

—conditions favourable to the homofermentative microorganisms,
—presence of the sour-dough yeasts,
—full-sours with acidity-indices 8–10, at pH 3·6–3·8.

Since the first stage takes place in a sealed tube, which is provided with a cooling facility for reducing the sour temperature to below 10°C, infections are prevented, and no surface drying-out can occur. Experience to date has shown that the single-stage processes, as used extensively by the craftsman-baker, are ideally flexible for the production of a very diverse variety of breads on a discontinuous batch or 'charge' basis. However, a more rational solution to the problem would be to produce sour-doughs continuously. Currently, pilot-plants are being used based on the VNI-Reimelt design model, using a fermentation-tube and tank for stage 1 and 2 respectively. In order to optimize such important process parameters as: microflora, substrate-effects-on-reproduction-rate, and lactic/acetic acid production ratio, laboratory-fermenters can be utilized to establish a basis for the construction of industrial-fermenters, Using these fermenters, the aim is to produce a wide range of sour-doughs from the various types of flour as substrates.

In the USSR, sour-dough processes for the production of the various types of rye and mixed rye/wheat-breads are also diverse. For the production of 100% rye-wholemeal bread, either a firm sour-dough (*golowka*), or a softer one (*kwass*) is normal practice. '*Golowka*' is prepared in four stages: initial-sour, intermediate-sour, basic-sour and full-sour. The production cycle is shortened by preparing the full-sour (*golowka*) on a continuous basis, thus involving two stages only: (1) full-sour, and (2) final-dough, i.e. full-sour to full-sour.

For batch processing, using machine-bowls, the *golowka* is divided into 3–4 parts. One part is mixed with more flour and water to prepare more *golowka*, and the

TABLE 59

|  | Initial-sour | Intermediate-sour | Basic-sour | Full-sour |
|---|---|---|---|---|
| Starter (*golowka* from previous batch) | 1·0 kg | — | — | — |
| Initial-sour | — | 6·5 kg | — | — |
| Intermediate-sour | — | — | 18·0 kg | — |
| Basic-sour | — | — | — | 56·0 kg |
| Rye-wholemeal flour | 2·8 | 6·5 | 22·2 | 68·0 |
| Water | 2·6 | 5·0 | 15·8 | 46·0 |
| Baker's yeast | 0·1 | — | — | — |
| Initial temperature | 25–26°C | 26–27°C | 27–28°C | 28–30°C |
| Standing-time | 3·5–4·5 h | 4·0–4·5 h | 4·0–4·5 | 3·5–4·0 |
| Final acidity-index | 9–11 | 11–13 | 13–15 | 13–16 |

Sources: *Production of Baked-products: A Technological Handbook*, I. M. Roiter, Technika, Kiev, 1966 (in Russian). *Technology of Breadmaking*. L. J. Auerman. VEB-Fachbuchverlag, Leipzig, 1977 (in German).

TABLE 60

|  | *Full-sour* | *Dough* |
|---|---|---|
| Full-sour | 15 kg | 46 kg |
| Rye-wholemeal flour | 18 | 74 |
| Water | 13 | adjusted |
| Salt | — | 1·5 |
| Initial temperature | 28–29°C | 30–31°C |
| Standing-time | 3·5–4·0 h | 1·5–1·75 h |
| Final acidity-index | 13–16 | 10–12 |

Sources: *Production of Baked-products: A Technological Handbook,* I. M. Roiter, Technika, Kiev, 1966, USSR (in Russian). *Technology of Breadmaking*, L. J. Auerman, VEB-Fachbuchverlag, Leipzig, 1977 (in German).

residual two or three parts used to prepare two or three portions of dough. The formulation given in Table 59 for a *golowka* is based on 100 kg rye-wholemeal flour. The water content of the *golowka* full-sour is 50%, equivalent to a sour-dough yield of 150.

The final dough is then prepared from the *golowka* full-sour on reaching maturity, according to the scheme given in Table 60. The procedure for the propagation of the soft sour-dough version, known as '*Kwass*', is detailed in Table 61. This scheme is based on 100 kg rye-wholemeal flour in the full-sour. The final dough is then prepared from the *Kwass* mature full-sour according to the scheme of Table 62. This formulation is based on 100 kg rye-wholemeal flour in the dough. Two-thirds of the *Kwass* full-sour is used to prepare the dough, and the residual one-third used to prepare a new batch of *Kwass* full-sour. The water content of the '*Kwass*'-type full-sour is about 1·5–2·0% higher than that of the *golowka* full-sour. In many Soviet bakeries, full-sours of much higher water content are utilized to reduce the acidic flavour of the bread, and facilitate transport of the sour-dough through pipelines.

TABLE 61

|  | *Initial-sour* | *Intermediate-sour* | *Full-sour (Kwass)* |
|---|---|---|---|
| Starter (*Kwass* from previous batch) | 2·0 kg | — | — |
| Initial-sour | — | 15·0 kg | — |
| Intermediate-sour | — | — | 50 kg |
| Rye-wholemeal flour | 7·0 | 20·0 | 72 |
| Water | 6·5 | 15·0 | 54 |
| Baker's yeast | 0·15 | — | — |
| Initial temperature | 27–28°C | 27–28°C | 28–29°C |
| Standing-time | 4·0–4·5 h | 3·5–4·0 h | 3·0–3·5 h |
| Final acidity-index | 10 | 10–11 | 11–12 |

TABLE 62

|                        | *Kwass*    | *Dough*   |
|------------------------|------------|-----------|
| *Kwass*                | 20·0 kg    | 76·0 kg   |
| Rye-wholemeal flour    | 32·0       | 56·0      |
| Water                  | 24·0       | adjusted  |
| Salt                   | —          | 1·5       |
| Initial temperature    | 28–29°C    | 28–30°C   |
| Standing-time          | 3·0–3·5 h  | 0·8–1·2 h |
| Final acidity-index    | 11–12      | 9–12      |

The quantities of *golowka* and *Kwass* full-sour used to prepare doughs can only be taken as guidelines, in practice these will be varied for the following reasons:

—time of year
—ambient temperature
—activity of the *golowka* or *Kwass*
—flour quality
—production-stoppages

The Saratower group of the baking industry developed a special process, due to E. A. Gladkova and co-workers known as the 'liquid-sour S-1 process'. This process uses a liquid sour, to which has been added pure cultures of group B lactic acid bacteria, according to the classification of Seliber. These are the so-called beta-bacteria, heterofermentative types producing volatile acids, alcohol and carbon dioxide, as well as lactic acid. The beta-bacteria vary in length somewhat, and show an optimum reproduction-temperature of 25–30°C. Using strains 1, 2 and 7 in the liquid reproductive sour-dough cycle, 50% of the mature sour is used for dough preparation, and to the remaining 50% is added the following nutrients: 40% of a saccharified, pregelatinized flour/water suspension (PFWS) (prepared by heating a flour/water mixture 1:3 to 1:2 for 2–4 h at a constant temperature of 62–65°C), water 51%, and flour 9%. The sour is then allowed to ferment at 33–35°C for 60–75 minutes, after which the sour reaches an acidity-index of 8–11. Again 50% of the mature sour is used to prepare a fresh sour. The sours are prepared in batch containers which remain stationary. To prepare the dough, 50% sour, based on flour weight, and the necessary amounts of flour, water and salt-solution are mixed. The dough is set at 34–35°C, and bulk-fermented for 120–140 minutes. Using rye-wholemeal flour, the dough reaches an acidity-index of 8·5–9·0. In most industrial bakeries, the doughmaking aggregate ChTR, system I. L. Rabinowitsch, is used for this process. Whereas, earlier, rye-flour doughs were always made, whether of a firm or soft consistency, on a batch or 'charge' basis using mobile machine-bowls, new machine aggregates operating both on a batch and continuous system were introduced in the Soviet baking industry during the 1960s. For rye-flour dough preparation using firm sours, the bunker-type doughmaking aggregate due to N. P. Gatilin found wide application. One bunker-aggregate is used to prepare the full-sour, and another the dough. The mature full-sour is metered, then diluted before being pumped into the dough-mixer. An appropriate amount of mature sour-dough is always dispensed into the mixer of the sour-dough preparation aggregate in order to make a fresh sour. Many Soviet bakeries preparing rye-bread from sours of firm consistency use the doughmaking aggregate due to A. M. Chrenow, which utilizes stationary bunkers for both the sour-dough and final dough fermentation. Where liquid intermediate products, such as: liquid yeast, liquid sour or liquid sours containing salt or other PFWS-products are utilized, stationary containers are also used, and the sour transported by either gravity or pumping through pipelines. For the purpose of mixing these intermediate products, the same machine aggregate, type ChSM-300, used for the preparation of PFWS-products has been successfully utilized.

Continuous rye dough preparation with liquid intermediate products involve the application of the mixing-aggregate type Ch-12, coupled with the sloping, trough-shaped fermentation-conveyor from the ChTR of the Rabinowitsch system, which is used for the production of wheat-flour doughs by the indirect sponge-dough procedure. Most of these aggregates are illustrated in Section 1.7.7.

The application of the technical solutions described have eliminated most of the manual operations of rye-bread production, similarly to those applied to wheat-bread production.

## Some Common Faults Encountered in Rye-bread Production

Flavour deficiencies can be due to a lack of total flavour acids, in which case insufficient of the rye-flour content of the dough has been subjected to the dough-souring process. Depending on the ratio of the blend rye-flour/wheat flour, a good guideline is 50–60% of total rye-flour content should be subjected to the souring process. If there is evidence of an unbalanced flavour, this is usually due to excess acetic acid production (sharp after-taste), resulting from an overripe sour. Often brought about by a combination of factors such as: excess temperature, too soft sour consistency (excessive dough-yield), prolonged standing-time, choice of too low a reproductive-factor. The ratio 75% lactic acid to 25% acetic acid in the sour is a guide to a strongly aromatic flavour. The acidic intensity of sour-doughs can be categorized according to the following ranges: mild-flavoured 8–10, medium-flavoured 10–13, and strongly aromatic 13–20, measured as the acidity-index according to the method of Neumann.

Poor symmetry (excessive spread) may be due to:

—dough mixed too warm, optimum range 23–27°C
—dough consistency too soft, dough-yield should be reduced
—overproofing, shorten proof-time (approximately 50 minutes final-proof)
—underproofing, extend proof-time
—hot/wet proof, hearth-breads require drier proof (approximately 70% RH) at approximately 35°C
—front of oven zonal-temperature too low, raise to achieve earlier crust-set
—overmixed dough tending to stick, reduce mix-time and dust adequately
—loose moulding, hearth-breads need tight moulding
—weakened top-surface due to cutting before final-proof, always cut after proofing and before oven-setting

Poor crumb-elasticity may be due to:

—sprouted rye-grain flour
—rye-flour deficient in swelling ability (pentosan-deficient)
—heat-damaged rye flour
—sour-dough ripening time either too short or too long
—sour-dough temperature too low
—sour-dough standing-time either too short or too long
—sour-dough content either too low or too high
—excessive use of stale-bread crumb
—dough temperature too low
—dough ripening-time too long
—final-proofing time too short
—dough temperature too low
—insufficient baking-time
—initial oven temperature too low, or too high
—oven-humidity insufficient, exposure time either too long or too short
—inefficient cooling, condensation

Owing to variations in rye-grain quality from year to year, the quality profile of each flour delivery should be evaluated in order to make a decision how best to blend with other deliveries received in the bakery. Initial Falling-number checks, complemented with Amylograms will reveal any extreme quality profiles. In extreme cases useful information can be obtained from an acid/Falling-number retest, involving the adjustment of the flour/water suspension used in the standard method to pH 4·0 with dilute lactic acid. Rye-flours with standard Falling-numbers around 90 (sprouted-grain samples) should respond positively to such adjustments, owing to enzymic inhibition. Standard Falling-number ranges from 90 to 110 must be regarded as suspect for raw material processing without appropriate corrective measures; and raw materials falling in the range 75 to 90 require blending with rye-flour of Falling-number values between 150 and 250. Additional information can be obtained by determination of the water-binding capacity using the Farinograph and a dilute lactic/acetic acid solution. Test-bakes are carried out using both lactic/acetic acid solution, and mature biological sour-dough. Bread evaluation after 24 h includes the following: pH and acidity-index of the bread-crumb; bread-volume; total measure of loaf cross-section (height + width); and sensory evaluation.

Processing corrective measures which can be taken in the case of excessively enzyme-active raw material are:

—use of shortened sour-dough process; increase the percentage of sour-dough; increase the full-sour temperature to produce more acids;
—reduction of the dough-yield, and reduce floor-time and final-proof time;
—in the case of mixed rye/wheat bread (*Mischbrot*), increase the wheat-flour content;
—increase the salt level in the dough;
—increase the temperature during the first-stage of baking.

A reduction of the pH down to 4·0 is the most important measure to take to improve the gelling-properties of sprouted rye-flours.

Other corrective additives which can be superimposed on a pH adjustment include:

—hydrocolloids, e.g. Quellmehl 1–3%, CMC 0·5%, xanthan gum 0·2–0·5% based on flour weight;

—suitable emulsifier at 0·5%, e.g. calcium stearoyl lactylate, sodium stearoyl lactylate, DATEM, mono- and diglycerides.

The other extreme rye quality profile encountered is when the grain shows a highly polymerized starch structure, and deficient alpha-amylase activity. In such cases the standard Falling-number rises to 250–300, and amylogram gelatinization maxima exceed 700. When processed into bread, such samples result in close crumb texture, cracked-crumb, poor volume and symmetry, and rapid-staling. In such cases, corrective processing measures for raw material with a hard, intact starch structure, and insufficient amylosis, are the following:

—blending with raw material of average Falling-number (110–150), or with smaller quantities of sprouted material of Falling-number 90–110;

—increase dough-yield as much as possible;

—increase the floor-time and final proof-time;

—intensify the dough-souring processes;

—appropriate corrective additives are: malt-flour 0·1%, Quellmehl 2·0%, stale bread-crumb paste up to 5·0%, combinations of bacterial and fungal amylases, proteinases and pentosanases (the optimal concentrations being established with the help of the Amylograph standard and swelling-curves);

—the preparation of a 'scalded' or pregelatinized portion of the flour (about 5%), made by treating one part flour with one part boiling water.

# 2.5   Formulation and Processing Techniques for Specialty-Breads

This sector of the bread-market has shown rapid growth in most European countries over the past 5–10 years. The greatest demand being for wholegrain and meal/grit-containing bread varieties. Even in the FRG, which has always been a classical 'black-bread' country, 20% of total bread consumption being of that type of product, demand has risen by about 15% over the past 5 years. In the UK, traditionally a white-bread country, the wholemeal percentage of all milled-products used for bread production has risen from about 3% to almost 15% over the past 10 years. Similar trends have been reported from the Scandinavian countries. The same trends have also taken place in the COMECON countries of Eastern Europe. In the GDR, specialty and dietary bread consumption has shown a growth rate of almost 20% from 1967 to the 1980s, while the consumption of rye, and mixed rye/wheat breads has fallen with improvement in living standards. Demands for freshly baked small fermented products, e.g. rolls, baguettes, kipfel and croissants, and confectionery products are also rising. Similar trends have taken place in Hungary, where the consumption of fibre-rich breads and rolls have increased from 1980 to 1988 from zero to almost 5%, and further increases up to 20% are considered a possibility. Such increases are considered generally desirable on nutritional grounds, in order to ensure that the population receive adequate amounts of the digestion-promoting fibre and the essential vitamins and minerals in the long term. COMECON nutritionalists have generally recommended the consumer to concentrate on fibre-rich menus consisting of foods containing naturally higher levels of fibre, such as: bread and grain products, fruit and vegetables.

In the USA, many special types of dietary and nutritionally changed breads have been formulated in recent years. The emphasis is placed on the requirements of specific age-groups in society. Typical specialization categories include: fibre-rich, protein-enriched, calcium-enriched, salt-reduced and special formulations for dietetic consumers. Further growth potential in specialized formulations for age-groups are anticipated in the future. The sensory evaluation, i.e. taste, aroma, of these special types of bread, however, is not always found to be acceptable. Textural problems and mastication properties due to overloading with high-density raw materials also require further research.

In most countries the standards and legal definition of specialty-breads are laid down, for the protection of the consumer. For example, in the GDR, according to Standard TGL 3067: 'A specialty-bread may contain rye and/or wheat-flours, or rye and/or wheatmeals, and/or specialty-flour products—often differing in the method of preparation from normal procedures—including, optionally, specific raw materials, which change the nutritional and/or flavour of the bread by comparison with the basic-bread-varieties'. In the above regulation, and its practical interpretation, it must be borne in mind that 'basic-bread-varieties' in the GDR already contain 50% rye-flour or rye-meal, and indeed more in most cases. A similar situation exists in the FRG, another 'black-bread' country, which already has at least 200 different bread varieties to offer the consumer.

In the formulation of specialty-breads, in order to ensure that the baked products do not suffer loss of volume and leavening, and remain appetizing in texture and flavour, particularly if only leavened with yeast, certain processing aids are often essential. These processing aids can be in the form of raw materials such as: vital-gluten, sugar in solid or liquid form, salt and shortening, which effect a rebalance, or dough-conditioners, e.g. stearoyl-2-lactylates or ethoxylated monoglycerides, and other emulsifier-type additives. The inclusion of the latter 'additives' will, however, depend on local regulations.

### 2.5.1 MULTI-GRAIN BREADS

When formulating for multi-grain or mixed grain breads, a 'soak-stage' is necessary, blending the grits, meals and other cereal flours to form a mash with most of the water, standing for about 6–8 h. The following formula indicates the average composition of multi-grain bread, and the production procedure involved.

*Multi-grain bread*

| Soak stage: | | Dough-stage: | |
|---|---|---|---|
| Wholewheat kernels | 15 parts | Mash from soak-stage | |
| Coarse rye meal | 15 | Gluten-rich flour | 37 parts |
| Buckwheat flour | 10 | Salt | 2·5 |
| Ground millet | 6 | Potato-flour | 4·0 |
| Ground oats | 8 | Defatted soy-flour | 3·0 |
| Ground barley | 4 | Margarine/butter | 4·0 |
| Sunflower-seed meal | 1 | Molasses (unsulphured) | 4·0 |
| Toasted wheatgerm | 2 | Lecithin | 0·4 |
| Vital gluten | 3 | Yeast | 4·0 |
| Water | 60 | Water | adjust accordingly |

| Procedure: | Procedure: | |
|---|---|---|
| Mix together at low speed | Mix 5 min at low speed, and | |
| for about 5 min. | 5 min at high-speed. | |
| Soak 8 h | Temperature: | 27°C |
| | Floor-time: | 30 min |
| | Bake: | 35 min at 200°C |
| | Scaled at: | 515 g units |

Other cereal flours, grits or meals can be combined with a gluten-rich flour, but vital gluten is an essential ingredient to give adequate leavening and gas-retention properties to an otherwise dense structure, dark-crumb and coarse texture. Many of these breads are formulated by the baker from commercially available concentrates containing balanced combinations of the various components. Alternatively, a multi-cereal bread can be prepared by substituting from 5% to 30% wheat-flour by mixtures of rye, corn and barley flours, adding vital-gluten to provide the gas-retaining structure. An appropriate combination would be the following: gluten-rich wheat-flour 60 parts, barley-flour 20, rye-flour 10, corn-flour 5 and vital-gluten 5.

## 2.5.2 HIGH-FIBRE BREADS

Dietary-fibre is the food fraction which is not enzymically degraded within the human alimentary digestive tract. The main components are cellulose and lignin, but will also include the hemicelluloses, pectins, gums and other carbohydrates not digestible by man. Analytically, it is determined as so-called 'crude-fibre'. This involves treatment with boiling acid and alkali, or neutral detergent solvents, the dried residue being reported as 'crude-fibre'. In fact, this residue is composed of about 97% cellulose and lignin, which comprises from about 13% to 70% of actual 'dietary-fibre'. Sources of fibre normally used are: wheat-bran, corn-bran, soy-bran, oat-hulls, rice-bran, and powdered cellulose. The range of dietary-fibre contents of these sources range from about 45% for rice-bran to 99·5% for cellulose (insoluble dietary-fibre).

Insoluble fibre is defined as including celluloses and lignins, and some hemicelluloses. Soluble fibre comprises pectins, gums, pentosans and some hemicelluloses, which is of physiological interest in possible control of cholesterol and diabetes. Total dietary-fibre consists of both soluble and insoluble fibre. The total dietary-fibre of a food is determined by the A.O.A.C. (TDF) procedure (Prosky method), which is based on an enzymic digestion, intended to simulate the human digestive tract. Other useful sources of fibre are: spent Brewer's grains (mash) or '*Malz-treber*' (German), and washed orange-pulp, which contains total dietary-fibre 46·5% (insoluble dietary-fibre 21·9%, soluble-fibre 24·6%). Roasted commercial fibre sources, such as oat-fibre, corn-fibre, and soy-fibre contain appreciable amounts of both insoluble and soluble dietary-fibre, compared with raw milled products.

When formulating for bread containing more than 10% added fibre, a high-protein, gluten-rich flour (16% protein) is necessary, and to ensure maximum gluten-swelling, 60–70% of the flour should be used in a sponge-stage. For breads containing 10–20% added fibre source, water absorption must be increased above 100% to ensure adequate hydration of the gluten proteins and the added fibre source. In the latter case, 6–10% vital-gluten will be required to ensure an acceptable cell-structure, about 5% of the vital-gluten being added to a sponge-and-dough process procedure for best results. Yeast levels must be increased to 3–5%, and 1–2% kept back for the dough-stage. Salt also needs adjustment up to about 3%, and the addition of high-fructose corn syrup at 8–12% should give adequate balance. Where wheat fibre sources are being utilized, honey or molasses should be used to replace part of the sugar. The addition of shortening should be avoided on caloric-content grounds, but sodium stearoyl-2-lactylate added at 0·5% will improve bread volume. Owing to inevitable higher bread moisture-contents, the addition of mould-inhibitors becomes a necessity. The sponge should be mixed for 1 minute at low-speed, and 4 minutes high-speed, being set at 24°C for about 3 h. The final dough is mixed for about 20 minutes, 2 minutes at low-speed and 18 minutes at high-speed, holding the salt back until one-half the mixing time has elapsed. Dough temperature should be 26–27°C, and given a floor-time of 10–15 minutes only before scaling. The dough-pieces are rounded and given a first proof of 10 minutes before sheeting with the head-rolls gap-setting slightly less than for normal white-bread doughs, and the long-roller board pressure lighter than that used for normal white bread. Final-proof times of 55 minutes at 43°C are appropriate for optimal volume, and baking should be at 220–230°C for 18–20 minutes.

Since fibre is essentially inert in terms of energy contribution (calories), certain sources are useful for the formulation of low-calorie yeasted breads. The FDA, in 1984, authorized the exclusion of non-digestible dietary-fibre when determining the calorie content of reduced calorie foods. Thus, for every 1·0% of total dietary fibre added, the product calorie content declaration can be reduced by 4 units/100 g. Calories/100 g are calculated as being equal to four times the percentage of protein, plus nine times the percentage of fat and four times the percentage total carbohydrates, minus the percentage of total non-digestible dietary-fibre.

To prepare a low-calorie (50 cal/28 g), high-protein (approximately 12%) yeast leavened bread, the following procedure is appropriate. To a high-protein spring flour (protein 16%), 10–20% washed citrus pulp flakes (orange, grapefruit, lemon) are added to increase the dietary-fibre content by about 10%. The total dietary-fibre content of pulp-flakes is about 46%, of which some 22% is insoluble dietary-fibre, and 24% is soluble dietary-fibre. The flour is mixed with yeast and water, and the citrus pulp flour added together with a premix of the other ingredients (approximately six parts by weight vital-gluten, salt, whey-powder, high-fructose syrup, vegetable oil, 0·5% sodium stearoyl-2-lactylate, and calcium propionate as a preservative). The dough should be mixed for 20 minutes, 2 minutes at low-speed and 18 minutes high-speed. Dough temperature should be 26–27°C, and given a floor-time of 2 h, degassed, and scaled and final-proofed for about 40 minutes. Baking at 210°C for 20 minutes. The final product should have the appearance, texture and taste of conventional white bread, and contain: protein approximately 10%, dietary-fibre approximately 10%, at a moisture level of approximately 46%.

## 2.5.3 HIGH-PROTEIN BREADS

High-protein breads have long been the most widely distributed of all dietary-breads. Fortification of wheat-bread with various soy-flours was successfully researched during the 1940–1950 decade. Full-fat soy flours have proved to be the most favourable raw materials for breadmaking owing to the presence of natural lecithins and glycolipids, which enhance baking-properties. Any lack of excess processing in the form of heat-treatment is also desirable to retain enzyme-activity, and preserve baking properties. Groundnut protein isolates exhibit similar properties, the lipoprotein isolate showing superior functional properties compared with the higher protein defatted products. Soy-bean products are the most functional and economic source of vegetable protein supplement available, and fortification levels of 15–20% produce breads of acceptable volume, flavour and general quality. To obtain the best results, an appropriate sugar rebalance is recommended, and an optimum level of oxidant in the form of potassium bromate, or ascorbic acid if the former is not permitted. The use of a dough-conditioner is also advisable to obtain perfection, 0·5% sodium stearoyl-2-lactylate being the usual choice. A typical formulation for high-protein bread would be the following: flour 100 parts, full-fat soy flour 10, yeast 3·0, salt 2·0, potassium bromate (optimized level), sodium stearoyl-2-lactylate 0·5, and sugar (high-fructose syrup) 5 parts; water variable.

The protein and gluten-forming properties of the flour are not as critical as is the case with other raw material supplements, although 11% protein (basis 14%) is a lower limit safety margin for good results. Appropriate oxidants and selected surfactants, together with a modified procedure for adding the soy-flour ensures acceptable results.

One procedure is to mix the dough to full-development, holding back the soy-flour for later addition and further development, having made it into a paste with a portion of the water. The dough is then scaled, rounded and rested for 40 minutes at 30°C and 85% relative humidity. Shaping, panning and final-proof to the required height is at 36°C and 92% relative humidity. Baking is at 220°C for 25 minutes.

A so-called 'Triple Rich Cornell Formula', containing soy-flour, wheatgerm and non-fat dry milk powder, often used with unbleached flour, raises the nutritional standard of the bread considerably, and is very popular in the USA among the natural-ingredient and healthful-eating consumers.

In the German-speaking countries, where rye-flours as well as wheat-flours are used for the basic bread varieties, the same rule applies to the specialty-breads (*Spezialbrote*). However, in addition to containing rye- and wheat-flour, many other milled products and special additions are utilized. These will include: oats, barley, maize, and special passages (streams) and mill-fractions, e.g. various grain-germs, grain outer-layers and brans. In addition, grain-starches, leguminosae-flours, and potato products. Other raw materials of vegetable origin include, fruit and vegetable products, spices; those of animal origin include, e.g. milk and milk-products and meat and meat-products. In some cases the methods of production are distinctive, e.g. baking in a steam-chamber, as for pumpernickel.

The requirement for a specialty-bread is either an improvement in the nutritional effect or a change in taste compared with the basic bread varieties. Dietary-breads also come into the specialty-bread category, their composition and properties being directed to meet the requirements of a particular medical condition, environmental circumstance or age-group, registered by the Ministry of Health. In the GDR, there are about 65 specialty-breads, about 10 of which can be classed as dietary-breads. These 65 varieties, owing to their method of preparation, can be manufactured in all bakeries. In addition, a limited assortment of breads baked in a steam-chamber which are sliced and wrapped in 200-g and 500-g units, can only be produced in bakeries with the necessary plant and equipment. Many formulations date back to the late 1960s, and many too were developed in the early 1980s. Every industrial bakery within a district collective has developed at least one new specialty-bread, often bearing a regional name. However, most have been developed centrally, e.g. *Mecklenburger Landbrot*, and *Wriezener Landbrot*, and the more recent ones have been based on high-extraction grain-products. The specialty-breads can be divided into two main groups: those containing mainly rye milled products and those containing mainly wheat milled products.

In the GDR, about 70% belong to the first group, and 30% to the second. Additionally, specialty-breads can be sub-divided into further groups, as follows:

—35% containing processed grain-products, e.g. wheatgerm, wheatbran, Nafa-Roggen, whole-wheat flours, rolled-rye or Kraftma-Special flour.
—For 50% of specialty-breads, the high-extraction grain-products, whole-wheat flours, meals, half or whole grains, are processed.
—Other specialty-breads, e.g. *Rehbrücker* special-toast, and fine wheat bread, are based on the use of low-extraction wheat, i.e. top-patent flour, and gluten-rich wheat-flour.
—About 75% of specialty-breads contain specific raw materials.

Depending on the formulae, about 27 materials of vegetable origin, and 10 of animal origin are utilized. Those of vegetable origin include: potato-semolina, syrup, margarine, sugar, linseed, Sauerkraut-juice, and various spices. Those of animal origin include: whole-egg powder, full-milk powder, whey, butter, and buttermilk. Most of these materials exert a flavour contribution to the bread, e.g. spices, syrup, sugar, fruit-concentrates, buttermilk, etc. Some raw materials and processed grain products are intended to increase the bread-yield, viz. vital-gluten, potato-semolina, and special-flours. Although about 35% of the specialty-breads are designed to be produced as panned varieties, in fact, most are produced as oven-bottom varieties, since this fits in with the line-production methods used for the basic range of bread varieties. It is acknowledged that specialty-breads which contain larger amounts of whole-grain material, binding increased amounts of water, represent an economy in raw material compared with the basic breads made from lower extraction products. The grain-input for specialty-bread varies considerably from 700 kg to 950 kg/t bread, owing to regional variations in consumption from 20 kg/head/year in Berlin to about 5 kg in more rural regions. Total consumption in the GDR of specialty-breads is about 10%, including dietary-breads, mainly contributed by the industrial bakeries. The smaller craft bakeries output potential is wide open. (Source: 'Specialty-breads'—a requirement from the baking industry, R. Schneeweiss and G. Heinrich, *Bäcker und Konditor*, **5**, 1984, 132 (in German).)

In the GDR, the selection of raw materials which best serve the nation's health has been analysed according to current dietary requirements. The main current requirement are raw materials which exert a stimulating effect on the stomach and intestinal tracts. This involves specialty-breads with fibre-enrichment; but they are not yet regularly consumed on a broad enough basis. Consumers at present dependent on various medications, could instead make use of a basic food to alleviate their problems. The raw material emphasis is therefore on: wheatgerm, brans, brewer's mash (*Bier-treber*), fruit-pulps, and processed products from oats, buckwheat, rye, barley, bean-flours, and maize. Other possible dietary supplementation could be iodized-salt and iron-enrichment. Enrichment of bread with either vegetable or animal proteins is regarded as superfluous, and the need for fat addition is ruled out. The requirement is for specialty-breads with a shelf-life of about 7 days, which only attain their peak flavour acceptability after 48–72 h.

Sliced and wrapped units of 250 g or 500 g have a prolonged shelf-life of up to 9 days for rye and wheat wholemeal-flour breads, and 4 days in the case of rye-flour bread made from type 997 and type 550 patent wheat-flour. The following is an overview of the production procedures and formulations for specialty-breads in the GDR.

### 2.5.4  *MECKLENBURGER LANDBROT*

This bread variety can be produced on a discontinuous or continuous basis, using a sour-dough process. Continuous production, however, is only purposeful when the plant can be used for long runs, and the rye-meal can be processed in the sour. Frequent changeovers from one variety to another are time-consuming. The following scheme is therefore batch-orientated, prepared from a basic-sour.

Basic-sour for two full-sours:

| | | |
|---|---|---|
| Starter | 30·000 kg | |
| Rye-meal, coarse | 82·000 | |
| Water | 68·000 | temperature: 25–27°C; standing-time: 5 h |

Full-sour:

| | | |
|---|---|---|
| Basic-sour | 90·000 kg | and 90·000 kg for a second full-sour |
| Whole rye flour | 17·500 | |
| Water | 17·500 | |
| | 125·000 kg | temperature: 28–30°C; standing-time: 2 h |

Dough:

| | | |
|---|---|---|
| Full-sour | 125·000 kg | |
| Whole rye flour | 16·000 | |
| Rye-flour 997 | 32·000 | |
| Wheat-flour 550 | 24·000 | |
| Yeast | 2·600 | |
| Salt | 2·600 | |
| Caraway seed milled | 0·100 | |
| Sugar (crystal) | 0·450 | dissolve in part of the water |
| Caramel colour | 0·950 | disperse in part of the water |
| Water | 27·000 | |
| | 230·700 kg | temperature: 30°C |

Procedure:
Mixing is in two stages, 2 minutes slow-speed and 8 minutes fast-speed using the high-speed vertical one-arm mixer S 250, which is described and illustrated in Section 1.9.5. The dough can be processed through the normal bread-line, scaling at 1150 g. Final-proof is at 32–34°C for 40–50 minutes, as appropriate. Baking is at initial temperature 280°C falling to 200°C at outlet for 50 minutes. The standing-times of the sour-doughs and final-dough must be adhered to in order to ensure the typical flavour of this variety. Each loaf is labelled according to the Technical Standards as '*Mecklenburger Landbrot*' 1000 g, selling at a fixed-price of 0·85 marks.
(Source: 'Recipes and production-methods for specialty-breads', G. Heinrich and R. Schneeweiss, *Bäcker und Konditor*, **5**, 1984, p. 143.)


### 2.5.5 *MALFA-KRAFTMA-BROT*

This is a long-established specialty-bread, 10% of the flour being in the form of Kraftma special-flour, which is prepared from malted barley, giving the bread a malt-aroma and dark crumb-colour. It can either be produced as an oven-bottom or panned bread weighing 1 kg.

Basic-sour:
Starter            6·500 kg
Rye-flour 997      18·500
Water              12·000

                   37·000 kg   temperature: 26–28°C; standing-time: 6–7 h

Full-sour:
Basic-sour         37·000 kg
Rye-flour 997      42·000
Water              42·000

                   121·000 kg  temperature: 28–30°C; standing-time: 2·5–3 h

Dough:
Full-sour          121·000 kg
Rye-flour 997      44·500
Wheat-flour 550    19·000
Kraftma special-
    flour          14·000
Salt               2·500
Water              31·000

                   232·000 kg  temperature: 29–31°C; floor-time: 15–25
                               minutes

Procedure:
Mixing in two stages, 2 minutes slow-speed and 7 minutes fast-speed using the high-speed mixer S 250.

Both sour-doughs and final-dough can be prepared using continuous plant, in which case the process parameters of the sour-doughs, especially dough-yield must be carefully controlled. The dough can otherwise be processed through a normal bread line, scaled at 1150 g. Final-proof time is 35–45 minutes. Baking-time for oven-bottom bread 45–50 minutes, and for panned-bread 60–70 minutes. Baking temperatures should be 250–270°C at the start, falling to 220 and 180°C. All standing-times for sours and dough must not be exceeded, in order to retain the malt flavour and aroma of this bread variety. *Malfa-Kraftma*-special bread flour is best mixed with water, and allowed to swell before adding to the dough.

Other specialty-breads are '*Trendbrot*' and '*Driftbrot*', which are products enriched with dietary-fibre, produced from rye-flour 997, gluten-rich wheat-flour, wheat-bran and Telavit-trend special concentrate. A full-sour is prepared using a basic-sour enriched with rye-flour 997 and water to stand for 2–2·5 h at 28–30°C. The final-dough is made by adding gluten-rich wheat-flour, wheat-bran, yeast, salt, Telavit-trend, an emulsion and water. The dough is given a floor-time of 20–30 minutes, scaled at 1160 g, and then a final-proof of 30–40 minutes at 32–35°C and relative humidity 80–90%. Baking is at 280, falling to 210°C for 70 minutes. *Hagenower Spezialbrot* is a light variety containing both whey and whole-egg powder, which gives it a mild taste and a tender crumb. A sponge is prepared with gluten-rich wheat-flour, yeast, whey, whole-egg powder and water, to stand 60–90 minutes at 26–28°C. The final-dough is then made by adding the mature sponge and a portion of mature basic-sour to rye-flour 997, more wheat-flour, more yeast, salt, a little prepared dry-sour and water. The dough has a floor-time of 15–25 minutes at 28–30°C, scaling at 1680 g for a 1·5 kg loaf, and 1150 g for a 1 kg loaf. Final-proof is for 40–50 minutes. Baking for 1 h for the 1·5 kg loaf, and 50 minutes for the 1 kg unit. Baking commences at about 270, falling to 180°C. The approximate composition of the Hagenower dough is: rye-flour 997 47·7%, wheat-flour 47·8%, whole-egg powder 0·5%, whey-powder 4·0%, salt 1·6%, yeast 3% (1% in the sponge, and 2% in the dough) and water approximately 65 parts (variable). Enough basic-sour from the sour maintained for normal rye and mixed rye/wheat bread production is added at doughmaking to acidify the rye-flour used (approximately 1 part basic-sour: five parts rye-flour).

## 2.5.6 PUMPERNICKEL

This is the most popular of specialty-breads, known for many decades. It is made from rye-wholemeal, salt and water, using a sour-dough or '*Quellstück*', which involves soaking part of the rye-meal. Optional ingredients are yeast at 0·3–0·5%, and sweeteners up to 3%. Baking-time is at least 16 h at about 100°C in a saturated steam chamber, the baked bread having a dark, dense crumb. Typical German procedure for pumpernickel:

Formula:

| | |
|---|---|
| Whole rye-meal (fine, medium, coarse ground) | 100 parts |
| Salt | 1·5 |
| Yeast | 0·6 |
| Water (variable) | 60·0 |

Procedure:

About 25% of the rye-meal is made into a sour-dough, using the classical three-stage process. Another 25% of the rye-meal is made into a *Quellstück* with warm water, and allowed to swell for about 7 hours at 22°C. When the above fractions have reached maturity, the dough is prepared by adding both fractions to the residual 50% rye-meal, more water, salt and yeast. The whole being intensively mixed to give a dough-yield of 160%. The dough-pieces are then worked-off with the help of dusting-flour and an emulsion to reduce adhesion, shaped, and deposited in closed, rectangular, covered baking-pans, or processed as an elongated dough-strip divided into sections. A full proof is given for 50–60 minutes with a strict control of relative humidity. Baking is in a special baking-chamber; or regular ovens may be utilized so long as they can provide enough steam-saturation, which is the essential feature of pumpernickel baking. Baking temperatures are held low at 100°C for at least 16 hours, under a saturated-steam atmosphere.

*Typical American Procedure for Pumpernickel*

Although German in origin, pumpernickel is now produced from a wide range of formulae. The ingredients used in the USA include: medium coarse ground whole rye-meal, salt, residual roasted pumpernickel bread, soaked overnight in warm water, and a properly processed sour-dough. Additional options are yeast 0·3–0·5%, sweeteners up to 3%, and '*Brühstück*' type of pre-ferment made by soaking up to 10% of the total rye-meal in water at 60°C for about 3 h to effect hydration and some autolysis. Sours for pumpernickel are produced by various processes, containing on average about 12% of a sour-dough; up to 20% residual-bread and whole rye-meal, water, yeast and salt completes the recipe. A simplified version for an American-style pumpernickel formulation would be:

*Brühstück*/sponge:

Clear flour 24 parts, dark rye flour 38, yeast 2 and water (variable) 32.

Dough:

Rye-meal 38 parts, sour-dough 10, salt 2·5, and water (variable) 39. Often, malt (non-diastatic) or other sweetener 1 part and shortening at 1 part are added to the dough.

The sponge is set at 24°C, with a standing-time of 3 hours. Dough temperature is set at 27°C. The dough is scaled, rounded, and allowed to rest for 15 minutes

*Handbook of breadmaking technology*

before final-shaping. A full-proof is given for 50 minutes with humidity control, and baking is at the higher temperature of 220°C, in saturated-steam.
(Source: J. G. Ponte, Jr., Production technology of variety breads, in *Variety Breads in USA*, edited by B. S. Miller, AACC, St Paul, MN, 1981.)

### 2.5.7 WHEAT OR RYE WHOLEMEAL

When processing wheat and rye wholemeal products into bread, the milling processes used in the preparation of the raw materials influence their baking properties, as demonstrated by Zwingelberg, Seibel, and Stephan (1984) and Seibel, Stephan and Zwingelberg (1984). Products prepared by roller-milling showed a lower specific weight (g/litre) than those prepared with a hammer-mill. Also, the bread volume and leavening of the roller prepared meals were superior to those prepared with a hammer-mill. For the processing of both wheat and rye wholemeal products into bread the above workers recommend the following procedure:

Two-stage sour-dough for wheat or rye wholemeal bread
Total amount of
    wholemeal/dough             10 kg
Sour-dough percentage           10%
Meal granulation:                 fine

First sour-dough stage

| | | |
|---|---|---|
| Wholemeal | 250 parts | Process parameters: |
| Starter | 1 | Dough-yield    300% |
| Water | 500 | Temperature    24–26°C |
| | ——— | Standing-time   15–20° h |
| Total without starter | 750 | |

Second sour-dough stage

| | | |
|---|---|---|
| Wholemeal | 750 parts | Dough-yield 300% |
| First sour-dough stage | 750 | Temperature    26–28°C |
| Water | 1 500 | Standing-time   3 h |
| Total | 3 000 | |

Dough

| | | | |
|---|---|---|---|
| Wholemeal | | 9 000 parts | Temperature    30°C |
| Second sour-dough stage | | 3 000 | Standing-time   2 h |
| Salt | | 120 | |
| Water: | | | |
|   Wheat wholemeal | approx. | 5 000 | Theoretical yield 170% |
|   Rye wholemeal | approx. | 5 500 | Theoretical yield 175% |

Total dough:

| | | |
|---|---|---|
| Wheat wholemeal | 17 120 | Final-proof: |
| Rye wholemeal | 17 620 | Temperature    40–45°C |
| | | Time (min)     60–90 |

The use of only 0·1% starter (based on soured meal weight) gives a mild acidification, but amounts of up to 10% can be used if desired. Both sour stages are made soft (dough-yield 300), the first overnight at 24–26°C, the second for 3 h at 26–28°C. This gives a reproductive factor of 4.

The important features of this process are:

—limiting the sour-dough percentage of the total meal content to 10–15%
—use of a 2-h dough floor-time, allowing an extra 'souring'
—limiting the final-proof time to 60–90 minutes
—the omission of yeast (added) permits the use of a dough temperature of 30°C, which stimulates fermentation during final-proof at 40–45°C
—limiting of salt content to 1·2% to improve the flavour-note

Both the Berliner and Detmold one-stage sour-dough processes can also be successfully applied, using 10–15% sour-dough, dough-yields of 200–300, and 2-h floor times.

For the production of specialty or variety-breads generally, the SEKOWA-special ferment can be applied at 0·05% total flour weight. This starter, grown on honey and natural seeds, contains both homofermentative and heterofermentative Lactobacilli (*L. plantarum*, *L. farciminis*, *L. fermentum*, *L. fructivorans*, and sour-dough yeasts). Its average pH is 4·3 and acidity-index 14, the microorganisms being cultured on a substrate of essentially natural components. The use of wheat-flour type 550/wholemeal 1:1 in the first sour-dough stage should limit the acidity-index to 14 on completion of fermentation.

Another valuable raw material with a wide reputation, owing to its active components, is linseed. These flattened, egg-shaped seeds, capable of forming a viscous, cellulosic slime, and containing 38–44% oil (mainly linolic and linoleic acids) in the endosperm, are very popular for the digestion. Linseed wholemeal bread is produced as sliced-bread, packed in the same way as pumpernickel and other wholemeal breads, usually 250 g and 500 g packs. A typical blend for this type of variety bread could be: wheat-meal (fine, medium or coarse) 75% and rye-meal (fine, medium, coarse) 25% with 8–25% linseed.

Other specialty-breads produced from vegetable sources include: wheat-germ bread, buckwheat bread, malt-bread, sesame-bread, soy-bread, raisin-bread, sunflower-seed bread, maize-flake bread and onion-bread. All added at 8–15% on average. Specialty-breads from raw materials of animal origin include: milk-bread, milk-protein bread, sour-milk bread, buttermilk bread, yoghurt-bread, kefir-bread, quark-bread, whey-bread. Average additions, based on milled cereal product in the formulation are 2–5 kg/100 kg for dry products and 10–15 litres for liquid products. A good average basic cereal blend for dairy-product addition is wheat-meal 75% and rye-meal 25%.

The conception and formulation of specialty-breads is only limited by the imagination.

A series of specialty toast-breads can be formulated, using a long patent grade type 550 wheat-flour by adding the following raw materials:

—cream-cheese 24%, margarine 2%, full-milk powder 2% and sugar 2% (based on flour weight), yeast 4%, salt 2% and water (variable) 52%
—tomato-paste 15%, sugar 1%, margarine 5%, full-milk powder 3%, paprika powder (hot) 0·5% (based on flour weight), yeast 3%, salt 2%, water, variable
—whey-powder 8%, sugar 6%, margarine 4%, lecithin 2% (based on flour weight), yeast 4%, salt 2%, water (variable) 60%.

Procedure:
Using a conventional mixer, mix for at least 12 minutes at 30°C. Stand for 30 minutes, remix, and allow to stand for further 30 minutes. Scale, round and mould for pan bread, final-proof for about 40 minutes. Bake at 240°C at start, falling to 210°C for 35 minutes 750-g units, or 30 minutes for 500-g units.

Owing to their friable texture, these breads should be cooled down to 40°C over a 2 h period in a chilling-cabinet or stored for 24 h before despatch.

### 2.5.8 BRAN-BREAD WITH RAISINS

This type of specialty-bread is very popular in the USA, Californian raisins ranking as a health food among consumers. Before adding to the dough, the raisins have to be conditioned by soaking in equal weight of water for at least 4 h. The raisins can be used as the only source of sweetening, or honey and brown sugar can be added to sweeten the dough. Typical formulae are the following:

Raisin-sweetened:
Enriched high-protein first clear flour 80 parts, wheat-bran 20, wheat-gluten 4, yeast 3, salt 2, shortening 3, dough conditioner 0·4, conditioning-water 12, additional water 52. The conditioned raisins and water are added at the start of mixing. Since raisins contain on average, 70% sugars, the amount of raisins can be up to 18% of the flour weight.
Added-sweetening: bran-bread basic formula.
Enriched high-protein first clear flour 80 parts, wheat-bran 20, wheat-gluten 4, yeast 3, salt 2, brown-sugar 4, honey 3, shortening 3, dough conditioner 0·4, water 64.

The addition of the raisins, which contain both natural tartaric acid and propionic acid is effective in retarding mould and effects a saving in added sweeteners.

Malted-grain products are particularly useful in formulating specialty- or variety-breads, since, in addition to carrying a high fibre and protein content (average 20% and 45% respectively), they impart a sweet, malty flavour, and a roasted grain aroma. In conclusion, it is generally predicted that this will be the most important growth sector in bread consumption towards the year 2000 and beyond. Since the consumer is constantly seeking new taste experiences, but insists on ingredients with a minimum of processing, and formulated with a minimum of processing-aids or non-essential additives, the potential sales-life of some specialty- or variety-breads may be short. Product loyalty may be short-lived, and substitution more frequent within the baked-product sector of the food industry.

# PART 3
## The Baking Process

# 3.1  Aims and Requirements of the Baking Process

In the previous chapters, the basic processes of raw material and additive selection, control and preparation, dough-mixing; and the various types of fermentation and dough leavening have been discussed. In paragraph 1.1.3 certain aspects and features of some baking-ovens utilized in craft and industrial bakeries have been touched upon as part of the total breadmaking operation. This paragraph deals with the objectives of the process generally, before considering the elements in greater detail. In this final step of breadmaking, the actual baking-process, the raw dough-piece undergoes transformation under the effect of heat application into a light-textured, and porous, digestible product of pleasant flavour and aroma. The diverse reactions between the dough polymers and the surface-active components, which result in irreversible structural changes in the dough components, are complex— physical, chemical and biochemical in nature. However, they must be carefully controlled by the rate of heat-transfer, amount of heat input, humidity level and time of exposure within the chamber. Assuming all previous dough preparation and processing procedures have been optimized (upstream), the chemical and physical/ colloidal state of the dough-piece should be fully matured and ready for the oven.

On setting in the oven, the dough-piece initially undergoes the final phases of swelling and solubilization under the influence of a limited heat-transfer situation. Progressive heat-transfer also produces gas-development within each dough-bubble, and gas expansion, and final gas-retention, as the solidification of the elastic-film surrounding each gas-bubble occurs. The collective result of these developments determines the final loaf-volume. Any estimation of total loaf-volume would require a knowledge of the number of bubbles in a dough. This could be determined experimentally by initial microscopic examination of the dough. When a proofed dough-piece is placed in the oven, the rheological properties developed during fermentation and proofing will vary with subsequent temperature changes. The most important rheological properties governing the condition of the dough-piece during baking are elastic extensibility and tensile strength. The immediate effect of heat on the dough surface is the formation of a film, the extensibility of which will depend on the humidity of the oven-chamber. The rapid increase in total volume of the entrapped gases results in dough expansion and so-called 'oven-spring'. Assuming gas-production is adequate, doughs with a high viscosity/rigidity modulus ratio, often called the 'relaxation-time', tend to give high oven-rises or 'oven-spring'. During the initial swelling and solubilization stage, the dough must retain sufficient viscosity and elastic-extension until the swelling of the starch occurs to increase the strength of the dough-structure.

539

Modern ovens are usually designed to convey the loaf either on trays or a travelling-band (hearth), through a series of zones, exposing it to various conditions of temperature and humidity for defined time periods. The initial stage on setting would be at a temperature of about 203°C for a duration of about 7 minutes, amounting to 25% of the total baking-time of 27 minutes for panned-bread. During this stage, the outer layers of the crumb increase in temperature at an average rate of 5°C/minute, until they reach a level of about 60°C to 70°C. The dough-surface reaches a temperature of 100°C within about 3 minutes, which progressively increases at a rapid rate, reaching about 130°C after 20 minutes residence-time. In contrast, the heat-transfer to the centre of the crumb is relatively slow. This is due to the rapid heating up of the dough-surface to 100°C, resulting in a temperature gradient between dough-surface and crumb-centre of 60–70°C. Since the heat-transfer proceeds from the outside to the inside, the unbound dough-moisture evaporates in so-called 'evaporation zones', as soon as the dough-layers attain a temperature of 100°C and over. This evaporation zone moves with increasing heat-transfer from the outer layers towards the centre. Moisture on reaching 100°C, evaporates and leaves the loaf via pores in the crust for the most part, thus increasing the humidity of the baking-chamber. However, a smaller amount of steam remains trapped within the loaf, and influences the swelling and gelatinization processes of the dough-layers located near the centre of the loaf. The processes described are closely linked with the consumption of heat, and a more exact knowledge of heat-transfer processes enables the oven operator to save energy. The first visual change brought about by the oven-heat is the rapid formation of the thin extensible surface-skin. As the temperature within the dough-piece progressively increases from 30°C to 70°C, a series of changes take place within narrow temperature ranges with time. The rise in temperature from 30 to 40°C accelerates swelling, enzymic activity and yeast growth. All fermentable sugars continue to be fermented by zymase, until a temperature of about 60°C is reached. Schneeweiss (1965) also confirmed the increase in the percentage of starch granules with a damaged outer layer depending on the temperature of the interior of the dough-piece during baking. Isolated starch granules were treated with china-blue, which only colours damaged granules. Within the temperature range 57 to 59°C there is a steep rise in coloured starch granules from about 20% to 100% on the completion of the baking process. Schneeweiss further points out, that the dynamics of the production of soluble carbohydrates during baking shows a rapid increase between 43 and 60°C from 14·5% up to about 24% after 55 minutes baking-time. This focuses attention on the activity of the amylolytic enzyme complex during baking. According to Schneeweiss, the enzyme dynamics of alpha-amylase commence at 30°C, reaching a maximum between 50 and 60°C; at 70°C its activity begins to fall off. These findings agree with those of Perten (1965), and other workers, but earlier work by S. F. Falunina (1952) indicated that alpha-amylase remains active up to 78–80°C. Complete inactivation of beta-amylase occurs when the dough-piece reaches the temperature range 64 to 66°C.

Rye-starch begins to gelatinize at 45°C, and becomes complete at about 55°C, whereas wheat-starch begins at 55°C continues up to about 65°C. During the starch swelling and gelatinization stages, water migrates from the other dough components to the starch-granule, and a partial dehydration of the gluten–protein complex takes place. This in turn increases the rigidity of the gluten molecular strands, the fibres becoming increasingly viscous and elastic. The temperature range for protein coagulation is 50 to 70°C, depending on molecular weight and colloidal state. At 55 to 60°C, the yeast enzymes become thermally inactivated, the zymase becoming inactivated at about 65°C. The visual process of 'oven-spring' continues during and after primary swelling and terminates during gluten coagulation, the plastic and elastic expansion of the dough-piece thus continues after starch swelling has started to change dough characteristics appreciably. As heat-transfer proceeds, dough mobility decreases, especially between 62 and 67°C, the dough finally assuming a more or less rigid state owing to protein coagulation and partial gelatinization of the starch. It must be stressed that all temperature data quoted refer to events taking place zonally at a specified temperature or temperature-range—whether the various layers or zones attain that temperature, and at which point in time during baking, will depend on the efficiency of heat-transfer from heat-source to each layer or zone of the dough-piece.

The effect of alpha- and beta-amylase on the gelatinizing starch increases the water-soluble fraction in the crumb of the dough-piece. The extent of this increase is both crucial and variable, depending on the dynamics of the amylolytic-enzymes. It depends on various factors: enzyme activity, the speed of heat-transfer to the inner layers of the dough-piece, degree of starch-damage, and dough pH. These degradation products from starch are dextrins and reducing-sugars, but other water-solubles include pentosans and hexosans, derived from the cellulosic components of flour. Such compounds have been found to vary in the water-soluble crumb fraction, after precipitation with ethanol, from about 0·4% to 1·2% of the crumb-weight. The total content after ethanol precipitation varies from 3·3% to 4·3% of the crumb-weight. The composition of this complex is xylose, arabinose and glucose in the ratio 5:4:3, these pentoses and hexoses having a strongly branched chain, similar to the structure of flour-pentosans. In fact, about 40% of all dough hydrolysates are pentosans. Also, during baking, the dextrin fraction of a dough made from a normal flour increases by about 15%, whereas that from a sprouted wheat-flour often increases by 50% or more.

On reaching 50 to 60°C, all the carbon dioxide becomes liberated and contributes to dough-piece expansion. As the surface-skin thickens, and loses its elasticity, the first signs of browning take place. This stage begins at 100°C, when a rather sudden loaf expansion of approximately one-third of its original volume takes place, which the baker refers to as 'oven-spring'. The second and third stages could amount to about 14 minutes, roughly 50% of the total baking-time. Oven temperature is held constant at 240°C and the crumb-temperature increases at the rate of about 5·5°C/minute during the second stage, until it acquires a level of 98 to 99°C at the commencement of stage three, thereafter remaining constant. This temperature satisfies the maximum rate of moisture evaporation, starch gelatinization and coagulation of the dough proteins. Also, at this temperature, ethanol evaporates, which contributes to the leavening process. The interior of the dough-piece is progressively transformed into a baked crumb-structure from its outer layers through to its inner layers, and ultimately the crumb-centre by heat penetration. When the crust temperature reaches 150 to 206°C the crust acquires the typical brown colouration due to caramelization. The latter process involving various residual sugars, of which about 2–3% must remain unfermented for normal crust formation. Other thermal reactions contributing to crust colour are reaction products from sugar/protein interactions, known as melanoidin compounds, which begin to form from about 100°C. These reactions involve amino acids and carbohydrate degradation products. The dextrins, formed from starch degradation also contribute to the non-enzymic browning reactions.

Under the influence of temperatures of 100 to 180°C, and simultaneous with the browning-reactions, the bread-aroma substances are formed. These are made up of volatile carbonyl compounds (aldehydes and ketones), produced mainly by the Maillard reaction in the crust region. However, on cooling and during storage they diffuse into the adjacent crumb-layers, and in dough, bread and oven-volatiles, over 200 aroma and flavour development substances have been identified. In white bread volatile organic acids are retained, the most prolific being lactic and acetic acids at concentrations of 10–50 mg, and 2–5 mg/100 g respectively. Rye-breads containing much higher levels owing to the lactic fermentation.

The final oven temperature zone is maintained at a constant temperature of 221–240°C, serving to firm-up the cell walls, and produce the desired crust colour intensity. This stage amounts to 25% of total baking-time, and results in the loss of organic volatiles, which constitutes the 'bake-out loss'.

The 'zoning' temperatures and times of the individual baking-stages described are representative of standard baking practice for the production of panned white bread using the conventional tunnel-oven conveyed either on trays or a travelling-band or hearth. In practice, considerable variations are encountered for both temperature and time. Important factors for consideration are: oven-design, weight and volume of the dough-piece, pan-shape and base-dimensions, type and thickness of crust required, level of residual crumb moisture acceptable and, above all, the type of product being baked, i.e. whether panned, hearth (oven-bottom), wheat-flour or rye-flour dough, or specialty (variety) bread, and its formulation components. Assuming the production schedule is for panned white bread, using an appropriately regulated temperature, a 500-g loaf may require only 18 to 20 minutes to obtain an adequate bake, whereas a 1 kg unit may require on that basis 22–24 minutes for the optimum degree of bake. A characteristic of continuously mixed white bread doughs is that they require baking with a rising temperature-gradient, starting as low as 200 to 228°C, and finishing at 230°C. When using a zoning system, each successive zone temperature must be higher than the preceding one, otherwise the sidewalls become weak and may collapse, resulting in a 'keyhole'-shaped cross-section. The correct temperature control, at constant baking-time for quality bread has been investigated by Prouty (1965). He established that zonal variations at constant time between feed-end, zone 2, zone 3 and the discharge-end, resulted in differences of up to 1·7% in bake-out loss, and 1·3% in bread-moisture loss. For panned white bread, unsliced, a bread-crumb moisture content of 37–38%

is considered desirable for the ultimate moistness at consumption. Whereas, the sliced version, owing to packaging, should be one per cent lower in moisture, at about 36%, although mould-inhibitors are usually added. The baking-process is a very important phase of the production schedule in the baking industry, since it results in a metamorphosis of an indigestible mass of nutrients, in the form of a visco-elastic fluid, into a solid with a light texture, which is easily digested and has an interesting flavour, stimulating the appetite.

Furthermore, the output capacity of a bakery will depend on the throughput or baking-surface of its ovens.

The processes which take place in the oven-chamber from the practical baker's viewpoint are: oven-proof, crumb-formation and crust-formation. These proceed simultaneously in the dough-piece, are complex in nature, and to some extent exert an influence on one another. The intensity with which the three processes proceed will depend on the interrelationship of the dough-piece residence-time and the following controllable variables: dough-piece weight and shape, its composition, temperature-gradient, efficiency of heat-transfer and humidity in the oven-chamber. Oven-proof, which is referred to as 'oven-spring' by the baker, manifests itself as a high degree of gas-development and 'leavening', and good elastic-extensibility for efficient gas-entrapment within the gas-cells or bubbles. Compared with the dough-piece, the baked loaf should show a volume increase of 15% to 25%, and when baked on the hearth, its height should increase at the expense of its width. The resulting crumb-porosity must be adequate for good digestibility.

Since the crumb forms 70–80% of the loaf, its leavening, porosity and elasticity are vital for bread quality. Its moisture content at baking is normally within the range 47–49%, depending on the raw materials and type of product. This level is then reduced on bake-out to 37–38%, giving an average 'bake-out loss' of 10–11%.

## Optimization of Baking-conditions According to Production Requirements

Although a trained and experienced baker can judge when a dough-piece has reached optimum final-proof, during the production process it is not always possible to adhere to the best time for setting in the oven. Therefore, some adjustments have to be made in the baking-temperature and relative humidity in the oven-chamber. If the dough-pieces show signs of underproof, the feed-end temperature should be reduced, and the humidity increased correspondingly. The reason for this being that, in the closed chamber, the extra steam increases the pressure, which is utilized by the under-proofed dough-piece. The external pressure acting counter to the pressure developed by gas-development within the dough-piece, thus hindering any excessive initial rising. Furthermore, since the surface of the dough-piece is cooler than the surroundings in the oven-chamber, condensation takes place on dough-surface. In this way, the surface of the dough-piece remains longer in the elastic state, until such time as gas-development and expansion is complete. In the case of over-proof, the feed-end temperature should be increased, and the humidity decreased accordingly. This will reduce the prolonged elasticity of the surface of the dough-piece, and reduce oven-proof to a minimum. In this way, loaf symmetry will be retained, and the formation of large vacuoles in the crumb avoided. Since, apart from doughmaking and fermentation errors, baking errors are the most prolific, the flexible and precise control of both baking-temperature and humidity prevent many bread-faults.

For the baking of hearth-breads, the optimum relative humidity in the oven-chamber during the initial baking stage, about 20% of the total baking-time, is between 70% and 80%. It is difficult for technical reasons to accurately measure the humidity within the baking chamber, and in practice the baker must rely on experience with his particular oven-system. Also, there are no exact values or guidelines for the level of steam necessary for the various baked-products. In deciding on the amount of steam required, the consistency of the dough-piece, and the degree of proof must be taken into account. In the case of hearth-breads, after the initial humid baking stage, most of the steam is released by opening the vents of the baking-chamber. Oven humidity is regulated by either injecting steam into the chamber, for products with a high percentage of crust and which are small in size, e.g. bread-rolls, or gradually releasing it by controlling the chamber-vents. As an approximate guideline, 100 kg of hearth-bread would require about 3000 kg of steam. In the case of modern ovens, this is supplied from outside the oven-chamber in the form of saturated steam under pressure of about 30 kPa; typical examples are the larger multideck-ovens with natural gas or oil firing, using hot-air circulation systems, and the travelling-band tunnel ovens.

Where older types of oven are in use, such as steam-tube ovens, and side-flue (fired with solid-fuel, or converted to gas or oil firing), steam can be produced within the chamber by using 'steamers', or steel-boxes containing water. Otherwise, the dough-pieces can be painted or sprayed with water just before setting in the oven. This latter technique can be used whether or not a steam-injection system is in use to supplement the humidity of the dough-surface.

## Utilization of Steam as an Aid to the Baking Process

Although most modern ovens are equipped with steam lines for the injection of saturated steam in the first or second heat-zones of the chamber, its application has diminished in recent years for the baking of panned-bread. The reason being that experience has shown that sufficient steam is generated during the initial stages of baking to produce a satisfactory crust for panned loaves. In the case of ovens used for various hearth-breads and crusty-roll types, additional steam is provided by lateral steam-headers installed in the first heat-zone. These will provide low-pressure steam at the approximate rate of 9 ft$^3$/1 lb of dough. Since the volume occupied by steam depends on pressure, it is more accurate to refer to steam by weight, i.e. the amount of water which the boiler must evaporate to produce it. The capacity of a boiler is often quoted in 'boiler hp/lb of dough', thus, 1·0 to 1·5 boiler hp/100 lb of dough is considered adequate for optimum steaming. Therefore, an oven capacity of 1000 lb dough/hour would require a 10–15-hp boiler.

Steam should be injected over the top of the dough-pieces during the first 2–3 minutes of baking, and the velocity of injection controlled at about 300 ft/min to keep turbulence to a minimum. The prime purpose of injecting steam being to effect condensation on the cool dough-surface to enable it to retain its elasticity long

Aims and requirements of the baking process 545

enough to attain maximum expansion without a ragged break and shred. This condensation results in the dough-piece actually gaining in weight during the first 3–5 minutes owing to humidity levels of 70–80%. However, the dough soon loses moisture to the oven-chamber atmosphere, and progressively loses weight almost linearly up to the end of the baking process.

The use of so-called 'primary' live-steam, injected into the oven-chamber before setting also improves the 'bloom' of the bread and helps to control evaporation losses. Bread baked in an oven without steam often has a dull greyish-brown crust, with irregular break and shred, dependent on the dryness of the chamber atmosphere. Primary steam results in the gelatinization of the starch on the dough-surface, giving it a glazed appearance, but once the crust has formed, no further surface gelatinization can occur. Instead, dextrinization takes place, depending on the humidity of the oven-chamber. Secondary-steam is that produced during baking by evaporation from the dough-piece.

In the case of travelling-band tunnel ovens, steam injection is only necessary at the feed-end and maybe zone 2; if used at any later stage, it may produce a tough, leathery crust by boiling the gluten on the dough-surface instead of coagulating it in a less humid atmosphere. Retention of too much steam in the oven after about 15 minutes baking-time will result in tough crusts, removal before the final stages of baking will ensure that the bread is baked-off in a relatively dry atmosphere. Removal of excessive steam is effected by dampers suitably positioned along the oven length. The swing-tray type of oven, of which the Baker Perkins Uniflow was an example, owing to their design provided an excellent steam-trap, reducing steam consumption to a minimum, once the oven had been in operation for about 1 hour.

All steam lines from the boiler to the oven must be completely lagged (insulated) to save energy and steam-quality. Uninsulated lines both lose heat and cause steam condensation, necessitating a steam-trap to remove the condensate before it reaches the injection points. Otherwise, the steam will carry moisture-droplets into the oven, causing surface blisters on the loaves, which result in a mottled top-crust. Steam pressure can be measured by a gauge mounted in-line near the oven, thus ensuring a stable pressure, and constant moisture content in the steam. Low-pressure steam is ideal for baking, since its temperature approximates to dough-temperature rather than oven-temperature. At 100 to 115°C, steam is in water-saturated state, and will condense on the cool surface of the dough-piece as it is set in the oven. If the steam is at a higher temperature, it is less humid, and the temperature gradient between oven and steam is such that no condensation on the surface of the dough-piece can occur. This prevents the dough surface from remaining flexible during the initial stage of baking, retards starch gelatinization, and hinders heat-transfer into the loaf and the formation of a gloss or glaze in the crust.

The multi-deck hot-air ovens are usually provided with a steam-trap at the mouth of the oven, which can be lowered at will, thus lowering the steam escape line. This set-up, combined with steam injection from a boiler producing sufficient steam, will ensure that the steam input at least balances that escaping, thus fulfilling the requirement of any oven for the baking of hearth, Vienna and crusty-rolls, and breads. Purpose-built Vienna ovens are constructed with a sloping-sole, sloping down from the back towards the mouth of the oven. Since the sole is higher at the back than at the top part of the oven-mouth, the steam can be retained in the oven during baking and setting. Thus allowing the production of products with the characteristic gloss or glaze and a crisp, hard crust. The sole at the oven-mouth rises more sharply than that within the oven-chamber, thus facilitating the retention of steam as the door is opened by sliding upwards. These ovens are fitted with vents, allowing the steam to be removed after it has served its purpose.

The quality of the steam used in both the proofer and oven have a significant influence on bread quality. For both applications, the steam should be at low pressure and saturated, produced with a boiler gauge pressure of 10–15 psi, which becomes reduced to about 5 psi by the injector. Steam at a pressure of 10 psi, in the saturated state, has a temperature of 115°C, and at atmospheric pressure (760 mm) 100°C. Thus, as the steam pressure is reduced, its temperature also falls, and its water-holding capacity. Therefore, when pressurized steam is allowed to expand to atmospheric pressure, the consequent heat loss results in moisture loss, as condensate or in the form of an aerosol or mist. When this form is injected into the oven-chamber, and contacts with the freshly loaded, cooler dough-pieces, its moisture condenses on the dough-surface, and produces the desirable effects already described. However, once the dough-surface temperature reaches 100°C, its moisture can no longer condense. The use of steam at higher pressures (and temperature) would not allow sufficient time for its temperature to fall to the point of condensation ('dew-point') before the effect of the oven-heat has raised it again to that of the dough-surface. Such high pressure/temperature steam would also have a depressing effect on loaf-volume, especially in the case of rye-doughs. In summary, the aims and requirements of the baking process are to produce from the raw rounded and shaped dough-piece, a baked product which is flawless in appearance, taste and flavour, light and easily digestible and with an acceptable shelf-life. This demands, in the first place, the transfer of heat within an optimum time-period to complete: crust and crumb formation, browning processes, evaporation of moisture and the elimination of all microorganisms.

If the transfer of heat is too rapid, the outer dough-layer becomes burnt, and the internal layers remain incompletely baked-out, owing to the heat-transfer from the outside to the inside not having been correctly synchronized. If the transfer of heat proceeds too slowly, the crust formation is unacceptable, and the internal layers become completely dried out.

A determination of the optimum baking-time and throughput capacity of the baking-system being utilized is essential for the maintenance of consistently good bread quality, and for the economic operation of the oven. In the case of discontinuous or batch loading of ovens, bread quality is usually controlled by the baking-time and checking the weight-loss. Where production is continuous, using travelling-band tunnel-ovens, the baking-time is regulated by setting the speed of the travelling-band. The speed of the band-drive is measured by means of dynamometers, and the electrical output measured by a voltmeter or potentiometer. The relationship between band-speed and energy output is then calibrated in terms of baking-time.

The resulting instrument, giving a direct read-out in minutes of the baking-time is referred to as a tachometer. These are either direct or alternating-current instruments, whose rotating-element is directly connected to the oven conveyor-band drive-shaft. They provide a daily read-out record of baking time, if desired, on connection to a recorder-chart. With solid-state technology, such instruments have been miniaturized and mounted integrally in control-panels of reduced dimensions.

**Wheat- and Rye-flour Breads—Differences in Baking Requirements and Conditions**
The requirements in terms of oven conditions of wheat and rye doughs differ in: baking-temperature, humidity, heat-transfer, and baking-time. This is due to the following:

(1)  Differences in composition and properties of wheat and rye doughs.
(2)  The lower density of wheat breads compared with rye breads.
(3)  The difference in the leavening characteristics, wheat (yeast) and rye (bacterial).

Any differences in the baking requirements between wheat bread and bread-roll products is due to the extreme difference in size and weight.

The injection of steam into the oven-chamber for wheat bread and roll products should be done immediately after the products have been set in the chamber, and should be ideally completed within 100 s. The weight of steam required increases with decreasing product weight, due to the smaller products having a larger surface-area, and percentage of crust than the larger units. An approximate guide is 130 to 150 g/m$^2$ of dough surface. Table 63 gives comparative data, showing the required optimal baking-times, temperatures, and amounts of steam for various types of wheat breads and wheat bread rolls. For baking bread, the oven temperature should have a falling tendency generally. This requirement, which is both technically and economically justified, is, however, only possible where ovens are under automatic temperature control.

TABLE 63

|  | *Mixed wheat/rye bread 1·500 kg hearth* | *Wholemeal bread 1·000 kg panned* | *Wheat bread 1·000 kg hearth* | *Wheat bread 0·500 kg hearth* | *Wheat rolls 0·045 kg* |
|---|---|---|---|---|---|
| Baking-time (minutes) | 60 | 60 | 40 | 35 | 18 |
| Baking-temperature (°C) | 280 → 210 | 270 → 220 | 240 → 220 | 230 → 220 | 220 |
| Litres of steam/ 100 kg product | 1 500 | 1 500 | 2 000 | 2 500 | 3 000 |

Source: 'Special processes in baking', Werner Schwate and Udo Ulrich, *A Process Manual for Baking*, 2nd edn, (in German), 1983, VEB-Fachbuchverlag, Leipzig.

In order to produce an acceptable baking-effect, the loss in weight during the baking-process and the thickness of crust must be under strict control. In many East European countries this is a legal obligation, and is laid down as a standard, together with the baking-times. Table 64 gives an example of the baking-temperatures and baking-times laid down in the GDR for the basic bread varieties under TGL 3067. A 1·5 kg hearth-baked mixed rye/wheat loaf must show a baking-loss of 11–12%, which requires under normal circumstances a baking-time of 50–60 minutes.

TABLE 64

**Baking-temperatures and baking-times for basic rye-bread varieties including the baking-times specified in TGL 3067 for use in the GDR**

| Bread type | Baking mode | Bread weight (kg) | Baking-temperatures (°C) | | Optimal baking-time (min) | Deviation tolerance (min) |
|---|---|---|---|---|---|---|
| | | | Start | Finish | | |
| Wholemeal | Hearth | 2 000 | 240 | 200 | 70 | −5 |
| (rye-flour) | Hearth | 1 500 | 250 | 210 | 60 | −5 |
| | Hearth | 1 000 | 250 | 210 | 50 | none |
| | Panned | 2 000 | 240 | 210 | 110 | −10 |
| | Panned | 1 500 | 240 | 210 | 100 | −10 |
| | Panned | 1 000 | 240 | 210 | 80 | −10 |
| Wholemeal | Hearth | 2 000 | 240 | 210 | 80 | −5 |
| (ryemeal) | Hearth | 1 500 | 250 | 210 | 70 | −5 |
| | Hearth | 1 000 | 250 | 210 | 60 | −5 |
| | Panned | 2 000 | 240 | 210 | 120 | −10 |
| | Panned | 1 500 | 240 | 210 | 100 | −10 |
| | Panned | 1 000 | 240 | 200 | 80 | −10 |
| Mixed | Hearth | 2 000 | 280 | 210 | 65 | −5 |
| rye/wheat | Hearth | 1 500 | 280 | 210 | 60 | −10 |
| (rye-flour) | Hearth | 1 000 | 280 | 200 | 50 | −10 |
| | Panned | 2 000 | 250 | 210 | 95 | −10 |
| | Panned | 1 500 | 250 | 210 | 80 | −10 |
| | Panned | 1 000 | 250 | 210 | 65 | −5 |

Source: 'Special processes in baking', Werner Schwate and Udo Ulrich, *A Process Manual for Baking*, 2nd edn (in German), 1983, VEB-Fachbuchverlag, Leipzig. (Adaptation from original.)

The amount of heat-energy required to bake a 1 kg loaf of bread is approximately 500–600 kJ, according to the type of oven, type of heating, and degree of automatic control of the heat-transfer operation. The heat-transfer during the baking-process takes on three forms, viz. convection, conduction and radiation. Depending on the oven type and design, and energy source, one or other of the three forms of heat-transfer will predominate.

The conductive heat-transfer will depend on the type of material used for the oven-sole, i.e. steel-plate, steel-mesh or fire-brick. This is actuated by the molecular oscillation within the material, and is often expressed in terms of the coefficient of conductivity (kJ m$^{-1}$ h$^{-1}$ K$^{-1}$). Heat-transfer by convection involves the atmosphere within the baking-chamber (a variable mixture of air, steam and fermentation volatiles), and its speed of circulation. The convection heat-transfer coefficient is expressed in kJ m$^{-2}$ h$^{-1}$ K$^{-1}$; under static atmospheric oven-chamber conditions this is of the order of 21 kJ m$^{-2}$ h$^{-1}$ K$^{-1}$, and at a velocity of 10 m s$^{-1}$ increases to 168 kJ m$^{-2}$ h$^{-1}$ K$^{-1}$ (Kolb, *Oven-construction and Heat Engineering* (in German), Manager of the Technical Department for Large-oven Construction, Werner and Pfleiderer, Stuttgart, 1958, p. 6). Since all bodies with a higher temperature than their surroundings radiate heat, the heating-elements of the oven-chamber are no exception and emit heat in the form of radiation. However, this type of radiation can only penetrate a few millimetres into the dough-piece, and heat-transfer then proceeds by conduction. The radiation-coefficient is expressed as kJ m$^{-2}$ h$^{-1}$ K$^{-4}$, and depends on the temperature of the radiating-element. Where the temperature is increased by a factor of 2, the radiant heat-energy must be increased by a factor of 16.

Although the chemical, physical and biological processes which take place during the baking of wheat, rye and mixed wheat/rye doughs are similar, there are appreciable differences in their intensity, and in certain specific aspects of reactions taking place. The most obvious difference is in the crust-colour, which is predominantly chestnut in the case of rye-bread; in the case of wheat-bread the crust is golden-yellow to dark-brown. The influence on bread-crumb and crust-colour becomes obvious when baking mixed rye/wheat breads 100% to 0% at 10% increments. The reason for these colour differences are the presence of more free amino acids and types of sugar in rye-doughs. These compounds become involved in the melanoidin-forming Meyer reaction, which improve bread flavour. During the baking of rye-bread, significant decreases in the levels of free lysine, leucine, valine, arabinose, xylose, galactose, and glucose are apparent owing to these reactions.

A comparison of minimum baking-times for hearth-breads of the same dough scaling-weight but made with various percentage mixtures of rye and wheat flours, show decreasing baking-times with increasing percentage of wheat flour in the mixture.

Baking-temperatures for wheat-bread are lower than for rye-bread of equal weight and shape. The higher gluten content of wheat-bread dough results in a higher coefficient of thermal conductivity, on account of its superior leavening properties. Therefore, the time required to reach the evaporation temperature at the centre of the loaf-crumb is much shorter than in the case of rye-bread dough.

In the case of wheat-flour bread rolls, the centre of the dough-piece reaches evaporation temperature within about 6 minutes. However, the dough-pieces must remain at 220°C for 18 minutes to ensure that the baked-quality of the crust and crumb is acceptable.

The use of higher temperatures to shorten the baking-time, result in loss of bread-quality. To obtain adequate starch gelatinization, and protein denaturation, the centre of the dough-piece must be maintained at 98°C for at least 10 minutes to avoid a damp, inelastic crumb, and premature staling. The ratio of crumb to crust varies with the type of product, but should be maintained for the sake of bread quality at the following average levels:

| | |
|---|---|
| Wheat-bread 1·000 kg | crumb 80%, crust 20% |
| Bread-rolls 0·045 kg | crumb 65%, crust 35% |
| Mixed wheat/rye bread 1·000 kg | crumb 80%, crust 20% |

(Source: 'Special processes in baking', Werner Schwate and Udo Ulrich, *A Process Manual for Baking*, 2nd edn 1983 (in German), VEB-Fachbuchverlag, Leipzig. (Adapted from Fig. 109, p. 179.)

These aspects are of great importance for the successful sale of bread as a basic food, and, coupled with freshness, constitute the sensory appreciation of the consumer. The baking-time has a profound effect on aroma and flavour build-up of all baked products. Amongst those that form the aroma profile are the following substance groups: acids, alcohols, esters and lactones, carbonyls, pyrazine, sulphur-compounds and phenols. They also produce the final taste sensations of acidity, sweetness, saltiness and bitterness, which must be optimally balanced. These, together with the thermally modified lipids, starch and proteins, with ethanol, water and gas-development form the product consistency, which also contributes to the final sensation of taste on the tongue, and general mouth-feel.

There is an increasing preference for bread varieties with a rustic character, having a strong aroma, and well-developed aromatic and 'nutty' flavour with a crust 3–4 mm thick. Most of these are based on rye or mixed wheat/rye milled products. Whereas wheat and wheat/rye bread varieties and bread-rolls must have a larger percentage of crust, freshly baked and crisp, since their shelf-life is restricted to 2–3 days for bread, and about 6 hours in the case of bread-rolls.

In countries where more wine than beer is consumed, e.g. France, and white bread predominates, the percentage crust varies between 32% and 36% (baguettes), whereas an average loaf of white-bread would only have about 20–25% crust. Crusty white bread is the only suitable accompaniment, without spreading, for the pleasant acid taste of wine.

The delicious aroma of the oven-fresh baguette, is a French creation which neutralizes the wine acids, and increases its consumption without any ill-effects due to excess stomach acid build-up. The bakeout loss of the baguette is about 20%, which is twice that of an average white panned loaf. However, it is not only the bread which becomes the deciding factor for taste and flavour from the consumer's viewpoint. The choice of the bread variety will depend on the food with which it is intended to be eaten, to produce the ideal harmonization of the taste and flavour components. Nevertheless, bread flavour is critically influenced by the baking-process, and the crust-thickness is crucial for the taste and flavour of the crumb.

There is no reason to suggest that the introduction of multi-deck and travelling-band tunnel-ovens are responsible for the lack of flavour development in certain types of bread. Of paramount importance for the taste and flavour direction of bread is the control of temperature and the temperature/time curve within the dough-piece during the baking-process rather than the type, construction or energy-source utilized for heating the oven. During the baking-process, the control of heat output must be such that the heat-transfer from oven to dough-piece reaches a 'steady-state' condition, i.e. an amount sufficient to be able to penetrate into the dough mass and evaporate the water, leaving the crumb structure intact. Any excess build-up of heat at the crust-surface must be avoided, since this results in a wasteful 'burning' of the crust instead of a controlled bake-out and browning. Thus a 1 000 kg loaf on bake-out requires an average of 40 minutes when set at 240°C, falling to 220°C. This is valid for white bread of the hearth variety, a relatively short baking-time at a higher temperature being better than a prolonged baking-time at a lower temperature. In the case of rye and mixed wheat/rye bread varieties, a relatively high initial temperature of 250°C, falling to 210°C over 50 to 60 minutes for a 1 000 kg baked weight, gives the best flavour development. Doughs containing ryemeals and panned rye doughs require baking at similar temperatures, but with reduced top-heat and stronger bottom-heat, baking times being up to 120 minutes for a 1 500 kg dough-weight. This allows adequate swelling of the coarser meals, thus avoiding splitting along the sides. The baking-loss will depend on the duration of the baking-time, the scaling-weight and the loaf-weight. Large loaves of 1000 kg or more give baking-losses of the order of 10–13%, whereas bread-rolls lose from 18–20% during baking. The larger the unit, the smaller the net baking loss. For the formation of Meyer-reaction products and caramelization in the crust, which perfect bread flavour, the steam-vents must be opened after the first baking-zone to remove all steam for two-thirds of the baking-time, and so produce a loaf with a crisp, firm crust. A technique often applied to the baking of rye-bread to improve flavour, crumb-elasticity, porosity and shelf-life, is to effect a so-called 'pre-bake'. This involves exposing large units of 1500 kg to temperatures of 420°C for about 2–3 minutes, this gives the load an extra 1 mm crust using the normal baking-time. The crumb retains more moisture than normal, and bread yields increase from 135 to 138 kg bread/100 kg flour for mixed rye/wheat breads (*Mischbrot*). Such bread is claimed to be more aromatic, of improved flavour, and have an improved shelf-life.

The temperature within the dough-piece rises, depending on its size and the baking-time lapsed, as well as degree of leavening, slowly up to 98°C in the centre of the crumb, and the crust-surface can reach about 160°C. When the crumb reaches 98°C, the optimal baking-time has been reached. Owing to the difference in the specific heat of wheat- and rye-doughs, the same dough-piece weight of wheat-dough will reach the critical temperature of 98°C in the crumb centre about 10–12 minutes before that of a rye-dough or mixed rye/wheat dough-piece. This explains the differences in baking conditions and techniques applied to the two diverse products. The lower specific heat of wheat-dough is partially due to its lower moisture content (45–46%), compared with 53–55% for a rye-dough, and especially the difference in its material structure, brought about by the formation of an elastic gluten network, which becomes more porous on leavening. Although, the latter aerated state, which increases during the initial baking stages, probably favours thermal-conductivity through the layers of the dough-piece.

Specific heat or specific-heat capacity of a material is defined simply as the quantity of heat required to raise the temperature of unit mass of that material from $t$°C to $(t+1)$°C, at any temperature, $t$. The unit of specific-heat capacity is

$J kg^{-1} K^{-1}$ expressed in SI (or international units), and $cal g^{-1} K^{-1}$ in the CGS system, the SI system now being generally accepted.

For any calculations concerning the heat requirements of dough and bread, the following basic data is useful when considering temperature differences 1 degree Celsius $(C) = 1 K$:

Specific-heat capacity of leavened wheat-dough approx. $2 \cdot 80 kJ kg^{-1} K^{-1}$
                                                                       rye-dough approx.    $3 \cdot 02$
                                                                       average value          $3 \cdot 0$
(variation with dough water-content, average range 45–55% of total composition)

| | |
|---|---|
| Specific-heat capacity of hearth baked | |
| wheat-bread (crumb-moisture 42%) | $2 \cdot 7 kJ kg^{-1} K^{-1}$ |
| Specific-heat capacity of panned rye-bread | $2 \cdot 8 kJ kg^{-1} K^{-1}$ |
| (crumb-moisture 45%) | |
| Specific-heat capacity of crust | $1 \cdot 68 kJ kg^{-1} K^{-1}$ |
| Specific-heat capacity of water | $4 \cdot 2 kJ kg^{-1} K^{-1}$ |
| Specific heat of evaporation of water | $2\,208 kJ kg^{-1}$ |
| Average crust-temperature | $150°C$ |
| Average crumb-temperature at centre | $98°C$ |
| Weight of water as steam required/100 kg bread | $3\,000 kg$ |
| A 1 500 kg bread unit, baked from a dough-piece | |
| weight 1 680 kg requires 0·045 kg of water as steam | |
| Average dough-piece temperature | $30°C$ |
| Boiling-point of water | $100°C$ |

With such data at hand, it is possible to calculate the theoretical heat requirement to bake a bread of any desired weight. The component parts of such a calculation would demand a computation of the following data:

—heat requirement to heat the dough-piece up to 100°C
—heat requirement to remove excess dough water content
—heat required to form the crust
—heat required to produce enough steam to humidify the oven-chamber

The cumulative total of these separate data will provide the theoretical heat-requirement for a loaf of any desired weight. Average data for a 1 kg loaf would be 500 to 600 kJ, which could comprise:

237 kJ to heat dough-piece to 100°C
247 kJ to remove excess dough-water content
17 kJ to form the crust
76 kJ to produce enough steam to humidify the oven-chamber

The theoretical heat-requirement, however, would not be enough to completely bake the loaf out, since no account has been taken of technical heat-losses, which include: exhaust-gases, unburnt-gas, heating-up, radiation, steam-evaporation, incorrect secondary-air feed at oil-burner.

# 3.2 Elements of the Baking Process and Their Control

Since the baking process constitutes the most energy-intensive operation of bread production, and 30–40% of all bread faults are due to deficiencies during baking, the optimization of the controllable variables merits some consideration. The central object for such considerations is the shaped and proofed dough-piece, and in order to understand what changes the dough-piece undergoes, and how they can be influenced by the baker, the baking process must be analysed into the various aspects involved in its technical management. These can be broadly analysed as: heat-energy input and temperature; heat-transfer; humidity; and baking-time.

**Heat-energy Input and Temperature**
The internal energy of a system is the sum of the kinetic energy of chaotic motion of the molecules, the potential energy of their interaction, and the intramolecular energy. This energy can be transferred from one body to another in two ways. One is by mechanical interaction, when work is done by mechanical or electromagnetic forces; and the other is by thermal interaction, whereby energy is transferred by the chaotic motion of the molecules to produce heat-conductivity or thermal radiation. The quantity of heat-energy produced by the thermal interaction of bodies is referred to as the 'quantity of heat' or simply as 'heat'. The unit of heat in the SI system is the joule, expressed in the CGS system as calories. 1 calorie $= 4.1868$ joules (J). The quantity of heat required to increase the temperature of a body of unit mass from $t_0$ to $t = t_0 + dt$, and is referred to as $dQ$. The mean specific-heat within the temperature interval $(t - t_0)$ is the ratio $dQ/dt$, the limiting value of this ratio is defined as the true specific heat at the temperature $t_0$, the true specific heat depending on temperature. However, this dependence on temperature is normally disregarded, and specific heat is defined as the quantity of heat required to raise the temperature of a body of unit mass from $t°C$ to $(t + 1)°C$ at any temperature $t$. The quantity of heat-energy $dQ$ absorbed by a body of mass $m$, when its temperature is increased by $dt$ is given by:

$$dQ = sm\, dt$$

where $s$ is the specific heat.

The specific heat also depends on the conditions under which the heating takes place, when under constant pressure, it is referred to as 'specific heat at constant pressure', if there is no volume change, then it is termed 'specific heat at constant volume'. The former always being greater than the latter. This concept is applicable to substances in the gaseous state, for solid state substances, the difference is

553

insignificant. The units of specific heat are $J\,kg^{-1}\,K^{-1}$ (SI), and $cal\,g^{-1}\,K^{-1}$. The sum of the quantity of heat $dQ$ which a body absorbs when heated, and the work $dW$ which it performs in so doing, represents the change in the internal energy, $dI$, which is the First Law of Thermodynamics, viz.

$$dQ + dW = dI$$

Any change in the internal energy, $dI$, depends on the final and initial states, and is independent of the heating process, whereas both $dQ$ and $dW$ depend on the transition process. The Second Law of Thermodynamics states that heat cannot be spontaneously transferred from a colder body to a hotter one without a change in the system. The fact that the specific heat of a body approaches zero as its temperature approaches absolute zero on the thermodynamic temperature scale, is the Third Law of Thermodynamics. The point of 'absolute-zero' on the thermodynamic (kelvin) scale being $273 \cdot 15\,K$ below the melting-point of ice at standard pressure. This is the baseline for temperature measurement in kelvins. Temperature is a measure of the average chaotic motion or kinetic energy of the molecules within a substance, and the velocity of their oscillation or movement. This energy of movement can be increased by applying an adequate amount of heat-energy, which raises the temperature of the body, unless the body loses heat simultaneously. Temperature, as an expression of heat-intensity, during the baking process, can be reduced or increased by the control of the heat-energy-input. On setting a batch of bread in the oven, after a short time the temperature falls, unless more heat input is applied, which confirms that in the transformation of the dough-piece into a loaf of bread a specific quantity of heat input is expended. As already stated, this quantity will vary with oven-design, heat-loss management, dough-piece weight and dimensions, efficiency of heat-transfer mechanisms, loaf-spacing in the chamber, humidity-control, and use or non-use of pans. Average data for a 1 kg loaf is 500 to 600 kJ. Most of this heat-energy is transferred to the dough-piece in the form of radiated heat, which means that the walls of the oven-chamber have to attain a temperature between 300 and 400°C to bake bread. The heat transferred onto the dough-piece by radiation, conduction and convection-currents, and condensation of steam, is further transferred by conduction from the peripheral

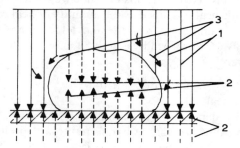

Fig. 90.  Heat-transfer in the dough-piece. 1. Radiation. 2. Conduction. 3. Convection currents. (Source: *Basic Processes in Baking*, A. Hirsekorn and W. Nehrkorn, GDR, 1982, p. 180.)

layers into the centre of the crumb in the manner illustrated (Figs 90–92). The graph (Fig. 93) shows the response of the moisture-content of the dough-piece to the rise in temperature; and the ideal temperature/time gradient during the baking-cycle (Fig. 93). During bread-baking, the oven temperature should ideally have a falling tendency, as the illustration shows. This technologically and economically justifiable approach is, however easier to realize when using an oven fitted with automatic temperature control. The curve illustrated (Fig. 94) is valid for rye and mixed rye/wheat doughs, an ideal curve for a 1 kg wheat-flour bread should show a

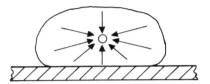

Fig. 91.    Thermal conduction within dough-piece. (Source: *Special Processes in Baking*, W. Schwate and U. Ulrich, VEB-Fachbuchverlag, Leipzig, 1986, p. 61.)

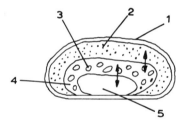

Fig. 92.    Water-evaporation zones. 1. Crust. 2. Firm-crumb. 3. Steam. 4. Instantaneous evaporation zone. 5. Warmed dough. (Source: *Special Processes in Baking*, W. Schwate and U. Ulrich, VEB-Fachbuchverlag, Leipzig, 1986, p. 61.)

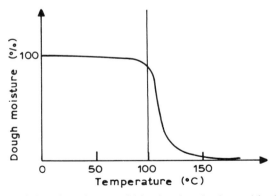

Fig. 93.    Reduction of dough-moisture within the dough-piece with rise in temperature. (Source: *Basic Processes in Baking*, A. Hirsekorn and W. Nehrkorn, GDR, 1982, p. 180.)

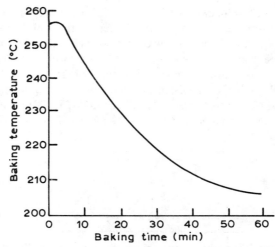

Fig. 94.   Ideal temperature/time curve for bread (hearth). (Source: *Special Processes in Baking*, W. Schwate and U. Ulrich, VEB-Fachbuchverlag, Leipzig, 1986, p. 61.)

sharp rise to a peak at 230°C, after about 5 minutes baking, followed by a more gradual temperature fall-off gradient up to the 40-minute baking-time. This is in sharp contrast with the illustrated curve for rye-bread, which shows a setting temperature of over 250°C with a much steeper fall-off gradient up to the 60-minute baking-time. Although it is not possible to generalize on the operation and control of ovens, owing to differences in design and source of energy used, certain basic requirements during their operation can be mentioned as guidelines:

(1)   The maximum capacity of an oven should be utilized to the full.
(2)   Final-proof and the optimum temperature of the oven should be synchronized.
(3)   Thermometers should be checked, and calibrated where necessary once a year.
(4)   Large ovens, e.g. travelling-band tunnel ovens, should be subjected to checks at regular intervals including: humidity in the chamber, gas and air consumption, also an exhaust-gas analysis. Where built-in continuous measurements are in use, the charts and read-outs should be monitored and interpreted.
(5)   All openings such as vents, steam-valves and loading-doors, should only be left open for as long as technologically necessary.
(6)   Draughts in the immediate vicinity of the ovens should be avoided.
(7)   Regular planned oven-maintenance is essential to limit breakdowns.

The theoretical heat-energy requirement for completion of the baking process, without consideration of inherent heat-losses, can be determined mathematically by calculating the quantity of heat, $dQ$, for the various elements of the process, viz. quantity of heat to heat the dough-piece to 100°C; quantity of heat to evaporate excess dough water; quantity of heat to form the crust; quantity of heat to produce enough steam to humidify the oven. The elements are calculated as follows:

*Dough-piece to 100°C*

Where: $dQ$ is the quantity of heat to raise temperature to 100°C in kJ

$m$ is the mass of the dough in kg

$s$ is the specific-heat of the dough in $kJ\,K^{-1}\,kg^{-1}$ $(3\cdot0\,kJ\,K^{-1}\,kg^{-1})$

$t_{100}$ dough temperature 100°C

$t_{30}$ dough temperature before baking

$dQ = ms\,(t_{100} - t_{30})$

*Evaporation of excess dough-water*

Where: $dQ$ is the quantity of heat to evaporate excess dough water in kJ

$m_2 - m_1$ is the difference in weight between the dough-piece and the baked loaf in kg

$r_{water}$ is the heat of evaporation of water $kJ\,kg^{-1}$ $(2208\,kJ\,kg^{-1})$

$dQ = (m_2 - m_1)r_{water}$

*Crust-formation*

Where: $dQ$ is the quantity of heat to form the crust in kJ

$m_{crust}$ is the weight of crust in kg

$s_{crust}$ is the specific-heat of the crust $kJ\,K^{-1}\,kg^{-1}$ $(1\cdot68\,kJ\,K^{-1}\,kg^{-1})$

$t_{crust}$ is the average crust-temperature °C, 150°C

$t_{100}$ is the boiling-point of water °C, 100°C

to calculate $m_{crust}$, the percentage crust for a bread of weight 1000 to 2000 kg is about 20%, and for bread-rolls 35% by weight

$dQ = m_{crust}s_{crust}(t_{crust} - t_{100})$

*Oven-humidity*

Where: $dQ$ is the quantity of heat to produce enough steam to humidify the oven in kJ at 100°C

$m_{water}$ is the weight of water in kg, 3000 kg/100 kg bread

$s_{water}$ is the specific heat of water $kJ\,K^{-1}\,kg^{-1}$ $(4\cdot2\,kJ\,K^{-1}\,kg^{-1})$

$t_{100}$ is the boiling-point of water °C, 100°C

$t_{20}$ is the initial water temperature, about 20°C (average)

$r_{water}$ is the heat of evaporation of water $kJ\,kg^{-1}$ $(2208\,kJ\,kg^{-1})$

$dQ = m_{water}s_{water}\,(t_{100} - t_{20})\,m_{water}r_{water}$

The summation of individual quantities of heat, $dQ$, required for each element gives the theoretical total heat-energy requirement for baking a loaf of any desired weight. However, the quantity arrived at would not be sufficient to bake the loaf of bread. This is due to the various sources of heat-loss, the most significant of which are: exhaust-gases, steam, radiation, and firing-losses. In the case of a travelling-band tunnel oven, of a total energy input expressed as 100%, some 45% would be utilized in baking the bread, 35% lost as exhaust-gases, 12% lost in the form of steam, and the residual 8% lost by radiation.

The 'energy-utilization index' of an oven is expressed as the ratio of the total theoretical heat-energy requirement to bake a loaf of bread of a defined weight to the total energy input, i.e.

$$\text{energy-utilization index} = \frac{\text{theoretical heat-energy input}}{\text{total energy input expended}}$$

This index value can be expressed as a percentage by multiplication by a factor of 100.

The total energy input expended, can be deduced as the product of the weight of fuel or energy-source in kg and the heat-capacity (calorific value) in kJ/kg of the energy-source, i.e.

$$Q_{\text{input}} = W_{\text{fuel}} \times H_{\text{value-fuel}}$$

In addition to the oven-capacity, baking-surface area, and loading characteristics, the energy-utilization index is a critical parameter for bakery ovens. It will depend upon the type of oven, the energy-source utilized, as well as the sense of responsibility and technical expertise of the person in charge of oven operation. This decisive stage of production, which used to be a work-intensive manual and unhealthy task because of the high-temperatures, has been made considerably easier through new oven design and developments. Newer oven-loading and setting systems, together with improved bake-out control techniques and climatic control within the oven-chamber have reduced the manual work, but in turn demand more technical know-how in terms of applied physics, machine construction and function, and measurement and control techniques. The essence of oven-operation is conscientious observation and a quick-reaction to correct unfavourable developments immediately. Typical routines are exact temperature readings taken before loading or 'setting' the oven, and throughout the baking-cycle; check actual readings against target values and readjust as discrepancies occur. Oven-humidity is often judged by experience, and lateral steam-headers or jets adjusted accordingly. Any excess steam is released by activation of the vents. Points to observe are crust-colour, loaf cross-sectional symmetry, volume and any signs of splitting at the sides in hearth-breads, or break and shred in panned breads. Any abnormal developments are countered by changes in temperature, humidity or baking-time. The energy-utilization index, expressed as a percentage, of two common oven-types: the steam-tube oven, and the travelling-band tunnel oven, heated by coal, natural gas and electricity are given in Table 65. The values suggest that electro-energy has the best utilization percentage, and is therefore the most economic energy source. Although this data is valid for the heat-energy-input to the oven alone, other factors which are external to the bakery itself, based on national economic issues such as availability and cost of raw material and resources to provide energy from them, will override all others. Where natural gas is readily available, this has proved to be popular, providing the gas:air mixture is automatically adjusted, since it is hygienic, requires no storage-space, and makes baking temperature adjustments easier. Multi-deck-ovens with fan-assisted cyclotherm hot-air or gas circulation, compared with Perkins steam-tube batch ovens, give an economy of up to 40% in energy.

## TABLE 65

| Oven-type | Energy-utilization (%) | | |
|---|---|---|---|
| | Coal | Natural gas | Electricity |
| Steam-tube batch-oven | 16–20 | 25–28 | 30–40 |
| Travelling-band tunnel-oven | — | 50–70 | 60–70 |

Source: *Special Processes in Baking*, W. Schwate and U. Ulrich, 2nd edn, 1983, VEB-Fachbuchverlag, Leipzig, p. 74 (in German).

These multi-deck ovens can be heated with diesel oil, gas or coal, and, independent of the form of energy, each deck can be individually regulated for baking-temperature. This allows the simultaneous baking of either an assortment of products, or the use of one deck to give a 'sharp pre-bake' and another to complete the baking-process. The average heating-up rate is about 3 K/min, and the heating options are oil, gas, electricity or coal. Roll-up oven-setters allow fast and efficient oven-loading. The relative energy-input requirements of an average 4–5 deck oven for heating-up and during continuous utilization are as given below.

*Heating-up*

| | Energy-consumption/h | Heating-up time (min) |
|---|---|---|
| Gas | approx. $32.0 \, m^3$ | 75 |
| Diesel oil | approx. $11.0 \, kg$ | 75 |
| Coal | approx. $45.0 \, kg$ | 120 |

*Continuous utilization*

| | Energy-consumption/h |
|---|---|
| Gas | $13 \, m^3$ |
| Diesel oil | $5.7 \, kg$ |
| Coal | $15.0 \, kg$ |

Source: 'Higher productivity through utilization of the Cyclo-therm deck-oven' (in German), *Bäcker und Konditor*, **20**(26), 1972, 5, p. 134.

For warm-water production the requirement is two Boilers/100 litres. The average baking-surface for a four-deck oven is a total of $10 \, m^2$.

### Control and Measurement of Heat-energy Input and Temperature

Gas is a popular source of heat-energy for baking-ovens. Natural gas (German: *Erdgas*) is fast replacing the traditional town-gas (German: *Stadtgas*) supply. The volume of gas-flow/unit time is measured by a cylinder-type digital-counter, designed for various throughputs. In addition the gas:air ratio must be checked, for which a minimum of air is necessary for efficient burning. This value is determined from the gas analysis; the actual quantity of air is either determined from the working-pressure of the measuring diaphragm, or the air velocity at the air-intake, and based on the cross-sectional area. The ratio of the actual quantity of air to the minimal quantity of air is referred to as the air-number. In the case of travelling-band tunnel oven BN-series, used extensively in the GDR and manufactured by the VEB-Kyffhäuserhütte, Artern, the 'air-number' or 'burner-number' should be between 1·1 and 1·3, to avoid energy-losses, according to Schneeweiss and Klose (*Technology of Industrial-baking*, VEB-Fachbuchverlag, Leipzig, 1981). Parallel with these measurements, a determination of the 'air-number', is also carried out by means of an exhaust-gas analysis. For this purpose, both discontinuous and continuous gas-analysers are available in both a stationary and transportable form. Reduced pressure in the combustion-chamber of tunnel-ovens can be measured with either a U-tube type manometer, or one functioning on the bent-tube principle coupled with a cogged-wheel and pointer system of scale-readout. An average combustion-chamber pressure for a travelling-band tunnel-oven should be regulated to within 40 down to 20 Pa by means of exhaust-gas adjustment.

Electro-energy can be measured with a meter, which registers the energy used in kilowatt-hours (kWh), and is defined as the energy equivalent of 1000 watts acting for 1 hour.

Temperature in both loading-type deck-ovens and travelling-band tunnel-ovens are measured by pyrometers. These instruments operate on the expansion principle, consisting of a bimetallic-strip of, for example, iron and copper of different expansion coefficients. The deformation of the strip at different temperatures is measured by registration on a calibrated-scale of the movement of a cogged-strip, which keys into a cogged-wheel, attached to a rigid pointer. Such instruments depend on conduction for their operation, whereas, true pyrometers measure radiated heat either by changes in electrical resistance, or millivolt differences of the order of 5–6 mV per 100 K. Thermocouples for resistance-thermometers consist of nickel and platinum; and the thermo-elements of instruments which depend on thermal-voltage differences, measured in millivolts, consist of Iron–Constantan couples, connected to the pyrometer by copper wires. In the case of continuously working ovens, temperatures are not only measured within the various oven-zones of the baking-chamber, but also within the burner-space, and the circulation-system. This involves temperature gradients of 150 to 600°C, which are measured remotely by means of either thermo-elements or electrical resistance thermometers. The read-out is usually performed by a six-colour dot-printer. By means of the electrical output-signal, the gas-feed and hence the temperature at the burner can be controlled. However, the normal two-point control-system is subject to temperature differences of the order of $\pm 20$ K, therefore for uniform temperature-control a continuous, or at least three-point, control-system is applied.

## Heat-transfer

During the baking-process, heat-transfer from energy source to dough-piece takes place in three forms. Depending on the type of oven and its design, the form of heat-transfer will vary, one or other form tending to predominate. Through the oven-bottom or 'sole', the dough-piece will receive heat from the area of contact by conduction, the material of the oven-sole being normally either steel sheeting or refractory tiles. Conduction results from the oscillation of the molecules within a material, brought about by the intensity and rate of heat-energy input. It is expressed as the coefficient of thermal conductivity, the units of which are either kJ $m^{-1} h^{-1} K^{-1}$, $J m^{-1} s^{-1} K^{-1}$, $W m^{-1} K^{-1}$ in SI units, or kcal $m^{-1} h^{-1} K^{-1}$ in CGS terms. The coefficient of thermal conductivity $(\lambda)$ is defined as the rate of flow of heat $dQ/dt$ through a surface of area $(A)$ of a medium or material, where the temperature-gradient in a direction normal to the surface is $dt/dx$. The relationship is mathematically expressed as:

$$dQ/dt = -\lambda A \, dt/dx$$

This constant, which depends on the moisture content of the material, is 45·4 for steel, 209·3 for aluminium, 0·683 for water at 100°C, but 0·648 at 50°C owing to temperature dependence, all values being expressed as W (watts)/metre/K $(W m^{-1} K^{-1})$. In the case of gases, which are relevant to the atmosphere within the baking-chamber, the thermal conductivity of air at standard-pressure (760 mm) and 20°C is $257 \times 10^{-4} W m^{-1} K^{-1}$, and that of carbon dioxide at 20°C $162 \times 10^{-4} W m^{-1} K^{-1}$.

Through the medium of the atmosphere within the baking-chamber, heat is transferred by another form or mode known as convection. This process involves the movement of the heated medium (air) from one place to another in the form of convection currents, and occurs within most types of oven. This transported heat-energy is transferred to the dough-piece by the movement of steam and air-currents when the vents are opened or by the assistance of fans. The convection heat-transfer coefficient in $kJ m^{-2} h^{-1} K^{-1}$ will depend on the speed of the air-currents, increasing from about $20 kJ m^{-2} h^{-1} K^{-1}$ when static to $168 kJ m^{-2} h^{-1} K^{-1}$ at air-speeds of $10 m s^{-1}$.

The third form of heat-transfer is by radiation, which is generally the most significant of the three modes of heat-transfer, although the heat so transferred can only penetrate a few millimetres into the dough-piece, after which the conduction mode takes over. The heating of the dough-piece in the baking-chamber is brought about by the heat-exchange with the heat-emitting elements of the baking-chamber and the existing steam/air mixture within it. At temperatures of 300 to 400°C, the heat-emitting oven surfaces give rise to electromagnetic radiation waves. The ideal thermal radiator is called a 'blackbody', such a body absorbing completely any radiation falling on it and, for a given temperature, emitting the maximum amount of thermal radiation possible. At temperatures of 300 to 400°C, the wavelength which corresponds to maximum radiant energy falls within the range 5 to 4·3 micrometres. This range of wavelength of electromagnetic oscillation lies within the range of the invisible infrared spectrum between 0·77 and 340 micrometres. The visible range of the spectrum is quite narrow, between 0·3 and 0·72 micrometres. The infrared spectrum generally is defined as the range from 0·72 to about 1000 micrometres. Bordering on the visible spectrum on the low-wavelength side are the ultraviolet rays, while microwaves border the infrared spectrum on the high side. Radiation-temperature sensing devices generally utilize some part of the range 0·3 to 40 micrometres. The law governing the ideal type of blackbody radiation is Planck's law. This allows the calculation of the distribution of radiant intensity with wavelength, since a blackbody at a given temperature emits some radiation per unit wavelength at every wavelength from zero to infinity, but not the same amount at each wavelength. The quantity $Q_\lambda$, the amount of radiation emitted from a flat surface into a hemisphere, per unit wavelength at the wavelength $\lambda$, is given by the equation:

$$Q_\lambda = \frac{C_1}{\lambda^5 (e^{c_2/(\lambda T)} - 1)}$$

where $Q =$ hemispherical spectral radiant intensity, $Q/(cm^2 \, \mu m)$, $C_1 = 37\,413$, $Q \, \mu m^4 \, cm^{-2}$, $c_2 = 14\,388 \, \mu m \, K$, $\lambda =$ wavelength of radiation, $\mu m$, $T =$ absolute temperature of blackbody K.

If $T$ is fixed at various values, e.g. 200, 250, 300, 350 and 400, and plots are made of $Q$ versus $\lambda$, the peaks appear at specific wavelengths, which become longer as the temperature decreases. The area under these curves represents the total emitted power, which increases rapidly with temperature. The power radiated at all wavelengths by a blackbody ($e$) is proportional to the fourth power of the absolute temperature, i.e. by doubling the temperature, the radiated power increases 16 times. This is known as Stefan–Boltzmann law, and allows the total energy, $e$, of all wavelengths radiated per second per square metre by a full radiator (blackbody) at temperature $T_1$ to surroundings at $T_0$ to be determined as follows:

$$\text{total energy } e = \sigma(T_1^4 - T_0^4)$$

where $\sigma =$ Stefan's constant $5 \cdot 6697 \times 10^{-8} \, W \, m^{-2} \, K^{-4}$. Although the concept of a blackbody is a mathematical abstraction, real physical bodies are constructed to approximate closely blackbody behaviour. Such sources are necessary to calibrate radiation thermometers, and usually consist of a blackened conical cavity of about 15 degrees cone-angle. The temperature being adjustable, automatically controlled for constancy, and measured by an accurate sensor such as a platinum resistance thermometer. While it is possible to construct a near perfect blackbody, the bodies whose temperatures are to be measured with some radiation-type instrument usually deviate considerably from such ideal conditions. This deviation from blackbody radiation is then expressed in terms of the emittance of the measured body. Various types of emittance have been defined for specific applications. The most fundamental form of emittance is the hemispherical spectral emittance $e_{\lambda,T}$. If $Q_{\lambda a}$ is the actual hemispherical spectral radiant intensity of a real body at temperature $T$, and assuming that it can be measured by using optical bandpass-filters, then:

$$e_{\lambda,T} = \frac{Q_{\lambda a}}{Q_\lambda}$$

where $Q_\lambda$ is the blackbody intensity at temperature $T$.

Emittance is therefore dimensionless, and always less than $1 \cdot 0$ for real bodies. In general it varies with both $\lambda$ and $T$.

The most insignificant practical part of total heat energy transfer during the baking-process is that transferred from the gaseous medium in the baking-chamber by convection to the dough-piece. However, its significance can be increased by the introduction of heat circulation fans within the chamber.

During the initial period of baking, the most significant mode of heat-transfer to the dough-piece is provided by heat of condensation of the steam onto the dough-surface and surface-layers. When the steam condenses on the surface, the latent heat of steam-formation is released as heat energy. This process accelerates the warming-up of the dough-layers near the crust, and changes the character of the temperature/baking-time curve in the central zone of the crumb. Therefore, in fixing a value for the optimum humidity within the baking-chamber, its importance as a heat-transfer medium onto the dough-piece should not be underestimated.

Having defined the fundamental principles of heat-transfer from one body to another in the direction from hot to cold by means of the three modes of transfer: (1) conduction in solids and liquids in the static state, (2) convection in gases and liquids in a state of motion, and (3) radiation which involves electromagnetic waves and no material carrier, it is necessary to consider how they apply to baking procedures.

The major role of conduction in baking is the transfer of heat from source to the hearth or oven-sole, and walls of the baking-chamber. It will represent the primary mode of heat-transfer to all flat products, which are baked by contact with a hot-surface, and those baked on a hot-plate. Its involvement in the baking of larger volumed products will be limited to the transfer of heat from the hearth or pan-walls to the sidewalls and bottom of the baking product. However, in modern indirect-fired ovens, where the heat from the fire-box is transferred to the chamber by flues and radiators, being transformed into radiant-heat, conducted-heat plays a major role initially.

Convection, which involves the transfer of heat by fluids or gases in a constant state of motion, is essential for water-heating and the creation of convection-currents, and equalizes heat distribution in tanks. In the oven a mixture of air and steam forms the main media for heat transmission, but passive-convection relies completely on natural air currents due to temperature differences of adjacent masses of air. Forced convection, however, which is created with the help of fans or blowers, results in a more rapid air movement and diffusion of heat. Thus, nearly all modern types of oven make use of forced convection to dissipate convection-heat. It is universally acknowledged, that rate of heat-transfer and uniformity of bake are both greatly enhanced by the application of forced-convection. The most effective method of air agitation and the volume and velocity to be applied for maximum heat-transfer are all subjects for open debate. Obviously, as air-velocity is increased, a point is reached beyond which any further increase would result in no further increase in heat-transfer from the source. The volume of air involved must also be calculated to achieve optimum heat-transfer by this technique.

Forced convection systems for ovens normally utilize a chamber at the bottom of the unit, from which air can be forced into the chamber at a measured rate, through slots to ensure an even distribution by agitation. Another technique is to recycle the oven atmosphere by using a fan system which removes air from the top of the chamber, passing it through perforated tubes placed above and below the baking product.

Heat-transfer by radiation is the most diverse mode of transfer. Energy being emitted by the heated body as electromagnetic waves, travelling with the speed of light in a straight line to the point of absorption. The range of wavelengths utilized practically for baking fall primarily within the infrared spectrum, those generally referred to as the 3 to 9 micrometre band. These infrared rays are absorbed by opaque bodies, heating them up; otherwise they are either deflected, or refracted or pass through transparent material. The intensity of these rays diminish depending on the square of the distance from their source. Owing to the direct mode of heat-transfer, commercially developed radiant heating-panels for ovens and cooking equipment can be applied. These consist of heating-coils embedded in ceramic refractory material, which absorbs heat uniformly when electrical energy is applied. Such ceramic radiant-panels provide a very energy-efficient heating-system for heat-transfer to the dough-piece.

It is generally accepted that direct-firing is the simplest and most economic method of heating ovens, being up to 25% more efficient than indirect systems. Since natural gas remains the most popular fuel, because of its availability, cost and ignition characteristics, ribbon-burner systems with a flame-band distributor have found acceptance. These produce a constant and efficient gas:air mixture, and a more homogeneous heat distribution in the oven-chamber. The natural convection currents set up by the heat from the burnt gas mixture usually produces adequate turbulence to ensure uniform heat distribution around the dough-piece.

During the baking-process, it is desirable that the moisture emerging from the dough-piece as steam is rapidly removed by air turbulence, thus facilitating heat penetration. The thermodynamics at the surface of the dough-piece are such that on entering the oven-chamber, moisture starts to migrate from the moist dough-piece interior layers towards the surface where evaporation can take place. This process requires considerable latent energy, and will delay the attainment by the dough-piece of the ambient oven temperature, since it will continue until crust-formation prevents further evaporation. Furthermore, prolonging this stage will result in a thick, tough crust. Adequate air turbulence around the dough-piece at this stage, will accelerate the evaporation process, by removing the protective steam-layer surrounding it. Ovens are now being fitted with air-circulating systems, which remove air from the chamber at several points, and recirculate it through suitably placed ducts using blowers. Ovens used in the biscuit industry, force the heated air through slotted-tubes running laterally across the oven-chamber. Such ovens are referred to as 'impingement-ovens'.

Improved heat-utilization continues to be the subject of research by oven manufacturers, which usually involves improved heat-transfer techniques.

Convection plays an important part for heat-transfer in the baking of panned-bread, ensuring a fast and uniform bake. By the circulation of 10 to 26 cubic feet of preheated air per square foot of oven-hearth at a velocity of 2000 to 4000 feet per minute around and over the top of the dough-piece, heat-transfer can be increased from less than $2 \cdot 0 \, \text{Btu} \, \text{h}^{-1} \, \text{ft}^{-2} \, {}^{\circ}\text{F}^{-1}$ up to about $20 \, \text{Btu} \, \text{h}^{-1} \, \text{ft}^{-2} \, {}^{\circ}\text{F}^{-1}$. Such dramatic changes in heat-transfer rates can only be realized, however, when a more complex system of turbulence is imposed on the dough-piece surface by vertical hot-air impingement. For such purposes, impingement-modules with tapered-jet nozzles must be directed both downwards and upwards in close proximity to the oven band-conveyor.

Brunson (*Baker's Digest*, **40**(1), 64, 1966) has described the improvement in efficiency of recirculation in indirect-fired band ovens, using air-circulation velocities within the baking-chamber of 200 to 400 ft/min. In air-impingement ovens, air-velocity rates of 3000 ft/min can be achieved in close proximity of the nozzle-orifices. These nozzles are designed to produce narrow jet air streams, which produce a maximum of turbulence around the dough-piece. Impingement air ovens allow the use of lower baking temperatures, and shorter baking-times, yielding products with higher moisture contents and prolonged staling-rates, at improved energy efficiency, according to test-bakes performed by Smith at Texas A and M University, College Station, Texas, USA, in 1983. These tests were compared with a reel oven as a controlled experiment.

Heat-transfer by conduction becomes important when baking hearth-breads, which involves a slower bake. In such cases, a solid-steel, refractory-tile, or wire-mesh reduces bottom-heat of radiation and cuts off the free-flow of convected-heat going completely around the dough-piece. Heavy, lower specific-volume, white-breads and variety breads such as rye and pumpernickel demand this slower baking technique, as mentioned previously. The technique and technological parameters are quite distinct from those for panned, high specific volume bread during the baking-process. For example, the enriched high specific volume/low density panned bread standard product produced on the North American continent allows the baking of a 16-ounce unit within 18–20 minutes, within a temperature range of 350 to 480°F (177 to 249°C). However, the speed of this bake will depend upon the rapid application of heat-transfer by forced convection, which in turn requires the oven-grids to be widely spaced, and the pan spacing on the grids adequate to allow free circulation of heated air around the pans. Also, the presence of a high percentage of added sugars demand a lower baking temperature.

A new approach to heat-transfer was introduced in 1965 by Rhodes in the USA, under Patent No. 4,072,762, Feb. 7, 1978. This device, with the trade-name of 'Acceletron', is capable of accelerating the rate of heat-transfer of both heat and steam in ovens. This effect is achieved by the creation of a type of 'corona wind' within the baking-chamber. An electrostatic field of high potential and constant polarity is set up between the oven-crown and the hearth by an electrode grid. The device is installed within the oven parallel to the steam-headers, and ionizes the gases near the electrodes, and the ionized gas molecules then become repelled by the electrodes, resulting in a downward air current towards the product on the hearth or on trays. It is reported that the use of this equipment can reduce the amount of steam required, and the baking-time by up to 20% in certain cases. The electrode-grid is installed above the hearth or trays, within the first 25% of baking-surface. A high-voltage direct current discharge through the electrodes results in the set-up of conditions for an improved convective heat-transfer to the dough-piece within the first 4–5 minutes of baking. The process gives rise to the generation of traces of ozone, as is normal with such discharges. Work done by Sievers (*Proc. Am. Soc.*

*Bakery Engineers*, 1978, 98), showed that the best results could be obtained by reducing the temperature in the first two heat zones by 50 to 80°F (28 to 45°C), using an adequate steam supply, and a corresponding reduction in the final zone of 25°F (14°C). Sievers' conclusions, from commercial trials using both direct- and indirect-fired ovens, and a wide product range, indicated that the important advantages to be derived from accelerated heat-transfer were (1) a 10% increase in oven-capacity, (2) an energy saving of 11·8%, due to a reduction of 40 Btu, down to 300 Btu/lb to complete a bake. This value comparing with a theoretical value (no heat-losses) of 235 Btu, and actual practical values for conventional modern ovens within the range 325 to 400 Btu/lb of bread, (3) a reduction in bake-out loss, which allows a reduction of 0·5 ounce in the scaling-weight of a 24-ounce unit. Average savings due to reduced bake-out loss are from 0·5 to 1·0 ounce/lb of dough.

Ionized-atmosphere or electrohydrodynamic heat-transfer enhancement in some oven installations, aimed at improved steam application, appear to enhance heat-transfer by radiation rather than by convection. This set-up would permit the pan-strap spacing on the oven-grid to be closed, thus improving oven efficiency still further.

For the modulation and control of temperature and heat supply in any zone of the oven, a modulation control valve to ensure the correct air:gas mixture is linked to a temperature control unit, which senses the temperature within the zone of burner operation, and signals when more or less heat is required. A single-lap direct-fired gas oven can have up to eight heat-zones. A typical link-up controller, actuating an electric modulating motor on the combustion air-line is illustrated in Fig. 95. The controller has a mercury-filled thermal sensing element, which is

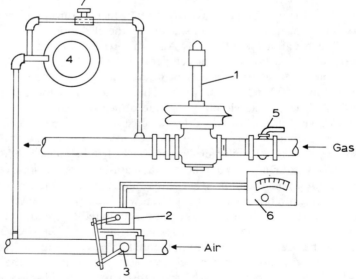

Fig. 95.   Burner-control system (gas), (direct-fired oven.) 1. Gas regulator. 2. Modulating electric motor. 3. Air-control valve. 4. Gas-burner. 5. Mains-gas valve. 6. Temperature-controller. 7. Gas-tap.

installed within the baking-chamber and moves an indicator on the instrument up and down the scale in response to the expansion and contraction of the mercury. This also actuates a contact-finger along a potentiometer-coil within the modulating temperature range. The coil itself moving with a red set-pointer, which can be set at the control point by turning a setter-knob positioned on the instrument housing. The potentiometer coil constitutes half of a Wheatstone-bridge circuit, and the other half of the bridge is formed by the potentiometer, which is built into the motor being driven by the motor-shaft. A relay sensor is incorporated into the motor, which detects any Wheatstone-bridge imbalance brought about by a change in temperature, and this in turn drives the motor in the appropriate direction to rebalance the bridge circuit. The motor-shaft is connected by a linkage to a device controlling the amount of air entering the gas mixture valve. Since the temperature controller is accurate to within 1·0% of the full scale range, temperatures in the various oven-zones are maintained within a few degrees of the instrument setting.

Using conventional baking-process heat transfer techniques, the internal loaf temperature must reach 96 to 98°C before it can be considered properly baked. However, it has been established that during the baking of a 22-ounce square pullman loaf (pan with cover), utilizing oven-atmosphere ionization equipment, which only took 17 minutes, the temperature at the loaf-centre remained below 43°C up until the 7th minute of baking. After which it rose rapidly during the next 7 minutes, levelling out between 88 and 93°C during the remaining 2·5 minutes in the oven.

Among the main factors which contribute to a more rapid and efficient bake-out are the following:

(1) Rapid and accurate temperature controls.
(2) Adequate natural and forced convection.
(3) Optimal air:gas mixture.
(4) Full burner modulation.
(5) Good lateral heat-control.
(6) Adequate pan-spacing.
(7) Use of aluminized-steel pans.
(8) Ionization of the oven atmosphere.
(9) Prevention of cold air entering the feed and discharge points.
(10) Formulation and enrichment ingredients, e.g. sugars and milk-solids.

For more accurate temperature control, the electronic potentiometer is the preferred instrument, capable of accuracies of about ±0·25% of total scale range. Pyrometers, based on the millivoltmeter, are slower and less sensitive in response. Thermometers of the pressure type, whether of the coiled Bourdon or capillary-tube design, have scale range accuracies of about ±1%.

The electronic potentiometer, available in several forms is the most accurate temperature control instrument for multi-zone ovens. In general it is advisable to have an instrument to serve each individual heat-zone of an oven, and complement each controller with a read-out chart to record changes in temperature within each zone during the baking-cycle.

Thermoelectric control systems, involving thermocouples, permit any number of thermocouples to be connected to one instrument, thus monitoring temperatures in several oven areas.

Associated burner-control systems can be of the electrical-contact, or pneumatic-proportional type (narrow-band throttler), depending on the burner type, the chosen control modes being high–low, on–off, or modulating. In the modulating mode, the controls operate the air-control valve of the burners in sharp response to preset temperatures, thus preventing any large temperature fluctuations.

**Humidity**

The presence of steam at an optimal concentration within the oven chamber is important for the quality of certain products such as: hearth-breads, bread-rolls and certain types of cake. The degree of saturation of steam depends on pressure and temperature, and as the temperature increases the capacity of air to hold moisture will increase. At normal atmospheric pressure (760 mm of mercury) water boils at 100°C, with a heat of evaporation (the quantity of heat per unit mass that must be supplied at its boiling-point to convert it completely into the gaseous state at that temperature) of 2260 J/g, whereas, if the pressure is reduced to 0·5 atm, it boils at 82°C, with a heat of evaporation of 2310 J/g. As the air reaches saturation point, it cannot hold any more steam, and it drifts as a fog in the air. At 20°C, the ratio of steam to air (kg) is 0·015, and at 90°C 1·56, but at 99°C it becomes 200. At temperatures in excess of 100°C, the water-holding capacity of air is such that steam can be mixed with it as required. At every temperature, air is capable of holding a specific maximum of steam (kg water/kg of dry air), and on exceeding this amount, condensation results. Conversely, water-saturated air will absorb increasing amounts of water with increasing temperature. The point of saturation of steam in air is often referred to as the 'dew-point', and every amount of steam in the air, depending on pressure, has a dew-point temperature. During the baking of bread, this dew-point temperature is between 70 and 90°C, which corresponds to a steam content of 0·29 to 1·56 kg/kg of dry-air. As long as the surface temperature of the dough-piece during the initial stage of baking remains below the appropriate dew-point temperature, moisture will condense on its surface, and in so doing 2270 kJ/kg of steam is released. Most of this free heat-energy is either absorbed by the dough-piece by conduction, or is carried directly onto the dough-piece as the water condenses. The effect of this condensation-heat is limited to the initial stage of baking only, since the surface temperature of the dough-piece will exceed the dew-point temperature after about 1 to 3 minutes baking-time. The ideal type of steam is at low pressure, produced with a boiler-gauge pressure of about 10–15 psi, reducing to 5 psi through the injector-nozzle. Steam at 10 psi pressure, in the saturated state has a temperature of 115°C, and at atmospheric pressure (760 mm mercury), 100°C. As steam pressure is reduced, its temperature falls, and its water-holding capacity, which results in condensation onto the relatively cold dough-piece at 30–35°C. Owing to the porous structure of the dough-piece, both sorption and condensation take place both on the surface and within the surface layers of the dough. Moisture condensation proceeds until the dough surface temperature exceeds the dew-point temperature, which will depend on the steam/air atmosphere within the oven chamber. This retains the surface in a moist hydrated state, allowing it to expand without excessive cracking or splitting at the sides, up until the surface temperature reaches 100°C. The use of steam at higher pressures (and temperatures) would not allow sufficient time for its temperature to fall to the dew-point, and condense, before the influence of the oven heat had raised it again to that of the dough-surface. High temperature/pressure steam also has a depressing effect on loaf-volume generally, and on rye-bread in particular. The presence of humidity during baking accelerates the heat-transfer to the crust area, surface layers and central zones of the loaf. The temperature/baking-time curves of all loaf-zones are changed, compared with those without humidification, comparable stages in baking with humidity showing occurrence at elevated temperatures up until the completion of the process (see Fig. 96(a)).

Increases in dough-moisture at doughmaking also increase the speed of heat-transfer, especially in those layers in closest proximity to the surface of the dough-piece.

During the baking of rye-bread on the hearth, the injection of steam at the right time and in the correct quantity is critical for bread quality. The optimal humidity injection and adjustment of top and bottom heat during the humidification stage often fall into a narrow range, depending on the type of oven and design. During the baking of mixed rye/wheat bread, if the initial-zone temperature is too low, and the humidity too low, the crust will develop both side and top splitting or even deep cracks. Other factors, however, which determine the requisite temperature and

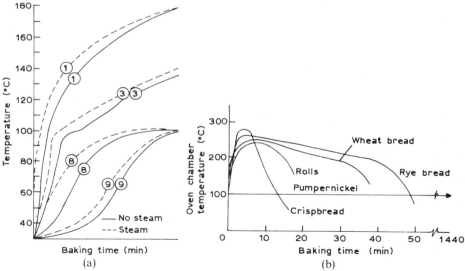

Fig. 96. (a) Temperature/time curves of the various dough layers during baking under humidified and unhumidified baking conditions. (Source: *Technology of Breadmaking*, L. Ja. Auerman (Ed.), VEB-Fachbuchverlag, Leipzig, 1977, p. 244.) (b) Temperature profiles in the oven chamber for various baked products. (Source: *Technology of Industrial Baking*, R. Schneeweiss and O. Klose, VEB-Fachbuchverlag, Leipzig, 1981, p. 206.)

humidity level are: recipe, mixing, proofing-conditions, size and shape and spacing of the dough-pieces inside the oven-chamber. Where no steam-injection facility is available, one would have to rely on spraying or moistening the dough-pieces with the help of a large brush prior to setting in the oven.

As an illustration of the use of humidity as a corrective measure for dough-pieces which have to be baked in a state of underproof, in such cases, plenty of steam is injected into the oven-chamber. Inside the closed chamber, this causes a pressure increase, which counteracts the increasing pressure within the dough-piece itself, due to gas-development and expansion. Thus suppressing excessive and uneven oven-proof of the dough-piece. Also, since the surface of the dough-piece is cooler than its surroundings, water condenses on the dough-piece, thus keeping its surface longer moist and elastic up until gas-development and expansion is complete.

During the baking of rye and mixed wheat/rye breads, using a batch-type oven, the bread is set at 250 to 280°C, and the steam allowed to remain in the oven for up to 1 minute. Depending on the amount of steam injected and temperature, as well as dough-piece shape and spacing, a variable amount of steam will condense on the dough surface. When this process is complete, the steam-channel vent is opened. This sets up a strong influx of colder air, replacing the hot, damp air. This incoming air is cool and up to 50 times drier than the previous atmosphere. At this stage radiated heat from the oven-walls takes over, the dry, warm air removing moisture from the crust rapidly, and effecting a general drying-out of the dough-piece. The remainder of the baking process should then proceed at a lower temperature, otherwise excessive browning of the crust will occur, assuming the baking-time is strictly adhered to. In the case of travelling-band continuous tunnel ovens, this reduction in temperature is easily achieved by reducing the heat in the subsequent zones, whereas, in batch ovens the bread is often transferred to a cooler oven-chamber.

The baking of wheat-breads and rolls presents much simpler circumstances, due to the greater elasticity of both crust and crumb, and more uniform temperatures and humidities can be used throughout the baking-process (Fig. 96(b)).

The longer wavelength infrared rays are partially absorbed by the steam, which is particularly significant for the effect of humidity, since these wavelengths predominate within the oven-chamber. During the injection of steam, initially, up to 3% of dough-piece weight can condense, this contributes about 57 kJ to the heat-transfer (about 5%) theoretically necessary to bake a 1-kg loaf. This is an efficient and direct method of heat-transfer to the dough-surface, which penetrates into the loaf interior by conduction. The condensation increases the surface moisture, and owing to this difference in concentration, water permeates through the porous structure into the interior of the loaf. This then becomes reversed as soon as the surface moisture evaporates, and the evaporation-zone, throughout the baking-process migrates further into the loaf interior. The surface temperature of the loaf, depending on the dew-point, is normally attained within 3 minutes, after which the heat-transfer effect ceases. At this stage any increase in the humidity of the oven-chamber would inhibit heat-transfer by absorption of the longer wavelength infrared rays. Air saturated with steam will not allow evaporation from the dough-surface, and will prolong drying out and crumb-firming. Increasing the dew-point from 70 to 80°C will require approximately twice as much steam, and in so doing the duration of condensation is not significantly prolonged. The heating of the dough-piece together with condensation of the humidified-air effects both chemical and mechanical changes to the dough-surface. The surface starch granules become gelatinized, and some proteins coagulate, the 'skin' so formed giving the dough its structure and exterior firmness.

An illustration of the change in weight of the dough-piece within the gas-medium of the oven chamber, which had been intensively humidified during the initial stage of baking, is graphically shown as a plot of dough-piece weight versus baking-time in Fig. 97. This work was carried out by Micheljew in the USSR, and published in

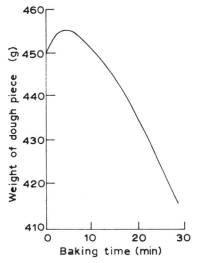

Fig. 97.   Change in weight of dough-piece baked in an oven-chamber gas medium which has been intensively humidified during the initial baking stage. (Source: *Technology of Breadmaking*, L. J. Auerman, Moscow, 1977.)

1943 in *Special Aspects of Bread-baking Heat-technology*, Alma-Ata, 1943. The maximum weight gain took place between 3 and 5 minutes baking-time, and the weight of the dough-piece increased by 1·3%, compared with its scaling-weight. This initial condensation does not only remain on the loaf surface, but also migrates into the adjacent dough-layers. The higher the humidity of the gas-medium within the oven-chamber, and the lower its temperature, and that of the dough-surface and adjacent layers, the greater the intensity and duration of the increase in weight of the dough-piece. However, as already stated, as soon as the dough-surface temperature exceeds that of the dew-point, no further condensation takes place. Instead, moisture evaporation immediately begins. Evaporation first takes place at the dough-surface, and adjacent layers, then, as soon as the surface-layers have reached the equilibrium moisture level and the crust forms, the evaporation-zone

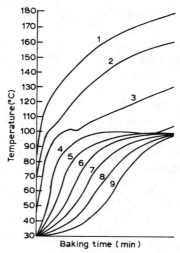

Fig. 98.   Temperature changes in specific layers of the dough-piece during baking. 1. Bread surface. 2, 3 and 4. Layers within the bread at 1/4, 1/2 and 3/4 of the crust thickness distant. 5. Layer dividing crust and crumb. 6, 7 and 8. Layers within the crumb at 1/4, 1/2 and 3/4 distant from the crust and crumb centre. 9. Point at the centre of the crumb. (Source: *Technology of Breadmaking*, L. J. Auerman, Moscow, 1977.)

immediately under the crust loses its moisture. This so-called evaporation-zone then continuously migrates, depending on crust build-up, farther into the crumb, forming the boundary-layer between the crust and the crumb. Crust formation results in lower porosity than that existing in the bread crumb, therefore the crust forms a resistant barrier to moisture seeking to escape from the evaporation-zone

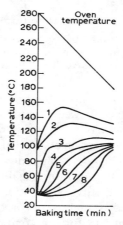

Fig. 99.   Temperature changes in specific layers of the dough-piece, baked under a falling temperature regime from 280°C down to 180°C. 1 and 2. Outer crust layers. 3 and 4. Central crust zones. 5, 6, 7 and 8. Crumb layers. (Source: *Technology of Breadmaking*, L. J. Auerman, Moscow, 1977.)

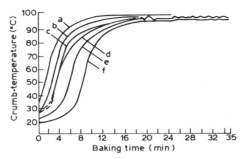

Fig. 100. Temperature changes in bread-crumb using various baking-temperatures (weight 750 g). (a) 250°C, (b) 240°C, (c) 230°C, (d) 220°C, (e) 210°C, (f) 200°C. (Source: *Technology of Breadmaking*, L. J. Auerman, Moscow, 1977.)

into the oven-chamber. Part of this steam then migrates towards the inner crumb layers, and on reaching the cooler layers near the centre will condense. This increases the moisture of these layers, and forms an inner condensation-zone for migrating steam. For any internal migration of moisture to take place, there must be differentials in both moisture concentration, and temperature at various zones within the dough-piece. Moisture-concentration gradients give rise to a 'moisture-flux' from points of high concentration to points of low concentration within the loaf. The same rule applies to temperature, which will assist the moisture-flux from points of higher temperature to points of lower temperature. Moisture-flux in moist-material is referred to as 'thermal-moisture-conduction'. The temperature/baking-time plots illustrated (Figs 98–100), show clearly the large difference in moisture content of the crust and crumb generally, and also large differences in temperature between the outer and inner crumb-layers, especially during the initial baking stages. The temperature influence or differential between the outer and inner crumb-layers is, however, the most significant. On this account, a situation of moisture-flux from the outer crumb layers towards the loaf centre takes place This phenomenon was confirmed experimentally by Ginsburg in the USSR, on whose data the moisture-migration diagram illustrated in Fig. 101 is based. This diagram shows clearly that the crumb-moisture during the baking-process, compared with the initial moisture of the dough-piece, increases by about 2%. The most rapid build-up of moisture takes place during the initial baking-stage within the outer crumb-layers, owing to the important part played by thermal-moisture-conduction during this period, which is in turn triggered by the temperature differential existing within the crumb. As a follow-up to the work described, Auerman (1977) reports an investigation concerning the migration of moisture within the individual layers of the dough-piece, baked at constant temperature and without any humidification of the oven-chamber medium. The conclusions of this work were:

(1) The moisture content of the surface-layer of the dough-piece falls very rapidly, reaching the equilibrium moisture level, as determined by the temperature and the relative humidity of steam/air medium of oven-chamber.

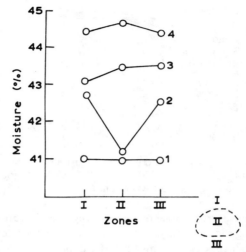

Fig. 101. Moisture migration within the various dough-piece crumb zones after (1) 0 min, (2) 5 min, (3) 10 min, and (4) 23 min baking-time (experimental data due to A. S. Ginsburg, USSR). (Source: *Technology of Breadmaking*, L. J. Auerman, Moscow, 1977.)

(2) More deep-seated layers, which later form part of the crust, attain the equilibrium moisture level at a much later stage, depending on their distance from the crust-thickness. In fact, these layers, which eventually help to form the crust, actually increase in moisture content during the initial baking-stage. The magnitude of this increase depends on the depth of the layer within the dough-piece, becoming greater with increasing depth in the dough-piece, first forming crumb and then part of the crust.

(3) The evaporation-zone, which shows an average moisture level during the baking-process (about 35%), has a moisture content on its inner surface (facing the crumb) made up of the initial moisture of the dough-piece (about 50%), plus the additional amount due to moisture migration, i.e. $M_0 + dM$. Whereas the surface bordering on the crust has a moisture corresponding to the equilibrium moisture level. The final moisture content of this zone being the average of $(M_0 + dM)$ and the equilibrium moisture level.

(4) The moisture content of the crumb-layers at 1/4, 1/2 and 3/4 distant from the crust, and the centre of the crumb all increase during the baking process. The initial build-up taking place in the outer layers of the crumb, then spreading deep into the innermost layers.

(5) As a result of the thermal-moisture-conduction, the moisture in the outer crumb-layers, close to the evaporation-zone, begin to diminish. However, the moisture level of these layers always remain higher than the initial moisture content of the dough-piece before the baking-process.

(6) The moisture content of the centre of the crumb increases at the slowest rate, and its final moisture content can be a little below that of the adjacent crumb-layers.

(7)  Repeated tests have shown that, at the end of the baking-process, the crumb moisture in total has risen by 1·5% to 2·5%.

To regulate the quantity of steam for injection into a travelling-band tunnel oven, a pressure manometer is installed between the regulating-valve and the jet-nozzle. In this case it is assumed that for various valve settings, the pressure read-out corresponds to the quantity of steam injected. This, however, is only valid when the cross-section of the jet-nozzles are uniform. The steam injected into the steaming-zone can become unevenly distributed, owing to the variations in draughts (air-currents) and temperatures prevailing. For the measurement of the relative humidity within the oven-chamber, at temperatures over 100°C, the dew-point method and psychrometry remains the classical procedure (see Fig. 102(a) and (b)). Direct determination of the baking atmosphere in the steaming-zone, or that of moisture variations and losses, are best carried out by the psychrometric procedure, since this method gives an almost instant reaction to moisture variations. This method involves a measurement of the temperature difference between a wet and a dry-bulb thermometer. From this, the dew-point temperature, and from this, the absolute moisture content of the baking atmosphere in kg steam/kg of dry-air can be calculated either from a chart or by using psychrometric formulae.

One, the dry bulb thermometer, reads the air temperature; the other, the wet-bulb is intended to read the temperature of adiabatic saturation. The measurement is carried out by sucking the air/steam medium out of the oven-chamber with an extractor-pump at a speed of 3 to 5 m/s or 1000 ft/min (see Fig. 102(a)). This is essential, since the wet-bulb must remain moist, and a suitably high air-velocity must be maintained over it. Also, to avoid any condensation before the measurement, it must not be allowed to fall below $(T + 10\,K)$, the dew-point temperature. In order to

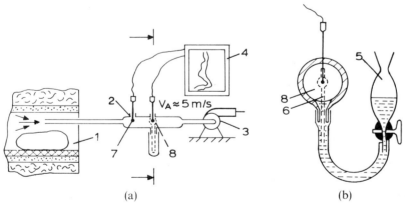

Fig. 102.   Principle of the continuous psychrometric measurement of humidity in the steam/air baking-medium. (a) Assembly. (b) Humidification principle. 1. Oven-chamber. 2. Measurement-chamber. 3. Exhaust-pump. 4. Registration-chart. 5. Water-reservoir. 6. Humidification-vane. 7. Dry thermometer. 8. Wet thermometer. (Source: *Technology of Industrial Baking*, R. Schneeweiss and O. Klose, VEB-Fachbuchverlag, Leipzig, 1981, p. 366.

maintain the psychrometric effect during continuous measurement, the wet-bulb must be maintained with an even stream of humid oven-chamber medium. To ensure that the flow of water in the exhaust and evaporation tube is maintained, the dry-air in the exhaust-pipe should be cooled below 150°C. The use of thermocouples or resistance-thermometers, allows the temperatures of both wet- and dry-bulbs to be recorded. Although the operations described may be automated to a degree, the complexity of the calculations equivalent to the psychrometric chart hinders development of this technique into a continuous-reading instrument. For continuous recording and/or control of humidity, electrical transducers of the Dunmore-type find wide application. These are based on a resistance element which changes resistance with relative humidity. The resistance element consists of a dual winding of noble-metal wires on a plastic form with a definite spacing between them. When these windings are coated with a lithium chloride solution, a conducting path is formed between the windings. The electrical resistance of this path varies reproducibly with the relative humidity of the surrounding medium, and thus may be used as a sensing element. Bridge-type resistance-measuring circuitry with a.c. excitation is usually applied. The resistance/relative humidity relationship is, however, quite nonlinear, and a single transducer can cover only a narrow range of relative humidity of the order of 10%. Where larger ranges are required, e.g. 5% to 99%, seven or eight transducers, each designed for a specific part of the total range, can be combined in a single package. Each sensing element is only about 2·5 cm in diameter, and 5 cm in length, and since these elements are also sensitive to temperature, some form of temperature compensation may be required. No moisture or heat is added or taken away from the environment, therefore, the sensors can be used in closed areas within the range −40 to +150°F.

   The lithium chloride sensor can also be modified to give a signal related to dew-point temperature, the dual wire windings being supplied with a.c. power, resulting in a heating of the lithium chloride film. Lithium chloride shows a very sharp

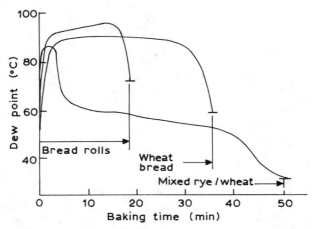

Fig. 103.   Dew-point profile in the oven-chamber. (Source: *Technology of Industrial Baking*, R. Schneeweiss and O. Klose, VEB-Fachbuchverlag, Leipzig, 1981, p. 206.)

decrease in electrical resistance when its relative humidity exceeds 11%. Thus, when surrounded by moist air, the lithium chloride momentarily absorbs moisture, and its resistance falls, allowing more current to flow, and more heat to be generated. This in turn, raises the temperature, driving off excess moisture, thus increasing the resistance, and reducing the heat. Therefore, this sensor regulates its own temperature, so that the relative humidity of the lithium chloride element remains near 11%, irrespective of the moisture content of the surrounding medium. Although the temperature attained by the lithium chloride element is not equal to the dew-point, it is related directly to it. Thus, by measuring this temperature with an appropriate sensor, the dew-point temperature may be established. This type of probe cover dew-point ranges of $-50$ to $+160°F$ with an error of the order of 1 or $2°F$. This covers dew-point ranges up to $71°C$. The illustrated graphical plot of dew-point *v.* baking-time shows the dew-point profile in the oven-chamber for bread-rolls, wheat-bread and mixed rye/wheat-bread (see Fig. 103).

The humidity distribution situation within a travelling-band tunnel-oven, both under conditions of correct external-draught exclusion and where an influx of cold-air exists, is illustrated in Fig. 104(a) and (b). A classical psychrometric measurement set-up for oven-chamber humidity is shown in Fig. 102(a) and (b).

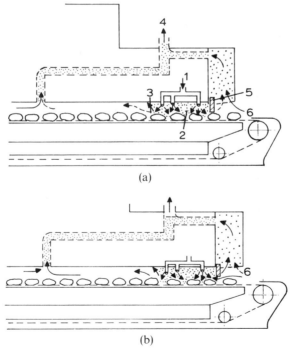

(a)

(b)

Fig. 104.   Humidity distribution in the steaming zone of a travelling-band tunnel-oven. A. Good protection. B. Influx of cold air. 1. Steam. 2. Steaming zone. 3. Steam trap. 4. Steam outlet vent. 5. Oven-feed apron. 6. Cold air. (Source: *Technology of Industrial Baking*, R. Schneeweiss and O. Klose, VEB-Fachbuchverlag, Leipzig, 1981, p. 366.)

## Baking-time

The minimum baking-time for a specific baked-product is determined by the minimal time required to give adequate starch gelatinization, protein coagulation, and formation of flavour compounds. In order to ensure acceptable crumb-elasticity, and good product digestibility, all products must be exposed to gelatinization temperature ranges for a sufficient time-period within the oven-chamber. The gelatinization-temperature of starches, depending on the cereal-source, cover a wide range from about 40 to 90°C. Furthermore, if starch gelatinization takes place slowly, in a controlled time sequence, the internal structure of the product will be less dense, of even texture and have an elastic crumb. If the gelatinization temperature range is traversed too rapidly, many starch granules have insufficient time to swell or gelatinize, since they become encapsuled by other gelatinized granules, and so cannot absorb enough moisture. The resultant baked products will show a crumbly, dense, doughy and inelastic crumb structure.

The upper limit of the baking-time is usually determined by economic considerations, such as production-rationalization and energy consumption, within the limits of quality standards for crust-thickness for the various bread varieties produced. As already mentioned in Section 3.1, the baking industry in many East European countries specify standards for both baking-times and temperatures for the basic bread varieties, as well as crust-thickness. However, in the western countries, this would require either national or community legislation to maintain bake-out quality standards for the consumer in this way. Instead, it is left to internal company quality standards and any lobbying by the consumer in the marketplace. Where panned bread is the standard bread, whether sliced or unsliced, baked-loaf moisture and potential shelf-life have always been important criteria. Against a background of conflicting interests, in the GDR under Standard TGL 3067 the baking-times for all basic bread varieties have been laid down according to weight and form in which they are baked (panned, hearth or '*angeschoben*', i.e. within the pan-depth). In so doing, both initial and final oven temperatures are specified, in addition to permissible baking-time tolerances (see tabulated data of TGL 3067 in Section 3.1). Such standardization is very desirable as a contribution to bread-quality, and is made simpler by the existence of standardized oven construction, whether in the form of travelling-band tunnel ovens or multi-deck batch ovens. Permissible baking-time variations are necessary to accommodate such parameters as:

—dough properties, flour type used, weight and shape of dough-piece, state of proof (density and degree of leaven) of dough-piece
—baking-regime, temperatures, humidity and forced-convection air-speed
—dough-piece spacing on the baking-surface

Bread quality is, however, heavily dependent on adherence to the optimal baking-time. On this depends not only the flavour and aroma of fresh bread but also such internal qualities as crumb-elasticity, texture and mastication properties, that ability to excite the appetite of the consumer.

For every chosen oven temperature-regime and heat-transfer situation, there is an appropriate residence-time in the oven for optimal bread-quality. For example, 1 kg long-rolled hearth wheat-flour bread can be baked at temperatures of 230°C, falling to 220°C for 35 to 50 minutes, provided the final crust-thickness achieved is at least 3 mm. The author's experience has shown that for hearth-baked units of 1 kg baked weight, 40 minutes at 240°C, falling to 230°C gives the required 3 mm crust-thickness, without losing the typical mild wheat-bread aroma. Baking-times of 30 minutes are inadequate for full flavour development, and 45 minutes resulting in some loss of flavour and aroma. The difference in baking-loss being 12·5% at 40 minutes, and 11·5% at 30 minutes, and although the difference represents a financial sacrifice calculated on the 100 kg flour price, which must be made good, the qualitative achievement more than compensates for the disadvantage. Using the same temperature regime a panned version should be adequately baked within 27–30 minutes, when made from a lean, low-enrichment formulation.

A comparable panned bread, made with an enriched North American formulation, and baked at 227°C, rising to 239°C at the discharge-end, should be adequately baked after 23 minutes, showing a baking-loss of about 10·2%, and a crumb-moisture of about 36·4% in the final loaf. The average dough-moisture percentage composition being 46·6%. In the case of a 1 kg rye or rye/wheat hearth-baked long-rolled unit, a temperature-regime of 250°C, falling to 200°C for 50 to 60 minutes would be appropriate, giving a final crust-thickness of 4·5–5·0 mm. The same weight-unit baked as '*angeschoben*' (batched adjacent), using the same temperature-regime would require 65–70 minutes for complete baking-through.

Often, in practice, these regimes quoted are not adhered to, but the most common bread-fault, generally, is due to the use of too short a baking-time, resulting in a lack of crust-thickness, and flavour and aroma. The author ventures to suggest that in the interests of bread-quality and increased sales, a baking-loss of up to 12% can be tolerated by the baker.

### 3.2.1 INFLUENCE OF THE ELEMENTS ON THE DOUGH-PIECE AND ITS COMPONENTS

During the process of baking many changes take place as a result of heat-transfer to the dough-piece, the physical effects of the elements temperature, heat-transfer, humidity and time have already been discussed in some detail. The dough-piece is converted into a loaf of bread in three stages:

(1)  Penetration of heat to the interior of the dough-piece until the boiling temperature of the dough liquid-phase is reached.
(2)  Removal of excessive liquid by evaporation.
(3)  Browning and flavour component formation in the form of dextrins, carbonyls from starch and sugars, and melanoidins with the help of proteins.

At the first stage the temperature differential 30 to 100°C has to be overcome. However, during this stage, the moisture content of the dough-piece does not decrease, but instead increases, owing to steam condensation onto its surface. The moisture is forced to migrate towards the centre of the loaf by the heat penetration, but the crust-forming layers lose their moisture. Then the internal moisture begins to migrate outwards, in the opposite direction to the heat-transfer flux. After reaching the boiling-point at the centre, the heat-flux from the outside is utilized to evaporate the residual moisture, which results in a slowing down of any temperature rise during this period. When the moisture is reduced to a specific level, the temperature of the dough-piece can continue to rise. The residual moisture is bound in the capillaries of the dough and can only be removed at higher temperatures by evaporation. The speed at which the three stages, i.e. heat-transfer flux, evaporation, and browning, take place in the loaf will depend on the formulation-enrichment, weight and shape of the loaf, and its degree of leaven. The main part of this section considers the effect on the colloidal and chemical properties of the various dough components.

**Changes in the Starch**
This macromolecular polysaccharide is made up of a complex network of two components: the straight-chained amylose, and the branched-chained amylopectin molecules. The amylose does not form a strong gel, whereas the amylopectin fraction in the presence of heat and sufficient moisture, swells strongly to form a gel. In baking one becomes involved with two main types of starch, viz. wheat-starch and rye-starch, both types beginning to swell at 40 to 50°C. This is followed by gelling, when they absorb from about 25% to 50% of their volume of water, forming a highly viscous suspension (see Fig. 105). In so doing, water penetrates the starch granule, first suspending the amylose, then the amylopectin. This causes the granule to swell, and the pressure built-up within by the water results in a splitting of the outer membrane of the granule. Schoch (*Baker's Digest*, **39**(2), 48, 1965) suggests that during this swelling and gelatinization, part of the amylose becomes dissolved and diffuses out of the granules into the aqueous dough medium, where it becomes concentrated as the interstitial water becomes reduced by continuous swelling. The dissolved amylose forms a gel on cooling, which tends to retrograde, thus influencing bread-staling. Sandstedt (*Baker's Digest*, **35**(3), 36, 1961), points out that a large proportion of the granules remain intact during baking, owing to the limited supply of water. The extent of mechanical starch damage sustained by the flour during milling, is also important during baking. Although these damaged granules

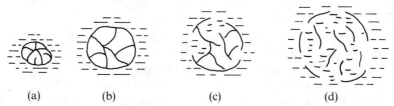

(a)            (b)                (c)                (d)

Fig. 105.  Gelatinization stages of starch-granule. (a) Dry-granule. (b) First-degree swelling. (c) Second-degree swelling. (d) Gelatinized. Gelatinization temperature range, 60–70°C. (a)–(d) Loss of opacity and crystallinity, increasingly amorphous with loss of birefringence under polarized light.

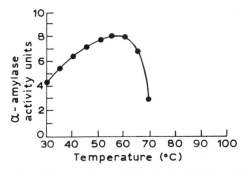

Fig. 106.   Changes in alpha-amylase activity during baking. (Schneeweiss, 1965.)

exhibit increased water absorption, they are also susceptible to enzymic attack. Within the range 43 to 60°C, alpha-amylase is most active, and attacks the granule before gelatinization can take place between 55 and 90°C (see Fig. 106). Furthermore, Schneeweiss (Conference Report 2: *International Problems of Modern Cereal Processing and Cereal Chemistry*, Vol. 2, 1965) established that, at temperatures between 57 and 79°C, within the dough-piece the number of damaged granules increased dramatically. This situation occurred after about 20 minutes baking-time, the isolated starch being coloured by china-blue, which only colours damaged granules (see Figs 107 and 108). Water-soluble carbohydrates correspondingly increased during the baking-process from 14·5% at oven-setting, steadily rising to 24% after 60 minutes baking-time (see Fig. 108). At 70°C, the activity of alpha-amylase appears to tail-off, the optimum range being 50–60°C (see Fig. 106). However, Falunina (1952) considered from experimental data that the enzyme could remain active up to 78–80°C. Other factors which determine the alpha-amylase activity during baking are dough pH, and the presence of salt. At pH values 3·0–4·0, which corresponds with a sour-dough pH level, alpha-amylase is

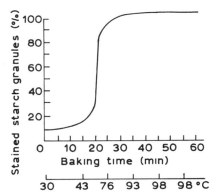

Fig. 107.   Changes in the damaged-starch content (stained granules) during baking. (Schneeweiss, 1965.)

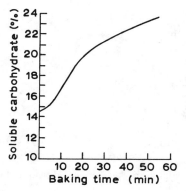

Fig. 108. Changes in the content of water-soluble carbohydrates during baking. (Schneeweiss, 1965.)

rapidly inhibited. Apart from pH, the most important determinant during baking is the exposure-time factor over the optimal temperature-range of activity of the enzyme. In the case of panned loaves, about 20 minutes baking-time is needed for the crumb-temperature to 'attain' a temperature of 99°C, which only allows about 6–10 minutes for total exposure at that level. Therefore, during this crucial 20 minutes, both the susceptibility of the starch-granule and the level of alpha-amylase activity must be contained, in order to maintain dough viscosity and structure (see Fig. 109). In the case of smaller units, of smaller cross-section, i.e. rolls and baguettes, the crumb reaches 99°C within about 8–10 minutes, leaving a further 20-minute exposure, resulting in almost complete starch gelatinization. By comparison, the 1 to 2 kg panned unit shows only limited starch pasting, hence most of the starch granules remain intact, the rapid crust-formation reducing the exposure/reaction-time. Therefore, many granules, apart from slight thermal deformation, remain in the crystalline, micellular state. This is evidenced by the

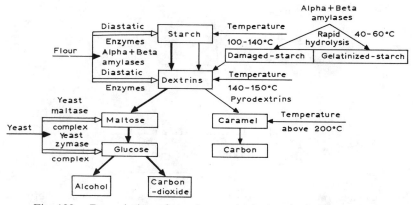

Fig. 109. Degradation of starch-granule during the baking process.

retention of the birefringent (German: *doppelbrechenden*) interference-cross, when viewed under polarized-light microscopy.

In the case of the high-frequency baking process, the dough-piece remains in the gelatinization-zone for about one-third of the time required in the case of conventional baking, therefore any defects normally encountered with amylolytic enzymes are avoided.

This baking technique involves the use of VHF (17·7 MHz) energy as a two-phase baking process. The dough-piece being subjected to the VHF energy for about 6 minutes, followed by high-temperature conventional heating for 5 minutes to give crust formation. Thus, the total baking-time is 10–11 minutes, compared with about 30 minutes using the conventional method. The use of VHF baking will permit the processing of flour with abnormally high alpha-amylase activity without the usual crumb defects. Using this technique, wheat-flours with Falling numbers as low as 95 can be baked without showing the characteristic sticky-crumb encountered with conventional baking.

If the total baking-time is to be shortened, it must be done during the initial-phase, since the period after passage through the gelatinization-zone (55–80°C), cannot be shortened, since this is essential for the formation of the flavour substances and firming of the crust. The amylases begin to hydrolyse the starch with the onset of swelling, their reaction rate increasing by about 50% for every 10°C rise in temperature, remaining active up to about 80°C, as indicated by Falunina. Starch is degraded either by amylolysis or temperature (100 to 140°C), forming dextrins and then maltose, the dextrin fraction increasing by about 15% during baking with normal flour. Flours milled from sprouted-wheat show increases of 30% and more in dextrin content. The rapid inactivation period for alpha-amylase (cereal), within the temperature range 68–83°C, is generally traversed in about 5 minutes baking. Assuming the dough has been given a bulk-ferment or floor-time of at least 1 hour, the maltose level will have increased during fermentation, and will have further increased during baking.

Beta-amylase, which is the saccharogenic enzyme, capable of degrading damaged susceptible starch, will have increased flour total sugars during mixing from about 1·5% to 2·0–2·5%, and will continue to degrade damaged starch and saccharify during baking up until the inactivation range of 57 to 72°C is reached. This latter range lasts not more than 3 minutes. The starch components of a typical type 550 (0·55% ash) flour, with 90% particles less than 90 $\mu$m in diameter, consist of: free secondary-starch (less than 10 $\mu$m), clusters of secondary-starch (10–30 $\mu$m), free prime-starch (10–30 $\mu$m), free prime-starch (30–50 $\mu$m), endosperm-cell fragments consisting of many prime-starch granules or many secondary-starch granules (30–50 $\mu$m), which constitutes a wide selection of damaged and undamaged granules with a wide range of accessibility and susceptibility. The dextrins are formed mainly in areas of the dough-piece where moisture is limited (crust-zone), from which the oven-heat constantly removes moisture. As a result of the lack of water, when the temperature reaches 140–150°C, some of these dextrins take part in non-enzymic, non-oxidative caramelization reactions, helping to form the crust. When the evaporation of moisture from the crust becomes too intense, carbonization occurs and the crust will become inedible. Caramelization, is the name given to the transformation of sugar-like substances and sugars, under the influence of heat, from colourless sweet substances into compounds which vary from yellow to brown. Their flavour varying from a pleasant caramel, to bitter and carbonized. Some of these products have been identified as complex polymers formed at temperatures of 150 to 190°C; they include: isomaltose, laevoglucosan, cellobiose, isomaltotriose, panose, oligosaccharides. Aldose sugars undergo enolization, forming 2-ketoses and 5-hydroxymethyl-2-furaldehyde; pentoses forming 2-furaldehyde. Hydrolytic splitting results in the formation of formic and laevulinic acids from furaldehydes; dihydroxyacetone, glyceraldehyde and carbonyls from 2-ketones; dehydration of trioses yielding acetol, maltol and pyruvaldehyde. Condensation reactions including, aldehydes and ketones with active hydrogen. Aldoses and ketoses also revert to di-, tri- and oligosaccharides (formed from three monosaccharide residues). Anhydrides are formed by dimerization, e.g. fructose to difructose anhydride. All caramelization reactions require high temperatures for their initiation.

## Changes in Protein

The most important part played by proteins during baking is that the molecules have unique colloidal properties, such as: water imbibition, hydrophilic and hydrophobic bonding, binding enough water for starch gelatinization. Simultaneously, their molecules provide rheological function, for structural formation and stability. As the temperature of the dough-piece reaches the range 50 to 70°C, the atoms and atomic groups of the protein molecules go into progressive oscillation. This results in the breaking of the weaker bonds of the molecular network, and the helical structure begins to uncoil. In so doing, the spatial arrangement within the protein molecules becomes changed, and water which had previously been imbibed and bound during the swelling stages of dough-mixing, fermentation and proof, is released. The extent of gluten-protein binding of water is estimated at about 30% of total dough absorption water, the hydrated proteins forming a dough structural matrix in which small starch granules are embedded. As the crumb temperature reaches about 60–70°C, progressive protein moisture loss and denaturation takes place, leading to eventual coagulation (see Fig. 110). The

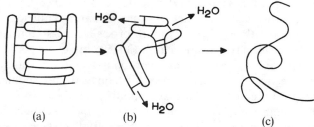

(a)                    (b)                                        (c)

Fig. 110.   Schematic of protein-denaturation. (a) Natural state of protein molecule (hydrated). (b) Opening up of molecule (50–70°C) owing to the application of heat-energy (moisture-loss). (c) Denatured molecule 40°C. (Source: S. M. Rappaport. In *Medical Biochemistry*, 8th edn, VEB-Verlag Volk und Gesundheit, Berlin, 1984.)

water released becomes transferred to the starch during its gelatinization process. At about 74°C, thermal denaturation transforms the gluten-network films, which enclose the gas vacuoles, into a more rigid structure, and together with the swollen starch, a firm but elastic crumb is formed. With the thermal expansion of the gas cell membranes, the viscous starch granules within the wall of the membranes elongate, and the gluten films become progressively thinner, eventually rupturing. However, by this time starch swelling, and rigidity has taken over the structural role.

Protein denaturation takes place within the temperature range 50–70°C, during which time the swollen molecules release most of their bound water. In so doing, the molecules become transformed from a viscoelastic state into that of a plastoelastic solid, a change which is very important for the solidification of the dough-piece. Popaditsch and Falunina (*Chlebopekarnaja i konditerskaja promyschlennost*, **2** (1962), 1, in Russian), during their studies of the influence of heat on fermented dough and proteolytic enzyme activity, determined the relative concentrations of water-soluble, and 0·1N acetic acid soluble proteins as the dough-piece temperature rose from 20 to 90°C. Their conclusions were that the water soluble fraction of dough proteins changed little from dough to bread, being 4·0 mg/g in dry matter and 3·10 mg/g in dry matter respectively at 20 and 90°C. By contrast, the 0·1N acetic acid soluble nitrogen decreased from 14·2 mg/g dry matter at 20°C, to 6·10 mg/g dry matter at 90°C. Within the temperature interval 60 to 70°C, the water-soluble fraction increased from 5·4 mg/g dry matter to 7·1 mg/g dry matter. Only on reaching 90°C did the water solubles show a decrease to 3·1 mg/g dry matter from the initial level of 4·0 mg/g dry matter. When the dough-piece reaches between 60 and 70°C, the acetic acid soluble fraction decreases significantly. This temperature range coincides with the onset of gluten denaturation, which reaches a maximum at 90°C. This was not only confirmed by the reduction in the acetic acid soluble fraction, but also in an increase in susceptibility to proteolytic enzymes. Up to 70°C and over, heated dough becomes increasingly prone to papain hydrolysis compared with the initial dough at ambient temperatures. In the case of proteolytic enzyme activity, during warming up and baking, the critical temperature range also fell between 60 and 70°C for inactivation, but traces of proteolysis could also be detected in the crumb as high as 85–90°C, indicating the tolerance of these enzymes. The processes of the protein coagulation and starch gelatinization form the porous structure of the dough-piece and the final development of the loaf-crumb. The rheological properties of the crumb are quite different from the dough from which it was transformed. These aspects of the baking process have been studied by Telegdy-Kovats and Lasztity during the period 1959–62 in Hungary (Fig. 111), Auerman and Melkina (1967) in USSR, and by Tunger and Thomas (1966) (Fig. 112) and Tunger and Erhard (1969) in the GDR. These workers have used specialized equipment to measure the general deformation of the crumb of the loaf during baking; the plastic deformation, and the elastic deformation as the difference between the total and residual deformation; and the relative-elasticity. The last quantity is expressed as a percentage of total deformation. Also, changes in the rheological properties of loaf formation during baking of rye and wheat breads of various weights, and from flours of various extractions are reported. Changes in the elastic deformation of the

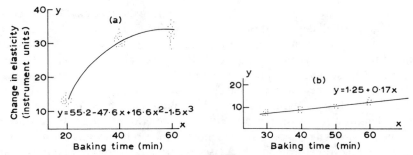

Fig. 111.   Changes in crumb-elasticity of bread made from 60% extraction rate flours. (a) Wheat-flour. (b) Rye-flour. (Telegdy-Kovats and Lasztity, 1959.)

crumb during the baking of rye and wheat bread from flours of the same extraction-rate (60%) are illustrated as graphical plots of baking-times, *v.* change in elastic-deformation of the crumb (instrument-units). These plots (Figs 111 and 112) indicate a certain correlation between flour extraction and relative crumb-elasticity of the resultant bread for both rye and wheat flours. The use of higher extraction flours results in the relative elasticity rising to a maximum limiting value and then falling-off, whereas when lower extraction flours are used, the relative crumb-elasticity continues to rise during the baking-process. The reason for this is considered to be due to the higher amylolytic activity of the higher-extraction/higher-ash flours.

The protein–proteinase complex of the baking dough-piece also produces some changes as a result of the progressive warming-up process. The extent of these changes will depend on the water-content of the dough, its pH, and the rapidity and temperature of the heat exchange. The extent of proteinase-activity can be measured as any increase in water-soluble protein nitrogen extracted from the dough. Work done by Popaditsch and Falunina has indicated that for a wheat-flour dough with a water-content of 48% at pH 5·8, the temperature optimum for wheat proteinases lies between 60 and 70°C, depending on the rapidity of the warming-up process. Increasing the water-content of the dough to 70% had the effect of lowering the temperature optimum of the enzymes to 50°C. In general, since the inactivation temperature of the enzymes in bread-dough during baking depends on the speed of

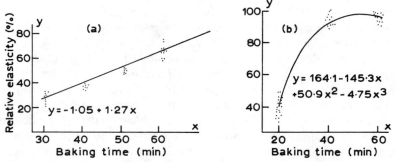

Fig. 112.   Changes in relative crumb-elasticity made from 60% extraction rate flours. (a) Wheat-flour. (b) Rye-flour. (Thomas and Tunger, 1966.)

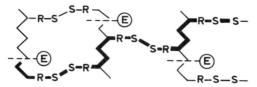

Fig. 113.   Random cleavage of peptide bonds of dough proteins by dough proteinases during the baking process at sites (E). Residual transverse and diagonal strands being held together by —SS— bonds. Activity in dough-piece depends on: dough-moisture level, dough pH and speed plus temperature of heat application.

warming-up, the more rapid the warm-up, the higher becomes the inactivation-temperature (see Fig. 113). The denaturation process of the dough-proteins can be followed at different temperatures by a determination of the nitrogen content of the fraction soluble in $0.1\,\text{N}$ acetic acid. The work of Popaditsch and Falunina shows a progressive decrease in acetic acid solubles from $55.9\%$ at $20°C$ to $24.01\%$ at $90°C$.

The formation of the loaf crust and so-called 'browning' effect involves both enzymic and non-enzymic processes. The enzymic ones have already been mentioned, viz. the formation of the light-yellow to black coloured dextrins. The non-enzymic browning results from the Maillard and melanoidin reactions, which are of great importance for the synthesis of colour and aroma substances. These reactions involve complex pathways with carbonyl and amino groups, which form the high molecular-weight brown coloured melanoidins. Melanoidins are formed from reactions between sugars and nitrogen-containing substances. The sources of nitrogen can be amino acids, polypeptides, proteins, amines or ammonium salts. The sugars can also be replaced by similar compounds with reactive aldehyde groups. The reaction proceeds according to the following equation:

$$R-\underset{\underset{\text{sugar}}{|}}{\overset{\overset{}{\|}}{C}}{=}O \;+\; \underset{\text{N-group}}{H_2N-R'} \;\longrightarrow\; R-\underset{\underset{\substack{\text{melanoidin}\\\text{(coloured)}}}{|}}{\overset{\overset{}{\|}}{C}}{=}N-R' \;+\; \underset{\text{water}}{H_2O}$$

The speed of the melanoidin reaction depends on several conditions, which the technologist can influence and control to a degree, i.e.

(1)   By control of temperature, pH and moisture-content of the dough, the speed of melanoidin formation can be controlled.

(2)   By ensuring the presence of free amino-groups ($H_2N$), aldehyde (CHO), in a reactive form.

(3)   The more reactive groups present and available, the quicker the melanoidin reactions take place.

(4)   Melanoidin formation can also be induced by the mixing process, since a longer and more intensive mixing-time can open up the protein molecule, exposing the free $H_2N$ and CHO reactive groups.

(5)   By careful control of the baking-time, and heat distribution and heat-transfer the melanoidin formation and loaf browning can be optimized.

Browning of the crust is mainly due to reactions between sugars and nitrogen-containing compounds, the formation of caramel having little significance for the browning. Therefore, any deficiency in browning is indicative of inadequate amounts of sugar and protein. The adequate presence of humidity (injected steam) also plays an important role in browning, therefore, provision for adequate steam, painting of the dough-pieces with water, and exposure·to controlled temperatures over 100°C are essential. In addition, the dextrins and caramel formation help to complenent the browning process. Parallel to their influence on browning, the melanoidins also contribute to the formation of flavour and aroma substances in the loaf. The formation of melanoidins is referred to generally as the 'Maillard reaction', and specifically involves the interaction of the free amino groups of amino acids, peptides or proteins with free reducing sugars. The reaction was first described by Maillard in 1912, but the complex pathways and intermediate compounds involved in the conversion of the starting materials into the wide range of compounds responsible for bread flavour have been the subject of intensive research in recent years. Hodge (*Agr. Food Chem.*, **1**, 928, 1953, and *Advances in Carbohydrate Chemistry*, **10**, 169, 1955), and (Namiki and Hayashi, in *The Maillard Reaction in Foods and Nutrition*, edited by G. R. Waller and M. S. Feather, ASC Symposium Series 215, ASC, Washington, DC, 1983; Lane and Nursten, ASC Symposium Series 215, 1983), have all made valuable contributions to the clarification of the melanoidin reactions of Maillard systems. Hodge's reaction scheme is the following:

> *I. Initial-stage (colourless)*
> (a) Sugar-amine condensation (b) Amadori rearrangement
> *II. Intermediate-stage (colourless or yellow)*
> (c) Sugar dehydration (d) Sugar fragmentation (e) amino acid degradation
> *III. Final-stage (highly coloured)*
> (f) Aldol condensation (g) Aldehyde-amine polymerization, and formation of heterocyclic nitrogen compounds.

Initially, reducing sugars and amines condense forming addition compounds from the amine group and the carbonyl group from the sugar. This is followed by the removal of a water molecule to form an N-substituted glycosylamine. Then follows the so-called Amadori rearrangement, whereby the N-substituted glycosylamines become isomerized to 1-amino-1-deoxy-2-ketoses, which remain in equilibrium with their enol forms (1,2-enol forms). All reaction products up to this point are reversible and colourless. At the intermediate stage, the products of Amadori rearrangement can then undergo a variety of fates, depending on the conditions of the reaction. They can, under neutral to acid conditions, either lose water and form a ring compound of the Schiff's base of hydroxy-methyl-furfural or furfural, or eliminate an amine to form a methyl alpha-dicarbonyl intermediate. The first type are colourless reductones and the second brown fluorescent substances. Another possibility is the splitting off of carbon dioxide from amino acids to form an aldehyde, which is known as the Strecker degradation process. These reaction products can either be colourless or have a yellow colouration.

During the final stage, dehydration pathways result in the formation of unsaturated dicarbonyls, which either condense or become polymerized to form various malanoidins, which vary in colour, taste and aroma intensity. Namiki and Hayashi have proposed another mechanism for the initial stage of the Maillard reaction involving the cleavage of the sugar molecule into a very reactive two-carbon fragment before Amadori rearrangement. The abundance of free radicals at the initial stage of the browning reaction, led these workers to conclude the following sequence: Stage 1: formation of glycosylamino compounds; Stage 2: fragmentation to $C_2$ and $C_4$ products, with simultaneous formation of reducing substances, free radicals and Amadori products; Stage 3: formation of glucosones. Lane and Nursten, during 1983, evaluated the odours formed from over 400 model systems of mixtures of 21 amino acids with 8 sugars, heated under various conditions of temperature and humidity. Odours associated with bread were synthesized when glucose was heated with the amino acids arginine, glutamine, histidine, lysine, proline, serine, theonine and tyrosine in a 1:1 ratio at temperatures of 100 to 140°C for periods ranging from 0·5 to 4 minutes. Basic reaction systems, between glucose and the simplest amino acid glycine, yield about 24 compounds, most of which are very reactive carbonyl derivatives. Although the mechanism of the Maillard reaction is not yet fully clarified in all its stages, Thomas and Rothe (Thomas, B. and Rothe, M., *Baker's Digest*, **34**(4), 50, 1960) found in model systems involving the Maillard reaction the relationships between starting materials and end-products shown in Table 66. The aldehydes are either formed by trans-amination through reduction, or by oxidation with furfural. Reactivities of the monosaccharides with amino acids by the Maillard reaction are variable, and are detected by the increase in furfural content and crust-browning. The most reactive sugar is the pentose, xylose, followed by arabinose, glucose, fructose, sorbose and rhamnose. The amino acids reactivity, in decreasing order of magnitude, is isoleucine, leucine, valine, methionine, alanine, and phenylalanine. According to Rothe, the amino acids arginine, lysine, and serine do not show any aldehyde formation by the Maillard reaction. The Maillard reactions are much more significant for bread-aroma formation than the caramelization reactions, since the

TABLE 66

| Parent substance (*starting material*) amino groups | Derivative carbonyl-group |
| --- | --- |
| Leucine | 3-Methylbutanol (beta-methylbutyl aldehyde) |
| Isoleucine | 2-Methylbutanol (alpha-methylbutyl aldehyde) |
| Valine | 2-Methylpropional (Isobutyl aldehyde) |
| Alanine | Acetaldehyde |
| Methionine | Methional |
| Phenylalanine | Phenylacetaldehyde |
| Threonine | 2-Hydroxypropional, transforming to methylglyoxal |

latter only take place at elevated temperatures. The formation of pyrazines as a bread-aroma component has been widely reported. In this case, possible reaction partners could be amino acids, diketones and ketoaldehydes. Although the aldehydes are formed in the crust, they diffuse from the crust into the crumb interior of the loaf on cooling, and are absorbed. It is also likely that these and other bread flavour substances become locked into the amylose starch fraction during cooling, since many higher alcohols are readily absorbed into the helical amylose structure of the starch molecule, and are released on the application of heat. The content of aroma precursors, in the form of monosaccharides and amino acids represents a basic potential for the formation of bread-aroma components. The availability of these reactive amino acids, sugars and alcohols will depend largely on the duration and activity of the enzymes, and the degree of polymerization and hydration capacity of their substrates from mixing to baking. The presence of sufficient amounts of amino acids and reducing sugars will depend on the balance of the starch/amylase and protein/proteinase activities, especially in the case of wheat-flours. When flours produce inadequate concentrations of these essential monomers, objective measures can be undertaken to increase their concentration, resulting in intensified aroma formation. This is one example where the basic technology of biotechnology can be applied to bread flavour, with the addition of specific enzymes. Similar results are obtained by the addition of malt-products, dry-milk products, and sugars.

**Changes in Volatile and Non-volatile Flavour and Aroma Components**
The evaporation of the volatile aroma substances during the baking-process progressively increases. These substances are mainly compounds formed during fermentation, and include: alcohol, diacetyl and iso-aldehydes. Less volatile flavour substances include: iso-alcohols, acetic acid, pyruvic acid, acetoin, butylene glycol, ethyl lactate, acetaldehyde and isovaleric aldehyde; also furfural, which is formed during baking with pyruvic aldehyde, the latter being relatively more volatile. Substances of relatively low volatility include: the melanoidins formed in the crust during baking, dihydroxyacetone, ethyl succinate, lactic acid and succinic acid. During the first 20 minutes there is a rapid loss of alcohol within the baking-chamber, falling from over 500 mg/100 g bread to about 400 mg/100 g. Rothe and co-workers (Shortened-process for wheat-flour bread and its influence on bread-aroma. I. Carbonyl compounds and alcohol, *Nahrung*, **16**, 1972, 5, 507 (in German) compared the influence of different dough-processes on the alcohol content of the resultant bread. They found that using conventional mixing and fermentation, 31% of the alcohol produced during fermentation evaporated during baking, whereas, in using an intensively mixed dough, about 60% was lost. Other volatile alcohols such as *n*-propanol and iso-butanol would also disappear at similar rates depending on dough-process employed. The importance of the length of fermentation and utilization or non-utilization of pre-ferments was confirmed by Coffman (Aroma of continuously-mixed bread, *Cereal Science Today*, 9, 1964, 7, 306). Coffman established that longer- process doughs emitted higher levels of amyl alcohol in the baking-chamber gases than continuous or short-time doughs. The average

distribution of the compounds continuously condensed from the oven-chamber gases in 1400 ml total condensate was as follows: *n*-propanol 2·3, isobutanol 18·3, amyl alcohol (isoamyl alcohol) 60·8, gamma-lacto-butyric acid 1·0, beta-phenyl-ethanol 5·7 and 2,5-furan-dialdehyde. This condensate of pH 3·5–3·6, when atomized gave the typical odour of baking-bread, and of the 64 compounds identified by gas-chromatography, only six could be identified by mass-spectroscopy and IR-spectroscopy, but these six compounds made up about 90% of the total condensate. The extremely complex nature of bread flavour was later made clear by Coffman in a later (1967) summary prepared by him in 'The chemistry and physiology of flavors', under Coffman, J. R., *Bread Flavor*, edited by W. H. Schultz, E. A. Day, and L. W. Libbey, AVI Publishing Co., Westport, CT, 1967. Coffman summarized the various categories of flavour compounds found in American white-bread as follows:

Alcohols: ethanol, isobutanol, *n*-propanol, isoamyl alcohol and *d*-amyl alcohol.
Acids: acetic, propionic, butyric, isobutyric, valeric, lactic, isovaleric, caproic, heptanoic, octanoic, nonanoic, capric, pyruvic, hydrocinnamic, benzilic, itaconic and laevulinic.
Esters: ethyl acetate, ethyl lactate, ethyl succinate, ethyl itaconate, ethyl pyruvate, ethyl laevulinate, and ethyl hydrocinnamate.
Aldehydes: formaldehyde, propionaldehyde, *n*-valeraldehyde, 2-methyl-butanal, 2-ethylhexanal, benzaldehyde, furfural, 2-butanal, acetaldehyde, isobutanalde-hyde, isovaleral, *n*-hexanal, crotonaldehyde, pyruvaldehyde, and hydroxymethyl furfural.
Ketones: acetone, methyl *n*-butyl, ethyl *n*-butyl, diacetyl, acetoin, and maltol.

However, the main contribution to the formation of bread-aroma is provided by the amino acid aldehydes, and furfural; and with increasing baking-time, the content of both aldehydes and furfural will increase. Bread flavour formation is basically governed by the raw material used, and the method of leavening, i.e. yeast, sour-dough, or chemical system. Furthermore, although a longer baking-time favours the browning and formation of flavour and aroma substances, it also reduces the nutritive-value of the proteins, since part of the protein compounds will be involved in the melanoidin-reactions, and lose their nutritive value. Destruction of vitamins can also occur, in particular vitamin B1. Kretovich and co-workers at the Bach A. N. Institute of Biochemistry in Moscow (*Prikl. Biokhim. Mikrobiol.*, 1988, **24**(2), 207–210, in Russian) have observed appreciable losses in the levels of free lysine, leucine, valine; and arabinose, xylose, galactose, and glucose during the baking of rye bread, due to involvement in melanoidin-reactions, improving bread flavour and aroma. Enrichment of rye-doughs with condensed fermented whey at 9·0%, increased bread amino acid levels by about 20%. The enriched bread containing much higher levels of isopentanal and isobutanal, which are formed from leucine and valine respectively, as well as other aroma compounds, than was the case with non-enriched bread. Johnson, and Linko (Analysis of bread-flour components, in *Qualitas Plantarum et Masteria Vegatabilis*, **11**, 1964, 2, 256) points out the significance of the bread-processing procedure on the formation of carbonyl-compounds in the crust and crumb. These results indicate that the use of a pre-ferment-sponge provide the highest level of carbonyls, and a liquid-pre-ferment slightly less, in both bread crust and crumb generally. However, the levels of furfural, hydroxymethylfurfural, acetone and proprionaldehyde were all significantly higher in the pre-ferment-sponges. Other factors influencing carbonyl aroma intensities are flour type and ash content. Certain products of yeast fermentation, such as acetyl-methyl carbinol, on oxidation yield diacetyl, a flavour component of fresh bread. The amount of the precursor acetyl-methyl carbinol formed depends on the process used, straight-doughs being deficient in acetyl-methyl carbinol by comparison with sponge and liquid pre-ferments, assuming adequate amounts of sugar and oxidants are present. Another product of yeast fermentation, dihydroxyacetone, is reported by Wiseblatt and Zournut (*Cereal Chemistry*, **40**, 1963, 116) to react thermally with proline, yielding substances associated with a crust-like odour. Organic acids formed during fermentation, in particular acetic and lactic acids, help to catalyse the rearrangement of the condensation products of amines and reducing sugars during baking for the Maillard reactions.

Equipment has been developed in the USSR by Telitschkun and co-workers (*Coloration of the Bread-surface during Baking—New Developments in Oven-technology*, Publ. Zinti, Moscow, 1968, p. 47) to continuously monitor and record with a potentiometer, the change in intensity of light reflected from the bread-crust during baking. The principle is that the beam of light impinging on the surface of the dough-piece is reflected, and is measured by the scale of a photocell. Changes in colour-intensity of the loaf-surface are measured as percentage reflection during baking, concurrent with loaf surface-temperature, both parameters being plotted against baking-time. On setting the dough-piece in the oven, the reflectivity rapidly falls, owing to the gelatinization of the surface starch-granules. This fact is confirmed by the fall reaching a maximum, i.e. lowest reflectivity coefficient (%) when the surface temperature reaches 60°C, i.e. the gelatinization range of the starch. Further temperature rises result in progressive increases in the reflectivity coefficient, owing to the dough-surface becoming lighter in colour as it dries out. When the temperature reaches 100°C the melanoidin-reactions begin to take place, and the crust takes on a darker colour, the reflectivity coefficient falls sharply. On reaching 130 to 170°C, the loaf-surface changes colour from light-yellow to dark-brown, and above 170°C the loaf-crust becomes progressively carbonized. The application of this equipment allows the optimal conditions for the ideal crust-colour to be determined, which is a contribution towards the full automatic control of the baking-process. In this way, the intensity of crust browning can be regulated and controlled.

Rooney, Salem and Johnson (Studies of the carbonyl-compounds produced by sugar-amino acid reactions. I. Model-systems, *Cereal Chemistry*, **44**, 1967, 539), working with model sugar/amino acid systems in buffered solutions at pH 5·5, heated at 95°C for 12 hours, identified the various carbonyl-compounds produced from the amino acids with the sugars glucose, maltose and xylose. Also the extent of their participation in the melanoidin colour reactions, measured as colour-absorption at

500 nm with the help of a Beckman spectrophotometer. In general, it was found that xylose, not glucose, gave the most intensive colour intensities with amino acids. This was established both in solution and model systems. Gas chromatograms showed the volatile carbonyl compounds formed from glucose, and the following amino-acids: alanine, valine, leucine, phenylalanine and methonine, to be principally: acetaldehyde, isobutyric acid, isovaleraldehyde, 2-methylbutanal, phenylacetaldehyde and methional, with trace amounts of acetone, formaldehyde, propionaldehyde with a few unidentified carbonyl-compounds. The amino acids lysine, arginine, histidine and tryptophane, on the other hand did not form any appreciable amounts of definable carbonyl compounds, although the first three coloured the solution intensely brown. Glutamic acid and proline neither formed any definable carbonyls nor produced much colouration. The strong participation of xylose, compared with glucose in melanoidin colour reactions is likely to be due to its non-fermentability by baker's yeast. The aldehydes formed during these reactions relate to the added amino acids, whereas, the type of sugar added has no influence on the aldehyde formed, but exerts a strong influence on crust-colouration. Tests carried out by the same research group at Kansas State University, Manhattan, KS, USA, in bread formulation systems with 3·0% saccharose (sucrose), utilized the following formula:

Flour 100 parts (12·0% protein, 14% moisture), Arkady-type improver 0·3, yeast 2·0, sugar 3·0, salt 2·0, shortening 3·0, water 49·0.

The straight-dough procedure involved fermenting the dough for 1·8 hours in bulk, punching, dividing after a further 50 minutes, and allowing to recover for 20 minutes. Moulding and proof followed for 50 minutes, and baking at 425°F (220°C) for 30 minutes. The amino acid and sugar were dissolved in water (0·02M sugar: 0·02M amino acid), and added to the formula before mixing. Crust-colour was measured with a Photovolt reflectometer, fitted with a green filter, expressed as an average of six readings on each of two loaves. Paper, thin-layer, and gas–liquid chromatography methods were used to isolate and identify the carbonyls as 2,4-dinitrophenylhydrazine (DNPH) derivatives. The derivatives were then estimated quantitatively by extraction and UV-absorption, data being expressed as milligrams of free carbonyl/100 g of bread crust or crumb on an as-is basis. Furfural contents of the crust and crumb were determined on 2-ml aliquots of the steam-distillate of 50 g of ground crust, or 100 g of crumb, according to the method of Linko (*Anal. Chem.*, 33, 1400–1403, 1961). Hydroxymethylfurfural contents were determined in 10 g samples of ground crust or crumb by extraction five times with a total volume of 25 ml of benzene and 15 minutes centrifugation. The combined furfural and hydroxymethylfurfural contents being determined in 5-ml aliquots of the benzene extract of crust or crumb, also by the method of Linko.

The test-bake formula used was a lean one, omitting dry-milk solids and malt-flour, and reducing the sugar content to 3%, thus allowing a better differentiation of crust-colour changes. Average reflectance (%) values on the top and bottom crust of bread with added amino acids and glucose or xylose were measured. Additions of amino acids increased the intensity of the crust colour generally, and xylose with amino acids produced darker crusts than did glucose, due to more reactive browning and non-fermentation. The lowest individual values for reflectance (darkest crust-colour) was shown by methionine and arginine, both reacting with glucose. The addition of amino acids to the formula increased the total carbonyl content of the bread-crust, with only small increases in the bread-crumb. Contents of furfural and hydroxymethylfurfural in bread-crust with added amino acids decreased in comparison with non-supplemented controls, with the exception of methionine, which showed an increase in hydroxymethylfurfural content (HMF). This result can be explained by the fact that, although furfural and HMF are products of the Maillard reaction, they are only intermediates in the presence of free amino acids, and undergo further condensation to form polymers. Therefore, lower HMF and furfural are to be expected where free amino groups are present to further react with furfural and HMF. Furthermore, furfural and HMF concentrations tend to be lowest in bread-crust containing the largest amounts of the other carbonyl compounds. During the Maillard browning reactions, many compounds are formed, e.g. furfural, HMF, reductones, Strecker degradation products, and, although many are lost as volatiles in oven-gases, these retained in the crust influence bread aroma. However, although bread flavour and aroma can be modified by the addition of amino acids to formulations, the determination of the optimal balanced ratios are of paramount importance. The total bread flavour experience is derived from the complex blend of aromatic substances, and their simultaneous effect on the sensory organ receptors, which in turn convey the external impulses received to the central nervous system.

The practical evaluation of bread-flavour tends to fall into four sensations: sweet, salty, sour and bitter. Sour tastes in bread are due to the presence of various organic acids. Lactic acid is the most prolific contributor to taste, followed by malic, tartaric, acetic and citric acids. Parallel with taste, the greatest contribution to an acidic odour is provided by acetic acid, followed by lactic, malic, and tartaric acids. Acetic acid gives an intensive and pungent impulse to the sensory organs of smell, and requires careful control where wheat- and rye-bread is prepared from liquid-yeast cultures, firm or liquid sours, and pre-ferments, especially with prolonged maturation-times. Appreciable differences in taste or smell can be detected by a trained taste-panel using the Triangle-test, where the panellists are asked to identify the odd sample from three, two being identical. The success of this method depends on the sensory amplitude of detection or sensitivity of the members of the panel. For more accurate analysis of the components of taste and smell, the product is placed in a flask and frozen with liquid nitrogen at $-196°C$, under a vacuum of 1 mm mercury. The sample is then allowed to warm up to room-temperature, and the evaporated liquid condensed. The condensate is then subjected to gas-chromatography for identification and quantitative determination of the volatiles.

In order to confirm the advantages of the various processing methods, e.g. solid-sponge, liquid-pre-ferment, sour-doughs *v.* straight and short-time mechanical/chemical maturing procedures, it is essential to determine whether any derived improvement in flavour and aroma substances occur during the pre-ferment, dough or baking stages. An example of the application of these research procedures is to be found in the work of Johnson, Rooney and Salem (Chemistry of bread-flavour, *Advances in Chemistry*, **56**, 1966, 152), and in that of Rothe (Influence of technology on the flavor of rye-bread, Report 5: *International Grain and Bread Congress*, 1970, 5, 203; also 'Bread-aroma', *Handbook of Aroma Research*, Akademie-Verlag, Berlin, GDR, 1974 (in German).

# 3.3  Energy Sources, Types of Oven and Oven Design

### 3.3.1 ENERGY SOURCES

The thermal treatment during the baking process is provided by energy derived from the following sources: wood, brown coal brickettes, fuel oil, natural gas, manufactured gas (town or city gas), anthracite, coke, bituminous coal (hard/mineral or pit-coal); or is applied as electroenergy. The heating effect of the individual energy sources during the baking process can be regarded as being similar, and the choice will depend on availability and cost. Wood, peat or mineral fuels could be the most economic choice in areas remote from industry. The energy is usually transmitted to the dough-piece via an intermediate medium, i.e. air, metal or brickwork. The relative efficiency or exploitation of each energy source varies considerably. Coal heating is about 20%, gas about 50%, and maximal in the case of a directly heated electrical oven, which is 85% efficient. The final choice of the energy source will depend on which resource is naturally available, and economically viable, as well as its relative efficiency and ease of control in the bakery. After taking into consideration all aspects, for example, in many European countries, natural gas has become the final choice for both the tunnel-ovens in industrial bakeries and the various batch-ovens used in the smaller bakeries. Amongst the advantages cited by users of natural gas-fired ovens are:

(1) The simplicity in regulation of the gas/air mixture, which is usually automated.
(2) Good temperature control and flexibility.
(3) Clean and hygienic in operation.

However, oven-design is important, and in the case of batch-ovens, the energy saving realized in installing a multi-deck gas-fired hot-air circulation oven, instead of a steam-tube oven with Perkin's tubes, could be 35% or more.

In the case of all energy-sources, which produce heat by combustion, the calorific-value is a measure of their heating capacity. It is defined as quantity of heat in kJ or Btu (British thermal unit (FPS-system) = quantity of heat required to raise temperature of 1 lb water by $1°F$, the therm = $10^5$ Btu) liberated by burning 1 kg or 1 lb of a solid/liquid fuel, $1 m^3$ or $1 ft^3$ of gaseous fuel. The heat of combustion, $H_{max}$, and calorific-value, $H_c$, are differentiated from one another by the liberated heat of condensation when steam condenses, i.e.

$$H_c = H_{max} - 2260 \, WA \, (kJ \, kg^{-1})$$

596

where WA is the water vapour in the exhaust-gas in $kg\,kg^{-1}$ of fuel, 2260 heat of condensation in kJ water/$kg^{-1}$ (nearest whole number).

The calorific-value and heat of combustion of some commonly utilized industrial gases and solid fuels are quoted in Table 67.

Heat production for baking, using the sources of energy already discussed, will depend on one of the following basic principles.

1. Exothermic reactions involving the burning of coal, natural gas, manufactured gas or fuel oil. The oven-chamber walls dissipate the heat-energy as radiant heat and conductive heat into the chamber atmosphere.

TABLE 67

| Fuel | Calorific-value | | Heat of combustion | |
|---|---|---|---|---|
| | $in\,kJ\,m^{-3}\,(N)$ | $in\,kJ\,kg^{-1}$ | $in\,kJ\,m^{-3}$ | $in\,kJ\,kg^{-1}$ |
| Gas mixtures: | | | | |
| Natural gas 90% methane | 33 600 | | 37 400 | |
| City gas | 14 250 | | 15 950 | |
| Gas from brown-coal | 15 500 | | 16 800 | |
| Solid fuel: | | | | |
| Anthracite | | 31 900–35 200 | | |
| Brown-coal | | 11 750–21 000 | | |
| Wood | | 10 500–14 900 | | |
| Coke | | 22 250–30 650 | | |
| Bituminous pit-coal | | 25 200–33 600 | | |
| Liquid fuels: | | | | |
| Fuel oil | | 39 900–42 000 | | 42 000–44 100 |
| Light gas oil | | 42 000–43 600 | | 44 100–45 350 |
| Heavy gas oil | | 41 000–42 800 | | 44 500–47 000 |

Source: *Technology of Industrial-baking*, R. Schneeweiss and O. Klose, Table 57, p. 355, VEB-Fachbuchverlag, Leipzig, 1981 (in German).

2. Mains power-system heating an electrical resistance, controlled by a temperature sensor, emitting diverse spectral waves extending into the bright-infrared range, which have a maximum emissive energy at a wavelength of $1\cdot3 \times 10^{-4}$ cm, amounting to over 40% relative to solar energy at 100%.

3. Under the influence of high-frequency electromagnetic waves, involving wavelengths within both the bright and dark infrared range between $3 \times 10^{11}$ and $4 \times 10^{14}$ Hz, the maximum emissive energy of the latter being at about $3 \times 10^{-4}$ cm, being only at the 20% energy level relative to solar energy, owing to their longer wavelength. The short wavelength bright-infrared frequencies are capable of much deeper penetration into the dough-piece than are the longer wavelength dark-infrared frequencies. This deeper penetration results in an improved evaporation of the water in the dough-piece, and shortens the baking-time.

4. The application of high-frequency electromagnetic field, which involves the principle of dielectric-heating, whereby, the dough-piece is subjected to a rapidly

alternating electrical field, generated by a magnetron (HF-generator). The dough-piece being placed as dielectricum between HF-electrodes, acting as capacitor-plates. When placed within an electric field a conductor acquires charges of opposite polarity, distributed over its surface, resulting in a field of zero intensity. When the dough-piece, which acts as an insulator or dielectricum, is placed within the electric field, it becomes polarized, i.e. the molecules become charged and displaced so that their electric field resembles that of two unlike point charges of equal magnitude. Such a system is referred to as an electric dipole.

In the rapidly alternating electrical field, a high-frequency oscillation is set-up of the dipolar water molecules. This causes the molecules to vibrate, rotate and oscillate more or less evenly, causing the product to heat up uniformly, without the normal temperature-gradient from outside to inside during conventional heating modes. Equipment for dielectric heating operate within the ranges 10 kHz to 500 kHz or 13 to 27 MHz. The lower electrode can form the mesh- or steel-band of a tunnel-oven, thus allowing a continuous baking process. Since the HF/dielectric baking process cannot produce a crust, it is necessary to combine it with a conventional process. The most effective combination is HF/IR radiation, as suggested by Maes (Automatic Bread-baking using combined HF and convection-heating, *Fr. Fermentativ*, 26, 1964, 5, in *Bulletin de l'Ecole Official Meunerie Belge*, 1964, p. 203), using the HF/IR oven type 'Reforma'. Maes found certain advantages over conventional ovens such as: 50% reduction in the baking-time compared with conventional ovens; a 6–7% increase in loaf-volume; desired crust-colour can be put on by using IR-radiation units. Certain constraints, however, have prevented the process from being commercially acceptable. Doughs must be extremely homogeneous, the water molecules being uniformly distributed to avoid localized desiccation and carbonization; dough-piece symmetry and surface must be even and smooth; metal bread-pans cannot be used, since metal reflects HF waves (alternatives are ceramic or Jena-glass); steam-injection poses problems due to the danger of 'shorting'; bread flavour and aroma tend to be somewhat deficient.

Although attempts to apply dielectric heating to bread-baking have not achieved commercial success to date, research continues in this field, and recent research by Salovarra *et al.* in Helsinki, Finland (*Acta Alimentaria*, 1988, 17(1), 67–76) details its application to high-amylase flours. Utilizing VHF frequency 17·7 MHz during the first baking-phase for 6 minutes, followed by high temperature conventional baking at 350°C for 5 minutes, avoided the conventional defects in baking bread with flours of Falling numbers 94 to 197. However, these workers point out that exposure-time control is critical to regulate baking-loss to the normal 10–12% level.

The most widely used fuels, whether in the solid, liquid or gaseous form, require combustion in the presence of an optimal volume of air to provide the maximum amount of heat or calorific value. The composition of a fuel determines its heat efficiency. Most fuels contain carbon and hydrogen as their main components. Coal contains carbon and hydrogen in a complex bituminous combination; coal gas contains hydrogen, carbon monoxide and other carbon and hydrogen compounds; coke contains a large amount of carbon and mineral salts; fuel oils contain carbon and hydrogen organically combined as hydrocarbons, e.g. methane; natural gas is about 90% methane.

A good sample of coal will burn, leaving a small ash residue, while an inferior sample will contain large amounts of incombustible slate or stone. The former sample, when burnt with adequate oxygen will produce more heat per unit weight than the latter. The weight of carbon and hydrogen in a fuel, however is the important criterion, and not merely the total weight of the fuel. In general, the greater the hydrogen content of a fuel, the greater its calorific value, which accounts for the higher calorific value of natural gas and fuel oils.

When carbon is burnt in an adequate supply of oxygen, it forms carbon dioxide, this being an exothermic reaction, yielding 8080 heat units thus:

$$C + O_2 = CO_2 + 8080 \text{ heat units}$$

However, if there is only a limited supply of air, only carbon monoxide is produced, and the heat produced is reduced to 1934 heat units, thus:

$$2C + O_2 = 2CO + 1934 \text{ heat units}$$

By burning this carbon monoxide to produce carbon dioxide, an additional 6146 heat units are produced, thus:

$$2CO + O_2 = 2CO_2 + 6146 \text{ heat units}$$

This illustrates the importance of oxygen in the efficient utilization of fuels, the introduction of oxygen being achieved by constant and skilled 'stoking' in the case of solid-fuel ovens; this requires the even distribution of the fire over the fire-bars at all times. In the case of gases and fuel-oils, the optimum gas:air or fuel oil:air ratios are controlled by specialized units, which control fuel:air feed-rates with the help of valve/membrane systems which depend on pressure differentials. Oil fuel-burners, unlike gas-burners, require the oil at the correct viscosity and temperature, this involves the use of electrical preheaters, to ensure the correct feeding of oil to the oil-pump, and finally to the atomizing jets. Fuel oils vary in density from $0.90 \text{ g ml}^{-1}$ for light-fractions to 0·95–0·98 for medium to heavy grades. The maximum viscosity at 50°C is $20 \text{ mm}^2 \text{ s}^{-1}$ for oven oil-burners, and the feed and feedback pipelines to the burners must be warmed electrically, and the oil fed to these lines. These heaters must in turn be controlled by control thermostats, therefore oil-fired ovens, especially travelling-ovens, demand a relatively high electroenergy consumption. The sulphur content of fuel-oils should not exceed 2·5% for light-grade, 3·2% for medium-grade and 4·2% for heavy-grade oils. Minimum calorific values should be taken at $37\,800 \text{ kJ kg}^{-1}$.

Hydrogen, when burnt, combines with oxygen to form water, producing simultaneously 32 000 heat-units. Thus any fuel containing hydrogen will have a higher calorific value than one containing only carbon, and the greater the hydrogen content of a fuel the higher its calorific value. The gas and liquid fuels are normally sold with a declared calorific value, whereas the solid fuels are not. Some typical declared fuel calorific values for Europe and USA are given in Table 68.

The objective is to ensure that a maximum of heat-energy produced during firing is transferred to the baking-chamber and evenly distributed. However, all these operations are associated with heat losses, and it is the duty of the bakery technologist to reduce these to a minimum. This can only be accomplished when the

TABLE 68

|  | Europe | USA |
|---|---|---|
| Fuel oil | 42 000 kJ/kg | 140 000 Btu/gal |
| Natural gas | 21 500 kJ/m³ | 1 000 Btu/ft³ |
| Manufactured gas | 16 800 kJ/m³ | 546 Btu/ft³ |
| Wood (no declaration) | 12 600 kJ/kg | |
| Brown-coal brickettes (no declaration) | 20 000 kJ/kg | |
| Electricity | 3 612 kJ/kWh | 3 412 Btu/kWh |

Sources: (1) *Special-processes in Baking: Process Manual*, W. Schwate and U. Ulrich, VEB-Fachbuchverlag, Leipzig, 1986. (2) 'Relative heat utilization in bread-baking', in *Conversion Factors and Technical Data for the Food Industry*, ed. Harrel, C. G. and Thelen, R. J., Burgess Publishing Co., Minneapolis, 1959.

potential loss-sources, and methods of containing them are known. Energy losses during the firing operation can be ascribed to either (1) defects in the constructional status of the combustion-area, e.g. cracks in fireplace or oven-chamber brickwork, which must be regularly checked, or (2) incorrect or inefficient heating techniques. Inefficient heating techniques, can be due to poor quality coal and/or inadequate combustion due to infrequent stoking operations. In the latter case, the ashes contain unburned coals. Where fuel oil or gas is used, inefficient combustion can be detected in the composition of the exhaust-gases. Where the oxygen level is too high, the excess air removes heat, and too little oxygen results in an incomplete combustion. The air and oil/gas pressures must be regulated to produce a flame of optimum calorific value. An average oil pressure for a Baker Perkins swing-tray travelling-oven is 12 lb/in², which is mixed with air at about 18 lb/in², to produce a flame of optimum calorific value. The secondary-air intake on each burner ensures more complete combustion, and can be adjusted accordingly. Too much secondary-air produces sparks, and too little a 'smokey' flame. The predominant constituents of fuel oils are carbon and hydrogen, and to extract the maximum heat-energy from it, the hydrogen must be burnt with the help of secondary-air. In coal-fired ovens the air uptake is controlled by the fireplace doors and exhaust-gas vents, at first the quantity of air must be greater than towards the end of firing.

Radiation-losses can be from the oven-walls, or baking-surface. The former is due to the oven-walls being at a higher temperature than the bakery surroundings; the best insurance against this is good insulation of the oven-walls, and the elimination of any gaps or draughts. Radiation-losses from the baking-surface are due to open feed and discharge openings of tunnel-ovens, or oven-doors in the case of batch-ovens. The only way this loss can be avoided is by opening as little as possible. Another source of heat-energy loss occurs due to the overheating of the humid air of the oven-chamber. This humidity heat-loss can be minimized by restricted opening of the steam vents and the feed and discharge openings. The various sources of heat-loss, and total energy requirements are always variable, depending on: oven-type and maintenance condition, fuel utilized, baking process, type of product, and the efficiency of the firing technique. Final-proof, oven-temperature, and the maximum oven-capacity must be synchronized. Thermometers must be correctly calibrated, and optimal temperatures maintained at all times. Regular fixed-period checks on oven-humidity, gas and air consumption, and exhaust-gas analysis should be carried out, where continuous monitoring facilities are not provided. All data should be recorded, preferably on recording-charts. All draughts in the close vicinity of the oven must be eliminated. Examples of average percentage heat-utilization and loss data for a wire-band tunnel oven are the following: heat-energy input 100%, heat-energy utilization 45%, exhaust-gas losses 35%, humidity heat-loss 12%, radiation-losses 8%, according to Schwate and Ulrich (*Special-processes in Baking: Process Manual*, Leipzig, 1986). A more detailed breakdown is given by Schneeweiss and Klose (*Technology of Industrial-baking*, Leipzig, 1981) as follows: effective heat-energy input 1200 kJ kg$^{-1}$, humidity-input 200 kJ kg$^{-1}$ humidity heat-loss 180 kJ, radiation-losses 120 kJ, exhaust-gas losses 480 kJ, dough water evaporation 340 kJ, heating the crumb to 98°C 160 kJ, heating the crust to 180°C 100 kJ. The weight-loss which occurs during baking, mainly due to the evaporation of water from exposed surfaces, amounts to about 97% for wheat-bread, and about 95% for rye-bread, the remainder is contributed by the evaporation of ethanol 1·5%, carbon dioxide 3·3% (0·3% volatile organic acids, 0·1% aldehydes). The actual baking-loss depends on the product, shape and weight, varying from 6% to 14%, according to the baking-regime.

TABLE 69

|  | *Fuel/100 lb of bake* | *Btu/lb of bake* |
|---|---|---|
| *Natural gas* (1000 Btu/ft$^3$) | | |
| Tray-ovens (direct-fired) | 52 ft$^3$ | 520 |
| Tray-ovens (indirect-fired) | 80 ft$^3$ | 800 |
| Tunnel-ovens | 82 ft$^3$ | 820 |
| Reel-ovens | 100 ft$^3$ | 1 000 |
| Peel-ovens (brick) | 142 ft$^3$ | 1 420 |
| *Manufactured gas* (546 Btu /ft$^3$) | | |
| Tray-ovens (direct-fired) | 82 ft$^3$ | 445 |
| Tray-ovens (indirect-fired) | 112 ft$^3$ | 610 |
| Tunnel-ovens | 125 ft$^3$ | 685 |
| Peel-ovens (brick) | 264 ft$^3$ | 1 440 |
| *Fuel oil* (140 000 Btu/gal) | | |
| Tray-ovens (indirect-fired) | 0·76 gal | 950 |
| Tunnel-ovens | 1·00 gal | 1 250 |
| Peel-ovens (brick) | 1·03 gal | 1 280 |
| *Electricity* (3 412 Btu/KWh) | | |
| Tray-ovens | 13·9 kWh | 475 |
| Rotary-ovens (cake) | 14·2 kWh | 485 |
| Tunnel-ovens | 15·0 kWh | 512 |
| Peel-ovens (brick) | 18·5 kWh | 563 |

Source: *Conversion-factors and Technical Data for the Food Industry*, ed. C. G. Harrel and R. J. Thelen, Burgess Publishing Co., Minneapolis, 1959.

Theoretical data concerning the heat-input required to bake a 1 kg or 1 lb loaf have been variously cited as 530 to 630 kJ kg$^{-1}$, and 325 to 400 Btu respectively. However, it is impossible to quote very accurate data, as heat requirements will depend on dough specific-heat levels, which could be 2·6–3·0 kJ kg$^{-1}$ K$^{-1}$ for lean dense European-type doughs, or 0·65 to 0·88 in the case of richer, lower density North American formulations. Oven temperature regimes, and rates of moisture evaporation will, in turn, depend on oven-type, fuel used and method of heat application, including steam where utilized.

Actual measurements have indicated that the heat input required per pound of bread in modern ovens falls within the wide range of 330 to 400 Btu, compared with 1000–2000 Btu in the older coal-fired brick ovens using a peel. The theoretical or calculated heat required to fully bake a dough of 0·75 specific-heat and 1 lb weight is 236 Btu, which emphasizes the variability of all factors involved, with the exception of the specific-heat of water at 4·2 kJ kg$^{-1}$ K$^{-1}$, and its heat of vapourization at 2260 kJ kg$^{-1}$ (970 Btu/lb). Bakers express oven capacity as pounds or kilograms of bread per hour, whereas, the oven-manufacturer will often express it as pounds of dough per square foot of baking-surface or hearth per hour, or more simply as baking-surface in square metres. A useful guideline of relative heat utilization expressed as fuel/100 lb bake and Btu/lb of bake for various types of oven, using various energy sources, has been compiled by Harrel and Thelen, as shown in Table 69.

### 3.3.2 TYPES OF OVEN AND OVEN DESIGN

The baking of the fermented raw dough-piece, so that it emerges from the oven in a palatable and appetizing form demands the control of its residence-time in the oven, the amount and intensity of the heat acting upon it, and the humidity of the oven-chamber atmosphere during that period. Technical advances in heat-transfer concepts, automation, electronics, and computer application, all demand from the oven-operator increasing expertise in exercising control over them. Such expertise involves an application of basic knowledge of physics, measurement and control technology, machine-construction, coupled with responsible, diligent and quick response to sudden changes in conditions, and their correction.

The choice of any particular type of oven will be decided after such considerations as required bakery production-throughput capacity, product diversity, available floor-space, energy-source, economy in operation, construction and maintenance.

Early commercial ovens were known as side-flue peel-ovens or peel brick ovens. These consisted of large chambers of brick construction, with a floor or sole of large square refractory-tiles. The crown of the chamber was arched, and the fireplace or furnace, long and narrow in shape, was built obliquely to feed the flames and products of combustion over the crown, and around the chamber before passing out through the flue at the left of the oven-mouth. Coal or coke could be utilized, or a combination of both fuels, coke requiring deeper fire-bed and more draught or secondary-air supply. Since these ovens were built into the bakery as integral structures, the space behind both the crown and sole were packed with sand. Owing to their massive construction, these ovens were capable of storing and delivering a solid bottom heat, and steady radiant top heat, which gave baking conditions conducive to an excellent bake, producing bread with an appetizing crust and crumb flavour and aroma. The operation of these ovens demanded skill and experience coupled by good judgement on the part of the master-baker, to consistently achieve an evenly baked product. Regular stoking and raking of the fire-bed was essential to ensure efficient and maximum combustion of solid fuels, without clumping and the build-up of vacuoles, which would result in the oven 'blowing-cold'. When the pyrometer-temperature reached 550°F (288°C), the batch was set using peels of various sizes (long-handled wooden shovels), as rapidly as possible to minimize heat-loss. After setting, the pyrometer would register about 500°F (260°C), owing to the presence of the relatively cold load. After a residence-time of up to 1 hour, the bread was unloaded with the peels. The oven temperature by this time was 'solid' enough (steady-state with stored heat) to bake the fermented sweetened products, soft-rolls and layered-pastry products, followed by smaller confectionery items and pound-cakes. Therefore, the baker had some control over product quality, crude as it was, provided he used his experience and judgement to exploit falling temperatures. The loading work with the peel-ovens was then made easier and faster by the introduction of the draw-plate-oven, in which the hearth could be withdrawn from the chamber on wheels running in tramlines. These draw-plates were constructed of steel-plate or tile, supported by a steel-frame. The main advantage of these ovens, which were sometimes double-decked, was that all products entered and left the chamber at the same time, instead of the first-loaf-in/last-loaf-out situation with the peel-oven. However, the large working area required for the draw-plate, and the large heat-loss involved when loading and unloading, limited its acceptance.

The early German ovens were similar to the side-flue peel-ovens, depending on the direct-heating principle; oven-heating and baking could not be done at the same time. These ovens were heat-storage ovens, the temperature in the tiles and brick walls falling during the course of production. After heating up, these ovens had to be cleaned out, and allowed to attain a steady-state of heat-distribution by allowing a resting period before loading. To improve steam-retention, the hearth had a slight gradient. Because of the high initial temperature, and even heat-transfer, they were ideal for baking rye and mixed rye/wheat-bread.

**Modern Ovens**
Modern ovens can be classified into batch-ovens, which are suitable for small to medium-size bakeries or to industrial bakeries with a specialized range of products, and larger-capacity continuous industrial baking ovens.

*Batch-ovens*
The heating-principle of these ovens depends either on the Perkins steam-tube, hot-gas circulation (cyclothermal), or electrically heated elements. Whichever heating-principle is involved, most batch-ovens tend to be of the multi-deck design. Those

using the Perkins patent tube are usually gas- or oil-fired, but formerly solid-fuels were applied. The solid drawn-steel tubes (filled with distilled water) extend into the chamber and under the hearth, being mounted at an angle so that the water can be heated at the firing-end, forming steam within with a pressure up to about 22 MPa. The mode of firing can be coal, oil or gas in the firing-chamber, and the Perkins system can be applied to single-chamber, multi-chamber designed ovens whether of brickwork or steel-frame construction. Oven-chamber temperatures must not exceed 500°C, because of the danger of explosion, and the tubes must be evenly heated at the firing-end. In the case of gas-firing, the gas-jets play directly onto the steam-tubes, since the flame temperature is much lower than the oil-flame, but regular cleaning of the tube-ends is essential to ensure even heat-transfer. Oil-burner flames are much too hot to come into direct contact with the steam-tubes, therefore heating is indirect, baffle-bricks being inserted between the burner-jets and the tubes.

Heat-circulation (cyclothermal) ovens are heated indirectly (Fig. 114), the hot gases or hot air (electrical heating) being circulated around the various flues either by chimney-draughts or by ventilators within radiator-channels between the oven-hearths. To improve heat-transfer, the channels or ducts and hearths are constructed of steel-sheeting, but the oven-sole is covered with tiles or steel-plates. Owing to their ease of control, gas or oil are the most popular; but electrically heated hot-air systems use a closed circulation system, since no exhaust gases are involved. Where coal, gas or oil are used the burnt gases must be disposed of, but part of the heating-gases from the ducts can be mixed with the hot combustion-gases, and become recirculated. In the case of multi-deck ovens with the oven-chambers closely stacked, the heating-ducts can be reduced in diameter when a more powerful fan is installed. Multi-deck ovens with heating-gas circulation, compared with steam-tube ovens, show an average energy saving of 35%. They normally have 4–5 decks, the heating-gas circulation type usually having 4 decks, with a baking-surface of $2 \cdot 4\,m^2$ each, making a total baking-surface of $9 \cdot 6\,m^2$. The temperature of each deck is independently controllable, enabling the baking of various products simultaneously, or one deck to be set at a 'sharp' higher temperature for 'pre-baking' of rye or variety-breads, and another for completion of baking. Using full-width oven doors, loading can be rationalized by using oven 'setter' equipment (German: *Abrollapparate*). Using all 4 decks, the average throughput is 120 units at 1 kg, or 860 bread-rolls at 45 g. The average heating-up time, using oil or gas is 70 minutes, which consumes about 11 kg diesel oil, or about $32\,m^3$ of gas per hour. When in continuous use, the consumption reduces to about $5 \cdot 7$ kg of diesel oil, and $13\,m^3$ of gas per hour respectively.

The typical start-up procedure for gas-firing is as follows. Actuate mains-power switch; switch on circulation-fan and allow to run for 5–10 minutes to aerate the system; set required temperature using the thermostat, taking into account that the thermostat measures and controls the heating-gas temperature, which is about 50°C higher than the oven-chamber temperature; open the mains gas tap; ignite the pilot-flame and adjust; burner selector-switch set either to 'manual' or 'automatic'; actuate the steam-line main-tap, and check that the injection-nozzles are free from mineral deposit.

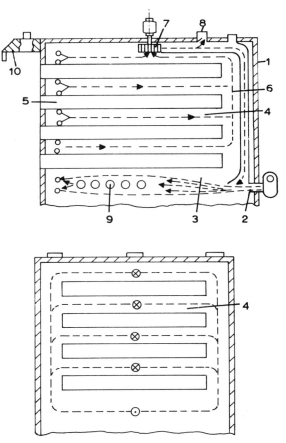

Fig. 114.  Forced heat circulation system (cyclothermal) of a multi-deck batch-oven (side-elevation). 1. Mineral-wool insulation. 2. Burner. 3. Channel or duct housing. 4. Heating zone. 5. Oven-chambers. 6. Channels or ducts. 7. Thermoflow fan. 8. Exhaust-gas vent. 9. Steam source (humidifier) for each deck and proofer unit. 10. Steam cowling and ventilator. (Source: *Machine Handbook (Baking)*, 2nd edn, B. Schramm, VEB-Fachbuchverlag, Leipzig, 1978.)

The loading procedure when using setter-equipment is as follows. Dough-pieces are carefully transferred from final-proof containers onto the setter-band; inject steam into chamber, immediately closing the door; place the setter-frame on the baking-surface, avoiding excessive shaking of the dough-pieces, but operate rapidly; firmly hook the guide-rails behind the oven-door; withdraw the setter-band by gripping the handle, taking care to control the speed of withdrawal, thus fixing the spacing of the dough-pieces; remove the empty band from the oven, and ensure the oven-door is closed; remove any steam coming from the oven into the bakery by using the ventilator-fan above the oven-door cowling.

The reel-oven, the first mechanical oven, is an American creation, and can still be found in some retail bake-shops in the USA. It was built like a ferris-wheel, a reel structure revolving vertically about a horizontal axis within the baking chamber, supporting the baking trays hung from hanger-pins protruding from large wheels inside the oven. A high baking-chamber is required to accommodate the reel, and fuel consumption is high relative to production capacity. Reel-ovens are normally directly fired with gas or electricity, the heating elements being located centrally across the floor of the chamber. A baffle positioned above the gas-burners provides some radiated-heat to balance the predominantly convection-heat of the burners. When fired indirectly, heat is generated in a separate fire-box beneath the oven chamber, the hot gases being then passed through radiators positioned across the oven floor, and vertically at the rear, between the oven-chamber and an insulated rear wall. The exhaust-gases then escape through a flue. The box-like structure provided a good steam-dome, and proved ideal for products requiring a high, moist atmosphere. The main aim in its design was to solve problems of loading and unloading deck-ovens, and to cope with mixed baking-times. Loading and unloading products is at a constant height from a full-width door, all products are readily accessible, the continuously rotating wheel being stopped when the appropriate carrier is positioned at the oven-mouth. Products with different baking-times can be baked simultaneously, assuming they require similar baking temperatures. Reel-oven carriers are designed for baking-trays of various dimensions, and will also take pan-straps. It is also possible to bake hearth-breads directly on the carriers, and also bread-rolls, proofing the doughs on dusted boards, and then slipping onto the carriers. Reel-ovens are ideal for a wide range of fermented-products and confectionery, providing an even temperature distribution that is due to the paddle-effect of the carriers, and being quick and simple to load and unload. The introduction of the rack-oven has, however, provided a more expedient solution for the batch production of a diverse range of products in the smaller bakeries, and also in the ever-increasing number of in-store bakeries of the grocery chain-stores. Were it not for this latter trend, the use of reel-ovens would probably have increased. The disadvantages of the reel-oven are: the large floor area compared with its baking-capacity, the large amount of moving parts involved in its construction, and the difficulty of steam distribution and control for certain products.

The electric multi-deck oven remains popular with the craft-baker, the in-store bakery and the hot-bread shop, owing to its flexibility and ease in operation. They are suitable for all types of product, with a high output relative to floor area, and are economic when correctly managed. The standard electric multi-deck unit is designed for one-tray oven depth, accommodating 2–3 trays across its width; the number of decks can be varied from 1 up to 6. Electronic temperature control is accurate, with a digital read-out, and top and bottom heat are controlled independently. Heating and insulation of each deck is also independent, and full-width doors allow the use of setter-equipment. The sheathed elements make heat distribution easier since they can be positioned above and below each deck, which is more difficult in the case of small gas-fired multi-deck ovens.

Rack-ovens are a relatively recent development, consisting of a vertical chamber into which the special racks are wheeled, carrying up to about 100 trays of product.

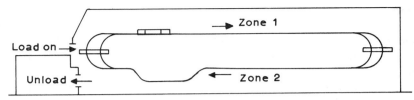

Fig. 115.   Single-lap tray oven—path of product. (Source: *Baking Science and Technology*, E. J. Pyler, Sosland Publ. Co., Kansas, USA, 1988.)

The rack stands on a turntable, which rotates during the baking cycle, thus ensuring uniform convection heat-transfer. Forced convection can also be used, hot air being blown over the products through slits or holes down one side of the chamber, passing through ducts in the other side, and being either recirculated through a heat-exchanger or electric elements to reheat. They may be heated electrically or with gas or fuel oil burners, often incorporating computerized programmable controls for the regulation of temperature, humidity and baking cycles. Single, or pairs of, racks can be placed in one chamber, or two racks can be placed in two single chambers heated from the same central source. 6–8-rack models are also available, either as single- or double-chambers. The very large tunnel rack-ovens will take from 10 to 30 racks, the racks being fed in at one end of the tunnel onto a variable speed floor-conveyor that moves them through the chamber to the discharge end. The most widely used are the single-, double-, and 3- or 4-rack models, since these offer more flexibility in baking-temperatures and utilization. These ovens are easily loaded and unloaded, with a minimum of handling involved, making possible the use of unskilled labour. They find general application within in-store bakeries, and freeze–thaw–proof–bake operations.

*Industrial Baking Ovens*
The travelling-tray ovens, of single-lap (Fig. 115) or double-lap (Fig. 116) design were an evolution of the reel-oven in the USA, being built to increase oven capacity and improve baking efficiency generally. The baking chamber was much lower, the reel being replaced by two parallel endless chains carrying trays, a third chain stabilizing the horizontal position of the trays during their passage through the oven. Single-lap models provide a single passage through the oven's two heat zones

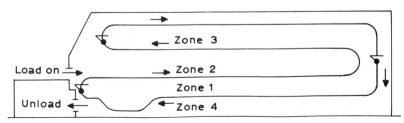

Fig. 116.   Double-lap tray oven—path of product. (Source: *Baking Science and Technology*, E. J. Pyler, Sosland Publ. Co., Kansas, USA, 1988.)

as illustrated in the schematic. Loading and unloading of these ovens takes place at the front end, and when automatic loading and unloading were added this type of oven increased in popularity. The single-lap design is generally considered to offer the best operational performance, owing to the simplicity in design with one horizontal run and lower crown for steaming. Modulation heat control is desirable to ensure lateral heat control and rapid response to loading changes. The number of trays varies from 18 to 74, and production capacities range from 2500 to 8000 lb bread/hour depending on the number of trays. About 38 trays, each measuring 148 × 32 inches, are adequate for a throughput of 7500 lb dough/hour. The double-lap tray oven is based on the same principle as the single-lap, however, in this case the trays travel through four heat zones instead of two. The vertical ceiling height is twice that of the single-lap, but the floor area required is much less, which is important where floor space is limited. Construction and ductwork is more complex in the double-lap models, and optimum steaming and zonal heat control is more difficult owing to stacking. Most models of this type of oven are direct gas-fired, with automatic oven-loader, the double-lap oven model accommodating about 48 trays.

On the European continent, a similar basic design principle to the lap tray oven is incorporated into the double wire-band return oven (Fig. 117), or *Doppelnetzband-Rücklauföfen*, as it is known in German. As the name suggests, it consists of two wire-bands in the form of endless-conveyors running parallel within two separate oven-chambers. The advantage of this design is the saving of labour with the aid of mechanical transport. The dough-pieces are transported from the mouth of the oven onto a desired position on the travelling-band, which transports them the full length of the bands within the oven-chambers, the baked products then being returned to the mouth of the oven, by which time the baking-cycle is complete. This latter operation is achieved by reversing the rotation of the motor which drives the wire-bands. On completion of the full distance in both directions, the motor automatically switches off; but overriding button-switches allow the bands to be halted in any position. In contrast to the multi-deck oven, the baking-surface is

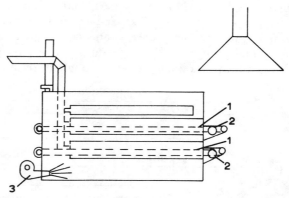

Fig. 117.   Double-wire-band return-oven (long section). 1. Wire-bands. 2. Motor. 3. Burner. (Source: *Machine Handbook (Baking)*, 2nd edn, B. Schramm, VEB-Fachbuchverlag, Leipzig, 1978.)

mobile, and part of the travelling-band protrudes outside the oven-mouth, allowing the dough-pieces to be deposited directly onto the band. Double wire-band return ovens can be installed in bakeries of any size, provided the production capacity demands it, combining the advantages of the multi-deck oven, viz. limited floor area and good heat-utilization, with ease of loading of the wire-band tunnel oven. Such ovens are used to bake bread-rolls, variety-breads baked on the hearth and sweetened products of most types. Most of these ovens are gas-fired, incorporating the same principle as the multi-deck ovens, i.e. indirect hot-gas cyclothermal heat circulation. The baking-surface of each oven-chamber is $6\cdot60\,m^2$, making a total of $13\cdot20\,m^2$, which is more than the $9\cdot6\,m^2$ of the four-deck multi-deck batch oven described. The throughput is normally about $216\,kg$ bread/hour, which can be increased to a maximum of $258\,kg$ bread/hour. Bread-roll output is 4000 units/hour maximum. Both outputs provide improved capacities compared with the four-deck multi-deck oven. The procedure for heating up is similar to the multi-deck oven with forced circulation (cyclothermal) heating system, but exact details are to be found in the instruction manual. The loading procedure is as follows. Inject steam according to product requirement, avoiding vibration. Load the part of the wire band which protrudes with dough-pieces. Switch the wire-band into 'forward' movement, and move the loaded part into the chamber. Load the next part and repeat the former operation. Repeat the same procedure until the wire-band is full. When this stage is reached the motor automatically cuts out. Inject more steam. The temperature is automatically controlled, and the oven humidity can be controlled by means of vents. The wire-band is then switched into 'reverse' movement, and the baked products either manually unloaded and placed on boards, or mechanically unloaded by placing a right-angled transport-band (conveyor) against the discharging oven wire-band. A typical example of this oven is the DNRO with $13\cdot2\,m^2$ baking-surface, manufactured by VEB Kombinat Fortschritt, Bäckerei-maschinenbau, Halle, GDR.

The most technologically, and economically advanced bread baking oven to date is the wire-band tunnel-oven, which has a good overall thermal efficiency. It distinguishes itself from all the other oven types by being completely continuous in operation, dough-pieces being fed in at one end of the chamber and discharged at the other. Such ovens are intended for line-production systems, where the baking stage must be synchronized with all previous and following production stages. In order to avoid considerable loss of production capacity and bread quality, the technologist must acquaint himself with all possible corrective control measures in the event of breakdowns and flow disturbances.

However, in spite of its many advantages, its economic justification for the baking of panned bread and rolls has been considered questionable in the USA, mainly on grounds of floor space. As Anderson (*Bakers Digest*, **47**(2), 28, 1973) points out, a tunnel oven with an output capacity of $8000\,lb$ of baked product/hour requires about $1000\,ft^2$ more floor area than a single-lap oven of comparable capacity, assuming that both are equipped with automatic loading and unloading equipment. Nevertheless, where a mixed production range of hearth-, pan- and variety-breads and rolls are baked, the many advantages of the tunnel-oven cannot be overlooked.

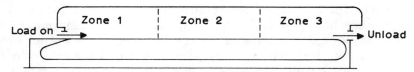

Fig. 118.   Tunnel-oven divided into heat zones.

The general features of the tunnel-oven are a long, low baking chamber, within which a motor-driven conveyor carries the oven hearth in a straight line for stretches from 100 to 400 ft. The length is divided by baffles into one or more steaming zones (Fig. 118), and up to eight heat-zones of 30–60 ft each, depending on length and manufacturer's specification. Each zone has its own individual control, which is made easier by provision for separate controls for top and bottom heat, and lateral heat balance. Steaming conditions can be easily optimized, and any problems with swing-tray stabilization become non-existent. Tunnel ovens are more or less standard within the biscuit industry in Europe and the cookie and cracker industry in the USA. The choice of baking-surface for tunnel ovens is woven-wire mesh, which is the commonest, or perforated-steel bands. In special cases, the band can be covered with fire-brick tiles.

As one example of an automatic continuous tunnel-oven of this type widely used in Western Europe, the Duotherm wire-band oven type DUO-NU, manufactured by Werner and Pfleiderer, 7000 Stuttgart 30, FRG, is described. This unit can be applied to the baking of all kinds of bread, whether panned with high side-walls, or free-spaced hearth- or variety-breads placed end-to-end or on all sides. It is also suitable for bread-rolls, cakes, bases, rusks, pizza and panettoni. This model incorporates the Duotherm forced-convection heat-transfer system in the rear part of the chamber, which can be utilized as desired for specific products. This provides even browning and crust formation on the sides of panned products, shorter baking times, and a thicker crust on free-spaced hearth products. The oven is in all-steel design, the oven front, steam hoods, top cover, feed and discharge frames being made of V2A stainless steel. Both feed and discharge doors are adjustable in height. A heating gas thermometer for each burner, and a baking-chamber thermometer for each control zone are fitted on the operating side. Heating-gas slides for top and bottom heat, the steam slides, and the observation-doors, are all fitted on the operating side. Exhausts are provided above the observation doors to collect the steam emitted, and release it through connecting pipes on the upper edge, thus preventing condensate from collecting on the casing sheets and forming rust. The front part of the oven, the burner zone, is fitted with the cyclothermal heating system of the Universal wire-band oven model NU. The rear section of the chamber being equipped with the Duotherm forced-convection system. Heating-gases for top and bottom heat are fed into numerous ducts, distributed across the width of the Duotherm radiators. The baking-chamber atmosphere is circulated in a vertical direction by a reversible fan through the spaces between the ducts. In this way, improved heat-transfer on a convective basis is achieved by the air flowing rapidly around the products, especially the side-walls of pans or loaves, which are not easily

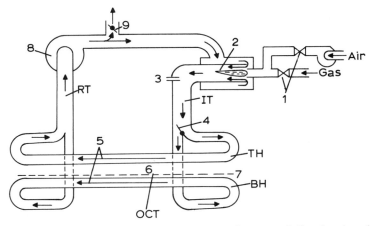

Fig. 119. Schematic of cyclothermal system. 1. Control systems. 2. Combustion chamber. 3. Partial-pressure measurement. 4. Control vent. 5. Heating duct. 6. Oven chamber. 7. Wire-band. 8. Circulation fan. 9. Exhaust-gas vent. IT, inlet temperature. TH, top-heat. BH, bottom-heat. OCT, oven-chamber temperature. RT, return temperature. (Source: *Technology of Industrial Baking*, R. Schneeweiss and O. Klose, VEB-Fachbuchverlag, Leipzig, 1981.)

reached by the infrared radiation. The direction of flow can be switched up or down or terminated by operating a switch on the control-panel.

With the indirect cyclothermal heating system (Fig. 119), the combustion gases are separate from the baking-chamber (partial vacuum system), having no contact with the products to be baked. The oven is divided into individual baking temperature control zones, which allow control of top and bottom independently of one another (Fig. 120).

The heating gases flow through heating radiators, emitting heat to the baking-chamber. The hot gases are conveyed by a circulating fan back to the combustion and gas-mixing chambers, where they are mixed with the combustion gases, heated up, and conveyed back into the radiators.

Burners for oil, natural gas, town gas or liquid gas can be connected to the heating system. Combined oil/gas burners for rapid conversion from oil to gas is also possible. A control unit with various safety devices simplify handling and ensure safety in operation. Optical and acoustic signals communicate any defects in the system, and the oven is protected against overheating by a safety contact in the baking-chamber thermometer. If the circulating-fan breaks down, a pressure sensor in the system shuts the burner off. A heat modulation unit in the heating gas circuit, adjusts the output of the burner to the heat requirements of the oven. Burner equipment design conforms to Standards 4755 and 4787 for oil, and DFGW Work Sheet E 663, and DIN Standards 4756 and 4788 for gas heating.

Steaming is by individual pipes, fitted one behind the other in the oven-feed station. The first steam pipe (spray) is round with boreholes, which ensure even distribution of the steam laterally. The direction of the spray can be changed by turning the pipe. All other steam pipes have a square cross-section, being jacketed,

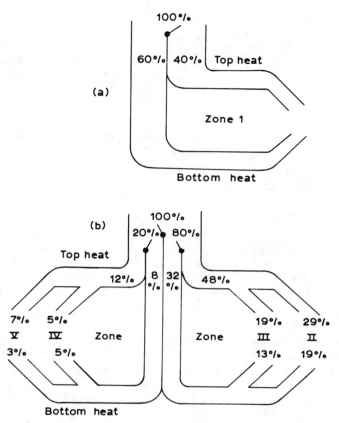

Fig. 120. Heat distribution of wire-band tunnel-oven, BN50, manufactured by VEB-Kyffhauserhütte, Artern, GDR. (a) Burner I. (b) Burner II. (Source: *Technology of Industrial Baking*, R. Schneeweiss and O. Klose, VEB-Fachbuchverlag, Leipzig.)

with openings pointing downwards, while the discharge openings of the round inner brass pipe are at the top. The amount of steam discharged by the first and next two steam-pipes can be regulated. The steaming zone can be separated from the baking-chamber by an adjustable screen. Steam is discharged from the baking chamber through slides, ventilation-flaps being fitted on both sides of the steaming zone to ensure complete discharge of the steam after the required reaction time has elapsed. The entire oven is insulated with mineral and glass wool to keep heat loss to a minimum.

The wire-band is driven by a variable-speed gear, driven by a three-phase motor, with a chain drive between the gear and the drive-drums. The required baking-time is set with a hand-wheel, and read off from a mechanical indicator. The wire-band moves over sheet-steel, ensuring smooth running and freedom from lateral wander and vibration, also reducing heat loss. Design of the wire-band tightening device is according to load/products baked.

Oven capacity is a function of the band conveyor width and length, and the dimensions of the bread-pan straps, hearth-loaf spacing or tray dimensions. Wire-band widths can be 1·65, 2·0, 2·5, 3·0 or 3·75 metres; and band length 9 to 45, or more, metres.

Example:
Wire-band conveyor: length 23 metres, width 2·5 metres; bread pan straps 570 mm × 240 mm, loaded end-on, containing 800 g × 4 loaves per strap
Output/hours using 30-minute baking-time 3228 × 800 g units
Output/hours using 26-minute baking-time 3725 × 800 g units

The output of hearth bread is more difficult to estimate due to spacing necessary for a crusty bake, but could be only 50% of that of panned bread in the same size oven, baking time is also longer, e.g. 60 minutes, to ensure crust crispness and thickness.

Also in the industrial bakeries of the Eastern European countries, the wire-band is the most utilized oven for the continuous production of basic breads, variety breads and rolls. Such ovens can be loaded manually or by mechanization and automation for line production. In the GDR, these ovens are manufactured by VEB Kyffhäuserhütte Artern, designed for both right- and left-side loading/unloading, and with a choice of town gas, natural gas or fuel oil as energy source. The most widely used models in the industrial bakeries of the GDR are the BN 50 and BN 72 wire-band tunnel-ovens (Figs 121–123) with baking-surfaces of 50 and 72 m² respectively. The output of the BN 50 is 835 to 1050 kg/h mixed rye/wheat hearth

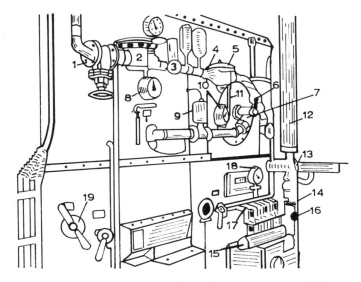

Fig. 121.   Heating and humidification system of BN 50. 1. Gas mains. 2. Pressure regulator. 3. Magnetic valve. 4. Gas-tap. 5. Mixture regulator. 6. Burner-tap. 7. Burner. 8. Manometer. 9. Magnetic valve. 10. Manometer. 11. Pipeline. 12. Steam feed-pipe. 13. Steam-valve. 14. Condensate pipeline (narrow gauge). 15. Steam-valve. 16. Regulator valve. 17. Steam-injector pipes. 18. Steam manometer. 19. Fresh air/exhaust gas.

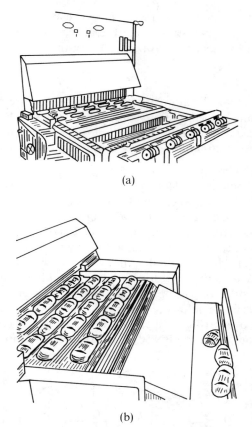

(a)

(b)

Fig. 122. BN 50 wire-band tunnel-oven. (a) Loading. (b) Unloading. (Source: *Special Processes in Baking—Process Manual*, W. Schwate and U. Ulrich, VEB-Fachbuchverlag, Leipzig.)

bread, or 665 to 685 kg/h bread-rolls. The energy requirements of the BN 50 are town-gas 100 m³ h⁻¹ (18 kW); natural gas 40 m³ h⁻¹ (18 kW); and heavy grade fuel-oil 32 kg h⁻¹ (30 kW). Steam consumption is 80–160 kg h⁻¹, depending on oven size and product. The most popular energy source is gas, which after combustion is circulated through ducts above and below the oven chamber. However, after leaving the heating channels, only a small volume of the heating gases is discharged through the chimney. Instead, the greater volume of the 200–400°C gases are returned to the burner chamber by means of a circulator-fan. They then become mixed with the hot ignition gases at 1200°C, and produce heating gas temperatures below 600°C. The combustion chamber must then be constructed of suitably temperature-resistant material, and since temperatures in excess of 600°C trigger off the melting of safety fuses, the burners are so constructed that the feedback gases cool the combustion chamber from the outside. As a result of the feedback of the heating-gases, these cyclo-

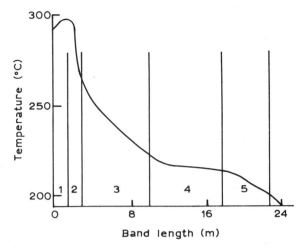

Fig. 123. Optimum baking temperatures for 1·5 kg rye wheat breads in the BN 50 oven heat-zones. (Source: *Special Processes in Baking—Process Manual*, W. Schwate and U. Ulrich, VEB-Fachbuchverlag, Leipzig.)

thermal tunnel-ovens can utilize up to 50% of the energy input. The heating-gases which emerge from the combustion and mixing chamber are conveyed, depending on the temperature-gradient, through vents in the feed pipes of the top and bottom heating system, and then become distributed throughout the heating-channels of the various heating-zones. To achieve improved heating-gas throughput and stability, the oven is divided lengthwise into heating-zones. Depending on the distribution of the heating-gases, and placement of the heating channels or ducts, the gases can be made to flow either with or against the direction of movement of the products. Since the heating-gases are continuously transferring heat energy to the oven-chamber, there exists along the whole length of each heat-zone, depending on the velocity of flow of the gases, and the difference in temperature between the heating ducts and the oven chamber, a temperature-gradient. This temperature-gradient can have either a positive or negative influence on the temperature curve. The circulation-fan creates a partial vacuum of 20–40 Pa within the combustion-chamber, and the flow velocity of the gases in the heating ducts reaches 5–6 m/s. Owing to the partial vacuum in the system, it is important that the system is completely sealed, otherwise cold-air draughts will be sucked into the system resulting in considerable energy losses. Other sources of heat loss result from inadequate protection of the feed and discharge ends of the oven-chamber from cold air streams entering the bakery. Radiation losses are minimized by insulation of the tunnel-housing with glass or mineral wool. For the removal of excess steam from the chamber, steam extraction pipelines are provided from the various zones of the chamber. The endless wire-band is driven by a cylindrical roller of about 500 mm diameter. The speed of the wire band, and hence the baking-time of the BN series ovens, is regulated by a three-phase motor with a variable-speed drive over the drive-roller, mounted on the discharge side of the oven. The selected motor-rpm determines the rotation-time of the band, and can be read off a tachometer calibrated in minutes as follows:

|  |  |
|---|---|
| 3 000 rpm | 17 minutes |
| 1 500 rpm | 35 minutes |
| 750 rpm | 70 minutes |

For smooth running, the straight path and tension of the band is very important. This can be corrected either by adjusting the tension on the cylindrical roller (drum) on the drive side, or manual adjustment of the tension-balance (with weights) on the corresponding drum at the oven-feed end. When necessary, the band must be shortened by a service-engineer. Most of the controls and temperature-recorders are located on the control-panel or desk. Figures 120 and 121 show the layout and cyclothermal heating system of the BN wire-band tunnel-oven. The main procedures to follow in actual operation are the following. Start the oven, set up the bridge-conveyor between final-proofer and oven-feed with the dough-cutting equipment, and switch on. Check for even gaps between the dough-pieces and spray them evenly with the spray equipment. Check the oven-development of the dough-pieces by visual observation through the side-windows. Check the baking temperatures and humidity in each heat-zone. Make sure that the spray equipment at the discharge end is correctly adjusted for even, moderate moistening of the baked loaves on leaving the oven. Check samples for quality, and the baked-weight continuously. Control the further transport of the loaves, and avoid any blockages. Utilize a counting machine to record production data.

The main operations required with the wire-band tunnel-oven (Fig. 121) is control and adjustment, which demands an exact knowledge of the oven functions. However, this must be backed up by a quick reaction, application of the appropriate corrective procedures, and a sense of responsibility. The illustrations also show the graphical relationship between the optimal temperatures in the five heat zones of the BN 50 wire-band tunnel-oven for 1·5 kg rye/wheat bread, and the distance travelled along the baking-surface in metres (Fig. 123).

Another highly efficient, high-capacity, indirectly fired tunnel-oven is the Simplex 2000, manufactured by Baker Perkins (APV-Baker), Westfield Road, Peterborough, UK. This unit incorporates a recirculating heating system, matching available heat to heat requirements, and utilizes forced convection to increase the rate of heat-transfer to the product. Automatic heat modulation control burners with preheated combustion air, and microprocessor temperature control, ensure a consistent bake for panned-bread, hearth-bread, rolls, cakes, pastries, pies, etc. Available conveyor widths are 2·5 and 3·0 metres, with lengths up to 36 metres; lengths from 30 m requiring two heat-circulation systems. Average heat input when running is approximately 70% of the maximum demand on heating up, which is 44 00 kcal $m^{-1} h^{-1}$ for the 3·25 m band width, and 52 000 kcal $m^{-1} h^{-1}$ for the 3·85 m band width.

In any bakery, the production output is controlled by the capacity of its ovens. Oven output is usually expressed on an hourly basis, and is often referred to as 'throughput'. In the case of a wire-band tunnel-oven it can be calculated as follows:

$$\text{throughput (kg/h)} = \frac{60 \times L \times n \times W \times 1000}{d \times t}$$

where:

$L$ = length of baking-surface
$n$ = number of loaves in each row
$W$ = weight of each loaf (kg)
$d$ = loaf spacing (mm)
$t$ = baking-time (minutes)
$60$ = unit conversion factor: minutes to hours
$1000$ = unit conversion factor: mm to m

In the case of batch ovens, working on a discontinuous basis, the calculation of the throughput is more difficult, since additional time factors will include: loading, baking, further heating, product transfer, and unloading. Comparative measurements of throughputs using a multideck gas-fired cyclothermal system, and a double wire-band return oven, with total baking-surfaces of 9·6 m and 13·2 m respectively, indicate values of 144 kg/h for 1·5 kg rye/wheat loaves for the multideck, and 220 kg/h for the double wire-band return ovens. (Special processes in baking, W. Schwate and U. Ulrich, *Process Manual*, VEB-Fachbuchverlag, Leipzig, 1986.)

A useful estimation of the comparative 'effective productivity' using various ovens can be calculated as follows:

$$\text{effective oven productivity} = \frac{\text{throughput in kg}}{\text{number of operators} \times \text{hours of work}}$$

Using this relationship, any increase in effective oven productivity can be determined, and expressed as a percentage increase, by comparison with that of the operation of another oven system. The selection of the correct type and capacity of oven for any bakery is a difficult task, since so many important factors have to be taken into account. Since the oven determines the throughput rate of products, the personnel required, the quality of the bake, and its costs, an in-depth evaluation before making the final choice merits the utmost attention of the baker. The author suggests that the best initial step is to visit bakeries of similar size and producing a similar product range to that of the prospective buyer. In this way, the ovens can be seen in operation, and the following points observed and noted on the spot: (1) quality of the bake, (2) floor space demanded, (3) mode of operation, (4) operator costs, (5) installation and fuel supply, steam provision and burner exhaust flues. When the decision on the best type of oven has been arrived at, it is a question of comparing models from the various manufacturers for baking effect and reliability, as well as after sales service versus price.

For the industrial bakery, the priorities influencing choice of oven may be quite different. The oven which demands the most floor space is the tunnel-oven, but it is also the most straightforward in design, construction and operation, as well as being the most flexible for different pan sizes and hearth products. Furthermore, temperature control of the oven-chamber is made easier and more accurate by dividing the chamber into up to five heat-zones, and the application of computer-controlled thermostats. The lower oven crown facilitates steaming, and loading and unloading are located at opposite ends of the oven.

The single-lap tray oven has many of the features of the tunnel-oven, and its greatest advantage is that it requires about 50% of the floor area of a tunnel-oven of the same capacity. Heat-zoning is more difficult than with a tunnel oven, but top and bottom heat supply and steam, are comparable. Product loading and unloading can be done automatically at the one end, which could be a planning advantage. Where pan bread is the main product, the swing-tray type of oven could be the best all-round choice for a bakery with limited floor area. The double-lap tray oven is usually installed where floor space dictates, but its complex design and difficult zonal heat control, coupled with less efficient humidity control, make it a less attractive choice. Both these swing-tray ovens have been popular in the USA, and remain so, owing to economy in space and suitability for pan-straps.

A conveyorized oven system must usually be integrated with a conveyorized final-proofer. The grid pan conveyor carries the panned dough through the final-proofer and oven without any transfer breaks. Residence times in the proofer and oven usually are 60 and 20 minutes respectively. Apart from their simplicity of design, their main advantage is absolute uniformity of bake, owing to each product taking the same path through the oven chamber. Heat-zoning with these ovens is non-existent, but facilities for steaming are possible. Another point to consider with an integrated proofing/baking conveyorized system is the difference in size between the proofer unit and the oven unit. This is due to original design being for the conveyor speed to be uniform throughout, necessitating a longer proofer conveyor length. However, in some later modifications of Lanham's original system (Lanham, *Bakers Digest*, **44**(6), 54, 1970), the proofer conveyor and oven conveyor are powered by separate drives. The oven is an enclosure formed from insulated panels with steel frame supports, housing an endless conveyor arranged in tier or spiral fashion, the panned product entering through a small aperture at the top or bottom, depending on the plant layout. Its chief advantage is a uniform bake, achieved essentially by convection-heat. These ovens are fired directly with gas ribbon-burners, located directly under the conveyor-grids, grouped on individual gas and air manifolds, giving a degree of zonal control. However, because of the open interior, complete zonal control of temperature is very difficult, and when steaming is required, this has to be done by enclosing a length of the conveyor from the point of entry in a tunnel, and injecting steam into it. Temperature adjustment is more rapid than for the lap-tray ovens, and an air distribution and recirculation system with some venting of hot gases is provided. Throughput capacities of the larger systems exceed 16 000 lb/h for panned-bread production, and can be integrated with the Tweedy high-speed mixing system, forming a continuous mixing, proofing and baking production system (Tweedy-Lanham).

In view of the high cost of energy in all its various forms, ovens must be evaluated in terms of their relative energy utilization efficiency, as well as their design and operation features. Manufacturers of ovens have optional burner systems to offer that allow the use of alternative fuels to natural gas. Ribbon-burners fitted with oil vapourizers or nozzle burners performing atomizing functions, can utilize either gas or fuel oil as desired. The oil is vapourized by a heat source, and the vapour mixed with air for combustion in the correct ratio. Nozzle or gun-type burners used in recirculation ovens can readily be adapted for both gas and oil. These burners have a very wide flame adjustment range, and the stability of the flame has been improved (Flanagan, *Bakers Digest*, 52(2), 42, 1978). Direct firing remains the simplest and most energy-efficient method of heating ovens, being on average 15–25% more efficient than the indirect method (Craig, *Bakers Digest*, **51**(5), 131, 1977). Multiple banks of ribbon-burners, located within the baking-chamber, lateral to the direction of product movement on the conveyor, both above and below the baking surface, give reasonable control of top and bottom heat. However, such ovens are normally limited to a clean-burn gas fuel, and although exhaust flues limit the products of combustion and moisture within the chamber by the creation of free convection, such ovens are much improved by the introduction of an additional forced-air circulation system. This is due to the fact that the thermodynamics at the surface of the dough-piece are such that heat penetration can be greatly increased by rapid removal of the steam emerging at its surface as a result of air turbulence.

Indirect firing systems offer the options of gas, fuel oil and other fuels; no combustion-products enter the oven-chamber and, in addition, each burner-chamber forms an oven zone, making temperature control easier, especially with the help of fans, ducts and temperature controllers. Design constraints for the installation of radiant surfaces in ovens result in less efficient energy utilization in indirect-fired ovens compared with direct-fired units.

Current research into improved heat-transfer methods for baking are being directed towards the use of combined techniques, making use of VHF energy sources of about 17 MHz, and high-temperature conventional heating of various modes. With such techniques, very careful control of the mixed-mode is required to keep baking-loss within conventional, acceptable limits. The VHF mode offers a rapid bake-out, coupled with a rapid transition of sensitive enzyme-activity ranges of temperature, especially that of alpha-amylase. This could allow the processing of flours with hitherto prohibitively high alpha-amylase activity levels.

# 3.4 Control Technology and Energy Recovery

Any attempt to evaluate the amount of heat energy required to bake a fixed weight of dough gives rise to wide variations owing to such variables as dough structure and formulation, mode of bake, e.g. hearth or panned, temperature control gradients and rate of removal of moisture, which in turn, for the most part, involve oven design. Anderson (*Bakers Digest*, **40**(6), 60, 1966) found that by increasing the final-proof temperature, and lowering the baking temperature, thus limiting moisture loss, the energy requirement was reduced to 162·2 Btu from 256·6 per pound of bread under controlled conditions. Johnson and Hoover (*Bakers Digest*, **51**(5), 58, 1977), analysed total energy requirements for baking, and estimated the following Btu requirements per pound of bread. Oven start-up 77·4; heating during baking 77·0; evaporation of moisture 179·7; starch gelatinization 2·0; heating of pan 34·8; insulation loss 45·4, and flue-gas losses 122·4. These results are based on industrial data obtained from a daily production of 80 000 lb of bread, the total energy input for 1 lb of bread being 528·9 Btu. Therefore, the two main loss sources in the total balance are: moisture evaporation 34·3%, and flue-gas losses 23·2%, which together constitute more than 50% of Btu requirements.

Similar research by Anderson (*Bakers Digest*, **47**(4), 40, 1973), and Skarin (*Proc, Am. Soc. Bakery Engineers*, 1964, p. 88), but using an efficient direct-fired oven using gas, and good insulation with forced convection, concluded that heat inputs of 300 to 400 Btu/lb of bread were adequate. In comparing these sets of data, it should be accepted that indirect-fired ovens require an average of 20% additional fuel to allow for a lower efficiency in fuel utilization. Thus taking a figure of 400 Btu/lb of bread, and assuming a gas fuel efficiency of 85%, the actual heat input in net terms into the oven-chamber would be 340 Btu/lb. Of this amount, 250 Btu/lb would be used to bake the dough, 40 Btu to heat the pan, and the residual 50 Btu being lost in the form of exhaust-gases and through the oven walls.

## Burner Control

Direct gas-fired ovens are fitted with so-called ribbon-burners (Fig. 124), installed within the oven-chamber. These generate a strip of flame continuously along their entire length, which can be adjusted laterally with the help of a lateral heat equalizer. This latter equipment is capable of lateral flame adjustment at any level across the width of the oven. Fig. 124 shows the various lateral flame patterns produced by ribbon burners, fitted with a flame band distributor.

Indirectly fired ovens, or recirculating heating systems, use large, higher-capacity

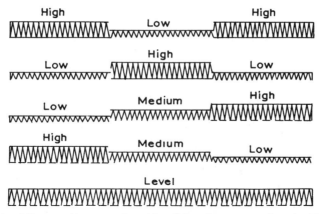

High　　　　　　　　　　　　　High
Low
High
Low　　　　　　　　　　　　Low
Low　　　　Medium　　　　High
High　　　Medium　　　Low
Level

Fig. 124.　Lateral flame patterns produced by ribbon-burners equipped with a flame band distributor (Flynn Burner Corp., USA). (Source: *Baking Science and Technology*, 3rd edn, E. J. Pyler, Sosland Publishing Co., Kansas, USA.)

burners, installed either singly or in pairs. These generate the hot gases of combustion in their own separate combustion chambers or boxes.

It is very important that the correct ratio of gas:air is maintained, to ensure correct combustion, resulting in the hottest possible flame, which is blue in colour and burns cleanly with no carbon deposition.

Using the older aspiration system, the gas is conveyed to the burner tube under a pressure of 1 to 5 psi. The design of the burner is such that the gas flow produces a venturi effect, drawing in enough air to form a combustible mixture. Manual heat control is then maintained on each gas control valve. However, since the amount of air drawn through depends on the gas flow-rate, at low settings, insufficient air is drawn in, and efficiency diminishes. The aspirator system also creates operating hazards, since the flame tends to blow out. A safer and more efficient system, involves mixing the air and gas in a zonal premixer unit, before it is fed through a

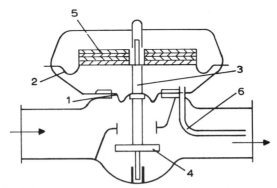

Fig. 125.　Principle of the gas-pressure regulator. 1 and 2. Membranes. 3. Valve axle. 4. Valve plate. 5. Loaded plates. 6. Air-pressure equalizer pipe. (*Source: Machine Handbook* (*Baking*), 2nd edn, B. Schramm, VEB-Fachbuchverlag, Leipzig, 1978.)

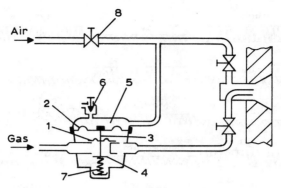

Fig. 126. Principle of the self-actuated gas/air mixture regulator. 1 and 2. Membranes. 3. Valve axle. 4. Valve plate. 5. Air-space. 6. Adjustable valve. 7. Spring. 8. Tap. (Source: *Machine Handbook* (*Baking*), 2nd edn. B. Schramm, VEB-Fachbuchverlag, Leipzig, 1978.)

common header to a number of ribbon-burners, which may be 15 or more (Flynn and Flynn, *Bakers Digest*, **39**(3), 63, 1965). Air at 1 psi pressure, after filtering, is fed into the mixer unit by blowers, and mixes with gas which has been reduced to zero pressure by a gas pressure regulator (Fig. 125) before being fed to the mixer. The air: gas mixture is then controlled by an air control valve (Fig. 126), which takes control of the burners. The air control valve is actuated by the temperature controller, which, on sensing a drop in oven temperature, signals the valve motor to increase the flow of air into the mixing chamber. Thus, more gas is drawn in, resulting in an increase in flame size and heat supply. When the oven temperature exceeds the preset temperature level, the air flow rate is correspondingly reduced, as is the flow of gas, thus automatically reducing the heat supply. The mixer unit is also provided with

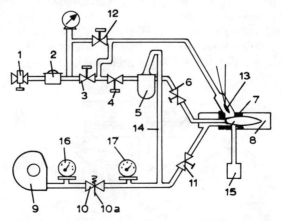

Fig. 127. Heat development system (gas). 1. Mains tap. 2. Pressure regulator. 3. Magnetic valve. 4. Cut-off tap. 5. Mixture regulator. 6. Tap. 7. Burner brick. 8. Burner pipe. 9. High-pressure ventilator. 10. Magnetic valve. 10a. Adjusting screw. 11. Setting-tap. 12. Magnetic valve. 13. Electrodes 14. Connecting pipeline. 15. Flame electrode. 16. Manometer. 17. Manometer. (Source: *Machine Handbook* (*Baking*), 2nd edn. B. Schramm, VEB-Fachbuchverlag, Leipzig, 1978.)

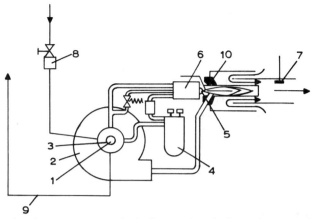

Fig. 128. Heat development system (fuel oil). 1. Drive-shaft. 2. Blower. 3. Oil-pump. 4. Oil pre-heater. 5. Atomizer. 6. Nozzle holder. 7. Photoelectric cell. 8. Tap. 9. Return pipeline. 10. Burner brick. (Source: *Machine Handbook (Baking)*, 2nd edn, B. Schramm, VEB-Fachbuchverlag, Leipzig, 1978.)

appropriate safety switches, which automatically release hazardous pressure levels, as well as gas lockout valves and either intermittent or continuous spark-ignition (Figs 127, 128). Where a higher level of control is required, separate proportional mixer units for each burner can be fitted, each with low-pressure air and zero-pressure gas lines.

In general, burner control systems can be classified into three types: (1) a simple on/off mode, in which the burners are fully on or off; (2) high/medium/low mode systems, whereby the air/gas controls vary between high and low flame settings, depending on heat demand; and (3) a modulating (proportional control) control valve to supply the correct amount of air and gas mixture to the burners, linked to a temperature controller to sense the temperature in the zone of burner operation, and signal when more or less heat is required.

An illustration of a typical modulation control set-up, actuating an electric modulating motor on the combustion air-line, is depicted in Fig. 95 and described in detail in Section 3.2. The controller has a mercury-filled thermal sensing element installed in the baking-chamber, causing an indicator on the instrument to move up and down the scale in response to the expansion and contraction of the mercury in the element. This in turn moves a contact along a potentiometer coil within the modulating temperature range. The potentiometer coil moves with a red pointer, which is set at the control point by turning a 'setter-knob' on the instrument cover. The potentiometer coil forms half of a Wheatstone-bridge circuit, and the other half of the bridge is formed by a potentiometer of similar electrical characteristics which is built into the proportioning motor, being driven by the motor-shaft. The proportioning motor also has a built-in detector relay, which detects any imbalance in the Wheatstone bridge due to a change in control temperature, then driving the proportioning motor in such a direction to regain a balanced bridge circuit. The motor-shaft is connected through a linkage to a device controlling the quantity of air entering the gas mixture valve. Such a temperature controller is accurate to less than 1·0% of the full scale, which allows oven-zone temperatures to be held within a few degrees of target setting on the instrument. Additionally, modern ovens are equipped with various safety devices, to prevent hazardous circumstances in the event of burner malfunction. For example gas or fuel oil may not ignite, because the pilot-light has gone out, or the electronic ignition has failed to provide an adequate spark. Blockages may occur in fuel lines, or flames may be extinguished by sudden draughts. Unless the oven is equipped with appropriate detectors and alarm systems, flame failures can result in underbaking of products.

The most fundamental safety measure is to have automatic shut-down of the fuel supply to the burner in the event of flame failure.

**General Objectives in the Application of Measurement and Control Equipment**
With increasing mechanization and automation of production processes, the application of measurement and control equipment becomes increasingly necessary. This ranges from simple indicating devices to more sophisticated monitoring, switching and functional control systems which confer a high degree of automation and accuracy to oven operation. Modern ovens are well provided with many types of control instruments to ensure safety and precision in their economic use of energy sources. Earlier electromechanical timers, switches and relays have now been replaced by solid state modules, and programmable control units, capable of performing all the functions of full automation through a central processor. Important criteria which influence the choice of appropriate control technology are:

—simple, centralized and well-organized for operation
—realization of an optimal control of the baking process
—simplification of repairs and servicing
—an increase in the degree of plant automation
—positive and reliable circuitry
—reduced energy consumption

As in other branches of industry, such objectives can only be made possible with the help of microelectronics. In the application of electronic control systems, the two basic possibilities are: control by information-storage programming, which allows the use of the same basic construction unit for various control functions, merely changing the programmes of their algorithms; or programming by wire, which involves working out a fixed programme by linking up specific groups of components. Since the control of ovens involves the use of a fixed functional sequence, programming by wire is the obvious choice.

The oven measurement and control system is divided into two areas: (1) the various equipment elements, viz. all recording instrumentation, fuel supply, and drive-motors for the wire-band and their speed control, coupled with a tachometer, and also the power supply for the bread-cooling plant, and (2) the control panel, which centrally controls the actual baking process.

Considering the oven equipment elements, these include the drive motors for the heating-gas circulation fan, and the steam ventilator. For the control of the correct function of the heating system, manometers are used to monitor gas pipeline pressures; an air-speed sensor controls the heating-gas circulation, and an excess temperature sensor safeguards the combustion-chamber; sensors within the combustion chamber and exhaust-gas vent perform the required temperature controls. The current tendency is to build a complete gas-burner in one unit, which includes the power units for gas and air regulation, ignition and flame controller units, and combustion air pressure monitor. The gas feed pipes together with the magnetic valves, and gas-pressure regulator are also built as a self-contained unit. The temperature sensors and illumination circuit for the various heat-zones are also grouped in one unit. Another unit houses the control mechanisms for the smooth running of the wire-band, which includes a revolution-counter monitor that detects any slackening of band-tension, signalling any slipping of the drive-drum. Mechanisms are also provided for the detection and signalling of any lateral wander of the wire-band, and equipment for integral brushing of the band conveyor. Since the grouping of these various block-units are both mechanically and electrically complementary, they can be assembled ready-wired at the point of oven installation, and connected by the use of clips attached to strips. This method of construction is also more efficient and economic in material.

The actual control and regulation of the baking process is performed centrally on the control panel, which is also built up from grouped electrical components. The measurement and read-out aggregate records the temperatures in the individual heat-zones of the oven-chamber, as well as the exhaust-gas temperature. The control-aggregate regulates the initial temperature of the heating-gas in the individual combustion-chambers, and incorporates a tachometer recording the baking-time relative to the velocity of the band-conveyor. The automatic control of the whole baking cycle proceeds from the control panel, but certain control functions can be made independent of the automatic cycle sequence.

A typical functional sequence for a wire-band oven with the control facilities described would be the following. On initiation of the automatic-start command, the first situation is acoustically activated. In order to activate the next situation, certain transition steps must be satisfied, which signal the correct function of previously actuated elements and power units. If one of these steps is omitted in time-order, an interruption of the programme sequence occurs, and an appropriate signal is given. With the aid of an appropriate logical interpretation, the interruption is reduced; then, depending on the source of the interruption, various cut-offs of the elements with appropriate signals take place, which facilitate the rapid detection of the source of the problem.

The ignition-programme of the burner is also carried out using the procedure described, allowing a minimum time period to elapse after switching on in the case of each situation, i.e. a fixed time-sequence for each successive situation. In this manner the burner ignition sequence takes place automatically, which makes the start-up operations considerably easier. A typical timed burner start-up programme is the following: Switch burner on; flush the unit for 3 minutes; switch to burner ignition

(30 s); check pressure (5 s); ignition (3 s); open main gas-tap (4 s); formation of flame, with complete ionization (7 s); programme complete, now ready for temperature settings. The same programme sequence is then followed for the other burners in the burner-block unit (usually two burners in each block).

It is also possible to switch the individual burners on and off, without interfering with the automatic programme function; a fixed period function facility can also be incorporated. In this case, the temperature-control mode is switched off, and the burner operated in the ignition situation. This facility is very useful for short breaks, e.g. change of shift, interruptions in the dough-fermentation schedule, etc. This avoids large heat losses due to re-ignition of the burner, and consequent flushing of the circulation.

This case study of some technical solutions with the application of micro-electronics for the control programme, illustrates the high degree of automation possible in bakery ovens. Summarizing, these include:

—Electrohydraulic powering of the flush-vent for the heating gas circulation system, which fixes the flush operation within the automatic burner start-up programme, thus eliminating manual operations.
—Introduction of the block gas-burner, which contains all power and control mechanisms for ignition, flame-maintenance and its control. This, combined with temperature regulators, which give an impulse-modulated signal, provide a continuous adjustment for the air and gas supply to the burner. This ensures a continuous control of the heating-gas temperature from the burner.
—To achieve the optimum heat-energy from the combustion process, the exhaust-gas vent is regulated in step with the air and gas supply to the burner. This makes sure that no more heat is lost from the exhaust-gas vent than is necessary, which reduces the gas consumption of the burners.
—To ensure a constant wire-band speed, and detection of any slippage of the drive-drum, or lateral wander of the band, the revolution counter provides a signal related to band-tension.

In spite of the increase in the number of hook-ups using microelectronics, the result is more exact than a traditional relay control system can offer. Also, the use of contact-free signal processing is more reliable, and wear-and-tear are almost eliminated. Savings in electro-energy, compared with non-electronic-controlled ovens are of the order of 20%. The use of the block gas-burner achieves savings of 16–24% in gas consumption, depending on dough formulation and product-spacing on the band, as well as on other factors.

**Standard Measurement and Control Equipment for Baking-ovens**
Basically, all measurement and control equipment perform a comparison of specific unknown physical parameters with known international calibrated 'norms'. If these are associated with a control-system, and the measured value deviates from the target-value, a related signal is passed to an appropriate control unit. The processing of this signal usually involves a conversion into either electrical or pneumatic units of measurement.

In the case of control instrumentation, the important question is not the measurement of an isolated value, but whether the measured values lie within the specified tolerance range. Control equipment frees personnel from routine control work and in certain areas can contribute to safe and hazard-free working conditions by linking to visual and acoustic alarm systems.

The illustrations show the principle and components of a gas-pressure regulator, and self-actuated gas/air mixture regulator for the gas-fired cyclothermal wire-band tunnel-oven Type BN 50, manufactured by Kyffhäuserhütte Artern VEB, in the GDR. Another illustration shows lateral flame patterns produced by ribbon-burners fitted with a variable flame-band distributor, as applied to direct gas-fired ovens with free convection heat distribution.

*Gas-pressure Regulator* (Fig. 125)
Gas pressure lifts the membranes (1) and (2), and the valve-axle (3) moves the valve-plate (4) upwards resisting the weight of the loaded plates (5). With increasing gas-pressure, the valve is raised, and falls as it reduces. The size of the valve aperture is fixed by the number of load-plates. For greater control accuracy, the air-pressure equalizer pipe (aspirator) (6) connects the two chambers, thus allowing an even pressure in spite of a changing inlet gas pressure.

*Self-actuated Gas/Air Mixture Regulator* (Fig. 126)
Owing to the gas pressure, the membranes (1) and (2), and hence the valve-axle (3) of the valve-plate (4) is raised. This movement compresses the air in the space (5), the magnitude being adjustable by means of valve (6). A preliminary setting is carried out making use of the spring (7), which carries the weight of the valve-plate, valve-axle and membranes. When the air-pressure in the feed pipe falls (e.g. by changes in the setting of the tap (8)), the gas pressure forces the valve-plate upwards, thus reducing the gas volume fed to the burner. Thus, the gas/air ratio fed to the burner is maintained at the desired constant level.

*Temperature Control System*
The temperature of the heating-gas is often measured by a thermocouple (nickel/chromium/nickel), which is hooked up to a control unit in the control panel. When the maximum of 650°C is reached, the magnetic valve on the high-pressure forced-air line falls, and thus closes the air line. However, by means of a spring-loaded adjusting screw, enough air to permit combustion on a low flame can be provided for on the forced-air line. This results in a fall in the heating-gas temperature in the oven-chamber, until the thermocouple on the two-point regulator of the control unit for the minimum tolerance limit gives its signal. At this moment, the magnetic valve opens again, allowing air to pass freely through the forced-air line. The temperature control is therefore achieved over the connecting pipeline to the air pressure within the gas/air mixture regulator, which controls the correct gas volume fed to the burner.

The main advantage of thermocouple measurement and control systems is that any number of thermocouples can be connected to the same instrument. However, for heat-zone temperature measurement, it is better to have one instrument for each zone, and record the temperature during the baking-cycle by means of a chart.

The most accurate temperature control for multi-zone ovens is obtained with electronic potentiometers, obtainable in several forms with various special features.

Temperature controllers associated with the burner control system are either of the electrical contact or pneumatic, narrow-band throttler type, depending on the type of burner in use. The mode of control can·be on/off, high/low or modulating. With the last system, already described in detail, the burner air control valve operates in rapid response to preset temperature limits, thus eliminating large temperature fluctuations.

*Tachometers*
Oven band speeds are measured and controlled by electric tachometers. These can be either direct-current or alternating-current generators, being connected to the oven band-conveyor drive shaft. Their output is measured with either a potentiometer or a voltmeter, and is usually calibrated in terms of baking-time in minutes. The selected three-phase motor-rpm driving the band determines the baking-time of the products. Any band-slip over the drive-drum can be detected by the resultant change in band-rpm by the tachometer, and an appropriate alarm signal given. This is important, since baking-time depends on band-rpm.

*Flame Control*
The lighting of ovens usually follows a programmed sequence to ensure that all elements of the ignition system are functioning properly. These are also part of the combustion safety precautions built into all modern ovens.

Instruments used for the detection of flame failures include flame rods, infrared and ultraviolet detectors, flame electrodes and photoelectric cells. These are all linked to various devices which will cut off the fuel supply. A shut-down of fuel supply may be triggered off by variations in the gas or fuel oil pressure, ignition failure, overheating in the combustion chamber or flue, or electrical power failure.

The most widely used instruments are the flame electrode and photoelectric cell. The principle of the flame electrode depends on the fact that a flame, owing to ionization, acts as an electrical conductor forming a circuit. As long as the flame is burning, a relay holds the burner feed valve in the open position, which reverts to the closed position, shutting off the fuel supply when fuel or power interruption occurs. The photoelectric cell and light sensitive sensors, depend on the closure of a relay as long as light-emitting flame prevails. If the flame is extinguished, the relay opens and actuates the relative safety devices, in a similar manner to the flame electrode. Re-ignition systems are fitted when the shut-off valve also functions as the temperature control system. The layout of the heat development systems of a typical tunnel oven using gas and fuel-oil is illustrated in Figs 127 and 128.

*Automatic and Computer Control Systems for Rack-Ovens*
In view of the rapid growth of the rack-oven for in-stores, freeze–thaw–proof–bake or 'bake off' operations, the optional control systems now available deserve

special mention. In the case of rack-ovens, the control systems are often integrated into the entire freeze–thaw–proof–bake cycle systems. In general the aim of computer control is to set automatically the optimum conditions for baking each product group. Typical factors which have to be varied are air flow and restart time after loading, since some products benefit from a delay in air flow restart for some minutes. The ability to vary oven temperature profiles during the baking cycle, steaming level and steam release is also desirable for many products. Rack-ovens equipped with heat exchangers of ample dimensions will ensure low exhaust-gas temperatures, thus making efficient use of energy-input. Use can also be made of exhaust-gas residual heat to preheat the combustion air, as applied to the industrial tunnel-ovens, and some multi-deck batch ovens. The rack-oven, Rototherm RE, manufactured by Werner and Pfleiderer, D-7000 Stuttgart 30, FRG, incorporates the rotating-rack design and the advantages of the block-building principle. Burner, fan and heat exchanger form one circuit, independent of the baking-chamber. As already described, the block-building unit design technique means that the electrics are supplied ready for plugging in. The installation electrician only has to make the mains connections via cable to the control panel. The burner, fan and heat exchanger can be heated up during the steam exposure-time. This reduces the temperature drop normally encountered when loading the oven, and thereafter continuous batch baking is possible. The heat exchanger, burner and steaming device are designed for continuous baking. The Rototherm rack-oven is fitted with an automatic baking control system as standard, but the computer control system can be programmed and then left to complete all the baking operations in sequence. When the loaded racks have been wheeled in the oven-chamber and the door closed, baking starts by selecting the programming and pushing the programme-start button. After the set steam exposure-time has elapsed, the air-circulation fan starts and the baking cycle proceeds until complete, which is signalled acoustically by a bell or a gong. The Rototherm is one of the most compact rack-ovens with the rotating turntable, giving a baking surface per 21 trays of 6·6 to 12·6 m², depending on the dimensions of the oven-chamber chosen. The power-rating of these ovens range from 60 to 105 kW/h for gas and fuel oil heating, and 42 to 60 kW for electricity.

**Energy Recovery**
Much research work has been done in recent years on heat-recovery systems, and most of this has been applied to industrial tunnel-ovens. However, before considering these techniques in more detail, it is appropriate to formulate some ideas on the most economic use of batch ovens with the aim of saving actual heat consumption within the smaller and medium-sized bakery. Some elementary considerations are the following:

—Ensure that the oven is consuming energy for the minimum time required only, by planning the production schedule to the most economical usage, i.e. make full use of falling temperatures for appropriate products, where a diverse range of products are being produced.

—Since multi-deck ovens have independently heated decks, they should be switched off immediately a deck is no longer required for the day's production. Oven insulation, however efficient, will not compensate for the energy necessary to maintain a deck at a certain temperature when empty.

—Where electricity is used, avoid periods of maximum demand tariffs by making full use of off-peak by staged switch-on. Control equipment making full use of electronic-based systems should be applied for this purpose. Negotiate terms for bake-schedule versus time.

—Always aim for maximum oven-loading, especially where pan-straps are utilized for bread. Strapping must be appropriate to the oven dimensions.

—Make full use of timers for baking-times to avoid opening oven doors, and ensure that steam dampers are closed after steam release, thus avoiding excessive re-heating, e.g. rack-ovens.

—Regularly maintain burners (gas and fuel oil) to ensure an efficient, clean burn, correct fuel/air mixture, and flame-profile. Failure to do so results in carbon deposition, and unburnt fuel losses with the exhaust-gases.

—Evaluate the cost of replacing an oven against the energy losses incurred with the existing oven, e.g. insulation efficiency, heat transfer and temperature control, quality of the bake generally.

Since all gas and fuel oil ovens have an exhaust-gas flue-stack for the removal of toxic gaseous by-products of combustion, a large volume of hot gases escape depending on the efficiency and dimensions of the heat exchanger. The more efficient the heat exchanger, the lower the temperature of the exhaust-gases.

In large industrial bakeries, where tunnel-ovens are standard plant, up to about 70% of heat can be recovered from the flue-gases, which can reduce heat-input to the oven by 5–6%. The first step in the design of a construction solution for such ovens is to measure the appropriate exhaust-gas parameters, viz. temperature, volume flow rate/hour, and weight/hour. Some typical data for an oven of average throughput 1000 kg/hour could be of the following order: temperature, 300 to 320°C; volume flow, 2000 m$^3$/hour; weight, 1220 kg/hour. These figures indicate that there is potentially a large amount of heat available for recovery. Heat-recovery units are of two main forms: (1) pre-heaters for burner combustion air; and (2) pre-heating of boiler water for steam, water-heating, space-heat, or proofer-heating.

*Burner Combustion Air Pre-heater*
The exhaust-gas duct is built into the flue-stack immediately above the burner-chamber, and exhaust gas is drawn through a heat exchanger with a counterflow system by an exhaust-fan. It then passes through an inlet-filter, from which it is fed by the combustion air blower to the burners. In the case of gas-fired ovens, a straight-through system is adequate, but for oil-fired installations using oil with a viscosity up to 38 mm$^2$ s$^{-1}$ at 50°C, a by-pass and wash-down system is necessary for periodic cleaning. Where fuel oils heavier than this are in use, this form of heat recovery cannot be recommended, owing to clogging by stack solids.

After prolonged use with controlled fresh air feeding of the burners, an exhaust-gas analysis must be carried out. Gas should give values of 10 to 12% carbon dioxide, and the carbon monoxide content zero. Such tests will give information concerning degree of burn of the fuels used. In the case of fuel oil, the deposition of carbon must be checked, and the content of sulphur-compounds determined. This is important to protect the oven from corrosion. If the exhaust-gas temperature falls below 180°C, which is a favourable trend from an energy-saving viewpoint, corrosion of the flue-stack and exhaust-gas duct sets in. Also, the actual flue temperature must not be reduced below the dew-point of the water vapour in the combustion gases, otherwise condensate is drawn back into the burner-chamber. Therefore, there are limits to the amount of heat recoverable from the burner-flue.

*Exhaust-gases for the Production of Low-pressure Steam*
This type of heat recovery unit is made up of the following functional elements: a steam-generator; preheater/presteamer; ventilator and exhaust-gas ducts. The steam generator is in the form of a low-pressure boiler, which is followed by a spiral of narrow pipeline serving as a heat-transfer unit, which acts as presteamer and preheater. This is followed by a medium-pressure radial ventilator. The exhaust-gas duct is connected to the exhaust-stack immediately above the burner-chamber. An integrated adjustable two-way vent allows the optional passage of the exhaust-gases either over the steam-generator or direct to the atmosphere. All functional elements of the plant are insulated throughout with 160 mm lagging. Electrical services effect an automatic control and adjustment of the steam-generator and the water-feed. When interruptions occur, the plant automatically shuts down, and the fault is indicated both optically and acoustically. The plant electrics are housed in a panel in front of the plant mounted on the plant-framework at eye-height. The whole plant is mounted within a framework on top of the wire-band tunnel-oven, being positioned between the second and third steam discharge vents. It serves as an independent auxiliary organ, and cannot cause any interruption in baking line function.

When the plant is in operation, the exhaust-gases from the burner-chamber are fed over the exhaust-gas vent, low pressure steam-generator, presteamer (evaporator), preheater, ventilator and additional exhaust-stack into the atmosphere. The low-pressure steam produced is conveyed through a system of pipelines into the steam-duct immediately in front of control valves of the steam-injection pipes. The steam-generator is fed with chemically treated water at 10–15°C, and can produce about 130 kg of low-pressure steam, at 20 kPa, per hour. This provides the 50 m$^2$ wire-band tunnel-oven with enough steam for baking, independently. The exhaust-gases become cooled down to about 120°C at the exhaust-stack exit. The plant requires minimal control, and this can be done by the oven-operator, owing to the automated control and adjustment system.

This type of exhaust-gas/steam building-block can be integrated into most types of band tunnel-ovens, rendering them self-sufficient in steam at low pressure and energy-effective in operation.

*Exhaust-gases for the Production of Hot Water*
Various designs are available, but the basic principle of their operation is a stainless-steel heat-exchanger coil, over which the exhaust-gases flow, water being pumped through the coil, thus extracting some heat from the exhaust-gases. Heat-exchanger

design varies considerably, e.g. air-blast, shell and tube exchangers, depending on application for condensing and evaporating functions. A crossflow system of heat exchange consists of a series of finned tubes through which the water flows, the warm exhaust-gases passing over the tubes and transferring heat to the water. Since the water is quite cool initially, condensation of water vapour contained in the exhaust gases occurs, thus some of the latent heat of evaporation is recovered. The amount of latent heat far exceeds sensible heat in the exhaust-gases, therefore relatively large quantities of hot water are produced. The maximum temperature of the water so produced is about 65°C.

Similar systems are designed for space-heating, preheating of boiler water and proofer-heating.

*Heat Recovery from Steam Liberated from Oven Feed and Discharge Ends and Steam Extraction Ducts*

This waste heat is at a lower temperature than the burner exhaust gases, and involves the processing of large volumes of steam-laden air. This makes any recovery system more complicated. The steam-laden air is collected from the steam extraction ducts and oven canopies, extracts the dust by filtering, condenses off the water vapour, and recycles the residual heat extracted for other uses. A wide range of information and advice is usually available concerning energy conservation schemes from government energy departments. The efficiency of boilers, baking ovens and any combustion process should be maximized by regular measurement of the four flue-gas parameters: oxygen, carbon monoxide, total combustibles and temperature. Analysers such as the Teledyne MAX portable unit, manufactured by Teledyne Analytical Instruments, Southall, Middlesex, UK, then calculate automatically net combustion efficiency; a theoretical calculation of carbon dioxide content is also provided. The MAX instrument has a three-digital LC display for all measurements and calculations. When connected to a printer, permanent data can be stored. Operating instructions are also stored in a permanent memory, and a temporary memory can handle up to 20 sets of data and all calibration settings. For *in situ* exhaust or flue gas analysis, the modular designed WGD *in situ* analyser manufactured by Thermox, and marketed by Fluid Data, Crayford, Kent, UK, is a state-of-the-art compact unit.

## Some Useful Instruments for Oven Heat and Humidity Monitoring in the Bakery
*Energy-loss*

For the detection of insulation defects, and insulation voids for the preventative maintenance of all heating equipment, the Emmaflex C-700C non-contact instrument measures both temperature, and heat flow (energy loss or gain). These instruments, with various measurement ranges are manufactured by Emmaflex, Milford, Stafford, UK.

*Non-contact Temperature Measurement*

Accurate measurement of inaccessible surfaces, and rapid scanning of surfaces for the location of hot-spots, can be performed with the Redpoint THI-300 infrared thermometer. Accuracy is $\pm 1\%$ with a 2 s response. Read-out functions are maximum/minimum, average, incorporating a dual Fahrenheit/Centigrade display. Read-out is digital, and the sensor probe is separate from the read-out unit.

### *Dew-point/Relative Humidity Measurement*
The full-blown, chilled-mirror dew-point meter, DP 383 R, marketed by Protimeter plc, Marlow, Bucks, UK., is a solid state instrument of robust design, with digital read-out of dew-point, temperature (ambient), and percentage relative humidity. The sensor probe and the read-out unit are separate, and it can be used for environmental control and monitoring in ovens and proofers and of water activity in foodstuffs, and for monitoring storage areas and production areas.

### Computer Monitoring of Energy Consumption per Process Unit or Plant
Computer software marketed by Stark Associates, Salfords, Surrey, UK, for the IBM PC, allows energy to be treated in the same way as any other raw material. By accurately monitoring the use of electricity, gas, oil and water throughout the production process, a more precise production cost calculation system can be established, both per process unit and in total for a production line. VDU plots of unit consumption per hour for electricity (in kWh), gas (in $m^3$), and water (in °C) are displayed for daily comparison. Information stored on the hard disc can be retrieved in time order as required.

### Mechanized Control of Oven-loading, Unloading and Depanning
In small bakeries with limited capacity, loading and unloading of ovens is still frequently done manually. Nevertheless, since this rather laborious task should be completed as rapidly as possible, it is being increasingly performed with the help of automatic loading and unloading devices (Fig. 129). These devices are known as 'setters' and have already been described. Such webbed bands are intended for hearth-baked products; for pan-straps another principle is applied. When the pan-straps with dough emerge from the final-proofer, they are moved by a cross-conveyor into position at the over-loaded platform. When enough straps have assembled to fill the oven width, or tray width in the case of a swing-tray oven, a pusher bar synchronized with the speed of the oven transfers the pan-straps onto a

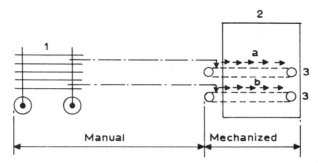

Fig. 129. Oven-loading set-up with a double wire-band return-oven. 1. Rack trolley. 2. Double wire-band return-oven. 3. Wire bands a and b for loading/unloading.

tray placed in the loading position, or onto the band-conveyor. This avoids the accumulation of pan-straps on the oven-feed conveyor. The pusher-bar can also be driven by a variable-speed motor, which allows slower movement on loading, and a rapid return to the starting position. This prevents damage to the friable, proofed dough and too many cripples. Even travel of trays in the oven during the loading of swing tray ovens can be ensured by the use of an interior load platform with hinges. This allows it to tilt downwards to accept the ascending empty tray, and follow its upward movement as it is being loaded with pan-straps. This avoids the need for stopping the trays to receive the pans. Actuation of the pusher-bar can be achieved by various mechanisms, which determine its return to the start position. In some cases, the bar moves to and fro in a horizontal plane, and in others the return movement involves lifting the bar over the pans back to the loading position. Another system moves the bar back under the loading-platform, and back up around the front of the loading plate. This set-up avoids the possibility of the bar descending or pans which are out of alignment.

Pan unloaders for tray ovens depend on either a tipping mechanism, or a pusher-bar. In the former case, the tray for unloading is tipped down just before it reaches the loading position, when the pans slide onto a discharge conveyor at a lower level, which transports them from the oven through an opening at the side. The oven tray remains level when the pusher-bar is transferring the pans onto the discharge conveyor. Uniform alignment of the pans on the discharge conveyor is maintained by intermittent stopping of the conveyor for the duration of pan transfer.

Bread pans are normally produced in sets strapped together, combining 3 to 5, but not often 6, the straps being rigid metal bands. The distance between each pan is normally fixed by spacer wires attached to the pans' rims lengthwise at regular intervals. The minimum distance between the sets of strapped-pans is fixed by lugs or offsets mounted on the actual straps. These are important for keeping the same spacing between the pan-strap sets at oven loading as that of each pan within the set, thus ensuring an even bake, and correct loading. The best material for baking-pans is aluminized or aluminium-coated steel. Aluminium-steel laminate is capable of withstanding the forming, shaping and seaming involved in their fabrication without any damage to the aluminium coating, which is about 25 $\mu$m in thickness. Aluminium gives the pan corrosion-resistance, and also makes very good pan material in its own right, finding wide application for cake pans and pie plates. Aluminium has a high thermal conductivity of 209·3 W m$^{-1}$ K$^{-1}$, is light in weight and corrosion-resistant, but is relatively expensive and needs to be strengthened with a backbone of steel, hence the wide use of aluminized steel. Stainless steel on the other hand, although tough and non-corrosive, has a lower thermal conductivity.

Originally, baking pans had to be 'burnt-in' or conditioned by first firing to over 200°C to in effect oxidize the tinned surface. Nowadays, pans are already oxidized or chemically etched. The majority of pans today are precoated with a high-molecular-weight silicon polymer resin. Such compounds, consisting of silicon and hydrocarbons show high thermal stability, electrical insulation, and releasing properties. Coated in this way, they will withstand about 500 bakes before requiring stripping and recoating with silicon.

In countries where hearth-breads predominate, industrial bread production (after a fermentation time of 38 to 70 minutes, depending on scheduled fermentation and baking-time, as well as oven-spacing), the dough-pieces leave the proofer and fall onto an endless conveyor-band, usually driven by a chain-drive from the proofer unit. This conveyor transfers the dough-pieces onto a bridge-band forming part of the oven-loading equipment, which is normally independently powered by a small drive motor of 0·8 kW and 125 rpm. After being loaded with a number of dough-pieces, the bridge-band passes in the direction of the oven, being often driven over a cable and pulley drive system from the motor. Shortly before reaching the wire-band of the oven, however, the lower track of the loading-band contacts a buffer stop, and moves the dough-pieces onto the band in the direction of the oven. The bridge-band moving on and over the wire-band of the oven. At this point, a switch is actuated by a stop-contact on the band, reversing the rotational direction of the motor. The bridge-band is then moved back by the cable. However, the band becomes held by a contact stopping mechanism, rolling it out and depositing the dough-piece on the oven wire-band. On moving back, the lower track of the bridge-band hits the buffer stop, and causes the band to roll back. Shortly afterwards, the bridge-band reaches another buffer stop, which shuts the motor off. On leaving the proofer and being deposited on the bridge-band, each dough-piece is cut in two places laterally at its extremities. This is carried out by a special cutting-aggregate, which is made up of 10 circular cutting wheels, mounted on a bar running the full width of the bridge-band. The height of this bar is adjusted to give a shallow or deep cut as desired, when the dough-pieces pass beneath it. The 10 cutting-wheels are driven by a separate motor. Often, another bar of band-width is used to impart a special embossed pattern on the dough-pieces. This is accomplished by pairs of ribbed wheels with embossed grooves, which are adjusted to impart a series of parallel lateral indentations on the dough-pieces as they pass underneath. Usually two such wheels serve each dough-piece, but the embossed profile can be changed by changing the wheels. No power is required to drive the wheels, as the band moves the dough-pieces forwards towards the oven under their path. The design and working principle of a typical oven-loading aggregate of this type is illustrated in Fig. 130.

*Depanning*
It is important that all panned bread is immediately removed from the pans on withdrawal from the oven, and allowed to cool before slicing and packaging. Formerly, depanning was carried out manually at the oven discharge apron by inverting the pan-straps, and striking the straps on a bumping-bar onto a table, then packing on racks to cool. This operation entailed fast, hot and hazardous work for those involved. Mechanical depanners were developed, but these were not a great success, since they could not cope with modern production rates, and frequently caused damage to both the product and the pans. Eventually the problem was overcome when vacuum depanning was developed. This technique involves the application of a vacuum to lift the baked product from the pan, thus causing minimum damage to both product and the pans. Where pans with lids are used, a magnetic lifter or rollers can either be placed in front of the depanner, or be built into

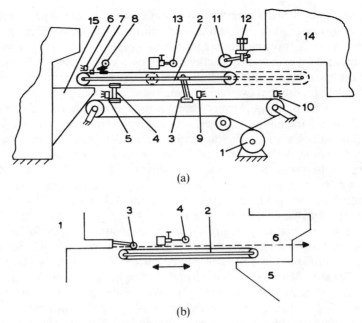

(a)

(b)

Fig. 130. (a) Oven-loading aggregate (final-proofer/wire-band tunnel-oven) for hearth-bread. 1. Motor. 2. Bridge band-conveyor. 3. Cable/pulley system. 4. Lower track rider. 5, 6, 9 and 10. Buffer-stops. 7. Switch. 8. Curved contact switch. 11. Cutter wheels. 12. Height adjustment. 13. Grooved roller wheels for embossed profile. 14. Final-proofer. 15. Oven feed. (Source: *Machine Handbook (Baking)*, 2nd edn, B. Schramm, VEB-Fachbuchverlag, Leipzig, 1978.) (b) Simple set-up for oven-loading (hearth-bread). 1. Final-proofer. 2. Transfer-band. 3. Cutter wheel. 4. Roller wheel for profile. 5. Oven feed. 6. Path of dough-piece.

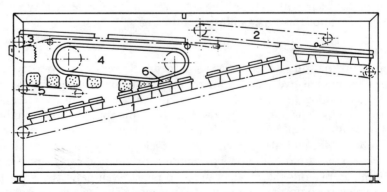

Fig. 131. Mechanics of a vacuum bread depanner. Components include: 1. Pan conveyor. 2. Magnetic lid conveyor. 3. Lid take-out conveyor. 4. Suction head. 5. Product take-out conveyor. 6. Air jets. (Source: *Baking Science and Technology*, 3rd edn, E. J. Pyler, Sosland Publishing Co., Kansas, USA.)

the front part of the depanner unit. Vacuum depanners vary in design, but a basic design incorporating a magnetic delidder is illustrated in Fig. 131. The lidded pans enter the unit on a conveyor above which are positioned powerful permanent magnets. These lift and hold the lids, then transfer them to a discharge conveyor. The delidded pans then proceed down a gradient on a flat-top conveyor, provided with a magnetic rail to hold them in place, towards the suction belt. This belt is fitted with a series of vacuum cups made of silicone rubber, but before reaching this vacuum belt, powerful cold air jets are directed between the pan walls and the baked loaves to facilitate separation of the loaves from the pans. The suction cups then lift the loaves from the pans, and transfer them onto the take-out conveyor to the cooler. All depanners have to be designed to suit specific products, i.e. bread, rolls and buns, and the equipment must be adjusted correctly, and also the pans must be treated with an efficient pan-release agent. The depeanning speed averages about 250 loaves/minute, and about 30 bun pans/minute.

The empty pans are removed from the unit by a conveyor, returning either to the moulder/panner area or to pan storage. In any event the pans must cool down to about 40°C for re-use at the moulder.

It is also possible to combine the functions of depanning and packaging in one unit for rolls and buns. One side of the unit accepts the cupped pans holding the rolls, the suction cups on the vacuum mechanism being aligned to coincide with those of the pan. On the other side the cartons or trays are placed to receive the rolls. The vacuum head sucks the rolls from the pan in any desired block number from 9 to 20, lifts, then moves over the empty carton depositing the rolls into it; the empty pans and filled cartons are removed by the conveyor; then another roll-filled pan and empty carton is conveyed in position for the next automatic transfer. Similar concepts have been applied in the robotic loading of trays and transport containers. Vacuum actuated heads move bagged loaves or packs from despatch lines in numbered blocks laterally, depositing them into the transport containers in a predetermined pattern. Such operations are fully automated, various menu patterns being programmed into a system for adaptation to container dimensions, all at the push of a button.

More recently, fully automated units have been designed for the transfer of products to the oven sole, known as 'oven sole setters'. These are dual-purpose machines, transferring both hearth products, e.g. rolls, sticks, variety-bread and buns; and backing sheets and pan-straps of specific dimensions.

Microprocessor oven-controllers with VDU and keyboard operator interface can be fed with oven-baking programs for rapid and accurate adjustment of oven temperature profiles to suit different products.

# 3.5  Bread Cooling and Setting

In general, the maturation of fermented baked products, which involves both the processes of cooling and setting, can take place in the dispatch area or during transport. For cooling in the dispatch area, adequate fresh air is essential, preferably making use of forced-air turbulence, such that the loaves can cool down gradually to about 35°C. Actual transportation is not recommended until loaves of bread of 1 kg and over have attained this temperature, assuming they are not for slicing, and packaging. During the cooling of bread there is always a potential hazard of microbial infection, whether in the small craft bakery with a short distribution chain to the point of sale, or in the large industrial bakery where distribution chains are always much longer. The main cause of microbial infections is the inevitable high humidity and incidence of condensation. All storage and dispatch areas must be kept clean, well-aerated and free from any contaminating foreign smells. Where forced convection is used, the air intake should ideally be filtered before entering the cooling area. Other aids to general hygiene in these areas are: the presence of UV-based radiation; insect-o-cutor' (Insect-o-cutor Inc., Georgia, USA) discharge units, for the control of insects; climatic control (temperature/humidity) and ozonization plants.

Traditionally, loaves were cooled by placing them hot from the oven on shelves of stationary racks, and simply exposing them to the ambient air temperature until their internal temperature had reached acceptable levels. This method of uncontrolled cooling is still widely used in small bakeries, but is time-consuming, and requires extra space and labour. Furthermore, the rate of actual cooling is non-uniform, owing to variations in weather and climate. In such circumstances, depending on the type of product and storage conditions, cooling would take between 8 and 24 hours.

However, for a product such as bread-rolls, it is essential that they reach the consumer within about 8 hours of baking. Ideal storage conditions for bread rolls would be about 22°C and 70% relative humidity, away from strong daylight.

Larger units such as loaves of bread must not be reduced in temperature too rapidly, since the condensation of the internal water vapour into localized smaller volumes of liquid can cause shrinkage and loss of quality.

Industrial bakeries utilize a process of controlled cooling, which ensures optimal product maturation after removal from the oven. For panned bread the optimal conditions are a temperature within the range 20 to 24°C, a relative humidity of around 85%, and an air-speed adjusted to produce a temperature rise above ambient at the exhaust point of approximately 10°C. Such conditions will allow a loaf of 500 g to cool to 40°C within the crumb during a residence-time in the cooler unit of approximately 1 hour.

## 3.5.1 TYPES OF BREAD COOLER

**Continuous Conveyor or Belt Coolers**

The introduction of this cooler design represented a major advance in the efficiency of the cooling process and large savings in floor space, since they could be suspended from the ceiling. The basic design consists of endless multi-tier overhead conveyors, travelling in straight stretches with U-turns at the end of each stretch. The cooling cycle is on average 60 to 90 minutes, and the number of cycles applied will depend on the type of product. Products formulated for a high specific volume (low density) will cool more quickly than leaner, more dense products, e.g. rye- and variety-breads. Sometimes these units are open all round, with an overhead fan to extract the radiant heat, and others are enclosed with panels to form a tunnel, a countercurrent air turbulence being used to accelerate cooling. The actual conveyor consists of either a wire-grid or metal-rod capable of flexing as required during travel. In addition to the popular overhead racetrack belt path design, spiral designs with either single or double helices are possible. In the case of the double helix design, the outer helix forms the ascending path and the inner one the descending product path. The height of these coolers can vary from about 4 up to 25 tiers, with track belt widths of 18 to 32 inches. In addition to the conservation of floor space, these double-helix or tower-in-tower designs allow greater flexibility in conveyor belt length to accommodate diverse cooling-time requirements. An example of this type of system which is applied in the GDR, is the Universal Transport System 'Regotrans', which can either be floor-mounted, or be suspended from the ceiling. This system is a wire-grid type conveyor, which finds general use for the transport of products and merchandise in the baking industry. The curve sections can provide an angle of 45 degrees, giving an outer radius of 1249 mm, and an inner radius of 708 mm. The maximum length of the straight stretches can be 19 500 mm, and the minimum length of 2600 mm, with a width of 578 mm, giving a usable width of 500 mm. The straight lengths can be used for ascending and descending slopes, the thickness of grid conveyor being 130 mm. The Regotrans system is very adaptable, and can be floor- or ceiling-mounted on a frame. The speed of travel is variable, and can be adjusted to suit any application. In the GDR, in order to satisfy bread slicing temperatures of about 40°C, the following cooling/maturing times are necessary: rye, rye/wheat, and wheat/rye breads 3 hours; wheat bread and toast bread 2 hours, according to Schneeweiss and Klose (*Industrial Baking Technology*, VEB-Fachbuchverlag, Leipzig, 1981, in German). The Universal 'Regotrans' system is applied generally in the GDR baking industry for the transport of a wide range of products in line production. In the USA, these conveyor coolers are manufactured extensively by Stewart Systems. In the UK, the average cooling time in the bread cooler is about 3 hours, the cooler temperature being kept at 20°C, and the relative humidity above that of the surrounding atmosphere.

On leaving the oven, the drop in temperature of a loaf of bread is quite rapid and uniform irrespective of atmospheric conditions. Almost the entire heat loss due to radiation is complete within about 40 minutes, the loaf surface having cooled to ambient temperature. However, the cooling of the crumb is far from complete, and unbound moisture within the crumb migrates slowly towards the crust. This moisture must be allowed to escape to avoid internal condensation and possible spoilage. Initial radiation losses account for only about 50% of the total moisture loss taking place during cooling, the residual 50% will take about 2–3 hours under normal circumstances, i.e. without any changes in the external environment.

### Tray Coolers

In this case, the conveyor is replaced by trays in the form of stainless steel grills mounted in steel frames. These frames being attached to parallel roller chains, drive sprockets being positioned at the conveyor loops. The loaves from the depanner have to be grouped and aligned for transfer onto the trays. Often, this type of cooler is fitted with high-pressure sanitizing facilities for use with hot detergent and water rinsing, as required.

### Rack Coolers

This operates in a manner similar to the rack-proofer. Since the loaves are usually densely packed in this type of unit, aimed at saving space, forced air circulation is desirable to expedite the cooling rate. In this case the racks must be enclosed with panelling. Where the process relies on natural convection currents, the sides are left open. The loading procedure is the same as with the tray cooler, at the loading/ unloading station a rack with about 10 or more shelves is dispatched upwards until all the shelves are full. During this time, the rack reaches the uppermost runway and proceeds towards the back of the cooler, leaving space for subsequent racks. On reaching the end of the upper runway, the loaves are then lowered to the bottom runway, returning to the loading/unloading station for removal. The vacant shelves are reloaded with more hot loaves. This type of cooler can be a solution where space is limited.

### Vacuum Coolers

The principle of this relatively new idea is the application of a modulated vacuum system, which can give a rapid reduction in the temperature of the product. The cooler consists of a conveyor, which runs through a vacuum chamber tunnel, both ends of which are provided with vertically operating doors, forming an air-tight seal when closed. A bridge conveyor is applied at the inlet and discharge ends to transfer the product into and out of the chamber. The vacuum system is made up of one or more vacuum pumps, which remove the gases and water vapour from the cooler tunnel and create a vacuum. The gases are entrapped by steam which is also injected into the tunnel. These then enter a water-jacketed condenser, where the water is condensed off, and the more volatile gases become drawn off by the pumps. Moisture removal from the baked product is very rapid under the influence of the partial vacuum within the cooling tunnel, which in turn results in a rapid decrease in product temperature. For example, bread rolls can be reduced from 70°C down to about 40°C within about 1 minute.

Since moisture losses are much higher than is the case with the atmospheric cooling methods described, the baking time of products to be cooled by this method should be reduced by 15–20% compared with standard baking times. This reduction being best applied at the final baking stage, when adjustments to crumb moisture can be achieved more easily.

The application of a vacuum to hot products results in moisture removal at a much more rapid rate, and that loss of moisture is more evenly distributed throughout the crumb of the loaf. The effect of the vacuum being analogous to that obtained during the freeze-drying of products. Therefore, the baking and cooling processes must be synchronized to give the desired final crumb moisture in the bread, and ensure an adequate shelf-life.

### 3.5.2 MOISTURE MOVEMENTS DURING THE COOLING AND MATURATION OF BREAD

On removal from the oven, a loaf of bread loses heat to the surrounding atmosphere, and simultaneously loses weight due to moisture transfer to it, depending on the atmospheric saturation surrounding it.

The temperature of the crust of the loaf on withdrawal from the oven falls within the range 130 to 180°C, and at the interface between crust and crumb the temperature reaches at least 100°C, at which moment in time the moisture content of the crust has been reduced to almost zero. Although the temperature of the crumb reaches 98 to 100°C, its moisture content exceeds the initial dough moisture by about 12%.

On placement in a bread storeroom or cooler at about 18 to 24°C, without forced convection, the loaf of bread begins to cool rapidly and loses weight owing to loss of moisture. This cooling process commences at the surface layers of the loaf, and gradually extends into the centre of the bread crumb.

During bread cooling there exists a considerable temperature gradient between the crust and crumb, e.g. 14°C, at the initial stage. With progressive cooling down this gradient is eventually reduced to zero. On the other hand, the moisture or humidity gradient becomes rapidly reduced during the first few minutes after withdrawal from the oven, but shows little change during further cooling and storage. One of the main factors responsible for the rapid loss of moisture during the initial cooling and storage period, apart from the surrounding temperature, is the high crumb temperature, which creates the temperature gradient between crust and crumb. This temperature gradient gives rise to the moisture migration or movement towards the crust. On reaching the temperature of the store surroundings, this gradient is reduced to zero. At this point, the thermal/moisture conductivity, and rapid moisture loss cease, in spite of the fact that a considerable moisture gradient between crumb and crust still exists. However, it is not only the thermal/moisture conductivity that is responsible for the acceleration of the moisture loss during the cooling process. This is quite clear from the fact that the rate of moisture movement, depending on its concentration, and triggered by the moisture gradient, also depends on temperature. The higher the temperature of the product, the more rapid the concentration-dependent moisture diffusion proceeds, which is important in the case of products like bread where moisture-diffusion can prove difficult. Another obstacle to the rate of moisture loss during bread cooling is the thin boundary layer of still air, which surrounds the external surface of the baked product. This thin layer impedes the free external diffusion of moisture into the atmosphere, depending on its temperature. This same phenomenon is well known in the technology of drying generally, and in such cases measures are taken to reduce the thickness of this boundary of still air by using turbulence, so increasing 'external moisture diffusion'. This diffusion process, at constant evaporative surface and external air speed, will depend on the difference in partial pressure of the air which becomes saturated owing to the product temperature, and that prevailing in the air immediately circulating around the product on the other side of the boundary layer. Partial pressure of the steam increases considerably with a rise in temperature of the product. After prolonged storage, the temperature of the cooling loaf of bread falls slightly below that of the surrounding bread-store. The reason for this is that the moisture evaporation from the loaf continues after the loaf temperature has assumed the bread-store temperature, in spite of some delay in its onset. The necessary heat for the evaporation process stems from the crumb zone that is adjacent to the crust, and not from the air which is separated from the crumb by the crust. The thermal-conductivity of the crust is much less than that of the crumb (Auerman, *Sowjetskoje mukomolje i chlebopetschenija*, **3**, 12, 708, 1929).

Figure 132 shows the temperature changes within the various layers of white bread during cooling after removal from the oven; however, this refers to a single loaf allowed to cool under ideal test conditions. Under production conditions,

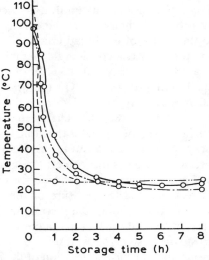

Fig. 132.   Temperature changes in the various layers of white bread during cooling after removal from oven. ——, temperature at crumb-centre. —·—, temperature in crumb layers adjacent to crust. ----, temperature of the crust. —··—, ambient storage temperature. (Source: *Technology of Breadmaking*, L. J. Auerman, Moscow, 1972 (in Russian); Leipzig, 1977 (in German).)

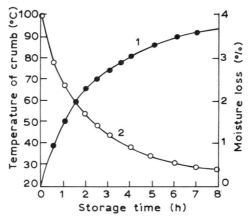

Fig. 133. Moisture loss and temperature changes at the centre of the crumb of rye bread during storage under production conditions. 1. Moisture loss. 2. Temperature of crumb. (Source: *Technology of Breadmaking*, L. J. Auerman, Moscow, 1972 (in Russian); Leipzig, 1977 (in German).)

without a cooling-plant providing controlled conditions, the rate of bread cooling is prolonged considerably. Figure 133 shows the moisture-loss and temperature changes taking place in rye bread of 1·5 to 1·9 kg made from wholemeal rye-flour, baked in pans. The data used for this graphical interpretation stems from production conditions, derived from the average of 120 trials conducted at five production centres in the USSR by Professor Auerman and co-workers at the Moscow Technological Institute for the Food Industry (MTIPP). Under the conditions of this investigation, the temperature of the bread-stores varied from 19 to 27°C, and the average crumb moisture of the bread was 53%. All the graphical illustrations (Figs 132–137) concerning 'Bread cooling and moisture-loss' originate

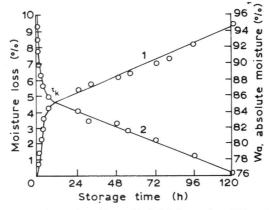

Fig. 134. 1. Drying out moisture loss ($\Delta g = ft$). 2. Moisture loss ($Wa = ft$) if bread stored 5 days after baking. [1] Moisture percentage based on weight of total solids. (Source: *Technology of Breadmaking*, L. J. Auerman, Moscow, 1972 (in Russian); Leipzig, 1977 (in German).)

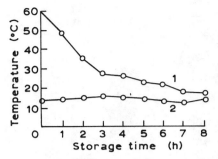

Fig. 135.   Comparison of air temperatures between loaves stored. 1. Boxes. 2. Bread-store.
(Source: *Technology of Breadmaking*, L. J. Auerman, Moscow, 1972 (in Russian); Leipzig,
1977 (in German).)

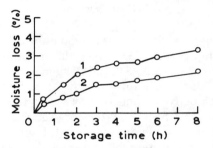

Fig. 136.   Comparison of moisture-loss between loaves stored. 1. Boxes. 2. Bread-store.
(Source: *Technology of Breadmaking*, L. J. Auerman, Moscow, 1972 (in Russian); Leipzig,
1977 (in German).)

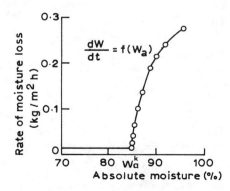

Fig. 137.   Rate of moisture loss versus absolute moisture percentage during bread storage.
(Source: *Technology of Breadmaking*, L. J. Auerman, Moscow, 1972 (in Russian); Leipzig,
1977 (in German).)

from Professor Auerman's reference book entitled *The Technology of Breadmaking*, first published in Russian in 1972, then published in German by VEB-Fachbuchverlag, Leipzig, GDR, in 1977, for the benefit of students and scientists in that country. This reference work is to be recommended for all students of baking technology in the English-speaking world, especially for its 'in-depth' treatment of most technological aspects, including bread baking and cooling. Immediately on withdrawal from the oven, bread starts to lose its moisture content, but simultaneously a movement or migration of moisture takes place within the loaf itself. For example, the crust is practically moisture-free on withdrawal from the oven, but since it cools rapidly, the moisture within the crumb migrates towards the crust owing to the concentration and temperature gradients which are set up. Therefore, its moisture will increase on cooling by about 12% to 14% within about 2 to 4 hours storage, depending on the weight of the bread, temperature of the storeroom and storage conditions. This moisture level of 12% to 14% for the crust, which approximates to its equilibrium relative humidity value (erh), remains constant during the remainder of the storage period. In contrast, the moisture of the breadcrumb continues to fall during storage.

After baking, the crumb layers immediately next to the crust have a moisture level slightly above that of the centre of the crumb. This is due to both the thermal/moisture conductivity away from the area bordering on the crust, and the migration of some of the water vapour away from the evaporation zone and into the cooler crumb layers, where it condenses.

Within the first hour of storage, the moisture level of crumb layers bordering on the crust falls, owing to moisture migration to drier areas of the crumb, and subsequent evaporation to the air. In this way, the outer layers and the internal crumb layers assume an equilibrium moisture level, which is about 1·5% lower than that immediately after baking. During further storage, the crumb layers nearest the crust lose their moisture much more rapidly than those near the centre of the loaf. Prolonged storage over several days results in the crumb layer immediately below the crust becoming quite firm. According to research work at the All-Union Research Institute for the Baking Industry (WNIIChP) in Moscow, the moisture of these layers near the crust fall from 44% after baking, to 18% after 3 hours storage; the thickness of these layers reaches 4 to 5 mm.

The conclusions which can be drawn from the research work described are:

—Temperature and temperature gradients during the cooling of bread are important factors in determining both the moisture evaporation by diffusion through the surface, and the water movement or migration by thermal-diffusion and concentration differentials within the bread crumb. These in turn determine the rate of the loss of moisture.

—When the bread acquires the temperature of its surroundings, the temperature factor ceases to affect moisture loss, and it therefore proceeds at a reduced rate.

—For studies concerning the moisture loss in bread, one can utilize graphical plots applied to drying-technology, such as percentage weight loss, and drying-rate curves. The drying-rate either being expressed as a function of initial moisture or the drying time. For examples of such curves see Figs 133–137.

—During the natural cooling process, the rate of moisture loss falls rapidly with time until a constant level is reached, as shown in Fig. 137, the point $t_k$ being easily read-off from the curve, as shown. Normally, this point coincides with that at which the bread temperature reaches ambient room temperature.

—The total time period of moisture loss can be divided into two phases, the first phase being $t_0$ to $t_k$, and the second where $dW/dt$ remains constant. Thus any shortening of the first phase will reduce evaporation moisture loss. This is best achieved by a rapid cooling of the bread from the oven down to ambient bread store temperature.

## Influence of Air Temperature, Relative Humidity and Air Speed on the Moisture Loss of Bread

### Air Temperature

The temperature of the bread store has a great influence on the rate of cooling and moisture loss. The lower the surrounding air temperature in the bread store, the more rapidly bread surface temperature falls, thus shortening the first phase of moisture loss, within which the rate of moisture loss is greatest (see Fig. 138). A low air temperature also slows down the moisture loss during the second phase, when the rate of moisture loss remains constant. The lower the air temperature, the lower the bread temperature during the second phase, and consequently the lower the

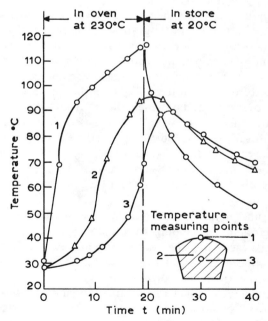

Fig. 138.    Temperature profile of loaf during baking and after removal from oven to cool at 20°C. (Source: *Technology of Breadmaking*, L. J. Auerman, Moscow, 1972 (in Russian); Leipzig, 1977 (in German).)

partial pressure of the water vapour at the surface, thus reducing the moisture loss from the bread. In general, the higher the temperature of the surrounding air on withdrawal of the bread from the oven, the more intensive the moisture loss (see Fig. 138). Auerman reported in 1932 (*Weight Loss of Bread during Cooling and Storage*, Moscow, Snabtechisdat) that the moisture loss of bread increases with increasing temperature of the surrounding air on removal from the oven. For example, the moisture loss of Ukrainian bread of 1·2 kg after 8 hours storage at 43–50°C amounted to 5%, and at 11·5–19°C only 2%. The storage of bread in the frozen state reduces moisture loss to a minimum.

*Relative Humidity*
Moisture loss is reduced with increasing relative humidity, owing to its effect on the evaporation process at the surface of the loaf. High humidities result in smaller differences in the partial pressure at the loaf surface compared with that of the air, thus reducing moisture loss rate. However, this effect of humidity is not very significant during the first phase of cooling. The higher the bread temperature, the greater the partial vapour pressure above the surface of the loaf, the difference between this and the partial vapour pressure of the air being too small to be of significance. During the second phase of moisture loss, when the temperature of the bread does not exceed the ambient temperature, the influence of the relative humidity on the rate of moisture loss increases. However, for normal cooling times of 3 to 6 hours under production conditions the second phase of moisture loss is relatively short.

*Air Speed*
During the first phase of bread cooling and moisture loss, it is expedient to surround the loaf with turbulent forced air at a speed of 0·3 to 0·5 m/s. This will speed up the cooling process, and effect a shortening of the first phase of moisture loss. Such conditions will simultaneously reduce the weight loss of the bread due to evaporation. Fig. 139 shows the influence of various air speeds at 20°C on the moisture release of the bread. The graphs show the effect of increased air speeds on the cooling rate, reducing the moisture loss of the loaf by about 0·5–0·7%. This data was obtained by Professor Auerman at the WNIIChP Research Institute in Moscow; and further investigations in the laboratory, using an air-conditioning plant based on an industrial plant, confirmed the advantage of such a cooling system for accelerating the cooling process and reducing moisture losses by about 0·5% to 0·9%.

**The Influence of Loaf Moisture and Baking Loss on Moisture Loss due to Evaporation**
Work done at the WNIIChP in Moscow has shown that the higher the loaf moisture, the greater the weight loss under identical conditions. For example, an increase in the crumb moisture content of rye bread made from wholemeal flour by 2%, results in an increase in the moisture loss of the bread over 4 hours of 0·26% to 0·42%, and over a period of 7 hours of 0·42% to 0·50%. These researchers further established that between the baking loss and the moisture loss of the bread due to

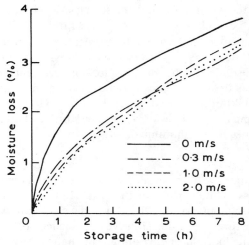

Fig. 139. Influence of air-speed on the moisture loss of bread during cooling. (Source: *Technology of Breadmaking*, L. J. Auerman, Moscow, 1972 (in Russian); Leipzig, 1977 (in German).)

cooling and evaporation there exists a reciprocal relationship. The greater the baking loss, the lower the loss due to cooling and evaporation, and vice versa.

### The Influence of Baking in Pan or on Hearth and Loaf Volume on the Moisture Loss of Bread

It is well known that hearth-baked breads show a greater baking loss and a lower crumb moisture than pan-baked breads of the same weight. Therefore, the moisture loss due to evaporation in the case of panned-bread is generally greater than that of hearth-breads. The data of Table 70 was obtained by Auerman in the case of rye bread made from wholemeal flour, weighing 1·1 kg, baked in pans and as round hearth-bread.

The internal structure of the dough, which will depend on the raw materials

TABLE 70

| Parameter (%) | Mode of baking | |
|---|---|---|
| | Panned | Hearth |
| Crumb moisture | 52·8 | 49·2 |
| Baking loss | 11·6 | 13·3 |
| Moisture loss | | |
| after 4 hours | 2·03 | 0·86 |
| after 6 hours | 2·19 | 1·04 |

Source: *Technology of Breadmaking*, Auerman, VEB-Fachbuchverlag, Leipzig, 1977, p. 286 (in German).

included in the formulation and their specific properties, determine the final texture and porosity of the bread crumb. Breads with a high specific volume and low density, due to their open crumb textures show greater moisture losses due to baking and cooling. Typical examples of this type of product are enriched formulations, containing NFMS and hydrolysed sugars, or those containing higher levels of gluten for low-calorie bread. The leaner and denser formulations such as rye breads and other breads based on dough-sours show appreciably less moisture loss due to both baking and cooling. The relative dimensional size of the loaf also determines the moisture loss during baking and cooling. Generally, the larger the loaf the lower the moisture loss.

### 3.5.3 BREAD QUALITY CHANGES DUE TO STORAGE

On removal from the oven, the temperature gradient which exists between crust and crumb is about 32°C, the crust being at about 130°C and the crumb about 98°C. At this stage the crust is dry and crisp, but during the first few hours of storage moisture migration from within the crumb results in the crust moisture increasing by 12–14%, as mentioned earlier. This movement of moisture gives the crust a tough, leathery and elastic character. These properties become extreme when the crust is thin and very moist. Longer storage of unpackaged bread results in the crust again becoming dry and flaky, which eventually also spreads to the crumb layers under the crust. In the larger loaves of 1 kg or more, the bread crumb takes much longer to cool than the crust, and after about 1 to 3 hours after removal from the oven, the crumb centre often still has a temperature of 50 to 60°C. Therefore, during this period many processes taking place during the baking process will continue. This is found to be particularly the case with the crumb of rye breads made with wholemeal flours, where the contents of sugars, dextrins and total water soluble carbohydrates, amongst other substances, increase. In addition, the gelatinization of starch and crumb stickiness increase. These observations were first made by Gogoberidse, Auerman, and Stscherbatenko in 1956–58, and reported in publications of both the MTIPP and WNIIChP in Moscow. The author can confirm these observations in rye bread, particularly where excessive sprouting has already produced an accumulation of amylase degradation products within the bread crumb during baking. It must be stressed, however, that these observations are not due to the amylases, which are completely destroyed during baking (cereal amylases), but rather by acid hydrolysis of the starch, and only proceeds as long as crumb temperature remains at 60°C or over. During the heat sterilization of bread to prolong shelf-life, in which case the centre of the crumb reaches temperatures of up to 85°C, Romanov (*Bread-storage*, Moscow, Pistschepromisdat, 1953) also reported an increase in crumb sugars and total water-solubles.

Changes in the composition and properties of wheat bread during storage in polyethylene packaging material have been reported by Gasiorowski and Jankowski (*Brot und Gebäck*, **21**, 7, 137, 1967). Storage for periods of up to 15 days at temperatures of −23 to +65°C were utilized to study the effects on the following bread parameters: moisture of the crumb and crust; starch, soluble starch; dextrins with a defined chain-length contained in the crumb; also the reducing sugars and total sugars in the crumb. Additional analyses carried out were: titratable acidity, pH, X-ray spectrum, and properties such as colour, compressibility and swelling. Sensory evaluation of freshness, flavour and aroma of the bread were also included. The results showed that the crumb moisture, its compressibility and swelling capacity, as well as the contents of sugar, soluble starch and dextrins, are all reduced during storage. However, the starch content of the crumb increased. The rate at which these changes took place were greatly affected by temperature. At the elevated temperatures, within the range 45 to 65°C, in the bread store, the acidity-index of the crumb rose considerably, and its colour became increasingly darker. Storage of bread under these conditions, resulted not only in flavour and aroma deterioration of the freshly baked bread, but also the development of an unpleasant by-taste.

Bread on removal from the oven is, however, not in a palatable condition. Until its crumb temperature has dropped, and part of the physically bound water diffused to stabilize the crumb structure, palatability has not been attained. Since these processes are progressive, and cannot be determined by experimentation, the digestibility of the bread must be determined sensorically. Characteristics important to the consumer are: appearance, smell, taste, crispness and digestibility. These are determined by the quality factors: starch, protein, concentration and type of aroma substances, condition of the pentosans and lipids, and the concentration and moisture binding. Additional factors include: volume, pH, crumb colour, its compressibility and elasticity as well as swelling.

This maturation or ripening process after baking must not be confused with the process of 'ageing' or 'staling', which sets in at a much later stage, resulting in negative changes in the palatability and freshness of bread. The time necessary for the latter process to set in will vary depending on the product. In the case of bread rolls, this is about 7 hours after baking, and in the case of a 1-kg wheat bread about 22 hours; rye bread can remain palatable for as long as 40 to 60 hours without packaging. Packaged breads normally remain palatable for periods up to 6 days. The maturation or ripening processes involve a series of complex physical-chemical reactions, which are interrelated, and relate to both the crust and crumb. The binding capacity between starch, protein and water play a major part in this reaction complex. After baking, the bread, the bread crumb cell walls are in a swollen state, the water being partly physically and partly chemically bound by adsorption. At temperatures above 70°C, the bread is in the form of an elastic gel, but on cooling the crumb forms a solid mass, during which phase chemical reactions take place. At 60°C, and below, intramolecular hydrogen bonding occurs, which cumulatively represents high levels of energy. Bonds form within the starch and protein molecules, as well as between the amylose and amylopectin intermolecularly, which form the firm crumb structure of the loaf. In this manner, the bonding between the molecules becomes progressively closer, and more hydrogen bonds are formed, the liberated energy giving rise to dehydration. Some of the water released is taken up by the proteins, and the rest is lost by diffusion. Therefore, water plays an important part in the processes of both maturation and staling after baking. The physical changes of loss of heat and moisture during cooling are interrelated, water loss being brought about by the high water vapour partial pressure, depending on the temperature. The maturing process also involves the loss of labile, highly volatile aroma components which decompose, the more stable aroma components diffusing into the crumb on solidification (see Fig. 140). Physical and chemically bound water is either lost or condensed to produce the optimal eating properties. The chemical processes involved in ripening are intermolecular bondings, whereas the physical ones involve mainly moisture loss, and the diffusion of the highly volatile aroma substances.

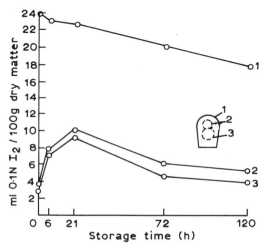

Fig. 140.   Changes in content of aldehyde–bisulphite addition compounds in bread during storage after baking. (Source: *Technology of Breadmaking*, L. J. Auerman, Moscow, 1972 (in Russian); Leipzig, 1977 (in German).)

On removal from the oven, the crumb temperature in the middle of the loaf is about 98°C, and at 1 cm below the crust about 110°C. The temperature falls most rapidly at first in this area under the crust, the temperature at the centre of the loaf only falling very slowly (see Fig. 141). The overall rapidity of the cooling process depends on the ambient storage temperature and relative humidity, and will proceed until a temperature equilibrium with the surroundings is attained. The actual relative humidity within the crumb of the loaf is about 100%, and that of the surroundings at 40% to 60%. Therefore, a diffusion process takes place, which removes about 2263 J/g of heat at boiling point and atmospheric pressure, which gives rise to further cooling. The approximate rate of cooling of the loaf being calculated from the basic Newtonian law of cooling equation:

$$\frac{dQ}{dt} = (t_{\text{loaf}} - t_{\text{surroundings}})Amst$$

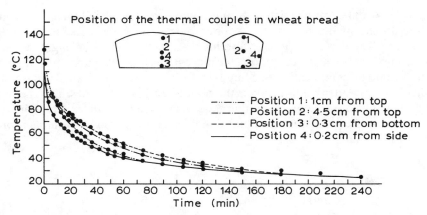

Fig. 141. Fall in temperature of 1kg wheat bread after baking. (Storage conditions: temperature 24°C; relative humidity 40%.) (Source: *Technology of Industrial Baking*, R. Schneeweiss and O. Klose, VEB-Fachbuchverlag, Leipzig, 1981 (in German).)

where:

$$A = \text{is the surface area of the loaf}$$
$$m = \text{loaf mass}$$
$$s = \text{specific heats of crumb and crust}$$
$$t = \text{crust thickness}$$

The total amount of heat lost during bread cooling can be determined approximately from the general heat equation:

$$Q = ms \, dt$$

For which the specific heat of crumb and crust and the crust thickness must be known. For a 1-kg white wheat loaf, the total heat loss amounts to about 180 kJ. This heat loss simultaneously results in condensation of water vapour within the loaf, a 1-kg wheat loaf of 2000 ml and water content 45%, contains about 0·5% moisture in the vapour form, according to Schneeweiss and Klose (*Technology of Industrial Baking*, VEB-Fachbuchverlag, Leipzig, 1981, p. 424, in German).

During the cooling process, part of this water vapour is lost, but most of it condenses, forming a partial vacuum, causing the air in the immediate surroundings of the loaf to be sucked in. Evidence of this can be obtained by the tightly sealed packaging of a loaf immediately after baking, the partial vacuum formed leading to a deformation of the loaf. Schneeweiss and Klose studied the volume change of a panned white loaf, which was packaged immediately after removal from the oven, and measured after various time intervals. They found that the volume change was greatest during the first 15 minutes after the baking process. After 45 minutes, a physical equilibrium between the partial vacuum and the partial water pressure (see Fig. 143) became established, after which only insignificant volume changes take place. Loaf volume was measured by placing the hot loaf in a sealed container, e.g. glass desiccator. The partial vacuum developed was measured with a U-tube

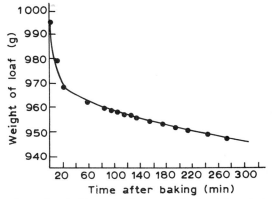

Fig. 142. Moisture loss after baking of 1 kg panned wheat bread stored at 25°C and a relative humidity of 34%. (Source: *Technology of Industrial Baking*, R. Schneeweiss and O. Klose, VEB-Fachbuchverlag, Leipzig, 1981 (in German).)

mercury manometer. The largest reduction in pressure was recorded during the first 60 minutes of cooling. In the case of unwrapped bread, this pressure difference due to the sucking in of surrounding air becomes equalized; it was measured with a gasometer. The same researchers also draw attention to the significance of these findings for the potential microbial infection of bread by mould (see Fig. 144). Since the amount of atmosphere gas sucked in or 'adsorbed' is greatest during the first minutes after removal from the oven, the microbial infection will commence at this

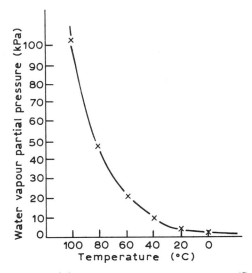

Fig. 143. Water vapour partial pressure versus temperature. (Source: *Technology of Industrial Baking*, R. Schneeweiss and O. Klose, VEB-Fachbuchverlag, Leipzig, 1981 (in German).)

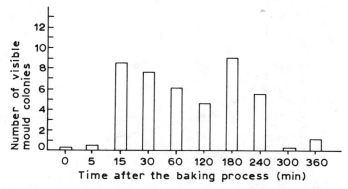

Fig. 144.   Number of visible mould colonies 3 days after baking of wheat bread, depending on the time at which packaging took place. (Source: *Technology of Industrial Baking*, R. Schneeweiss and O. Klose, VEB-Fachbuchverlag, Leipzig, 1981 (in German).)

stage. However, owing to the initial high temperature of the loaf surface, most of the spores become killed, therefore, a greater infection hazard can be expected after about 5 minutes exposure to the atmosphere has elapsed. Bread samples were packed in UV-irradiated plastic bags after sampling at various times after the baking process. These were stored in an incubator at 30°C, to test for mould development. It was found that if the bread was packed about 5 to 6 hours after baking, the moisture content of the crust was lower, and the spores did not find growth conditions as favourable. In this case, a delay in the microbial spoilage is achieved, although the spore contamination stands at the same level. On completion of cooling, the partial vacuum becomes equalized, and the spores which then settle on the dry top crust present less danger for the mould-free shelf-life of the bread. This emphasizes the need to cool bread in a storage atmosphere which is as low as possible in spore content. Table 71 shows the relationship between quality and storage-time of various baked products stored at 70–80% RH, and 18–22°C.

TABLE 71

| Product | Still fresh at | Still edible after | Staling influences quality after |
|---|---|---|---|
| Bread-rolls | 4 h | 6 h | 7 + h |
| Rolls with shortening or milk products | 6–8 h | 12–16 h | 16 h |
| Wheat-flour bread | 16 h | 20 h | 22 h |
| Rye/wheat bread | 19 h | 23 h | 26 h |
| Rye bread | 35 h | 40 h | 41 h |
| Rye wholemeal | 12–24 h | 36–60 h | 60 h |
| Packaged toast | 8–24 h | 6 days | 6 days |

Source: *Technology of Industrial Baking*, Schneeweiss and Klose, VEB-Fachbuchverlag, Leipzig, 1981 (in German).

Closely related to temperature and heat loss is the moisture loss of a loaf of bread. Figure 142 clearly shows that the water diffusion during the first few hours of cooling is greatest, but on reaching the ambient temperature of 25–26°C, the moisture loss per unit time tends to approach a constant value. The large loss of moisture during the first few hours is due to the large difference in water pressure between surrounding atmosphere and the loaf of bread. The dependence of the partial water pressure on the temperature is made clear in Fig. 143. A comparison of the cooling curve and moisture diffusion, together with the partial pressure of the water, and their co-dependence on temperature clearly confirms a close correlation between these variables. The importance of the cooling time lapse factor before packaging bread is clearly indicated in Fig. 144, expressed as the number of mould colonies visibly growing on samples stored over a 3-day period.

The structural solidity, and suitability for slicing depends on the transition of the crumb from a structureless gel. This is achieved by molecular bonding at temperatures below 60°C as the loaf cools. Hydrogen bonding plays a major part in this solidification process, in particular between the amylose and amylopectin molecules of the starch, thus forming the porosity of the bread crumb. This process of crumb firming starts when the loaf is removed from the oven, and finishes after 6–8 days, when the loaf becomes hard, dry and crumbly. This process can be monitored by measuring the response of the crumb to pressure deformation with time, referred to as 'compressibility'. For this purpose, the Instron pressure and stretching test equipment is often applied. Characteristics which are important to the consumer are: appearance, smell, taste and digestibility. Fresh bread is identified by its crispness and pleasant smell, the aroma profile consisting of a complex mixture of volatile and less volatile compounds as described earlier. The aldehydes, which form the bulk of the aroma substances, are chemically unstable and readily oxidized to carbonic acid. Hence, their stability depends on the formation of addition compounds with the amylose or amylopectin molecule. Schoch (*Bakers Digest*, **39**, 2, 48, 1965) suggested that the amylose spiral-chain is most likely to assume this role, since on reheating stale bread these aromatic aldehydes become released. Aroma substances are initially most prolific in the crust of the loaf, but during a 20-hour storage period, they migrate into the crumb zone immediately under the crust, then towards the central zone. During further storage for periods in excess of 20 hours, the content of aroma substances in the crumb progressively decreases. It is unlikely, however, that these are lost by evaporation. A more likely explanation for their disappearance is chemical or physical-chemical transformation into unknown non-aromatic compounds. Whilst the crust progressively loses its aroma substances by diffusion, the carbonyl content of the crumb will increase up to the second or third day of storage, reaching a maximum. This is mainly due to the relatively less volatile aldehydes condensing with the moisture under the influence of a partial vacuum into the crumb area. Immediately after baking, furfural can only be detected in the crust, but after about 24 hours it can be detected in the crumb, and gives the crumb its characteristic smell and taste. The exchange of aroma substances between crumb and crust are little changed by the packaging of bread; but packaging too soon after baking must be avoided, to ensure adequate cooling, moisture loss, and the evaporation of highly volatile aroma components which are unstable and would give rise to undesirable aroma contributions. In the crumb, during the cooling process many chemical and physical changes take place, which contribute to an aromatic balanced flavour profile, whereas the crust changes relatively little, being of low moisture and crisp. However, eventually some moisture from the crumb will migrate into the crust, causing it to become tough and leathery, and many aroma substances formed during baking within the crust are lost or decompose after diffusing into the crumb of the loaf. The summation of all these chemical and physical changes after baking with storage-time vary depending on the type of product and the raw materials and process applied. The relationship between product quality and storage-time of various baked products under average conditions are quoted earlier in this section. However, the perception of 'freshness' remains essentially a subjective judgement, and should remain the privilege of the consumer, regardless of the hours elapsed since product removal from oven.

In the case of industrially produced panned white bread, the crust after baking contains about 12% moisture, and the crumb about 44% to 45% moisture. After 4 days' storage at 22°C, without any moisture loss to the atmosphere, the crust moisture increases to about 29%, with a corresponding moisture loss from the crumb. This redistribution of moisture from crumb to crust is the best general indicator of freshness, since it tends to parallel crumb firming and loss of sensory attributes. The practice of packaging bread in moisture-proof material will promote crust staling, since this restricts any evaporation, resulting in the retention of larger amounts of moisture within the crust. A similar situation exists where bread is packaged before adequate cooling, more moisture being retained within the crumb, which eventually diffuses and accumulates in the crust. Any increase in moisture level of the crust is accompanied by a flavour devaluation. The exact mechanism of this flavour reversal is little understood.

### Bread Ageing and Staling Processes

The term 'staling' refers to the progressively decreasing consumer acceptance of bakery products due to physical-chemical changes in the constituents of the crust and crumb, and the flavour and aroma compounds formed during baking, but excludes any subsequent microbial spoilage. The result of these processes is a product which the consumer no longer considers 'freshly baked'.

Since losses which result from bread staling are of great economic importance, especially under conditions of industrialized and centralized production, considerable attention has been focused on this problem. Much research effort has been expended on the development of methods of retarding the staling process or minimizing its effect. These include modifications to fermentation and processing procedures, and/or the use of antifirming/antistaling additives, and the inclusion of moisture-retaining substances in the dough formulation.

Although the process of bread staling has been studied for over 100 years, the processes and their mechanisms remain obscure. The suggestion that bread staling is due to moisture loss, was proven incorrect by Boussingault over 100 years ago (Boussingault, M., *Ann. Chim. Phys.*, **38**, 490, 1853), who demonstrated that the storage of bread under conditions which ruled out moisture loss still became stale. The practical method of reheating bread in the oven to recover its freshness confirms these findings, since in spite of additional moisture loss, the crumb recovers its physical properties of freshness. Also, any reduction in crumb compressibility and increase in crumbliness on staling remain the effect and not the cause of the staling processes. Nevertheless there is a limit to refreshing of bread by reheating, as confirmed by von Bibra (von Bibra, E., *Getreidearten und das Brot*, Nuremberg, 1861), who showed that products whose moisture content was reduced to below 30%, could no longer be refreshed.

Current knowledge of the bread staling process indicates that the carbohydrates, proteins and water as the main components of bread are responsible in a complex manner. However, the role of the glycolipids, glycoproteins, pentosans and lipids cannot be overlooked. Owing to the diversity of the changes, few theories find general acceptance. During the staling process, many processes can become either accelerated or inhibited, and it is very difficult to isolate any one process for study. Therefore, before discussing the individual theories of staling processes, the observed changes in the microstructure of the bread crumb reported by Auerman (*Technology of Breadmaking*, 1st German edn, Leipzig, 1977, p. 291; original Russian edn, Moscow, 1972) offers a sound basis for the complex changes taking place within the crumb. The above author states that bread-crumb structure depends on the formation of the cell walls, which form the crumb porosity, in the form of a spongy framework. A microscopic examination of these pore cell walls reveals that they consist of a mass of coagulated gluten proteins. However, within them are embedded swollen, partially gelatinized starch granules. These granules expand somewhat, but remain parallel and surrounded on all sides by coagulated protein, only a few individual granules being in direct contact with one another. Therefore, the continuous phase of the spongy framework of the crumb is coagulated protein. In fresh bread the starch granules lie close to the surface of the coagulated protein throughout, and no sharply visible boundary layer is visible between them.

Whereas, within the crumb of staling bread the granularity of the partially gelatinized starch particles is clearly visible, owing to the formation of a thin layer of air around part of their surface. The older the bread, the more clearly visible the air layers become. This confirms the volume shrinkage of the starch. Any structural changes in the protein of the cell walls were not discernible with the microscope. The build-up of these thin layers of air around the changed granules are considered responsible for the increasing crumbliness of staling bread. Nevertheless, it is pointed out that the volume change of the starch granules is not the primary process of staling, but probably its effect. Numerous test results confirmed that during bread staling, the hydrophilic properties of the crumb change. The swelling capacity and water absorption capacity of the crumb become reduced, together with the solubility of the colloids and other components. The total water solubles are reduced, and the water solubility of the starch.

Changes in the hydrophilic properties of the crumb undoubtedly effect its physical properties, but these changes cannot be regarded as the cause of the staling process but rather an effect. Lindet (*Bull. Soc. Chem. de Paris*, **27**, 634, 1902) appears to have been the first to associate starch 'retrogradation' with bread staling. The term 'retro' meaning 'back or return', which implies in this case the increasing crystallization of the starch components, amylose and amylopectin. During baking, the starch is partially gelatinized and absorbs water from the proteins, as a result the starch changes from its initial crystalline form to becoming amorphous. In so doing, the starch absorbs a considerable amount of water, which the proteins release on coagulation, during the baking process. During subsequent storage, the starch within the crumb retrogrades, i.e. the starch tends to gradually revert to the crystalline form in which it existed within the dough before baking. In so doing the internal structure of the starch becomes firmer, it loses its solubility, and part of the water absorbed during gelatinization is released. This released moisture is then considered to be taken up by the gluten proteins in the crumb. This concept of retrogradation gave impetus to a promising new approach to research into the staling reactions. Katz in a series of studies (*Z. für physik. Chemie*, Abt. A, **150**, 1, 60; 1, 67; 2, 81; 2, 90; 2, 100, 1930; **155**, 3/4, 199, 1931; **158**, 5/6, 321, 337, 346, 1932) established that the initial native starch of wheat and flour gave an X-ray spectrum which was typical for a crystalline structure. This pattern persists in the dough up to the baking stage. During baking the starch is partially gelatinized, but only the first stage of gelatinization took place. This was due to the amount of available water in the dough being insufficient for complete gelatinization. As a result, the crumb gave a so-called V-pattern X-ray, which differed from the original pattern. The V-pattern of partial crumb gelatinization characterized the combination of the amorphous element and the crystalline structural element. However, the crystalline structure of this partially gelatinized starch was quite different from that in the flour and dough before baking. The crumb of stale bread gave an X-ray pattern designated as B-pattern, which was a combination of the initial pure crystalline pattern and the V-pattern. Moreover, with increasing staling, the B-pattern became increasingly like the pure crystalline pattern in character. This X-ray pattern B of staled bread Katz designated as the 'starch retrogradation spectrum'.

Katz regarded the starch in the bread crumb as a thermodynamic equilibrium system, of alpha- and beta-starch forms, the alpha form being typical for fresh bread, and the beta form for staled bread. The alpha form was stable at temperatures above 60°C. Thus, at this temperature, no retrogradation of starch takes place and no staling of the crumb. At lower temperatures from 60°C down to −10°C, the system equilibrium shifts in favour of the beta-starch form, which is characteristic for staled bread. Thus, Katz considered retrogradation of starch as a change in the ratio of alpha and beta-starch forms, staling resulting in a transition from the more soluble and less structural alpha form into the beta form. Katz also emphasized the importance of both temperature and moisture in these changes. He found that when the loaf moisture fell below 16·4%, or if it was moistened with excess water, the physical effects of staling did not occur. Breads stored under controlled conditions of temperature and moisture from 24 to 48 hours at 60°C or above remained fresh; became semi-stale at 40°C; almost stale at 30°C; stale at 17°C, and very stale at 0°C. It remained quite fresh between −7 and −184°C.

Schoch (*Bakers Digest*, **39**(2), 48, 1965) and Schoch and French (*Cereal Chem.*, **24**, 231, 1947) considered that the starch granules only undergo partial gelatinization and limited swelling during baking, owing to the limited amount of water available in dough. With increasing swelling, the linear amylose fraction becomes more soluble, and diffuses outwards into the aqueous phase. As a result of continued swelling, the concentration of amylose in the interstitial water increases, forming a concentrated solution. Since the amylose tends towards retrogradation, this concentrated solution sets to a gel by the time the loaf has cooled. Fresh bread thus contains swollen, elastic starch granules, embedded in a firm gel of amylose. This gel is considered to remain stable during further storage, and not to participate in the staling processes. Crumb firming is attributable to changes in the physical orientation of the branched amylopectin molecules of starch within the swollen granule. In fresh bread, the branched chains of amylopectin are spread out, within the limits of available water, thus forming a concentrated system. However, the outer branches of the amylopectin gradually aggregate, aligning with one another by various types of bonding. This system is less stable than retrogradation, being an intramolecular association, which solubilizes with moderate heat. This system gives rise to an increasing rigidity of the internal structure of the swollen starch granules. This physical transformation explains the effect of crumb hardening with staling (see Fig. 145). Schoch's original concept has undergone some enhancement by Lineback ('The role of starch in bread staling' in *International Symposium on Advances in Baking Science and Technology*, Dept. of Grain Science, Kansas State University, Manhatten, KS, USA, 1984), in order to explain the mechanical functions of starch. Whilst amylopectin plays the most important role, amylose is considered to be also

Fig. 145. Reactions of starch fractions during baking and ageing of normal bread. ○, starch granule. ⅄, amylopectin 3/4. ⚹, amylose 1/4. (Source: T. J. Schoch, *Bakers Digest*, 37(2), 48, 1965.)

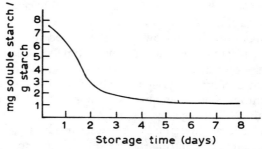

Fig. 146. Decrease in soluble amylose during the staling process. (Source: *Technology of Industrial Baking*, R. Schneeweiss and O. Klose, Leipzig, 1981.)

involved in the staling process (see Fig. 146). Portions of both amylose and amylopectin molecules extend from the swollen starch granule, and associate with other carbohydrate chains present in the interstitial aqueous phase, and with those extending from granules in close proximity of one another. This postulate introduces the inter-granular function of both amylopectin and amylose to structural firming. Cornford (*Cereal Chem.*, **41**, 216, 1964) established that the crumb compressibility (elastic modulus) increased as the storage temperature fell to 0°C, and was proportional to the concentration of crystallized starch present during staling. Data from these workers indicated that the mechanism of starch retrogradation is one of instantaneous nucleation and crystalline growth. Senti and Dimler (*Bakers Digest*, **34**(1), 28, 1960) also postulated a change of state to explain crumb staling. Starch is at a higher energy level in the amorphous form than in the crystalline form. Therefore, when no external heat energy is applied, the gelatinized starch tends to retrograde from the amorphous state to the lower energy crystalline state. This involves a reorientation of the linear molecules into a more orderly structure of lateral bonding between chains. Crystallization may account for about 15% of the starch, and renders it insoluble and rigid. Crystallized regions strengthen the bread crumb, being interlinked by long linear amylose chains, extending through the amorphous regions. This forms a three-dimensional network, within which the rigid retrograded areas form a backbone, depending on the number or crystallized molecules. Hollo (*Stärke*, **12**, 106, 1960) postulated retrogradation in three stages: (1) breaking of intramolecular bonds by stretching of the chain, thus uncoiling its helical form; (2) loss of bound water, and molecular reorientation; (3) hydrogen bonding between adjacent chains forming crystalline areas. (1) and (2) require the input of energy, but (3) releases energy in the form of heat, sufficient to make the complete process exothermic. If heat is applied to the retrograded gel, the hydrogen bonds joining the linear molecules in the crystalline areas are broken. This allows free kinetic movement for the chains, which could revert to their normal helical conformation. This is the assumed mechanism of retrogradation reversal experienced on reheating staled bread (Figs 145 and 147).

The most comprehensive staling model has been presented by Knjaginičev, M. J. ('Bread staling and the maintenance of its freshness' *Russ. Journal of Chem. All-*

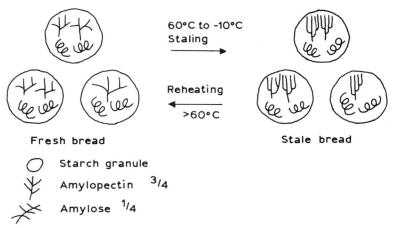

Fig. 147. Reactions of starch fractions during ageing of 'crumb-softened' (emulsifer-treated) bread. ○, starch granule. ¥, amylopectin 3/4. ✳, amylose 1/4. (Source: T. J. Schoch, *Bakers Digest*, 37(2), 48, 1965.)

*Union Mendelejew-Institute, Moscow*, **10**, 3, 1965, pp. 277–286). The essence of his postulates are that staling is caused by the change in distribution of the water (see Figs 148–151), the crumb changing to a solid structured form; hydration and water migration plays a major role. According to Knjaginičev, bread, depending on the type, begins to lose its original softness at 6–10 hours after baking, when stored at

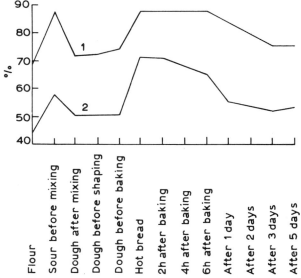

Fig. 148. Changes in content of bound water during the preparation of wheat bread. (2) and rye bread made with meal (1). Ordinates show the amount of water bound by 100 g of absolute dry product. (Source: M. J. Knjaginičev (USSR).)

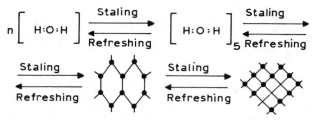

Fig. 149.   Proposed scheme for the state of water in the crumb (starch plus protein) during staling and refreshing of bread. (Source: M. J. Knjaginičev (USSR).)

normal temperatures of 15 to 20°C. The crumb becomes friable, and the bread loses its aroma. This rapid loss of consumer acceptance and flavour restricts the long-term storage of large quantities produced in advance, and their transport over long distances. The enlargement of industrial bakeries, and product specialization is also limited by this economic factor. Bread has always been evaluated for its freshness organoleptically, but now objective rheological methods are also available. During the first 3 days, the moisture content in the crumb centre remains constant, but bread staling proceeds independently (Auerman, L. J., *Technology of Baking*, 6th edn, Piscepromizdat, Moscow, 1956; and Scerbatenko, V. V., Gogoberidse, N. I., and Zelman, G. S., in *Maintenance of the Freshness of Bread*, CINTIPiscepromizdat, Moscow, 1962). The perimeter of the crumb loses moisture, forming a dense layer of 3–5 mm thickness. This layer restricts the compressibility of the crumb when pressed with the fingers, which is how the consumer tests freshness. A detailed analysis of stale bread showed that, in spite of equal moisture contents, the crumb of staling bread differed from that of fresh bread in respect of many measurable parameters.

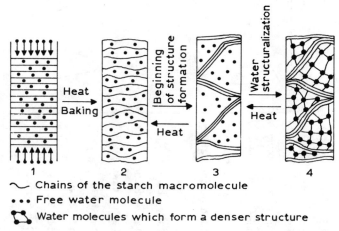

Fig. 150.   Proposed scheme for swelling gel-formation of the starch and proteins and the state of the water molecules during baking, staling and refreshing of bread. 1. Weak swollen gel. 2. Strong swollen gel (freshly baked bread). 3. Gel with initial structural formation (beginning of staling). 4. Structured gel (staled bread). (Source: M. J. Knjaginičev (USSR).)

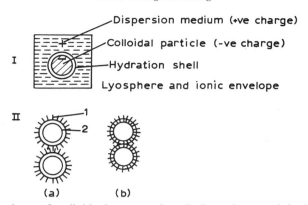

Fig. 151. Behaviour of colloids in suspension. I. Lyosphere and ionic envelope. II. Attraction of two colloidal particles. (a) Prevention from contact by hydration or solvation shell. (b) Attraction due to Brownian motion deformation of the ionic envelope and lyosphere. 1. Ionic envelope. 2. Lyosphere.

Amongst these are the following: compressibility of the crumb (elasticity) is sharply reduced, and the mechanical firmness of the cell walls increase; its content of bound-water is reduced; the relative viscosity of an aqueous extract of the crumb, prepared by rubbing and filtering, together with its speed of filtration, both become reduced considerably. The crumbliness markedly increases, and the water solubility of the crumb components become reduced (Auerman, L. J., Nikolaev, B. A., and Kulman, A. G., *Colloids during Baking*, Moscow, 1940, 1956). See Fig. 152, according to M. J. Knjaginičev, Leningrad.

It was also established that with increasing loss of freshness, the viscosity of the crumb suspension, measured with the Amylograph is reduced (Emikeeva, N. G., and Auerman, L. J., *Moscow Tech. Inst. Food Industry*, **4**, 105, 1956; Bechtel, W. J., Meisner, D. F., and Bradley, W. F., *Cereal Chem.*, **31**, 3, 171, 1954). Also, the ability of methylene blue to become bleached by the crumb decreases with staling (Auerman, L. J., and Rachmankulova, R. G., *Bread and Confectionery Industry*, **22**, 22, 1957; Nazarov, V. I., Sacharov, V. G., and Tichomirova, T. P., *High School News Food Technology*, No. 4, 113, 1958). The resistance of the crumb starch of staled bread to beta-amylase is also less than in the case of fresh bread (Rachmankulova, R. G., and Falunina, Z. F., *High School News Food Technology*, No. 2, 63, 1960). The influence of staling on crumb suspension viscosity is illustrated in Fig. 153.

Knjaginičev draws attention to the established changes in the dough and bread coloids, in particular their relative contents of bound water. Figure 148 shows the changes in bound water content of wheat and rye dough and bread during the various technological phases of production and storage. The illustration clearly shows that an increase in content of bound-water takes place at two stages during processing: in the sour/sponge before final dough mixing, and during baking, while it remains hot. The observed increase in bound-water during the pre-ferment stage is due to the increased hydration capacity of the proteins, coupled with the increase in acidity-index on ripening. The role of starch at this stage is secondary. During

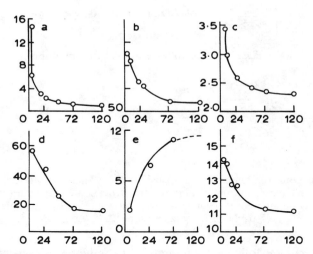

Fig. 152.   Changes in physical-chemical properties of the bread crumb during staling versus time in hours (a to f). a. Compressibility (firmness) in units of Katz-apparatus. b. Water binding, percentage. c. Relative viscosity of water suspension. d. Filtrate from 25 ml of suspension of the bread crumb in 1 min. e. Friability of the crumb, percentage. f. Water solubility of the crumb, percentage. (Source: *Staling of bread and maintenance of freshness*, M. J. Knjaginičev (USSR), *Die Stärke*, **22**, 12, 1970 (in German).)

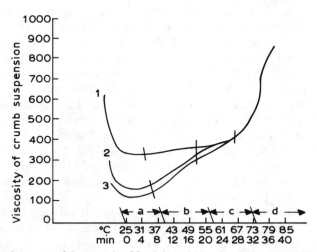

Fig. 153.   Amylograms of the crumb of fresh bread (1), after 24 h storage (2), and 48 h (3), at room temperature. (Source: *Staling of bread and maintenance of freshness*, M. J. Knjaginičev (USSR), *Die Stärke*, **22**, 12, 1970 (in German).)

baking the bound-water increases, but falls when the proteins become denatured. However, when the starch gelatinizes, the bound-water again increases by virtue of the starch, at the expense of the proteins. The general pattern of the curves for rye and wheat processing is similar, but the rye curve (1) is appreciably higher in overall water-binding capacity. This difference is due to the considerably higher pentosan content of rye, and the colloidal properties of its proteins. In bread, 80% to 85% of its content consists of starch and water, the sugar, cellular substances, lipid and mineral contents being relatively small, their chemical structure being such that they have little influence on the hydrophilic properties of the bread. The hydration capacity of the proteins, amounting to 8% or more, have a great influence on the hydrophilic properties of the dough, and determine bread volume and the porosity of the crumb. The cellular content of rye meal, consisting in the main of arabinans and xylans, have a great influence on both the colloidal properties of the rye dough, and the bread. During baking, the pentosans become denatured; and as far as the physical-chemical properties of the flour and bread components are concerned, whenever the nature of the staling process is under investigation, only the changes in the starch and protein have attracted attention.

The refreshment of bread of moisture content 30% and above by heating has been long known, and if the moisture is below 30%, it can be refreshed by first dipping in water for a short time. The influence of temperature on the maintenance of loaf freshness after baking, has been extensively studied. In so doing, in addition to the organoleptic evaluation of the bread, the swelling of the crumb was also measured during a 24 hour period. The data obtained are detailed in Table 72. These data confirm the importance of the hydrophilic high-molecular-weight polymers (colloids), the crumb losing its ability to swell with staling.

TABLE 72

| Temperature (°C) | Sensory freshness evaluation | Crumb swelling (in ml) during 24 hours |
|---|---|---|
| 85–92 | fresh | 50·0 |
| 70 | fresh | 50·5 |
| 60 | fresh | 51·0 |
| 50 | almost fresh | 49·0 |
| 40 | signs of staling | 43·5 |
| 30 | semi-stale | 40·0 |
| 17 | stale | 34·5 |
| 0 | completely stale | 30·0 |
| −2 | completely stale | 34·0 |
| −6 | stale | 39·0 |
| −7 | semi-stale | 41·0 |
| −20 | fresh | |
| −30 | completely fresh | 49·0 |
| −190 | completely fresh | |

Source: M. J. Knjaginičev, *The staling of bread and the maintenance of its freshness*, Die Stärke, **22**, 12, 1970.

The storage of freshly baked bread within the temperature range 60 to 90°C, or −20 and −190°C, result in the retardation of the staling processes. Temperatures between +50 and −7°C, including the room or ambient range are unfavourable in this respect. The technological application of temperature control to dough and bread, as well as other baked products, is discussed in detail in Section 3.6.3. Although the part played by temperature in the maintenance of freshness of the bread crumb has not yet led to a full clarification of staling, investigations of the changes in physical-chemical properties parallel with the development of processes for the retardation of staling have yielded information concerning bread components, especially the starch.

Bread and baked products contain about 50% to 60% starch, which in turn is made up of about 20% amylose and 80% amylopectin. These components differ in both their molecular weight and chemical behaviour, being arranged differently within the starch granule. Some areas within the granule are discretely organized showing a crystalline space-lattice X-ray pattern spectrum; and other areas are randomly organized, according to Badenhuizen ('Structure and formation of the starch granule', in *Starch Handbook*, Paul Parey, Berlin and Hamburg, 1971). The interference pattern of the rays, in fact show restricted sharpness in the contours of the rings in the X-ray pattern. Rye bread shows a wide, diffuse ring, similar to completely gelatinized wheat starch. The water content of the starch is also reflected in the X-ray spectrum, starch at atmospheric moisture levels conforming to that of native starch, while water-free starch exhibits a pattern similar to that of the amorphous form. The interference pattern of the X-rays show a strong dependence on the water absorption of the starch. The water in crystalline starch is built in the lattice in the form of hydrogen bonds. In flour, the starch particles only bind water at the surface by adsorption at 20°C, hydration only starting at the damaged areas of the granule. The energy input at baking, however, results in the swelling of the whole starch granule. The water rests between the amylose and the amylopectin molecules, both the organized and randomly organized areas becoming hydrated. However, the spatial network remains intact during this swelling process. These network bondings, being typically hydrogen bridges, and van der Waal bonds, become partially broken during hydration. This hydrated starch molecule at temperatures above 60°C is the energy-stable form. On cooling, the system becomes thermo-dynamically unstable, and the staling process progressively commences. The term 'retrogradation' is now applied also to reactions taking place in the aqueous starch suspension, and includes the ageing of both amylose and amylopectin. Of crucial importance are the physical-chemical reactions taking place immediately after baking. The amylose is mainly dispersed colloidally, whereas the amylopectin is in a swollen globular state. During the cooling period, a process of partial crystallization begins (gel with initial structural formation). Crystals of amylose, amylopectin, and mixed crystals form, which progressively form an increasingly firm mechanical structure. This speed of crystallization, will depend on the ratio of amylose to amylopectin and the molecular arrangement of the components. Wheat starch crystallizes more rapidly than potato starch, hence the use of the latter in formulations, to retard the entire staling process.

The stability of colloidal systems are governed by processes of particle aggregation, which is also applicable to the starch granule. Therefore, amylose and amylopectin become surrounded by thick water layers, known as 'hydration-shells', 'solvation-shells' (Fig. 151). In this manner, mutual contact is prevented, since on account of surface charge, repulsion of the shells determines their spacing. Repulsion depends on the size of the hydration-shell, and the magnitude of the electrical charge. The thicker the hydration-shell, the stronger the electrical charge, and any Brownian motion will depend on the kinetic energy of the system. A build-up of such movement leads to structural deformation of the hydration-shells, resulting in them merging to form flattened particles in place of the spherical ones. These colloidal processes explain the technological effects of staling, and the important part played in it by hydration and water movement. The speed at which staling proceeds depending on the magnitude of the charge on the hydration-shells, and the morphology of the particles involved. The presence of ions from electrolytes also exert a strong influence on the hydration-shells, forming an 'electric double-layer'. Any double layer so formed, consists of two parts, a fixed double-layer adjacent to the charged colloidal particles at ionic equilibrium, and a diffused double-layer between the fixed double-layer and the bulk of the surrounding medium. Both parts are free to move as separate entities. In such cases, the fall in potential across the diffused double-layer is measured as the 'zeta or electrokinetic potential', which has a value of 0·05 volts for many aqueous dispersions. If this value falls to about 0·02 to 0·03 volts, the system becomes unstable.

Molecular aggregation is mainly due to the formation of intermolecular forces. In the case of starch, hydrogen bridge formation has the greatest significance. The ability to form hydrogen bonds will depend on the steric formation of the molecule, and amylose is capable of assuming various helical conformations in solution as pointed out by French (French, A. D., *Bakers Digest*, **53**, 1, 39, 1979). The formation of hydrogen bridges in amylose and amylopectin is schematically illustrated in Fig. 154. Retrogradation results in the formation of a space lattice structure from amylopectin, into which the retrograding amylose becomes engulfed, which explains why a reduction in the soluble amylose occurs after about 2 days storage. Depending on the conformation of the amylose, structural changes take place. As mentioned earlier, according to Hollo, the spiral-shaped amylose helix uncoils, and loses its hydration-shell becoming dehydrated. The energy required for these reactions being derived from the hydrogen bondings. These hydrogen bonds give rise to the formation of the crystalline areas which appear in the X-ray structural analysis mentioned in Knjaginičev's staling model. According to the X-ray spectrum pattern, grain starches are classified under type A starches, and potato starch is assigned to type E. On retrogradation, wheat starch no longer shows the crystallinity of type A, but instead is similar to the type B pattern, which was confirmed by Badenhuizen ('Structure and formation of the starch granule', in *Handbook of Starch*, Paul Parey, Berlin and Hamburg, 1971, in German). X-ray patterns of fresh and aged wheat starch show that the retrogradation process becomes apparent at about 55°C on cooling the gels, and on storage over several days, the crystalline areas increase considerably. The reversion process (refreshing),

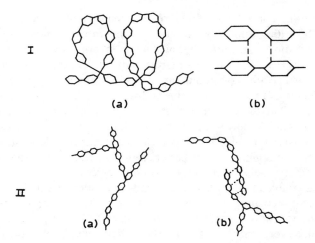

Fig. 154. I. Amylose retrogradation. (a) Amylose helix. (b) Hydrogen bonding between stretched amylose molecules. II. Amylopectin retrogradation. (a) Amylopectin chain. (b) Formation of hydrogen bonds between adjacent molecules.

involving conversion to the hydrated state requires a considerable energy input. The hydrogen bonds of the amylose molecule requiring temperatures of 150°C, and the amylopectin about 60°C to break them. Therefore, a complete reversion of stale bread to the fresh state is not possible, and partial refreshing results in a rapid change back to the retrograded state. Maximum staling occurs at about −2°C, showing a slow decline up to about 60°C, but a rapid decline due to the 'freezing' of retrogradation between −10 and −35°C.

Knjaginičev has stressed the importance of the dough colloids, and their relations with the water molecule within the bread-crumb during staling and refreshing. His proposed model scheme is illustrated in Fig. 150. It postulates that the individual water molecule during staling is subjected to a molecular reorientation. The water molecule, depending on its four charges, orientates four further water molecules, forming an aggregate from five such molecules. These aggregates form a hexagonal lattice (as in the case of ice), characterized by loose spacing. The latter can switch to a denser spacing, through a reorientation of the hexagonal lattice to form a tetragonal lattice, or through the filling of vacant spaces with more water molecules. Figure 148 shows in schematic form that during dough processing, (mixing, fermentation of the pre-ferment, followed by dough-mixing and maturing) the flour forms with the water, under limited swelling, a viscoelastic gel, in which not all the dough water takes part in the swelling of the polymers. Mechanically bound water is depicted by dots and arrows. The swollen gel of the flour/dough polymers, can be initially regarded as an unstructured gel, forming a one-phase system. The heating of the dough during baking of the bread favours the diffusion of the water in the intermolecular spaces of the starch and proteins. In so doing, the monomers of the molecules, i.e. glucose and amino acid residues, increase in mobility, becoming flexible and losing their compact posture. This results in the formation of micro and macro hollow cavities. During preparation of the pre-ferment, only a limited amount of water is available, which only permits a limited swelling. Under the influence of the higher temperature in the oven, the proteins coagulate, forming a framework and fixing the porous volume of the bread. The cell walls of the crumb, which consist of starch and protein, represent a swollen system in which one part of the water molecule is thermodynamically bound, and the other part distributed in the intermolecular spaces of the denatured proteins, and the swollen, partially gelatinized starch. This system, together with that of the dough, can be regarded as a swollen, elastic, structureless gel. During cooling and storage of the bread, owing to the flexibility of the starch monomers, the molecular chains move closer together, and intermolecular van der Waals force bonding results in the formation of a mechanically firm network, i.e. a structural gel. The formation of this network is followed by the bread staling processes, which continue to a maximum state of mechanical firmness. Knjaginičev postulates that the water in the macropores is probably not in the free state, but exists in an ordered form owing to the high polarity of the molecules, and the electrostatic forces at the surface of the micropores. The walls of these micropores consist of the macromolecular starch and proteins, as schematically illustrated in Figs 149 and 150. Eventually, the water, starch and proteins form a uniform structured system. As a result of heating to refresh the bread, the water structure within the pores of the crumb is destroyed, and the macromolecular chains revert to the freshly baked state, as shown schematically in 2, 3 and 4 of Fig. 150, depicting the stages of swelling and gel-formation depending on the state of the water molecules.

These models, based on diverse collective research carried out by Knjaginičev and co-workers, illustrated in Figs 148–150 represent a revision of former postulates in order to present a more realistic approach to the questions posed by bread staling. The basis of his considerations is the formation of a structure within the micropores of the starch and protein polymers of the bread as a result of the reorientation of the water molecules. The resulting aggregates forming an hexagonal space-lattice similar to that of ice, but loosely packed. As staling proceeds, the spatial density becomes compacted, either because of reorientation to form a tetragonal lattice, or as a result of water molecules entering the empty spaces within the lattice. The crumb is thus transformed into a firm, structured state, the main participant being the water molecule, its mobility and colloidal activity. As a result of storage, within the temperature range $+50$ and $-7°C$, both the macromolecules of the starch, and the water molecule lose their mobility, giving rise to a firm structured system. Additional tests involving the addition of starch granules to white bread before baking confirmed that similar moisture losses occurred after 4 hours and 8 days, both in the added starch and the bread-crumb. Therefore, there is no appreciable migration of water from the water-rich crumb to the water-free starch. Knjaginičev consequently concludes that the crumb consists of a system of pores with cell-walls of starch and denatured proteins in a swollen state, in which the water is partly thermodynamically bound and partly distributed within the intermolecular spaces of the proteins and swollen, partially gelatinized starch. Thus, the molecules of water, starch and proteins form a uniform structural system. During the refreshing of staled bread by heating, the structure of the water within the micropores of the crumb is disturbed, and the starch and protein macromolecules revert to a state similar to that of fresh bread.

There appears to be considerable general acceptance that the retrogradation of starch is a crystallization process, which can be represented by the so-called modified Avrami equation (Avrami, M. J., *J. Chem. Phys.*, **8** (1940) 212; Avrami, M., *J. Chem. Phys.*, **9** (1941) 177; and Cornford, S. J., Axford, D. W. E., and Elton, G. A. H., *Cereal Chem.*, **41** (1964), 216), written in the following form:

$$\text{amount of non-crystalline material at time } t = \frac{E - E_t}{EE_0} = e^{-kt^n}$$

where $E_0$, $E_t$ = elasticity modulus at time zero and time $t$; $E_\infty$ = elasticity modulus after unlimited storage, i.e. the final stage of crystallization; $k$ = speed constant; and $n$ = Avrami exponential.

The reciprocal of the speed constant, $k$, is designated as the time-constant $1/k$; $n$ is a function of the directional growth of the crystals, and the time required to form the crystal nucleus (Sharples, A., *Introduction to Polymer Crystallization*, Edward Arnold, London, 1966). For experimental purposes, the value of $n$ for measurements on bread and starch gels was found to be 1 (Kim, S. K., and D'Appolonia, B. L., *Cereal Chem.*, **54**, 1977, 216 and 225; Elton, G. A. H., *Bakers Digest*, **43**, 3, 1969, 24; Axford, D. W. E. *et al.*, *J. Sci. Food Agric.*, **19**, 1968, 95). This indicates that nucleation is spontaneous, i.e. independent of time, and that growth of the crystals is rod-shaped and unidimensional. For a more detailed explanation of Avrami analysis, the reader is recommended to the following literature: Cornford, S. J. *et al.*, *Cereal Chem.*, **41**, 1964, 216; and McIver, R. G., *et al.*, *J. Sci. Food Agric.*, **19**, 1968, 560. Krüsi and Neukom, *Die Stärke*, **36**, 1984, 2, 40–45) who used this technique to study retrogradation in concentrated wheat gels. The influence of starch concentration, and mode of preparation of the gels were investigated by determination of gel-strength (Avrami analysis), and changes in cold water soluble starch. An increase in gel concentration from 40% to 50%, resulted in corresponding increases in gel strength when fresh, and during storage, with consequent reductions in soluble starch fractions. Temperature increases from 100 to 130°C in gel preparation, resulted in significant increases in water-soluble starch. The amylose content in the gel extracts indicates that both starch components take part in retrogradation. Since water-binding of the insoluble fraction decreases during storage, retrogradation also occurs in the insoluble starch, thus increasing gel strength. Retrogradation of concentrated gels occurs both on the surface of the swollen granule, and within the soluble starch which becomes leached out during heating. Since crystallized starch is mainly insoluble, the amount of soluble starch extracted from bread-crumb provides an index to the degree of retrogradation (staling). Kim and D'Appolonia (*Cereal Chem.*, **54**, 207 and 216, 1977), found that the amylose content in the soluble starch extracted from the bread crumb at 10 minutes after baking was small, and decreased further during a 5-hour cooling period, therefore confirming Schoch's observation that amylose retrogradation occurs during baking and cooling. Soluble starch (mainly amylopectin), decreases from 3·34% to 1·22% over 5 days storage of bread made with flour of 11% protein (14% moisture basis). Decreases in soluble starch were less in bread stored at 30°C than at 21°C, confirming less crystallization of starch at higher temperatures.

Amperometric determinations of the amylose content of gelatinized starch solution and freshly prepared gels by Krüsi and Neukom yielded about 25% for a 1% solution, and 12% to 16·5% in the extracts of the fresh gels. This confirms that the amylose retrogrades mainly after cooling of the gels, or remains insoluble under the extraction conditions used (20 g starch dry matter/200 g deionized water at 25°C mixed in a homogenizer for 2 minutes, then re-extracted with a further 100 g water at 25°C for a further 30 minutes, made up to 370 g and centrifuged at 4000 rpm for 10 minutes. The residual weight in the centrifuge-beaker is used to calculate the water-binding of the insoluble starch, g water/g starch dry matter. The extraction yield is determined by drying an aliquot of the supernatant overnight at 105°C. The cold water soluble starch is expressed as a percentage of the 20 g of starch dry matter in the gel). The iodine-binding capacity of the soluble starch (amylose content) is determined by amperometric titration, the binding of pure amylose being taken as 19·5%. Kim and D'Appolonia (*Cereal Chem.*, **54**, 1977, 216) obtaining similar results for soluble starch in bread crumb. Krüsi and Neukom found that in the case of concentrated wheat gels, the largest fall in amylose content of the extract occurs during the first 24 hours of storage. The amylose:amylopectin ratio changes increasingly in favour of the amylopectin, the reduction in content of the branched polymer proceeding more slowly. In contrast, Neukom and Rutz (*Lebensm. Wiss. und -Technologie*, **14**, 1981, 292), found from determinations of the amylose: amylopectin ratio in bread-crumb, that amylose and amylopectin retrograde at approximately the same rate. Owing to retrogradation of the starch, the gel structure becomes firmer, as the gel strength increases, and starch solubility falls. However, these two gel properties only have a limited relationship, since the fall in cold water soluble starch, and especially the fall in the soluble amylose, proceed more rapidly than the increase in gel strength. From freshly prepared 40% and 50% dry matter gels, only a small part of the starch could be extracted (5·35% and 2·56% respectively in dry matter). This suggests that changes in the insoluble, non-extractable starch are responsible for the ageing of the gels. Reduced water-binding capacity of the insoluble starch confirms this view. Most of the research work into bread staling discussed hitherto has concentrated on the changes taking place in starch and the mode of binding of the water molecule. Since the gluten proteins constitute the continuous phase of the bread structure, their role in staling cannot be completely overlooked, since they are at least responsible for the formation of the framework and porosity of the crumb before they undergo denaturation during baking. Recently, the protein content of the starch granule itself has attracted attention. Starch granule protein represents from 0·2% to 0·35% of the starch isolated by water-washing. On extraction with sodium dodecylsulphate (SDS), and polyacrylamide gel electrophoretic (PAGE) separation, the 10 protein bands obtained show one dominant band, which is characteristic for soft wheats. This protein has been named 'friabilin', since it is considered to be responsible for the friable structure of the soft wheat endosperm, certain genes being instrumental in its formation. This illustrates the significance of even trace amounts of components for functionality during processing.

Many researchers have considered that the gluten proteins participate in the staling processes, but their exact role remains open. Most researchers look upon the proteins as a vehicle for moisture redistribution within staling bread. One notable exception to this type of postulate is that of Erlander and Erlander (*Die Stärke*, 21, 1969, 12, 305), who consider the protein:starch ratio important; protein–starch exchange reactions and the formation of protein/starch complexes, coupled with the significance of the ionic series. The complete removal of water from proteins is very difficult, if not impossible without complete decomposition. Therefore, changes due to heating and adsorption may be partly or completely reversible or irreversible. Transformation of hydrophilic to hydrophobic sol can result from a change in composition, brought about by the rupture of linkages and a partial combination between ionizable protein groups. Reversible changes involve drying and rehydration during cooling, and salting-out by salts of alkaline metals; irreversible changes occur as a result of heating, adsorption on the surface by metal salts, or by the action of non-electrolytes such as alcohol. During freezing of colloidal systems mutual absorption of colloids can occur. In all cases the effect of the lyotropic series is important. Lyophilic colloids often form reversible gels, and lyophobic colloids irreversible gels. The term 'synerisis' refers to the ageing of gels at constant temperature and without evaporation, resulting in the expulsion of only a small amount of the liquid component, owing to internal rearrangement. Hydrated gluten proteins in dough undergo a mild transformation during baking, resulting in denaturation and modification, the hydration water being transferred in part to the starch. Recent studies concerning the effect of storage on the physical chemical properties of wheat bread proteins by Kolpakova, Nazarenko, Zharinov and Burak, USSR (*Mlyn-Pek. Prum. Tech. Skladovani Obili*, 33, 11, 1987, 321–323, in Czech) revealed that within the first 14 hours of storage, bread proteins underwent denaturation with separation of free-water. During the following 24 hours, renaturation took place and water was re-absorbed, then during the third day of storage, denaturation returned. Ageing of bread from gluten-rich strong flour was due mostly to ionic bond interactions, since ionic bonds prevail in the gluten proteins of the flour, whereas hydrophobic–ionic interactions were found to be characteristic for the bread from weaker low-gluten flours.

Some conflicting views have been put forward by Cluskey *et al.* (*Cereal Chem.*, 36, 236, 1959) and Willhoft (*Bakers Digest*, 47, 6, 1973, 14). While the former claims a gain in moisture by gluten during staling, the latter proposes an irreversible modification in the hydrated structure of gluten during staling, water being released and absorbed by the starch.

Bread made from strong flours, forming gluten of adequate quantity and good quality, has high volume and a slower rate of staling than is the case with weak flours, forming inadequate quantity and quality of gluten. Adequate fermentation both in bulk and after shaping also reduces the staling rate. Maleki *et al.* (*Cereal Chem.*, 57, 1980, 138) also confirm that breads from different flours with various levels of protein, staled at differing rates. Increasing protein levels progressively prolonging the rate of staling. Data obtained by Cluskey *et al.* indicated that moisture was transferred from starch to gluten in the crumb during staling. A determination of the moisture sorption capacity of starch and gluten gels at 97–98% RH, as in the crumb, showed a decrease from 58% to 55% for starch after 1 day, and then a slower fall to 53% at 4 days, reducing to 51% at 7 days. The water sorption capacity for gluten, however, remained constant over the same period. Senti and Dimler (*Bakers Digest*, **34**, 1, 1960, 28) established 2% loss of moisture from starch during ageing, and point out that if all this water were taken up by the gluten, it would show a moisture gain of 12% (assuming the normal starch:protein ratio in flour of 6:1). Furthermore, this relatively small contribution to crumb rigidity could hardly account for the increase in firmness of the bread-crumb which takes place at a constant moisture content during ageing. The enrichment of wheat-flour doughs with such sources of protein as: soy, gluten, groundnut, and non-fat milk solids reduce staling rates and crumb-firming, but has little effect on the ageing of the crust, which loses its crispness, becoming elastic and tough. This latter trend is due to moisture exchange with the crumb, as is the case with non-enriched bread.

The non-starch polysaccharides, which include: cellulose, pentosans, hemi-cellulose, beta-glucans, mannans, gluco- and galactomannans, as well as glycoproteins, all constitute a group of substances which exert an effect on dough and baked products, right through to the staling processes due to their chemical, physical and rheological properties. Wheat-flours can contain 3% to 4% of this group, depending on extraction rate, some of which are cold water soluble, 20–25%. Rye-flours contain about the same levels at low extraction, but the cold water soluble fraction is about 40%. Kim and D'Appolonia (*Cereal Chem.*, **54**, 1977, 225) in studies of the effect of pentosans on staling rates, concluded that they slowed down the rate of crumb-firming, especially the water-insoluble fraction. The water-soluble fraction reacting with the starch amylopectin only, but the insolubles retarding retrogradation by interacting with both the amylose and amylopectin. Gilles *et al.* (*Cereal Chem.*, **38**, 1961, 229) found pentosans in the soluble starch fraction of bread crumb. Of the 4·3% soluble starch extracted from fresh bread crumb, 11·7% were pentosans. This compared with 3·3% soluble starch, of which 19·3% were pentosans in the case of stale bread crumb. This increase is attributed to amylose and amylopectin crystallization; and since pentosans retard the retrogradation of amylose, the fall in soluble starch on staling is deemed to be the result of amylopectin aggregation. Kim and D'Appolonia (*Cereal Chem.*, **54**, 1977, 150), in further studies on the effect of wheat-flour pentosans on starch retrogradation, observed a distinct retarding effect by the pentosans on the rate of retrogradation of starch gels on ageing, the water-insoluble pentosans being the most effective. The water-soluble fraction appeared to interact with the amylose, and the insoluble fraction with both the amylose and amylopectin. The presence of the pentosans tends to reduce the exposure of the starch components to crystallization.

Any complete study of the staling mechanisms should include, in addition to measurements of moisture redistribution between components of bread, swelling capacity, starch hydration and enzymic breakdown, saccharide analysis, pentosan analysis, protein extraction and resolution to monitor denaturation, soluble starch and amylose. Instrumental techniques should include rheological tests in elasticity and oscillatory modes; differential scanning calorimetry (DSC) endotherms; and Fourier transform IR (FTIR) spectroscopy to study biopolymer functionality in aqueous sols and gels. Changes in IR-spectra can be monitored during storage, and partly related to crystallinity development in materials. The time course of FTIR measurements during retrogradation of starches is similar to those produced from other techniques, e.g. shear modulus. This technique is applicable to starch gels, and real foods, e.g. commercial bread. DSC techniques are frequently applied to the study of the effects on bread staling and starch-gel ageing of various emulsifiers and crumb softeners. An example of this type of DSC application is that of Eliasson at the Chemistry Centre of the University of Lund, Sweden ('New approaches to research in cereal carbohydrate', *Prog. Biotechnol.*, **1**, 1985, 93–98). DSC endotherms of stale bread or aged starch gels gave amylopectin melt read-outs. The size of the endotherm increasing with increasing storage temperature and time. The addition of sodium stearoyl lactylate to the starch gel decreases the staling endotherm. Amylose/lipid complexes affect the crystallization of amylopectin. It was also found that the amount of unfreezable water in starch gel increases from 40% to 53% of total water content during storage at room temperature for 7 days, suggesting that water plays a role in starch retrogradation. It must be stressed, however, that DSC measurements do not monitor the same changes in ageing starch gels as those measured during gel-strength tests using compression techniques in the elasticity mode (Instron Universal Model TM-M). For the measurement of water concentrations in the free and bound state, specific NIR bands for water at 1940 and 1450 nm can be utilized. For example, the band at 1450 nm can be resolved into two discriminant spectral patterns at 1410 and 1460 nm for proportions of free and bound water.

As far as protein structural changes during staling are concerned, hydrogen bonding is of less significance, owing to the lack of free hydroxyl groups; instead, intermolecular bonds, due to dipole and van der Waals forces predominate. It is doubtful whether the primary peptide chains between the amino acids are influenced by the baking process, unlike the secondary and tertiary structure. Therefore, protein staling will depend on the degree of denaturation, which can be monitored by extraction and resolution, as suggested earlier. Auerman has established that changes in protein only take place at one-quarter to one-eighth of the speed of starch retrogradation. Banecki ('Influence of wheat and rye proteins on staling', Congress Report 5, the World Wheat and Bread Congress, Dresden, 1970, Vol. 5, p. 179) studied the physicochemical changes of the crumb structure during staling. He found that the extensibility of the gluten isolated from the crumb became reduced by more than 50% during 12 hours storage. No further changes taking place between the 12th and 48th hour. Both Hüttinger (*Gordian*, **72**, 7/8, 1972, 261) and Erlander and Erlander (*Die Stärke*, **21**, 12, 1969, 305) consider that the gluten protein/starch complex has an important role in staling, and retards the starch retrogradation.

The lipids of flour amounting to about 1–2% in wheat and rye, which contain the non-polar fraction, i.e. mono-, di-, and triglycerides, stearins and stearin esters, also free fatty acids; and the polar lipid fraction, sometimes known as 'lipoids', comprising phosphatides, and sterin glycosides and tocopherols. According to Acker and Hamza (*Getreide Mehl und Brot*, 7, 1974, 181–186), the non-polar (neutral) lipids amount to 35·8% of total lipids in wheat; and the polar fraction consists of: digalactosyl diglyceride 13·4%, monogalactosyl diglyceride 6·0%, phosphatidyl choline (lecithin) 2·1%, and lysophosphatidyl choline 9·2%. Acker *et al.* (*Getreide und Mehl*, **18**, 6, 1968, 45), consider that the polar lipids interact with the gelatinized starch, forming inclusion complexes with the amylose. However, both non-polar and polar fractions prolong the staling of bread, and supplementation with many of these have found industrial application. Additional compounds, which complex with starch, proteins, pentosans, and lipids, are known as: glycoproteins, lipoproteins, glycolipids, and phospholipids. These have also been found to possess valuable functional properties both in dough and final bread quality. Such substances as dextrins, sugars, enzymes and mineral salts, all contribute to crumb softness and the postponement of the staling processes.

Gels form from amylose suspensions during cooling, resulting from phase separation, giving rise to a three-dimensional polymer network; crystallization, which is detected by X-ray diffraction patterns is a slower process originating in the polymer-rich phase.

The retrogradation process is usually studied by a combination of X-ray diffraction, DSC, and shear modulus techniques. Essentially, starch gels are composites with gelatinized granules embedded in an amylose matrix. In the short term, the gel structure develops and is dominated by irreversible gelation and crystallization below 100°C within the amylose matrix. Longer-term increases in the shear modulus of starch gels are linked to reversible crystallization due to amylopectin within the granule on storage. Crystallization increases the rigidity of the granules and reinforces the amylose matrix.

Breads made with rye-flour, and those containing various percentages of rye-flour (*Mischbrot*) enjoy a much longer edible shelf-life than those made from wheat-flour alone. Whilst a wheat-flour loaf is considered to be still fresh at 16 hours, and still edible at 20 hours after leaving the oven, a loaf made with 100% rye-flour is considered to be still fresh at 35 hours, and still edible after 40 hours. A *Mischbrot* is judged to be still fresh at 19 hours, and still edible at 23 hours. Such differences are due to properties of rye-flour and its components:

—Higher water-binding capacity, due to the higher pentosan content, especially the water-insoluble pentosans.
—Greater susceptibility of the rye starch to amylolytic breakdown, owing to its lower degree of condensation and lower threshold temperatures of gelatinization. At 55°C about 40% of starch granules become gelatinized, at 60°C 75–80%, and at 65°C and above 90% plus.
—The proteins of rye have a lower gliadin and glutenin content than wheat generally, and a much higher water-soluble fraction. This is due to lower-molecular-weight proteins with lower polymerization. These forming water

soluble complexes with the pentosans resulting in the structural build-up of the crumb, the water insoluble pentosans being responsible for the water binding functionality.
—The presence of the sour-dough bacteria and lower dough pH of about 4·0, resulting in increased enzymolysis and therefore more saccharide and peptide degradation products.

### Flavour Changes in White Bread during Storage

The effect of fermentation conditions and storage-time on the sensory properties and volatile compounds of white bread have been investigated by Stoellman ('Shelf-life of foods and beverages, *Dev. in Food Science*, **12**, 1986, 293–301). Bread was baked after fermentation at different times and dough temperatures, then subjected to sensory and physicochemical analysis. Both freshly baked loaves, and loaves stored for 1, 2, 4, and 7 days were tested. The loaves being wrapped in aluminium foil, and stored at 20°C. The volatile compounds were analysed by gas-chromatography, using a headspace sampling technique, and identified by mass spectrometry. The results showed that varied fermentation conditions, gave small differences, while storage times were of great significance for the sensory properties of the bread. During the first 24 hours, there was a dramatic decline in freshly baked flavour, while sensory properties described as 'old' and 'stale' began to appear. Instrumental analysis of the bread crust revealed that the concentration of many volatile compounds, viz. aldehydes and ketones, decreased during storage. The concentration of the headspace gas of certain substances such as alkyl pyrazines appeared to have increased temporarily after 24 hours storage. The sensory properties of the crumb were weaker than those of the crust, but the changes showed similar profiles. The decline of the freshly baked character was not as sharp, whilst the appearance of properties such as old/stale became more pronounced in the crumb. The concentration of many volatile compounds increased during the first 2 days of storage, owing to diffusion from the crust. After that they decreased again. Analysis of the aroma profile of bread, performed by sniffing the effluent from the gas chromatograph were carried out to determine which compounds were associated with fresh bread flavour. The results show that this odour was associated with substituted pyrazines, present in the crust. The odour of freshly baked crumb could not be associated with a particular group of compounds.

Changes in the physicochemical properties of the crust represent a significant indicator of the staling process. Various crust zones show differences in moisture and mechanical firmness. Crust formation is a drying out process, offering protection for the crumb from microbial contamination and rapid drying out. With the drying of the crust zone, its mechanical firming becomes complete, and coupled with the setting of the crumb, a stable structure is formed. The maximum temperature within the crumb during baking is about 98°C, whereas the crust reaches about 190°C, which represents a considerable temperature gradient. Within the crust, dextrins, caramel and browning reaction products are formed, the high temperature resulting in irreversible changes in the starch and protein. The crust then loses its crispness on staling, becoming elastic and tough, due to part of the water diffusing from the crumb into the crust. The remaining moisture finding a state of equilibrium between the crust and the surrounding atmosphere.

If the crust of a loaf is separated from the crumb immediately after baking, and stored under conditions whereby moisture changes are ruled out, the crust retains all its freshly baked physical properties, therefore confirming that any changes which occur in the crust are a function of its state of hydration with time.

The processes of staling can be summarized by analysis into several phases, which occur as a result of a basically unstable thermodynamic state, the staling processes being essentially physicochemical in nature. These changes can only be influenced by regulation of the technological process.

The first phase of physicochemical processes is the formation of intermolecular bonding forces, loss of the hydration-shells in part (see Fig. 151), and molecular aggregation. This phase can be prolonged by the addition of antistaling/crumb-softening agents to a degree, choice of packaging material, and controlled storage temperature/RH conditions. The second phase of physicochemical processes involve structural re-arrangements of the molecules from energy-rich bonds (covalent and ionic) to low energy bondings (hydrophobic and van der Waals forces). The addition of antistaling/crumb-softening agents, and controlled storage conditions are the main retarding possibilities.

The summation of the physicochemical effects on loaf quality are: diffusion of moisture both in to the atmosphere and from crumb to crust, resulting in crumb firming, changes in solubility of the starch components, and proteins; and progressive loss of crumb swelling capacity.

The best technological control measure which is applicable to date to effect a retardation of these processes is to rapidly freeze the baked bread after about 4 hours cooling. Before freezing, the freezer unit must be pre-cooled down to 2–4°C, and the bread temperature reduced to −18 to −20°C rapidly, and maintained at this temperature. This constant final temperature prevents separation of crust from crumb, and excess ice-crystal sublimation in the crumb area under the crust. For longer-term storage, freezing to −25 to −30°C can be applied. In all cases, forced air circulation is required, and only products of the same dimensions should be placed in the same freezer unit. In this way bread can be maintained fully fresh at −30°C, the swelling capacity of the crumb approximating to that of oven-fresh bread. However, the flavour changes described above cannot be retarded or prevented by freezing, as any sensory evaluation will reveal. The description 'fully fresh' refers to the physicochemical properties of the frozen product only.

Measures should be undertaken to limit bread faults during processing of doughs, which in turn improve general quality and prolong staling, viz.

—Maintain fermentation times and temperatures as appropriate.
—When processing flours of poor swelling capacity, use a pre-ferment and an appropriate improving agent, increasing final proof-time.
—Avoid excess mechanical abuse of doughs, and adjust initial oven temperatures accordingly.
—Maintain baking times, and aim at same-day sale; store relative humidity

should not be above 60%, or under 30°C. Use of acidic additives, e.g. lactic acid lowers crumb pH and prolongs mould-free shelf-life of bread.

### Evaluation of Staling

As yet, it is not possible to measure the freshness of bread directly. Instead, one has to rely on a number of chemical or physical methods, and/or sensory evaluations of the crumb and crust. Sensory evaluation remains the most important from a practical viewpoint, and includes an assessment of the texture of the crust and crumb, especially the elasticity and porosity of the latter, and its colour. In addition, palatability, flavour and aroma of the bread is assessed.

Chemical tests are restricted to research work and often include a measurement of the iodine-binding intensity of the amylose and amylopectin. The starch being extracted from the crumb with a formamide ammonium sulphate-sulphosalicyclic acid solution, according to Hampel (*Brot und Gebäck*, **23**, 6, 1969, 106). Other tests are: crumbliness, soluble starch, soluble amylose and swelling tests. However, these tests do not always correlate with the organoleptic evaluation of a trained tasting-panel, with the help of statistical analysis.

Crumb firmness can be evaluated with good reproducibility with the Instron Universal instrument, when the method is standardized. This instrument consists of a drive mechanism, and a load-sensing and recording system. The recorded force/distance curves offering a wide range of measured parameters, which can be correlated with specific sensory properties. Among the instruments used for measurement of the rheological properties of the crumb are: the Penetrometer, the Amylograph, the Farinograph and viscometers.

Crumb swelling tests, that test the capacity to absorb water, which decreases with ageing, form the basis for several methods of estimating the degree of staling. Procedures essentially involve maceration of the crumb in water, then measuring the crumb volume, followed either by sedimentation, centrifugation or a determination of the increase in weight of the crumb sediment after centrifugation.

Enzymatic procedures include testing the sensitivity of the crumb to beta-amylase, and to the proteinase papain.

# 3.6   Dough and Bread Preservation

### 3.6.1 RETARDATION OF STALING PROCESSES

Currently, cryogenic freezing appears to be the most effective method for the preservation of the essential qualities of freshness, apart from 'oven-freshness' of baked products over prolonged periods. Long before the application of cryogenics, bakers were experimenting with production processes, formulations, and additives which would either prevent or delay the onset of staling characteristics, during the normal shelf-life of their products. On account of food safety regulations the choice of anti-staling agents has been restricted to existing food substances and their modifications. These substances can be grouped as follows:

*Protein basis:* vital wheat gluten; milk products; soy proteins; protein hydrolysates.

*Carbohydrate basis:* sugars; starch hydrolysates; modified starches; dextrins; pregelatinized starches and flours (*Quellmitteln/Quellmehl*), including guar-flour and locust or carob bean flour (German: *Johannisbrotkernmehl*); Sorbitol and Xylitol for specialty products; amylose.

*Cellulose basis:* pentosans; powered cellulose; xanthan gum; carboxy methyl cellulose.

*Enzymes and enzyme-containing substances:* alpha- and beta-amylases; malt products; amyloglucosidases; pullulanase; proteinases; pentosanases (for rye breads); cellulases (for specialty products); enzyme-active full fat soy flour.

*Shortenings and emulsifiers:* vegetable and animal shortening blends with melting points within the range 40 to 50°C; lipoprotein (protein 70% and lipoid 30%); glycolipids; lecithins and hydroxylated lecithins; glycerol fatty acid esters (glycerol mono- and diglycerides, polyglycerides, sugar and sugar alcohol fatty acid esters); propylene glycol monoesters; diacetyl tartaric acid esters (DATEM) of mono- and diglycerides.

*Aliphatic and cyclic aldehydes:* acetaldehyde, paraldehyde, aldol and furfuraldehyde (with increasing chain-length the anti-staling function increases); practical application is not yet allowed, due to toxicity. Their action is due to location between the hydroxyl groups of the starch and protein, thus blocking the hydroxyl groups and retarding staling processes.

*Dough improver/conditioner mixtures*: these are manufactured in powder, paste, dispersion or tablet form, and consist of blends of emulsifiers, enzyme preparations, shortenings, malt flours, sugars, hydrocolloids, and carrier material.

Often, ascorbic acid is added with or without other oxidants, to form a complete processing aid. The application of most of these substances have been discussed in Section 1.5.

The most widely accepted emulsifiers are the phosphatides, of which lecithin is the commonest. These polar lipids, of natural origin, form complexes with the amorphous starch fraction, which reduces starch hydration, or prolongs the process. The presence of non-polar lipids prolong hydration of the crystalline regions of the starch granule. Also, according to Grosskreutz, the sheet-like phospholipid molecule becomes bound to the proteins by a salt-type bonding. Apart from retarding staling, the phosphatides have a valuable function when used in combination with vegetable oil and water in the ratio 5:5:90 in emulsion form. This emulsion at 0·05% flour weight is then combined with 0·3% enzyme active full fat soy-flour (flour weight), as a source of active lipoxygenase, and made into a dough with wheat-flour, yeast, salt and water in the usual manner. The soy-flour is ideally first solubilized in part of the dough water to release the lipoxygenase into solution. In this way, the polyunsaturated fatty acids of the vegetable oil and the phosphatide emulsion concentrate are oxidized by the lipoxygenase and the entrapped oxygen. The resulting hydroperoxides act as active oxidants, reacting with the —SH— groups of the protein–proteinase complex of the wheat-flour. This reaction improves the gas and structure retaining capacity of the dough, resulting in bread with improved volume, symmetry, texture, and crumb elasticity. Also, the peroxides of the fatty acids oxidize and bleach the carotenoid pigments, giving the crumb a brighter colour. This technique is a cost-effective method of processing untreated/unbleached wheat-flour with successful results.

The other cost-effective and safe anti-firming agent, of quality and long standing, is enzyme-active, full-fat soy-flour. The addition of 0·5% based on flour weight, of this enzyme active material reduces the firming rate of bread by reducing the rate of moisture interchange between starch and gluten proteins in the crumb. The average composition of the full-fat soy-flour is: protein 41%, fat 20% and moisture 5%, with about 3% pentosans,which contribute to the water-binding capacity of the dough. The most commonly used soy-flour for baking is the defatted variety, which has a reduced beany flavour and inactivated enzymes as a result of heat treatment. This material can be used at levels of 3–5% flour weight. However, owing to the heat treatment, its protein dispersion index (PDI) becomes reduced from about 90 to 70, which reduces water dispersibility, and its desirable functional properties are less apparent. Detailed information concerning the numerous improver/conditioner agents is given in Section 1.5. Many of these have anti-staling properties, but owing to current trends, both from the consumer and international food legislatory organizations, the inclusion of most of these substances does not satisfy the 'clean label' requirement.

## Packaging
Staling processes can be appreciably retarded by the application of appropriate packaging materials, and the products protected from external contamination.

Retardation of staling is due to the packaging material hindering dehydration by preventing the partial water pressure from causing further moisture loss to the atmosphere. Therefore, the degree of staling retardation depends on the choice of packaging material. In the case of sliced bread, the requirements from the material are the following:

—non-toxic, and without smell or flavour token,
—as good an insulator as possible,
—good sealing capability, preferably by heat,
—essentially water-tight, but partially pervious to steam from within,
—essentially translucent,
—resistant to alkali, acid and organic solvents,
—cost-effective versus properties.

In practice, all requirements cannot be fulfilled by one material, therefore a compromise must be found between price and functionality. Also, the material must be compatible with the packaging-machine design, as intended by the manufacturer, otherwise frequent interruptions occur.

Waxed paper was the only wrapping material used for bread at the turn of the century, and prevailed for many decades. At first this was coated with paraffin-wax, but later this was blended with micro crystalline wax and polyethylene resin to improve its appearance, elasticity and sealing properties. The use of paper having the added advantage of ease of imprinting for labelling purposes.

The introduction of cellophane films in the 1930s soon spread to the baking industry, and gained wide acceptance. Originally, it was in the form of a film or sheet of a defined thickness, which determined its relative strength. This was improved by the application of coatings of either nitrocellulose or polyvinylidene chloride, which improve its resistance, gas permeability, heat-sealability, and moisture vapour transmission rate (MVTR), and is also an effective oxygen barrier. It is an excellent material, applicable to virtually all baked products, but is relatively expensive. During the 1950s, the plastic polyolefin films replaced cellulose films for certain product applications. Polyethylene and polypropylene, as examples of these transparent materials, differ in tensile strength, rigidity, permeability and temperature-sensitivity. Various methods of manufacture provide a wide range of films of different gauges and orientations (cast/unbalanced) and coatings, and laminate combinations. Their special properties and suitability for application must be carefully evaluated. The following is a summary of the manufacture and properties of the basic materials.

Cellophane is prepared from cellulose by hydrolysis, followed by formation into foil, drying and bleaching. It is transparent, gas-, aroma-, fat- and oil-proof, but porous to steam. It can be sealed by sticking, and can be coated with nitrocellulose or polyvinylidene chloride. It can be made weather-proof and impervious to steam, and is heat-sealable, but remains prone to splitting.

Polyethylene is prepared by polymerization from ethers under pressures of $10^2$ to $2 \times 10^2$ N mm$^{-2}$ at 200°C, yielding high-pressure polyethylene, under normal pressure and vacuum conditions with a catalyst. High-pressure polyethylene is transparent, water-proof and impervious to steam, being stable within the temperature range $-50$ to $+80°C$. It is less impervious to gas, aromas, fat and oil, but can be heat-sealed.

Polyethylene film, available in high, medium and low densities is the most popular packaging material for bread and rolls, the standard low density being the basic material for perforated bags. This film satisfies normal packaging requirements, but where good heat stability is necessary, a medium density film is appropriate. Polypropylene has similar properties but is more rigid and stronger at the same gauge or thickness. It is used as a component in laminated films to give stability at heat-sealing. Special inks and printing techniques provide a range of possible colours and designs for marketing purposes.

Hoffmann *et al.* (*Extending the Freshness of Bread, Rolls and Specialty Products*, Part I: *Bread and Rolls*, Research report IGA, A4, Bergholz-Rehbrücke, 1976) in the GDR, found that the use of different packaging material gave different improvements in the retention of freshness, as follows:

| Packaging material | Improvement in retention of freshness (%) |
|---|---|
| Aluminium–polyethylene bags | maximum 23 |
| Polyethylene coated paper | maximum 16 |
| Polyethylene film | maximum  9 |

The rapidity of staling was also found to depend on storage temperature and relative humidity. In the case of *Mischbrot* (rye/wheat), when stored at 20°C, the crumb firmness within 24 hours was equivalent to that at 28 hours when stored at 30°C; and at 60°C the crumb firmness at 48 hours was comparable with that at 16 hours at 20°C. The greatest crumb-firming was noted at 4°C.

The principal criteria which the baker must apply when making a decision on packaging are:

—degree of product protection,
—performance on existing packaging machines,
—attractiveness of the package to the customer, to keep ahead of competition,
—does it justify its cost?

Since polyolefin films have become established as the predominant packaging material for bread, plastic films can be expected to grow in popularity. Their application for bun, and orientated polypropylene for sweet goods, confirm this. Coated, orientated polypropylene find increasing application in overwrapping bakery products in trays and cartons. For specialty products where a low moisture transmission vapour rate (MTVR) and oxygen transmission rate (O2TR) is required, aluminium foil gives the lowest values, the cost being about 2·6 times greater, relative to polyethylene.

## 3.6.2 PREVENTION OF MICROBIAL INFECTION

Since the possibility of infection under industrial baking conditions cannot be ruled out, certain safety precautions must be met to limit the conditions favourable to the growth of moulds and bacteria. General maintenance of cleanliness and hygienic practices by personnel within production, storage and dispatch areas is a prerequisite. All transport containers, operators' clothing and wash areas must be regularly cleaned and disinfected.

Any infection of the products can be reduced by reducing the air-borne spore count within the bakery. This can be achieved by UV-irradiation of the air drawn into the bakery for ventilation, and irradiation of the internal atmosphere by the use of wall-mounted units at a height of about 2 metres. As a result of convection currents, the spores are destroyed when they pass within about 3 cm of the radiation source, and are met at right-angles by the rays. Equipment can also be UV-irradiated. Personnel must be provided with green-filter goggles, or the irradiation carried out in their absence. In order to obtain the best effect from air-filters, the areas must be hermetically sealed, and the air should be cooled before entering the filter. However, filter design, size and electrical charge of the particles, their temperature and RH, and also air pollution, all influence their effectiveness. In the bread slicing and wrapping department a strict hygiene regime must prevail, which must be adhered to at all times. Only people who work therein should have access; temperature of 18–20°C and RH 65% should be automatically regulated; personal hygiene and regular medical check-ups for all personnel working in the slicing and wrapping are essential disciplines.

Careful control during processing can also help to prevent any microbial infections or premature onset of staling processes. Optimal fermentation exerts a significant influence on the keeping properties of bread. Using normal processing methods with ample levels of yeast (about 2% flour weight), dough temperatures should be held within the range 22–26°C, and doughs allowed to mature for a maximum of time, depending on the fermentation tolerance of the flour quality in use. The limiting factor is always flour quality.

In order to prevent any incidence of *Bacillus mesentericus* (potato bacillus) or *B. subtilis* (hay bacillus), which are broadly distributed in air, soil and plants, and which can be found in grain and flour at times, the baker can also take preventative measures. Since these bacilli, when present in the dough, also survive the baking process and liquefy and destroy the crumb structure, their optimal conditions of reproductive growth must be avoided. This is best achieved by ensuring that the dough pH is always kept within the range 4·8 to 5·0 in the case of wheat-flour doughs, and in the case of doughs from rye wholemeal flours, the final bread should have an acidity-index of 12 (*Standard Methods AGF*, Detmold, FRG, p. 79, 5th edn). In the case of dough sours, this can be done by increasing the souring intensity of the basic sour, or in both yeasted wheat and soured rye doughs by the addition of an appropriate amount of lactic or acetic acid. In this respect a target pH of 3·8 in the final dough should solve the problem.

In many countries, chemical preservatives or mould inhibitors are widely applied, depending on the food regulations. These include salts of propionic and acetic acids and sorbic acid and sorbates. In the USA, a maximum of 0·32% sodium propionate, 0·4% sodium acetate, or 0·75% calcium monophosphate are applied, all based on flour weight. Sorbic acid used at the 0·1% flour weight level is a very effective fungistat, owing to unsaturation in molecular terms, and lower dissociation constant in solution. Where shortenings are used, it can be added to the dough after premixing with the shortening to avoid yeast inhibition. Hickey (*Bakers Digest*, **54**, 4, 1980, 20) has reported that a spray application of 1·0–1·5% potassium sorbate solution on hot freshly baked bread, rolls and partially baked products is effective in doubling or tripling the mould-free shelf-life of the products. This gives sorbate residuals of 0·02% based on flour.

Schulz (*Brot und Gebäck*, 13, 7, 1959, 141) has suggested the use of a sour combining lactic and propionic fermentation, by using special cultures of propionic acid bacteria. This resulted in a prolongation of bread moulding to 14 to 19 days, without affecting either the flavour or volume of the bread.

The continental European practice of using relatively high initial baking temperatures, known as '*vorbacken*', resulting in early crust formation, is recommended for hearth products. This is then followed by extended baking at a lower temperature.

### Pasteurization/Sterilization of Bread and Baked Products

The heat sterilization of specialty sliced breads, e.g. pumpernickel, is carried out either at high temperature, or by energy-rich ionized radiation. Sterilization within the packaging is the best guarantee for prolonged mould-free shelf-life, providing the packaging material has been sterilized, and no further contamination takes place. The process depends on temperature and time. At 150°C, a sterilization time of 15 minutes is normally adequate; but at 98°C an exposure time of 12 hours is necessary to obtain the same degree of sterilization. Where high-frequency microwave heating is utilized, exposure times of 10 s to 2 minutes are sufficient. The choice of conditions depends on the type of product, its dimensions, and the packaging material used. An average schedule could be 2 hours at 90°C for a 250-g sliced bread pack. Owing to the effect of temperature the package shrinks, and the overlapping film is immediately sealed from external contamination. The heat treatment is normally carried out in a tunnel heated with either gas or electricity, positioned downstream of the packaging machine and adjacent to it. There is a temperature gradient between the inlet and outlet end of the shrink-tunnel, the packages being transported through on a conveyor-band. The heat-stability of the packaging material is often the limiting factor for the choice of the sterilization temperature. When a certain temperature is exceeded, most materials become brittle or melt, no longer protecting the bread from contamination. The application of microwave heating will not allow the use of metallic packaging materials, instead, coated polyethylene films are more appropriate.

These techniques are routinely only applied to specialty products such as pumpernickel and whole-grain meal breads, e.g. *Vollkornbrot*, which have a dense, close texture. Breads from low extraction wheat-flours with high volume, porous textures could suffer damage from heat sterilization. Burg (*Brot und Gebäck*, **22**, 58, 1968) describes a tunnel microwave sterilizer of 400 to 600 lb bread capacity, sterilization taking place by conveying the packaged bread through the unit with a residence time of 45–90 s. In this case a moisture permeable wrapping material must be used, since the shock treatment results in some moisture release within the packs.

Alternatively, a moisture-permeable material can be used, followed by an airtight material; otherwise, opened plastic bags can be used, the bags then being sealed on completion of sterilization and moisture evaporation. Zboralski (*Getreide Mehl und Brot*, **27**, 6, 1973, 213–216) describes the application of IR-sterilization for bread and baked products, which offers considerable technological and economic advantages. The method was first applied in France in 1965 to sliced bread and cake. The process utilizes IR rays in the middle range of the spectrum, about 2500 Ångström degrees for bread and baked products. The radiation is absorbed by the product, and transformed into heat, reaching temperatures of 160–170°C. Important for the success of this process is the packaging material, a high-temperature-stable special polyamide foil (nylon-type) being most suitable, together with a low vacuum. The packaging material must satisfy certain technical specifications, viz. have heat-stability up to 200°C for the duration of sterilization; be steam-proof, gas-proof and impervious to spores; have good heat-sealing capability; allow penetration of IR-rays without loss of energy.

These properties could only be satisfied by a stabilized film from Polyamide 11 ('Rilsan'), available in tubular form of 40–60 $\mu$ strength. The product is placed in the bag by means of special equipment, and then packaged under a low vacuum with a vacuum sealing machine, the latter being adjusted to avoid product deformation.

The generated IR-rays from parabolic lamps (220 V/250 W) are chemically and biologically neutral, causing no unfavourable reactions within the product, unlike ionizing or UV radiation. The tungsten alloy elements are capable of withstanding temperatures up to 2200°C, with a life of 3000–4000 hours, and the lamps are coated internally with silver to ensure maximum radiation. Compared with standard illumination lamps, which only utilize 4% of energy input, these IR-lamps can utilize 96% of total radiated energy. The short to medium IR-rays provide a much greater penetration than the long wavelength IR-rays, which are emitted from ceramic or iron rods or surfaces. The great advantage of the IR-sterilization process is the extremely short time required compared with the conventional heat-convection process in a drying-chamber or baking oven. This is due to a two-phase system of heat-transfer, which proceeds as follows. (1) The IR-rays of optimal wavelength penetrate a few millimetres into the product, and are transformed almost instantaneously into heat energy. This nascent, *in statu nascendi*, liberated energy achieves levels which cannot be accurately determined. One can estimate temperatures of about 300°C, which become spontaneously distributed by conduction throughout the interior of the product. On this basis, the temperature at the centre of the product rapidly attains the desired 70–80°C. (2) IR-sterilization results in much greater energy transfer onto the product surface than in the case of hot-air convection, thus providing a rapid product penetration. This allows a reduction of the sterilization-time by 50%, compared with that of hot-air sterilization. The choice of packaging material for IR-radiation of bread and baked products must satisfy a number of technical specifications, viz. (a) heat stability up to 200°C for the duration of the sterilization; (b) water-vapour and gas-proof, as well as impervious to microbial spores; (c) good and reliable heat-sealing capacity of the film; and (d) penetration of the material by the IR-rays without energy loss.

Polyamide film is slightly pervious to water-vapour; therefore, over a period of one month, a weight loss of 2–4% per pack occurs. Therefore, either a 60 μm film, or a 40 μm film with an overwrap of Cellophane treated for water-vapour proof properties is recommended. This allows printing, and improves the appearance.

The packed product is transferred to a grill-profile band conveyor, proceeding into the IR-tunnel of about 10 m length, moving at a speed of 80 cm/min. The sterilization-time can be regulated by controlling the speed of the conveyor. The tunnel is equipped with 250 W IR-lamps positioned both above and below the conveyor-band, 1 m² being provided with about 30 lamps, with an output of 7 kW. To maintain a tunnel temperature of 160–170°C, the lamps are connected via a mercury contact relay switch, which switches on and off with a defined rhythm. The products have a spacing of 6–7 cm on the band-conveyor so that the whole bread surface is exposed to the rays. The side-walls of the tunnel are covered with reflective metal to obtain maximum IR-radiation. The IR-lamps within the tunnel are positioned in such a manner that three zones of different heat-intensity and length are created. The first zone consists of 1 to 1·5 m, with a temperature of 160–170°C and an 80% output, providing a rapid heating up of the product to sterilization temperature. The second and third zones are each 3·5 m long, and are maintained at temperatures of 160 and 155°C respectively. These represent 60% and 50% of plant capacity. In these zones, the attained temperature is maintained or reduced slightly as necessary.

As a result of the radiation, a very rapid heating-up of the product surface takes place, resulting in considerable expansion of the residual air, and water-vapour development within the packs. This can result in the packs becoming inflated, but this does not affect the sterilization intensity. On leaving the IR-tunnel, the products are transferred to another band-conveyor and allowed to cool gradually; externally applied cooling should be avoided since this results in condensation on the inside of the packaged product. The temperature at the centre of the product on leaving the IR-tunnel is 40–55°C, and first attains the desired centre temperature level of 70–80°C after 5–20 minutes into the cooling phase. The required sterilization-time to adjust the centre temperature depends on the texture of the product, its porosity and crumb-density. The most rapid heat conductivity occurs with light, high volume products, e.g. toast and white breads. The more dense and close the crumb-porosity, the longer it takes to reach the desired centre temperature. Other factors which influence the required sterilization-time are the dimensions of the product and whether or not it is sliced. Sliced products generally require shorter sterilization-times. The comparisons given in Table 73 illustrate the relative sterilization-times required for a diverse range of products, using a tunnel temperature of 170°C for all products.

In the case of all products, the final temperature in the centre is determined by heating up time in the IR-tunnel, and the heating up which occurs during the cooling phase outside the tunnel. In this connection, it is surprising to note that time of heating during the cooling phase is only marginally less than that during the actual heating-up time in the IR-tunnel. Measurements have confirmed that the centre temperature after the heating-up time in the tunnel only reach 40 to 55°C, therefore, the cooling phase is very important for the attainment of the desired centre

TABLE 73

| Product | Dimensions ($l \times b \times h$) | Volume ($cm^3$) | Weight (g) | Density | Time required (minutes) | Centre temperature (after time in minutes) |
|---|---|---|---|---|---|---|
| Toast | $90 \times 90 \times 120$ | 960 | 254 | 0·26 | 9·0 | 65°C (18) |
| sliced | | | | | 9·5 | 70°C (16) |
| Wheat/rye | | | | | | |
| bread 70/30 | $100 \times 130 \times 90$ | 1 380 | 500 | 0·38 | 15·0 | 65°C (25) |
| sliced | | | | | 17·0 | 75°C (25) |
| Rye/wheat | | | | | | |
| bread 70/30 | $100 \times 110 \times 105$ | 1 150 | 505 | 0·44 | 15·0 | 65°C (29) |
| sliced | | | | | 17·0 | 75°C (28) |
| Graham bread | $90 \times 115 \times 100$ | 900 | 503 | 0·55 | 15·0 | 65°C (27) |
| sliced | | | | | 19·0 | 75°C (30) |
| Linseed bread | $80 \times 115 \times 90$ | 820 | 487 | 0·59 | 17·0 | 65°C (30) |
| sliced | | | | | 19·0 | 75°C (29) |

Source: Zboralski, U. (Budenheim) *Getreide Mehl und Brot*, **27**, 6, 1973, 214 (in German).

temperature. This emphasizes the need to determine first by experimentation the necessary heating up time for each type of product. The best correlation appears to exist between product density and heating up time.

Storage tests at ambient and incubation temperatures of IR-sterilized products versus non-sterilized controls, and some parallel samples additionally infected with mould-spores, showed that on average non-sterilized samples often had signs of mould growth after a few days. Whereas, sliced bread packages, sterilized at centre-temperatures of 65°C remained mould-free for up to 2 months. The use of centre temperatures above 70°C resulted in zero mould growth, even in the case of the additionally infected packages. Entire unsliced bread products, as well as cakes required only 8–9 minutes sterilization-time to ensure an elimination of mould growth. Cakes packed in aluminium pans and sterilized for 8–9 minutes gave similar results. However, the sealing efficiency of the package is equally important to maintain a mould-free shelf-life. It has been demonstrated generally that the IR-sterilization of bread and baked products is very effective and reliable in the preservation of their mould-free shelf-life, thus confirming industrial experience gained in France using IR-sterilization and Polyamide-11 ('Rilsan') for cakes packed in aluminium-pans, and sliced honey cake (*pain de miel*). In France, preservation times of up to 1 year have been guaranteed to customers. Such techniques are ideal for long wholesale distribution chains marketing specialty types of bread in delicatessens and health shops, and packaged cakes sold from vending machines and kiosks in cities and railway stations worldwide. In such cases as export and camping services, where fresh deliveries are impossible, IR-sterilization offers distinct advantages with distribution problems.

Other preservation techniques successfully applied to bread include, X-ray, beta- and gamma-radiation. Warming of rye/wheat breads to 50°C, and irradiation with

0·05 Mrad, resulted in a preservation of the bread for several weeks, without any sensory devaluation (Stehlick, G., 'Gamma-rays on bread', *Atompraxis* (Karlsruhe), **14**, 4/5, 1968, 1, in German). For the application of ionized radiations for preservation, specially treated polyethylene has proved a suitable material. Preservation by ionized radiation has not yet become widely accepted, since their potential health hazards are not yet fully proven.

Although the packaging of bread has no effect on the chemical staling processes, it provides a hygienic barrier for subsequent handling, and reduces the rate of crumb-firming compared with unwrapped bread. This is largely due to a reduced moisture loss, as Martin (*Food*, **13**, 1944, 129) demonstrated in samples of 2-lb loaves of white bread, stored in the wrapped and unwrapped state for 6 days. Wrapped loaves only lost 0·85% in weight, whereas comparable weight losses in the unwrapped state were about 10·25% during the winter, and about 12·4% in summer. Cathcart (*Cereal Chem.*, **17**, 1940, 100), in his studies, also found that sliced bread packaged in moisture-proof material only lost about 2% over a 72-hour period, due to a state of equilibrium being reached between crumb and crust. In the case of unwrapped bread, moisture is lost mainly from the outer 0·5 in of the crumb after about 24 hours, forming a hard, dry crumb adjacent to the crust. Berg (*Am. Soc. Bakery Engrs Bulletin*, **39**, 1929) investigated the influence of the wrapping temperature on the keeping properties and sensory characteristics. Temperatures of 45, 38 and 31°C were compared for wrapping suitability, and the samples allowed to cool to ambient temperature, one sample after wrapping at 31°C being placed in the refrigerator. The sample breads were evaluated by a panel for crumb-softness, aroma and taste, and also any mould contamination, after 36 hours storage. Warm wrapping at 45°C gave the best crumb-softness, but scored the worst for aroma and taste. Conversely, cool wrapping at 31°C gave the greatest loss in crumb-softness, but the best retention of aroma and taste. Rapid cooling in the refrigerator scored second-best on all counts. However, these wrapping temperatures refer to panned bread of high specific volume, and would not be valid for hearth-breads of lower specific volume, especially if they are made from rye/wheat-flour blends. If such products are to be sliced warm, the maximum temperature in the centre of the crumb should be 40°C; and when cold slicing is the chosen procedure, crumb centre temperatures should fall within the range 20–25°C. European experience indicates optimal bread maturing times of 2 hours for warm slicing at 40°C in the case of wheat-flour bread, and 3 hours for rye and rye/wheat bread types. For wrapping hot crusty bread in the unsliced form, a 360 μm perforated polypropylene film has proved to be a versatile material. It is obtainable as a centre folding film on the reel, or as bags. A new white pigmented, biaxially orientated polypropylene film 35 μm thick, and co-extruded on both sides with polyolefinic copolymers, has been found to be ideal for specialty-breads. It offers excellent water-vapour retention properties and UV-light transmission, and is heat-sealable on both sides.

Nevertheless, it must be acknowledged that under production conditions, it is impossible to package baked products under aseptic conditions, guaranteeing that no spores enter the packaging area. For the packaging of bread and rolls intended to be consumed fresh, or for consumption within 1 week, the slicing and wrapping room atmosphere must be kept as spore-free as possible; and so too must the packaging materials and the packaging-machines, their blades being kept clean and sterilized. The packaging of baked products should prolong their shelf-life by a factor of two to three times that of unpackaged products. This demands a rapid cooling of the products to room temperature, immediate packaging, and the stability of the packaging material to damage with safe and reliable sealing.

The introduction and implementation of an adequate quality control programme should minimize any structural or sensory faults in baked products, which can also limit their shelf-life. This programme must cover raw materials, formulation errors, environmental conditions, and adherence to correct times, temperatures and humidity requirements of processing and baking. Equipment must also be maintained in optimal operating condition. The cause of many bread faults are often difficult to diagnose definitively, and the close attention to raw material quality, and accurate control of the production processes will prevent any final product quality defects.

## 3.6.3 DOUGH PRESERVATION BY FREEZING

The low-temperature treatment of bakery foods provides a very rational method of temporarily halting or retarding the chemical and biological reactions which normally take place. Frozen bread, bread-rolls and cakes, and their intermediate products in dough form can be rendered stable for weeks or months at appropriate storage temperatures in the freezers of bakeries, or retail outlets. Whether defrosted and warmed up in the case of baked products; baked off in the home, or baked-off by a small retail outlet or in-store bakery in the case of the frozen dough-piece, the eating quality of the final product has the characteristic of 'freshness'. In the USA, the pioneering research work of Bailey, Cathcart and Luber in the 1930s, and that of Pence *et al.* in the 1950s, laid the foundation for the freezing concept within the baking industry. Pence's work also determined the optimal parameters for its commercial realization.

The freezing of prepared and shaped dough-pieces on a large scale for a wide variety of baked products attracted renewed attention with the introduction of the in-store bakeries within the multiple chain stores and supermarkets. The advantage for this type of operation is that the frozen dough-pieces, can be stored, thawed, proofed and baked as and when required by relatively unskilled, in-house trained personnel. Where the frozen merchandise is supplied by centrally located frozen dough manufacturers, minimal equipment is required to proof and bake off the various products according to demand. However, in order to provide the consumer with fresh products all day long, it is essential to build up an inventory. For this purpose adequate freeze-storage facilities providing a product core (centre) temperature of $-18°C$ will greatly increase supply flexibility.

Another market for frozen doughs is the domestic consumer who likes to do some baking at home, but wishes to save the time of preparation. Also, institutions responsible for catering for large numbers are tending to rely increasingly on frozen dough products owing to the lack of skilled labour, economy and product quality-uniformity. An additional frozen dough market sector is that of two-stage baking products or 'brown-and-serve', which originated in the USA in the 1950s. After proofing, the product is baked at a sufficiently high temperature to give adequate crumb rigidity, but no crust colour. This is achieved by allowing the core temperature to reach 75–80°C for a duration of 10 to 15 minutes only. This can be either completed in a conventional oven at 135 to 147°C, or by utilizing a microwave oven.

The importance of this freezing process technology is that one can separate in terms of time and place the dough-preparation and fermentation from the baking or 'bake-off' process. This separation, according to the author's experience, tends to have a positive effect on the quality of bread-rolls. Probably because of the reduced time-pressure during fermentation, and longer maturing times, the rolls acquire an improved aroma and flavour.

In the German-speaking countries of continental Europe, where a bread-roll is no longer considered fresh after 4–5 hours of leaving the oven, owing to a loss of crispness, this freeze–thaw–proof–bake cycle allows the industrial bakeries and in-store bake-offs to compete with the master-craftsman baker to a degree. Nevertheless, only the master-craftsman has the expertise to produce quality in this type of product area, where the consumer is quality-conscious and discerning. In Bavaria and Austria, where the *Kaisersemmel* (bread-roll) is the favourite partner for food and wine, these rolls have to be eaten within 2 hours to be fresh. Therefore, production is organized on a continuous basis in small two-hourly batches; for this purpose a freeze–store–thaw–proof and bake-off cycle permits supply to meet demand with a minimum of stress for the baker.

In France, although the bake-off concept is growing in popularity, the bake-off equipment is mainly to be found within the bakery, rather than being exposed in the retail outlets.

In The Netherlands, bake-off equipment is utilized by bakers to bake their own products, in outlets not necessarily attached to the main production bakery. Products usually include a local specialty in addition to the standard range of snack-products.

In Belgium, bake-off equipment is usually to be found in the supermarkets, often located in an in-store bakery, sited so that it is in full view of the public.

In the UK, bake-off equipment is utilized by the in-store supermarket chains, private baker, and smaller retail grocer and general-store owner on a reduced scale. Typical product ranges include: bread-rolls, doughnuts, croissants, Danish pastry, Chelsea buns, Belgian buns, white, wholemeal and brown loaves of 400 g, baps, and hamburger buns. Typical part-baked (brown-and-serve) products include croissants, puff pastry, short crust, French bread, Belgian buns and American cakes. The multiple industrial bakery groups also operate in this market through frozen-dough company subsiduaries, e.g. Speedibake of the ABF group, and Falconis, acquired by Northern Foods group in 1987.

**Freezing Systems Applied to Dough Products**

The two basic systems utilized for baked products are also applied to dough. Mechanical refrigeration, involving an air blast to cool the product is applied more extensively for baked products or dough-products generally than the cryogenic system, owing to the difference in the operational costs. Where throughputs of 1000 kg/hour and more are required, the application of the cryogenic system with either liquid nitrogen or carbon dioxide as the coolant becomes viable. Since this also involves spraying the refrigerant onto the products, while they are passing through a tunnel on a band-conveyor, the process expenditure on coolant becomes a continuous one.

Blast-freezing provides a means of spreading production so that the quieter periods during the week can be utilized to build up an inventory or stock to ease pressure during the busier peak periods. For example, products can be produced on a Monday and Tuesday ready for sale at weekends. Seasonal peaks such as Easter and Christmas can be organized well in advance, and products placed in the cold-store for withdrawal as required.

Blast-freezing involves the application of very cold air derived from evaporators of large output capacity, moving the air rapidly across the products to reduce the centre (core) temperature to about $-20°C$. Typical air temperatures fall into the range $-35°C$ to $-40°C$. Choice of air speeds shows some variation; on the continent of Europe air speeds of 2–4 m/s are a popular range; in the USA air speeds of 600 ft/min appear to be generally acceptable during transit through the freezing chamber; in the UK, air speeds of about 4000 cubic feet/minute are often suggested. In all cases, the limiting factor on air speed is the heat transfer coefficient produced at a given air speed. Where an air speed increase starts to reduce the heat-transfer coefficient, optimal working conditions progressively decline.

The commonest general procedure for the preparation of a frozen dough is to prepare a rather tight (stiff) no-time straight dough, using a high-speed mixer with a refrigeration jacket to keep the dough at 18 to 22°C. This produces a plastic dough, inhibits fermentation and reduces the freezing-time. Mixing is to optimal development, indicated by clean-up (when dough pulls away from mixer-wall). For direct freezing, only allow enough floor-time to recover from mixing, before machining—approximately 15 minutes. The dough is then divided and rounded using standard equipment. In some cases, the rounded dough-pieces are given a 3–5 minute intermediate proof for relaxation. The moulded and shaped dough-pieces are then deposited onto trays, which are then loaded into racks or onto conveyors, being either held within a freezer-cabinet for the required time, or slowly passed through a freezer-tunnel. Temperatures within the main freezer area are maintained within the range $-30$ to $-40°C$, with product exposure to an optimal air turbulence during residence or transit through the tunnel.

Freezing of the dough-piece proceeds from the exterior to the interior, the outer layers being frozen solid first of all. When the centre core of the product reaches $-20°C$, all fermentation activity has become suspended, and the product is solidly frozen.

According to Marston (*Bakers Digest*, **52**(2), 18, 1978), and the industrial experience of many others, rapid freezing of the outer dough layers, resulting in a solidly frozen layer of 3–4 mm minimizes moisture loss. Rapid freezing gives little opportunity for fermentation to proceed, which can result in surface-spotting, moisture loss and ice-crystal formation which can give rise to soggy textures with some products, e.g. gateaux. In addition, meat-containing products, e.g. pasties and sausage-meat rolls, must have a minimal time-lapse between shaping and cold storage to reduce microbiological hazards. Once the core of the products have reached zero degrees, further freezing can be completed in a holding-room at $-23°C$. Where possible, frozen dough-pieces are packaged in polyethylene bags, and packed in moisture-resistant containers before transfer to the cold-store. Although cold-store temperatures within the range $-15$ to $-20°C$ are usually regarded as standard, it is preferable to maintain an average of $-23°C$. Lehmann and Dreese (*Am. Inst. Baking Tech. Bull.*, **3**, No. 7, 1981) at the American Institute of Baking, during studies entitled, 'Stability of frozen dough—effects of freezing temperatures', observed that in the case of frozen bread doughs, the best results were obtained by blast-freezing at temperatures $-21$ to $-29°C$ to a core temperature of $-9$ to $-5°C$.

The important aspect for frozen dough products is stability, and the many factors which determine it. The ultimate dough, when thawed or defrosted, must respond to proof within an acceptable time, and bake out into a product of normal volume and characteristics.

During studies concerning the effect of dough fermentation-time, and freezing conditions on gluten quality properties, and final bread quality, Teshitel, at the Technological Institute at Odessa, USSR (*Khlebopek. Konditer. Prom-st*, **8**, 1987 23–24, in Russian) made the following significant observations. Bread dough, fermented with yeast for a period of less than 2 hours, and then held in frozen storage at $-12°C$ or $-18°C$ for a 4-week period, resulted in gluten quantity increases at 2 weeks and $-18°C$, but gave a decrease at $-12°C$. Rheological properties and dough hydration-state changes were similar at both storage temperatures, but less expressed at $-12°C$.

## Commercial Applications of Dough-freezing Systems and Equipment

The true value of freezing for the baker is that it enables him to plan ahead for the peaks and troughs of demand, utilizing quieter periods to build up his inventories for weekends and holiday periods. Catering and wholesale contracts can be satisfied, without the stress of trying to meet such extra demands when already under pressure. Products in heavy demand can be produced and stored in advance. Working on this basis, machinery can be used for long runs, e.g. bread-roll plant and pastry lines, with improved results. A week's turnover can be produced in one long production-run, instead of small batches of each product being produced daily. Moreover, instead of merely baking off frozen-dough products from external frozen-dough suppliers, the baker can prepare his own products for bake-off, e.g. meat-containing products, Danish pastry and bread-rolls.

A baker with a two-rack retarder/proofer, and a single-rack blast-freezer can

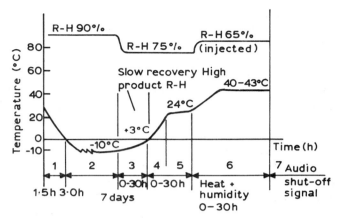

Fig. 155. Retarder-proofer applications. 1. Temperature pull down 1·5–3·0 h. 2. Retard/store, 3·0 h to 7 days (max.) (adjustable); defrost every 6 h. 3. Preheat automatic only, 0–30 h (variable). 4. Recovery (heat) zone. 5. Recovery (refrigerate and heat) zone, 0–30 h (variable). 6. Proofing zone (heat and humidity), 0–30 h (variable). 7. Shut-off audio signal. (Source: Foster Refrigeration Group, Kings Lynn (in-store and independent bakeries).)

freeze a rack of bread-rolls within 25 minutes, and the use of polystyrene insulated boxes will keep the products frozen during transit, where a refrigeration van is uneconomic. Within the shop, the equipment required is a freezer, a proofer and an oven—or for controlled thawing or defrosting, ensuring a quick start each morning, a retarder and/or retarder-proofer cabinet (see Fig. 155).

Freezing products in a cold-store compared with a blast-freezer gives loss of product quality. Owing to the rapidity of the freezing time in a blast-freezer, preventing fermentation, moisture loss and ice-crystal formation, no such loss of product quality is encountered. The intended use of a cold-store is to hold already-frozen products at $-18$ to $-20°C$, and not for the freezing of ambient temperature products. The ideal combination, is a blast-freezer for rapid freezing with minimal moisture loss to $-20°C$, followed by deposit in a cold-store to retain them in that condition until required.

Blast-freezers of various capacities are available, a small reach-in model suitable for freezing about 30 lb of product would achieve this in about 50 minutes. Such a unit would extract heat at about 12 000 Btu/hour (47·6 cal or 11·4 J) using a 3-hp compressor, the capacity being about eight trays. The next step up in capacity is a roll-in or trolley model, with one, two, three or more racks capacity. This would be capable of freezing a rack of bread rolls from ambient temperature to $-20°C$ within about 20 minutes, at an extraction rate (heat-transfer rate) of 21 000 Btu/hour (83·3 cal or 19·8 J). These roll-in models can be integrated into a cold-store complex, incorporating freezer storage, and a recovery room to thaw out the racks under controlled conditions of RH, temperature and air speed. A further development of standard retarder/proofer systems incorporates the latest microprocessor techniques. This gives the baker a high degree of control over the freeze–thaw–proof–bake cycle. Once programmed, such a unit will control temperature, air distribution, RH, and heating with precision for a wide product range. This minimizes loss of quality, producing doughs in perfect condition for baking off. The system can accommodate various loads, accommodating bread, bread-rolls and specialty-breads, of various weights and formulations, simultaneously. A 7-day programme can be set, with varying programme for each day. Weekends and extended weekends can also be accommodated by a simple modification. The programme also includes the required proofing-time for normal daily production. An oven-ignition contact ensures that ovens are at the correct temperatures to coincide with the end of the conditioning cycle.

The functions of proofing and retarding can be either carried out in separate units, or combined in one unit the retarder/proofer. Dough retarders are normally of two types: Type I, which has been on the market for many years, retards within the temperature range +2 to +5°C, being suitable for a wide range of small products on a day-to-day basis. The leaner formulated products have a shorter retarded storage life than enriched products. Type II has the facility of a retarding range of −5°C, or lower, up to +5°C. This unit offers more versatility, allowing longer retarded storage periods when the retarder is set at a lower temperature, e.g. day-to-day retardation at +2°C, or weekend or holiday retarding at −2 to −5°C. This type of unit normally has a heat-transfer capacity for retarding 400-g hearth breads.

The baker must decide which type of retarder unit suits his production needs. If only 60-g units have to be retarded on a daily basis, with Saturday to Monday being the maximum requirement, then the +2 to +4°C range is adequate when strictly applied. However, for larger dough units and periods longer than the daily schedule, e.g. weekly stock inventories, for subsequent proofing and baking to meet peak demands, a unit with a higher heat-transfer rating at a lower temperature must be utilized. Figure 156 provides an overview of the temperature/time systems for

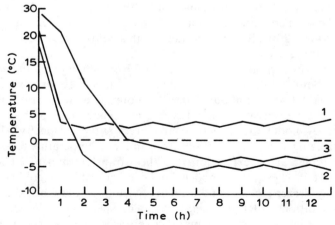

Fig. 156. Typical time/temperature diagram for dough-retarder systems. 1. High-temperature system over 24 h, e.g. 60 g (2 oz) bread rolls. 2. Low-temperature system over 72 h, e.g. 60 g (2 oz) bread rolls. 3. Low-temperature system over 72 h, e.g. 900 g (2 lb) loaf. (Source: Foster Refrigerator Group, King's Lynn (In-store and independent bakeries).)

dough-freezing, recovery and proof in in-store and independent bakeries, as marketed by Foster Refrigeration Group of Kings Lynn, UK.

Attention must be paid to the retarder design to ensure efficient operation and final product quality. The RH level within the retarder must be maintained at a value which prevents dough-skinning on the surface, at the appropriate air speed of the circulation system installed. The requirement is a low-turbulence, indirect movement over the products; lower-temperature retarders often have an automatic evaporator-coil defrost sequence to remove ice.

A disciplined operation regime is also recommended for the most efficient use of retarders. Excessive opening and closing of doors must be avoided to avoid temperature/RH fluctuations, and possible dough-skinning. Doughs should be placed in the retarder immediately after make-up to avoid skinning and partial-proof.

As a general guide to frozen-dough formulation and make-up procedures, the following fermentation tests conducted by Merritt, illustrate some of the parameters involved and their effect on final product quality.

*Formula:*
Flour 100 parts, water 65, dry yeast 4, bromated yeast-food 0·5, salt 1·75, dextrose 7·0, shortening 9·0, sweet-whey 5·0.

*Procedure:*
Combine all ingredients in a high-speed mixer with a refrigeration jacket. Mix to full development, and final dough temperature 18 to 21°C, allowing 10–15 minutes fermentation during make-up. Freeze at air temperature − 32°C, and store below − 12°C until required. This dough gave a shelf-life of 1 year, and slabs of dough scaled at 16 to 20 ounces show good performance after 16 months storage. Merritt's fermentation tests involved fermenting portions of the above formula doughs for 30–45 minutes before processing. Processing, apart from the fermentation periods before freezing, was identical to that given to the unfermented dough. All frozen samples, when removed from the freezer were proofed at 35°C and 80% RH. The samples were baked at 218°C for 17 minutes. The conclusions which Merritt drew from these tests were the following:

—Final product specific volume (ml/g) should not be less than 4·5, specific volumes of 4·0 ml/g or less give compact, heavy products.
—Storage times of up to 8 weeks resulted in doughs with normal proof-times; after 21 days storage, the baked rolls were judged to be unacceptable.
—When the proof-time exceeds, 3 hours, specific volume is low, and the structure poor.
—Rolls baked from frozen-dough without fermentation, developed a flavour during the proofing period, and continued to perform well with proof-time.

*Effect of storage on performance* (unfermented doughs) Merritt (1960)
Storage temperature: − 32°C

| Time (months): | 1 | 6 | 9 |
|---|---|---|---|
| Proof-time (minutes) at 35°C and RH 80% | 70 | 80 | 85 |
| Specific volume (ml/g) | 4·6 | 4·3 | 4·2 |

Flavour, grain and volume all remained high during the extended storage period.

Merritt warns that fermentation of dough before freezing has more effect on dough stability, and required proofing time than any other single factor. Stability of frozen dough is inversely related to the amount of fermentation before freezing. Fermentation times before freezing of not more than 1 hour give a stability of a few weeks only; 30 minutes gives 3–4 months; and zero time gives about 12 months. (Source: Merritt, P. P., *Bakers Digest*, **34**, 4, 1960, 57).

The routine application and operation of the separate dough retarder unit is typically the following. Doughs are made up according to basic formulations, which have often been modified or adjusted, and tested for performance and quality in the final products. The doughs are then made up and shaped, being placed onto trays for a short relaxation period. Common adjustments made to doughs are the use of lower temperatures of 18–21°C, to minimize any partial-proof, and to yeast levels, depending on the size or surface area of the product. When using separate retarders and proofers, the yeast levels in the larger dough-pieces can be reduced to minimize proofing tendencies initially. In some cases this could reduce the speed of recovery and final-proof later. Where combined retarder/proofer units are being used, this problem does not arise. After the required residence time in the retarder, the dough-pieces are removed from the unit as required, and left at ambient bakery temperature for about 10 to 60 minutes, depending on product and size, to gradually acclimatize to ambient, and avoid a large temperature gradient between dough surface and core temperatures. Large temperature gradients can result in poor quality products of uneven texture. The larger dough-pieces require a longer rest period than the smaller units, and those retarded at temperatures below zero, will require more time than those retarded at 2–4°C. Placing cold dough-pieces and trays direct into the proofer unit results in excess condensation on both products and trays, and should be avoided. Various types and sizes of retarder are obtainable, cabinet models, with capacities of 20–80 trays being a popular unit. Smaller units of the bench-retarder type, with up to five retarding sections, each taking 7–8 trays, depending on spacing, give both retarding and work top space where floor space is limited.

The larger wheel-in rack retarders are available in modular design from one rack up to 60 or more rack capacities. With the larger units, the important point is the refrigeration output capacity, which must satisfy the required hourly output, and an efficient air circulation system to ensure rapid and even cooling of all products. In addition, all rack retarders designed for operation below zero temperatures must be well insulated, including the floor.

The main function of the dough-proofer unit is to provide an atmosphere with sufficient humidity to prevent dough-skimming, and to maintain an even temperature for uniform proof and regular proof-times. Therefore, independent control of temperature and RH is vital, together with an efficient air-circulation system providing forced convection at an optimized air speed. Humidity is often controlled by water being sprayed onto a protected heat source at the command of a hydrostat set to the required RH level. Again, efficient insulation is important, to ensure that the influence of the bakery climate is excluded, which must also include the proofer-floor. Accurate control of temperature, RH, and air speed by the application of the use of reliable sensors, backed up by solid, well-insulated enclosures which are rust-free will repay their investment. Instrumentation is very important, allowing precise setting and read-out of target and actual data, preferably with a recorded print-out. Water hardness and pressure are other parameters, which must be optimized for local operation, and booster pumps installed where necessary. Proofing temperatures generally fall into the range 35°C to 45°C to give wide choice of cover for all products, but lower proof-temperatures and longer proof-times give the best results in most cases. Typical products benefiting from this treatment include: baguettes, croissants, wholemeal, rye and some specialty hearth varieties. Baguettes proofed at 30°C for 2 hours often produce a better result than those proofed at higher temperatures for a shorter time, thus for extensive production of hearth and variety products a proof-temperature range of 20 to 45°C is preferable.

The benefits derived from the use of a retarder/proofer system compared with the unit separation of the retarding and proofing operation can be summarized as follows:

(1)  Proofing can be performed before commencing work, the products being ready for baking either immediately or within a short time. Thus allowing a later start in the morning.

(2)  Proofing-times, as well as defrosting and recovery times can be programmed to fit in with work start and oven temperatures. Oven switch-on times can be synchronized with product retard and proof schedules.

(3)  A retarder/proofer can double as a proofer only for freshly produced products, until the retarder facility is required again.

The retarder/proofer is available in many types and capacities, with varying degrees of automatic control in-built. The choice will depend on the requirement, i.e. daily retarding and proofing of small products, or a complete range up to 800-g to 18kg for retardation over a holiday weekend. In the latter case, larger units of increased refrigeration capacity must be considered. Every baker must plan how best a retarder/proofer will complement his production schedule. This is best accomplished by visiting installations working *in situ*. As well as capacity, the degree of automatic control for the functional requirements of production schedules must be evaluated, and this set against engineering quality and price. The need for a defrost or thaw sequence is desirable for any unit with a retardation-temperature range of −5 to +4°C, −10 to +4°C and −18 to +4°C. The following are the commonest types of retarder/proofer on the market.

*Type 1:* For the daily retarding/proofing of small dough-pieces, and also suitable for a Saturday to Monday holding operation. In the latter case a lower yeast level and longer proof-time is appropriate. Retarding temperature range is 2 to 4°C, with a proofer temperature range 20–43°C.

*Type 2:* This offers a recovery sequence and the gradual warming up of the doughs, suitable for long and large dough-pieces. Owing to the relatively high retarding temperature, larger units must have a reduced yeast level, and a longer recovery and proof-time. Retarding temperature range is 2 to 4°C, with proofer temperature range 20–43°C, coupled with a recovery sequence.

*Type 3:* Retardation of small and large products, and over prolonged holiday weekends. Retardation temperature range is $-5$ to $+4$°C, with recovery sequence and proofer temperature range 20–43°C.

*Type 4:* This unit is classed as a freezer retarder/proofer, since it is capable of freezing doughs, allowing extended storage times at $-10$ to $-18$°C. However, it is available in several versions with different retarding temperature ranges, viz. $-5$ to $+4$°C, $-10$ to $+4$°C, and $-18$ to $+4$°C. All versions have a recovery sequence, defrost sequence, and proofer temperature range 20–43°C.

Retarder/proofers are also utilized by in-store bakeries, and non-production outlets to defrost overnight, recover and proof frozen dough-pieces supplied by specialist frozen-dough companies. These could include 500-g panned or hearth loaves, croissants, muffins, hamburger rolls, baguettes and fruit loaves, etc.

The compact retarder/proofers manufactured by Revent Equipment Limited, which are widely used in the in-store bakeries of at least one very large supermarket chain in the UK, will accommodate 1, 2, 3 or 4 bakery racks in various combinations. The roof and walls are lined with 40-mm polyurethane, and the floors with 15-mm. In both these and the larger modular room units (4 to 36 racks), the air is gently circulated by fans. Humidification is achieved by atomizing water into the recirculating air stream. The entire programme of proofing, retarding, and proofing again for early morning production can be preset for a whole week.

The Koma CDS dough-conditioning system, marketed in the UK by Bakery Conservation Systems International Limited, Maidenhead, Berks, is a further development of the standard retarder/proofer, using the latest microprocessor techniques. Once programmed, the CDS unit controls temperature, air-distribution, RH, and heating with precision over a wide product range. A 7-day programme can be set with different programmes each day; and weekends and extended weekend periods are covered by an appropriate modification. Daily production proofing times are also included, and oven-ignition is linked and synchronized with the conditioning cycle, ensuring that the requisite temperature for baking has been reached.

Williams (Bakery Division) of Downham Market, Norfolk, UK, offer a wide range of units, cabinets for 20, 40, and 60 trays, and 1 to 12 rack roll-in tunnel rooms. Stainless-steel construction with 75–90 mm polyurethane foam insulation ensures low running costs. Controls are kept simple but effective on all freeze/retard/recover systems, and the refrigeration unit is easily serviced, being either top- or remote-mounted to avoid dust and high ambient temperatures. The Leutenegger & Frei range of retarder/proofers, manufactured in Switzerland by Leutenegger & Frei of CH-9204 Andwil/St. Gallen, is of heavy-gauge stainless-steel interior construction. Each unit is designed to retard over extended periods, with a maximum temperature of $-10°C$ for long-term storage of up to 6 days. All operations are managed by a Sensomatic control unit, with an integrated microprocessor which will handle 16 different programmes. These units are marketed in the FRG and the UK through agents.

In the FRG, a cryogenic system of freezing has been developed and marketed by AGEFKO, Kohlensäure-Industrie GmbH, 4000 Düsseldorf 1. This system involves the application of liquid carbon-dioxide, which has an expansion cryogenic temperature of $-78·9°C$. This effect can be utilized in cooling, freezing or storage enclosures. The products, whether in the dough or baked state are 'shockfrosted' either by static residence on trays within an insulated enclosure, or by transit through a long tunnel, 5 to 13 m in length, with freezing capacities of 100 to 2660 kg/hour respectively. Another system utilizes a spiral tunnel of vertical design, for application where space is limited. The fundamental demarcation between all the systems described is whether they are of the discontinuous (batch), or continuous design. The numerous retarder/proofers described operate on a discontinuous system, although the freeze–thaw cycle can be automated and computerized to improve precision in operation. In all cases the freezer cabinets or roll-in rooms have to be batch loaded. The mechanical refrigeration system used in the batch systems utilize a vapour-compression unit, powered by an electric motor. This compresses a vapourized refrigerant of low density, e.g. 'freon' (halogenated hydrocarbon) into a high-pressure vapour at high energy and temperature. A condenser/heat-exchanger dissipates the hot vapour either by air or water-cooling, causing it to condense into the liquid state to be stored in a receiver. An evaporator produces the actual cooling, being equipped with a thermostatic expansion valve, which reduces the pressure on the refrigerant, causing it to change into the gaseous state, and in so doing it absorbs and extracts heat from its surroundings, which in this case is the insulated cabinet or cold-room. In the case of blast-freezers, the cooling-coil is mounted within the insulated enclosure or tunnel, and fans distribute the cold air over the products for rapid freezing.

Mechanical refrigeration units take the form of simple static insulated boxes, into which the products are placed on racks to be exposed to air blasts at $-30°C$.

Tunnel blast-freezers represent a more rational and continuous system of operation, the throughput speed on the band-conveyor depending on the time required to freeze the product.

Continuous spiral freezers for freezing bakery products comprise large enclosures containing one or two helices of conveyors for transportation of the products through the freezing cycle.

Another continuous system uses a multi-tray blast-freezer, whereby, open-ended wire trays loaded with the product are automatically stacked to a fixed height. These pass through the freezing enclosure on a bed-conveyor, being discharged after an adequate residence time for automatic stacking. The versatility and flexibility of this system allows the simultaneous freezing of products of various dimensions and freeze-time requirements, since the speed of each bed-conveyor can be controlled independently. However, all these primary freezing systems must be complemented by an adequate number of holding-freezer units.

**Freezing Conditions**

Although there appears to be some differences in experiences from one country to another, owing to equipment, packaging and product formulation, it is desirable to aim at some specification guidelines for the industry generally. Frozen products containing meat in particular must be processed with a minimum of time lapse at bacterial incubation temperatures and humidities, and without direct handling by operatives. Meat-containing products constitute a very real bacteriological hazard, especially in a finely divided state. In the USA, R. Bamford of the Am. Soc. Bakery Engineers (*Bakers Digest*, **49**, 3, 1975, 40) suggested the conditions given in Table 74 for blast-freezing, using a standardized air flow of 600 ft/min.

TABLE 74
**Blast-freezing conditions for products at an air-flow of 600 ft/min**

| Product | Packaging | Freezing-time to reach a centre (core) temperature of −12°C (h) |
|---------|-----------|-------------------------------------|
| Bread 1-lb loaf | None | 2·75 |
| | Single-wrap | 3·25 |
| | Double-wrap | 3·50 |
| 2-lb loaf | None | 3·00 |
| | Single-wrap | 3·50 |
| | Double-wrap | 4·00 |
| Dinner rolls | Foil pan in plastic bag | 1·00 |
| Bread dough | Plastic bag | 3·00 |
| Danish pastry | Plastic bag sealed or overwrapped box | 2·00 |

In all the above product freezing (Table 74), the air temperature was −36°C. Thompson (*Bakers Digest*, **50**, 2, 1976, 28) recommends blast-freezing at temperatures of −35 to −40°C, using air velocities of 600 to 700 ft³/min, and for such time that the core temperature of the product reaches −18°C. In addition, holding-freezer units should be maintained at air temperatures between −23 and −29°C, not being allowed to rise above −18°C. The air should also be held at the highest RH level practically possible, and should be circulating freely. If these conditions are strictly adhered to, product quality can be preserved for several weeks or months.

The extremely rapid freezing rates of cryogenic freezers, which are 10 to 30 times faster than any blast-freezer, are to be recommended where outputs justify their use. Countercurrent tunnel freezers can freeze up to 5000 lb/hour on a 70-ft belt conveyor, liquid nitrogen being introduced into the freezing zone by atomized spray headers at a controlled rate. The resultant low-temperature gas is circulated at speeds of up to 7000 ft/minute on a zonal basis, the entry zone effecting a pre-cool of the product at −18°C, the freezing zone rapidly freezing at −196°C, and a tempering final zone before exit of −108°C. The net result is a significantly better retention of product quality than is the case with mechanical freezers. In addition, cryogenic freezing offers simplicity in design of equipment, lower initial cost and subsequent maintenance costs, lower energy costs and considerably less floor space.

Where product capacities of around 10 000 lb/hour are involved, and space is limited, cryogenic spiral freezers offer the best solution.

**Thawing Conditions**
On removal from the holding-freezer units, the core temperature of the products is on average −18 to −21°C. In order to avoid too great a temperature gradient between the core temperature and the dough-surface temperature, the products must be placed in a defroster to thaw gradually. According to Bamford (ASBE, 1975), the optimal defroster temperature is 37 to 49°C, at RH 50% to 60%, using an air velocity of 200 to 500 ft/minute. However, initially, airflow rates of 150 to 200 ft$^3$/minute are more appropriate to raise the product temperature from −18°C up to 21°C. The time required to reach a core temperature of 21°C will depend upon the size, type and packaging of the product. Table 75 gives some examples of thaw-times from −18 up to 21°C.

TABLE 75
**Relative thawing times utilizing various packaging materials**

| *Product* | *Packaging* | *Thawing-time required to raise centre (core) temperature from − 18 to 21°C (h)* |
|---|---|---|
| Bread 1-lb | Double-wrapped | 3·50 |
| 2-lb | Double-wrapped | 3·75 |
| Dinner rolls | Plastic bag | 1·25 |
| Bread dough 1-lb | Poly bag | 4·50 |
| Danish pastry | Plastic bag | 1·50 |

Defroster air temperature: 37–49°C; air flow: 150–200 ft$^3$/min

Source: Bamford, R., *Am. Soc. Bakery Engineers*, 1975.

The holding of bread doughs in a retarder for up to 24 hours before proofing considerably reduces the final proof-time, and often results in loaf volume improvements, but no improvements in texture or flavour of significance were detected by the author. Nevertheless, direct freezing of the dough after preparation appears to be the best protection for yeast viability, providing it is graduated.

**Influence of Freezing and Thawing Rates**
The rate of freezing has a considerable influence on the life-span of yeast cells. Rapid freezing resulting in greater cell-wall damage than is the case with a graduated freezing process.

Mazur and Miller (*Cryobiology*, **3**, 1967, 365) established that yeast-cells cooled rapidly to −30°C and below showed a survival rate of less than 0·01%, but, when cooled more slowly to the same temperature, viability can be as much as 65% of the total cell count. The explanation for this destruction by rapid freezing is the formation of intracellular ice crystals, which damage the cellular protoplasm structure sufficiently to terminate its life-span. Where cooling rates are slower, it is considered that the yeast cells are in a position to transfer sufficient intracellular water to external ice, thus preventing ice crystal formation within the actual cell.

Excessively low freezing temperatures have been confirmed to have an effect on yeast viability by Hsu *et al.* (*Cereal Chem.*, **56**, 1979, 419). These workers found that in a dough, yeast cells freeze at about −35°C. Doughs frozen at −10°C and −40°C, showed respective proof-times of 72 and 132 minutes. Also, doughs frozen at −78°C showed minimal viability on thawing. Lorenz (*Bakers Digest*, **48**, 2, 1974, 14) found that improved frozen-dough stability was obtained by freezing only to a temperature of −10°C, freezing being carried out slowly, and subsequent thawing rapidly. Many of these recommendations deviate from current bakery practice.

Bruinsma and Giesenschlag (*Bakers Digest*, **58**, 6, 1984, 6), studied the influence of 12 weeks' storage, with daily freeze–thaw cycles on dry and compressed yeast viability. They found that the concentration of dry yeast was the most significant variable in the determination of proof-time. Increasing levels of dry yeast gave consistently shorter proof-times than compressed yeast. Proof-time and gas-production remained reasonably constant over a seven consecutive freeze–thaw cycle programme for both types of yeast. However, crumb structure rapidly deteriorated over the seven cycles. In addition, with each successive freeze–thaw cycle, dough stability became reduced, the dough showing friability and stickiness on the outer surface.

### Formulation and the Use of Dough-conditioners and Other Additives

The usual technique for the preparation of frozen doughs is to make a firm no-time straight dough by mixing at a temperature of 18–21°C, preferably in a water-jacketed high-speed mixer. The dough is worked off immediately, and in some cases the rounded dough-pieces are given a 5-minute relaxation period. The shaped pieces are then deposited on trays, which are either loaded onto racks or conveyors to be held within a blast-freezer for the requisite time, or slowly transported through a freezing tunnel or enclosure. The freezer air temperature is kept at −30 to −40°C, under an air turbulence of 600 ft/minute.

According to the experience of De Stefanis *et al.* ('Oxidation requirements for the preparation of frozen-dough', Paper No. 19, 71st Annual Meeting of AACC, Toronto, Canada, 5–8 Oct. 1986), in an investigation concerning factors which cause negative changes during freezing, found that the general practice of mixing doughs at lower than ambient temperatures resulted in inferior oven-spring, and a loss in loaf volume. However, he points out that this can be compensated for by adding more oxidant. De Stefanis even found that dough mixing-temperature was inversely related to the optimal oxidation level. Adequate oxidation is generally desirable in dough for freezing, and the omission of any reducing agents. This precaution will ensure fullest dough maturity, since floor-times are ruled out. Varriano-Marston *et al.* (*Bakers Digest*, **54**, 1, 1980, 32) found that when ascorbic acid was added together with potassium bromate, proof-times of frozen doughs stored for 2 months were reduced by 23 minutes. In formulating for frozen doughs, the most important criterion is the ability of the thawed dough to give an even proof within an acceptable time, baking off to a normal-volumed loaf with acceptable texture and flavour, as well as crust colour.

The choice of yeast and its quality is important for frozen doughs. Both compressed and dry yeasts are satisfactory for use in doughs to be frozen, but freshness and vitality of the cells is essential. Fresh compressed yeast has a light grey to yellow colour, and a pleasant, fresh, faintly acid taste and smell, with a minimum dry matter content of 25%. The presence of any dead cells can be identified by mixing a water suspension of the cells with an equal volume of a solution containing: methylene blue 200 mg, potassium dihydrogen phosphate 27·2 g, disodium hydrogen phosphate 0·07 g per litre of distilled water. The dead cells stain red, allowing counting under a microscope. The relative gassing-power of yeasts can be evaluated by the procedure described in Section 1.8.1. The quality of active dry yeasts should be compared for the presence of denatured, hydrolysed cell material. Evaluation tests should include: glutathione-content, trihalose content, and the amount of low-molecular-weight material released on hydration at 8°C (an index of fermentation power changes).

Yeast stability and proofing power decreases sharply in a frozen straight dough, where a substantial fermentation is allowed before moulding and freezing. Fermentation times of 1 hour or longer could result in an 80–90% loss in the viable yeast cell content during prolonged storage, and a proofing-time on thawing out exceeding 5 hours duration.

Freshly prepared doughs fermented in the conventional manner do not exhibit this instability, since cell reproduction remains constant as fermentation time proceeds. It is during the stage of reproductive growth, or 'budding', that the yeast cells are unstable and vulnerable to destruction during the frozen storage.

It is important to keep the yeast cells in a dormant state within frozen doughs, discouraging any budding. This is best achieved by streamlining the processes of freezing, packaging, storage, transport and distribution to avoid temperature fluctuations.

During formulation, the amount of yeast used should be double that used conventionally, combined with a minimal fermentation-time. The usage range for compressed yeast is 6% to 10% flour weight, and for active dry yeast 3% to 5% flour weight. The actual amounts used within each range will depend on desired speed of proofing, bearing in mind that active dry yeast requires an induction period for the initiation of metabolism compared with compressed yeast. Although this might be regarded as an advantage in unfermented frozen doughs, the presence of dead cell material in dry yeasts will weaken the dough (glutathione release), compared with compressed yeast. The use of 6% compressed yeast, with a fermentation time of 15 minutes, will provide a substantial yeast cell count after 15 weeks storage at −18°C, coupled with a respectable proof-time of about 1 hour. A minimal recovery time of 15 minutes at 27–28°C is necessary to render the dough sufficient workability (elasticity and plasticity). This stability is, however, only achieved at the expense of product flavour and aroma, and to some extent loaf volume.

The use of longer fermentation times at lower dough temperatures are a possibility, and would allow the dough to be introduced into the freezing step at a lower temperature, simultaneously allowing flavour development. Frozen dough stability is difficult to achieve using a sponge and dough method. However, one possibility is to shorten the sponge time, using NFMS milk solids as a buffer, then chilling the sponge and dough at the doughing-up stage and adding extra yeast. This has been reported to provide good retention of yeast cell viability and proof-times of about 1·5 hours after 15 weeks storage at $-18°C$. Nevertheless, most frozen unbaked bread dough is produced using straight doughs with increased yeast levels and shortened fermentation times. Such limitations are necessary for the retail market, where a shelf-life of more than 2 months is desired. Frozen doughs for institutional baking, requiring frozen storage holding for less than 3 weeks can be fermented for periods of up to 1 hour before freezing. The dating of all frozen doughs before storage is strongly recommended to avoid prolonged and hazardous storage times.

A high-quality medium to strong flour of the patent type is recommended for frozen doughs. Good gluten-forming quality is more important than actual quantity, and the use of dough oxidants in the form of bromate or ascorbic acid, depending on local legislation, is a useful aid. Bromated yeast-foods can be added at the 0·5% level. The levels of salt used are within the range 1·5% to 2·0%, flour weight. Sugars are used at the 5% to 10% flour weight level for richer formulations, but in the case of leaner, bread-doughs 5% to 7% flour weight is adequate. Sucrose and dextrose give similar results, but higher levels will increase proof-times.

Shortening, in the form of a good quality lard or vegetable shortening based on palm-kernel oil is added at the 5% flour weight level. This ensures a fine grain and tender texture, and an emulsifier of the mono-/diglyceride type, added at about 3% of the shortening weight, will improve both volume and crumb quality.

The use of NFMS milk solids will improve crust colour in many products. Dough water absorption must be carefully adjusted within the range 52% to 64%, depending on flour protein quality, to give a dough which is slightly stiffer or firmer than the conventional to compensate for mechanical handling methods. A good average absorption is about 60%. Slack doughs are to be avoided, as these will have poor gas retention during proofing. The addition of higher oxidant levels, especially bromate at 70–75 ppm flour weight, will counter the effect of the release of any reducing substances from dead yeast cells, but has no direct effect on yeast stability during freezing. Variations in thawed product quality can be obtained by minor adjustments to basic raw materials, viz. shortening, sugar, type of yeast, oxidants, use of milk solids, and remixing the dough after thawing. But the overriding criterion is the maintained vitality of the yeast cells during the freezing stage. Research may succeed in culturing a yeast strain with an improved tolerance to prolonged low temperature storage conditions, suitable for frozen doughs. Such a development would allow increasing flexibility during processing, and shorter proof-times, leading to market expansion and a wider acceptance of frozen-dough technology.

Certain fungal and bacterial amylolytic enzymes make useful supplementations in low sugar doughs, and at relatively high concentration are more effective in increasing loaf volume than fungal alpha-amylase. Amyloglucosidase, for example, will split the limit dextrins resulting from starch degradation by alpha-amylase into glucose. Also known as limit-dextrinase, amyloglucosidase is capable of cleaving the alpha-1,6- linkage of amylopectin, and low-molecular-weight saccharides, thus producing a more complete hydrolysis of branched carbohydrates. Amyloglucosidase is classified as an isoamylase.

The addition of the surfactants/dough conditioners sodium stearoyl-2-lactylate (SSL), and diacetyl tartaric acid esters of mono- and diglycerides (DATA), and mono- and diglycerides, as well as their ethoxylated derivatives (EOM), all improve dough rheological properties, but have no significant effect on proof-time. SSL improves both volume and crumb softness, owing to its ability to complex with both protein and starch. In considering the functionality of surfactants, the most important aspect is the hydrophilic/hydrophobic balance (HLB). According to their ionic activity, they can be divided into three main groups:

(1) Anionically active surfactants, which are water-soluble and become ionically dissociated carrying a negative charge, e.g. calcium and sodium stearoyl-2-lactylate (CSL and SSL).

(2) Non-ionic surfactants, which do not dissociate into ions, e.g. mono- and diglycerides of fatty acids, sugar alcohol esters of fatty acids (sorbitol mono-esters), and the so-called 'fat-sugars', which include esters of sucrose (saccharose) and glucose with various fatty acids, also the numerous polyoxyethylene derivatives.

(3) Ampholytic or dipolar surfactants, with mixed ionic functionality, e.g. the phosphatides, of which phosphatidyl choline or lecithin is an example. These lecithins form a group of allied compounds, differing in chemical structure, the phosphoryl-choline structural moiety functioning as a dipolar or 'zwitterion'.

In formulating some frozen doughs, the inclusion of certain sucro-glycerides, especially those based on palm oil (palmitic acid 47·5%, oleic acid 40·0%, linoleic acid 7·5%, stearic acid 4·0% and myristic acid 1·0%), show useful functionality.

Another useful group of substances for inclusion into frozen doughs are the hydrocolloids, whose function is to stabilize water movement within the dough during freezing, using any number of freeze–thaw cycles. Locust bean (carob), and guar gum both present excessive ice crystal formation during freezing, the active component in both cases being galactomannans, derived from the seed endosperms. Both gums produce a high viscosity in aqueous solutions of low concentration. Therefore, only low concentrations should be included; amounts of up to 0·1% flour weight are normally sufficient. Karaya gum is also applicable when used at low concentration of 0·1% flour weight, and exerts a hydrocolloidal synergistic effect when used with monoglycerides at about 0·3%. The rheology of the polysaccharide gums can be adjusted to give either extensible or short textured gels, depending on the application.

In addition, many native and modified starches have extremely good freeze–thaw properties in frozen dough products, preventing 'synerisis', or loss of water. Waxy sorghum shows relatively low water separation over about three freeze–thaw cycles; some cross-linked amylopectin starches yield, after phosphorulation mono- and di-starch phosphates, which also exhibit good freeze–thaw stability. Native starches which give long, cohesive textured gels include: tapioca, potato and amioca. Pregelatinized potato starch is also cold-swelling, and is stable at low temperatures. Many companies which market bases and conditioners, also include in their product programme specialized products for addition to frozen doughs, giving longer frozen shelf-life and improved recovery. Many of these will include the components already mentioned.

### Examples of Frozen Dough Preparation
The application of the Reddi-sponge process (R. G. Henika) patented by Foremost Foods, Dublin, California, USA, doubles the shelf-life, and allows the use of lower dough temperatures, 18 to 21°C. The use of direct biochemical action on the flour proteins prolongs dough storage potential.

*Reddi-sponge process*:
Flour 100 parts, water 62, yeast 4·0, yeast food 0·5, sugar 8·0, salt 2·0, shortening 3·0, mono- diglyceride emulsifier 0·4, Reddi-sponge 3·0, total oxidant 70–80 ppm.

Procedure:
(1) Mix at low and high speeds to full development. If the mixing-time exceeds 12 minutes, hold the yeast and add during the last 4–5 minutes of the mixing-time.
(2) Adjust the water-jacket temperature to give a final dough temperature of 21 to 24°C.
(3) Allow to relax for 10 minutes.
(4) Divide, round, giving minimum overhead proof, and mould.
(5) Freeze in a blast-freezer at −24°C or lower.

It has been established that the sequence in which the ingredients are added during mixing has a significant effect on final proof-time and yeast activity. Doughs held at retardation temperatures for up to 24 hours show reduced final proof-times, and often improved volume.

Baguette formulation:
Flour (protein 12% dry matter, gluten, minimum, 25%), 100 parts, yeast (compressed) 6, ascorbic acid 50–70 ppm, amyloglucosidase preparation (adjusted to optimum concentration), salt 2·0 (added at final stage of mixing).

Croissant formulation:
Flour (medium strength) 100 parts, sucrose 10, invert sugar 1·0, sucroglyceride derived from palm oil 2·0, yeast (compressed) 4·0, water 50, lemon juice 5, whole egg 2, roll-in butter 44, salt 2·0.

Brioche formulation:
Flour 100 parts, sugar 10, sucroglyceride derived from palm oil 3·0, butter 40, eggs (fresh) 1.

Freeze–store–thaw–proof procedure:
Use a low temperature and rapid freezing to minimize ice crystal size (blast-freezer). Doughs are best covered with polyethylene film or bags during freezing to prevent moisture loss and skinning. Liquid nitrogen flash or jet streams offers the fastest and best procedure (average freeze-time 6 minutes).

Storage at $-18°C$ after rapid freezing will keep the dough in the completely frozen state, and the yeast inert. Many producers use $-23°C$ to allow a safety margin against errors and any temperature fluctuations.

Thawing is best carried out in a warm and draught free situation at 27 to 29°C, either in the bakery covered with a damp cloth, or within an enclosure at RH 60%.

Final proofing is best carried out at 30 to 45°C, depending on the product.

**Partially Baked or 'Brown and Serve' Products**
This concept is applied to rolls, breads and pastries, which are preformed and prebaked to the final size and shape, requiring only crust browning and fresh flavour development in a domestic oven. The aim is to bake the product to the point of maximum rigidity and volume, without any crust formation. This can be done by reducing the oven temperature to the range 120 to 150°C. Proof has to be so controlled that the oven-spring which takes place at the lower temperature range is maintained within limits.

Dough water absorption must be reduced to give a stiff dough, which will remain rigid on removal from the oven.

Formulation modifications require straight doughs to be set at 32 to 35°C, but sponge-doughs at the normal lower temperatures. Levels of yeast and yeast-food must be reduced moderately to avoid excess oven-spring. Enriched formulations perform best for eating quality, and final-proof should be rapid at 38 to 40°C. Baking is between 120 and 150°C, being for as long as possible without any crust browning, usually about 10 to 15 minutes at a solid 140°C. The core temperature of the product must reach 82°C, to avoid shrinkage or collapse, on cooling. Product cooling and packaging must be carried out under sterile conditions to avoid any mould infection, or product damage.

**Marketplace**
Frozen foods generally represent a growth market, projected 1990 sales figures for the EC being valued at $12 billion at 1985 price levels. In percentage terms, the three fastest growth sectors of this market are: ready-meals at 53%, bakery products at 48%, and desserts at 41%. The ready-meal sector, however, having been recently subjected to a 'bacteriological scare' in the form of *Listeria* contamination, could face an uncertain growth pattern. The bakery product sector, provided meat and dairy raw materials are carefully monitored, should not become exposed to any adverse publicity in this way. The important aspect for the baking industry is diligent environmental sanitation control and hygienic handling regimes.

The ownership of home freezing facilities in the form of freezers and fridge-freezers can be closely correlated with frozen food consumption. Within the EC, about 50% of households now have these facilities, whereas the household penetration of microwave ovens remains relatively small. In the bakery frozen product sector there is still considerable scope for new product innovation and growth, wherever a convenience aspect can be introduced. In general terms, certain socio-economic trends must be taken into account. The number of 1 to 2 person households is on the increase, the population average age is increasing, and increasing leisure time has become available to the consumer. The type of foods, fast food snacks and complete meals consumed have changed dramatically since the 1960s, and will continue to do so into the year 2001 and beyond. Health and nutritional aspects have now come to the forefront, and will form an increasing part in food consumption trends. The large supermarket chains have published brochures to inform the consumer about the various types of food, viz. bread and baked products, fruit and vegetables, meat and meat-products, cheese and dairy products, wines, fish and poultry, and about the nutritional significance of salt, fats and fibre in the diet. Nutritional labelling information printed on the packaging includes: a typical analysis, the energy contribution to the diet of a given weight, and in some cases the contribution to the 'recommended daily amounts' of each nutrient (RDAs). Much of this information can confuse the consumer; but the message is that a healthy diet does not involve slimming or being confined to a boring diet. Instead, it involves making minor adjustments to individual diets, in order to make them more healthy and better balanced.

For the baking industry, the success of the frozen-product sector, whether in the form of dough or baked products, is due to flexibility and product freshness. The concept of 'bake-off' according to consumer demand is a worthwhile technical innovation, which is a benefit to the producer and consumer of baked products.

The demand for 'oven-fresh' products is also a growth market, especially in the bread-roll and specialty-bread sector. The health and nutritional factor has resulted in the development of more interesting specialty products of added value for the consumer. These products are showing steady sales increases, and are expected to make further inroads into the conventional white panned bread market. The pathway to success is innovation, and the production of a wide product range of freshly baked varieties.

## 3.6.4 BREAD PRESERVATION BY FREEZING

The preservation of bread and baked products by freezing technology has been known and applied both in the USA and Europe for several decades. Although many refinements in equipment and technology have taken place within the past two decades, the choice of temperature ranges for the retardation of staling of baked products falls within the wide basic range $-20$ and $-190°C$; the most unfavourable range being $+50$ to $-7°C$.

For its successful application, the products must be frozen within 4 to 8 hours of leaving the oven. Freezing can take place either with or without packaging, or with a double-layer of wrapping material. Although packaging prior to freezing reduces the rate of freezing to reach the required loaf centre temperature of $-30°C$, the time required being approximately twice as long under identical conditions, the wrapped products remain fresh two to three times longer, owing to improved moisture retention. The choice of wrapping often depends on the product, but polyethylene films remain a popular material. In a series of tests using commercial freezers, Cathcart (*Cereal Chem.*, **18**, 1941, 771), and Cathcart and Luber (*Ind. Eng. Chem.*, **31**, 1939, 362), found that bread remained fresh in flavour and aroma for up to 30 days when held at $-35°C$, and in a saleable condition for up to 345 days, the limiting factor being the ultimate development of an off-odour.

Baked products are normally frozen by using a forced air current, air speeds of 2 to 4 m/s being found optimal to reach air temperatures of $-25$ to $-35°C$. This air speed would provide a heat-transfer coefficient of 18 to 24 W m$^{-2}$ K$^{-1}$. Any increase in air speed above this 2 to 4 m/s level will progressively reduce the heat-transfer coefficient. This is due to the relatively poor thermal conductivity of baked products generally, although the air-cells in products with increasing shortening contents are correspondingly less. In order to maintain the pre-frozen quality level of the product, by utilizing the freezing process to reduce staling and any physical changes, a rapid freezing is essential, passing through the unfavourable temperature zone of $+50$ to $-7°C$ as rapidly as possible. However, the limiting factor here is air speed versus the heat-transfer coefficient from the products. As a guide, in the case of the average 1-kg loaf, cooled down to 20°C after baking, the time required to reach a temperature of $-18°C$ should not exceed 4 to 5 hours.

Bread and baked products frozen in the unwrapped state will show excessive drying out within 1 to 2 weeks, owing to the low RH, in storage-freezers.

Freezer storage temperatures also have a significant effect on the staling rate; for example, bread held within the range $-10$ to $-7°C$, which corresponds to its freezing range, will show significant crumb-firming and flavour-loss within a period of 1 week. At storage temperature of $-18°C$, crumb-softness will remain relatively stable and constant for about 1 month. Storage for longer periods at $-18°C$ will progressively devalue bread quality. Even if the temperature is allowed to rise to $-12°C$, which is only just outside the freezing-range, crumb-firming becomes apparent within a few days. Temperature fluctuations above $-7°C$ give rise to moisture migration within the crumb of the frozen bread. Even at temperatures around $-10°C$ for 2 weeks, signs of sublimation of ice crystals can be observed as white rings forming in the crumb layers immediately under the crust. This is indicative of crumb-drying beneath the crust, moisture having been transferred by sublimation and diffusion from the high-moisture crumb centre to the relatively low-moisture crust zone.

Since bread staling proceeds rapidly within the temperature range $+60$ to $-2°C$, cryogenic preservation of bread and baked products must be carried out by so-called 'shock-freezing', which avoids any subsequent separation of the crust from the crumb. This phenomenon is due to redistribution of moisture within the layers under the crust, the internal crumb layers losing moisture to the outer regions, resulting in a separation from the crumb.

Baked products are usually frozen within a tunnel by air-current flow, the heat-transfer taking place by convection according to Newton's law as follows:

$$dQ = Aa(t_1 - t_2)\,dT$$

where:

| | |
|---|---|
| $dQ$ | heat-transfer output (in W) |
| $A$ | heat-transfer surface (in m²) |
| $a$ | heat-transfer coefficient (in $W\,m^{-2}\,K^{-1}$) |
| $t_1$ | surface temperature of product (in °C) |
| $t_2$ | air temperature (in °C) |
| $dT$ | time (in hours). |

The temperature gradient $(t_1 - t_2)$ becomes less as the freezing-time increases. The value of the heat-transfer coefficient will depend mainly on the air speed within the tunnel, which is optimally about 2–4 m/s. If this range is exceeded, its influence on heat-transfer steadily decreases. Average tunnel temperatures are within the range −25 to −35°C. Such freezing systems are generally known as 'blast-freezers'.

In the USA, blast-freezing is also used for bread products, using standard air flow (air speed) of 600 ft/min, reaching final temperatures of −29 to −30°C in the case of fermented goods. Baked cakes and ready-to-bake products are taken down to −40°C, packaged in fully sealed packages with a small headspace, or in foil pans with overlay and box. Freezing times depend on the type and extent of the packaging material used. Table 76 gives examples of times required for products to reach −12°C.

TABLE 76
**Freezing times for various products to reach −12°C at an air-flow of 600 ft/min**

| Product | Packaging | Freezing time (h) for −12°C | Air-flow (ft/min) | Final target temperature (°C) |
|---|---|---|---|---|
| Bread 1-lb | None | 2·75 | 600 | −29 |
| | Single-wrap | 2·25 | 600 | −29 |
| | Double-wrap | 3·50 | 600 | −29 |
| 2-lb | None | 3·00 | 600 | −29 |
| | Single-wrap | 3·50 | 600 | −29 |
| | Double-wrap | 4·00 | 600 | −29 |
| Dinner rolls | Plastic bag & foil pan | 1·00 | 600 | −29 |
| Bread dough | Plastic bag | 3·00 | 600 | −29 |
| Danish pastry | Foil pan & sealed box | 2·00 | 600 | −29 |
| Fruit pies | Foil pan overlay & box | 4·50 | 600 | −29 |

Source: *Reference Source 1987.* Sosland Publishing Co., Kansas, USA.

## Freezing Applications for Baked Products

The application of cryogenic technology (freezing) to baked products can be divided into (1) chilled storage, and (2) longer-term preservation by freezing. Chilled storage techniques are only intended for short-term preservation of baked products, using temperatures of 0 to +4°C, which will inhibit the growth of moulds and liquefying bacteria, e.g. *B. subtilis*, and *B. mesentericus*, and retard spoilage to a degree. This method is suitable for: sweetened, fermented baked products, puff-pastry baked products and puff-pastry dough for domestic shaping and bake-off, short-paste dough for domestic shaping and bake-off, sliced and assietted cakes, cream gateaux, desserts, meat pasties, etc. The permissible residence-time of these products will depend on (a) whether or not they are packaged, (b) product type and formulation, and (c) whether the chill-cabinet is open or closed. In the case of open chill-cabinets set at +4°C, residence-times must be limited to days; but if the cabinets are closed and set for −18°C many products can be retained for weeks, or months if adequately packaged or sterilized before packaging.

Longer-term preservation by freezing is intended for setting up a production-inventory programme, in order to rationalize and better cope with peak demand periods, e.g. week-ends, and holidays. It prevents waste, and retards the onset of staling, but can only preserve the quality of a product as it leaves the oven. Therefore, only top-quality raw materials should be used in formulation, and hygienic processing procedures maintained. Depending on the size, technical and technological resources, distribution radius and personnel situation, there are various processes for the longer-term preservation by freezing. However, irrespective of their special objectives, preservation by freezing takes place in the following phases:

—cooling of the product and packaging where appropriate,
—freezing of the product,
—product storage,
—product thawing.

A typical technological system for the preservation by freezing of bread-rolls or bread can be analysed as shown in Table 77. Figures 157 and 158 show in schematic form a system suitable for the blast-freezing of bread-rolls, and the equipment and technological parameters involved. Products which tend to retrograde quickly, e.g. bread-rolls must be thawed out rapidly, since retrogradation soon sets in again after temporary suspension. Since bread-rolls are already considered stale 4 hours after baking, they should ideally be thawed out shortly before consumption. Since the consumer prefers crispness in rolls, these are best thawed out in an oven at 150 to 200°C with moderate steaming, requiring about 5 minutes residence time.

In order to exploit preservation by freezing to the full for product freshness, it is better to despatch baked products in the frozen state, and either thaw them out during transit, or at the point of sale. Where distribution chains are relatively long, and many wholesale and retail outlets have to be supplied, freezer vans or insulated containers must be utilized.

Bread-rolls and other lean-formulated products often are subject to a separation

TABLE 77

**Technological system for the freeze-preservation and thawing of bread and bread rolls**

| Process phase | Equipment | Technological parameters |
|---|---|---|
| Cooling | Bucket-elevator or band-conveyor Bread cooler Forced air | Oven-fresh products cooled by fresh air circulation to about 35°C |
| Packaging | Slicing/wrapping machine Suitable material | Products wrapped in Cellophane, polyethylene, polypropylene or aluminium film |
| Freezing | Freeze-tunnel (blast) Two-stage freezers Band-conveyors and drive-belts | Rapid cooling, with an air speed of 3·5 m/s, and air temperature −35°C, for crumb centre temperature −18°C |
| Storage | Freezer store-rooms adjacent to, and loaded from freeze-tunnel directly | Temperature about −20°C, and air speed 0·5 m/s |
| Defrosting (thawing) | Band-conveyors Product containers Ideally a defroster unit with controlled temperature, RH, and air-speed flow | Alternatives: (1) Thawing at room-temperature 20–25°C for 1–2 hours (2) Infrared lamps (3) Defroster with air-temperature 30–40°C, RH, 50–70% and air speed 0·5–2·0 m/s. For 1-kg loaves, air temperature 60–70°C can be applied |

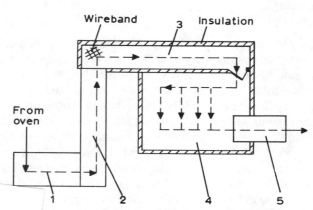

of a system for blast-freezing of bread-rolls. 1. Band conveyor. 2. Steel-
3. Blast-freeze tunnel. 4. Freeze store. 5. Conveyor. (Source: *Special
J.* Schwate and U. Ulrich, VEB-Fachbuchverlag, Leipzig, 1986, p. 193.)

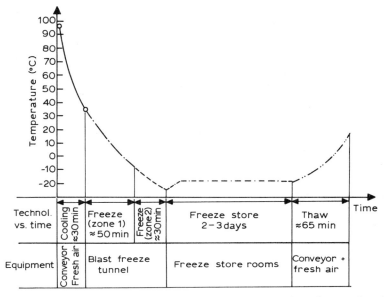

Fig. 158.   Ideal temperature curve, technological processes and equipment for the blast-freezing of bread-rolls. (Source: *Special Processes in Baking*, W. Schwate and U. Ulrich, VEB-Fachbuchverlag, Leipzig, 1986, p. 194.)

of crust from crumb during freezer storage, owing to drying out and shrinkage. These defects can be controlled to a degree in several ways. A reduction of the baking time by 20%, coupled with an increase in baking temperature by about 15°C and simultaneous addition of 2% corn syrup (HFCS), tends to postpone this crust/crumb separation from the third day of storage to about day 10. In the case of bread with added shortening at 1–2%, the addition of such surfactants as phosphatides and monoglycerides tend to hinder this separation process, through improved lipid distribution between the dough polymers.

All the freezing applications described hitherto rely on mechanical refrigeration techniques involving either static air or an air blast to cool the product.

For industrial bakeries specializing in frozen bakery products, or frozen-food chains, with throughputs of up to 1000 kg/hour, the direct contact application of refrigerant, often referred to as 'cryogenic' can be economically applied. Furthermore, since the liquid coolants come into direct contact with the product, cryogenic freezing rates are far more rapid than is the case with mechanical refrigeration techniques. The cryogenic system uses liquid nitrogen or liquid carbon dioxide. The boiling point of liquid nitrogen is −196°C, therefore the cryogenic system allows the application of very much lower temperatures than the air-blast system. Cryogenic systems use tunnels of up to 12 m with widths of about 1·8 m, the products for freezing passing under liquid nitrogen spray, which evaporates on contact. By means of fans, cold evaporated gas flows in a countercurrent direction. Equilibrium at −18°C is achieved by retaining the product in the tunnel. However, this system involves the continuous expenditure of the coolant which increases the operational cost very significantly, depending on the potential throughput on a 24 hour basis. The main advantages of the cryogenic system are: the relatively small floor area required, low maintenance costs, improved product quality in many cases, and a high degree of flexibility.

Although the application of freezing technology to baked products remains important for the baking industry, of far greater significance is its application to doughs. This technology has revolutionized the working schedules in production, and made the forward planning of inventories and their distribution to wholesale and retail outlets much easier. This sector of the industry now includes the large and small baked-product manufacturer, and the ever-increasing number of non-manufacturers in the wholesale and retail market, who are described as 'bake-off' bakeries or retail outlets.

# 3.7 A Preview of the 1990s and Changes in Product Demand and Supply

Within Europe, the removal of barriers to trade will inevitably result in commercial opportunities and adversities for all member states. The removal of these barriers will also affect the technological aspects of baking, since a list of additives acceptable to all member states will have to be incorporated into European Community legislation. Since the cost-effectiveness of producing certain types of product depend on the inclusion of specific dough additives, within some member states, their retention in the final list is likely to be the subject of much discussion and negotiation. Most continental member states have, however, produced good quality products with untreated, unbleached flour, and a minimum of permitted additives, e.g. oxidants and emulsifiers, for a period of more than 30 years. The situation regarding the import/export of bakery products will be determined by consumer demand and cost/retail-price structures. Product preservation techniques are now available which allow the inter-state transport and distribution of both baked and unbaked products without deterioration. The most significant technique is preservation by freezing, the freezing of shaped dough-pieces permitting bake-off into the fresh-product state at any desired location. If large supermarket chains import frozen cakes, or frozen dough-pieces for rolls, baguettes, croissants, specialty-breads or pastries, the sale of such imported products would represent a corresponding loss of business for home-based raw material suppliers, i.e. milling industry, etc., craft-bakeries and plant bakers alike. To a limited extent this is already happening, but may well escalate when the barriers are removed, at least in the short-term. At the present time, frozen, sliced and wrapped bread is being sent from the UK to the continent; and the UK imports regular supplies of rye-based wholegrain breads, e.g. pumpernickel, *Vollkornbrot*, and wrapped rye/wheat breads with extended shelf-life, from West Germany and Holland. Aluminium-foil packed cakes from West Germany and Denmark can be found on UK supermarket shelves; and Bahlsen of Hannover have a long-established market in the UK for biscuits, *Pfefferkuchen, Stollen, Sandkuchen,* and cocktail snacks. Packed individual cakes are also increasing in frequency on UK grocery shelves, imported from Holland— quality at a competitive price.

The UK prepacked cake manufacturer, whose market is valued at about £500 million per year, relies on chlorine-treated high-ratio cake flour at the present time. If chlorine is removed from the European Community permitted list, the type of cake and formulation would have to change, unless an alternative flour technology can be found.

The requirement to meet the challenge of 1992 and beyond for small and large bakers in all member states, is to study the bakeries and products made by their Community neighbours, and exchange recipes and ideas. In this way, each can extend his product range and satisfy the demand of the consumer for specialties, often from other countries, which he or she has taken to while on vacation. Therefore, it is possible that the export/import rates will not show large increases in the long term. However, there are already distinct product trends taking place in central Europe, as a result of diverse baking cultures and traditions. The Scandinavian/German 'black-bread' product trend for meal and wholegrain products has extended far beyond the borders of these countries, and will make steady sales progress in most 'white-bread' consuming countries. Such traditional French products as baguettes and croissants, which offer freshness and consumer satisfaction, will be found to be part of the product assortment throughout the Community. The delicately light-eating brioche, as produced traditionally in Spain and France, represents another quality product with general appeal and potential. The crusty Italian variety breads, of high volume and open crumb porosity, coupled with the aromatic flavour, also have great potential Community appeal for the consumer. The Mediterranean pizza and pita breads, originally a part of restaurant meals, have now found a place in the households of most Europeans and North Americans.

In West Germany, a country with over 200 types of product, there has long been an exchange of regional product recipes, the Westfalian pumpernickel being as well known in Bavaria as the *Bretzel* in Hannover and Hamburg. This development is now expected to extend to the European level, extending across national frontiers.

The German baking industry is starting from a position of strength within the Community, owing to its large choice of products and their quality. The technological expertise of the German baker and the standard of his training is acknowledged worldwide. Without the completion of an apprenticeship, known as the '*Lehre*', and attainment of the '*Meisterprüfung*', he or she cannot open a bakery business. Since the craftsmanship and technical standards are maintained at a high level, for the processing of wheat- and rye-flours into bread, the product standard is also high, and held in high esteem by the consumer. For the year 1988/89, the consumption per head of bread and bread-rolls in the FRG stands at 80·3 kg, having risen by an average of 17% over the previous 15 years. Mixed wheat/rye breads represent about 52% of all breads sold. Furthermore, it has been established by the Emnid Institute in Bielefeld, FRG, as a result of a consumer census, that West Germans on vacation abroad are not satisfied with the bread products on offer, and miss in particular the quality of their mixed rye/wheat and wholegrain breads. In most 'black-bread' countries, up to 50% of all bread produced contains rye in some form. It now remains for the 'white-bread' countries to acquire the rye processing technology, and offer their consumers an improved range and quality of breads and baked products.

It must also be acknowledged that the bread quality standards within the COMECON countries compare favourably with those in the West. Product standards found in the GDR compare favourably with those available in the FRG. In the GDR, machinery and equipment standardization in design and construction has considerable advantages. Also, baking standards have been laid down under

TGL-3067 for all basic bread varieties. These include baking-times, initial and final baking-temperatures and crust-thickness, according to weight, shape, and mode of baking. Permissible baking-time tolerances are also specified for each variety and mode of baking under TGL-3067, as detailed in Section 1.7.

In the FRG, where a night-baking ban has been in force since 1915, industrial bakeries in particular are seeking a solution to this restriction on production between 22.00 and 04.00 hours, which only increases their cost base, where baker's wages are already high. Such restrictions could sharpen competitition from certain products from neighbouring countries, in particular France. Therefore, there is considerable incentive for automating the whole production process, pre-programming mixing, proofing and baking schedules, in a similar way to that applied to frozen-dough/bake-off cycles. A computer solution can be utilized to integrate the 'islands' of automation. However, such a solution demands manufacturing experience, technology and systems integration in evaluating a particular process; and will comprise such elements as software, hardware, robotics and control systems. Such a fully automated, centrally controlled processing concept, 'Project CIM 2000', computer-integrated manufacture is under development by Dietrich Reimelt KG.

On the bakery product vending (selling) side, automation is also making a valuable contribution. Product exposure can be increased to 24 hours per day, and 7 days per week by installing a modern vending machine. Service points can be outside a baker and confectioner's shop, or located in a busy shopping mall. The 'Quickbread' day and night service, '*Pain frais nuit et jour*', is manufactured by Dataa, 12, rue de la Blanchardière, BP 513-49305 Cholet Cedex, France, and marketed by Eurofours Baking Systems. Such a 'Quickbread' vending unit has a capacity of about 80 loaves, being easy to load and simple in operation, backed by after-sales service and promotional advice. The unit is loaded with loaves of the same type and price, fed to the dispenser down a spiral vibrator chute when the coin mechanism is actuated by the correct amount of money being inserted. The coin units are made by national companies to suit the currency. The Dataa 'Quickbread' is widely sold in France and other EC countries to promote bread sales outside normal shop hours, and in locations where staffing is minimal, but customer potential high, e.g. garage forecourts and holiday parks. Such machines are also available on lease for 5 years, allowing a precise calculation of individual break-even and pay-back rates from price/profit product structures. These machines will also accept rolls in a prepacked form.

For industrial-bakeries Dietrich Reimelt KG (Frankfurt/Main, FRG) are developing a Computer Integrated Manufacturing (CIM) system 'CIM 2000', which is a step towards fully automated, centrally controlled processing. Important elements for such automation are:

—consistent preparation and processing
—immediate and unrestricted availability of data
—immediate adaptation of parameters
—quality engineering
—computer-controlled maintenance and cleaning schedule (CIP)

For the Handcraft-bakery, Winkler GmbH & Co. KG, Villingen/Black Forest, FRG, represented in UK by Atlas Equipment (London) Ltd, are working on a similar project incorporating computer linked and guided systems to integrate: ingredient addition, mixing, dividing, make-up, proofing, and baking, thus reducing personnel, and quality variations.

# PART 4
## Notes and References

# 4.1 Notes and References for Part 1

## CHAPTER 1.1 THEORETICAL MODEL TO EXPLAIN THE DOUGHMAKING PROCESS and CHAPTER 1.2 APPLICATION OF FUNDAMENTAL DOUGH-MIXING PARAMETERS

*Tschernych, W. Ja., Putschkowa, I. J., Ljaskowskij, D. I., Salapin, M. B., and Tscheuschner, H.-D.* (1985) Investigations concerning the influence of mixing-intensity and mixing-time during the discontinuous preparation of wheat-flour dough (German). Collaborative research USSR/GDR. *Bäcker und Konditor*, **1**, p. 22–4 (VEB-Fachbuchverlag, Leipzig, GDR).
Fig. 1 Dough viscosity versus mixing-time at various mixer-rpm.
Fig. 2 Specific mixing-intensity versus mixing-time.
Fig. 3 Specific mixing-intensity versus mixing-time at various rpm.
Fig. 4 Specific energy input versus mixing-time at various rpm.

*Stear, C. A.* (1986) Dynamics of the discontinuous mixing process and its rationalization. *Getreide Mehl und Brot*, **40**(10), p. 294–7.
Fig. 5 Dough viscosity versus mixing-time at various rpm.
Fig. 6 Mixing torque versus mixing-time at various rpm.
Fig. 7 Specific mixing-intensity versus mixing-time at various rpm.
Fig. 8 Specific energy input versus mixing-time at various rpm.

## CHAPTER 1.3 FUNDAMENTAL CONSIDERATIONS CONCERNING DOUGH RHEOLOGICAL ELEMENTS AND DYNAMIC MIXING PARAMETERS

*Halton, P., and Scott-Blair, G. W.* (1936–1937).
*J. Physical Chemistry*, **XI**, p. 561, 1936.
*J. Physical Chemistry*, **XI**, p. 811, 1936.
*Cereal Chemistry*, **XIV**, p. 201, 1937 (English).

*Lerchental, C. H., and Müller, H. G.* (1967) Research in dough rheology at the Israel Institute of Technology. *Cereal Science Today*, **12**(5), p. 185–92 (English).

*Hlynka, I., and Anderson, J. A.* (1952) Relaxation of tension in stretched dough. *Canadian Journal Technology*, **30**, p. 198 (English).

*Dempster, C. J., Hlynka, I., and Anderson, J. A.* (1953) *Cereal Chemistry*, **30**, p. 492 (English).

*Frazier, P. J., Brimblecombe, F. A., Daniels, N. W. R., and Russell-Eggitt, P. W.* (1979) Better bread from weaker wheats—rheological considerations. *Getreide Mehl und Brot*, **33**, 10, p. 268–71 (German).
Fig. 3 Comparison of rheological data and baking quality of mechanically developed doughs.

*Lutsishina, E. G., and Matyash, A. I.* (1984) Deposited Doc. 1984 VINITI 3547-84 10 pp. (Russian).

*Örsi, F., Pallagi-Bânkfalvi, E., and Lástity, R.* (1985) Analysis of the correlation between flour quality and electrophoretic protein spectra. *Acta Alimentaria,* **14**, 1, p. 49–57 (German).

*Kawamura, E. G. and co-workers* (1985) Selective reduction of inter-polypeptide and intra-polypeptide SS-bonds of wheat glutenin from defatted flour. *Cereal Chemistry,* **62**, p. 279–83 (English).

*Moonen, J. H. E., Scheepstra, A., and Graveland, A.* (1985) Biochemical properties of some high molar weight sub-units of glutenins. *Journal of Cereal Science,* **3**(1), p. 17–27 (English).

## CHAPTER 1.4 WATER-BINDING CAPACITY OF DOUGH COMPONENTS AND DOUGH CONSISTENCY CONTROL

*Jazuba, W. I., Siderowa, O. G., Putschkowa, L. I., and Tscheuschner, H.-D.* (1985) Influence of mixing-intensity and mixing-time on the water-binding capacity of the components of wheat flour dough. *Bäcker und Konditor,* **3**, p. 90–2 (VEB-Fachbuchverlag, Leipzig, GDR) (German). Collaborative research USSR/GDR.

*Baker, J., Parker, H., and Mize, M.* (1946) Supercentrifugates from dough. *Cereal Chemistry,* **1**, p. 16.

*Auerman, L. Ja.* (1972) *Breadmaking Technology,* 7th edn, Food Industry Publishers, Moscow, USSR, 1972, p. 511 (Russian).

*Kosmina, N. P.* (1978) *Biochemistry of Breadmaking,* Food Industry Publishers, Moscow, USSR, 1978, p. 277 (Russian).
Table 1 Distribution of moisture in the dough components, Jazuba, W. I., Siderowa, O. G., Putschkowa, L. I., and Tscheuschner, H.-D. *Bäcker und Konditor,* **3**, 1985, p. 92.
Table 2 Analysis of the dough fractions after ultracentrifugation, Jazuba, W. I., Siderowa, O. G., Putschkowa, L. I., and Tscheuschner, H.-D. *Bäcker und Konditor,* **3**, 1985, p. 92. (Collaborative research USSR/GDR.)

## CHAPTER 1.5 EFFECTS OF DOUGH ADDITIVES

*Marston, P. E.* (1971) *Bakers Digest,* December issue, p. 16–19 and 62. Chemical activation of dough development under slow mixing conditions. (Australian Bread Research Institute, N. Ryde, NSW, Australia.)

*Henika, R. G., and Rogers, N. E.* U.S. Patent 3,053,666 Sept. 11, 1962. Foremost Dairies, Inc., Dublin, California, USA.

*Allen, W. G., and Spradlin, J. E.* (1974) Properties of commercial amylases for bakery applications. *Bakers Digest,* **48**(3), 14.
Reference source 1985, Proteolytic activity of malt and fungal enzymes supplements. *Bakers Digest,* Kansas 66201, USA.

*Pfeilsticker, K., and Marx, F.* (1986) Gas-chromatographic and Mass-spectrographic studies on the redox reaction kinetics of L-ascorbic acid and L-dehydro-ascorbic acid in wheat-flour doughs. Rheinischen Friedrich-Wilhelms University, Bonn, FRG. *Z. Lebensmittel Untersuchung und Forschung,* **182**(3), p. 191–5 (German).

*Vangelov, A., Naidenova, R., and Karadzhov, G.* (1985) Comparative evaluation of the effect of ascorbic acid and iso-ascorbic acid on bread quality. Higher Technological Institute for the Food Industry, Plovdiv, Bulgaria. *Nauchui Tr-Vissh. Khranit. Vkusova Prom.-st, Plovdiv,* **32**(1), p. 133–40 (Bulgarian).

*Kuninori, T., and Matsumoto, H.* (1963) L-Ascorbic acid oxidizing system in dough and dough improvement. *Cereal Chemistry,* **40**, p. 647–57.

*Prihoda, J., Hampl, J., and Holas, J.* (1971) Effects of ascorbic acid and potassium bromate on the viscous properties of dough measured with a Höppler-consistometer. Chemical and Technological University, Prague, Czechoslovakia. *Cereal Chemistry*, **48**, 1, p. 68–74.

*Giacanelli, E.* (1972) Mechanical and chemical development of bread doughs III. *Ind. Alimentari*, **11**(5), p. 93–8.

*Jørgensen, H.* (1935) Contribution to the explanation of the inhibitory effect of oxidants on proteolytic enzyme activity: the nature of the effect of potassium bromate and similar substances on the baking properties of wheat-flour. *Biochemische Zeitschrift*, **280**, p. 1–37 and **283**, p. 134–145 (German).

*Jørgensen, H.* (1935) Nature of the bromate-effect. *Mühlenlaboratorium*, **5**, p. 113–125.

*Jørgensen, H.* (1939) Further investigations into the nature of the action of bromates and ascorbic acid on the baking strength of wheat-flour. *Cereal Chemistry*, **16**, p. 51–60.

*Elion, E.* (1945) The role of proteolysis in bread dough. *Bakers Digest*, April issue, p. 15–17 and 27–28.

*Coventry, D. R., Carnegie, P. R. and Jones, I. K.* (1972) The total glutathione content of flour and its relation to the rheological properties of dough. *J. Sci. Food Agric.*, **23**, p. 587–94.

*Karpenko, V. I.* (1985) Relation of dough properties and bread quality to the separation of glutathione from yeast. *Khlebopek. Konditer Prom-st*, **9**, p. 40–2. Voronezh. Tekhnol. Inst., Voronezh, USSR (Russian).

*Grunert, S., Möhr, B., and Kroll, J.* (1986) Model techniques for investigation of the interaction between proteins and emulsifiers. *Lebensmittelindustrie*, **1**, p. 17–20 (German).

*Lucny, M.* Wheat and rye/wheat products with the addition of monoacylglycerol diacetyltartrate. Czech patent, CS 210,528, 15 July 1983 (Czech).

*Kawka, A., and Gasiorowski, H.* (1985) Effect of sodium stearoyl-2-lactylate on the overall quality of wheat/rye mixed bread (1:1), enriched with skimmed milk powder. *Getreide Mehl und Brot*, **39**, 12, p. 369–73. Institute of Food Technology, Agric. University, Poznan, PL-60624, Poznan, Poland (German).

*Moore, W. R., and Hoseney, R. C.* (1986) Influence of shortening and surfactants on retention of carbon dioxide in bread doughs. *Cereal Chemistry*, **63**, 2, p. 67–70. Kansas State University, Manhatten, Kansas, USA.

*Ofelt, C. W., and Smith, A. K.* (1954) Baking behavior and oxidation requirements of soy-flour. I. Commercial full-fat soy-flours. II. Commercial defatted soy-flours. *Cereal Chemistry*, **31**(1), p. 15–22.

*Pollock, J. M., and Geddes, W. F.* (1960) Soy-flour as a white-bread ingredient. I. Preparation of raw and heat-treated soy-flours, and their effects on dough and bread. II. Fractionation of raw soy-flour and effects of the fractions in bread. *Cereal Chemistry*, **37**(1), p. 19–29 and 30–54.

*J. Rank Limited*, partly in collaboration with J. G. Hay. British patents: 646311, 706942, 771361, 880182. French patent: 1,070174. Methods for the improvement of the baking properties and crumb color of bread by high-speed, or 2–7 times extended mixing-time.

*Daniels, N. W., Frazier, P. J., and Wood, P. S.* (1971) Flour lipids and dough development. *Bakers Digest*, August issue, p. 20–26.

*Pashchenko, L. P., Mazur, P. Ya., Serbulov, Yu. S., and Zolotykh, I. N.* (1984) Dependence of the rheological properties of liquid semi-products on a surfactant and oxygen. *Khlebopek. Konditor Prom-st*, **11**, p. 24–7 (Russian).

Model of a phospholipid (lecithin) molecule. *Kleine Enzyklopädie-Natur* (1971) p. 189. VEB-Bibliographisches Institut, Leipzig, GDR (German).

*Sullivan, B.* (1954) Proteins in flour—Review of the physical characteristics of gluten and reactive-groups involved in changes in oxidation. *J. Agr. Food Chem.*, **2**, 1231–4.

*Grosskreutz, J. C.* (1960) The physical structure of wheat protein.
*Biochim. biophys. Acta.*, **38**, 400–9.
A lipoprotein model of wheat gluten structure. *Cereal Chem.*, **38** 336–49.

*Hess, K.* Zwickel—und Haftprotein (Wedge and Adherent Protein) *Kolloid Z.*, **136** (1954) 84.
*Kolloid Z.* **141** (1955) 61.

*Rohrlich, M. and Müller, K.* (1968) Investigations concerning the Fat/Protein Complex in Cereals. *Die Mühlle*, **41**.

Kyowa Hakko Kogo Co Ltd (Patent) Euro. Patent Appl. 134,658 March 20, 1985.
'Improvement of Performance of Vital Wheat Gluten'.
Schematic diagrams Fig. 15 Amino acids in Wheat Gluten protein (Dimler).
Fig. 16 Model of gluten-sheet showing double-layer of Phospholipid in the Lipoprotein slip-plane (Grosskreutz).

*Stepanova, O. N. and Akimova, A. A.* (1986) (Russ) Khlebopek, Konditer Prom-st, 2, 32–4.

*Farrand, E. A.* (1969) Starch damage and alpha-amylase as a basis for mathematical models relating to flour water absorption. *Cereal Chem.*, **46**, March issue, p. 103–16.

*Schulz, A.* (1962 and 1963) The influence of starch-degrading enzymes of rye and wheat of the chloride ion. *Brot und Bebäck*, **16**, 8, p. 101–43, 1962 (German).
The salt-effect during the processing of sprouted rye-flours. *Brot und Gebäck*, **17**, 6, p. 112, 1963 (German).

*Stephan, H. and Schulz, A.* (1960 and 1961) Investigations concerning an expedient processing of sprouted rye flours. *Brot und Gebäck*, **14**, 12, p. 240–5, 1960 (German).
A contribution to the processing of sprout-damaged flours. *Brot und Gebäck*, **15**, 8, p. 162–5, 1961 (German).

*Huber, H.* (1961, 1962 and 1964) Influence of acid and salt on the baking properties of rye flour. *Industriebackmeister*, **9**, No. 4, August issue, 1961 (German).
Influence of acid and salt on the baking properties of rye flour. *Brot und Gebäck*, **16**, 5, p. 88–95, 1962 (German).
The salt-effect during the processing of wheat and rye flour. *Brot und Gebäck*, February issue, 1964 (German).

*Mecham, D. K., and Weinstein, N. E.* (1952) Lipid-binding in doughs—effects of dough ingredients. *Cereal Chem.*, **29**, 6, p. 448–55.

*Fullerton, J. G.* (1969) Lipid–protein interaction. *Bakers Digest*, December issue, 1969.

## CHAPTER 1.6 CHEMICAL BONDING DURING DOUGHMAKING

*Laignelet, B., and Dumas, C.* (1984) Lipid oxidation and distribution of oxidized lipids during the mixing process of bread flour wheat. *Lebens. Wissenschaft und Technologie*, **17**, 4, p. 226–30.

*Jackson, G. M.* (1983) The effect of endogenous phenolic acids on the rheology of wheat flour doughs. *Dissertation Abstracts International*, 1983, B 44, No. 5, 1439 B.

*Kolesov, V. M.* (1951) The amino-nitrogen content of grain-prolamines. *Biochimja*, **16**, p. 346–9 (Russian).

*Reznichenko, M. S.* (1951) The length of the polypeptide-chains in gliadin. *Biochimija*, **16**, p. 579–83 (Russian).

*Ewart, J.* (1985) Blocked thiols in glutenin and protein quality. *J. Sci. Food Agric.*, **36**, 2, p. 101–12.

*Hess, K.* (1954) Protein, gluten and lipoid in wheat grain and flour. *Kolloid-Z.*, **136**, 2/3, p. 84–99 (German).

*Traub, W., et al.* (1957) X-ray studies of the wheat protein complex. *Nature (London)*, **179**, p. 769–70.

*Grosskreutz, J. C.* (1961) A lipoprotein model of wheat gluten structure. *Cereal Chem.*, **38**, 4, p. 336–49.

*Vercouteren, R., and Lontie, R.* (1954) The solubilization of the gluten of wheat using dimethyl formamide. *Arch. Intern. Physiol.*, **62**, p. 579–80 (French).

*Bloksma, A. H.* (1958) The significance of thiol- and disulfide-groups in gluten for the baking properties. *Getreide und Mehl*, **8**, 6, p. 65–9 (German).

*Bushuk, W.* (1961) Accessible sulfhydryl-groups in dough. *Cereal Chem.*, **38**, 5, p. 438–48.

*Axford, D. W. E., et al.* (1962) Disulfide groups in flour proteins. *J. Sci. Food Agric.*, **13**, 2, 5, p. 177–181.

*Agatova, A. I., and Proskurjakov, N. I.* (1962) Sulfhydryl and disulfide bonds in flour proteins. *Biochimija*, **27**, 1, p. 88–93 (Russian).

*Lee, C. C., and Samuels, E. R.* (1962) A radiochemical method for the estimation of sulfhydryl/disulfide ratio in wheat gluten. *Can. J. Chem.*, **40**, 5, p. 1040–2.

*Lutsishina, E. G., and Matyash, A. I.* (1984) Wide-band proton NMR-spectrographic study of gluten proteins of wheat. Deposited Doc. 1984 VINITI 3547–84, 10 pp. (Russian).

*Ponte, J. G. Jr.* (1968) Modification of dough properties by organic solvents. *Cereal Sci. Today*, **13**, 10, p. 364–94.

*Popineau, Y.* (1985) Fractionation of acetic acid soluble proteins from wheat gluten by hydrophobic interaction chromatography; evidence for different behaviour of gliadin and glutenin proteins. *J. Cereal Sci.*, **3**, 1, p. 29–38.

*Zawistowska, U., and Bushuk, W.* (1986) Electrophoretic characterization of a low molecular weight wheat protein of variable solubility. *J. Sci. Food and Agric.*, **37**, 4, p. 409–17.

*Ma Ching Yung, Oomah B. Dave, and Holme, John* (1986) Effect of deamination and succinylation on some physicochemical and baking properties of gluten. *J. Food Sci.*, **51**, 1, p. 99–103.

*Bushuk, W., Bekes, F., McMaster, G. J., and Zawistowska, U.* (1985) Carbohydrate and lipid complexes with protein in gluten. *Veröff. Arbeitsgem. Geteideforsch.*, 1985, 198, 3–9 (German). (*Berliner Tag. Getreidechem.*, **35**, 1984.)

*Wall, J. S.* Review of wheat protein; Bietz, J. A., Huebner, F. R., and Wall, J. S. Schematic of configuration differences between gliadin and glutenin and the hydrated visco-elastic structures which form gluten. *Bakers Digest*, **47**, 1, 26, 1973.
Schematics of complex interactions during breadmaking.
(A) Starch–lipid–protein complex of flour, according to Hess, K.
(B) Phospholipid–protein bonding.
(C) Lipoprotein model, according to Grosskreutz, J. C.
(D) Gliadin–glutolipid–glutenin complex, according to Hoseney, R. C. *et al.*
(E) Starch–glycolipid–gluten complex, according to Wehrli, H. P.

*Hoseney, R. C., Finney, K. F., Pomeranz, Y.* (1970) Functional (breadmaking) and biochemical properties of wheat flour components VI. Gliadin–lipid–glutenin interaction in wheat gluten. *Cereal Chem.*, **47**, 2, p. 135–9.

*Wehrli, H. P., and Pomeranz, Y.* (1970) A note on the interaction between glycolipids and wheat flour macromolecules. *Cereal Chem.*, **47**, 2, p. 160–6.

*Pomeranz, Y.* (1985) Wheat flour lipids—what they can and cannot do in bread. *Cereal Foods World*, **30**, 7, p. 443–6.

*MacRitchie, F.* (1985) Studies of the methodology for fractionation and reconstruction of wheat flours. *J. Cereal Sci.*, **3**, 3, p. 221–30.

*Pallagi-Bânkfalvi, E.* (1983) Analysis by polyacrylamide-gel electrophoresis (PAGE) of proteins in wheat flours and doughs. *Acta Alimentaria,* **12,** 1983 (English).

*Bogdanov, V. P., et al.* (1985) Difference in polypeptide composition of glutenins from wheats with different baking qualities. *Prikl. Biokhim. Mikrobiol.,* **21,** 6, p. 838–42 (Russian).

*Graveland, A. et al.* (1985) A model of the molecular structure of glutenins from wheat flour. *J. Cereal Sci.,* **3,** 1, p. 1–16.

*Hoseney, R. C. et al.* (1970) *Cereal Chem.,* **47,** 2, p. 135–139.

*Daniels, N. W., et al.* (1971) Flour lipids and dough development. *Bakers Digest,* August issue, p. 20–6.

*Wehrli, H. P., and Pomeranz, Y.* (1970) *Cereal Chem.* **47,** 2, p. 160–6.

*Tortosa, E., et al.* (1985) Chemical changes during bread dough fermentation I. Lipids of bread dough. *Rev. Agroquim. Tecnol. Aliment.,* **25,** 3, p. 417–27 (Spanish).

*Ma Ching Yung, Oomah B. Dave, and Holme John* (1986) *J. Food Sci.,* **51,** 1, p. 99–103.

*Woychik, J. H., Boundy, J. A., and Dimler, R. J.* (1961) Hydrophobicities of gliadin and glutenin. *J. Agr. Food Chem.,* **9,** 4, p. 307–10 (under title: 'Amino acid composition of proteins in wheat flour').

*Cluskey, J. E., and Wu, Y. V.* (1966) Structure of gliadin and glutenin. *Cereal Chem.,* **43,** p. 119.

*Lenard, J., and Singer, J. S.* (1966) Alpha-helices stabilization by hydrophobic amino acids. *Proc. Nat. Acad. Sci. USA,* **56,** p. 1828.

*Vakar, A. B. et al.* (1985) Influence of $D_2O$ on the physical properties of gluten and wheat dough. *Prikl. Biokhim. Mikrobiol.,* **1,** p. 5–24 (Russian).

*Bloksma, A. H.* (1975) *Cereal Chem.,* **52** (3, Part II). 170r.

*Ewart, J. A. D.* (1977) *J. Sci. Food Agric.,* **28,** p. 191.

*Ewart, J. A. D.* (1978) *J. Sci. Food Agric.,* **21,** p. 551.

*Ewart, J. A. D.* (1985) Blocked thiols in glutenin, and protein quality. *J. Sci. Food Agric.,* **36,** 2, p. 101–12.

*Tsen, C. C., and Bushuk, W.* (1963) Dough mixing in air vs. in nitrogen. *Cereal Chem.,* **40,** p. 399–408 (under title: Changes in SH and SS contents of doughs during mixing under various conditions).

*Sullivan, B., et al.* (1963) The oxidation of wheat flour—labile and non-labile SH groups. *Cereal Chem.,* **40,** 5, p. 515–31.

*Mamaril, F. P., and Pomeranz, Y.* (1966) The isolation and characterization of wheat flour proteins. IV. Effects on wheat flour proteins of dough mixing and of oxidizing agents. *J. Sci. Food Agric.,* **17,** p. 339–43.

*Hguyen-Brem, P. T., et al.* (1983) Oxidation/reduction effects on flour proteins—molecular weight distribution. *Getreide Mehl und Brot,* **37,** p. 35 (German).

*Dirndorfer, M., et al.* (1986) Changes in glutens of various wheat varieties by oxidation. *Lebensmittel Untersuch. und Forsch.,* **183,** 1, p. 33–8 (German).

*Kosmina, N. P.* (1958) Proteins of rye and rye gluten. *Getreidemühle,* **2,** 6, p. 122–3 (German).

*Vones, E., et al.* (1964) Significance of wedge proteins of rye flour for dough properties. *Cereal Chem.,* **41,** p. 456–64.

*Anger, H., et al.* (1986) Molecular weight and limit-viscosity of arabinoxylan (Pentosan) from rye (*Secale cereale*); formulation of the Mark–Honwink relationship. *Die Nahrung*, **30**, 2, p. 205–8 (German).

*Neukom, H., and Markwalder, M.* (1978) Oxidative gelling of wheat flour pentosans—a new possibility of polymeric linkage. *Cereal Foods World*, **55**, 7, p. 374–6.

*Holas, J., et al.* (1973) Grain polysaccharides of the pentosan type VI. Investigation of the protein component of the pentosans and hemicelluloses. *Mlynsko-pekárensky prùm. Praha*, **19**, 1, p. 11–14 (Czech.).

*Kolesov, V. M.* (1951) The amino-nitrogen content of grain prolamines. *Biochimija*, **16**, 579–83 (Russian).

*Perlin, A. S.* (1951) Structure of the soluble pentosans of wheat flours. *Cereal Chem.*, **28**, 5, p. 382–93.

*Montgomery, R., and Smith, F.* (1955) The carbohydrates of the gramineae V. The constitution of a hemicellulose of the endosperm of wheat (*Triticum vulgare*). *J. Am. Chem. Soc.*, **77**, p. 2834–7.

*Cole, E. W.* (1967) Isolation and chromatographic fractionation of hemicelluloses from wheat flour. *Cereal Chem.*, **44**, 4, p. 411–16.

*Medcalf, D. G., D'Appolonia, B. L., and Gilles, K. A.* (1968) Comparison of the chemical composition and properties between HRS and durum wheat endosperm pentosans. *Cereal Chem.*, **45**, 6, p. 539–49.

*Kosmina, N. P. The Biochemistry of Breadmaking*, VEB-Fachbuchverlag, Leipzig, 1977 (German).

*Holas, J. et al.* (1974) Grain polysaccharides of the pentosan type XII. Changes in the pentosan fractions during breadmaking. *Mlýnsko-pekárenský prùm. Praha*, **20**, 2, p. 46–8 (Czech.).

*Schmieder, W.* (1977) Problems concerning the evaluation of rye for breadmaking from analytical data, with special reference to breeder's sample material. Dissertation, 1977 (German).

*Drews, E.* (1973) Variations in the quality of flour type 997, depending on the properties of the rye grain. *Getreide Mehl und Brot*, **27**, 10, p. 305–311 (German).

*Drews, E., and Zwingelberg, H.* (1977) Results from many years of rye investigations concerning varieties, location, and crop year with particular reference to flour properties. *Getreide Mehl und Brot*, **33**, 2, p. 34–7 (German).

*Drews, E.* (1979) Milling properties as a function of grain properties, and their effect on flour quality. *Getreide Mehl und Brot*, **33**, 2, p. 29–34 (German).

*Meuser, F., et al.* (1986) Chemical/physical properties of rye pentosans. *Veroff. der Arbeitsgem. für Getreideforsch. in Berlin*, **203**, p. 73–89 (German).

*Kühn, M. C., and Grosch, W.* (1985) The fractionation and reconstruction of rye flour—special technique to study the non-starch polysaccharide hydrolases in rye bread production. *Getreide Mehl und Brot*, **39**, 11, p. 340–344 (German), and in *Veroff. Arbeitsgem. Getreideforsch. Berlin*, 1986, **203**, p. 21–29 (German).

*Simmonds, H., and Orth, R. A.* (1973) Structure and composition of cereal proteins, as related to their potential utilization. In: *Industrial Uses of Cereals*, Y. Pomeranz, editor, American Association of Cereal Chemists, St Paul, MN, USA.

*Ewart, J. A. D.* (1968) Action of glutaraldehyde, nitrous acid or chlorine on wheat proteins. *J. Sci. Food Agric.*, **19**, p. 370.

## CHAPTER 1.7 TYPICAL FORMULATION AND PROCESS SCHEDULES (INCLUDING CASE STUDIES) FOR WHEAT AND RYE BREADS EMPLOYED IN WESTERN AND EASTERN EUROPE AND NORTH AMERICA

### 1.7.1 France

*Calvel, R.* (1972) Technical requirements for flours used for bread and baked-products in France. *Getreide Mehl und Brot*, **26**, 3, p. 75–8 (German).

### 1.7.2 Spain

*Barber, S., et al.* (1983) Microflora of the sour dough of wheat flour bread I. Identification and functional properties of microorganisms of industrial sour doughs. *Rev. Agroquim. Tecnol. Aliment.*, **23**, 4, p. 552–62 (Spanish). Laboratorio de Cereales y Proteaginosas, Instituto de Agroquimica y Technologia de Alimentos (CSIC), Valencia 10, Spain.
II. Functional properties of commercial yeasts and pure strains of *S. cerevisiae* in sugar solutions. *Rev. Agroquim. Technol. Aliment.*, **25**, **3**, 1985, p. 436–46 (Spanish).
III. Functional properties of commercial yeasts and pure strains of *S. cerevisiae* in wheat flour. *Rev. Agroquim. Tecnol. Aliment.*, **25**, 4, 1985, p. 447–57 (Spanish).

*Galli, A., and Ottogalli, G.* (1973) Microflora of the sour dough of panettone cake. *Annali di Microbiologia ed Enzimologia*, **23**(1/2/3), p. 39–49 (Italian).

*Azar, M., et al.* (1977) Microbiological aspects of Sangak bread. *J. Food Sci. and Technol.*, **14**, 6, p. 251–254.

### 1.7.5 Federal Republic of Germany (FRG)

Quality of the German wheat crop 1986, AGF Detmold, published in *Die Mühle und Mischfuttertechnik*, 1986, **42**, p. 572–5 (German).

### 1.7.6 German Democratic Republic (GDR)

Quality standards of flours and meals TGL 27424/01.
*Technology of Industrial Baking* by Schneeweiss, R., and Klose, O. VEB-Fachbuchverlag, Leipzig, 1981, p. 394, Table 66 (German).
*Special Processes in Baking* by Schwate, W., and Ulrich, VEB-Fachbuchverlag, Leipzig, 1986 (German). Authoritative reference for the preparation of special breads.
National bread and roll types produced in the GDR, and flour types used to produce them. Adaptation from: *Technology of Industrial Baking* by Schneeweiss, R., and Klose, O. Leipzig, 1981, Table 3, p. 21, Definition of Bread Varieties (German).
*Maschinenlehre Backwaren* by B. Schramm, 2nd edn, 1978, VEB-Fachbuchverlag, Leipzig, GDR (German), a machinery handbook for the baking industry recognized by the Food Ministry as suitable for professional education.

### 1.7.7 USSR

Reference sources include:
*The Technology of Breadmaking* by L. Ja. Auerman, Moscow, 1972 (Russian); German language version published by VEB-Fachbuchverlag, Leipzig, GDR (German), 1977.
*The Biochemistry of Breadmaking* by N. P. Kosmina, Moscow, 1971 (Russian); German language version published by VEB-Fachbuchverlag, Leipzig, GDR (German), 1977.
*The Problem of Baking Properties* by Natalie P. Kosmin (Professor of Grain Chemistry at the High School for Milling at Tomsk, USSR), published by Moritz Schäfer, Leipzig, GDR.
*Biochemistry of Grain and Breadmaking* (Biokhimiya zerna i Klebo-pecheniya) by V. L. Kretovich, published by the Academy of Sciences of USSR, 1958.

*Lutsishina, E. G., and Matyash, A. I.* (1984) Wide-band proton NMR-spectrographic study of gluten proteins of wheat. Deposited Doc. 1984 VINITI 3447-84, 10 pp. (Russian) available from VINITI.

*Pashchenko, L. P., et al.* (1984) Dependence of the rheological properties of liquid semiproducts on a surfactant and oxygen. *Khlebopek. Konditer Prom-st*, **11**, p. 24–7 (Russian).

*Mazur, P. Ya.* (1986) Influence of oxidizers on the change in binding-energy of moisture in flour and dough. *Izv. Vyssh. Uchebn. Zaved. Pisheh. Tekhnol.*, **5**, p. 33–7 (Russian).

*Losa, A. I.* (1939) Use of PFWS soured with *L. delbrückii* instead of a dough sponge. *Bulletin Chlebopekarnaja promyschlennost* (WNIIChP publication 1939–1940), **4–5**, 46.

*Chernaya, L. S., et al.* (1986) A biologically-active mixture for improving the quality of wheat bread. *Khlebopek. Konditer Prom.-st*, **5**, p. 28–9 (Russian).

*Karpenko, V. I.* (1986) Relation of dough properties and bread quality to the separation of glutathione from yeast. *Khlebopek. Konditer Prom.-st*, **9**, p. 40–2 (Russian).

*Belova, L. D., et al.* (1985) Use of starch-syrup to obtain baker's yeast. *Khlebopek. Konditer Prom.-st*, **4**, p. 40–1 (Russian).

### 1.7.8 Hungary
*Case study:* The Vârpalota Bakery—a modern production unit in the District of Veszprém (Lake Balaton), PRH. *Bäcker und Konditor*, **1**, 1987, p. 11–12.

### 1.7.9 Czechoslovakia
International Exhibition SALIMA-85 in Brno, Czechoslovakia.
Baked-product range, Mixing aggregate T 457·0, Flour feed aggregate T 437·0, as components of the continuous sour-dough and doughmaking plant T 995·0 of Topos, Sluknov, Czechoslovakia.
Wire-band tunnel-ovens PPC series of the milling-machinery manufacturers at Pardubice, Czechoslovakia; 3-deck oven type 30 of Merkuria, Czechoslovakia.
Complete Doughmaking-lines KVT 1000, 1500 and 1800 for Full-sour to Full-sour production of bread on a continuous basis. Bread-roll lines t 940 and t 985.

*Leipzig Fair 1985 (GDR)* KVT Type 995·0 for continuous doughmaking line for rye and mixed rye/wheat doughs with outputs 1800 to 2500 kg/h dough, exported by Technopol, Bratislava, Czechoslovakia.
Mixer T 457·0 for the KVT doughmaking plant.

### 1.7.10 Poland
*Case study:* Newest and largest bakery and confectionery in Warsaw, Krakowiak Street (near Warsaw Airport). First stage of construction began 1976 for confectionery production, second stage construction of the bakery with six production lines and tunnel-ovens, 2 laboratories, administration and cultural areas, commissioned in 1982. Further construction projects are in hand. *Bäcker und Konditor*, **1**, 1985, p. 19–21.

### 1.7.12 United Kingdom (UK)
*Case study:* Two 1950s comparisons of a large multiple-chain London plant bakery, and that of the author's rural family bakery located in a South Devon market town on the coast.

*Todd, J. P., et al.* (1954) *Chem. and Industry*, **50**.

*Hawthorn, J., and Todd, J. P.* (1955) *J. Sci. Food Agric.*, 501.
'The Blanchard batter process'—no-time dough with less power, article in *Milling*, **146**, p. 520–21, 1965.
'Blanchard batter, process for bread', article in *Milling*, **147**, p. 519, 1966.

'Process for the improvement of wheat-flour bread', State Committee for Inventions of USSR No. 164860 July 7, 1963. Auerman, L. J., Kretowitsch, W. L., and Polandowa, R. D.

*Meredith, P.* (1966) Combined action of ascorbic acid and potassium bromate as bread dough improvers. *Chem. and Industry*, p. 948–9.

*Barret, F. F., and Joiner, R. R.* (1967) Dough developer combination ADA and potassium bromate. *Bakers Digest*, **41**, 6, p. 46.

*Menger, A.* (1977) *Getreide Mehl und Brot*, **31**, p. 48.

*Menger, A.* (1978) *Getreide Mehl und Brot*, 32, p. 13.

### 1.7.13 North America

*Kilborn, R. H., and Tipples, K. H.* (1968) Sponge-and-dough type bread from mechanically developed doughs. *Cereal Sci. Today*, **13**, 1, p. 25–30.
Fermi-tech pre-ferment process, developed by Fermitech Company, Rogers Group Inc., Oberlin, Ohio 44074, USA.
Nutrient slurries for pre-ferment systems.

*Case study:* Automatic flour-pre-ferment system at Flowers Baking Company, Miami, Florida, USA.

*Johnson, J. A.* (1974) Precursors of bread flavor. Department of Grain Science, Kansas State University, USA.
Proceedings of the 8th National Conference on Wheat Utilization Research at Denver, Colorado, USA.

*Ponte, J. G. Jr.* (1985) Short-time doughs, *Bakers Digest* (Technical Source), May issue.

*Freed, R. J.* (1963) Continuous dough-developer and extruder. *Bakers Digest*, **37**, 3, p. 55.

*Kline, L., Sugihara, T. F., and McCready, L. B.* (1970) *Bakers Digest*, **44**, p. 48. Nature of the San Francisco sour dough French bread process. I. Mechanics of the process.
*Bakers Digest*, **44**, p. 51. II. Microbiological aspects.

*Sugihara, T. F., Kline, L. and Miller* (1971) Microorganisms of the San Francisco sour dough bread process. I. Yeasts responsible for the leavening action. *Applied Microbiology*, **21**, p. 456. II. Isolation and characterization of undescribed bacterial species, responsible for the souring activity. *Applied Microbiology*, **21**, p. 459 (Kline and Sugihara).

## CHAPTER 1.8 MEASUREMENT AND CONTROL TECHNIQUES FOR RAW MATERIALS AND PROCESS VARIABLES

*Auerman, L. Ja.* (1972) *Technology of Breadmaking*, Moscow; VEB-Fachbuchverlag, Leipzig, 1977 (German), under: Complex parameters which reflect the quantity of gluten and its quality in flour, p. 62–3.
Gluten evaluation using the $H_{pek}$ and $H_d$ Parameters, refer to: *The Laboratory Handbook for Baking Technology* by L. I. Putschkova, Moscow, 1971, published by Pistschewaja promyschlennost (Russian), which contains the calculation tables.

*Wünsche, R. and Tscheuschner, H.-D.* (1984) The cell-structure of wheat-flour dough and the bread-crumb and possibilities for their automatic analysis. *Wiss. Zeitschrift der TU Dresden*, **33**, 3, p. 115–18 (Germany).

*Bloksma, A. H.* (1981) Effect of surface-tension in the gas/dough interface on the rheological behaviour of dough. *Cereal Chem.*, **58**, 6, p. 481–6.

*Carlson, T. L.-G., and Bohlin, L.* (1978) Free surface energy in the elasticity of wheat-flour dough. *Cereal Chem.*, **55**, 4, p. 539–44.

*Kuzminskij, R. V., Scerbatenko, V. V., and Vassin, M. I.* (1977) Dynamics of the cell structure of the bread-crumb. *Chlebopek. i kond. prom.*, **21**, 5, p. 21–2.

*Kuzminskij, R. V., Scerbatenko, V. V., and Vassin, M. I.* (1977) Model for the evaluation of bread-crumb cell structure. *Chlebopek. i kond. prom.*, **21**, 6, p. 19–20.

*Zimmermann, R., and Schmidt, S.* (1977) The microscopic characteristics of the baked-crumb. *Bäcker und Konditor*, **25**, 9, p. 262–4 (German).
USSR laboratory wheat-flour baking test GOST 9404-60 Standard, paragraph 55–64, Flour and Bran Test Methods: Moscow Isdatelstwo Standardow, 1963.
*Laboratory Manual for Technology of Baked Products* by Putschkowa, L. I., Moscow, published by Pistschewaja promyschlennost, 1971.
AACC standard method for assessment of the baking quality of bread-flour by the sponge-dough pound-loaf method in USA—AACC Method 10–11.
Canadian Remix Baking Test, 1960. Irvine, G. N., and McMullan, M. E. *Cereal Chem.*, **37**, 5, p. 603, 1960.
Rye baking-test: sour-dough baking test, *Standard Methods for Grain Flour and Bread*, 4th edn, 1964, p. 151 (German).

## Flour Pentosan Determination

*Cerning, J., and Guilbot, A.* (1973) *Cereal Chem.*, **50**, p. 176.

*Petzold, H., and Volkmer, M.* (1980) *Bäcker und Konditor*, **28**, 1980, p. 282.

## Yeast

Standard test for the baker's yeast industry (proof-time). GDR Standard TGL-25111/03, GDR, May 1971.

*Jelitzki and Semichatowa:* Production-based method for the enzyme-activity of yeast, All-Union Research for the Baking Industry of USSR.

*Bergander, K., and Bahrmann, K.* (1957 and 1058) Shelf-life of yeast by electrometric measurement of pH and redox-potential. *Nahrung*, **1**, p. 74, 1957; and *Nahrung*, **2**, p. 500, 1958. Standard Test Method GDR Standard TGL-25111/04 Shelf-life of Baker's Yeast.

*Josić, D.* (1980) Criteria for the determination of the quality of active dry yeast. *Getreide Mehl und Brot*, **34**, 7, p. 179–81 (German).

## Sour-dough Starter Cultures and Sour-dough

*Stephan, H.* (1982) Characteristics of various sour-dough processes and resultant bread quality. *Getreide Mehl und Brot*, **1**, p. 16–19 (German).

*Schulz, A.* (1941) Development and physiological activity of the bacteria and yeasts during the sour-dough stages of whole-rye bread production, and their influence on flavour and quality of the bread. *Mehl und Brot*, **41**, p. 377–9 (German).
*Fortschrittliche Bäckerei*, **71**, Berlin, 1941 (German).

*Schulz, A.* (1954) *Zeitschrift für das gesamte Getreidewesen*, **31**, 7–9, p. 51.

*Gerstenberg, H.* (1978) Detection of dough-acidulants in bread from the citric acid content. *Z. Lebensmittelchemie und gerichtlichen-chemie*, **32**, p. 125–6.

*Rabe, E.* (1980 and 1981) Organic acids in breads from various processes. Part I: Methods for the determination of the mode of acidulation. *Getreide Mehl und Brot*, **34**, 4, p. 90–4, 1980.

Part II: Various methods for the determination of the types of acidity. *Getreide Mehl und Brot*, **6**, p. 146–50, 1981.

*Stephan, H.* (1982) *Getreide Mehl und Brot*, **1**, p. 16–19 (German).

*Pelshenke, P., and Schulz, A.* (1941) Berliner short-sour process. *Mühlenlaboratorium*, **11**, 11, p. 105.

### Wheat Sour-dough

*Barber, S., Báguena, R., Martinez-Anaya, M. A., and Torner, M. J.* (1983) The microflora of the sour dough of wheat flour bread. I. Identification and functional properties of microorganisms of industrial Sour-doughs. *Rev. Agroquim. Tecnol. Aliment*, **23**, 4, p. 552–62 (Spanish).

*Spicher, G., and Schöllhammer, K.* (1977) Comparative investigations concerning the yeasts of pure sour-dough cultures and spontaneous sour-doughs. *FSTA*, **12**, M 1431.

*Lorenz, K.* (1983) Sour-dough processes—methodology and biochemistry. *Bakers Digest*, **57**, 4, p. 41–5.

*Spicher, G., Schröder, R., and Schöllhammer, K.* (1979) The microflora of sour-dough VIII. Yeast composition of sour-dough starters. *Z. Lebensm. Untersuch. und Forschung*, **169**, p. 77–81 (German).

*Spicher, G., and Stephan, H.* (1982) *Handbook of Sour-dough Biology, Biochemistry and Technology*. BBV Wirtschafts-information GmbH, Hamburg (German).

*Sugihara, F.* (1977) Non-traditional fermentation in the production of baked-goods. *Bakers Digest*, **51**, 5, p. 76, 78, 80 and 142.

*Calvel, R.* (1980) Fermentation and processing of natural sour (French). *Industries des céréales*, No. 7 p. 27–35. *Industries des céréales*, No. 5 p. 31–6.

*Galli, A., and Ottogalli, G.* (1973) Fermentation and processing of the sour-dough of panettone cake. *Annali di Microbiologia ed Enzimologia*, **23**(1,2,3), p. 39–49.
*Difco Laboratories Inc.*, Detroit 1, Michigan, USA. Manufacturers of culture media and reagents for microbiological and clinical laboratory procedures.

*Man, J. C. de, Rogosa, M., and Sharpe, M. E.* (1960) A medium for the cultivation of Lactobacilli. *J. Appl. Bact.*, **23**, p. 130.

*Gibson, T., and Abd-el-Malek, Y.* (1945) The formation of carbon-dioxide by Lactic acid Bacteria and B. licheniformis, and a cultural method of detecting the process. *J. Dairy Research*, **14**, p. 35.

*Stamer, J. R., Albury, M. N., and Pederson, C. S.* (1964) Substitution of Manganese for Tomato Juice in the Cultivation of Lactic acid bacteria. *Applied Microbiology*, **12**, p. 165.

*Breed, R. S., Murray, E. G. D., and Smith, N. R.* (1957) 'Bergey's Manual of Determinative Bacteriology', 7th Edition, Baltimore: Williams and Wilkins.

*Gibbs, B. M., and Skinner, F. A.* (editors) (1966) 'Identification Methods for Microbiologists', 2 vols. London: Academic Press.

## 1.8.2 Process Variables

*1988 Reference Source*—A statistical reference manual and specification guide for wholesale baking, published by Sosland Publishing Company, P. O. Box 29155, Shawnee Mission, KS 66201, USA. *Temperature Calculation Equations*, p. 8.

Intensive mixing-aggregate IMK 150, made in the GDR—*Winter and Summer dough-water temperature adjustments for Wheat and Rye-flour processing. Machine Manual Baking* by B. Schramm, VEB-Fachbuchverlag, Leipzig, GDR (1978), p. 112 (German).

## Application of NIR Spectroscopy

In-line monitoring of flour-protein, -moisture, -damaged starch, ash, and particle-size, using the Infra Alyzer 250 of Technicon Instruments Corporation, 511 Benedict Ave, Tarrytown, NY 10591-5097. Dough moisture determination at 2045 nanometres within 1 minute, using the Garner/Neotec 3000 or 7000 units. Gardner Neotec Division, Pacific Scientific, Silver Spring, MD, USA.

Final baked-product analysis includes: protein, moisture, fat, sugar, fibre, and SSL levels, using selected wavelengths.

*TKO plate-dryer for the determination of dough-yield,* GDR Standard TGL-29066 (Food Engineering School, Dippoldiswalde, GDR).

*Nomogram application* to relate flour-weight to dough-yield, dough piece yield and baked product yield.

*Mixing and mixing regimes* versus optimization of dough-yield.

*Efficient use of electric-motors as a power source.*

*Microelectronics, biotechnology and information technology* application for process measurement and control systems, leading to CAM and CIM.

*Data acquisition at each production phase.*

*Fixation of responsibilities, work-flow, and information-flow.*

*Training, and constant updating of qualified personnel.*

*Coordination of problems involving measurement by formation of working-groups, within corporate organizations.*

*In-line monitoring of dough viscosity and rheology in the oscillatory and relaxation modes, time programmed and computer interfaced.*

*Optical in-line torque transducers can also be applied to measure viscosity; mounted in-line with the drive shaft and mixer element, both speed and torque during dough mixing can be continuously monitored.*

*Measuring systems based on dough deformation in the oscillatory and relaxation modes:*
System R2 of Rheotec AB, S-22370 Lund, Sweden, for both 'in-line' and 'on-line' measurement of dough rheological deformation in the oscillatory mode, at chosen optimal material response frequencies, within the range $10^{-3}$ to 20 Hz.

*IR-absorption spectroscopy* can be applied to the measurement of hydrogen-bonding, monitoring changes taking place during storage of both wheat and flour (maturation), under specified conditions of time, temperature and humidity.

*Fourier transform infrared (FTIR)* spectroscopy, in combination with attenuated total reflectance (ATR) sample presentation, will now allow improved quality spectra from aqueous solutions of food biopolymers, thus overcoming the hitherto limited use of water as an IR solvent, owing to strong band absorption within the range 5000-400 cm$^{-1}$.

*Wilson, R. H., Goodfellow, B. J., and Belton, P. S.* (1988) Fourier transform infrared spectroscopy for the study of food biopolymers (includes a complete IR ATR spectrum of bread). *Food Hydrocolloids*, **2**, 2, p. 169–78.

*McClure, W. F., and Davies, A. M. C.* (1987) Fourier self-deconvolution in the analysis of NIR spectra of chemically complex samples. *Mikrochim. Acta*, **1**(1–6), p. 93–6 (published 1988).

*Wilson, R. H., and Belton, P. S.* (1988) A Fourier transform infrared study of wheat starch gels. *Carbohydrate Research*, **180**, p. 339–44.

*NIR spectroscopy* (reflectance) has also been applied to the measurement of the cold pasting viscosity of extruded starches and flours. The amount of reflected light increases with increasing cold pasting viscosity. Correlations between measured and NIR-reflectance spectra calculated viscosities, from 200 to 1600 mPa s were good, indicating possible process control applications.

*Meuser, F., et al.* (1987) Application of NIR spectroscopy for measurement of the viscosity of extruded starch and flours (German). *Veröff. Arbeitsgem. Getreideforsch.*, **208**, p. 61–7 (*Berliner Tagung für Getreidechemie*, **37**, 1986).

*NMR instrumental techniques* apply powerful magnetic fields from superconducting magnets, which are capable of resolving and identifying minor, but significant structural differences in complex molecules, e.g. polymer composition, chain-branching, tacticity and residual monomer content. High resolution pulsed proton NMR can be used to measure spin–lattice, and spin–spin NMR relaxation-times, thus allowing a determination of free and bound water in materials, and water mobility on storage. Changes in the colloidal state of water with time can be monitored under specific storage conditions, e.g. starch, wheat proteins, and non-starch polysaccharides. The advantage of NMR spectroscopy is that it is a rapid and non-destructive method of food analysis. The most widely applied techniques are based on proton NMR spectroscopy, i.e. water content of foods, solid/liquid ratio of fats and emulsions, lipid–protein exchange reactions in emulsions, monitoring the freezing-processes of various foods. More recently, pulsed NMR spectroscopy has gained in importance for the investigation of food components.

*Korn, M.* (1983) *Deutsche Lebensmittelrundschau*, **1**, p. 1 (German).

*Lutsishina, E. G., and Matyash, A. I.* (1984b) Wide-band proton NMR spectroscopic study of gluten proteins of wheat. Deposited Doc. 1984 VINITI 3547-84, USSR, 10 pp. Available VINITI USSR (Russian).
Proton NMR spectroscopy is also being applied to the study of the mechanism of tripeptide glycosylation (glycoproteins), and their conformations.

*UV spectroscopy* has been applied to baking quality determination in Hungary, using 2M urea as a resolution medium for protein, gluten-forming proteins, correlating well with dough water absorption and baking quality.

*Kovacs, E., and Selmeczy, A.* (1985) Application of UV-values in the determination of baking quality. *Elelmez. Ipar*, **39**, 7, p. 272–5 (Hungarian).

## Cleanliness and Sanitation (hygiene)

*American Baking Industry Sanitation Standards Committee* (BISSC), represented institutions are: American Bakers Association, Bakery-equipment Manufacturers, American Society of Bakery Engineers, Associated Retailers of America (Bakers), Biscuit and Cracker Manufacturers Association, American Institute of Baking, also Consultants from nationally well-known sanitarians. As a result the FDA publishes lists of equipment and certified manufacturers which meet BISSC standards.

*Hazard Analysis and Critical Control Point System* (HACCP) concept accepted in USA by the FDA, as accepted procedure for Food Process Inspection (Denver National Conference of Food Protection, 1971) Issue of '*FDA Compliance Program Guide Manual*'.

*USA concept of 'good manufacturing practices' (GMPs)*
*FDA Inspectorate in the USA* empowered by Congress to examine food producer's quality assurance programs and records to assess daily operations.

*Federal Food and Cosmetic Act* requires the processor to *identify control points, and hazards of contamination* from unsanitary environment, and *accept total responsibility legally* for its entire programme.

*Good Manufacturing Practices (GMPs) Manual* is a guideline as to how a food plant is maintained in a clean state.

*Duty of food industry* to demonstrate to the public, government and regulatory-agencies that it is producing clean, wholesome and safe foods.

*Swab test-kits* for surface-sanitation, Millipore Corporation, Bedford, MA 01730, USA.

*UV-liquid sterilizers, PWS Sterilizer Bulletin*, available from: Pure Water Systems, 343 Boulevard, Hasbrouck Heights, NJ 07604, USA.

*Instant spray-nozzle hot/cold water and steam cleaner units*, available from: Strahman Valves, Inc., 3 Vreeland Road, Florham Park, NJ 07932, USA.

*Birco nocorrode foamer*, available from: Birco Chemical Corp., P.O. Box 1315, Denver, CO 80201, USA.

*Morton Biocidal System* includes static and mobile high-pressure washing, cleaning and sanitizing equipment, and *in situ* production of low pH gaseous chlorine, and sodium hypochlorite from just salt, water and low voltage electricity, available from: Morton Salt Company, Division of Morton Norwich Products Inc., 44 Dock Street, St. Louis, Missouri 63147, USA.

*Quarternary based combined* cleaner/disinfectant/sanitizer/virucide, marketed by: Robert Langer Co., Inc., USA.

*Water-activity measurement* for prediction of microbial growth support, available from: American Instrument Company, Travenol Laboratories, Inc., Silver Spring, MD 20910, USA.

*Automatic insect control units*, USDA approved, marketed by Virginia Chemicals, Dept 116, Portsmouth, VA 23703, USA.

*High-pressure/low-volume, compact, stationary or mobile, cleaning-station systems*, marketed under the 'Economiser' trademark, are available from: Chemidyne Corporation, 8679 Freeway Drive, Macedonia, OH 44056, USA.

*UHF rodent scarer*, operating on compressed air, at low cost, and maintenance-free, is marketed by: Rat-X, 325 W. Huron, Chicago 60610, USA, being international leading rodent-control specialists.

*Critical Control Point*, total-systems approach, considering each unit operation. '*Reference Source*', *Bakers Digest*, 1986, p. 40.

*Falcon swubes*, and Falcon disposable plastic equipment, available from: Falcon Plastics, Division of BioQuest, Baltimore Biological Laboratories, Cockeysville, MD, USA.

*Atomic Energy Agency, Vienna*, responsibile authority for the application of ionizing radiation to food.

*Official Journal of European Communities No. C99*, 13:4:87 lists current radiation clearances obtained by national governments.

*UK food irradiation* currently based on Food (Control of Irradiation) Regulations Amendment 1972.

*Baltimore Biological Laboratories (BBL)* culture media for Microbiological studies, available from: BBL, Cockeysville, MD, USA.

*Difco Laboratories Inc.*, Detroit 1, Michigan, USA, suppliers of culture media for microbiological studies.

*Public Health Service Publication*, paragraph 729 Fluorescent Antibody Techniques, US Government Printing Office, Washington, D.C., 1960, USA.

*Spicher, G.* (1969) Investigations concerning the occurrence of aflatoxin-forming microflora, and aflatoxin found in bread. *Brot und Gebäck*, **23**, p. 149–52.

*Frank, H. K.* (1966) Aflatoxins in food *Arch. Lebensmittelhyg.*, **17**, p. 237–42 (German).

*Frank, H. K., and Eyrisch, W.* (1968) The detection of aflatoxins, and the occurrence of pseudo-aflatoxins in food. *Z. Lebensm. Unters, Forsch.*, **138**, p. 1–11 (German).

*Bösenberg, H., and Eberhardt, E.* (1969) Studies concerning the contamination of food in supermarkets by moulds. *Med. und Ernährung*, **10**, p. 12–13 (German).

*Spicher, G.* (1970) Investigations of the occurrence of aflatoxin in bread. *Zbl. f. Bakt.*, **II**, Abt. 124, p. 697–706.

*Spicher, G.* (1984) The toxins of mould growth on baked products, 2nd Part: Mycotoxins occurring in mould contaminated sliced-bread. *Lt. Lebensm. Rdsch.*, **80**, 2, p. 35–8.

*Mycotoxin quantitative estimations* using chromatography with UV-detection.

*Gorst-Allmann, C. P., and Steyn, P. S.* (1979) (TLC method) *J. Chromatography*, **175**, p. 325.

*Duracková, Z., Betina, V., and Nemec, P.* (1976) (TLC method) *J. Chromatography*, **116**, p. 141.

*Howell, M. V., and Taylor, P. W.* (1981) (TLC method) *J. Assoc. off, analyt. Chemists*, **64**, p. 1356.

*Lee, K. Y., Poole, C. F., and Zlatkis, A.* (1980) (HPTLC method) Simultaneous multi-mycotoxin analysis by HPTLC with continuous multiple development and two solvent systems of different polarity. In: Bertsch, W., and Kaiser, R. E. (editors): *Instrumental HPTLC*, Published by Hüthig, New York, 1980, p. 263–73.

*Scott, P. M.* (1981) (HPLC method) Liquid chromatography in the analysis of mycotoxins. In: *Trace Analysis*, Vol. I, edited by Lawrence, J. F., Academic Press, New York, p. 193 *et seq.*

*Castegnaro, M., and O'Neill, I. K.* (1982) *Environmental Carcinogens—Selected Methods of Analysis, Some Mycotoxins.* Vol. 5, IARC, Lyon, 1982.

## CHAPTER 1.9 WEIGHER/MIXER FUNCTIONS AND DIVERSE TYPES OF MIXERS AND MIXING-REGIMES

*Integrated electronic and microprocessor-based control systems for ingredient handling*: Dietrich Reimelt KG, 6074 Rödermark-Urberach, FRG.

*Reimelt-Atlas SP-80 computer-interfaced* ingredient-batching/weighing, checking/ optimization, production-line control and formulation cost control systems, tailor-made from core intelligence software elements and appropriate hardware.

*Pneumatic transport* of raw materials of variable density, e.g. brans and meals, also brown flours.

*One-step weigh and mix systems*
Lift and sift system marketed by Russel Finex Limited, London WC2H 7EQ, UK.
*Negative vacuum* raw material transport systems.
*Travelling-weigher* aggregates.
*Tweedy weighing and mixing systems* for CBP brews and straight doughs.
*Kerry Flour Handling Systems Limited*, Crawley, Sussex RH10 2PX, UK. Automatic pneumatic transport, load-cell censor weighing, and centrifugal sifting aggregates for flours of diverse type and composite bulk density and particle-sizes flours, e.g. brown flours, specialty flours.

*Weigher/mixer function concepts* and their relative merits.
*Types of mixer and mixer-regimes:*

*Single-arm conventional discontinuous (batch) mixers*, operating at 15 to 50 rpm, with mixing times of up to 20 minutes.

*Single-arm high speed discontinuous (batch) mixers,* operating at 70 to 100 rpm, with a choice of 2 speeds, e.g. Diosna/Dierks, Osnabrück, FRG; or types HLK-50, S-125, and S-250 manufactured in the GDR by VEB-Bäckereimaschinenbau, Halle, GDR. Extremely robust and versatile machines for large and small bakeries, hotels and restaurants, with a high torque at a reduced rpm. A mixer for the craftsman, who believes in fermentation, and who has a mixed trade.

*Twin-arm, elliptical-orbit, discontinuous (batch) mixers*, with the unique cyclodynamic action, of which the Artofex is the prototype. Of Swiss origin, these machines are marketed in the UK by: Artofex Limited, Enfield, Middlesex. The Record Bakery Equipment Co. Ltd of St Albans, Herts AL1 1JF, UK, market a machine of similar design.

*Spiral mixers* are ideal for use with the CBP, activated dough development (ADD), short-time, or soy/untreated flour processes. Although intended as a discontinuous (batch) machine, it can be integrated into a quasi-continuous system, using interchangeable mixing bowls with a full-length centre column, and non-wrapping spiral. The short mixing-cycle allowing doughs to be worked off on a line system, with a throughput of 3 tonnes/h. Such a system with computer control is available from: All Foods Engineering Services Limited, Newcastle, Staffordshire ST5 6NR, UK. Spirals have gained in popularity in EC countries, owing to their versatility, short mixing-cycle, simplicity and reliability. Design options are: spiral arm only, spiral and central pillar or plinth, fixed/removable bowls, bowl-tippers, etc., from Dierks 'Diosna' range (FRG), Kemper (Austria), Thomas Collins (UK), Eberhardt (FRG), Werner and Pfleiderer (FRG), APV-Baker, formerly Baker Perkins (UK), Sottoriva (Italy), Sancassiano (Italy), Boku (FRG), Oase (FRG), Record (UK), and Artofex (Swiss), etc.

*Wendel mixers*, are multifunctional, with an advanced complex wave path, resulting in alternative short, frequent dough stretching and compression actions, forming a developed three-dimensional gluten network. Ideal for all wheat and rye doughs; the action is especially suited to coarse meal doughs, and soy/untreated flour doughs for 'clean-label' products. This machine is also ideal for biscuit doughs and cake batters. The prototype of these high technology mixers is the Diosna Wendel, which has two wendel-elements, rotating in opposite directions.

*Type A2-ChTM* for 45 kg flour batches, and the *A2-Ch-TB* for up to 100 kg, manufactured in the Ukrainian SSR of the USSR, are batch mixers with adjustable mixing elements, and fixed mixing-bowl, which can be removed and transported on tyred-wheels.

*Type SM85 spiral-kneader*, manufactured in Yugoslavia by GOSTOL of Nova Gorica, is a heavy-duty machine for doughs of 135 kg. *Machine type SM 125* for 200 kg flour has a work-input of 11 to 15 kW, and the smaller SM 85 delivers 7 to 11 kW. The spiral mixing element rotates at 110 to 220 rpm within one half of the machine-bowl, and the ploughshare mounted within the machine-bowl axis, distributes the dough evenly. The machine-bowl rotating in the opposite direction to the mixing-arm, being independently powered by a 2·2-kW motor, its direction of rotation being reversed during mixing. On completion of mixing, the mixing-arm, ploughshare and drive are raised electromechanically to allow removal of the bowl.

*Intensive discontinuous (batch) mixers*, which can be either vertical or horizontal in design, can often be operated on a quasi-continuous basis, owing to the short mixing cycle, thus allowing integration into line-production systems. Those used for the CBP in the UK include the following machines: Tweedy model range 35, 70, 140, 160, 200, 4400, 300 and 6600, giving outputs of 216 to 3300 kg dough/h; Gilbert 35 and 70, or 140; Mono 28, 70 and 100, all introduced with the CBP during the 1960s. These machines delivering energy-inputs of 8 to 11 W h kg$^{-1}$ dough, at 400 to 420 rpm, over 90 to 120 s. However, much energy is dissipated as heat, resulting in dough temperature increases. This, coupled with the 'clean-label' requirement and use of only organic-based additives, has resulted in a current trend away from these machines.

*The Stephan TK 150, top-loading, horizontal mixer,* can be used for all mixing requirements, including: doughs, pizza-doughs, pastry, cake-batters, and grinding stale-bread crumb, by changing mixing-heads. It is engineered in the FRG by A. Stephan und Söhne GmbH & Co., D-3250 Hameln 1, mixing speeds being 750, 1500 or 3000 rpm, depending on the product, utilizing a power-rating of 15·5 kW. Maximum dough-loading is 90 kg per batch, rapid mixing and unloading cycles allowing throughputs of about 800 kg/h. This is a high-technology intensive batch mixer with hydraulic discharge, direct into dough-trolley, or onto conveyors.

*The Collette intensive mixers,* engineered in Belgium, represent prototype intensive mixers for the baking and food industries, pioneered during the 1970s. The wheel-in bowls eliminate handling and tipping, with capacities of 5 to 600 litres, and the use of additional bowls provides a quasi-continuous operation.

*The IMK 150,* engineered in the GDR by VEB Bäckereimaschinenbau, Halle, GDR, is a very-heavy-duty intensive-mixer, mounted within a twin-standing, which houses the water-dosage unit and the control panel. Driven by a 60-kW motor, transmission is direct via 10 drive-belts, reducing the rpm from 1470 to 415 to 420, delivering an energy-input of $8–10\,W\,h\,kg^{-1}$ dough, within 2–3 minutes, which is recorded on an amperometer. Raising and lowering of the mixer-bowl is achieved hydraulically, the oil-pump being driven by a separate 3-kW motor. This machine is extremely flexible for wheat and rye flour doughs, allowing throughputs of up to 3000 kg dough/h. Using eight machine-bowls, one IMK 150 at maximum output of about 1900 kg/h, is capable of feeding two lines continuously for an oven of $50\,m^2$ baking-surface.

*The horizontal-bar and drum high-speed mixers* used in the USA and Canada, range in capacity from 100 lb to 2500 lb, operating at two speeds, 35 rpm in the slow mode, and 70 rpm maximum. Power-ratings ranging from 5 hp up to 125 hp. Horizontal mixers can be either of the stationary or tilting bowl design. In the former case, the bowl drum is solidly fixed to the mixer frame, the dough being removed by lowering the electrically operated sliding-door and slowly revolving the mixer arms. The tilting machine bowl design ejects the dough by tilting the bowl hydraulically at a 90 to 140 degree angle. Horizontal mixers are heavy-duty, high-torque machines with transmission ratios of 2:1. The mixer drum is trough-like with a curved bottom and flat sides and ends, of stainless steel construction, with rounded joints and corners. The drum has a refrigerated jacket, running around it, with external insulation to prevent condensation and refrigeration loss. Dough mixing and development is effected by a rolling, kneading and stretching action imparted by cylindrical bars, which are mounted along the length of the bowl-drum on 2, 3, or 4 opposite arms of a cradle attached to the agitator shaft, a 3-arm cradle having a Y-profile, and a 4-arm cradle that of a cross. The profile of the bars can vary from being smooth and rounded to being curved or sigma-shaped, thus improving traction and mixing intensity. The 3-arm cradle usually also has a breaker-bar positioned at the top rear to absorb the impact, and fold the dough when tossed towards the rear of the bowl-drum. Although these mixers have quoted capacities of 1000, 2000 and 2500 lb of dough, these are only maxima, and experience has shown that much improved mixing efficiency in terms of mixing intensity is achieved by reducing dough loads by about 15%. Since these mixers are also utilized for the preparation of sponges for the sponge and dough processes on the North American continent, a reduction in weight loading of about 60% for sponge mixing becomes mandatory, owing to the firmer, lower absorption of these doughs. These mixers are marketed by Baker Perkins (APV-Baker); Peerless Machinery Corporation; and Oshikiri HM Series marketed by Gemini, Philadelphia, PA 19115, USA; the American Machine and Foundry Company of Richmond, Virginia 23227, USA, is also a long-established manufacturer of horizontal-bar mixers.

*The UniPlex and BiPlex* intensive batch mixing systems were introduced by Baker Perkins in

1984. The UniPlex is an intensive mixer with a similar mixing concept to the Stephan TK150, intended for mechanically developed doughs (MDD), ADD, and bulk-fermented systems. However, the UniPlex is intended for the larger operation, requiring more ingredient-feed sophistication, coupled with a choice of three levels of automation. Wheat and rye flours can be accommodated as well as enriched doughs of most types. Four maximum output capacities are available: 1600 kg/h, 2200 kg/h, 3000 kg/h and 4000 kg/h, using power-ratings of 35, 45, 55 and 75 kW respectively. These output capacities are based on mixing times of 2·75 to 3·25 minutes. The BiPlex utilizes a two-stage mixing concept, the spiral mixer-element describing a planetary orbit within a horizontal fixed cylindrical bowl at high speed for about 4 minutes, thus effecting hydration, dispersion, and atmospheric oxygen uptake; the second stage, of about 1 minute duration, involves a high-speed rotation about a fixed shaft, thus completing the development stage. Oxidants synergistic to oxygen, e.g. ascorbic acid and azodicarbonamide perform best with this mixer, yielding short-time doughs. However, untreated flour with a soy-lipoxygenase source to effect oxidation could be utilized. In the case of sponge-and-dough processing, or straight doughs, involving bulk fermentation, the planetary mode only is necessary. Owing to the efficient dynamics, minimal heat is dissipated, reducing undesirable dough temperature rises and the need for cooling or refrigeration. The BiPlex 200 with 2 litres capacity provides 106 kg batches at the rate of 15/h, equivalent to a throughput of 1590 kg dough/h. The high volume output systems have a fully integrated and automated mixer, ingredient weighing and feed facility under microprocessor control, which stores formulations and provides diagnostic information. The main motor power-rating is 22 kW, and the planetary motor 4 kW. The BiPlex process and mixing system has been developed, patented and manufactured by APV-Baker (formerly Baker Perkins Limited) of Peterborough, UK.

*Continuous dough-mixing systems* demand exact and systematic operating procedures, the dough being assembled, mixed and developed in distinct stages, leaving the developer aggregate by extrusion. Processing operations comprise two stages: raw material preparation and dosage, and continuous dough mixing. Sieved and blended flour is transported into the daily-silos and fed to the weigh-head (with load-cells), of the mixing-aggregate. Yeast, salt, shortening and dough improvers/conditioners being prepared in either slurry or solution form, as appropriate, at the preparation-station using special containers and agitation-aggregates. The exact amount of yeast and salt are automatically metered according to the flour weight measured by the weigh-head load-cells. Usually, if the weight of the flour is changed, a synchronization-device adjusts all the liquid ingredients accordingly. However, the operator must check this function before delivery to ensure that the formulation is adhered to, and correct as necessary. Rotational speeds of continuous mixing-aggregates can usually be regulated from 25 to 170 rpm, dough residence times being about 2–4 minutes, depending on the rpm and extrusion outlet adjustment. Mixer-housings are usually cylindrical, with baffle-plates or pins often mounted on the inner surface, to increase resistance during mixing. Dough throughput capacities, depending on chosen rpm, average 500 to 1500 kg dough/h, larger units for 1800 to 3000 kg/h being also available. The use of an active pre-ferment with this mixing system improved its feasibility.

*Do-Maker and Amflow systems*, as commercially practised in the USA, produce an unorientated cell structure in the bread, and a very soft crumb texture, fine grain and high specific volume. The excessive soft, gummy crumb texture, and flavour and aroma reduction being ameliorated by: passing through a conventional make-up process; use of 50% total flour in the pre-ferment; air injection into the developer. Some loss of operational rationalization resulting. Essential processing steps being: an active pre-ferment containing yeast, inorganic nitrogen, flour or fermentable sugars and pH adjustment; continuous production of an assembled dough in a pre-mixer/incorporator from the mature pre-ferment, residual ingredients, high oxidant levels; intensive mixing, at high speed and pressure to full development within a small chamber, anaerobically; direct scale/deposit into baking pans. This latter development-head being a counter-rotating double-arc element at speeds of up to

290 rpm, over about 1 minute, at energy-input levels of 0·3 to 0·4 hp min$^{-1}$ lb$^{-1}$ of dough. Production capacity ranges being 4000 to about 7000 lb h$^{-1}$. Automatic divider/panning rates approximate to 80 pieces/min.

*European continuous systems*, had to be designed or modified to satisfy the consumer demand for variety- and hearth-breads from wheat- and rye-flours, mixed wheat/rye breads, specialty-breads, producing products of lower specific volumes, higher densities, employing much leaner formulations on average. Whereas, the American systems perform ingredient incorporation or premixing, and actual mechanical dough development in two distinct mixing modes, the European systems perform these two functions within the same machine aggregate. Furthermore, in many cases, European systems do not rely entirely on purely mechanical development, variable bulk fermentation floor-times being allowed to attain full dough maturity, even when pre-ferments are being utilized.

*The Konpetua continuous system* was developed in the FRG in 1967, and built by Werner and Pfleiderer of Stuttgart, the first commercial installation on a large scale being in 1968. The main feature of this aggregate is its mixing flexibility, and ease of adaptation to changing production requirements. The twin rotating mixer shafts, equipped with variously shaped mixing elements are housed in a cylindrical chamber. Production capacities range from 300 to 5400 lb/h (136 to 2454·5 kg/h), depending on chamber diameter. When a small and a large unit are utilized as a dual system, hourly dough outputs can be increased to around 8500 lb/h (3864 kg/h). Dough consistency is controlled by feedback from a wattmeter, which automatically corrects fluctuations in metered dough water. The developed dough is extruded through an adjustable outlet onto a belt conveyor, and proceeds to make-up aggregates.

*The Kontinua continuous system*, also manufactured by Werner and Pfleiderer of Stuttgart, suitable for standard type bread production from wheat- and rye-flours, and wheat/rye mixtures, offers high volume outputs (see Figs 71 to 74). The two types FMS 400 and FMS 500 deliver output capacities of up to 2000 and 3500 kg dough/h, respectively of mixed rye/wheat-flour doughs; and 1500 and 2400 kg/h of wheat-flour dough respectively. Variable speed motors of 42 kW provide an rpm range of 600 to 2200.

*The Buss-Ko-System List*, is a Swiss continuous mixer, consisting of a cylindrical housing, enclosing a mixing worm designed as a helix with interruptions. Sets of mixing protrusions, fixed to the housing, project into the gaps in the helix, so that as the worm rotates, the dough is forced towards the extrusion end. Mixing and development of the dough being achieved simultaneously.

*The Ivarsson system*, commercially introduced in Sweden in 1957, has metering and feeding aggregates for liquid and dry ingredients feeding into the cylindrical mixing-aggregate of horizontal design. The mixing elements are two straight shafts, and one spiral ribbon kneader rotating eccentrically. The ingredients are fed in at one end of the cylinder, being mixed, and then forced by the spiral towards the discharge end, thus effecting dough development. The residence time being controlled by an adjustable extruder outlet is on average 90 s. The average throughput is 100 lb/minute, the mixed dough being given a 30–60-minute maturation-time, before passing through a conventional make-up schedule.

*The Strahmann mixer*, of German design and manufacture, also attracted interest in the UK during the early 1960s. Modified patents were taken out by Simon of Stockport under the joint name Simon–Strahmann. These modifications included a degree of control over shear and pressure rates during mixing, and the retention of conventional moulding techniques, thus providing products with either accepted standards or finer-grained 'texturized' crumb porosities.

Elias, D. G., and Wragg, B. H. (1963) *Cereal Sci. Today*, **8**, p. 271.
Eggitt, P. W. R., and Coppock, J. B. M. (1965) *Cereal Sci. Today*, **10**, p. 406.
The mixer shaft, rotating at 120–300 rpm, was provided with a series of impellers, with a static shear disc positioned in front of each. The shear discs have apertures of varying dimensions,

depending on their location along the shaft. The dough is forced through these apertures by the preceding impellers, about 20 in number, driven by a 35 hp motor. The first four sections had open discs with coarsely pitched impellers, thus effecting a premix function. Subsequent disc apertures were reduced, giving pressures up to 30 psi, and resulting in dough development with considerable air occlusion, leading to texturization. Mixing intensity could be controlled by shaft-speed, shear-disc aperture, and feed-rate. Energy-input levels were about 5 W h lb$^{-1}$ for fine texture, and 4 W h lb$^{-1}$ for irregular, conventional dough porosity. The whole system is enclosed within a horizontally mounted cylindrical housing. An additional feature of the Simon–Strahmann system was a special patented rotary-divider. This was fitted to the extruder outlet, the dough being extruded into a circular chamber with a 6-segmented scaling rotor, controlled by a pressure-sensitized microswitch, which rotates the rotor as each segment is filled with dough. The dough being discharged from the segments, rounded, given an intermediate proof before sheeting and moulding. Such treatment resulting in an open-textured conventional type loaf.

*The Oakes continuous mixer/modifier*, designed by E. T. Oakes Limited in the UK, was the only continuous system which could be recorded as a commercial success in the UK baking industry. First introduced commercially in 1964, this unit was compact, flexible in product adaptation, and easily cleaned. The dough was extruded as a continuous ribbon, being either put through a standard divider, or an Oakes continuous automatic divider, designed to give better uniformity, and dough-weight accuracy. The dough-pieces then passed through a conventional make-up procedure. A system was designed for the CBP with complete ingredient facilities, dough development, followed by dividing and conventional make-up. Direct panning never found favour in the UK baking industry. Throughput capacities of 1800 to 4950 lb dough/h were possible with the Oakes mixer/modifier, and a specially designed automatic divider eliminated variations in dough density and consequent scaling weight variations. Continuous feeding of the divider with a constant head of dough from the mixer/modifier in an inert state gave reduced quality difference between bread from bulk fermented and mechanically developed doughs. However, the UK Baking Industry concluded that the disadvantages of a continuous system, viz. reduced flexibility of product types; synchronization problems with conventional make up machine-aggregates; reduced water absorption; and floor space and line integration, compared with the use of CBP batch mixers, could not be justified during the 1960s. Therefore, continuous processing progressively diminished in favour of the quasi-continuous CBP batch system.

*Socialist community countries* (COMECON) applied continuous breadmaking systems of diverse processing techniques and machine-aggregates with much success, after pioneering work carried out in the USSR during the 1950s. After that time, mixing-aggregates suitable for various output requirements for wheat and rye-flour, as well as mixed rye/wheat-flour doughs were developed in most of the COMECON countries as a result of special trade arrangements to avoid excessive competition.

In the USSR, during the 1930s work was already under way concerning continuous breadmaking technology under such pioneers as Woronkow, Beljakow, Molodych and Proworichin. In 1944, an engineering and scientific collective in the mechanics laboratory of the WNIIChP in Moscow under the direction of G. E. Nudelman, began research on intensive mechanical methods of continuous mixing, and the first production installation was realized in 1947. Then followed the development of a continuous two-phase system for wheat-flour dough, the first phase being a liquid pre-ferment, the final doughs being fermented in a carousel system.

Nudelman, G. E. (1952) Automatic production line for the manufacture of bread-roll doughs. Collective papers from scientific lectures No. 178 Moscow, Pistscherpromisdat, 1952, (Russian).

*The N. F. Gatilin bunker-aggregate*, was first installed in Soviet bread factories in 1946 for the preparation of rye bread from firm sour-doughs, together with the AZCh oven, for the production of 100 t of bread per day. However, since then its application has been extended to the production of whole wheatmeal bread from firm sour-doughs, and wheat-flour bread from second grade wheat-flour using a sponge-dough process. For factories with outputs of 20 t per day, a smaller version, the BAG-20 was introduced for wheat-flour sponge-doughs, and rye doughs from firm sour-doughs at up to 30 t daily output. These systems eliminated the use of machine-bowls for sours, sponges and doughs, but the preparation of the pre-ferments (sours and sponges) was on a batch basis. During the early 1970s, Bunker aggregates with continuous-mixing and reduced bulk fermentation-times were introduced.

*Tschernjakow, B. I. Chlebopekarnaja i konditerskaja promyschlennost (ChKP)*, **12**, 9, 33, 1968 (Russian).

Such aggregates were designed for outputs of about 30 t of 1 kg hearth-breads per day, using sponges prepared on a large scale from 70% of total flour, and 66% of total dough water. The sponges were fermented for 4·5–5·0 h at 28°C. The final doughs, set at 32°C, remained in a stationary-bunker for 20–25 minutes maturity-time before make-up. This aggregate is fully automated. A similar concept using aggregate L4-ChAG was applied to the production of both wheat and rye bread at 15 t per day.

*The I. L. Rabinowitsch continuous aggregate ChTR*, for wheat-flour doughs, utilizes a direct fermentation process, using either 1·5% compressed yeast and 5% to 20% liquid yeast, or 50% to 60% liquid yeast. The dough temperature being set at 27 to 29°C for a bulk fermentation time of 4·5 to 5·0 h.

*Roiter, I. M. Modern Breadmaking Technology in Industrial Bakeries*, Kiev, Published by Technika, 1971 (Russian).

*The W. M. Dontschenko continuous aggregate ChTU-D*, utilizes a Ch-12 mixing-aggregate (Fig. 24, item 5) with discharge into a worm-pump, which further develops the dough before it is transported upwards through a pipeline into the divider.

*The A2-ChTT continuous mixing-aggregate*, exhibited at the 1987 Leipzig Fair, is intended for throughputs of 1300 kg/h at a power-rating of only 3 kW.

*The Czechoslovakian manufactured continuous doughmaking plant series KVT 1000, 1500 and 1800*, producing 1000, 1500 and 1800 kg dough/h, illustrated in Fig. 75, although intended as a complete line assembly, can also be adapted for non-line systems. This system is utilized for mixed wheat/rye-, or rye-flour processing, at a power-rating of 12 kW, the KVT 1000 throughput ranging from 850 to 1000 kg dough/h. This is a well-engineered plant, which has found very wide acceptance in the GDR baking industry, as well as other baking industries of the COMECON countries.

*The FTK 1000 continuous doughmaking plant*, which is also well engineered, is manufactured in Hungary, and also exported throughout COMECON. It is capable of outputs of 1000 kg of rye-flour dough/h, or 810 kg/h of wheat-flour dough. This plant has centralized control systems, but offers control units on individual aggregates, utilizing a building block elemental design. The illuminated central control panel comprises the grouped control mechanisms, automatic systems making full use of transistors and thyristors for motor control. Grouped control-mechanisms are provided for the following functions: energy-distribution; programming; instrumentation; switches and preset; malfunction warning lights; switchover from manual to automatic control; safety devices for electrics; illuminated display for all functions. Sequence numbering on the control panel clarifies the construction and schematic layout of the FTK 1000 plant.

*In the Federal Republic of Germany (FRG)*, mechanization and automation of the sour-dough acidification and perpetuation processes are described by D. Thörner of Düsseldorf.
*Thörner, D. (1982) Getreide Mehl und Brot*, **2**, p. 44–7 (German).
Patented as the 'Isernhäger natural dough-souring process', this process can be applied to the production of a complete range of rye-flour bread, providing a uniform supply of sour-dough

from Monday to Saturday with flexibility. The maturing time of the sour is 18 h, and the required amount of sour, about 36 kg, for each batch, is directly metered into the mixer. In the case of the 'salt/sour process', the amount of sour-dough for a given flour mix is read off from tables. Any mature sour not processed within 6 h of the full maturity-time, is transferred to the starter-machine, and then flows into the dosage tank after 24 h, which contains enough sour-dough for a day's production. The plant is intended for sour-dough batches of 1500 kg, but can be increased to 2000 kg. When two starter-machines are coupled with one dosage-tank, outputs of up to 4000 kg of sour-dough/day are possible.

## CHAPTER 1.9 WEIGHER-MIXER FUNCTIONS AND DIVERSE TYPES OF MIXERS AND MIXING-REGIMES

*A joint design and construction between VNI in Vienna and Reimelt of Frankfurt/FRG*, has produced a patented system for the continuous preparation and storage of natural sour-dough.
*Foramitti, A.* (1982) *Getreide Mehl und Brot*, **2**, p. 47–50 (German).
This is a two-stage process, using a fermentation-tube and a tank. The first stage can be stored for 4 days, under control, without excess acid formation or damage to the organisms, thus permitting stoppages as and when necessary. The second stage provides a uniform quality sour for bread production. Selection of process parameters; fermentation power of sour-dough yeasts are maintained to give a balanced lactic:acetic acid content. The first plant is in operation at the VNI, one of Europe's largest bakeries in Vienna, producing up to 4·8 t sour-dough/h, commissioned in 1979. This fully mechanized, continuous, automated plant can be operated by just one person and guarantees homogeneous sour-dough and consistent bread quality.

*Short-time (no dough-time) processes*, involving either totally mechanical, or a combined mechanical/chemical development process, have found favour in some countries. However, the elimination of fermentation results in a bread flavour, and crust and crumb structure which is unacceptable in many countries with discerning consumers and bakery technologists.

*Sheeting methods for dough development*, using sheeting-rolls and repeated passage of doughs through them are utilized in countries where energy resources are limited, and the cost of expensive mixers prohibitive. In countries such as Peru, parts of Spain, south-east Asia, and parts of Africa, labour-intensive methods remain efficient owing to the plentiful supply of manual labour. Where bread is produced from composite-flours, e.g. wheat/maize, wheat/sorghum, wheat/millet, or wheat/triticale, in various ratios, low-power pressure sheeting remains an ideal form of dough development.

*Bushuk, W., and Hulse, J. H. (1974) Cereal Science Today*, **19**, 8, p. 424–7. produced bread of acceptable volume, grain, texture and colour from composite flours of 80% CWRS (Manitou) wheat-flour mixed with up to 20% non-wheat flour, using 100% CWRS wheat-flour as control raw material. The doughs being mixed by hand, and rested for 30 minutes before dividing and passing, in 10 small pieces of equal weight, through three sets of sheeting-rolls, using 10 passes in the case of each set. Thus effecting mechanical dough development. After a 10-minute first proof at 35°C and 80% RH, the doughs were again sheeted successively through three sets of rolls adjusted to increasing roll-gap. The dough-pieces were then moulded either manually or mechanically, and panned. Final-proofing was 55 minutes at 35°C and 80% RH, baking at 221°C for 25 minutes. Results were also compared with the same flours processed by the CBP, and Canadian remix-test bake procedures. The conclusions were that simple dough development by pressure sheeting, gives results comparable with those obtained by high-energy mixing procedures.

*Kilborn, R. H., and Tipples, K. H.* (*1974*) Cereal Chemistry, **51**, 5, p. 648–57, using a Canadian CWRS wheat-flour of 12·5% protein, confirmed the efficient development work performed by sheeting-rolls at much reduced energy input levels, compared with high-speed mixers. The actual rate of energy input is only slightly less than that imparted by an intensive pin-mixer. This confirms the importance of the mixing parameters 'specific mixing-intensity' and the type or mode of work-input, and dough structuralization, i.e. compression (pressure), elongation (stretching), or shear forces. Excessive sheeting of doughs gives rise to symptoms of loss of resilience, and permanent deformation. However, structuralization by sheeting at optimal intensity (about 20 folds), will create over 1 millions layers, and simultaneously, the dough becomes reorientated through a 90 degree angle at each folding-action, thus creating a two-dimensional (cross-hatched) versus a unidirectional network formation. Research has indicated that the sheeting-procedure can effect energy savings of up to 70%, compared with mixer energy-inputs, and produce bread of equal quality to that produced by the conventional, latter technique. This concept has potential, but requires line-flow mechanization/automation, in order to fix process control parameters. Integration into a continuous mixing system is indicated, allowing reduced or no-fermentation-time doughmaking.

*The Reddi-sponge system*, can completely eliminate the fermentation process stage, enabling the baker to save about 4 hours in bread and roll production. On completion of mixing, and a 15-minute rest period, doughs can be divided, proofed and baked.

*Pre-ferment systems*, however, protect bread quality and flavour. Pre-ferments can be varied in composition to include soluble nutrients for the yeasts, and sours for both wheat- and rye-flour processing; up to 25% of total dough salt content; and pregelatinized/partially hydrolysed biopolymer-rich raw materials. Their viscosity can be varied by adding from 10% to 60% of total flour content; temperature control, depending on maturation-time, being varied from 20°C up to about 40°C as required. Such procedures allow floor-times of 10–15 minutes, or direct dividing, proofing and baking.

*Maes, E.* (1959 and 1960) Automatic bread machine.
*Bäcker und Konditor*, **13**, 19, p. 9, 1959 (German).
*Brot und Gebäck*, **14**, 1, p. 1, 1960 (German).
*Maes, E.* (1964) *Brot und Gebäck*, **18**, 6, 1964 (German).

Other innovations include the use of liquid-yeast, and the application of biotechnology in the form of specific enzymes, e.g. amylases, proteinases, amyloglucodisases, pentosanases, lipoxygenases (soy source, marketed in the USA as 'Wytase W-52' of J. R. Short Milling Company, Chicago, IL 60606), cellulases, and oxidoreductases, etc.
Yeast species, other than *S. cerevisiae*, which can be included in pre-ferments include the following strains: *S. delbrückii*, for rapid flavour production; *Candida lusitaniae*, for increased flavour development; *Candida krusei*, *Pichia saitoi* and *Torulopsis holmii* as starters for sour-doughs: *Candida milleri* and *S. exiguus* for wheat-sours and pannetoni bread (Italy); *S. rouxii* for sweet-dough ferments; *Torulopsis cellulosa*, and *T. candida* used as starters for sangak bread (Iran); also *S. rosei* for all frozen-dough baked products.
The basic control parameters of yeasts, sours and their pre-ferments are: nutrient food supply (up to 6% is optimum); water content (100% of flour absorption gives the most rapid digestion rate); dough temperature (30 to 40°C is appropriate for short pre-ferment-times); medium pH range 4·0 to 6·0 (for enzymic activity); flour content (10% allows rapid digestion, 25% is a good average concentration); salt concentration (25% of total dough salt content controls cell osmosis).

# Notes and References for Part 2

## CHAPTER 2.1 FERMENTATION OF WHEAT AND RYE-FLOUR DOUGHS

*Yeast-forms for breadmaking include:* fresh compressed; fresh cream liquid; dry active; dry instant active. Water contents are, 68% to 75%, 80% to 82%, 6% to 8% and 4% to 6% respectively. Relative storage life is approximately 3 weeks, 10 days, 2 to 12 months, 1 year plus, respectively.

*Fresh compressed* requires dispersal in water before mixing.

*Fresh cream liquid* can be metered directly.

*Dry active* requires rehydration in water at 40°C for 15 minutes before addition.

*Dry instant active* can be dry blended with dry raw materials, or added at a later stage.

*Compared to fresh compressed* used at 1 part by weight, fresh cream liquid requires 1·5 to 2·0 parts, dry active 0·5 parts with extra water, and dry instant active 0·4 parts with extra water added.

*Frozen-dough yeast,* is a strain engineered to become desensitized to sub-zero temperatures. This permits a normal proof-time after about 90 days storage at −20°C of the frozen dough products. Other strains of yeast require extended proof-times, with no guarantee of consistent results.

Compared with the use of fresh compressed at 1 part, frozen-dough yeast requires only 0·4 parts with the appropriate 0·6 parts water added extra, to adjust the weight difference. This yeast requires no rehydration, the frozen yeast product being transferred from sachet to mixer bowl. It is viable for about 1 year when stored at −18°C. Its relatively slow fermentation rate makes it also ideal for bulk fermentation and Viennoiserie work. Also, whereas about 5% fresh compressed yeast is required for frozen-doughs (flour weight), the frozen-dough yeast can be used at the 2% level for lean doughs.

*Liquid yeast* is gaining in popularity for industrial bakeries, owing to the advantages of bulk-handling systems, being delivered in refrigerated tankers depending on climatic temperature. Liquid yeast has been used in the USSR for many decades in the large city industrial bakeries. The average dry matter content of liquid yeast is 20%, compared with about 30% for fresh compressed yeast.

## CHAPTER 2.2 INDUSTRIAL PROPAGATION AND PRODUCTION OF YEAST FOR THE BAKING INDUSTRY

*Yeast reaction products:* 100 parts of monosaccharide reacts with the yeast enzymes to yield 49 parts of ethanol and 47 parts of carbon dioxide, together with such other organic by-products as glycerol; lactic, acetic, malic and succinic acids; aldehydes; fusel oils; and numerous flavour compounds.

*Belova, L. D., et al.* (1985) at the WNIIChP (All-Union Research Institute for the Baking Industry) Moscow, USSR, have shown that hydrolysed starch syrup, derived from potato or wheat can be used instead of molasses for the production of baker's yeast, at biomass yields 61–63%.

*Khlebopek. Konditer Prom.-st.,* **4**, p. 40–1 (Russian).

The yeasts by S. Burrows, in A. H. Rose and J. S. Harrison (editors), Academic Press, New York, Vol. 3, 1970, p. 349–420.

*Harrison, J. S.*, in D. J. D. Hockenhull (editor), *Progress in Industrial Microbiology*, Vol. 10, Elsevier, Amsterdam, The Netherlands, 1971, p. 162–3.

Microbial technology. In H. J. Peppler and D. Perlman (editors), Academic Press, New York, Vol. 1, 1979, p. 157–85.

*Reed, G.* In G. Reed (editor), *Prescott and Dunn's Industrial Microbiology*, 4th edn, AVI Publishing Co., Westport, CT, 1982, p. 593–633.

*Trivedi, N. B., Cooper, E. J., and Buinsma, B. L.* (1984) *Food Technology*, **38**, p. 51–7.

## 2.2.1 Yeast Physiology

*Wang, H. Y., et al.* (1977) *Biotechnology and Bioengineering*, **19**, p. 69–86.

*Grossman, M. K., and Zimmermann, F. K.* (1979) *Mol. Gen. Genet.*, **175**, p. 223–9

*Cohen, J. D., et al.* (1984) *Mol. Gen. Genet.*, **196**, p. 208–16.

## 2.1.2 Improvement of Industrial Yeasts

*Trivedi, N. B., and Jacobson, G.* (1986) Recent advances in baker's yeast. In M. R. Adams (editor), *Progress in Industrial Microbiology*, Vol. 23, Elsevier Applied Science.

*Entain, K. D., and Frohlich, K.-U.* (1984) *J. Bacteriology*, **158**, p. 29–35.

*Carlson, M., et al.* (1983) *Molecular Cell Biol.*, **3**, p. 439–47.

*Toda, K., and Yale, I.* (1979) Critical glucose concentration for catabolic repression. *Biotechnology and Bioengineering*, **21**, p. 487–502.

*Rogers, D. T. and Szostak, J. W.* (1985) of the US Genetics Institute, US Application 796,551 dated Nov. 8, 1985. Method of yeast-biomass production suppression and increased leavening-power (carbon dioxide and ethanol production).

*Societé Industrielle Lesaffre*, UK Patent 1,593,211, 1979, Method of industrial hybridization of baker's yeast meiotic spore clones from strains with desirable properties (rapid and slow rates of fermentation). European applications for leaner doughs demand more rapid fermenting strains.

*Commercial yeast strains* can either be constructed by the classical hybridization techniques, or by protoplast fusion after cell-wall digestion with specific enzymes.

*van Solingen, P., and van der Pratt, J. B.* (1977) *J. Bacteriology*, **130**, p. 946–7 is an example of fusion between yeasts achieved by a commercial yeast company.

*Spore clones* are not always haploid (*n* chromosomes), and often exhibit sporulation; resultant crosses can give hybrids which exceed the diploid (2*n* chromosome) number. Digestion is carried out in an osmotically stabilized medium to protect the protoplasts from lysis, the two desired strain protoplasts being mixed in the presence of polyethylene glycol and calcium ions. This results in spheroplast-agglutination followed by fusion. Numerous hetero- and homologous fusion ratios are possible, multinucleate protoplasts being formed. Sexual reproduction involves three phases: plasmogamy (plasmic-fusion), karogamy (nuclear-fusion), and meiosis (reduction division). The plasmic-fusion mixture is then plated on isotonic recovery agar, where karogamy (nuclear-fusion), cell-wall regeneration and growth takes place. Owing to participation of several nuclei in this process, a genetically unstable hybrid nucleus may result. At the subsequent mitotic events of cell to colony growth, chromosomes are randomly lost, therefore each colony recovered from a protoplast fusion experiment can consist of many variants, which differ in chromosome content. In the case of

*Baker's yeast* (*S. cerevisiae*), the haploid (*n* chromosome) stage is rare and short-lived. It occurs during meiosis (reduction division), and is normally restricted to the formation of the ascospores, which copulate immediately on release from the ascus. The diploid (*2n* chromosome) phase is characterized by the formation of large cells, after plasmogamy and karyogamy have yielded the zygote, which sooner or later reproduces by budding on a large scale. The budding process results from mother cells forming small bladder-like protrusions, which result from a combination of cytoplasmic and nucleic material, the resulting daughter-cells eventually being pinched off as independent living entities.

*Protoplast-fusion experimentation* can provide problems in the selection of a heterologous fusion product. This can be effected by utilizing a number of techniques:

—the industrial strain can be converted into a mitochondrial DNA-less petite, which can be fused to a laboratory-cultured strain with auxotrophic markers and the desired gene properties, e.g. maltose production. The resulting heterologous fusant can then be selected on a small amount of medium, containing glycerol as a non-fermentable source of carbon;

—another procedure, as described by Putrament, A., Baramowska, H., and Prazmo, W. in *Molec. Gen. Genet.*, **126**, 1973, p. 357–66, two industrial strains can be treated with manganese, in order to create mitochondrial point mutations, resulting in respiration deficiency or resistance to mitochondrial specific antibiotics (e.g. oligomycin, chloramphenicol and erythromycin). The heterologous fusant is then selected depending on complementary petite mutations that restore growth on non-fermentable substrates, or a double antibiotic resistance.

Both techniques depend on the complementation of cytoplasmic markers, and there are no guarantees of karogamy (nuclear fusion).

*Karogamy* requires the presence of a dominant selectable marker from each of the two strains to be fused. Selectable markers are few in *S. cerevisiae*, but marker genes known as CUP1 (copper resistance), and ROC (resistance to the disinfectant roccal) have been reported by *Broach, J. R.* in J. N. Stratern, E. W. Jones and J. R. Broach (editors), *The Molecular Biology of the Yeast Saccharomyces—Life Cycle and Inheritance*, Cold Spring Harbor Laboratory, Cold Spring Harbor, New York, 1981, p. 653–727.

*Protoplast fusion* can now be induced electrically, thus eliminating the use of polyethylene glycol. An alternating current of chosen frequency, producing a microsecond pulse, induces an electric dipole in the spheroplasts, resulting in a binding of the protoplasts to one another, and the formation of protoplast chains between the electrodes. The microsecond pulse causes localized disruption at the point of protoplast/protoplast interface. This technique gives up to 100% efficiency in a large mass fusion chamber, although fusants from microliter fusion chambers are difficult to recover.

*A baker's yeast strain* was produced by protoplast-fusion by Universal Foods Corporation, being the subject of U.S. Patent Appl. 503,323, 1983, due to Jacobson, G. K., and Trivedi, N. However, this strain was constructed by utilizing complementary spontaneously occurring petites found in commercial strains. The fusants were selected on hypertonic glycerol, and then streaked on non-selective medium. The clones from this medium were then screened on glycerol, acetate and sucrose media. Colonies showing good growth on these media, indicate stable regeneration. Clones showing at least 75% of parent strain yields can then be evaluated in various dough systems, containing flour with no additives as control; flour with 2% flour weight of salt; and flour with 2% added salt and 20% flour weight of added sucrose. In each case doughs are made by adding 15 ml of water, containing 50 mg of yeast solids. In each case, total carbon dioxide over a 4 hour period, recorded every 30 minutes, is measured barometrically; plots being made as total gas evolved, and rate of gas liberated during each 30-minute period. Such tests are an indicator of the organism's resistance to high osmotic pressures, and relative independence from nitrogen sources from yeast-foods, often added in bakeries. Activity in lean doughs, and in the presence of salt are also important criteria.

*Maltose* is the predominant sugar in flour, only trace amounts of glucose and sucrose being present, and maltose fermentation is quite sensitive to the addition of salt.

*Sucrose*, used at high concentration in excess of a dough's leavening requirements, has an inhibitory effect on fermentation, repressing the maltose utilization system of the yeast. Therefore, strains which are designed for sweet doughs do not require an active maltose utilization system.

*Yeast strain vitality*, is measured as total gas evolved and linear rate of gas production over a 4 h period, using the three dough systems described above.

*Yeast* is classed as a unicellular eukaryote (Greek: eu = normal/typical) since, in common with plants and animals, it has a well-defined nucleus with chromosomes, which divides by mitosis. Bacteria have no nuclear membrane, or endoplasmic reticulum and mitochondria and do not show sexual reproduction. Therefore, yeast can be utilized in eukaryotic molecular biology. This involves 'recombination' or the crossing-over between parts of chromosomes, and the reassortment of chromosomes. A typical haploid yeast has a single set of 15 or more chromosomes, two such cells of opposite mating types, a and b, can fuse, their nuclei fusing to form a diploid nucleus with two complete sets of chromosomes. During meiosis (a phase of sexual reproduction), the chromosomes become double structures, forming two chromatids. Homologous chromosomes then pair and exchange parts of their chromatids by crossing-over, thus four haploid sexual spores form, each spore can then have a new gene combination different from the parent cells. Genes on the same chromosome recombine by cross-over, and genes on different chromsomes become shuffled as chromosome pairs reassort.

*Biological yeast activity*, can be enhanced by mixing with other ingredients which promote the fermentation process. Such mixtures can be dried down to 10% moisture, and stored for up to 3-month periods with improved leavening power, compared with the use of the control yeast. Dough fermentation times can reduce by 50–80 minutes, and bread quality be improved.

*Chernaya, L. S., et al.* (1986) *Klebopek. Konditer Prom.-st*, **5**, p. 28–9 (Russian), at WNIIChP (All-Union Research Institute of USSR).

*Chernaya, L. S., et al.* (1986) *Klebopek. Konditer Prom.-st*, **7**, p. 25–8 (Russian), at the Vses Nauchno-Issled. Institute Khlebopek. Prom., Moscow, USSR.

*Waste bread-crumb* can also be hydrolysed with either acid at pH 1·6 and 100°C, or alpha-amylases followed by amyloglucosidase to produce low-molecular-weight polymers at 60°C. Such hydrolysates act as useful substrate starters for both yeasts and lactobacilli in wheat and rye bread production.

*Pashchenko, L. P., et al.* (1986) *Klebopek. Konditer Prom-st*, **8**, p. 81–3 (Russian), at Voronezh. Tekhnol. Institute, Voronezh, USSR. Use of bread-crumb hydrolysates in liquid rye-starters. *Berchtold, J., and Meuser, F.* Eur. Pat. Appl. EP 154,135 Sept. 11, 1985. Lieken-Batschneider Mühlen und Bäckereibetrieb GmbH: Amylolytic flour/crumb hydrolysate for bread dough.

*A perpetual liquid yeast system* for preparation in the bakery was developed by A. I. Ostrowski in the USSR. A gelatinized flour/water suspension is initially fermented at 48–54°C with thermophilic lactobacilli. The resultant high lactic content mash is then cooled to 28–30°C, transferred to another container and used as a substrate for yeast propagation using various strains. After maturation, a definite amount of the mature liquid yeast is taken for doughmaking, and replaced with fresh substrate. This maintains the propagation cycle of the liquid yeast. Such a product is economic in use and produces a flavourful baked product, but its preparation must be under careful aseptic control to avoid wild-culture contamination.

## 2.2.3 Post-fermentation Yeast Technology

*Fresh compressed or cake yeast*, quality standards should include a sensory evaluation of colour, consistency, smell and taste. Other quality parameters should include: moisture content (maximum 75%), time required to reach a standard 70 mm gassing height should not exceed 75 minutes, and the shelf-life at 35°C not less than 48 hours. Samples should be evaluated for technical culture purity, and glutathione content (proteinase activation). For bakeries, fermentation-speed is the first priority. A standard dough proof height of 70 mm in a cylinder, should be attained within 45–52 minutes.

*Dry-yeast*, whether in pellet or granular form, should contain a maximum of 10% dust material, and a maximum moisture limit of 8%. A storage-life of not less than 12 months, at a moisture maximum of 8% should be a realistic standard. The 70 mm cylinder proof height should be realized within 90 minutes.

*Instant active dry yeast (IADY)*, was introduced in the 1960s by Gist-Brocades to overcome the problem of rehydration. Porous, cylindrical shaped particles providing a larger surface area, and instantization.

*Frozen-dough yeast*, for use in doughs for frozen products, as well as *liquid yeast*, with about 20% solids for use in industrial bakeries, are latter-day developments in Western Europe.

# CHAPTER 2.3 CHEMICAL CHANGES IN YEASTED DOUGHS DURING FERMENTATION

*Sponge, pre-ferment, sour-dough and final-dough* convert an initially dense, inelastic dough-mass into a physical state which can be leavened by the gaseous products of fermentation, simultaneously producing flavour and aroma substances. The final result is a fully 'ripened or matured' dough for baking.

*Dough-maturation*, can only be achieved when the following prerequisites proceed in a systematic manner: adequate gassing immediately after mixing; optimal physical dough properties for machining: adequate elasticity for gas-retention; adequate protein and starch hydrolysis products for crust-colouration, flavour and aroma.

*Dough fermentation is essentially an anaerobic process*, carbohydrate being utilized by the yeast cell to produce energy, the by-products being alcohol and carbon dioxide, in the absence of oxygen.

*Many enzymes* participate in the series of intermediate stages of anaerobic breakdown, known as the tricarboxylic acid (TCA) cycle.

*The collective zymase–enzyme complex* in yeast, hydrolyses the monosaccharide molecule to alcohol and carbon dioxide (two molecules of each).

*Baker's yeast (S. cerevisiae)* is capable of fermenting most types of sugar in dough, e.g. glucose, fructose, saccharose (sucrose) and maltose.

*Glucose and fructose* are fermented first, almost spontaneously, followed by saccharose, which can be hydrolysed extracellularly to glucose and fructose by the periplasmic invertase. The activity of this external invertase, and its concentration depends on cultural conditions, the appropriate gene having been sequenced to encode a 1·9 kilo base pair (kb) and a 1·8 kb mRNA, the former being under glucose regulation, encoding the external invertase.
*Chu, F. K., Watorek, W., and Maleg, F.* (1983) *Arch. Biochem. Biophys.*, **223**, p. 543–55.
*Taussig, R., and Carlson, M.* (1983) *Nucl. Acids Res.*, **11**, p. 1943–54.
*Carlson, M., and Botstein, D.* (1982) *Cell*, **28**, p. 145–54.
*Carlson, M., Taussig, R., Kustu, S., and Botstein, D.* (1983) *Molec. Cell. Biol.*, **3**, p. 439–47.

*Invertase*, is encoded by any one of six genes of a *SUC* polymeric system, comprising both a cytoplasmic, and a periplasmic enzyme. The exact number of *SUC* genes in industrial yeast strains is not certain, and can vary depending on which ones are present, and their respective copy-number (ploidy of the chromosome on which they reside). Gene dosage has been shown to influence yeast activity, as reported by *Grossman, M. K., and Zimmermann, F. K.* (1979) *Molec. Gen. Genet.*, **175**, p. 223–9. The cytoplasmic (internal) invertase concentration of the cell remains relatively stable. The products of invertase action, glucose and fructose then diffuse into the cell with the help of phosphorylation.
*Van Steveninck, J.* (1969) *Arch. Biochem. Biophys.*, **130**, p. 244–52.
Shortly after mixing of the dough, all saccharose (sucrose) molecules have been converted by the invertases to glucose and fructose.

*Carbohydrate fermentation* by yeasts in general follow a definite scheme according to Kluyver, viz. glucose is the most widely fermented sugar by all yeast species; if a yeast ferments glucose, it will also ferment fructose and mannose by the Embden–Meyerhof–Parnas pathway (EMP); raffinose is only fermented when saccharose (sucrose) is also present in the anaerobic system. *S. cerevisiae* is only capable of splitting off the fructose molecule from raffinose, the disaccharide melibiose not being digested by baker's yeast strains. Some yeast species can ferment di- and trisaccharides, and others, e.g. *S. diastaticus*, even ferment starch.

*Yeasts* cannot utilize as many types of sugar under anaerobic conditions of metabolism as they can over the aerobic respiratory route.

*Maltose* fermentation is important in lean doughs, its metabolism also being controlled by a polymeric gene system, five unlinked genes having been identified; as with the *SUC* system, no single strain carries all of them, but the presence of any one of these results in the ability to ferment maltose. The *MAL* genes are more complex than the *SUC* system, being made up of three factors: a regulatory gene, the enzyme maltase (alpha-glucosidase), and maltose permease. Both permease and maltase are induced by the presence of maltose, and repressed by glucose.
*Naumov, G. I.* (1971) *Genetika*, **7**, p. 141–8.
*Naumov, G. I.* (1976) *Genetika*, **12**, p. 87–100.
*Cohen, J. D. et al.* (1984) *Mol. Gen. Genet.*, **196**, p. 208–16.

*Maltase repression by glucose*, manifests itself in three forms: (1) enzyme inhibition, (2) enzyme inactivation, and (3) complete repression of enzyme synthesis.
*Siro, M., and Lovgren, T.* (1979) *Eur. J. Appl. Microbiol. Biotechnol.*, **7**, p. 59–66.

*Galactose*, is fermented by all strains of baker's yeast, and some strains can also ferment the trisaccharide melezitose, the disaccharide trehalose, and alpha-methyl glucoside. Where hydrolysed whey is used as an ingredient, the lactase enzyme system becomes of interest taxonomically. In lean doughs, which often contain isomaltose, the ability to utilize alpha-methyl glucoside could be useful for cleavage of the alpha-1-6 bond.
*Frankel, D. G.*, in the J. N. Strathern, E. W. Jones, and J. R. Broach (editors), *The Molecular Biology of the Yeast Saccharomyces—Metabolism and Gene Expression*, Cold Spring Harbour Laboratory, Cold Spring Harbor, New York, 1982, p. 1–37.

*Alcoholic fermentation by yeasts* can be analysed into the following stages:
*Glycolysis*
Activation of the intracellular hexoses by phosphorulation.
Splitting of the hexose-diphosphate to form a phosphorulated triose.
Formation of pyruvic acid.
*Conversion of pyruvic acid* to carbon dioxide and alcohol.

*Good baking activity*, is dependent upon genes and gene-systems which feed carbohydrates into glycolysis, actual glycolysis, and the TCA cycle.

*Genes* involved in nitrogen metabolism and storage carbohydrate synthesis are also important.

*Rate of carbon flow* through the glycolytic pathway is crucial to both yeast production, and its baking application. Comparing the short times of leavening with production propagation, i.e. 2–4 hours *v.* 12–18 hours.

*Pre-ferments and yeast-foods* are desirable to improve the biological activity of the substrate, since leavening is an entirely anaerobic process, and the rate at which the organism can feed hexoses through glycolysis is of paramount importance.

*Yeast invertase activity*, can be evaluated from the amount of reducing sugar formed within 10 minutes from a 15% saccharose (sucrose) solution at pH 5·2 and 60°C, by a yeast suspension containing 6·2 mg dry-cells/ml.

*Carbohydrate utilization* of the yeast cell depends on the conditions; during fermentation (anaerobic), only 10% is utilized for cell growth, whereas during propagation 50–60% is utilized for cell growth (aerobic).

*Thermophilic yeasts*, compared with the mesophils (e.g. baker's yeast), have fermentation temperature optima of 39–40°C, thriving in acid media. Thermophils can be used for wheat-flour doughs and sponges, fermenting at 40°C, producing carbon dioxide and alcohol at a rapid rate. Final proof-times are also much reduced.

*Bread dough*, being a biological filled-polymer system containing many enzymes is best regarded as a 'redox-process', involving the simultaneous transfer of electrons for ionic or covalent bonding. During fermentation, this oxidation-reduction system is controlled by the desmolase group of enzymes, which are flavoproteins with a prosthetic group. The hydrolase enzyme group, controlling hydrolysis and resynthesis of saccharides, amides, proteins and esters, have no such prosthetic group in the molecule.

*The redox-potential, or rH scale*, can be divided into 0–15 at the strongly reducing end, and 25–41 at the oxidation end. Wheat-flours show rH values within the range 15–19, with the higher extraction flours around 15. Wheatgerm, being strongly reducing has rH values between 6 and 7. Rye-flours tend to be between 17 and 18. After 2 hours fermentation, rH values can fall to 17 in the case of wheat doughs, owing to progressive reduction.

*The rH value*, represents the electrical potential developed when the oxidants and reducing components of the dough system reach equilibrium, its magnitude depends on pH and temperature. Since the changes in pH are relatively narrow, the rH scale provides a broader spread for more meaningful interpretation. On the basis of measurements of aerobic and anaerobic rH potential and pH of fermenting bread dough, mathematical models of autolysis and fermentation can be formulated.

Zlobin, L. A., et al. (1986) *Vses. Zaochn. Inst. Pishehevoi Prom.-st*, **9**, p. 22–23.
This is an important contribution to the biotechnological control and CIM of industrial baking.

*Conventional baking processing techniques* require considerable time and floor-space, and are labour-intensive on an industrial scale. Therefore, techniques are constantly being researched to rationalize the various process stages involved: mixing, ripening (maturing), make-up (work-off), proofing, baking and cooling.

*Mixing innovations* include increased mixer energy input levels to effect a mechanical dough development. This is complemented by mixer design, involving bowl-shape, mixing-element contour, and path swept out by the rotation.

*Dough-ripening (maturing) innovations* are represented by: Liquid pre-ferments and sponges containing 10–25% of total flour weight, with appropriate temperature, yeast and salt adjustments; application of biotechnological techniques involving fungal amylases, protinases, and hemicellulases, etc.; the addition of various oxidants and reducing agents (where legislation permits) to accelerate physical changes; substrate supplementation in the form of yeast-foods.

*Make-up and Proofing innovations* comprise 2–6 measuring-pocket volumetric dividers with good scaling-accuracy due to servo-motors controlled from in-line checkweighers. Supertex moulder/panner for tightly moulded four-piece, and single-piece pan breads; and cross-grain moulders, which mould the dough in two directions to improve texture; marketed by APV-Baker. Air conditioner final-proofers either free-standing or integral with tunnel-ovens, followed by automatic lidding units.

*Baking and cooling innovations:* indirect fired tunnel-ovens with forced convection, with a choice of gas or oil firing, automatic heat-modulating burners allowing rapid and precise zonal temperature control; automatic depanning units, and travelling band conveyor coolers with air-conditioning.

## CHAPTER 2.4 WHEAT- AND RYE-SOURS AND SOUR-DOUGH PROCESSING

*Wheat-flour sour-doughs*, although not yet widely accepted for leavening, owing to the lack of standardized manufacturing techniques, are applied in Spain, San Francisco (USA), Central and Eastern European countries, and in Scotland in the form of 'barms'.

*Sour-dough standardization and development*, begins with the careful choice of raw materials and starter-cultures during the initial or first stage. Type and species of microflora, temperature, dough-yield (dough-firmness), and standing or maturing-time, will all determine the acidity-index, flavour and aroma, and crust/crumb characteristics of the final bread.

*Wheat sour-dough cultures*, consist of various species of lactobacilli, and yeasts which ferment under acidic conditions, e.g. pH 3·8.

*Mother-doughs*, 'masa madre' (Spanish), or 'Mutterteige' (German), contain the sour-dough cultures in very firm flour/water doughs. Their exact preparation is usually a closely guarded secret, perpetuated over generations with both individualism and idealism. Mother-doughs can be stored overnight at 2–8°C for up to 8 hours, acidity depending on temperature and storage-time. Their application in latter-day breadmaking is very appropriate for the upgrading of wheat bread from a bland, yeast-raised, mundane food commodity. It involves a three-stage process over about 3 days to obtain the desirable flavour intensity and acidity-index (*Säuregrad*) of about 24, at final dough pH 3·8–4·0. Ideal flour ash contents for this fermentation are 0·55% to 0·65%, i.e. flour types 550 to 650.

*Sour-dough bacteria*, depending on type and species, have the capability of converting citric and malic acids in flour into lactic and acetic acids, which enhances bread flavour considerably.

*San Francisco sour-dough French bread*, continuously made for over 100 years, contains the yeast *S. exiguus* and *Lactobacillus sanfrancisco*, which coexist within the pH range 3·8–4·5, the lactobacillus bacteria preferring to ferment maltose, the yeast not competing with the bacteria for this sugar.

*Scottish barms*, are prepared from starter-cultures containing hop-infusion, malt flour, and a portion of old-barm. This is fermented for 3–4 days. On reaching maturity, the barm can be added to 25% flour sponge or a 20% flour pre-ferment.

*Wheat- and rye-sours* can be dried and milled to fine powders for reactivation. Freeze-dried products offer the best protection for the microorganisms from heat-damage during manufacture. Such products allow a one-step process of about 3 hours duration for the production of rye bread, compared with a 3-day/3-step traditional fermentation schedule.

These products give consistent quality, and uniform dough souring at acidity-indices 20–24 for wheat-sours, and 40–44 for rye-sours. These biologically produced sours are used at about 10% of total flour weight. In the USA, such sours are manufactured by the Caravan Products Company Inc., NJ 07512, for wheat and rye doughs, marketing and shipping to Europe being handled by Richards International Ltd, 40 East 34th Street, New York, NY 10016. In the FRG, similar product ranges are marketed by: Dr. Otto Suwelack GmbH & Co., 4425 Billerbeck; Böcker Sauerteigprodukte, Minden 4950; Ulmer-Spatz, Ulm (Donau); and Böhringer & Söhne, Ingelheim-am-Rhein, amongst others.

*Rye-sour starters*, contain lactic acid bacteria and yeasts existing in symbiosis. The lactobacilli are classified as either homofermentative or heterofermentative, according to their enzyme systems. The homofermentative species ferment glucose by the fructose disphosphate pathway, converting it with a 90% lactic acid yield. The heterofermentative species ferment it over the hexose monophosphate route, yielding about 50% lactic acid, carbon dioxide, ethanol, acetic acid, and other organic acids. An average starter composition is homofermentative lactobacilli 54%, and heterofermentative lactobacilli 46%. The most significant strains internationally are: *L. brevis, L. fermentum, L. plantarum*, and *L. casei*. The most significant yeast strains found in starter cultures are: *S. cerevisiae, Pichia saitoi, Candida krussei*, and *Torulopsis holmii*. The composition and amount of starter-culture used will determine the sour-dough ripening-time.

*The ripening- or maturing-time of the sour-dough* always depends on the rhythm of the growth curve of the microflora, i.e. cell-count *v.* time (hours). During the first 2 hours, a doubling of the cell-count (generation-time) takes place owing to exponential reproduction. At about 4 hours, the state of exponential reproduction lapses into stagnation, equilibrium existing between newly formed cells and dying cells. This coincides with optimum maturity, detected by measuring acidity-index, pH, gas-development, and cell-count. An experienced baker can detect maturity from the aroma, and cracks which appear in the flour-strewn top surface of the sour-dough. In practice, the baker coordinates the controllable variables to optimize the chosen technological process.

*Controllable process variables* include, temperature, standing-time, dough consistency (dough yield), type and amount of flour added at each process-stage (dough substrate nutrients, i.e. reproductive-factor or *Vermehrungsfaktor* to the German baker), aeration, and amount of initial starter-culture used (*Anstellgut* to the German baker).

*Sour-dough processes* can either be discontinuous (batch), or continuous, the number of process-stages involved showing some variation from bakery to bakery. At each sour-dough stage, appropriate control variables are chosen to favour either lactobacilli or yeasts. Thus ensuring an optimal balance of yeast and bacteria at the final doughmaking stage.

*The three-stage discontinuous or batch sour-dough process*, often used in small to medium sized bakeries, involves the preparation of initial (*Anfrischsauer*), basic (*Grundsauer*), and full (*Vollsauer*) sours before the final dough stage. The basic-sour stage is stood overnight for 4–8 hours at 22–28°C, with a dough-yield of 145–160%. This stage is utilized to propagate the lactobacilli, and so intensify acid and aroma build-up, thus suppressing the sour-dough yeasts. By scaling up the quantities of each stage accordingly, enough basic-sour can be produced for four full-sours on reaching maturity. These, in turn will produce four final doughs at maturity. The basic-sour is self-preserving, when the acidity-index reaches about 20 (approximately 2% acid concentration), owing to lactobacilli activity-stagnation but retention of vitality. Therefore, the basic-sour stage can be not only perpetuated from day to day, but also stored over the week-end or for a whole week's production. This can, however, only be done by preparing a basic-sour of firm consistency (dough-yield 150, or 2 parts flour to 1 part water). Following this procedure, the aroma-strength and preservation capacity of the basic-sour allows it to be utilized as a flavour concentrate, adding it at 5–15% of total flour weight depending on the type of bread required. Leavening is supplemented by the addition of yeast. However, the orthodox procedure is to proceed from basic-sour to full-sour by adding a further 32–35% of total rye-flour content, and 55–60% more water. Adjusting the dough yield to 180–200%, and the temperature to 30–32°C, allowing a ripening-time of 2–4 hours. These conditions favour the production of both lactic and acetic acids, as well as activating the yeasts. This influences both bread-crumb properties and flavour later. The final dough is prepared by adding 48–52% of total rye/wheat-flour, and 40–45% water with 1·6–1·8% salt (total flour weight), the final dough-yield being adjusted back to 145–160%. Final dough temperature is set at 28–32°C. Mixing times are 20 minutes for conventional mixers, and 6 and 2 minutes for high-speed and intensive mixers respectively. In the case of rye-meals mixing times of 30 minutes (conventional), and 20 minutes (high-speed/intensive) are appropriate. Starting from scratch, this typical three-stage system would require the preparation of a starter-sour by mixing 0·5% total flour weight of dry starter with 1–2% of total flour, and 2–3% of water, which is then added to the initial–sour as 'seed'. The percentage of total rye-flour which has been subjected to the souring process is an important technological parameter for each type of bread, valid guidelines are the following: 100% rye-flour bread 40–50%; mixed rye/wheat bread 40–60%; and mixed wheat/rye bread 50–100%. This ensures bread quality and flavour intensity.

*The quasi-continuous sour-dough process*, although prepared on a discontinuous (batch) basis, is regarded as being continuous since part of the mature full-sour is used to prepare a fresh initial-sour, thus perpetuating the sour-dough preparation and maturing system. The KVT 1000, 1500 and 1800 are examples of plants using this system. The process is perpetuated from one mature full-sour to another, two-thirds of the mature-sour being used to prepare the final dough, and the residual one-third used to prepare the fresh initial-sour. The full-sour dough-yield in this process is 230% at 28–29°C, requiring a ripening- or maturation-time of 3·5 to 5·0 hours. The full-sours are mixed, and then allowed to mature in segmented compartments of a carousel-bunker. After degassing, they are mixed into a final dough using a continuous-aggregate, from which extrusion takes place. Since this system is routinely used for the production of mixed rye/wheat doughs, wheat- and rye-flour, water, salt solution, and the mature full-sour are fed into the mixing-aggregate.

*The Isernhäger sour-dough process* provides sour-dough of constant quality, and in the required quantity on a flexible basis, to satisfy the demands of the essentially craft-operated bakery. This patented process gives a pumpable product, which can be integrated into a modern computer-controlled doughmaking system, and which can be stored and still give reproducible bread quality for a day's production. Simultaneously, this patent allows the processing of stale-bread paste, which improves crumb-elasticity and avoids wastage.

*The Berliner short sour-dough process* propagates the heterofermentative lactobacilli by using a dough-yield of 180–200% and a temperature of 35°C, which gives an adequate acidity-index of 10–13 within a maturing-time of 3 hours. This gives the favourable ratio of acetic acid to lactic acid 30:70, but some loss of crumb properties could present itself.

*Quality rye-bread preparation* demands a uniform control over the sour-dough processing-stages, and the elimination of external influences. The diverse types of sour-dough processes confront the microorganisms with specific conditions, which control their metabolic output, and result in differences in both dough acidity-indices and pH. The type and species of the lactobacilli are typical for each process. The relative amounts of lactic and acetic acids produced by fermentation can be influenced by the sour-dough temperature; even the ratio of D- to L-lactic acid can be influenced by the choice of lactobacilli. Control of dough-yield (and consistency), also influence the acid build-up from the various types of bacteria used, e.g. *L. casei* and *L. brevis* var. *lindneri*. In addition, the total microflora content of the starter-culture influences the quality of the sour-dough in which it is used. The lactobacilli exist in symbiosis with the various yeast types and strains, benefiting from the proteolytic activity of the latter. Without a complete knowledge of the sour-dough microflora, and the exact understanding of their mode of growth, and the factors which influence them, any changes in process parameters cannot be correlated with their effects on sour-dough quality. Starter-cultures for rye and rye/wheat flour breads can contain the following microorganisms: *L. brevis* var. *lindneri, L. brevis, L. fructivorans, L. plantarum,* and *L. farciminis,* amongst others. The prime function of which is to lower the dough pH, and produce a balanced ratio of lactic to acetic acids, referred to as the 'fermentation-quotient'.

*Maturing-times* vary considerably between 2 and 5 hours, with the exception of overnight, or longer term basic-sours made with a firmer consistency with more flour to slow down the microbial activity, and excess souring. Standing times are the prime consideration in practice, and such process variables as: temperature, dough-yield (consistency), starter-culture concentration, reproductive-factor, and rye-flour type are adapted and adjusted accordingly. More bread-faults result from immature doughs than over-ripe ones, the latter sour-dough situation being capable of correction by the addition of appropriate amounts of potassium carbonate.

*Rye-flour quality* shows wide variations due to climatic conditions during growth and at the time of harvest, as well as post-harvest treatment and storage. However, contrary to common belief, rye quality is often more influenced by deficiencies in the degree of polymerization of the grain storage materials, than by actual increases in enzymic activity. In many cases, rye-flour deliveries with Falling-number (Hagberg-Perten) values less than 100 show quite satisfactory or even good processing properties. However, values in excess of 150, but not over 220 will provide better tolerance to a wider process range. Amylogram viscosity at the maximum gelatinization temperature should be over 300 units, at a temperature of at least 64°C, gelatinization commencing at about 53°C. Rye wholemeal samples should be at least 150 Falling-number to produce flours of adequate tolerance. These values are based on wholemeal and flour sample weights of 9 g/25 ml water, *v.* 7 g/25 ml for wheat samples. Amylogram values are based on 80 g/450 ml water. Rye-flour water absorptions, measured at 350 Farinograph units, range from about 64% to 73%, depending on crop-year and extraction rate. Any reduction in flour water absorption suggests less swelling capacity initially, but this is normally realized during the sour-dough stages. Test dough yields of 168% (based on total flour weight 100 parts), with dough consistency about 325 Farinograph units, should give bread volumes of 520–550 ml/250 g. Loaf-symmetry (width and height measure), and sensory evaluation of crust/crumb structure, colour and flavour, are most important attributes for all rye breads. The loaf width:height ratio should be ideally 1·5–1·6, with little tendency to dough flow-out.

*The multi-stage processing* of rye-flour doughs, allows the optimization of the process for diverse bread varieties, using rye-flour of variable characteristics. By using a batch process, and various process-stages of appropriate temperatures, consistencies, standing-times, and reproductive-factors to suit the microorganisms, large quantities of dough can be leavened and acidified with a relatively small quantity of starter-culture (*Anstellgut*). Processing objectives are: sour-dough yeast reproduction; bacterial reproduction; inhibition of wild microflora; achievement of optimal pH and acidity-index; and controlled swelling of the flour hydrocolloid components. Process control is effected by precise fixation of temperature, weight, and time. A facility for testing the sour-dough acidity-index at each stage is desirable for consistent bread quality. Many bakeries no longer use multi-stage processes, but in so doing compromises have to be made.

*The basic-sour, or basic-sour without full-sour* is an example of a rationalized batch process schedule for the production of rye breads. This is achieved by separation and specialization of the functions of the various sour-dough stages of multi-stage procedures. The basic-sour being utilized to produce the aroma substances, and natural acidification, any deficiency in acidity being supplemented by the addition of dough-acidulants containing lactic, citric or adipic acids. Deficiencies in leavening effect are supplemented by the addition of baker's yeast at the 0·8 to 1·5% (flour-weight) level. Initially, a basic-sour is prepared with an amount of starter-culture depending on the desired standing-time, i.e. 5% to 50% of total rye-flour content. Using 50% starter, the basic-sour matures in about 4 to 6 hours. The use of 500 g starter/10 kg flour (5·0%) in the sour, adjusting dough yield to 150% (2 kg rye-flour/litre water) and maintaining dough temperature at 24°C, should give a basic-sour of pure aroma and flavour over 11–12 hours. The temperature latitude being 20–30°C for acid-formation and fermentation. Once enough mature basic-sour has been produced, the process is progressively continued from one basic-sour to the next in one stage. For quasi-continuous production, two-thirds mature basic-sour are processed into the final dough, and the residual one-third is used to prepare the next basic-sour. For convenience, salt is added as a 25% solution, and up to about 3% of stale-bread crumb (flour weight), finely milled and soaked in water at 24°C, can be added to the final dough.

*The Berliner short sour-dough process*, can reach maturity within 3 hours, the hetero-fermentative bacteria producing an acidity-index of 10–13. This is achieved by using a flour content five times that in the starter-culture (reproductive-factor), dough yield being set at 180% to 200%, and sour-dough temperature 35°C. As a general guideline, at least 50% of total rye-flour content of rye/wheat doughs should be subjected to the sour-dough process.

*KVT 1500 continuous doughmaking plant* (CSSR) Dough-preparation rates and feed-rates. Source: *Special Bakery Processes* by Werner Schwate and Udo Ulrich, VEB-Fachbuchverlag, Leipzig, GDR, 2nd edn, 1986, p. 38.

*Progressively produced rye-sour* for continuous automated production of rye bread varieties is the subject of a patented system of joint design and construction by the Vereinigte Nahrungsmittelindustrie (VNI) in Vienna, and Reimelt of Frankfürt, FRG.

*Foramitti, A.* (1982) *Getreide Mehl und Brot*, **2**, p. 47–50 (German).

*In the USSR sour-doughs for rye breads* are either made with firm consistency sours (*Golowka*), involving a four-stage build-up, or a softer sour (*Kwass*). The *golowka* production cycle can be shortened by working from full-sour to full-sour, on a quasi-continuous basis. The water content of the *golowka* full-sour being 50%, i.e. a sour-dough yield of 150%. In this case it becomes a two-stage process: full-sour and final dough. Using a batch system, the mature *golowka* full-sour can be divided into 3–4 parts. One part being mixed with more flour and water to prepare more *golowka*, and the residual 2 or 3 parts used to prepare 2 or 3 portions of dough. The build-up of the softer sour or *kwass* involves three stages, and the water content of the resulting mature full-sour is on average 2·0% higher than the *golowka* full-sour. On a batch system, two-thirds of the mature *kwass* full-sour is used to prepare the dough, and the residual one-third used to prepare a new batch of *kwass* full-sour. In some bakeries much higher levels of water are added to the full-sours to adjust flavour, and facilitate its transport through pipelines. Also the amounts of mature full-sour used to prepare doughs will be varied to suit production conditions, such as: ambient temperatures, *golowka* or *kwass* activity, flour quality, and production-stoppages.

*The S-1 liquid-sour process*, due to E. K. Gladkova and co-workers, utilizes a lower viscosity sour, containing so-called beta-bacteria, pure cultures of group B lactic acid bacteria (Seliber classification). These heterofermentative types produce many volatile flavour acids in addition to lactic acid, alcohol and carbon dioxide. By using selected strains of these bacteria, with a reproduction temperature optimum of 25–30°C, 50% of the mature sour is used for dough preparation, and the residual 50% supplemented with the following nutrients: 40% of a saccharified, pregelatinized flour/water suspension (PFWS), prepared by heating a 1:3 or 1:2 flour/water mixture for 2–4 hours at a constant 62–65°C, 51% water and 9% flour. This enriched sour is fermented at 33–35°C for 60–75 minutes, the sour reaching an acidity-index 8–11. Then, 50% of this mature sour is utilized to prepare a fresh sour. Sours are prepared in stationary batch containers. To prepare the final dough, 50% sour (based on flour weight), and the requisite amount of flour, water and salt-solution are adequately mixed. The dough is set at 34–35°C, and bulk-fermented for 120–140 minutes. A 100% rye wholemeal dough would attain an acidity-index of 8·5–9·0. The doughmaking aggregate ChTR of the I. L.

Rabinowitsch system is often utilized in Soviet bakeries for this purpose. For processing the *golowka* (firm consistency sours), the bunker-type system of N. P. Gatilin became widely used during the 1960s. This involved the use of one bunker-aggregate for preparation of the full-sours, and another for the final doughs. About one-third of a mature full-sour being fed back into the sour-dough mixer aggregate to inoculate the fresh sour. Stationary bunker-systems, such as that of A. M. Chrenow can also be utilized for both sour-dough and final dough preparation, and by placement of a rotary sectioned-bunker aggregate between them, a quasi-continuous doughmaking system can be constructed. The sectioned-bunker is used for sour-dough or sponge (wheat-flour) maturation. For continuous rye dough preparation, using liquid-sours of the *kwass* or S-1 type described, a Ch-12 mixing-aggregate, coupled with a sloping trough-shaped fermentation-conveyor ChTR (as used in the Rabinowitsch system), is utilized. This being also applicable to wheat-flour doughs, where indirect pre-ferment systems are in use. The various aggregates mentioned are illustrated in Section 1.7.

*Rye bread faults* can be categorized into: flavour deficiencies; poor loaf symmetry; and poor crumb-elasticity, after 24 hours setting and cooling.

*Rye-flour delivery quality profiles* with standard Falling-number (9 g/25 ml water) values within the range 90 to 110 must be regarded as suspect for processing without appropriate corrective measures. A retest using water acidified to pH 4·0 with dilute lactic acid will provide supplementary information, since such flours often respond positively to enzymic inhibition adjustments, and produce quite acceptable bread when correctly soured. Test-bakes are carried out using both lactic and acetic acid solutions, and mature biologically produced sour-dough. Other corrective measures which can be superimposed on a pH adjustment include the use of one of the following hydrocolloids: *Quellmehl* 1–3%, CMC 0·5%, and xanthane gum 0·2–0·5%, all based on flour weight. These substances regulate the dough water absorption and its distribution, when flour component polymerization is deficient due to rye growth conditions in the field. Simultaneous addition of one of the following emulsifiers, calcium stearoyl lactylate, sodium stearoyl lactylate, DATEM, or mono- and diglycerides, at the 0·5% flour weight level, exerts a synergistic effect on hydrocolloid function. The other rye-flour quality profile-extreme results from a highly polymerized grain structure, which shows resistance to enzymic breakdown. Such flour samples have standard Falling-number values of 250 to 300, and Amylogram gelatinization-maxima of 700 or more. When processed into bread without corrective measures, the baked loaf shows a close, cracked crumb, with poor volume, which stales rapidly. This is due to the hard, intact starch structure, and insufficient amylosis, and can be corrected either by blending with a smaller percentage of sprouted rye-flour, or by the use of corrective additives. Appropriate corrective additives are malt-flour at 0·1%, stale bread-crumb paste at up to 5·0%; or it is possible to prepare a 'scald' (pregelatinate) of about 5% of the flour, treating 1 part flour with 1 part boiling water. Optimum concentrations of bacterial and fungal amylases, proteinases, and pentosanases, can also have the desired effect, but correct treatment levels must be determined with the help of Amylograph standard and swelling-curves.

# CHAPTER 2.5 FORMULATION AND PROCESSING TECHNIQUES FOR SPECIALTY-BREADS

This sector of the bread market has shown promising growth in most countries during the 1970 and 1980 decades, and in some countries growth was already under way during the 1960s. In general, the greatest demand has been for wholegrain and meal/grit-containing bread varieties. In traditional white-bread consuming countries such as the UK, continental-style rye and other cereal- and seed-containing breads have shown growth rates exceeding 300% between 1986 and 1989. The innovative potential for the British baker to market an improved variety of high-quality continental breads remains very considerable. From the regulatory viewpoint, a specialty-bread can contain rye and/or wheatmeals and/or certain specialty-flour products (often differing in the method of preparation from normal procedures), containing specific raw materials (often in the form of a premixed concentrate), which change the nutritional value and/or flavour of the bread when compared with basic bread varieties. The interpretation of such regulations, however, will depend on the country. For example, in 'black-bread' consuming countries such as: GDR, FRG, Austria, Czechoslovakia, Hungary, Poland and the USSR, basic bread varieties may already contain 50% or more rye-flour or rye-meal.

*Specialty-bread formulation*, particularly when yeast leavened, often requires the addition of suitable processing aids and/or a formula rebalance. Typical processing aids in the form of raw materials are: vital gluten, sugar in solid or liquid form, salt and shortening, which can effect a rebalance. Typical dough-conditioners are: stearoyl-2-lactylates, ethoxylated monoglycerides, and other emulsifier-type additives, depending on local legislation.

*Multi-grain or mixed-grain breads*, require a 'soak-stage' to effect complete swelling of grain components. The grits, meals or other cereal flours are blended together to form a mash with most of the water, and allowed to stand for 6 to 8 hours. The increasing demand for wholegrain bread varieties is due to nutritionalists generally recommending the consumer to concentrate on fibre-rich menus.

*Specialization categories* for formulation include: fibre-rich, protein-enrichment, calcium-enrichment, salt-reduced, and special formulations for dietetic consumers.

*Multi-cereal breads* can be formulated by substitution of 5–30% wheat-flour by mixtures of rye-, corn-, oat- and barley-flours, with the addition of vital-gluten to provide the structure.

*High-fibre breads* containing the fibre fraction not enzymically degraded within the human digestive system, i.e. cellulose, lignin, hemicelluloses, pectins, gums and other non-starch carbohydrates not digested by man (loosely referred to as 'crude-fibre'), are formulated with such raw materials as: wheat-bran, corn-bran, soy-bran, oat-hulls, rice-bran, and powdered cellulose. In fact, the so-called 'crude-fibre residue' is made up of about 97% cellulose and lignin, 13% to 70% of which is actual 'dietary-fibre'. The range of dietary-fibre contents of the above raw material sources is about 45% (rice-bran) to 99·5% for powdered cellulose (insoluble dietary-fibre). Insoluble-fibre is defined as including celluloses and lignins, and some hemicelluloses. Soluble-fibre comprises pectins, gums, pentosans, and some hemicelluloses, which has physiological significance in the control of cholesterol and diabetes. Total dietary-fibre (TDF) consists of both soluble and insoluble fibre, and is determined by the A.O.A.C. Prosky-TDF procedure, which is an enzymic digestion intended to simulate human digestion. Other sources of fibre include: spent brewer's grains ('mash') or '*Malztreber*' (German), and washed orange-pulp (TDF 46·5%, insoluble dietary-fibre 21·9% and soluble-fibre 24·6%). Roasted commercial fibre sources, e.g. oat-fibre, corn-fibre, and soy-fibre, contain significant amounts of both insoluble and soluble dietary-fibre, compared with raw milled-products. Formulations containing more than 10% added fibre require a high-protein, gluten-rich flour (16% protein), and 60–70% of the flour should be used in a sponge-stage. Where 10–20% added fibre is used, water absorption must be increased in excess of 100% to ensure complete hydration of the gluten proteins, and the added fibre. In this case, 6–10% vital-gluten is necessary to provide an acceptable cell-structure, about 5% of which is best added to a sponge-and-dough process procedure. Yeast levels should be increased to 3–5%, 1–2% retained for the dough stage. A salt adjustment of up to about 3%, and the addition of high-fructose corn-syrup at 8–12% should provide an adequate rebalance. Honey or molasses can replace part of the sugar where wheat-fibre is utilized. Shortening additions should be avoided on calorific grounds, but sodium stearoyl-2-lactylate, added at 0·5% does improve bread volume. Mould-inhibitors usually become necessary, owing to higher bread moisture levels. Since fibre is inert in terms of energy contribution (calories), some sources can be utilized for the formulation of low-calorie yeasted breads. In 1984, the FDA in the USA, allowed the exclusion of non-digestible dietary-fibre when calculating the calorie content of reduced calorie foods. Therefore, for every 1·0% of total dietary fibre added, the product calorie content declaration can be reduced by 4 units/100 g. Calories/100 g are calculated as being equal to four times the protein percentage, plus nine times the fat percentage, and four times the percentage total carbohydrates, minus the percentage of total non-digestible dietary-fibre.

*High-protein breads*
This can be achieved by using a diverse source of high-protein raw materials such as: full-fat soy-flours, groundnut protein isolates, soy protein isolates, cottonseed protein, gluten-flour, vital-gluten, milk protein fractions, and composite mixtures of these to form premix concentrates. To obtain satisfactory results, a sugar rebalance is desirable, and an optimal level of potassium bromate or ascorbic acid, if the former is not permitted. The additional use of a dough-conditioner, such as 0·5% sodium stearoyl-2-lactylate, will give an even better end-product. Improved results are also obtained with selected surfactants of the sucro- or glycolipid type, and first preparing a sponge, to which is added on maturity the 10–12% soy-flour in paste form at the doughmaking stage. In countries where rye-flours as well as wheat-flours are used for basic bread varieties, the same general regulations for specialty-breads (*Spezialbrote*) apply. However, many other milled-products and special fractions are utilized, e.g. oats, barley, maize, and their mill-stream (passage) products, grain-germs, outer layers and brans. Other raw materials of vegetable origin are: fruit and vegetable products, spices, oil-bearing seeds (sunflower, linseed, cottonseed, etc.), buckwheat, sesame-seed and carob-bean flour. Those of animal origin include: milk and milk-products, and meat and meat-products, such as full-milk powder, whey, buttermilk, whole-egg powder and butter. Certain partially cooked extruded cereal products also find application. Many of these materials increase bread yield, e.g. vital-gluten, potato-semolina (flour), and soy-based products. In developed countries, the need for protein-enrichment is limited, the only case being to supplement the lysine deficiency of wheat flours. More urgent requirements are for increased dietary-fibre, reduced fat and sodium intake. There may also be a case for iodized potassium, iron, selenium and other trace elements in certain countries. Specialty-breads, containing larger amounts of whole-grain material, binding increased amounts of water, represent an economy in raw material, compared with the basic breads made with lower extraction products. Total specialty-bread consumption in the GDR, including dietary-breads, amounts to about 10%, which is mainly produced in the industrial-bakeries, but the potential for craft-bakeries remains open.

*Schneeweiss, R., and Heinrich, G.* (1984) *Bäcker und Konditor*, **5**, p. 132 (German).
In the GDR, raw materials are selected which best serve the nation's dietary requirements. Current analysis indicates that specialty-breads with fibre-enrichment are not yet being consumed on a broad enough basis, especially as a regular part of the diet. Many consumers who are dependent on various medications, could instead use a basic food to alleviate their problems. In the above-cited report, entitled 'Specialty-breads—a requirement from the baking industry', the authors state that the required shelf-life of such breads is about 7 days, and that they should ideally attain their peak flavour acceptability after 48–72 hours. When sliced and wrapped, 250 g and 500 g units show a prolonged shelf-life of up to 9 days, when made from rye and wheat wholemeal flours, compared with only 4 days when they contain rye-flour type 997 and patent wheat-flour type 550. Some well-known specialty-breads from the GDR are *Mecklenburger Landbrot*, and *Malfa-Kraftma-Brot*.

*Mecklenburger Landrot*, can be produced either discontinuously or continuously, using a sour-dough process. However, continuous production requires long runs with the ryemeal processed in the sour. On a discontinuous basis, enough basic-sour is prepared for the maturation of two full-sour batches, the basic-sour standing for 5 hours at 25–27°C. The full-sour matures for 2 hours at 28–30°C, having been enriched with whole rye-flour. The final dough consisting of the mature full-sour, a blend of whole rye-flour, rye-flour type 997, and wheat-flour type 550, yeast, salt, caraway-seed (milled), sugar, caramel colour and water to give a dough temperature of 30°C. Final proof is at 32–34°C for 40–50 minutes for units scaled at 1150 g. Baking is set at 280°C initially, falling to 200°C at the outlet for a period of 50 minutes.

*Heinrich, G., and Schneeweiss, R.* (1984) *Bäcker und Konditor*, **5**, p. 143. Recipes and production methods for specialty breads—Part I (German). *Mecklenburger Landbrot* was sold as 1000 g loaves at 0·85 MDM (fixed price), which is equivalent to 30p at 1989 exchange-rates (based on a DM:MDM exchange-rate of 1:1). The baking-loss allowed for is 13% for *Mecklenburger Landbrot*.

*Malfa-Kraftma Brot*, is a long-established specialty-bread in the GDR, containing 10% Kraftma special-flour, prepared from malted barley, thus giving the bread a malt-aroma and dark crumb-colour. It can be produced either as a hearth (oven-bottom), or panned bread weighing 1 kg, and continuous plant can be utilized for both sours and final doughs. In this case, dough-yields must be controlled. The basic-sour is prepared from rye-flour type 997 instead of coarse rye-meal, including the appropriate amount of starter-culture and water. The temperature of the sour is set at 26–28°C, and allowed to stand for 6–7 hours. On reaching maturity, the full-sour is prepared by enrichment of the mature basic-sour with more rye-flour type 997 and water. This is set at 28–30°C, and stood for 2·5–3·0 hours for maturity. The final dough is then made up of the mature full-sour, and a blend of rye-flour type 997, wheat-flour type 550, and Kraftma special-flour, salt and water to give a dough temperature of 29–31°C. This receives a floor-time of 15–25 minutes. Final-proof is for 35–45 minutes at 32–34°C for units scaled at 1150 g. Baking temperatures are set at 250–270°C initially, falling to 220 and 180°C, the baking-time for hearth bread being 45–50 minutes, and 60–70 minutes for panned-bread. As with all coarse-textured cereal meals and flours, the Malfa-Kraftma special bread-flour is best mixed with water, and allowed to swell before addition to the dough.

*Trendbrot and Driftbrot*, are GDR specialty-breads enriched with dietary fibre, made from rye-flour 997, gluten-rich wheat-flour, wheat-bran, and Telavit-trend concentrate. A Full-sour is prepared from a mature basic-sour, enriched with rye-flour 997 and water to stand for 2–2·5 hours at 28–30°C. The final-dough is made by the addition of gluten-rich wheat-flour, wheat-bran, yeast, salt, Telavit-trend concentrate, an emulsion and water. The dough is given a floor-time of 20–30 minutes, scaled at 1160 g per unit, and then given a final-proof of 30–40 minutes at 32–35°C and RH 80–90%. Baking is at an initial temperature of 280°C, falling to 210°C for at least 70 minutes.

*Driftbrot*, in contrast to low extraction wheat-bread, has a reduced energy or calorific value. Instead, it has a high protein and dietary-fibre content, especially suitable for overweight subjects, by preventing the formation of excess adipose tissue. It is prepared by a direct or straight dough procedure, but the rye-flour should be added in a soured form, having undergone a three-stage souring process to the mature full-sour stage. The mature full-sour is added to the final dough, which is made up of the following materials: gluten-rich wheat-flour, special *drift* flour, yeast, salt, and an emulsifier in paste form. The dough receives a floor-time of 60 minutes, with a punch or remix at 20-minute intervals. For 1500 kg baked unit weight, the baking-time must be at least 80 minutes, at an initial temperature of 280°C, falling to 210°C with steam injection.

*Hagenower Spezialbrot*, is a light-eating bread containing both whey and whole-egg powder, which gives it a mild taste and tender crumb. It is produced as a hearth-loaf of 1 kg or 1·5 kg baked weight. A sponge is prepared with gluten-rich wheat-flour, yeast, whey, whole-egg powder and water, to stand 60–90 minutes at 26–28°C. The final dough is made by adding the mature sponge and a portion of mature basic-sour to rye-flour 997, more gluten-rich wheat-flour, more yeast, salt, a small amount of prepared dry-sour and water. The dough is given a floor-time of 15–25 minutes at 28–30°C, scaling at 1680 g for a 1·5-kg loaf, and 1150 g for a 1-kg loaf. Final-proof is for 40–50 minutes, baking the 1·5 kg units for 60 minutes, and the

1-kg units for 50 minutes. Baking-temperatures commencing at about 270°C, falling to 180°C.

*Heinrich, G., and Schneeweiss, R.* (1984) *Bäcker und Konditor*, **6**, p. 175. Recipes and production methods for specialty-breads II (German).
The floor-time should not be exceeded, otherwise the characteristic aroma is lost.

*Pumpernickel* is an old-established specialty-bread made from rye-wholemeal, salt and water, using a sour-dough or *Quellstück*, which means soaking part of the rye-meal. Yeast at 0·3–0·5% and up to 3% sweeteners can also be added. Baking-time must be at least 16 hours, at about 100°C in a chamber saturated with steam. The baked bread has a dark, dense crumb. About 25% of the rye-meal is made into a sour-dough, applying the traditional three-stage process, another 25% of rye-meal being used to prepare a '*Quellstück*' with warm water, and allowed to stand to swell for about 7 hours at 22°C. On reaching maturity, both the sour-dough and *Quellstück* are added to the residual 50% rye-meal, more water, salt and yeast. The final dough-yield is adjusted to 160%, with intensive mixing. The dough-pieces are then worked-off by depositing them into baking-pans; or they can be processed in one elongated dough-strip, divided into sections. Proof-time is 50–60 minutes with exact relative humidity control. Baking must be carried out with steam-saturation, either in a special baking-chamber, or in a standard oven providing adequate steam-saturation. Baking-temperatures are maintained at 100°C for at least 16 hours, under a saturated-steam atmosphere. A simplified American-style pumpernickel is made from a *Brühstück*/sponge with clear flour, dark rye-flour, yeast and water, being set at 24°C for 3 hours. The final dough is then made by the addition of ryemeal, sour-dough, salt and water. Additionally, malt, sweeteners and shortening are used at 1·0% flour weight. The final dough is set at 27°C, scaled, rounded, and rested for 15 minutes before final-shaping. Final-proof requires 50 minutes with humidity control, baking at 220°C, in saturated-steam.

*Ponte, J. G. Jr.* In *Production Technology of Variety Breads*, edited by B. S. Miller, AACC, St. Paul, MN, USA, 1981.

*Wheat and rye wholemeal* products for bread processing are best prepared by roller-milling as opposed to the use of hammer-milling.

*Zwingelberg, H., Seibel, W., and Stephan, H.* (1984) *Getreide Mehl und Brot*, **38**, 3, p. 69–76. Influence of milling wheat and rye wholemeals on their baking-performance (German).

*Seibel, W., Stephan, H., and Zwingelberg, H.* (1984) *Getreide Mehl und Brot*, **38**, 11, p. 339–45. The manufacture of whole-grain-meal alternative breads (German).
The above researchers established that the mode of grinding/milling of wheat and rye into wholemeals had a great influence on the specific weight (g/litre), as well as the relative kernel-hardness of the grain from which it is ground/milled. The roller-milled products are lighter in weight for the same volume than the corresponding hammer-milled products. In general, the softer kernel structured grains give lower g/litre (lighter) meals than the harder grains. These workers also developed certain suitable dough-processing methods involving a two-stage sour-dough for wheat or rye wholemeal bread, as well as the application of both the Berliner and Detmold one-stage sour-dough processes, and the SEKOWA special ferment method. For the two-stage process, to achieve a mild souring, only 0·1% starter, based on total meal for souring, is necessary. If stronger souring is required, up to 10% of the total meal for souring can be added. The first sour-dough stage for a 10 kg wholemeal dough requires 250 g wholemeal (fine), 1 g starter, 500 ml water, which gives a dough-yield of 300%. This is allowed to stand at 24–26°C for 15–20 hours. To the mature first-stage sour (750 g) is added a further 750 g of wholemeal and 1500 g (ml) of water. This is equivalent to a dough-yield of 300%, being allowed to stand for 3 hours at 26–28°C. The final dough is prepared by adding 9000 g of wholemeal (fine) to the 3000 g of mature second sour-dough stage, 120 g of salt, and approximately 5000 g water for wheat wholemeal dough or 5500 g for a rye wholemeal dough. Respective yields being 170% and 175% of final dough. Final-dough proof is at 40–45°C for

60–90 minutes. The SEKOWA special ferment procedure utilizes a three-stage sour, using a total of 0·05% of the special ferment, which contains hetero- and homofermentative lactobacilli, and yeasts cultured under wild and natural conditions. The first stage consists of 100 g of wholemeal, 100 g wheat-flour type 550, 20 g special ferment and 220 g water for a 10 kg wholemeal batch, giving a dough-yield of 210%. This is allowed to mature for 24 hours at 26–30°C. The second stage is made up of the 400 g of mature first sour, 300 g wholemeal, and 100 g water, adjusting to a dough-yield of 160%. This stage matures for a further 24 hours at 26–30°C. On reaching maturity, only 100 g of the 800 g total of this material is used to make the final dough. The residual 700 g is placed in a cold-room for further batches, being stable for several months. 100 g is appropriate for 10 kg of wholemeal. The pre-ferment is 100 g of mature second-stage sour, 30 g of SEKOWA special ferment, 3·5 kg wholemeal, and 3·5 kg water. This adjusts the dough-yield to 200. This pre-ferment is allowed to mature for a further 24 hours at 26–30°C, yielding a total pre-ferment weight of 7·130 kg. The final-dough is then made up of about 7·000 kg of mature pre-ferment, 6·500 kg wholemeal, 150 kg salt, and 3·500 to 4·000 kg water, giving a dough-yield of 170–175%. This is allowed 45 minutes at 30°C. The scaled dough-pieces are then final-proofed for 60–90 minutes at 35–40 minutes.

*Whole-grain bread*, containing whole grains, whole crushed grains or very coarse meals or grits, are becoming a favourite specialty-bread with the consumer. Typical characteristics of whole-grain specialty-breads are: a soft crumb; good keeping-properties; moist crumb, but no signs of dough-like structure; mild, slightly sour taste; good slicing properties. The grain-kernels added to the dough must be readily masticated, and have good digestibility, but be clearly visible on slicing. The baking procedure for this type of bread containing whole rye grains can be one of the following:

—in a steam-chamber as panned bread, with or without a sour-dough stage;
—in a conventional oven as panned bread made with sour-dough either as sliced and wrapped portions or whole loaves;
—in a conventional oven as hearth-bread (oven-bottom) dispatched in the form of a whole loaf.

Irrespective of the baking procedure, the grains must be allowed to swell beforehand, since the amount of water added during sour or dough preparation is not adequate, and the time-interval from doughmaking to baking inadequate for proper grain swelling. The pretreatment of the whole or crushed grains is carried out in the form of a soaking in water between 70 and 100°C (*Brühstück*), or by a lower temperature soak (*Quellstück*) at 15 to 70°C. The *Quellstück* is preferable owing to ease of preparation. Possible variations in baked-products include: whole-grain panned bread with 50% of the milled product as rolled rye grains; light whole-grain containing 25% rolled rye grains; dark whole-grain containing 25% rolled rye grains. The preparation of the *Quellstück* can be rationalized for quasi-continuous production and processing on a bread-line. This is achieved by using a mixer unit with only three mixing-cycles/4 hours, in which is placed the crushed grain, salt and water. After a 4-hour swelling period, the product is drained through a mesh-grill, and added to the mature full-sour in the machine bowl in the desired quantity. After mixing, the dough is processed in the normal manner with a floor-time of 30 minutes, followed by rounding, final-proofing, and baking. The swelling process can be accelerated somewhat by using a dilute solution of lactic or acetic acid instead of just water when preparing the *Quellstück*. The use of a mixer to prepare the *Quellstück*, avoids the presence of unswollen white kernels in the cut bread.

Whole-grain breads represent a considerable efficiency in the nutrition of a nation, owing to the high extraction rate of the grain consumed, and a reduced processing cost for the grain employed. The qualitative value of the end-product is extremely high.

*Bran-bread with raisins* was originally an American specialty, where Californian raisins are considered a health food by the consumer. Before being added to the dough, raisins must be conditioned by soaking in an equal weight of water for at least 4 hours. The raisins can be used as the only source of sweetening, or honey and brown sugar can be added as extra sweeteners. Raisins contain an average of 70% sugars, and the amount added can be up to about 18% of the flour weight. A typical bran-bread formulation is the following: high-protein, enriched, first clear flour 80 parts, wheat-bran 20, wheat-gluten 4, yeast 3, salt 2, brown sugar 4, honey 3, shortening 3, dough-conditioner 0·4, water 64.

*Malted-grain products* are particularly useful raw materials for the formulation of specialty- or variety-breads, owing to their high fibre and protein content, as well as the roasted grain flavour and aroma token.

The specialty/variety-bread sector of the bread market respresents the greatest potential for expansion and innovation generally. The consumer seeks new taste experiences, whilst insisting on ingredients with a minimum of processing, and a minimum of processing-aids or additives. This is also the trend in food legislation towards the year 2000.

# Notes and References for Part 3

## Chapter 3.1 AIMS AND REQUIREMENTS OF THE BAKING PROCESS

*Raw dough-piece* undergoes transformation under the influence of heat application to form a light-textured, porous, digestible product of pleasant flavour and aroma. The irreversible structural changes in the dough components involve physical, chemical and biochemical reactions, complex in nature. The reaction rate must be carefully controlled by the amount of heat input, rate of heat-transfer, level of humidity, and time of exposure within the oven chamber. However, this assumes that all 'upstream' process procedures have been optimized to produce a fully mature dough-piece ready for the oven. On setting, the dough-piece undergoes the final phases of swelling and solubilization due to limited heat-transfer. As heat-transfer progresses, gas-expansion and then gas-retention and solidification of the elastic-film surrounding the gas-bubbles take place. The net effect of these developments determine loaf-volume, resulting from dough expansion or 'oven-spring'.

*Rheological attributes* of the proofed dough-piece include elastic-extensibility and tensile-strength, and doughs with a high viscosity/rigidity-modulus ratio, or 'relaxation-time', provide large oven rises. During initial swelling and solubilization, the dough must retain sufficient viscosity and elastic-extension, until the starch can swell and gelatinize, thus contributing to the strength and rigidity of the dough-piece.

*Modern industrial baking ovens* are designed to convey the dough-pieces on trays or a travelling-band (hearth), through a series of controlled heat-zones, each subjecting the dough-piece to strict conditions of temperature and humidity. For panned-bread, the initial setting temperature would be about 203°C, for a duration of about 7 minutes (about 25% of total baking-time 27 minutes). At this stage the outer crumb layers show temperature increases of on average 5°C/minute up to 60–70°C. The dough-surface reaching 100°C within about 3 minutes, progressively increasing at a rapid rate, reaching about 130°C within 20 minutes of setting. However, heat-transfer to the centre of the crumb proceeds relatively slowly. This is due to the large temperature-gradient of 60–70°C between the dough-surface and the crumb-centre. The heat-transfer proceeds from the outside to the inside, any unbound moisture evaporating in 'evaporation-zones' on attaining a temperature of 100°C or more. This zone then moves with increasing heat-transfer from the outer layers towards the centre. Moisture, on reaching 100°C evaporates by diffusing through the pores in the crust for the most part, increasing the humidity of the baking-chamber. A smaller amount of steam entrapped within the loaf contributes to the swelling and gelatinization of the dough-layers located near the centre of the loaf. Such processes are linked with heat consumption, and an exact knowledge of heat-transfer mechanism will enable the oven-operator to save energy.

*Formation of a thin extensible surface-skin*, represents the first visual change during the baking process.

*Temperature/time* changes within the dough-piece between 30 and 70°C take place within narrow temperature ranges. At 30 to 40°C, swelling, enzymic activity and yeast growth progress, all fermentable sugars continuing to be fermented by the zymase enzyme-complex, until a temperature of about 60°C is reached. The percentage of starch granules with a damaged outer-layer increases, depending on the interior temperature of the dough-piece during baking.

*Schneeweiss, R.* (1965) Report of 2nd Conference 'International Problems of Modern Cereal Processing and Cereal Chemistry', Vol. 2, p. 260. The temperature changes during the baking process and their influence on bread quality (German).
Within the temperature range 57 to 79°C, a steep rise in damaged-starch granules (stained by china-blue dye) from 20% to 100% on completion of baking takes place. Schneeweiss, R. (1965) also established that the rate of soluble carbohydrate formation during baking increases rapidly between 43°C and 60°C, from 14·5% up to about 24% after 55 minutes baking-time (hearth-bread). According to Schneeweiss, *alpha-amylase enzyme kinetics* begin at 30°C, reaching a maximum between 50 and 60°C, beginning to fall off at 70°C. Falunina, S. F. claims that alpha-amylase can remain active up to 78–80°C, and that some residual activity remains at 97–98°C. Even in baked bread, some activity remains at the crumb-centre, providing the crumb pH remains above about 4·3.

*Falunina, S. F., and Popaditsch, I. A.* (1951) *DAN* (Reports of the USSR Academy of Sciences), **78**, 1, 103 (Russian).

*Falunina, S. F.* (1954) MTIPP Report 3, 60 (collective work of Moscow Technological Institute for the Food Industry).

*Rye-starch* starts to gelatinize at 45°C, and becomes complete at about 55°C.

*Wheat-starch* starts to gelatinize at 55°C, and will continue up to about 65°C.

*Starch-swelling and gelatinization* is made possible by the migration of water from the other dough components to the starch-granule, resulting in partial dehydration of the gluten-complex. This causes the gluten proteins to become more viscous and elastic, before coagulation becomes complete at about 70°C, depending on molecular weight and colloidal state.

*Yeast-enzyme inactivation* occurs between 55 and 60°C, but the zymase-complex remains active up to about 65°C.

*Oven-spring* continues after starch-swelling has begun, due to plastic and elastic dough expansion, but terminates during gluten coagulation.

*Dough structural rigidity*, due to protein coagulation and partial starch gelatinization, takes place between 62 and 67°C, as heat-transfer proceeds.

*Temperature data cited* assume events taking place zonally within the dough-piece at the temperature specified or range cited. At which point in time the various layers attain these temperatures will depend on the efficiency of the heat-transfer system from heat-source to each layer or zone of the dough-piece.

*Alpha- and beta-amylase* act on the gelatinizing starch within the crumb of the dough-piece, increasing the water-soluble fraction. Its extent is both crucial and variable, depending on enzyme kinetics and degree of polymerization of the native starch in the flour used. Both heat-transfer, and dough pH also exert an influence on the speed of liquefaction.

*Pentosans and hexosans*, derived from the cellulosic flour components, are also present in the water-soluble dough fraction. Their content varying from about 0·4% to 1·2% of crumb-weight. These branched-chain compounds are mixtures of xylose, arabinose and glucose, in the approximate ratio 5:4:3.

*Dextrin content* of doughs made from standard flours increase by about 15% during baking, whereas sprouted wheat-flours show increases of 50% or more.

*Carbon dioxide liberation* is complete at 60°C, and contributes to dough expansion. With dough-surface skin-drying and thickening, its elasticity is reduced, and crust-browning sets in, at 100°C.

*Oven-spring or loaf expansion*, approximating to about one-third of the proofed-loaf volume, takes place at about 100°C.

*Second and third stages* of panned-bread baking amount of approximately 14 minutes, roughly 50% of total baking-time, oven-temperature being held at a constant 240°C. The crumb temperature rises at about 5·5°C per minute during stage two, until it acquires a level of 98 to 99°C at the onset of stage three, after which it reaches a steady-state constant level for the remainder of the baking-process. This level of temperature provides the maximum rate of moisture evaporation, starch gelatinization, and dough protein coagulation. Ethanol evaporates, contributing to leavening, and the loaf interior is transformed into a baked-crumb structure, from its outer layers through to its inner layers, thence to the crumb-centre by heat penetration. On reaching 150 to 206°C, the crust acquires the typical brown colour due to caramelization.

*Unfermented sugars* prior to baking need to be about 2–3% to ensure normal crust formation. Other thermal reactions between sugars and proteins yield melanoidin compounds at 100°C and above. Amino acid, and carbohydrate monomers are the reactants. The dextrins from starch degradation are also involved in non-enzymic browning reactions.

*Bread-aroma substances* are formed at temperatures of 100 to 180°C, parallel with the browning reactions. These are volatile carbonyl compounds (aldehydes and ketones), resulting from Maillard reactions in the crust area. During cooling and storage, these diffuse into adjacent crumb-layers. In dough, bread and oven-volatiles, more than 200 aroma- and flavour-developing substances have been identified. In white bread, volatile organic acids are retained, lactic acid at concentrations of the order of 10–15 mg/100 g and acetic acid at 2–5 mg/100 g. Rye breads contain much higher levels of these acids, owing to lactic fermentation.

*Fourth and final oven-temperature zone* for panned bread is maintained constant at 221–240°C, which is intended to firm up the cell-walls of the crumb-porosity, and adjust the desired crust-colour intensity. The duration of this stage is approximately the same as that of the initial stage, about 7 minutes (25% of total baking-time). This results in a loss of organic volatiles, which contributes to the 'bake-out loss'. The latter amounts to 10–12%, showing differences of up to 2·0%, owing to temperature/time variations, product weight/volume, formulation, ratio of crumb:crust, oven-design, crust-texture, and efficiency of heat-transfer.

*Prouty, W. W.* (1965) *Bakers Digest*, **39**, 4, p. 69. Effect of temperature variations at constant baking-time on bake-out and moisture-losses in the finished bread.

*Unsliced panned white bread*, crumb-moisture is considered ideal for ultimate consumption when the cooled moisture content is 37–38%, but the sliced/wrapped version should be nearer 36%, although a permitted mould-inhibitor is normally added.

*Output-capacity* of an industrial bakery depends on the baking-surface of its ovens.

*Visual perception by the baker*, during baking, identifies: oven-proof, crumb-formation and crust-formation. Oven-proof or 'oven-spring', is the result of a high degree of gas-development and efficient gas-entrapment within the gas-cells. This should give a volume increase of 15% to 25%, and in the case of hearth-bread a height increase at the expense of its width. Crumb-porosity determines potential digestibility, forming 70–80% of the loaf. Its moisture at setting lies within the range 47–49%, which reduces on bake-out to about 37–38%, depending on the mode of baking.

*Underproofed dough-pieces*, on setting, require a reduction of the oven feed-end temperature, and an increase in humidity input. The latter, within a closed chamber, increases the pressure, thus countering the pressure developed by gas-production within the dough-piece. This hinders excessive initial rising, and dough-surface condensation retains it longer in the elastic state until gas-development and expansion is complete.

*Overproofed dough-pieces* on setting, require an increase in feed-end temperature, and the humidity input decreased accordingly, both of which exert the opposite effects on the dough-piece. Thus loaf symmetry can be retained, and the formation of large vacuoles in the crumb avoided.

*Hearth-breads* require the input of 70–80% relative humidity during the initial baking stage, which represents about 20% of the total baking-time. However, since the accurate measurement of humidity within the baking-chamber presents technical problems, the baker must rely on experience with his particular oven-system. Also, no exact guidelines for steaming-levels to apply to individual products are possible.

*Optimal steam-input levels* will depend on the consistency of the dough-piece, and its degree of proof. In the case of hearth-breads, after the initial humid stage, most of the steam is released by opening the oven-chamber vents. Oven humidity is regulated by either injecting steam into the chamber, e.g. for products with a high percentage of crust such as bread-rolls, or release through the chamber-vents. An approximate guideline is 100 kg of hearth-bread require about 3000 kg of steam. With modern ovens, this is supplied from outside the oven-chamber as saturated-steam under pressure at about 30 kPa. Where older ovens are in use, 'steamers' (steel-boxes containing water) can be placed within the chamber, or dough-pieces can be either painted or sprayed with water before setting. The latter technique supplements dough-piece surface humidity in all situations.

*Steam-injection* is widely used in Europe for hearth and crusty products, injection taking place in the first or second heat-zones of the chamber by lateral steam-headers. These provide saturated, low-pressure steam over the top of the dough-pieces during the first 2–3 minutes of baking, at a velocity of about 300 ft/min. Since steam volume varies with pressure, steam is more accurately measured by weight, i.e. the amount of water which the boiler must evaporate to produce it. Boiler capacity often being quoted as 'boiler hp/lb of dough', e.g. 1·0 to 1·5 boiler hp/100 lb of dough is regarded as optimal. Oven-capacities of 1000 lb dough/hour would require 10- to 15-hp boiler. The prime purpose of steam injection is to effect condensation on the cool dough-surface, enabling it to remain elastic long enough to expand to a maximum without a ragged break and shred. Other functions of primary live-steam before setting or firming of the loaf are improvement of 'bloom', and evaporation loss control. Secondary-steam is that produced during baking by evaporation from the dough-piece. The use of steam for panned-bread production is now less common, since the secondary-steam generated is considered adequate. With travelling-band tunnel-ovens, steam-injection is only used at the feed-end and zone 2, since, when used at a later stage, it can produce a tough, leathery crust. This is due to gluten boiling on the dough-surface instead of being coagulated in a less humid atmosphere. Prolonged retention of steam, after about 15 minutes baking-time results in tough crusts. Bake-off in a relatively dry atmosphere imparts crispness to the crust. Steam removal is controlled by dampers, positioned along the oven length. All steam-lines from boiler to oven must be insulated to retain steam-quality, and save energy; this prevents condensation, and any carry over of moisture droplets into the oven, causing surface blisters on the products. Low-pressure steam is ideal for baking, its temperature approximating to dough-temperature rather than oven-temperature. At 100 to 115°C steam is in a water-saturated state, and condenses on the cool dough-piece. Steam at 10–15 psi pressure from the boiler, reduces to about 5 psi through the injector. If the steam were at a higher pressure, and temperature, insufficient time would be available for it to fall to the point of condensation (dew-point) before the oven temperature had raised it again to dough-surface temperature. Also, high pressure/temperature steam would exert a depressing effect on loaf-volume, especially rye-doughs.

*Heat-transfer* from source to dough-piece must be accomplished within an optimum time-period to complete crust and crumb formation, browning processes, evaporation of moisture, and elimination of all microorganisms. If too rapid, the outer dough-layer becomes burnt, and the internal layers remain incompletely baked-out, the heat-transfer from outside to inside not having been correctly synchronized. When heat-transfer proceeds too slowly, the crust formation is inferior, and the internal-layers become dried out.

*Optimum baking-time and throughput-capacity* of the baking system utilized is essential for consistent bread-quality, and the economic operation of the oven. In the case of discontinuous or batch oven loading, bread quality is usually controlled by the baking-time and checking weight-loss. For continuous industrial baking, using travelling-band tunnel-ovens, baking-time is regulated by setting the speed of the travelling-band. This is measured with dynamometers, the electrical output being measured either by a voltmeter or a potentiometer. The relationship between band-speed and energy-output is then calibrated in terms of baking-time. The resulting instrument, giving a direct read-out in minutes of the baking-time is referred to as a tachometer.

*Tachometers* are either direct- or alternating-current instruments, whose rotating-element is directly connected to the oven conveyor-band drive-shaft. These instruments provide a daily read-out record of baking-time when connected to a recorder-chart. Solid-state technology now miniaturizes these instruments, allowing them to be mounted integrally in central control-panels of reduced dimensions.

*Wheat- and rye-flour breads* differ in baking requirements and conditions, i.e. baking-temperature gradient, humidity, heat-transfer, and baking-time. Reasons for such differences are: composition and properties of wheat and rye doughs; lower density of wheat *v.* rye doughs: leavening procedures, wheat (yeast) *v.* rye (bacterial).

*Wheat-bread and bread-roll products* differ in baking requirements and conditions owing to the extreme difference in size and weight and also to the relative ratios of crust:crumb.

*Weight of steam required* increases with decreasing product weight, owing to smaller products having a larger surface-area, and crust percentage than the larger units. An approximate guide is 130 to 150 g/m$^2$ of dough surface. 1 kg hearth-baked wheat-bread, baked at 240°C, falling to 220°C over 40 minutes, would require about 2000 litres of steam/100 kg product. Whereas, 45 g wheat-rolls, baked at 220°C over 18 minutes, would require 3000 litres of steam/100 kg product.

*Bread-baking* requires oven temperature programming such that the temperature-gradient is a falling one. This procedure is both technically and economically justified, but can only be successful where ovens are under automatic temperature control. Baking-loss and the thickness of crust must be under strict control to produce products of acceptable baked quality. In many Eastern European countries this is regulated, and laid down as a standard, together with the baking-time. For example, in the GDR, a basic bread variety such as a 1·5-kg hearth-baked mixed rye/wheat loaf, must show a baking-loss of 11–12%, which demands a baking-time of 50–60 minutes at 280°C falling to 210°C.

*Heat-energy*, required to bake a 1-kg loaf of bread is approximately 500–600 kJ, depending on oven type, type of heating, and degree of automatic control of the heat-transfer system.

*Heat-flux and heat-transfer* during baking can take three different modes, i.e. convection (movement through gaseous media); conduction (movement through solid material); and radiation (emission from surfaces). Which mode predominates will depend on oven-type and design and energy-source. Convective heat-transfer, involves the atmosphere within the baking-chamber, a variable mixture of air, steam and fermentation volatiles, and its circulation speed. The convection heat-transfer coefficient is expressed in kJ m$^{-2}$ h$^{-1}$ K$^{-1}$. Under static atmospheric chamber conditions, this is of the order of 21 kJ m$^{-2}$ h$^{-1}$ K$^{-1}$; but at a velocity of 10 m s$^{-1}$ this increases to 168 kJ m$^{-2}$ h$^{-1}$ K$^{-1}$.

*Korb, W.* (1958) *Oven Construction and Heat Engineering* (German), p. 6. Technical Department for Large Oven Construction, Werner and Pfleiderer, Stuttgart, FRG.
Conductive heat-transfer depends on the type of material used for the oven-sole, i.e. steel-

plate, steel-mesh or fire-brick. It is the result of molecular oscillation within the the the material, and is expressed in terms of the coefficient of conductivity (kJ m$^{-1}$ h$^{-1}$ K$^{-1}$). Radiation results when there is a temperature-gradient between a body and its surroundings. The heating-elements of the oven chamber emitting heat in the radiant mode. However, this type of non-ionized radiation can only penetrate a few millimetres into the dough-piece, after which heat-transfer proceeds by conduction. The radiation coefficient is expressed as kJ m$^{-2}$ h$^{-1}$ K$^{-4}$, depending on the temperature of the radiating-element. A temperature increase by a factor of 2, requires a radiant heat-energy increase by a factor of 16.

*Wheat, rye and mixed wheat/rye doughs*, although similar in baking behaviour, show certain specific differences in reactants and reactions taking place in the oven. Visual differences are apparent as differences in crust-colour, and bread-crumb character. Rye-bread crust-colour is predominantly chestnut, whereas that of wheat-bread is golden-yellow to dark-brown. The influence of these differences become apparent when blends of rye- and wheat-flour from 100% to 0% are made at 10% increments. Rye doughs have more free amino acids and types of sugar than wheat doughs. These compounds become involved in the melanoidin-forming Meyer reaction, which considerably enhances rye bread flavour. Significant decreases in free lysine, leucine, valine, xylose, arabinose, galactose and glucose occur during rye-bread baking as a result of such reactions. Minimum baking-times for hearth-breads of the same scaling weight, containing various percentage mixtures of rye- and wheat-flours, show decreases with increasing percentage of wheat-flour in the mixture. The higher gluten content of wheat-bread dough results in a higher thermal conductivity, on account of superior leavening properties. The time necessary to reach evaporation temperature at the centre of the loaf-crumb is less than in the case of rye-bread dough. Bread-roll products may reach evaporation temperature within about 6 minutes of setting in the oven, but the dough-pieces must remain at 220°C for 18 minutes to ensure an acceptable baked-quality of the crust and crumb.

*Baking-times and temperatures* for specific products must be adhered to for baked-out quality. To control starch gelatinization, and protein denaturation, the centre of the dough-piece must be maintained at 98°C for at least 10 minutes to avoid a damp, inelastic crumb and premature staling. The ratio of crumb:crust varies with the type of product.

*Schwate, W., and Ulrich, U.* (1983) *Special Processes in Baking*, 2nd edn, (German), published by VEB-Fachbuchverlag, Leipzig, GDR (see Fig. 109, p. 179—Average crumb-crust %ages of bread and rolls).

*Sales success of bread* as a basic food commodity depends on strict control of the baking process. Baking-time has a profound effect on aroma and flavour build-up of all baked products. These, and product freshness, constitute the sensory appreciation of the consumer. Substances amongst those forming the aroma profile of bread are: acids, alcohols, esters and lactones, carbonyls, pyrazine, sulphur-compounds and phenols. These produce the final taste sensations of acidity, sweetness, saltiness and bitterness, all of which must be optimally balanced. These, together with the thermally modified lipids, proteins and carbohydrates, combined with ethanol, water, fermentation by-products and gas-development make up the product consistency. The latter contributing to the taste sensation on the tongue and general mouth-feel.

*Rustic products* with a 3–4-mm crust thickness are increasing in popularity. Such breads have a strong aroma, and well-developed aromatic and 'nutty' flavour. Most of these are based on rye or mixed wheat rye milled products, but the addition of oat milled products, e.g. oat-bran and oat-germ, and malted cereals, will add value both nutritionally and from a flavour viewpoint. By comparison, standard wheat and wheat/rye bread varieties and bread-rolls require a larger percentage of crust, and to be freshly baked and crisp. Bread-rolls having a limited shelf-life of 6 hours, and 2–3 days in the case of breads.

*The choice of bread variety or bread-roll type* will depend on the food or drink with which it is consumed, thus producing an ideal harmonization of the various taste and flavour components. However, bread flavour is critically influenced by the baking-process, and crust-thickness. In countries where more wine than beer is consumed, e.g. France, white-bread predominates in the form of the baguette, which has 32–36% crust. Whereas, the average loaf of white-bread has only about 20–25% crust. Crusty white-bread is the ideal accompaniment to complement the pleasant fruity/acid flavour notes of good wine. Bread neutralizes the wine-acids, allowing increased consumption without any adverse effects due to excess stomach-acid build-up. The bake-out loss of a baguette is about 20%, which is twice that of an average white panned-loaf. In Germany, the '*Salzstange*' is very popular with both beer and wine, and is found all over Germany in the restaurants. The dough is given a short proof before being rolled out in an elliptical form and rolled-up. After being painted with water, they are covered with salt and caraway-seed before setting in the oven. Most of such rolls are prepared from a lean 'water-dough', with no shortening or sugar enrichment. Another type is the '*Mohnsemmel*', made in the form of a rosette, wreath or straight, from a lean dough. Before setting, the dough-piece is moistened and strewn with blue poppy-seed.

*Bread-flavour development* does not depend on the use of an oven-fired with solid fuel, and the entry of the flame into the oven-chamber. Excellent quality and flavour breads can be baked in multi-deck and travelling-band tunnel-ovens. Temperature and temperature/time control within the dough-piece during the baking cycle are more important than oven-type, construction or energy-source. Heat-transfer from oven to dough-piece must reach a 'steady-state' condition, the heat-flux being sufficient to penetrate the dough mass and evaporate the water, leaving the crumb structure intact and elastic. Any excessive heat build-up at the crust-surface must be avoided, since this results in wasteful 'burning' of the top-crust instead of a controlled bake-out and browning. A 1-kg hearth-baked white loaf requires an average of 40 minutes to bake out, when set at 240°C, falling to 220°C. Whereas, a 1-kg rye or mixed wheat/rye variety would require a higher initial temperature of 250°C, falling to 210°C over 50 to 60 minutes, to give optimum flavour development. Doughs made from rye-meals, and panned rye doughs require similar temperatures, but with reduced top-heat, and stronger bottom-heat, baking-times being up to 120 minutes for a 1·5 kg scaling weight.

*Baking-loss* is a function of baking-time, scaling-weight, and loaf-surface exposure. Loaves of 1 kg or more give baking-losses of 10–13%, whereas bread-rolls lose from 18–20% during baking. In general, the larger the unit the smaller the net baking loss.

*Pre-baking (Vorbacken)*, is a technique often applied to rye-bread baking to improve flavour, crumb-elasticity, porosity and shelf-life. The larger 1·5-kg units are exposed to temperatures of 420°C for about 2–3 minutes, which gives the loaf an extra 1 mm of crust, using the normal baking-time and temperature subsequently. This increases the crumb-moisture retention, and bread-yield increases from 135 to 138 kg bread/100 kg flour for mixed rye/wheat breads.

*Crumb-centre temperature* must reach 98°C, and when this has been reached the optimal baking-time has almost lapsed, but the crust-surface can reach about 160°C. Differences in specific heat between wheat and rye doughs, mean that the same dough-piece weight of wheat dough will reach the critical 98°C at the centre on average 10–12 minutes before that of a rye dough or mixed rye/wheat dough-piece.

*Specific heat* differences between wheat and rye doughs are due to differences in moisture content, i.e. 45–46% *v.* 53–55%, their material structural porosity, which in turn favours thermal conductivity through the layers in the case of a wheat dough. Specific heat is defined as the quantity of heat required to raise the temperature of unit mass of material from $t$ to $(t + 1)$ °C, at any temperature $t$. The unit of specific heat capacity is $J\,kg^{-1}\,K^{-1}$ SI units, or cal $g\,K^{-1}$ in the CGS system. Temperature differences of $1°C = 1\,K$ (kelvin). Approximate specific heat capacities of leavened dough range from 2·80 to 3·0 kJ $kg^{-1}\,K^{-1}$, depending on dough water-content, ranging from 45% to 55%.

*Theoretical heat requirement* to bake a 1-kg loaf has been calculated to be 500 to 600 kJ, but such values take no account of such technical heat-losses as: exhaust-gases, unburnt-gas, heating-up, radiation, steam-evaporation, incorrect secondary-air feed at the burner.

# CHAPTER 3.2 ELEMENTS OF THE BAKING PROCESS AND THEIR CONTROL

*Baking process* is the most energy-intensive operation of bread production, and 30–40% of all bread-faults are due to baking inadequacies. Therefore, optimization of the controllable variables is very desirable.

*The fate of the shaped and proofed dough-piece* during the baking-process under the influence of the following variables must be carefully analysed: heat-energy input and temperature; heat-transfer; humidity; and baking-time. The unit of heat in the SI system is the joule, which can also be expressed in non-SI system CGS units as calories. 1 calorie = 4·1868 joules (J). The quantity of heat-energy $dQ$, absorbed by a body of mass $m$, when its temperature is increased by $dt$ is: $dQ = sm\,dt$, $s$ being the specific heat of the body. The sum of the quantity of heat a body absorbs on heating, and the work $dW$ performed in so doing, represents the change in internal energy, $dI$, i.e. $dQ + dW = dI$, the First Law of Thermodynamics. Any change in internal energy, $dI$, will depend on the final and initial state of the dough-piece, and is independent of the actual heating process, whereas both $dQ$ and $dW$ depend on the transition process. The fact that heat cannot be spontaneously transferred from a colder body to a hotter one without a change in the system is known as the Second Law of Thermodynamics. The Third Law of Thermodynamics states that the specific heat of a body approaches zero as its temperature approaches absolute zero on the thermodynamic temperature scale. The point of 'absolute-zero' on the thermodynamic (Kelvin) scale being 273·15 K below the melting-point of ice at standard pressure. This being the baseline for temperature measurement on the Kelvin scale. Temperature is a measure of the average chaotic motion or kinetic energy of the molecules within a substance, and their velocity of oscillation or movement. This energy of movement can be increased by the application of an adequate amount of heat-energy, which raises the temperature of the body, unless it loses heat simultaneously. Temperature, as an expression of heat-intensity, can be reduced or increased by the control of the heat-energy input. After setting a batch of bread in the oven, the temperature falls, unless more heat is applied. Therefore, in transforming the dough-piece into a loaf of bread, a specific quantity of heat-input is expended. This quantity varies with oven-design, heat-loss management, dough-piece weight and dimensions, efficiency of the heat-transfer systems, loaf-spacing within the chamber, humidity-control, and the use or non-use of pans. The average range for a 1-kg loaf being 500–600 kJ, most of this heat-energy is transferred to the dough-piece in the radiated form, which means that the oven-chamber walls must attain temperatures between 300 and 400°C to bake the bread. Heat transferred onto the dough-piece by radiation, conduction and convection currents, and steam condensation, must be further transferred by conduction from the peripheral layers into the centre of the crumb, as shown in Figs 87, 88 and 89. The graphs shown in Figs 90 and 91, show the ideal temperature/time curve for hearth-bread, and the reduction of dough-piece moisture as the dough-piece rises in temperature.

*Oven operation and control* should take into account the following: utilization of oven capacity to the full; synchronization of final-proof with oven-temperature; regular calibration of thermometers; industrial bakery ovens require regular checks for humidity, gas/air ratios and exhaust-gas analysis, chart-recorders should be monitored and interpreted; vents and loading and discharge points should be kept closed after use; draught sources near the oven must be eliminated; regular oven-maintenance is essential.

*Energy-utilization calculations* are more meaningful than theoretical heat requirement calculations, owing to the numerous sources of heat-loss inherent in all ovens, e.g. exhaust-gases, steam, radiation, and firing-losses. Travelling-band tunnel ovens show the following average energy-utilization percentages: total energy-input base 100%, of which, some 45% is utilized in actually baking the bread, 35% lost as exhaust-gases, 12% lost as steam, and a residual 8% lost by radiation.

*Energy-utilization index* is expressed as the ratio of the total theoretical heat-energy requirement to bake a loaf of defined weight, to the total energy input. When multiplied by a factor 100, it is expressed in percentage terms. Alongside such data as: oven-capacity, baking-surface area, and loading characteristics, the energy utilization index is a critical parameter for bakery ovens. In effect, it is a measure of the summation: oven type and design, the energy source, and the efficiency of the operator.

*Improved oven-design and control techniques* have reduced manual work, and improved bake-out control, but the demand for technical know-how from the operator has increased accordingly. Oven-operation requires dedicated observation, and a quick-reaction to correct any unfavourable developments. Exact temperature readings taken before loading and during the baking-cycle for checking against target values are typical routines. This enables readjustments to be made manually. If the oven temperature-control system utilizes a servo or feedback capability, providing modulated heat control, this can be computer interfaced. Oven-humidity is often judged by experience using manual control of the lateral steam-headers or jets, but this results in about 20–33% loss of steam for bread or rolls. An automatic steam-control system can reduce energy consumption by up to 10%, being linked to the hygrostat. This facility is also available on many increasingly high-technology equipped multi-deck ovens, together with check control instrumentation for the precise diagnosis of electrical and burner faults. Exhaust-gas monitors keep excess air and soot-formation limits to a minimum. Burner monitors allow comprehensive checks to be made on oil-consumption, in-built computers on the instrument-panel inform on burner operation-time, the number of switch-ons, oil-consumption, oil consumed since the last reset, as well as oil supply and tank-level, thus ensuring heating-times and baking characteristics are stabilized.

*Baking dough-piece characteristics* demanding close observation are: crust-colour, loaf cross-sectional symmetry, volume and any signs of splitting at the sides (hearth-bread), or break and shred (panned bread). Any abnormal developments being countered by adjustments to temperature, humidity or baking-time.

*Energy-utilization indices* are the highest in the case of travelling-band tunnel-ovens, whether fired by natural gas or electricity, and electricity provides the most efficient energy-utilization index compared with other conventional energy sources. However, national economic issues, such as natural resources available to produce energy, will override such bakery internal evaluations.

*Schwate, W., and Ulrich, U., Special Processes in Baking*, 2nd edn, 1986, p. 74 (German).

*Multi-deck ovens* offer great flexibility when fan-assisted, cyclotherm hot-air or gas circulation is utilized. Each deck is under individual temperature control, allowing the simultaneous baking of a product assortment, or one deck to give a 'sharp pre-bake', and another to complete the baking-process. Average heating-up rates are about 3 K/min, with the heating options oil, gas, electricity or coal. Fast and efficient loading with roll-up oven-setters is now widely applied. Relative energy-input requirements of a 4–5-deck oven for heating-up and continuous operation respectively are: gas $32 \cdot 0 \, m^3/h$ and $13 \, m^3/h$ at 75 minutes heating-up time; diesel oil $11 \cdot 0 \, kg/h$ and $5 \cdot 7 \, kg/h$ at 75 minutes heating-up time; coal $45 \cdot 0 \, kg/h$ and $15 \cdot 0 \, kg/h$ at 120 minutes heating-up time. This data is approximate, and reflects relative energy-input differences in requirements. The average baking-surface of a four-deck oven being a total of $10 \, m^2$, but can be $19 \cdot 2 \, m^2$ with a large deck width of 3100 mm, and a capacity of $240 \times 1$-kg hearth-loaves. Ovens fitted with fire-brick tiles on the sole, known as 'chamotte', are capable of storing so much heat that there is no need to heat up again for baking, the heat being emitted more slowly than the rapid heat-transmission from steel-plates.

*Vogel, W.* (1972) Higher productivity through utilization of the cyclotherm deck-oven (German), *Bäcker und Konditor*, **20**, 26, issue 5, p. 134.

*Heat-energy input and temperature control and measurement* is the most important routine operation in oven management. Close attention, and logging of measured data will reduce running cost to a minimum. Where natural gas is in use, the volume of gas-flow per unit time is measured with a cylinder type digital-counter, designed for various throughputs. The gas:air ratio must also be routinely checked, a minimum of air being essential for an efficient combustion. This parameter is determined by gas analysis, the quantity of air being either determined from the working pressure of the measurement-diaphragm, or the air velocity at the intake point, based on cross-sectional area. The ratio of the actual quantity of air to the minimal quantity required is known as the 'air-number', or sometimes the 'burner-number'. For travelling-band tunnel-ovens of the BN-series, manufactured in the GDR by VEB-Kyffhäuserhütte, Artern, the 'air-number', or 'burner-number' should be 1·1 to 1·3, in order to avoid energy-losses, according to:

*Schneeweiss, R., and Klose, O.* (1981) in *Technology of Industrial-Baking*, VEB-Fachbuchverlag, Leipzig, GDR.

Parallel checks to determine the 'air-number', are carried out by an exhaust-gas analysis, gas-analysers being available in both a stationary and transportable form. Any fall in pressure in the combustion-chamber of tunnel-ovens can be measured with a U-tube manometer or Bourdon-gauge scale read-out. This should be regulated to fall within the range 40 down to 20 Pa by exhaust-gas adjustment. Electro-energy is measured with a meter reading energy consumed in kWh or kilowatt-hours (1000 watts acting for 1 hour). Temperatures in both multi-deck and travelling-band tunnel-ovens are measured by pyrometers. These instruments operate on the expansion principle, consisting of a bimetallic strip, often iron and copper of differing coefficients of expansion. The resulting deformation of the bimetallic-strip at different temperatures is then measured by registration on a calibrated scale of the movement of a cogged-strip, which keys into a cogged-wheel attached to a rigid pointer. Such instruments depend on conduction, whereas true pyrometers measure radiated-heat (infrared), either by changes in electrical resistance, or millivolt differences of the order of 5–6 mV per 100 K. Thermocouples for resistance-thermometers consist of nickel and platinum, and those depending on thermal-voltage differences, measured in millivolts, consist of iron–constantan couples connected to the pyrometer by copper wires. Continuously working ovens require temperature measurements not only within the various zones of the baking-chamber, but also within the burner-space and circulation-system. This involves temperature gradients of 150 to 600°C, which are measured remotely with either thermo-elements or electrical resistance thermometers. An electrical output-signal controls the gas-feed and the burner temperature. Uniform temperature-control requires the application of either continuous, or at least a three-point control system. A new type of temperature sensor, which can float on an air-cushion, like a mini-hovercraft, just 2 mm from the dough-piece, depends on convection instead of radiation or conduction. This floating sensor rides on convection currents drawn over the surface of the dough-piece, a thermocouple measuring the air temperature. Requirements for measurement of localized convective, radiative, or total heat-transfer rates have resulted in the development of several types of heat-flux sensors.

*The slug-type sensor* consists of a slug of metal embedded in, but insulated from, the surface across which the heat-transfer rate is to be measured, a thermocouple being connected at the rear interface between the slug and surrounding insulation.

*The Gardon Gauge* represents a 'steady-state', or asymptotic type of sensor.

*Gardon, R.* (1953) *Rev. Sci. Instrum.*, May issue, p. 366. An instrument for the direct measurement of intense thermal radiation.

It consists of a thin constantan disk connected at its edges to a large copper heat sink, while a very thin, 0·005 in diameter, copper wire is connected at the centre of the disk, thus forming a differential thermocouple between the disk centre and its edges. When the disk is exposed to a constant heat-flux, an equilibrium temperature gradient is established, which is proportional to the heat-flux. The thermocouple signal, being directly proportional to the heat-flux, no corrections, or thermocouple reference junction is necessary. With the Gardon gauge, there is a radial temperature gradient over the disk, which if excessive, causes a variation in the local convection-coefficient and thus an error. However, when only the radiation component of the total flux is required, the front of the sensor can be covered with a thermally isolated sapphire window, which passes the radiation-flux, but blocks the convective-flux.

*Heat-transfer* from energy source to dough-piece occurs in three forms, depending on oven design, viz. conduction, convection and radiation. Conductive transfer takes place from the oven-bottom of 'sole', depending on its area of contact with the surface, and the material of the oven-sole, i.e. steel-sheeting or refractory (chamotte) tiles. It results from molecular oscillation within the material as a result of an intense heat-energy input, being expressed as the coefficient of thermal conductivity, the units in the SI system are kJ or $J\,m^{-1}\,s^{-1}\,K^{-1}$, $W\,m^{-1}\,K^{-1}$, and kcal $m^{-1}\,h^{-1}\,K^{-1}$ in CGS units. This constant for any given material, is expressed as the rate of flow of heat through a surface of defined area, where the temperature-gradient normal to the surface within a fixed time is measured. Its value will depend on both moisture content and temperature. In the case of gases within the baking-chamber, the thermal-conductivity, expressed in $W\,m^{-1}\,K^{-1}$, will depend on both pressure and temperature. The oven atmosphere also transfers heat by convection, which involves atmospheric movement in the form of convection currents. This involves the movement of steam and volatiles either as the result of opened vents or by fan-assistance. Convection heat-transfer coefficients, expressed in $kJ\,m^{-2}\,h^{-1}\,K^{-1}$, depend on the air-speed within the oven-chamber, increasing from about $20\,kJ\,m^{-2}\,K^{-1}$ when static, to $160\,kJ\,m^{-2}\,K^{-1}$ at air-speed $10\,m\,s^{-1}$. The third form of heat-transfer, radiation, can only penetrate a few millimetres into the dough-piece, but the heat-emitting elements of the oven effect a heat-exchange with the chamber atmosphere, and the resultant heat-flux initiates the conduction mode of transfer. However, the total heat energy transfer to the dough-piece by convection, remains quite insignificant unless some form of forced heat circulation is introduced within the chamber. During the initial period of the baking cycle, a very significant mode of heat-transfer to the dough-piece is the heat of condensation of the steam onto the dough-surface and adjacent surface-layers. The latent heat of formation is released as heat-energy. The optimum adjustment of baking-chamber humidity by steaming emphasizes the importance of steam as a heat-transfer medium in baking. Warming up of the dough-layers in the crust-forming zone changes the temperature/baking-time curve in the crumb-centre zone, therefore, care should be taken to fix the optimal level of humidity for products.

*Forced-convection heat-transfer* techniques are designed to provide maximum heat-transfer by adjustment of air-speed and volume of air in circulation. Often air is circulated through slots into the chamber being forced from a chamber located at the base of the oven. Alternatively, the oven atmosphere is recycled by a fan system, removing hot air from the top of the chamber, and inserting it through perforated tubes located above and below the baking-product.

*Direct-firing* is the most economic and the simplest method of heating ovens, giving up to 25% more efficiency compared with indirect methods.

*Natural gas* is the most popular source of energy in many countries, owing to availability and cost. The introduction of the ribbon-burner systems, with a flame-band distributor having found general acceptance. The resultant gas:air mixture remains constant and efficient, and provides good heat distribution, even with natural convection turbulence.

*Heat-penetration* and migration of moisture from the loaf-centre demand considerable latent energy. Therefore, moisture emerging from the dough-piece as steam should be removed by air-turbulence as soon as possible, thus promoting heat-penetration. All evaporation must be completed before crust-formation takes place, since this prevents it. Also, any prolongation of this stage will result in a thick, tough crust.

*Panned-bread baking* demands efficient convective heat-transfer systems ensuring a fast and uniform bake. This is best achieved by a complex system of turbulence, provided by impingement-modules, equipped with tapered-jet nozzles directed both downwards and upwards, carrying hot air onto the dough-piece surface, in close proximity to the oven band-conveyor. With such a system of air-impingement, air-velocity rates of 3000 ft/min are possible near the nozzle-orifices.

*Impingement air ovens* permit the use of lower baking-temperatures, and shorter baking-times. Thus yielding products of higher moisture contents, and prolonged staling-rates, with lower net energy inputs. This technique is appropriate for white panned bread, especially in the sliced and bagged form.

Smith, D. P. (1985) AIB Paper: Update on Oven Technology Seminar, Orlando, Florida, September 21, 1985, Jet accelerated convection baking.

Smith, D. P. (1986) *Food Technology*, **40**, p. 112.

*Improved air circulation* in indirectly fired band-ovens can be achieved by the recirculation of air within the chamber at 200 to 400 ft/minute.

*Hearth-bread baking* requires another technique involving a slower bake. Heavier, lower specific-volume white and variety-breads such as rye and pumpernickel demand this procedure. The solid-steel, refractory (chamotte) tile, or wire-mesh bottom-heat reduces radiation, and cuts off the free flow of convected heat around the dough-piece. Therefore, heat-transfer by conduction becomes predominant, involving a slower bake. The chamotte stores heat, and releases it slowly and progressively. Baking-times approximate to 60 minutes for a 1-kg loaf *v.* 27–30 minutes in the case of an equivalent panned white-loaf.

*Accelerated heat-transfer* under the trade name 'Acceletron', introduced by I. Rhodes, in 1965, depends on the development of a 'corona-wind' within the oven-chamber. An electrode-grid sets up an electrostatic field of high potential and constant polarity between the oven-crown and the hearth. The device is installed parallel to the steam-headers, ionizing the gases near the electrodes, the ionized gas molecules then becoming repelled by the electrodes, giving rise to a downward current of air towards the product. Reduction in baking-times of up to 20% have been achieved with this equipment when installed within the front 25% of baking-surface, using a high-voltage direct current discharge during the first 4–5 minutes of baking. Ozone is simultaneously liberated in trace amounts, as is normal with electrostatic discharges.

Sievers, R. S. (1978) *Proc. Am. Soc. Baking Engineers*, p. 98, obtained the best results by reducing the temperature within the first two heat-zones by 28 to 45°C with adequate steam input, and reducing the final heat-zone by 14°C. Commercial trials indicated the most important advantages of accelerated heat-transfer to be: (1) reduced spacing of the pan-straps, allowing a 10% increase in oven-capacity; (2) an average energy saving of 11·8%; (3) a reduction in bake-out loss, allowing a scaling-weight reduction of 0·5-ounce (14 g) for a 24-ounce unit. Reduced bake-out loss, giving average savings of 0·5 to 1·0 ounces (14–28 g)/lb (454 g) of dough.

*Electrohydrodynamic heat-transfer* in some oven applications, are aimed at improved steam application, enhance transfer by radiation versus convection, thus allowing even closer pan-strap spacing on the oven grid.

*Modulation heat and temperature-control units* for any oven-zone ensure that the correct air: gas mixture is linked to a temperature-control unit, sensing the temperature in the zone of burner operation, feeding back a signal when more or less heat is required. Fig. 95 shows the link-up controller, actuating an electric modulating motor on the combustion air-line. Such devices allow oven-zone temperature to be maintained within a few degrees of instrument preset, the controller being accurate to within 1·0% of the scale range.

*Conventional heat-transfer techniques* must ensure that the centre-temperature of the loaf reaches 96 to 98°C, in order to attain the fully baked state.

*Oven-atmosphere ionization* can bake a 22-ounce (616 g) square Pullman-loaf (pan with cover) within 17 minutes, the loaf-centre temperature remaining below 43°C up to the 7th minute, then rising rapidly over the next 7 minutes, levelling out at 88–93°C during the last 2·5 minutes in the oven.

*Accurate temperature control* can be obtained with the electronic potentiometer, which gives readings accurate to ±0·25% of total scale-range. Pyrometers based on the millivoltmeter being slower and less sensitive in response. Thermometers based on the pressure principle, whether of the coiled Bourdon- or capillary-tube design, have scale range accuracies of about ±1% maximum. Thermoelectric temperature and control systems find the widest application, since any number of thermocouples can be connected to one instrument. However, it is advisable to have one instrument serving each individual heat-zone of an oven, complementing each indicating controller with its own recording-chart to monitor temperature changes within each zone during the baking-cycle. Electronic-pyrometers can be adapted to give modulating heat control, keeping oven temperature within narrow limits.

*Optimal humidity concentration* within the oven-chamber is important for end-product quality of hearth- or variety-breads, bread-rolls, and specific types of cake. At every temperature, air is capable of holding a specific maximum of steam (kg water/kg of dry air), and when this amount is exceeded, condensation results. At 20°C, under atmospheric pressure, the ratio of steam:air (kg) is 0·015, at 90°C 1·56, but at 99°C it reaches 200. At temperatures in excess of 100°C, the water-holding capacity of air is such that steam can be mixed with it in increasing amounts as the temperature is increased. The point of saturation of steam in air is referred to as the 'dew-point', every amount of steam in air, depending on the prevailing pressure, has a dew-point temperature. During baking, this dew-point temperature falls between 70 and 90°C, which corresponds to a steam content of 0·29 to 1·56 kg per kg dry-air. As long as the dough-piece surface temperature during the initial stage of baking remains below the appropriate dew-point temperature, moisture will condense onto its surface, releasing 2270 kJ/kg of steam. Most of this free heat-energy is either absorbed by the dough-piece owing to conduction, or is deposited onto it as the water condenses. However, the effect of this condensation-heat is limited to the initial stage of baking only, since after about 1 to 3 minutes baking-time the dough-surface temperature will exceed the dew-point temperature. The ideal type of steam is at low pressure, produced with a boiler-gauge pressure of about 10–15 psi ($68·94 \times 10^3$ Pa to $103·41 \times 10^3$ Pa), reducing to 5 psi ($34·47 \times 10^3$ Pa) through the injection-nozzle. Steam at 10 psi pressure, in the saturated state, has a temperature of 115°C, whereas at atmospheric pressure (760 mm mercury) only 100°C. With the reduction in pressure, the steam temperature falls with its water-holding capacity, resulting in condensation onto the relatively cold dough-piece at 30–35°C. The porous structure of the dough-piece allows both sorption and condensation on the surface and within the dough-layers. Condensation retains the dough surface in a moist, hydrated state, allowing expansion without cracking or splitting up until the surface temperature reaches 100°C. The presence of humidity accelerating heat-transfer to the crust, surface-layers and central zones of the loaf at the same time. The curves of Fig. 98 illustrate the effect of the temperature/baking-time response of the various loaf-zones both with and without humidity, up until the completion of baking. Dough-moisture increases at doughmaking also contribute to the speed of heat-transfer, especially the surface-layers.

*Rye-bread baked on the hearth* requires the injection of steam at the right time, and in the correct quantity for bread quality. Often, depending on oven design, the optimal adjustment of humidity and top and bottom heat during this stage of humidification becomes critical, falling into a narrow range. If the initial-zone temperature is too low, and the humidity also too low, the crust will develop both side and top splitting. The required levels of temperature and humidity depend generally on the recipe, mixing, proofing-conditions, size and shape, as well as spacing of the dough-pieces within the oven. Where no steam-injection facility is available, one has to rely on spraying, or moistening the dough-pieces with the help of a large brush before setting in the oven. Dough-pieces, which have to be baked in a state of underproof, in an emergency situation, can be given an ample supply of steam. This sets up a pressure increase within the oven-chamber, which counteracts the increasing pressure within the dough-piece itself owing to gas-development and expansion. As a result uneven oven-proof and spring of the dough-piece becomes suppressed, and the dough-surface is kept moist and elastic up until expansion is complete. In baking rye and mixed wheat/rye breads in a batch-type oven, the bread is set at 250 to 280°C, and the steam kept in the oven for up to 1 minute. This allows enough time for the steam to condense on the dough-surface, after which the steam-vent is opened, allowing colder air to enter, replacing the hot, humid air. Radiated heat then takes over, removing moisture from the crust more rapidly and effecting a general drying-out of the dough-piece. The remainder of the baking-time should then proceed at a lower temperature to prevent excess crust browning over the average 55–60-minute total residence time. In travelling-band tunnel-ovens, this can be precisely controlled, whereas in batch ovens, the bread is often transferred to a cooler oven-deck chamber.

*Weight-changes* in the dough-piece within the oven-chamber atmosphere upon intensive initial-stage humidification were studied by Micheljew, A. A., in the USSR, and published in: Special aspects of breadbaking heat-technology, Micheljew, A. A., Alma-Ata, 1943 (Russian). The plot of dough-piece weight *v.* baking-time in Fig. 99 shows that the maximum weight gain occurred between the third and 5th minute of baking-time; the dough-piece increasing by 1·3% over its scaling-weight.

*Moisture-migration or 'flux'*, often referred to as 'thermal-moisture-conduction', results from the movement of moisture from the outer crumb-layers progressively towards the loaf centre, owing to the large temperature gradient between the outer and inner crumb-layers, especially during the initial baking stages. This phenomenon was confirmed experimentally by A. S. Ginsburg in the USSR (see Fig. 100). Conclusions of subsequent work carried out by L. J. Auerman under constant baking temperature, and without added humidification are detailed in Section 3.2.

*Oven-chamber humidity or dew-point* measurement at temperatures in excess of 100°C are carried out with the wet and dry-bulb psychrometric procedures, since these methods give an instantaneous response to moisture variations (see Fig. 101). This involves the measurement of the temperature diffence between a wet and dry-bulb thermometer, from which the 'dew-point' temperature, and absolute moisture content of the baking chamber atmosphere in kg steam/kg of dry-air can be deduced. This is calculated either from a chart, or by using psychrometric formulae. The chamber atmosphere is sucked out by means of an extractor-pump at a speed of 3 to 5 m/s (approximately 1000 ft/min), as shown in Fig. 101A. The high velocity is essential to maintain the wet-bulb-surface moist; also, to avoid any condensation before the measurement, the bulb must not be allowed to fall below the dew-point temperature. During continuous measurement, the psychrometric effect must be maintained with an even stream of chamber-atmosphere. Dry-air in the exhaust-pipe should be cooled to below 150°C to ensure a flow of water in the exhaust and evaporation tube is maintained. Thermocouples are used to record the temperatures of both wet and dry-bulbs. For continuous recording and control of humidity, electrical transducers of the Dunmore type now find wide application. The lithium chloride sensor is very sensitive to humidity over 11%, showing a sharp fall in electrical resistance. Modification gives a signal related to dew-point temperature. Fig. 103 shows how the various products respond over the total baking-time *v.* dew-point temperature. Fig. 102 shows the difference in humidity distribution within a travelling-band tunnel-oven, under conditions of correct external-draught exclusion (A) and where an influx of cold-air exists (B).

*Product baking-time*, can be defined as the minimal time required to provide adequate starch gelatinization, protein coagulation, following oven-proof; and the formation of colour and flavour compounds. However, these processes must take place within a controlled time sequence in order to ensure that the internal structure of the product is well leavened, of even texture and has an elastic crumb. When the starch gelatinization temperature is traversed too rapidly, many starch granules have insufficient time to swell or gelatinize, but instead become encapsulated by other granules already gelatinized, and therefore cannot absorb enough moisture. The resultant products will show a crumbly, dense, doughy and inelastic crumb structure. In many industrial bakeries, the upper limit of the baking-time is determined by economic factors as a matter of priority, i.e. potential after-sales shelf-life with crumb-moisture levels of 35–38%, oven-throughput rationalization and maximum energy economy. This approach often results in a reduction in quality standards of the baked product, owing to inadequate bake-out. Without appropriate consumer protection legislation, coupled with sharp competition, this situation is unlikely to change. In contrast, the baking industry in many East European countries (COMECON) specify standards for both baking-times and temperatures for all basic bread varieties, as well as crust-thickness. In spite of conflicting interests, in the GDR, under Standard TGL-3067, the baking-times and initial and final baking-temperatures are specified, according to weight, variety and form or mode of baking, i.e. panned, hearth or *Angeschoben* (baked adjacently batched). Permissible baking-time tolerances for each basic variety and mode of baking are also provided. Examples of baking standards included in TGL-3067 are tabulated in Section 1.7. They originate from: *Schwate, W., and Ulrich, U., Special Processes in Baking* (Process Manual), VEB-Fachbuchverlag, Leipzig, GDR, 1986 edition, p. 62, Table 7 (German).

Such standardization is very desirable as a contribution to bread-quality, and is facilitated by oven construction standardization. Bread quality depends heavily on adherence to optimal baking-times. This embodies, not only flavour and aroma, but also crumb-elasticity, texture, and general mastication properties. Parameters which can influence the baking-time include: baking-regime, temperatures, humidity, forced-convection, atmospheric-ionization, and dough-piece spacing on the baking-surface, as well as its weight and shape. Dough-piece compositional parameters include: flour type, dough properties, and state of proof before setting. For a chosen temperature-regime and oven heat-transfer situation, there is an appropriate oven residence-time for optimal bread quality. A 1-kg long-rolled hearth white loaf can be baked at 230°C, falling to 220°C for 35–50 minutes, providing the final crust-thickness reaches at least 3 mm. The author's experience has shown that for 1-kg hearth white bread (baked-weight) 40 minutes at 240°C falling to 230°C, gives the required 3 mm crust-thickness, without loss of the mild wheat-bread flavour. A baking-time of 30 minutes being inadequate for full flavour development, and 45 minutes giving rise to some loss of flavour and aroma. The difference in baking-loss of 12·5% at 40 minutes, and 11·5% at 30 minutes, although representing a financial sacrifice, based on 100-kg flour price, provides a qualitative improvement of the product which more than compensates for the disadvantage. On the other hand, using the same temperature regime, a panned white loaf should be adequately baked within 27–30 minutes, when made from a lean formulation. Using an enriched North American formulation, baking at 227°C, rising to 239°C at the discharge-end, a comparable 1-kg panned loaf would be adequately baked after 23 minutes, with a baking-loss of about 10·2%, leaving a crumb moisture in the final loaf of about 36·4%. By comparison, a 1-kg rye or rye/wheat hearth-baked long-rolled unit, baked under a temperature-regime of 250°C, falling to 200°C for 50 to 60 minutes would give an optimal crust-thickness of 4·5–5·0 mm. The same weight-unit, baked as '*angeschobene Brot*' (adjacently batched), using the same temperature-regime, would require 65–70 minutes for complete bake-through. These practical examples illustrate the importance of the adherence to the optimal baking-time. The most common bread-fault encountered is too short a baking-time, resulting in a lack of crust-thickness, flavour and aroma. The author ventures to suggest that in the interests of bread quality, and increased sales, a baking-loss of up to 12% can be tolerated by the baker.

### 3.2.1 Influence of the Elements on the Dough-piece and its Components

The physical effects of the elements temperature, heat-transfer, humidity and time on the dough-piece, convert it into a baked-loaf in three stages:

(1) Heat penetration into the interior of the dough-piece, until the temperature of the dough liquid-phase reaches boiling point.
(2) Progressive evaporation of excess liquid.
(3) Browning and flavour component formation, in the form of dextrins, and carbonyls from starch and sugars, and melanoidins with the help of protein degradation products.

During the initial phase of baking, the temperature differential of about 30 to 100°C has to be bridged. However, at the initial stage, the moisture content of the dough-piece cannot decrease, but rather increases due to steam condensation onto its surface. Instead the moisture is forced to migrate towards the centre of the loaf owing to progressive heat penetration. However, the crust-forming layers lose their moisture. After a considerable concentration of moisture has taken place at the central zones, this internal moisture begins to migrate outwards, in the opposite direction to the heat-transfer flux. Once the boiling-point at the loaf centre has been reached, the heat-flux from the outside is utilized to evaporate excess moisture. This process is accompanied by a slowing down of any temperature rise. When the moisture level has been reduced to a specific degree, the temperature of the dough-piece will continue to rise. Since residual moisture is bound within the capillaries of the dough, this can only be removed at higher temperatures of evaporation. The rate at which the three stages: heat-transfer flux, evaporation, and loaf-browning take place, will depend on formulation-enrichment, weight and shape of the loaf, and its degree of leaven.

*Colloidal and chemical changes* in the properties of the dough components are the most important aspects.

*Wheat and rye starches*, both begin to swell at 40 to 50°C, absorbing from 25% to 50% of their volume of water forming a viscous suspension (see Fig. 104). On penetration, the amylose starch component is the first to go into suspension, followed by the amylopectin fraction. This causes swelling, and the pressure built-up within by the water, results in a splitting of the granule outer membrane.

*Schoch, T. J.* (1965) *Bakers Digest*, **39**, 2, p. 48, postulates that during this swelling and gelatinization, part of the amylose becomes dissolved, and diffuses out of the granules into the aqueous dough medium. Here, it becomes concentrated as the interstitial water is reduced by swelling, forming a gel on cooling, which retrogrades, thus influencing bread-staling.

*Sandstedt, R. M.* (1961) *Bakers Digest*, **35**, 3, p. 36, points out that a large proportion of the starch granules remain intact during baking, owing to the limited supply of water. Such granules are now considered to form the so-called 'resistant-starch', which contribute to the non-starch polysaccharide content.

*Degree of polymerization* of the starch in grain and the resultant flour is a function of environmental conditions during growth. Differences in molecular weight distribution of polymers give rise to variations in gelatinization behaviour (melting), as well as differences in depolymerization behaviour with enzymes. The gelatinization range of wheat starch is 56 to 64°C.

*Degree of mechanical starch-damage* sustained during milling is also important during the baking process. Although exhibiting increased water absorption, damaged starch granules are also susceptible to enzymic attack, and within the range 43 to 60°C alpha-amylase is most active. The enzyme attacking the granule between 55 and 90°C before gelatinization is complete (see Fig. 108).

*Schneeweiss, R.* (1965) also found that at temperatures between 57 and 79°C within the dough-piece, the number of damaged granules increased dramatically. This situation occurred after about 20 minutes baking-time, the isolated starch being coloured by china-blue, which only colours damaged granules (see Figs 106 and 107). Fig. 107 indicates how the water-soluble carbohydrates increase during baking from 14·5% on setting in the oven, steadily rising to 24% after 60 minutes baking-time. Fig. 108 indicates that alpha-amylase is most active within the temperature range 50–60°C, tailing-off abruptly at 70°C.

*Wheat and rye starches* differ in their gelatinization response. At 55°C, 40% of rye starch has already gelatinized, whereas wheat starch has only just begun; at 60°C, up to 80% rye starch has gelatinized, compared with about 20% in the case of wheat; at 65°C, rye starch has reached about 90% gelatinization, compared with about 45% for wheat. Even at 70 to 75°C, the wheat starch gelatinization level only reaches about 80% maximum, whereas rye starch reaches over 90% at that temperature range. Fig. 105 shows in schematic form the importance of the starch granule, and the diverse reactions in which it is involved during the baking process.

*Alpha-amylase activity* is influenced during baking by dough pH and salt concentration in addition to temperature. Both gelatinized and damaged starch are very susceptible to attack by alpha-amylase, resulting in the formation of low-molecular-weight dextrins and only small amounts of maltose, accompanied by trace amounts of glucose and low molecular weight saccharides, e.g., amylotriose, amylotetraose, and amylopentose.

*Pazur, J. H., Sandstedt, R. M.* (1954) *Cereal Chemistry*, **31**, 5, p. 416.
At pH 4·3 to 4·5 (acidity-index 10–11) in rye doughs, and temperature 60°C, and pH 4·7 to 4·8 (acidity-index 4–6) within temperature range 73 to 77°C, alpha-amylase becomes almost completely inhibited. Rye doughs at pH 4·9 (acidity-index 4·4) show alpha-amylase activity in the centre of the bread-crumb up to temperatures of 96°C. Inactivation temperatures for the amylases for both wheat and rye breads depend on the rapidity and duration of the baking process.

*Susceptibility of flour starch* to alpha-amylase attack, depends on the particle-size distribution (granularity) of the flour, the size of the starch granules, and their degree of mechanical damage during the milling process. The greater the molecular surface exposure of the starch granule and its fragments, the greater the enzymic susceptibility.

*Salt* delays the beginning and reaching of the peak gelatinization of flours by 5 to 10°C.
*Huber, H.* (1964) *Brot und Gebäck*, February issue (German).

*Salt and acidification* in the correct quantities gives rye-flour similar properties to that of wheat-flour, allowing optimal baking properties.
*Huber, H.* (1964) *Brot und Gebäck*, February issue (German).

*Products with smaller cross-section*, e.g. bread-rolls and baguettes, reach 99°C at the crumb-centre within about 8–10 minutes, leaving a further 20 minute exposure, resulting in almost complete starch gelatinization.

*Panned units of 1 to 2 kg* experience limited starch-pasting only, hence, many granules, apart from slight thermal deformation, remain in the crystalline form. This is confirmed by the retention of the birefringent (German: *doppelbrechenden*) interference-cross, when viewed under polarized-light microscopy.

*High-frequency (HF) baking techniques*, using VHF (17·7 MHz) energy as a two-stage baking process, results in the dough-piece remaining in the gelatinization-zone for about one-third of the time required for conventional baking. Thus, any defects normally encountered with amylolytic enzymes are avoided. VHF baking permits the processing of flour with abnormally high alpha-amylase activity without the usual crumb-defects. Flours with Falling numbers as low as 95 can be baked without showing the conventionally baked sticky-crumb symptoms. Any shortening of the total baking-time must take place during the initial-phase, since the period after passage through the gelatinization-zone (55–80°C) is essential for the formation of flavour substances and crust-firming.

*Starch-swelling*, allows the amylases to start hydrolysing the substrate, the reaction rate increasing by about 50% every 10°C rise in temperature, remaining active up to 75°C in some cases.

*Starch degradation*, can be due to amylolysis (enzymic), or temperature within the range 100 to 140°C. In both cases dextrins and some maltose are formed, the dextrin fraction increasing by about 15% during baking with flour of normal quality. The rapid inactivation period for cereal alpha-amylase, within the temperature range 68–83°C, is usually traversed within about 5 minutes baking-time.

*Beta-amylase*, is the saccharogenic enzyme, capable of degrading susceptible damaged starch substrates whose free surface has been exposed to attack. Under comparable conditions, the effect of beta-amylase on various starch-substrates yields the following amounts of maltose (in mg): wheat starch (intact) 0·43; dextrin 144·0; gelatinized wheat starch 148; coarse starch fraction 8·4; medium starch fraction 18·0; fine starch fraction 44·0; starch damaged by grinding 127.

*Glasunow, I. W.* (1938) *Biochimija Chlebopetschenija (BCh)*, **1**, p. 51 (Russian).

Glasunow demonstrated that beta-amylase produces 335 times more maltose from dextrin than from intact wheat starch, and even more from gelatinized wheat starch. The relative influences of starch particle-size, and damage by grinding is also clear from his data. Beta-amylase is capable of degrading damaged, susceptible starch during mixing, increasing flour sugars from about 1·5–2·0% to 2·5%. If the dough is bulk fermented (given a floor-time), the maltose level will increase further. During baking, starch degrading and saccharification will continue up until the inactivation range of 57 to 72°C, which lasts for about 3 minutes only.

*Starch components* of a typical type 550 (0·55% ash) flour, with 90% particles less than 90 μm diameter, consist of a wide selection of damaged and undamaged granules with equally wide ranges of accessibility and susceptibility.

*Dextrins* are formed within the dough-piece mainly in areas of limited moisture, i.e. the crust-zone, from which, moisture is constantly being removed. Some of these dextrins take part in non-enzymic, non-oxidative caramelization reactions at 140–150°C, contributing to crust formation.

*Caramelization* results from the transformation of sugar-like substances and sugars, under the influence of heat into compounds which vary in colour from yellow to brown. Their flavour ranges from a pleasant caramel, to bitter and carbonized. Some are complex polymers, formed at 150 to 190°C, condensation products and anhydrides.

*Proteins during baking* on reaching temperatures of 50 to 70°C lose their helical structures, beginning to uncoil because of the weaker bonds of the network being broken. Owing to the colloidal nature of protein molecules, they imbibe water, form hydrophilic and hydrophobic bonds, binding sufficient water for starch gelatinization. This gluten-protein binding of water represents about 30% of total dough absorption water, the resulting hydrated protein matrix having small starch granules embedded in it. When the crumb temperature reaches about 60–70°C, protein denaturation takes place, leading to eventual coagulation at about 90°C (see Fig. 109). At about 74°C, thermal denaturation already transforms the gluten network films, enclosing the gas vacuoles, into a more rigid structure, and this, together with the swollen starch, forms a firm but elastic crumb. Thermal expansion of the gas-cell membranes results in the viscous starch granules within the membrane-wall elongating, the gluten films becoming progressively thinner, eventually rupturing. However, before this occurs, starch-swelling and rigidity has taken over the structural role. At the protein denaturation temperature range 50–70°C, the molecules release most of their bound-water, and in so doing, become transformed from a viscoelastic state into that of a plastoelastic solid.

*Protein solubility* changes during the dough-piece temperature rise from 20°C to 90°C were researched by Popaditsch and Falunina in the USSR in 1962.

*Popaditsch, I. A., and Falunina, S. F.* (1962) *Chlebopekarnaja i konditerskaja promyschlennost (ChKP)*, **2,** 1. Influence of heating-up of fermenting wheat-dough on the activity of the proteolytic enzymes.

Their conclusions were that the water-soluble fraction of dough-proteins change little from dough to bread, between 20 and 90°C. However, the fraction soluble in 0·1N acetic acid, expressed as soluble nitrogen, decreased from 14·2 mg/g dry matter at 20°C to 6·1 mg/g dry matter at 90°C. This is a measure of the protein denaturation process of the gluten-forming proteins. Significantly, within the temperature interval 60 to 70°C, the water-soluble fraction increased from 5·4 mg/g dry matter to 7·1 mg/g dry matter, owing to the activity of the proteolytic enzymes. Within the same temperature interval, the 0·1N acetic acid soluble dough-protein fraction decreased from 13·8 mg/g dry matter to 8·9 mg/g dry matter, falling to 6·1 mg/g dry matter on reaching 90°C. Thus confirming a progressive denaturation of the gluten-forming proteins. Although the critical temperature range for proteolytic enzyme inactivation is between 60 and 70°C, traces of its activity were detected in the crumb as high as 85–90°C. The Fig. 111 schematic indicates the dough-protein cleavage sites, designated ---E, at which the dough-proteinases become active during the baking process.

*Protein-coagulation and starch-gelatinization* together form the porous structure of the baking dough-piece, and finally that of the loaf-crumb. The rheological properties of the bread-crumb are quite different from that of the dough from which it was transformed.

*Deformation modes of the baking loaf-crumb* of wheat and rye breads of various weights, from flours of various extractions have been studied by several researchers during the period 1959–69 in Hungary, the USSR and the GDR (see Figs 110 and 112).

*Auerman, L. J., and Melkina, G. M.* (1967) *NTInf.,* **15,** 6 (Russian).

These workers used specialized equipment to study the changes in bread-crumb elasticity at 10- or 20-minute intervals during baking for a total time of 60 minutes. Other parameters measured were: plastic deformation; and elastic deformation expressed as the difference between the total and residual deformation; relative-elasticity being expressed as a percentage of total deformation. Bread samples of various weights, and from flours of various extraction-rates were included in these studies. Graphical plots of relative-elasticity% versus baking-time, indicate certain correlations between flour extraction rates and relative crumb-elasticity of both wheat- and rye-flour breads. The higher extraction flours show relative-elasticity rising to a maximum limiting value, and then falling-off, whereas the lower extraction flours show continuous rises in relative-elasticity of the crumb during the baking process. This phenomenon is considered to be a function of enzymic content of the flours, involving both the starch/amylase, and the protein/proteinase complexes during the warming-up period. Other covariables are: dough water content, dough pH, and the temperature and rapidity of the heat-exchange process.

*Dough water absorption*, when increased, tends to lower the temperature optimum of the flour enzymes.

*Speed of warming-up* of the dough-piece is crucial to enzyme-inactivation. The more rapid the warm-up, the higher the inactivation temperature of the enzyme complexes.

*Popaditsch, I. A., and Falunina, S. F.* (1962) *ChKP,* **2,** 1 (Russian). The influence of the heating-through of the fermenting wheat dough on the activity of the proteolytic enzymes.

Results obtained by these workers indicated that for a wheat dough of 48% water-content, at pH 5·8, the temperature optimum for the wheat proteinases lies between 60 and 70°C, always depending on the dough warming-up process. Increasing the water-content of the dough to 70% had the effect of lowering the temperature optimum of the enzymes to 50°C. Popaditsch and Falunina found that the nitrogen fraction soluble in 0·1N acetic acid, indicative of dough-protein denaturation at different temperatures, decreased progressively from 55·9% at 20°C to 24·0% at 90°C.

*Proteinases* are proteolytic enzymes capable of degrading true, complex protein molecules into lower-molecular-weight derivatives, i.e. polypeptides, peptides and peptones, attacking the CO—NH peptide linkages. Peptidases then complete the hydrolysis to the amino acid stage. Both proteinases and peptidases occur naturally in wheat, being mainly concentrated in the germ and bran fractions. Endopeptidases attack the internal sites of the protein chain, whereas exopeptidases attack peptide linkages joining terminal amino acid residues, thus producing amino acids as end-products. Cereal proteinases appear to be most active at pH 3·8 to 4·5, but this varies with temperature and the substrate-structure. Their activity is difficult to measure with any accuracy, but the most widely used procedure is to prepare an extract and a suspension from about 10 g of sample with a 0·1M acetate buffer at pH 3·8. After centrifugation, the proteinase-activity is determined in both the extract and the residual suspension by reaction with a suitable substrate, e.g. haemoglobin, azo-dye derivatives of gluten or casein. After precipitation with trichloroacetic acid, the soluble-nitrogen released can be determined according to Kjeldahl, or determined colorimetrically using Folin–Ciocalteau reagent. A suitable calibration-curve being prepared with appropriate dilutions of a standard tyrosine solution. Proteinase activity is expressed in tyrosine-equivalents per gram of sample. Reagent-blanks and about 8–10 parallel samples of the test material are recommended for precision-improvement.

*Loaf-crust formation* due to 'browning' effects are both enzymic and non-enzymic processes. Enzymic ones are due to the formation of the light-yellow to black coloured dextrins. Non-enzymic browning results from the Maillard and melanoidin reactions, of great importance for the synthesis of colour and aroma substances, from carbonyl and amino acid groups. The high-molecular-weight melanoidins forming from reactions between sugars and nitrogenous substances. Typical nitrogen sources can be amino acids, polypeptides, proteins, amines or ammonium salts. Sugars can also be replaced by similar compounds with reactive aldehyde groups. Reaction speed of melanoidins will depend on the control of temperature, pH, moisture-content of the dough, and the adequate presence of free amino-groups and aldehydes in a reactive form. Longer and more intensive mixing-times will open up and expose the free reactive groups of the protein molecule. Careful control of baking time, heat distribution and heat-transfer, ensure the optimization of melanoidin formation and loaf browning. For the formation of a good crust colour adequate sugars and protein, coupled with the presence of humidity (injected-steam) are essential. These conditions are ideally achieved by dough fermentation with careful control of temperature, time, dough-consistency, dough-pH, and balanced formulation of ingredients.

*Bread flavour and aroma*, is also the result of the formation of melanoidins, often referred to generally as the 'Maillard reaction'. The many possibilities of reactions between the various nitrogenous groups and free reducing sugars have been the subject of much research, mainly concerning the complex pathways and intermediates involved in forming the actual flavour compounds.

Hodge, J. E. (1953) *Agr. Food Chem.*, **1**, p. 928.
Hodge, J. E. (1955) *Carbohydrate Chemistry*, **10**, p. 169.
Namiki, M., and Hayashi, T. A. (1983) In *The Maillard Reaction in Foods and Nutrition*, edited by G. R. Waller and M. S. Feather, ASC Symposium Series 215, ASC, Washington, DC, USA.
Lane, M. J., and Nursten, H. E. (1983) ASC Symposium Series 215, Washington, DC, USA.

*Melanoidin reactions of the Maillard system*, according to Hodge, involve initially reducing sugars and amines condensing to form addition compounds, i.e. amine-group and carbonyl-group from the sugar. This is followed by removal of one molecule of water (dehydration), to form an N-substituted glycosylamine. Isomerization then occurs (Amadori rearrangement), forming 1-amino-1-deoxy-2-ketones, which remain in equilibrium with their enol forms (1,2-enol forms). All reaction products up to this point are reversible, and colourless. The intermediate products of Amadori rearrangement then undergo various reactions depending on prevailing conditions. Under neutral or acid pH, they can either lose water to form a ring compound (Schiff's base) hydroxy-methyl-furfural or furfural, or eliminate an amine group to form a methyl alpha-dicarbonyl compound. The former are colourless reductones, and the latter brown fluorescent substances. Alternatively, carbon dioxide can become split off from amino acids to form an aldehyde (a Strecker degradation process), the latter being either colourless or yellow. Final reactions involve dehydration, resulting in the formation of dicarbonyls, which then condense or become polymerized, forming various melanoidins varying in colour, taste and aroma intensity.

*Thomas, B., and Rothe, M.* (1960) *Baker's Digest*, **34**, 4, p. 50, studied model systems involving the Maillard reactions to establish a correlation between starting materials and end-products. Reactivity of the monosaccharides and amino acids in the Maillard reaction are variable, and can be detected by the relative increase in furfural content and crust-browning. The most reactive sugar is the pentose xylose, and the most reactive amino acid isoleucine and leucine. Maillard reactions are more significant for bread-aroma formation than caramelization-reactions, the latter only taking place at elevated temperatures. Pyrazines, as bread-aroma components have been widely reported. Possible reactants for their formation are amino acids, diketones and ketoaldehydes. Many aldehydes formed in the crust diffuse into the crumb interior of the loaf during cooling, and become absorbed. Flavour substances can also become locked into the amylose starch fraction during cooling.

*Full bread flavour development* depends on the availability of free, reactive amino acids, sugars and alcohols. Their presence is a function of enzyme activity, and the degree of polymerization and hydration of their substrates from mixing to baking. A good balance between starch/amylase, and protein/proteinase activities, especially in the case of wheat-flours will produce an end-product of appetizing flavour. Intensification of flavour and aroma can be achieved by: 2 hours or more final-dough fermentation in bulk; the use of solid or liquid pre-ferments (sponges or brews); the application of biotechnology, by the careful addition of specific enzymes; or the use of a biologically active mixture to stimulate yeast activity and fermentation. Most important of all, however, is oven control, and the formation of at least 3 mm of crust in the case of a 1-kg long-rolled hearth white loaf, and at least 4·5 mm of crust for a 1-kg rye or rye/wheat, hearth-baked, long-rolled loaf.

*Volatile flavour and aroma components*, such as alcohol, diacetyl and iso-aldehydes tend to progressive evaporation during baking. Less volatile flavour substances, e.g. iso-alcohols, acetic acid, pyruvic acid, acetoin, butylene glycol, ethyl lactate, acetaldehyde, iso-valeric aldehyde and furfural (formed during baking with pyruvic-aldehyde, the latter being more volatile), tend to be retained, and diffuse through the crumb during cooling, becoming either absorbed, or locked into the starch amylose fraction.

*Low-volatility flavour and aroma components*, which include the melanoidins formed within the crust, dihydroxyacetone, ethyl succinate, lactic acid and succinic acid, are components which form the basis of the typical bread flavour and aroma.

*Fermentation duration and the use of pre-ferments* encourage the formation of flavour and aroma substances in bread.

*Rothe, M., et al.* (1972) *Nahrung*, **16**, 5, p. 507. Shortened process for white bread and its influence on bread aroma I. Carbonyl compounds and alcohol (German).
Rothe *et al.* compared the influence of different dough-processes on the alcohol content of the resultant bread. In general, during the first 20 minutes of baking, there is a rapid loss of alcohol within the chamber, falling from over 500 mg/100 g bread to about 400 mg/100 g. Using conventional mixing and fermentation procedures, about 31% of the alcohol produced during fermentation was found to have evaporated during baking, compared with about 60% when using intensive-mixing and shortened processing. Other volatile alcohols such as: *n*-propanol and Iso-butanol also disappear at similar rates, depending on the dough-process.
*Rothe, M.* (1966) Bread aroma determination, *Brot und Gebäck*, **20**, 10, p. 189 (German).
*Rothe, M.* (1974) Bread aroma. In *Handbook of Aroma Research*, Berlin: Akademie-Verlag, Berlin, 1974 (German).

The significance of longer fermentation-times was confirmed by Coffman, who isolated much higher levels of amyl alcohol from bread made by a straight (direct) processed dough, than could be obtained from bread prepared from a continuous, sponge or short-time dough process.

*Coffman, R.* (1964) *Cereal Science Today*, **9**, 7, p. 306. The aroma of continuously mixed bread.

It is the author's own experience that straight-processed doughs made to stand for 10–12 hours overnight, compared with those processed over 2–3 hours, show much higher concentrations of aroma substances in general, amongst which the predominance of amyl alcohol in the fermentation gases is very marked. However, on baking into bread, the sensory perception of both flavour and aroma appears to change in favour of volatile organic acids, acetic acid predominating, with background notes of lactic and succinic acids. The diversity of bread flavour and its complexity is made clear by Coffman in a 1967 publication.

*Coffman, J. R.* (1967) Bread flavour. In *The Chemistry and Physiology of Flavours*, edited by W. H. Schulz., E. A. Day and L. W. Libbey, AVI Publishing Co., Westport, CT, 1967.

*Kretovich, V. L., et al.* (1988) *Prikl. Biokhim. Mikrobiol.*, **24**, 2, p. 207–10, at the Bach A. N. Institute of Biochemistry in Moscow, observed appreciable losses in free lysine, leucine, valine, and the monosaccharides arabinose, xylose, galactose and glucose during the baking of rye bread, owing to involvement in melanoidin reactions to improve bread flavour and aroma. Enrichment of the doughs with condensed, fermented whey at 9% increased bread amino acid levels by about 20%. The enriched bread containing much higher levels of isopentanal and isobutanal, which are formed from leucine and valine respectively, together with other aroma compounds, than was the case with non-enriched bread.

*Johnson, J., and Linko, Y.* (1964) Analysis of bread-flour components. *Qualitas Plantarum et Materia Vegatabilis*, **11**, 2, p. 256, draw attention to the significance of the breadmaking process on the formation of carbonyl compounds in the crust and crumb. Their results indicated that the use of a pre-ferment-sponge provided the highest level of carbonyls generally (crust 29·8, crumb 3·0 mg/100 g). The liquid pre-ferment produces 25·7 and 4·0 mg/100 g in the crust and crumb respectively. However, the levels of furfural, hydroxymethyl-furfural, acetone and proprionaldehyde were all significantly higher in the pre-ferment-sponges.

*Carbonyl compound formation* in bread is strongly influenced by the baking procedure. The bread variety pumpernickel, which has a very long maturing-time, and a prolonged baking-time in pans at relatively low temperatures, contains far more furfural in the crumb than other bread varieties baked from the same flour.

*Flour-type and extraction rate* also influence the formation of furfural, acetaldehyde, and other aldehydes, as established by Rothe and Thomas.

*Rothe, M., and Thomas, B.* (1965) Differences in aroma profile of rye-bread, wheat-bread and mixed wheat/rye bread. *Ref. Konf. Miedz. Zytn.*, **4** (German).

*Less volatile alcohols*, amyl and iso-amyl alcohols, iso-butanol, iso-propanol and propanol, although only present in trace amounts, have a potent influence on the flavour profile of wheat breads.

*Products of yeast fermentation*, such as acetyl methyl carbinol and dihydroxyacetone, act as precursors in the formation of flavour components of fresh bread. On oxidation, the former yields diacetyl, but straight doughs are deficient in acetyl methyl carbinol compared with sponge and liquid pre-ferments, and diacetyl formation depends on the presence of adequate concentrations of sugar and oxidants. Dihydroxyacetone reacts thermally with proline yielding substances associated with crust-odour.

*Wiseblatt, L., and Zournut, H.* (1963) *Cereal Chem.*, **40**, p. 116.

*Organic acid formation* during fermentation, especially acetic and lactic acids help to catalyse the rearrangement of the condensation products formed from amines and reducing sugars in the Maillard reactions during baking.

*Measurement of bread-crust surface colouration* continuously during the baking-cycle has been developed in the USSR.

Telitschkun, W. I., et al. (1968) Colouration of the bread-surface during baking. *New Developments in Oven Technology*, Zinti, Moscow, p. 47 (Russian).

The change in light intensity reflected from the bread-crust during baking is continuously monitored and recorded with a potentiometer. |A beam of |light impinging on the dough-surface is reflected, and is measured on the scale of a photocell. Colour-intensity changes are measured as percentage reflection during baking, concurrently with loaf-surface temperature, both parameters being plotted against baking-time. On setting in the oven, the reflectivity of the dough-piece falls rapidly owing to surface starch-granule gelatinization, reaching a minimal reflectivity coefficient (%) when the surface-temperature reaches 60°C, i.e. the starch gelatinization range. Further temperature rises result in progressive increases in the reflectivity coefficient, as the surface becomes lighter in colour owing to drying out. On reaching 100°C, the melanoidin reactions commence, the crust taking on a progressively darker colour, causing the reflectivity coefficient to fall sharply. On reaching the range 130 to 170°C, loaf surface colour changes from light-yellow to dark-brown, and above 170°C progressive carbonization occurs. The application of this equipment allows the optimal conditions for an ideal crust-colour to be ascertained, and then controlled, which contributes to the full automatic control of the baking-process.

Rooney, K., Salem, A. and Johnson, J. (1967) *Cereal Chem.*, **44**, p. 539. Studies of the carbonyl compounds produced by sugar–amino acid reactions I. Model systems.

Working with sugar/amino acid mixtures in buffered solutions at pH 5·5, heating at 95°C for 12 hours, these researchers identified various carbonyl compounds produced from the sugars: glucose, maltose and xylose. It was found that xylose gave the most intensive colour with amino acids in the melanoidin reactions, measured at 500 nm. This correlates with the more intensive crust and crumb colour development in rye bread versus wheat bread, owing to the presence of larger concentrations of pentoses. Also, pentose sugars are not fermented by baker's yeast strains. Aldehyde carbonyls formed during these reactions relate to the amino acid present, rather than the type of sugar participating. However, the type of sugar has a strong influence on crust-colouration.

*Research into Maillard-browning*, as applied to bread formulations, requires careful experimental design on the part of the researchers. In the case of wheat bread, a straight-dough process with added saccharose at 3·0% is appropriate to provide adequate basic substrate for the yeast, coupled with an Arkady-type improver. The dough can then be fermented for 2 hours in bulk, allowing a 20-minute recovery, followed by final moulding and proofing for 55 minutes, baking at 220°C for 30 minutes. The amino acid variable can be premixed with the standard sugar in solution form (0·02M amino acid:0·02M standard sugar), being added to the formula prior to mixing. Changes in crust-colour can be measured with a reflectometer, similar to that designed by Telitschkun *et al.*, cited previously. A green filter can be utilized, and readings taken from three loaves of a sample batch, using an average of 10 readings from each loaf. Isolation and identification of the carbonyls, as 2,4-dinitrophenyl-hydrazine (DNPH) derivatives, can be carried out using GLC, TLC or PC techniques. Quantitative estimation can be accomplished by extraction and UV-absorption, expressing data as milligrams of free carbonyl/100 g of crust or crumb. Furfural concentrations in both crust and crumb, can be determined by the procedure of Linko.

Linko, P. (1961) *Anal. Chem.*, **33**, p. 1400–3.

Hydroxymethylfurfural concentrations can be determined by extraction from the ground crust or crumb with benzene, followed by centrifugation. Combined concentrations of these aldehydes can also be determined by the method of Linko. These aldehyde Maillard reaction products, however, only represent intermediates, which, in the presence of available free amino acids, undergo further condensation forming polymers. Therefore, reduced levels of the aldehydes can be expected where free amino groups persist, and their concentrations are lowest in bread-crust with otherwise large amounts of carbonyl compounds. Although bread flavour and aroma can be modified by amino acid addition to formulations, the optimal end effect of the total flavour experience requires a balanced ratio of a complex blend of aromatic compounds, their simultaneous effect on the sensory organo-receptors, and the transmission of their impulses to the central nervous system.

*Routine bread flavour evaluations* tend to fall into the sensations of being: sweet, salty, sour, and bitter. Sour tastes are due to the presence of various organic acids. Lactic acid is the most important contributor to taste, followed by malic, tartaric, acetic and citric acids. Acidic odours are due to acetic acid, followed by lactic, malic and tartaric acids. Acetic acid gives a pungent impulse to the sensory organs of smell, requiring careful control when wheat and rye bread is prepared from liquid-yeast cultures, firm or liquid sours, and pre-ferments. Appreciable differences in taste and smell can be detected by a trained taste-panel using the Triangle-test. This requires the panellists to identify the odd sample from three, two of which are identical. Its success depends on the sensory-amplitude of detection or sensitivity of the members of the panel.

*Accurate quantitative analysis* of the components of taste and smell are conducted by placing the product in a flask and freezing with liquid-nitrogen at $-196°C$, under a vacuum of 1 mm mercury. The sample is then allowed to warm up to room-temperature, and the volatile liquid condensed. The condensate is then subjected to GLC for identification, and quantitative determination of the volatile components.

*Breadmaking process evaluations versus bread flavour and aroma* demand an exact determination whether any derived improvement in flavour and aroma substances take place during the pre-ferment, dough or baking stages. Typical process variables are: solid-sponge, liquid pre-ferment, straight-dough, short-time mechanical/chemical maturing procedures; rye sour-dough processes using three-stage sour-builds, Berliner short-sour, or continuous liquid sour-dough processes. Examples of research procedures involving such studies are the following:

*Johnson, J., Rooney, L., and Salem. A.* (1966) *Advances in Chemistry*, **56**, p. 152. The chemistry of bread flavour.

*Rothe, M.* (1970) Influence of technology on the flavour of rye bread. Report 5: World Grain and Bread Congress 1970, Vol. 5, p. 203 (German).

*Rothe, M.* (1965) Differences in the aroma content of rye bread, wheat bread and mixed rye/wheat bread. *Ref. Konf. Miedz. Zytn.*, **4** (German).

*Johnson, J., and Miller, B.* (1956) *J. Agric. and Food Chem.*, **4**, p. 82. The effect of fermentation time on certain chemical components of pre-ferments used in breadmaking.

*Rubenthaler, G., Pomeranz, Y., and Finney, K. F.* (1963) *Cereal Chem.*, **40**, 6, p. 658–65. Effect of sugars and certain free amino acids on bread characteristics.

*Hunter, J., et al.* (1961). *J. Food Sci.*, **26**, 6, p. 578–80. Volatile organic acids in pre-ferments for bread.

*Tokarewa, R. R., and Kretovich, V. L.* (1963) *Tr. Tsentr. Nauchn. Issled. Inst. Khlebopekar Prom*, **9**, p. 95–9. Effect of enzyme preparations on the accumulation of reducing sugars during dough-handling and on the aromatic complex of bread.

*Tokarewa, R. R., and Kretovich, V. L.* (1963) *Acid. Sci. USSR, Moscow Proc. Internat. Congr. Biochem.*, 5th Moscow, 1961. The use of concentrated enzyme preparations from fungi in breadmaking.

# CHAPTER 3.3 ENERGY SOURCES, TYPES OF OVEN AND OVEN DESIGN

*Heat-energy sources* for the baking of bread can be derived from: wood, brown coal briquettes, peat, fuel oil, natural gas, manufactured-gas (town or city gas), anthracite, coke, bituminous-coal (hard/mineral or pit coal), or applied as electro-energy. The net heating effect of these energy-sources is similar, choice depending on local availability and cost. In areas remote from industry, wood, peat or mineral-fuels could be the most economic choice. The final choice will depend on which resource is naturally available, and economically viable, also its relative efficiency and ease of control in the bakery. In many European countries, natural gas has become the final choice for both the tunnel-ovens in industrial bakeries and the batch-ovens used in smaller bakeries.

*The relative efficiency, or exploitation, of energy sources* varies considerably. Coal-heating is about 20%, gas 50%, being maximal in the case of a directly heated electrical oven, which is 85% efficient.

*Calorific value*, is a measure of the heating-capacity of any energy-source which produces heat of combustion. It is defined as the quantity of heat in kilojoules, liberated by burning 1 kg of a solid/liquid fuel, or 1 m$^3$ of gaseous fuel.

*Heat of combustion and calorific value*, are differentiated by the liberated heat of condensation when steam condenses.

$$\text{calorific value} = \text{maximum heat of combustion} - 2260\,WA$$

where WA is the water-vapour in the exhaust-gas in kg/kg of fuel, 2260 being the heat of condensation in kJ water/kg$^{-1}$ (nearest whole number). The appropriate units are: kJ m$^{-3}$ for gaseous fuels, and kJ kg$^{-1}$ for solid fuels. Examples of calorific values and heats of combustion are: natural gas (90% methane) 33 600 and 37 400; city gas 14 250 and 15 950; gas from brown-coal 15 500 and 16 800; fuel oil 42 000 and 44 100.

*Heat production for baking*, depends on the following basic principles:

*Exothermic reactions* involving the burning of coal, natural gas, manufactured gas, or fuel oil, the oven-chamber walls dissipating heat-energy as radiant-heat and conductive-heat into the chamber atmosphere.

*A mains power-system* heating an electrical resistance, controlled by a temperature sensor, emitting diverse spectral waves, extending into the bright infrared range, amounting to over 40% relative to solar-energy at 100%.

*High-frequency (HF) electromagnetic waves*, including wavelengths within the bright and dark infrared range between $3 \times 10^{11}$ and $4 \times 10^{14}$ Hz, maximum emissive-energy of the latter being at about $3 \times 10^{-4}$ cm, representing only 20% compared with solar-energy, owing to the longer wavelength. The bright, shorter wavelength infrared frequencies penetrate deeper into the dough-piece, shortening the baking-time.

*Dielectric heating* involving the application of a high-frequency electromagnetic field, whereby the dough-piece is subjected to a rapidly alternating electrical field, generated by a magnetron (HF generator). The dough-piece acts as dielectricum between HF-electrodes, which act as capacitor-plates, thus setting up an electric-dipole system. The rapidly alternating field results in a high frequency oscillation of the dipolar water molecules, causing the product to heat up uniformly from within, instead of relying on an external heat-flux and penetration. Equipment for dielectric heating operate within the frequency ranges 10 kHz to 500 kHz, or 13 to 27 MHz. The oven-mesh or steel-band of a tunnel-oven can serve as the lower electrode for continuous-baking processes. Since HF/dielectric baking cannot produce a crust, it must be combined with conventional radiation. The most effective combination is HF/IR radiation, as suggested by Professor E. Maes.

Maes, E. (1964) *Fermentativ*, **26**, 5. In *Bulletin de l'Ecole Official Meunerie Belge*, 1964, p. 203 (French).

Using the HF/IR oven type 'Reforma', Maes found certain advantages compared with conventional ovens, viz. 50% reduction in baking-time, a 6–7% increase in loaf-volume, and crust control by using IR-radiation units. However, certain inherent disadvantages have to be overcome. Recent research by Salovarra *et al.* in Finland has shown the versatility of its application for baking high-amylase flours.

*Salovarra, H., et al.* (1988) *Acta Alimentaria,* **17**, 1, p. 67–76 (English). High-amylase flour in baking with high-frequency (VHF) energy (EKT Dept. Food Tech. Univ. Helsinki, SF-00710 Helsinki, Finland).

*Fuel-efficiency* depends on its purity, the relative weight of carbon and hydrogen it contains (hydrogen content increases its calorific value), and an adequate supply of oxygen to ensure optimal combustion.

*Heat-energy utilization,* can be maximized by careful oven design, and control of unnecessary sources of heat-loss, e.g. unrestricted feed and discharge-point opening. Typical examples of net heat-utilization for a wire-band tunnel-oven are: heat-energy input 100%, net heat-utilization 45%; exhaust-gas losses 35%, humidity heat-loss 12%, radiation losses 8%, according to Schwate and Ulrich.
*Schwate, W., and Ulrich, U.* (1986) In *Special Processes in Baking—A Process Manual,* VEB-Fachbuchverlag, Leipzig, GDR.
A more detailed breakdown of heat losses is given by Schneeweiss and Klose.
*Schneeweiss, R., and Klose, O.* (1981) In *Technology of Industrial Baking,* VEB-Fachbuchverlag, Leipzig, GDR.

*Theoretical input levels of heat* required to bake a 1-kg loaf have been cited as 530 to 630 kJ kg$^{-1}$, and for a 1-lb loaf as 325 to 400 Btu. Such data is only a guideline, and will depend, amongst other variables, on the specific-heat levels of the dough, e.g. 2·6–3·0 kJ kg$^{-1}$ K$^{-1}$ for lean doughs, and 0·65–0·88 for enriched lower density formulations.

*Actual heat-input* measurements indicate that ranges of 330 to 400 Btu per pound of bread are more realistic, compared with 1000–2000 Btu for the older coal-fired brick-ovens using a peel.

*Oven capacity* is often expressed by the baker as pounds, or kilograms of bread per hour, whereas the oven-manufacturer may express it as pounds of dough per square foot of baking-surface per hour, or more simply as baking-surface in square metres.

*Relative oven heat-utilization* guidelines, expressed in fuel/100 lb bake and Btu/lb of bake for the various types of oven have been compiled by Harrel and Thelen.
*Harrel, C. G., and Thelen, R. J.* (1959) In *Conversion Factors and Technical Data for the Food Industry,* edited by Harrel and Thelen, Burgess Publ. Co., Minneapolis, MN.
Their data clearly indicate that direct-fired tray-ovens produce the highest heat-utilization per pound of bake, irrespective of the choice of fuel, and that electro-energy offers the lowest Btu/lb of bake for all common oven-types. Manufactured gas and natural gas being the next best all round choice of energy-source for most ovens.

### 3.3.2 Types of Oven and Oven Design
*Choice of oven type* will depend on such considerations as: bakery production-throughput capacity, product diversity, available floor-space, energy source, economy in operation, construction and maintenance.

*Early commercial oven designs* were known as side-flue peel-ovens, or peel brick-ovens, consisting of large chambers of brick construction. The floor or 'sole' being of large square refractory tiles, and the crown arched, with the fire-place or furnace (long and narrow in shape) positioned obliquely to feed the flames over the crown, and around the chamber, before passing out through the flue to the left of the oven-mouth. A combination of coal and coke gave the best heating effect, the latter requiring a deeper fire-bed, and more draught or secondary-air supply. Their massive construction, integral with the bakery, allowed the storage and delivery of a solid bottom heat, and a steady radiant top heat. Such conditions were conducive to an excellent bake, producing bread with an appetizing crust and crumb flavour and aroma. Operation of these ovens, however, demanded skill and experience, coupled with good judgement on the part of the master-baker, to achieve consistently even-baked products. Regular stoking and raking of the fire-bed ensured an efficient combustion, without any cold-air blowing through. The baking technique was to attain a temperature of 550°F (288°C) by constant firing, setting the batch in the oven with long-handled wooden 'peels' of various sizes as rapidly as possible. After setting, the pyrometer would normally register about 500°F (260°C), owing to the presence of the relatively cold load. After a residence-time of about 1 hour, the bread was unloaded with the peels. During this time, the oven conditions had reached a 'steady-state' due to the stored heat, having a temperature suitable for baking smaller fermented products and confectionery items. The early German ovens were similar in design, depending on direct heat-transfer from the fire-place to the oven-chamber and brickwork. Such ovens depended on heat-storage, the temperature of the tiles and brick walls falling as the products are baked. Steam retention was often achieved by building the hearth or sole with a slight gradient. The high initial temperature, and even heat release made these ovens ideal for baking all types of rye and wheat breads. The introduction of the drawplate-oven, whereby the hearth could be withdrawn from the chamber on wheels running on tramlines, allowed all products to enter and leave the chamber simultaneously, instead of the first-in/last-out situation of the side-flue peel-oven. However, these ovens required a large work area, and suffered a large heat-loss at loading and unloading. The invention of the Perkins steam-tube led to the development of the steam-oven. Such ovens depended on heat-transfer through a liquid medium, with which the tubes or pipes were filled. The tubes were located beneath each chamber floor (sole), running parallel, about 10 tubes serving each deck. The tubes were usually heated by solid fuel or gas, extending into the furnace to acquire the initial conductive heat.

*Modern ovens* can be classified into batch-ovens, which are suitable for small to medium size bakeries, or industrial bakeries with a specialized product range, and larger capacity continuous industrial baking ovens.

*Batch-ovens* depend on a hot-gas circulation system (cyclothermal) from electrically heated elements. Very few ovens are still directly fired with solid fuels, or indirectly heated with the Perkins steam-tube. However, one should not assume that older oven-heating technologies are completely obsolete, since the end-product quality from such ovens was excellent, although more variable in character. Independent of the heating principle involved, most batch-ovens (Fig. 113) tend to be multi-deck in design, with the exception of rack-ovens, which employ racks with trays, either fitted with wheels or running on tracks. Typical multi-deck ovens have 4–5 decks, with total baking-surfaces of 9 to 19 m², the temperature of each deck being independently controllable, enabling the baker to bake various products simultaneously. Such ovens allow throughput capacities of 120 to 240 1-kg long loaves, or 800 to 1600 round bread-rolls at 50 g. Using oven 'setter' equipment (German: *Abrollapparate*), 1-setter width 600 mm, 2-setter width 1200 mm, 3-setter width 1800 mm, and 4-setter width 2400 mm, loading can be rationalized.

*Reel-ovens*, the first mechanical oven design, are of American origin being still found in bake-shops in the USA. They were built like a ferris-wheel, the reel-structure revolving vertically about a horizontal axis within the baking-chamber, supporting baking-trays, hung from hanger-pins protruding from the large wheels inside the oven.

*Rack-ovens* offer a more expedient solution for the batch production of a diverse range of products. They consist of a vertical chamber into which the special racks are wheeled, carrying up to 100 trays of product. Usually, the racks are clamped onto a turntable, which rotates during the baking cycle, thus ensuring convective heat-transfer to the products. The Static Radiant Rack Oven manufactured by Gouet of France (marketed in the UK by EPP Ltd of Banstead SM7 2NT, UK), produces very solid heat characteristics, and much less heat loss, thus combining the baking quality of a deck-oven with the ease of operation of a rack-oven. This oven is energy efficient, the oil-filled radiators above and below each tray of products, providing a unique convection system, heating from cold to bake within about 20 minutes. The thermal heating fluid is circulated and heated by remote heat-exchanger and pump unit. Steaming facilities are also provided. Such ovens are very popular for in-store bakeries, and craft bakeries with a mixed trade.

*Industrial-baking ovens*, are designed for continuous production, and are often referred to as 'travelling-ovens'. The travelling *tray-ovens of single-lap or double-lap design*, are illustrated in Figs 114 and 115 respectively, being a development of the reel-oven in the USA.

*Single-lap models* provide one passage through the oven's two heat-zones, as shown in the schematic, loading and unloading taking place at the front end. With automatic loading and unloading equipment added, this oven increased in popularity, the single-lap design offering a simple operational performance with one horizontal run, and a lower crown for steaming. Modulation heat control ensures lateral heat control and rapid response to loading changes. The number of trays vary from 18 to 74, production capacities ranging from 2500 to 8000 lb bread/hour.

*Double-lap models* are based on the same design principle as the single-lap, but the trays travel through four heat-zones instead of two. The vertical crown height is twice that of the single-lap, but the floor-area required is much less, which is important where floor-space is limited. Construction and duct work is more complex with the double-lap models, steaming and zonal heat control being more difficult owing to stacking. Most models are direct gas-fired, with automatic oven-loader, accommodating about 48 trays on average.

*Double wire-band return-ovens* (German: *Doppelnetzband-Rücklauföfen*), to be found on the European continent, are also based on a double-lap principle, but consist of two wire-bands in the form of endless conveyors, running parallel, but within two separate oven-chambers (see Fig. 116). This design offers a saving in labour, with the help of mechanical transport of the dough-pieces from the oven-mouth onto a desired position on the travelling-band wire conveyors. This transports the dough-pieces the full length of the wire-band conveyors within the oven-chambers, the baked products then being returned to the mouth of the oven, by which time, the baking cycle is complete. The latter operation being achieved by reversing the rotation of the motor which drives the wire-bands. On completion of the full distance in both directions, the motor automatically switches off, but button-switches can override, allowing the bands to be halted in any position. This combines the advantages of the multi-deck oven, i.e. limited floor-area and good heat utilization, with the ease of loading of the wire-band tunnel-oven. The baking-surface is mobile, and part of the travelling-band protrudes outside the oven-mouth, allowing the dough-pieces to be deposited directly onto the band. Most of these ovens are gas-fired, using the indirect hot-gas cyclothermal heat circulation system. The baking-surface of each chamber is $6·60 \, m^2$, making a total of $13·20 \, m^2$. Throughput is normally about 216 kg bread/hour, which can be increased to a maximum of 258 kg bread/hour. Bread-roll output is 4000 units/hour maximum, bread and roll outputs providing improved capacities compared with the four-deck multi-deck oven. The best engineered example of this unique design concept is the DNRO, with $13·2 \, m^2$ baking-surface, manufactured by VEB Kombinat Fortschritt, Bäckereimaschinenbau, Halle, GDR.

*Wire-band tunnel ovens* represent the most technologically and economically advanced baking ovens to date, having a good general thermal efficiency. They are completely continuous in operation, dough-pieces being fed in at one end of the chamber and discharged at the other. Such ovens are eminently suitable for line-production systems, where the baking stage must be synchronized with all previous and following production stages. For European bakery requirements, where a mixed production range of hearth-, panned-, and variety-breads and rolls are produced, and baked in the same oven, the flexibility of these ovens is ideal. However, on the North American continent, for the baking of panned bread and rolls, its economic justification is questionable, on grounds of floor-space versus output capacity. *Anderson, R. C.* (1973) *Baker's Digest*, **47**, 2, p. 28.
Anderson points out that a tunnel-oven of output capacity 8000 lb of bread/hour requires about $1000 \, ft^2$ more floor area than a single-lap oven of comparable capacity, when both are

equipped with automatic loading and unloading equipment. General features of wire-band tunnel ovens are a long, low baking-chamber, within which a motor-driven conveyor carries the oven-hearth in a straight line for stretches of 100 to 400 ft (30·5 to 121·9 m). The length being divided by baffles into one or more steaming zones, and up to eight heat-zones of 30–60 ft (9–18 m) each, depending on specification (see Fig. 118). Each zone has its own individual control, made easier by separate controls for top and bottom heat, and lateral heat balance. Steaming conditions are easily optimized, without the stabilization problems of swing-tray ovens. An example of such an oven is the *Duotherm wire-band oven type DUO-NU*, manufactured by Werner & Pfleiderer, 7000 Stuttgart 30, FRG. This oven unit is applied to the baking of all types of product: panned-bread, hearth- and variety-breads, bread-rolls, cakes, bases, rusks, pizza and panettoni. The Duotherm forced convection heat-transfer system in the rear part of the chamber can be applied for specific products, providing even browning and crust formation on the sides of panned products, shorter baking-times, and a thicker crust on free-spaced hearth products. At the feed-end—the location of the burners—the heat circulation system is cyclothermal, the rear section of the chamber being fitted with the Duotherm forced convection system. This involves feeding the heating-gases for top and bottom heat into the numerous ducts, distributed across the width of the Duotherm radiators. The baking-chamber atmosphere being circulated in a vertical direction by a reversible fan through the spaces between the ducts. In this way, improved heat-transfer, on a convective basis is achieved by rapid air circulation around the products, e.g. side-walls of pans or hearth products not easily reached by infrared radiation. The direction of flow being switched up or down, or terminated by a switch on the control-panel.

*The indirect cyclothermal heating system* depends on the combustion-gases, or hot air from electrical heating, being circulated through ducts around the baking chamber, having no contact with the products. The ducts are of steel-sheeting, and the oven sole of chamotte or brick-tiles, steel-plates, or a wire-mesh band. Owing to ease of regulation, gas or oil are mainly used, but with electrically heated air, a closed circulation system can be used. Cyclothermal heating systems are used in tunnel-ovens, multi-deck ovens, double wire-band return ovens (DWRO), and rack-ovens (see Fig. 118).

*Wire-band tunnel-oven heat distribution* is achieved by dividing the chamber into individual baking temperature control zones, allowing control of top and bottom heat independently of one another (see Fig. 119) The conveyor element within the chamber can be endless wire-bands, steel-bands, metal-plates or metal-plates covered with chamotte, which permits the user to label his bread 'brick-oven baked'. The source of energy is gas, which after combustion, is distributed by means of the cyclothermal system through heating-ducts positioned above and below the oven-chamber, as shown in Fig. 119 (a) and (b). These heating-ducts acting like a network of radiators. Only part of combustion gases, on leaving the duct network, pass out of the chimney. Instead, most of the 200 to 400°C hot gases are returned to the combustion-chamber, where the burner flame gases mix with it at a temperature of about 1000°C or more. However, the gases for heating the oven-chamber must not exceed 600°C, otherwise the security fuses melt, extinguising the burner. The burner is so constructed that the recirculated gases simultaneously cool the combustion chamber from the outside. As a result of the partial recycling of the heating-gases, the cyclothermal tunnel oven can utilize up to 50% of the energy-input. Oven-capacity depends on the band conveyor width and length, dimensions of the bread-pan straps, hearth-loaf spacing, or tray dimensions. Wire-band width options are 1·65, 2·0, 2·5, 3·0 or 3·75 metres, and band length 9 to 45 or more metres. An oven with a wire-band conveyor 23 metres long, and 2·5 metres wide, loading pan-straps 570 mm × 240 mm end-on with 4 × 800 g dough-pieces per strap, would provide an output capacity of 3228 units using a 30-minute baking-time, and 3725 units using a 26-minute baking-time. Hearth-bread yields are more difficult to predict, owing to the spacing necessary for a crusty bake, but may be only 50% of panned bread in the same oven, baking-time being also longer, e.g. 60 minutes to ensure crust crispness and thickness. In the industrial bakeries of the COMECON countries, the wire-band tunnel-oven is the most utilized oven, for basic breads, variety breads, and rolls. In the GDR, these are manufactured by VEB Kyffhäuserhütte, Artern, the most widely used models being the BN 50 and BN 72, with baking-surfaces of 50 and 72 m² respectively. Figs 120 and 121 show the external features of the BN 50, and baking-temperature/band-length relationship within the five heat-zones.

*The BN 50 wire-band tunnel-oven output* is 835 to 1050 kg/h of mixed rye/wheat-flour hearth bread, or 665 to 685 kg/h of bread-rolls. Energy requirements are: town-gas 100 m³ h⁻¹ (18 kW); natural-gas 40 m³ h⁻¹ (18 kW); and heavy grade fuel oil 32 kg h⁻¹ (30 kW). Steam consumption is 80–160 kg h⁻¹, depending on oven size and product baked. Wire-band speed, and hence baking-time of the BN series ovens, is regulated by a three-phase motor with a variable-speed drive over the drive-roller, mounted at the discharge end of the oven. The selected motor-rpm determining the rotation-time of the band, which is read off a tachometer calibration in minutes, e.g. 3000 rpm/17 minutes, 1500 rpm/35 minutes, etc.

*The Simplex 2000*, is another highly efficient, high-capacity tunnel-oven, manufactured by APV-Baker (formerly Baker Perkins) of Peterborough PE3 6TA, UK. This oven also incorporates a recirculating heating system, relocating the heat-flux as required, and utilizing forced convection to increase heat-transfer. Automatic heat modulation burner control, preheated combustion-air, and microprocessor temperature control ensure a consistent bake for panned-, and hearth-breads, rolls, cakes, pastries and pies, etc. Available conveyor widths are 2·5 and 3·0 metres, with lengths up to 36 metres, lengths from 30 m requiring two heat circulation systems in the case of the Simplex 2000E oven. The Simplex 2000 version has conveyor widths of 3·25 and 3·85 metres, with the same length options. Average heat-inputs when running is approximately 70% of the maximum heating up demand, which is 20 200 kcal m⁻¹ h⁻¹ for the 2·5 m band width, and 24 200 kcal m⁻¹ h⁻¹ for the 3·0 m band width (Simplex 2000E). Comparable heating-up demand for the Simplex 2000 model are: 3·25 m band width 44 000 kcal m⁻¹ h⁻¹, and 3·85 m band width 52 000 kcal m⁻¹ h⁻¹. Total gross heat-input is calculated in each case by multiplying these figures by the relevant oven-conveyor length in metres. Both the Simplex 2000 and the 2000E incorporate the single constant mass recirculating heating system, which matches heat available to the product requirement profile, and high turn-down automatic modulating burners. The main difference in the models described is the lighter weight grid conveyor for fuel economy with panned and sheeted products, in the case of the Simplex 2000; the Simplex 2000E being designed for European product ranges.

*Bakery production output* is controlled by the capacity of its ovens. This is expressed on an hourly basis, and is often referred to as 'throughput'. In the case of tunnel-ovens, it can be calculated in kg/h from: baking-surface length, number of loaves/row, weight of each loaf (kg), loaf-spacing (mm), baking-time (min), taking into account a conversion factor of 60 for minutes to hours and a factor of 1000 for conversion of millimetres to metres. Such calculations for discontinuous batch ovens are more difficult, owing to additional factors, e.g. loading, baking, further heating, product transfer, and unloading. Comparative throughput measurements between a multi-deck gas-fired cyclothermal system, and a double wire-band return-oven (DWRO), with total baking surfaces of 9·6 m and 13·2 m respectively, indicate values of 144 kg/hour, and 220 kg/hour respectively for the baking of 1·5 kg mixed rye/wheat-flour bread.

Schwate, W., and Ulrich, U. (1986) In *Special Processes in Baking—Process Manual* (German), VEB-Fachbuchverlag, Leipzig, GDR.

*Effective productivity of various ovens* can be calculated by dividing throughput (in kg) by the product of the number of operators and hours of work. Thus, any increase in effective oven productivity can be determined, and expressed as a percentage increase compared with that of the operation of another oven system. The choice of the best oven type and capacity is a difficult task for any bakery, since so many important factors have to be taken into account. Since oven capacity determines throughput of baked products, the personnel requirement, quality of the bake, and production costs, an in-depth evaluation before making a final choice merits the baker's time. Visits to bakeries with comparable outputs and product ranges to that of the prospective buyer will allow the following points to be assessed on the spot: (1) quality of the bake; (2) floor-space demand; (3) mode of operation and degree of maintenance required; (4) operator costs; (5) installation and fuel supply, steam provision, and burner exhaust flues. Once the type of oven has been decided upon, the various models from manufacturers have to be compared for baking effect and reliability, as well as after sales service versus price. For the industrial bakery, the priorities influencing choice may be quite different. Although the tunnel-oven demands the most floor-space, it is the most straightforward in design, construction and operation, as well as being the most flexible for different pan sizes and hearth products. Oven-chamber temperature control is also made easier and more accurate by dividing into five heat-zones, and the application of a 'feedback' system (servo-system) with computer-controlled thermostats. The lower crown facilitates steaming, and loading and unloading are located at opposite ends of the oven. The single-lap swing-tray oven, such as the Series 440 oven of APV-Baker, Peterborough, UK, has many of the advantages of the tunnel-oven, and for the production of panned products, offers about 50% economy in floor-space compared with the tunnel-oven of the same output capacity. Heat distribution is divided into four zones, and the design allows superior fuel utilization efficiency compared with tunnel ovens. The 440 Series oven, has a gentle steaming gradient, with still air and bottom radiation, for good oven-spring, followed by a mild convection-zone, then a high convection-zone whereby adjustable jets direct air flow between the pans for rapid baking. Finally, a mild convection-zone provides even browning on the return run of the trays. Automatic loading and discharge mechanisms have been perfected at the one service-end, and integral in-oven lidding systems added.

*Conveyorized oven systems* are normally integrated with a conveyorized final-proofer, the grid pan conveyor carrying the panned dough through the final-proofer and oven without any transfer breaks. Residence-times in the proofer and oven are usually 60 and 20 minutes respectively. Such systems offer design simplicity and uniformity of bake, each product taking the same path through the oven chamber. Heat-zoning with these ovens is non-existent, but facilities for steaming are possible. Since the original design of the conveyor speed was to be uniform throughout, a longer proofer-conveyor was necessary to accommodate the longer proofing time versus baking-time. However, this problem was overcome in later modifications to Lanham's original system by powering the proofer and oven conveyors with independent drives.

*The Lanham proof-and-bake system*, was developed in 1967 by E. Lanham, bread and roll plant installations being utilized worldwide. The Atlanta-based company now operates under the APV-Lanham title, having been acquired by APV-Baker in 1989. The endless chain conveyor conducts products through proofing, baking and cooling. Pan-straps of six abreast are locked in position on the conveyor, passing evenly through proofer and oven to give a consistent bake.
Lanham, W. E. (1970) *Baker's Digest*, **44**, 6, p. 54.
An important feature of the system is the relative ease of maintenance. In the UK alone, there are currently 14 Lanham plants producing various breads and small fermented products for national groups, independent plants, and burger bun specialists. Lanham have a licence agreement with Tweedy of Burnley for all markets outside North America and Japan involving Tweedy in the design and manufacture of components such as conveyors, oven temperature controls, belt-speeds and pan-spacing. The Lanham system has a good growth potential as a roll production plant.

*Ribbon-burners* utilize either gas or fuel oil; in the latter case they are fitted with oil vapourizers or nozzle burners to perform an atomizing function. The oil is vapourized by a heat source, and the vapour mixed with air for combustion in the correct ratio. The nozzle and gun-type burners used in recirculation ovens can be readily adapted for both gas and oil. These burners have a very wide flame adjustment range, and stability on low turn-down.

Flanagan, P. (1978) *Baker's Digest*, **52**, 2, p. 42.
Direct firing still remains the simplest and most energy efficient method of oven heating, being about 15–25% more efficient than an indirect method.
Craig, S. (1977) *Baker's Digest*, **51**, 5, p. 131.
Multiple banks of ribbon-burners within the oven-chamber, positioned laterally to the direction of product movement on the conveyor, both above and below the baking-surface, are capable of providing reasonable top and bottom heat control. However, ovens using them normally employ a clean-burn gas fuel, and although exhaust flues limit the combustion products, and moisture within the oven-chamber by free convection, some form of forced convection improves the performance of these ovens considerably. This phenomenon is explained by the thermodynamics at the surface of the dough-piece, which are such that heat flux and penetration can be greatly increased by rapid removal of steam emerging at its surface by air turbulence.

*Indirect-firing-and-heating systems* offer the options of gas, fuel oil and other fuels, with no combustion-products entering the oven-chamber, and each burner-chamber forming an oven zone, thus facilitating temperature control, especially with the aid of fans, ducts, and temperature controllers. Indirectly fired ovens, owing to design constraints for the installation of radiant panels, are less efficient in energy utilization than directly fired ovens.

*Current research into improved heat-transfer* is being directed towards the use of combined techniques, making use of VHF energy sources of about 17 MHz, and high-temperature conventional modes of heating. These techniques demand very careful control of the mixed heating modes to keep baking-loss within conventional, acceptable commercial limits. The two-phase process (HF and high temperature) takes about 10 minutes, compared with 30 minutes for conventional baking. The HF mode is applied for about 6 minutes, followed by supplementary high temperature conventional heating for 4–5 minutes at 350°C for crust formation. The HF mode provides a rapid bake-out, and a rapid transition of sensitive enzyme-activity temperature optima, in particular that of alpha-amylase. This permits the use of flours with relatively high alpha-amylase activities (Falling number values for wheat-flour below 200), the lower limit of Falling number values for wheat-flours being about 70, even with two-phase baking.
Salovarra, H., et al. (1988) *Acta Alimentaria*, **17**, 1, p. 67–76 (English).

*The air radio frequency assisted (ARFA) oven system,* manufactured by Greenbank-Darwen Engineering, part of the Greenbank Engineering Group, of Blackburn BB1 3AJ, UK, uses a combination of radio frequency (RF), and forced convection of hot air. Applied simultaneously in the ARFA oven, the baking-time becomes dramatically reduced. The RF element of the process generates heat uniformly within the product, reducing the limiting factors of heat and mass transfer, thus allowing free migration of moisture to the product surface. The air system is then used to remove surface moisture from the product, and impart texture, colour, and develop flavour. This balance being the most efficient use of energy. Metallic conveyors, baking-pans and foil containers can now be tolerated in the RF field, the forced convection nozzles or perforated plates being utilized as the electrodes. The forced convection hot air system is applied from above and below the product, either through an air nozzle system, or through perforated plates. The latter gives lower density RF waves, which can be an advantage for some products. The conveyor material can be: steel-mesh, steel-slats, or non-metallic. The choice of energy source for the hot air system is open, heating being either direct or indirect.

## CHAPTER 3.4 CONTROL TECHNOLOGY AND ENERGY RECOVERY

*Heat-energy requirements per unit weight of dough per bake* show wide variations, depending on dough formulation, structure, mode of baking, and oven-design. Anderson found that by increasing final-proof temperature, and reducing the baking temperature, thus reducing moisture loss, the energy requirement could be reduced from 256·6 per pound of bread to 162·2 per pound (Btu), under control.

*Anderson, R. C. (1966) Baker's Digest,* **40**, 6, p. 60.

Johnson and Hoover, analysed total energy requirements for baking, estimating the following Btu data per pound of bread: oven start-up 77·4; heating during baking 77·0; moisture evaporation 179·7; starch gelatinization 2·0; heating of pan 34·8; insulation loss 45·4; and flue-gas losses 122·4. These results are based on industrial data from a daily production of 80 000 lb of bread, the total energy input per pound of bread being 528·9 Btu. Thus, the two main loss sources in the overall balance are: moisture-evaporation 34·3%, and flue-gas losses 23·2%, together making up more than 50% of Btu consumption.

*Johnson, L. A., and Hoover, W. J. (1977) Baker's Digest,* **51**, 5, p. 58.

Later research by Anderson and Skarin, using an efficient direct-fired oven with gas, good insulation and forced-convection, indicated heat-inputs of 300 to 400 Btu/lb of bread as adequate. However, it must be borne in mind that an indirect-fired oven would demand an average of 20% more fuel to compensate for the lower fuel utilization efficiency. Therefore, using 400 Btu/lb bread as a base, and assuming a gas fuel efficiency of 85%, the actual heat-input in net terms entering the oven-chamber would be 340 Btu/lb. Of this amount, 250 Btu/lb is necessary to bake the dough, 40 Btu to heat the pan, and the residual 50 Btu being lost as exhaust-gases and through the oven walls.

*Anderson, R. C. (1973) Bakers Digest,* **47**, 4, p. 40.

*Skarin, R. (1964) Proc. Am. Soc. Bakery Engineers,* 1964, p. 88.

*Burner control systems* for direct-fired gas ovens are usually achieved by the fitting of so-called 'ribbon' burners equipped with a flame-band distributor, as shown in Fig. 122. These are installed within the oven-chamber, and generate a strip of flame continuously along their entire length, which can be adjusted laterally by using a lateral heat equalizer. The latter equipment is capable of lateral flame adjustment at any desired level across the oven width. The illustrations show various lateral flame patterns produced by the burners, fitted with a flame band distributor. Indirect-fired ovens, using recirculating (cyclothermal) heating systems, are fitted with larger higher capacity burners, installed singly or in pairs. These generate the hot gases of combustion within their own separate combustion chambers or boxes. Correct gas:air ratios must be maintained to produce the hottest possible flame, burning with a blue colour, with no carbon deposition. The older aspiration burner systems involved conveying the gas to the burner under a pressure of 1 to 5 psi, the burner design being such that the gas flow produces a venturi-effect, drawing in sufficient air to form a combustible mixture. Manual control was then maintained on each gas control valve. Owing to the quantity of air drawn through depending on the gas flow rate, at low settings insufficient air is drawn in, and efficiency diminishes. Also, the aspirator system is potentially hazardous, should the flame blow out.

*The zonal premixer unit* avoids these problems by mixing air and gas before being fed through a common header to a number of ribbon-burners, which may be 15 or more.

*Flynn, J. H., and Flynn, E. S. (1965) Baker's Digest,* **39**, 3, p. 63.

Filtered air at 1 psi, is fed into the mixer unit by blowers, and mixes with gas, which has been reduced to zero pressure by a gas pressure regulator, shown in Fig. 123—I, before being fed to the mixer unit.

*The air:gas mixture* is controlled by an air-control valve, which effectively takes over control of the burners. Actuation of this valve is by the temperature controller, which, on sensing a fall in oven temperature, signals the valve motor to increase the air flow into the mixture chamber. More gas is then drawn in, resulting in a flame-size increase to supplement the heat supply. When the oven temperature exceeds the preset temperature level, the air flow rate is reduced, together with the flow of gas, thus reducing the heat supply. These mixer units, which are self-actuated gas/air mixture regulators, are also provided with safety switches, which automatically release hazardous pressure levels, lock-out valves for gas supply, and intermittent or continuous spark-ignition. Fig. 123—II shows the design features of a self-actuated gas/air mixture regulator. Higher levels of control can be obtained by using separate proportional mixer units for each burner fitted, also with low-pressure air and zero-pressure gas lines.

*Burner control systems generally* can be classified into three types according to their degree of sophistication, as follows: (1) a simple on/off mode, the burners being either fully on or off; (2) high/medium/low mode systems, the air/gas controls varying between high and low flame settings, according to heat demand; (3) a modulating, proportional control, valve, which supplies the correct amount of air and gas mixture to the burners, being linked to a temperature controller, which senses temperature in the zone of burner operation, and signals when more or less heat is required. A typical example of a modulation control hook-up, actuating an electric modulating motor on the combustion air-line is illustrated schematically in Fig. 95. This controller has a mercury thermal sensing element installed in the oven-chamber, causing an indicator on the instrument to move up and down the scale in response to the expansion and contraction of the mercury. In turn this moves a contact along a potentiometer coil within the modulating temperature range. The potentiometer coil follows a red-pointer, which can be set at the control point by turning a setter-knob located on the instrument cover. The potentiometer-coil forms one half of a Wheatstone-bridge circuit, and the other half of the bridge is formed by a potentiometer of similar electrical characteristics, built into the proportioning motor and driven by the motor-shaft. The proportioning motor also has an inbuilt detector, detecting any imbalance in the Wheatstone-bridge due to a change in control temperature. This imbalance is then rectified in the form of a balanced bridge circuit by the proportioning motor being driven in an appropriate direction. The motor-shaft is connected via a linkage to a device controlling the amount of air entering the gas-mixture valve. This type of control hook-up between oven temperature and burner, provides accuracies of less than $1.0\%$ of the full scale reading, which is sufficient to maintain oven zone temperatures to within a few degrees of target settings on the instrument.

*Safety devices* are also fitted to modern ovens to prevent hazardous conditions in the event of burner malfunction. Gas or fuel oil ignition may fail owing to an extinguished pilot-light; blockages can occur in fuel lines; a flame may become extinguished by a sudden draught. Unless ovens are equipped with suitable detection and alarm systems, flame failures can result in product under-baking. However, the most fundamentally important safety measure is to have automatic shut-down of the fuel supply to the burner in the event of flame failure (see Fig. 123—III).

*Diagnostic-check display instruments* are now provided on many multi-deck ovens, whereby, in the event of a fault, the type of breakdown is indicated by lights and symbols, i.e. whether the fault is in the electrical system or the burner system. Thus, faults can be diagnosed by the baker, before contacting the service-engineer.

*Exhaust-gas (soot) monitors* reduce oil consumption and maintenance costs by monitoring the exhaust-gases from combustion, keeping the gap between excess air and the soot limit to a minimum.

*Oil-consumption monitors with inbuilt computer* update the baker on burner operating time, the number of times the burner is switched on, current oil consumption, amount of oil used since the last reset, and current oil supply versus minimum tank level. Changes in oil flow are immediately apparent for perusal, thus ensuring constant heating-up times, and stable baking conditions.

*Automated steaming control* reduces energy demands when baking products such as bread-rolls in rapid succession. Average losses of steam for rolls are $33\%$, and about $22\%$ for 1-kg hearth long-loaves. These losses can be reduced by up to $10\%$ using an automatic steam system.

*Oven control technology* must be simple, centralized and well organized for operation, provide an optimal control over the baking-process, simplify repairs and servicing, increase the degree of automation, consist of positive and reliable circuitry, and result in reduced energy consumption. Oven measurement and control systems are divided into two areas: (1) equipment elements, i.e. recording instrumentation, fuel-supply, drive-motors for the conveyor band and speed controls, coupled with the tachometer, and power supply for the bread-cooling plant; (2) control panel, which centrally controls the actual baking-process. As in other branches of industry, microelectronics and information technology are emerging in the baking industry as key technologies. Computerized baking and computer-guided production, applied also in craft-bakeries, reduces stress-induced operational errors and quality fluctuations. In this respect, the two basic possibilities are: control by information-storage programming, which allows the use of the same basic construction unit for various control functions, merely changing the programs of their algorithms; or programming by wire, which involves working out a fixed program, by linking up specific groups of components. Since oven control involves the use of a fixed functional sequence, programming by wire is the obvious choice. Control of the correct function of the heating-system is achieved by pipeline pressure manometers, and an air-speed sensor controls the heating-gas circulation. The combustion chamber is protected by an excess temperature sensor; temperature sensors are also placed within the combustion chamber and exhaust-gas vent. Equipment elements tend to be constructed as complete units, e.g. a complete gas burner includes power units for gas and air regulation, ignition and flame control units, and combustion air pressure monitor. Gas feed pipes, magnetic-valves, and gas-pressure-regulator are also built as a complete unit. Temperature sensors and illumination circuits for the various heat-zones are also grouped in one unit. Since the grouping of all these 'block-units' are mechanically and electrically complementary, they can be assembled ready-wired at the point of installation, being connected by clips attached to strips.

*Actual control and regulation* of the baking process is performed centrally from the control-panel, which is also built up from grouped electrical components. Measurement and read-out aggregates record the temperature in the individual heat-zones of the oven-chamber, and exhaust-gas temperature. Initial temperature of the heating-gas in individual combustion-chambers are regulated by a control-aggregate, which incorporates a tachometer, recording the baking-time relative to the velocity of the band-conveyor. Automatic control of the whole baking cycle proceeds from the control panel, but certain control functions can be made independent of the automatic control sequence. In spite of the number of hook-ups and servo (feedback)-systems using microelectronics, the result is more exact than a traditional relay control system can offer. Also, contact-free signal processing is more reliable, and wear-and-tear is almost eliminated. Savings in electro-energy, compared with non-electronic-controlled ovens are of the order of 20%, and savings achieved by the use of 'block' gas burner units range from 16% to 24% of gas consumption. Actual economies depend on the type of product, spacing on the band, and other factors.

*Measurement and control equipment* generally performs a comparison of specific unknown physical parameters with defined internationally calibrated norms. When associated with a control-system (servo- or feedback), and the measured value deviates from the target value, a related signal is passed to an appropriate control unit. The processing of this signal usually involves a conversion into either electrical or pneumatic units of measurement. Control instrumentation is not provided to measure isolated values, but rather to determine whether measured values lie within a specified tolerance range. Control equipment frees personnel from routine control work, and contributes to safe, hazard-free working by linking to visual and acoustic alarm systems. Figs 123—I and 123—II, show the principle and components of a gas-pressure regulator, and self-actuated gas/air mixture regulator for the gas-fired cyclothermal wire-band tunnel-oven Type BN 50, manufactured by Kyffhäuserhütte, Artern, VEB, in the GDR. Fig. 122 shows the lateral flame patterns developed by ribbon-burners fitted with a variable flame-band distributor, which are applied to direct gas-fired ovens with free convection heat distribution.

*Temperature control systems* for ovens, consist of a thermocouple (nickel/chromium/nickel) as the temperature sensing device wired to a calibrated instrument on the control-panel. When a maximum of 650°C is reached, the magnetic-valve on the high-pressure forced-air line falls, thus sealing the air-line. However, a spring-loaded adjusting screw allows enough air to maintain combustion on a low flame on the forced-air line. Thus, the heating-gas temperature in the oven-chamber falls until the thermocouple on the two-point regulator of the control unit for the minimum tolerance limit gives its signal. At this moment, the magnetic-valve opens again, allowing air to pass freely through the forced-air line. In this way, temperature control is achieved via the connecting pipeline to the air pressure within the gas/air mixture regulator, which controls the gas volume feed to the burner. Although any number of thermocouples can be connected to the same instrument, for heat-zone temperature measurement, one instrument for each zone is desirable, the temperature record during the whole baking-cycle being maintained on a chart. The most accurate temperature control for multi-zone ovens is obtained with electronic potentiometers, obtainable in several forms and with special features. Temperature controllers associated with burner control systems are either of the electrical contact or pneumatic, narrow-band throttler type, depending on the type of burner used. The control mode can be on/off, high/low, or modulating. The latter system involves the burner control valve operating in rapid response to preset temperature limits, eliminating large temperature fluctuations.

*Oven band-conveyor speeds* are measured and controlled by electric tachometers (rpm-counters), which can be either direct-current or alternating-current generators, being connected to the band-conveyor drive-shaft. Their output is either measured with a potentiometer or a voltmeter, and is calibrated in terms of baking-time in minutes. The selected three-phase motor-rpm driving the band determines the baking-time of the products. Any band-slip over the drive-drum is detected by the resultant change in band-rpm by the tachometer, and an appropriate alarm signal given. Since baking-time depends on band-rpm, the detection of band-slip is important.

*Flame ignition and the detection of flame-failure* are part of the combustion safety precautions built into all modern ovens. The lighting of ovens follows a programmed sequence to ensure that all elements of the ignition system are functioning properly. Instruments used for the detection of flame failures include flame rods, infrared and ultraviolet detectors, flame electrodes and photoelectric cells. These instruments are all linked to various devices which will cut off the fuel supply. A fuel supply shut-down may be triggered off by variations in the gas or fuel oil pressure, ignition failure, overheating in the combustion-chamber or flue, or an electrical power failure. The most widely used instruments are the flame-electrode and the photoelectric cell. The positioning of these instruments within the heat-development systems of a gas-fired oven, is at (13) in Fig. 123—III, and at (7) in the case of an oil-fired oven respectively, as shown in Fig. 123—III(b). The principle of the flame electrode depends on the fact that a flame, due to ionization, acts as an electrical conductor, forming a circuit. As long as the flame burns, a relay holds the burner feed-valve in the open position, reverting to the closed position and shutting off the fuel supply when a fuel or power interruption occurs. The photoelectric cell, and light sensitive sensors, depend on the closure of a relay as long as a light-emitting flame prevails. If the flame is extinguished, the relay opens, and actuates the relative safety devices, similar to the flame electrode. Re-ignition systems are fitted when the shut-off valve also functions as the temperature control system.

*Rack-oven automatic and computer-control systems* have become important with the rapid growth of rack-ovens for in-stores, freeze–thaw–proof–bake and 'bake-off' operations generally. Their control systems are often integrated into the entire freeze–thaw–proof–bake cycle system. The aim of computerized control is to automatically set-up the optimum conditions for baking each product group. Typical process variables are air-flow and restart time after loading, some products benefiting from a delay in air-flow restart time. The facility to vary oven temperature profiles during the baking-cycle, steaming levels and steam release are also desirable for some products. When equipped with heat exchangers of adequate dimensions, rack ovens can maintain low exhaust-gas temperatures, thus making efficient use of energy-imput. Exhaust-gas residual heat can also be utilized to preheat the combustion air, in the same manner as applied to tunnel-ovens in industrial bakeries, and multi-deck batch-ovens. The Rototherm RE, manufactured by Werner & Pfleiderer, D-7000 Stuttgart 30, FRG, incorporates the rotating-rack design, and the advantages of the block-building principle. Burner, fan and heat-exchanger form one circuit, independent of the baking-chamber. The block-building unit design technique enables the electrics to be supplied ready for plug-in, the installation electrician only having to make the mains connections via cable to the control-panel. Burner, fan and heat-exchanger can be heated up during the steam exposure-time, thus reducing the temperature-drop normally encountered when loading the oven. Thus making continuous batch-baking possible afterwards. The heat exchanger, burner and steaming device are designed for continuous baking. The Rototherm rack oven is fitted with an automatic baking control system as standard, but the computer control system can be programmed and left to complete all the baking operations in sequence. Once the loaded racks have been wheeled into the oven-chamber, and the door closed, the baking cycle starts by selecting the programme, and pushing the programme start button. After the preset steam exposure-time has elapsed, the air circulation fan starts, and the baking-cycle proceeds until complete, which is signalled acoustically by a bell or gong. The Rototherm is one of the most compact rack-ovens with the rotating turntable, providing a baking-surface per 21 trays of 6·6 to 12·6 m$^2$, depending on the dimensions of the oven-chamber chosen. Power-ratings of these ovens range from 60 to 105 kW/hour for gas and oil-fuel heating, and 42 to 60 kW/hour for electricity.

*Energy recovery systems* have attracted much research in recent years, most of this having been applied to industrial tunnel-ovens. However, with planning and discipline, considerable savings can be made in the economic usage of energy with batch-ovens in the smaller and medium sized bakery. Some elementary considerations are the following:

—planning of production schedules to make full use of falling temperatures for a diverse product range,
—switching off unwanted multi-deck oven decks when no longer needed,
—making full use of off-peak tariffs by staged switch-on of electricity, making full use of electronically based systems, negotiating special terms for bake-schedule versus time,
—always aim for maximum oven-loading, especially where pan-straps are utilized for bread, strapping being appropriate to oven dimensions,
—making full use of timers for baking to avoid opening oven doors, ensuring steam dampers are closed after steam release, thus avoiding reheating, e.g. rack-ovens,
—ensuring regular maintenance of burners to give efficient combustion, correct fuel/air mixture, and flame profiles, thus minimizing carbon deposition,
—evaluate current replacement cost against energy losses incurred with existing oven equipment.

Since all gas and fuel oil ovens have an exhaust-gas flue-stack for removal of toxic gaseous by-products of combustion, a large volume of hot gases escape. However, by passing these gases through an efficiently designed and adequately dimensioned heat-exchanger, the exhaust-gas temperature can be reduced to a minimum, and the heat so extracted used for generating both hot air and hot water. Thus, extra heat can be utilized for heating water, proofing-cabinets, or the bakery. In large industrial bakeries, where tunnel-ovens are standard, up to 70% of heat can be recovered from the flue-gases, which can reduce heat-input to the oven by 5–6%. The first step is to evaluate the potential amount of heat available for recovery, appropriate exhaust-gas parameters for measurement are: temperature, volume flow rate/hour, and weight/hour. If these figures indicate that a large amount of heat is available for recovery, the installation of a heat exchanger of appropriate dimensions is justified. Heat recovery equipment includes: burner combustion air preheaters; preheaters for boiler water for steam, water-heating; space-heating; heat and steam for proofers. Other sources of waste-heat, at lower temperature and high volume, e.g. steam liberated from ovens, and steam-extraction ducts, are more difficult and expensive in equipment.

*Advice and information* concerning energy conservation schemes can be obtained from government energy departments, and the large oven-construction companies.

*Heat-exchangers* can generate both hot air and hot water, and all oven types from multi-deck to tunnel-ovens can be equipped with these energy saving devices.

The usefulness of such systems is described in detail in the study: 'Opportunities for improving energy utilization in bakery ovens' carried out by the Fraunhofer Institute, München, FRG. The Winkler Bakery Oven and Machinery manufacturer, D-7730 VS-Villingen, Black Forest, FRG, have applied these studies in the construction of their Columbus multi-deck ovens. These ovens, having received a design award and approval of merit in 1985 by the Baden-Württemburg State Office of Trade, are compatible with computer-aided production in craft bakeries.

*Combustion-process efficiency* of boilers, baking-ovens, etc. should be maximized by regular measurement of the four flue-gas parameters: oxygen, carbon monoxide, total combustibles and temperature. Portable gas analysers, such as the Teledyne MAX, manufactured by Teledyne Analytical Instruments, Southall, Middlesex, UK, also automatically calculate combustion efficiency in net terms. A theoretical calculation of carbon dioxide content being also provided. When connected to a printer, permanent data can be stored. For *in situ* exhaust or flue gas analysis, the modular designed WGD *in situ* analyser, manufactured by Thermox, and marketed by Fluid Data, Crayford, Kent, UK, is a state-of-the-art, compact unit.

*Energy-loss through insulation defects* can be detected with the Emmaflex C-700C non-contact instrument, which measures both temperature, and heat flow (energy loss or gain). Insulation voids can be detected during preventative maintenance of all heating equipment. These instruments, with various measurement ranges, are manufactured by Emmaflex, Milford, Stafford, UK.

*Non-contact temperature measurement*, of inaccessible surfaces, and scanning for surface location of hot-spots, can be performed with the Redpoint THI-300 infrared thermometer. Accuracy is $\pm 1\%$, with a 2-s response. Read-out functions are maximum/minimum and average, incorporating a dual Fahrenheit/Centigrade display. Read-out is digital, and the sensor probe is separate from the read-out unit.

*Dew-point/relative humidity measurement* can be carried out with the full-blown, chilled mirror dew-point meter DP 383 R, marketed by Protimeter plc, Marlow, Bucks SL7 1LX, UK. This is a solid-state instrument of robust design, with digital read-out of dew-point, temperature (ambient), and percentage relative humidity. The sensor probe and read-out units are separate, and it can be used for environmental control and monitoring in ovens and proofers, also water-activity in food, and monitoring storage, and production areas. Lee-Integer Ltd, of Kettering, Northants N18 7QW, UK, market a Dewpoint Signal Converter DP800, a block-format electronic module for surface or rail mounting, which converts %RH and temperature signals into a 4-20 mA output dew-point temperature signal, thus providing an alternative method of measuring dew-point, without the need for chilled-mirror techniques. Dew-point temperatures can be measured in a wide range of industrial environments, using rugged probes. These probes incorporate thin film chromium-mosaic capacitive sensors, which is the most advanced relative-humidity sensor currently available.

*Computer monitoring of energy consumption per process unit or plant* allows energy to be treated as any other raw material. By monitoring electricity, gas, oil, and water accurately throughput the production process, a more precise production cost calculation system can be established, both per process unit, and in total for a production line. Computer software for this task is marketed by Stark Associates, Salfords, Surrey, UK, for use with the IBM-PC. VDU plots of unit consumption per hour for electricity (in kWh), gas (in m³), and water (in °C), are displayed for daily comparison. Stored information on the hard-disk can be retrieved in time order, as required.

*Mechanized control of oven loading, and unloading and depanning* is rapidly replacing the rather slow and laborious task of manual loading and unloading. In the case of batch-ovens, devices known as 'setters' are employed for the loading of hearth-baked products. These consist of webbed band-conveyors, which are pushed into the oven-chamber with the products deposited on them. The setter band rollers are then slowly rotated in a counter-clockwise direction, and the band withdrawn from the chamber. This deposits the dough-pieces onto the oven sole at the required spacing. This set-up is only appropriate for use with static sole batch-ovens. In the case of the double wire-band return-oven (DWRO), the loading device can deposit the dough-pieces onto the moving band-conveyors allowing appropriate spacing (see Fig. 126).

*Pan-strap loading* involves the application of a different technique. On emerging from the proofer, the straps are moved by a cross-conveyor into position at the oven-loading platform. When sufficient straps have assembled to fill the oven width, or swing-tray width in the case of the swing-tray oven, a pusher bar, synchronized with the speed of the oven, transfers the pan-straps onto the band-conveyor or the tray in the loading position.

*Bridge-bands* form part of the oven-loading equipment in countries where hearth bread predominates production. These are driven independently by a motor of about 0·8 kW and 125 rpm, over a cable and pulley drive system. On leaving the proofer, and being deposited on the bridge-band, each dough-piece is cut in two places laterally at its extremities by a special cutting-aggregate, made up of 10 circular cutting-wheels mounted on a bar, which is the full width of the band. The height of this bar is adjusted to give the desired depth of cut, as the dough-pieces pass beneath it. These bars are often interchangeable, to allow the use of another bar of band width, which imparts a special embossed pattern on the dough-pieces. The design and working principle of this bridge-band conveyor aggregate is shown in Fig. 124.

*Depanning*, which is hot hazardous work when performed manually, was initially replaced by mechanical depanners, but these aggregates had only limited success. The problem was overcome by vacuum depanning. This technique involves the application of a vacuum to lift the baked loaves from the pans, thus causing minimum damage to both product and pans. Designs vary, but a basic design incorporating a magnetic delidder is illustrated schematically in Fig. 125.

*Depanning/packaging*, can be combined in one unit for rolls and buns. One side of the unit accepting the cupped-pans holding the rolls, suction cups on a vacuum mechanism being aligned to coincide with those of the pan. On the other side, the cartons or trays are placed to receive the rolls. The vacuum head sucks the rolls from the pan in any desired block number from 9 to 20, lifts, then moves over the empty carton, depositing the rolls into it. The empty pans and filled cartons being removed by the conveyor, then another roll-filled pan and empty carton is conveyed in position for the next automatic transfer.

*Robotic loading of trays and transport containers* involves the application of similar concepts. Vacuum-actuated heads moving bagged loaves or packs from dispatch lines into number blocks laterally, then depositing them into the transport containers in a predetermined pattern. Such an operation is fully automated, appropriate menu patterns being programmed into a system according to container dimensions.

*Automated oven-sole setters* transfer hearth products, baking-sheets, and pan-straps of specific dimensions to the oven sole of multi-deck batch-ovens.

*Microprocessor-based oven-controllers*, with VDU and keyboard/operator interface can be fed with oven-baking programme-menus for rapid and accurate adjustment of oven-temperature profiles to suit different products.

## CHAPTER 3.5 BREAD COOLING AND SETTING

*Maturation of fermented baked products*, involves the processes of cooling and setting, which can take place on racks in the dispatch area, with air turbulence, or within purpose-built continuous coolers. For cooling in the dispatch area, adequate fresh air is essential, which should be filtered from dust particles before it enters the cooling area. The loaves should be allowed to cool down gradually to about 35°C, with adequate spacing, on wire-racks. 1-kg loaves should not be transported until they have cooled to this temperature, assuming they are not for slicing, and packaging.

*Microbial infections during cooling* result from contaminated air during cooling, and the presence of excess humidity resulting in condensation. All storage and dispatch areas must be kept clean, well-aerated and free from any contaminating foreign smells. Other aids to general hygiene in bread cooling areas are: the presence of UV-based radiation; 'Insect-o-cutor' discharge units for insect control (Insect-o-cutor Inc., *Georgia* 30083, USA), and climatic control (temperature/humidity), combined with ozonization.

*Bread-rolls* should reach the customer within about 8 hours of baking, and should be stored at about 22°C and 70% RH, away from strong sunlight.

*Loaves of bread* must not be reduced in temperature too quickly, since internal condensation of water vapour in localized zones can cause shrinkage and loss of quality. Industrial bakeries utilize a process of controlled cooling, ensuring optimal product maturation. For panned-bread optimal conditions are a temperature within the range 20 to 24°C, a relative humidity of around 85%, and an air speed adjusted to produce a temperature rise above ambient at the exhaust point of about 10°C. These conditions will allow a 500-g loaf to cool to 40°C within the crumb during a residence-time in the cooler unit of about 1 hour.

*Continuous conveyor or belt coolers* consist of endless multi-tier overhead conveyors, travelling in straight stretches and U-turns at the end of each stretch. Cooling cycles are on average 60 to 90 minutes, the number of cycles applied depending on the type of product. Formulation and product density will determine this factor. The construction of these coolers are sometimes open all round, with an overhead fan to extract the heat radiated, and others are enclosed with panels, forming a tunnel, a counter-current turbulence being used to accelerate the process. The travelling conveyor consists of either a wire-grid or metal-rod capable of flexing during travel. Spiral designs with a single or double helices are also possible, in the latter case the outer helix forms the ascending path, and the inner one the descending product path. These tower-in-tower designs, save floor-space, and provide greater flexibility in conveyor-belt length to accommodate diverse cooling-times for products. An example of this type of system, applied in the GDR is the Universal Transport System 'Regotrans', which can be either floor- or ceiling-mounted. In the GDR, in order to satisfy bread slicing temperatures of about 40°C, the following cooling/maturing times are necessary: rye, rye/wheat, and wheat/rye breads, 3 hours; wheat bread, and toast bread, 2 hours.

Schneeweiss, R., and Klose, O. (1981) In *Industrial Baking Technology*, VEB-Fachbuchverlag, Leipzig, GDR (German).

In the UK, average cooling time within the cooler is about 3 hours, the cooler temperature being kept at 20°C, and the RH above that of the surrounding atmosphere.

*Heat loss of a loaf of bread* on leaving the oven is rapid and uniform, independent of atmospheric conditions. Almost the entire heat loss due to radiation is complete within 40 minutes, the loaf surface having cooled to ambient temperature. However, crumb cooling is far from complete.

*Initial moisture evaporation loss* accounts for only about 50% of the total moisture loss during cooling. Unbound moisture within the crumb migrates slowly towards the crust, and must be allowed to escape to avoid condensation and possible spoilage. The residual 50% will take about 2–3 hours to evaporate under normal circumstances, i.e. no change in external environment. Other types of bread coolers include, tray coolers, rack coolers, and vacuum coolers.

*Vacuum coolers* are a relatively recent innovation, involving the use of a modulated vacuum system, giving a rapid reduction in the product temperature. The cooler consists of a conveyor, which runs through a vacuum chamber tunnel fitted with vertically operating doors at both ends, forming an air-tight seal when closed. Vacuum pumps remove the gases and water vapour from the tunnel, creating a vacuum. Gases become entrapped in the steam, which is also injected into the tunnel. A water-jacketed condenser, condenses the water off. Moisture removal is rapid, due to the partial vacuum within the tunnel, and product temperature decreases. Bread rolls can be reduced from 70°C to 40°C within about 1 minute. Owing to the higher moisture loss, compared with atmospheric cooling, baking times should be reduced by 15–20% to synchronize the final crumb moisture as desired in the final bread, thus ensuring an adequate shelf-life.

*Moisture content* of a loaf of bread on leaving the oven reaches almost zero in the crust when the crust/crumb interface reaches about 100°C; but although the crumb temperature reaches 98 to 100°C, its moisture content exceeds the initial dough moisture by about 12%. On placement in a cooler at 18 to 24°C, with forced convection, the loaf cools rapidly, losing weight in the form of moisture. During cooling there is an initial temperature gradient between crust and crumb of about 14°C, which progressively reduces to zero as cooling becomes complete. This gradient is created by the relatively high crumb temperature, resulting in moisture movement towards the crust. On reaching the temperature of the cooler-atmosphere, the gradient is reduced to zero. However, thermal/moisture conductivity is not the only factor responsible for moisture-loss acceleration during cooling, it is also temperature-dependent. The higher the product temperature, the more rapid the concentration-dependent moisture diffusion proceeds, especially in bread, where diffusion of moisture is often difficult. Still air surrounding the external surface layer of the loaf impedes free external diffusion of moisture into the atmosphere, depending on its temperature. This phenomenon is overcome by the use of forced-air or turbulence.

*Moisture diffusion* from the hot, cooling loaf, at constant evaporative surface, and external air-speed turbulence, depends on the difference in partial pressure of air which becomes saturated as a result of product temperature, and that prevailing in the air immediately circulating around the product on the other side of the boundary layer. The partial pressure of the steam increasing considerably with a rise in temperature of the product. The effect of turbulence being to progressively reduce the thickness of the boundary-layer of still air.

*Prolonged storage* results in the temperature of the cooling loaf falling slightly below that of the surrounding bread-store or cooler. This occurs, since the moisture evaporation from the loaf will continue after the loaf temperature has assumed the temperature of its surroundings, in spite of the delay in its onset. The necessary heat for the evaporation process stems from the crumb-zone adjacent to the crust, and not from the air separated from the crumb by the crust. The thermal-conductivity of the crust is much less than that of the crumb.

Auerman, L. J. (1929) *Sowjetskoje mukomolje i chlebopetschenija*, **3**, 12, p. 708 (Russian).

The graphical illustrations Figs 127 to 132, concerning 'bread cooling and moisture loss', originate from L. J. Auerman's reference book, *The Technology of Breadmaking*, published in Russian in 1972, then published in German by the VEB-Fachbuchverlag, Leipzig, GDR, in 1977 for the benefit of students and scientists in that country. This reference work, encyclopaedic in its coverage, is to be strongly recommended for all students of baking technology in the English-speaking world, including, as it does much 'in-depth' research data. Data used to produce the graphical interpretation shown in Fig. 128 was obtained by

Auerman and co-workers at the Moscow Technological Institute for the Food Industry (MTIPP). This illustration shows the moisture loss due to drying out, and temperature changes with storage-time, which take place at the centre of the crumb under production conditions. The data was obtained from 1·5 to 1·9 kg rye wholemeal bread baked in pans, being an average of 120 investigations, conducted in five production centres throughout the USSR. The temperature of the bread-stores was 19 to 27°C, and the average crumb-moisture of the bread 53%. Immediately on removal from the oven, the moisture loss or 'drying-out' begins. Simultaneously, a redistribution of moisture within the loaf of bread takes place. At the moment of oven withdrawal, the crust is almost moisture-free, but on rapid cooling, the moisture in the crumb, owing to differences in concentration and temperature between the internal and external layers, migrates into the crust. The cooling, and simultaneous water absorption of the crust to levels of 12% to 14%, will depend on the temperature of the bread-store, the weight of the loaf, and the storage conditions, taking place over the first 2 to 4 hours after baking. The crust moisture level of 12–14%, approximates to its equilibrium-relative-humidity (e.r.h.), remaining constant during the remaining storage period. In contrast, the moisture of the bread-crumb continues to fall during storage. Data used for the graphical interpretation shown in Fig. 129, was obtained by the All-Union Institute for Research to the Baking Industry of the USSR (WNIIChP). The moisture-loss of 467 mixed rye/wheat wholemeal flour breads (70% rye : 30% wheat) were studied over a period of 5 days by means of changes in their weight taking place on bread-rack trolleys under production conditions. The total surface area of the 467 loaves was 46·15 m², and the absolute initial moisture of the loaves 94·7% (moisture in percentage based on dry matter of the material). On this basis, the drying-out and absolute moisture curves shown in Fig. 129; and the curves for drying-speed, Fig. 132, were obtained. As Fig. 132 shows, during the process of natural moisture loss of the bread, the drying-out rate falls, depending on the cooling process, until a point $t_k$ is rapidly reached, then it remains constant. The point $t_k$ is easily read off the drying-out rate curve, in this case the $W_a^k$ being equivalent to 85%. Normally this $t_k$ value coincides with the moment in time when the loaf temperature acquires the ambient temperature. Therefore, the time period of moisture loss can be divided into two phases, the first phase being one of a rapidly changing drying-out rate from $t_0$ to $t_k$, and the second phase one of a constant rate of drying-out, and moisture loss. During the first phase of moisture loss, the rate of drying-out falls with falling bread-temperature, and the temperature-gradient within the bread. During the second phase, the bread-temperature approximates to that of the surrounding atmosphere, remaining practically constant. As a result, the moisture-loss (drying-out) occurs at a constant rate, largely determined by the hydrophilic properties of the bread, its dimensions, shape, and the environmental parameters (temperature, RH, and air speed). Fig. 132 clearly indicates that the rate of moisture loss during the first phase of the cooling/drying-out time period is greatest, and during the second phase is considerably less. It therefore follows that any shortening of the first phase of the total time period is the best method of reducing evaporation losses during cooling. In practice, this is achieved by rapid cooling of the bread on removal from the oven, down to the ambient temperature of bread-stores.

*Auerman, L. J.* (1977) In *Technology of Breadmaking*, VEB-Fachbuchverlag, Leipzig, GDR (German), paragraph 9·1, p. 282–4.

*Initial changes in bread moisture* during the first 30–60 minutes after baking involve a reduction in those crumb-layers adjacent to the crust, owing to migration of moisture into the dried-out crust and subsequent evaporation towards the exterior. This results in the moisture of the outer layers and the inner layers of the crumb reaching the same level of moisture, which is 1·0% to 1·5% less than on withdrawal from the oven.

*Subsequent cooling and storage* results in these crumb-layers near the crust losing moisture much more rapidly than those nearer the loaf centre.

*Prolonged storage* for several days often results in the crumb-layers immediately below the crust becoming quite hard, due to moisture loss. Even light pressure at the surface fails to deform the crust. According to research carried out by the WNIIChP, Moscow, USSR, the moisture level of crumb-layer adjacent to the crust falls within 3 days after baking from 44% (3 hours after leaving the oven) to 18%, and the thickness of this hardened layer amounts to 4–5 mm.

*Auerman, L. J.* (1977) In *Technology of Breadmaking*, VEB-Fachbuchverlag, Leipzig, GDR (German), paragraph 9.1, p. 281.

*Air-temperature of the bread-store* influences the rate of cooling, and the moisture loss of the bread. The higher the temperature of the atmosphere surrounding the bread during storage, after removal from the oven, the more intensive the moisture loss.

*Auerman, L. J.* (1932) Snabtechisdat, Moscow. Weight loss of bread during cooling and storage (Russian).

Moisture-loss of Ukrainian bread of 1·2 kg during storage for 8 hours at 43–50°C was 5%, compared with only 2% at 11·5 to 19°C. The storage of bread in the frozen state reduces moisture loss to a minimum (see Fig. 133).

*Relative humidity of the bread-store* also influences the rate of moisture loss. Moisture loss is reduced with increasing relative humidity of the bread-store atmosphere. Higher humidities result in smaller differences in partial pressure at the loaf-surface and in the air, thus reducing the rate of moisture loss. However, this effect is not very significant during the first phase of cooling, since the higher bread-temperature increases the partial vapour pressure above the loaf-surface, any difference between this and the partial vapour pressure of the air being too small for significance. During the second phase of moisture loss, when the bread-temperature does not exceed the ambient bread-store temperature, the influence of relative humidity on the intensity of moisture loss increases. Nevertheless, during normal cooling/storage times of 3 to 6 hours under production conditions, the second phase of moisture loss is very short.

*Forced-air turbulence* is also expedient during the first phase of bread cooling and moisture loss. Fig. 134 illustrates the effect of various air-speeds on the moisture loss of bread during cooling. When the bread is surrounded with turbulent forced-air at speeds of 0·3 to 0·5 m/s, this shortens the time period for the first cooling phase, simultaneously reducing the weight loss due to evaporation by about 0·5% to 0·7%. Research carried out at the WNIIChP in Moscow, placing bread-rack trolleys within a chamber cooled by an air-conditioning plant, has shown that the cooling process is speeded up, and the moisture loss reduced by 0·5% to 0·9%.

*Loaf-moisture and baking-loss* influence moisture loss due to evaporation. Further research work carried out at the WNIIChP in Moscow, has confirmed that the higher the loaf-moisture, the greater the weight loss under otherwise identical conditions. An increase in the crumb-moisture content of rye bread made from wholemeal flour, of 2%, results in an increase in the moisture loss of the baked bread over 4 hours of 0·26% to 0·42%, and over a period of 7 hours of 0·42% to 0·50%. WNIIChP research also confirmed that there exists a reciprocal relationship between baking-loss and moisture loss due to drying-out as a result of cooling and evaporation. The greater the baking-loss, the lower the drying-out loss, due to cooling and evaporation, and vice versa.

*Mode of baking (i.e. in pan or on hearth) and loaf-volume* also influence bread moisture loss. Hearth-baked breads always show larger baking-losses, and lower crumb-moisture levels than pan-baked breads of the same weight. Therefore, moisture loss due to evaporation in the case of panned bread is generally greater than that of hearth breads. Auerman established that in the case of rye bread made from wholemeal flour weighing 1·1 kg, baked in pans, and as round hearth loaves, significant loss differences occurred. Crumb moistures were 52·8% and 49·2% for panned- and hearth-breads, respectively; baking-losses were 11·6% and 13·3% respectively; and moisture losses after 4 hours storage were 2·03% and 0·86% respectively; after 6 hours, storage moisture losses were 2·19% and 1·04% respectively.

*Auerman, L. J.* (1977) In *Technology of Breadmaking*, VEB-Fachbuchverlag, Leipzig, GDR, p. 286 (German).

Raw materials and formulation determine final loaf texture and crumb-porosity. Breads of low density and high specific volume will show larger moisture losses, due to baking and cooling, than those made from leaner, denser formulations generally. Dimensional size of the loaf also determines moisture losses, larger units showing lower moisture losses.

*Bread quality changes* due to storage after baking are essential for bread ripening or maturation, and must not be confused with the processes of 'ageing' or 'staling', which set in later, resulting in diminished freshness and palatability. These latter changes commence after about 22 hours in the case of a 1-kg wheat-bread, whereas rye-bread can remain palatable for up to 40 to 60 hours without packaging. Bread-rolls reach this stage at about 7 hours after baking. Packaged loaves remain palatable for periods up to 6 days. The maturation processes involve a series of complex physical–chemical reactions, which are interrelated, and include both crust and crumb. Loaves of 1 kg or more take longer to cool, the crumb taking much longer than the crust. After about 1 to 3 hours after removal from the oven, the crumb centre often still has a temperature of 50 to 60°C, therefore many processes taking place during the baking process will continue. Depending on the type of bread, contents of sugars, dextrins, and total water-soluble carbohydrate, amongst other substances, increase during the cooling period. In addition, the gelatinization of starch, and crumb stickiness increase. Such observations were made by Gogoberdise, N. I., Auerman, L. J., and Stscherbatenko, W. W., during the period 1956–58, and reported in publications of both the MTIPP, and WNIIChP in Moscow, USSR. As long as crumb temperatures remain at 60°C or over, increases in crumb sugars, and total water-solubles can occur. During the heat-sterilization of bread to prolong shelf-life, whereby, the centre of crumb can rise to 85°C, sugars and water-solubles increase.

*Romanov, A. N.* (1953) Bread storage. In *Pistschepromisdat*, Moscow, 1953.

Changes in the composition of Wheat bread during storage in polyethylene packaging have been studied by Gasiorowski, H., and Jankowski, S.

*Gasiorowski, H., and Jankowski, S.* (1967) *Brot und Gebäck*, **21**, 7, p. 137 (German).

This study comprises storage periods of up to 15 days, at temperatures of $-23$ to $+65°C$, measuring effects on such bread parameters as: crumb and crust moisture; starch; soluble starch; dextrins with a defined chain-length in the crumb; reducing sugars and total sugars in the crumb; titratable acidity; pH; X-ray spectrum; and properties such as colour, compressibility and swelling. Sensory evaluations for freshness, flavour and aroma of the bread were also included. The binding capacity between starch, protein and water play a major part in bread ripening or maturation after baking. After baking, the crumb cell walls are in a swollen state, water being partly physically and partly chemically bound by adsorption. At 70°C, and above, the crumb is in the form of an elastic gel, but on cooling forms a solid mass, during which phase chemical reactions take place. At 60°C and below, intramolecular hydrogen bonding takes place, cumulatively representing high levels of energy. These form within the starch and protein molecules, as well as between the amylose and amylopectin intermolecularly, thus forming the firm crumb structure of the loaf. In this manner, bonding between molecules become progressively closer, and more hydrogen bonds are formed, the energy so liberated giving rise to dehydration. Some of the water released is taken up by the proteins, and the rest is lost by diffusion. Water plays an important part in both maturation and staling, after baking. Loss of heat and moisture during cooling are interrelated, water loss being brought about by the high water vapour partial pressure, which depends on temperature. Maturation also involves the loss of labile highly volatile aroma compounds, which decompose, the more stable aroma compounds diffusing into the crumb on solidification (see Fig. 135). Physically and chemically bound water becomes either lost, or condensed to produce the optimal eating properties. Chemical ripening processes are intermolecular bondings, whereas the physical ones involve mainly moisture loss and the diffusion of highly volatile aroma substances.

*Physical changes* exert an important influence on bread cooling and storage. After baking, loaf crumb temperature at the centre is about 98°C, and at 1 cm below the crust about 110°C. The temperature falls most rapidly at first in this area below the crust, the temperature at the centre of the loaf only falling very slowly. Fig. 136 clearly shows the relative temperature-falls with time taking place at various locations within a 1-kg wheat-bread during storage under specified conditions of 24°C and RH 40%. Rapidity of the cooling process, will depend on ambient storage temperature and RH, and air turbulence, and proceeds until a temperature equilibrium with the surroundings is attained. Owing to an RH gradient between the loaf crumb, at about 100% RH, and the external surrounds at 40% to 60% RH, a process of diffusion takes place, removing about 2263 J/g of heat at boiling point, and atmospheric pressure, resulting in further cooling. An approximate rate of cooling of the loaf can be estimated from the basic Newtonian law of cooling equation, and the total heat lost during cooling from the general heat equation:

$$Q = \text{loaf-mass} \times \text{specific heats} \times \text{temperature drop}$$

taking into consideration the specific heat of crumb and crust, and crust-thickness. The total heat loss of a 1-kg white loaf amounts to about 180 kJ. This loss of heat simultaneously results in water vapour condensation within the loaf. A 1-kg wheat loaf of 2000 ml, and water content 45%, contains about 0·5% moisture in the vapour form according to Schneeweiss and Klose.

*Schneeweiss, R., and Klose, O.* (1981) In *Technology of Industrial Baking*, VEB-Fachbuchverlag, Leipzig, GDR, 1981, p. 424 (German).

Although, during cooling, part of the water vapour is lost, most of it condenses, forming a partial vacuum, causing the air immediately surrounding the loaf to be sucked in. Proof of this fact can be obtained by the tightly sealed packaging of a loaf immediately after baking, th partial vacuum formed leading to a deformation of the loaf. Schneeweiss and Klose studied the volume change of a panned white loaf of 1 kg, which was packaged immediately after removal from the oven, by measuring after various time intervals (see Fig. 138). Volume changes were found to be greatest during the first 15 minutes after baking. After 45 minutes, a physical equilibrium between the partial vacuum and the partial water pressure became established, after which only insignificant volume changes take place. Loaf-volume was measured by placing the hot loaf in a glass desiccator, perfectly sealed. The developed partial vacuum was measured with a U-tube mercury manometer. The largest reduction in pressure was recorded during the first 60 minutes of cooling. In the case of unwrapped bread, this pressure difference due to the sucking-in of the surrounding air became equalized, which was measured with a gasometer, directly (see Fig. 137). These researchers also draw attention to the significance of partial pressure/suction effects for the potential microbial infection of bread by mould. Since the amount of atmospheric medium adsorbed is greatest during the first few minutes after removal from the oven, microbial infection will commence at this stage. However, owing to the initial high temperature of the loaf surface, most of the spores become killed, a greater infection hazard being expected after about 5 minutes exposure to the atmosphere (see Fig. 139). It was found that bread packed about 5–6 hours after baking, owing to its lower crust moisture, showed reduced spore growth, although the actual spore contamination level was the same. When cooling is complete, the partial vacuum becomes equalized, and any spores then settling on the dry top-crust present less danger to the mould-free shelf-life of the bread. This emphasizes the need to cool bread in a storage atmosphere, which is as low as possible in spore content. For each type of baked product there exists a relationship between quality and storage-time, assuming the products are stored at 18–22°C and 70–80% RH. Specific storage-time limits can be placed on products, in order to categorize them as: 'still-fresh'; 'still-edible'; and 'onset-of-staling'. Bread rolls can be described as 'fresh' at up to 4 hours after baking; at up to 6 hours they can be described as 'still-edible'; but at 7 hours or more, there exists the 'onset-of-staling' (see Table 70).

*Structual solidification* of the loaf crumb involves its transition from a structureless gel. This process commences on its removal from the oven, and finishes after 6–8 days, when the loaf becomes hard, dry, and crumbly. Crumb solidification is achieved by molecular bonding at temperatures below 60°C as the loaf cools. Hydrogen bonding plays an important part in this process, particularly between the amylose and amylopectin molecules of the starch, forming crumb porosity. Physical covariables which influence this process are: rates of temperature and heat loss, moisture-loss and diffusion, and surrounding air turbulence. The large moisture loss during the first few hours of cooling is due to the large difference in water pressure between the surrounding atmosphere and the loaf. Comparisons between the cooling curve and moisture diffusion, water-vapour partial pressure, and the latter's codependence on temperature (Fig. 137) confirm a close relationship between these physical variables. On reaching the ambient temperature of 25–26°C, moisture loss per unit time tends to approach a constant value.

*Dynamic/mechanical thermal analysis*, involving a torsion rheometer system, applied in the testing of polymer-melts in polymer research, can also be applied to dough during the 'heat-up' and 'cool-down' stages of processing. Heat-up and cool-down measurements are performed in the torsion-head, consisting of parallel plates, one being a disk-rotor, driven by a continuous rotating motor at the top; and the bottom plate, the oscillating-plate, driven by an electromechanical oscillating force transducer from the base. Small oscillations of controlled amplitude, using frequency sweeps of 0·02 to 50 Hz, with about four points per decade, can be utilized to obtain viscoelastic data. Measurements at low strain amplitude, under isothermal conditions, with temperature increased, and then decreased in steps, can be applied. A 10 minute equilibrium is allowed at isothermal temperature, and the frequency sweep is completed in 20 minutes. Intervals of 10°C during heat-up, and 4°C during cool-down can be taken for isothermal temperature measurements. Relaxation modulus peaks obtained depend on the measurement frequencies used. At high frequencies the peaks indicate polymer chain entanglement, and at low frequencies peaks are interpreted as being due to the association of hydroxyl chain ends by hydrogen bonding. In the preparation of viscoelastic dough, the water content is the most critical factor. The storage (elastic) modulus, designated $G'$; and the loss (viscous) modulus, designated $G''$, both decrease with increasing dough water content, there being no observed interactions between frequency and water content. This dynamic test method is ideal for monitoring flour-processing properties, and dough-baking physics continuously during the whole baking process. However, since strain sweep tests show a sharp fall in $G'$ at strain rates above 0·2%, baking sweeps have to be carried out at very low strain, so that the dough structure is not disturbed during the experiment. The baking experimental sweep shows an increase and a decrease of both $G'$ and $G''$, as a result of starch gelatinization, and molecular interactions between the flour components and water during the baking process. The maxima or peaks obtained at starch pasting and the final values of $G'$ and $G''$ are highly correlated with flour quality parameters. Therefore, the baking sweep curves are a good procedure for monitoring and interpreting structural/textural changes in dough during processing. The RFS II Fluids Spectrometer, marketed in Europe by Rheometrics Europe GmbH, Hahnstrasse 70, D-6000 Frankfurt 71, FRG, and Rheometrics, Inc., One Possumtown Road, Piscataway, NJ 08854, USA, offers dynamic and steady testing, with independent control of oscillation frequency, strain, and temperature. Frequency and strain sweeps, and a stress relaxation mode are standard options. Menu-driven software, provided with the computer-controlled system, allows temperature-dependent studies to be made, as well as sequenced test routines for quality control testing. Measurements can be made with the RFS II in steady or dynamic shear, using parallel plate, cone and plate, or couette geometries. Properties evaluated include such viscoelastic data as: steady shear viscosity ($n$), complex viscosity ($n'$), complex modulus ($G$), elastic modulus ($G'$), loss modulus ($G''$), and damping of low viscosity materials (tan $d$).

*Weipert, D.* (1987) *Getreide Mehl und Brot*, **41**, 11 (German).
*Faubion, J. M.* (1985) In *Rheology of Wheat Products*, AACC, St. Paul, MN, p. 91 (English).

*Bread staling* can be measured by monitoring the crumb response to pressure deformation with time, referred to as 'compressibility'. For this purpose, the Instron test equipment is often applied. Important consumer characteristics are: appearance, smell, taste, and digestibility. Freshness is identified by crispness, a pleasant smell, and an aroma profile consisting of a complex mixture of volatile and less volatile compounds. Aldehydes, forming the bulk of the aroma substances, are chemically unstable and readily oxidized to carbonic acid. Their stability depends on the formation of addition compounds with the amylose or amylopectin molecule of the starch.

*Schoch, T. J.* (1965) *Bakers Digest*, **39**, 2, p. 48.

Schoch considers that the spiral amylose chain is most likely to assume this role, since on reheating stale bread, aromatic aldehydes become released. Aroma substances initially prolific in the crust of the loaf, migrate into the crumb zone immediately under it during a 20-hour storage period, then towards the crumb centre. However, after 20 hours, the content of aroma substances within the crumb decrease, owing not to evaporation, but to chemical or physical transformation into unknown non-aromatic compounds. Although the crust loses its aroma substances by diffusion, the carbonyl content of the crumb will increase to a maximum up to the second or third day of storage, owing to the condensation of less volatile aldehydes with moisture into the crumb area, under the influence of a partial vacuum. An exchange of aroma substances between crust and crumb is evident after about 24 hours, when furfural becomes detectable in the crumb, whereas, immediately after baking it can only be detected in the crust. This exchange of aroma substances between crumb and crust also takes place in packaged bread; too rapid packaging after baking must be avoided to allow adequate cooling, moisture loss and evaporation of unstable, highly volatile aroma components, which would give rise to undesirable aroma contributions. Within the crumb, many chemical and physical changes take place during cooling, which contribute to an aromatic balanced flavour profile. The crust, however, being of low moisture and crisp, changes relatively little. During prolonged storage, some moisture from the crumb migrates into the crust, resulting in it becoming tough and leathery. This movement of moisture from crumb to crust is the best general indicator of freshness, since it parallels crumb-firming, and loss of sensory attributes. The packaging of bread in moisture-proof material promotes crust-staling, since evaporation is restricted, and the crust retains larger amounts of water.

*Processes and mechanisms of the staling of bread* have been studied for well over 100 years, but their exact nature remain obscure. The idea that bread staling is due to moisture loss was disproved by Boussingault.

*Boussingault, M.* (1853) *Ann. Chim. Phys.*, **38**, p. 490.

Storage under conditions of zero moisture loss, still resulted in staling. Also, the well-known method of reheating bread to recover its freshness confirms this fact, since, in spite of additional moisture loss, the crumb recovers its physical properties of freshness. Increases in crumbliness, and reduction in crumb compressibility, are symptoms of staling, and not its cause. However, there is a limit to refreshing bread by reheating, as confirmed by von Bibra.

*von Bibra, E.* (1861) *Getreidearten und das Brot*, Nuremburg, FRG (German).

von Bibra showed that products whose moisture content were reduced to less than 30%, could no longer be refreshed. Owing to the diversity of the changes taking place, isolation of any one process for study is very difficult, and few theories have found general acceptance.

*Microstructural changes* in the bread crumb, observed by Auerman, reported in his reference work *Technology of Breadmaking* form a sound basis for further research.

*Auerman, L. J.* (1977) In *Technology of Breadmaking*, VEB-Fachbuchverlag, Leipzig, GDR, p. 291 (German). Original Russian edition, 1972, Moscow, USSR (Russian).

Auerman states that bread-crumb structure depends on cell wall formation, which, on baking, build crumb porosity in the form of a spongy framework. Microscopic examination of these pore cell walls, reveal a mass of coagulated gluten proteins, with swollen, partially gelatinized starch granules embedded in them. Although these granules expand somewhat, they remain parallel, and surrounded on all sides by coagulated protein, only a few isolated granules being in direct contact with one another. Coagulated protein forms the continuous phase of the spongy crumb framework. In fresh bread, the starch granules lie close to the coagulated protein surface throughout, and no sharply visible boundary layer can be seen between them. In staling bread, the granularity of the partially gelatinized starch is clearly visible, owing to the formation of a thin layer of air around part of the granule surface. The older the bread, the more clearly visible the air layers become, thus confirming the volume shrinkage of the starch. Structural changes in the protein of the cell walls were not discernible under the microscope. The build-up of these thin layers of air around these changed granules are considered responsible for the increasing crumbliness of staling bread. It is pointed out, however, that the volume change of the starch granules is not the staling process, but probably its effect. Numerous further tests confirmed that during bread staling, the hydrophilic properties of the crumb change. The swelling capacity and water absorption capacity of the crumb become reduced, together with the solubility of the colloids and other components. Total water solubles are reduced, and the water solubility of the starch. However, it is acknowledged that although changes in hydrophilic properties of the crumb affect its physical properties, these changes represent effects of staling rather than their direct cause.

*Retrogradation of starch* was first associated with bread staling by Lindet in 1902.
*Lindet, L.* (1902) *Bull. Soc. Chem. de Paris*, **27**, p. 634.
As the term implies, 'retrogradation' involves a process of reversion, in this case a reversion to the crystalline state, after becoming transformed into the amorphous state as a result of partial gelatinization, and absorption of water from the gluten proteins and other hydrocolloids, during coagulation at baking. On subsequent storage, the starch within the crumb tends to gradually revert to the crystalline form in which it existed within the dough before baking. This process results in the starch structure becoming firmer, less soluble, and part of its absorbed water taken up by the proteins of the crumb. With the aid of the X-ray spectrograph, Katz during 1930–32 was able to gain more fundamental information on starch crystallization, and the effects of temperature and moisture on any changes taking place.
*Katz, J. R.* (1930, 1931, 1932) *Z. für physik. Chemie*, Abt. A, **150**, 1, p. 60 and 67; 2, p. 81, 90 and 100; **155**, 3/4, p. 199; **158**, 5/6, p. 321, 337 and 346.
Katz established that the initial native starch of wheat and flour gave an X-ray spectrum typical for a crystalline material, which pattern persisted in the dough up to the baking stage. During baking, limited gelatinization took place, depending on the amount of available water in the dough, but being insufficient for complete gelatinization. The pattern of the X-ray so produced being referred to as a V-pattern, differing from the original native pattern. This characterized the presence of an amorphous element, and a crystalline element within the structure. The crystalline element, however, being quite different from that in the flour and dough before baking. The crumb of stale bread gave an X-ray pattern designated as B-pattern, which was a combination of the initial pure crystalline pattern, and the V-pattern. Moreover, with increasing staling, the B-pattern became increasingly like the pure crystalline pattern in character. This X-ray pattern-B of staled bread Katz designated as the 'starch-retrogradation-spectrum'. Katz regarded the starch in the bread crumb as a thermodynamic system in equilibrium, consisting of alpha- and beta-starch forms. The alpha form being typical for fresh bread, and the beta form for staled bread. The alpha form being stable at temperatures above 60°C. At this temperature, no retrogradation can take place, and the crumb remains in a non-staled state. At temperatures from 60°C down to −10°C, the system equilibrium shifts in favour of beta-starch, characteristic for staled bread. Katz also emphasized the importance of temperature and moisture for these changes. When loaf moisture fell below 16·4%, or when it was moistened with excess water, the physical effects of staling did not occur. Breads stored under controlled conditions of temperature and moisture from 24 to 48 hours at 60°C or above remained fresh; becoming semi-stale at 40°C; almost stale at 30°C; stale at 17°C; and very stale at 0°C. Remaining quite fresh between −7 and −184°C. These temperature ranges form the basis of recommended storage conditions valid today.
*Schoch, T. J.* (1965) *Bakers Digest*, **39**, 2, p. 48.
*Schoch, T. J., and French, D.* (1947) *Cereal Chem.*, **24**, p. 231.

Schoch and French also considered that the starch granule only underwent partial gelatinization and limited swelling during baking, owing to the limited availability of water in the dough. With increasing swelling, the linear amylose fraction becomes more soluble, and diffuses outwards into the aqueous phase, as illustrated in Fig. 140. Continuous swelling results in increasing concentration of amylose in the interstitial regions, forming a concentrated solution. This concentrated solution setting to a gel by the time the loaf has cooled, Schoch considering that amylose tends to retrograde. Fresh bread is considered to contain swollen, elastic starch granules embedded in a firm amylose gel matrix. This gel remaining stable during further storage, and not participating in the staling processes. Crumb firming is attributed to changes in the physical orientation of the branched amylopectin molecules of starch, within the swollen granule.

*In fresh bread*, according to Schoch, the branched chains of the amylopectin are outspread, within the limits of available water, forming a concentrated system. However, the outer branches of the amylopectin gradually aggregate, aligning with one another by various types of bonding. This system, being less stable than retrogradation, can be regarded as an intramolecular association, which solubilizes again under moderate heating at 60°C plus.

*In staling bread*, at temperatures between 60°C and −10°C, this system gives rise to increasing rigidity of the structure of the swollen granule internally, thus explaining the effects of crumb hardening. (see Fig. 140). This original concept of Schoch has undergone some enhancement by Lineback in 1984.

Lineback, D. R. (1984) The role of Starch in Bread Staling. In *International Symposium on Advances in Baking Science and Technology*, published by Department of Grain Science, Kansas State University, Manhatten, KS, USA.

The intention of this concept enhancement is to explain the mechanical functions of starch.

*Amylose* is also involved in the staling process, although amylopectin plays the most important role. Schneeweiss and Klose have established a decrease in soluble amylose during the staling process, falling from about 8 mg/g starch to about 3 mg/g starch, after 2 days' storage (see Fig. 147). As a result of the retrograding (crystallizing) amylopectin, the retrograding amylose can occupy vacant space within the spatial network structure, which could account for the reduction in soluble amylose during staling. The formation of intermolecular bonds result in structural changes, amongst which, a stretching of the starch helix takes place. This in turn will depend on the conformation of the amylose. The necessary energy required to stretch the helix, and for dehydration, is supplied by the formation of hydrogen bonds according to Hollo.

Hollo, J., et al. (1959) *Periodica Polytechnica*, **3**, 3, p. 163. The process of amylose retrogradation (English).

Hollo considers that the whole process of retrogradation of starch after the hydration during the baking process, can be explained in three stages:

—the helical macromolecules become stretched as a result of the absorption of energy,
—after loss of the hydration-shell, a reorientation takes place,
—hydrogen bonds form between the hydroxyl groups of the amylose, and a crystalline
   structure emerges. (See Fig. 148.)

*An intergranular function* of both amylopectin and amylose in structural firming is the subject of another postulate, whereby portions of both amylose and amylopectin molecules extend from the swollen starch granules. These then associate with other carbohydrate chains present in the interstitial aqueous phase, and with those extending from granules in close proximity to one another.

*Crumb-compressibility*, or elastic modulus, increases as the storage temperature falls to 0°C, and is proportional to the concentration of crystallized starch present during staling, according to Cornford and co-workers.

Cornford, S. J., et al. (1964) *Cereal Chem.*, **41**, p. 216.

Data obtained by these workers indicated that the mechanism of starch retrogradation is one of instantaneous nucleation and crystalline growth.

*Senti, F. R., and Dimler, R. J.* (1960) *Bakers Digest,* **34**, 1, p. 28.

Senti and Dimler also postulate a change of state to explain crumb staling. Starch is at a higher energy level in the amorphous form than in the crystalline form. In the absence of external energy, the gelatinized starch tends to retrograde from the amorphous state to the lower-energy crystalline state. This involves a structural reorientation of the linear molecules into a more orderly structure of lateral bonding between chains. Crystallization can account for about 15% of the starch, forming rigid, insoluble regions. These chains, extending through many intervening amorphous regions, interlink the crystallized regions to form a three-dimensional network, within which, the rigid retrograded areas form a backbone for polymer chain entanglement.

*Hollo, J., et al.* (1960) *Stärke,* **12**, p. 106, postulates the three-stage process of retrogradation:

(1)  Breaking of intramolecular bonds, and uncoiling of the helix.
(2)  Loss of bound water (hydration-shells), and molecular reorientation.
(3)  Hydrogen bonding between adjacent chains forming crystalline areas.

If heat is applied to the retrograded gel, the hydrogen bonds joining linear molecules in the crystalline areas are broken, allowing free kinetic movement for the chains, permitting reversion to the normal helical conformation. This explains the mechanism of retrogradation reversal when stale bread is reheated, as illustrated in Figs 140 and 141.

*The most comprehensive staling model* has been presented by Knjaginičev in the USSR.

*Knjaginičev, M. J.* (1965) *Russ. Journal of Chem. Allunion Mendelejew Institute, Moscow,* **10**, 3, p. 227–86 (Russian).

The essence of his postulates are that staling is caused by the change in distribution of water, as illustrated in Fig. 144. According to Knjaginičev, bread, depending on the type, begins to lose its original softness at 6–10 hours after baking, when stored at normal temperatures of 15 to 20°C. The crumb becoming friable, and the bread loses its aroma. During the first 3 days, the moisture content of the crumb centre remains constant, but staling proceeds independently, as confirmed by the following researchers.

*Auerman, L. J.* (1956) In *Technology of Baking,* 6th edn, Piscepromizdat, Moscow.

*Scerbatenko, V. V., Gogoberdize, N. I., and Zelman, G. S.* (1962) In *Maintenance of the Freshness of Bread,* CINTIPisceprom., Moscow.

The crumb perimeter loses moisture, forming a dense layer of 3–5 mm thickness, restricting compressibility when pressed with the fingers, as practised by the consumer. A comparison between the crumb of a staling loaf with that of a fresh one, in spite of equal moisture contents, showed many differences in measurable parameters, as follows: compressibility (elasticity) of the crumb sharply reduced, and the mechanical firmness of the cell walls increased; content of bound-water is reduced; the relative viscosity of an aqueous crumb extract, prepared by rubbing and filtering, together with its relative speed of filtration, both become considerably reduced. Stale bread crumbliness markedly increases, and the water solubility of the crumb components become reduced according to Auerman and Nikolaev and Kulman (*Colloids during Baking,* Moscow, 1940 and 1956). (See Fig. 142.)

*Viscosity of the Crumb-suspension,* measured with the Amylograph is measurably reduced, with increasing loss of freshness. (See Fig. 143.)

*Enikeeva, N. G., and Auerman, L. J.* (1956) *Moscow Tech. Inst. Food Industry,* **4**, p. 105 (Russian).

*Bechtel, W. J., Meisner, D. F., and Bradley, W. F.* (1954) *Cereal Chem.,* **31**, 3, p. 171.

*Ability of the bread-crumb to bleach methylene blue* also decreases with staling.

*Auerman, L. J., and Rachmankulova, R. G.* (1957) *Bread and Confectionery Industry,* **2**, p. 22 (Russian).

*Nazarov, V. I., Sacharov, V. G., and Tichomirova, T. P.* (1958) *High School News Food Technology,* No. 4, p. 113.

*Resistance of crumb-starch to beta-amylase* in stale bread becomes reduced compared with that of fresh bread.

*Rachmankulova, R. G., and Falunina, Z. F.* (1960) *High School News Food Technology*, No. 2, p. 63.

*Changes in the dough and bread colloids,* in particular the bound-water content of wheat and rye doughs during the various technological phases of production and storage are presented in Fig. 144. This shows that an increase in bound-water content occurs at two stages during processing: in the sour/sponge before final-dough mixing, and during baking, whilst the loaf remains hot. Knjaginičev observes that the increase in bound-water during the pre-ferment stage is due to the increased hydration capacity of the proteins, coupled with the increased acidity-index on ripening (lower pH). Initially during baking, the bound-water increases, but decreases as the proteins become denatured. However, when the starch gelatinizes, the bound-water increases again by virtue of the starch, at the expense of the proteins. The general pattern of the curves (Fig. 144) for wheat and rye processing is similar, but the rye curve (1), is appreciably higher in overall water-binding capacity. This difference is due to the much greater pentosan content of the rye, and the colloidal properties of its proteins. In bread, starch and water make up 80–85% of its contents, the sugars, cellular material, lipid and mineral salts being relatively small. The chemical structure of these minor components are such that they exert little influence on the hydrophilic properties of the bread. Whereas, the hydration capacity of the proteins, amounting to 8% or more, have a great influence on the hydrophilic properties of the dough, determining bread volume, crumb porosity and elasticity. The cellular content of the rye meal, consisting mainly of arabinans and xylans, have a great influence on both the colloidal properties of the rye dough, and bread. During baking, these pentosans become denatured, but as far as the physical–chemical properties of the flour and bread components are concerned, whenever the nature of the staling process is under investigation, only changes in the starch and protein have received attention.

*Refreshment of bread* with a moisture content of 30% and above by heating has been long known, and if the moisture is below 30%, it can be refreshed by first dipping in water for a short time.

*Evaluation of the degree of staling* usually only involves a sensory appraisal of the organoleptic properties, but Knjaginičev also studied the changes in crumb-swelling over a 24-hour period at various temperatures. His results confirmed the importance of the hydrophilic, high-molecular-weight polymers (colloids), the crumb losing its ability to swell with staling. Crumb swelling over 24 hours was found to be a maximum at 60°C, when the bread was sensorically judged as 'fresh', and reached a minimum at 0°C, when it was judged as being 'completely stale'. Bread stored at −30°C was judged as completely fresh, and the crumb swelling almost reached the 'fresh' state maximum value. Table 72 shows the crumb-swelling data at various storage temperatures during 24 hours, and the corresponding sensory evaluations.

*Freshly baked bread,* stored within the temperature range 60 to 90°C, or −20 to −190°C remains in the fresh state due to a successful retardation of the staling processes. Temperatures between 50°C and −7°C, including room or ambient temperature range, are unfavourable in this respect. Although the part played by storage temperature in the maintenance of freshness has not yet provided a full clarification of staling, investigations of changes in physical–chemical properties conducted parallel with any retardation-process-development can yield further information concerning bread components and their response.

*Starch,* which contributes 50% to 60% of the composition of bread and baked products, consists of about 20% amylose and 80% amylopectin. These components differ in both their molecular weight and their chemical behaviour, being arranged differently within the starch granule. Some areas within the granule are discretely organized with a crystalline space-lattice X-ray pattern spectrum; other areas are randomly organized, according to Badenhuizen.

*Badenhuizen, N. P.* (1971) Structure and formation of the starch granule. In *Starch Handbook*, Paul Parey, Berlin and Hamburg.

Interference patterns of the rays show restricted sharpness in the contours of the rings of the X-ray pattern. Rye-bread shows a wide, diffuse ring, similar to completely gelatinized wheat starch. Water content is also reflected in the X-ray spectrum, starch at atmospheric moisture levels conforming to the native starch spectrum; water-free starch exhibits a pattern similar to amorphous starch. Interference patterns of the X-rays show a strong dependence on the starch water absorption, in the case of crystalline starch, water is built into the lattice in the form of hydrogen bonds. In flour, starch particles only bind water at the surface by adsorption at 20°C, any hydration only starting at the damaged areas of the granule. During baking, the energy input results in granule swelling, water resting between the amylose and the amylopectin molecules, both organized and randomized areas becoming hydrated. Spatial arrangements remain intact during this swelling process. Network bondings being typically hydrogen bridges, and van der Waals force bonds, which become partially broken during hydration; the hydrated starch molecule at temperatures above 60°C is the energy-stable form.

*Starch cooling* results in the system becoming thermodynamically unstable, and staling or 'retrogradation' progressively commences. Retrogradation is applied to reactions taking place in the aqueous starch suspension, in addition to those in the crystalline state, including the ageing of amylose and amylopectin. Immediately after baking, the amylose is mainly colloidally dispersed, and the amylopectin in the swollen globular state. Cooling involves a process of partial crystallization, the formation of a structured gel; then crystals of amylose and amylopectin, and mixed crystals form, which progressively form an increasingly firm mechanical structure. Speed of crystallization depends on the ratio of amylose to amylopectin, and the molecular orientation of the components. Wheat starch crystallizes more rapidly than potato starch, hence the use of the latter to retard the entire staling process. The stability of colloidal systems depend on particle aggregation, as is the case with the starch granule. Both amylose and amylopectin become surrounded by thick water layers known as 'hydration-shells', or 'solvation-shells'. Mutual contact is prevented by surface charges, shell repulsion determining their spacing. The repulsive force depends on the dimension of the hydration-shell, and the magnitude of the electrical charge. The ability to form hydrogen bridges is very significant in the case of starch, depending on the steric formation of the molecule; amylose is capable of assuming various helical conformations in solution, as pointed out by French.

*French, A. D.* (1979) *Bakers Digest*, **53**, 1, p. 39.

Hydrogen bridge formation in amylose and amylopectin is schematically illustrated in Fig. 148.

*Retrogradation* results in the formation of a space lattice structure from amylopectin, inside which the retrograding amylose becomes engulfed, which explains why a reduction in the soluble amylose occurs after about 2 days' storage. Also, according to Knjaginičev, during staling, the water molecule becomes subjected to a molecular reorientation, due to its four charges. One water molecule, depending on its four charges, can orientate four other water molecules, forming an aggregate of five molecules as an hexagonal lattice with loose spacing (as in ice formation). A denser spacing is made possible by a reorientation of the hexagonal lattice to form a tetragonal one, or owing to filling of vacant spaces with more water molecules. Fig. 146 shows the proposed Knjaginičev model in schematic form, depicting how, during dough processing, the flour forms with water, under limited swelling, a viscoelastic gel. Only part of the dough water participates in the swelling of the polymers. Mechnically bound water is depicted as dots and arrows. Initially, the dough polymers in the form of a swollen gel are unstructured, forming a one-phase system.

*Dough-heating during baking* favours diffusion of water into the intermolecular spaces of the starch and proteins. This results in the macromolecular monomers, viz. sugars and amino acid residues, becoming mobile, losing their initial compact posture. Thus, macro and micro hollow cavities develop, and under the influence of higher temperature, the proteins coagulate, forming a framework, and fixing the porous volume of the bread. The crumb cell-walls, consisting of starch and protein, constitute a swollen system in which one part of the water is thermodynamically bound, and the other part distributed in the intermolecular spaces of the denatured proteins and the swollen partially gelatinized starch. In effect, both the dough stage and the baking dough-piece constitute a swollen, elastic, structureless gel.

*On removal from the oven*, owing to the flexibility of the starch monomers, the molecular chains move closer together, and intermolecular van der Waals force bonding results in the formation of a mechanically firm network, i.e. a structured gel. This completes cooling and setting of the loaf.

*Bread staling processes* result in this network progressively reaching a maximum state of mechanical firmness. Knjaginičev postulates that the water in the macropores is probably not in the free state, but rather exists in an ordered form owing to the high polarity of the molecules, and the electrostatic forces at the surface of the micropores. The walls of these micropores consist of the macromolecular starch and proteins, as schematically illustrated in Figs 145 and 146. Eventually, the water, starch and proteins form a uniform structured system.

*Refreshing the bread by heating* results in the water structure within the pores of the crumb being destroyed, according to Knjaginičev, and the macromolecular chains reverting to the freshly baked state, as shown schematically in Fig. 146, 4 to 3, then to 2. Thus depicting the stages of swelling and gel formation, depending on the state of the water molecules.

*Knjaginičev's bread-crumb staling model* represents a revision of former postulates, in order to present a more realistic approach to the questions posed by bread staling. The basic concept is the formation of a structure within the micropores of the starch and protein polymers of the bread as a result of the reorientation of the water molecules. The resulting aggregates forming an hexagonal space-lattice, similar to ice, but loosely packed. As staling proceeds, the spatial density becomes compacted, either owing to reorientation of water molecules to form a tetragonal lattice, or as a result of water molecules entering the empty spaces within the lattice. The crumb is thus transformed into a firm structured state, the main participant being the water molecule, its mobility and colloidal activity. Storage of bread within the temperature range 50 to $-7°C$ results in both the starch macromolecules, and the water molecule losing their mobility, giving rise to a firm structured system. Additional tests, involving the addition of starch granules to white bread before baking confirmed that similar moisture losses occurred after 4 hours and 8 days, both in the added starch and the bread-crumb. Therefore, there is no appreciable migration of water from the water-rich crumb to the water-free starch. Knjaginičev consequently concludes that the crumb consists of a system of pores with cell-walls of starch and denatured proteins in a swollen state; within which the water is in part thermodynamically bound, and partly distributed within the intermolecular spaces of the proteins, and swollen, partially gelatinized, starch. The molecules of water, starch and proteins forming a uniform structural system. During the refreshing of staled bread by heating, the structure of the water within the micropores of the crumb is disturbed, and the starch and protein macromolecules revert to a state similar to that of fresh bread.

*Polymer crystallization* principles have been applied to the study of starch retrogradation, since it has been widely accepted that it involves a crystallization process.
Avrami, M. (1940) *J. Chem. Phys.*, **8**, p. 212.
Avrami, M. (1941) *J. Chem. Phys.*, **9**, p. 177.
Cornford, S. J., Axford, D. W. E., and Elton, G. A. H. (1964) *Cereal Chem.*, **41**, p. 216.
Sharples, A. (1966) *Introduction to Polymer Crystallization*, Edward Arnold, London.
Kim, S. K., and D'Appolonia, B. L. (1977) *Cereal Chem.*, **54**, p. 216 and 225.
Elton, G. A. H. (1969) *Bakers Digest*, **43**, 3, p. 24.

*Axford, D. W. E., et al.* (1968) *J. Sci. Food Agric.*, **19**, p. 95.
*McIver, R. G., et al.* (1968) *J. Sci. Food Agric.*, **19**, p. 560.
*Krüsi, H., and Neukom, H.* (1984) *Die Stärke*, **36**, 2, p. 40–5 (German).

*Soluble starch extraction from bread crumb* has also been applied to the measurement of the degree of retrogradation (staling).
*Kim, S. K., and D'Appolonia, B. L.* (1977) *Cereal Chem.*, **54**, p. 207 and 216.

*Amylose/amylopectin ratios*, and their rates of retrogradation have also been the subject of research.
*Neukom, H., et al.* (1981) *Lebensm. Wiss. und -Technologie*, **14**, p. 292 (German).

*Protein/starch ratios*, and their exchange reactions in the presence of ions, have also been considered to participate in bread staling processes.
*Erlander, S. R., and Erlander, J. G.* (1969) *Die Stärke*, **21**, 12, p. 305.

*Protein physical–chemical changes* during bread storage, as a result of ionic bond and hydrophobic–ionic interactions, have also been linked with the staling processes.
*Kolpakova, V. V., Nazarenko, Y. A., Zharinov, V. I., and Burak, I. A.* (1987) (USSR) *Mlyn-Pek. Prum. Tech. Skladovani Obili*, **33**, 11, p. 321–3 (Czech).

*Gluten/starch moisture exchange mechanisms*, have been variously proposed by some researchers.
*Cluskey, J. E., et al.* (1959) *Cereal Chem.*, **36**, p. 236.
*Willhoft, E. M. A.* (1973) *Bakers Digest*, **47**, 6, p. 14.

*Flour gluten-protein quality* and adequate fermentation in bulk, and after shaping, have also been found to influence staling rates.
*Maleki, M., et al.* (1980) *Cereal Chem.*, **57**, p. 138.

*Starch energy levels* in the amorphous and crystalline states, and the availability of external heat energy, has been linked with gelatinized starch retrogradation.
*Senti, F. R., and Dimler, R. J.* (1960) *Bakers Digest*, **34**, 1, p. 28.

*Dough-enrichment raw materials*, containing various protein sources such as soy, gluten, ground-nut, and non-fat milk solids, will reduce crumb-firming rates, but have little effect on the ageing of the crust, which loses its crispness, becoming elastic and tough. This latter trend is due to moisture exchange with the crumb, as is the case with unenriched bread.

*Non-starch polysaccharides* such as cellulose, pentosans, hemicellulose, beta-glucans, mannans, gluco- and galactomannans (pectins), also glycoproteins, all constitute a group of substances which have an effect on dough and baked products, right through to the staling processes, on account of their chemical, physical and rheological properties. Wheat-flours can contain 3–4% of these substances, depending on the extraction rate, some of which are 20–25% cold-water soluble. Rye-flours contain about the same levels at low extraction, but the cold-water soluble fraction is about 40%. Studies of the effect of pentosans on staling rates, concluded that they slowed down the rate of crumb firming, especially the water insoluble fraction.
*Kim, S. K., and D'Appolonia, B. L.* (1977) *Cereal Chem.*, **54**, p. 225.
The water-soluble fraction of the pentosans reacts with the starch amylopectin only, but the insoluble fraction retarded and retrogradation rate by interaction with both the amylose and amylopectin.
*Kim, S. K., and D'Appolonia, B. L.* (1977) *Cereal Chem.*, **54**, p. 150.
The presence of the pentosans tends to reduce the exposure of the starch components to crystallization during staling, acting like a protective-colloid. Protective-colloids owe their stability and high viscosity to their hydration-shells, giving rise to the formation of large particles.
Gilles and co-workers found pentosans in the soluble-starch fraction of the bread-crumb; the 4·3% soluble starch extracted from fresh bread-crumb contained 11·7% pentosans. This

compared with 3·3% soluble starch extracted from stale bread-crumb, of which 19·3% was pentosan. This increase in relative amount of pentosans in soluble starch with bread staling is attributed to the crystallization of amylose and amylopectin. Since pentosans have been shown to retard amylose retrogradation, the decrease in soluble starch due to staling is considered to be essentially the result of the aggregation of amylopectin.

*Gilles, K. A., Geddes, W. F., and Smith, F.* (1961) *Cereal Chem.,* **38**, p. 229.

*Instrumental physical techniques* applied to the study of mechanisms of starch gelation and retrogradation (staling) include: dynamic/mechanical thermal analysis, involving a torsion rheometer system, providing viscoelastic data, viz. steady shear viscosity, complex viscosity, complex modulus, elastic modulus, loss modulus, and damping of low viscosity materials; differential scanning calorimetry (DSC); X-ray diffraction; Fourier transform IR (FTIR) spectroscopy. The time course of FTIR measurements during storage retrogradation of starches is similar to that produced from shear modulus tests. This method being applicable to starch gels, and real foods, e.g. commercial bread. DSC techniques are frequently used to study the effects of various emulsifiers and crumb softeners on bread staling and starch-gel ageing.

*Eliasson, A. C.* (1985) *Prog. Biotechnol.,* **1**, p. 93–8. New approaches to research in cereal carbohydrate.

The DSC endotherms of stale bread or aged starch-gels provide amylopectin melt read-outs. The size of the endotherm increasing with increasing storage temperature and time. The addition of sodium stearoyl lactylate to starch gel decreases the staling endotherm. Amylose/lipid complexes affect the crystallization of amylopectin. DSC endotherm measurements do not monitor the same changes in ageing starch gels as those measured during gel strength tests using compression methods in the elasticity mode. Measurement of water concentrations in the free and bound state is possible with NIR spectroscopy, utilizing specific bands at 1940 and 1450 nm. The band at 1450 nm can be resolved into two discriminant spectral patterns at 1410 and 1460 nm for proportions of free and bound water. The application of nuclear magnetic resonance (NMR) techniques are also becoming increasingly important for the study of aggregate and binding states in solid/liquid food systems. Nuclear resonance impulse spectroscopy can be applied to the measurement of diffusion coefficients during drying-out processes, and molecular exchange reactions. Moisture redistribution between bread components during staling is a possible application for NMR studies.

*Chemical procedures* applied to studies of the staling processes often include the following: soluble starch/amylose determination/g starch with time; saccharide analysis; protein extraction and resolution; pentosan extraction and resolution; presence of gluten/starch complexes; presence of starch/lipid inclusion complexes.

*Banecki, H.* (1970) Congress Report No. 5, World Wheat and Bread Congress, Dresden, Vol. 5, p. 179.

*Hüttinger, R.* (1972) *Gordian,* **72**, 7/8, p. 261. Emulsifiers in baking.

*Erlander, J. G.* (1969) *Die Stärke,* **21**, 12, p. 305. Explanation of the ionic in connection with various phenomena. Protein/Carbohydrate exchange reactions, and the mechanism of bread staling (English).

*Acker, L., et al.* (1968) *Getreide und Mehl,* **18**, 6, p. 45. Wheat lipids (German).

*Acker, L., et al.* (1974) *Getreide Mehl und Brot,* **7**, p. 181–6 (German).

*Effects of fermentation schedules and storage-time* on sensory evaluation and volatile compounds in white bread crust and crumb, were investigated by Stoellman.

*Stoellman, U. M.* (1986) Shelf-life foods and beverages. *Developments in Food Science,* Vol. 12, p. 293–301.

*Iodine-binding intensity* of amylose and amylopectin have been used to monitor the staling of baked products.

*Hampel, G.* (1969) *Brot und Gebäck,* **23**, 6, p. 106. Investigations of the staling-processes in baked-products (German).

## CHAPTER 3.6 DOUGH AND BREAD PRESERVATION

*Retardation of staling processes* has been most successfully achieved by freezing and the application of cryogenics, which represents the closest approximation to 'oven-freshness' in the case of most baked products produced by fermentation.

*Anti-staling agent application* is restricted by food safety regulations, and has been increasingly restricted to natural raw materials, and their modifications. These substances can be grouped as follows: protein, carbohydrate, cellulosic-based; enzymes and enzyme-containing substances; shortenings and emulsifiers; aliphatic and cyclic aldehydes, which locate between the hydroxyl groups of the starch and protein, anti-staling function increasing with chain-length.

*Dough Improver/conditioner mixtures* are manufactured in powder, paste, dispersion or tablet form, consisting of blends of emulsifiers, enzyme preparations, shortenings, malt-flours, sugars, hydrocolloids, with a carrier material. Often, ascorbic acid is added, with or without other permitted oxidants to form a complete processing-aid. See Section 1.5. The most widely accepted emulsifiers are the phosphatides, of which the lecithins are the commonest. These polar lipids form complexes with the amorphous starch fraction, the presence of non-polar groups prolonging hydration of the crystalline regions of the starch granule.

*Phosphatides* are also highly functional when used in combination with vegetable oil and water in the ratio 5:5:90 in emulsion form. This emulsion, at 0·05% flour weight is then combined with 0·3% of enzyme-active full-fat soy-flour (flour weight), as a source of active lipoxygenase, and made into a dough with wheat-flour, yeast, salt and water in the usual manner. The soy-flour is ideally first dispersed in a portion of the dough water to release the lipoxygenase into solution. The mechanism of the functional reactions is the oxidation of the polyunsaturated fatty acids of the vegetable oil, and the phosphatide emulsion concentrate by the lipoxygenase and the entrapped oxygen. The resulting hydroperoxides act as nascent or 'active-oxidants', reacting with the SH-groups of the protein–proteinase complex of the wheat-flour. This reaction improves the gas- and structure-retaining capacity of the dough, resulting in bread with improved volume, symmetry, texture and crumb elasticity. Also, the peroxides of the fatty acids oxidize and bleach the carotenoid pigments, giving rise to a brighter crumb colour. This technique is a cost-effective method of processing untreated/unbleached wheat-flour with successful results, and satisfying a 'clean-label' legislation requirement. For the preparation of the so-called 'liquid oxidation phase', which can be on a batch or continuous basis, about 50–75% of the total dough water is utilized; and in the case of a pre-ferment sponge, all the water is normally used in its preparation.

Auerman, L. J., Kretowitsch, W. L., Polandowa, R. D. Process for the quality improvement of wheat bread. USSR State Committee for Inventions and Discoveries, Patent No. 164860, with priority from 8:7:1963 (Russian).

Auerman, L. J., Kretowitsch, W. L., Polandowa, R. D. *PrBchMb*, 1, 1, 66, 1965 (Russian). This process results in a considerable improvement in dough rheology, improved water binding, and a more rapid dough ripening owing to acid build-up. This process is similar in conception to the Blanchard process, but not in practical execution.

*Enzyme-active full-fat soy-flour* is a cost-effective, safe, anti-firming agent of quality and long-standing. The addition of 0·5%, based on flour weight, reduces the crumb-firming rate of bread by reducing the rate of moisture interchange between starch and gluten proteins. The average composition of full-fat soy-flour is: protein 41%, lipid 20%, moisture 5%, and about 3% pentosans, which also contribute to dough water-binding capacity. Defatted soy-flour has a reduced beany flavour, and inactivated enzymes, as a result of heat treatment. This material can be used at levels of of 3–5% flour weight, and as a nutritional supplement, especially for the amino acid lysine, deficient in wheat-flour. Detailed information concerning dough improver conditioners can be found in Section 1.5.

820		*Handbook of breadmaking technology*

*Packaging materials* with the appropriate properties can offer considerable retardation of staling by hindering dehydration and preventing the partial water pressure from causing progressive moisture loss to the atmosphere. The degree of staling retardation depends on the choice of packaging material; however, since no material can satisfy all requirements, a compromise is found between price and functionality. Waxed paper was the only material used for bread for many decades, being first coated with paraffin-wax, but later this was blended with microcrystalline wax and polyethylene resin to improve appearance, elasticity and sealing properties. The advantage of paper was ease of imprinting for labelling purposes. Cellophane films gained wide acceptance in the late 1930s, first being a film or sheet of a defined thickness and hence strength.

*Cellophane* was improved by the application of coatings of either nitrocellulose or polyvinylidene chloride, which contributed to its resistance, gas permeability, heat-sealability, and moisture vapour transmission rate (MVTR), also acting as an oxygen barrier. It is an excellent material with wide application, but relatively expensive.

*Plastic polyolefin films*, e.g. polyethylene and polypropylene, replaced Cellophane (cellulose) films during the 1950s for certain product applications. They differ in tensile strength, rigidity, permeability, and temperature-sensitivity. Various manufacturing methods provide a wide range of films of various gauges, orientations (cast/unbalanced), coatings, and laminate combinations.

*Polyethylene film*, available in high, medium and low densities, is the most popular packaging material for bread and rolls, the standard low-density film being the basic material for perforated bags. For greater heat stability, a medium density film is appropriate.

*Polypropylene film* has similar properties, but is more rigid and stronger at the same gauge or thickness. It can also be used as a component in laminated films to provide stability at heat-sealing. Special inks and printing techniques provide a range of colours and designs for marketing. Hoffmann and co-workers in the GDR, found that the use of different packaging material gave different improvements in moisture retention.
*Hoffmann, R., et al.* (1976) Extending the freshness of bread, rolls, and specialty products. Part I: Bread and rolls. Research report IGA, A4, Bergholz-Rehbrücke, GDR (German).

*Improvement in retention of freshness* with aluminium–polyethylene bags was maximum 23%, polyethylene-coated paper maximum 16%, and polyethylene film maximum 9%. Rapidity of staling was also found to depend on storage temperature and RH. In the case of mixed rye/wheat bread (*Mischbrot*), storage at 20°C for 24 hours gave crumb-firmness equivalent to that at 28 hours when stored at 30°C; and at 60°C, crumb firmness at 48 hours was comparable with that at 16 hours at 20°C. Crumb firming reaching a maximum at 4°C.

*Packaging material criteria for the baker* are: degree of product protection; performance on existing packaging-machines: attractiveness of the package to the consumer, to keep ahead of competition; justification of cost. Polyolefin films have now become the established material for bread, and the orientated polypropylene for sweetened products. Coated orientated polypropylene find application in overwrapping products in trays and cartons.

*Aluminium foil* is used for specialty products, where a low moisture transmission vapour rate (MTVR) and oxygen transmission rate (O2TR) is required. The cost is, however, about two to three times greater relative to polyethylene.

*Prevention of microbial infection* in all storage and dispatch areas is a prerequisite, and general maintenance of cleanliness and hygiene practices by personnel within production, storage and dispatch is essential to avoid cross-contamination. Transportation containers, operative's clothing and wash areas must be regularly cleaned and disinfected.

*Airborne spore-counts* within the bakery must be reduced to a minimum. Where air-filters are installed, areas must be hermetically sealed, and the air cooled before entering the filter. Filter efficiency also depends on filter design, size and electrical charge of particles, their temperature, RH, and degree of air pollution. Air drawn into the bakery should be UV-irradiated by wall-mounted units at a height of about 2 m. Spores passing within about 3 cm of the irradiation source, as a result of convection currents are destroyed, provided the rays strike them at right-angles. Equipment can also be UV-irradiated, but personnel must be provided with green-filter goggles, or the irradiation carried out in their absence.

*Bread-slicing and -wrapping departments* must be kept under a strict hygiene regime at all times. Only people working therein should have regular access; temperature 18–20°C and RH 65% should be automatically regulated; personal hygiene and regular medical check-ups for personnel are essential disciplines.

*Optimal fermentation and process control* can also help to prevent microbial infections, and premature onset of staling processes. Adherence to adequate yeast levels (about 2% flour weight), and use of dough temperatures within the range 22–26°C, allows doughs to mature for a maximum time, depending on flour quality, and will hinder the establishment of airborne fungal or bacterial spores. Maintenance of dough pH within the 4·8 to 5·0 range for wheat-flour, and final bread acidity-index of 12 for rye wholemeal flour doughs, will hinder any incidence of *B. mesentericus* (potato bacillus) or *B. subtilis* (hay bacillus). These species are broadly distributed in air, soil and plants, and can be found in grain and flour on occasion. Since these bacilli survive the baking process, and liquefy and destroy the crumb structure, their optimal conditions of reproductive growth must be avoided. In the case of dough-sours, this can be done by increasing the souring intensity of the basic-sour, or in both yeasted wheat and soured rye doughs by the addition of an appropriate amount of lactic or acetic acid. In this respect, a target pH of 3·8 in the final dough should solve the problem.

*Chemical preservatives or mould inhibitors* are widely applied in countries where legislation permits. They include salts of propionic and acetic acids, and sorbic acid and sorbates. In the USA, a maximum of 0·32% sodium propionate, 0·4% sodium acetate, or 0·75% calcium monophosphate are applied, all based on flour weight. Sorbic acid used at the 0·1% flour weight level is a very effective fungistat, owing to unsaturation in molecular terms and lower dissociation constant in solution. Where shortenings are used, it can be added to the dough after premixing with the shortening to avoid yeast inhibition. Hickey has suggested a spray application of 1·0–1·5% potassium sorbate solution on hot, freshly baked bread, rolls, and partially baked products, as effective in doubling or tripling the mould-free shelf-life of the products. This gives sorbate residuals of 0·02%, based on flour weight.
*Hickey, C. S.* (1980) *Bakers Digest*, **54**, 4, p. 20.
Schulz has suggested the use of dough sours combining lactic and propionic fermentation, by using special cultures of propionic acid bacteria. This resulted in a prolongation of bread moulding to 14–19 days, without affecting either the flavour or volume of the bread.
*Schulz, A.* (1959) *Brot und Gebäck*, **13**, 7, p. 141 (German).
The continental European practice of using relatively high initial baking temperatures, known as '*vorbacken*' (German), resulting in early crust formation, is recommended for hearth breads. This is then followed by extended baking at a lower temperature.

*Pasteurization/sterilization of bread and baked products.*
Heat sterilization of specialty sliced breads, such as pumpernickel, is achieved either at high temperature, or by the utilization of energy-rich ionized radiation. Sterilization within the packaging gives the best guarantee for prolonged mould-free shelf-life, providing the packaging material has been previously sterilized, and no subsequent contamination takes place. The process depends on temperature and time. At 150°C, a sterilization time of 15 minutes is normally sufficient; but at 98°C, an exposure-time of 12 hours is necessary to obtain the same degree of sterilization.

*High-frequency microwave heat sterilization* permits exposure-times of 10 seconds to 2 minutes, depending on the type of product, dimensions, and packaging materials used. An average schedule could be 2 hours at 90°C for a 250-g sliced bread pack. Owing to the effect of temperature the package shrinks, and the overlapping film must be immediately sealed from external contamination. Heat treatment is normally carried out in a tunnel, heated either by gas or electricity, positioned downstream of the packaging-machine, but adjacent to it. A temperature gradient always exists between the inlet and outlet end of the shrink-tunnel, the packages being transported through on a band conveyor. The heat-stability of the packaging material is often the limiting factor for the choice of the sterilization temperature. On exceeding a certain temperature, most materials become brittle or melt. The application of microwave heating will not allow the use of metallic packaging materials, coated polyethylene films being more appropriate. Such techniques are routinely only applied to specialty products such as pumpernickel and whole-grain meal breads, e.g. *Vollkornbrot*, which have a dense, close texture. Breads made from low extraction wheat-flours with higher volumes, and porous textures could suffer damage from heat sterilization processing. Burg has described a tunnel microwave sterilizer designed for 400 to 600 lb bread capacity.

Burg, F. (1968) *Brot und Gebäck*, **22**, p. 58 (German).

Sterilization takes place by conveying the packaged bread through the tunnel with a residence-time of 45–90 seconds. In such cases, a moisture-permeable wrapping material has to be used, since the shock treatment results in some moisture release within the packs. Another solution is to use a moisture-permeable material, followed by an airtight packaging material; or opened plastic bags can be used, the bags then being sealed on completion of sterilization and moisture evaporation.

*IR-sterilization* for bread and baked products, has been described by Zboralski; it offers technical and economic advantages, being first applied in France in 1965, for sliced bread and cake. The process utilizes IR rays in the middle range of the spectrum (2500 Ångström degrees), the radiation being absorbed by the product, and transformed into heat, temperatures of 160–170°C being attained. Important for the success of this process is the packaging material, a high-temperature stable special polyamide foil (nylon-type) being most suitable, together with a low vacuum. The packaging material must be heat-stable up to 200°C for the duration of sterilization; it must be steam-proof, gas-proof and impervious to spores; have good heat-sealing capability; and allow penetration of the IR-rays without loss of energy.

Zboralski, U. (1973) *Getreide Mehl und Brot*, **27**, 6, p. 213–16 (German).

The necessary requirements were only satisfied by a stabilized film from Polyamide 11 ('Rilsan'), available in tubular form of 40–60 $\mu$m strength. The product is placed in the bag by special equipment, and then packed under a low vacuum with a low vacuum sealing machine, adjusting to avoid product deformation. The advantage of the IR-sterilization process is the extremely short time required compared with heat-convection processes. This is due to a two-stage process of heat-transfer. The IR-rays penetrate a few millimetres into the product, and become almost instantaneously transformed into heat energy. This *in statu nascendi* liberated energy can achieve temperatures of about 300°C, but rapid conduction distributes heat throughout the product interior, the centre rapidly attaining 70–80°C. The product surface also receives a rapid energy transfer, resulting in a reduction in sterilization-time of 50% compared with hot-air sterilization. To facilitate heat control, the tunnel is normally divided into three heat-zones (See Table 73.)

*X-ray, beta- and gamma-radiation*, techniques have also been successfully applied to bread preservation. Warming of rye/wheat breads to 50°C, and irradiation with 0·05 Mrad, permits preservation for several weeks, without sensory devaluation.

Stehlick, G. (1968) *Atompraxis, Karlsruhe*, **14**, 4/5, p. 1. Gamma-rays on bread (German).

For the application of ionized radiations for preservation, specially treated polyethylene has proved a suitable packaging material. Although preservation by ionized radiations has not yet become widely acceptable, owing to the potential health hazards—mainly due to induced radioactivity—the energy levels used are too low to result in radioactivity in the product. The maximum irradiation dose likely to be permitted generally, as a result of the WHO

recommendation published in 1981, is 10 kGy (0·005 kGy = 0·5 Gy = 50 rad). Extent of penetration depends on product density, and electron beam energy. In The Netherlands, all foods can be irradiated up to the 10 kGy dosage level, which substantially reduce the microbial load, and the number of non-sporing pathogens. For sterilization, higher doses can be used commercially. Clearances by the Atomic Energy Agency in Vienna, have been granted for rye bread sterilization in The Netherlands and USA, and for wheat and wheat-products in USSR, Canada, Brazil, Spain, Chile, Thailand, and Bangladesh.
*Official Journal of European Communities*, No. C99, 13:04:87.
Various claims against irradiated foods on grounds of safety hazards, and carcinogenic residues have been investigated by the WHO/IAEA/FAO, and rejected. For food products, the accepted radiation unit is the 'gray' (Gy), which is defined as the amount of radiation absorbed by 1 kg food, equivalent to 1 joule of energy; 1 kGy = 1000 Gy. Considerable confusion exists between 'radioactivity' and food preservation by the technique of 'irradiation'. Irradiation is a physical method of food preservation, using ionizing radiation, ionizing in the form of short wavelength electromagnetic energy, e.g. X-rays. Resulting free radicals lead to the destruction of microbial DNA, but the dosage levels are too low to induce radioactivity in the product. Also, in the practical execution of these techniques, the food never contacts the radioactive source, e.g. $^{60}$Co, or $^{137}$Cs, therefore, there is no chance of the food becoming radioactive. Currently, there are two main forms of ionizing radiation commercially applied. Gamma-rays emitted by radioactive sources such as cobalt-60 or caesium-137, in which case the food never contacts the source, hence no chance of the food becoming radioactive; and the use of high energy electrons in the form of electron beam irradiation, which does not require a permanent radioactive source, and is therefore environmentally attractive. Its application is, however, limited by the relatively weak penetrating power of the electrons, being only suitable for foods up to about 10 cm thickness. The application of X-rays is under current research, but no commercial X-ray irradiators are presently available.

*Packaging of bread*, although having little effect on the chemical staling processes *per se*, does reduce the rate of crumb-firming compared with unwrapped bread, and also provides a hygienic barrier for handling.
*Martin, G. W.* (1944) *Food*, **13**, p. 129.
*Cathcart, W. M.* (1940) *Cereal Chem.*, **17**, p. 100.
Sliced bread packaged in a moisture-proof material only lost about 2% during 72 hours, owing to a state of equilibrium being reached between crumb and crust. Unwrapped bread loses moisture mainly from the outer 13 mm of crumb, already at 24 hours, forming a dry crumb adjacent to the crust.

*Wrapping-temperature*, is an important parameter both for keeping properties, and for sensory characteristics. Berg concluded that warm wrapping at 45°C gave the best crumb softness after 36 hours storage, but resulted in severe loss of aroma and taste. Conversely, cool wrapping at 31°C gave the greatest loss of crumb-softness, but the best retention of aroma and taste. These results, however, refer to high-specific-volume panned bread, and cannot be applied to hearth-breads.
*Berg, I. A.* (1929) *Am. Soc. Bakery Engrs Bulletin*, 39.
Hearth breads should have a maximum temperature at the crumb-centre of 40°C, and where cold slicing is the chosen procedure, crumb-centre temperatures should fall within the range 20–25°C. European experience indicates optimal bread maturing-times of 2 hours for warm slicing at 40°C in the case of wheat bread, and 3 hours for rye and rye/wheat bread types.

*Wrapping of hot, crusty hearth-breads in the unsliced form* can be successfully achieved by using a 360 micrometre, perforated polypropylene film, which is available as a centre-folding film on the reel, or as bags. Alternatively, a new white-pigmented, biaxially orientated polypropylene film, 35 micrometres thick, and co-extruded on both sides with polyolefinic copolymers, has been found ideal for specialty-breads. This is due to excellent water-vapour retention properties and UV-light transmission, being heat-sealable on both sides. Since many compounds from packaging materials can migrate into food products, the legal consequences of resultant interactions for food quality and consumer protection should be evaluated.

*Piringer, O.* (1988) *Chem. Ing. Tech.,* **60**, 4, p. 255–60 and 265 (German).

*Interactions between Food and Packaging.* Fraunhofer Inst. Lebensmittel-technologie und Verpackung, D-8000 Munich 50, FRG.) Review of 14 references.

For the packaging of bread and rolls, intended for fresh consumption, or consumption within 1 week of baking, the slicing-room atmosphere must be kept as spore-free as possible, and packaging materials, machines, and their blades kept clean and sterilized. If these basic regimes are adhered to, the packaging of baked products should prolong their shelf-life by a factor of two to three times that of unpackaged products.

### 3.6.3 Dough Preservation by Freezing

Although the application of freezing technology to baked products remains important for the baking industry, of far greater significance is its application to doughs. This technology has revolutionized the working schedules of production, and facilitated forward-planning of inventories, and their distribution to wholesale and retail outlets. This sector of the industry now includes the large and small frozen-dough product manufacturer, and the ever-increasing number of non-manufacturers in the wholesale and retail market, who are described as 'bake-off' bakeries or retailers. Whether defrosted and warmed up in the case of baked products, baked-off domestically, or baked-off by a retailer or in-store bakery as a frozen dough-piece, the eating quality of the final product has the 'fresh' characteristics.

*Prepared and shaped dough-pieces* for a wide variety of baked products, can be blast-frozen at high volume output with a core temperature of $-20°C$ within 20 to 60 minutes, depending on product mass. For smaller output requirements, tray-loading freezers offer similar results at about one-third of the cost of a tunnel-freezer. Such heavy-duty units are suitable for fermented products, gateaux, and pastries with mass up to 400 g. The frozen products are then transferred to a low-temperature holding cold-room. Once frozen, the dough-pieces must be maintained at $-18$ to $-21°C$, until required for the retarder-proofer, for gentle adjustment to the retardation temperature state, and subsequent proofing. Distribution to receiving bakery outlets is by refrigerated vehicle, and once the frozen items are received, they are held either in a freezer cabinet/room or retarder-proofer, depending on the operation. The advantage of a retarder-proofer unit is that the dough-pieces can be stored and gently brought to a retarded state until required, at which point, proofing can be initiated ready for bake-off. Retarder cold-rooms can be employed, set at $2°C$ for overnight storage, or $-4°C$ for 72-hour storage. Retarder cabinets are used in smaller outlets, which combine both temperatures, selected at the flick of a switch. The important concept of freezing-process technology in the bakery is that one can separate in terms of time and place the dough preparation and fermentation from the bake-off process.

*Retarder-proofers* combine the benefits of dough-retardation with an automatic pre-heat, recovery and proofing facility, thus minimizing early-morning preparation work. The baker arrives to a cabinet full of ready proofed dough-pieces ready for the oven. To enable the baker to process products of diverse weight and variety in pre-prepared, and shaped form, retarder-proofers have a built-in manual control option for the adjustment of retard, recover, and proof periods, as well as temperatures. These functions are operated from a central control-panel, covering such typical ranges as: $+2$ to $-4°C$ for retarding; $-8$ to $+3°C$ for automatic pre-heat; 3 to $24°C$ for recovery; and 24 to $43°C$ for proofing, with adjustable humidity.

*Dough retarders* permit the baker to produce dough-pieces in advance, and retard ready for proofing and baking off, either the following day or at a later date, thus eliminating early-morning preparation and setting up an inventory of products which can be drawn on as demand arises. Product inventories are built up during off-peak periods. The baker is not limited to overnight or short-term retardation, both rich and lean formulated doughs can be retarded for up to 72 hours, thus bridging weekends and short vacational periods. Adjustable thermostats allow the baker to reduce dough-piece temperature from 27°C to −4°C within 1·5 to 3·0 hours, and for overnight storage, retardation at +2°C is adequate. Retarding to a stabilizing temperature of −4°C, which is still above the dough freezing-point, allows extended storage for up to 72 hours; and at −10°C, up to 7 days storage is possible.

*Storage freezers* are intended for extended storage of dough or baked products, thus allowing the build-up of a back-up inventory to cope with unexpected or variable sales demands. These units operate at temperatures of −18 to −21°C, and are normally fitted with an automatic defrost plug-in facility. Storage times can be days or weeks as required.

*Blast freezers* are based on a modular (block) construction principle, to meet individual requirements for a larger capacity operation. Multi-tunnel or multi-trolley arrangements are possible to suit most freezing schemes and schedules, thus enabling bakers and frozen-dough manufacturers to meet the expanding market potential for high quality frozen products at competitive prices. Mechanical blast-freezers with a multi-trolley loading facility provide flexibility, economic running costs, and the ability to handle all dough-piece shapes and sizes. Figs 155 and 156 provide an overview of the various temperature/time applications for dough-freezing, retarding/storing, preheating, recovery and proofing.

*Proofing schedules of retarder-proofers*, can be usually varied between 0 and 30 hours, at variable heat and humidity. Average temperature ranges being 40 to 43°C, at 85% RH injected humidity.

*Brown-and-serve, or two-stage baking*, which originated in the USA in the 1950s, represents another frozen-dough market sector. Products such as bread-rolls, croissants, baguettes, buns, Danish pastries, etc., are proofed, then baked at a sufficiently high temperature to give adequate crumb rigidity, but no crust colour. This is achieved by allowing the core temperature to reach 75–80°C for a duration of 10 to 15 minutes only. This operation can be either completed in a conventional oven at 135 to 147°C, or by utilizing a microwave oven.

*Bake-off facilities* provide small and medium-sized bakers with value-added sales potential, both for freshly baked products, and snack take-away products. This new market sector has been identified and exploited in continental Europe, both at manufacturing and retail shop level. Most bakery and confectionery retail shops now have a 'snack-corner', which serves fresh bake-off products, snacks, and savoury/salad-filled baked products, which can be consumed either with coffee at the 'snack-corner' bar, or packed for taking away. This facility is normally run alongside a high-class restaurant or tea-rooms with table-service, in addition to the baked product retail counter sales. In some countries, the bake-off facility is located on display in the shop, and in others adjacent to the shop within the production area. Where freezer capacity and production capability coexist, bake and freeze, and make and freeze and store, can all become complementary integrated operations. The existence of a bake-off 'show-bakery' (German *Schaubäckerei*) or in-store (German *Kaufhallenbäckerei*) gives the sales areas an active focal point of interest, and a freshly baked or 'home-baked' atmosphere.

*In the German-speaking countries* where a bread-roll is no longer considered fresh after 4–5 hours of leaving the oven, owing to loss of crispness, a freeze–thaw–proof–bake cycle allows both industrial bakeries and in-store bake-offs to compete with the master-craftsman baker. Although, the master-craftsman baker also makes full use of dough-freezing processes to plan production, and provide freshly baked products round the clock. Only the master-craftsman baker (*Bäckermeister*) has the expertise to produce the quality required in this important bread-roll market, where the consumer is most discerning. In Bavaria and Austria, where the *Kaisersemmel* is the favourite partner for food and wine, these rolls have to be consumed within 2 hours of leaving the oven to be fresh. Therefore, production is organized on a continuous basis in small 2-hourly batches. A freeze–store–thaw–proof and bake-off cycle permits supply to meet demand with a minimum of stress for the baker.

*In France* the freeze and bake-off concept is applied within the bakery rather than being on show in the shop.

*In The Netherlands* bake-off equipment is utilized by bakers as a general procedure to make their own products, often in outlets not attached to the main production bakery. Products usually also include a local specialty, in addition to the standard range of snacks.

*In Belgium* bake-off equipment is to be found in the supermarkets, often located in an in-store bakery, sited in full view of the public.

*In the UK* bake-off equipment for savoury products, e.g. pies, pasties, sausage-rolls, has been used in baker's shops for several decades, but its application by the baker for fermented products is a relatively recent innovation. The supermarket-chain in-store bakeries first applied the concept to fermented products in the UK. With the continuing growth of the bake-off concept, many frozen-dough companies have entered the market, often marketing bake-off equipment and the necessary advice and expertise. The multiple industrial bakery groups also operate in the market through frozen-dough company subsidiaries. Many innovative bakers with retail outlets now manufacture from scratch those items in great demand, and purchase in frozen dough form, items which are either uneconomic to produce, or which have only limited demand. In both cases the dough-pieces can be frozen and stored to provide an inventory of products to meet demand through subsequent thawing, proofing and bake-off.

*Dough-freezing* presents new technological problems, which demand considerable research in order to apply the appropriate solutions. On freezing, life processes in a product become suspended, and ice crystals form. Subsequent application of thawing temperatures, recovery temperature, then proofing temperatures, and finally baking temperatures, all result in changes in dough structure, and the colloidal behaviour of the individual dough components relative to the concentrations of free and bound water present. Structural stability of the gas cell-wall membranes, which form eventual crumb-porosity, during these changes in temperature and state, coupled with the ability of the yeast cells to live through such changes, and retain their vitality, are crucial to final product quality. Therefore, yeast stability and proofing-power after the hazards of freezing, are as important as the dough structural integrity itself. Wolt and D'Appolonia, have defined frozen-dough stability in practical terms as the 'ability of a thawed dough to proof in an acceptable period of time, and to bake into a loaf with normal volume and characteristics'.

*Wolt, M. J., and D'Appolonia, B. L.* (1984) *Cereal Chem.*, **61**, p. 209.

Aspects of frozen-dough processing, which determine its stability will include product formulation, yeast genotype and properties, dough fermentation period before freezing, duration and storage conditions, and freeze/thaw conditions. When freezing, it is important to achieve a core-temperature of $-15°C$ as rapidly as possible, average product temperature then being under $-18°C$. Owing to the presence of salt and sugars, and other soluble dough components, the dough does not begin to freeze at $0°C$, but instead within the range $-3$ to $-8°C$, depending on the concentration of soluble matter in the dough water (depression of the freezing-point). The pure water molecule will crystallize out, and the residual solution becomes more concentrated. With increasing concentration, its freezing-point is depressed, requiring progressively lower temperatures for the subsequent freezing process. The effective freezing-time is the sum of the 'pull-down'-time (ambient to zero), the solidification-time (0 to $-10°C$), and the residual cooling-time. Within the solidification-time most of the free water becomes frozen. As an index of the correct freezing of a food, the 'effective-rate-of-freezing' is the accepted norm, expressed in $cm\ h^{-1}$, being the quotient of the shortest distance of the food surface from its centre (cm) and the effective freezing-time (hours). However, the 'nominal-rate-of-freezing' is more generally employed, which is the quotient of the shortest distance of the food surface from its centre (cm) and solidification-time (hours). According to Gutschmidt, the nominal-rate-of-freezing allows a classification of relative rates for food materials, as follows: under $0.2\ cm\ h^{-1}$ = very slow freeze; $0.2$ to $0.9\ cm\ h^{-1}$ = slow freeze; 1 to $5\ cm\ h^{-1}$ = rapid freeze; and over $5\ cm\ h^{-1}$ = very rapid freeze.

*Gutschmidt, J.* (1963) *Tiefkühl-Praxis*, **4**, 10, p. 8. Fundamentals of freezing and the frozen storage of foods (German).
The more rapid the freezing-rate, the less the damage the microstructure of the food undergoes. The formation of small, evenly distributed ice crystals prevails, and the separation of the liquid phase is therefore prevented. Thus, the nominal-rate-of-freezing must be at least 1 cm h$^{-1}$. This rate can normally be achieved with mechanical blast-freezers, and cryogenic spraying with liquid nitrogen results in freezing-rates of more than 15 cm h$^{-1}$. However, in the case of baked-products the technological advantages are not reflected in the end-product quality.

*Frozen storage of dough*, which is carried out at $-18°C$ or below, must ensure that the optimal structural condition achieved by rapid freezing is maintained. Temperature fluctuations should be avoided, since they only encourage recrystallization. Recrystallization is the progressive diffusion of water-vapour from the small ice crystals to the larger ones. This process, which depends on the size of the crystals and the surface pressure which they exert, has the same net effect on the structure of the product as a slow freeze. Recrystallization increases with temperature, since the pressure difference of the water-vapour around the large and small crystals is greater. Also, the smallest ice crystals melt first with each temperature rise, which only increases the vapour pressure difference still further. According to Marston, once a solidly frozen layer of some 3–4 mm has been formed by rapid freezing of the outer shell of the dough-piece, and the core temperature has attained 0°C, further freezing can take place in a storage-freezer maintained at $-20$ to $-23°C$. Excessive moisture loss is prevented, by packaging the dough-pieces in polyethylene bags and placing in corrugated cases prior to transfer to the storage-freezer. The storage-freezer temperature is maintained at $-23·3°C$.
*Marston, P. E.* (1978) *Bakers Digest*, **52**, 2, p. 18.

*Freezing-temperatures for bread doughs*, amongst other dough-freezing variables, were studied at the American Institute of Baking by Lehmann and Dreese. They concluded that bread doughs have the most consistent quality bread when blast frozen at $-21$ to $-29°C$ to a core temperature of $-9$ to $-5°C$. The inclusion of 62 D.E. corn syrup or a 42% high fructose corn syrup (HFCS) instead of sucrose, at the 8% and 10% levels, reduced final proof-times of frozen-dough stored at $-24·4°C$ for 26 weeks.
*Lehmann, T. A., and Dreese, P.* (1981) *Am. Inst. Baking Tech. Bull.*, **3**, No. 7. Stability of frozen dough—effects of freezing temperatures.
Further work on frozen white bread dough by Dubois and Dreese at the same Institute, established that doughs containing corn syrup gave in general, lower-volumed bread than the other sweeteners employed.
*Dubois, D. K., and Dreese, P.* (1984) *Am. Inst. Baking Tech. Bull.*, **6**, No. 7. Frozen white bread dough—effects of sweetener type and level.

*Frozen-dough fermentation stability* is the most important aspect of frozen-dough technology. The final dough, when thawed must respond to proofing temperatures within an acceptable time, and bake-out to a product of normal volume and characteristics. In studies concerning the effects of the length of dough fermentation and conditions of freezing on gluten properties, and quality of the finished products, Teshitel, at the Technological Institute of Odessa, USSR, made the following significant observations. Bread dough fermented with yeast for periods not exceeding 2 hours, and then held in frozen storage at $-12°C$, or $-18°C$ for 4 weeks, resulted in increases in the amount of gluten in the dough during storage for 2 weeks at $-18°C$, but showed a slight decrease at $-12°C$ during that period. Dough rheological properties and hydration changes appeared similar at both storage temperatures, but were less expressed at $-12°C$.
*Teshitel, O. V.* (1987) *Klebopek. Konditer. Prom-st*, **8**, p. 23–4 (Russian).
There is considerable research evidence to confirm that fermentation before freezing reduces yeast viability.
*Sugihara, T. F., and Kline, L.* (1968) *Bakers Digest*, **42**, 5, p. 51.

Merritt considers that dough fermentation before freezing has more effect on dough stability, and the proofing-time required, than any other single variable. The stability of frozen dough is inversely related to the amount of fermentation before freezing. Fermentation times before freezing of not more than 1 hour gave dough stability of a few weeks only; 30 minutes gave 3–4 months; and zero time gave about 12 months. Merritt's tests refer to a dough containing: flour 100 parts, water 65, dry yeast 4, bromated yeast-food 0·5, salt 1·75, dextrose 7·0, shortening 9·0, sweet-whey 5·0, mixed at high-speed with a refrigeration jacket, mixing to full development to give a final dough temperature of 18 to 21°C, allowing 10–15 minutes for fermentation during make-up. Freezing was at air temperature −32°C, and frozen storage at below −12°C until required. This dough could be stored for 1 year, 16–20-ounce slabs showing good proof after 16 months storage. On removal from the freezer, proofing was at 35°C and 80% RH, baking at 218°C for 17 minutes.
*Merritt, P. P. (1960) Bakers Digest,* **34**, 4, p. 57.
Merritt attributes the greater yeast stability in unfermented frozen doughs to its dormant state at the mixing stage.

*Yeast management for frozen-dough preparation* can present problems. Both compressed- and dry-yeasts are satisfactory for doughs to be frozen. Recommended amounts, based on flour weight, are 6–10% for compressed, and 3–5% for active dry-yeast, depending on the desired speed of final-proof. Where active dry-yeast is used, it must be free of significant amounts of dead cell material, which weakens the dough. Only fresh yeast should be used, and storage fluctuations avoided. Dead cells can be identified by staining a cell suspension 1:1 with an equal volume of methylene blue 200 mg, potassium dihydrogen phosphate 27·2 g, and disodium hydrogen phosphate 0·07 g/litre distilled water. Dead cells stain red, allowing counting under a microscope. Hsu and co-workers also observed that the quality of the yeast greatly affected frozen dough stability. Finding those with protein contents in excess of 57% performing best.
*Hsu, K. H., Hoseney, R. C., and Seib, P. A. (1979) Cereal Chem.,* **56**, p. 419.
Wolt and D'Appolonia found that fresh compressed yeast performed slightly better than either active dry-yeast or instant-active dry-yeast over a storage period of 20 weeks.
*Wolt, M. J., and D'Appolonia, B. L. (1984) Cereal Chem.,* **61**, p. 213.
Bruinsma and Giesenschlag conducted tests to establish the effects of 12 weeks' frozen-storage, with daily freeze/thaw cycles on the viability of dry-yeast and compressed-yeast. They found that higher levels of active dry-yeast gave consistently shorter final proof-times than compressed-yeast. Proof-times and gas-production showed little change over seven consecutive freeze/thaw cycles for either type of yeast, but crumb structure became rapidly devalued over the seven cycles; the dough became progressively weaker and friable, making handling difficult.
*Bruinsma, B. L., and Giesenschlag, J. (1984) Bakers Digest,* **58**, 6, p. 6.
Mazur and Miller found that the freezing-rate had a profound effect on the survival rate of yeast cells. Rapid freezing to −30°C and below, reducing the cell survival rate to less than 0·01%; whereas slow cooling to the same temperature results in up to 65% remaining viable.
*Mazur, P., and Miller, R. H. (1967) Cryobiology,* **3**, p. 365.
Yeast cell destruction by rapid cooling is due to the formation of intracellular ice crystals which destroy the structure of the cell protoplasm. Slower cooling rates allow the yeast cells to transfer sufficient intracellular water to the outside, thus preventing ice crystal formation inside the cells. Freezing at excessively low temperatures also diminishes yeast viability. In dough, yeast cells freeze at about −35°C, and are unable to reproduce much below −12°C. Hsu and co-workers established proof-times of 72 and 132 minutes for doughs frozen at −10°C and −40°C respectively. Doughs taken down to −78°C show only minimal activity on thawing.
*Hsu, K. H., Hoseney, R. C., and Seib, P. A. (1979) Cereal Chem.,* **56**, p. 424.
Lorenz has suggested that freezing the dough slowly down to −10°C only, followed by rapid defrosting, results in improved frozen dough stability.

*Lorenz, K.* (1974) *Bakers Digest,* **48**, 2, p. 14.
However, it must be acknowledged that many of these suggestions are contrary to current practice. In order to seek solutions to such problems it is essential to consider the basic requirements of yeast, as a microorganism, for life and growth. In the vegetative (reproductive) state, the yeast cell contains 85% water, since for the processes of nutrient uptake, transport of the metabolites, water is essential. When this water is removed, metabolism ceases together with growth and reproduction. Most vegetative cells die as a result of drying-out, but spores can survive considerable drying-out. However, yeast cell development depends less on the absolute moisture content of the substrate (dough) than on the content of 'available' or 'active-water'. The unit of measurement of the available water of a substrate is referred to as its 'water-activity' ($a_w$), which is expressed as the ratio of the vapour pressure of the substrate, i.e. its surrounding atmosphere ($p$), and the saturated vapour pressure of pure water ($p_0$) at the same temperature, i.e. $a_w = p/p_0$. Pure water has an $a_w$ value of 1·000 but with increasing content of dissolved particles, this falls to $a_w$ values below 1. Each soluble particle has an envelope of water molecules surrounding it, which are no longer available to the yeast. A completely water-free substrate having an $a_w$ value 0. Most microorganisms can only develop in substrates with $a_w$ values within the range 0·85 to about 1·00, but some specialized ones can grow at 0·62. Most species of yeast have a minimum $a_w$ value for growth between 0·88 and 0·91.

*Burcik, E.* (1950) *Arch. Mikrobiol.,* **15**, p. 203–35. Concerning the relationship between the hydration-state and growth of bacteria and yeasts (German).
The resistance of microorganisms to cold depends on the chemical composition of the substrate (dough) within which they are subjected to low temperatures. Resistance against cold increases with increasing concentrations of soluble solids. Anions and cations also have an influence on resistance to cold; also unfavourable pH conditions lower their resistance. The presence of colloids generally, and protein and lipids exert a protective influence against the cold. Protective colloids like gelatine, which are hydrophilic (water-binding), considerably reduce the death rate due to cold of yeasts and bacteria during freezing. Bound water cannot be easily frozen, therefore no ice crystal formation can take place. In general, high moisture contents result in the death of considerable numbers of yeast cells, whereas lower levels offer better protection against the cold, hence the general use of freeze-drying (lyophilization) for long-term storage of culture collections of microorganisms. Yeast cells maintained in the stationary reproductive phase remain much more resistant to cold than those in the vegetative phase. Spores in particular showing greater resistance. Rapid freezing of cells with adequate bound water is better tolerated than slow freezing, which is also true for thawing. The critical temperature range for the cell liquid is freezing temperature range $-4$ to $-10°C$, and repeated freeze/thaw cycles results in a cumulative loss of living cells. Mechanical damage as a result of the volume increase due to ice formation is not the only problem. During freezing, the water component of the solubles is first frozen, resulting in a concentration of the dissolved solids. This results in an osmotic pressure increase, changes in pH, and a concentration of mineral salts, which causes irreversible damage to the physical and chemical properties of the colloidal structure of the cell protoplasm. Such changes also give rise to toxicity, resulting from concentration at freezing, thus inhibiting cell function. Damage to the cytoplasmic membrane results in a disturbance of the metabolic equilibrium, and irreplaceable loss of energy. Prolonged storage of yeasted doughs in the frozen state results in progressive loss of living cells, owing to free-radical formation. Microorganisms generally are best protected from freezing damage by freezing with liquid nitrogen at about $-196°C$, in this way, their biochemical properties remain unchanged for years.

*Stevenson, K. E., and Graumlich, T. R.* (1978) *Adv. Appl. Microbiol.,* **23**, p. 203–17. Injury and recovery of yeasts and moulds.

*Farrell, J., and Rose, A. H.* (1965) *Adv. Appl. Microbiol.,* **7**, p. 335–78.

*Farrell, J., and Rose, A. H.* (1967) *Ann. Rev. Microbiol.,* **21**, p. 101–20.

Proceedings of low-temperature microbiology symposium 1961, Camden, New Jersey. Campbell Soup Co., 1962. Collective authorship.

*Precht, H., et al.* (1973) *Temperature and Life*, Springer-Verlag, Berlin, Heidelberg, New York. Unless new yeast strains are developed with higher tolerance to prolonged low temperature storage in dough products, the only option is to modify the formulation, processing, and freeze/thaw procedures to protect the viability of the yeast cells enough to result in an acceptable final proof-time, and crumb porosity.

*Frozen-dough production procedures* currently involve the preparation of a rather stiff straight dough on a no-time schedule, using 4–6% yeast, shortening about 5%, sugar 5%, and non-fat dry milk at 4% (as necessary), all based on flour weight. The dough should be well developed and homogeneous, mixing temperature being kept at about 22°C, by chilling if necessary. The dough can then either be transferred immediately to make-up, as suggested by Rosenholtz and Boyd.

*Rosenholtz, S.* (1985) *Proc. Am. Soc. Bakery Engrs*, p. 141.

*Boyd, W. E.* (1985) *Proc. Am. Soc. Bakery Engrs*, p. 38.

or given a maximum 15-minute relaxation-time to render more workable. Where the strict no-time schedule is utilized, the rounded dough-pieces are given an intermediate proof of about 5 minutes to relax. Flour quality should be above average, with gluten elasticity and resilience. A patent flour, milled from medium-strength spring/winter wheat grists, with added bromate or ascorbic acid is suggested, gluten quality rather than quantity being the criterion. A good water absorption starting level being 60%, instead of the normal 64% for a conventional dough. Dough salt levels should be 1·5–2·0% flour weight. The moulded and shaped dough-pieces are then deposited on trays, and loaded on racks or conveyors and either placed in a blast-freezer for the required time, or slowly passed through a freezing-chamber or tunnel. Primary freezer area temperature should be maintained at −29 to −40°C, being exposed to air currents of 600 ft/min during transit. Doughs are best covered with polyethylene film or bags during freezing to prevent surface moisture loss and crusting. A liquid nitrogen flash or jet stream is the most rapid freeze rate in use, taking an average of 6 minutes. A blast-freezer of the staionary design will take about 2–3 hours to reach a core temperature of −12°C in the case of a 2-lb loaf. Packaging will add about 30 min/unit to freezing-times, depending on unit mass. In the USA, R. Bamford of the American Society of Bakery Engineers has suggested the conditions shown in Table 74 for blast-freezing with a standardized air flow of 600 ft/min. During the freezing of all the listed products, the air temperature was maintained at −36°C.

*Thompson, D. R.* (1976) *Bakers Digest*, **50**, 2, p. 28, recommends blast-freezing at temperatures of −35 to −40°C, using air velocities of 600 to 700 ft³/min, and for such time that the core temperature of the product reaches −18°C. Further stipulations are that the holding-freezer units should be maintained at air temperatures between −23 and −29°C, not allowing it to rise above −18°C at any time. Furthermore, the air should be held at the highest RH possible in practice, and freely circulating. Under these strict conditions, product quality could be preserved for several weeks or months. Many producers use a temperature of −23°C, although −18°C is often quoted as satisfactory after rapid cryogenic-freezing; this provides an insurance against mistakes and temperature fluctuations.

*Cryogenic freezing-rates* are 10 to 30 times faster than any blast-freezer, and a 70-ft belt conveyor will freeze up to 5000 lb/hour in a countercurrent tunnel system. Liquid nitrogen being introduced into the freezing zone by atomized spray headers at a controlled rate. The low temperature gas circulating at speeds of up to 7000 ft/min zonally. Entry zones effect a precool of the product to −18°C. The actual freezing zone freezes at −196°C, with a tempering final zone prior to exit of −108°C. A significantly better retention of product quality, compared with a mechanical freezer is obtained. Equipment simplicity in design, lower initial and subsequent maintenance costs, lower energy costs, and less floor-space are other attractions. Where space is limited, and capacities of around 10 000 lb/hour are involved, spiral cryogenic freezers offer the best solution.

*Thawing conditions* are important, and represent a long, frustrating step, unless integrated into the production planning on a programmed basis. On removal from the holding-freezers, the core temperature of the products is $-18$ to $-21°C$. In order to avoid too great a temperature gradient between the core temperature and the dough surface temperature, the products are placed in a defroster to thaw gradually. According to R. Bamford of the American Society of Bakery Engineers (1975), the optimal defroster temperature is 37 to 49°C, at RH 50% to 60%, using an air velocity of 200 to 500 ft/min. However, initially, air flow rates of 150 to 200 ft$^3$/min 150 to 200 ft$^3$/min are required to raise the product temperature from $-18°C$ up to 21°C. The time required to reach a core temperature of 21°C will depend on the size, type and packaging of the product. Table 75 provides some examples of thaw-times from $-18°C$ up to 21°C.

*Potassium bromate* when supplemented with ascorbic acid results in a proof-time reduction of 23 minutes, after a 2-month frozen-dough storage period.
*Varriano-Marston, E. et al.* (1980) *Bakers Digest*, **54**, 1, p. 32.

*Oxidant type and amount* will depend on local regulations, and processing conditions. Both ascorbic acid and potassium bromate improve frozen-dough stability, either used alone or in combination. The level of treatment will depend on flour quality and response, and the inverse relationship between oxidant level and dough mixing temperature. Lower dough temperatures require more oxidation to optimize response to final-proof. Bromated yeast foods can be added at the 0·5% level.

*Sugars* are used at the 5% to 10% level, based on flour weight, in rich formulations, but in the case of leaner bread doughs 5% to 6% is adequate. Sucrose or dextrose give similar results, or maltodextrin can be used to reduce calorific values.

*Shortenings* are included in the form of good quality lard, or vegetable shortening based on palm-kernel oil, at the 5% flour weight level, depending on the formulation.

*Mono-/diglyceride*-type emulsifiers added at about 3% of shortening weight improves product volume and crumb quality. The distilled 90% monoglyceride is the most active source on a weight-for-weight basis.

*Datem, or diacetyl tartaric esters of monoglycerides*, owing to their protein complexing, make the dough more tolerant and gas-retaining; but the yeast vitality must be also present at final-proof.

*Ethoxylated derivatives of mono- and diglycerides* (EOM) also improve dough rheology, but requires yeast vitality at final-proof.

*Sodium stearoyl-2-lactylate* (SSL) strengthens the dough, producing a greater oven-spring and improved crumb grain; but yeast vitality is a prerequisite for final-proof.
*Dubois, D. K., and Blockcolsky, D.* (1986) *Am. Inst. Baking Tech. Bull.*, **8**, No. 4. Frozen bread dough—effect of additives.

*Surfactant functionality* depends on hydrophilic/hydrophobic balance. According to ionic activity, surfactants can be divided into three groups. Anionically active ones are water-soluble, and become ionically dissociated, e.g. calcium or sodium stearoyl-2-lactylates (CSL and SSL); non-ionic ones, which do not dissociate into ions, e.g. mono- and diglycerides of fatty acids, and sugar alcohol esters of fatty acids (sorbitol monoesters, and the 'fat-sugars' such as sucrose and glucose esters), also their polyoxyethylene derivatives; ampholytic or dipolar ones with mixed ionic functionality, e.g. the phosphatides of which the lecithins form a group of allied compounds. The phosphoryl-choline moiety functioning as a dipolar, or 'zwitterion'.

*Frozen-dough formulation* in certain cases, can utilize the functionality of sucroglycerides, especially those based on palm oil. Formulations containing sugars, shortening and eggs benefit in particular from their inclusion at the 2–3% flour weight level. The functionality of the sucroglycerides derived from palm-nut oil, containing palmitic acid 47·5%, oleic acid 40·0%, linoleic acid 7·5%, stearic acid 4·0%, and myristic acid 1·0%, are particularly effective.

*Hydrocolloids* also show good frozen dough functionality. Their precise function is to stabilize water movement within the dough during freezing, when it is being subjected to a number of freeze/thaw cycles. Locust bean (carob), and guar gum, both prevent excessive ice crystal formation during freezing, the active component in both cases being the galactomannans, derived from the seed endosperms. Both gums produce high viscosities at low concentration, therefore, only concentrations of up to 0·1% flour weight are normally included. Karaya gum, used at 0·1% flour weight, often exerts a hydrocolloidal synergistic effect when used with monoglycerides at about 0·3% flour weight in doughs. The rheology of the polysaccharide gums can be adjusted to provide either extensible or short-textured gels as appropriate.

*Native and modified starches* also exert good freeze/thaw tolerance properties in frozen-dough products, preventing moisture movement within the dough system. Waxy sorghum starch has a relatively low moisture release over about three freeze/thaw cycles; cross-linked amylopectin starches yield mono- and distarch phosphates, on phosphorylation, which are freeze/thaw stable. Oxidatively modified maize (corn) starch, producted by treating the starch with 0·05–0·07% potassium bromate prior to drying at 160°C, improves dough stability, fermentation response, and product porosity. Native starches, which give long, cohesive-textured gels are: tapioca, potato and amioca. Pregelatinized potato starch also provides cold-swelling, and is stable at low temperatures.

*Dough base and conditioner manufacturers* also market composites for addition to frozen-doughs, providing longer frozen storage life, and improved recovery. Many of these products include the functional components already discussed.

*Frozen-dough formulations* based on the Reddi-Sponge Process (R. G. Henika) patented by Foremost Foods, Dublin, California, USA, allow a doubling of the frozen storage shelf-life, and permits the use of lower dough temperatures, e.g. 18 to 21°C. The application of direct biochemical action on the flour proteins prolongs dough storage potential. Typical dough formulations contain: yeast 4·0%, yeast-food 0·5%, salt 2·0%, shortening 3·0%, mono-/diglyceride emulsifier 0·4%, Reddi-sponge 3·0%, using total oxidant level 70–80 ppm, with added sugar up to 8%, all based on flour weight 100 parts. Doughs should be mixed to full development at low and high speeds, and if the mixing-time exceeds 12 minutes, it is better to withhold the yeast and add over the last 4–5 minutes of mixing for subsequent batches. Mixer-jacket temperature should be adjusted to aim at dough temperatures between 21 and 24°C. The dough is allowed a 10-minute relaxation period, after which it is divided, rounded, and given a minimum intermediate-proof before shaping. The dough-pieces are then blast-frozen at −24°C or lower. The sequence in which the dough ingredients are added during mixing has a significant effect on final proof-time and yeast activity. Doughs held at retardation temperatures for up to 24 hours before proofing, show reduced (shortened) final-proof times, and improved product volume, although other quality criteria may not show an improvement. These trends were also confirmed by Dubois and Blockcolsky at the American Institute of Baking, Manhatten, Kansas, USA.
*Dubois, D. K., and Blockcolsky, D. (1986) Am. Inst. Baking Tech. Bulletin, 8, No. 6.*

*Amyloglucosidase supplementation* is more effective in increasing loaf volume in lean sugar-free doughs than fungal alpha-amylases, and is a useful ingredient for frozen-dough work. Limit-dextrinase, present in fungal enzymes, is capable of hydrolysing starch or dextrins to glucose. This enzyme was shown by Pomeranz to belong to a group of amyloglucosidases which specifically hydrolyse starch or dextrins to glucose; earlier work by Pomeranz and co-workers, using combinations of alpha-amylase and amyloglucosidase from fungal sources, confirmed that the improving effect of amyloglucosidase depends on the sugar level in the formulation. Leaner low-sugar doughs require higher levels of added amyloglucosidase to yield loaf volumes comparable to doughs with higher sugar and lower levels of enzyme.
*Pomeranz, Y., and Finney, K. F. (1975) Bakers Digest, 49, p. 20.*
*Pomeranz, Y., Rubenthaler, G. L., and Finney, K. F. (1964) Food Technol., 18, p. 138.*

These workers established that at higher dosage levels, amyloglucosidase was more effective than fungal alpha-amylase in sugar-free doughs for the purpose of increasing loaf volume. For a frozen-dough baguette formula: flour (protein 12% dry matter, gluten minimum 25%) 100 parts, compressed-yeast 6, ascorbic acid 50–70 ppm, amyloglucosidase preparation added at optimal concentration (determined by baking-tests), salt 2·0 (added at final mixing-stage), water adjusted to flour absorption.

*Frozen-dough procedure*, as a guideline to the freeze–store–thaw–proof sequence, should be carried out as follows. Use a rapid freeze to a low temperature to minimize ice crystal size utilizing a blast-freezer technique. A liquid nitrogen flash, or jet streams, offer the fastest and best procedure (average freeze-time 6 minutes), otherwise a mechanical blast system is preferred. Doughs are ideally covered with polyethylene film or bags during freezing to prevent moisture loss and skinning. A blast-freezer will reduce the core-temperature of products to $-20°C$ within about 30 minutes; a large 16-rack roll-in unit freezing a rack of rolls within 20 minutes to $-20°C$ from ambient. Freeze-storage is maintained at $-20$ or $-23°C$ to allow a safety margin against temperature fluctuations, being part of a cold-store complex, with attached recovery room for controlled thaw-out of products (humidity, air-flow, and temperature). A slow recovery/thaw-out system is placed under automatic control, preheating from sub-zero temperatures up to about 3°C. This can be varied in duration to meet requirements, maintaining humidity at 70–75%. The actual recovery phase takes place between 3°C and about 24°C at 70–75% RH. The final-proof phase ranges from 24°C to 43°C, with RH maintained at 85%. In the absence of a cold-store recovery complex, thawing is best carried out in a warm draught-free location at 27 to°C to 29°C, ideally within an enclosure at RH 60–70%. If carried out in the open bakery, the products must be covered to prevent moisture loss and skinning. Final-proof is then completed at 30 to 45°C, depending on the product.

## 3.6.4 Bread Preservation by Freezing

Preservation of bread and baked products by freezing technology has been applied in the USA and Europe for several decades. Although many refinements in technology and equipment have taken place over the past two decades, the choice of temperature ranges for the retardation of staling of baked products falls between $-20$ and $-190°C$. The most unfavourable range is $+50$ to $-7°C$. For successful application, the products must be frozen within 4 to 8 hours of leaving the oven. Freezing can take place either with or without packaging, or with a double-layer of wrapping material. Although packaging prior to freezing reduces the freezing-rate to reach a loaf centre-temperature (core) of $-30°C$—the time required being about twice as long under identical conditions—the wrapped products remain fresh two to three times longer, owing to improved moisture retention. The choice of wrapping material often depends on the product, but polyethylene films remain the popular choice on economic grounds. Cathcart and co-workers, in commercial freezers, found that bread remained fresh in flavour and aroma for up to 30 days when held at $-35°C$, and in a saleable condition for up to 345 days, the limiting factor being the ultimate development of an off-flavour.

*Cathcart, W. H. (1941) Cereal Chem.,* **18**, p. 771.
*Cathcart, W. H., and Luber, S. V. (1939) Ind. Eng. Chem.,* **31**, p. 362.

*Baked products* are normally frozen by using a forced air current. Air speeds of 2 to 4 m/s are found optimal to reach air-temperatures of $-25$ to 35°C. This air speed provides a heat-transfer coefficient of 18 to 24 $W\,m^{-2}\,K^{-1}$. Any increase in air speed above this 2 to 4 m/s level, will progressively reduce the heat-transfer coefficient. This is due to the relatively poor thermal conductivity of baked products generally, although the air-cells in products with increasing shortening contents are correspondingly fewer.

*Pre-frozen quality levels* of products can only be maintained, when utilizing the freezing process to retard staling and physical changes, by rapid freezing, which means passing through the unfavourable temperature zone of +50 to −7°C as rapidly as possible. The limiting factor here is air speed *v.* the heat-transfer coefficient from the products. As a guideline, a 1-kg loaf cooled down to 20°C after baking should not require longer than 4 to 5 hours to reach a core temperature of −18°C. The time taken to attain a specific core temperature will depend on product mass, under any specified conditions of blast-freezing. By moving very cold air across products at high velocity, fast-freezing is achieved without the formation of large ice crystals, which damage texture, and leave a dry end product, depending on their water content and water-binding capacity. This is particularly important when freezing doughs.

*Unwrapped bread and baked products,* when frozen, show excessive drying out within 1 to 2 weeks, owing to the lower RH, in storage freezers.

*Freezer storage-temperature* has a significant effect on the staling rate. Bread held within the range −10 to −7°C, which corresponds to its freezing-range, will show considerable crumb-firming, and flavour loss within a period of 1 week. At a storage temperature of −18°C, crumb-softness will remain relatively stable and constant for about 1 month. However, prolonged storage at −18°C results in a progressive devaluation of bread quality. Even if the temperature is allowed to rise to −12°C, only just outside the freezing range, crumb-firming becomes apparent within a few days. Temperature fluctuations above −7°C result in moisture migration within the crumb of the frozen bread. Temperatures around −10°C even give rise to sublimation of the ice crystals after about two weeks' storage, which can be identified as white rings forming in the crumb layers immediately under the crust. This is indicative of crumb drying beneath the crust, moisture having been transferred by sublimation and diffusion from the high-moisture crumb-centre to the relatively low-moisture crust-zone.

*Blast-freezing or 'shock-freezing'* is the most efficient method of bread and baked product preservation, whether in the form of a high-velocity, very cold air movement around the product, or by the use of a liquid cryogenic, e.g. nitrogen or carbon dioxide. This procedure avoids any subsequent separation of the crust from the crumb, which is brought about by a moisture redistribution within the layers under the crust, the internal layers losing moisture to the outer layers resulting in a separation of crust from crumb. Bread staling processes proceed most rapidly within the temperature range +60°C and −2°C; hence the importance of a blast-freezing system to preserve freshly baked quality. On a batch freezing basis, this is achieved with a trolley-loading blast-freezer, coupled with a low-temperature holding cold-room. This provides a good output/time ratio. Tunnel blast-freezing is, however, a more rational operation, the products being passed through a tunnel on a band conveyor, within which a very cold forced-air flow prevails. Heat-transfer takes place by convection, according to Newton's law, the temperature gradient becoming less as the freezing-time increases. The value of the heat-transfer coefficient depending on the air speed within the tunnel, which is optimally set at about 2–4 m/s. If this range is exceeded, its influence on heat-transfer steadily decreases. Average tunnel temperatures are within the range −25 to −35°C. In the USA, blast-freezing is also used for bread products, using standard air-flow rates of 600 ft/min, reaching final temperatures of −20 to −30°C for fermented goods.

*Freezing-time* depends on the type and extent of the packaging material used. Table 76 shows relative times required for products to reach −12°C in hours. A 2-lb unwrapped loaf of bread would require 3·0 hours, a single-wrapped loaf 3·5 hours, and a double-wrapped loaf 4·0 hours to reach −12°C at the centre (core), at the standard 600 ft/min air-flow. In the case of the blast-freezing of baked bread-rolls, the temperature/time sequence is as follows:

—rolls cooled from 98°C to 35°C, taking about 30 minutes;
—rolls cooled from 35°C down to about −10°C over a period of about 50 minutes, in the first zone of the blast-freezer tunnel;
—rolls further cooled down from −10°C to −20°C over a further 30 minutes, in the second zone of the blast-freezer tunnel;

—freeze-store at $-18°C$ to $-21°C$ for about 2–3 days in freeze-store rooms;
—thawing from $-18°C$ to $-21°C$, up to ambient temperature, 20°C plus, requiring about
65 minutes.
See Figs 157 and 158.

*Freezing applications for baked products* can be divided into:

(1)   Chilled storage, intended for short-time preservation only, using temperatures of 0 to
$+4°C$, e.g. chill-cabinets with residence-times limited to days. This will inhibit the
growth of moulds and liquifying bacteria, e.g. *B. subtilis*, to a degree. In such cases,
fermented products, baked-pastries, and pastry-dough intended for domestic bake-
off, gateaux, desserts, and meat pasties should be suitably wrapped. Permissible
residence-times of products will depend on whether they are wrapped, on their type
and on their formulation, and on whether the cabinet is open or closed. Recent draft
food hygiene regulations in the UK stipulate, for same-day sale, that sandwiches be
stored at a maximum of 8°C, and cream cakes at a maximum of 5°C.
(2)   Longer-term preservation, extending to weeks or months, with adequate packaging or
sterilization before packaging, can be carried out in closed cabinets set at $-18°C$.

Longer-term preservation by freezing is intended for setting up a production-inventory
programme, aimed at rationalizing and better coping with peak-demand periods.
Table 77 shows the various phases, equipment and technological parameters involved in the
preservation by freezing of bread-rolls and bread.
Figure 157 shows the technological system for the freeze-preservation and thawing of bread
and bread-rolls.

*Preservation of baked product freshness* is best achieved during transit by dispatching in the
frozen state and allowing to thaw, either during transit, or at the point of sale, in the case of
relatively short distribution chains. For longer distribution chains, and where many
wholesale and retail outlets have to be supplied, insulated freezer-vans or containers must be
utilized.

*Industrial bakeries or frozen-food chains* specializing in bakery products, with throughputs of
1000 kg/hour, utilize a direct contact refrigerant, known as a 'cryogenic'. Direct contact of the
cryogenic in the gaseous state with the product provides very rapid freezing rates. Cryogenic
systems use either liquid nitrogen or liquid carbon dioxide. Since the boiling point of liquid
nitrogen is $-196°C$, a cryogenic system allows the application of very much lower
temperatures than the air blast-freezer. Average tunnel lengths are 12 m with widths of about
1·8 m, the products passing under liquid nitrogen spray, which evaporates on contact; fans
move the cold evaporated gas in a countercurrent direction, equilibrium at $-18°C$ being
achieved by retaining the product in the tunnel. Coolant cost on a continuous basis must be
justified by a 24-hour production output. Advantages are the relatively small floor-space
required, low maintenance, improved product quality, and flexibility.

*Partially baked or 'Brown-n-serve' products* are preformed and prebaked to the final size and
shape, requiring only crust browning and fresh flavour development in a domestic oven. This
concept is applied to rolls, breads and pastries, the aim being to bake the product to a point of
maximum rigidity and volume, without actual crust formation. This is achieved by reducing
the oven temperature to the range 120 to 150°C. Proof has to be controlled so that the oven-
spring which takes place at the lower temperature is maintained within limits. Dough water
absorption must be reduced to give a stiff dough, which will remain rigid on removal from the
oven. Formulation modifications require straight doughs to be set at 32°C to 35°C, and
sponge-doughs at the usual lower temperatures. Levels of yeast and yeast-food must be
moderately reduced to contain oven-spring. Enriched formulations perform best for eating
quality, and final-proof should be rapid at 38–40°C. Baking between 120°C and 150°C is
maintained for as long as possible without crust browning, usually for about 10 to 15 minutes
at a solid 140°C. The product core-temperature must reach 82°C to avoid shrinkage or

collapse on cooling. Product cooling and packaging must be carried out under sterile conditions to avoid mould infection, or product damage.

*Market trends*, projected into the 1990s, indicate a growth market for frozen foods generally, with bakery products representing about 48%, being only second to ready-meals as the fastest growth sector. Where meat and dairy raw materials are utilized, the baking industry must adhere to safe storage-temperatures and handling procedures to avoid microbiological hazards, and the adverse publicity resulting from 'health-scares'. In the bakery frozen product sector there is also considerable scope for new product innovation and growth, wherever a convenience aspect can be introduced. Socio-economic trends show that the number of 1- to 2-person households is on the increase, that the population average age is increasing, and that the consumer has more leisure time at his or her disposal. Fast-food consumption has shown dramatic increases, and bake-off concepts are part of this trend. Product variety, and their health and nutritional attributes are now in the limelight. The large supermarket chains have taken the lead in publishing brochures and applying nutritional-labelling to the various types of food. The nutritional significance of salt, fats and fibre in the diet, the energy contribution to the diet of a given weight, and the contribution to the recommended daily amounts (RDAs) of each nutrient, have all been included in consumer information.

For the baking industry, the success of the frozen product sector, whether in the form of dough or baked products, is due to greater flexibility in production, and the sale of fresh products round the clock. The 'bake-off' concept, according to consumer demand has proved to be a valuable technical innovation allowing the baker to plan production, and better cope with the peaks and troughs of demand. The demand for 'oven-fresh' products is also a growth market, especially the bread-roll and specialty- or variety-bread sector. The health and nutritional factor has resulted in the development of more interesting specialty products with added value for the consumer. These products are already showing steady sales increases and are expected to take part of the conventional white panned-bread market, which has hitherto commanded 70% of total bread sales in the UK. The secret of sales expansion in the baking industry is innovation, and the production of a wide range of freshly baked products. In-shop baking and product promotion, including special offers and features from other countries, resulting from recipe exchanges, will stimulate demand into the 1990s and beyond.

# Index

A2-ChTB/M, 422
A-2 ChTT, 445
A 12 assembly 120, 124
Acceletron, 565
Accurist dough divider, 436
Acetic acid, 59, 66, 73, 103, 342, 343–4,
    347, 350–1, 484–5, 505
Activated dough development process
    (ADD), 25, 27, 231–233, 235, 490
Active dry yeast, 467
*Aerobacter* species, 326, 388
  *cloaceae*, 326
Aflatoxins, 389, 390–392
Agatova and Proskurjakov, 66
Agene, 182, 216, 222
Albumins, 3, 68, 73
Alcohol, 479, 480
Allied Bakeries, 244, 248, 250, 253
Allied Mills, 237, 242, 249, 257
Alpha-amylase, 30, 54, 235
Amadori rearrangement, 588, 589
America. *See* United States of America
American Machine and Foundry
    Company, 281, 431
Amflow units, 28, 236, 280–283, 285, 443,
    490
Amino acids 51, 53, 61, 63, 64, 68, 75, 76,
    78, 482, 483, *see also* under
    individual names
Ammonium hydroxide, 234
Amylase, 55, 84, 234
Amylograph paste viscosity, 58, 59
Amylopectin, 38, 82
Amylose, 38, 40, 82
Anger *et al.*, 80
Aniline, 335
AP-4/1, 308, 313
AP-4/2, 308, 313
APV-Baker, 262, 264, 265, 266, 420, 436,
    *see also* Baker-Perkins
Arabinoxylan, 80, 83
Arkady Company, 236, 241, 250
Armenia, 483
Artofex mixer, 93, 261, 411, 417

Ascorbic acid, 31–4, 36, 42, 66, 85, 230
L-Ascorbic acid, 78, 235, 236, 320–1
*Aspergillus*
  *awamori*, 173, 234
  *candidus*, 326
  *niger*, 30
  *oryzae*, 30, 234
  species, 390
Associated British Foods, 236
Atlas Equipment, 400, 718
Auerman, L., 52, 161, 162, 169, 179, 181,
    190, 191, 313, 573, 648, 674
Auerman *et al.*, 158, 225, 643, 645
Australia, 27, 36, 72
Austria: bread-making in, 94–5
Autoclave cookers, 55
AWB-100, 180
Axford, D. W. E., 66
AZCh oven, 167
Azodicarbonamide, 29, 33, 34, 66, 230,
    231, 233, 268

*Bacillus cereus*, 382, 383
*Bacillus mesentericus*, 389, 683, 711
*Bacillus subtitils*, 30, 389, 683, 711
Bacto-Wort Agar, 354
BAG-20, 167
Baker *et al.*, 21–2
Baker-Perkins-Invarsson system, 226
Baker-Perkins
  mixers, 182, 207, 224, 261, 262, 430
  ovens, 211–13, 215, 600
  *see also* APV-Baker
Baker's percent, 48
Baking
  aims, 539–52
  aroma components, 590–5
  caramelization, 583, 584, 588
  elements, 553–95
  flavour components, 590–5
  heat-transfer, 560–8
  humidity, 544, 558, 568–77
  loss, 601

Baking—*contd.*
  moisture and, 555, 564, 573, 579
  optimization of, 543
  requirements, 539–22
  starch and, 580–4
  steam, 544–7, 558, 569, 570, 575
  temperature, 548, 550–2, 553–60, 559–60,
    568, 570, 572, 573, 579
  times, 548, 569, 578–9, 583
  weight loss, 548
  *see also* Ovens
Baking tests, 327–35
Baltimore Biological Laboratories, 379
Barber, S., 91
Barber *et al.*, 348
Barms, 493
Barnes and Blakeney, 316
Barret and Joiner, 233
Bath processing, 37
Bean-flour, 88
Belova, L. D., 187
Belova *et al.*, 469
Benzoyl peroxide, 42, 43, 216
Berliner, Dr, 313, 317, 318
Berliner and Koopman, 52, 80
Berliner short-sour process, 333, 345, 346,
  533
Berlin short-sour, 106, 507
Beta-amylase, 54, 174
Bezenchukskaya, 148
Bezostaya, 148–9
BIAKS, 187, 477
Biotin, 245
'Biplex' mixer, 262, 263, 264, 435–8
BK oven, 158
Blanchard, G., 179, 182, 183, 224, 225
Blanchard process, 43, 179, 224–5
Bloksma, A. H., 66, 76, 328
BM-2, 158
BN 25 oven, 196, 204
BN 40 oven, 133, 134
BN 50 oven, 119, 120, 139, 204, 459, 627
BN 72 oven, 204
Bogdanov *et al.*, 73
Böhringer, 346
Boku mixers, 260
Boku Spiral, 421
Bolling and Zwingelberg, 96
Bösenberg and Eberhardt, 390
Brabender Amylograph, 82, 83, 315, 500–1
Bradender-Do-corder, 10, 313
Brabender Extensograph, 18, 36, 313
Brabender Farinograph, 3, 148, 313, 501
Brabender Glutentester, 307

Brabender Maturograph, 40–1
Bran-bread, 534–5
Bread
  ageing, 656–76
  air temperature and, 646–7
  aroma components, 656
  cooling, 213–14, 638, 639–41;
    moisture, 641–9, 656
  freezing, 679, 708–10, 711, 712
  humidity and, 647
  microbial infection, prevention of, 682–9
  mould and, 653–4
  nutritional value, 245–7
  packaging, 656, 680–2, 688
  pasteurization, 684–9
  refreshing, 657, 665, 667
  slicing, 213, 214, 235, 655
  staling, 656–78
    retardation, 40, 679–82
  storage, 649–78
    flavour changes, 676–8
    moisture and, 655, 662
    temperature gradient, 676
  wrapping, 213, 214, 681
British Arkady Co. Ltd, 42
British Bakeries, 248, 249, 253
British Baking Industries Research
    Association, 225, 227
British Cellophane, 214
British Soya Products, 240
Brookfield, 9
Bühler Brothers, 216
Burkitt, D., 254
Bushuk, W., 66
Bushuk and Hulse, 461
Bushuk *et al.*, 69
Butanol, 74


Calcium, 247, 248
Calcium bromate, 33, 35
Calcium iodate, 33, 34, 66
Calcium peroxide, 33, 34
Calcium stearoyl-2-lactylate, 39
Canada
  baking tests, 332–3
  bread-making in, 294–8
Canadian Engineering Research Service,
  18
Canadian Grain Research Laboratory,
  149
Canadian Hard Red Spring, 77, 296, 461
Canadian Prairie spring wheat, 295
Canadian Utility wheat, 295

Canadian Western Red Spring, 51, 52, 229, 266, 294, 462
Canadian Western Red winter wheat, 295
*Candida* species, 353, 355, 478, 495
Carbohydrates, 3, 40
Carbon dioxide, 479, 480, 481, 486, 541, 599
Carbon monoxide, 599
Carboxyl methyl cellulose (CMC), 56, 57
Carlson and Bohlin, 328
Carlson *et al.*, 473
Carob bean flour, 55
Cathcart and Luber, 709
Cavel, R., 351
Cavenham Foods, 240
Cellophane, 681, 686
Cellulose and derivatives, 55–8
Central Soy, 44
Ch-12 mixer, 175, 194
Charlock, 324–5
Chefarox, 216
Chernaya *et al.*, 187
Chlorine dioxide, 42, 43, 216
Chopin Alveograph, 41, 90, 147–8
Chorleywood Bread Process (CBP), 25, 182, 183, 225, 226, 228, 230, 231, 267, 317, 491
Chrenow, A. M., 194, 519
Chr. Hansen's Laboratorium, 352
ChSM-300, 172, 176, 185, 194, 519
ChTR, 170, 171, 176, 179, 519
ChTU-D unit, 176, 177
Citric acid, 59
*Claviceps purpurea*, 323, 324
Clement wheat, 78
*Clostridium perfrigens*, 382, 383, 385
Cluskey and Wu, 76
Cluskey *et al.*, 672, 673
Coal, 598, 600
Cohen *et al.*, 472
Cole, E. W., 81
Combination bakeries, 305
COMECON, collaboration in, 203–6
Computers, 369, 402, 463, 470, 488, 717
Continuous bread process, gluten addition, 49–53
Convenience food, 255–6
Coppock, Dr, 183
Corn syrup, 338
Coventry *et al.*, 36
Crabtree effect, 470
Cresta Doughmaster, 230
Crest Catering Equipment Limited, 252
Cysteic acid, 36

Cysteine, 27–9, 36, 65, 76
L-Cysteine-HCl, 27, 28, 232, 235, 236, 490
Cystine, 36
Czechoslovakia, 197–201

Dal'nevostochnaya, 148
Daniels *et al.*, 43, 74
Dark Northern Spring wheat, 266, 269
Dataa, 717
Deamination, 75
Deck-ovens, 98
Deep-freezing, 99
L-Dehydroascorbic acid, 32, 33
Dehydro-D-isoascorbic acid, 33
Dempster *et al.*, 18
Detmold baking test, 333
Detmolder process, 346, 508
Detmold one-stage process, 106, 533
Detmold two-stage process, 106
Deuterium oxide, 76
Dextrin, 31
Diacetyl tartaric acid, 39, 40, 79, 235
Diasoy, 42
Dierks, 260
Dietary breads, 45, 522
Dietrich Reimelt, 207, 400, 401, 717
Diosna, 260, 261, 414, 419, 420
Dirndorfer *et al.*, 78
Disinfectants, 380–1
Disulphide bonding, 63–4
Divider-moulders, 122
Do-corder, 10
Do-maker, 28, 226, 280–1, 283, 443, 490
Donelson and Yamazaki, 316
Donezk Bread-factory No. 2, 177
Dontschenko, W. M., 171, 175, 176
Do-Soy, 42
Dough
    acidity, 484, 485
    additives, effects of, 27–59
    chemical changes in, 479–91
    components, main, 22
    conditioners, 42
    consistency, 27–29, 32–34, 82
    deformation, 371–2
    elasticity, 67
    fermentation, 479–91
    freezing, 689–708, 715
    gluten addition, 45–53
    measuring systems, 371–2
    pH, 581–2
    proofer unit, 696–7
    proofing time, 39

Dough—*contd.*
  relaxation time, 43
  retardation, 98
  retarders, 694, 695, 696
  rheological elements, 11–20, 539
  ripening, 18
  structure deformation, 11
  temperature and, 27, 28, 359–62
  testing apparatus, 313
  viscosity, 9, 10, 11, 13, 17, 21, 29, 30
  water and, 24, 32, 35, 47
  yield, 363–6
Dough-making:
  chemical bonding, 60–85
  consistency control unit, 25–6
  model explaining, 3–8
  stress relaxation, 28
  temperature, 26
  water and, 16, 17, 21–6
Dough-mixing:
  control of, 366–71
  discontinuous, 10
  high-speed, 57
  intensity, 9–10, 13, 29
  mixers, 11
  mixer types, 408–64
  mixer torque, 13–16
  parameters, 9–10, 11–20
  rpm, 10
  speed, 29
  time, 9, 10, 13, 26
Doughnuts, 301–5
Dough-souring, 57, *see also* sour-doughs
DP 383P, 633
Drews, E., 81, 82, 97, 316
DUO-NU oven, 610–17
Durum wheat, 269


E221, 29
E223, 29
E300, 42
E466, 57
E471, 38
E472e, 39
E481, 39
E482, 39
E-924, 229
Eberhardt, 260
EC
  additives and, 29, 34, 35, 38, 39, 42, 715
  tariffs, 49, 266
Elion, E., 36
Elsworth and Philipps, 66

Emmaflex C-700C, 632
Emulsifiers, 40, 268
Emulsions, 180, 181
Emulthin, 45
Entain and Frohlich, 473
*Escherichia*
  *coli*, 326, 327, 385, 388
  species, 388
Ethanol, 83, 481, 541
E.T. Oakes Limited, 226, 443, 444–5
Ewart, J., 64, 76, 85
Extensigraph, 18, 19


Falling number, 31, 83, 84, 96, 315
Falunina, S. F., 540, 581, 583
FAO, 149
Farinogram, 67
Farinograph, 16, 34, 76, 77, 148
Farrand, E. A., 54, 229, 317
Fatty acids, 180, 181, 224
Fermentation
  control of, 488
  time for, 489, 490, 491
Ferulic acid, 60, 80, 82
Fibre, 56, 254, 255, 256
*Flavobacterium*, 326
Flavol, 278, 279
Flour
  acidity, 319–20, 350, 484, 485
  additive tests, 320–1
  animal contaminants, 322–3
  catalase, 322
  colour, 319
  components, 21
  evaluating, 327–33
  gluten content, 306–9
  improvers, 42
  measurement and, 306–35
  mineral contaminants, detection of, 322
  particle size, 318–19
  pH, 320
  preparation, 362–3, 397
  protein-nitrogen measurement, 314
  storage, 394, 395, 396, 402
  strength, 485
  structural elements, 60
  thermal treatment, 327
  vegetable contaminants, 323–5
  vital wheat gluten and, 47, 52
Flowers Bakery, 276, 277
Folic acid, 245
Food poisoning, 382–3, 387
Foramitti, A., 460

Foremost Foods, 290, 706
Foremost Whey Products, 279
France
  bread-making in, 86–90
  dough mixing, 25
Frank, H. K., 390
Frank and Eyrich, 390
Frazier *et al.*, 18
Freed, R. J., 282
Friabilin, 671–2
Fructose, 195, 337, 473, 480, 481, 482
FTK 1000, 126, 127, 204, 451–60
FTK 1500, 127, 204
Fuchs, 133
Fuchs, K., 499
Fullington, J. G., 59
Fumigants, 380
Fungicides, 391
Furfural, 655

Galactolipids, 75
Galactose, 40, 55, 70, 83, 472
Gatelin process, 168
Gatilin, N. P., 194
Gatilin, W. F., 167
Gélinas *et al.*, 470
General Mills, 374
German Democratic Republic
  bread-making in, 108–43, 522, 526–8,
    716–17
  dietetic breads, 115–16
  Grain Processing Institute, 58, 135, 137
  national bread and rolls, 117–19
  production lines in, 119–43
  quality control in, 108–10, 112–13
  rolls, 108, 117–19, 140
  special breads, 111–15, 136–43
  Technical Control Organization, 109,
    110
  TGL numbers, 108
    TGL 27242/01, 110, 112–13
    TGL 2742/01, 110
    TGL-3067, 108, 522, 548, 578, 717
    TGL 8029, 313
    TGL 26972, 117
  wheat breads, 111
  wholewheat breads, 111
  WTOZ, 142, 143
Germany, Federal Republic of:
  bread-making in, 96–108, 522, 523, 716
  Grain Research Institute, 59
  rye-flours, 102–8
  wheat-flours, 97–102

Gerstenberg, H., 344
Giacenelli, E., 34
Gilbert mixers, 422
Gill and Duffus, 244
Ginsberg, A. S., 161
Ginsburg, A. G., 183
Girrek spirals, 261
Gladkowa, E. A., 192
Gliadin, 29, 39, 51, 60, 61, 68–72, 74–76,
    79, 80
Globulins, 3, 68, 73
Glucanases, 335
Glucose, 337, 473, 480, 481–483, 541
Glutamic acid, 35, 75, 79
Glutamine, 67
Glutaraldehyde, 85
Glutathione, 33, 35–7, 76, 187, 483
Glutelins, 3, 61
Gluten
  analysis, 60
  coagulation, 541
  composition, 65
  dough formation, 6, 19
  drying, 45
  quality, 313
  solubility, 76
  stretching, 65
  structure, 20
  viscoelasticity, 39, 40
  water and, 23, 24–5
Glutenin, 20, 29, 39, 40, 60, 61, 67, 69–73,
    76, 78
Glutomatic, 210, 307
Glycerol, 481, 482
Glycine, 35, 77
Glycolipids, 16, 20, 71, 74, 75
Glycoprotein, 20
Goldsmith, James, 253
Goodman Fielder, 244, 253
Gorky, Maxim, 144
Goroschenko, M. K., 167
GOSTOL, 422
Graham flour, 111
GRAS, 57
Graveland *et al.*, 73
Great Britain. *See* United Kingdom
GRL-1000 mixer, 462, 463
Grosskreutz, J. C., 50, 51, 65, 71
Grossman and Zimmermann, 472
Grunert *et al.*, 39
GSH, 36, 37
GSSG, 36, 37
Guar gum, 705
Gupta, N. K., 161

Haake, 9
Hagberg-Pertent, 315
Halton and Scott-Blair, 17
*Hansenula subpelliculosa*, 91, 349, 353
Hard Red Spring, 37, 269
Hard Red Winters, 269
Harrel and Thorn, 602
Hawthorne and Todd, 222
Hay, J. G., 42
Hemicellulase, 335
Hemicellulose, 55, 56, 80, 81, 103, 335
Henika and Rogers, 28
Hess, K., 50, 51, 65, 71
Hexose, 337, 480, 541
High-fibre breads, 524–5
High fructose corn syrup (HFCS), 41, 488, 713
High-protein breads, 525–8
HLK 50 mixer, 140, 414
Hoeppler Consistometer, 33
Holas *et al.*, 80
Hookean element, 17
Horizontal mixers, 262, 263, 428–38
Hoseney *et al.*, 70, 74
Hovis, 242, 250
Huber, H., 59
Humboldt-Universität, 39
Hungary, 196–7, 522
Hydrocolloids, 53–4, 78, 84, 705
Hydrogen bonding, 62, 63, 66, 67, 71, 75, 76, 655
Hydrogen peroxide, 44
Hygiene, 372–93

IADY (instant active dry yeast), 476, 477
ICC-Standard, 107, 108, 315, 316
IDK-1, 148
Ilyinych, K. E., 162
IMK 150 mixer, 119, 120, 124, 361, 368, 426–8
Improver-wheat, 149, 150
India, 149
Instant active dry yeast, 468
In-store bakeries, 689
  GDR, 140
  UK, 241, 242, 244
  USA, 287–91, 292, 305
Instron, 18, 19, 655, 678
International Society of Cereal Chemists, 206
Ionic binding, 63
Ionowa, W. W., 178
Ireks-Arkady, 346

IR spectroscopy, 70
Isernhäger natural dough-souring, 460
Isoascorbic acid, 33

Jackson, G. M., 60
Jaroslawer bread-factory, 193
Jazuba *et al.*, 21
Jegorowa *et al.*, 185
Jelitzki and Semichatowa, 337
Johnson, J. A., 278
Jørgensen, H., 36
J. Rank Limited, 42
J. R. Short Milling Company, 42

Karaya gum, 705
Karl-Marx-Stadt Bakery Cooperative, 136
Karpenko, F. P., 158
Karpenko, V. I., 37, 187
Kawamura *et al.*, 20
Kawka and Gasiorowski, 41
Kazanskaya, L., 194
Kelesov, I. M., 61
Kemerovo Technical Institute, 53
Kiev Bread-factory No. 6, 177
Kilborn and Tipples, 277, 296, 462
Kilborn *et al.*, 463
Kim and D'Appolonia, 670, 671, 673
Kjeldahl method, 314
Kline *et al.*, 291
Kogan, M. A., 169
Kolesov, V. M., 80
Koma CDS system, 698
Konova *et al.*, 53
Kontinua mixer, 439, 441
Kosmina, N. P., 80
Kosmina and Holas, 81
Kovbasa, V., 194
Krasnodar bread-making factory, 175, 178
Kretowich and Vakar, 76
Krijaginičev, M. J., 660–5, 667–669
Kühn and Grosch, 83
Kuninori and Matsumoto, 33
Kusmenko, W. W., 176
Kuzminskij *et al.*, 328
KVT 1000, 127, 205, 345, 445–51, 509
KVT 1300, 202
KVT 1500, 127, 129, 130, 131–133, 135, 136, 168, 198, 205, 345, 445, 509, 510
KVT 1800, 127, 202, 205, 345, 445, 509
*Kwaß*, 190, 191, 517, 518
Kyowa Hakko Isogyo Company Limited, 53

Lactic acid, 44, 52, 55, 58, 59, 83, 103, 104, 172, 313, 334, 340, 342–4, 350, 484–5, 505
Lactobacilli, 494, 495, 533
  *Lactobacillus acidophilus*, 355, 356
  *Lactobacillus alimentarius*, 355
  *Lactobacillus brevis*, 91, 340, 341, 348, 349, 352, 353, 355, 356, 495, 496
  *Lactobacillus bulgaricus*, 356
  *Lactobacillus casei*, 355, 356, 495
  *Lactobacillus cellobiosus*, 348
  *Lactobacillus delbrückii*, 104, 173, 186, 187, 350, 352, 485, 496
  *Lactobacillus farciminis*, 355
  *Lactobacillus fermentum*, 104, 355, 356, 495
  *Lactobacillus fructivorans*, 355
  *Lactobacillus leichmanii*, 496
  *Lactobacillus plantarum*, 92, 348, 349, 351, 352, 355, 356, 495, 496
  *Lactobacillus sanfrancisco*, 291, 348, 493
Laignelet Dumas, 60
Lagansk bread-factory, 177, 178
Langroller B 750, 90
Lecithin, 43–4
Lee and Samuels, 66
Lee *et al.*, 391
Lehmann and Dreese, 692
Lenard and Singer, 76
Leningrad bread-factories, 185
Lerchental and Müller, 17
Leucine, 79
*Leuconostoc mesenteroides*, 348
Leutenegger and Frei, 698
Lignin, 103
Linoleic acid, 42
Lipids, 40, 43, 49, 59, 61, 70–6, 674–5
Lipoprotein, 16, 20
Lipoxygenase, 42, 60, 78, 179, 180, 182, 224
Liquefaction number, 83, 84
Liquid oxidation phase (LOP), 177, 179, 180–1, 225
Liquid pre-ferment, 181
Liquid-sour, 172, 185, 186–95
Liquid yeast, 172, 173, 175, 176, 185, 186–95
List, 443
*Listeria*, 707
Locust bean, 705
Longroller ELR 680, 98
Lorenz, K., 349
Losa, A. I., 187
Lucas Meyer, 45

Lucny, M., 41
Lutescens, 148
Lutsishina, E. G., 20
Lutsishina and Matyash, 67, 150
LVF viscometer, 9
Lyons and Company, 217
Lysine, 41, 45, 79, 85

McCarrison, Sir Robert, 256
Ma Ching Yung *et al.*, 68, 75
MacDonalds, 267, 298
Machkamow, G. M., 162
MacRitchie, F., 72
Maillard reaction, 542, 587, 588, 589, 592, 594
Malfa-Kraftma-Brot, 529–30
Maltase, 336, 337, 478, 483
Maltose, 54, 58, 84, 314–17, 337, 472, 480, 583
Mamaril and Pomeranz, 77
Mannose, 473
Maple Leaf Mills, 297, 298
Maris Huntsman, 78
Marsakow, G. P., 167
Marssakov, G. P., 144
Marston, P. E., 27–8
Martin process, 45
Maslow, I. N., 161
Maslow and Nikolajew, 163
*Mass madre panaria*, 91, 348, 492
Maturox, 233
Mauri Bros, 234
Maximat-Spiralknetautomat, 420
Maxwell's relaxation time, 17
Mazur, P., 181
Mazur and Miller, 701–2
MB III mixer, 140
Mecham and Weinstein, 59
Mecham *et al.*, 61
Mechanically developed doughs, 262
*Mecklenburger Landbrot*, 138, 139, 528–9
Medcalf *et al.*, 81
Meissel, M. N., 183
Meredith, P., 229
Merritt, P. P., 695, 696
Metallic ions, 59
Meuser *et al.*, 83
Mayerhof coefficient, 473
Mayer reaction, 549, 551
Micheljew, A. A., 167, 169, 571
Microbiological examination, 325–7
Miljukow, P. M., 174
Minel production lines, 189

MOCKIP, 313
Molasses, 187, 353, 470, 472
Monhamer process 346, 508
Monod, 353
Monoglyceride, 40
Mono mixers, 422, 423
Montgomery and Smith, 81
Moonen *et al.*, 20
Moore and Hoseney, 41
Moscow bread factory No. 2, 159
Moscow bread-factory No. 3, 169
Moscow bread-factory No. 4, 179, 184
Moscow bread-factory No. 5, 185, 186
Moscow bread-factory No. 10, 169, 189
Moscow bread-factory No. 12, 167, 169
Moscow Technological Institute, 53
MRS broth, 356, 357
Muffins, 298, 299
Multi-grain breads, 523–4
Mycotoxins, 389, 390, 391

Nessler's reagent, 356
Netherlands: Institute of Cereals, Flour
    and Bread, 73
Neukom and Markwalder, 80, 82
Newton, 8
Newton's law, 710
Nguyen-Brem *et al.*, 77
Nikolajew, B. A., 312
Nitrogen, 66, 353
NMR spectroscopy, 20, 37, 67, 70, 150
Novadelox, 216
Novoukrainka, 149
NU oven, 610
Nutrex, 207

Oase-Pumpen, 421
Odessa bakeries, 188, 189
Odesskaya, 149
Ofelt and Smith, 41
Ofentriebgerät, 41
Oils, 598, 599, 600, 601
Orla-Jensen, 356
Örsi and Pallagi-Bánkfalvi, 20
Osborne *et al.*, 68
Osborne fractions, 67, 68–9
Ostrowski, A. I., 172, 185, 477
Ovens
    batch, 603–7
    burner control, 620–4
    control of, 620–9
    depanning, 635–7

Ovens—*contd.*
    design, 602–17
    dielectric, 597–8
    electric, 596–7, 601
    energy recovery, 629–32
    energy sources, 596–602
    energy utilization, 558, 559
    gas and, 559–60, 598–601, 619
    industrial, 607–19
    loading, 633–7
    multi-deck, 545, 558–9, 603, 604, 606, 617
    rack, 606–7
    reel, 605–6
    requirements, 556–8
    ribbon burners, 620, 622
    temperature, 98
    thermometers, 601
    travelling-band tunnel, 211–13, 545, 546,
        550, 558, 577
    tray, 607–8, 618
    types of, 602–19
    unloading, 633–7
    wire-band tunnel, 601, 609, 617, 625–6,
        627
Oven spring, 28, 29, 541, 543
OW-204, 308
Oxidants, 6, 19, 34–5, 78, 268, *see also*
    under names of
Oxygen transfer, 353

Pallagi-Bánkfalvi, E., 73
Pan-o-mat Roll Plant, 267
Papain, 31
*Paracolobacterium aerogencides*, 326
Paschchenko *et al.*, 43, 181
Patt and Stscherbatenko, 163
PAU-1, 308
Pauling and Corey, 61
Pawperow, A. A., 167
Pawperow and Tschernjakow, 167
*Pediococcus cerevisiae*, 348
PEK-3A, 308
Pekar test, 319, 320
Pelshenke and Schulz, 345
Penetrometer, 308, 313, 338
*Penicillium*
    *citrinum*, 390
    *expansum*, 390
Pentosanases, 55, 82, 83, 234, 484
Pentosans, 3, 19, 22–3, 25, 45, 79–83, 335,
    484, 541
Perkins ovens, 558
Perlin, A., 81

Pest control, 374–93
Petroleum ether, 74
Pettri dish, 52
Pfeilsticker and Marx, 32
PFWS (pre-gelatenized flour/water
    suspension), 185–8
Pharmacia AB, 316
Phenolic acid, 60
Phosphate, 353
Phosphatic acid, 43
Phosphatidyl choline, 43, 44
Phospholipids, 49, 59, 71, 72, 74, 75
Phytin, 247
*Pichia polymorpha*, 91, 349
*Pischia saitoi*, 349, 353, 355, 495
Pitta bread, 251
Pizzas, 251–4
Plastometer, 309
Plastometer AB-1, 151
Plate-counts, 384, 385
Plotnikow, P. M., 193
Poland, 201–3
Polydextrose, 56
Polyethylene, 681–2
Polypeptides, 61–5, 69, 73, 79, 80
Polypropylene, 681
Polythene, 214
Pomeranz, Y., 72
Ponte, J. G., 67, 279
Popaditsch and Falunina, 585, 586
Poponeau, Y., 67
Potassium bromate, 27, 28, 33–35, 41, 66,
    77, 79, 85, 149, 232, 235
Potassium carbonate, 136
Potassium chloride, 115
Potassium iodate, 28, 33, 34, 35, 66, 177,
    233
Potato semolina, 139
Potato starch, 31, 667
Potentiometer, 567, 592
PPC ovens, 198
Pressed-yeast, 172, 467, 476, 478
Pressure sheeting, 461–3
Prihoda *et al.*, 33
Procea, 207, 240
Prolamines, 3, 61
Proline, 51, 67, 75, 79
Proofer, automatic, 210–11
Protein
    baking, changes in, 584–90
    denaturation, 549, 585, 587, 669, 672
    disaggregation, 29
    mixers and, 5
    molecular surface, 17

Protein—*contd.*
    structure, 61, 65, 69
    *see also* under names
Protein Digestibility Index, 41
Proteolipids, 51
*Pseudomonas*, 387
PTC-24 oven, 196
PTK 1000, 345
PTK 1500, 345
Pumpernickel, 115, 500, 530–2, 684
Puratos Corporation, 279
Putschkowa, L. I., 309
Pyrometers, 567
Pyruvic acid, 481–483

Quasi-continuous process, 508–12
Quellkurve, 82–84, 97
Quellmehl, 344, 346
Quellzahl, 313, 314
Quellzahl-flasks, 52

Rabe, E., 344
Rabinowitsch, I. L., 169, 170
Raffinose, 472
Rank Hay process, 42
Rank-Hovis, 217, 222
Rank Hovis McDougall, 236, 237, 242,
    243, 248, 253
Rank-Pullen meter, 367
Rebinger, P. A. 309, 312
Reddi-Sponge, 279, 290, 706–7
Redox potential, 67
Redox-systems, 321
Redpoint THI-300 thermometer, 632–3
Red Spring, 273
Red Winter, 273
Regotrans, 639
Rehrlich *et al.*, 71
Reimelt, 515
Resistance oven, 41
Retarder cabinet, 259
Retarder/proofers, 692, 693, 696, 697–9
Reznichenko, M. S., 61
Rheo-Tech VER, 370
Rheotest 2, 9
Riplex mixer, 432
Rogers and Szostak, 475
Rogers Fermitech Company, 275, 280
Rohrlich and Müller, 50
Roiter, I. M., 163, 190, 191, 193
Roiter *et al.*, 174
Rosedowns, 230
Rothe *et al.*, 590

Rototherm 90, 629
Rotovisco RV-3, 9
Rumsey-Ritter, 317
Russel Finex Limited, 402
Russia. *See* Union of Soviet Socialist Republics
Rye proteins, 79
Rye-bread
  baking, 569
  USSR, 190–5
Rye-flour
  properties, 675–6
  wheat flour and, 86
Rye-sours, 492–7, 500, 501, 502, 512–13, 516, 519–21
  faults, 519–21


S 70 divider moulder, 204, 459
S-125, 414
S-250, 414, 423
*Saccharomyces*
  *cerevisiae* 91, 193, 335, 340, 348, 349, 351, 355, 467, 471, 472, 495
  *exiguus*, 291, 348, 493
  *fructuum*, 91, 349
  *minor*, 194, 340, 353
Saizew, N. W., 159, 169
Saizew *et al.*, 185
Salmonella, 382, 385, 392
*Salmonella enteritidis*, 382
Salt, 28, 43, 58, 73
Salt-sour process, 508
Sancassiano range, 420
Sandstedt *et al.*, 31
S. and W. Berisford, 244
Sanger and Edman, 61
Saratovskaya, 148
Saratower system, 109
Sarrubra, 148, 149
Schneeweiss, R., 206, 540, 581
Schneeweiss and Klose, 639, 652
Schneider, W., 81
Schoch, T. J., 580, 655, 659, 670
Schulz, A., 59, 222, 340
Schulz and Stephan, 59
Scott, P. M., 391
SDS-electrophoresis, 20
Seibel *et al.*, 532
Semolina, 110, 269
Senti and Dimler, 660, 673
Sephadex G-100, 77
SH groupings, 32, 34, 36, 65, 66
—SH— groups, 77, 181, 226, 233

Sievers, R. S., 565–6
Simmonds and Orth, 85
Simplex 2000, 616
*Sinapis arvensis*, 324–5
Slimcea, 240
Smith, D. P., 565
SM mixers, 425
Sodium bisulphate, 66
Sodium metabisulphite, 29–30
Sodium stearoyl-2-lactylate, 39, 674
Sodium sulphite, 29–30
Soest, P. J. van, 56
Soft Red Winter, 269
Sorbic acid, 683–4
Sotscha divider, 189
Sottoriva, 419
Sour-dough, 340–58
  acidity, 506, 511
  bacteria, 104
  baking-process, 513
  consistency, 498
  drying, 347–8
  flour quality, 500–1
  maturation, 504
  microflora, 355, 494, 495
  nutrients, 499
  processing, 492–521
  processing stages, 501–21
  process variables, 497–501
  standing time, 499–500
  starter, 340
  starter culture, 103, 104, 495
  temperature, 497–8
  USSR, 516–19
Sandstedt, R. M., 580
Soy additives, 41–5, *see also* Soy flour
Soya foods, 240
Soy flour, 41–43, 75, 180, 223–225, 240
SP-80 systems, 400
Spain, 90–2, 258, 348, 351
SP-ALH 261, 419
Specialty breads, techniques for, 522–35
Spicher, G., 355, 495
Spicher and Schöllhammer, 349
Spillers, 183, 217, 236, 240, 244, 250, 253, 257
Spindle-moulder, 209–10
Spiral mixers, 93, 259–62, 417–23
SPP, 240, 241, 250
—SS— bridges, 65, 233
—SS— groups, 66, 77
Staling, retarding, 40, 679–82
*Staphylococcus aureas*, 326, 327, 383, 385–88

Starch
    baking, changes during, 580–4
    damaged, 53–4, 316–17, 484, 580
    gelatinization, 40, 58, 540–1, 549, 578, 665, 669
    granules, changes, 23, 31
    particle size, 314–15
    products added to dough, 54–5
    retrogradation, 670, 671, 674
    swelling, 3, 5
    water and, 3, 23, 24, 25
Stepanova, O. N., 53
Stephan, H., 345
Stoljarowa *et al.*, 173, 490
Stoll and Bouteville, 324
*Streptococci*, 385, 386, 387,
*Subtilis mesentericus*, 326
Suchumier bread factory No. 2, 169
Sucrose, 195, 337, 338, 472, 480, 482
Sullivan, B., 49, 60
Sullivan *et al.*, 77
Sulphuric acid, 31
Sulphydryl, 6, 34, 66
Supertex, 208, 230
Surfactants, 37–41
SU-US 854349, 194
Sverdlovsk bread factory, 167
Switzerland, bread-making in, 92–4

t940, 121, 198
t985, 121, 122, 124, 198
TCA, 474, 480, 482
T. Collins Ltd, 218, 219
Technoplast, 419–20
Teltoma sour, 346
Tesco, 242
Thiol, 76, 77
Thomas Collins, 416, 417, 419
Thomson Pty Ltd, 234
Thörner, D., 460
TK-150, 424
TLC, 75
TM-M, 674
Toda and Yale, 474
Todd *et al.*, 222
Tortosa *et al.*, 75
*Torulopsis candida*, 348
*Torulopsis colluculosa*, 348
*Torulopsis holmii*, 353, 355, 495
Transducers, 13, 16
Traub *et al.*, 65
Treloar's equation, 19, 20
*Trichosporon margaritiferum*, 91, 349

*Trichosporon penicillatum*, 349
Tscheuschner, H.-D., 161
Tscheuschner and Auerman, 308
Tschtscherbatenko, 174
Tsen and Bushuk, 77
Turbo-radiant oven, 211, 212
Tweedy mixers, 182, 225, 227, 228, 361, 368, 402–3, 407, 422, 423

UAD-15, 184
UAD-60, 184
Uniflow oven, 211, 212, 215
Union of Soviet Socialist Republics
    Academy of Sciences, 76
    All-Union Institute for Grain and
        Grain-processing Research
        (WNIIS), 148
    All-Union Research Institute for the
        Baking Industry (WNIIChP), 158,
        160, 170, 173–5, 179, 188, 193, 477,
        478, 645, 647
    All-Union Scientific Research Institute
        of Applied Molecular Biology and
        Genetics, 73
    bakeries, number of, 145
    baking tests in, 146–7
    bread, chemical composition, 165
    bread, energy content, 166
    bread-making in, 143–95
    bread varieties, 154–66, 195
    dough-processing in, 166–83
    flour, digestibility coefficients, 165
    flour-milling, 151, 152
    GOST, 146, 148, 151, 328, 490
    Moscow Technology Institute for the
        Food Industry, 161
    State Standards, 150, 153
Uniplex mixer 262, 264, 265, 432–4
United Kingdom
    Bakery and Allied Trades Association,
        230
    Bread and Flour (Amendment)
        Regulations (1972), 231, 490
    Bread and Flour Regulations (1984), 52,
        229, 235, 241
    bread-making in, 206–522
    bread quality, 217, 222, 239
    bread varieties, 248
    Food (Control of Irradiation)
        Regulations Amendment (1972), 381
    Food Labelling Regulations (1984), 52,
        229
    Home-Grown Cereals Authority, 237–8

United Kingdom—*contd.*
  milling, 216, 217, 236, 237, 244
  National bread, 42, 208
  National flour, 206, 207
  nutritional labelling, 254, 255
United States of America
  baking tests, 328–32, 335
  bread-making in, 268–94, 534
  enrichment standards, 270
  FDA, 31, 33–5, 39, 55, 525
  milling, 272
  nutritional labelling, 270–2
  pre-ferments, 274–7
  short-time dough processes, 277–94
  sour-dough French bread, 291–2, 493,
      494
  sour rye bread, 272–3
  sponge-dough in, 274, 277, 279
Urea, 76, 77, 353
URK-2, 152


Van der Waals force, 64, 65, 67, 71, 666,
      669, 674
Vangalev *et al.*, 33
VATW-4 divider-moulder, 140
VATW 515, 124
VEB-Bäckereimaschinenbau, 414, 423, 426
VEB Backwarenkombinat, 346
VEB Bakery Collective, 140
VEB Bakery Cooperative, 142
VEB Metal-works, 142
VEB Prüfgeräte, 9
Vercouteren and Lontie, 65
Veron AV/AC, 84
Viscometer, 9, 10, 82, 83
Vital wheat gluten, 45–53
Vitbe, 250
VNI, 460, 515
Vogl, A. E., 323
Vones *et al.*, 80
Voronezh Technological Institute, 181, 187


Wall, J. S., 69
Walsh *et al.*, 78
Wedge-protein, 51
Wehrli and Pomeranz, 70
Wehrli *et al.*, 74

Weigher-mixer functions, 394–9
Weighing methods, 399–408
Weisblatt, 327
Wendel-type mixer, 93
Werner and Pfeiderer, 90, 95, 98, 420, 439,
      548, 610
Wheat-flour, rye-flour and, 86
Wheat-sours, 492–7, 503
Whey, 28, 29
Whey/cystein, 283, 285
Winkler, 300, 301
Winkler GmbH, 718
Wohlgemuth procedure, 31
Wolarowitsch, M. P., 309, 310, 312
Woltman counter, 362
Wood-pulp, 56
Woychik *et al.*, 76
Wünsche and Tscheuschner, 328


Xanthan, 84
Xylose, 55, 81, 541


Yeast
  acidity, 342–3
  cell structure, 183
  dead cells, 335
  fermentation, 479–91
  food, 275
  gassing power, 336, 337, 339, 467
  glutathione, 37, 338–9, 478
  measurements and, 335–40
  physiology, 471–4
  post-fermentation technology, 476–8
  pre-activation, 183–6
  propagation and production, 469–78
  shelf-life, 338
  types of, 467–8
  water content, 338, 339
*Yersinia enterocolitis*, 382–3, 392


Zeleny, 313
ZGL proofer, 98
Zimmermann and Schmitt, 328
ZNK proofer, 98
Zwingelberg *et al.*, 532
Zwitterions, 43